ENGINEERING THERMODYNAMICS

SEVENTH EDITION

AF650551

About the Authors

Late P K Nag had a distinguished 50-year teaching career, including nearly 40 years at IIT Kharagpur (1962–2002). Post-retirement, he served as an AICTE Emeritus Fellow at Jadavpur University and later taught at IISWBM, Government College of Engineering and Leather Technology, and Heritage Institute of Technology in Kolkata. He was also Guest Professor and Head of Mechanical Engineering at Aliah University for three years and Visiting Professor at the Technical University of Nova Scotia (now Dalhousie University), Canada, during 1985–86 and 1993–94.

A pioneer in 'Circulating Fluidised Bed Combustion and Heat Transfer' in India, his research interests included thermodynamic optimization, combined cycle power generation, and waste heat recovery. He published over 150 papers and several books, and presented them internationally. He was a recipient of the 'President of India' medal in 1995. Dr. Nag was a Fellow of FNAE and the Institution of Engineers (India), and a member of the New York Academy of Sciences.

Sudipta De holds a BE in Mechanical Engineering from Bengal Engineering College (now IIEST), Shibpur and M.Tech and Ph.D from IIT Kharagpur. He is currently a Professor in the Mechanical Engineering Department at Jadavpur University. He has been a guest researcher at Lund University, Sweden, and participated in international academic exchanges, including those at the Technical University of Munich (INSA Fellowship) and KTH Stockholm (India4EU Program), focusing on clean and sustainable energy.

He has also delivered lectures as a guest faculty at several prominent institutions. His research has been supported by UGC, DST (India), EU, Swedish Research Council, DFG-Germany, SIU, and DIKU-Norway. Prof. De has authored over 150 publications, including journal and conference papers and textbooks. He is a Fellow of both the West Bengal Academy of Science and Technology and the International Society for Energy, Environment and Sustainability. His awards include the INSA Teacher Award and the 'Shiksharatna' by the Government of West Bengal.

ENGINEERING THERMODYNAMICS

SEVENTH EDITION

P K NAG
Former Professor
Indian Institute of Technology
Kharagpur

SUDIPTA DE
Professor
Jadavpur University
Kolkata

McGraw Hill Education (India) Private Limited

Published by McGraw Hill Education (India) Private Limited,
Registered Office: Anjana Complex No: 5/90 A, Butt Road, St. Thomas Mount, Chennai – 600016

Engineering Thermodynamics, 7e

Copyright © 2026, by McGraw Hill Education (India) Private Limited.

This book is protected under copyright law. Any unauthorized copying, reproduction, distribution, adaptation, conversion, transmission, storage or use of this publication or any part or extract of it, through any means or any medium, including electronic, mechanical, photocopied, recorded, or otherwise, whether for any direct/indirect commercial gain or not, and without the prior written consent of the publisher, is explicitly prohibited and will result in severe legal consequences. Such restrictions apply to social media channels/groups, YouTube videos, audiobooks, AI applications, and any other means of copying, reproducing, distributing, adapting, converting, transmitting, storing the content of this book, using current or future technologies.

This edition can be exported from India only by the publisher,
McGraw Hill Education (India) Private Limited.

Print Edition:
ISBN (13): 978-93-6444-432-3
ISBN (10): 93-6444-432-9

E-Book Edition:
ISBN (13): 978-93-6444-439-2
ISBN (10): 93-6444-439-6

1 2 3 4 5 6 7 8 9 7093309 29 28 27 26 25

Printed and bound in India

Information contained in this work has been obtained by McGraw Hill Education (India), from sources believed to be reliable. However, neither McGraw Hill Education (India) nor its authors guarantee the accuracy or completeness of any information published herein, and neither McGraw Hill Education (India) nor its authors shall be responsible for any errors, omissions, or damages arising out of use of this information. This work is published with the understanding that McGraw Hill Education (India) and its authors are supplying information but are not attempting to render engineering or other professional services. If such services are required, the assistance of an appropriate professional should be sought.

Typeset at CoKrator Content Solutions Private Limited, Delhi-110092
and printed at Rajkamal Electric Press, Kundli, Haryana-131 028.

Cover Designer: Vipin Kumar Raghav
Cover Image Source: Shutterstock

Visit us at: www.mheducation.co.in
Write to us at: support.india@mheducation.com
CIN: U80302TN2010PTC111532
Toll Free Number: 1800 103 5875

Dedicated to the memory of my parents
Late Ananta Lal Nag
Late Jyotsna Rani Nag
And with blessings and love to my grandchildren
Madhurima Basu Thakur
Abhiroop Guha

- P K Nag

Dedicated to the memory of my father
Late Syama Prasad De

- Sudipta De

Preface to the Seventh Edition

I vividly recall my first class at IIT Kharagpur, taught by Prof. P K Nag, whose textbook was our primary reference during undergraduate studies. It was through that course that my deep interest in thermodynamics was first ignited. Later, during my Ph.D, I had the opportunity to revisit and clarify many fundamental concepts of the subject. It is, therefore, a matter of great privilege and personal significance for me to be associated with this book, which has long been regarded as one of the most widely used texts on Engineering Thermodynamics in India.

Thermodynamics is a fundamental subject that explains the laws governing natural, and spontaneous processes. Its foundational nature makes it essential across nearly all branches of engineering and science, and even relevant in certain aspects of social sciences. While basic engineering thermodynamics focuses on core principles, applied thermodynamics translates those into real-world systems and devices. This book covers both dimensions comprehensively.

The content has been carefully structured to align with the syllabi of most undergraduate courses on basic and applied thermodynamics offered by leading institutes in India. It also includes select topics relevant to postgraduate studies. In line with the nationally adopted Outcome-Based Education (OBE) framework, this revised edition helps students clearly understand learning objectives and assess their own progress through review questions and problem-solving exercises. Each problem is tagged with a difficulty level to aid structured learning.

The chapters in this book are structured to facilitate progressive learning. It begins with fundamental concepts and gradually advancing to more complex topics and real-world applications. This arrangement allows undergraduate students to build a strong conceptual foundation before exploring advanced material. Depending on specific course requirements, certain chapters can be selectively skipped without affecting overall understanding. The book will be a useful resource for postgraduate students as well. Moreover, industry professionals will find selected chapters useful for refreshing core concepts or exploring practical applications in engineering systems.

Organization of the Book

Chapter 1 introduces the foundational concepts of thermodynamics, providing essential background and historical context. It also defines the fundamental entities and terminologies that form the basis of the subject.
Chapter 2 is dedicated entirely to the concept of temperature and its measurement principles, with a clear connection to the Zeroth Law of Thermodynamics.

Chapter 3 focuses exclusively on the fundamental concepts of heat and work, exploring their definitions, differences, and roles in thermodynamic processes.

Chapter 4 explores the fundamental principles of energy conservation by examining the interactions between a thermodynamic system and its surroundings - formally known as the First Law of Thermodynamics. This chapter specifically focuses on the application of the First Law to closed systems only.

Chapter 5 extends the discussion of the First Law of Thermodynamics to open systems, covering both steady-state and unsteady-state processes.

In addition to the First Law, natural processes must also comply with the Second Law of Thermodynamics. **Chapter 6** introduces this fundamental principle, along with key associated concepts such as reversible processes and absolute thermodynamic temperature.

Chapter 7 is fully dedicated to entropy, one of the most fundamental thermodynamic properties that governs the direction of spontaneous natural processes.

Exergy is another fundamental concept in thermodynamics, representing the maximum work potential of a system relative to a defined reference environment. **Chapter 8** is dedicated to the analysis of exergy for both closed and open systems, providing insight into the quality and usability of energy.

To apply thermodynamic principles in practical devices, working fluids are essential. **Chapter 9** provides a detailed discussion on the properties of pure substances, with a specific focus on steam. It covers the development and use of steam property tables and charts, which are critical for real-world thermodynamic analysis.

Gases and gas mixtures are also commonly used as working fluids in thermodynamic systems. **Chapter 10** is exclusively devoted to their analysis, covering both ideal and real gas behaviors in detail.

Not all thermodynamic properties can be measured directly; therefore, thermodynamic property relations are used to calculate these from measurable quantities. **Chapter 11** provides a detailed discussion on these property relations and highlights their importance in thermodynamic analysis.

Vapour power cycles remain the predominant method for power generation in steam power plants. **Chapter 12** focuses on the thermodynamics of steam power plants and also covers key aspects of binary vapour power cycles and cogeneration systems.

Gas power cycles offer an alternative for bulk power generation and are now widely used in aviation and transportation sectors. **Chapter 13** covers these applications, including gas-steam combined cycles, which achieve high efficiency.

Cooling is a fundamental requirement in many applications. **Chapter 14** explores the thermodynamic principles underlying cooling processes.

Psychrometrics deals with the thermodynamics of dry air mixed with moisture, requiring special consideration beyond the separate properties of steam and air. This topic is covered comprehensively in **Chapter 15**.

Chemical reactions play a vital role not only in producing useful products but also in applications such as heat release through fuel combustion. The thermodynamics of chemical reactions is discussed in detail in **Chapter 16**, highlighting its significance in both industrial and energy-related processes.

A fluid is considered compressible when its specific volume (or density) varies with both pressure and temperature. This behavior requires separate treatment due to its unique characteristics and wide range of applications. **Chapter 17** is dedicated to the study of compressible fluid flow, covering its fundamental principles and key applications.

Chapter 18 provides a detailed discussion on the thermodynamics of various types of gas compressors, covering their principles, performance, and real-world applications.

Steam turbines are the primary prime movers in large-scale power generation, while condensers serve as essential heat sinks in steam power plants. **Chapter 19** discusses the thermodynamic principles governing both devices.

Large-scale steam generation is carried out in steam generators or boilers, which operate based on multiple engineering principles. The two core aspects involved are combustion and heat transfer. **Chapter 20** *(available online)* provides an in-depth discussion on the thermodynamics of boilers, along with key engineering considerations related to their design and operation.

What is New in this Edition?

This edition incorporates several updates to enhance its relevance and usability. Key revisions include:

- Revised and updated Learning Objectives across all chapters
- New figures added to support newly introduced content
- Additional problems and questions included to reinforce updated material
- Inclusion of two new chapters:
  - Chapter 19: Steam Turbine and Condenser
  - Chapter 20: Steam Generators *(Available Online)*

Additionally, this edition also includes newly added content across the chapters, enhancing the text with updated and expanded information.

- New section on the working of boilers
- Enhanced discussion on real gases
- New section on Supercritical and Ultra supercritical Steam Cycle
- New section on Deviation in Actual Cycle from Ideal Brayton Cycle
- Additional content on pulse jet
- Expanded content on Refrigerants, including a discussion on modern green refrigerants
- New content on Carrier's Equation
- New sections on Stoichiometric Calculation and Exhaust Gas Analysis and Thermodynamics of Combustion in Internal Combustion Engines
- Additional content on Combustion in SI engines and CI engines

Online Resources

The online resources provided for both instructors and students are designed to enhance the effectiveness of teaching and learning. For instructors, the comprehensive resource materials consolidate various teaching tools, making it easier to integrate the textbook into classroom instruction. For students, the resources aim to improve learning outcomes and concept retention by reinforcing key topics and interlinking related concepts. Below is a detailed list of resources for readers.

Instructor Resources

- Solution Manual
- PowerPoint Presentations

Student Resources

- Chapter 20: Steam Generators
- Additional Reading Material on Steam Turbine and Condenser
- Appendix C: Multiple-Choice Questions
- Appendix D: Miscellaneous Solved Examples

Students can download these resources from *www.mheducation.co.in* using the scratch code mentioned on the back inner cover. Instructors can access the resources on *https://connect.mheducation.com* following the steps mentioned on the front inner cover of the book.

The book would serve as a comprehensive resource, offering insights and guidance tailored to diverse needs. Extensive efforts were dedicated to refining and revising the content to ensure clarity, relevance, and engagement in every chapter. These enhancements aim to provide an informative and enriching experience for readers.

– Sudipta De

Publisher's Note

McGraw Hill Education (India) invites suggestions and comments from you, all of which can be sent to *support.india@mheducation.com* (kindly mention the title and author name in the subject line). Piracy-related issues may also be reported.

Preface to the First Edition

The purpose of this book is to provide a mature approach to the basic principles of classical thermodynamics which is one of the few core subjects for the undergraduate students of almost all branches of engineering. The system-surroundings interactions involving work and heat transfer with associated property changes, and the system-control volume approaches of the first law have been emphasized. The second law has been elaborated upon in considerable detail. Except for some physical explanations, a statistical or microscopic analysis of the subject has not been made.

The first eight chapters of the book are devoted to a thorough treatment of the basic principles and concepts of classical thermodynamics. The second law and entropy have been introduced using the concept of heat engine. Chapters 9 and 10 present the properties of substances. Chapter 11 gives the general thermodynamic relationships among properties. A detailed analysis of power and refrigeration cycles is given in Chapters 12 to 14. Chapter 15 deals with air-water vapour mixtures, and reactive systems are analyzed in Chapter 16. To increase the utility of the book, Chapters 17 and 18 dealing with compressible fluid flow and heat transfer respectively, have been added.

Many illustrative examples are solved and many problems are provided in each chapter to aid comprehension and to stimulate the interest of the students.

Throughout the text SI units have been used. Tables and charts given in the Appendix are also in SI units.

This book is based mainly on the lecture notes prepared for classes on the subject at IIT Kharagpur. I am grateful to the authors of the books that I used in preparing the notes, a list of which is given in the bibliography. I am thankful to my colleagues in the mechanical engineering department of IIT Kharagpur, for many stimulating discussions, and for encouragement. I am indebted to all those who have helped in the preparation of the book.

I would very much appreciate criticisms, suggestions for improvement, and detection of errors from my readers, which will be gratefully acknowledged.

P K Nag

Acknowledgements

The authors gratefully remember the contributions made by Late Prof. Bindu Madhav Belgaumkar, the Head of the Department of Mechanical Engineering (1954–1970), at Indian Institute of Technology (IIT), Kharagpur, in building up his teaching career during his earlier formative years, by providing the right academic environment in the campus. They also humbly acknowledge the contribution of Prof. R G Mokadam for maintaining high standard of teaching Thermodynamics in the decades of 1950s and 1960s at IIT, Kharagpur, the first IIT which played the pioneering role in technical education in the country.

The authors are indebted to erstwhile colleagues at IIT for many positive interactions, some of whom taught the M.Tech course, viz., Prof. C N Lakshminarayan, Late Prof. H R Narayan, Prof. Jayanand H. Hiranandani, Prof. H V Rao, Prof. B G Ghosh, Prof. A K Roy, Late Prof. P K G Pannikar, Prof. Darshan Lal, Prof. N S Murthy and Prof. S K Som.

The authors sincerely acknowledge the students of Jadavpur University from various engineering disciplines, whose questions and interactions during teaching significantly deepened the understanding of thermodynamics. The curiosity and engagement of these students have been invaluable in shaping the approach to the subject.

Heartfelt thanks are extended to retired colleague Dr. Sandip Das, whose innovative and collaborative teaching methods enriched the authors' perspective on thermodynamics.

Special acknowledgments are also due to esteemed colleagues: Prof. Himadri Chattopadhyay, Dr. Snehangshu Mondal, Prof. Sudip Simlandi, and Prof. Nilkanta Barman—for their insightful discussions during the co-teaching of this subject.

– Authors

Nomenclature

The symbols listed here are not exhaustive, but are the main ones used in the text. Other symbols have been defined at appropriate places in the book.

Symbols

Symbol	Meaning
a	Activity: Specific availability or exergy, kJ/kg
a_f	Exergy of a stream per unit mass, kJ/kg
A	Availability or exergy kJ
A	area, m^2
b	Specific Keenan function, kJ/kg
B	Keenan function, kJ
c	Velocity of sound, m/s
c_p	Specific heat at constant pressure, kJ/kgK
c_v	Specific heat at constant volume, kJ/kgK
C_p	Heat capacity at constant pressure, kJ/K
C_V	Heat capacity at constant volume, kJ/K
COP	Coefficient of performance
d, D	Diameter
e	Specific energy, kJ/kg
E	Total energy, kJ
f	Fugacity; Specific Helmholtz function, $u - Ts$, kJ/kg
F	Force N
F	Helmhotz function, kJ
g	Gravitational acceleration, m/s^2
g	Specific Gibbs function, $h - Ts$, kJ/kg
G	Total Gibbs function, kJ
h	Specific enthalpy, kJ/kg
h_f	Enthalpy of formation, kJ/kg mol
H	Total enthalpy, $U + pV$, kJ
H_P	Enthalpy of products, kJ
H_R	Enthalpy of reactants, kJ
h_{RP}	Enthalpy of combination, kJ/kg fuel
HHV	Higher heating value, kJ/kg fuel
i	Specific irreversibility, kJ/kg
I	Total irreversibility, kJ
I	Electric current, A
K	Boltzmann constant, J/molecule–K.
K	Indicator spring constant, N/cm^3
k	Thermal conductivity, W/mK
k_T	Isothermal compressibility, K^{-1}
k_s	Adiabatic compressibility, m^2/N
K	Equilibrium constant
KE	Total kinetic energy, kJ

l	Latent heat, kJ/kg
L	Length, m
LHV	Lower heating value, kJ/kg fuel
m	Mass, kg
$\dot{m}$	Mass flow rate, kg/s
M	Mach number
M	Molar mass, kg/kg mol
MEP, mep	Mean effective pressure, N/m^2
m_f	Mass fraction
n	Polytropic index, number of moles
N	Number of molecules
p	Pressure, N/m^2 (Pa)
p	Partial pressure, Pa
p_c, p_{cr}	Critical pressure, kPa
p_o	Atmospheric pressure, kPa
p_r	Reduced pressure
p_v	Vapour pressure, kPa
PE	Total Potential Energy, kJ
Q	Total heat transfer, kJ
Q	Heat transfer rate kJ/s
Q_H, Q_1	Heat transfer from high temperature body, kJ
Q_L, Q_2	Heat transfer to low temperature body, kJ
Q_{rev}	Heat transferred and reversibly, kJ
r_c	Cut-off ratio
r_e	Expansion ratio
γ_k	Compression ratio
γ_p	Constant volume pressure ratio
R	Characteristic gas constant, kJ/kg – K
R, R_u	Universal gas constant, kJ/kg mol – K
s	Specific entropy, kJ/kg – K
S	Total entropy, kJ/K$^-$
S_{iso} the	Siso, Suniv Entropy of isolated system or universe, kJ/K
t	Time, and; temperature, °C
t_f	Final temperature, °C
T	Temperature K
T_{as}	Adiabatic saturating temperature, °C or K.
T_{cr}, T_c	Critical temperature, K
T_{db}	Dry bulb temperature, °C or K
T_{dp}	Dew point temperature, °C
T_{wb}	Wet bulb temperature, °C
T_H, T_1	Temperature of hot reservoir or source, K
T_L, T_2	Temperature of cold reservoir or sink, K
T_0	Temperature of the surroundings, K
T_r	Reduced temperature
u	Specific internal energy, kJ/kg
U	Total internal energy, kJ
v	Specific volume, m^3/kg
v_c, v_{cr}	Critical specific volume, m^3/kg
v_r	Reduced volume
V	Total volume, m^3
W	Total work, kJ
$\dot{W}$	Power, kW
W_{max}	Maximum work, kJ
W_{rev}	Reversible work, kJ
W_u	Useful work, kJ
x	Quality, dryness fraction
x	Mass fraction, Mole fraction
z	Elevation, m
Z	Compressibility factor

Greek Letters

β	Volume expansivity, K^{-1}
Δ	Finite change in quantity
ε	Effectiveness
η_{th}	Thermal efficiency
η_{I}	First law efficiency

η_{II}	Second law efficiency	*fg*	Difference in property between saturated liquid and saturated vapour
μ	Chemical potential, kJ/kg	*g*	Saturated vapour
μ_J	Joule-Thomson coefficient, K/kPa	*gen*	Generation
ν	Stoichiometric coefficient	*H*	High temperature (as in T_H and Q_H)
ρ	Density, kg/m^3	*i*	Inlet conditions
σ	Stefan-Boltzmann coefficient	*i*	*i*-th component
σ	Surface tension, N/m	*L*	Low temperature (as in T_L and Q_L)
τ	Torque, Nm	*m*	Mixture mean
τ	Time, *s*	*r*	Reduced
ϕ	Relative humidity	*rev*	Reversible
Φ	Availability or energy of a closed system, kJ/kg	*s*	Isentropic
ψ	Availability or exergy of a steady flow system, kJ/kg	sat	Saturated
ω	Specific humidity or humidity ratio, kg water vap/kg dry air	surr	Surroundings
e	Exit conditions	sys	system
f	Saturated liquid	*v*	Water vapour
		0	Dead state
		1	Initial or inlet state
		2	Final or exit state

Subscripts

a	Air
abs	Absolute
act	Actual
atm	Atmospheric
av	Average
cr	Critical point
CV	Control volume

Superscripts

. (over dot) Quantity per unit time
‾ (over bar) Quantity per unit mol
° (circle) Standard reference state
′ (one prime) Solid phase
″ (two prime) Liquid phase
‴ (three prime) Vapour phase

Contents in Brief

Contents

Visual

LEARNING TOOLS

LO4.1 State first law for a closed system executing a cycle
LO4.2 Derive first law for a closed system for a change of state
LO4.3 Recognize total energy as a property
LO4.4 Differentiate between stored energy and energy in transition
LO4.5 State specific heat of a substance at constant volume
LO4.6 Discuss enthalpy and compare it with heat transfer
LO4.7 State specific heat of a substance at constant pressure
LO4.8 Analyze total energy of an isolated system
LO4.9 Discuss perpetual motion machine of the first kind (PMM1)

Learning Objectives

Each chapter is organized into multiple Learning Objectives (LOs) that provide a quick reference to the chapter's key aspects. These help students better anticipate what they will be studying and help instructors measure the students' understanding.

Pedagogy

The pedagogy is arranged as per the levels of difficulty which are derived from Bloom's Taxonomy.

E indicates Level 1 and Level 2, i.e., Remember and Understand based easy-to-solve problems.

M indicates Level 3 and Level 4, i.e., Apply and Analyze based medium-difficulty problems.

D indicates Level 5 and Level 6, i.e., Evaluate and Create based high-difficulty problems

Levels of Difficulty

(E) Low complexity: Level 1 and Level 2 Category
(M) Medium complexity: Level 3 and Level 4 Category
(D) High complexity: Level 5 and Level 6 Category

SUMMARY

The first law of thermodynamics is a special statement of the law of conservation of energy. It may be written for a cycle

$$(\Sigma Q)_{cycle} = (\Sigma W)_{cycle}$$

or for a process.

$$\begin{aligned} Q &= \Delta E + W \\ &= \Delta E_K + \Delta E_p + \Delta U + W \\ &= \frac{1}{2} mV^2 + mgZ + \Delta U + W \end{aligned}$$

where E is the amount of energy stored in a system and it is a property of the system. The stored energy can be either in macroscopic mode like kinetic and potential energy (E_K and E_p) and also in microscopic mode called intergy energy (U) where energy is stored in the molecules of a system.

Summary

A detailed chapter-end summary is provided for a quick review of the important concepts.

Walkthrough

INTEGRATED ASSESSMENT TOOLS

SOLVED EXAMPLES

Example 4.1 *A stationary mass of gas is compressed without friction from an initial state of 0.3 m³ and 0.105 MPa to a final state of 0.15 m³ and 0.105 MPa, the pressure remaining constant during the process. There is a transfer of 37.6 kJ of heat from the gas during the process. How much does the internal energy of the gas change?* (E) **[LO 4.2]**

Solution First law for a stationary system in a process gives

$$Q = \Delta U + W$$

or
$$Q_{1-2} = U_2 - U_1 + W_{1-2} \quad \text{(i)}$$

Example 9.1 *Find the saturation temperature, the changes in specific volume and entropy during evaporation and the latent heat of vapourisation of steam at 1 MPa.* (E) **[LO 9.7, 9.8]**

Solution At 1 MPa, from Table A.1.2 in Appendix A

$$t_{sat} = 179.91°C$$
$$v_f = 0.001127 \text{ m}^3/\text{kg}$$
$$v_g = 0.19444 \text{ m}^3/\text{kg}$$
$$\therefore \quad v_{fg} = v_g - v_f = \mathbf{0.1933 \ m^3/kg}$$
$$s_f = 2.1387 \text{ kJ/kg K}$$
$$s_g = 6.5865 \text{ kJ/kg K}$$
$$\therefore \quad s_{fg} = s_g - s_f = \mathbf{4.4478 \ kJ/kg \ K}$$
$$h_{fg} = h_g - h_f = \mathbf{2015.3 \ kJ/kg}$$

In-Text Examples

The text has been integrated with practice questions linked with the Learning Objectives. This offers great retention through looping mechanism.

Chapter-end Exercises

Inclusion of comprehensive exercise questions to help students consolidate their understanding and apply key concepts, reinforcing learning through practice.

REVIEW QUESTIONS

(E) **4.1** State the first law for a closed system undergoing a cycle. **[LO 4.1]**

(M) **4.2** What was the contribution of J.P. Joule in establishing the first law? **[LO 4.1]**

(M) **4.3** What is the caloric theory of heat? Why was it rejected? **[LO 4.1]**

(E) **4.4**

(E) **4.5**

(M) **4.6**

(M) **4.7**

PROBLEMS WITH HINTS

(E) **4.1** A closed system of mass 2 kg undergoes a process in which there is a heat transfer of 25 kJ from the system to the surroundings. The amount of work done on the system is 100 kJ. The specific internal energy decreases by 15 kJ/kg and the elevation of the system increases by 1000 m. The acceleration of gravity is 9.6 m/s². Determine the change in K.E. of the system. **[LO 4.4]**

Hint: By first law,

$$\therefore \quad \Delta KE + mg\,\Delta Z + m\,\Delta U = \dot{Q} - W$$

PROBLEMS

(E) **4.1** An engine is tested by means of a water brake at 1000 rpm. The measured torque of the engine is 10,000 mN and the water consumption of the brake is 0.5 m³/s, its inlet temperature being 20°C. Calculate the water temperature at exit, assuming that the whole of the engine power is ultimately transformed into heat which is absorbed by the cooling water. **[LO 4.1]**

(E) **4.2** In a cyclic process, heat transfers are + 14.7 kJ, – 25.2 kJ, – 3.56 kJ and + 31.5 kJ. What is the net work done for this cyclic process? **[LO 4.1]**

1

Introduction

Learning Objectives

LO 1.1 Recall the background of thermodynamics
LO1.2 Recall the application of thermodynamics
LO1.3 Discuss two views of thermodynamic analysis
LO1.4 State thermodynamic systems
LO1.5 Discuss thermodynamic state, property, process and cycle
LO1.6 Compare homogeneous and heterogeneous systems
LO1.7 Appraise thermodynamic equilibrium
LO1.8 Analyze quasi-static or quasi-equilibrium process
LO1.9 Describe the concept of continuum
LO1.10 Discuss SI units and dimensions of basic thermodynamic quantities
LO1.11 Discuss pressure and its various measuring devices
LO1.12 Describe history of engineering thermodynamics

LO 1.1
Recall the background of thermodynamics

Thermodynamics is the science of energy transfer and its effect on the physical properties of substances. It is based upon observations of common experiences which have been formulated into thermodynamic laws. These laws govern the principles of energy conversion. The applications of the thermodynamic laws and principles are found in all fields of energy technology, notably in steam and nuclear power plants, internal combustion engines, gas turbines, air conditioning, refrigeration, gas dynamics, jet propulsion, compressors, chemical process plants and direct energy conversion devices.

One may be curious to know about the founders of engineering thermodynamics. A list of individuals whose lives were vital to the founding of the subject is given on the next page with some information in regard to their contributions.

One thermodynamicist, *Antoine Laurent Lavoisier* (1743–1794), was the father of modern chemistry and was particularly interested in the compounding of gunpowder. It was Lavoisier's unfortunate fate that he lived during the French Revolution (1789). Before the revolution, he was a recognised scientist and contributed to the analysis of the combustion process. He led in developing the concept of elements; he proposed the name of 'oxygen' (discovered by Joseph Priestley) and described its function in the combustion process. He was born in a prosperous middle-class family and pursued his education in the sciences. He was active in politics before the revolution, seeking greater social reforms. He pursued various scientific topics including the development of the metric system. However, his role as a tax collector did not endear him to the public and he was beheaded by the guillotine.

Another foremost contributor was *Nicolas Leonard Sadi Carnot* (1796–1832), whose observations led to the naming of a theoretical thermodynamic cycle after him. He lived his youth in Napoleon's France, and his father, one of the Napoleon's military organisers, was able to provide Carnot, with a sound technical education. He left the army at the age of 24 and moved to Paris to do work in thermodynamics and wrote the famous monograph, '*Reflections on the Motive Power of Fire*' (1824) and established the basic principles governing the conversion of heat to work that led to the second law of thermodynamics.

James Prescott Joule (1818–1889), a British scientist of the mid-nineteenth century, performed careful and thoughtful experiments to establish heat as a form of energy refuting the calorific theory of heat. It led to the founding of the first law of thermodynamics with experimental proof of heat as a form of energy.

William Thomson (1824–1907), who later became *Lord Kelvin*, a renowned British physicist, developed the thermodynamic temperature scale. The absolute temperature scale was named after him. From his experiments he deduced the doctrine of available energy, and along with Joule expounded the Joule–Kelvin effect.

A German professor, *Rudolph J. Clausius* (1822–1888), another famous contributor in thermodynamics, used some of Carnot's reflections and spent the mid-1860s developing ideas about matter and publishing papers dealing with an abstract quantity called 'entropy'.

The last thermodynamics pioneer we will discuss is *J. Willard Gibbs* (1839–1903), a comparatively unknown American, who contributed tremendously in various areas of thermodynamics, statistical mechanics, chemistry and mathematics. He studied and taught at Yale University, and being disreputed for his odd ideas worked for nine years without pay. Eventually his worth was recognised in academic circles, and he was paid his salaries later.

The strength of thermodynamics lies in its ability to analyse and design a wide range of systems using only a few tenets—the four laws of thermodynamics. Some common applications of work-producing and work-consuming plants are being described in the following sections.

1.1 SIMPLE STEAM POWER PLANT

LO 1.2
Recall the application of thermodynamics

A schematic diagram of a simple steam power plant is shown in Fig. 1.1. High pressure and high-temperature superheated steam leaving the steam generator or boiler does work in the turbine which drives the electric generator to produce electricity. The low-pressure steam from the turbine enters the condenser, where heat is transferred from the steam to the cooling water circulating through the tubes. Steam is condensed to water which is then pumped back to the boiler.

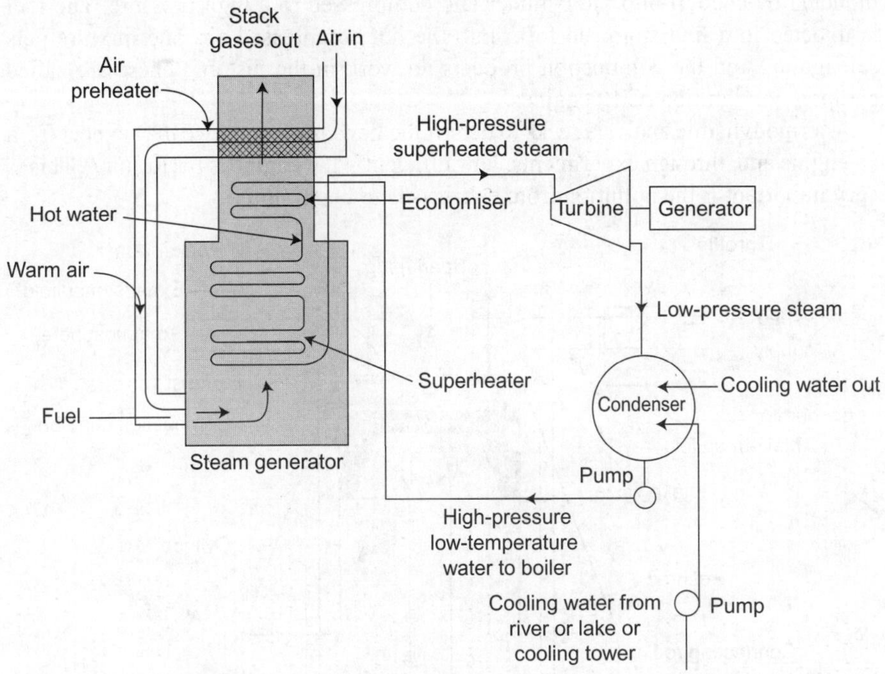

Fig. 1.1 Schematic diagram of a simple steam power plant

A fossil fuel (coal, fuel oil or natural gas) is used to release heat by combustion, and the products of combustion or fuel gases heat water to generate steam and then superheat it. The exhaust gases, before leaving through stack or chimney, heat the feedwater in a heat exchanger called economizer before the feedwater enters the steam drum and then heats the air for combustion in another heat exchanger called *air preheater*. The exhaust gases leaving the stack have a considerable amount of greenhouse gases (CO_2) and cause thermal pollution and global warming. A large power plant has many other pieces of equipment, some of which will be considered later.

In a nuclear power plant, the nuclear reactor replaces the furnace of a conventional power plant. The turbine, condenser and pump are the three other basic components. Safety considerations and the disposal of the radioactive wastes are important in the design and operation of a nuclear plant.

1.2 INTERNAL COMBUSTION (I.C.) ENGINES

These are largely used in transport vehicles like cars, buses, trucks, motor cycles and so on. Here, a fuel is burned, and the energy from the burning fuel is transferred to the pistons, which through gears turn the wheels, thus moving the automobile (Fig. 1.2).

In *petrol engines*, the fuel-air mixture after being compressed is ignited by an electric spark, hence the name spark ignition (S.I.) engine, and the products of combustion do work on the piston, and through a crank-and-connecting-rod mechanism, power is transferred to the crank shaft. In *diesel engines*, only air is inducted by suction into the cylinder and compressed to a high pressure. The fuel is injected in a fine atomised form into the hot compressed air. The mixture gets self-ignited and the combustion products do work in the pistons. These are called compression ignition (C.I.) engines.

Thermodynamic analysis seeks to determine how much work we may expect from an engine and, through experiments, how efficiently the engine is performing. This is very important if the pollution from exhausts is to be minimised.

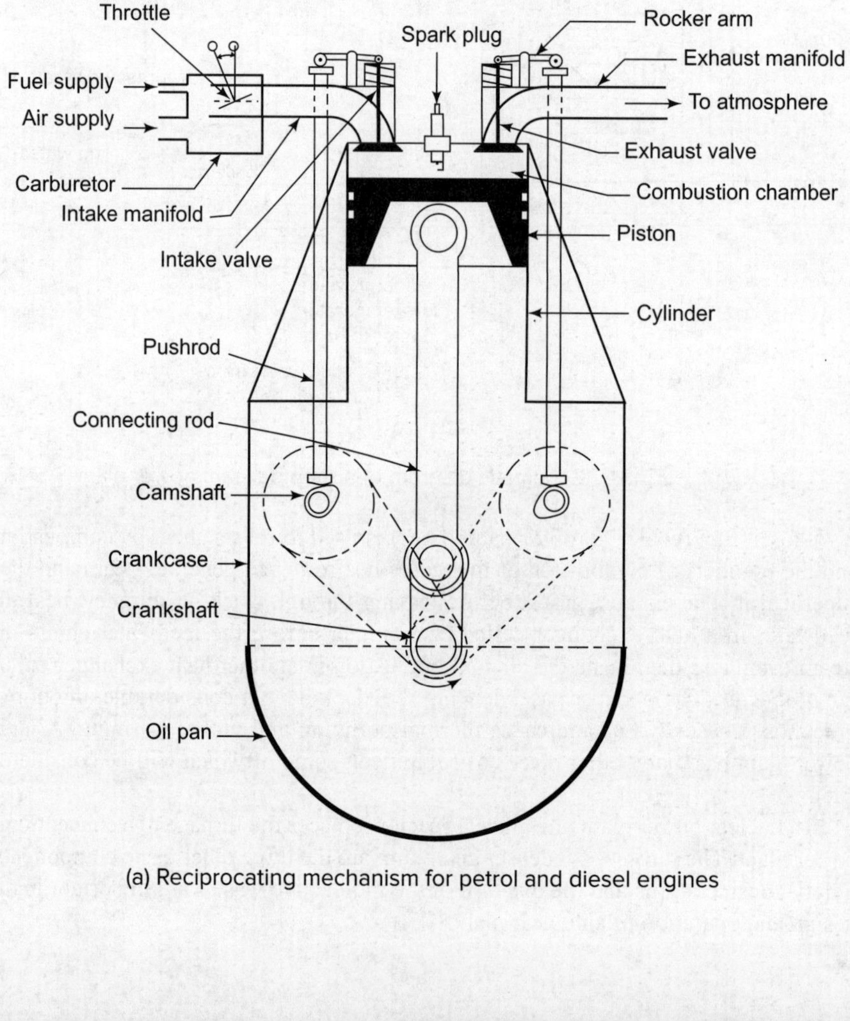

(a) Reciprocating mechanism for petrol and diesel engines

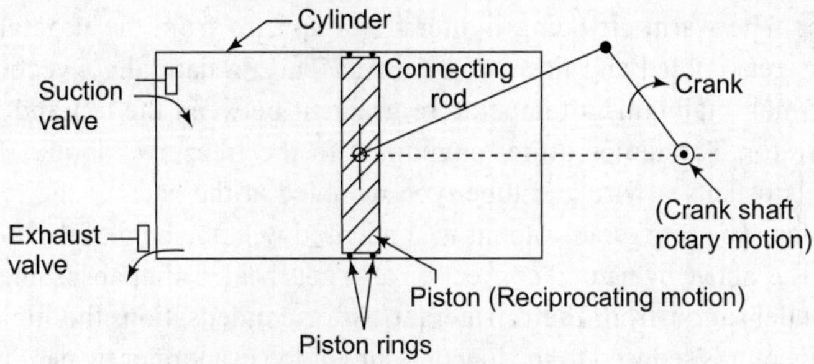

(b) An automotive engine

Fig. 1.2 Internal combustion engine

The gas turbine (Fig. 1.3) is another automotive power source, more commonly found in jet planes. There is an upsurge in the development of gas turbine plants in both electric power generation and ship propulsion. Air is compressed and energy is added to it by burning fuel in a combustion chamber; this mixture, viz. the products of combustion, expands through a turbine, doing work, which drives the electric generator or the ship. The analysis is similar to that of most power plants, and all these analyses have a common purpose, which is to consider how efficiently the chemical energy of the fuel is converted into mechanical energy. The processes of converting the energy are different, but the principle of energy conversion remains the same.

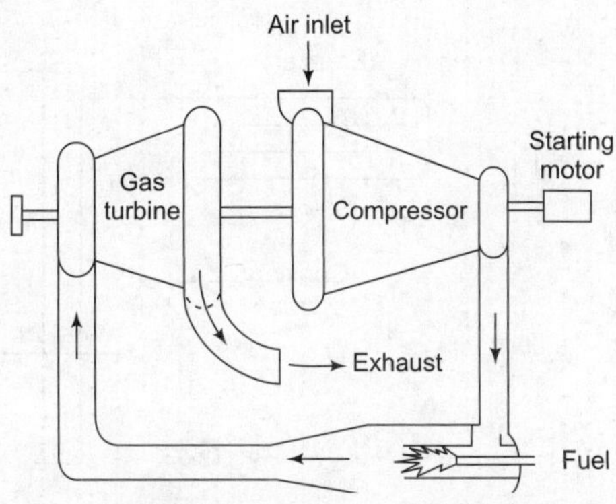

Fig. 1.3 A gas turbine unit

1.3 DOMESTIC REFRIGERATOR

A schematic diagram of the refrigerator is shown in Fig. 1.4. The basic components are (a) Evaporator, (b) Compressor, (c) Condenser and (d) Expansion device. The evaporator where the refrigerant (working fluid) evaporates absorbing the latent heat of vaporisation is the coldest part of the *freezer* cabinet, where it is located. In modern *frost-free refrigerators*, the evaporator is located outside the cabinet. A fan circulates air from the evaporator to the freezer. Just below the freezer, there is a chiller tray. Further below are compartments with progressively higher temperatures. The bottom-most compartment meant for vegetables is the least cold one. The cold air being heavier flows down from the freezer to the bottom of the refrigerator. The warm air being lighter flows upward from the vegetable box to the freezer, gets cooled and flows down again. Thus, a natural convection current is set up which maintains a temperature gradient between the top and bottom of the refrigerator. The temperature maintained in the freezer is about –15°C. The condenser is usually a wire-and-tube type mounted at the back of the refrigerator, having no fan. The refrigerant vapour is condensed with the help of the surrounding air which rises above by natural convection as it gets heated after absorbing the latent heat of condensation from the refrigerant. After condensation, the high pressure liquid refrigerant is reduced to the low pressure of the evaporator by passing through an expansion device (throttle valve or capillary tube) and the cycle is completed.

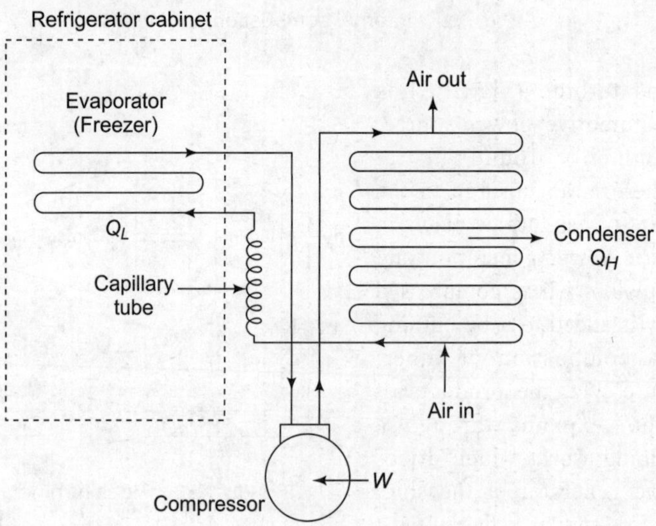

Fig. 1.4 Schematic diagram of a domestic refrigerator

Earlier, the refrigerant used was R-12 and because of the ozone layer depletion it is now isobutene C_4H_{10}), propane (C_3H_8) or tetrafluoroethane ($C_2H_2F_4$ or R-134a).

1.4 ROOM AIR CONDITIONER

The schematic diagram of a typical window-type air conditioner is given in Fig. 1.5. Let us consider a room to be maintained at a constant temperature of 25°C. The air from the room is drawn by a fan and is made to pass over a *cooling coil* of the evaporator, the surface of which is maintained at a temperature of 10°C. After passing over the coil, the air is cooled to around 10°C before being supplied to the room at 25°C by the fan. In the cooling coil, the refrigerant, R–22 or R–134a enters at, say, 5°C and evaporates absorbing the latent heat of vaporisation from the room air.

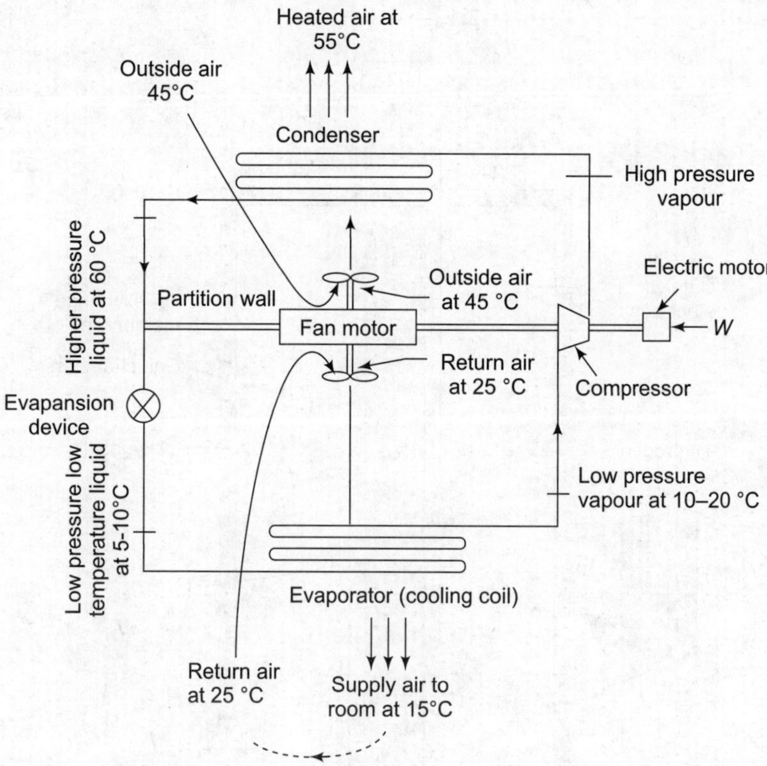

Fig. 1.5 Schematic diagram of a room air conditioner

The refrigerant vapour from the evaporator is compressed to a high pressure before entering the condenser where the atmospheric air at about say, 45°C, in summer is circulated by a fan. After picking up the latent heat of condensation from the refrigerant, the air is thrown back to the atmosphere, say, at 55°C. The high-pressure liquid refrigerant from the condenser is reduced to the low evaporator pressure by passing through the expansion device (capillary tube) before entering the evaporator. The cycle repeats itself.

1.5 FUEL CELLS

In a conventional power plant, fuel and air enter the power plant and products of combustion leave the unit. There is also a transfer of heat to the cooling water and work is done in the form of electrical energy. The fuel cell accomplishes this objective converting the chemical energy of fuel directly to electricity. Figure 1.6 shows a schematic arrangement of a fuel cell of ion-exchange membrane type. In this fuel cell, hydrogen and oxygen react to form water. The flow of electrons in the external circuit is from anode to cathode. Hydrogen enters at the anode side, and oxygen enters at the cathode side. At the surface of the ion-exchange membrane, the hydrogen is ionised according to the reaction

$$2H_2 \rightarrow 4H^+ + 4e^-$$

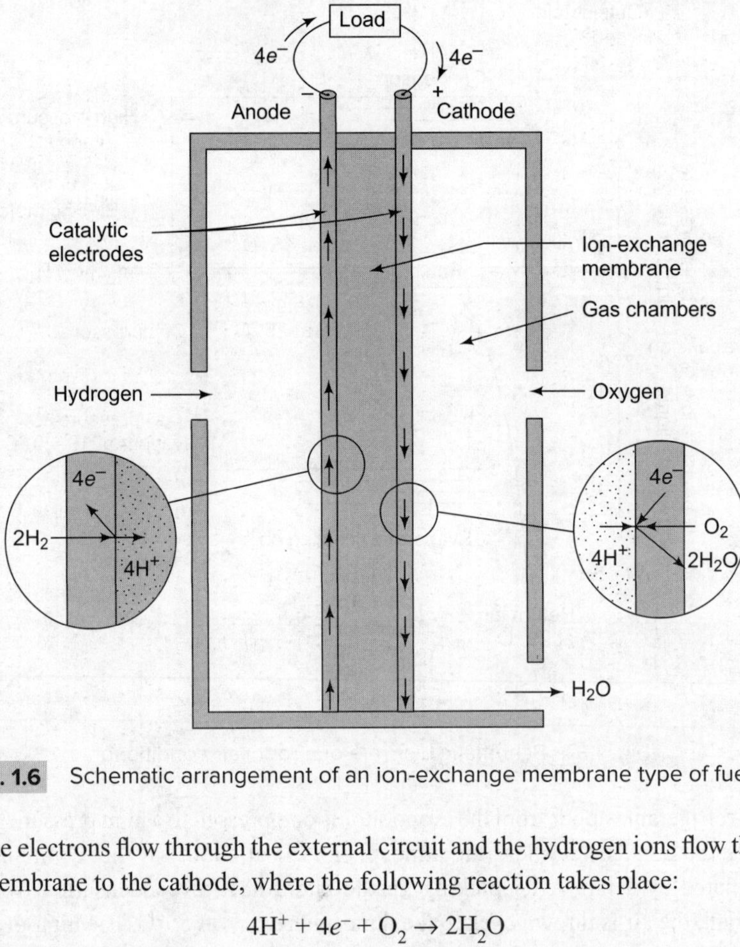

Fig. 1.6 Schematic arrangement of an ion-exchange membrane type of fuel cell

The electrons flow through the external circuit and the hydrogen ions flow through the membrane to the cathode, where the following reaction takes place:

$$4H^+ + 4e^- + O_2 \rightarrow 2H_2O$$

There is a potential difference between the anode and the cathode, and thus there is a flow of electricity.

At present, the fuel used in fuel cells is either hydrogen or a mixture of gaseous hydrocarbons and hydrogen. The oxidiser is usually oxygen. However, current development is directed towards the production of fuel cells that use hydrocarbon fuels and air. The fuel cell is already being used to produce power for space and other special applications.

Thermodynamics plays a vital role in the analysis, development and design of all power-producing systems, including I.C. engines and gas turbines as well as power-absorbing systems like refrigerators and air conditioners. Considerations such as the increase of efficiency, improved design, optimum operating conditions, environmental pollution and alternative methods of power generation involve, among other factors, the careful application of the fundamentals of thermodynamics.

1.6 MACROSCOPIC VERSUS MICROSCOPIC VIEWPOINT

> **LO 1.3**
> Discuss two views of thermodynamic analysis

There are two points of view from which the behaviour of matter can be studied: the macroscopic and the microscopic. In the macroscopic approach, a certain quantity of matter is considered, without the events occurring at the molecular level being taken into account. From the microscopic point of view, matter is composed of myriads of molecules. If it is a gas, each molecule at a given instant has a certain position, velocity and energy, and for each molecule these change very frequently as a result of collisions. The behaviour of the gas is described by summing up the behaviour of each molecule. Such a study is made in *microscopic* or *statistical thermodynamics*. *Macroscopic thermodynamics* is only concerned with the effects of the action of many molecules, and these effects can be perceived by human senses. For example, the macroscopic quantity, pressure, is the average rate of change of momentum due to which all the molecular collisions made on a unit area. The effects of pressure can be felt. The macroscopic point of view is not concerned with the action of individual molecules, and the force on a given unit area can be measured by using, e.g. a pressure gauge. These macroscopic observations are completely independent of the assumptions regarding the nature of matter. All the results of classical or macroscopic thermodynamics can, however, be derived from the microscopic and statistical study of matter.

1.7 THERMODYNAMIC SYSTEM AND CONTROL VOLUME

> **LO 1.4**
> State thermodynamic systems

A thermodynamic *system* is defined as a quantity of matter or a region in space upon which attention is concentrated in the analysis of a problem. Everything external to the system is called the *surroundings* or the *environment*. The system is separated from the surroundings by the system boundary (Fig. 1.7). The boundary may be either *fixed* or *moving*. A system and its surroundings together comprise a *universe*.

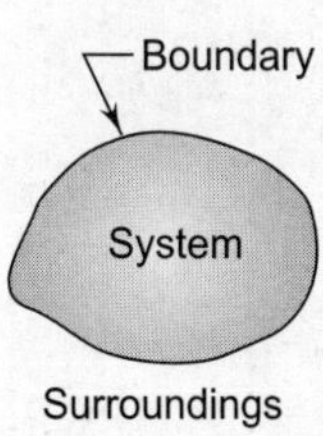

Fig. 1.7 A thermodynamic system

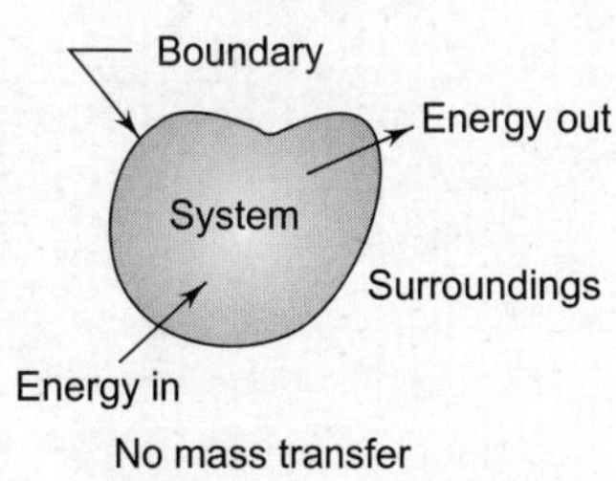

Fig. 1.8 A closed system

There are three classes of systems: (a) closed system, (b) open system and (c) isolated system. The *closed system* (Fig. 1.8) is a system of fixed mass. There is no mass transfer across the system boundary. There may be energy transfer into or out of the system. A certain quantity of fluid in a cylinder bounded by a piston constitutes a closed system. The *open system* (Fig. 1.9) is one in which matter crosses the boundary of the system. There may be energy transfer also. Most of the engineering devices are generally open systems, e.g. an air compressor in which air enters at low pressure and leaves at high pressure and there are energy transfers across the system boundary. The *isolated system* (Fig. 1.10) is one in which there is no interaction between the system and the surroundings. It is of fixed mass and energy, and there is no mass or energy transfer across the system boundary.

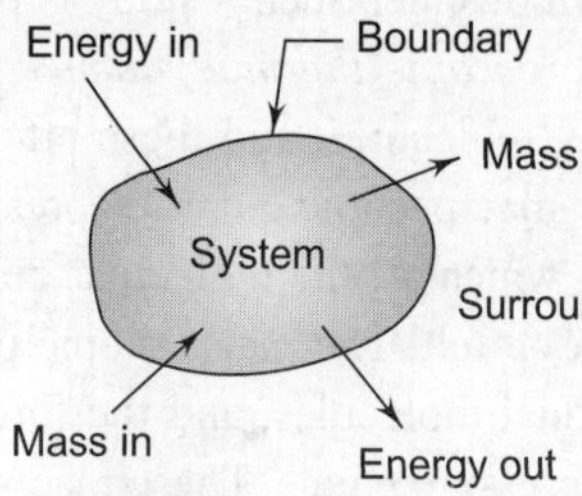

Fig. 1.9 An open system

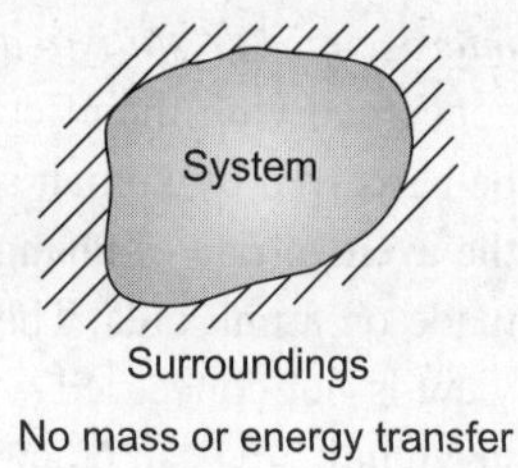

Fig. 1.10 An isolated system

If a system is defined as a certain quantity of matter, then the system contains the same matter and there can be no transfer of mass across its boundary. However, if a system is defined as a region of space within a prescribed boundary, then matter can cross the system boundary. While the former is called a closed system, the latter is an open system.

For thermodynamic analysis of an open system, such as an air compressor (Fig. 1.11), attention is focused on a certain volume in space surrounding the compressor, known as the *control volume*, bounded by a surface called the *control surface*. Matter as well as energy crosses the control surface.

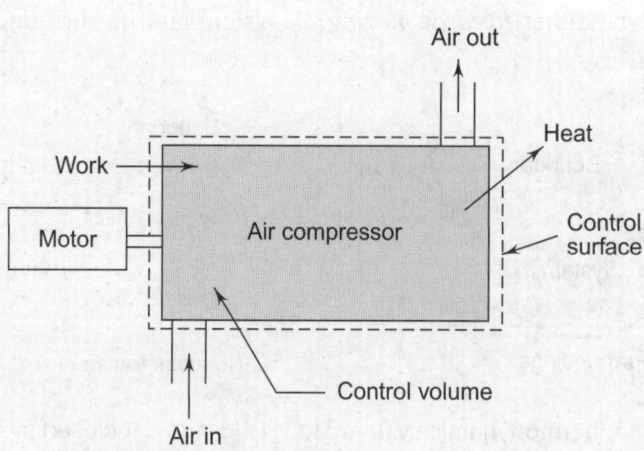

Fig. 1.11 Control volume and control surface

A closed system is a system closed to matter flow, though its volume can change against a flexible boundary. When there is matter flow, then the system is considered to be a volume of fixed identity, the control volume. There is thus no difference between an open system and a control volume.

1.8 THERMODYNAMIC PROPERTIES, PROCESSES AND CYCLES

LO 1.5 Discuss thermodynamic state, property, process and cycle

Every system has certain characteristics by which its physical conditions may be described, e.g. volume, temperature, pressure and so on. Such characteristics are called *properties* of the system. These are all *macroscopic* in nature. When all the properties of a system have definite values, the system is said to exist at a definite *state*. Properties are the coordinates to describe the state of a system. They are the state variables of the system. Any operation in which one or more of the properties of a system change is called a *change of state*. The succession of states passed through during a change of state is called the *path* of the change of state. When the path is completely specified, the change of state is called a *process*, e.g. a constant pressure process. A *thermodynamic cycle* is defined as a series of state changes such that the final state is identical with the initial state (Fig. 1.12).

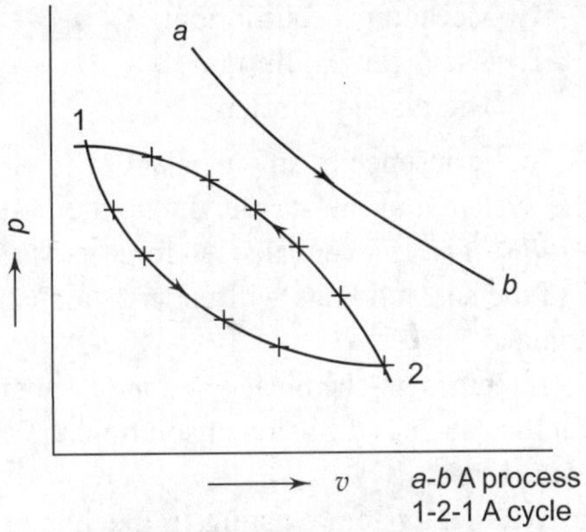

Fig. 1.12 A process and a cycle

Properties may be of two types. *Intensive properties* are independent of the mass in the system, e.g. pressure, temperature, and so on. *Extensive properties* are related to mass, e.g. volume, energy and so on. If mass is increased, the values of the extensive properties also increase. Specific extensive properties, i.e. extensive properties per unit mass, are intensive properties, e.g. specific volume, specific energy and so on.

1.9 HOMOGENEOUS AND HETEROGENEOUS SYSTEMS

LO 1.6 Compare homogeneous and heterogeneous systems

A quantity of matter homogeneous throughout in chemical composition and physical structure is called a *phase*. Every substance can exist in any one of the three phases, viz. solid, liquid and gas. A system consisting of a single phase is called a *homogeneous system*, while a system consisting of more than one phase is known as a *heterogeneous system*.

1.10 THERMODYNAMIC EQUILIBRIUM

LO 1.7
Appraise thermodynamic equilibrium

A system is said to exist in a state of *thermodynamic equilibrium* when no change in any macroscopic property is registered, if the system is isolated from its surroundings.

An isolated system always reaches in course of time a state of thermodynamic equilibrium and *can never depart from it spontaneously*.

Therefore, there can be no *spontaneous change in any macroscopic property* if the system exists in an equilibrium state. Thermodynamics studies mainly the properties of physical systems that are found in equilibrium states.

A system will be in a state of thermodynamic equilibrium, if the conditions for the following three types of equilibrium are satisfied:

1. Mechanical equilibrium
2. Chemical equilibrium
3. Thermal equilibrium

In the absence of any unbalanced force within the system itself and also between the system and the surroundings, the system is said to be in a state of *mechanical equilibrium*. If an unbalanced force exists, either the system alone or both the system and the surroundings will undergo a change of state till mechanical equilibrium is attained.

If there is no chemical reaction or transfer of matter from one part of the system to another, such as diffusion or solution, the system is said to exist in a state of *chemical equilibrium*.

When a system existing in mechanical and chemical equilibrium is separated from its surroundings by a diathermic wall (diathermic means 'which allows heat to flow') and if there is no spontaneous change in any property of the system, the system is said to exist in a state of *thermal equilibrium*. When this is not satisfied, the system will undergo a change of state till thermal equilibrium is restored.

When the conditions for any one of the three types of equilibrium are not satisfied, a system is said to be in a *non-equilibrium state*. If the non-equilibrium state is due to an unbalanced force in the interior of a system or between the system and the surroundings, the pressure varies from one part of the system to another. There is no single pressure that refers to the system as a whole. Similarly, if the non-equilibrium is because of the temperature of the system being different from that of its surroundings, there is a non-uniform temperature distribution set up within the system and there is no single temperature that stands for the system as a whole. It can thus be inferred that when the conditions for thermodynamic equilibrium are not satisfied, the states passed through by a system cannot be described by thermodynamic properties which represent the system as a whole.

Thermodynamic properties are the macroscopic coordinates defined for, and significant to, only thermodynamic equilibrium states. Both classical and statistical thermodynamics study mainly the equilibrium states of a system.

1.11 QUASI-STATIC PROCESS

LO 1.8
Analyze quasi-static or quasi-equilibrium process

Let us consider a system of gas contained in a cylinder (Fig. 1.13). The system initially is in an equilibrium state, represented by the properties p_1, v_1 and t_1. The weight on the piston just balances the upward force exerted by the gas. If the weight is removed, there will be an unbalanced force between the system and the surroundings, and under gas pressure, the piston will move up till it hits the stops. The system again comes to an equilibrium state, being described by the properties p_2, v_2 and t_2. But the intermediate states passed through by the system are non-equilibrium states which cannot be described by thermodynamic coordinates.

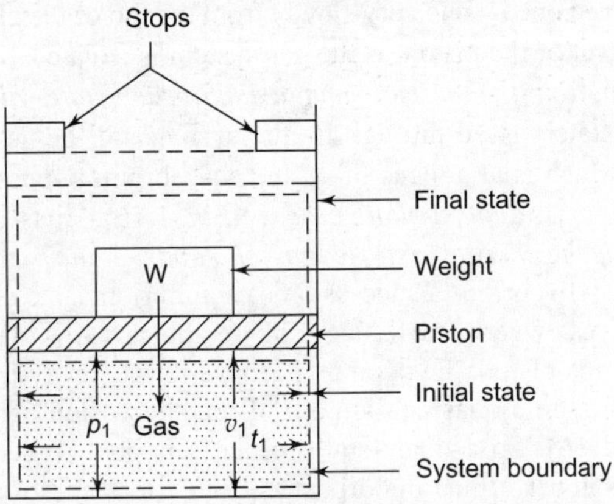

Fig. 1.13 Transition between two equilibrium states by an unbalanced force

Figure 1.14 shows points 1 and 2 as the initial and final equilibrium states joined by a dotted line, which has got no meaning otherwise. Now if the single weight on the piston is made up of many very small pieces of weights (Fig. 1.15), and these

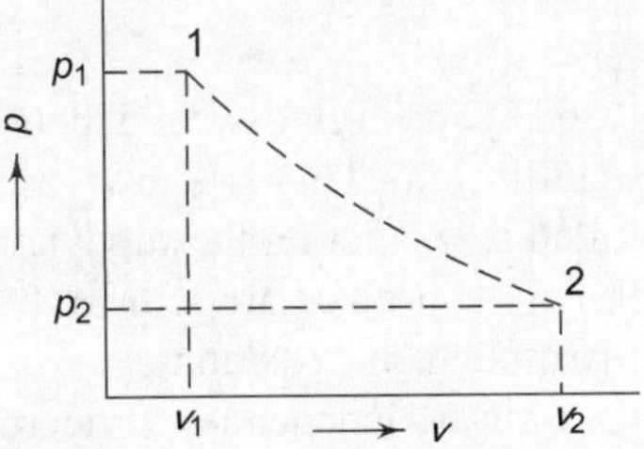

Fig. 1.14 Plot representing the transition between two equilibrium states

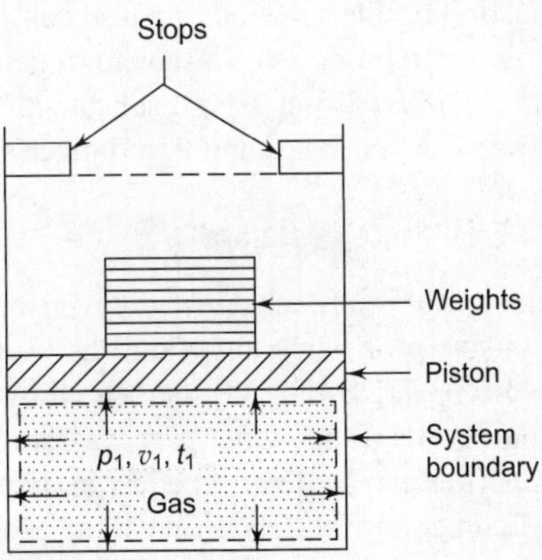

Fig. 1.15 Infinitely slow transition of a system by infinitesi-mal force

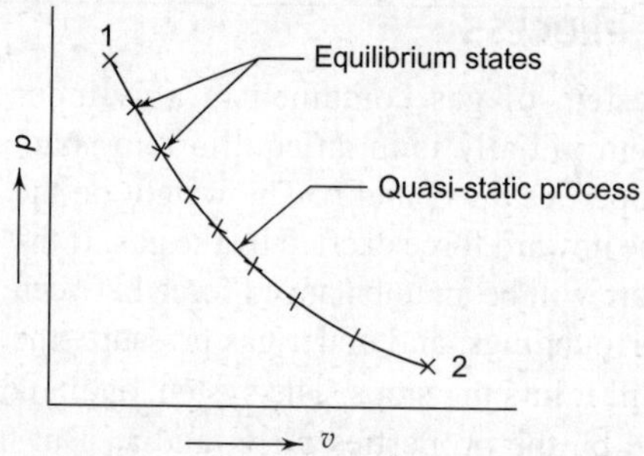

Fig. 1.16 A quasi-static process

weights are removed one by one very slowly from the top of the piston, at any instant of the upward travel of the piston, if the gas system is isolated, *the departure of the state of the system from the thermodynamic equilibrium state will be infinitesimally small*. So every state passed through by the system will be an equilibrium state. Such a process, which is at a locus of all the equilibrium points passed through by the system, is known as a *quasi-static process* (Fig. 1.16), 'quasi' meaning 'almost'. *Infinite slowness is the characteristic feature of a quasi-static process*. A quasi-static process 1–2 (Fig. 1.16) is thus a succession of equilibrium states.

If the same slices or very small pieces of weights are now placed slowly one by one on the top of the piston, the piston will move down slowly from state 2 to state 1 by following the same quasi-static path of the succession of equilibrium states (Figs. 1.15 and 1.16). So a quasi-static process is also a *reversible process*. A reversible process is performed in both ways from 1 to 2 and from 2 to 1 involving *infinite time* in executing each of the processes.

1.12 PURE SUBSTANCE

A pure substance is defined as one that is homogeneous and invariable in chemical composition throughout its mass. The relative proportions of the chemical elements constituting the substance are also constant. Atmospheric air, steam-water mixture and combustion products of a fuel are regarded as pure substances. But the mixture of air and liquid air is not a pure substance, since the relative proportions of oxygen and nitrogen differ in gas and liquid phases in equilibrium.

1.12.1 State Postulate

The state of a pure substance of given mass can be fixed by specifying two independent intensive properties, provided the system is in equilibrium. This is known as the '*two-property rule*'. The state can thus be represented as a point on thermodynamic property diagrams. Once any two properties of a pure substance are known, other properties can be determined from the available thermodynamic relations.

Temperature and specific volume, for example, are always independent properties, and together they can fix the state of a simple compressible system. Temperature and pressure, however, are independent for single-phase systems, but are dependent for two-phase systems.

1.13 CONCEPT OF CONTINUUM

LO 1.9

Describe the concept of continuum

From the macroscopic viewpoint, we are always concerned with volumes which are very large compared to molecular dimensions. Even a very small volume of a system is assumed to contain a large number of molecules so that statistical averaging is meaningful and a property value can be assigned to it. Disregarding the behaviour of individual molecules, matter is here treated as continuous. Let us consider the mass δm in a volume δV surrounding the point P (Fig. 1.17). The ratio $\delta m/\delta V$ is the average mass density of the system within the volume δV. We suppose that at first δV is rather large, and is subsequently shrunk about the point P. If we plot $\delta m/\delta V$ against δV, the average density tends to approach an asymptote as δV increases (Fig. 1.18). However, when δV becomes so small as to contain relatively few molecules, the average density fluctuates substantially with time as molecules pass into and out of the volume in random motion, and so it is impossible to speak of a definite value of $\delta m/\delta V$. The smallest volume which may be regarded as continuous is $\delta V'$. The density ρ of the system at a point is thus defined as

$$\rho = \lim_{\delta V \to \delta V'} \frac{\delta m}{\delta V} \tag{1.1}$$

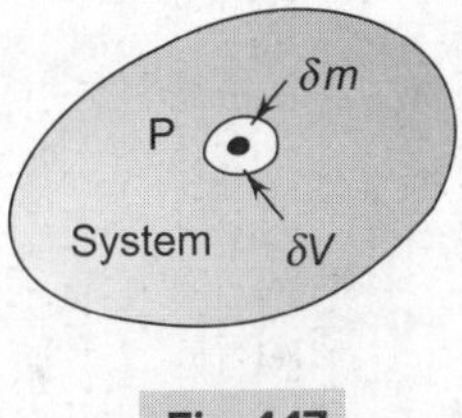

Fig. 1.17

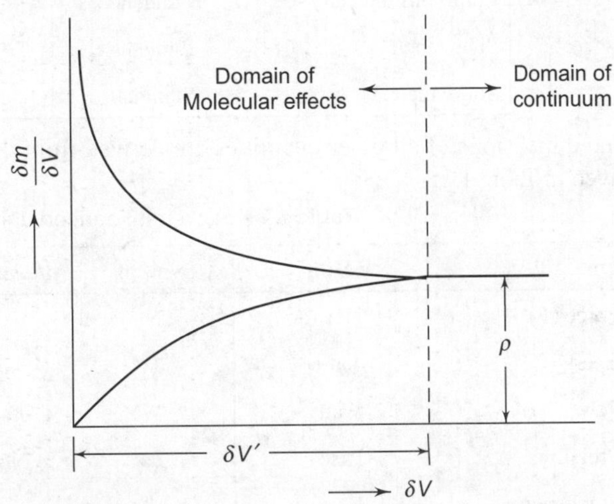

Fig. 1.18 Definition of the macroscopic property, density

Similarly, the fluid velocity at a point P is defined as the instantaneous velocity of the centre of gravity of the smallest continuous volume $\delta V'$.

The concept of continuum loses validity when the mean free path of the molecules approaches the order of magnitude of the dimensions of the vessel, as, for instance, in highly rarefied gases encountered in high vacuum technology, in rocket flights at high altitudes and in electron tubes. In most engineering applications, however, the assumption of a continuum is valid and convenient, and goes hand in hand with the macroscopic point of view.

1.14 UNITS AND DIMENSIONS

In the present text, the SI (System International) system of units has been used. The basic units in this system are given in Table 1.1.

LO 1.10 Discuss SI units and dimensions of basic thermodynamic quantities

Table 1.1 System: basic units

Quantity	Unit	Symbol
Length (*L*)	Metre	m
Mass (*M*)	Kilogram	kg
Time (*t*)	Second	s
Amount of substance	Mole	mol
Temperature (*T*)	Kelvin	K
Electric current	Ampere	A
Luminous intensity	Candela	cd
Plane angle	Radian	rad
Solid angle	Steradian	sr

The dimensions of all other quantities are derived from these basic units which are given in Table 1.2.

Table 1.2 SI system: derived units

Quantity	Unit	Symbol	Alternative unit	In basic units
Force (*F*)	Newton	N		kg m/s^2
Energy *E*)	Joule	J	Nm	kg m^2/s^2
Power	Watt	W	J/s	kg m^2/s^3
Pressure	Pascal	Pa	N/m^2	kg/(ms^2)
Frequency	Hertz	Hz		s^{-1}
Electric charge	Coulomb	C		As
Electric potential	Volt	V	W/A = J/C	kg m^2/(s^3 A)
Capacitance	Farad	F	C/V	s^4 A^2/(kg m^2)
Electrical resistance	Ohm	Ω	V/A	kg m^2/(s^3 A^2)
Magnetic flux	Weber	Wb	Vs	kg m^2/(s^2 A)
Magnetic flux density	Tesla	T	Wb/m^2	kg/(s^2 A)
Inductance	Henry	H	Wb/A	kg m^2/(s^2 A^2)

It is often convenient and desirable to use multiples of various units, the standard list of which is given in Table 1.3.

Table 1.3 SI system: standard multipliers

Factor	Prefix	Factor	Prefix
10^{12}	tera, T	10^{-3}	milli, m
10^{9}	giga, G	10^{-6}	micro, μ
10^{6}	mega, M	10^{-9}	nano, n
10^{3}	kilo, k	10^{-12}	pico, p

1.14.1 Force

The force acting on a body is defined by Newton's second law of motion. The unit of force is the newton (N). A force of one newton produces an acceleration of 1 ms^{-2} when applied to a mass of 1 kg.

$$1 \text{ N} = 1 \text{ kg m/s}^2$$

The weight of a body (W) is the force with which the body is attracted to the centre of the earth. It is the product of its mass (m) and the local gravitational acceleration (g), i.e.

$$W = mg$$

The value of g at sea level is 9.80665 m/s^2. The mass of a substance remains constant with elevation, but its weight varies with elevation.

1.14.2 Pressure

Pressure is the normal force exerted by a system against unit area of the bounding surface. If δA is a small area and $\delta A'$ is the smallest area from continuum consideration, and δF_n is the component of force normal to δA (Fig. 1.19), the pressure p at a point on the wall is defined as

LO 1.11

Discuss pressure and various measuring devices of it.

$$p = \lim_{\delta A \to \delta A'} \frac{\delta F_n}{\delta A}$$

The pressure p at a point in a fluid in equilibrium is the same in all directions.

The unit for pressure in the SI system is pascal (Pa), which is the force of one newton acting on an area of 1 m^2.

$$1 \text{ Pa} = 1 \text{ N/m}^2$$

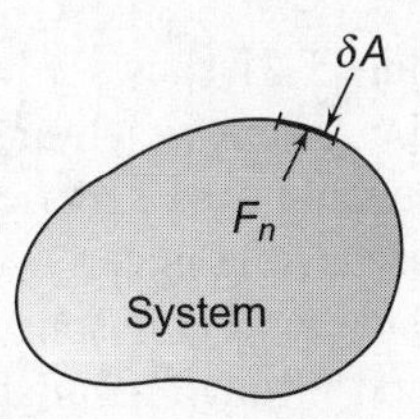

Fig. 1.19 Definition of pressure

The unit of pascal is very small. Very often kilo-pascal (kPa) or mega-pascal (MPa) is used.

Two other units, not within the SI system of units, continue to be widely used. These are the *bar*, where

$$1 \text{ bar} = 10^5 \text{ Pa} = 100 \text{ kPa} = 0.1 \text{ MPa}$$

and the standard atmosphere, where

$$1 \text{ atm} = 101.325 \text{ kPa} = 1.01325 \text{ bar}$$

One bar is thus close to 1 atm.

Most instruments indicate pressure relative to the atmospheric pressure, whereas the pressure of a system is its pressure above zero, or relative to a perfect vacuum (Fig. 1.20). The pressure relative to the atmosphere is called *gauge pressure*. The pressure relative to a perfect vacuum is called *absolute pressure*.

$$\text{Absolute pressure} = \text{Gauge pressure} + \text{Atmospheric pressure}$$

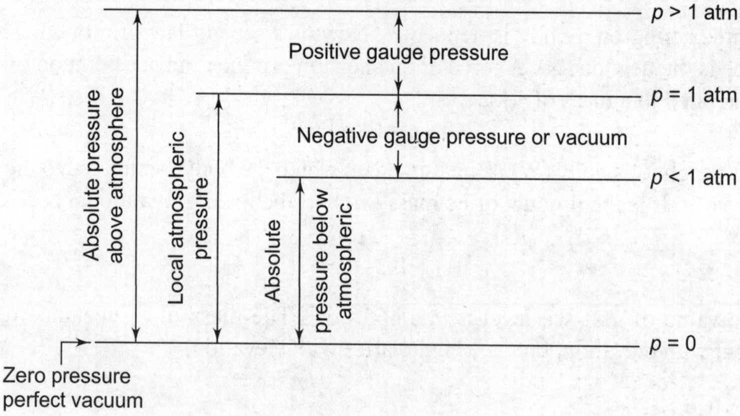

Fig. 1.20 Relationship between pressures

When the pressure in a system is less than the atmospheric pressure, the gauge pressure becomes negative, but is frequently designated by a positive number and is called *vacuum*. For example, 16 cm Hg vacuum will be

$$\frac{76-16}{76} \times 1.013 = 0.80 \text{ atm}$$

Thus,

$$p_{\text{abs}} = p_{\text{atm}} - p_{\text{vac}}$$

Figure 1.21 shows a few pressure measuring devices. Figure 1.21(a) shows the Bourdon gauge which measures the difference between the system pressure inside the tube and atmospheric pressure. It relies on the deformation of a bent hollow tube of suitable material which, when subjected to the pressure to be measured on the inside (and atmospheric pressure on the outside), tends to unbend. This moves a pointer through a suitable gear-and lever mechanism against a calibrated scale. Figure 1.21(b) shows an open U-tube indicating gauge pressure, and Fig. 1.21(c) shows an open U-tube indicating vacuum. Figure 1.21(d) shows a closed U-tube indicating absolute pressure. If p is atmospheric pressure, this is a *barometer*. These are called U-tube manometers.

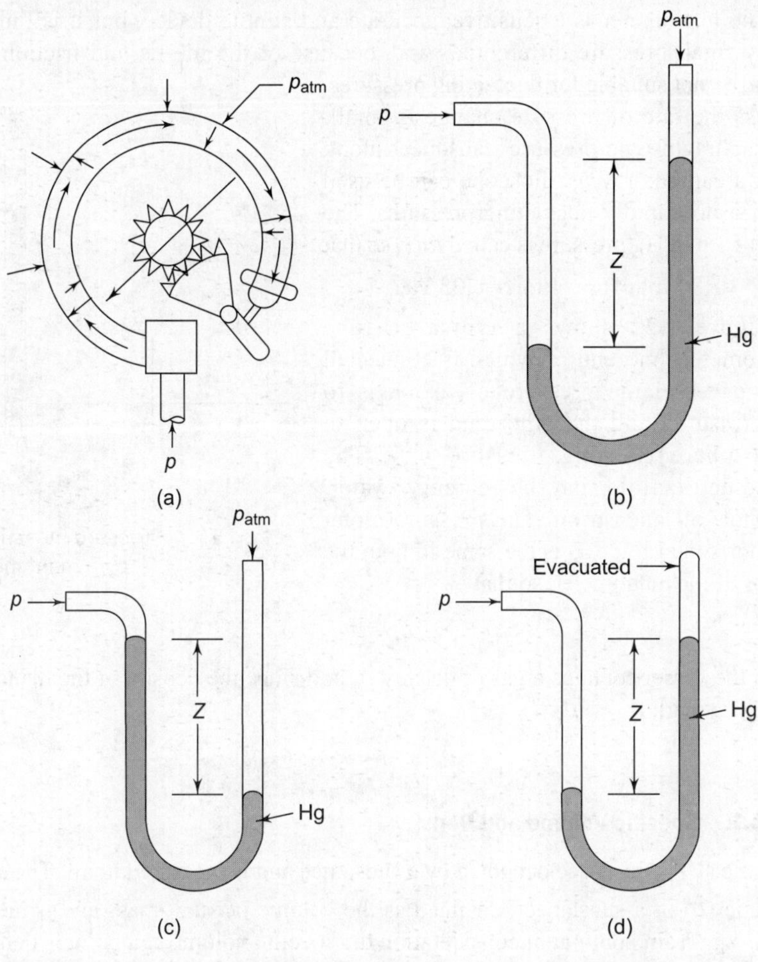

Fig. 1.21 Pressure gauges: (a) Bourdon gauge, (b) Open U-tube indicating gauge pressure, (c) Open U-tube indicating vacuum, (d) Closed U-tube indicating absolute pressure

If Z is the difference in the heights of the fluid columns in the two limbs of the U-tube [Figs. 1.21(b) and (c)], ρ is the density of the fluid and g is the acceleration due to gravity, then from the elementary principle of hydrostatics, the gauge pressure p_g is given by

$$p_g = Z\rho g \left[\text{m} \cdot \frac{\text{kg}}{\text{m}^3} \frac{\text{m}}{\text{s}^2} \right] = Z\rho g \text{ N/m}^2$$

If the fluid is mercury having $\rho = 13{,}616$ kg/m^3, one metre head of mercury column is equivalent to a pressure of 1.3366 bar, as shown below.

$$1 \text{ m Hg} = Z\rho g = 1 \times 13616 \times 9.81$$

$$= 1.3366 \times 10^5 \text{ N/m}^2 = 1.3366 \text{ bar}$$

The manometer is a sensitive, accurate and simple device, but it is limited to fairly small pressure differentials and, because of the inertia and friction of the liquid, is not suitable for fluctuating pressures, unless the rate of pressure change is small. A diaphragm-type pressure transducer along with a cathode ray oscilloscope can be used to measure rapidly fluctuating pressures. The unit of 1 mm Hg pressure is called *torr*, so that

$$1 \text{ mm Hg} = 1 \text{ torr} = 133 \text{ Pa}$$

Figure 1.22 shows a typical U-tube manometer, one end of which is connected to the vessel the pressure of which is to be measured and the other end is open to atmosphere exerting pressure p_{atm}. The manometric fluid may be mercury, water, alcohol, oil and so on. The pressure along the horizontal line AB is the same in both the limbs of the manometer, so that

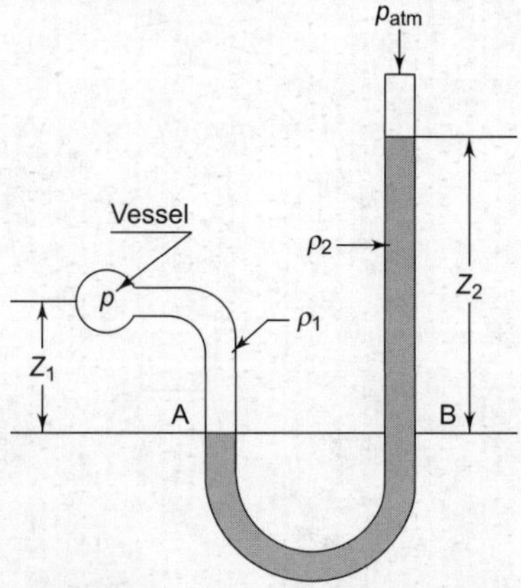

Fig. 1.22 Pressure measurement by a manometer

$$p + \rho_1 g Z_1 = p_{atm} + \rho_2 g Z_2$$

If the vessel contains a gas of density ρ_1 and since the density of the manometric fluid (say Hg), $\rho_2 >> \rho_1$,

$$p - p_{atm} = \rho_2 g Z_2 = p_{gauge}$$

1.14.3 Specific Volume and Density

Volume (V) is the space occupied by a substance and is measured in m^3. The specific volume (v) of a substance is defined as the volume per unit mass and is measured in m^3/kg. From continuum consideration the specific volume at a point is defined as

$$v = \lim_{\delta V \to \delta V'} \frac{\delta V}{\delta m}$$

where dV' is the smallest volume for which the system can be considered a continuum.

Density (ρ) is the mass per unit volume of a substance, which has been discussed earlier, and is given in kg/m^3.

$$\rho = \frac{m}{v}$$

In addition to m^3, another commonly used unit of volume is the litre (l).

$$1\ l = 10^{-3} \text{ m}^3$$

The specific volume or density may be given either on the basis of mass or in respect of mole. A *mole* of a substance has a mass numerically equally to the molecular weight of the substance. One gram mol of oxygen has a mass of 32 g and 1 kg mol (or kmol) of nitrogen has a mass of 28 kg. The symbol $\bar{v}$ is used for molar specific volume (m^3/kmol).

1.14.4 Energy

Energy is the capacity to exert a force through a distance, and manifests itself in various forms. Engineering processes involve the conversion of energy from one form to another, the transfer of energy from place to place and the storage of energy in various forms, utilising a working substance.

The unit of energy in the SI system is Nm or J (joule). The energy per unit mass is the specific energy, the unit of which is J/kg.

1.14.5 Power

The rate of energy transfer or storage is called *power*. The unit of power is watt (W), kilowatt (kW) or megawatt (MW).

$$1 \text{ W} = 1 \text{ J/s} = 1 \text{ Nm/s}$$
$$1 \text{ kW} = 1000 \text{ W}$$

1.15 HISTORY OF ENGINEERING THERMODYNAMICS

> **LO 1.12**
> Describe history of engineering thermodynamics

The latter half of the eighteenth century ushered man into the modern world of machinery and manufacture, and brought about cataclysmic change in the social, economic and political life of people. The historians have called it the *Industrial Revolution*. It began in England due to a fortuitous combination of many factors. There was bustling creative activity in science and technology during this period in England, with the appearance of a galaxy of same brilliant individuals. The invention of the steam engine gave an impetus to this activity, and for the first time made man free from the forces of nature. The name of Savery, Newcomen and notably James Watt are associated with this invention. Watt brought about considerable improvement in the performance of the steam engine, which began to be widely used in coal mines, iron metallurgy and textile mills. George Stephenson introduced steam engine for rail transport, and Robert Fulton used it in steam boats. A variety of industries grew up, and man gradually entered into the modern machine age. The advent of steam engine also gave stimulus to the birth of thermodynamics. Thermodynamics is said to be the 'daughter' of the steam engine (D. Bradley, *Thermodynamics–the Daughter of Steam*, Engineering Heritage, Vol.2, I.Mech.E., London, 1966).

There was once a young inventor who thought that he could produce energy out of nothing. 'It is well known', said he, 'that an electrical motor converts electrical energy into mechanical energy and that an electric generator converts mechanical energy into electrical energy. Why not then, use the motor to run the generator and the generator to run the motor, and create thereby an endless supply of energy'? But this is never going to happen. A hypothetical device which creates energy from nothing is called a perpetual motion machine of the first kind, a PMMI. Like the proverbial touchstone which changes all metals into gold, man attempted to find such a PMMI for several years, but it turned out to be a wild goose chase. In fact, the development of the principle of conservation of energy has been one of the most significant achievements in the evolution of physical science. The first recognition of this principle was made by Leibnitz in 1693, when he referred to the sum of kinetic energy and potential energy in a gravitational force field. Energy is neither created nor destroyed. Energy manifests in various forms and transforms the chemical energy of the food and the oxygen into heat, sound and useful work. Work was always considered a form of energy. The concept of heat was, however, a very actively debated scientific topic. Until the middle of the nineteenth century, heat was regarded as an invisible colourless, weightless, odourless fluid that flowed from a body of higher calorie to a body of lower calorie. This was known as the caloric theory of heat, first proposed in 1789 by Antoine Lavoisier (1743–1794), the father of modern chemistry. When an object becomes full of caloric, it was then said to be saturated with it. This was the origin of the terms 'saturated liquid', 'saturated vapour' and so on, which we use in thermodynamics today. The caloric was said to be conserved and it was indestructible. The caloric theory was, however, refuted and heat was confirmed as a form of energy in the middle of the nineteenth century leading to the formulation of the first law of thermodynamics. The names which stand out in the establishment of the first law were Benjamin Thompson (1753–1814), James Prescott Joule (1818–1889) and Julius Robert Mayer (1814–1878).

Benjamin Thompson, an American born in Massachusetts, did not support the revolt against the British during the American war of independence, and in 1775, he left for England where he took up a government service. On a trip to Germany, he met the prince of Bavaria who offered him a job. He introduced many reforms in the government for which the title of Count Von Rumford was conferred on him. While boring brass cannon hole, Count Rumford noticed that there was a continuous heat release. How could the caloric fluid be conserved, when it was being produced continuously by a mechanical friction? From the established principle of conservation of mass, a true fluid can be neither created nor destroyed, so heat could not be a fluid if it could be continuously created in an object by the mechanical friction. Rumford conceived that heat was 'a kind of motion' and the hotness of an object was due to the vibrating motion of the particles in the objects. On his return to England, he become a member of the Royal Society, and later founded the Royal Institution for the Advancement of Science. Rumford married the widow of Lavoisier and lived in Paris for the rest of his life.

In the early forties of the nineteenth century, James P Joule and Julius R Mayer almost simultaneously set forth the idea that heat transfer and mechanical work were simply different forms of the same quantity, which we now recognise as energy in transit. In some modern treatments of engineering thermodynamics, Joule's name alone is attached to the establishment of the equivalence of 'heat' and 'work'. The published record, however, shows that the idea of convertibility of heat into work was published independently by Mayer in May, 1842 and Joule in August, 1843. For an important aspect in the history of the first law, is the fact that both Mayer and Joule had difficulty in getting their papers published and in being taken seriously by their established contemporaries.

Robert Mayer was a doctor in a ship in the East Indies and from physiological observations, he believed in a principle of conservation of energy. He derived theoretically, the mechanical heat equivalent based on the calorimetric data of Joseph Black of Glasgow University. Mayer tried to publish his paper but remained unsuccessful for a long time. His despair was so great that he attempted suicide by jumping from a window, breaking his two legs. He was placed in an asylum for some time. In later years, however, he was given some measure or recognition and honoured equally with Joule in establishing the mechanical theory of heat.

Mayer argued that an amount of gas needs to be heated more at constant pressure than at constant volume, because at constant pressure it is free to dilate and do work against the atmosphere, which in today's notations becomes

$$mc_p \Delta T - mc_v \Delta T = p_{atm} \Delta V \tag{1.2}$$

Using the c_p and c_p/c_v constants that were known in his time, he estimated the left-hand side of the equation in calories, while the right-hand side was known in mechanical units. He thus established numerically the equivalence between these units. If the relation

$$pv = RT \tag{1.3}$$

is used in Eq. (1.2), Mayer's argument reduces to

$$c_p - c_v = R \tag{1.4}$$

This classic relationship between the specific heats of an ideal gas in called Mayer's equation: while the ideal gas equation of state, Eq. (1.3), was first derived by Chaperon (Bejan, 1988)*.

Joule was the ultimate experimentalist. His experiments seem to be the direct continuation of those of Rumford and the gap of some 40 years between the two investigations appeared puzzling to some authors. Joule's first discovery from his measurements was that the flow of current in a resistance is accompanied by the development of heat proportional to the resistance. He concluded that caloric was an indeed created by the flow of current. He was firmly convinced that there existed some conservation law of a general nature and hence set out to investigate whether the conversion of the various forms of energy is governed by definite conversion factors. He considered the conversion of chemical, electric, caloric and mechanical energy forms in all combinations. The determination of the mechanical equivalent of heat forms the central part of his experiments, the result of which can be summed up in the general relation:

$$W = JQ \tag{1.5}$$

where J is the mechanical equivalent of heat. Joule's experiments suggested that this relation may have a universal validity with the same numerical value of J under all conditions.

Joule communicated the results of his experiments to the British Association for the Advancement of Science in 1843. It was received with entire incredulity and general silence. In 1844, a paper by Joule on the same subject was rejected by the Royal Society. To convince the skeptics, he produced a series of nakedly simple experiments whose message proved impossible to refute. From the point of view of mechanical engineers, the most memorable among these experiments was the heating of a pool of water by an array of paddle wheels driven by falling weights. In 1847, he discussed his experimental results before the British Association at Oxford, in which he suggested that the water at the bottom of the Niagara waterfall (160 feet high) should be warmer than at the top (by 0.2 °F). From the thermal expansion of gases, Joule deduced that there should be a

* Adrian Bejan, 'Research into the Origins of Engineering Thermodynamics', ***Int. Comm. Heat Mass Transfer***, Vol. 15, No. 5, 1988, pp. 571–580.

'zero of temperature', 480 °F below the freezing point of ice. This was the first suggestion of absolute zero. Although these results failed to provoke further discussion, it created interest in a young man who only two year ago had passed from the University of Cambridge with the highest honour. The young man was William Thomson, who later became Lord Kelvin. He somewhere stated that it was one of the most valuable recollections of his life, Michael Faraday was also present in the 1844 Oxford meeting, and he communicated Joule's paper, 'On the Mechanical Equivalent of Heat' to the Royal Society in 1849.

The paper ultimately appeared in its Philosphical Transactions in 1850. Even while Joule was perfecting the experimental basis of the energy law now called the Mayer–Joule principle, Herman Ludwig von Helmohltz (1821–1894) published in 1847, his famous essay on the conservation of force. In this work, he advanced the conservation of energy as a unifying principle extending over all the branches of physics. Helmholtz, like Mayer, was a physician by profession and self-taught in Physics and Mathematics. He also faced great difficulties in getting his paper published in professional journals.

In the history of classical thermodynamics, one thinks of only the closed system formulations of the first law which were deliberated by the pioneers as stated above. In engineering thermodynamics, however, open system formulations are of prime interest. The first-low-for open systems was first stated by Gustave Zeuner, as a part of the analysis of flow systems that operate in the steady state. Zeuner's formula for the heat transfer rate to a stream in steady flow and without shaft work in present notations is given to be

$$\delta\dot{Q}\,/\dot{m} = d(u + pv + V^2/2 + gz) \tag{1.6}$$

The reference of this formula is found in Stodola's classic treatise on steam turbines, first published in the German language in 1903.

The first person to invent a theory simultaneously involving the ideas of conservation and conversion of energy was the young French military engineer Nicolas Leonard Sadi Carnot (1796–1832). The strikingly original ideas of Carnot's work make it among the most brilliant new departures in theoretical physics. Sadi Carnot was the son of Napoleon's general, Lazare Carnot, and was educated at the famous Ecole Polytechnique in Paris. Between 1794 and 1830, Ecole Polytechnique had such famous teachers as Lagrange, Fourier, Laplace, Ampere, Cauchy, Coriolis, Poisson, Gay-Lussac and Poiseuille. After his formal education, Carnot chose a career as an army officer. Britain was then a powerful military force, primarily as a result of the industrial revolution brought about by the steam engine. French technology was not developing as fast as Britain's. Carnot was convinced that France's inadequate utilisation of steam power had made it militarily inferior. He began to study the fundamentals of steam engine technology and in 1824; he published the results of his study in the form of a brochure, '*Reflection on the Motive Power of Heat and on Machines Fitted to Develop that Power*'. Carnot was trained in the basic principles of hydraulics, pumps and water wheels at the Ecole Polytechnique. During Carnot's time, caloric theory of heat was still persisting, and the water wheel as the major source of mechanical power was gradually getting replaced by the steam engine. Carnot conceived that the power of a steam engine was released, as the heat fluid or caloric fell from the high temperature of the boiler to the lower temperature of the condenser, in much the same way that water falls through a water wheel to produce mechanical shaft work output. Carnot stated, 'The motive power of a water wheel depends on its height and the quantity of liquid. The motive power of heat also depends on the quantity of caloric used and on the height of its fall, i.e. the difference of temperatures of the bodies between which the exchange of caloric is made.'

Till Carnot's time, thermodynamics was developed primarily on an empirical basis provided by chemistry. Carnot approached an engineering problem, the efficiency of heat engines, in terms of entirely new concepts with the steam engine serving as the stimulus. Carnot observed that the existence of temperature differences is a necessary condition for producing mechanical work by means of a heat engine. He simplified the problem to its bare essentials and stipulated that this system, consisting essentially of a working substance, should exchange heat with its surroundings only at two fixed temperatures. In order to conceptualise such a situation, he introduced the idea of heat reservoirs. Two important conclusions emerged from Carnot's work:

1. No one could build a water wheel that would produce a continuous work output unless water actually entered and exited the wheel. If water with a certain kinetic and potential energy entered the wheel, then the same amount of water with a lower energy must also exit the wheel. It is thus impossible to make a water wheel that converts all the energy of the inlet water into shaft work output. There must be an outflow of water from the wheel.
   If this idea is extended to a steam engine by replacing the water by heat fluid caloric, it can be concluded that when caloric at a certain energy level (temperature) enters a work producing heat engine, it must also exit the heat engine at a low energy level (temperature). Thus a continuously operating heat engine that converts all of its caloric (heat) input directly into work output is not possible. This is very close to the Kelvin–Planck statement of second law as it is known today.
2. The maximum efficiency of a water wheel was independent of the type of the liquid and dependent only on the inlet and outlet flow energies. The maximum efficiency of the steam engine (or any heat engine) depends only on the temperatures of the high and low temperature thermal reservoirs of the engine and is independent of the working fluid. To achieve the maximum efficiency there must not be any mechanical friction or other losses of any kind.

At the age of 36, Sadi Carnot succumbed to an attack of cholera and scarlet fever. The significance of 'Carnot's work was not recognised until 1850, when Rudolf Clausius (1822–1888) and William Thomson (1824–1907) worked out a clear formulation of the conservation of energy principle. Carnot's second conclusion was to develop the concept of an absolute temperature scale. Thermodynamics is thus said to have originated from the 'clumsy puffing of the early steam engines' and is often called 'the daughter of steam engine'.

Carnot's ideas were so revolutionary that they were largely ignored. Soon after Carnot's death, Emile Clapeyron (1799–1864), a French mining engineer, strengthened Carnot's ideas by using more precise mathematical derivation. Clapeyron constructed its thermodynamic cycle by deducing that it must be composed of two reversible isothermal processes and two reversible adiabatic processes. It is now known as Carnot's cycle. It was the first heat engine cycle to be conceptualised. No other heat engine can equal its efficiency.

Clapeyron was later able to derive a relation for the enthalpy change of the liquid to vapour phase (h_{fg}) in terms of pressure, temperature and specific volume. This provided the first equation, now called the Clausius–Clapeyron equation, representing the first-order phase transition, which could be used to estimate a property that is not directly measurable in terms of properties that are directly measurable. Clapeyron's equation is now most easily derived from one of Maxwell's equations.

William Thomson (1824–1907), who become a professor of natural philosophy at the University of Glasgow in 1848 at the age of 24, rejected the caloric theory of heat and for the first time used the terms 'thermodynamics' and 'mechanical energy'. Apart from the deduction of the absolute temperature scale, Thomson worked with Joule from 1852 to 1862 in a series of experiments to measure the temperature of gas in a controlled expansion and propounded the Joule–Thomson effect for real gases.

Rudolf Julius Emanuel Clausius (1822–1888) realized that there were two distinct laws at work, the first law due to Joule and Mayer and the second law as expounded by Carnot. He defined the internal energy *U*. Although both Kelvin and Clausius used the function Q_{rev}/T for some years, Clausius recognised the value of this function and to describe it, he coined the word 'entropy' from the Greek work '*trope*' meaning 'transformation' and assigned it the symbol S. Clausius in 1865, summarised the first and second law of thermodynamics in the following words:

"Die Energie der Welt ist constant.
Die Entropie der Welt strebt einem Maximum zu"

This is translated as

"The energy of the world is constant.
The entropy of the world tends toward a maximum"

The world here means the universe, the system and the surroundings together. These statements made a strong impression upon a young student, Max Karl Ernst Ludwig Planck (1858–1947). He was educated at the universities of Munich and Berlin. In his autobiography he stated, 'One day I happened to come across the treatises of Rudolf Clausius, whose lucid style and enlightening clarity of reasoning made an enormous impression on me, and I become deeply absorbed in his articles, with an ever-increasing enthusiasm. I appreciated especially his exact formulation of the two laws of thermodynamics, and the sharp distinction, which he was the first to establish between them'. In 1897, Planck* demonstrated the close connection between the second law and the concept of reversibility. He stated the second law as the impossibility of a cyclic device which produces positive work and exchanges heat with a single reservoir. Similar statement was

* M. Planck, Treatise on Thermodynamics (1897), translated by A. Ogg, Longman and Green; London, 1927.

also made by Kelvin, and is now recognised as Kelvin–Planck statement of the second law. Poincare* in 1908 extended the work of Planck and prescribed a complete structure of classical thermodynamics.

The property, entropy, plays a stellar role in thermodynamics. It was introduced via the concept of heat engines. In 1909, the Greek mathematician Caratheodory proved the existence of entropy function mathematically without the aid of Carnot engines and refrigerators. Caratheodory's statement of second law may be stated as: 'In the neighbourhood of any arbitrary state P_0 of a physical system, there exist neighbouring states which are not an accessible form P_0 along quasi-static adiabatic paths'. From the standpoint of the engineer and physicist, it is entirely mathematical in form and devoid of physical insight.

William John Macquorn Rankine (1820–1872) defined the thermodynamic efficiency of a heat engine and showed the usefulness of p–v diagrams as related to work. He wrote the first textbook on thermodynamics**, and was the first to work out the thermodynamic cycle for the adiabatic cylinder steam engine, now known as Rankine cycle for a vapour power cycle.

In 1862, the cycle used in modern gasoline-powered I.C. engines was proposed in a patent issued to Alphonse Beau de Rockas (1815–1893). The first practical engine was, however, built by Nilkolous Auguest Otta (1832–1891) which was demonstrated at the Paris Exposition in 1878. Otto fought many legal battles with Beau de Rockas for the production of these engines, but finally lost to him.

Captain John Ericsson (1803–1889) was a Swedish engineer who marketed small solar-powered and coal-fired hot air engines. Rev. Robert Stirling (1790–1879), an English parish minister, patented a practical heat engine in 1816 that used air as the working fluid. In theory, the cycle used the Stirling engine approaches the ideal cycle later proposed by Carnot (1824).

Gerorge Bailey Brayton (1830–1892), an American engineer, marketed an I.C. engine with a separate combustion chamber, where combustion of fuel occurred at about constant pressure. This cycle later formed the basis for modern gas turbine plants.

Gottieb Daimler (1834–1900) obtained a patent in 1879 for a multi-cylinder automotive engine, which was commercially successful. Dr. Rudolf Charistian Karl Diesel (1858–1913) studied at Technische Hochschule in Munich. He designed large steam engines and boliers. He later developed in 1897 an I.C. engine with fuel injection which resembled the modern diesel engine. Flailing health, continuing criticism and serious financial setbacks beset Diesel who in 1913 disappeared from a boat crossing the English Channel on a moonlit night. Josiah Willard Gibbs (1839–1903) is often regarded as the most brilliant thermodynamicist in the USA. He received the first doctorate degree in engineering in Yale University, USA. He contributed significantly in many areas of thermodynamics like heterogeneous systems, phase rule, physical chemistry and statistical thermodynamics. Some of his very important papers were published in obscure journals like Connecticut Academy of Sciences and remained unknown to most scientists. These were discovered posthumously.

---

* H. Poincare, Thermodynamique, Guthier-Villars, Paris, 1908.

** W.J.M. Rankine, "Manual of the Steam Engine and other Prime Movers", 1859 going through 17 editions, as mentioned by Robert Balmer in "Thermodynamics", West Publishing Co., 1990, page 399.

SOLVED EXAMPLES

Example 1.1 *The pressure of gas in a pipeline is measured with a mercury manometer having one limb open to the atmosphere (Fig. 1.23). If the difference in the height of mercury in the two limbs is 562 mm, calculate the gas pressure. The barometer reads 761 mm Hg, the acceleration due to gravity is 9.79 m/s², and the density of mercury is 13,640 kg/m³.* E [LO 1.11]

Solution At plane *AB*, we have

$$p = p_0 + \rho g z$$

Now

$$p_0 = \rho g z_0$$

where z_0 is the barometric height, ρ the density of mercury and p_0 the atmospheric pressure.

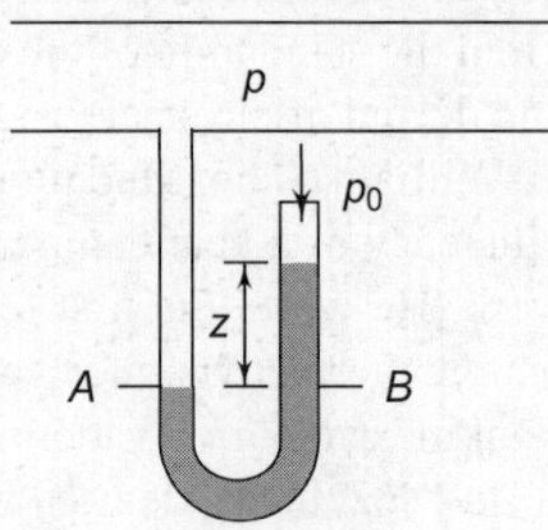

Fig. 1.23

Therefore,

$$p = \rho g (z + z_0)$$
$$= 13{,}640 \text{ kg/m}^3 \times 9.79 \text{ m/s}^2 (0.562 + 0.761) \text{ m}$$
$$= 177 \times 10^3 \text{ N/m}^2 = 177 \text{ kPa} = 1.77 \text{ bar} = \mathbf{1.746 \text{ atm}}$$

Example 1.2 *A turbine is supplied with steam at a gauge pressure of 1.4 MPa. After expansion in the turbine the steam flows into a condenser which is maintained at a vacuum of 710 mmHg. The barometric pressure is 772 mmHg. Express the inlet and exhaust steam pressures in pascals (absolute). Take the density of mercury as 13.6 × 10³ kg/m³.* E [LO 1.11]

Solution The atmospheric pressure p_0

$$= \rho g z_0 = 13.6 \times 10^3 \text{ kg/m}^3 \times 9.81 \text{ m/s}^2 \times 0.772 \text{ m}$$
$$= 1.03 \times 10^5 \text{ Pa}$$

Inlet steam pressure

$$= [(1.4 \times 10^6) + (1.03 \times 10^5)] \text{ Pa}$$
$$= 15.03 \times 10^5 \text{ Pa} = \mathbf{1.503 \text{ MPa}}$$

Condenser pressure

$$= (0.772 - 0.710) \text{ m} \times 9.81 \text{ m/s}^2 \times 13.6 \times 10^3 \text{ kg/m}^3$$
$$= 0.827 \times 10^4 \text{ Pa} = \mathbf{8.27 \text{ kPa}}$$

Example 1.3 *Convert the following readings of pressure to kPa, assuming that the barometer reads 760 mm of Hg.*

(a) *40 cmHg vacuum*
(b) *90 cmHg gauge*
(c) *1.2 m of H_2O gauge* E [LO 1.11]

Solution

(a) $p_{vacuum} = h\rho g = (40 \times 10^{-2}) \times (13.6 \times 10^3) \times 9.8$

$= 53.31 \times 10^3 \text{ N/m}^2 = \mathbf{53.31\ kPa}$

$p_{absolute} = p_{atm} - p_{vac}$

$= (760 - 400) \times 9.8 \times 13.6 \times 10^3$

$= 48 \times 10^3 \text{ N/m}^2 = \mathbf{48\ kPa}$

Also, $p_{abs} = 101.325 - 53.31 = 48.015$ kPa

(b) $p_{gauge} = h\,\rho\,g$

$= (90 \times 10^{-2}) \times (13.6 \times 10^3) \times 9.8$

$= 120 \times 10^3 \text{ N/m}^2 = \mathbf{120\ kPa}$

Since $p_{atm} = 760$ mm Hg $= 101.325$ kPa

$p_{abs} = p_{gauge} + p_{atm}$

$= 120 + 101.325 = \mathbf{221.325\ kPa}$

(c) $p_{gauge} = h\,\rho\,g = 1.2\text{ m} \times 1000\ \frac{\text{kg}}{\text{m}^3} \times 9.81\ \frac{\text{m}}{\text{s}^2}$

$= 11.772$ kPa

$\therefore$ $p_{abs} = 11.772 + 101.325 = \mathbf{113.097\ kPa}$

Example 1.4 *The pressure and specific volume of the atmosphere are related according to the equation $pv^{1.4} = 2.5 \times 10^5$, where p is in N/m^2 and v in m^3/kg. What depth of atmosphere is required to produce a pressure of 1.033 bar at the earth's surface? Assume g = 9.81 m/s^2.* D **[LO 1.11]**

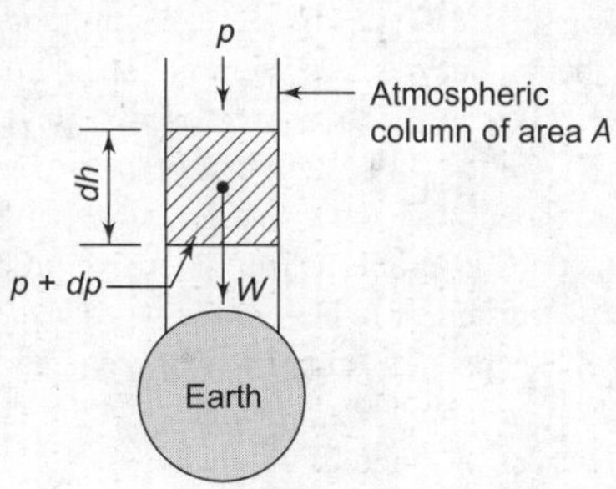

Fig. 1.24

Solution Let H be the depth of the atmosphere required to produce the pressure of 1.033 bar at the earth's surface. Considering an element of length dh (Fig. 1.24), by force balance:

$$A\,(p + dp) = mg + pA$$

$$pA + Adp = \rho A dhg + pA$$

or, $$dh = \frac{1}{g}\, vdp$$

or, $$\int_0^H dh = \int_0^{1.033\times10^5} \frac{1}{g}\left(\frac{2.5\times10^5}{p}\right)^{\frac{1}{1.4}} dp$$

$$H = \frac{1}{g}\,(2.5\times10^5)^{0.714} \int_0^{1.033\times10^5} p^{-0.714}\, dp$$

$$= 731.274\ \frac{\left(1.033\times10^5\right)^{0.286}}{0.286} = 69{,}462 \text{ m} = \mathbf{69.462\ km}$$

Example 1.5 *Acceleration is sometimes measured in g's, or multiples of the standard acceleration of gravity. Determine the net upward force that an astronaut whose mass is 68 kg experiences if the acceleration on lift-off is 10 g's.*

E **[LO 1.11]**

Solution Net vertical force

$$F_{net} = F = mg = ma$$

$$a = 10 \times 9.806 = 98.06 \text{ m/s}^2$$

$$F = 68 \times 98.06 = \mathbf{6668\ N}$$

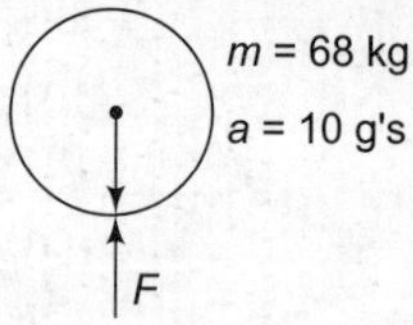

SUMMARY

A certain quantity of matter or a region in space upon which attention is focused in the analysis of a problem is called a *system*. Everything external to the system is called the *surroundings* which is separated from the system by a *boundary*. The system and its surroundings together is called a universe.

Thermodynamics studies the *energy interactions* like work and heat transfer and also mass transfer between a system and the surroundings and how these interactions affect the properties of the system. It is an empirical science drawing its material from certain observed facts of nature formulated into thermodynamic laws. Only the macroscopic behaviour of the system is considered, without taking into account the events occurring at the molecular or microscopic level. Properties which are the coordinates to describe the state of a system can be *intensive*, which are independent of mass, like pressure and temperature. *Extensive* properties like volume and energy depend on the mass. Specific extensive properties are intensive properties.

Systems can be of three types: a *closed system* can have only energy interactions and no mass transfer. An *open system* can have both mass and energy transfer across its boundary. An *isolated system* is one in which there is no interaction, either mass or energy transfer, across the boundary.

Any operation in which one or more of the properties changes is called a *change of state*. When the path of the change of state is specified, it is called a *process*. A *thermodynamic cycle* is a series of state changes such that the final state is identical with the initial state.

A system is said to exist in a state of *thermodynamic equilibrium* when no change in any macroscopic property is registered, if the system is isolated from its surroundings. If a system satisfies three types of equilibrium, viz. mechanical, chemical and thermal, it is said to be in thermodynamic equilibrium. A *quasi-static process* is a succession of equilibrium states, occurring infinitely slowly, so that the process takes infinite time to complete it across a finite gradient.

A *pure substance* is one that is homogeneous and invariable in chemical composition throughout its mass like a mixture of ice and water. The *two-property rule* holds that the state of a pure substance of given mass can be fixed by specifying two independent intensive properties.

REVIEW QUESTIONS

(D) **1.1** What do you understand by macroscopic and microscopic viewpoints? **[LO 1.3]**

(M) **1.2** How does the subject of thermodynamics differs from the concept of heat transfer? **[LO 1.13]**

(M) **1.3** What is the scope of classical thermodynamics? **[LO 1.3]**

(M) **1.4** What is a thermodynamic system? **[LO 1.4]**

(E) **1.5** What is the difference between a closed system and an open system? **[LO 1.4]**

(D) **1.6** An open system defined for a fixed region and a control volume is synonymous. Explain. **[LO 1.4]**

(E) **1.7** Define an isolated system. **[LO 1.4]**

(D) **1.8** Distinguish between the terms 'change of state', 'path' and 'process'. **[LO 1.5]**

(E) **1.9** What is a thermodynamic cycle? **[LO 1.5]**

(M) **1.10** What are intensive and extensive properties? **[LO 1.5]**

(M) **1.11** What do you mean by homogeneous and heterogeneous systems? **[LO 1.6]**

(D) **1.12** Explain what you understand by thermodynamic equilibrium. **[LO 1.7]**

(D) **1.13** Explain mechanical, chemical and thermal equilibrium. **[LO 1.7]**

(D) **1.14** What is a quasi-static process? What is its characteristic feature? **[LO 1.8]**

(D) **1.15** What is the concept of continuum? How will you define density and pressure using this concept? **[LO 1.9]**

(M) **1.16** What is vacuum? How can it be measured? **[LO 1.11]**

(M) **1.17** What is a pressure transducer? **[LO 1.11]**

PROBLEMS WITH HINTS

(E) **1.1** Determine the absolute pressure exerted on an object submerged 1000 m below the surface of the sea. The density of sea water is 1020 kg/m^3 and the acceleration due to gravity is 9.7 m/s^2. The local atmospheric pressure is 0.98 bar. **[LO 1.11]**
Hint: Use the following equation to find the absolute pressure: $P = P_{atm} + \rho g L$)

(M) **1.2** An inclined manometer illustrated in Fig. 1.25 is used to measure the pressure of a gas. The liquid inside the manometer has a density of 0.89 kg/cm^3 and the manometer reading is shown in the figure. If the atmospheric pressure is 101 kPa and the acceleration of gravity $g = 9.81$ m/s^2, what is the absolute pressure of the gas? **[LO 1.11]**
Hint: For the inclined manometer, $L = x \sin 30° = 0.125$ m
$p_g = \rho g L$ and $p_{abs^.} = p_g + p_{alm^.}$]

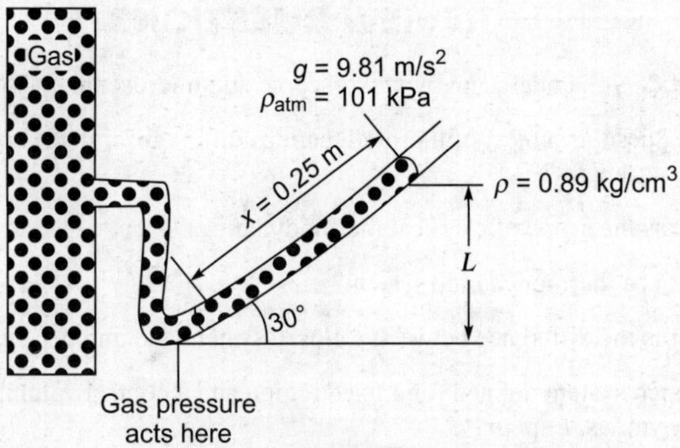

Fig. 1.25

PROBLEMS

Ⓔ **1.1** A pump discharges a liquid into a drum at the rate of 0.032 m^3/s. The drum, 1.50 m in diameter and 4.20 m in length, can hold 3000 kg of the liquid. Find the density of the liquid and the mass flow rate of the liquid handled by the pump. **[LO 1.10]**

Ⓜ **1.2** The acceleration of gravity is given as a function of elevation above sea level by

$$g = 980.6 - 3.086 \times 10^{-6}\ \mathrm{H}$$

where g is in cm/s², and H is in cm. If an aeroplane weighs 90,000 N at sea level, what is the gravity force acting upon it at 10,000 m elevation? What is the percentage difference from the sea-level weight? **[LO 1.10]**

Ⓓ **1.3** Prove that the weight of a body at an elevation H above sea-level is given by

$$W = \frac{mg}{g_0}\left(\frac{d}{d + 2H}\right)^2$$

where d is the diameter of the earth. **[LO 1.10]**

Ⓓ **1.4** The first artificial earth satellite is reported to have encircled the earth at a speed of 28,840 km/h and its maximum height above the earth's surface was stated to be 916 km. Taking the mean diameter of the earth to be 12,680 km, and assuming the orbit to be circular, evaluate the value of the gravitational acceleration at this height.

The mass of the satellite is reported to have been 86 kg at sea level. Estimate the gravitational force acting on the satellite at the operational altitude. **[LO 1.10]**

Ⓔ **1.5** Convert the following readings of pressure to kPa, assuming that the barometer reads 760 mm Hg:

(a) 90 cm Hg gauge, (b) 40 cm Hg vacuum,

(c) 1.2 m H_2O gauge and (d) 3.1 bar. **[LO 1.11]**

Ⓔ **1.6** A 30 m high vertical column of a fluid of density 1878 kg/m³ exists in a place where $g = 9.65$ m/s². What is the pressure at the base of the column? **[LO 1.11]**

Ⓜ **1.7** Assume that the pressure p and the specific volume v of the atmosphere are related according to the equation $pv^{1.4} = 2.3 \times 10^5$, where p is in N/m² abs and v is in m³ kg. The acceleration due to gravity is constant at 9.81 m/s^2. What is the depth of atmosphere necessary to produce a pressure of 1.0132 bar at the earth's surface? Consider the atmosphere as a fluid column. **[LO 1.11]**

(M) **1.8** The pressure of steam flowing in a pipeline is measured with a mercury manometer, shown in Fig. 1.26. Some steam condenses into water. Estimate the steam pressure in kPa. Take the density of mercury as 13.6×10^3 kg/m^3, density of water as 10^3 kg/m^3, the barometer reading as 76.1 cmHg and g as 9.806 m/s^2. **[LO 1.11]**

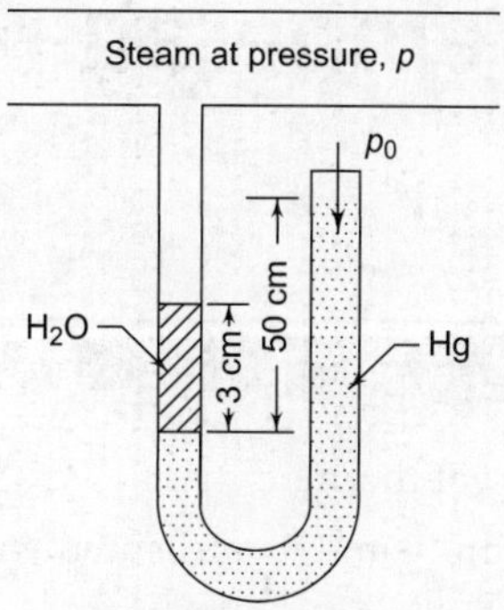

Fig. 1.26

(E) **1.9** A vacuum gauge mounted on a condenser reads 0.66 mHg. What is the absolute pressure in the condenser in kPa when the atmospheric pressure is 101.3 kPa? **[LO 1.11]**

ANSWERS TO PROBLEMS WITH HINTS

1.1 99.92 bar

1.2 101.98 kPa

ANSWERS TO PROBLEMS

1.1 12.934 kg/s

1.2 89,716.4 N, 0.315%

1.4 8.9 m/s^2; 765 N

1.6 544 kPa

1.7 64.8 km

1.9 13.3 kPa

2

Temperature

Learning Objectives

LO 2.1 State Zeroth law of thermodynamics
LO 2.2 Discuss temperature and its measurement principle
LO 2.3 Compare different thermometers
LO 2.4 Discuss ideal gas thermometers
LO 2.5 Recognize ideal gas temperature
LO 2.6 Compare ideal gas and Celsius temperatures
LO 2.7 Recognize the working principle of electrical resistance thermometer
LO 2.8 Explain the working principle of thermocouples
LO 2.9 Discuss international practical temperature scale

2.1 ZEROTH LAW OF THERMODYNAMICS

LO 2.1
State Zeroth law of thermodynamics

The property which distinguishes thermodynamics from other sciences is temperature. One might say that temperature bears as important a relation to thermodynamics as force does to statics or velocity does to dynamics. Temperature is associated with the ability to distinguish hot from cold. When two bodies at different temperatures are brought into contact, after sometime they attain a common temperature and are then said to exist in thermal equilibrium.

When a body A is in thermal equilibrium with a body B, and also separately with a body C, then bodies B and C will be in thermal equilibrium with each other.

This is known as the *zeroth law of thermodynamics*. It is the basis of temperature measurement.

In order to obtain a quantitative measure of temperature, a reference body is used, and a certain physical characteristic of this body which changes with temperature is selected. The changes in the selected characteristic may be taken as an indication of changes in temperature. The selected characteristic is called the *thermometric property*, and the reference body which is used in the determination of temperature is called the *thermometer*. A very common thermometer consists of a small amount of mercury in an evacuated capillary tube. In this case the extension of the mercury in the tube is used as the thermometric property.

There are five different kinds of thermometer, each with its own thermometric property, as shown in Table 2.1.

2.2 MEASUREMENT OF TEMPERATURE—THE REFERENCE POINTS

LO 2.2 Discuss temperature and its measurement principle

The temperature of a system is a property that determines whether or not a system is in thermal equilibrium with other systems. If a body is at, say, 70°C, it will be 70°C, whether measured by a mercury-in-glass thermometer, resistance thermometer or constant volume gas thermometer. If X is the thermometric property, let us arbitrarily choose for the temperature common to the thermometer and *to all systems in thermal equilibrium with it* the following *linear* function of X:

Table 2.1 Thermometers and thermometric properties

Thermometer	Thermometric property	Symbol
1. Constant volume gas thermometer	Pressure	p
2. Constant pressure gas thermometer	Volume	V
3. Electrical resistance thermometer	Resistance	R
4. Thermocouple	Thermal e.m.f.	ε
5. Mercury-in-glass thermometer	Length	L

$\theta(X) = aX$, where a is an arbitrary constant.

If X_1 corresponds to $\theta(X_1)$, then X_2 will correspond to

$$\frac{\theta(X_1)}{X_1} \cdot X_2$$

that is

$$\theta(X_2) = \frac{\theta(X_1)}{X_1} \cdot X_2 \tag{2.1}$$

Two temperatures on the linear X scale are to each other as the ratio of the corresponding X's.

2.2.1 Method Used Before 1954

The thermometer is first placed in contact with the system whose temperature $\theta(X)$ is to be measured, and then in contact with an arbitrarily chosen standard system in an easily reproducible state where the temperature is $\theta(X_1)$. Thus,

$$\frac{\theta(X_1)}{\theta(X)} = \frac{X_1}{X} \tag{2.2}$$

Then the thermometer at the temperature $\theta(X)$ is placed in contact with another arbitrarily chosen standard system in another easily reproducible state where the temperature is $\theta(X_2)$. It gives

$$\frac{\theta(X_2)}{\theta(X)} = \frac{X_2}{X} \tag{2.3}$$

From Eqs. (2.2) and (2.3)

$$\frac{\theta(X_1) - \theta(X_2)}{\theta(X)} = \frac{X_1 - X_2}{X}$$

or,
$$\theta(X) = \frac{\theta(X_1) - \theta(X_2)}{X_1 - X_2} \cdot X \tag{2.4}$$

If we assign an arbitrary number of degrees to the temperature interval $\theta(X_1) - \theta(X_2)$, then $\theta(X)$ can be calculated from the measurements of X, X_1 and X_2.

An easily reproducible state of an arbitrarily chosen standard system is called a fixed point. Before 1954, there were two fixed points: (*i*) *the ice point*, the temperature at which pure ice co-existed in equilibrium with air-saturated water at one atmosphere pressure, and (*ii*) *the steam point*, the temperature of equilibrium between pure water and pure steam at one atmosphere pressure. The temperature interval, $\theta(X_1) - \theta(X_2)$, between these two fixed points was chosen to be 100 degrees.

The use of two fixed points was found unsatisfactory and later abandoned, because of (*i*) the difficulty of achieving equilibrium stage/pt. between pure ice and air-saturated water (since when ice melts, it surrounds itself only with pure water and prevents intimate contact with air-saturated water), and (*ii*) extreme sensitiveness of the steam point to the change in pressure.

2.2.2 Method in Use After 1954

Since 1954 only one fixed point has been in use, viz. *the triple point of water*, the state at which ice, liquid water and water vapour co-exist in equilibrium. The temperature at which this state exists is arbitrarily assigned the value of 273.16 degrees Kelvin, or 273.16 K (the reason for using Kelvin's name will be explained later). Designating the triple point of water by θ_t, and with X_t being the value of the thermometric property when the body, whose temperature θ is to be measured, is placed in contact with water at its triple point, it follows that

$$\theta_t = aX_t$$

∴
$$a = \frac{\theta_t}{X_t} = \frac{273.16}{X_t}$$

Therefore,
$$\theta = aX = \frac{273.16}{X_t} \cdot X$$

or,
$$\theta = 273.16 \frac{X}{X_t} \tag{2.5}$$

The temperature of the triple point of water, which is an easily reproducible state, is now the *standard fixed point of thermometry*.

2.3 COMPARISON OF THERMOMETERS

LO 2.3
Compare different thermometers

Applying the above principle to the five thermometers listed in Table 2.1, the temperatures are given as follows:

1. Constant volume gas thermometer $\theta(p) = 273.16 \dfrac{p}{p_t}$
2. Constant pressure gas thermometer $\theta(V) = 273.16 \dfrac{V}{V_t}$
3. Electric resistance thermometer $\theta(R) = 273.16 \dfrac{R}{R_t}$
4. Thermocouple $\theta(\varepsilon) = 273.16 \dfrac{\varepsilon}{\varepsilon_t}$
5. Liquid-in-glass thermometer $\theta(L) = 273.16 \dfrac{L}{L_t}$

If the temperature of a given system is a measured simultaneously with each of the five thermometers, it is a found that there is a considerable difference among the readings. The smallest variation is, however, observed among different gas thermometers. That is why a gas is chosen as the standard thermometric substance.

2.4 IDEAL GAS

LO 2.4
Discuss ideal gas thermometers

It has been established from experimental observations that the $p - v - T$ behaviour of gases at a low pressure is closely given by the following relation:

$$p\bar{v} = \bar{R}T \tag{2.6}$$

where $\bar{R}$ is the universal gas constant, 8.3143 J/mol K and $\bar{v}$ is the molar specific volume, m^3/gmol. (see Sec. 10.3.). Dividing Eq. (2.6) by the molecular weight μ,

$$pv = RT \tag{2.7}$$

where v is specific volume, in m^3/kg, and R is the characteristic gas constant. Substituting $R = \dfrac{\bar{R}}{\mu}$ /μ J/kg K, we get in terms of the total volume V of gas,

$$PV = n\bar{R}T$$

$$PV = mRT \tag{2.8}$$

where n is the number of moles and m is the mass of the gas. Equation (2.8) can be written for two states of the gas,

$$\frac{p_1V_1}{T_1} = \frac{p_2V_2}{T_2} \tag{2.9}$$

Equation (2.6), (2.7) or (2.8) is called the *ideal gas equation of state*. At very low pressure or density, all gases and vapours approach ideal gas behaviour.

2.5 GAS THERMOMETERS

A schematic diagram of a constant volume gas thermometer is given in Fig. 2.1. A small amount of gas is enclosed in bulb B which is in communication via the capillary tube C with one limb of the mercury manometer M. The other limb of the mercury manometer is open to the atmosphere and can be moved vertically to adjust the mercury levels so that the mercury just touches the lip L of the capillary. The pressure in the bulb is used as a thermometric property and is given by

$$p = p_0 + \rho_M Zg$$

where p_0 is the atmospheric pressure, ρ_M is the density of mercury.

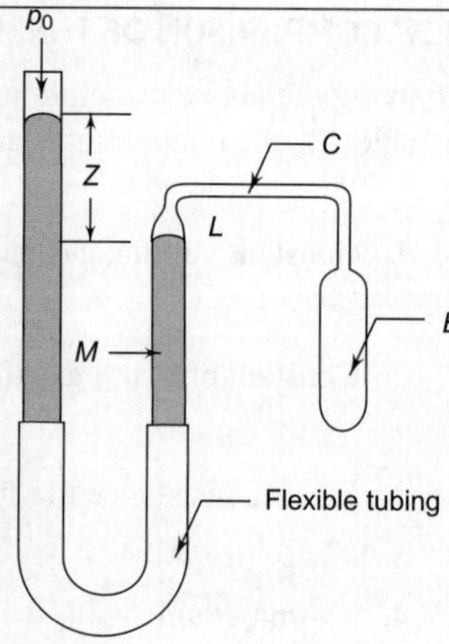

Fig. 2.1 Constant volume gas thermometer

When the bulb is brought in contact with the system whose temperature is to be measured, the bulb, in course of time, comes in thermal equilibrium with the system. The gas in the bulb expands, on being heated, pushing the mercury downward. The flexible limb of the manometer is then adjusted so that the mercury again touches the lip L. The difference in mercury level Z is recorded and the pressure p of the gas in the bulb is estimated. Since the volume of the trapped gas is constant, from the ideal gas equation,

$$\Delta T = \frac{V}{R} \Delta p \tag{2.10}$$

i.e. the temperature increase is proportional to the pressure increase.

In a constant pressure gas thermometer, the mercury levels have to be adjusted to keep Z constant, and the volume of gas V, which would vary with the temperature of the system, becomes the thermometric property.

$$\therefore \qquad \Delta T = \frac{p}{R} \Delta V \tag{2.11}$$

i.e. the temperature increase is proportional to the observed volume increase. The constant volume gas thermometer is, however, mostly in use, since it is simpler in construction and is easier to operate.

2.6 IDEAL GAS TEMPERATURE

> **LO 2.5**
> Recognize ideal gas temperature

Let us suppose that the bulb of a constant volume gas thermometer contains an amount of gas such that when the bulb is surrounded by water at its triple point, the pressure p_t is 1000 mm Hg. Keeping the volume V constant, let the following procedure be conducted:

1. Surround the bulb with steam condensing at 1 atm, determine the gas pressure p and calculate

$$\theta = 273.16 \frac{p}{1000}$$

2. Remove some gas from the bulb so that when it is surrounded by water at its triple point, the pressure p_t is 500 mm Hg. Determine the new value of p and then θ for steam condensing at 1 atm.

$$\theta = 273.16 \frac{p}{500}$$

3. Continue reducing the amount of gas in the bulb so that p_t and p have smaller and smaller values, e.g. p_t having, say, 250 mm Hg, 100 mm Hg and so on. At each value of p_t calculate the corresponding θ.
4. Plot θ vs. p_t and extrapolate the curve to the axis where $p_t = 0$. Read from the graph

$$\lim_{p_t \to 0} \theta$$

The graph, as shown in Fig. 2.2, indicates that although the readings of a constant volume gas thermometer depend upon the nature of the gas, *all gases indicate the same temperature as* p_t *is lowered and made to approach zero.*

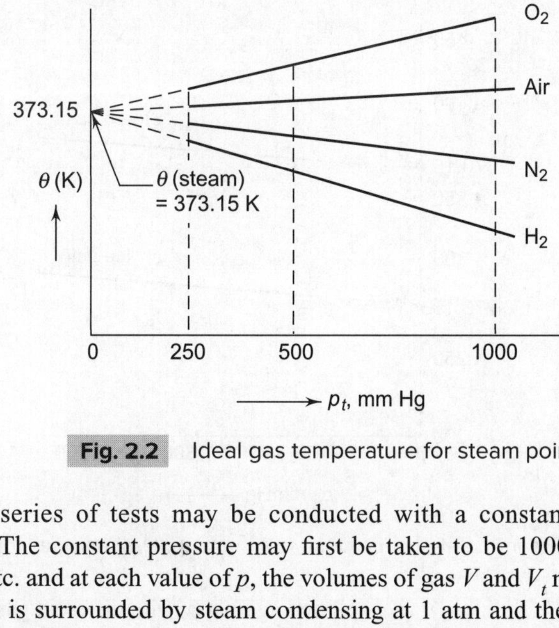

Fig. 2.2 Ideal gas temperature for steam point

A similar series of tests may be conducted with a constant pressure gas thermometer. The constant pressure may first be taken to be 1000 mm Hg, then 500 mm Hg, etc. and at each value of p, the volumes of gas V and V_t may be recorded when the bulb is surrounded by steam condensing at 1 atm and the triple point of water, respectively. The corresponding value of θ may be calculated from

$$\theta = 273.16 \frac{V}{V_t}$$

and θ vs. p may be plotted, similar to Fig. 2.2. It is found from the experiments that all gases indicate the same value of θ as p approaches zero.

Since a real gas, as used in the bulb, behaves as an ideal gas as pressure approaches zero (which would be explained later in Chapter 10), the *ideal gas temperature T* is defined by either of the two equations:

$$T = 273.16 \lim_{p_t \to 0} \frac{p}{p_t}$$

or,

$$T = 273.16 \lim_{p \to 0} \frac{V}{V_t} \tag{2.12}$$

where θ has been replaced by T to denote this particular temperature scale, the *ideal gas temperature scale*.

If p_s and p_t are the measured pressures at the steam point and the triple point, respectively, one gets the value of the steam point temperature T_s as

$$T_s = 273.16 \frac{V}{V_t} \tag{2.13}$$

which is equal to 373.15 K.

Similarly, the temperature T_i at the ice point is

$$T_i = 273.16 \lim_{p_t \to 0} \frac{p_s}{p_t} \tag{2.14}$$

which is equal to 273.15 K (Fig. 2.3).

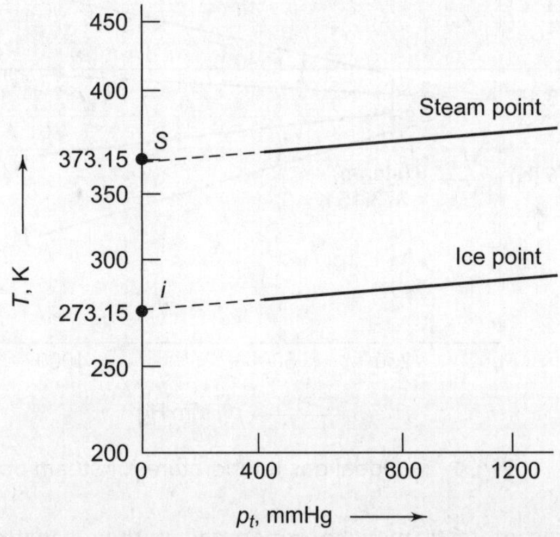

Fig. 2.3 Steam-point and ice-point from constant volume gas thermometer

Alternatively, if the ratio p_s/p_i is plotted against p_t with different gases in the bulb, one gets different curves as in Fig. 2.3. However, when extrapolated to zero pressure all curves converge, and the ratio p_s/p_i tends to a constant value giving

$$\frac{T_s}{T_i} = \lim_{p_t \to 0} \frac{p_i}{p_t} = 1.366099 \tag{2.15}$$

This value may be considered as a *universal constant*.

One may now decide to have a certain number of divisions between the steam point and the ice point, say 100 as in the Kelvin and Celsius scales so that

$$T_s - T_i = 100 \tag{2.16}$$

Solving Eqs. (2.15) and (2.16),

$$T_s = 373.15 \text{ K and } T_i = 273.15 \text{ K}$$

2.7 CELSIUS TEMPERATURE SCALE

> **LO 2.6**
> Compare ideal gas and Celsius temperatures

The Celsius temperature scale employs a degree of the same magnitude as that of the ideal gas scale, but its zero point is shifted, so that the Celsius temperature of the triple point of water is 0.01 degree Celsius or 0.01°C. If t denotes the Celsius temperature, then

$$t = T - 273.15°$$

Thus, the Celsius temperature t_s at which steam condenses at 1 atm. pressure

$$t_s = T_s - 273.15°$$

$$= 373.15 - 273.15 = 100.00°\text{C}$$

Similar measurements for ice points show this temperature on the Celsius scale to be 0.00°C. The only Celsius temperature which is fixed by definition is that of the triple point.

2.8 ELECTRICAL RESISTANCE THERMOMETER

> **LO 2.7**
> Recognize the working principle of electrical resistance thermometer

In the resistance thermometer (Fig. 2.4), the change in resistance of a metal wire due to its change in temperature is the thermometric property. The wire, frequently platinum, may be incorporated in a Wheatstone bridge circuit. The platinum resistance thermometer measures temperature to a high degree of accuracy and sensitivity, which makes it suitable as a standard for the calibration of other thermometers.

In a restricted range, the following quadratic equation is often used:

$$R = R_0 (1 + \text{A}t + \text{B}t^2)$$

where R_0 is the resistance of the platinum wire when it is surrounded by melting ice and A and B are constants.

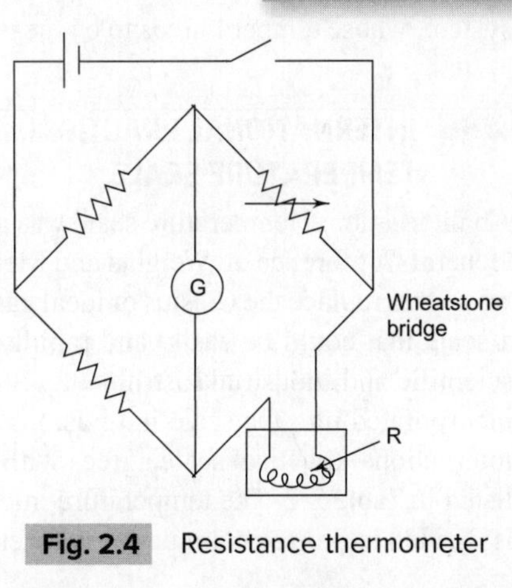

Fig. 2.4 Resistance thermometer

2.9 THERMOCOUPLE

LO 2.8

Explain the working principle of thermocouples

A thermocouple circuit made up from joining two wires *A* and *B* made of dissimilar metals is shown in Fig. 2.5. Due to the Seebeck effect, a net e.m.f. is generated in the circuit which depends on the difference in temperature between the hot and cold junctions and is, therefore, a thermometric property of the circuit. This e.m.f. can be measured by a microvoltmeter to a high degree of accuracy. The choice of metals depends largely on the temperature range to be investigated, and copper-constantan, chromel–alumel and platinum–platinum–rhodium are typical com-binations in use.

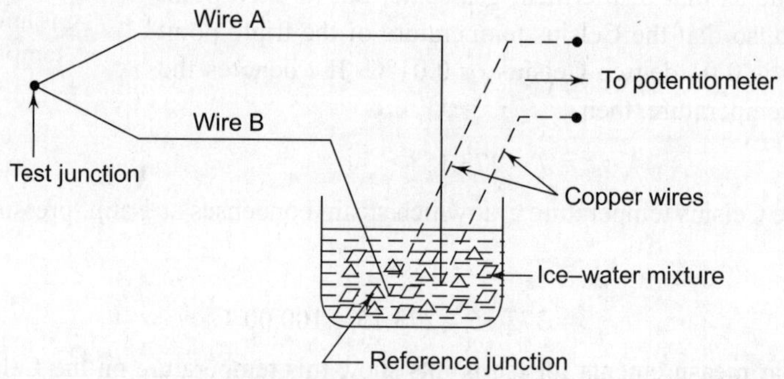

Fig. 2.5 Thermocouple

A thermocouple is calibrated by measuring the thermal e.m.f. at various known temperatures, the reference junction being kept at 0°C. The results of such measurements on most thermocouples can usually be represented by a cubic equation of the form:

$$\varepsilon = a + bt + ct^2 + dt^3$$

where ε is the thermal e.m.f. and the constants *a*, *b*, *c* and *d* are different for each thermocouple.

The advantage of a thermocouple is that it comes to thermal equilibrium with the system, whose temperature is to be measured, quite rapidly, because its mass is small.

2.10 INTERNATIONAL PRACTICAL TEMPERATURE SCALE

LO 2.9

Discuss international practical temperature scale

An international temperature scale was adopted at the Seventh General Conference on Weights and Measures held in 1927. It was not to replace the Celsius or ideal gas scales, but to provide a scale that could be easily and rapidly used to calibrate the scientific and industrial instruments. Slight refinements were incorporated into the scale in revisions adopted in 1948, 1954, 1960 and 1968. The international practical scale agrees with the Celsius scale at the defining fixed points listed in Table 2.2. The temperature interval from the oxygen point to the gold point is divided into three main parts, as given below.

1. From 0 to 660°C A platinum resistance thermometer with a platinum wire whose diameter must lie between 0.05 and 0.20 mm is used, and the temperature is given by the equation:

$$R = R_0 (1 + At + Bt^2)$$

where the constants R_0, A and B are computed by measurements at the ice point, steam point and sulphur point.

Table 2.2 Temperatures of fixed points

	Temperature °C
Normal boiling point of oxygen	– 182.97
Triple point of water (standard)	+ 0.01
Normal boiling point of water	100.00
Normal boiling point of sulphur	444.60
(Normal melting point of zinc-suggested	
as an alternative to the sulphur point)	419.50
Normal melting point of antimony	630.50
Normal melting point of silver	960.80
Normal melting point of gold	1063.00

2. From –190 to 0°C The same platinum resistance thermometer is used, and the temperature is given by

$$R = R_0 [1 + \mathrm{A}t + \mathrm{B}t^2 + C(t - 100)t^3]$$

where R_0, A and B are the same as before, and C is determined from a measurement at the oxygen point.

3. From 660 to 1063°C A thermocouple, one wire of which is made of platinum and the other of an alloy of 90 percent platinum and 10 percent rhodium, is used with one junction at 0°C. The temperature is given by the formula:

$$\varepsilon = a + bt + ct^2$$

where a, b and c are computed from measurements at the antimony point, silver point and gold point. The diameter of each wire of the thermocouple must lie between 0.35 and 0.65 mm.

An optical method is adopted for measuring temperatures higher than the gold point. The intensity of radiation of any convenient wavelength is compared with the intensity of radiation of the same wavelength emitted by a black body at the gold point. The temperature is then determined with the help of Planck's law of thermal radiation.

4. A revised international temperature Scale, ITS-90, was adapted in 1990 by the International Committee of Weights and Measures with further refinements defining more fixed points and conforming more closely the thermodynamic temperature scale based on the the second law of thermodynamics (*See Chapter 6*).

SOLVED EXAMPLES

Example 2.1 *Two mercury-in-glass thermometers are made of identical materials and are accurately calibrated at 0°C and 100°C. One has a tube of constant diameter, while the other has a tube of conical bore, 10 percent greater in diameter at 100°C than at 0°C. Both thermometers have the length between 0 and 100 subdivided uniformly. What will be the straight bore thermometer read in a place where the conical bore thermometer reads 50°C?* **D [LO 2.5]**

Solution The volume of mercury in the tube at t°C, V_t, is given by

$$V_t = V_0\,[1 + \beta\,(t - t_0)]$$

where V_0 is the volume of mercury at 0°C, β is the coefficient of volume expansion of mercury, and t_0 is the ice point temperature which is 0°C. The volume change of glass is neglected.

Therefore, $$V_t - V_0 = \beta\, V_0\, t$$

The temperature t is thus a linear function of volume change of mercury $(V_t - V_0)$.

Therefore, $$\Delta V_{0-100} = \beta\, V_0 \cdot 100$$

$$\Delta V_{0-50} = \beta\, V_0 \cdot 50$$

$$\therefore \qquad \frac{\Delta V_{0-50}}{\Delta V_{0-100}} = \frac{1}{2}$$

i.e. at 50°C, the volume of mercury will be half of that at 100°C, for the straight bore thermometer (Fig. 2.6a).

But if the bore is conical (Fig. 2.6b), mercury will fill up the volume *ACDB*, which is less than half of the mercury volume at 100°C, i.e. volume *AEFB*. Let t be the true temperature when mercury rises half the length of the conical tube (the apparent temperature being 50°C). Let *EA* and *FB* be extended to meet at G. Let l represent the length of the thermometers and l' the vertical height of the cone *ABG*, as shown in the figure. Now,

$$\frac{l'}{l + l'} = \frac{d}{1.1d} = \frac{1}{1.1}$$

$$l' = 10$$

and $$\frac{l'}{l' + l/2} = \frac{d}{CD}$$

$$\therefore \qquad CD = \frac{10.5}{10}\, d = 1.05d$$

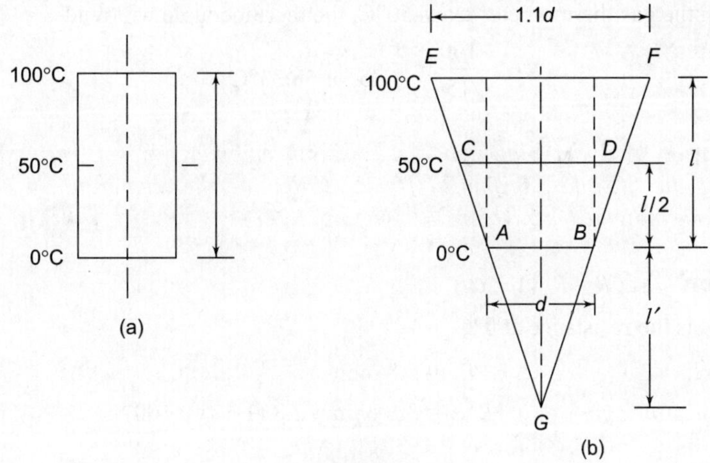

Fig. 2.6

Again

$$\Delta V_{0-100} = V_0 \cdot \beta \cdot 100$$

$$\Delta V_{0-t} = V_0 \beta t$$

$$\frac{\Delta V_{0-t}}{\Delta V_{0-100}} = \frac{t}{100}$$

or, $$\frac{\text{Volume } ACDB}{\text{Volume } AEFB} = \frac{t}{100}$$

or, $$\frac{\frac{1}{3}\frac{\pi}{4}(1.05d)^2 \times 10.5l - \frac{1}{3}\frac{\pi}{4}d^2 \cdot 10l}{\frac{1}{3}\frac{\pi}{4}(1.1d)^2 \times 11l - \frac{1}{3}\frac{\pi}{4}d^2 \cdot 10l} = \frac{t}{100}$$

or, $$\frac{1.05 \times 1.05 \times 10.5 - 10}{1.1 \times 1.1 \times 11 - 10} = \frac{t}{100}$$

$\therefore$ $$t = \frac{1.58}{3.31} \times 100 = \mathbf{47.7°C}$$

Example 2.2 *The e.m.f. in a thermocouple with the test junction at t°C on gas thermometer scale and reference junction at ice point is given by*

$$\varepsilon = 0.20\, t - 5 \times 10^{-4}\, t^2 \text{ mV}$$

The millivoltmeter is calibrated at ice and steam points. What will this thermometer read in a place where the gas thermometer reads 50°C? E **[LO 2.8]**

Solution At ice point, when $t = 0°C$, $\varepsilon = 0$ mV

At steam point, when $t = 100°C$, $\varepsilon = 0.20 \times 100 - 5 \times 10^{-4} \times (100)^2$
$= 15$ mV

At $t = 50°C$, $\varepsilon = 0.20 \times 50 - 5 \times 10^{-4} (50)^2 = 8.75$ mV

When the gas thermometer reads 50°C, the thermocouple will read

$$\frac{100}{15} \times 8.75, \text{ or } \mathbf{58.33°C}$$

Example 2.3 *A platinum resistance thermometer has a resistance of 2.8 ohm at 0°C and 3.8 ohm at 100°C. Calculate the temperature when the resistance indicated is 5.8 ohm.* Ⓔ **[LO 2.7]**

Solution Let $R = R_0 (1 + \alpha t)$

where R_0 is the resistance at 0°C.

Therefore, $R_0 = 2.8$ ohm

$$R_{100} = 3.8 \text{ ohm} = 2.8 (1 + \alpha \times 100)$$

$$\therefore \quad \alpha = \left(\frac{3.8}{2.8} - 1\right) \times 10^{-2} = 0.357 \times 10^{-2}$$

When $R = 5.8$ ohm,

$$5.8 = 2.8 (1 + 0.357 \times 10^{-2}\, t)$$

or,

$$t = \left(\frac{5.8}{2.8} - 1\right) \times \frac{100}{0.357} = \mathbf{300°C.}$$

SUMMARY

When two bodies at two different temperatures are brought into contact, after sometime they attain a common temperature and are then said to exist in *thermal equilibrium*.

When a body A is in thermal equilibrium with a body B and also separately with body C, then bodies B and C will be in thermal equilibrium with each other.

This is known as the *zeroth law of thermodynamics*. It is the basis of temperature measurement. A physical characteristic of an arbitrarily chosen body which changes with change in temperature is called *thermometric property* (X) and the reference body is called the *thermometer*. The common thermometers are mercury-in-glass, (length of mercury column, $L \propto t$), resistance ($R \propto t$), thermocouple ($\varepsilon \propto t$), constant volume gas thermometer ($p \propto t$) and constant pressure gas thermometer ($v \propto t$). Before 1954, two fixed points, the ice point and steam point, were used to quantify the temperature of a system. After 1954, only one fixed point, the *triple point of water (0.01°C)* is used and it is the standard fixed point of thermometry. A thermometer scale is set up by assuming a linear variation of temperature with the reading of some arbitrarily selected standard thermometer. The most suitable standard has been found to be the gas thermometer, which is used in laboratories to determine the reference points at which other thermometers are calibrated.

An absolute temperature scale may be constructed by taking the temperature directly proportional to the volume of the gas in a constant pressure gas thermometer. The scale, based upon a gas at zero pressure, is the physical realisation of a logically formulated absolute thermodynamic temperature scale, which will be introduced in connection with the Second Law of Thermodynamics. For ordinary purposes, absolute temperatures may be found from the relation (°C = 273.15 K).

REVIEW QUESTIONS

(M) **2.1** What is the zeroth law of thermodynamics? **[LO 2.1]**

(M) **2.2** Define thermometric property. **[LO 2.1]**

(E) **2.3** What is a thermometer? **[LO 2.1]**

(M) **2.4** What is a fixed point? **[LO 2.2]**

(D) **2.5** How many fixed points were used prior to 1954? What are these? **[LO 2.2]**

(D) **2.6** What is the standard fixed point in thermometry? Define it. **[LO 2.2]**

(D) **2.7** Why is a gas chosen as the standard thermometric substance? **[LO 2.3]**

(E) **2.8** What is an ideal gas? **[LO 2.4]**

(M) **2.9** What is the difference between the universal gas constant and a characteristic gas constant? **[LO 2.4]**

(D) **2.10** What is a constant volume gas thermometer? Why is it preferred to a constant pressure gas thermometer? **[LO 2.3]**

(D) **2.11** What do you understand by the ideal gas temperature scale? **[LO 2.5]**

(M) **2.12** How can the ideal gas temperature for the steam point be measured? **[LO 2.5]**

(M) **2.13** What is the Celsius temperature scale? **[LO 2.6]**

(M) **2.14** What is the advantage of a thermocouple in temperature measurement? **[LO 2.8]**

(M) **2.15** How does the resistance thermometer measure temperature? **[LO 2.7]**

(D) **2.16** What is the need of the international practical temperature scale? **[LO 2.9]**

PROBLEMS WITH HINTS

(E) **2.1** A new absolute temperature scale is proposed. On this scale the ice point of water is 150 °S and the steam point is 300 °S. Determine the temperatures in °C that correspond to 100 °S and 400 °S, respectively. What is the ratio of the size of °S to the size of °S to the Kelvin? **[LO 2.2]**

Hints: For the given data, 150 °S → 0 °C and 300 °S = 100 °C

Thus the relation between °S and °C is expressed as (°S) = $\frac{300-100}{100}$ °C + 150 = 1.5°C + 150

Using this relation, find 100 °S and 400 °S in terms of °C.
Using (°S) = 1.5 [K–273.15] + 150
The size of °S to the Kelvin is **1.5 °S/K.**

(E) **2.2** Over a limited temperature range, the relation between electrical resistance R and temperature T for a resistance temperature detector is

$$R = R_o[1 + \alpha(T - T_o)]$$

where R_o is the resistance in ohms (Ω), measured at reference temperature T_o (in °C). The following data are obtained for a particular resistance thermometer: 0 °C $\rightarrow$ 51.39 Ω, 91 °C $\rightarrow$ 51.72 Ω.
What is the resistance reading when the temperature is 50 °C? **[LO 2.7]**

Hints: Substituting the two test readings in the relation $R = R_o\,[1 + \alpha(T - T_o)]$, obtain $\alpha = 7.0565 \times 10^{-5}$ (°C^{-1}). When $T = 50$ °C,
$R = \mathbf{51.57\ \Omega}$

PROBLEMS

(E) **2.1** The limiting value of the ratio of the pressure of gas at the steam point and at the triple point of water when the gas is kept at constant volume is found to be 1.36605. What is the ideal gas temperature of the steam point? **[LO 2.5]**

(M) **2.2** In a constant volume gas thermometer the following pairs of pressure readings were taken at the boiling point of water and the boiling point of sulphur, respectively:

Water b.p.	50.0	100	200	300
Sulphur b.p.	96.4	193	387	582

The numbers are the gas pressures, mm Hg, each pair being taken with the same amount of gas in the thermometer, but the successive pairs being taken with different amounts of gas in the thermometer. Plot the ratio of $S_{b.p.}:H_2O_{b.p.}$ against the reading at the water boiling point, and extrapolate the plot to zero pressure at the water boiling point. This gives the ratio of $S_{b.p.}:H_2O_{b.p.}$ on a gas thermometer operating at zero gas pressure, i.e. an ideal gas thermometer. What is the boiling point of sulphur on the gas scale, from your plot? **[LO 2.3]**

(E) **2.3** The resistance of a platinum wire is found to be 11,000 ohms at the ice point, 15.247 ohms at the steam point and 28.887 ohms at the sulphur point. Find the constants A and B in the equation:

$$R = R_0\,(1 + \mathrm{A}t + \mathrm{B}t^2)$$

and plot R against t in the range 0 to 660°C. **[LO 2.7]**

(M) **2.4** When the reference junction of a thermocouple is kept at the ice point and the test junction is at the Celsius temperature t, and e.m.f. e of the thermocouple is given by the equation:

$$\varepsilon = at + bt^2$$

where $a = 0.20$ mV/deg, and $b = -5.0 \times 10^{-4}$ mV/deg^2

(a) Compute the e.m.f. when $t = -100$°C, 200°C, 400°C and 500°C, and draw graph of ε against t in this range.

(b) Suppose the e.m.f. ε is taken as a thermometric property and that a temperature scale t^* is defined by the linear equation.

$$t^* = a'\,\varepsilon + b'$$

and that $t^* = 0$ at the ice point and $t^* = 100$ at the steam point. Find the numerical values of a' and b' and draw a graph of ε against t^*.

(c) Find the values of t^* when $t = -100$°C, 200°C, 400°C and 500°C, and draw a graph of t^* against t.

(d) Compare the Celsius scale with the t^* scale. **[LO 2.8]**

(M) **2.5** The temperature t on a thermometric scale is defined in terms of a property K by the relation:

$$t = a \ln K + b$$

where a and b are constants.

The values of K are found to be 1.83 and 6.78 at the ice point and the steam point, the temperatures of which are assigned the numbers 0 and 100, respectively. Determine the temperature corresponding to a reading of K equal to 2.42 on the thermometer. **[LO 2.2]**

(M) **2.6** The resistance of the windings in a certain motor is found to be 80 ohms at room temperature (25°C). When operating at full load under steady state conditions, the motor is switched off and the resistance of the windings, immediately measured again, is found to be 93 ohms. The windings are made of copper whose resistance at temperature t°C is given by

$$R_t = R_0\,[1 + 0.00393\,t]$$

where R_0 is the resistance at 0°C. Find the temperature attained by the coil during full load. **[LO 2.7]**

(M) **2.7** A new scale N of temperature is divided in such a way that the freezing point of ice is 100°N and the boiling point is 400°N. What is the temperature reading on this new scale when the temperature is 150°C? At what temperature both the Celsius and the new temperature scale reading would be the same? **[LO 2.2]**

(E) **2.8** A platinum wire is used as a resistance thermometer. The wire resistance was found to be 10 ohm and 16 ohm at ice point and steam point, respectively, and 30 ohm at sulphur boiling point of 444.6°C. Find the resistance of the wire at 500°C, if the resistance varies with temperature by the relation.

$$R = R_0\,(1 + \alpha t + \beta t^2)$$

[LO 2.7]

ANSWERS TO PROBLEMS WITH HINTS

2.1 –33.33 °C and 167.67 °C

ANSWERS TO PROBLEMS

2.1 100°C

2.2 445°C

2.5 21.346°C

2.6 70.41°C

2.7 550°N, – 50°C

2.8 31.3 ohm

3

Work and Heat Transfer

Learning Objectives

LO 3.1 Discuss work transfer and its sign convention
LO 3.2 Explain displacement (pdV)-work
LO 3.3 Compare path and point functions
LO 3.4 Analyze indicator diagrams of engines
LO 3.5 Discuss works other than pdV-work
LO 3.6 Recognize free expansion
LO 3.7 Discuss heat transfer and its sign convention
LO 3.8 Analyze heat and work transfer as path function
LO 3.9 Demonstrate specific and latent heats

A closed system and its surroundings can interact in two ways: (*i*) by work transfer, and (*ii*) by heat transfer. These may be called *energy interactions* and these bring about changes in the properties of the system. Thermodynamics mainly studies these energy interactions and the associated property changes of the system.

3.1 WORK TRANSFER

LO 3.1
Discuss work transfer and its sign convention

Work is one of the basic modes of energy transfer. In mechanics, the action of a force on a moving body is identified as work. A force is a means of transmitting an effect from one body to another. But a force itself never produces a physical effect except when coupled with motion and hence, it is not a form of energy. An effect such as the raising of a weight through a certain distance can be performed by using a small force through a large distance or a large force through a small distance. The product of force and distance is the same to accomplish the same effect.

In mechanics, work is defined as: *The work is done by a force as it acts upon a body moving in the direction of the force.*

The action of a force through a distance (or of a torque through an angle) is called *mechanical work* since other forms of work can be identified, as discussed later. The product of the force and the distance moved parallel to the force is the magnitude of mechanical work.

In thermodynamics, work transfer is considered as occurring between the system and the surroundings. *Work is said to be done by a system if the sole effect on things external to the system can be reduced to the raising of a weight.* The weight may not actually be raised, but the net effect external to the system would be the raising of a weight. Let us consider the battery and the motor in Fig. 3.1 as a system. The motor is driving a fan. The system is doing work upon the surroundings. When the fan is replaced by a pulley and a weight, as shown in Fig. 3.2, the weight may be raised with the pulley driven by the motor. The sole effect on things external to the system is then the raising of a weight.

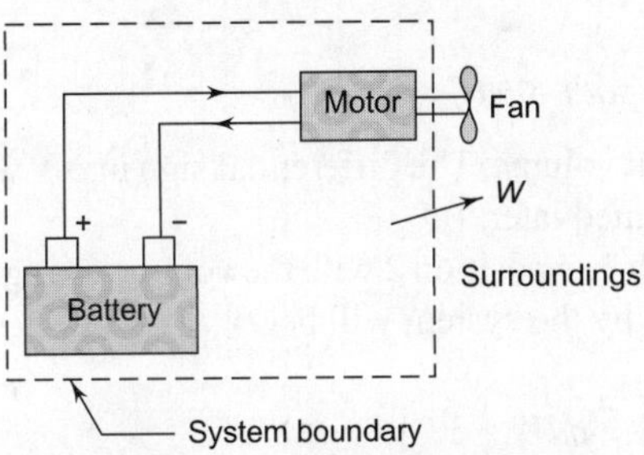

Fig. 3.1 Battery-motor system driving a fan

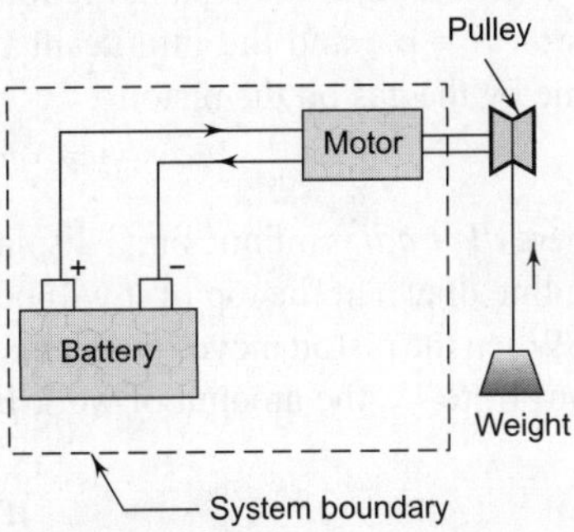

Fig. 3.2 Work transfer from a system

When work is done by a system, it is arbitrarily taken to be positive, and when work is done on a system, it is taken to be negative (Fig. 3.3). The symbol W is used for work transfer.

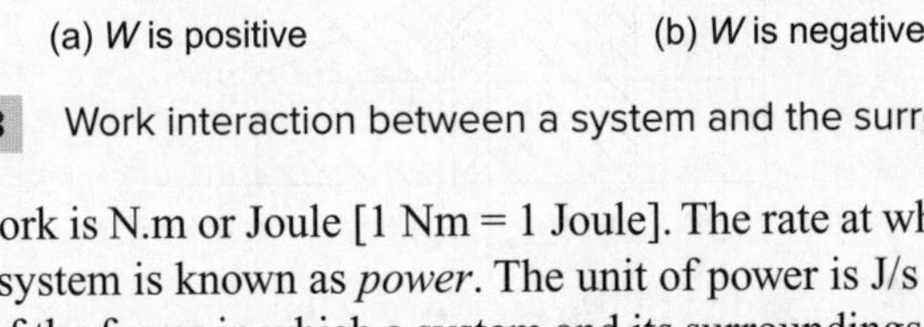

Fig. 3.3 Work interaction between a system and the surroundings

The unit of work is N.m or Joule [1 Nm = 1 Joule]. The rate at which work is done by, or upon, the system is known as *power*. The unit of power is J/s or watt.

Work is one of the forms in which a system and its surroundings can interact with each other. There are various types of work transfers which can get involved between them.

3.2 *pdV*-WORK OR DISPLACEMENT WORK

LO 3.2
Explain displacement (*pdV*)-work

Let the gas in the cylinder (Fig. 3.4) be a system having initially the pressure p_1 and volume V_1. The system is in thermodynamic equilibrium, the state of which is described by the coordinates p_1 and V_1. The piston is the only boundary which moves due to gas pressure. Let the piston move out to a new final position 2, which is also a thermodynamic equilibrium state specified by pressure p_2 and volume V_2. At any intermediate point in the travel of the piston, let the pressure be p and the volume be V. This must also be an equilibrium state, since macroscopic properties p and V are significant only for equilibrium states. When the piston moves an infinitesimal distance dl, and if 'a' be the area of the piston, the force F acting on the piston F = p.a and the infinitesimal amount of work done by the gas on the piston

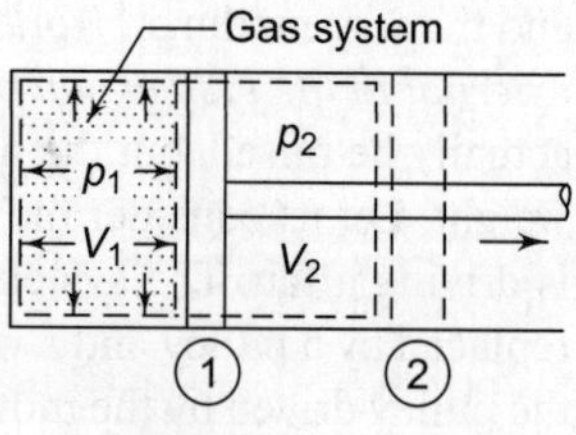

Fig. 3.4 *pdV* work

$$đW = F \cdot dl = padl = pdV \tag{3.1}$$

where $dV = adl$ = infinitesimal displacement volume. The differential sign in δW with the line drawn at the top of it will be explained later.

When the piston moves out from position 1 to position 2 with the volume changing from V_1 to V_2, the amount of work W done by the system will be

$$W_{1-2} = \int_{V_1}^{V_2} pdV$$

The magnitude of the work done is given by the area under the path 1–2, as shown in Fig. 3.5. Since p is at all times a thermodynamic coordinate, all the states passed through by the system as the volume changes from V_1 to V_2 must be equilibrium states, and the path 1–2 *must be quasi-static*. The piston moves infinitely slowly so that every state passed through is an equilibrium state. *The integration* $\int pdV$ *can be performed only on a quasi-static path.*

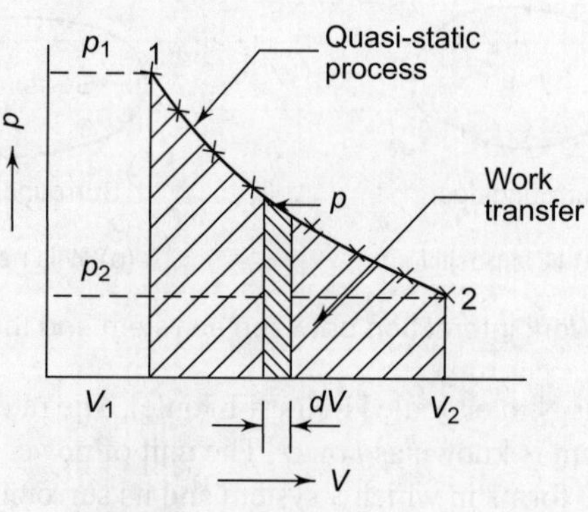

Fig. 3.5 Quasi-static *pdV* work

3.2.1 *pdV*-Work in Various Quasi-Static Processes

1. Constant pressure process (Fig. 3.6) (isobaric or isopiestic process)

$$W_{1-2} = \int_{V_1}^{V_2} pdV = p(V_2 - V_1) \tag{3.2}$$

2. Constant volume process (Fig. 3.7) (isochoric process)

$$W_{1-2} = \int pdV = 0 \tag{3.3}$$

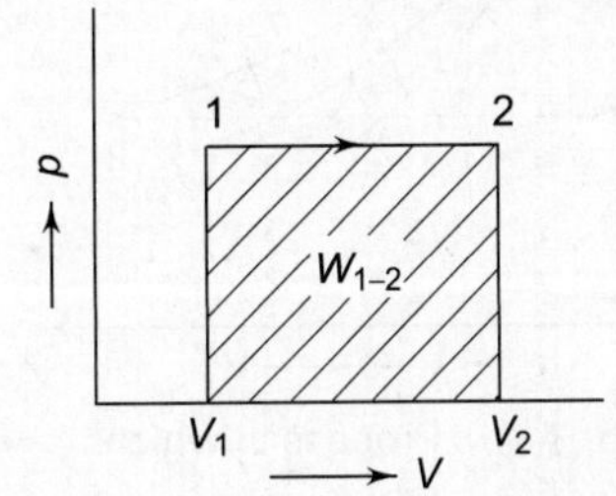

Fig. 3.6 Constant pressure process

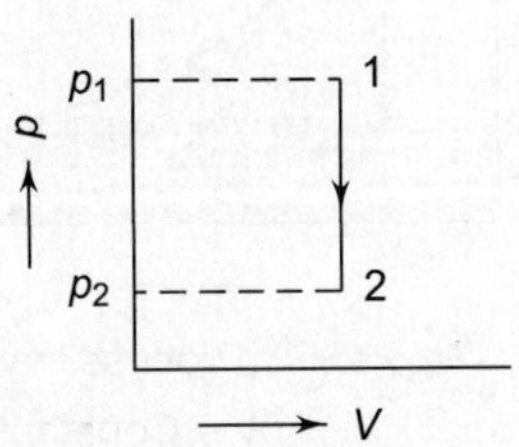

Fig. 3.7 Constant volume process

3. Process in which $pV = C$ (Fig. 3.8) (*isothermal process*)

$$\therefore \quad W_{1-2} = \int_{V_1}^{V_2} pdV \qquad pV = p_1 V_1 = C$$

$$p = \frac{(p_1 V_1)}{V}$$

$$W_{1-2} = p_1 V_1 \int_{V_1}^{V_2} \frac{dV}{V} = p_1 V_1 \ln \frac{V_2}{V_1}$$

$$= p_1 V_1 \ln \frac{p_1}{p_2} \tag{3.4}$$

4. Process in which $pV^n = C$, where n is a constant (Fig. 3.9) (polytropic process).

$$pV^n = p_1 V_1^n = p_2 V_2^n = C$$

$$\therefore \quad p = \frac{\left(p_1 V_1^n\right)}{V^n}$$

$$\therefore \quad W_{1-2} = \int_{V_1}^{V_2} pdV = \int_{V_1}^{V_2} \frac{p_1 V_1^n}{V^n} \cdot dV = (p_1 V_1^n) \left[\frac{V^{-n+1}}{-n+1}\right]_{V_1}^{V_2}$$

$$= \frac{p_1 V_1^n}{1-n} (V_2^{1-n} - V_1^{1-n}) = \frac{p_2 V_2^n \times V_2^{1-n} - p_1 V_1^n \times V_1^{1-n}}{1-n}$$

$$= \frac{p_1 V_1 - p_2 V_2}{n-1} = \frac{p_1 V_1}{n-1}\left[1 - \left(\frac{p_2}{p_1}\right)^{\frac{n-1}{n}}\right] \tag{3.5}$$

Similarly, for process in which $pv^{\gamma} = c$, where $\gamma = c_p/c_v$,

$$W_{1-2} = \frac{p_1 V_1}{n-1}\left[1-\left(\frac{p_2}{p_1}\right)^{\frac{\gamma-1}{\gamma}}\right] \tag{3.5a}$$

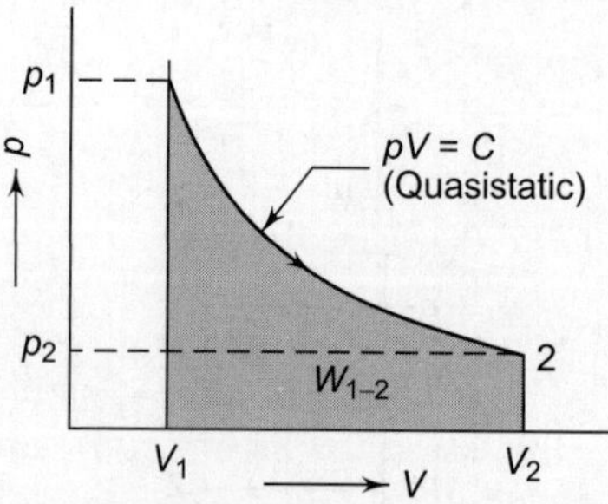

Fig. 3.8 Process in which pV = Constant

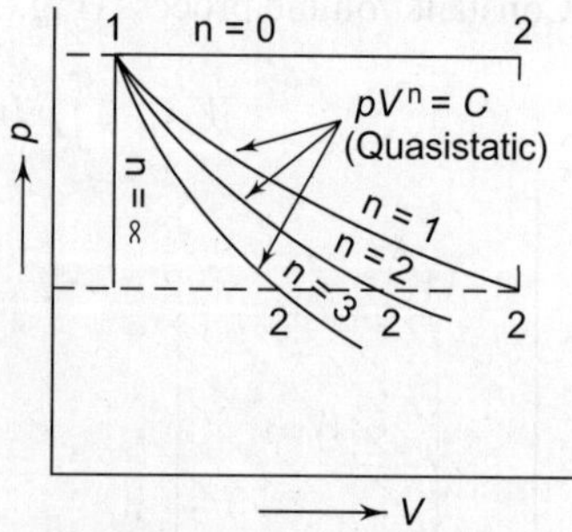

Fig. 3.9 Process in which pV^n = Constant

3.3 PATH FUNCTION AND POINT FUNCTION

LO 3.3 Compare path and point functions

With reference to Fig. 3.10, it is possible to take a system from state 1 to state 2 along many quasi-static paths, such as *A*, *B* or *C*. Since the area under each curve represents the work for each process, the amount of work involved in each case is not a function of the end states of the process, and it depends on the path the system follows in going from state 1 to state 2. For this reason, work is called a *path function,* and $đW$ is an *inexact or imperfect differential.*

Thermodynamic properties are *point functions*, since for a given state, there is a definite value for each property. The change in a thermodynamic property of a system in a change of state is independent of the path the system follows during the change of state, and depends only on the initial and final states of the system. The differentials of point functions are *exact or perfect differentials*, and the integration is implied.

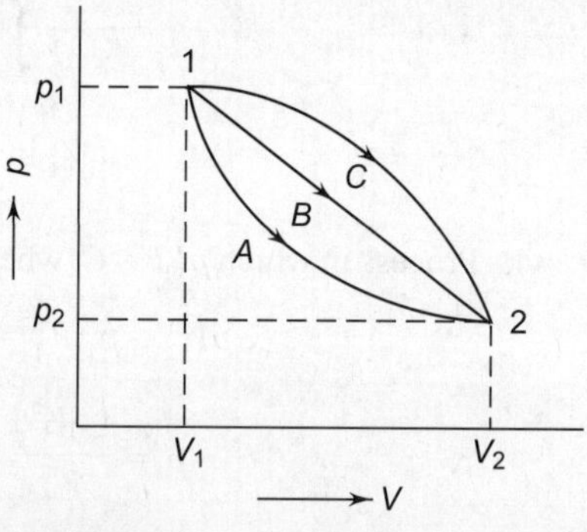

Fig. 3.10 Work—A path function

$$\int_{V_1}^{V_2} dV = V_2 - V_1$$

The change in volume thus depends only on the end states of the system irrespective of the path the system follows. realized

On the other hand, work done in a quasi-static process between two given states depends on the path followed.

$$\int_1^2 đW \neq W_2 - W_1$$

Rather,

$$\int_1^2 đW = W_{1-2} \quad \text{or} \quad {}_1W_2$$

To distinguish an inexact differential $đW$ from an exact differential dV or dp the differential sign is being cut by a line at its top.

Fom Eq. (3.1),

$$dV = \frac{1}{p} đW \tag{3.6}$$

Here, $1/p$ is called the *integrating factor*. Therefore, an inexact differential δW when multiplied by an integrating factor $1/p$ becomes an exact differential dV.

For a cyclic process, the initial and final states of the system are the same, and hence, the change in an property is zero, i.e.

$$\oint dV = 0, \oint dp = 0, \oint dT = 0 \tag{3.7}$$

where the symbol $\oint$ denotes the cyclic integral for the closed path. Therefore, *the cyclic integral of a property is always zero.*

3.4 INDICATOR DIAGRAM

LO 3.4

Analyze indicator diagrams of engines

A reciprocating engine (Figs. 1.2 and 13.4) where the reciprocating motion of the piston is converted to the rotary motion of the crank shaft can be (*i*) single-acting or double-acting, depending on whether the working fluid does work either on one side or both sides of the piston, (*ii*) two-stroke or four-stroke, depending on whether the engine cycle is completed by two or four strokes of the piston, and (*iii*) Single cylinder or multi-cylinder, depending on the number of cylinders in the engine. An indicator diagram is a trace made by a recording pressure gauge, called the indicator, attached to the cylinder of a reciprocating engine. This represents the work done in one engine cycle. Figure 3.11 shows a typical engine indicator.

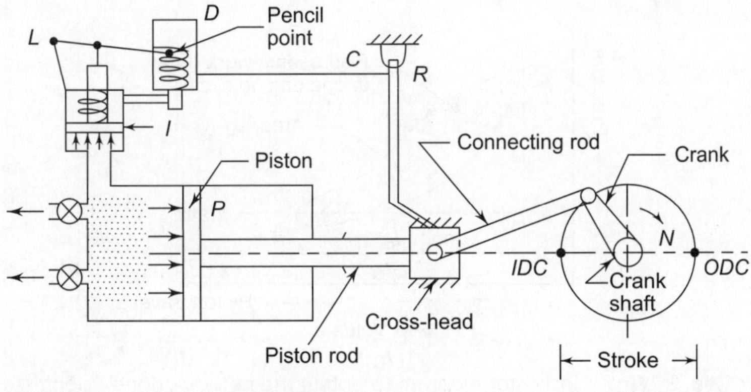

Fig. 3.11 Engine indicator

The same gas pressure acts on both, the engine piston P and the indicator piston I. The indicator piston is loaded by a spring and it moves in direct proportion to the change in pressure. The motion of the indicator piston causes a pencil held at the end of the linkage L to move upon a strip of paper wrapped around drum D. The drum is rotated about its axis by cord C, which is connected through a reducing motion R to the piston P of the engine. The surface of drum D moves horizontally under the pencil while the pencil moves vertically over the surface and a plot of pressure upon the piston vs. piston travel is obtained.

Since an engine normally operates in a succession of essentially identical machine cycles, the indicator diagram will be a closed curve which shows only relative pressures and volumes, there being no reference lines or zero axes (Fig. 3.12a).

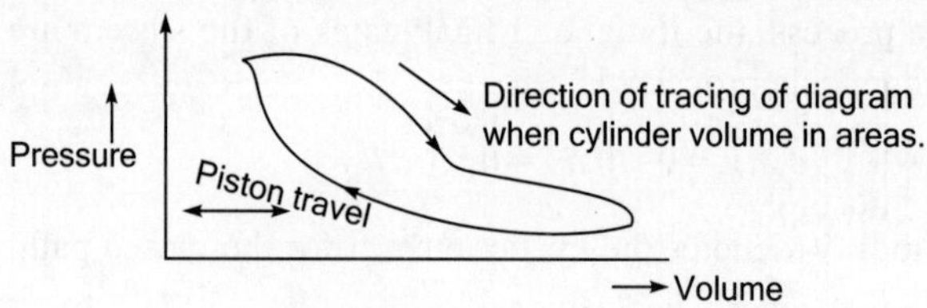

Fig. 3.12(a) Indicator plot

Before tracing the final indicator diagram, a pressure reference line is recorded by subjecting the indicator to the atmosphere and tracing a line at a constant pressure of one atmosphere.

The area of the indicator diagram represents the magnitude of the net work done by the system in one engine cycle. The area under the path 1–2 represents work done by the system and the area under the path 2–1 represents work done upon the system (Fig. 3.12b). The area of the diagram, a_d, is measured by means of a planimeter, and the length of the diagram, l_d, is also measured. The *mean effective pressure* (m.e.p.) p_m is defined in the following way:

$$p_m = \frac{a_d}{l_d} \times K$$

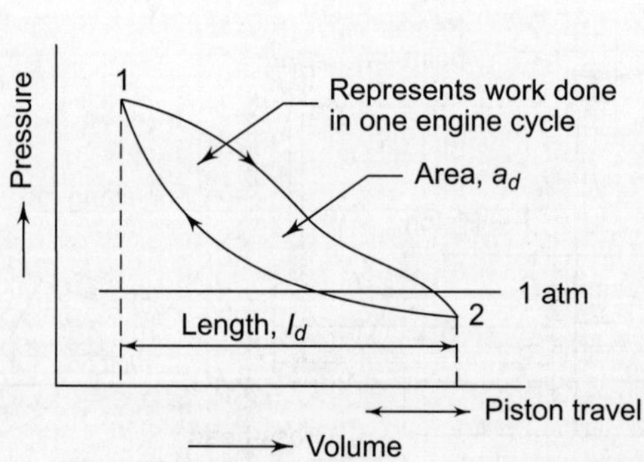

Fig. 3.12(b) Indicator diagram to obtain the network done upon the pistion during a complete machine cycle

where K is the indicator spring constant (N/cm^2 × cm travel). Work done in one engine cycle

$$= (p_m \cdot A) L$$

where A = cross-sectional area of the cylinder

$$= \frac{\pi}{4} D^2, \text{ where } D \text{ is the cylinder diameter}$$

and L = stroke of piston, or length of cylinder.

Let N be the revolutions per minute (r.p.m.) of the crankshaft. In a two-stroke cycle, the engine cycle is completed in two strokes of the piston or in one revolution of the crankshaft. In a four-stroke cycle, the engine cycle is completed in four strokes of the piston or two revolutions of the crankshaft.

For a two-stroke engine, work done in one minute = $p_m ALN$, and for a four-stroke engine, work done in one minute = $p_m ALN/2$.

The power developed inside the cylinder of the engine is called *indicated power* (IP),

$$\therefore \quad \text{IP} = \frac{p_m AL\left(N \text{ or } \dfrac{N}{2}\right) n}{60} \text{ kW} \tag{3.8}$$

where p_m is in kPa and n is the number of cylinders in the engine.

The power available at the crankshaft is always less than this value (IP) due to friction, etc. and is called the *brake power* (BP) or *shaft power* (SP). If w is the angular velocity of the crankshaft in radian/sec, then

$$\text{BP} = T\omega \tag{3.9}$$

where T is the torque transmitted to the crankshaft in mN.

$$\therefore \quad \text{BP} = \frac{2\pi TN}{60} \tag{3.10}$$

where N is the number of revolutions per minute (rpm).

The mechanical efficiency of the engine, η_{mech}, is defined as

$$\eta_{\text{mech}} = \frac{\text{BP}}{\text{IP}} \tag{3.11}$$

An engine is said to be *double acting*, if the working fluid is made to work on both sides of the piston. Such an engine theoretically develops twice the amount of work developed in a single-acting engine. Most reciprocating steam engines are double acting, and so are many marine diesel engines. Internal combustion engines for road transport are always single acting.

3.5 OTHER TYPES OF WORK TRANSFER

LO 3.5

Discuss works other than pdV-work

There are forms of work other than pdV or displacement work. The followings are the additional types of work transfer which may get involved in system-surroundings interactions.

1. Electrical work When a current flows through a resistor (Fig. 3.13), taken as a system, there is a work transfer into the system. This is because the current can drive a motor, the motor can drive a pulley and the pulley can raise a weight.

The current flow, I, in amperes, is given by

$$I = \frac{dC}{d\tau}$$

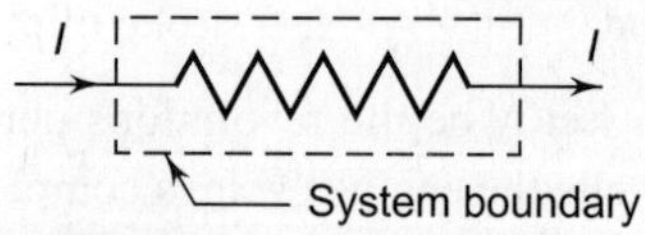

Fig. 3.13 Electrical work

where C is the charge in coulombs and τ is time in seconds. Thus dC is the charge crossing a boundary during time $d\tau$. If E is the voltage potential, the work is

$$đW = E \cdot dC$$

$$= EI\, d\tau$$

$$\therefore \quad W = \int_1^2 EI\, d\tau \tag{3.12}$$

The electrical power will be

$$\dot{W} = \lim_{d\tau \to 0} \frac{đW}{d\tau} = EI \tag{3.13}$$

This is the rate at which work is transferred.

2. Shaft work When a shaft, taken as the system (Fig. 3.14a), is rotated by a motor, there is work transfer into the system. This is because the shaft can rotate a pulley which can raise a weight. If T is the torque applied to the shaft and $d\theta$ is the angular displacement of the shaft, the shaft work is

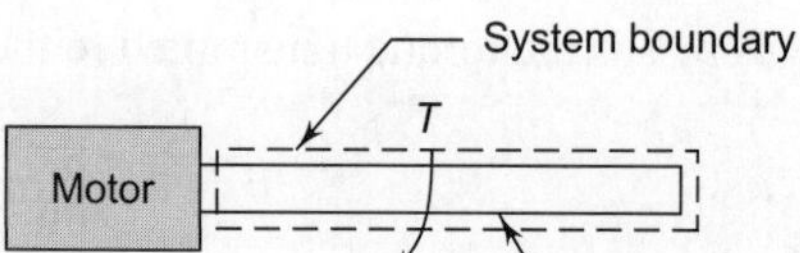

Fig. 3.14(a) Shaft work

$$W = \int_1^2 T d\theta \tag{3.14}$$

and the shaft power is

$$\dot{W} = \int_1^2 T \frac{d\theta}{d\tau} = T\omega \tag{3.14b}$$

where ω is the angular velocity and T is considered a constant in this case.

Some common shaft work machines are shown in Fig. 3.14(b).

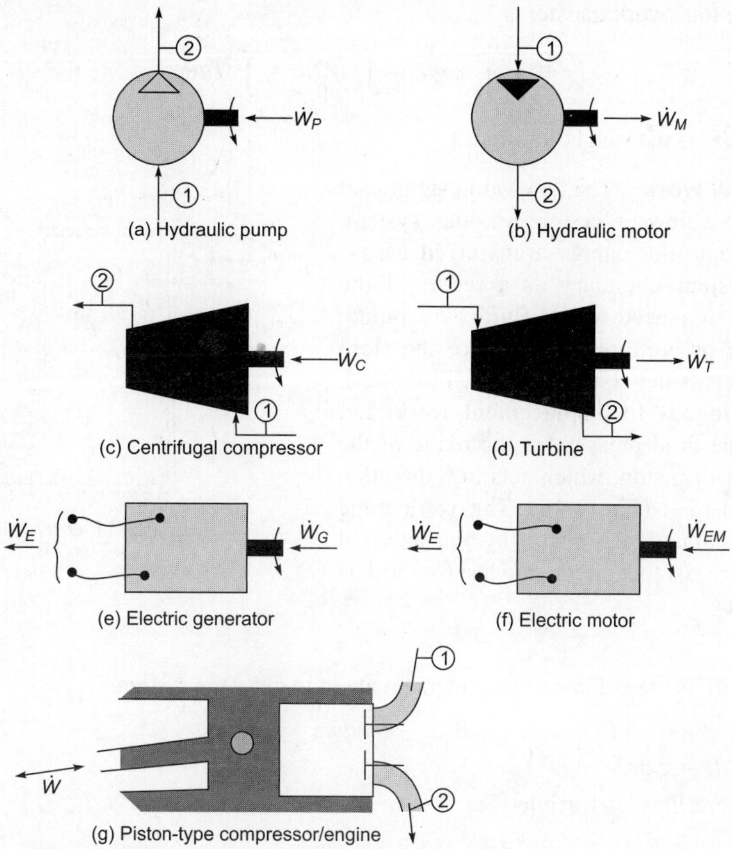

Fig. 3.14(b) Graphical symbols for common shaft work machines

3. Paddle-wheel work or stirring work As the weight is lowered, and the paddle wheel turns (Fig. 3.15), there is work transfer into the fluid system which gets stirred. Since the volume of the system remains constant, $\int pdV = 0$. If m is the mass of the weight lowered through a distance dz and T is the torque transmitted by the shaft in rotating through an angle $d\theta$, the differential work transfer to the fluid is given by

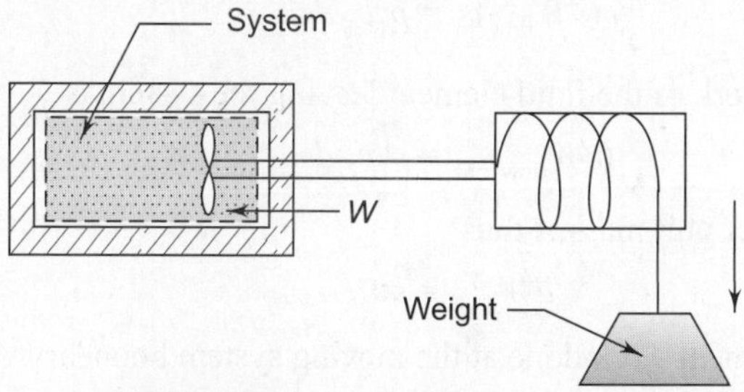

Fig. 3.15 Paddle-wheel work

$$đW = mgdz = Td\theta$$

and the total work transfer is

$$W = \int_1^2 mgdz = \int_1^2 W'\, dz = \int_1^2 Td\theta \tag{3.15}$$

where W' is the weight lowered.

4. Flow work The flow work, significant only in a flow process or an open system, represents the energy transferred across the system boundary as a result of the energy imparted to the fluid by a pump, blower or compressor to make the fluid flow across the control volume. Flow work is analogous to displacement work. Let p be the fluid pressure in the plane of the imaginary piston, which acts in a direction normal to it (Fig. 3.16). The work done on this imaginary piston by the external pressure as the piston moves forward is given by

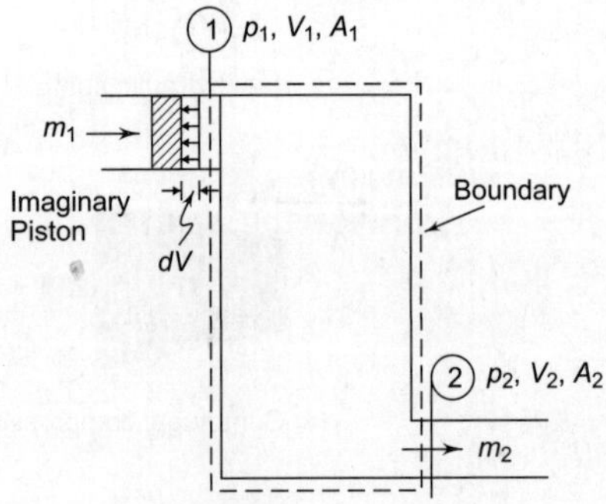

Fig. 3.16 Flow work

$$đW_{\text{flow}} = p\, dV, \tag{3.16}$$

where dV is the volume of fluid element about to enter the system.

$$\therefore \quad đW_{\text{flow}} = pv\, dm \tag{3.17}$$

where $dV = v\, dm$

Therefore, flow work at inlet (Fig. 3.16),

$$(đW_{\text{flow}})_{\text{in}} = p_1 v_1\, dm_1 \tag{3.18}$$

Equation (3.18) can also be derived in a slightly different manner. If the normal pressure p_1 is exerted against the area A_1, giving a total force $(p_1\, A_1)$ against the piston, in time $d\tau$, this force moves a distance $\mathbf{V}_1 d\tau$, where $\mathbf{V}_1$ is the velocity of flow (piston). The work in the time $d\tau$ is $p_1\, A_1\, \mathbf{V}_1\, d\tau$, or the work per unit time is $p_1\, A_1\, \mathbf{V}_1$. Since the flow rate

$$\omega_1 = \frac{A_1\, \mathbf{V}_1}{v_1} = \frac{dm_1}{d\tau}$$

the work done in time $d\tau$ becomes

$$(đW_{\text{flow}})_{\text{in}} = p_1 v_1\, dm_1$$

Similarly, flow work of the fluid element leaving the system is

$$(đW_{\text{flow}})_{\text{out}} = p_2 v_2\, dm_2 \tag{3.19}$$

The flow work per unit mass is thus

$$đW_{\text{flow}} = pv \tag{3.20}$$

It is the displacement work done at the moving system boundary.

5. Work done in stretching a wire Let us consider a wire as the system. If the length of the wire in which there is a tension $\mathcal{F}$ is changed from L to $L + dL$, the infinitestimal amount of work that is done is equal to

$$đW = -\mathcal{F}\,dL$$

The minus sign is used because a positive value of dL means an expansion of the wire, for which work must be done on the wire, i.e. negative work. For a finite change of length,

$$W = -\int_1^2 \mathcal{F}\,dL \tag{3.21}$$

If we limit the problem to within the elastic limit, where E is the modulus of elasticity, s is the stress, ε is the strain and A is the cross-sectional area, then

$$\mathcal{F} = sA = E\varepsilon A, \text{ since } \frac{s}{\varepsilon} = E$$

$$d\varepsilon = \frac{dL}{L}$$

$$đW = -\mathcal{F}\,dL = -E\varepsilon AL\,d\varepsilon$$

$$\therefore \quad W = -A\varepsilon L \int_1^2 \varepsilon d\varepsilon = -\frac{AEL}{2}(\varepsilon_2^2 - \varepsilon_1^2) \tag{3.22}$$

6. Work done in changing the area of a surface film A film on the surface of a liquid has a surface tension, which is a property of the liquid and the surroundings. The surface tension acts to make the surface area of the liquid a minimum. It has the unit of force per unit length. The work done on a homogeneous liquid film in changing its surface area by an infinitesimal amount dA is

$$đW = -\sigma\,dA$$

where σ is the surface tension (N/m).

$$\therefore \quad W = -\int_1^2 \sigma\,dA \tag{3.23}$$

7. Magnetisation of a paramagnetic solid The work done per unit volume on a magnetic material through which the magnetic and magnetisation fields are uniform is

$$đW = -H dI$$

and

$$W_{1-2} = -\int_{I_1}^{I_2} H dI \tag{3.24}$$

where H is the field strength, and I is the component of the magnetisation field in the direction of the field. The minus sign provides that an increase in magnetisation (positive dI) involves negative work.

The following equations summarise the different forms of work transfer:

Displacement work (compressible fluid) $\quad W = \int_1^2 pdV$

Electrical work $\quad W = \int_1^2 E\,dC = \int_1^2 EI\,d\tau$

Shaft work $W = \int_1^2 T\, d\theta$

Surface film $W = -\int_1^2 \sigma\, dA$ (3.25)

Stretched wire $W = -\int_1^2 \mathcal{F}\, dL$

Magnetised solid $W = -\int_1^2 H\, dI$

It may be noted in the above expressions that the work is equal to the integral of the product of an intensive property and the change in its related extensive property. These expressions are valid only for infinitesimally slow quasi-static processes.

There are some other forms of work which can be identified in processes that are not quasi-static, for example, the work done by shearing forces in a process involving friction in a viscous fluid.

3.6 FREE EXPANSION WITH ZERO WORK TRANSFER

LO 3.6

Recognize free expansion

Work transfer is identified only at the boundaries of a system. It is a boundary phenomenon, and a form of energy in transit crossing the boundary. Let us consider a gas separated from the vacuum by a partition (Fig. 3.17). Let the partition be removed. The gas rushes to fill the entire volume. The expansion of a gas against vacuum is called *free expansion*. If we neglect the work associated with the removal of partition, and consider the gas and vacuum together as our system (Fig. 3.17a), there is no work transfer involved here, since no work crosses the system boundary, and hence

$$\int_1^2 đW = 0, \quad \text{although} \int_1^2 pdV \neq 0$$

If only the gas is taken as the system (Fig. 3.17b), when the partition is removed there is a change in the volume of the gas, and one is tempted to calculate the work from the expression $\int_1^2 pdV$. However, this is not a quasi-static process, although the initial and final end states are in equilibrium. Therefore, the work cannot be calculated from this relation. The two end states can be located on the p–V diagram and these are joined by a dotted line (Fig. 3.17c) to indicate that the process had occurred. However, if the vacuum space is divided into a large number of small volumes by partitions and the partitions are removed one by one slowly (Fig. 3.17d), then every state passed through by the system is an equilibrium state and the work done can then be estimated from the relation $\int_1^2 pdV$ (Fig. 3.17e). Yet, in free expansion of a gas, there is no resistance to the fluid at the system boundary as the volume of the gas increases to fill up the vacuum space. Work is done by a system to overcome some resistance. Since vacuum does not offer any resistance, there is no work transfer involved in free expansion.

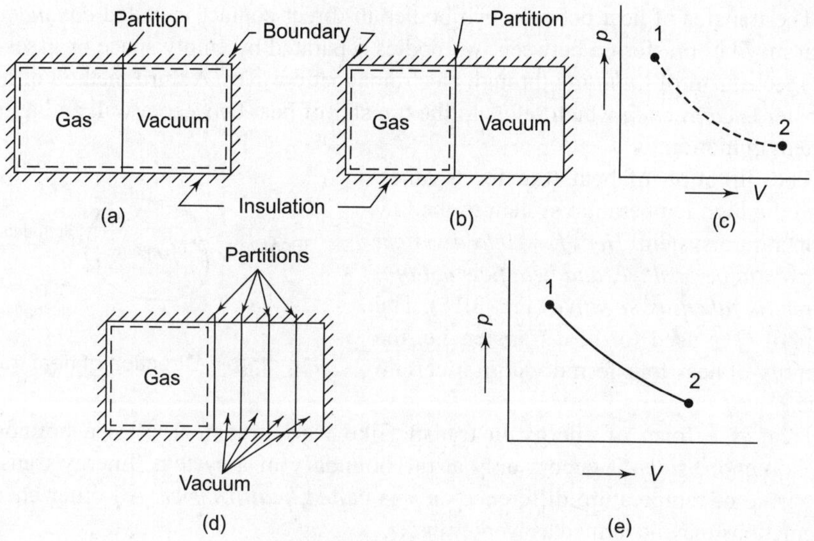

Fig. 3.17 Free expansion

3.7 NET WORK DONE BY A SYSTEM

Often different forms of work transfer occur simultaneously during a process executed by a system. When all these work interactions have been evaluated, the total or net work done by the system would be equal to the algebraic sum of these as given below:

$$W_{\text{total}} = W_{\text{displacement}} + W_{\text{shear}} + W_{\text{electrical}} + W_{\text{stirring}} + \ldots$$

3.8 HEAT TRANSFER

LO 3.7 Discuss heat transfer and its sign convention

If a system has a non-adiabatic boundary, its temperature is not independent of the temperature of the surroundings, and for the system between the states 1 and 2, the work W depends on the path, and the differential $đW$ is inexact. The work depends on the terminal states 1 and 2 as well as the non-adiabatic path connecting them. For consistency with the principle of conservation of energy, some other type of energy transfer besides work must have taken place between the system and surroundings. This type of energy transfer must have occurred because of the temperature difference between the system and its surroundings, and it is identified as 'heat'. Thus, when an effect in a system occurs solely as a result of temperature difference between the system and some other system, the process in which the effect occurs shall be called a transfer of heat from the system at the higher temperature to the system at the lower temperature.

Heat is defined as the form of energy that is transferred across a boundary *by virtue of a temperature difference*. The temperature difference is the 'potential' or 'force' and heat transfer is the 'flux'.

The transfer of heat between two bodies in direct contact is called *conduction.* Heat may be transferred between two bodies separated by empty space or gases by the mechanism of *radiation* through electromagnetic waves. A third method of heat transfer is *convection* which refers to the transfer of heat between a wall and a fluid system in motion.

The direction of heat transfer is taken from the high temperature system to the low temperature system. *Heat flow into a system is taken to be positive, and heat flow out of a system is taken as negative* (Fig. 3.18). The symbol Q is used for heat transfer, i.e. the quantity of heat transferred within a certain time.

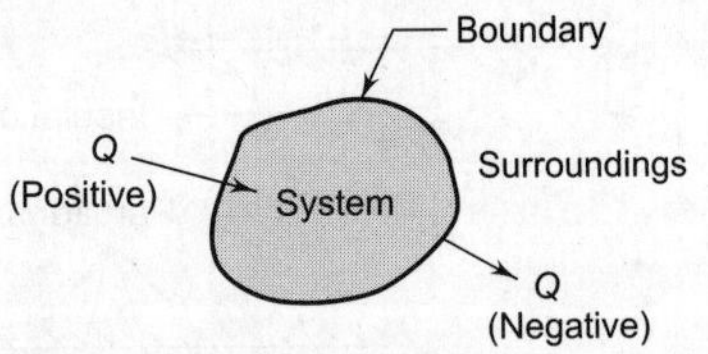

Fig. 3.18 Direction of heat transfer

Heat is a form of energy in transit (like work transfer). It is a boundary phenomenon, since it occurs only at the boundary of a system. Energy transfer by virtue of temperature difference only is called *heat transfer*. All other energy interactions may be termed as work transfer.

Heat is not that which inevitably causes a temperature rise. When heat is transferred to an ice-and-water mixture, the temperature does not rise until all the ice has melted. When a temperature rise in a system occurs, it may not be due to heat transfer, since a temperature rise may be caused by work transfer also. Heat, like work, is not a conserved quantity, and is not a property of a system.

A process in which no heat crosses the boundary of the system is called an *adiabatic process*.

Thus, an adiabatic process is one in which there is only work interaction between the system and its surroundings.

A wall which is impermeable to the flow of heat is an *adiabatic wall*, whereas a wall which permits the flow of heat is a *diathermic wall*.

The unit of heat is Joule in S.I. units.

The rate of heat transfer or work transfer is given in kW or W.

3.9 HEAT TRANSFER—A PATH FUNCTION

> **LO 3.8**
> Analyze heat and work transfer as path function

The heat flow Q in a process can be quantified in terms of the work W done in the same process between two given terminal states. Work W is different for different non-adiabatic processes between these two states and it is not conserved. However, from the principle of conservation of energy, the difference $(Q - W)$ is conserved for all paths between the same two states. The heat flow Q, like W, depends on the process and is path-dependent and not a property.

Heat transfer is a *path function*, that is, the amount of heat transferred when a system changes from state 1 to state 2 depends on the intermediate states through which the system passes, i.e. its path. Therefore, $đQ$ is an inexact differential and we write

$$\int_1^2 đQ = Q_{1-2} \quad \text{or} \quad {}_1Q_2$$

The displacement work is given by

$$W_{1-2} = \int_1^2 đW = \int_1^2 pdV$$

It is valid for a quasi-static process, and the work transfer involved is represented by the area under the path on *p-v* diagram (Fig. 3.19a). Whenever there is a difference in pressure, there will be displacement work. The pressure difference is the cause and work transfer is the effect. The work transfer is equal to the integral of the product of the intensive property, *p* and the differential change in the extensive property, *dV*.

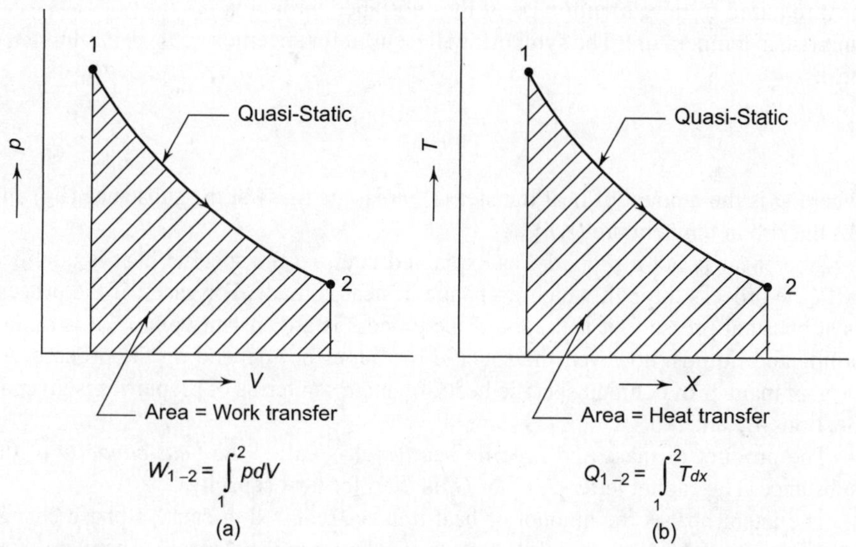

Fig. 3.19 Representation of work transfer and heat transfer in quasi-static processes on *p-v* and *T-x* coordinates

Likewise, whenever there is a difference in temperature, there will be heat flow. The temperature difference is the cause and heat transfer is the effect. Just like the displacement work, the heat transfer can also be written as the integral of the product of the intensive property *T* and the differential change of an extensive property, say *X* (Fig. 3.19b).

$$Q_{1-2} = \int_1^2 đQ = \int_1^2 TdX \tag{3.26}$$

It must also be valid for a quasi-static process only, and the heat transfer involved is represented by the area under the path 1–2 in *T-X* plot (Fig. 3.19b). Heat transfer is, therefore, a path function, i.e. the amount of heat transferred when a system changes from state 1 to state 2 depends on the path the system follows (Fig. 3.19b). Therefore $đQ$ is an inexact differential. Now,

$$đQ = T\,dX$$

where X is an extensive property and dX is an exact differential.

$$\therefore \qquad dX = \frac{1}{T}đQ \tag{3.27}$$

To make δQ integrable, i.e., an exact differential, it must be multiplied by an integrating factor which is, in this case, $1/T$. The extensive property X is yet to be defined. It has been introduced in Chapter 7 and is called 'entropy'.

3.10 SPECIFIC HEAT AND LATENT HEAT

LO 3.9 Demonstrate specific and latent heats

The *specific heat* of a substance is defined as the amount of heat required to raise a unit mass of the substance through a unit rise in temperature. The symbol c will be used for specific heat.

$$\therefore \qquad c = \frac{Q}{m \cdot \Delta t} \text{ J/kg K}$$

where Q is the amount of heat transfer (J), m is, the mass of the substance (kg) and Δt, the rise in temperature (K).

Since heat is not a property, as explained later, so the specific heat is qualified with the process through which exchange of heat is made. For gases, if the process is at constant pressure, it is c_p, and if the process is at constant volume, it is c_v. For solids and liquids, however, the specific heat does not depend on the process. An elegant manner of defining specific heats, c_v and c_p, in terms of properties is given in Sections 4.5 and 4.6.

The product of mass and specific heat (mc) is called the *heat capacity* of the substance. The capital letter C, C_p or C_v is used for heat capacity.

The latent heat is the amount of heat transfer required to cause a phase change in unit mass of a substance *at a constant pressure and temperature*. There are three phases in which matter can exist: solid, liquid and vapour or gas. The *latent heat of fusion* (l_{fu}) is the amount of heat transferred to melt unit mass of solid into liquid, or to freeze unit mass of liquid into solid. The *latent heat of vaporisation* (l_{vap}) is the quantity of heat required to vaporise unit mass of liquid into vapour or condense unit mass of vapour into liquid. The *latent heat of sublimation* (l_{sub}) is the amount of heat transferred to convert unit mass of solid to vapour or vice versa. l_{fu} is not much affected by pressure, whereas l_{vap} is highly sensitive to pressure.

3.11 POINTS TO REMEMBER REGARDING HEAT TRANSFER AND WORK TRANSFER

1. Heat transfer and work transfer are the *energy interactions*. A closed system and its surroundings can interact in two ways: by heat transfer and by work transfer. Thermodynamics studies how these interactions bring about property changes in a system.

2. The same effect in a closed system can be brought about either by heat transfer or by work transfer. Whether heat transfer or work transfer has taken place or not depends on what constitutes the system.
3. Both heat transfer and work transfer are boundary phenomena. Both are observed at the boundaries of the system, and both represent energy crossing the boundaries of the system.
4. *It is wrong to say 'total heat' or 'heat content' of a closed system*, because heat or work is not a property of the system. Heat, like work, cannot be stored by the system. Both heat and work are the energy in transit.
5. Heat transfer is the energy interaction due to temperature difference only. All other energy interactions may be termed as work transfer.
6. Both heat and work are path functions and inexact differentials. The magnitude of heat transfer or work transfer depends upon the path the system follows during the change of state.

SOLVED EXAMPLES

Example 3.1 *An object of 40 kg mass falls freely under the influence of gravity from an elevation of 100 m above the earth's surface. The initial velocity is directed downward with a magnitude of 100 m/s. Ignoring the effect of air resistance, what is the magnitude of the velocity, in m/s, of the object just before it strikes the earth? The acceleration of gravity is g = 9.81 m/s².* E [LO 3.2]

Solution Since the only force is that due to gravity,

$$\frac{1}{2}mV_2^2 + mgz_2 = \frac{1}{2}mV_1^2 + mgZ_1$$

$$V_2 = \sqrt{2gz_1 + V_1^2}$$

$$= \sqrt{2 \times 9.81 \times 100 + 100^2}$$

$$= 104.4 \text{ m/s} \qquad \textbf{Ans.}$$

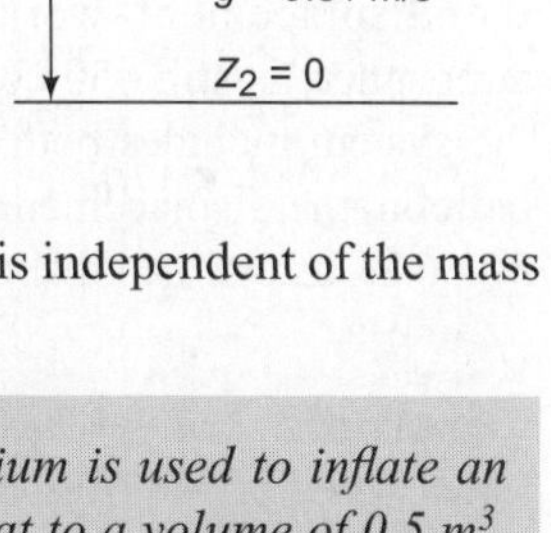

The velocity increases, as expected, and the magnitude is independent of the mass of the object.

Example 3.2 *Gas from a bottle of compressed helium is used to inflate an inelastic flexible balloon, originally folded completely flat to a volume of 0.5 m³. If the barometer reads 760 mm Hg, what is the amount of work done upon the atmosphere by the balloon? Sketch the system before and after the process.* E [LO 3.3]

Solution The firm line P_1 (Fig. 3.20) shows the boundary of the system before the process, and the dotted line P_2 shows the boundary after the process. The displacement work

$$W_d = \int_{\text{Balloon}} pdV + \int_{\text{Bottle}} pdV = p\,\Delta V + 0$$

$$= 101.325\ \frac{\text{kN}}{\text{m}^2} \times 0.5\ \text{m}^3 = \mathbf{50.66\ kJ}$$

This is positive, because work is done by the system. Work done by the atmosphere is –50.66 kJ. Since the wall of the bottle is rigid, there is no pdV-work involved in it.

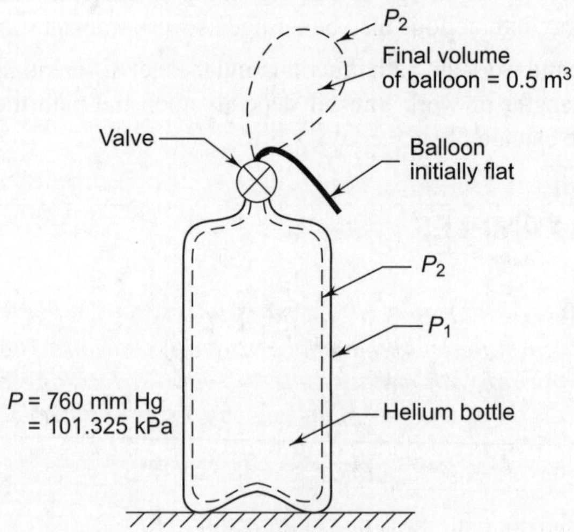

Fig. 3.20

It is assumed that the pressure in the balloon is atmospheric at all times, since the balloon fabric is light, inelastic and unstressed. If the balloon were elastic and stressed during the filling process, the work done by the gas would be greater than 50.66 kJ by an amount equal to the work done in stretching the balloon, although the displacement work done by the atmosphere is still –50.66 kJ. However, if the system includes both the gas and the balloon, the displacement work would be 50.66 kJ, as estimated above.

Example 3.3 *When the valve of the evacuated bottle (Fig. 3.21) is opened, atmospheric air rushes into it. If the atmospheric pressure is 101.325 kPa, and 0.6 m^3 of air (measured at atmospheric conditions) enters into the bottle, calculate the work done by air.* E [LO 3.2]

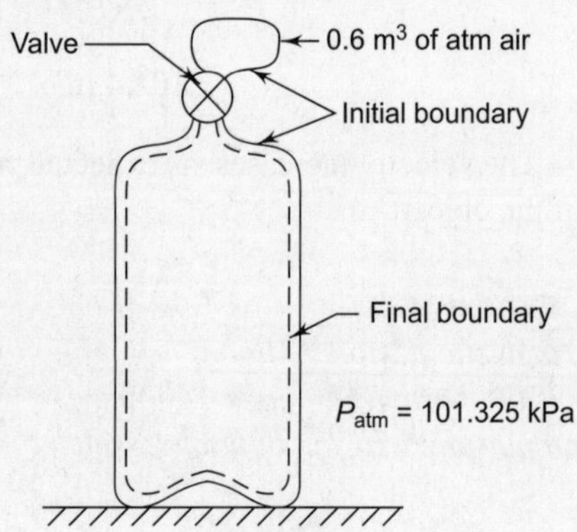

Fig. 3.21

Solution The displacement work done by air

$$W_d = \int_{\text{Bottle}} pdV + \int_{\substack{\text{Free-air}\\ \text{boundary}}} pdV$$

$$= 0 + p\,\Delta V$$

$$= 101.325 \text{ kN/m}^2 \times 0.6 \text{ m}^3 = \mathbf{60.8 \text{ kJ}}$$

Since the free-air boundary is contracting, the work done by the system is negative (ΔV being negative), and the surroundings do positive work upon the system.

Example 3.4 *A 2 kW, 25 litre storage water heater is used to raise the temperature of incoming cold water to a desired pre-set end use temperature ranging from 25°C to 75°C. Assume that all the electrical energy has gone into heating the water. If the cold water temperature is 15°C and the end use temperature is 50°C, determine the time required to reach the end use temperature after the heater is switched on. The heat capacity c_p for water is 4.2 kJ/kgK and density of water is 1000 kg/m³.* Ⓜ [LO 3.9]

Solution Mass of water in the heater is $25 \times 10^{-3} \times 1000 = 25$ kg

Heat required to raise the temperature of water from 15°C to 50°C is

$$Q = 25 \times 4.2(50 - 15) = 3675 \text{ kJ}$$

Energy input to the water per minute is $= 2 \text{ kJ/s} \times 60 \text{ s} = 120$ kJ

Time required to raise the temperature is

$$= \frac{3675}{120} \text{ minutes} = 30.625 \text{ minutes}$$

Example 3.5 *A piston and cylinder machine containing a fluid system has a stirring device in the cylinder (Fig. 3.22). The piston is frictionless, and it is held down against the fluid due to the atmospheric pressure of 101.325 kPa. The stirring device is turned 10,000 revolutions with an average torque against the fluid of 1.275 mN. Meanwhile the piston of 0.6 m diameter moves out 0.8 m. Find the net work transfer for the system.*

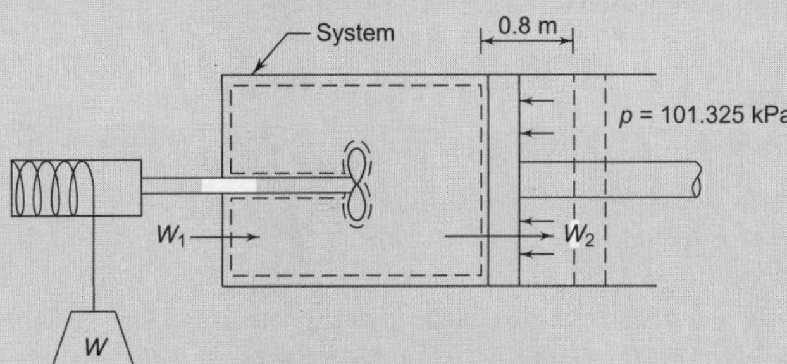

Fig. 3.22 Ⓜ [LO 3.5]

Solution Work done by the stirring device upon the system (Fig. 3.22)

$$W_1 = 2\pi TN$$

$$= 2\pi \times 1.275 \times 10{,}000 \text{ Nm} = 80 \text{ kJ}$$

This is negative work for the system.

Work done by the system upon the surroundings

$$W_2 = (pA) \cdot L$$

$$= 101.325 \frac{\text{kN}}{\text{m}^2} \times \frac{\pi}{4} (0.6)^2 \text{ m}^2 \times 0.80 \text{ m} = 22.9 \text{ kJ}$$

This is positive work for the system. Hence, the net work transfer for the system

$$W = W_1 + W_2 = -80 + 22.9 = -57.1 \text{ kJ}$$

Example 3.6 *The following data refer to a 12-cylinder, single-acting, two-stroke marine diesel engine:*

Speed—150 rpm
Cylinder diameter—0.8 m
Stroke of piston—1.2 m
Area of indicator diagram—5.5 × 10⁻⁴ m²
Length of diagram—0.06 m
Spring value—147 MPa per m

Find the net rate of work transfer from the gas to the pistons in kW. Ⓜ [LO 3.4]

Solution Mean effective pressure, p_m, is given by

$$p_m = \frac{a_d}{l_d} \times \text{spring constant}$$

$$= \frac{5.5 \times 10^{-4} \text{ m}^2}{0.06 \text{ m}} \times 147 \frac{\text{MPa}}{\text{m}} = 1.35 \text{ MPa}$$

One engine cycle is completed in two strokes of the piston or one revolution of the crank-shaft.

∴ work done in one minute

$$= p_m LAN$$

$$= 1.35 \times \frac{\pi}{4} (0.8)^2 \times 1.2 \times 150 = 122 \text{ MJ}$$

Since the engine is single acting, and it has 12 cylinders, each contributing an equal power, the rate of work transfer from the gas to the piston is given by

$$\mathbf{W} = 122 \times 12 \text{ MJ/min}$$

$$= 24.4 \text{ MJ/s}$$

$$= 24.4 \text{ MW} = \mathbf{24,400 \text{ kW}}$$

Example 3.7 *It is required to melt 5 tonnes/h of iron from a charge at 15°C to molten metal at 1650°C. The melting point is 1535°C and the latent heat is 270 kJ/kg. The specific heat in solid state is 0.502 and in liquid state (29.93/atomic weight) kJ/kg K. If an electric furnace has 70 percent efficiency, find the kW rating needed. If the density in molten state is 6900 kg/m³ and the bath volume is three times the hourly melting rate, find the dimensions of the cylindrical furnace if the length to diameter ratio is 2. The atomic weight of iron is 56.* Ⓓ [LO 3.8]

Solution Heat required to melt 1 kg of iron at 15°C to molten metal at 1650°C = Heat required to raise the temperature from 15°C to 1535°C + Latent heat + Heat required to raise the temperature from 1535°C to 1650°C

$$= 0.502\,(1535 - 15) + 270 + 29.93\,(1650 - 1535)/56$$
$$= 763 + 270 + 61.5 = 1094.5 \text{ kJ/kg}$$

Melting rate = 5×10^3 kg/h

So, the rate of heat supply required

$$= (5 \times 10^3 \times 1094.5) \text{ kJ/h}$$

Since the furnace has 70 percent efficiency, the rating of the furnace would be

$$= \frac{\text{Rate of heat supply per second}}{\text{Furnace efficiency}}$$

$$= \frac{5 \times 10^3 \times 1094.5}{0.7 \times 3600} = \mathbf{217 \times 10^3\ kW}$$

$$\text{Volume needed} = \frac{3 \times 5 \times 10^3}{6900} \text{ m}^3 = 2.18 \text{ m}^3$$

If d is the diameter and l the length of the furnace

$$\frac{\pi}{4} d^2\, l = 2.18 \text{ m}^3$$

or $$\frac{\pi}{4} d^2 \times 2d = 2.18 \text{ m}^3$$

$\therefore$ $$d = 1.15 \text{ m}$$

and $$l = \mathbf{2.30\ m}$$

Example 3.8 *If it is desired to melt aluminium with solid state specific heat 0.9 kJ/kg K latent heat 390 kJ/kg, atomic weight 27, density in molten state 2400 kg/m³ and final temperature 700°C, find out how much metal can be melted per hour with the above kW rating. Other data are as in the above example. Also, find the mass of aluminium that the above furnace will hold. The melting point of aluminium is 660°C.* M **[LO 3.8]**

Solution Heat required per kg of aluminium

$$= 0.9\,(660 - 15) + 390 + \frac{29.93}{27}\,(700 - 660)$$
$$= 580.5 + 390 + 44.3 = 1014.8 \text{ kJ}$$

$$\text{Heat to be supplied} = \frac{1014.8}{0.7} = 1449.7 \text{ kJ/kg}$$

With the given power, the rate at which aluminium can be melted

$$= \frac{217 \times 10^3 \times 3600}{1449.7} \mathbf{\ kg/h = 5.39\ tonnes/h}$$

Mass of the aluminium that can be held in the above furnace

$$= 2.18 \times 2400 \text{ kg} = \mathbf{5.23\ tonnes}$$

Example 3.9 *A cooling tower nozzle disperses water into a stream of droplets. If the average diameter of the droplets is 60 microns, estimate the work required for atomizing 1 kg of water isothermally at the ambient conditions. Given: surface tension of water in contact with air = 0.07 N/m, density of water = 1000 kg/m³. Water is assumed to enter the nozzle through a pipe of 15 mm diameter.* D **[LO 3.6]**

Solution As water passes through the nozzle and comes out in the form of a spray of droplets, there is a tremendous increase in the surface area. The process of atomization can be viewed as analogous to increase in the surface area of the film, and the work done on the film is

$$W = \int_i^f \sigma\, dA = \sigma (A_f - A_i)$$

where A_f = final area, A_i = initial area

If N = number of droplets with an average diameter of 60 μm,

$$N \times \frac{\pi}{6}\,(60 \times 10^{-6})^3 \times 1000 = 1 \text{ kg}$$

$$\therefore \qquad N = \frac{6}{\pi \times 216 \times 10^{-15} \times 1000} = 8.842 \times 10^9$$

Total surface area of all the droplets

$$A_f = \pi\,(60 \times 10^{-6})^2 \times 8.842 \times 10^9 = 100 \text{ m}^2$$

If we assume water enters the nozzle through a pipe of 15 mm diameter and 1 kg of water is contained in $L(m)$ length of the inlet pipe, then

$$\frac{\pi}{4} d^2 L \rho = 1 \text{ kg}$$

Surface area would be

$$S = \pi d L \text{ m}^2$$

Dividing, $$\frac{S}{L} = \frac{\pi\, dL}{\frac{\pi}{4} d^2 L \rho} = \frac{4}{15 \times 10^{-3} \times 1000} = 0.267 \text{ m}^2/\text{m}$$

Work done during atomisation

$$W = (0.07)\,(100 - 0.287) = \mathbf{6.98\ J}$$

Example 3.10 *An electric motor drives a stirrer fitted with a horizontal cylinder. The cylinder of 40 cm diameter contains a fluid restrained by a frictionless piston. During the stirring of fluid for 15 minutes the piston moves outward slowly by a distance of 30 cm against the atmospheric pressure of 1 bar. The current supplied to the motor is 0.5 amp. from a 24-V lead-acid accumulator. If the conversion efficiency from electrical work to mechanical work output is 90 percent, estimate the work done on the motor, stirrer and the atmosphere.*

M **[LO 3.5]**

Solution Work input to the motor

$$W_m = \int V\, dQ_c = 24 \times 0.5 \times 15 \times 60$$

$$= 10800 \text{ J} = \mathbf{10.8\ kJ}$$

Work input to the stirrer

$$W_s = 10.8 \times 0.9 = \mathbf{9.72\ kJ}$$

Work done by the fluid on the atmosphere

$$W = p \cdot \Delta V = 10^5 \times \frac{\pi}{4}\,(0.4)^2 \times 0.3 \text{ Nm}$$

$$= 3.77 \text{ kJ} \quad \textit{Ans.}$$

The fluid receives a work input of 9.72 kJ from the stirrer and does 3.77 kJ of work on the atmosphere at the moving foundary.

Example 3.11 *The authority of an engineering college has decided to install window air-conditioner(s) into a newly constructed seminar room of capacity 80 people with an objective to maintain the temperature inside the room as 22°C. The rate of heat transfer to the classroom through the walls, roofs and the windows because of the temperature difference between the ambient and the room is estimated to be 24000 kJ/h. There are 20 tubelights in the room, each with a rating of 40 W. The rate of dissipation of heat by a tubelight is 90 percent of the power it consumes. The average rate of heat dissipation by a person is approximated as 350 kJ/h. If the capacity of an air-conditioning unit is 2 tonnes, determine the number of air-conditioning units required. Note that 1 tonne cooling capacity is equivalent to heat removal at the rate of 14000 kJ/h. Assume that there is no computer or projector in the room.* E [LO 3.8]

Solution The total rate of heat that is to be transferred from the seminar room is

$$\dot{Q} = \dot{Q}_{\text{heat gain}} + \dot{Q}_{\text{tubelights}} + \dot{Q}_{\text{people}}$$

where $\dot{Q}_{\text{heat gain}}$ is the rate of heat transfer to the room from the ambient through the walls, roofs and the windows, $\dot{Q}_{\text{tubelight}}$ is the rate of heat generated inside the room by the tubelights and $\dot{Q}_{\text{people}}$ is the rate of heat dissipation by the persons sitting inside the room.

Here $$\dot{Q}_{\text{heat gain}} = 24000 \text{ kJ/h}$$

$$\dot{Q}_{\text{tubelights}} = 20 \times 40 \times 0.9 \text{ W} = \frac{720 \times 3600}{1000} \text{ kJ/h} = 2592 \text{ kJ/h}$$

and $$\dot{Q}_{\text{people}} = 80 \times 350 = 28000 \text{ kJ/h}$$

Therefore, $$\dot{Q} = 24000 + 2592 + 28000 = 54592 \text{ kJ/h}$$

The number of air-conditioning unit of cooling capacity 2 tonnes required is found to be

$$n = \frac{54592}{2 \times 14000} = 1.94 \rightarrow 2$$

Example 3.12 *One undergraduate student has taken a project on the estimation of misuse of energy on the academic complex of his institute. The academic complex has 200 faculty rooms, 70 classrooms and 30 laboratories. The classrooms and laboratories are equipped with 15 tubelights, each with a rating of 40 W. Each faculty room is equipped with four tubelights, each consuming 40 W. The institute is open 240 days per year for faculty members and 150 days for students. The laboratory classes are scheduled for 90 days a year. Although all the lights are kept on an average of 9 hours per day in the classrooms and laboratories, the students are not occupied average of 3 hours per day. Similarly, the faculty members are not in their office on an average 4 hours per day although the lights are kept on for 10 hours per day. If the lights in the classrooms, laboratories and faculty room are turned off during unoccupied periods, determine how much energy the institute can save in a year. Also find the energy saving percentage by turning off the lights during unoccupied periods.* E [LO 3.8]

Solution The energy consumed by the lights is found to be

$$\dot{W}_{\text{class room}} = 70 \times 15 \times 40 = 42000 \text{ W} = 42 \text{ kW}$$
$$\dot{W}_{\text{laboratory}} = 30 \times 15 \times 40 = 18000 \text{ W} = 18 \text{ kW}$$
$$\dot{W}_{\text{faculty room}} = 200 \times 4 \times 40 = 32000 \text{ W} = 32 \text{ kW}$$

The energy consumed by the lights per year is obtained as follows

$$\dot{W}_{\text{class room}} = 150 \times 9 \times 42 = 56700 \text{ kWh}$$
$$\dot{W}_{\text{laboratory}} = 90 \times 9 \times 18 = 14580 \text{ kWh}$$
$$\dot{W}_{\text{faculty room}} = 240 \times 10 \times 32 = 76800 \text{ kWh}$$

The total energy consumed by the lights per year is

$$\dot{W} = \dot{W}_{\text{class room}} + \dot{W}_{\text{laboratory}} + \dot{W}_{\text{faculty room}}$$
$$= 56700 + 14580 + 76800 = 148080 \text{ kWh}$$

The energy consumed by the lights per year during unoccupied period is

$$\dot{W}_{\text{unocc, class room}} = 150 \times 3 \times 42 = 18900 \text{ kWh}$$
$$\dot{W}_{\text{unocc, laboratory}} = 90 \times 3 \times 18 = 4860 \text{ kWh}$$
$$\dot{W}_{\text{unocc, faculty room}} = 240 \times 4 \times 32 = 30720 \text{ kWh}$$

The total energy consumed by the lights per year during unoccupied period is

$$\dot{W}_{\text{unoccupied}} = \dot{W}_{\text{unocc, class room}} + \dot{W}_{\text{unocc, laboratory}} + \dot{W}_{\text{unocc, faculty room}}$$
$$= 18900 + 4860 + 30720 = 54480 \text{ kWh}$$

The percentage saving in the energy due to turning off the lights during unoccupied periods is

$$= \frac{54480}{148080} = 0.3679 \text{ or } 36.79 \text{ percent}$$

Example 3.13 *A piston-cylinder device with air at an initial temperature of 30°C undergoes an expansion process for which pressure and volume are related as given below:*

p (kPa)	*100*	*37.9*	*14.4*
V (m³)	*0.1*	*0.2*	*0.4*

Calculate the work done by the system. E **[LO 3.3]**

Solution $p_1 V_1^n = p_2 V_2^n = p_3 V_3^n$

Here, $$n = \frac{\ln (p_1/p_2)}{\ln (V_2/V_1)} = \frac{\ln (100/37.9)}{\ln (0.2/0.1)} = 1.4$$

Also, $$n = \frac{\ln (p_2/p_3)}{\ln (V_3/V_2)} = \frac{\ln (37.9/14.4)}{\ln (0.4/0.2)} = 1.4$$

Therefore, n can be taken as 1.4 for the expansion process.

$\therefore$ $$W = \frac{p_1V_1 - p_3V_3}{n-1} = \frac{100 \times 0.1 - 14.4 \times 0.4}{1.4 - 1}$$
$$= \mathbf{10.6 \text{ kJ}}$$

Example 3.14 *A piston-cylinder device operates 1 kg of fluid at 20 atm. pressure. The initial volume is 0.04 m³. The fluid is allowed to expand reversibly following a process $pV^{1.45}$ = constant so that the volume becomes double. The fluid is then cooled at constant pressure until the piston comes back to the original position. Keeping the piston unaltered, heat is added reversibly to restore it to the initial pressure. Calculate the work done in the cycle.* E **[LO 3.3]**

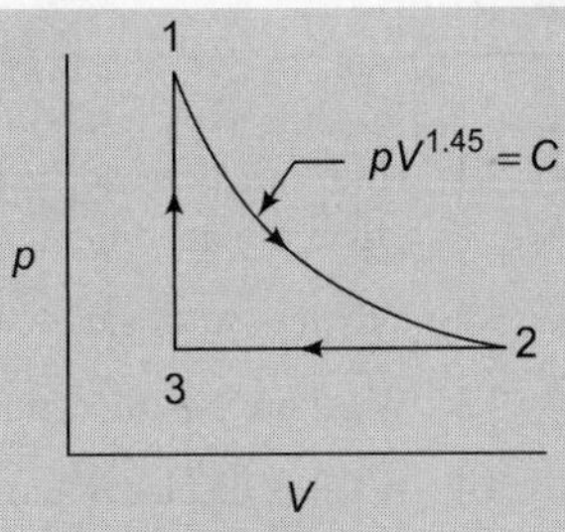

Fig. 3.23

Solution

$$p_1 = 20 \text{ atm} = 20 \times 1.01325 = 20.265 \text{ bar}$$

$$V_1 = 0.04 \text{ m}^3,\ V_2 = 2V_1 = 0.08 \text{ m}^3$$

$$p_2 = (V_1/V_2)^n \cdot p_1 = (1/2)^{1.45} \times 20.265$$

$$= 7.417 \text{ bar}$$

$$W_{1-2} = \frac{p_1V_1 - p_2V_2}{n-1} = \frac{100\,[20.265 \times 0.04 - 7.417 \times 0.08]}{0.45}$$

$$= 47.82 \text{ kJ}$$

$$W_{2-3} = p_2(V_2 - V_1) = -\,100 \times 0.04 \times 7.417 = -\,29.65 \text{ kJ}$$

Work in the cycle = 47.82 – 29.65 = **18.17 kJ**

SUMMARY

Work is done by a system if the sole effect on things external to the system could be reduced to the raising of a weight. The weight may not actually be raised, but the net effect external to the system would be the raising of a weight. Work done by a system is taken to be positive, if it is done upon the system it is negative. There are various types of work transfer crossing the boundary of a system:

(a) *pdV* work or displacement work, (b) electrical work, (c) shaft work, (d) stirring or paddle-wheel work, (e) flow work, etc. In a stationary fluid system, the work done by or upon the system is given by $\int pdV$. If dV is positive, W is positive and if dV is negative W is negative. An indicator diagram is similar to a pressure-volume plot, but it is a plot of pressure vs piston travel, and not a system pressure-volume diagram. The integration $\int pdV$ can only be done on a quasi-static path. Work done is a path function and δW is an inexact differential.

Any transfer of energy to a system by virtue of temperature difference is known as *heat*. Heat, like work, is a form of energy in transit and flows across a boundary. Heat flow to a system is arbitrarily taken to be positive and heat flow *out* of the system is regarded as negative. Heat is a path function and $đQ$ is an inexact differential.

The specific heat of a substance is defined as the amount of heat required to raise the temperature of unit mass of the substance by 1°C. For a compressible substance there can be two specific heats, c_p and c_v. The latent heat is the amount of heat transfer required to cause a phase change at constant pressure of unit mass of a substance like latent heats of fusion, vaporisation and sublimation.

REVIEW QUESTIONS

(M) 3.1 How can a closed system and its surroundings interact? What is the effect of such interactions on the system? **[LO 3.1]**

(M) 3.2 When is work said to be done by a system? **[LO 3.1]**

(M) 3.3 What are positive and negative work interactions? **[LO 3.1]**

(E) 3.4 What is displacement work? **[LO 3.2]**

(M) 3.5 Under what conditions is the work done equal to $\int_1^2 pdV$? **[LO 3.2]**

(D) 3.6 What do you understand by path function and point function? What are exact and inexact differentials? **[LO 3.3]**

(M) 3.7 Show that work is a path function, and not a property. **[LO 3.3]**

(D) 3.8 What is an indicator diagram? **[LO 3.4]**

(D) 3.9 What is mean effective pressure? How is it measured? **[LO 3.4]**

(M) 3.10 What are the indicated power and the brake power of an engine? **[LO 3.4]**

(D) 3.11 How does the current flowing through a resistor represent work transfer **[LO 3.5]**

(D) 3.12 What do you understand by flow work? Is it different from displacement work? **[LO 3.5]**

(E) 3.13 Why does free expansion have zero work transfer? **[LO 3.6]**

(M) 3.14 What is heat transfer? What are its positive and negative directions? **[LO 3.7]**

(E) 3.15 What are adiabatic and diathermic walls? **[LO 3.7]**

(M) 3.16 What is an integrating factor? **[LO 3.8]**

(M) 3.17 Show that heat is a path function and not a property. **[LO 3.8]**

(M) 3.18 What is the difference between work transfer and heat transfer? **[LO 3.8]**

(E) 3.19 Does heat transfer inevitably cause a temperature rise? **[LO 3.7]**

PROBLEMS WITH HINTS

(E) 3.1 An object of mass 40 kg falls freely under the influence of gravity from an elevation of 100 m above the earth's surface. The initial downward velocity is 100 m/s. Ignoring the effect of air resistance, what is the magnitude of velocity when the object just strikes the earth. Take $g = 9.81$ m/s². **[LO 3.1]**

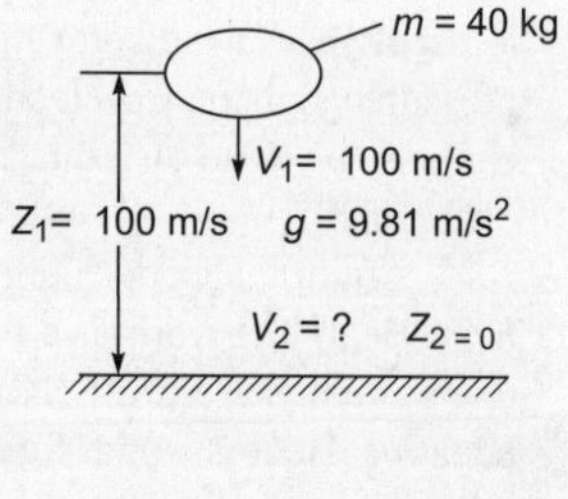

Hint:

Since the only force is that due to gravity $\frac{1}{2}mV_2^2 = \frac{1}{2}mV_1^2 + mgZ_1$

$$\therefore \quad V_2 = \sqrt{2gZ_1 + V_1^2}$$

Substituting, V_2 = **109.4 m/s**

(E) **3.2** A gas expands from an initial state where the pressure is 340 kPa and the volume is 0.0425 m³ to a final pressure of 136 kPa. The relationship between the pressure and volume of the gas is pv^2 = constant. Determine the work for the process. **[LO 3.1]**

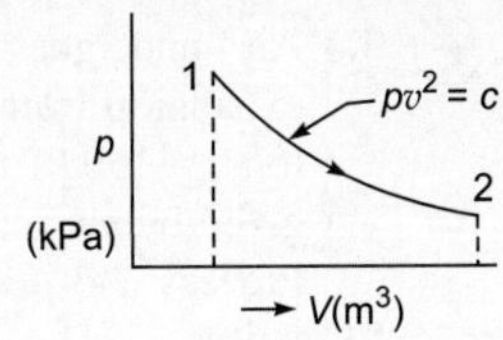

Hint:

From the relation,

$$PV^2 = C$$
$$P = CV^{-2}$$
$$\frac{V_2}{V_1} = \sqrt{\frac{P_1}{P_2}}$$

Final $V_2 = 0.0672\ \text{m}^3$

$$W_{1-2} = \int_1^2 pdv = (P_1V_1^2)\left(\frac{1}{V_1} - \frac{1}{V_2}\right) = \mathbf{5.31\ kJ.}$$

(M) **3.3** Air is trapped in a piston-cylinder machine (Fig. 3.24). Initially P_1 = 100 kPa, $V_1 = 2 \times 10^{-3}$ m³. The spring exerts no force on the piston in the initial

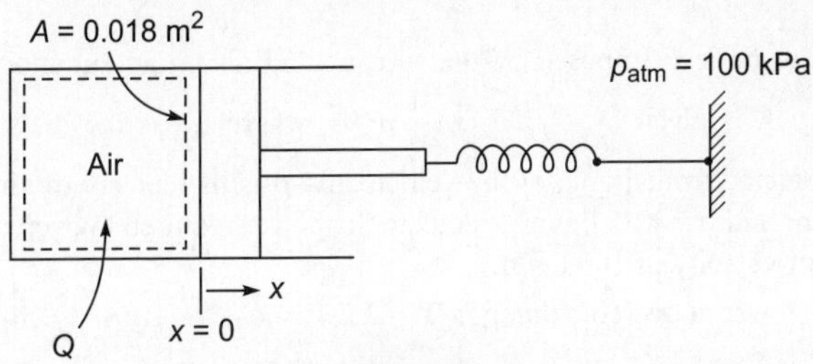

Fig. 3.24

position. The atmospheric pressure is 100 kPa and the area of the piston is 0.018 m². The air expands slowly till its volume is $V_2 = 3 \times 10^{-3}$ m³. During the process, the spring exerts a force on the piston $F = -kx$, where $k = 16.2 \times 10^3$ N/m. Neglecting any friction between piston and cylinder wall, determine the final pressure of air and the work done by the air on the piston. **LO 3.4]**

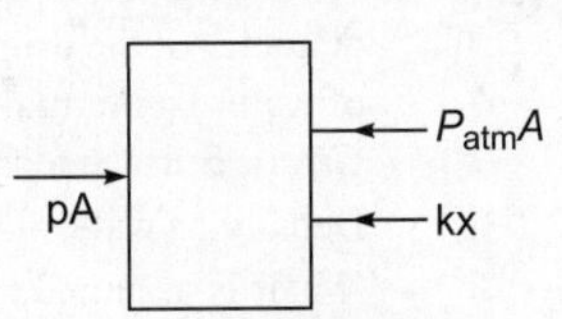

Fig. 3.25

Hint:

From the free body diagram on the left, $P_2A = P_{atm}A + kx_2$

Now, $V_2 - V_1 = A(x_2 - x_1) = 0$

$$\therefore \quad P_2 = P_{atm} + \frac{kx_2}{A}$$

$$x_2 = \frac{V_2 - V_1}{A} = 5.56 \times 10^{-2}\ \text{m}$$

$$\therefore \quad P_2 = \mathbf{150\ kPa}$$

The work done:
$$W = \int_{x_1}^{x_2} pAdx, \text{ where } P = P_{atm} + \frac{kx}{A}$$
$$= \mathbf{0.125\ kJ}$$

(E) **3.4** Air undergoes two processes: Process 1–2 expansion from $P_1 = 300$ kPa, $v_1 = 0.019$ m^3/kg to $P_2 = 150$ kPa during which the p-v relation is given by pv = constant. **[LO 3.2]**

Process 2-3 constant pressure compression to $v_3 = v_1$.

Sketch the processes on a p-v diagram and determine the work done per unit mass.

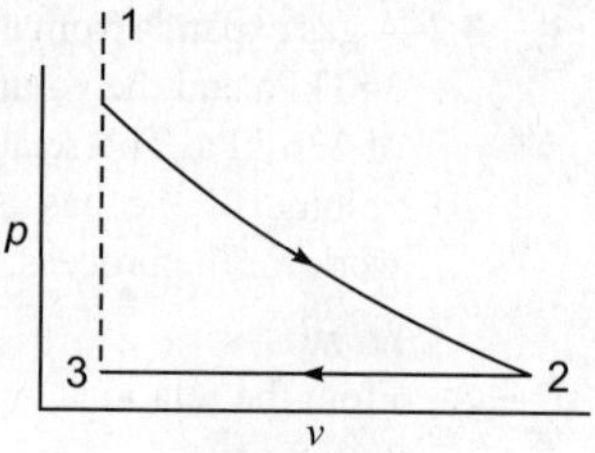

Fig. 3.26

Hint:

$$W_{1-2} = \int_1^2 pdv = P_1 v_1 \text{ In } \frac{v_2}{v_1}$$

where $$v_2 = \frac{p_1 v_2}{p_2} = 0.038 \text{ m}^3/\text{kg}$$

$$W_{1-2} = 3.95 \text{ kJ/kg}$$

$$W_{2-3} = P_2 (v_3 - v_2) = -2.85 \text{ kJ/kg}$$

$$\therefore \quad \frac{W}{m} = \mathbf{1.1\ kJ}$$

(M) **3.5** The drag force F imposed by the surrounding air on an automobile moving with velocity V is given by $F_d = C_d A \frac{1}{2} PV^2$, where C_D is the drag coefficient, A is the projected frontal area of the vehicle and ρ is the density of air. For $C_D = 0.42$, $A = 2$ m^2 and $\rho = 1.25$ kg/m^3, calculate the power required to overcome the drag at a constant velocity of 100 km/h. **[LO 3.5]**

Hint: Power to overcome drag is $\dot{W} = FD\ddot{V} = C_D A \frac{1}{2} \rho V^2$

$$= \mathbf{11.08\ kW}$$

(D) **3.6** A cooling tower nozzle disperses water into a stream of droplets. If the average diameter of the droplets is 60 microns, estimate the work required for atomising 1 kg of water isothermally at the ambient condition. **[LO 3.5]**

Given: Surface tension of water in contact with air = 0.07 N/m,

Density of water = 1000 kg/m^3.

Water is assumed to enter the nozzle through a pipe of 15 mm diameter.

Hint: Atomisation of water gives a tremendous increase in the surface area of the film and it can be viewed as the work done on the film, $W = \int^f \sigma dA = \sigma (A_f - A_i)$, where A_f = final area, A_i = initial area

$$1 \text{ kg of water} = N \times \frac{\pi}{6} (60 \times 10^{-6})^3 \times 1000,$$

N being the number of droplets of av. diameter 60 μm.

Find N and then surface area of all the droplets into which 1 kg of water is distributed.

$$A_f = \pi (60 \times 10^{-6})^2 N = 100 \text{ m}^2$$

To find A_i, let 1 kg of water is contained in $L(m)$ length of the inlet pipe of 15 mm diameter.

$$\frac{\pi}{6} d^2 LP = 1 \text{ kg}$$

Surface area, $S = \pi dL$ m^2

$$\therefore \quad \frac{S}{L} = \frac{\pi dL}{\frac{\pi}{4} d^2 LP} = \frac{4}{dP} = 0.267 \text{ m}^2 = A_i$$

$\therefore$ work done in atomisation, $W = 0.07\ (100 - 0.267)$

$= \mathbf{6.98\ J}$

(M) **3.7** An electric motor drives a stirrer fitted with a horizontal cylinder of 40 cm diameter which contains a fluid restrained by a frictionless piston. During the stirring of fluid for 15 minutes the piston moves outward slowly by a distance of 30 cm against the atmospheric pressure of 1 bar. The current supplied to the motor is 0.5 amp from a 24 V lead-acid accumulator. If the conversion efficiency from electrical work to mechanical work is 90 percent, estimate the work on the motor, stirrer and the atmosphere. **[LO 3.5]**

Hint: Work input to the motor, $W_m = \mathbf{0.5 \times 15 \times 60 \times 24\ J}$

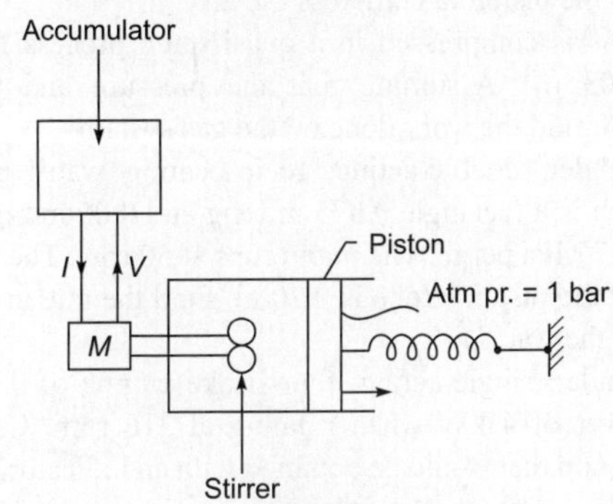

Fig. 3.27

Work input to stirrer = $W_m \times \eta$, where $\eta = 0.9$.

$= \mathbf{9.72\ kJ}$

Work done by the fluid on the atmosphere

$$W_{atm} = P\,\Delta V = \mathbf{3.77\ kJ}$$

The fluid receives a work input of 9.72 kJ from the stirrer and does 3.77 kJ of work on the atmosphere.

PROBLEMS

(E) **3.1** (a) A pump forces 1 m^3/min of water horizontally from an open well to a closed tank where the pressure is 0.9 MPa. Compute the work the pump must do upon the water in an hour just to force the water into the tank against the pressure. Sketch the system upon which the work is done before and after the process. **[LO 3.5]**

(M) (b) If the work done as above upon the water had been used solely to raise the same amount of water vertically against gravity without change of pressure, how many metres would the water have been elevated? **[LO 3.5]**

(M) (c) If the work done in (a) upon the water had been used solely to accelerate the water from zero velocity without change of pressure or elevation, what velocity would the water have reached? If the work had been used to accelerate the water from an initial velocity of 10 m/s, what would the final velocity have been? **[LO 3.5]**

(E) **3.2** The piston of an oil engine, of area 0.0045 m^2, *moves downwards* 75 mm, drawing in 0.00028 m^3 of fresh air from the atmosphere. The pressure in the cylinder is uniform during the process at 80 kPa, while the atmospheric pressure is 101.325 kPa, the difference being due to the flow resistance in the induction pipe and the inlet valve. Estimate the displacement work done by the air finally in the cylinder. **[LO 3.2]**

(E) **3.3** An engine cylinder has a piston of area 0.12 m^3 and contains gas at a pressure of 1.5 MPa. The gas expands according to a process which is represented by a straight line on a pressure-volume diagram. The final pressure is 0.15 MPa. Calculate the work done by the gas on the piston if the stroke is 0.30 m. **[LO 3.2]**

(E) **3.4** A mass of 1.5 kg of air is compressed in a quasi-static process from 0.1 MPa to 0.7 MPa for which pv = constant. The initial density of air is 1.16 kg/m^3. Find the work done by the piston to compress the air. **[LO 3.2]**

(E) **3.5** A mass of gas is compressed in a quasi-static process from 80 kPa, 0.1 m^3 to 0.4 MPa, 0.03 m^3. Assuming that the pressure and volume are related by pv^n = constant, find the work done by the gas system. **[LO 3.2]**

(E) **3.6** A single-cylinder, double-acting, reciprocating water pump has an indicator diagram which is a rectangle 0.075 m long and 0.05 m high. The indicator spring constant is 147 MPa per m. The pump runs at 50 rpm. The pump cylinder diameter is 0.15 m and the piston stroke is 0.20 m. Find the rate in kW at which the piston does work on the water. **[LO 3.4]**

(E) **3.7** A single-cylinder, single-acting, four-stroke engine of 0.15 m bore develops an indicated power of 4 kW when running at 216 rpm. Calculate the area of the indicator diagram that would be obtained with an indicator having a spring constant of 25×10^6 N/m^3. The length of the indicator diagram is 0.1 times the length of the stroke of the engine. **[LO 3.4]**

(E) **3.8** A six-cylinder, four-stroke gasoline engine is run at a speed of 2520 rpm. The area of the indicator card of one cylinder is 2.45×10^3 mm^2 and its length is 58.5 mm. The spring constant is 20×10^6 N/m^3. The bore of the cylinders is 140 mm and the piston stroke is 150 mm. Determine the indicated power, assuming that each cylinder contributes an equal power. **[LO 3.4]**

(M) **3.9** A closed cylinder of 0.25 m diameter is fitted with a light frictionless piston. The piston is retained in position by a catch in the cylinder wall and the volume on one side of the piston contains air at a pressure of 750 kN/m^2. The volume on the other side of the piston is evacuated. A helical spring is mounted coaxially with the cylinder in this evacuated space to give a force of 120 N on the piston in this position. The catch is released and the piston travels along the cylinder until it comes to rest after a stroke of 1.2 m. The piston is then held in its position of maximum travel by a ratchet mechanism. The spring force increases linearly with the piston displacement to a final value of 5 kN. Calculate the work done by the compressed air on the piston. **[LO 3.4]**

(M) **3.10** A steam turbine drives a ship's propeller through an 8 : 1 reduction gear. The average resisting torque imposed by the water on the propeller is 750×10^3 mN and the shaft power delivered by the turbine to the reduction gear is 15 MW. The turbine speed is 1450 rpm. Determine (a) the torque developed by the turbine, (b) the power delivered to the propeller shaft and (c) the net rate of working of the reduction gear. **[LO 3.6]**

(M) **3.11** A fluid, contained in a horizontal cylinder fitted with a frictionless leakproof piston, is continuously agitated by means of a stirrer passing through the cylinder cover. The cylinder diameter is 0.40 m. During the stirring process lasting 10 minutes, the piston slowly moves out a distance of 0.485 m against the atmosphere. The net work done by the fluid during the process is 2 kJ. The speed of the electric motor driving the stirrer is 840 rpm. Determine the torque in the shaft and the power output of the motor. **[LO 3.5]**

(M) **3.12** At the beginning of the compression stroke of a two-cylinder internal combustion engine the air is at a pressure of 101.325 kPa. Compression reduces the volume to 1/5 of its original volume, and the law of compression is given by $pv^{1.2}$ = constant. If the bore and stroke of each cylinder is 0.15 m and 0.25 m, respectively, determine the power absorbed in kW by compression strokes when the engine speed is such that each cylinder undergoes 500 compression strokes per minute. **[LO 3.4]**

(E) **3.13** Determine the total work done by a gas system following an expansion process as shown in Fig. 3.23. **[LO 3.2]**

(M) **3.14** A system of volume V contains a mass m of gas at pressure p and temperature T. The macroscopic properties of the system obey the following relationship:

$$\left(p + \frac{a}{V^2}\right)(V - b) = mRT$$

where a, b and R are constants.

Obtain an expression for the displacement work done by the system during a constant-temperature expansion from volume V_1 to volume V_2. Calculate the work done by a system which contains 10 kg of this gas expanding from 1 m^3 to 10 m^3 at a temperature of 293 K. Use the values $a = 15.7 \times 10$ Nm^4, $b = 1.07 \times 10^{-2}$ m^3 and $R = 0.278$ kJ/kg-K. **[LO 3.2]**

(E) **3.15** If a gas of volume 6000 cm^3 and at pressure of 100 kPa is compressed quasi-statically according to pV^2 = constant until the volume becomes 2000 cm^3, determine the final pressure and the work transfer. **[LO 3.2]**

(E) **3.16** The flow energy of 0.124 m^3/min of a fluid crossing a boundary to a system is 18 kW. Find the pressure at this point. **[LO 3.5]**

(E) **3.17** A milk chilling unit can remove heat from the milk at the rate of 41.87 MJ/h. Heat leaks into the milk from the surroundings at an average rate of 4.187 MJ/h. Find the time required for cooling a batch of 500 kg of milk from 45°C to 5°C. Take the c_p of milk to be 4.187 kJ/kg K. **[LO 3.8]**

(E) **3.18** 680 kg of fish at 5°C are to be frozen and stored at – 12°C. The specific heat of fish above freezing point is 3.182, and below freezing point is 1.717 kJ/kg K. The freezing point is – 2°C, and the latent heat of fusion is 234.5 kJ/kg. How much heat must be removed to cool the fish, and what per cent of this is latent heat? **[LO 3.8]**

(M) **3.19** A gas in a piston-cylinder assembly undergoes an expansion process for which the relationship between pressure and volume is given by pV^n = constant. The initial pressure is 0.3 MPa, the initial volume is 0.1 m^3 and the final volume is 0.2 m^3. Determine the work for the process in kJ if (a) $n = 1.5$, (b) $n = 1.0$ and (c) $n = 0$. **[LO 3.2]**

(E) **3.20** A gas expands from an initial state with $p_1 = 340$ kPa and $V_1 = 0.0425$ m^3 to a final state where $p_2 = 136$ kPa. If the pressure-volume relationship during the process is pV^2 = constant, determine the work in kJ. **[LO 3.2]**

(M) **3.21** The drag force, F_d, imposed by the surrounding air on an automobile moving with velocity V is given by $F_d = C_d A \frac{1}{2} \rho V^2$

where C_d is a constant called the drag coefficient, A is the projected frontal area of the vehicle and ρ is the air density.

For $C_d = 0.42$, $A = 2\text{m}^2$ and $\rho = 1.23$ kg/m^3, calculate the power required in kW, to overcome the drag at a constant velocity of 100 km/h. **[LO 3.5]**

(M) **3.22** An electric generator coupled to a windmill produces an average electrical power output of 5 kW. The power is used to charge a storage battery. Heat transfer from the battery to the surroundings occurs at a constant rate of 0.6 kW. Determine the total amount of energy stored in the battery, in kJ, in 8 hours of operation. **[LO 3.5]**

ANSWERS TO PROBLEMS

3.1 **a.** 5400 kJ/h
b. 91.74 m
c. 42.4 m/s, 43.6 m/s

3.2 27 J

3.3 29.7 kJ

3.4 251.62 kJ

3.5 −11.83 kJ

3.6 43.3 kW

3.7 505 mm^2

3.8 243.57 kW

3.9 3.07 kJ

3.10 **a.** $T = 98.84$ km N
b. 14.235 MW
c. 0.765 MW

3.11 0.08 mN, 6.92 W

3.12 17.95 kW

3.13 2.253 MJ

3.14 1742 kJ

3.15 900 kPa, −1.2 kJ

3.16 8709 kPa

3.17 2h 13 min

3.18 186.28 MJ, 85.6 percent

3.19 **a.** 17.6 kJ
b. 20.79 kJ
c. 30 kJ

3.20 5.31 kJ

3.21 11.08 kW

3.22 1.27×10^5 kJ

4

First Law of Thermodynamics

Learning Objectives

LO4.1 State first law for a closed system executing a cycle
LO4.2 Derive first law for a closed system for a change of state
LO4.3 Recognize total energy as a property
LO4.4 Differentiate between stored energy and energy in transition
LO4.5 State specific heat of a substance at constant volume
LO4.6 Discuss enthalpy and compare it with heat transfer
LO4.7 State specific heat of a substance at constant pressure
LO4.8 Analyze total energy of an isolated system
LO4.9 Discuss perpetual motion machine of the first kind (PMM1)

4.1 FIRST LAW FOR A CLOSED SYSTEM UNDERGOING A CYCLE

LO 4.1
State first law for a closed system executing a cycle

The transfer of heat and the performance of work may both cause the same effect in a system. Heat and work are different forms of the same entity, called energy, which is conserved. Energy which enters a system as heat may leave the system as work, or energy which enters the system as work may leave as heat.

Let us consider a closed system which consists of a known mass of water, m, contained in an adiabatic vessel having a thermometer and a paddle wheel, as shown in Fig. 4.1. Let a certain amount of work W_{1-2} be done upon the system by the paddle wheel. The quantity of work can be measured by the fall of weight which drives the paddle wheel through a pulley. The system was initially at temperature t_1, the same as that of atmosphere, and after work transfer let the temperature rise to t_2. The pressure is always 1 atm. The process 1–2 undergone by the system is shown in Fig. 4.2 in

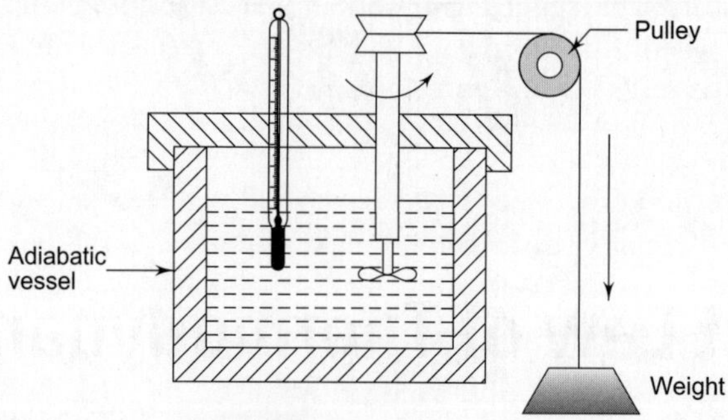

Fig. 4.1 Adiabatic work

generalised thermodynamic coordinates X, Y. Let the insulation now be removed. The system and the surroundings interact by heat transfer till the system returns to the original temperature t_1, attaining the condition of thermal equilibrium with the atmosphere. The amount of heat transfer Q_{2-1} from the system during this process, 2–1, shown in Fig. 4.2, can be estimated from $Q_{2-1} = mc_p (t_2 - t_1)$. The system thus executes a cycle, which consists of a definite amount of work input W_{1-2} to the system followed by the transfer of an amount of heat Q_{2-1} from the system. It has been found that this W_{1-2} is always proportional to the heat Q_{2-1}, and the constant of proportionality is called the Joule's equivalent or the *mechanical equivalent of heat*. In the simple example given here, there are only two energy transfer quantities as the system performs a thermodynamic cycle. If the cycle involves many more heat and work quantities, the same result will be found. Expressed algebraically,

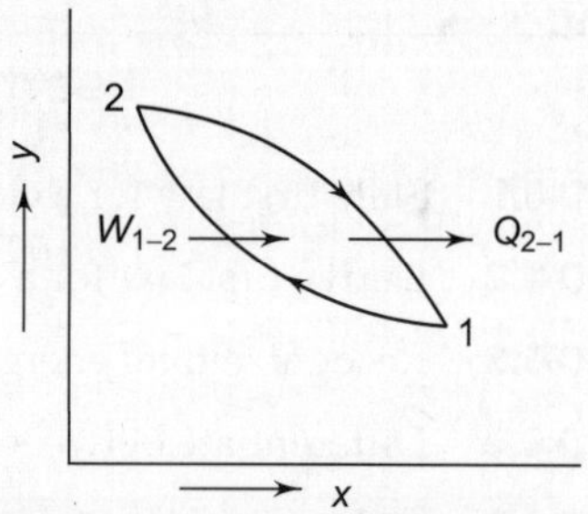

Fig. 4.2 Cycle completed by a system with two energy interactions: adiabatic work transfer W_{1-2} followed by heat transfer Q_{2-1}

$$(\Sigma W)_{\text{cycle}} = J(\Sigma Q)_{\text{cycle}} \tag{4.1}$$

where J is the Joule's equivalent. This is also expressed in the form

$$\oint đW = J \oint đQ$$

where the symbol $\oint$ denotes the cyclic integral for the closed path. This is the *first law for a closed system undergoing a cycle*. It is accepted as a *general law of nature*, since no violation of it has ever been demonstrated.

In the S.I. system of units, both heat and work are measured in the derived unit of energy, the Joule. The constant of proportionality, J, is therefore unity ($J = 1$ Nm/J).

The first law of thermodynamics owes much to J.P. Joule who, during the period 1840–1849, carried out a series of experiments to investigate the equivalence of work and heat. In one of these experiments, Joule used an apparatus similar to the one shown in Fig. 4.1. Work was transferred to the measured mass of water by means of paddle wheel driven by the falling weight. The rise in the temperature of water was recorded. Joule also used mercury as the fluid system, and later a solid system of metal blocks which absorbed work by friction when rubbed against each other. Other experiments involved the supplying of work in an electric current. $W_{1-2} = EIt = Q_{2-1} = mc_p(t_2 - t_1)$, where E is the voltage and I is the current flowing through the wire for t seconds. In every case, he found the same ratio (J) between the amount of work and the quantity of heat that would produce identical effects in the system, where J is unity in S.I. System.

Prior to Joule, heat was considered to be an invisible fluid flowing from a body of higher calorie to a body of lower calorie, and this was known as the *caloric theory of heat*. It was Joule who first established that heat is a form of energy, and thus laid the foundation of the first law of thermodynamics.

4.2 FIRST LAW FOR A CLOSED SYSTEM UNDERGOING A CHANGE OF STATE

LO 4.2 Derive first law for a closed system for a change of state

The expression $(\Sigma W)_{\text{cycle}} = (\Sigma Q)_{\text{cycle}}$ the applies only to systems undergoing cycles, and the algebraic summation of all energy transfer across system boundaries is zero. But if a system undergoes a change of state during which both heat transfer and work transfer are involved, the *net* energy transfer will be stored or accumulated within the system. If Q is the amount of heat transferred to the system and W is the amount of work transferred from the system during the process (Fig. 4.3), the net energy transfer ($Q - W$) will be stored in the system. Energy in storage is neither heat nor work, and is given the name *internal energy* or simply, the *energy* of the system.

Therefore,
$$Q - W = \Delta E$$
where ΔE is the increase in the energy of the system

or
$$Q = \Delta E + W \tag{4.2}$$

Here Q, W and ΔE are all expressed in the same units (in joules). Energy may be stored by a system in different modes, as explained in Section 4.4.

If there are more energy transfer quantities involved in the process, as shown in Fig. 4.4, the first law gives

$$(Q_2 + Q_3 - Q_1) = \Delta E + (W_2 + W_3 - W_1 - W_4)$$

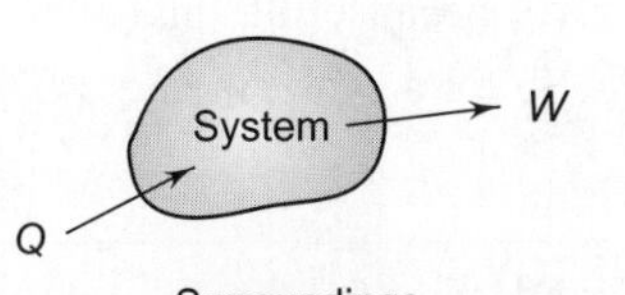

Fig. 4.3 Heat and work interactions of a system with its surroundings in a process

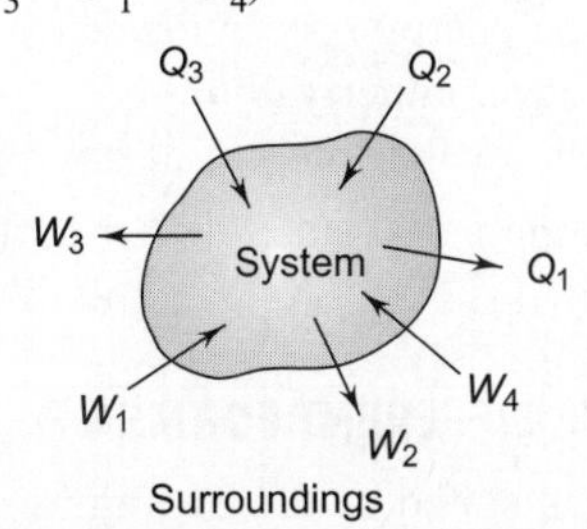

Fig. 4.4 System-surroundings interaction in a process involving many energy fluxes

Energy is thus conserved in the operation. The first law is a particular formulation of the principle of the conservation of energy. Equation (4.2) may also be considered as the definition of energy. This definition does not give an absolute value of energy E, but only the change of energy ΔE for the process. It can, however, be shown that the energy has a definite value at every state of a system and is, therefore, a property of the system.

4.3 ENERGY—A PROPERTY OF THE SYSTEM

LO 4.3

Recognize total energy as a property

Consider a system which changes its state from state 1 to state 2 by following the path A, and returns from state 2 to state 1 by following the path B (Fig. 4.5). So the system undergoes a cycle. Writing the first law for path A

$$Q_A = \Delta E_A + W_A \quad (4.3)$$

and for path B

$$Q_B = \Delta E_B + W_B \quad (4.4)$$

The processes A and B together constitute a cycle, for which

$$(\Sigma W)_{\text{cycle}} = (\Sigma Q)_{\text{cycle}}$$

or $$W_A + W_B = Q_A + Q_B$$

or $$Q_A - W_A = W_B - Q_B \quad (4.5)$$

From Eqs. (4.3), (4.4) and (4.5), it yields

$$\Delta E_A = -\Delta E_B \quad (4.6)$$

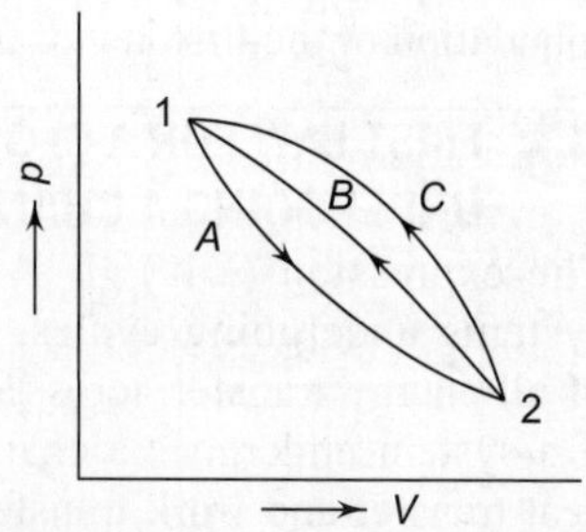

Fig. 4.5 Energy—A property of a system

Similarly, had the system returned from state 2 to state 1 by following the path C instead of path B

$$\Delta E_A = -\Delta E_C \quad (4.7)$$

From Eqs. (4.6) and (4.7)

$$\Delta E_B = \Delta E_C \quad (4.8)$$

Therefore, it is seen that the change in energy between two states of a system is the same, whatever path the system may follow in undergoing that change of state. If some arbitrary value of energy is assigned to state 2, the value of energy at state 1 is fixed independent of the path the system follows. Therefore, energy has a definite value for every state of the system. Hence, it is a *point function and a property of the system*.

The energy E is an extensive property. The *specific energy*, $e = E/M$ (J/kg), is an intensive property.

The cyclic integral of any property is zero, because the final state is identical with the initial state. $\oint dE = 0$, $\oint dV = 0$, etc. So for a cycle, the Eq. (4.2) reduces to Eq. (4.1).

4.4 DIFFERENT FORMS OF STORED ENERGY

LO 4.4

Differentiate between stored energy and energy in transition

It was stated at the beginning that thermodynamics is the science of energy transfer and it effect the physical properties of a substance. Energy, as we know, is the capacity of doing work. In thermodynamics, energy can be in two forms: (*i*) Energy in transit, (*ii*) Energy in storage. Work and heat intractions are the forms of energy in transit, observed at the boundaries of a system. They are not the properties of a system. They are path functions, their magnitudes depending upon the path the system follows during a change of state. Energy in storage, called internal energy, is a point or state function and hence a property of a system.

The symbol E refers to the total energy stored in a system. Basically there are two modes in which energy may be stored in a system:

1. Macroscopic energy mode
2. Microscopic energy mode

The macroscopic energy mode includes the macroscopic kinetic energy and potential energy of a system. Let us consider a fluid element of mass m having the centre of mass velocity or barycentric velocity $\bar{V}$ (Fig. 4.6). The macroscopic kinetic energy E_K of the fluid element by virtue of its motion is given by

$$E_K = \frac{m\bar{V}^2}{2}$$

If the elevation of the fluid element from an arbitrary datum is z, then the macroscopic potential energy E_p by virtue of its position is given by

$$E_p = mgz$$

The microscopic energy mode refers to the energy stored in the molecular and atomic structure of the system, which is called the *molecular internal energy* or *simply internal energy*, customarily denoted by the symbol U. Matter is composed of molecules. Molecules are in random thermal motion (for a gas) with an average velocity $\bar{\boldsymbol{v}}$, constantly colliding with one another and with the walls (Fig. 4.6). Due to a collision, the molecules may be subjected to rotation as well as vibration. They can have translational kinetic energy, rotational kinetic energy, vibrational energy, electronic energy, chemical energy and nuclear energy (Fig. 4.7). If ε represents the energy of one molecule, then

$$\varepsilon = \varepsilon_{\text{trans}} + \varepsilon_{\text{rot}} + \varepsilon_{\text{vib}} + \varepsilon_{\text{chem}} + \varepsilon_{\text{electronic}} + \varepsilon_{\text{nuclear}} \tag{4.9}$$

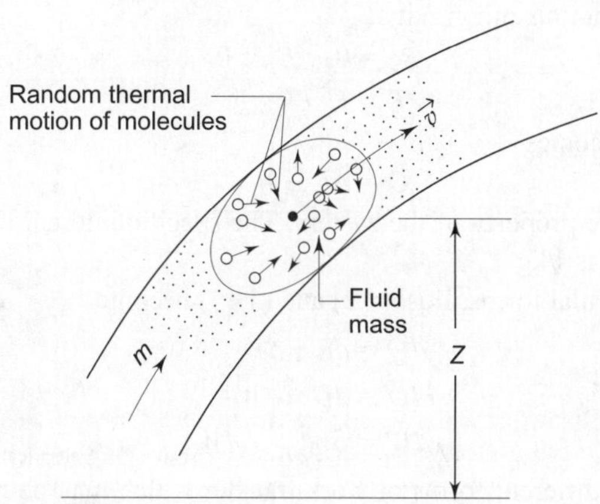

Fig. 4.6 Macroscopic and microscopic energy

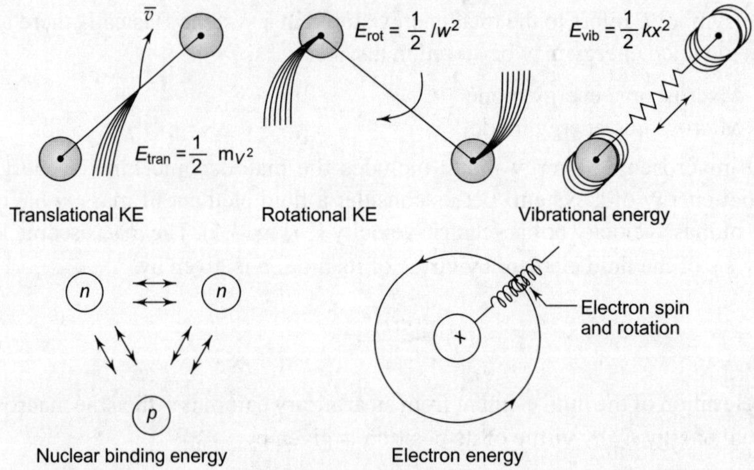

Fig. 4.7 Various components of internal energy stored in a molecule

If N is the total number of molecules in the system, then the total internal energy

$$U = N\varepsilon \tag{4.10}$$

In an ideal gas there are no intermolecular forces of attraction and repulsion, and the internal energy depends only on temperature. Thus,

$$U = f(T) \text{ only} \tag{4.11}$$

for an ideal gas.

Other forms of energy which can also be possessed by a system are magnetic energy, electrical energy and surface (tension) energy. In the absence of these forms, the total energy E of a system is given by

$$E = \underbrace{E_K + E_P}_{\text{macro}} + \underbrace{U}_{\text{micro}} \tag{4.12}$$

where E_K, E_P and U refer to the kinetic, potential and internal energy, respectively. In the absence of motion and gravity

$$E_K = 0, \quad E_P = 0$$
$$E = U$$

and Eq. (4.2) becomes

$$Q = \Delta U + W \tag{4.13}$$

U is an extensive property of the system. The specific internal energy u is equal to U/m and its unit is J/kg.

In the differential forms, Eqs. (4.2) and (4.13) become

$$đQ = dE + đW \tag{4.14}$$

$$đQ = dU + đW \tag{4.15}$$

where $$đW = \delta W_{pdV} + đW_{\text{shaft}} + đW_{\text{electrical}} + \ldots,$$

considering the different forms of work transfer which may be present. When only pdV work is present, the equations become

$$đQ = dE + pdV \tag{4.16}$$

$$đQ = dU + pdV \tag{4.17}$$

or in the integral form

$$Q = \Delta E + \int pdV \tag{4.18}$$

$$Q = \Delta U + \int pdV \tag{4.19}$$

4.5 SPECIFIC HEAT AT CONSTANT VOLUME

> **LO4.5**
> State specific heat of a substance at constant volume

The specific heat of a substance at constant volume c_v is defined as the rate of change of specific internal energy with respect to temperature when the volume is held constant, i.e.

$$c_v = \left(\frac{\partial u}{\partial T}\right)_v \tag{4.20}$$

For a constant-volume process

$$(\Delta u)_v = \int_{T_1}^{T_2} c_v \cdot dT \tag{4.21}$$

The first law may be written for a closed stationary system composed of a unit mass of a pure substance

$$Q = \Delta u + W$$

or

$$đQ = du + đW$$

For a process in the absence of work other than pdV work

$$đW = pdV$$

$$đQ = dU + pdV \tag{4.22}$$

When the volume is held constant

$$(Q)_v = (\Delta u)_v$$

$\therefore$

$$(Q)_v = \int_{T_1}^{T_2} c_v \, dT \tag{4.23}$$

Heat transferred at constant volume increases the internal energy of the system.

If the specific heat of a substance is defined in terms of heat transfer, then

$$c_v = \left(\frac{\partial Q}{\partial T}\right)_u$$

Since Q is not a property, this definition does not imply that c_v is a property of a substance. Therefore, this is not the appropriate method of defining the specific heat, although $(đQ)_v = du$.

Since u, T and v are properties, c_v is a property of the system. The product $mc_v = C_v$ is called the *heat capacity at constant volume* (J/K).

4.6 ENTHALPY

> **LO4.6**
> Discuss enthalpy and compare it with heat transfer

The enthalpy of a substance, h, is defined as

$$h = u + pv \tag{4.24}$$

It is an intensive property of a system (kJ/kg).

Internal energy change is equal to the heat transferred in a constant volume process involving no work other than pdV work. From Eq. (4.22), it is possible to drive an expression for the heat transfer in a constant pressure process involving no work other than pdV work. In such a process in a closed stationary system of unit mass of a pure substance

$$đQ = du + pdv$$

At constant pressure

$$pdv = d(pv)$$

$\therefore$ $$(đQ)_p = du + d(pv)$$

or $$(đQ)_p = d(u + pv)$$

or $$(đQ)_p = dh \tag{4.25}$$

where $h = u + pv$ is the *specific enthalpy*, a property of the system.

Heat transferred at constant pressure increases the enthalpy of a system.

For an ideal gas, the enthalpy becomes

$$h = u + RT \tag{4.26}$$

Since the internal energy of an ideal gas depends only on the temperature Eq. (4.11), the enthalpy of an ideal gas also depends on the temperature only, i.e.

$$h = f(T) \text{ only} \tag{4.27}$$

Total enthalpy $H = mh$

Also $$H = U + pV$$

and $$h = H/m \text{ (J/kg}$$

4.7 SPECIFIC HEAT AT CONSTANT PRESSURE

LO 4.7 State specific heat of a substance at constant pressure

The specific heat at constant pressure c_p is defined as the rate of change of enthalpy with respect to temperature when the pressure is held constant

$$c_p = \left(\frac{\partial h}{\partial T}\right)_p \tag{4.28}$$

Since h, T and p are properties, so c_p is a property of the system. Like c_v, c_p should not be defined in terms of heat transfer at constant pressure, although $(đQ)_p = dh$.

For a constant pressure process

$$(\Delta h)_p = \int_{T_1}^{T_2} c_p\, dT \tag{4.29}$$

The first law for a closed stationary system of unit mass

$$đQ = du + pdv$$

Again $$h = u + pv$$

$\therefore$ $$dh = du + pdv + vdp$$
$$= \delta Q + vdp$$
$$đQ = dh - vdp \tag{4.30}$$

$\therefore$ $$(đQ)_p = dh$$

or $$(Q)_p = (\Delta h)_p$$

$\therefore$ From Eqs. (4.19) and (4.20)

$$(Q)_p = \int_{T_1}^{T_2} c_p\, dT$$

c_p is a property of the system, just like c_v. The *heat capacity at constant pressure* C_p is equal to mc_p (J/K).

4.8 ENERGY OF AN ISOLATED SYSTEM

LO 4.8
Analyze total energy of an isolated system

An isolated system is one in which there is no interaction of the system with the surroundings. For an isolated system

$\delta Q = 0 \quad \delta W = 0.$

The first law gives

$$dE = 0$$

or

$$E = \text{constant}$$

The energy of an isolated system is always constant.

4.9 PERPETUAL MOTION MACHINE OF THE FIRST KIND—PMM1

LO 4.9
Discuss perpetual motion machine of the first kind (PMM1)

The first law states the general principle of the conservation of energy. *Energy is neither created nor destroyed, but only gets transformed from one form to another*. There can be no machine which would continuously supply mechanical work without some other form of energy disappearing simultaneously (Fig. 4.8). Such a *fictitious machine* is called a *perpetual motion machine of the first kind*, or in brief, PMM1. *A PMM1 is thus impossible.*

The converse of the above statement is also true, i.e. there can be no machine which would continuously consume work without some other form of energy appearing simultaneously (Fig. 4.9).

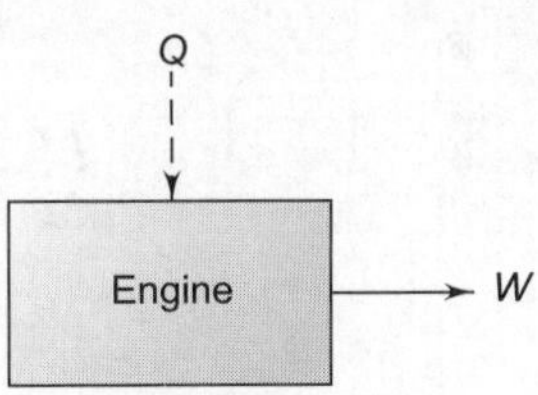

Fig. 4.8 PMM1

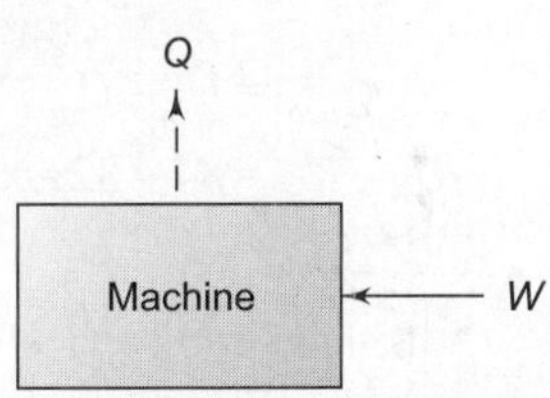

Fig. 4.9 Converse of PMM1

SOLVED EXAMPLES

Example 4.1 *A stationary mass of gas is compressed without friction from an initial state of 0.3 m³ and 0.105 MPa to a final state of 0.15 m³ and 0.105 MPa, the pressure remaining constant during the process. There is a transfer of 37.6 kJ of heat from the gas during the process. How much does the internal energy of the gas change?* E [LO 4.2]

Solution First law for a stationary system in a process gives

$$Q = \Delta U + W$$

or

$$Q_{1-2} = U_2 - U_1 + W_{1-2} \tag{i}$$

Here
$$W_{1-2} = \int_{V_1}^{V_2} pdV = p(V_2 - V_1)$$
$$= 0.105\ (0.15 - 0.30)\ \text{MJ} = -15.75\ \text{kJ}$$
$$Q_{1-2} = -37.6\ \text{kJ}$$

∴ Substituting in Eq. (i)

$$-37.6\ \text{kJ} = U_2 - U_1 - 15.75\ \text{kJ}$$

∴ $$U_2 - U_1 = \mathbf{-21.85\ kJ}$$

The internal energy of the gas decreases by 21.85 kJ in the process.

Example 4.2 *When a system is taken from state a to state b, in Fig. 4.10 along path acb, 84 kJ of heat flow into the system, and the system does 32 kJ of work. (a) How much will the heat that flows into the system along path adb be, if the work done is 10.5 kJ? (b) When the system is returned from b to a along the curved path, the work done on the system is 21 kJ. Does the system absorb or liberate heat, and how much of the heat is absorbed or liberated? (c) If U_a = 0 and U_d = 42 kJ, find the heat absorbed in the processes ad and db.*

M [LO 4.1, 4.2]

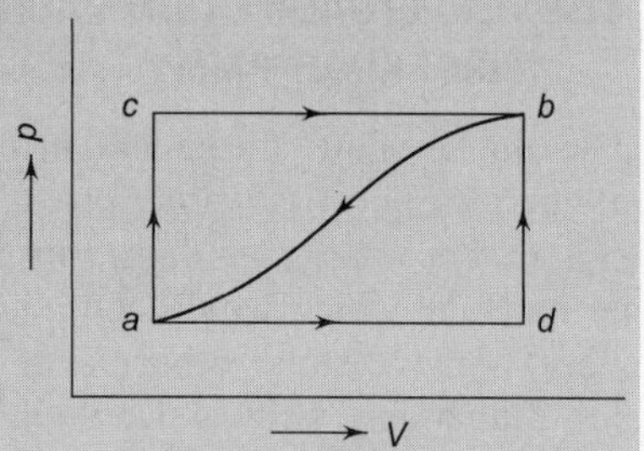

Fig. 4.10

Solution

$$Q_{acb} = 84\ \text{kJ}$$
$$W_{acb} = 32\ \text{kJ}$$

We have

$$Q_{acb} = U_b - U_a + W_{acb}$$

∴ $$U_b - U_a = 84 - 32 = \mathbf{52\ kJ}$$

(a) $$Q_{adb} = U_b - U_a + W_{adb}$$
$$= 52 + 10.5 = \mathbf{62.5\ kJ}$$

(b) $$Q_{b-a} = U_a - U_b + W_{b-a}$$
$$= -52 - 21 = \mathbf{-73\ kJ}$$

The system liberates 73 kJ of heat

(c) $$W_{adb} = W_{ad} + W_{db} = W_{ad} = 10.5\ \text{kJ}$$

∴ $$Q_{ad} = U_d - U_a + W_{ad}$$
$$= 42 - 0 + 10.5 = 52.5\ \text{kJ}$$

Now $$Q_{adb} = 62.5\ \text{kJ} = Q_{ad} + Q_{db}$$

∴ $$Q_{db} = 62.5 - 52.5 = \mathbf{10\ kJ}$$

Example 4.3 *A piston and cylinder machine contains a fluid system which passes through a complete cycle of four processes. During a cycle, the sum of all heat transfers is – 170 kJ. The system completes 100 cycles per minute. Complete the following table showing the method for each item, and compute the net rate of work output in kW.*

M [LO 4.1, 4.2]

Process	Q (kJ/min)	W (kJ/min)	ΔE (kJ/min)
a–b	0	2,170	—
b–c	21,000	0	—
c–d	–2,100	—	–36,600
d–a	—	—	—

Solution Process a–b:

$$Q = \Delta E + W$$
$$0 = \Delta E + 2170$$
$$\therefore \quad \Delta E = -2170 \text{ kJ/min}$$

Process b–c:

$$Q = \Delta E + W$$
$$21{,}000 = \Delta E + 0$$
$$\therefore \quad \Delta E = 21{,}000 \text{ kJ/min}$$

Process c–d:

$$Q = \Delta E + W$$
$$-2100 = -36{,}600 + W$$
$$\therefore \quad W = 34{,}500 \text{ kJ/min}$$

Process d–a: $\sum_{\text{cycle}} Q = -170 \text{ kJ}$

The system completes 100 cycles/min.

$$\therefore \quad Q_{ab} + Q_{bc} + Q_{cd} + Q_{da} = -17{,}000 \text{ kJ/min}$$
$$0 + 21{,}000 - 2100 + Q_{da} = -17{,}000$$
$$\therefore \quad Q_{da} = -35{,}900 \text{ kJ/min}$$

Now, $\oint dE = 0$, since cyclic integral of any property is zero.

$$\therefore \quad \Delta E_{a-b} + \Delta E_{b-c} + \Delta E_{c-d} + \Delta E_{d-a} = 0$$
$$-2170 + 21{,}000 - 36{,}600 + \Delta E_{d-a} = 0$$
$$\therefore \quad \Delta E_{d-a} = 17{,}770 \text{ kJ/min}$$
$$\therefore \quad W_{d-a} = Q_{d-a} - \Delta E_{d-a}$$
$$= -35{,}900 - 17{,}770 = \mathbf{-53{,}670 \text{ kJ/min}}$$

The table becomes

Process	Q (kJ/min)	W (kJ/min)	ΔE (kJ/min)
a–b	0	2170	– 2170
b–c	21,000	0	21,000
c–d	–2100	34,500	– 36,600
d–a	–35,900	– 53,670	17,770

Since $\sum_{\text{cycle}} Q = \sum_{\text{cycle}} W$

Rate of work output

$$= -17{,}000 \text{ kJ/min} = \mathbf{-283.3 \text{ kW}}$$

Example 4.4 *The internal energy of a certain substance is given by the following equation:*

$$u = 3.56\, pv + 84$$

where u is given in kJ/kg p is in kPa, and v is in m³/kg.

A system composed of 3 kg of this substance expands from an initial pressure of 500 kPa and a volume of 0.22 m³ to a final pressure 100 kPa in a process in which pressure and volume are related by $pv^{1.2}$ = constant.

(a) *If the expansion is quasi-static, find Q, ΔU and W for the process.*

(b) *In another process the same system expands according to the same pressure-volume relationship as in part (a), and from the same initial state to the same final state as in part (a), but the heat transfer in this case is 30 kJ. Find the work transfer for this process.*

(c) *Explain the difference in work transfer in parts (a) and (b).* Ⓜ [LO 4.2]

Solution

(a)

$$u = 3.56\, pv + 84$$

$$\Delta u = u_2 - u_1 = 3.56\,(p_2v_2 - p_1v_1)$$

∴ $$\Delta U = 3.56\,(p_2V_2 - p_1V_1)$$

Now $$p_1V_1^{1.2} = p_2V_2^{1.2}$$

∴ $$V_2 = V\left(\frac{p_1}{p_2}\right)^{1/1.2} = 0.2\left(\frac{5}{1}\right)^{1/1.2}$$

$$= 0.22 \times 3.83 = 0.845 \text{ m}^3$$

∴ $$\Delta U = 356\,(1 \times 0.845 - 5 \times 0.22) \text{ kJ}$$

$$= -356 \times 0.255 = \mathbf{-91\ kJ}$$

For a quasi-static process

$$W = \int pdV = \frac{p_2V_2 - p_1V_1}{1-n}$$

$$= \frac{(1 \times 0.845 - 5 \times 0.22)100}{1-1.2}$$

$$= 127.5 \text{ kJ}$$

∴ $$Q = \Delta U + W$$

$$= -91 + 127.5 = \mathbf{36.5\ kJ}$$

(b) Here $$Q = 30 \text{ kJ}$$

Since the end states are the same, ΔU would remain the same as in (a).

∴ $$W = Q - \Delta U$$

$$= 30 - (-91) = \mathbf{121\ kJ}$$

(c) The work in (b) is not equal to $\int pdV$ since the process is not quasi-static.

Example 4.5 *A rigid tank contains 1 m³ of air 200 kPa at 50°C. How much heat must be transferred to the air to raise the air pressure to 400 kPa? Air can be considered as an ideal gas and the value of constant volume specific heat (c_v) for air at 210°C is kJ/kg°C.* Ⓜ [LO 4.2]

Solution The mass of air is found to be

$$m = \frac{pV}{RT} = \frac{200 \times 10^3 \times 1}{287 \times 323} = 2.157 \text{ kg}$$

Since the tank is rigid, volume of the air will not change. The final temperature of the air can be found from the characteristic equation of an ideal gas as

$$\frac{p_1}{T_1} = \frac{p_2}{T_2}$$

or $$T_2 = \frac{p_2}{p_1} T_1 = \frac{400}{200} \times 323 = 646 \text{ K}$$

The average temperature during the heating process is $T_{avg} = \frac{323 + 646}{2} = 484.5 \text{ K} = 211.5°\text{C}$. Consider air in the tank as a system. The work transfer is zero for this constant-volume process. Consequently, the first law of thermodynamics for the system gives

$$Q_{1-2} = U_2 - U_1 = mc_v \Delta T = mc_v(T_2 - T_1)$$

$$Q = 2.157 \times 0.742 \times (746 - 323) = 677 \text{ kJ}$$

Note that although the specific heat of air varies with temperature, the specific heat is considered as constant in the above example and the constant value is evaluated at average temperature of air.

Example 4.6 *An insulated tank contains 10 litres of water at 100 kPa and 20°C. A 2 kW electric resistance heater is immersed in water. Considering all the electrical energy is used to raise the temperature of water, find the time required to raise the water temperature to 70°C. Water can be considered as a constant density fluid with density 1000 kg/m^3 and the specific heat is constant as 4.2 kJ/kg°C in the given range of temperature.* E [LO 4.2]

Solution Since the density of water is constant, the mass of water in the tank is also constant and equals to

$$m = \rho V = 1000 \times 10 \times 10^{-3} = 10 \text{ kg}$$

Consider water in the tank as a system. There is only work transfer on the system and as a result of which temperature of water increases. Applying first law of thermodynamics for the system, we have

$$W_{1-2} = U_2 - U_1$$

or $$\dot{W}_{1-2} \Delta t = mc(T_2 - T_1)$$

or $$2000 \,\Delta t = 10 \times 4200(70 - 20)$$

or $$\Delta t = 1050 \text{ s} = 17.5 \text{ minute}$$

Example 4.7 *A fluid is confined in a cylinder by a springloaded, frictionless piston so that the pressure in the fluid is a linear function of the volume ($p = a + bV$). The internal energy of the fluid is given by the following equation:*

$$U = 34 + 3.15\, pV$$

where U is in kJ, p in kPa, and V in cubic metre. If the fluid changes from an initial state of 170 kPa, 0.03 m^3 to a final state of 400 kPa, 0.06 m^3, with no work other than that done on the piston, find the direction and magnitude of the work and heat transfer. M [LO 4.2]

Solution The change in the internal energy of the fluid during the process.

$$U_2 - U_1 = 3.15\,(p_2V_2 - p_1V_1)$$
$$= 315\,(4 \times 0.06 - 1.7 \times 0.03)$$
$$= 315 \times 0.189 = \mathbf{59.5\ kJ}$$

Now

$$p = a + bV$$
$$170 = a + b \times 0.03$$
$$400 = a + b \times 0.06$$

From these two equations

$$a = -60\ \text{kN/m}^2$$
$$b = 7667\ \text{kN/m}^5$$

Work transfer involved during the process

$$W_{1-2} = \int_{V_1}^{V_2} p dV = \int_{V_1}^{V_2} (a + bV)\, dV$$

$$= a(V_2 - V_1) + \frac{V_2^2 - V_1^2}{2}$$

$$= (V_2 - V_1) + \left[a + \frac{b}{2}(V_1 + V_2) \right]$$

$$= 0.03\ \text{m}^3 \left[-60\ \text{kN/m}^2 + \frac{7667}{2}\frac{\text{kN}}{\text{m}^5} \times 0.09\,\text{m}^3 \right]$$

$$= \mathbf{8.55\ kJ}$$

Work is done by the system, the magnitude being 8.55 kJ.

∴ Heat transfer involved is given by

$$Q_{1-2} = U_2 - U_1 + W_{1-2}$$
$$= 59.5 + 8.55 = \mathbf{68.05\ kJ}$$

68.05 kJ of heat flow into the system during the process.

Example 4.8 *A stationary fluid system goes through a cycle shown in Fig. 4.11 comprising the following processes:*

(a) Process 1–2 isochoric heat addition of 235 kJ/kg.

(b) Process 2–3 adiabatic expansion to its original pressure with loss of 70 kJ/kg in internal energy.

(c) Process 3–1 isobaric compression to its original volume with heat rejection of 200 kJ/kg.

Prepare a balance sheet of energy quantities and find the overall changes during the cycle. M [LO 4.1, 4.2]

Solution

$$Q_{1-2} = 235\ \text{kJ/kg},\ W_{1-2} = 0,\ u_2 - u_1$$
$$= Q_{1-2} - W_{1-2} = 235\ \text{kJ/kg}$$
$$Q_{2-3} = 0,\ u_3 - u_2 = -70\ \text{kJ/kg};\ W_{2-3}$$
$$= Q_{2-3} - (u_3 - u_2) = 0 - (-70)$$
$$= 70\ \text{kJ/kg}$$

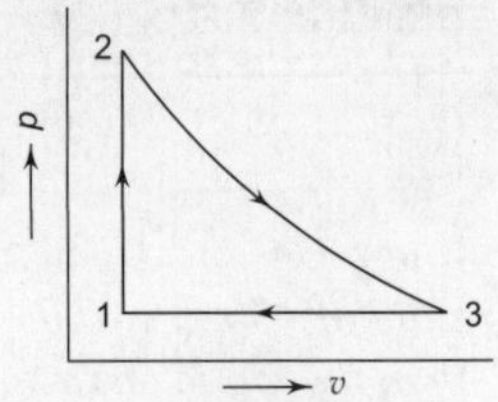

Fig. 4.11

$$Q_{3-1} = -200 \text{ kJ/kg}, u_1 - u_3 = (u_1 - u_2) - (u_3 - u_2) = -235 + 70$$
$$= -165 \text{ kJ/kg}$$
$$W_{3-1} = Q_{3-1} - (u_1 - u_3) = -200 - (-165) = -35 \text{ kJ/kg}$$
$$\sum_{\text{cycle}} Q = Q_{1\text{-}2} + Q_{2\text{-}3} + Q_{3\text{-}1} = 235 + 0 - 200 = \mathbf{35 \text{ kJ/kg}}$$
$$\sum_{\text{cycle}} W = W_{1\text{-}2} + W_{2\text{-}3} + W_{3\text{-}1} = 0 + 70 - 35 = \mathbf{35 \text{ kJ/kg}}$$
$$\therefore \quad \sum_{\text{cycle}} Q = \sum_{\text{cycle}} W$$

This cycle is known as *Lenoir cycle*, and the operation of pulse jet approximates to it.

Example 4.9 *A cylinder fitted with a piston as shown in Fig. 4.12 initially contains 0.2 m^3 of fluid at 200 kPa. Heat is transferred to the fluid and as a result of which the fluid is expanded in a constant-pressure quasi-equilibrium process to a volume thrice of initial one. The piston is then locked with a pin and heat transfer continues until the pressure of fluid reaches 100 kPa. The internal energy of the fluid is given by the following equation*

$$U = 3.4\, pV + 80$$

where U is in kJ, p is in kPa and V is in m^3.
Find the work and heat transfer in both the processes.

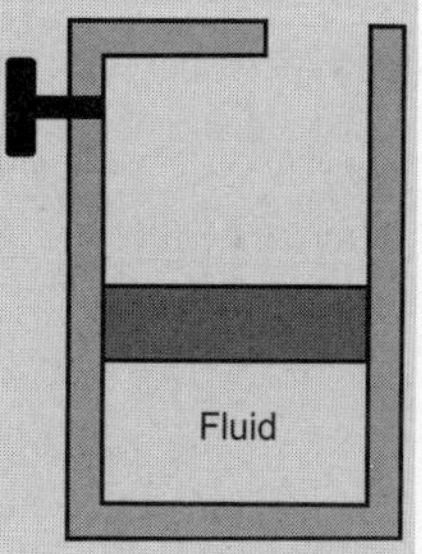

Fig. 4.12

M [LO 4.2]

Solution The fluid in the cylinder bounded by the piston is considered as system. During process 1-2, there is an expansion of the fluid from 0.2 m^3 to 0.6 m^3 at constant pressure of 200 kPa. The work transfer during this process can be found to be

$$W_{1\text{-}2} = \int pdV = p_1(V_2 - V_1) = 200\,(0.6 - 0.2) = 60 \text{ kJ}$$

The change in internal energy during the process 1-2 is

$$U_2 - U_1 = 3.4\,(p_2V_2 - p_1V_1)$$
$$= 3.4 \times 200\,(0.6 - 0.2) = 272 \text{ kJ}$$

The heat transfer during the process 1-2 can be found by applying the first law of thermodynamics for the system as

$$Q_{1\text{-}2} = U_2 - U_1 + W_{1\text{-}2} = 272 + 60 = 332 \text{ kJ}$$

When the piston is locked, there is no change in volume of the fluid and hence the work transfer during this process (2-3) is zero.

The change in internal energy during the process 2-3 is

$$U_3 - U_2 = 3.4\,(p_3V_3 - p_2V_2)$$
$$= 3.4 \times 0.6\,(100 - 200) = -204 \text{ kJ}$$

The heat transfer during the process 2-3 is then

$$Q_{2\text{-}3} = U_3 - U_2 = 204 \text{ kJ}$$

Negative sign indicates that the heat transfer takes place from the fluid and therefore the temperature of the fluid will decrease.

SUMMARY

The first law of thermodynamics is a special statement of the law of conservation of energy. It may be written for a cycle

$$(\Sigma Q)_{\text{cycle}} = (\Sigma W)_{\text{cycle}}$$

or for a process.

$$\begin{aligned} Q &= \Delta E + W \\ &= \Delta E_K + \Delta E_p + \Delta U + W \\ &= \frac{1}{2} mV^2 + mgZ + \Delta U + W \end{aligned}$$

where E is the amount of energy stored in a system and it is a property of the system. The stored energy can be either in macroscopic mode like kinetic and potential energy (E_K and E_P) and also in microscopic mode called intergy energy (U) where energy is stored in the molecules of a system.

The specific heat at constant volume is defined as $c_v = \left(\frac{\partial u}{\partial T}\right)_v$. The heat supplied to a system at constant volume increases its internal energy. The enthalpy of a substance h is defined as $h = u + pv$. The specific heat at constant pressure is defined as $c_p = \left(\frac{\partial h}{\partial T}\right)_p$. The heat supplied to a system at constant pressure increases its enthalpy. The heat capacities are equal to $C_v = mc_v$ and $C_p = mc_p$. The energy of an isolated system is always constant.

REVIEW QUESTIONS

(E) **4.1** State the first law for a closed system undergoing a cycle. **[LO 4.1]**

(M) **4.2** What was the contribution of J.P. Joule in establishing the first law? **[LO 4.1]**

(M) **4.3** What is the caloric theory of heat? Why was it rejected? **[LO 4.1]**

(E) **4.4** Which is the property introduced by the first law? **[LO 4.2]**

(E) **4.5** State the first law for a closed system undergoing a change of state. **[LO 4.2]**

(M) **4.6** Show that energy is a property of a system. **[LO 4.2]**

(M) **4.7** What are the modes in which energy is stored in a system? **[LO 4.4]**

(M) **4.8** Define internal energy. How is energy stored in molecules and atoms? **[LO 4.4]**

(E) **4.9** What is the difference between the standard symbols of E and U? **[LO 4.4]**

(M) **4.10** What is the difference between heat and internal energy? **[LO 4.4]**

(D) **4.11** Define enthalpy. Why does the enthalpy of an ideal gas depend only on temperature? **[LO 4.6]**

(M) **4.12** Define the specific heats at constant volume and at constant pressure. **[LO 4.5, 4.7]**

(D) **4.13** Why should specific heat not be defined in terms of heat transfer? **[LO 4.5, 4.7]**

(E) **4.14** Which property of a system increases when heat is transferred: (a) at constant volume, (b) at constant pressure? **[LO 4.5, 4.6]**

(D) **4.15** What is a PMM1? Why is it impossible? **[LO 4.9]**

PROBLEMS WITH HINTS

(E) **4.1** A closed system of mass 2 kg undergoes a process in which there is a heat transfer of 25 kJ from the system to the surroundings. The amount of work done on the system is 100 kJ. The specific internal energy decreases by 15 kJ/kg and the elevation of the system increases by 1000 m. The acceleration of gravity is 9.6 m/s^2. Determine the change in K.E. of the system. **[LO 4.4]**

Hint: By first law,

$$\therefore \quad \Delta KE + mg\,\Delta Z + m\,\Delta U = \dot{Q} - W$$

$$\therefore \quad \Delta KE = -mg\,\Delta Z - m\,\Delta u + Q - W = \mathbf{85.8\ kJ}$$

(E) **4.2** An electric generator coupled to a windmill produces an average electric power of 5 kW. The power is used to charge a storage battery. Heat transfer from the battery to the surroundings is 0.6 kW. Determine the total amount of energy stored in the battery in 8-hour operation. **[LO 4.4]**

Hint: The storage battery is charged for 8 hours.

$$\therefore \quad W = +5\text{ kW} \times 8\text{ h} = +40\text{ kwh},\ Q = -0.6 \times 8 = -4.8\text{ kWh}$$

$$\therefore \quad \Delta E = (+40 - 4.8) = 35.2\text{ kWh} = \mathbf{1.27 \times 10^5\ kJ}$$

(M) **4.3** A gas contained within a piston-cylinder machine is shown in Fig. 4.13. Initially, the piston face is at $x = 0$ and the spring exerts no force on the piston. As a result of heat transfer, the gas expands, raising the piston until it hits the stops. At this point the piston face is located at $\boldsymbol{x} = 0.05$ m, and the heat transfer ceases. The force exerted by the spring on the piston as the gas expands and varies linearly with $\boldsymbol{x}$ according to $F_{spring} = -kx$, where k = 10, 000 N/m. Kinetic and potential energy effects and friction between the cylinder and the piston are neglected. The acceleration of gravity is $g = 9.81$ m/s^2. **[LO 4.4]**

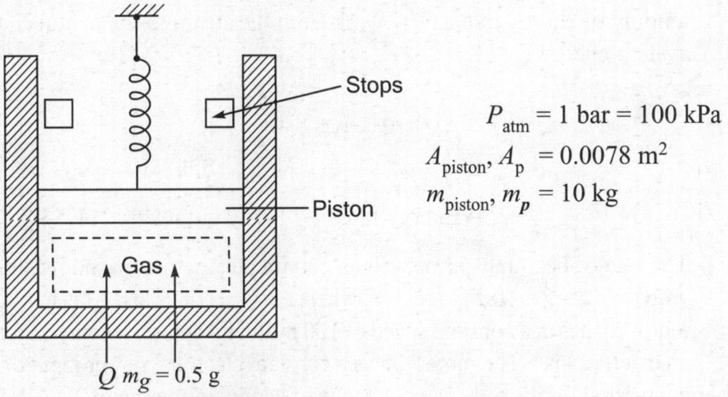

Fig. 4.13

(a) What is the initial pressure of the gas? (b) Determine the work done by the gas on the piston. (c) If the specific internal energies of the gas at the initial and final states are 214 and 337 kJ/kg, respectively, calculate the heat transfer. **[LO 4.4]**

Hint: (a) For the piston $\Sigma \boldsymbol{F_x} = 0$

$$P_{gas} A_{pist} = P_{atm} \cdot A_{pist} + F_{spring} + m_{pist}\, g$$

$$\therefore \quad P_{gas} = P_{atm} + \frac{m_{pist}.g}{A_{pist}} = \mathbf{112.6\ kPa.}$$

(b) Work done by the gas on the piston

$$W = \int_{x_1}^{x_2} P_{gas} A_{pist}\, dx = \int_{x_1}^{x_2} (P_{atm} A_{pist} + m_{pist} g + kx)\, dx$$

$$= \left[(p_{atm} A_{pist} dx + m_{pist} g) x + k\frac{x^2}{2} \right]_{x=0}^{x=0.05\text{ m}}$$

$$= \mathbf{56.4\ J}$$

First law gives,

$$Q = W + \Delta U$$

$$= m(u_2 - u_1) + W$$

$$= \mathbf{117.9\ J}$$

(E) **4.4** A gas of 4 kg is contained within a piston-cylinder machine. The gas undergoes a process for which $pV^{1.5}$ = constant. The initial pressure is 3 bar and the initial volume is 0.1 m^3, and the final volume is 0.2 m^3. The specific internal energy of the gas decreases by 4.6 kJ/kg. There are no significant changes in KE and PE. Determine the net heat transfer for the process. **[LO 4.4]**

Hint:

$$W_{1-2} = \frac{P_1V_1 - P_2V_2}{n-1} \text{ where } P_2 = P_1\left(\frac{V_1}{V_2}\right)^n = 3\left(\frac{0.1}{0.2}\right)^{1.5} = 1.06 \text{ bar}$$

$$= +\ 17.6 \text{ kJ}$$

$$Q = \Delta U + W = m(u_2 - u_1) + W = 4 \times (-4.6) + 17.6$$

$$= \mathbf{-0.8\ kJ}$$

(E) **4.5** A piston-cylinder assembly contains 5 kg of steam. The steam having an internal energy of 2709.9 kJ/kg expands to a state where the internal energy is 2659.6 kJ/kg. During the process, there is heat transfer of 80 kJ to the steam and also a paddle-wheel work transfer of 18.5 kJ. Neglecting *KE* and *PE* of steam, determine the amount of energy transfer by work from the steam to the piston and the decrease in system energy. **[LO 4.4]**

Hint:

$$Q = \Delta U + W = \Delta U + W_{pw} + W_{pav}$$

$$\therefore \quad W_{pdv} = Q - m(u_2 - u_1) - W_{pw} = \mathbf{350\ kJ}$$

$$\Delta E = \text{Energy input} - \text{Energy output} = \mathbf{-251.5\ kJ}$$

(M) **4.6** The rate of heat transfer between a certain electric motor and its surroundings varies with time as $\dot{Q} = -0.2\,[1 - e^{-0.65t}]$ where *t* is in seconds and $\dot{Q}$ is in kW. The shaft of the motor rotates at a constant speed of 100 rad/s and applies a constant torque of 18 Nm to an external load. The motor draws a constant electrical power input of 2 kW. Obtain an expression for the time rate of change of energy of the motor. **[LO 4.4]**

Hint: The time rate of change of system energy

$$\frac{dE}{dt} = \dot{Q} - \dot{W} = \dot{Q} - \left(\dot{W}_{shaft} + \dot{W}_{elect}\right)$$

$$\dot{W}_{shaft} = Tw = 0.08 \times 100 \times 10^{-3} = +1.8 \text{ kW}$$

$$\dot{W}_{elect} = -\ 2.0 \text{ kW}$$

$$\therefore \quad \frac{dE}{dt} = 0.2\,[1 - e^{-0.65\,t}] + 1.8 - 2$$

$$= \mathbf{0.2\, e^{-0.65\,t}}$$

PROBLEMS

(E) **4.1** An engine is tested by means of a water brake at 1000 rpm. The measured torque of the engine is 10,000 mN and the water consumption of the brake is 0.5 m^3/s, its inlet temperature being 20°C. Calculate the water temperature at exit, assuming that the whole of the engine power is ultimately transformed into heat which is absorbed by the cooling water. **[LO 4.1]**

(E) **4.2** In a cyclic process, heat transfers are + 14.7 kJ, – 25.2 kJ, – 3.56 kJ and + 31.5 kJ. What is the net work done for this cyclic process? **[LO 4.1]**

(M) **4.3** A slow chemical reaction takes place in a fluid at the constant pressure of 0.1 MPa. The fluid is surrounded by a perfect heat insulator during the reaction which begins at state 1 and ends at state 2. The insulation is then removed and 105 kJ of heat flow to the surroundings as the fluid goes to state 3. The following data are observed for the fluid at states 1, 2 and 3:

State	*V*(m^3)	*t*(°C)
1	0.003	20
2	0.3	370
3	0.06	20

For the fluid system, calculate E_2 and E_3, if $E_1 = 0$ **[LO 4.2]**

(E) **4.4** During one cycle the working fluid in an engine engages in two work interactions: 15 kJ to the fluid and 44 kJ from the fluid, and three heat interactions, two of which are known: 75 kJ to the fluid and 40 kJ from the fluid. Evaluate the magnitude and direction of the third heat transfer. **[LO 4.1]**

(E) **4.5** A domestic refrigerator is loaded with food and the door closed. During a certain period the machine consumes 1 kW h of energy and the internal energy of the system drops by 5000 kJ. Find the net heat transfer for the system. **[LO 4.2]**

(E) **4.6** 1.5 kg of liquid having a constant specific heat of 2.5 kJ/kg K is stirred in a well-insulated chamber causing the temperature to rise by 15°C. Find ΔE and W for the process. **[LO 4.2]**

(E) **4.7** The same liquid as in Problem 4.6 is stirred in a conducting chamber. During the process 1.7 kJ of heat are transferred from the liquid to the surroundings, while the temperature of the liquid is rising to 15°C. Find ΔE and W for the process. **[LO 4.2]**

(E) **4.8** The properties of a certain fluid are related as follows:

$$u = 196 + 0.718\, t$$
$$pv = 0.287\,(t + 273)$$

where u is the specific internal energy (kJ/kg), t is in °C, p is pressure (kN/m^2) and v is specific volume (m^3/kg). For this fluid, find c_v and c_p. **[LO 4.5, 4.7]**

(E) **4.9** A system composed of 2 kg of the above fluid expands in a frictionless piston and cylinder machine from an initial state of 1 MPa, 100°C to a final temperature of 30°C. If there is no heat transfer, find the net work for the process. **[LO 4.2]**

(M) **4.10** If all the work in the expansion of Problem 4.9 is done on the moving piston, show that the equation representing the path of the expansion in the *pv*-plane is given by $pv^{1.4}$ = constant. **[LO 4.2]**

(E) **4.11** A stationary system consisting of 2 kg of the fluid of Problem 4.8 expands in an adiabatic process according to $pv^{1.2}$ = constant. The initial conditions are 1 MPa and 200°C, and the final pressure is 0.1 MPa. Find W and ΔE for the process. Why is the work transfer not equal to $\int pdV$? **[LO 4.2]**

(E) **4.12** A mixture of gases expands at constant pressure from 1 MPa, 0.03 m^3 to 0.06 m^3 with 84 kJ positive heat transfer. There is no work other than that done on a piston. Find ΔE for the gaseous mixture.
The same mixture expands through the same state path while a stirring device does 21 kJ of work on the system. Find ΔE, W and Q for the process. **[LO 4.2]**

(E) **4.13** A mass of 8 kg gas expands within a flexible container so that the p–v relationship is of the form $pv^{1.2}$ = constant. The initial pressure is 1000 kPa and the initial volume is 1 m^3. The final pressure is 5 kPa. If specific internal energy of the gas decreases by 40 kJ/kg, find the heat transfer in magnitude and direction. **[LO 4.2]**

(M) **4.14** A gas of mass 1.5 kg undergoes a quasi-static expansion which follows a relationship $p = a + bV$, where a and b are constants. The initial and final pressures are 1000 kPa and 200 kPa, respectively and the corresponding volumes are 0.20 m^3 and 1.20 m^3. The specific internal energy of the gas is given by the relation,

$$u = 1.5\, pv - 85 \text{ kJ/kg}$$

where p is in kPa and v is in m^3/kg. Calculate the net heat transfer and the maximum internal energy of the gas attained during expansion. **[LO 4.2]**

(M) **4.15** The heat capacity at constant pressure of a certain system is a function of temperature only and may be expressed as

$$C_p = 2.093 + \frac{41.87}{t + 100} \text{ J/°C}$$

where t is the temperature of the system in °C. The system is heated while it is maintained at a pressure of 1 atmosphere until its volume increases from 2000 cm^3 to 2400 cm^3 and its temperature increases from 0°C to 100°C. (a) Find the magnitude of the heat interaction. (b) How much does the internal energy of the system increase? **[LO 4.2, 4.7]**

(M) **4.16** An imaginary engine receives heat and does work on a slowly moving piston at such rates that the cycle of operation of 1 kg of working fluid can be represented as a circle 10 cm in diameter on a p–v diagram on which 1 cm = 300 kPa and 1 cm = 0.1 m^3/kg. (a) How much work is done by each kg of working fluid for each cycle of operation? (b) The thermal efficiency of an engine is defined as the ratio of work done and heat input in a cycle. If the heat rejected by the engine in a cycle is 1000 kJ per kg of working fluid, what would be its thermal efficiency? **[LO 4.2]**

(M) **4.17** A gas undergoes a thermodynamic cycle consisting of three processes beginning at an initial state where p_1 = 1 bar, V_1 = 1.5 m^3 and U_1 = 512 kJ. The processes are as follows:
(a) Process 1–2: Compression with pV = constant to p_2 = 2 bar, U_2 = 690 kJ
(b) Process 2–3: $W_{23} = 0$, $Q_{23} = -150$ kJ and
(c) Process 3–1: $W_{31} = +50$ kJ. Neglecting KE and PE changes, determine the heat interactions Q_{12} and Q_{31}. **[LO 4.1, 4.2]**

(M) **4.18** A gas undergoes a thermodynamic cycle consisting of the following processes: (i) Process 1–2: Constant pressure p = 1.4 bar, V_1 = 0.028 m^3, W_{12} = 10.5 kJ, (ii) Process 2–3: Compression with pV = constant, $U_3 = U_2$ and (iii) Process 3–1: Constant volume, $U_1 - U_3 = -26.4$ kJ. There are no significant changes in KE and PE. (a) Sketch the cycle on a p-V diagram. (b) Calculate the net work for the cycle in kJ. (c) Calculate the heat transfer for process 1–2 (d) Show that $\sum_{\text{cycle}} Q = \sum_{\text{cycle}} W$. **[LO 4.1, 4.2]**

(M) **4.19** A fluid contained in a cylinder receives 150 kJ of mechanical energy by means of a paddle wheel, together with 50 kJ in the form of heat. At the same time, a piston in the cylinder moves in such a way that the pressure remains constant at 200 kN/m^2 during the fluid expansion from 2m^3 to 5m^3. What is the change in internal energy and in enthalpy? **[LO 4.2]**

ANSWERS TO PROBLEMS

4.1 20.5°C
4.2 17.34 kJ
4.3 $E_2 = -29.7$ kJ, $E_3 = -110.7$ kJ
4.4 – 6 kJ
4.5 – 8.6 MJ
4.6. $\Delta E = 56.25$ kJ, $W = -56.25$ kJ
4.7 $\Delta E = 54.55$ kJ, $W = 56.25$ kJ
4.8 0.718, 1.005 kJ/kg K)
4.9 100.52 kJ
4.11 $W = 217.35$, $\Delta E = -217.35$ kJ, $\int pdV = 434.4$ kJ
4.12 54 kJ, – 21 kJ, 33 kJ
4.13 + 2615 kJ
4.14 660 kJ, 503.3 kJ
4.15 (a) 238.32 J
(b) 197.79 J
4.16 (a) 2356.19 kJ/kg
(b) 0.702
4.17 74 kJ, 22 kJ
4.18 (b) – 8.28 kJ
(c) 36.9 kJ
4.19 – 400 kJ, + 200 kJ

5

First Law Applied to Flow Processes

Learning Objectives

LO 5.1 State flow process in thermodynamics

LO 5.2 State steady flow

LO 5.3 Demonstrate steady flow energy equation (S.F.E.E.)

LO 5.4 Apply S.F.E.E. to typical engineering devices

LO 5.5 Interpret Euler's and Bernoulli's equations from S.F.E.E.

LO 5.6 Discuss unsteady flow process

LO 5.7 Analyze unsteady flow using both system and control volume approach

LO 5.8 Analyze variable flow for charging and discharging a tank

5.1 CONTROL VOLUME

LO 5.1
State flow process in thermodynamics

For any system and in any process, the first law can be written as

$$Q = \Delta E + W$$

where E represents all forms of energy stored in the system.

For a pure substance,

$$E = E_K + E_P + U$$

where E_K is the K.E., E_P is the P.E. and U is the residual energy stored in the molecular structure of the substance.

$$\therefore \quad Q = \Delta E_K + \Delta E_P + \Delta U + W \tag{5.1}$$

When there is mass transfer across the system boundary, the system is called an open system. Most of the engineering devices are open systems involving the flow of fluids through them.

Equation (5.1) refers to a system having a particular mass of substance, and it is free to move from place to place.

Consider a steam turbine (Fig. 5.1) in which steam enters at a high pressure, does work upon the turbine rotor, and then leaves the turbine at low pressure through the exhaust pipe.

If a certain mass of steam is considered as the thermodynamic system, then the energy equation becomes

$$Q = \Delta E_K + \Delta E_P + \Delta U + W$$

and in order to analyse the expansion process in turbine the moving system is to be followed as it travels through the turbine, taking into account the work and heat interactions all the way through. This method of analysis is similar to that of Lagrange in fluid mechanics.

Although the system approach is quite valid, there is another approach which is found to be highly convenient. Instead of concentrating attention upon a certain quantity of fluid, which constitutes a moving system in flow processes, attention is focused upon a certain fixed region in space called a *control volume* through which the moving substance flows. This is similar to the analysis of Euler in fluid mechanics.

To distinguish the two concepts, it may be noted that while the system (closed) boundary usually changes shape, position and orientation relative to the observer, the control volume boundary remains fixed and unaltered. Again, while matter usually crosses the control volume boundary, no such flow occurs across the system boundary.

The broken line in Fig. 5.1 represents the surface of the control volume which is known as the *control surface*. This is the same as the system boundary of the open system. The method of analysis is to inspect the control surface and account for all energy quantities transferred through this surface. Since there is mass transfer across the control surface, a mass balance also has to be made. Sections 1 and 2 allow mass transfer to take place, and Q and W are the heat and work interactions, respectively.

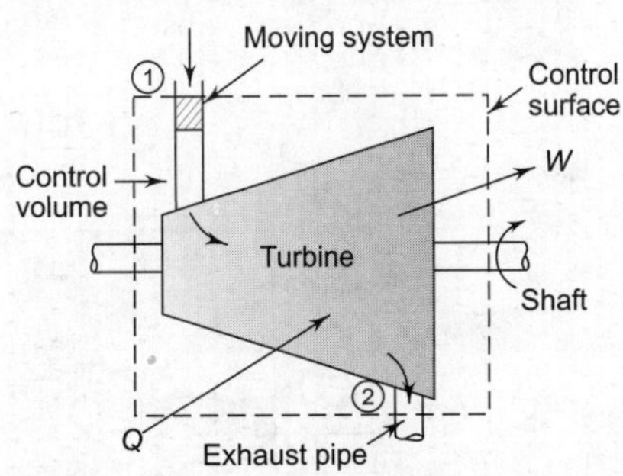

Fig. 5.1 Flow process involving work and heat interactions

5.2 STEADY FLOW PROCESS

> **LO 5.2**
> State steady flow

As a fluid flows through a certain control volume, its thermodynamic properties may vary along the space coordinates as well as with time. If the rates of flow of mass and energy through the control surface change with time, the mass and energy within the control volume also would change with time.

'Steady flow' means that the rates of flow of mass and energy across the control surface are constant.

In most engineering devices, there is a constant rate of flow of mass and energy through the control surface, and the control volume in course of time attains a steady state. *At the steady state of a system, any thermodynamic property will have a fixed value at a particular location, and will not alter with time.* Thermodynamic properties may vary along space coordinates, but do not vary with time. 'Steady state' means that the state is steady or invariant with time.

5.3 MASS BALANCE AND ENERGY BALANCE IN A SIMPLE STEADY FLOW PROCESS

> **LO 5.3**
> Demonstrate steady flow energy equation (S.F.E.E.)

In Fig. 5.2, a steady flow system has been shown in which, one stream of fluid enters and an another stream leaves the control volume. There is no accumulation of mass or energy within the control volume, and the properties at any location within the control volume are steady with time. Sections 1–1 and 2–2 indicate, respectively, the entrance and exit of the fluid across the control surface. The following quantities are defined with reference to Fig. 5.2:

A_1, A_2—cross section of stream, m^2

w_1, w_2—mass flow rate, kg/s

p_1, p_2—pressure, absolute, N/m^2

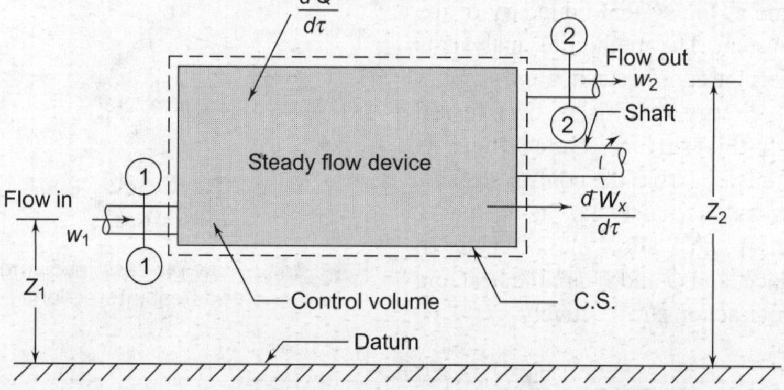

Fig. 5.2 Steady flow process

v_1, v_2—specific volume, m^3/kg

u_1, u_2—specific internal energy, J/kg

$\mathbf{V}_1, \mathbf{V}_2$—velocity, m/s

Z_1, Z_2—elevation above an arbitrary datum, *m*

$\frac{đQ}{d\tau}$—net rate of heat transfer through the control surface, J/s

$\frac{đW_x}{d\tau}$—net rate of work transfer through the control surface, J/s

exclusive of work done at Sections 1 and 2 in transferring the fluid through the control surface.

τ—time, s.

Subscripts 1 and 2 refer to the inlet and exit sections.

5.3.1 Mass Balance

By the conservation of mass, if there is no accumulation of mass within the control volume, the mass flow rate entering must equal the mass flow rate leaving, or

$$w_1 = w_2$$

or

$$\frac{A_1 \mathbf{V}_1}{v_1} = \frac{A_2 \mathbf{V}_2}{v_2} \tag{5.2}$$

This equation is known as the *equation of continuity*.

5.3.2 Energy Balance

In a flow process, the work transfer may be of two types: the *external work* and the *flow work*. The external work refers to all the work transfer across the control surface other than that due to normal fluid forces. In engineering thermodynamics, the only two kinds of external work of importance are *shear* (shaft or stirring) *work* and *electrical work*. In Fig. 5.2 the only external work occurs in the form of shaft work, W_x. The flow work, as discussed in Section. 3.4, is the displacement work done by the fluid of mass dm_1 at the inlet Section 1 and that of mass dm_2 at the exit Section 2, which are $(-\,p_1 v_1\, dm_1)$ and $(+\,p_2 v_2\, dm_2)$, respectively. Therefore, the total work transfer is given by

$$W = W_x - p_1 v_1 dm_1 + p_2 v_2 dm_2 \tag{5.3}$$

In the rate form,

$$\frac{đW}{d\tau} = \frac{đW_x}{d\tau} - p_1 v_1 \frac{dm_1}{d\tau} + p_2 v_2 \frac{dm_2}{d\tau}$$

or

$$\frac{đW}{d\tau} = \frac{đW_x}{d\tau} - w_1 p_1 v_1 + w_2 p_2 v_2 \tag{5.4}$$

Since there is no accumulation of energy, by the conservation of energy, the total rate of flow of all energy streams entering the control volume must equal the total rate of flow of all energy streams leaving the control volume. This may be expressed in the following equation:

$$w_1 e_1 + \frac{đQ}{d\tau} = w_2 e_2 + \frac{đQ}{d\tau}$$

Substituting for $\frac{đW}{d\tau}$ from Eq. (5.4)

$$w_1 e_1 + \frac{đQ}{d\tau} = w_2 e_2 + \frac{đW_x}{d\tau} - w_1 p_1 v_1 + w_2 p_2 v_2$$

$$\therefore \qquad w_1e_1 + w_1p_1v_1 + \frac{đQ}{d\tau} = w_2e_2 + w_2p_2v_2 + \frac{đW_x}{d\tau} \tag{5.5}$$

where e_1 and e_2 refer to the energy carried into or out of the control volume with unit mass of fluid.

The specific energy e is given by

$$e = e_k + e_p + u = \frac{\mathbf{V}^2}{2} + Zg + u \tag{5.6}$$

Substituting the expression for e in Eq. (5.5)

$$w_1\left(\frac{\mathbf{V}_1^2}{2} + Z_1g + u_1\right) + w_1p_1v_1 + \frac{đQ}{d\tau}$$

$$= w_2\left(\frac{\mathbf{V}_2^2}{2} + Z_2g + u_2\right) + w_2p_2v_2 + \frac{đW_x}{d\tau}$$

or

$$w_1\left(h_1 + \frac{\mathbf{V}_1^2}{2} + Z_1g\right) + \frac{đQ}{d\tau}$$

$$= w_2\left(h_2 + \frac{\mathbf{V}_2^2}{2} + Z_2g\right) + \frac{đW_x}{d\tau} \tag{5.7}$$

where $h = u + pv$.

And, since $\qquad w_1 = w_2$, let $w = w_1 = w_2 = \dfrac{dm}{d\tau}$

Dividing Equation (5.7) by $\dfrac{dm}{d\tau}$

$$h_1 + \frac{V_1^2}{2} + Z_1g + \frac{đQ}{dm} = h_2 + \frac{V_2^2}{2} + Z_2g + \frac{đW}{dm} \tag{5.8}$$

Equations (5.7) and (5.8) are known as *steady flow energy equations* (S.F.E.E.), for a single stream of fluid entering and a single stream of fluid leaving the control volume. All the terms in Equation (5.8) represent energy flow per unit mass of fluid (J/kg), whereas the terms in Equation (5.7) represent energy flow per unit time (J/s). The basis of energy flow per unit mass is usually more convenient when only a single stream of fluid enters and leaves a control volume. When more than one fluid stream is involved the basis of energy flow per unit time is more convenient.

Equation (5.8) can be written in the following form:

$$Q - W_x = (h_2 - h_1) + \frac{\bar{\mathbf{V}}_2^2 - \bar{\mathbf{V}}_1^2}{2} + g(Z_2 - Z_1) \tag{5.9}$$

where Q and W_x refer *to* energy transfer per unit mass. In the differential form, the SFEE becomes

$$đQ - đW_x = dh + \bar{\mathbf{V}}d\bar{\mathbf{V}} + gdZ \tag{5.10}$$

When more than one stream of fluid enters or leaves the control volume (Fig. 5.3), the mass balance and energy balance for steady flow are given as follows.

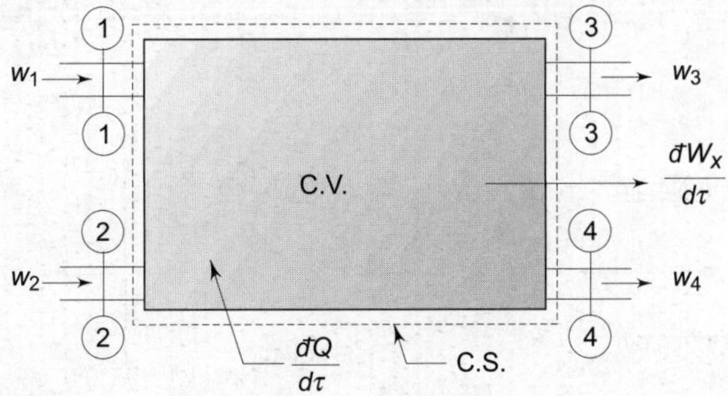

Fig. 5.3 Steady flow process involving two fluid streams at the inlet and exit of the control volume

Mass balance

$$w_1 + w_2 = w_3 + w_4 \tag{5.11}$$

$$\frac{A_1\mathbf{V}_1}{v_1} + \frac{A_2\mathbf{V}_2}{v_2} = \frac{A_3\mathbf{V}_3}{v_3} + \frac{A_4\mathbf{V}_4}{v_4} \tag{5.12}$$

Energy balance

$$w_1\left(h_1 + \frac{\mathbf{V}_1^2}{2} + Z_1 g\right) + w_2\left(h_2 + \frac{\mathbf{V}_2^2}{2} + Z_2 g\right) + \frac{đQ}{d\tau}$$

$$= w_3\left(h_3 + \frac{\mathbf{V}_3^2}{2} + Z_3 g\right) + w_4\left(h_4 + \frac{\mathbf{V}_4^2}{2} + Z_4 g\right) + \frac{đW_x}{d\tau} \tag{5.13}$$

The steady flow energy equation applies to a wide variety of processes like pipe line flows, heat transfer processes, mechanical power generation in engines and turbines, combustion processes and flows through nozzles and diffusors. In certain problems, some of the terms in steady flow energy equation may be negligible or zero. But it is best to write the full equation first, and then eliminate the terms which are unnecessary.

5.4 SOME EXAMPLES OF STEADY FLOW PROCESSES

LO 5.4
Apply S.F.E.E to typical engineering devices

The following examples illustrate the applications of the steady flow energy equation in some of the engineering systems.

5.4.1 Nozzle and Diffusor

A nozzle is a device which increases the velocity or K.E. of a fluid at the expense of its pressure drop, whereas a diffusor increases the pressure of a fluid at the expense of its K.E. Figure 5.4 shows a nozzle which is insulated. The steady flow energy equation of the control surface gives

$$h_1 + \frac{\mathbf{V}_1^2}{2} + Z_1 g + \frac{đQ}{dm} = h_2 + \frac{\mathbf{V}_2^2}{2} + Z_2 g + \frac{đW_x}{dm}$$

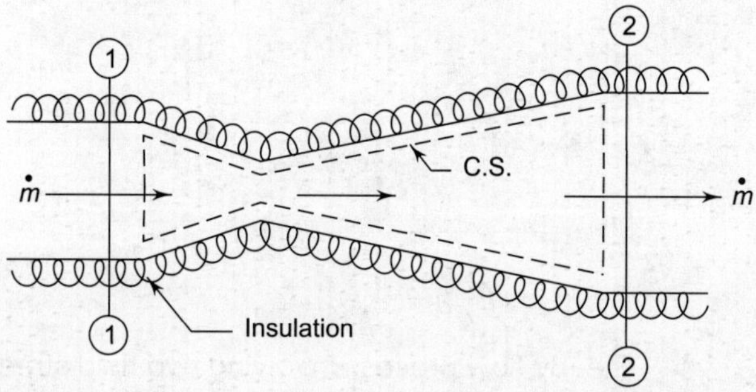

Fig. 5.4 Steady flow process involving two fluid streams at the inlet and exit of the control volume

Here, $\frac{đQ}{dm} = 0$, $\frac{đW_x}{dm} = 0$ and the change in potential energy is zero. The equation reduces to

$$h_1 + \frac{\mathbf{V}_1^2}{2} = h_2 + \frac{\mathbf{V}_2^2}{2} \tag{5.14}$$

The continuity equation gives

$$w = \frac{A_1\mathbf{V}_1}{v_1} = \frac{A_2\mathbf{V}_2}{v_2} \tag{5.15}$$

When the inlet velocity or the 'velocity of approach' $\mathbf{V}_1$ is small compared to the exit velocity $\mathbf{V}_2$, Eq. (5.14) becomes

$$h_1 = h_2 + \frac{\mathbf{V}_2^2}{2}$$

or

$$\mathbf{V}_2 = \sqrt{2(h_1 - h_2)} \text{ m/s}$$

where $(h_1 - h_2)$ is in J/kg.

Equations (5.14) and (5.15) hold good for a diffusor as well.

5.4.2 Throttling Device

When a fluid flows through a constricted passage, like a partially opened valve, an orifice, or a porous plug, there is an appreciable drop in pressure, and the flow is said to be throttled. Figure 5.5 shows the process of throttling by a partially opened valve on a fluid flowing in an insulated pipe. In the steady-flow energy Eq. (5.8),

$$\frac{đQ}{dm} = 0, \quad \frac{đW_x}{dm} = 0$$

and the changes in P.E. are very small and ignored. Thus, the S.F.E.E. reduces to

$$h_1 + \frac{\mathbf{V}_1^2}{2} = h_2 + \frac{\mathbf{V}_2^2}{2}$$

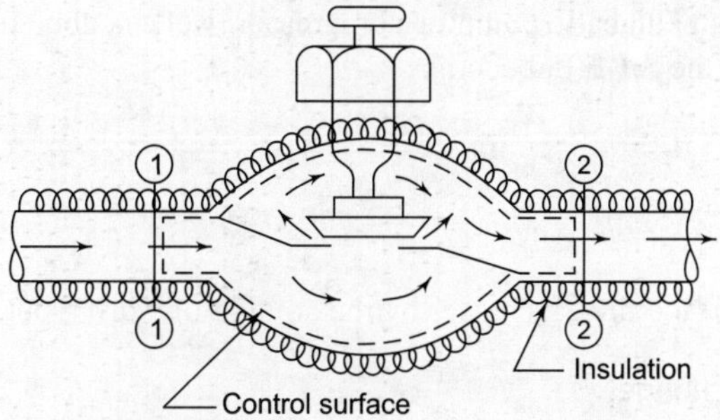

Fig. 5.5 Flow through a valve

Often the pipe velocities in throttling are so low that the K.E. terms are also negligible. So

$$h_1 = h_2 \tag{5.16}$$

or the enthalpy of the fluid before throttling is equal to the enthalpy of the fluid after throttling.

Enthalpy is the energy of a flowing fluid per unit mass. Since no energy is imparted to, or rejected from, the fluid during a throttling process, its enthalpy i.e., energy remains the same.

5.4.3 Turbine, Pump and Compressor

Turbines and engines give positive power output, whereas compressors and pumps require power input.

For a turbine (Fig. 5.6) which is well insulated, the flow velocities are often small, and the K.E. terms can be neglected. The S.F.E.E. then becomes

$$h_1 = h_2 + \frac{đW_x}{dm}$$

or

$$\frac{W_x}{m} = (h_1 - h_2)$$

It is seen that work is done by the fluid at the expense of its enthalpy.

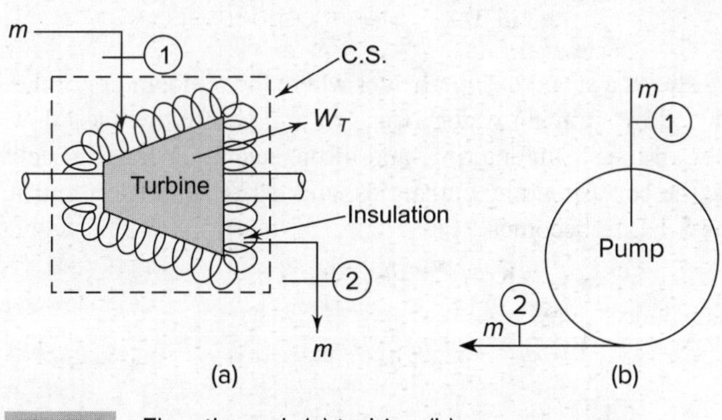

Fig. 5.6 Flow through (a) turbine (b) pump

Similarly, for an adiabatic pump or compressor, work is done upon the fluid and W is negative. So the S.F.E.E. becomes

$$h_1 = h_2 - \frac{W_x}{m}$$

or
$$\frac{W_x}{m} = h_2 - h_1$$

The enthalpy of the fluid increases by the amount of work input.

5.4.4 Heat Exchanger

A heat exchanger is a device in which heat is transferred from one fluid to another. Figure 5.7 shows a steam condenser, where steam condenses outside the tubes and cooling water flows through the tubes. The S.F.E.E. for the C.S. gives

$$w_c h_1 + w_s h_2 = w_c h_3 + w_s h_4$$

or
$$w_s (h_2 - h_4) = w_c(h_3 - h_1)$$

Here the K.E. and P.E. terms are considered small, there is no external work done, and energy exchange in the form of heat is confined only between the two fluids, i.e. there is no external heat interaction or heat loss.

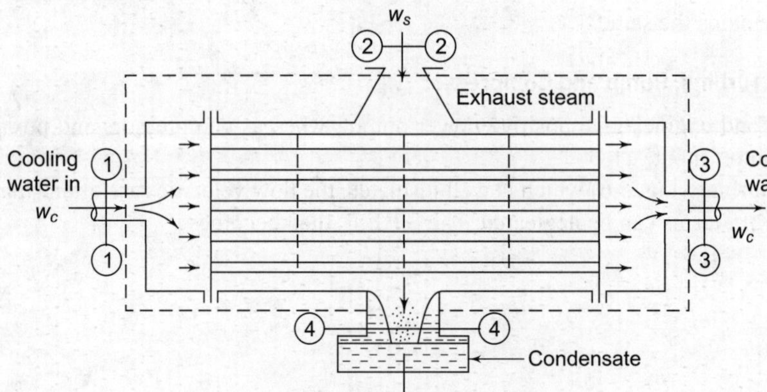

Fig. 5.7 Steam condenser

Figure 5.8 shows a steam desuperheater where the temperature of the superheated steam is reduced by spraying water. If w_1, w_2 and w_3 are the mass flow rates of the injected water, of the steam entering, and of the steam leaving, respectively, and h_1, h_2 and h_3 are the corresponding enthalpies, and if K.E. and P.E. terms are neglected as before, the S.F.E.E. becomes

$$w_1 h_1 + w_2 h_2 = w_3 h_3$$

and the mass balance gives

$$w_1 + w_2 = w_3$$

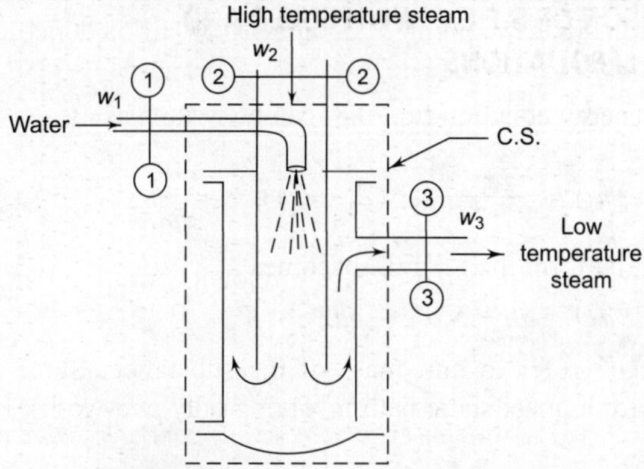

Fig. 5.8 Steam desuperheater

5.4.5 Boiler

A boiler is a device designed to generate steam by heating water on a large scale. Fig. 5.9 shows an illustration of a boiler. It receives air and fuel (primarily coal) for combustion, which releases the fuel's calorific value (CV_f). The flue gas produced during combustion transfers heat to the water, raising its enthalpy through both latent heat (during evaporation) and sensible heat (which includes liquid heating and, if applicable, steam superheating). Ash, a byproduct of solid fuels, is either deposited at the bottom of the boiler or carried over to the stack or chimney. The enthalpy change of the steam is provided by the calorific value of the fuel and the enthalpy of the incoming air. Some heat is inevitably lost to the atmosphere through the flue gas. Assuming complete combustion of the fuel and no heat loss, the S.F.E.E. of the boiler will be

$$m_a h_2 + m_f \mathrm{CV}_f + m_w h_3 = m_s h_5 + m_{fg} h_4$$

Neglecting, any loss of water and steam,

$$m_w = m_s$$

The mass of steam generated from the boiler

$$m_s = (m_a h_2 + m_f \mathrm{CV}_f - m_{fg} h_4)/(h_5 - h_3)$$

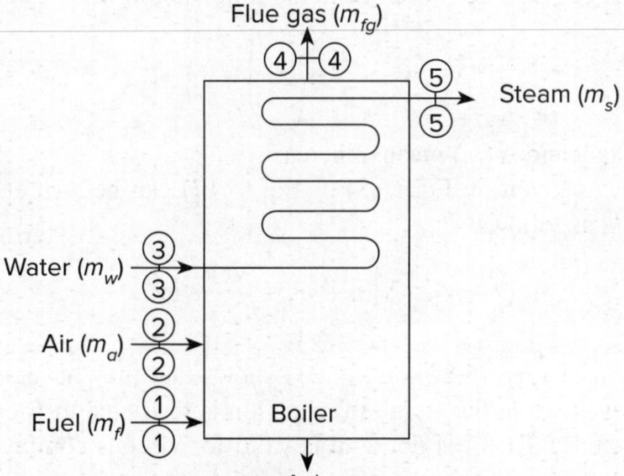

Fig. 5.9 Boiler

5.5 COMPARISON OF S.F.E.E. WITH EULER AND BERNOULLI EQUATIONS

LO 5.5 Interpret Euler's and Bernoulli's equations from S.F.E.E

The steady flow energy equation Eq. (5.8) can be written as

$$\frac{đQ}{dm} = (h_2 - h_1) + \frac{\bar{V}_2^2 - \bar{V}_1^2}{2} + (Z_2 - Z_1)\,g + \frac{đW_x}{dm}$$

In the differential form the S.F.E.E. becomes

$$đQ = dh + \bar{\mathbf{V}}d\bar{\mathbf{V}} + gdZ + đW_x \tag{5.17}$$

where $đQ$ and $đW_x$ refer to unit mass of the substance. Since $h = u + pv$ and $đQ = du + pdv$ (for a quasi-static path involving only pdv-work), Eq. (5.17) can be written as

$$du + pd\mathbf{v} = du + pd\mathbf{v} + \mathbf{v}dp + \mathbf{V}d\mathbf{V} + gdZ + đW_x$$

For an inviscid frictionless fluid flowing through a pipe, the equation reduces to

$$\mathbf{v}dp + \mathbf{V}d\mathbf{V} + gdZ = 0 \tag{5.18}$$

This is the *Euler equation*. If we integrate between two Sections 1 and 2 of the pipe

$$\int_1^2 \mathbf{v}dp + \int_1^2 \mathbf{V}d\mathbf{V} + \int_1^2 gdZ = 0$$

For an incompressible fluid, v = constant

$$\therefore \quad \mathbf{v}(p_2 - p_1) + \frac{\mathbf{V}_2^2}{2} - \frac{\mathbf{V}_1^2}{2} + g\,(Z_2 - Z_1) = 0 \tag{5.19}$$

Since the specific volume v is the reciprocal of the density ρ, we have

$$\frac{p_1}{\rho} + \frac{\mathbf{V}_1^2}{2} + Z_1 g = \frac{p_2}{\rho} + \frac{\mathbf{V}_2^2}{2} + Z_2 g \tag{5.20}$$

or

$$\frac{p}{\rho} + \frac{\mathbf{V}^2}{2} + Zg = \text{constant} \tag{5.21}$$

This is known as the *Bernoulli equation*, which is valid for an inviscid incompressible fluid. It can also be expressed in the following form:

$$\Delta\left(p\mathbf{v} + \frac{\mathbf{V}^2}{2} + gZ\right) = 0 \tag{5.22}$$

where v is constant and Δ (...) means 'increase in ...

The S.F.E.E. as given by Eq. (5.8) or Eq. (5.17) can be written with $(u + pv)$ substituted for h, as follows:

$$Q - W_x = \Delta\left(u + p\mathbf{v} + \frac{\mathbf{V}^2}{2} + gZ\right) \tag{5.23}$$

A comparison of Eqs. (5.22) and (5.23) shows that they have several terms in common. However, while the Bernoulli equation is restricted to frictionless incompressible fluids, the S.F.E.E. is not, and is valid for viscous compressible fluids as well. The Bernoulli equation is, therefore, a special limiting case of the more general steady flow energy equation.

5.6 VARIABLE FLOW PROCESSES

LO 5.6
Discuss unsteady flow process

Many flow processes, such as filling up and evacuating gas cylinders, are not steady. Such processes can be analysed by the control volume technique. Consider a device through which a fluid is flowing under non-steady state conditions (Fig. 5.10). The rate at which the mass of fluid within the control volume is accumulated is equal to the net rate of mass flow across the control surface, as given below.

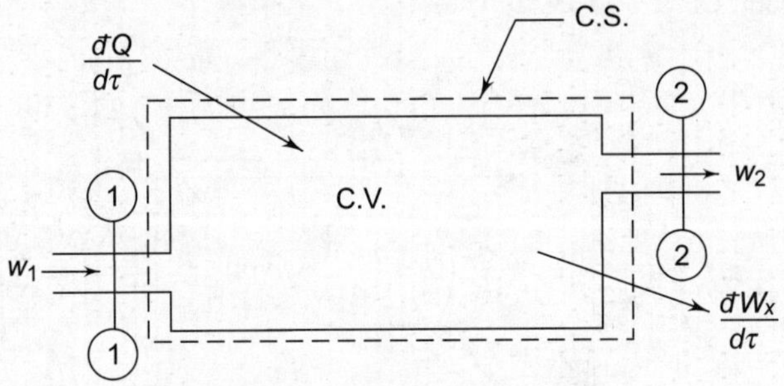

Fig. 5.10 Variable flow process

$$\frac{dm_V}{d\tau} = w_1 - w_2 = \frac{dm_1}{d\tau} - \frac{dm_2}{d\tau} \tag{5.24}$$

where m_V is the mass of fluid within the control volume at any instant.

Over any finite period of time

$$\Delta m_V = \Delta m_1 - \Delta m_2 \tag{5.25}$$

The rate of accumulation of energy within the control volume is equal to the net rate of energy flow across the control surface. If E_v is the energy of fluid within the control volume at any instant,

Rate of energy increase = Rate of energy inflow – Rate of energy outflow

$$\frac{dE_V}{d\tau} = w_1\left(h_1 + \frac{\mathbf{V}_1^2}{2} + Z_1 g\right) + \frac{đQ}{d\tau}$$

$$- w_2\left(h_2 + \frac{\mathbf{V}_2^2}{2} + Z_2 g\right) - \frac{đW_x}{d\tau} \tag{5.26}$$

Now $$E_V = \left(U + \frac{mV^2}{2} + mgZ\right)_V$$

where m is the mass of fluid in the control volume at any instant.

$$\therefore \quad \frac{dE_V}{d\tau} = \frac{d}{d\tau}\left(U + \frac{mV^2}{2} + mgZ\right)_V$$

$$= \left(h_1 + \frac{\mathbf{V}_1^2}{2} + Z_1 g\right)\frac{dm_1}{d\tau} + \frac{đQ}{d\tau}$$

$$- \left(h_2 + \frac{\mathbf{V}_2^2}{2} + Z_2 g\right)\frac{dm_2}{d\tau} - \frac{đW_x}{d\tau} \quad (5.27)$$

Figure 5.11 shows all these energy flux quantities. For any finite time interval, Eq. (5.27) becomes

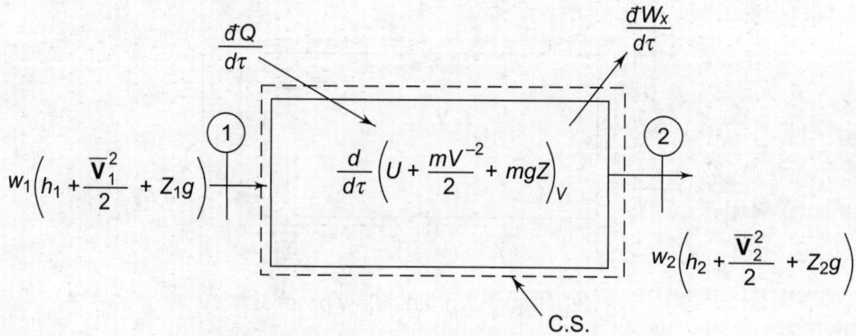

Fig. 5.11 Energy fluxes in an unsteady system

$$\Delta E_V = Q - W_x + \int \left(h_1 + \frac{\mathbf{V}_1^2}{2} + Z_1 g\right) dm_1$$

$$- \int \left(h_2 + \frac{\mathbf{V}_2^2}{2} + Z_2 g\right) dm_2 \quad (5.28)$$

Equation (5.26) is the *general energy equation*. For steady flow,

$$\frac{dE_V}{d\tau} = 0,$$

and the equation reduces to Eq. (5.7). For a closed system $w_1 = 0$, $w_2 = 0$, then from Eq. (5.26),

$$\frac{dE_V}{d\tau} = \frac{đQ}{d\tau} - \frac{đW_x}{d\tau}$$

or $$dE_V = đQ - đW_x \quad \text{or,} \quad đQ = đE + đW_x$$

as obtained earlier.

5.7 EXAMPLE OF A VARIABLE FLOW PROBLEM

Variable flow processes may be analysed either by the system technique or the control volume technique, illustrated as follows.

> **LO 5.7**
> Analyze unsteady flow using both system and control volume approach

Consider a process in which a gas bottle is filled from a pipeline (Fig. 5.12). In the beginning, the bottle contains gas of mass m_1 at state p_1, t_1, v_1, h_1 and u_1. The valve is opened and gas flows into the bottle till the mass of gas in the bottle is m_2 at state p_2, t_2, v_2, h_2 and u_2. The supply to the pipeline is very large so that the state of gas in the pipeline is constant at p_p, t_p, v_p, h_p, u_p and $\mathbf{V}_p$.

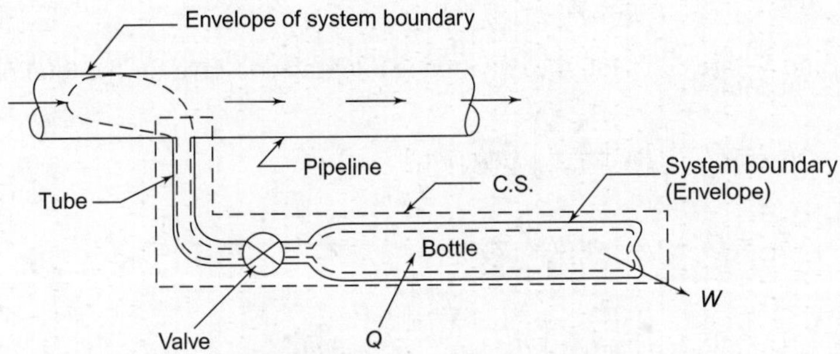

Fig. 5.12 Bottle-filling process

System technique Assume an envelope (which is extensible) of gas in the pipeline and the tube which would eventually enter the bottle as shown in Fig. 5.12.

Energy of the gas before filling

$$E_1 = m_1u_1 + (m_2 - m_1)\left(\frac{\mathbf{V}_p^2}{2} + u_p\right)$$

where $(m_2 - m_1)$ is the mass of gas in the pipeline and tube which would enter the bottle.

Energy of the gas after filling

$$E_2 = m_2u_2$$

$$\Delta E = E_2 - E_1 = m_2u_2 - \left[m_1u_1 + (m_2 - m_1)\left(\frac{\mathbf{V}_p^2}{2} + u_p\right)\right] \quad (5.29)$$

The P.E. terms are neglected. The gas in the bottle is not in motion, and so the K.E. terms have been omitted.

Now, there is a change in the volume of gas because of the collapse of the envelope to zero volume. Then the work done

$$W = p_p(\mathbf{V}_2 - \mathbf{V}_1) = p_p[0 - (m_2 - m_1)v_p]$$
$$= -(m_2 - m_1)p_pv_p$$

$\therefore$ using the first law for the process

$$Q = \Delta E + W$$
$$= m_2u_2 - m_1u_1 - (m_2 - m_1)\left(\frac{\mathbf{V}_p^2}{2} + u_p\right) - (m_2 - m_1)p_pv_p$$
$$= m_2u_2 - m_1u_1 - (m_2 - m_1)\left(\frac{\mathbf{V}_p^2}{2} + h_p\right) \quad (5.30)$$

which gives the energy balance for the process.

Control volume technique Assume a control volume bounded by a control surface, as shown in Fig. 5.12. Applying the energy Eq. (5.27) to this case, the following energy balance may be written on a time rate basis:

$$\frac{dE_V}{d\tau} = \frac{đQ}{d\tau} + \left(h_p + \frac{\mathbf{V}_p^2}{2}\right)\frac{dm}{d\tau}$$

Since h_p and $\mathbf{V}_p$ are constant, the equation is integrated to give for the total process

$$\Delta E_V = Q + \left(h_p + \frac{\mathbf{V}_p^2}{2}\right)(m_2 - m_1)$$

Now

$$\Delta E_V = U_2 - U_1 = m_2u_2 - m_1u_1$$

$$\therefore \quad Q = m_2u_2 - m_1u_1 - \left(h_p + \frac{\mathbf{V}_p^2}{2}\right)(m_2 - m_1)$$

This equation is the same as Eq. (5.30).

5.8 DISCHARGING AND CHARGING A TANK

LO 5.8 Analyze variable flow for charging and discharging a tank

Let us consider a tank discharging a fluid into a supply line (Fig. 5.13). Since $\delta W_x = 0$ and $dm_{in} = 0$, applying first law to the control volume,

$$dU_V = đQ + \left(h + \frac{\mathbf{V}^2}{2} + gz\right)_{out} dm_{out} \qquad (5.31)$$

Assuming K.E. and P.E. of the fluid to be small and $đQ = 0$

$$d(mu) = hdm$$

$$mdu + udm = udm + pv\,dm$$

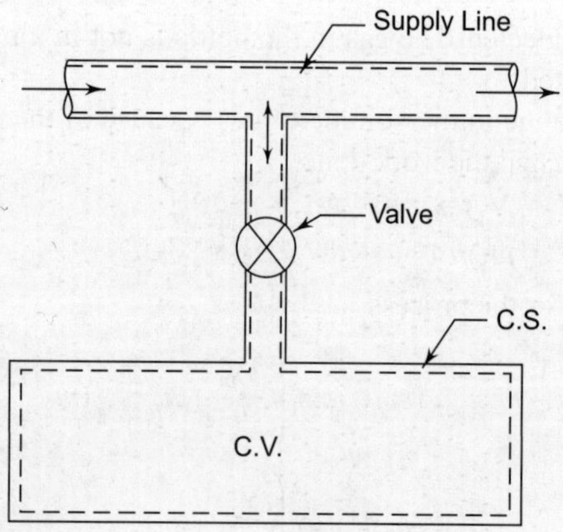

Fig. 5.13 Charging and discharging a tank

$$\frac{dm}{m} = \frac{du}{pv} \tag{5.32}$$

Again the total volume $V = vm = \text{const.}$

$$vdm + mdv = 0$$

or

$$\frac{dm}{m} = -\frac{dv}{v} \tag{5.33}$$

From Eqs. (5.32) and (5.33),

$$\frac{du}{pv} = -\frac{dv}{v}$$

$$d(u + pv) = 0$$

or

$$đQ = 0$$

which shows that the process is adiabatic and quasi-static.

For charging the tank

$$\int (hdm)_{\text{in}} = \Delta U_V = m_2u_2 - m_1u_1 \tag{5.34}$$

$$m_ph_p = m_2u_2 - m_1u_1$$

where the subscript p refers to the constant state of the fluid in the pipeline. If the tank is initially empty, $m_1 = 0$.

$$m_ph_p = m_2u_2$$

Since

$$m_p = m_2$$

$$h_p = u_2$$

If the fluid is an ideal gas, the temperature of the gas in the tank after it is charged is given by

$$c_pT_p = c_vT_2$$

or

$$T_2 = \gamma T_p \tag{5.35}$$

SOLVED EXAMPLES

Example 5.1 *In a steam power plant, steam flows steadily from the steam generator to the turbine. In order to increase the velocity of stream before striking the turbine blades, it passes through a horizontal nozzle having an inlet area of 0.1 m². Steam at 1000 kPa and 400°C enters the nozzle with a velocity of 70 m/s. The steam leaves the nozzle at 300 kPa and 300°C and there is a heat loss of30 kW while steam passing through it. Properties of steam are as follows:*

Pressure (kPa)	*Temperature (°C)*	*Enthalpy (kJ/kg)*	*Specific volume (m^3/kg)*
1000	400	3263.9	0.292
300	300	3069.3	0.875

Determine the mass flow rate of steam, the exit velocity and the exit area of the nozzle. E [LO 5.3]

Solution From the mass balance, the mass flow rate can be found as

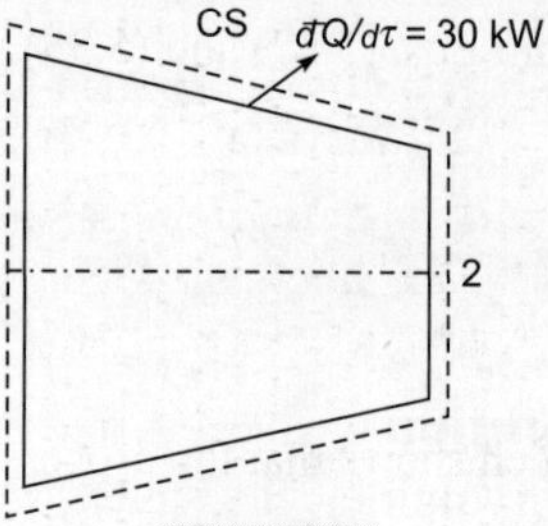

Fig. 5.14

$$w = \frac{A_1V_1}{v_1} = \frac{0.1 \times 70}{0.292} = \mathbf{23.97\ kg/s}$$

The steady flow energy equation for the control volume as shown in Fig. 5.14 gives

$$w\left(h_1 + \frac{\mathbf{V}_1^2}{2} + gz_1\right) + \frac{đQ}{d\tau} = w\left(h_2 + \frac{\mathbf{V}_2^2}{2} + gz_2\right) + \frac{đW}{d\tau}$$

Since the nozzle is horizontal, $Z_1 = Z_2$, and work transfer is zero $\left(\frac{đW}{d\tau} = 0\right)$. The energy equation reduces to

$$w\left(h_1 + \frac{V_1^2}{2}\right) + \frac{đQ}{d\tau} = w\left(h_2 + \frac{V_2^2}{2}\right)$$

Substituting the respective values, we get

$$23.97 \times \left(3263.9 \times 10^3 + \frac{70^2}{2}\right) - 30 \times 10^3 = 23.97 \times \left(3069.3 \times 10^3 + \frac{\mathbf{V}_2^2}{2}\right)$$

or $\mathbf{V}_2 = \mathbf{625.78\ m/s}$

For steady flow of steam, the mass flow rate can also be expressed as

$$w = \frac{A_2\mathbf{V}_2}{v_2}$$

Substituting the respective values, we get

$$23.97 = \frac{A_2 \times 625.78}{0.875}$$

or $A_2 = \mathbf{0.0335\ m^2}$

Example 5.2 *In a domestic refrigerator, throttle valve is placed in between the condenser and evaporator to reduce the pressure of the refrigerant. Refrigerant R-134a is throttled from the saturated liquid at 1000 kPa to a pressure of 140 kPa. Properties of refrigerant R-134a are as follows:*

Pressure (kPa)	Saturation temperature (°C)	Enthalpy for saturated liquid (kJ/kg)	Enthalpy for saturated vapour (kJ/kg)
1000	39.37	107.34	271.04
140	–18.77	27.06	239.19

At a given pressure, temperature at which liquid is converted into its vapour state is called saturation temperature and until the liquid is completely converted into vapour, there is a mixture of liquid and vapour. The enthalpy of liquid-vapour mixture at a given pressure is in between the enthalpy of saturated liquid and saturated vapour. Neglecting the changes in kinetic and potential energies, determine the temperature drop during the process. E [LO 5.3]

Solution There is one inlet and exit in the control volume. The steady flow energy equation for the control volume as shown in Fig. 5.15 gives

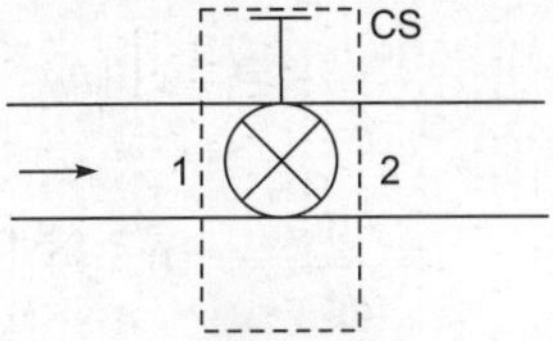

Fig. 5.15

$$w\left(h_1 + \frac{V_1^2}{2} + gz_1\right) + \frac{đQ}{d\tau} = w\left(h_2 + \frac{V_2^2}{2} + gz_2\right) + \frac{đW}{d\tau}$$

Both the heat and work transfer are zero because the throttling process is fast. The energy equation reduces to

$$h_1 = h_2$$

It is given that h_1 = 107.34 kJ/kg and from the above expression, we can write

$$h_2 = 107.34 \text{ kJ/kg}$$

Since the enthalpy of refrigerant after throttling is in between the values of the enthalpy of saturated liquid and saturated vapour of R-134a at 140 kPa, the state of the refrigerant after throttling will be a mixture of liquid and vapour and the temperature will be the saturation temperature at 140 kPa, i.e., –18.77°C.

Thus, the temperature drop during throttling is

$$\Delta T = -18.77 - 39.37 = \mathbf{-58.14°C}$$

Example 5.3 *Air flows steadily at the rate of 0.5 kg/s through an air compressor, entering at 7 m/s velocity, 100 kPa pressure and 0.95 m*^{3}*/kg volume, and leaving at 5 m/s, 700 kPa and 0.19 m*^{3}*/kg. The internal energy of the air leaving is 90 kJ/kg greater than that of the air entering. Cooling water in the compressor jackets absorbs heat from the air at the rate of 58 kW. (a) Compute the rate of shaft work input to the air in kW. (b) Find the ratio of the inlet pipe diameter to the outlet pipe diameter.* E [LO 5.3]

Solution Figure 5.16 shows the details of the problem.

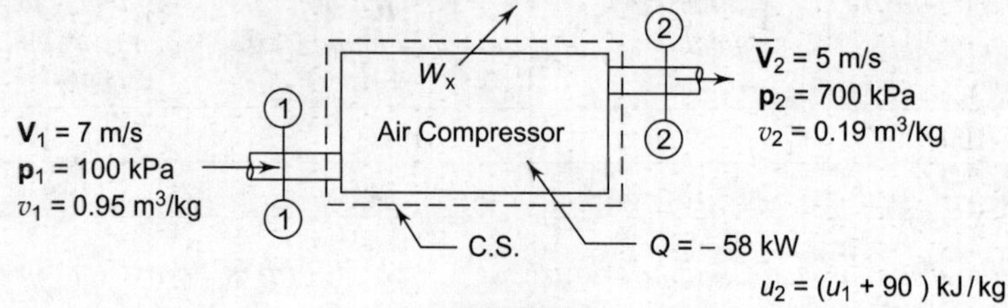

Fig. 5.16

(a) Writing the steady flow energy equation, we have

$$w\left(u_1 + p_1 v_1 + \frac{\mathbf{V}_1^2}{2} + Z_1 g\right) + \frac{đQ}{d\tau} = w\left(u_2 + p_2 v_2 + \frac{\mathbf{V}_2^2}{2} + Z_2 g\right) + \frac{đW_x}{d\tau}$$

$$\therefore \quad \frac{đW_x}{d\tau} = -w\left((u_2 - u_1) + (p_2 v_2 - p_1 v_1) + \frac{\mathbf{V}_2^2 - \mathbf{V}_1^2}{2} + (Z_2 - Z_1)g\right) + \frac{đQ}{d\tau}$$

$$\therefore \quad \frac{đW_x}{d\tau} = -\,0.5\frac{\text{kg}}{\text{s}}\left[90\frac{\text{kJ}}{\text{kg}} + (7 \times 0.19 - 1 \times 0.95)100\frac{\text{kJ}}{\text{kg}} + \frac{(5^2 - 7^2) \times 10^{-3}}{2}\frac{\text{kJ}}{\text{kg}} + 0\right] - 58 \text{ kW}$$

$$= -\,0.5\,[90 + 38 - 0.012] \text{ kJ/s} - 58 \text{ kW}$$

$$= -\,122 \text{ kW}$$

Rate of work input is **122 kW**.

(b) From mass balance, we have

$$w = \frac{A_1 \mathbf{V}_1}{v_1} = \frac{A_2 \mathbf{V}_2}{v_2}$$

$$\therefore \quad \frac{A_1}{A_2} = \frac{v_1}{v_2} \cdot \frac{\mathbf{V}_2}{\mathbf{V}_1} = \frac{0.95}{0.19} \times \frac{5}{7} = 3.57$$

$$\therefore \quad \frac{d_1}{d_2} = \sqrt{3.57} = \mathbf{1.89}$$

Example 5.4 *In a combined power and process plant steam is generated at 20 MPa and 500°C at a rate of 120 kg/s and is expanded through a turbine. While the steam is expanded in the turbine, 30 kg/s of steam at 5 MPa and 300°C is extracted for process work from the middle of the turbine. The remaining steam is further expanded and exits the turbine as liquid-vapour mixture with 90% quality (90% vapour in a mixture of liquid-vapour) at 75 kPa. There is a heat loss to the surroundings at the rate of 3 kW. Properties of steam are as follows:*

Pressure (kPa)	*Temperature (°C)*	*Enthalpy (kJ/kg)*
20000	500	3238.2
5000	300	2924.5

The specific enthalpy of water at 75 kPa and 90% quality is 2514.4 kJ/kg. Assuming there is no change in kinetic energy and potential energy, determine the total power output of the turbine. E [LO 5.3]

Solution There is one inlet and two exits for the control volume as shown in Fig. 5.17. From the mass balance, we can write

$$w_1 = w_2 + w_3$$

or $$w_3 = w_1 - w_2 = 120 - 30 = 90 \text{ kg/s}$$

The steady flow energy equation for the control volume as shown in Fig. 5.17 gives

$$w\left(h_1 + \frac{V_1^2}{2} + gz_1\right) + \frac{đQ}{d\tau} = w\left(h_2 + \frac{V_2^2}{2} + gz_2\right) + \left(h_2 + \frac{V_3^2}{2} + gz_2\right) + \frac{đW}{d\tau}$$

For negligible changes in kinetic and potential energy, the energy equation reduces to

$$w_1 h_1 + \frac{đQ}{d\tau} = w_2 h_2 + w_3 h_3 + \frac{đW}{d\tau}$$

Substituting the respective values, we get

$$120 \times 3238.2 \times 10^3 - 3 \times 10^3$$

$$= 30 \times 2924.5 \times 10^3 + 90 \times 2514.4 \times 10^3 + \frac{đW}{d\tau}$$

or $$\dot{W}_T = \frac{đW}{d\tau} = 74550000 \text{ W} = \mathbf{74.55\ MW}$$

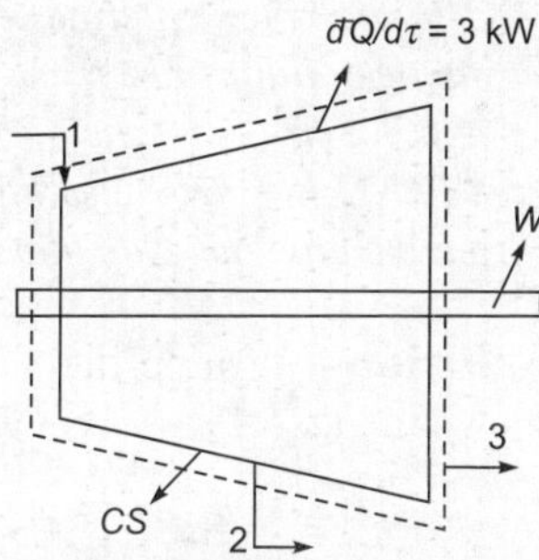

Fig. 5.17

Example 5.5 *In a steady flow apparatus, 135 kJ of work is done by each kg of fluid. The specific volume of the fluid, pressure and velocity at the inlet are 0.37 m³/kg, 600 kPa and 16 m/s respectively. The inlet is 32 m above the floor, and the discharge pipe is at floor level. The discharge conditions are 0.62 m³/kg, 100 kPa and 270 m/s. The total heat loss between the inlet and discharge is 9 kJ/kg of fluid. In flowing through this apparatus, does the specific internal energy increase or decrease, and by how much?* E [LO 5.3]

Solution Writing the steady flow energy equation for the control volume as shown in Fig. 5.18.

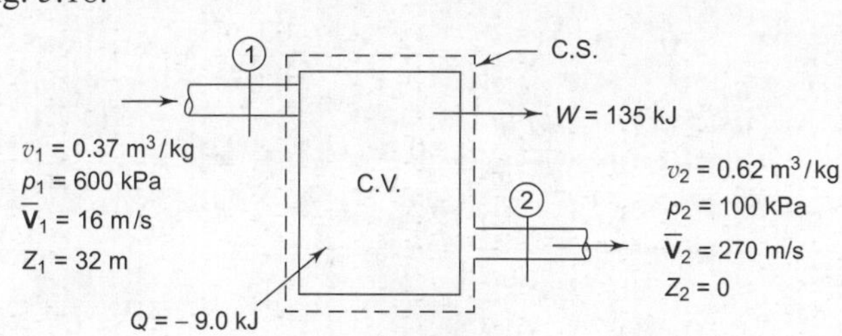

Fig. 5.18

$$u_1 + p_1 v_1 = \frac{\mathbf{V}_1^2}{2} + Z_1 g + \frac{\text{đ}Q}{dm} = u_2 + p_2 v_2 + \frac{\mathbf{V}_2^2}{2} + Z_2 g + \frac{\text{đ}W_x}{dm}$$

$$\therefore \quad u_1 - u_2 = (p_2 v_2 - p_1 v_1) + \left(\frac{\mathbf{V}_2^2 - \mathbf{V}_1^2}{2}\right) + (Z_2 - Z_1)g + \frac{\text{đ}W_x}{dm} - \frac{\text{đ}Q}{dm}\mathbf{V}W$$

$$= (1 \times 0.62 - 6 \times 0.37) \times 10^2 + \frac{\left(270^2 - 16^2\right) \times 10^{-3}}{2}$$

$$+ (-32 \times 9.81 \times 10^{-3}) + 135 - (-9.0)$$

$$= -160 + 36.45 - 0.314 + 135 + 9$$

$$= 20.136 \text{ kJ/kg}$$

Specific internal energy decreases by **20.136 kJ.**

Example 5.6 *In a steam power station, steam flows steadily through a 0.2 m diameter pipeline from the boiler to the turbine. At the boiler end, the steam conditions are found to be: p = 4 MPa, t = 400°C, h = 3213.6 kJ/kg and v = 0.073 m³/kg. At the turbine end, the conditions are found to be: p = 3.5 MPa, t = 392°C, h = 3202.6 kJ/kg and v = 0.084 m³/kg. There is a heat loss of 8.5 kJ/kg from the pipeline. Calculate the steam flow rate.* M **[LO 5.3]**

Solution Writing the steady flow energy equation for the control volume as shown in Fig. 5.19.

$$h_1 + \frac{\mathbf{V}_1^2}{2} + Z_1 g + \frac{\text{đ}Q}{dm} = h_2 + \frac{\mathbf{V}_2^2}{2} + Z_2 g + \frac{\text{đ}W_x}{dm}$$

Here, there is no change in datum, so change in potential energy will be zero.

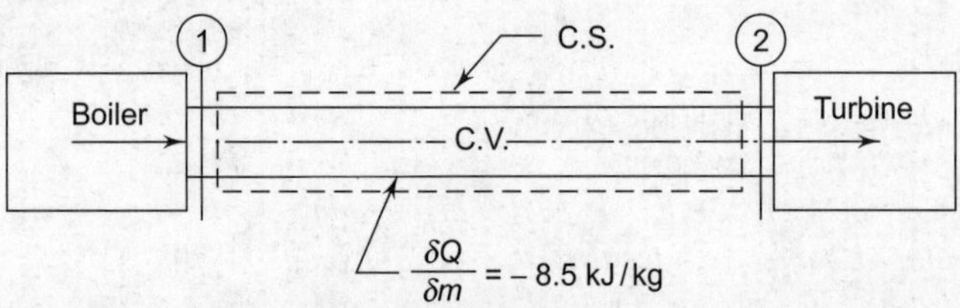

Fig. 5.19

Now
$$\frac{A_1 \mathbf{V}_1}{v_1} = \frac{A_2 \mathbf{V}_2}{v_2}$$

$$\therefore \quad \mathbf{V}_2 = \frac{A_1 \mathbf{V}_1}{v_1} \cdot \frac{v_2}{A_2} = \frac{v_2}{v_1} \cdot \mathbf{V}_1 = \frac{0.084}{0.073}\,\mathbf{V}_1 = 1.15\,\mathbf{V}_1$$

and
$$\frac{\text{đ}W_x}{dm} = 0$$

$$h_1 + \frac{\mathbf{V}_1^2}{2} + \frac{\text{đ}Q}{dm} = h_2 + \frac{\mathbf{V}_2^2}{2}$$

$$\therefore \quad \frac{\left(\mathbf{V}_2^2 - \mathbf{V}_1^2\right) \times 10^{-3}}{2} = h_1 - h_2 + \frac{\text{đ}Q}{dm}$$

$$= 3213.6 - 3202.6 + (-8.5) = \mathbf{2.5\ kJ/kg}$$

$$\mathbf{V}_1^2\,(1.15^2 - 1^2) = 5 \times 10^3$$

$$\mathbf{V}_1^2 = 15{,}650\ \text{m}^2/\text{s}^2$$

$$\therefore \quad \mathbf{V}_1 = 125.1\ \text{m/s}$$

$$\therefore \quad \text{Mass flow rate } w = \frac{A_1\mathbf{V}_1}{v_1} = \frac{\frac{\pi}{4} \times (0.2)^2\ \text{m}^2 \times 125.1\ \text{m/s}}{0.073\ \text{m}^3/\text{kg}} = \mathbf{53.8\ kg/s}$$

Example 5.7 *A certain water heater operates under steady flow conditions receiving 4.2 kg/s of water at 75°C temperature, enthalpy 313.93 kJ/kg. The water is heated by mixing with steam which is supplied to the heater at temperature 100.2°C and enthalpy 2676 kJ/kg. The mixture leaves the heater as liquid water at temperature 100°C and enthalpy 419 kJ/kg. How much steam must be supplied to the heater per hour?*

E [LO 5.3]

Solution By mass balance across the control surface (Fig. 5.20)

$$w_1 + w_2 = w_3$$

By energy balance

$$w_1\left(h_1 + \frac{\mathbf{V}_1^2}{2} + Z_1 g\right) + \frac{đQ}{d\tau} + w_2\left(h_2 + \frac{\mathbf{V}_2^2}{2} + Z_2 g\right)$$

$$= w_3\left(h_3 + \frac{\mathbf{V}_3^2}{2} + Z_3 g\right) + \frac{đW_x}{d\tau}$$

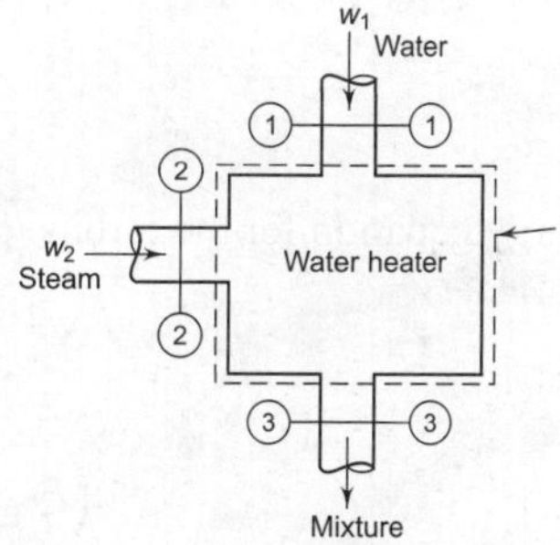

Fig. 5.20

By the nature of the process, there is no shaft work. Potential and kinetic energy terms are assumed to balance zero. The heater is assumed to be insulated. So the steady flow energy equation reduces to

$$w_1 h_1 + w_2 h_2 = w_3 h_3$$

$$4.2 \times 313.93 + w_2 \times 2676 = (4.2 + w_2)\ 419$$

$$\therefore \quad w_2 = 0.196\ \text{kg/s} = \mathbf{705\ kg/h}$$

Example 5.8 *Air at a temperature of 15°C passes through a heat exchanger at a velocity of 30 m/s where its temperature is raised to 800°C. It then enters a turbine with the same velocity of 30 m/s and expands until the temperature falls to 650°C. On leaving the turbine, the air is taken at a velocity of 60 m/s to a nozzle where it expands until the temperature has fallen to 500°C. If the air flow rate is 2 kg/s, calculate (a) the rate of heat transfer to the air in the heat exchanger, (b) the power output from the turbine assuming no heat loss and (c) the velocity at exit from the nozzle, assuming no heat loss. Take the enthalpy of air as $h = c_p t$, where c_p is the specific heat equal to 1.005 kJ/kg K and t is the temperature.*

M [LO 5.3, 5.4]

Solution As shown in Fig. 5.21, writing the S.F.E.E. for the heat exchanger and eliminating the terms that are not relevant,

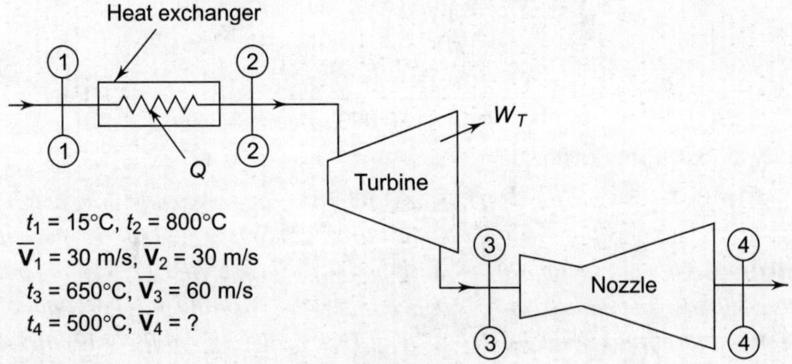

Fig. 5.21

$$w\left(h_1 + \frac{\mathbf{V}_1^2}{2} + Z_1 g\right) + Q_{1-2} = w\left(h_2 + \frac{\mathbf{V}_2^2}{2} + Z_2 g\right) + W_{1-2}$$

$$\therefore \quad wh_1 + Q_{1-2} = wh_2$$

$$\therefore \quad Q_{1-2} = w(h_2 - h_1) = w\, c_p(t_2 - t_1)$$

$$= 2 \times 1.005\ (800 - 15)$$

$$= 2.01 \times 785$$

$$= \mathbf{1580\ kJ/s}$$

Energy equation for the turbine gives

$$w\left(\frac{\mathbf{V}_2^2}{2} + h_2\right) = wh_3 + w\,\frac{\mathbf{V}_3^2}{2} + W_T$$

$$\frac{\mathbf{V}_2^2 - \mathbf{V}_3^2}{2} + (h_2 - h_3) = W_T/w$$

$$\frac{\left(30^2 - 60^2\right) \times 10^{-3}}{2} + 1.005\ (800 - 650) = W_T/w$$

$$\therefore \quad \frac{W_T}{w} = -1.35 + 150.75$$

$$= 149.4\ \text{kJ/kg}$$

$$\therefore \quad W_T = 149.4 \times 2\ \text{kJ/s}$$

$$= \mathbf{298.8\ kW}$$

Writing the energy equation for the nozzle

$$\frac{\mathbf{V}_3^3}{2} + h_3 = \frac{\mathbf{V}_4^2}{2} + h_4$$

$$\frac{\mathbf{V}_4^2 - \mathbf{V}_3^2}{2} = c_p\,(t_3 - t_4)$$

$$\mathbf{V}_4^2 - \mathbf{V}_3^2 = 1.005\,(650 - 500) \times 2 \times 10^3$$

$$= 301.50 \times 10^3 \text{ m}^2/\text{s}^2$$

$$\mathbf{V}_4^2 = 30.15 \times 10^4 + 0.36 \times 10^4$$

$$= 30.51 \times 10^4 \text{ m}^2/\text{s}^2$$

∴ velocity at exit from the nozzle

$$\mathbf{V}_4 = \mathbf{554 \text{ m/s}}$$

Example 5.9 *In a gas turbine the gas enters at the rate of 5 kg/s with a velocity of 50 m/s and enthalpy of 900 kJ/kg and leaves the turbine with a velocity of 150 m/s and enthalpy of 400 kJ/kg. The loss of heat from the gases to the surroundings is 25 kJ/kg. Assume for gas R = 0.285 kJ/kg K and c_p = 1.004 kJ/kgK and the inlet conditions to be at 100 kPa and 27°C. Determine the power output of the turbine and the diameter of the inlet pipe.* Ⓜ [LO 5.3]

Solution Steady flow energy equation for the C.V. (Fig. 5.22) gives

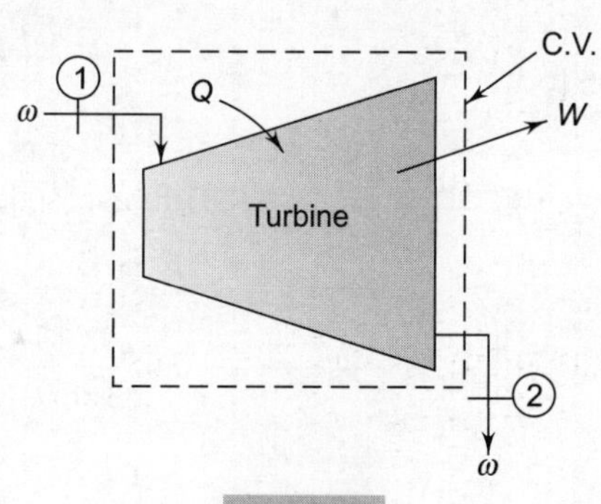

Fig. 5.22

$$w\left(h_1 + \frac{V_1^2}{2} + gZ_1\right) + \frac{đQ}{d\tau} = w\left(h_2 + \frac{V_2^2}{2} + gZ_2\right) + \frac{đW_x}{d\tau}$$

$$\therefore \quad \frac{đW_x}{d\tau} = w\left[(h_1 - h_2) + \frac{V_1^2 - V_2^2}{2}\right] + \frac{đQ}{d\tau}$$

$$= 5\left[\left(900 - 400 + \frac{50^2 - 150^2}{2} \times 10^{-3}\right)\right] - 25 \times 5$$

$$= \mathbf{2325 \text{ kW}}$$

Using ideal gas equation of state at pipe inlet,

$$p_1 \dot{V}_1 = w_1 R\, T_1$$

$$\dot{V} = \text{volume flow rate at inlet} = \frac{5 \times 0.285 \times 300 \times 10^3}{100 \times 10^3}$$

$$= 4.275 \text{ m}^3/\text{s}$$

Inlet area, $A_1 = \dfrac{4.275}{50}$

$$= 0.086 \text{ m}^2 = \frac{\pi}{4} D_1^2$$

$\therefore$ $D_1 = \mathbf{0.33\ m}$ or $\mathbf{33\ cm}$

Example 5.10 *The air speed of a turbojet engine in flight is 270 m/s. Ambient air temperature is – 15°C. Gas temperature of outlet of nozzle is 600°C. Corresponding enthalpy values for air and gas are, respectively 260 and 912 kJ/kg. Fuel-air ratio is 0.0190. Chemical energy of the fuel is 44.5 MJ/kg. Owing to incomplete combustion 5 percent of the chemical energy is not released in the reaction. Heat loss from the engine is 21 kJ/kg of air. Calculate the velocity of the exhaust jet.* Ⓜ **[LO 5.3]**

Solution Energy equation for the turbojet engine (Fig. 5.23) gives

$$w_a \left(h_a + \frac{\mathbf{V}_a^2}{2} \right) + w_f E_f + Q = w_g \left(h_g + \frac{V_a^2}{2} + E_g \right)$$

$$\left(260 + \frac{270^2 \times 10^{-3}}{2} \right) + 0.0190 \times 44500 - 21$$

$$= 1.0190 \left(912 + \frac{\mathbf{V}_g^2 \times 10^{-3}}{2} + 0.05 \frac{0.019}{1.019} \times 44500 \right)$$

$$260 + 36.5 + 845 - 21 = 1.019 \left(912 + \frac{\mathbf{V}_g^2 \times 10^{-3}}{2} + 42 \right)$$

$$\therefore \quad \frac{\mathbf{V}_g^2}{2} = 156 \times 10^3 \text{ m}^2/\text{s}^2$$

$$\mathbf{V}_g = \sqrt{3.12} \times 100 \text{ m/s}$$

Velocity of exhaust gas, $\mathbf{V}_g = \mathbf{560\ m/s}$

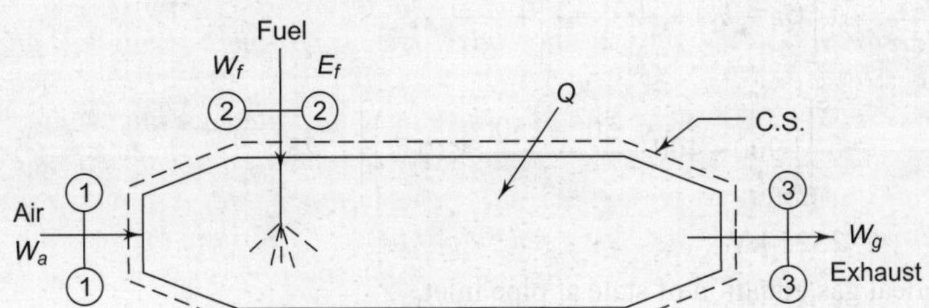

Fig. 5.23

Example 5.11 *In a reciprocating engine, the mass of gas occupying the clearance volume is m_c kg at state p_1, u_1, v_1 and h_1. By opening the inlet valve, m_f kg of gas is taken into the cylinder, and at the conclusion of the intake process the state of the gas is given by p_2, u_2, v_2 and h_2. The state of the gas in the supply pipe is constant and is given by p_p, u_p, v_p, h_p and V_p. How much heat is transferred between the gas and the cylinder walls during the intake process?* M [LO 5.6]

Solution Let us consider the control volume as shown in Fig. 5.24. Writing the energy balance on a time rate basis

$$\frac{dE_V}{d\tau} = \frac{đQ}{d\tau} - \frac{đW}{d\tau} + \left(h_p + \frac{\mathbf{V}_p^2}{2}\right)\frac{đm}{d\tau}$$

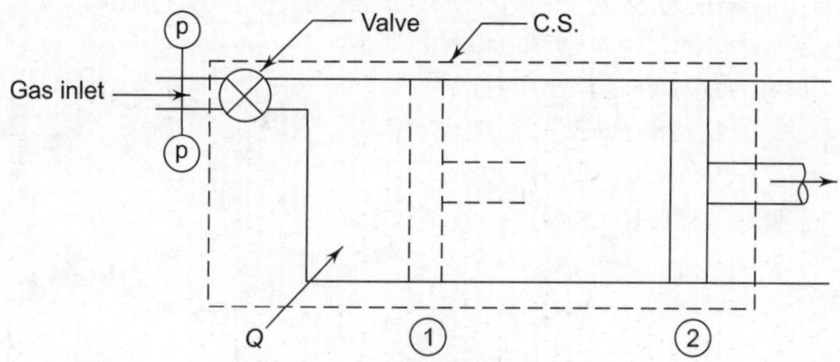

Fig. 5.24

With h_p and $\mathbf{V}_p$ being constant, the above equation can be integrated to give for the total process

$$\Delta E_V = Q - W + \left(h_p + \frac{\mathbf{V}_p^2}{2}\right) m_f$$

Now $$\Delta E_v = U_2 - U_1 = (m_e + m_f)\, u_2 - m_e u_1$$

$\therefore$ $$Q = (m_c + m_f)u_2 - m_c u_1 - m_f\left(h_p + \frac{\mathbf{V}_p^2}{2}\right) + W$$

Example 5.12 *The internal energy of air is given by*

$$u = u_0 + 0.718\, t$$

where u is in kJ/kg, u_0 is any arbitrary value of u at 0°C in kJ/kg and t is the temperature in °C. Also for air, pv = 0.287 (t + 273), where p is in kPa and v is in m³/kg.

A mass of air is stirred by a paddle wheel in an insulated constant volume tank. The velocities due to stirring make a negligible contribution to the internal energy of the air. Air flows out through a small valve in the tank at a rate controlled to keep the temperature in the tank constant. At a certain instant, the conditions are as follows: tank volume 0.12 m³, pressure 1 MPa, temperature 150°C and power to paddle wheel 0.1 kW. Find the rate of flow of air out of the tank at this instant. M [LO 5.6]

Solution Writing the energy balance for the control volume as shown in Fig. 5.25

$$\frac{dE_V}{d\tau} = \frac{đW}{d\tau} - (h_p)\frac{dm}{d\tau}$$

Since there is no change in internal energy of air in the tank,

$$h_p \cdot \frac{dm}{d\tau} = \frac{đW}{d\tau}$$

where $h_p = u + pv$.

Let $u = 0$ at $t = 0\ K = -273°C$

$$u = u_0 + 0.718\ t$$

$$0 = u_0 + 0.718\ (-273)$$

$$u_0 = 0.718 \times 273 \text{ kJ/kg}$$

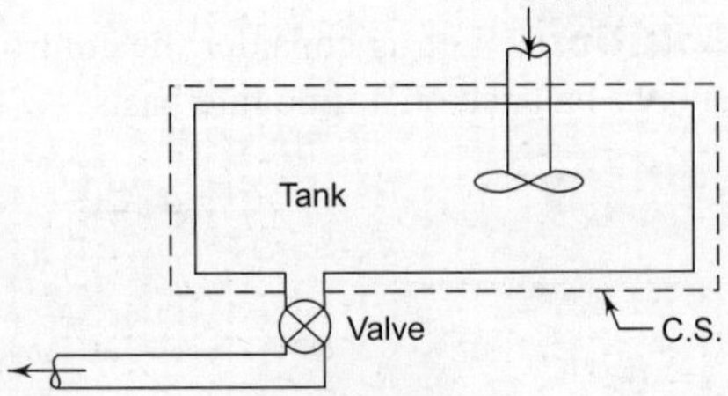

Fig. 5.25

At t°C

$$u = 0.718 \times 273 + 0.718\ t$$

$$= 0.718\ (t + 273) \text{ kJ/kg}$$

$$h_p = 0.718\ (t + 273) + 0.287\ (t + 273)$$

or $$h_p = 1.005\ (t + 273)$$

At 150°C

$$h_p = 1.005 \times 423$$

$$= 425 \text{ kJ/kg}$$

$$\therefore \quad \frac{dm}{d\tau} = \frac{1}{h_p}\frac{đW}{d\tau}$$

$$= \frac{0.1 \text{ kJ/s}}{425 \text{ kJ/kg}} = 0.236 \times 10^{-3} \text{ kg/s}$$

$$= \mathbf{0.845 \text{ kg/h}}$$

This is the rate at which air flows out of the tank.

Example 5.13 *A well-insulated vessel of volume V contains a gas at pressure p_0 and temperature t_0. The gas from a main at a uniform temperature t_1 is pumped into the vessel and the inflow rate decreases exponentially with time according to $\dot{m} = \dot{m}_0 e^{-a\tau}$, where a is a constant. Determine the pressure and temperature of the gas in the vessel as a function of time. Neglect the K.E. of the gas entering the vessel and assume that the gas follows the relation,*

$$pv = RT, \text{ where } T = t + 273$$

and its specific heats are constant.

(a) If the vessel was initially evacuated, show that the temperature inside the vessel is independent of time.

(b) Determine the charging time when the pressure inside the vessel reaches that of the main. D [LO 5.7]

Solution Since the vessel is well insulated, $\dot{Q} = 0$ and there is no external work transfer, $\dot{W} = 0$. Therefore,

$$\frac{dE_V}{d\tau} = h_1 \frac{dm}{d\tau} m = h_1 \dot{m}_0 e^{-a\tau}$$

where h_1 is the enthalpy of the gas in the main

On integration,

$$E = E_0 + \frac{h_1 \dot{m}_0}{a} (1 - e^{-a\tau})$$

where E_0 is the initial energy of the vessel at the beginning of the charging process, i.e. $E = E_0$ at $\tau = 0$. Neglecting K.E. and P.E. changes, by energy balance

$$Mu = M_0 u_0 + \frac{\dot{m}_0}{a} (1 - e^{-a\tau}) (u_1 + p_1 v_1) \quad \text{(i)}$$

Again,

$$\frac{dm}{d\tau} = \dot{m}_0 e^{-a\tau}$$

On integration,

$$M = M_0 + \frac{\dot{m}_0 \left(1 - e^{-a\tau}\right)}{a} \quad \text{(ii)}$$

where M_0 is the initial mass of the gas. Eliminating M from Eqs. (i) and (ii),

$$\left\{ M_0 + \frac{\dot{m}_0}{a}\left(1 - e^{-a\tau}\right) \right\} u - M_0 u_0$$

$$= \frac{\dot{m}_0}{a} (1 - e^{-a\tau}) (u_1 + RT_1)$$

$$M_0 c_v (T - T_0) = \frac{\dot{m}_0}{a} (1 - a^{-a\tau} \{c_v (T_1 - T) + RT_1\}$$

$$\therefore \quad T = \frac{M_0 c_v T_0 + \frac{\dot{m}_0}{a}\left(1 - e^{-a\tau}\right) c_p T_1}{\left\{ M_0 + \frac{\dot{m}_0}{a}\left(1 - e^{-a\tau}\right) \right\} c_v}$$

$$p = \frac{MRT}{V} = \frac{R}{V c_v} \left\{ M_0 c_v T_0 + \frac{\dot{m}_0}{a}\left(1 - e^{-a\tau}\right) c_p T_1 \right\}$$

$$= p_0 + \frac{\dot{m}_0 R}{aV} (1 - e^{-a\tau}) \gamma T_1$$

The above two equations show the temperature and pressure of the gas in the vessel as functions of time.

(a) If $M_0 = 0$, $T = \gamma T_1$, i.e. the temperature inside the vessel becomes independent of time and is equal to γT_1 throughout the charging process.

(b) The charging process will stop when pressure inside the vessel reaches that of the main. The charging time can be found by setting $p = p_1$ in the pressure relation

$$p_1 - p_0 = \frac{\dot{m}_0 R\gamma T_1}{aV} - \frac{\dot{m}_0 R}{aV} e^{-a\tau} \gamma T_1$$

By rearrangement,

$$e^{a\tau} = \frac{\dot{m}_0 R\gamma T_1/(aV)}{\left(\frac{\dot{m}_0 R\gamma T_1}{aV}\right) - (p_1 - p_0)}$$

$$\therefore \qquad \tau = \frac{1}{a} \ln\left[1 - (p_1 - p_0)\frac{aV}{\dot{m}_0 R\gamma T_1}\right]$$

Example 5.14 *An insulated piston-cylinder assembly has an initial volume V_1 and contains air at p_1 and T_1. Air is supplied to the cylinder from a pipeline, maintained to the constant p_p and T_p, through a valve fitted into the cylinder. The piston is restrained in such a manner that the pressure of the air in the cylinder remains constant at p_1 during the process of filling. The filling process is terminated when the final volume V_2 is twice the initial volume V_1. Show that the final temperature of the air T_2 in the cylinder is given by*

$$T_2 = \frac{2}{\frac{1}{T_1} + \frac{1}{T_p}}$$

If p_1 = 400 kPa, T_1 = 200°C, T_p = 150°C and V_1 = 1.2 m^3, estimate the mass of air that has entered the cylinder through the valve. Assume that air behaves as an ideal gas and use $pV = mRT$, $u = c_v T$ and $h = c_p t$, wher c_v = 0.718 kJ/kgK and R = 0.287 kJ/kgK. **D [LO 5.7]**

Solution

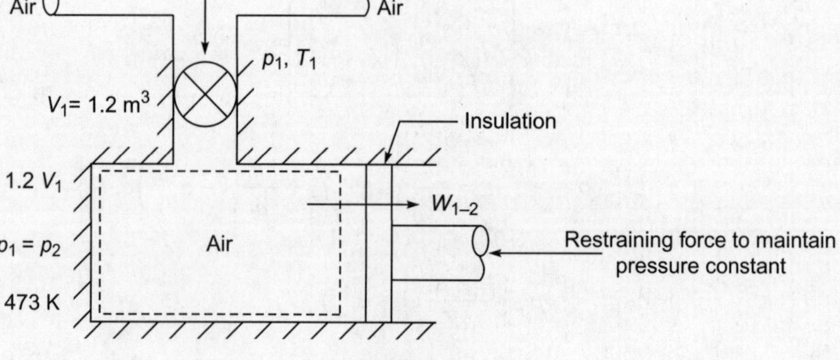

Fig. 5.26

Amount of air that has entered the cylinder = $m_2 - m_1$

Energy equation can be written as

$$h_p (m_2 - m_1) = m_2 u_2 - m_1 u_1 + W_{1-2}$$

Using $u = c_v T$, $h = c_p T$ and $m = \dfrac{pV}{RT}$, and since $W = p_1(V_2 - V_1)$

$$c_p T_p \left(\frac{p_2 V_2}{R T_2} - \frac{p_1 V_1}{R T_1} \right) = \frac{p_2 V_2}{R T_2}\, c_v T_2 - \frac{p_1 V_1}{R T_1}\, c_v T_1 + p(V_2 - V_1)$$

Since $p_1 = p_2 = p$,

$$V_2 - V_1 - \frac{c_p T_p}{R} \left(\frac{V_2}{T_2} - \frac{V_1}{T_1} \right) + \frac{c_v}{R} (V_2 - V_1) = 0$$

$$(V_2 - V_1) \left(1 + \frac{c_v}{R} \right) - \frac{c_p T_p}{R} \left(\frac{V_2}{T_2} - \frac{V_1}{T_1} \right) = 0$$

$$(V_2 - V_1) \frac{c_p}{R} - \frac{c_p T_p}{R} \left(\frac{V_2}{T_2} - \frac{V_1}{T_1} \right) = 0$$

Since $\dfrac{c_p}{R}$ cannot be zero,

$$(V_2 - V_1) - T_p \left(\frac{V_2}{T_2} - \frac{V_1}{T_1} \right) = 0$$

$$T_p \frac{V_2}{T_2} = V_2 - V_1 + T_p \frac{V_1}{T_1}$$

$$\therefore \quad T_2 = \frac{T_p V_2}{V_2 - V_1 + \dfrac{T_p V_1}{T_1}} = \frac{V_2}{\dfrac{V_1}{T_1} + \dfrac{V_2 - V_1}{T_p}}$$

Since $V_2 = 2 V_1$,

$$T_2 = \frac{2V_1}{\dfrac{V_1}{T_1} + \dfrac{V_1}{T_p}} = \frac{2}{\dfrac{1}{T_1} + \dfrac{1}{T_p}} \quad \textit{Proved.}$$

Given: $T_1 = 473$ K and $T_p = 150 + 273 = 423$ K

$$T_2 = \frac{2}{\dfrac{1}{473} + \dfrac{1}{423}} = 446.6 \text{ K}$$

The mass of air that has entered

$$m_p = m_2 - m_1 = \frac{p_2 V_2}{R T_2} - \frac{p_1 V_1}{R T_1}$$

$$= \frac{p}{R} \left(\frac{2V_1}{T_2} - \frac{V_1}{T_1} \right)$$

$$= \frac{pV_1}{R} \left(\frac{2}{T_2} - \frac{1}{T_1} \right) = \frac{400 \times 1.2}{0.287} \left(\frac{2}{446.6} - \frac{1}{473} \right)$$

$$= \mathbf{3.95\ kg}$$

Example 5.15 *Coal with a calorific value of 33,000 kJ/kg is fed to a boiler at a rate of 500 kg/h. It has 10% ash. Air feed rate is 2.72 kg/s with an inlet enthalpy of 400 kJ/kg. Flue gas exits from the boiler with an enthalpy of 550kJ/kg. If the feedwater to the boiler enters with an enthalpy of 170 kJ/kg and the generated steam in superheated state has an enthalpy of 2900 kJ/kg. Calculate the steam generation rate. Assume no mass or heat loss from the boiler.* Ⓔ [LO 5.4]

Solution From mass balance of the air, fuel, flue gas and ash,

$$m_{fg} = m_a + m_f - \text{m}_{\text{ash}} = 2.72 + (500/3600) - 0.1 \times (500/3600) = 2.845 \text{ kg/s}$$

From steady flow energy equation of the boiler,

$$m_s = (m_a h_2 + m_f CV_f - m_{fg} h_4)/(h_5 - h_3) = \{(2.72 \times 400) + (500/3600) \times 33{,}000 - 2.845 \times 550\}/(2900\text{–}170)$$

$$= 1.5 \text{ kg/s} = 5400 \text{ kg/h}$$

Example 5.16 *A boiler needs to generate steam at a rate of 6000 kg/h. Available feedwater at the local ambient condition has an enthalpy of 200 kJ/kg. Delivered steam has an enthalpy of 3000 kJ/kg. What should be the coal feed rate with a calorific value of 35,000 kJ/kg, if 10% of the heat in the boiler is lost through the flue gas, surfaces and others?* Ⓜ [LO 5.4]

Solution Rate of heat required to convert the feed water to steam $Q_s = (6000/3600) \times (3000\text{-}200) = 4666.67$ kW

Useful heat from the fuel in the boiler $Q_u = 0.9 \times m_f\ CV_f = 0.9 \times 35{,}000 \times m_f = 31{,}500 \times m_f$

From steady flow energy balance, $Q_u = Q_s$ or $mf = 4666.67/31{,}500 = 0.1481$ kg/s = 533.16 kg/h

SUMMARY

The first law may be applied to the flow processes. A control surface is the boundary of a control volume which is a fixed region in space upon which attention is concentrated in the analysis of a problem. A steady flow process is a process in which all conditions within the control volume remain constant with time. For a simple steady flow process involving one mass stream entering and one mass stream leaving the control volume, the following equations hold:

Mass balance or continuity equation

$$w = \frac{A_1V_1}{v_1} = \frac{A_2V_2}{v_2}$$

Energy balance or *steady flow energy equation* (S.F.E.E.) written on a unit mass basis

$$u_1 + p_1v_1 + \frac{V_1^2}{2} + Z_1g + \frac{đQ}{dm} = u_2 + p_2v_2 + \frac{V_2^2}{2} + gZ_2 + \frac{đW_e}{d\tau}$$

or on time basis

$$w_1\left(h_1+\frac{V_1^2}{2}+gZ_1\right)+\frac{đQ}{d\tau}=w_2\left(h_2+\frac{V_2^2}{2}+gZ_2\right)+\frac{đW_x}{d\tau}$$

For variable flow problems energy equations may be set up by means of a system technique or control volume technique. In these cases, account must be taken not only of mass and energy quantities crossing the control surface but also of accumulations within the control volume.

REVIEW QUESTIONS

(M) **5.1** Explain the system approach and the control volume approach in the analysis of a flow process. **[LO 5.1]**

(E) **5.2** What is a steady flow process? **[LO 5.2]**

(E) **5.3** Write the steady flow energy equation for a single stream entering and a single steam leaving a control volume and explain the various terms in it. **[LO 5.3]**

(E) **5.4** Give the differential form of the S.F.E.E. **[LO 5.3]**

(M) **5.5** Under what conditions does the S.F.E.E. reduce to Euler's equation? **[LO 5.5]**

(M) **5.6** How does Bernoulli's equation compare with S.F.E.E.? **[LO 5.5]**

(E) **5.7** What will be the velocity of a fluid leaving a nozzle, if the velocity of approach is very small? **[LO 5.4]**

(M) **5.8** Show that the enthalpy of a fluid before throttling is equal to that after throttling. **[LO 5.4]**

(D) **5.9** Write the general energy equation for a variable flow process. **[LO 5.6]**

(D) **5.10** What is the system technique in a bottle-filling process? **[LO 5.7]**

(D) **5.11** Explain the control volume technique in a variable flow process. **[LO 5.7, 5.8]**

PROBLEMS WITH HINTS

(E) **5.1** Air enters a compressor operating at steady state at a pressure of 1 bar, a temperature of 290 K and a velocity of 6 m/s through an inlet with an area of 0.1 m^2. At exit, the pressure is 7 bar, the temperature is 450 K and the velocity is 2 m/s. Heat transfer from the compressor to the surroundings occurs at the rate of 180 kJ/min. Employing the ideal gas model, calculate the power input to the compressor. Take c_p = 1.005 kJ/kgK.

Hint: **[LO 5.3]**

P_1 = 1 bar $\qquad$ P_2 = 7 bar

T_1 = 290 K $\qquad$ T_2 = 450 K

V_1 = 6 m/s $\qquad$ V_3 = 2 m/s

A_1 = 0.1 m^2 $\qquad$ $\dot{Q}$ = –180 kJ/min

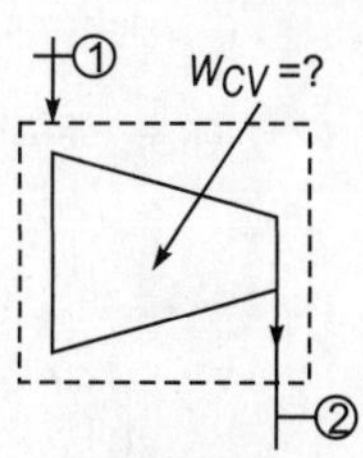

Fig. 5.27

First find the mass flow rate, $w=\frac{A_1V_1}{v_1}=\frac{AV_1p_1}{RT_1}$

where for air,

$$R=\frac{\bar{R}}{\mu}=\frac{8.3143}{28.96}=0.287\ \frac{\text{kJ}}{\text{kgK}}.$$

Then use the steady flow energy equation

$$\dot{W} = \dot{Q} + w\left[(h_1 + h_2) + \frac{V_1 - V_2^2}{2} + g(Z_1 - Z_2)\right]$$

$$= -\frac{180}{60} + w\,w\left[1.005(290 - 450) + \frac{6^2 - 2^2}{2} \times 10^{-3} + 0\right]$$

$$= \mathbf{-119.4\ kW}$$

(E) **5.2** Air enters an insulated diffuser operating at steady state with a pressure of 0.7 bar, a temperature of 5.7°C and a velocity of 200 m/s. At the exit the pressure is 1 bar. The exit flow area is 20% greater than the inlet flow area. Potential energy effects can be neglected. Determine the air exit temperature and the velocity. Take $c_p = 1.005$ kJ/kgK and $R = 0.287$ kJ/kgK. **[LO 5.3, 5.4]**

Hint: In the S.F.E.E.

$$\dot{Q} = \dot{W} + w\left[(h_1 + h_2) + \frac{V_2^2 - V_1^2}{2} \times 10^{-3} + g(Z_2 - Z_1)\right] = 0$$

$$\dot{Q} = 0,\ \dot{W} = 0,\ Z_1 = Z_2$$

$$\therefore \quad \left[c_p(T_2 - T_1) + \frac{V_2^2 - V_1^2}{2} \times 10^{-3}\right] = 0$$

$$w = \frac{A_1V_1}{v_1} = \frac{A_2V_2}{v_2} = \frac{1.2\,A_1V_2}{v_2}$$

$$V_2 = \frac{V_1}{v_1} \cdot \frac{v_2}{1.2} \quad \text{and} \quad V_1 = \frac{RT_1}{P_1} = \frac{0.287 \times (273 + 57)}{0.7 \times 100}\,\frac{\text{m}^3}{\text{kg}}$$

$$\text{and} \quad v_2 = \frac{0.287 \times T_2}{100}\,\frac{\text{m}^3}{\text{kg}}$$

From the above equations, get T_2 = **342 K** and V_2 = **120.9 m/s**

(E) **5.3** A pump steadily delivers water at a volumetric flow rate of 0.05 m³/s through a pipe of diameter 18 cm located 100 m above the inlet pipe which has a diameter of 15 cm. The pressure is nearly equal to 1 bar at both the inlet and the exit, and the temperature is nearly constant at 20°C throughout. Determine the power required by the pump. Take $g = 9.81$ m/s². **[LO 5.3, 5.4]**

Hint: The power is determined from the S.F.E.E.

$$0 = \dot{Q} - \dot{W} + w\left[(h_1 - h_2) + \frac{V_1^2 - V_2^2}{2} + g(Z_1 - Z_2)\right]$$

where $w_1 = w_2 = w$, $T_2 = T_1$ and $p_2 = p_1$. Since $h_1 = h_2$,

$$\therefore \quad \dot{W} = w\left[\frac{V_1^2 - V_2^2}{2} + g(Z_1 - Z_2)\right]$$

$$\text{where} \quad w = \frac{A_2V_2}{v} = \frac{0.05\ \text{m}^3/\text{s}}{0.001\ \text{m}^3/\text{kg}} = 50\ \text{kg/s}$$

$$V_2 = \frac{0.05\ \text{m}^3/\text{s}}{\frac{\pi}{4}(0.18)^2\ \text{m}^2} = 1.965\ \text{m/s}$$

$A_1V_1 = 0.05$ m³/s. Find V_1 and using the S.F.E.E. obtain.

$$\therefore \quad \dot{W} = \mathbf{-48.9\ kW}$$

(E) **5.4** Air is heated as it flows at steady state through a constant area duct. At the inlet, the temperature is 300 K, the pressure is 2 bar and the velocity is 87 m/s. At the exit, the pressure and velocity are 1.45 bar and 360 m/s, respectively. Neglecting P.E., determine the heat transfer per unit mass. **[LO 5.3, 5.4]**

Hint: The S.F.E.E gives

$$\frac{\dot{Q}}{w} = h_2 - h_1 + V_2^2\, V_1^2 + g(Z_2 - Z_1)$$

where $\quad Z_1 - Z_2,\ w = \rho_1 A_1 V_1 = \rho_2 A_2 V_2$

Solving for $\quad T_2,\ T_2 = \frac{P_2}{P_1}\frac{A_2}{A_2^1}\frac{V_2}{V_1}\cdot T_1 = 900 \text{ K}$

$$\frac{\dot{Q}}{w} = 1.005\,(900 - 300) + \frac{360^2 - 87^2}{2} \times 10^{-3}$$

$$= \mathbf{690\ kJ/kg}$$

(E) **5.5** Air as an ideal gas flows at steady state through the compressor and heat exchanger as shown in Fig. 5.25. Data for the flow streams are shown in the figure. Neglecting the K.E. and P.E. effects and heat loss to the surroundings, calculate the compressor power and the mass flow rate of cooling water. Take c_{pa} = 1.005 kJ/kgK. **[LO 5.3, 5.4]**

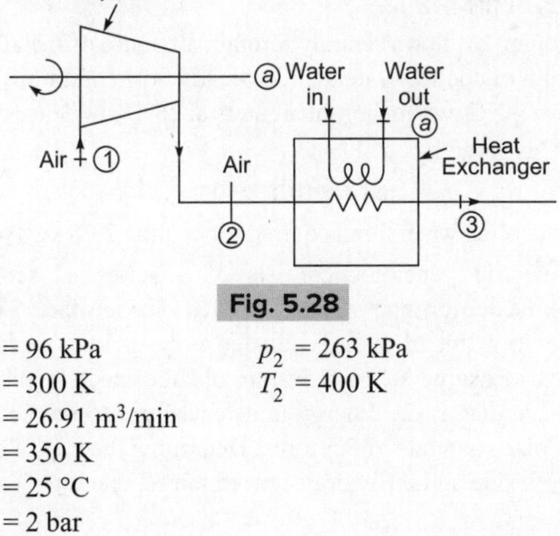

Fig. 5.28

P_1 = 96 kPa $\qquad p_2$ = 263 kPa
T_1 = 300 K $\qquad T_2$ = 400 K
$(AV)_1$ = 26.91 m³/min
T_3 = 350 K
T_a = 25 °C
p_a = 2 bar
T_b = 40 °C
p_b = 1.5 bar

Hint: The S.F.E.E. gives

$$O = \dot{Q} - \dot{W} + w\left[(h_1 - h_2) + \frac{v_1^2 - v_2^2}{2} + 9(Z_1 - Z_2)\right]\cdot$$

where $w_1 = w_2 = w_{air}$ since $V_1 = V_2$ and $Z_2 - Z_1$

$$\dot{W} = w(h_1 - h_2).$$

$$w_{air} = \frac{(AV)_1}{v_1} = \frac{p_1 (AV)_1}{RT_1} \text{ gives } \mathbf{-50.4\ kW}$$

$\therefore \qquad \dot{W} = 30 \times c_{pa}\,(300 - 400)$ gives **–50.4 kW**

By energy balance in the heat exchanger,

$$w_{air}\, c_{pa}\,(T_3 - T_2) = w_w\, c_{pw}\,(T_b - T_a), \text{ where } c_{pw} = 4.187 \text{ kJ/kgK}$$

which gives w_w = **0.403 kg/s**

PROBLEMS

(E) **5.1** A blower handles 1 kg/s of air at 20°C and consumes a power of 15 kW. The inlet and outlet velocities of air are 100 m/s and 150 m/s, respectively. Find the exit air temperature, assuming adiabatic conditions. Take c_p of air is 1.005 kJ/kg-K. **[LO 5.4]**

(E) **5.2** A turbine operates under steady flow conditions, receiving steam at the following state: pressure 1.2 MPa, temperature 188°C, enthalpy 2785 kJ/kg, velocity 33.3 m/s and elevation 3 m. The steam leaves the turbine at the following state: pressure 20 kPa, enthalpy 2512 kJ/kg, velocity 100 m/s and elevation 0 m. Heat is lost to the surroundings at the rate of 0.29 kJ/s. If the rate of steam flow through the turbine is 0.42 kg/s, what is the power output of the turbine in kW? **[LO 5.4]**

(E) **5.3** A nozzle is a device for increasing the velocity of a steadily flowing stream. At the inlet to a certain nozzle, the enthalpy of the fluid passing is 3000 kJ/kg and the velocity is 60 m/s. At the discharge end, the enthalpy is 2762 kJ/kg. The nozzle is horizontal and there is negligible heat loss from it. (a) Find the velocity at exit from the nozzle. (b) If the inlet area is 0.1 m^2 and the specific volume at inlet is 0.187 m^3/kg, find the mass flow rate. (c) If the specific volume at the nozzle exit is 0.498 m^3/kg, find the exit area of the nozzle. **[LO 5.4]**

(E) **5.4** In an oil cooler, oil flows steadily through a bundle of metal tubes submerged in a steady stream of cooling water. Under steady flow conditions, the oil enters at 90°C and leaves at 30°C, while the water enters at 25°C and leaves at 70°C. The enthalpy of oil at t°C is given by

$$h = 1.68\, t + 10.5 \times 10^{-4}\, t^2 \text{ kJ/kg}$$

What is the cooling water flow required for cooling 2.78 kg/s of oil? **[LO 5.3]**

(D) **5.5** A thermoelectric generator consists of a series of semiconductor elements (Fig. 5.29), heated on one side and cooled on the other. Electric current flow is produced as a result of energy transfer as heat. In a particular experiment, the current was measured to be 0.5 amp and the electrostatic potential at (1) was 0.8 volt above that at (2). Energy transfer as heat to the hot side of the generator was taking place at a rate of 5.5 watts. Determine the rate of energy transfer as heat from the cold side and the energy conversion efficiency.

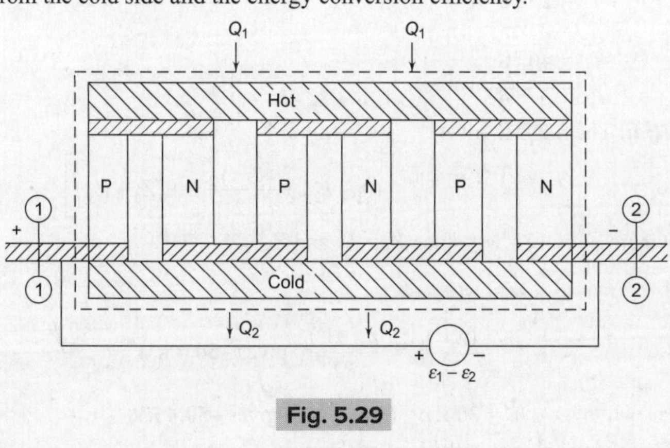

Fig. 5.29

[LO 5.3]

(E) **5.6** A turbo compressor delivers 2.33 m^3/s of air at 0.276 MPa and 43°C which is heated at this pressure to 430°C and finally expanded in a turbine which delivers 1860 kW.

During the expansion, there is a heat transfer of 0.09 MJ/s to the surroundings. Calculate the turbine exhaust temperature if changes in kinetic and potential energy are negligible. **[LO 5.4]**

(E) **5.7** A reciprocating air compressor takes in 2 m^3/min at 0.11 MPa, 20°C which it delivers at 1.5 MPa and 111°C to an aftercooler where the air is cooled at constant pressure to 25°C. The power absorbed by the compressor is 4.15 kW. Determine the heat transfer in (a) the compressor, and (b) the cooler. State your assumptions. **[LO 5.4]**

(E) **5.8** In a water cooling tower air enters at a height of 1 m above the ground level and leaves at a height of 7 m. The inlet and outlet velocities are 20 m/s and 30 m/s respectively. Water enters at a height of 8 m and leaves at a height of 0.8 m. The velocity of water at entry and exit are 3 m/s and 1 m/s, respectively. Water temperatures are 80°C and 50°C at the entry and exit, respectively. Air temperatures are 30°C and 70°C at the entry and exit, respectively. The cooling tower is well insulated and a fan of 2.25 kW drives the air through the cooler. Find the amount of air per second required for 1 kg/s of water flow. The values of c_p of air and water are 1.005 and 4.187 kJ/kg K, respectively. **[LO 5.3]**

(M) **5.9** Air at 101.325 kPa and 20°C is taken into a gas turbine power plant at a velocity of 140 m/s through an opening of 0.15 m^2 cross-sectional area. The air is compressed, heated, expanded through a turbine and exhausted at 0.18 MPa and 150°C through an opening of 0.10 m^2 cross-sectional area. The power output is 375 kW. Calculate the net amount of heat added to the air in kJ/kg. Assume that air obeys the law $pv = 0.287\ (t + 273)$, where p is the pressure in kPa, v is the specific volume in m^3/kg and t is the temperature in °C. Take c_p = 1.005 kJ kg K. **[LO 5.4]**

(M) **5.10** A gas flows steadily through a rotary compressor. The gas enters the compressor at a temperature of 16°C, a pressure of 100 kPa and an enthalpy of 391.2 kJ/kg. The gas leaves the compressor at a temperature of 245°C, a pressure of 0.6 MPa and an enthalpy of 534.5 kJ/kg. There is no heat transfer to or from the gas as it flows through the compressor. (a) Evaluate the external work done per unit mass of gas assuming the gas velocities at entry and exit to be negligible. (b) Evaluate the external work done per unit mass of gas when the gas velocity at entry is 80 m/s and that at exit is 160 m/s. **[LO 5.4]**

(M) **5.11** The steam supply to an engine comprises two streams which mix before entering the engine. One stream is supplied at the rate of 0.01 kg/s with an enthalpy of 2952 kJ/ kg and a velocity of 20 m/s. The other stream is supplied at the rate of 0.1 kg/s with an enthalpy of 2569 kJ/kg and a velocity of 120 m/s. At the exit from the engine the fluid leaves as two streams, one of water at the rate of 0.001 kg/s with an enthalpy of 420 kJ/kg and the other of steam; the fluid velocities at the exit are negligible. The engine develops a shaft power of 25 kW. The heat transfer is negligible. Evaluate the enthalpy of the second exit stream. **[LO 5.4]**

(M) **5.12** The stream of air and gasoline vapour, in the ratio of 14 : 1 by mass, enters a gasoline engine at a temperature of 30°C and leaves as combustion products at a temperature of 790°C. The engine has a specific fuel consumption of 0.3 kg/kWh. The net heat transfer rate from the fuel-air stream to the jacket cooling water and to the surroundings is 35 kW. The shaft power delivered by the engine is 26 kW. Compute the increase in the specific enthalpy of the fuel-air stream, assuming the changes in kinetic energy and in elevation to be negligible. **[LO 5.4]**

(E) **5.13** An air turbine forms part of an aircraft refrigerating plant. Air at a pressure of 295 kPa and a temperature of 58°C flows steadily into the turbine with a velocity of 45 m/s. The air leaves the turbine at a pressure of 115 kPa, a temperature of 2°C, and a velocity of 150 m/s. The shaft work delivered by the turbine is 54 kJ/kg of air. Neglecting changes in elevation, determine the magnitude and sign of the heat transfer per unit mass of air flowing. For air, take $c_p = 1.005$ kJ/ kg K and the enthalpy $h = c_p\, t$. **[LO 5.4]**

(E) **5.14** In a turbomachine handling an incompressible fluid with a density of 1000 kg/m³, the conditions of the fluid at the rotor entry and exit are as given below:

	Inlet	**Exit**
Pressure	1.15 MPa	0.05 MPa
Velocity	30 m/s	15.5 m/s
Height above datum	10 m	2 m

If the volume flow rate of the fluid is 40 m³/s, estimate the net energy transfer from the fluid as work. **[LO 5.4]**

(E) **5.15** A room for four persons has two fans, each consuming 0.18 kW power, and three 100 W lamps. Ventilation air at the rate of 80 kg/h enters with an enthalpy of 84 kJ/kg and leaves with an enthalpy of 59 kJ/kg. If each person puts out heat at the rate of 630 kJ/h determine the rate at which heat is to be removed by a room cooler, so that a steady state is maintained in the room. **[LO 5.3]**

(E) **5.16** Air flows steadily at the rate of 0.4 kg/s through an air compressor, entering at 6 m/s with a pressure of 1 bar and a specific volume of 0.85 m³/kg, and leaving at 4.5 m/s with a pressure of 6.9 bar and a specific volume of 0.16 m³/kg. The internal energy of the air leaving is 88 kJ/kg greater than that of the air entering. Cooling water in a jacket surrounding the cylinder absorbs heat from the air at the rate of 59 W. Calculate the power required to drive the compressor and the inlet and outlet cross-sectional areas. **[LO 5.3]**

(E) **5.17** Two streams of air, one at 1 bar, 27°C and velocity of 30 m/s and the other at 5 bar, 227°C and 50 m/s velocity, mix in equal proportion in a chamber from which heat at the rate of 100 kJ/kg is removed. The mixture is then passed through an adiabatic nozzle. Find the velocity of the stream issuing out of the nozzle. The temperature of air leaving the nozzle is 27°C, and its $c_p = 1.005$ kJ kgK. **[LO 5.4]**

(M) **5.18** Steam flowing in a pipeline is at a steady state represented by p_p, t_p, u_p, v_p, h_p and $\mathbf{V}_p$. A small amount of the total flow is fed through a small tube to an evacuated chamber which is allowed to fill slowly until the pressure is equal to the pipeline pressure. If there is no heat transfer, derive an expression for the final specific internal energy in the chamber, in terms of the properties in the pipeline. **[LO 5.7]**

(D) **5.19** The internal energy of air is given, at ordinary temperatures, by

$$u = u_0 + 0.718\, t$$

where u is in kJ/kg, u_0 is any arbitrary value of u at 0°C, kJ/kg and t is temperature in °C.

Also for air, $pv = 0.287\,(t + 273)$

where p is in kPa and v is in m³/kg.

(a) An evacuated bottle is fitted with a valve through which air from the atmosphere, at 760 mm Hg and 25°C, is allowed to flow slowly to fill the bottle. If no heat is transferred to or from the air in the bottle, what will its temperature be when the pressure in the bottle reaches 760 mm Hg?

(b) If the bottle initially contains 0.03 m³ of air at 400 mm Hg and 25°C, what will the temperature be when the pressure in the bottle reaches 760 mm Hg? **[LO 5.7]**

(D) **5.20** A pressure cylinder of volume V contains air at pressure p_0 and temperature T_0. It is to be filled from a compressed air line maintained at constant pressure p_1 and temperature T_1. Show that the temperature of the air in the cylinder after it has been charged to the pressure of the line is given by

$$T = \frac{\gamma T_1}{1 + \dfrac{p_0}{p_1}\left(\gamma \dfrac{T_1}{T_0} - 1\right)}$$

If $p_0 = 400$ mm hg, $p_1 = 760$ m Hg, $T_0 = T_1 = 298$ K and $\gamma = 1.4$, what is the value of T. **[LO 5.8]**

(D) **5.21** A small reciprocating vacuum pump having the rate of volume displacement V_d is used to evacuate a large vessel of volume V. The air in the vessel is maintained at a constant temperature T by energy transfer as heat. If the initial and final pressures are p_1 and p_2, respectively, find the time taken for the pressure drop and the necessary energy transfer as heat during evacuation. Assume that for air, $pV = mRT$, where m is the mass and R is a constant, and u is a function of T only. **[LO 5.8]**
[***Hint:*** $dm = -p(V_d \cdot dt)/(RT) = V\,dp/(RT)$].

(M) **5.22** A tank containing 45 kg of water initially at 45°C has one inlet and one exit with equal mass flow rates. Liquid water enters at 45°C and a mass flow rate of 270 kg/h. A cooling coil immersed in the water removes energy at the ate of 7.6 kW. The water is well mixed by a paddle wheel with a power input of 0.6 kW. The pressures at inlet and exit are equal. Ignoring changes in KE and PE, find the variation of water temperature with time.
If $p_1 = 760$ mm Hg, $V_1 = 0.3$ m³ and $V_d = 0.001$ m³/s, find the required time for the pressure to drop from p_1 to p_2 and the associated heat transfer to maintain the temperature constant. **[LO 5.4]**

(M) **5.23** A rigid tank of volume 0.5 m³ is initially evacuated. A tiny hole develops in the wall, and air from the surroundings at 1 bar and 21°C leaks in. Eventually, the pressure in the tank reaches 1 bar. The process occurs slowly enough that heat transfer between the tank and the surroundings keeps the temperature of the air inside the tank constant at 21°C. Determine the amount of heat transfer. **[LO 5.6]**

(M) **5.24** A well-insulated rigid tank with a volume of 10 m³ is connected to a large steam line through which steam flows at 15 bar and 280°C. The tank is initially evacuated. Steam is allowed to flow into the tank until the pressure inside is 15 bar. Calculate the amount of mass that flows into the tank. **[LO 5.8]**

ANSWERS TO PROBLEMS

5.1 28.38°C)
5.2 112.51 kW
5.3 (a) 692.5 m/s
(b) 32.08 kg/s
(c) 0.023 m²
5.4 1.47 kg/s
5.5 $Q_2 = 5.1$ watts, $\eta = 0.073$
5.6 157°C
5.7 – 0.17 kJ/s, – 3.76 kJ/s
5.8 3.16 kg/s
5.9 150.23 kJ/kg
5.10 143.3 kJ/kg, 152.9 kJ/kg
5.11 2402 kJ/kg
5.12 – 1877 kJ/kg mixture
5.13 + 7.96 kJ/kg
5.14 60.3 MW
5.15 1.92 kW
5.16 45.4 kW, 0.057 m², 0.0142 m²
5.17 51.96 m/s
5.19 (a) 144.2°C
(b) 71.6°C
5.20 344.64 K
5.21 $t = \dfrac{V}{V_d} \ln \dfrac{p_1}{p_2}, \quad Q = (p_1 - p_2)V$
5.22 $T = 318 - 22\,[1 - \exp(-6t)]$
$t = 19.26$ s, $Q = 1.44$ kJ
5.23 50 kJ
5.24 47.4 kg

6

Second Law of Thermodynamics

Learning Objectives

LO 6.1 Discuss second law of thermodynamics and the qualitative difference between heat and work

LO 6.2 State thermal and mechanical energy reservoirs

LO 6.3 Recognize heat engines for closed and open systems

LO 6.4 State Kelvin-Plank statement of the second law of thermodynamics

LO 6.5 Explain working principles of refrigerator and heat pump referring to Clausius statement

LO 6.6 Identify the equivalence of Kelvin-Plank and Clausius statements of second law

LO 6.7 Discuss reversible process and irreversibility

LO 6.8 Discuss types of irreversibility

LO 6.9 Demonstrate Carnot cycle

LO 6.10 Explain reversed heat engine and Carnot's theorem

LO 6.11 Demonstrate absolute thermodynamic temperature scale and third law of thermodynamics

LO 6.12 Demonstrate efficiency of reversible devices

LO 6.13 Relate ideal gas temperature with Kelvin temperature

6.1 INTRODUCTION TO THE SECOND LAW

> **LO 6.1**
> Discuss second law of thermodynamics and the qualitative difference between heat and work

The first law of thermodynamics states that a certain energy balance will hold when a system undergoes a change of state or a thermodynamic process. But it does not give any information on whether that change of state or the process is at all feasible or not. The first law cannot indicate whether a metallic bar of uniform temperature can spontaneously become warmer at one end and cooler at the other. All that the law can state is that if this process did occur, the energy gained by one end would be exactly equal to that lost by the other. *It is the second law of thermodynamics which provides the criterion as to the probability of various processes.*

Spontaneous processes in nature occur only in one direction. Heat always flows from a body at a higher temperature to a body at a lower temperature, fluid always flows from higher mechanical energy to mechanical energy. Material always diffuses from high concentration to low concentration. Time always flows in the forward direction. The reverse of these never happens spontaneously. The spontaneity of the process is due to a finite driving potential. These transfer processes can never occur spontaneously from a lower to a higher potential.

Figure 6.1 demonstrates the various transport processes. The process 1-2 (firm line) from a high to a low potential is spontaneous, whereas the process 2-1 (dotted line) can never happen spontaneously and is made possible only by energy addition.

The second law of thermodynamics puts a limitation on energy transformation other than that imposed by the first law.

Joule's experiments (Section 4.1) amply demonstrate that energy, when supplied to a system in the form of work, can be completely converted into heat (work transfer → internal energy increase → heat transfer). But the complete conversion of heat into work in a cycle is not possible. *So heat and work are not completely interchangeable forms of energy.*

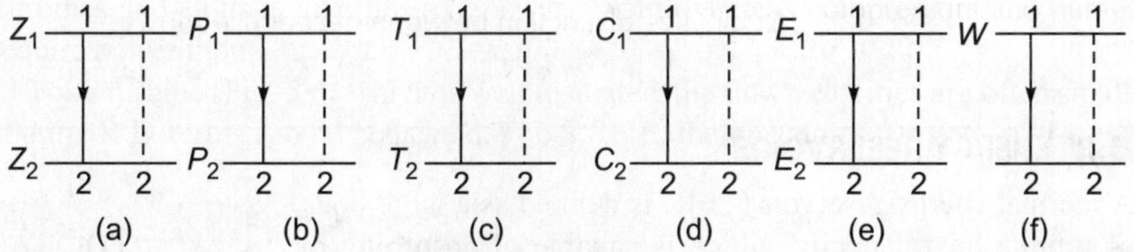

Fig. 6.1 (a) Flow from a high to allow elevation (Z), (b) Fluid flow from a high to a low pressure (p), (c) Heat transfer from a high to a low temperature (T), (d) Mass transfer of a constituent in a mixture from a high to a low concentration (C, gmol/m^3), (e) Flow of electric current from a high to a low electric potential (E), (f) Energy flux from work to heat transfer ($W \rightarrow Q$)

When work is converted into heat, we always have

$$W \equiv Q$$

but when heat is converted into work in a complete closed cycle process

$$Q \geq W$$

The arrow indicates the direction of energy transformation. This is illustrated in Fig. 6.2. As shown in Fig. 6.2(a), a system is taken from state 1 to state 2 by work transfer W_{1-2}, and then by heat transfer Q_{2-1} the system is brought back from state 2 to state 1 to complete a cycle. It is always found that $W_{1-2} = Q_{2-1}$.

But if the system is taken from state 1 to state 2 by heat transfer Q_{1-2}, as shown in Fig. 6.2(b), then the system cannot be brought back from state 2 to state 1 by work transfer W_{2-1}. Hence, heat cannot be converted completely and continuously into work in a cycle. Some heat has to be rejected. In Fig. 6.2(c), Q_{1-2} is the heat supply to the system, W_{2-3} is the work done by the system, Q_{3-4} is the heat rejected from the system and W_{4-1} is the work done upon the system to complete the cycle. The net work done in the cycle is $(W_{2-3} - W_{4-1})$ which is less than Q_{1-2} and some heat Q_{3-4} must be rejected in the cycle. This underlies the work of Sadi Carnot, a French military engineer, who first studied this aspect of energy transformation (1824). Work is said to be a *high grade energy* and heat a *low grade energy. The complete conversion of low grade energy into high grade energy in a cycle is impossible.*

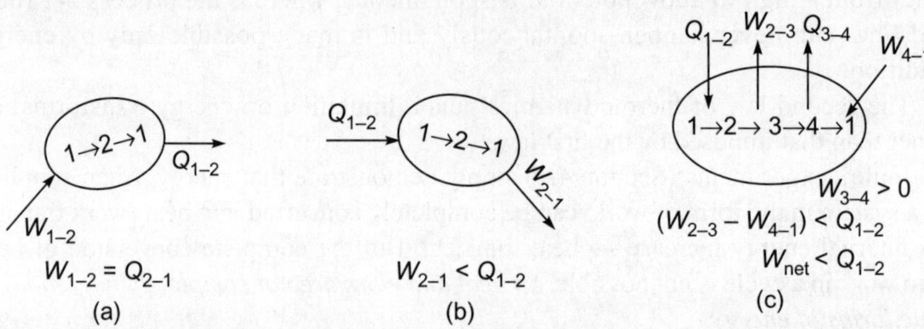

Fig. 6.2 Qualitative distinction between heat and work

6.2 ENERGY RESERVOIRS

> **LO 6.2**
> State thermal and mechanical energy reservoirs

A thermal energy reservoir (TER) is defined as a large body of infinite heat capacity, which is capable of absorbing or rejecting an unlimited quantity of heat without suffering appreciable changes in its thermodynamic coordinates. The changes that do take place in the large body as heat enters or leaves are so very slow and so very minute that all processes within it are quasi-static.

The thermal energy reservoir TER_H from which heat Q_1 is transferred to the system operating in a heat engine cycle is called the *source*. The thermal energy reservoir TER_L to which heat Q_2 is rejected from the system during a cycle is the *sink*. A typical source is a constant temperature furnace where fuel is continuously burnt, and a typical sink is a river or sea or the atmosphere itself.

A mechanical energy reservoir (MER) is a large body enclosed by an adiabatic impermeable wall capable of storing work as potential energy (such as a raised weight or wound spring) or kinetic energy (such as a rotating flywheel). All processes of interest within an MER are essentially quasi-static. An MER receives and delivers mechanical energy quasi-statically.

Figure 6.3 shows a cyclic heat engine exchanging heat with a source and a sink and delivering W_{net} in a cycle to an MER.

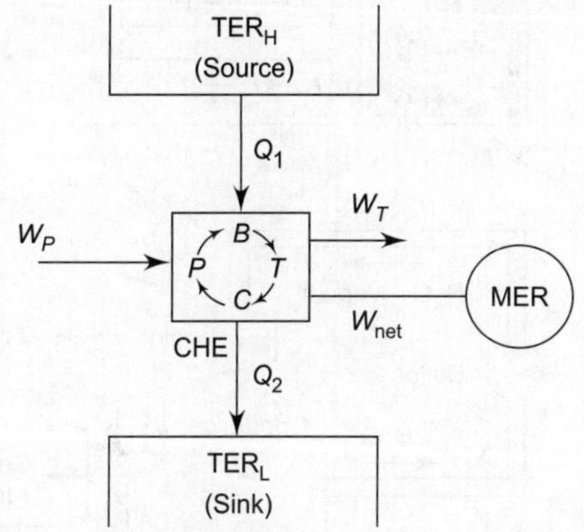

Fig. 6.3 Cyclic heat engine (CHE) with source and sink

6.3 HEAT ENGINES

LO 6.3

Recognize heat engines for closed and open systems

It has been already mentioned that work can easily be converted to other forms of energy; however, conversion of other forms of energy to work is not an easy task. From the foregoing discussion (Section 6.1), it has also been observed that work can be converted to heat directly and completely, but complete conversion of heat to work is not possible. For the purpose of conversion of heat into work, we require special devices, called *heat engines*.

A heat engine may be defined as a device, which operates in a thermodynamic cycle, converts part of the heat received from high temperature reservoir to useful work and rejects the remaining heat to low temperature reservoir. Thus, there is a net heat transfer to a heat engine and a network transfer from the heat engine. The thermodynamic cycle in which the heat engine works is called *heat engine cycle.* Heat engines usually involve a substance which receives energy from the source and rejects energy to the sink. This substance is called the *working fluid or working substance.*

A heat engine may be in the form of mass of gas in a cylinder bounded by a piston (Fig. 6.4a) or a mass of water moving in a steady flow through a steam power plant (Fig. 6.4b). In the cyclic heat engine, as represented in Fig. 6.4(a), heat Q_1 is transferred to the system, work W_E is done by the system, work W_C is done upon the system, and then heat Q_2 is rejected from the system. The system is brought back to the initial state through these four successive processes which constitute a heat engine cycle. In Fig. 6.4(b), heat Q_1 is transferred from the furnace to the water in the boiler to form steam which then works on the turbine rotor to produce work W_T, then the steam is condensed to water in the condenser in which an amount of heat Q_2 is rejected from the system, and finally work W_P is done on the system (water) to pump it to the boiler. The system repeats the cycle.

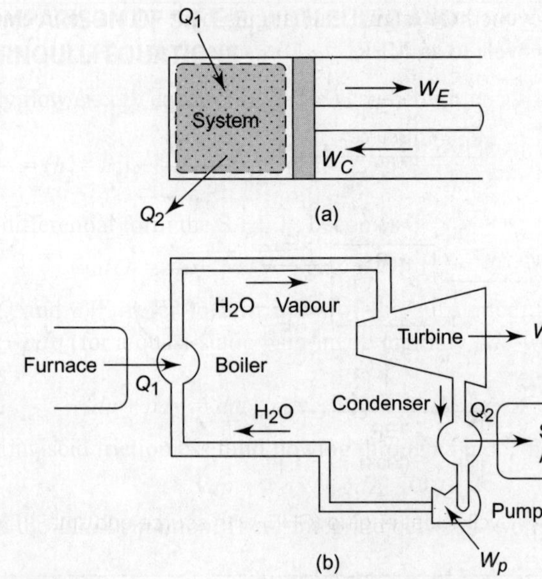

Fig. 6.4 Cyclic heat engine (a) Heat engine cycle performed by a closed system undergoing four successive energy interactions with the surroundings, (b) Heat engine cycle performed by a steady flow system interacting with the surroundings as shown

The net heat transfer in a cycle to either of the heat engines is given as:

$$Q_{net} = Q_1 - Q_2 \tag{6.1}$$

and the net work transferred in a cycle is given as:

$$W_{net} = W_T - W_P \tag{6.2}$$

$$(\text{or } W_{net} = W_E - W_C)$$

By the first law of thermodynamics for a cycle, we have:

$$\sum_{cycle} Q = \sum_{cycle} W$$

$$\therefore \quad Q_{net} = W_{net}$$

or

$$Q_1 - Q_2 = W_T - W_P \tag{6.3}$$

Figure 6.5 represents a cyclic heat engine in the form of a block diagram indicating various energy interactions during a cycle. Boiler (B), turbine (T), condenser (C), and pump (P), all four together constitute a heat engine. A heat engine here is a certain quantity of water undergoing the energy interactions, as shown, in cyclic operations to produce net work from a certain heat input.

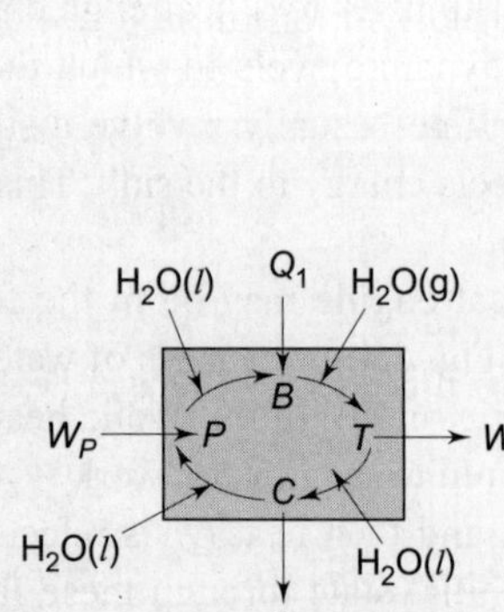

Fig. 6.5 Cyclic heat engine with energy interactions represented in a block diagram

The performance of a heat engine is expressed by the thermal efficiency, which is defined as the ratio of the desired output to the required input. The function of a heat engine is to produce work *continuously* at the expense of heat input to it, hence the net work W_{net} (desired output) and heat Q_1 (required input) are of primary interest. The *thermal efficiency* of a heat engine can be expressed as follows:

$$\eta = \frac{\text{Net work output}}{\text{Total heat input}} = \frac{W_{net}}{Q_1} \tag{6.4}$$

From Eqs. (6.1), (6.2), (6.3) and (6.4),

$$\eta = \frac{W_{\text{net}}}{Q_1} = \frac{W_T - W_P}{Q_1} = \frac{Q_1 - Q_2}{Q_1}$$

or

$$\eta = 1 - \frac{Q_2}{Q_1} \tag{6.5}$$

Since heat Q_1 transferred to a system cannot be completely converted to work in a cycle (Section 6.1), $W_{net} < Q_1$. Therefore, thermal efficiency of a heat engine is always less than unity. Alternatively, we can conclude that a heat engine can never be 100% efficient.

The thermal efficiency of a heat engine is a parameter that measures how efficiently it converts the heat received to work. The thermal efficiencies of real engines are relatively low. It is important to know in this context the typical values of the thermal efficiency of real engines. The usual values of the thermal efficiency are about 30–35% for spark ignition automobile engines, 35–40% for diesel engines, 35–50% for large power plants, and 60–65% for modern gas-steam power plants. Therefore, a significant amount of energy supplied to the engine is useless and this motivates the engineers to work on engines to improve theirthermal efficiencies.

6.4 KELVIN–PLANCK STATEMENT OF THE SECOND LAW

LO 6.4

State Kelvin-Plank statement of the second law of thermodynamics

It is clear from the previous discussion that no heat engine can convert all the heat it receives to useful work. In other words, a heat engine, operating in a cycle, must reject some heat to a low-temperature reservoir. To produce net work in a thermodynamic cycle, a heat engine has thus to exchange heat with two reservoirs, the source and the sink. This limitation forms the basis of the *Kelvin–Planck statement* of the second law of thermodynamics, which is expressed as follows: *It is impossible for a heat engine to produce net work in a complete cycle if it exchanges heat only with bodies at a single fixed temperature.*

If $Q_2 = 0$ (i.e., $W_{net} = Q_1$, or $\eta = 1.00$), the heat engine will produce net work in a complete cycle by exchanging heat with only one reservoir, thus violating the Kelvin-Planck statement (Fig. 6.6). Such a heat engine is called a *perpetual motion machine of the second kind*, abbreviated to PMM2. A PMM2 is impossible.

A heat engine has, therefore, to exchange heat with two thermal energy reservoirs at two different temperatures to produce net work in a complete cycle (Fig. 6.7). So long as there is a difference in temperature, motive power (i.e., work) can be produced. If the bodies with which the heat engine exchanges heat are of finite heat capacities, work will be produced by the heat engine till the temperatures of the two bodies are equalised.

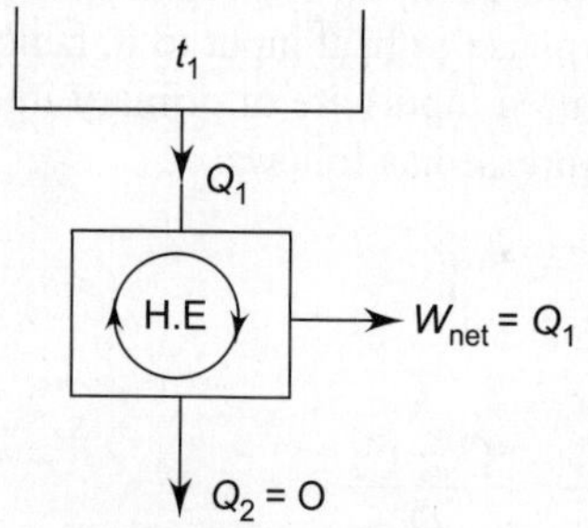

Fig. 6.6 A PMM2

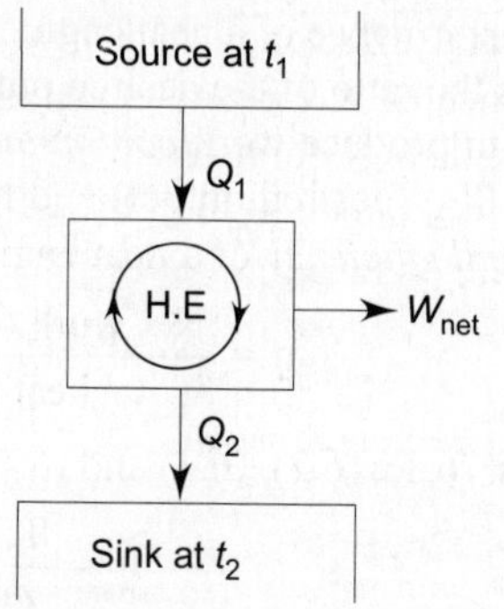

Fig. 6.7 Heat engine producing net work in a cycle by exchanging heat at two different temperatures

6.5 CLAUSIUS' STATEMENT OF THE SECOND LAW

Heat always flows from a body at a higher temperature to a body at a lower temperature.The reverse process never occurs spontaneously.

The above observation forms the basis of the Clausius statement of the second law of thermodynamics, which is as follows:It is impossible to construct a devicewhich, operating in a cycle, will produce no effect other than the transfer of heat froma cooler to a hotter body.

Heat cannot flow of itself from a body at a lower temperature to a body at a higher temperature. Some work must be expended to achieve this.

6.6 REFRIGERATOR AND HEAT PUMP

> **LO 6.5**
> Explain working principles of refrigerator and heat pump referring to Clausius statement

A *refrigerator* is a device which, operating in a cycle, maintains a body at a temperature lower than the temperature of the surroundings. Let the body *A* (Fig. 6.8) be maintained at t_2, which is lower than the ambient temperature t_1. Even though *A* is insulated, there will always be heat leakage Q_2 into the body from the surroundings by virtue of the temperature difference. In order to maintain body *A* at the constant temperature t_2, heat has to be removed from the body at the same rate at which heat is leaking into the body. This heat (Q_2) is absorbed by a working fluid, called the *refrigerant*, which evaporates in the evaporator E_1 at a temperature lower than t_2 absorbing the latent heat of vaporisation from the body *A* which is cooled or refrigerated (Process 4–1). The vapour is first compressed in the compressor C_1, driven by a motor, which absorbs work W_C (Process 1–2), and is then condensed in the condenser C_2 rejecting the latent heat of condensation Q_1 at a temperature higher than that of the atmosphere (at t_1) for heat transfer to take place (Process 2–3). The condensate then expands adiabatically through an expander (an engine or turbine) producing work W_E, when the temperature drops to a value lower than t_2 such that heat Q_2 flows from the body *A* to make the refrigerant evaporate (Process 3–4). Such a cyclic device of flow through E_1–C_1–C_2–E_2 is called a *refrigerator*. Just like thermal efficiency in a heat engine, the parameter that characterises the performance of a refrigerator is called the *coefficient of performance*, abbreviated to COP, which is defined as the ratio of the desired effect to the required input.

In a refrigerator cycle removal of heat Q_2 from a body A (desired effect) and work input W are of primary interest. The COP of a refrigerator can be expressed as

$$\text{COP} = \frac{\text{Desired effect}}{\text{Work input}} = \frac{Q_2}{W}$$

$$\therefore \qquad [\text{COP}]_{\text{ref}} = \frac{Q_2}{Q_1 - Q_2} \qquad (6.6)$$

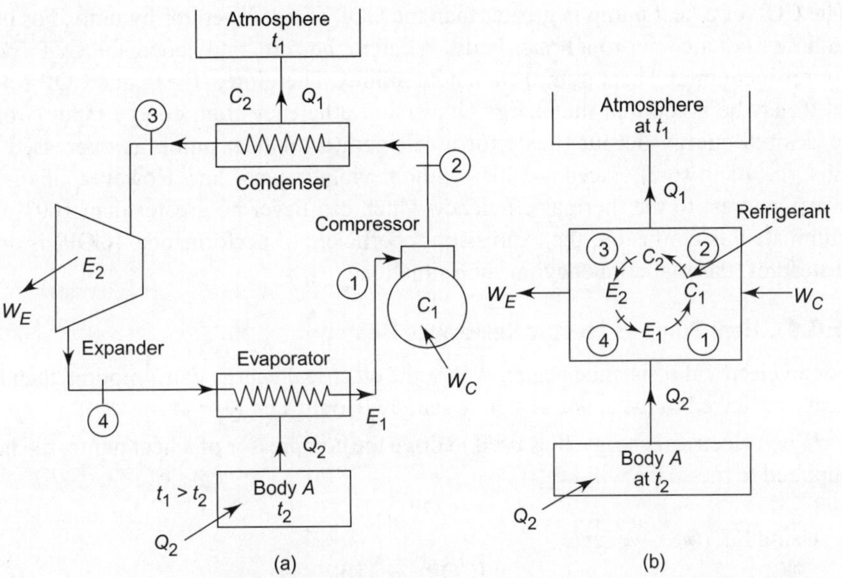

Fig. 6.8 A cyclic refrigeration plant

A *heat pump* is a device which, operating in a cycle, maintains a body, say *B* (Fig. 6.9), at a temperature higher than the temperature of the surroundings. By virtue of the temperature difference, there will be heat leakage Q_1 from the body to the surroundings. The body will be maintained at the constant temperature t_1, if heat is discharged into the body at the same rate at which heat leaks out of the body. The heat is extracted from the low temperature reservoir, which is nothing but the atmosphere, and discharged into the high temperature body *B*, with the expenditure of work *W* in a cyclic device called a *heat pump*. The working fluid

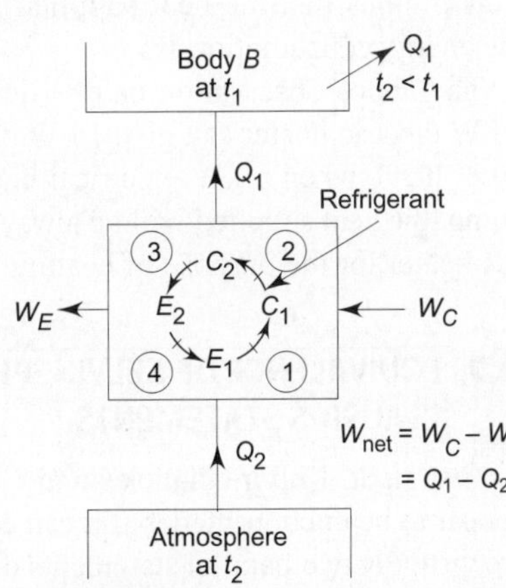

Fig. 6.9 A cyclic heat pump

operates in a cycle flowing through the evaporator E_1, compressor C_1, condenser C_2 and expander E_2, similar to a refrigerator, but the attention is here focussed on the high temperature body B. Here Q_1 and W are of primary interest, and the COP is defined as

$$\text{COP} = \frac{Q_1}{W}$$

$$\therefore \qquad [\text{COP}]_{H.P.} = \frac{Q_1}{Q_1 - Q_2} \tag{6.7}$$

From Eqs. (6.6) and (6.7), it is found that

$$[\text{COP}]_{H.P.} = [\text{COP}]_{\text{ref}} + 1 \tag{6.8}$$

The COP of a heat pump is greater than the COP of a refrigerator by unity. For heat to flow from a cooler to a hotter body, W cannot be zero, and hence, the COP (both for refrigerator and heat pump) cannot be infinity. Therefore, $W > 0$, and COP $< \infty$.

It may be noted that the energy conversion efficiency from energy input (work) to desired energy output (heat) for a refrigerator or heat pump as expressed by Eqs. (6.6) and (6.7), exceeds 100% without violating any law. However, this is in sharp contrast to the thermal efficiency which can never be greater than 100%. To eliminate this confusion, the expression 'coefficient of performance' (COP) is used instead of 'thermal efficiency' in such situations.

6.6.1 Heat Pump vs Electric Resistance Heater

For an electrical resistance heater, if W is the electrical energy consumption, then the heat transferred to the space at steady state is W only, i.e., $Q_1 = W$.

When electrical energy W is used to drive the compressor of a heat pump, the heat supplied to the space will be

$$Q_1 = [\text{COP}]_{\text{HP}} W$$

Using Eq. (6.8), we have

$$Q_1 = [\text{COP}_{\text{ref}} + 1]W \tag{6.9}$$

Since $[\text{COP}]_{\text{ref}}$ is a positive quantity, it is evident from Eq. (6.9) that $Q_1 > W$. This implies that the heat supplied by a heat pump Q_1 will always be more than the energy consumption W.

The above observation can be demonstrated through an example as follows: A 1 kW electric heater can give 1 kW of heat at steady state and nothing more. On the other hand, when 1 kW electrical energy is used to drive the compressor of a heat pump, the heat supplied will be always more than 1 kW. This shows that the use electric heater for the purpose of heating is thermodynamically inefficient.

6.7 EQUIVALENCE OF KELVIN–PLANCK AND CLAUSIUS STATEMENTS

LO 6.6

Identify the equivalence of Kelvin-Plank and Clausius statements of second law

At first sight, Kelvin–Planck's and Clausius' statements may appear to be unconnected, but it can easily be shown that they are virtually two parallel statements of the second law and are equivalent in all respects.

The equivalence of the two statements will be proved if it can be shown that the violation of one statement implies the violation of the second, and vice versa.

1. Let us first consider a cyclic heat pump P which transfers heat from a low temperature reservoir (t_2) to a high temperature reservoir (t_1) with no other effect, i.e., with no expenditure of work, violating Clausius statement (Fig. 6.10).

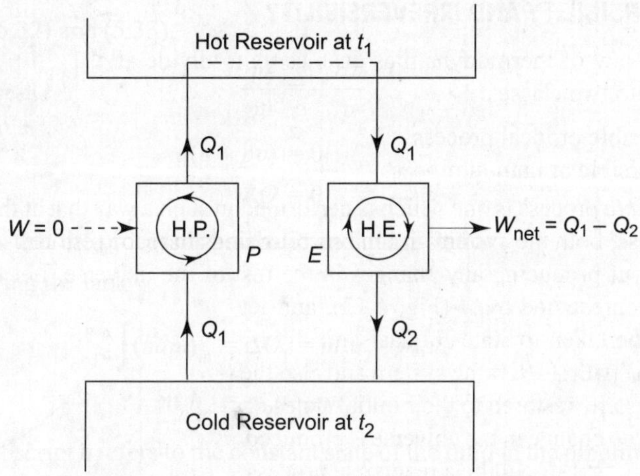

Fig. 6.10 Violation of the Clausius statement

Let us assume a cyclic heat engine E operating between the same thermal energy reservoirs, producing W_{net} in one cycle. The rate of working of the heat engine is such that it draws an amount of heat Q_1 from the hot reservoir equal to that discharged by the heat pump. Then the hot reservoir may be eliminated and the heat Q_1 discharged by the heat pump is fed to the heat engine. So we see that the heat pump P and the heat engine E acting together constitute a heat engine operating in cycles and producing net work while exchanging heat only with one body at a single fixed temperature (t_2). This violates the Kelvin–Planck statement.

2. Let us now consider a perpetual motion machine of the second kind (E) which produces net work in a cycle by exchanging heat with only one thermal energy reservoir (at t_1) and thus violates the Kelvin–Planck statement (Fig. 6.11).

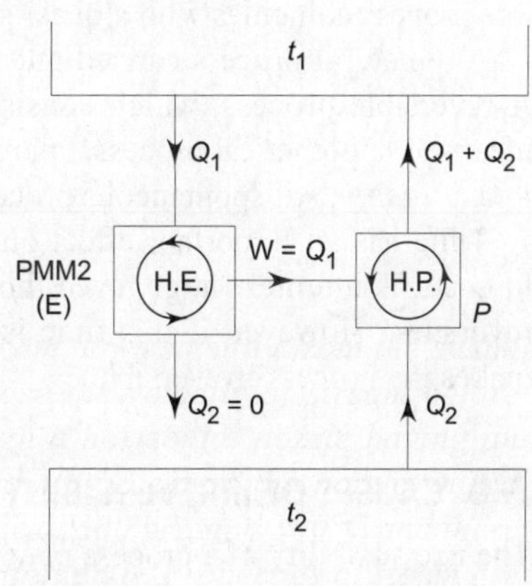

Fig. 6.11 Violation of the Kelvin–Planck statement

Let us assume a cyclic heat pump (P) extracting heat Q_2 from a low temperature reservoir at t_2 and discharging heat to the high temperature reservoir at t_1 with the expenditure of work W equal to what the PMM2 delivers in a complete cycle. So E and P together constitute a heat pump working in cycles and producing the sole effect of transferring heat from a lower to a higher temperature body, thus violating the Clausius statement.

6.8 REVERSIBILITY AND IRREVERSIBILITY

> **LO 6.7**
> Discuss reversible process and irreversibility

The second law of thermodynamics enables us to divide all processes into two classes:

1. Reversible or ideal process
2. Irreversible or natural process

A *reversible* process is one which is performed in such a way that at the conclusion of the process, both the system and the surroundings may be restored to their initial states, without producing any changes in the rest of the universe. Let the state of a system be represented by A (Fig. 6.12), and let the system be taken to state B quasistatically by following the path A–B. If the system and also the surroundings are restored to their initial states on reverse, and no change in the universe is produced, then the process A–B will be a reversible process. In the reverse process, the system has to be taken from state B to A by following the same path B–A. A reversible process should not leave any trace or relic behind to show that the process had ever occurred.

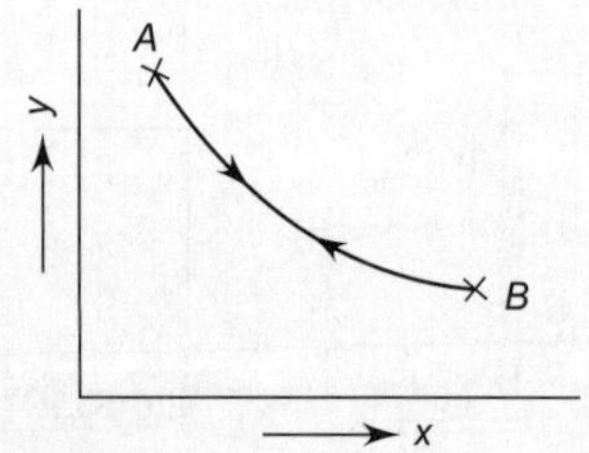

Fig. 6.12 A reversible process

A reversible process is carried out infinitely slowly with an infinitesimal gradient, so that every state passed through by the system is an equilibrium state. So a reversible process coincides with a quasi-static process.

Any natural process carried out with a finite gradient is an irreversible process. A reversible process, which consists of *a succession of equilibrium states*, is an idealised hypothetical process, approached only as a limit. It is said to be an *asymptote to reality*. All spontaneous processes are irreversible.

Time has an important effect on reversibility. If the time allowed for a process to occur is infinitely large, even though the gradient is finite, the process becomes reversible. However, if this time is squeezed to a finite value, the finite gradient makes the process irreversible.

6.9 CAUSES OF IRREVERSIBILITY

The irreversibility of a process may be due to either one or both of the following:

1. Lack of equilibrium during the process
2. Involvement of dissipative effects

6.9.1 Irreversibility due to Lack of Equilibrium

The lack of equilibrium (mechanical, thermal or chemical) between the system and its surroundings, or between two systems, or two parts of the same system, causes a spontaneous change which is irreversible. The following are specific examples in this regard:

1. Heat transfer through a finite temperature difference A heat transfer process approaches reversibility as the temperature difference between two bodies approaches zero. We define a reversible heat transfer process as one in which heat is transferred through an infinitesimal temperature difference. So to transfer a finite amount of heat through an infinitesimal temperature difference would require an infinite amount of time, or infinite area. All actual heat transfer processes are through a finite temperature difference and are, therefore, irreversible, and greater the temperature difference, the greater is the irreversibility.

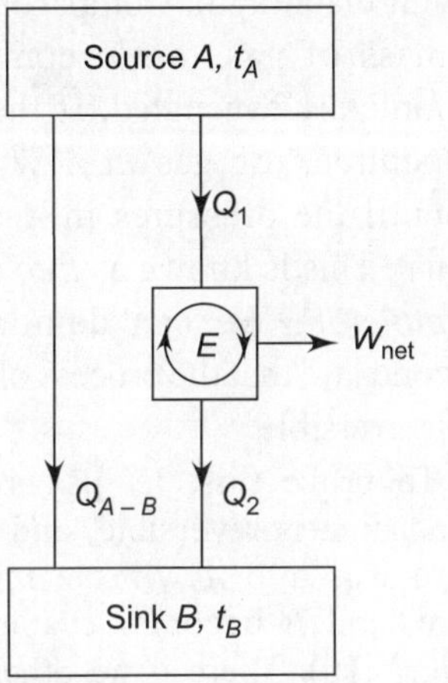

Fig. 6.13 Heat transfer through a finite temperature difference

We can demonstrate by the second law that heat transfer through a finite temperature difference is irreversible. Let us assume that a source at t_A and a sink at t_B ($t_A > t_B$) are available, and let Q_{A-B} be the amount of heat flowing from A to B (Fig. 6.13). Let us assume an engine operating between A and B, taking heat Q_1 from A and discharging heat Q_2 to B. Let the heat transfer process be reversed, and Q_{B-A} be the heat flowing from B to A, and let the rate of working of the engine be such that

$$Q_2 = Q_{B-A}$$

(Fig. 6.14). Then the sink B may be eliminated. The net result is that E produces network W in a cycle by exchanging heat only with A, thus violating the Kelvin–Planck statement. So the heat transfer process Q_{A-B} is irreversible, and Q_{B-A} is not possible.

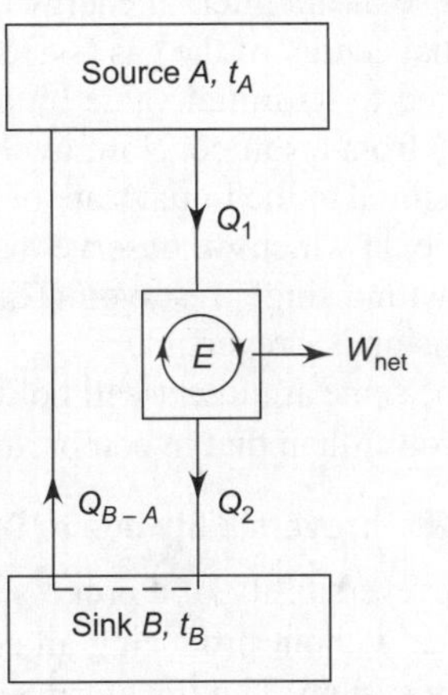

Fig. 6.14 Heat transfer through a finite temperature difference is irreversible

2. Lack of pressure equilibrium within the interior of the system or between the system and the surroundings When there exists a difference in pressure between the system and the surroundings, or within the system itself, then both the system and its surroundings or the system alone, will undergo a change of state which will cease only when mechanical equilibrium is established. The reverse of this process is not possible spontaneously without producing any other effect. That the reverse process will violate the second law becomes obvious from the following illustration.

3. Free expansion Let us consider an insulated container (Fig. 6.15) which is divided into two compartments *A* and *B* by a thin diaphragm. Compartment *A* contains a mass of gas, while compartment *B* is completely evacuated. If the diaphragm is punctured, the gas in *A* will expand into *B* until the pressures in *A* and *B* become equal. This is known as *free* or *unrestrained expansion*. We can demonstrate by the second law, that the process of free expansion is irreversible.

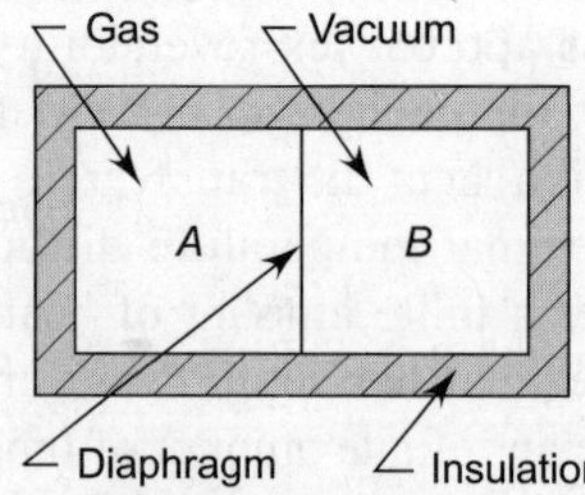

Fig. 6.15 Free expansion

To prove this, let us assume that free expansion is reversible, and that the gas in *B* returns into *A* with an increase in pressure, and *B* becomes evacuated as before (Fig. 6.16). There is no other effect. Let us install an engine (a machine, not a cyclic heat engine) between *A* and *B*, and permit the gas to expand through the engine from *A* to *B*. The engine develops a work output *W* at the expense of the internal energy of the gas. The internal energy of the gas (system) in *B* can be restored to its initial value by heat transfer *Q* (= *W*) from a source. Now, by the use of the reversed free expansion, the system can be restored to the initial state of high pressure in *A* and vacuum in *B*. The net result is a cycle, in which we observe that net work output *W* is accomplished by exchanging heat with a single reservoir. This violates the Kelvin–Planck statement. Hence, free expansion is irreversible.

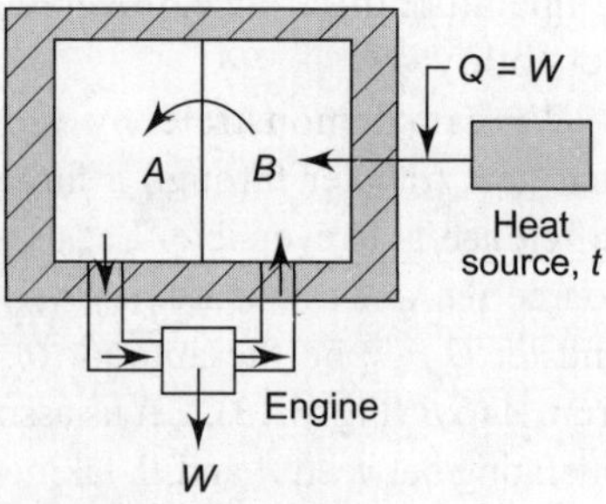

Fig. 6.16 Second law demonstrates that free expansion is irreversible

The same argument will hold if the compartment *B* is not in vacuum but at a pressure lower than that in compartment *A* (case b).

6.9.2 Irreversibility due to Dissipative Effects

The irreversibility of a process may be due to the *dissipative effects* in which work is done without producing an equivalent increase in the kinetic or potential energy of any system. The transformation of work into molecular internal energy either of the system or of the reservoir takes place through the agency of such phenomena as friction, viscosity, inelasticity, electrical resistance and magnetic hysteresis. These effects are known as dissipative effects, and work is said to be dissipated.

1. Friction Friction is always present in moving devices. Friction may be reduced by suitable lubrication, but it can never be completely eliminated. If this were possible, a movable device could be kept in a continual motion without violating either of the two laws of thermodynamics. The continual motion of a movable device in the complete absence of friction is known as *perpetual motion of the third kind.*

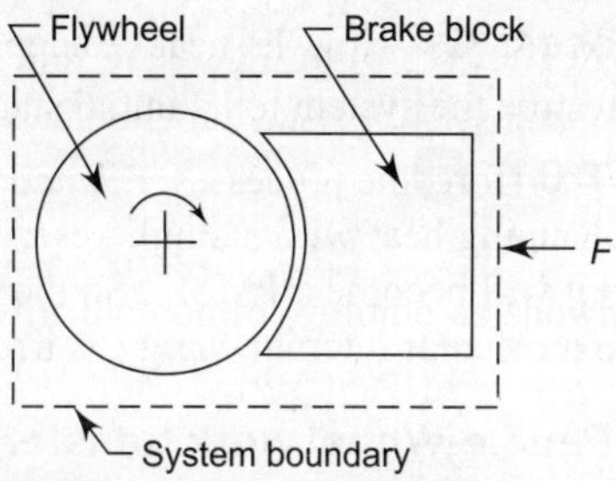

Fig. 6.17 Irreversibility due to dissipative effect like friction

That friction makes a process irreversible can be demonstrated by the second law. Let us consider a system consisting of a flywheel and a brake block (Fig. 6.17). The flywheel was rotating with a certain rpm, and it was brought to rest by applying the friction brake. The distance moved by the brake block is very small, so work transfer is very nearly equal to zero. If the braking process occurs very rapidly, there is a little heat transfer. Using suffix 2 after braking and suffix 1 before braking, and applying the first law, we have

$$Q_{1-2} = E_2 - E_1 + W_{1-2}$$

$$0 = E_2 - E_1 + 0$$

$$\therefore \quad E_2 = E_1 \tag{6.10}$$

The energy of the system (isolated) remains constant. Since the energy may exist in the forms of kinetic, potential and molecular internal energy, we have

$$U_2 + \frac{m\mathbf{V}_2^2}{2} + mZ_2g = U_1 + \frac{m\mathbf{V}_1^2}{2} + mZ_1g$$

Since the wheel is brought to rest, $V_2 = 0$, and there is no change in P.E.

$$U_2 = U_1 + \frac{m\mathbf{V}_1^2}{2} \tag{6.11}$$

Therefore, the molecular internal energy of the system (i.e. of the brake and the wheel) increases by the absorption of the K.E. of the wheel. The reverse process, i.e. the conversion of this increase in molecular internal energy into K.E. within the system to cause the wheel to rotate is not possible. To prove it by the second law, let us assume that it is possible, and imagine the following cycle with three processes:

Process A Initially, the wheel and the brake are at high temperature as a result of the absorption of the K.E. of the wheel, and the flywheel is at rest. Let the flywheel now start rotating at a particular rpm at the expense of the internal energy of the wheel and brake, the temperature of which will then decrease.

Process B Let the flywheel be brought to rest by using its K.E. in raising weights, with no change in temperature.

Process C Now, let heat be supplied from a source to the flywheel and the brake to restore the system to its initial state.

Therefore, the processes *A*, *B* and *C* together constitute a cycle producing work by exchanging heat with a single reservoir. This violates the Kelvin–Planck statement, and it will become a PMM2. So the braking process, i.e. the transformation of K.E. into molecular internal energy, is irreversible.

2. Paddle-Wheel work transfer Work may be transferred into a system in an insulated container by means of a paddle wheel (Fig. 6.18) which is also known as stirring work. Here work transferred is dissipated adiabatically into an increase in the molecular internal energy of the system. To prove the irreversibility of the process, let us assume that the same amount of work is delivered by the system at the expense of its molecular internal energy, and the temperature of the system goes down (Fig. 6.19). The system is brought back to its initial state by heat transfer from a source. These two processes together constitute a cycle in which there is work output and the system exchanges heat with a single reservoir. It becomes a PMM2, and hence the dissipation of stirring work to internal energy is irreversible.

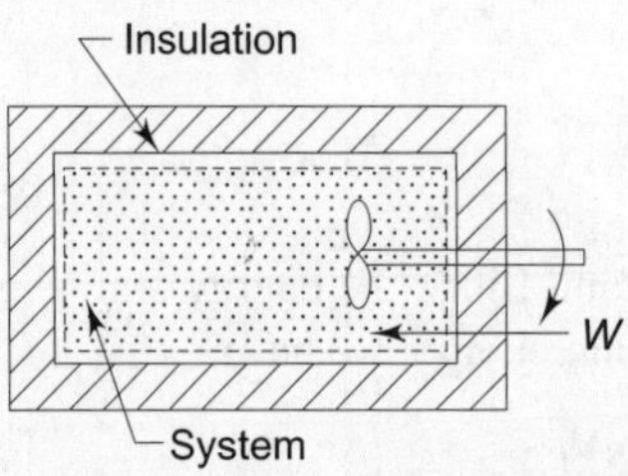

Fig. 6.18 Adiabatic work transfer

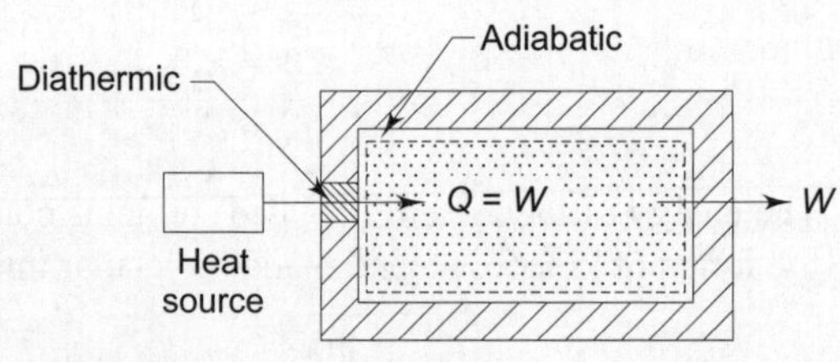

Fig. 6.19 Irreversibility due to dissipation of stirring work into internal energy

3. Transfer of electricity through a resistor The flow of electric current through a wire represents work transfer, because the current can drive a motor which can raise a weight. Taking the wire or the resistor as the system (Fig. 6.20) and writing the first law

$$Q_{1-2} = U_2 - U_1 + W_{1-2}$$

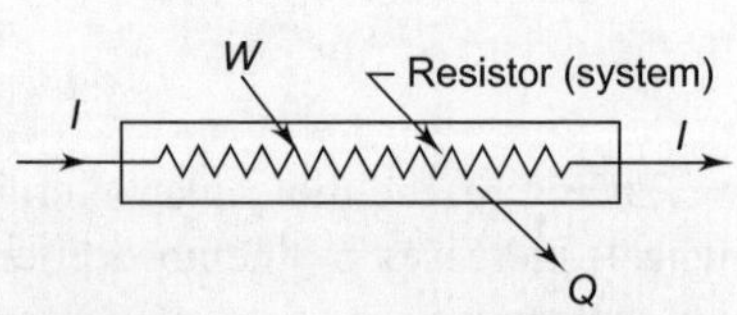

Fig. 6.20 Irreversibility due to dissipation of electrical work into internal energy

Here, both W_{1-2} and Q_{1-2} are negative.

$$W_{1-2} = U_2 - U_1 + Q_{1-2} \tag{6.12}$$

A part of the work transfer is stored as an increase in the internal energy of the wire (to give an increase in its temperature), and the remainder leaves the system as heat. At steady state, the internal energy and hence the temperature of the resistor become constant with respect to time and

$$W_{1-2} = Q_{1-2} \tag{6.13}$$

The reverse process, i.e. the conversion of heat Q_{1-2} into electrical work W_{1-2} of the same magnitude is not possible. Let us assume that this is possible. Then heat Q_{1-2} will be absorbed and equal work W_{1-2} will be delivered. But this will become a PMM2. So the dissipation of electrical work into internal energy or heat is irreversible.

6.10 TYPES OF IRREVERSIBILITY

LO 6.8

Discuss types of irreversibility

It has been discussed in Section 6.9 that a process becomes irreversible if it occurs due to a finite potential gradient like the gradient in temperature or pressure, or if there is dissipative effect like friction, in which work is transformed into internal energy increase of the system. Two types of irreversibility can be distinguished:

1. Internal irreversibility
2. External irreversibility

The *internal irreversibility* is caused by the internal dissipative effects like friction, turbulence, electrical resistance, magnetic hysteresis, etc., within the system. The *external irreversibility* refers to the irreversibility occurring at the system boundary like heat interaction with the surroundings due to a finite temperature gradient.

Sometimes, it is useful to make other distinctions. If the irreversibility of a process is due to the dissipation of work into the increase in internal energy of a system, or due to a finite pressure gradient, it is called *mechanical irreversibility*. If the process occurs on account of a finite temperature gradient, it is *thermal irreversibility*, and if it is due to a finite concentration gradient or a chemical reaction, it is called *chemical irreversibility*.

A heat engine cycle in which there is a temperature difference (*i*) between the source and the working fluid during heat supply, and (*ii*) between the working fluid and the sink during heat rejection exhibits external thermal irreversibility. If the real source and sink are not considered and hypothetical reversible processes for heat supply and heat rejection are assumed, the cycle can be reversible. With the inclusion of the actual source and sink, however, the cycle becomes externally irreversible.

6.11 CONDITIONS FOR REVERSIBILITY

LO 6.9

Demonstrate Carnot cycle

A natural process is irreversible because the conditions for mechanical, thermal and chemical equilibrium are not satisfied, and the dissipative effects, in which work is transformed into an increase in internal energy, are present. For a process to be reversible, it must not possess these features. If a process is performed quasistatically, the system passes through states of thermodynamic equilibrium, which may be traversed as well in one direction as in the opposite direction. *If there are no dissipative effects, all the work done by the system during the performance of a process in one direction can be returned to the system during the reverse process.*

A process will be reversible when it is performed in such a way *that the system is at all times infinitesimally near a state of thermodynamic equilibrium and in the absence of dissipative effect of any form*. Reversible processes are, therefore, purely ideal, limiting cases of actual processes.

6.12 CARNOT CYCLE

A reversible cycle is an ideal hypothetical cycle in which all the processes constituting the cycle are reversible. Carnot cycle is a reversible cycle. For a stationary system, as in a piston and cylinder machine, the cycle consists of the following four successive processes (Fig. 6.21):

1. *A reversible isothermal process* in which heat Q_1 enters the system at t_1 reversibly from a constant temperature source at t_1 when the cylinder cover is in contact with the diathermic cover A. The internal energy of the system increases.

   From first law,

$$Q_1 = U_2 - U_1 + W_{1-2} \tag{6.14}$$

   (for an ideal gas only, $U_1 = U_2$)

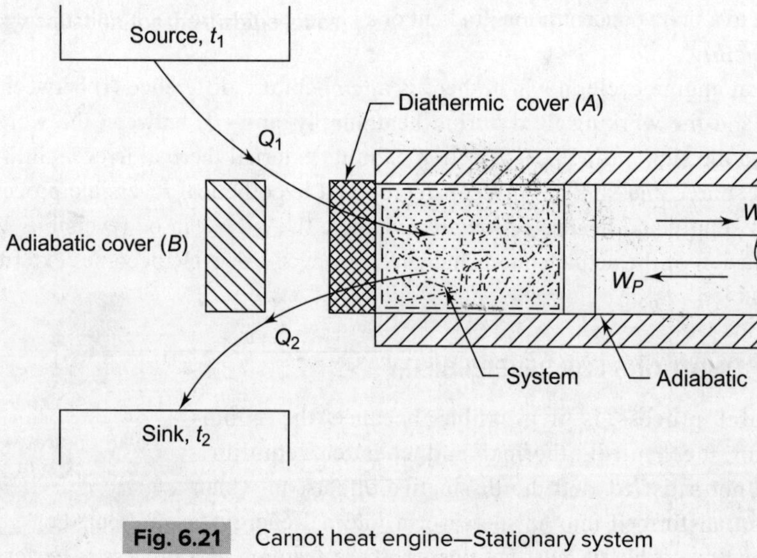

Fig. 6.21 Carnot heat engine—Stationary system

2. *A reversible adiabatic process* in which the diathermic cover A is replaced by the adiabatic cover B, and work W_E is done by the system adiabatically and reversibly at the expense of its internal energy, and the temperature of the system decreases from t_1 to t_2.

   Using the first law,

$$0 = U_3 - U_2 + W_{2-3} \tag{6.15}$$

3. *A reversible isothermal process* in which B is replaced by A and heat Q_2 leaves the system at t_2 to a constant temperature sink at t_2 reversibly, and the internal energy of the system further decreases.

   From the first law,

$$-Q_2 = U_4 - U_3 - W_{3-4} \tag{6.16}$$

   only for an ideal gas, $U_3 = U_4$

4. *A reversible adiabatic process* in which B again replaces A, and work W_p is done upon the system reversibly and adiabatically, and the internal energy of the system increases and the temperature rises from t_2 to t_1.

   Applying the first law,

$$0 = U_1 - U_4 - W_{4-1} \tag{6.17}$$

Two reversible isotherms and two reversible adiabatics constitute a Carnot cycle, which is represented in p-v coordinates in Fig. 6.22.

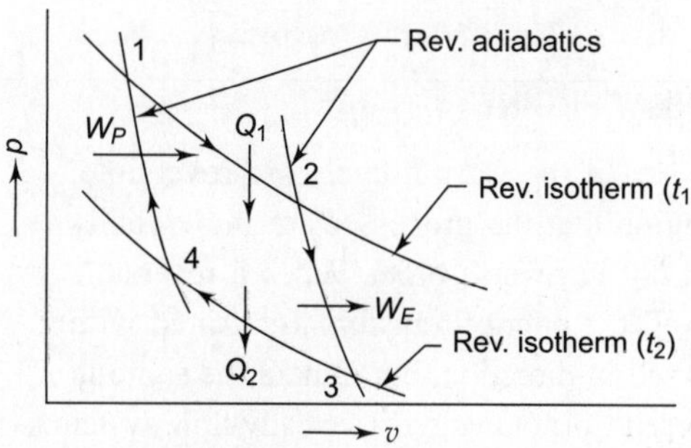

Fig. 6.22 Carnot cycle

Summing up Eqs. (6.14) to (6.17),

$$Q_1 - Q_2 = (W_{1-2} + W_{2-3}) - (W_{3-4} + W_{4-1})$$

or

$$\sum_{\text{cycle}} Q_{\text{net}} = \sum_{\text{cycle}} W_{\text{net}}$$

A cyclic heat engine operating on the Carnot cycle is called a Carnot heat engine.

For a steady flow system, the Carnot cycle is represented as shown in Fig. 6.23. Here heat Q_1 is transferred to the system reversibly and isothermally at t_1 in the heat exchanger A, work W_T is done by the system reversibly and adiabatically in the turbine (B), then heat Q_2 is transferred from the system reversibly and isothermally at t_2 in the heat exchanger (C), and then work W_p is done upon the system reversibly and adiabatically by the pump (D). To satisfy the conditions for the Carnot cycle, there must not be any friction or heat transfer in the pipelines through which the working fluid flows.

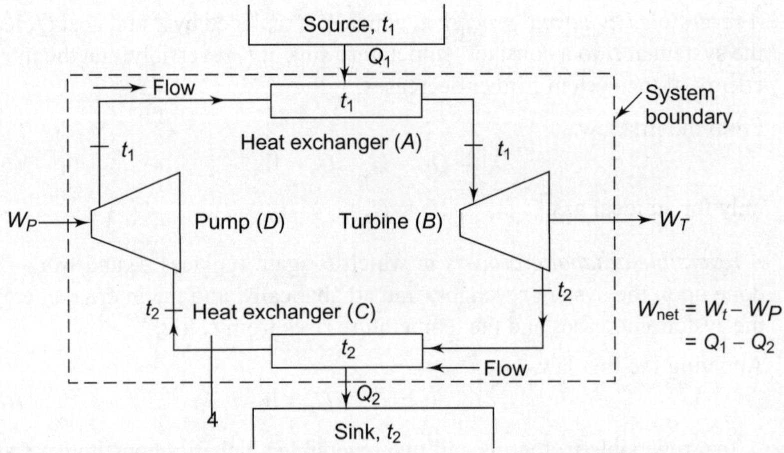

Fig. 6.23 Carnot heat engine—Steady flow system

6.13 REVERSED HEAT ENGINE

> **LO 6.10**
> Explain reversed heat engine and Carnot's theorem

Since all the processes of the Carnot cycle are reversible, it is possible to imagine that the processes are individually reversed and carried out in reverse order. When a reversible process is reversed, all the energy transfers associated with the process are reversed in direction, but remain the same in magnitude. The reversed Carnot cycle for a steady flow system is shown in Fig. 6.24. The reversible heat engine and the reversed Carnot heat engine are represented in block diagrams in Fig. 6.25. If E is a reversible heat engine (Fig. 6.25a), and if it is reversed (Fig. 6.25b), the quantities Q_1, Q_2 and W remain the same in magnitude, and only their directions are reversed. The reversed heat engine ∃ takes heat from a low temperature body, discharges heat to a high temperature body, and receives an inward flow of network.

The names *heat pump* and *refrigerator* are applied to the reversed heat engine, which have already been discussed in Section 6.5, where the working fluid flows through the compressor (B), condenser (A), expander (D) and evaporator (C) to complete the cycle.

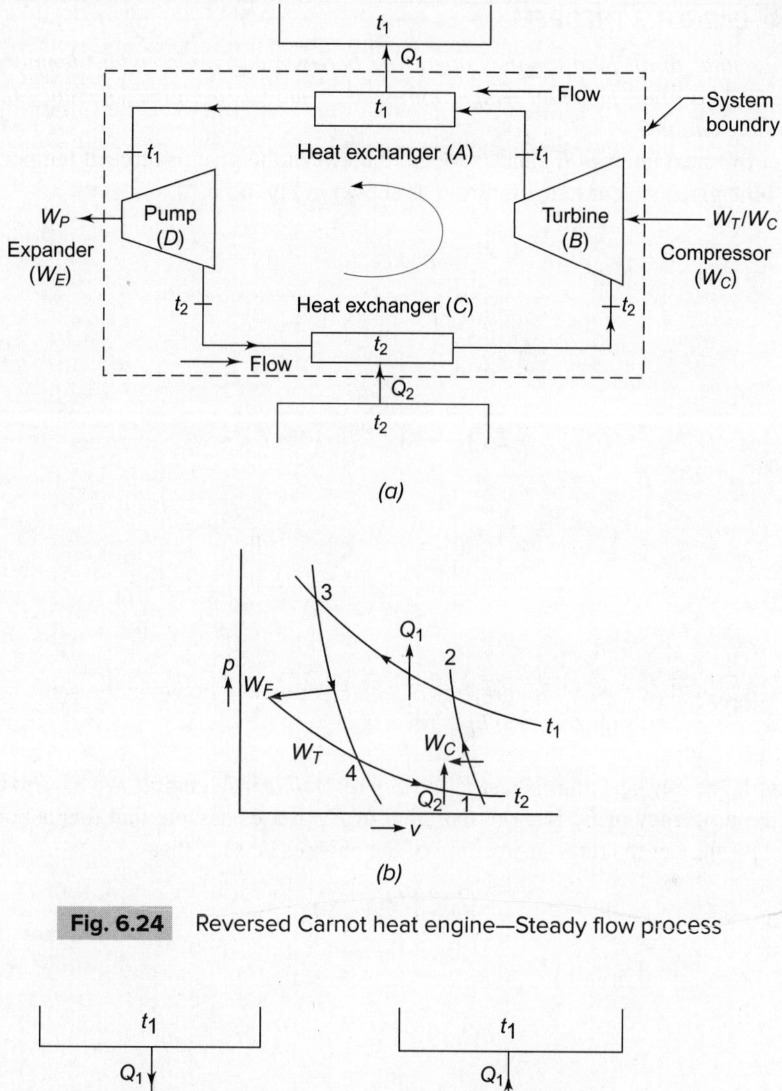

Fig. 6.24 Reversed Carnot heat engine—Steady flow process

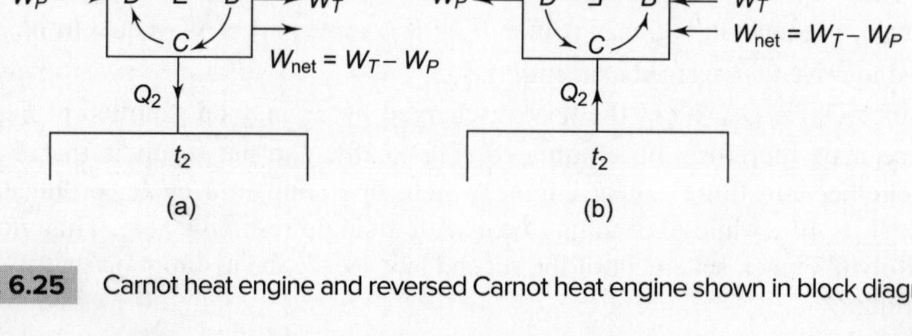

Fig. 6.25 Carnot heat engine and reversed Carnot heat engine shown in block diagrams

6.14 CARNOT'S THEOREM

It states that of all heat engines operating between a given constant temperature source and a given constant temperature sink, none has a higher efficiency than a reversible engine.

Let two heat engines E_A and E_B operate between the given source at temperature t_1 and the given sink at temperature t_2 as shown in Fig. 6.26.

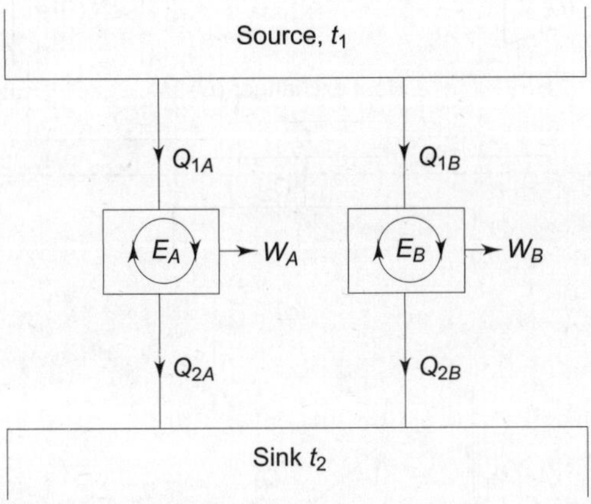

Fig. 6.26 Two cyclic heat engines E_A and E_B operating between the same source and sink, of which E_B is reversible

Let E_A be *any* heat engine and E_B be *any reversible* heat engine. We have to prove that the efficiency of E_B is more than that of E_A. Let us assume that this is not true and $\eta_A > \eta_B$. Let the rates of working of the engines be such that

$$Q_{1A} = Q_{1B} = Q_1$$

Since $$\eta_A > \eta_B$$

$$\frac{W_A}{Q_{1A}} > \frac{W_B}{Q_{1B}}$$

$\therefore$ $$W_A > W_B$$

Now, let E_B be reversed. Since E_B is a reversible heat engine, the magnitudes of heat and work transfer quantities will remain the same, but their directions will be reversed, as shown in Fig. 6.27. Since $W_A > W_B$, some part of W_A (equal to W_B) may be fed to drive the reversed heat engine $\exists_B$.

Since $Q_{1A} = Q_{1B} = Q_1$, the heat discharged by $\exists_B$ may be supplied to E_A. The source may, therefore, be eliminated (Fig. 6.28). The net result is that E_A and $\exists_B$ together constitute a heat engine which, operating in a cycle, produces net work $W_A - W_B$, while exchanging heat with a single reservoir at t_2. This violates the Kelvin–Planck statement of the second law. Hence the assumption that $\eta_A > \eta_B$ is wrong.

Therefore, $$\eta_B \geq \eta_A$$

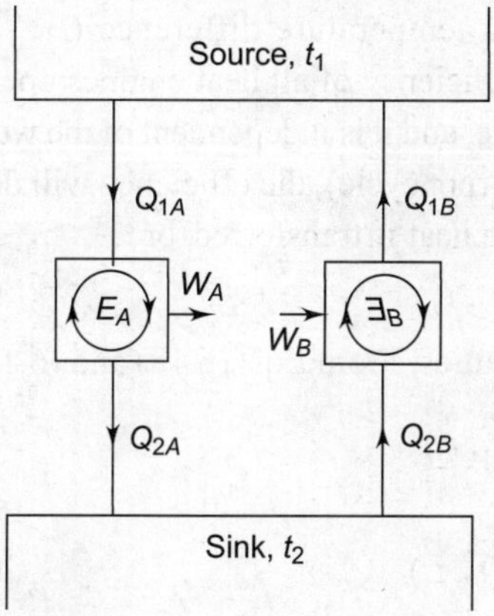

Fig. 6.27 E_B is reversed

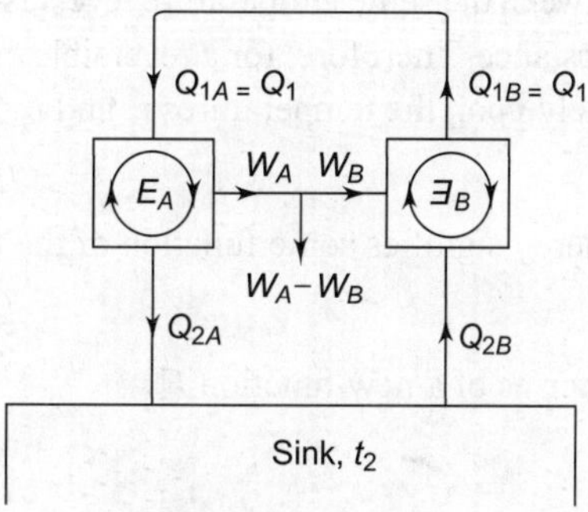

Fig. 6.28 E_A and $Ǝ_B$ together violate the K-P statement

6.15 COROLLARY OF CARNOT'S THEOREM

> **LO 6.11**
> Demonstrate absolute thermodynamic temperature scale and third law of thermodynamics

The efficiency of all reversible heat engines operating between the same temperature levels is the same.

Let both the heat engines E_A and E_B (Fig. 6.26) be reversible. Let us assume $\eta_A > \eta_B$. Similar to the procedure outlined in the preceding article, if E_B is reversed to run, say, as a heat pump using some part of the work output (W_A) of engine E_A, we see that the combined system of heat pump, E_B and engine E_A, becomes a PMM2. So η_A cannot be greater than η_B. Similarly, if we assume $\eta_B > \eta_A$ and reverse the engine E_A, we observe that η_B cannot be greater than η_A.

Therefore, $\qquad \eta_A = \eta_B$

Since the efficiencies of all reversible heat engines operating between the same heat reservoirs are the same, *the efficiency of a reversible engine is independent of the nature or amount of the working substance undergoing the cycle.*

6.16 THE THERMODYNAMIC TEMPERATURE SCALE

The efficiency of any heat engine cycle receiving heat Q_1 and rejecting heat Q_2 is given by

$$\eta = \frac{W_{\text{net}}}{Q_1} = \frac{Q_1 - Q_2}{Q_1} = 1 - \frac{Q_2}{Q_1} \tag{6.18}$$

By the second law, it is necessary to have a temperature difference $(t_1 - t_2)$ to obtain work for any cycle. We know that the efficiency of all heat engines operating between the same temperature levels is the same, and it is independent of the working substance. Therefore, for a reversible cycle (Carnot cycle), the efficiency will depend solely upon the temperatures t_1 and t_2, at which heat is transferred, or

$$\eta_{rev} = f(t_1, t_2) \tag{6.19}$$

where f signifies some function of the temperatures. From Eqs. (6.18) and (6.19)

$$1 - \frac{Q_2}{Q_1} = f(t_1, t_2)$$

In terms of a new function F

$$\frac{Q_1}{Q_2} = F(t_1, t_2) \tag{6.20}$$

If some functional relationship is assigned between t_1, t_2 and Q_1/Q_2, the equation becomes the definition of a temperature scale.

Let us consider two reversible heat engines, E_1 receiving heat from the source at t_1, and rejecting heat at t_2 to E_2 which, in turn, rejects heat to the sink at t_3 (Fig. 6.29a).

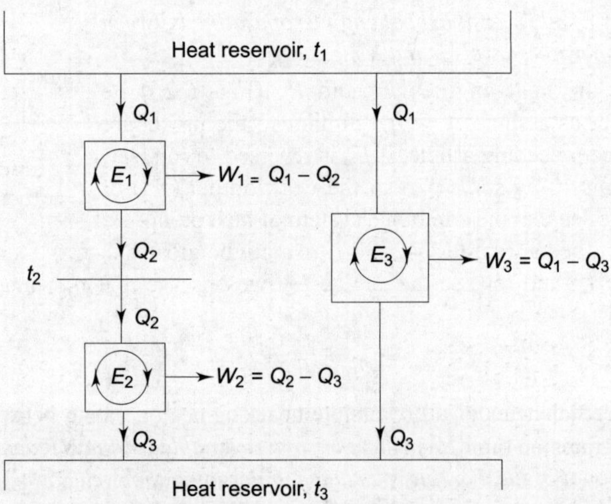

Fig. 6.29(a) Three Carnot engines

Now $$\frac{Q_1}{Q_2} = F(t_1, t_2); \quad \frac{Q_2}{Q_3} = F(t_2, t_3)$$

E_1 and E_2 together constitute another heat engine E_3 operating between t_1 and t_3.

$$\therefore \quad \frac{Q_1}{Q_3} = F(t_1, t_3)$$

Now $$\frac{Q_1}{Q_2} = \frac{Q_1/Q_3}{Q_2/Q_3}$$

or $$\frac{Q_1}{Q_2} = F(t_1, t_2) = \frac{F(t_1,t_3)}{F(t_2,t_3)} \tag{6.21}$$

The temperatures t_1, t_2 and t_3 are arbitrarily chosen. The ratio Q_1/Q_2 depends only on t_1 and t_2, and is independent of t_3. So t_3 will drop out from the ratio on the right in Eq. (6.21). After it has been cancelled, the numerator can be written as $\phi(t_1)$, and the denominator as $\phi(t_2)$, where ϕ is another unknown function. Thus,

$$\frac{Q_1}{Q_2} = F(t_1, t_2) = \frac{\phi(t_1)}{\phi(t_2)}$$

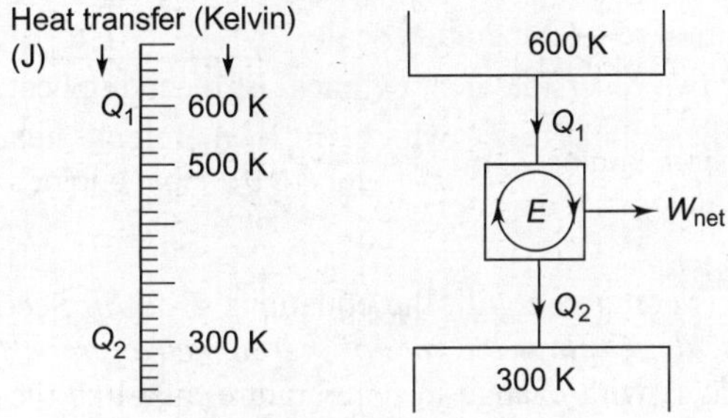

Fig. 6.29(b)

Since $\phi(t)$ is an arbitrary function, the simplest possible way to define the *absolute thermodynamic temperature* T is to let $\phi(t) = T$, as proposed by Kelvin. Then, by definition

$$\frac{Q_1}{Q_2} = \frac{T_1}{T_2} \tag{6.22}$$

The absolute thermodynamic temperature scale is also known as the *Kelvin scale*. Two temperatures on the Kelvin scale bear the same relationship to each other as do the heats absorbed and rejected respectively by a Carnot engine operating between two reservoirs at these temperatures Fig. (6.29 (b)). The Kelvin temperature scale is, therefore, independent of the peculiar characteristics of any particular substance.

The heat absorbed Q_1 and the heat rejected Q_2 during the two reversible isothermal processes bounded by two reversible adiabatics in a Carnot engine can be measured. In defining the Kelvin temperature scale also, the triple point of water is taken as the standard reference point. For a Carnot engine operating between reservoirs at temperatures T and T_t, T_t being the triple point of water (Fig. 6.30), arbitrarily assigned the value 273.16 K,

$$\frac{Q}{Q_t} = \frac{T}{T_t}$$

$$\therefore \qquad T = 273.16 \frac{Q}{Q_t} \tag{6.23}$$

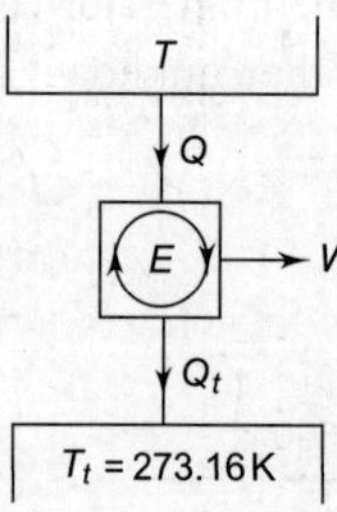

Fig. 6.30 Carnot heat engine with sink at triple point of water

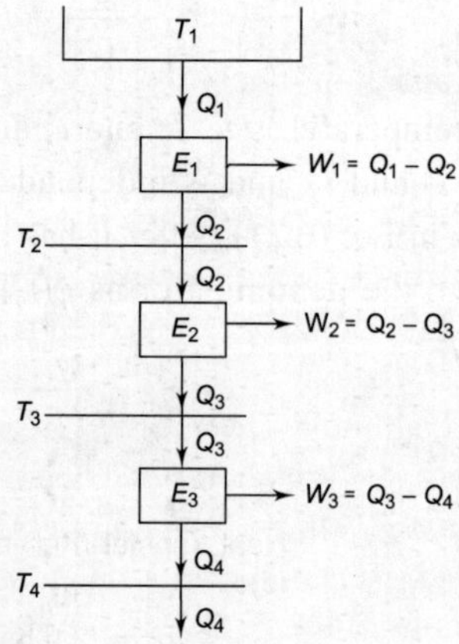

Fig. 6.31 Heat engines operating in series

If this equation is compared with the equations given in Section 2.3, it is seen that *in the Kelvin scale, Q plays the role of thermometric property*. The amount of heat supply Q changes with change in temperature, just like the thermal emf in a thermocouple.

That the absolute thermodynamic temperature scale has a definite zero point can be shown by imagining a series of reversible engines, extending from a source at T_1 to lower temperatures (Fig. 6.31).

Since

$$\frac{T_1}{T_2} = \frac{Q_1}{Q_2}$$

$$\therefore \qquad \frac{T_1 - T_2}{T_2} = \frac{Q_1 - Q_2}{Q_2}$$

or

$$T_1 - T_2 = (Q_1 - Q_2)\frac{T_2}{Q_2}$$

Similarly,

$$T_2 - T_3 = (Q_2 - Q_3)\frac{T_3}{Q_3}$$

$$= (Q_2 - Q_3)\frac{T_2}{Q_2}$$

$$T_3 - T_4 = (Q_3 - Q_4)\frac{T_2}{Q_2}$$

and so on.

If $T_1 - T_2 = T_2 - T_3 = T_3 - T_4 = ...$, assuming equal temperature intervals

$$Q_1 - Q_2 = Q_2 - Q_3 = Q_3 - Q_4 = ...$$

or $$W_1 = W_2 = W_3 = ...$$

Conversely, by making the work quantities performed by the engines in series equal ($W_1 = W_2 = W_3 = ...$), we will get

$$T_1 - T_2 = T_2 - T_3 = T_3 - T_4 = ...$$

at equal temperature intervals. A scale having one hundred equal intervals between the steam point and the ice point could be realised by a series of one hundred Carnot engines operating as in Fig. 6.31. Such a scale would be independent of the working substance.

If enough engines are placed in series to make the total work output equal to Q1, then by the first law the heat rejected from the last engine will be zero. By the second law, however, the operation of a cyclic heat engine with zero heat rejection cannot be achieved, although it may be approached as a limit. When the heat rejected approaches zero, the temperature of heat rejection also approaches zero as a limit. Thus it appears that a definite zero point exists on the absolute temperature scale but this point cannot be reached without a violation of the second law.

It follows from Eq. (6.23),

$$T = 273.16 \frac{Q}{Q_t}$$

that the heat transferred isothermally between two given adiabatics decreases as the temperature decreases. Conversely, the smaller the value of Q, the lower the temperature T. The smallest possible value of Q is zero, and the corresponding T is zero. *Thus if a system undergoes a reversible isothermal process without transfer of heat, the temperature at which this process takes place is called absolute zero.* In other words, at absolute zero, an isotherm and an adiabatic are identical.

Also, since $Q_2 = 0$, the isothermal process at $T_2 = 0$ would also be adiabatic rendering Carnot cycle ambiguous.

Thus any attainable value of absolute temperature is always greater than zero. This is also known as the *third law of thermodynamics* which may be stated as follows: *It is impossible by any procedure, no matter how idealised, to reduce any system to the absolute zero of temperature in a finite number of operations.*

This is what is called the Fowler–Guggenheim statement of the third law. The third law itself is an independent law of nature, and not an extension of the second law. The concept of heat engine is not necessary to prove the non-attainability of absolute zero of temperature by any system in a finite number of operations.

6.17 EFFICIENCY OF THE REVERSIBLE HEAT ENGINE

> **LO 6.12**
> Demonstrate efficiency of reversible devices

The efficiency of a reversible heat engine in which heat is received solely at T_1 is found to be

$$\eta_{rev} = \eta_{max} = 1 - \left(\frac{Q_2}{Q_1}\right)_{rev} = 1 - \frac{T_2}{T_1}$$

or $$\eta_{rev} = \frac{T_1 - T_2}{T_1} \tag{6.24}$$

It is observed here that as T_2 decreases, and T_1 increases, the efficiency of the reversible cycle increases.

Since η is always less than unity, T_2 is always greater than zero and positive.

The COP of a refrigerator is given by

$$(\text{COP})_{\text{refr}} = \frac{Q_2}{Q_1 - Q_2} = \frac{1}{\dfrac{Q_1}{Q_2} - 1}$$

For a reversible refrigerator, using

$$\frac{Q_1}{Q_2} = \frac{T_1}{T_2}$$

$$[\text{COP}_{\text{refr}}]_{\text{rev}} = \frac{T_2}{T_1 - T_2} \tag{6.25}$$

Similarly, for a reversible heat pump

$$[\text{COP}_{H.P.}]_{\text{rev}} = \frac{T_1}{T_1 - T_2} \tag{6.26}$$

6.18 EQUALITY OF IDEAL GAS TEMPERATURE AND KELVIN TEMPERATURE

LO 6.13 Relate ideal gas temperature and Kelvin temperature

Let us consider a Carnot cycle executed by an ideal gas, as shown in Fig. 6.32.

The two isothermal processes *a–b,* and *c–d* at θ_1 and θ_2 are represented by equilateral hyperbolas whose equations are, respectively

$$pV = nR\,\theta_1$$

and

$$pV = nR\,\theta_2$$

For any infinitesimal reversible process of an ideal gas, the first law may be written as

$$đQ = C_v\, d\theta + pdV = pdV$$

Applying this equation to the isothermal process *a–b*, the heat absorbed is found to be

$$Q_1 = \int_{V_a}^{V_b} pdV = \int_{V_a}^{V_b} \frac{nR\theta_1}{V}\, dV = nR\theta_1 \ln \frac{V_b}{V_a}$$

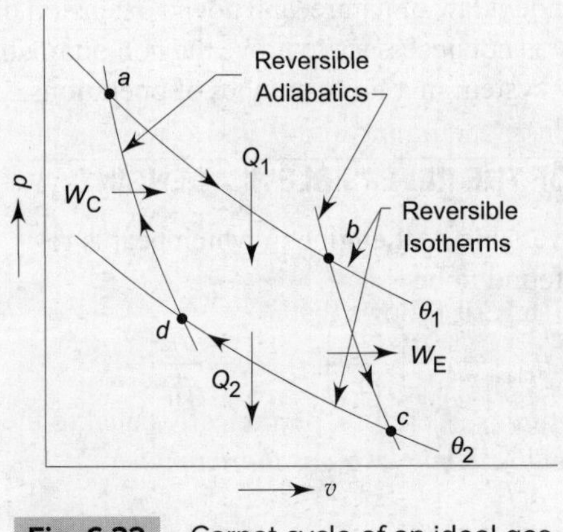

Fig. 6.32 Carnot cycle of an ideal gas

Similarly, for the isothermal process *c*–*d*, the heat rejected is

$$Q_2 = nR\theta_2 \ln \frac{V_c}{V_d}$$

$$\therefore \qquad \frac{Q_1}{Q_2} = \frac{\theta_1 \ln \dfrac{V_b}{V_a}}{\theta_2 \ln \dfrac{V_c}{V_d}} \tag{6.27}$$

Since the process *b*–*c* is adiabatic, the first law gives

$$-C_v\, d\theta = p dV = \frac{nR\theta}{V}\, dV$$

$$\frac{1}{nR}\int_{\theta_2}^{\theta_1} C_v \frac{d\theta}{\theta} = \ln \frac{V_c}{V_b}$$

Similarly, for the adiabatic process *d*–*a*

$$\frac{1}{nR}\int_{\theta_2}^{\theta_1} C_v \frac{d\theta}{\theta} = \ln \frac{V_d}{V_a}$$

$$\therefore \qquad \ln \frac{V_c}{V_b} = \ln \frac{V_d}{V_a}$$

or

$$\frac{V_c}{V_b} = \frac{V_d}{V_a}$$

or

$$\frac{V_b}{V_a} = \frac{V_c}{V_d} \tag{6.28}$$

Equation (6.27) thus reduces to

$$\frac{Q_1}{Q_2} = \frac{\theta_1}{\theta_2} \tag{6.29}$$

Kelvin temperature was defined by Eq. (6.22)

$$\frac{Q_1}{Q_2} = \frac{T_1}{T_2}$$

If θ and T refer to any temperature, and θ_t and T_t refer to the triple point of water,

$$\frac{\theta}{\theta_t} = \frac{T}{T_t}$$

Since $\theta_T = T_t = 273.16$ K, it follows that

$$\theta = T \tag{6.30}$$

The Kelvin temperature is, therefore, numerically equal to the ideal gas temperature and *may be measured by means of a gas thermometer.*

SOLVED EXAMPLES

Example 6.1 *A cyclic heat engine operates between a source temperature of 800°C and a sink temperature of 30°C. What is the least rate of heat rejection per kW net output of the engine?* E [LO 6.13]

Solution For a reversible engine, the rate of heat rejection will be minimum (Fig. 6.33).

$$\eta_{max} = \eta_{rev} = 1 - \frac{T_2}{T_1}$$

$$= 1 - \frac{30 + 273}{800 + 273}$$

$$= 1 - 0.282 = 0.718$$

Now $\frac{W_{net}}{Q_1} = \eta_{max} = 0.718$

$\therefore \quad Q_1 = \frac{1}{0.718} = 1.392 \text{ kW}$

Now $Q_2 = Q_1 - W_{net} = 1.392 - 1 = \mathbf{0.392 \ kW}$

This is the least rate of heat rejection.

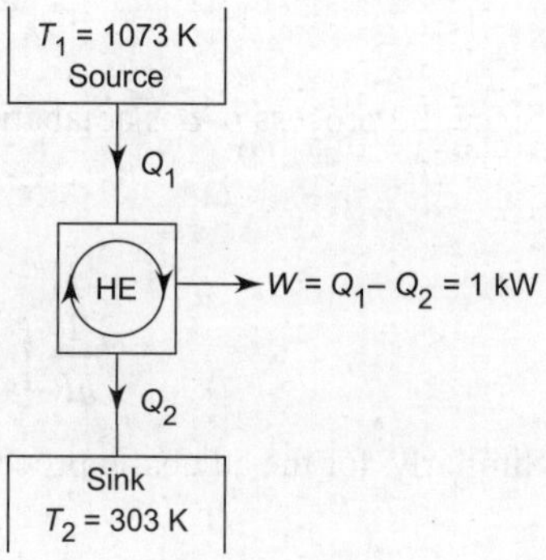

Fig. 6.33

Example 6.2 *A Carnot engine absorbs 200 J of heat from a reservoir at the temperature of the normal boiling point of water and rejects heat to a reservoir at the temperature of the triple point of water. Find the heat rejected, the work done by the engine and the thermal efficiency.* E [LO 6.12]

Solution

$$Q_1 = 200 \text{ J at } T_1 = 373.15 \text{ K}$$

$$T_2 = 273.16 \text{ K}$$

$$Q_2 = Q_1 \frac{T_2}{T_1} = 200 \times \frac{273.16}{373.15} = \mathbf{146.4 \ J}$$

$$W = Q_1 - Q_2 = \mathbf{53.6 \ J}$$

$$\eta = \frac{W}{Q_1} = \frac{53.6}{200} = \mathbf{0.268}$$

Example 6.3 *An inventor comes to an industrialist and claims to have developed a device that operating in a thermodynamic cycle receives 800 kJ of heat from a source at 450 K and produces 280 kJ of net work while rejecting the waste heat to the atmosphere at 300 K. What would you advise to the industrialist, he should invest or not?* M [LO 6.12]

Solution Thermal efficiency of the device is

$$\eta = \frac{W}{Q_1} = \frac{280}{800} = 0.35 \text{ or } 35\%$$

For a reversible heat engine operating between the same two reservoirs, the efficiency will be maximum and the maximum value can be obtained as

$$\eta_{\text{max}} = 1 - \frac{T_2}{T_1} = 1 - \frac{300}{450} = 0.33 \text{ or } 33\%$$

Since the thermal efficiency of the device is higher than the maximum possible efficiency, the inventor's claim is not reasonable.

Example 6.4 *In a power plant steam, steam is continuously extracted from the turbine and in a heat exchanger there is a condensation of the extracted steam. The energy released during condensation of steam is used to operate a heat engine. A heat engine receives thermal energy at a rate of 1250 kJ/min from condensing steam and rejects waste heat to a lake at 20°C at a rate of 800 kJ/min. Determine the lowest possible condensing steam temperature.* E [LO 6.12]

Solution Thermal efficiency of the heat engine is

$$\eta = 1 - \frac{Q_2}{Q_1} = 1 - \frac{800}{1250} = 0.36$$

The thermal efficiency must be less than or equal to the thermal efficiency of a reversible heat engine operating between the same two reservoirs. Therefore,

$$\eta \le \eta_{rev}$$

or $$0.36 \le 1 - \frac{T_2}{T_1}$$

where T_1 is the condensing steam temperature and T_2 is the temperature of lake.

∴ $$0.36 \le 1 - \frac{293}{T_1}$$

or $$\frac{293}{T_1} \le 0.64$$

or $$T_1 \ge 457.8 \text{ K} = \mathbf{184.8°C}$$

The lowest possible steam temperature is 184.8°C

Example 6.5 *Which is the more effective way to increase the efficiency of a Carnot engine: to increase T_1, keeping T_2 constant; or to decrease T_2, keeping T_1 constant?* M [LO 6.12]

Solution The efficiency of Carnot engine is given by $\eta = 1 - \frac{T_2}{T_1}$

Let T_2 be decreased by ΔT with T_1 remaining the same

$$\eta_1 = 1 - \frac{T_2 - \Delta T}{T_1}$$

If T_1 is increased by the same ΔT, T_2 remaining the same

$$\eta_2 = 1 - \frac{T_2}{T_1 + \Delta T}$$

Then

$$\eta_1 - \eta_2 = \frac{T_2}{T_1 + \Delta T} - \frac{T_2 - \Delta T}{T_1}$$

$$= \frac{(T_1 - T_2)\Delta T + (\Delta T)^2}{T_1(T_1 + \Delta T)}$$

Since $T_1 > T_2, (\eta_1 - \eta_2) > 0$

The more effective way to increase the cycle efficiency is to decrease T_2.
Alternatively, the efficiency of a Carnot engine is given by

$$\eta = 1 - \frac{T_2}{T_1}$$

If T_2 is constant

$$\left(\frac{\partial \eta}{\partial T_1}\right)_{T_2} = \frac{T_2}{T_1^2}$$

As T_1 increases, η increases, and the slope $\left(\frac{\partial \eta}{\partial T_1}\right)_{T_2}$ decreases (Fig. 6.34). If T_1 is constant,

$$\left(\frac{\partial \eta}{\partial T_2}\right)_{T_1} = -\frac{1}{T_1}$$

As T_2 decreases, η increases, but the slope $\left(\frac{\partial \eta}{\partial T_2}\right)_{T_1}$ remains constant (Fig. 6.35).

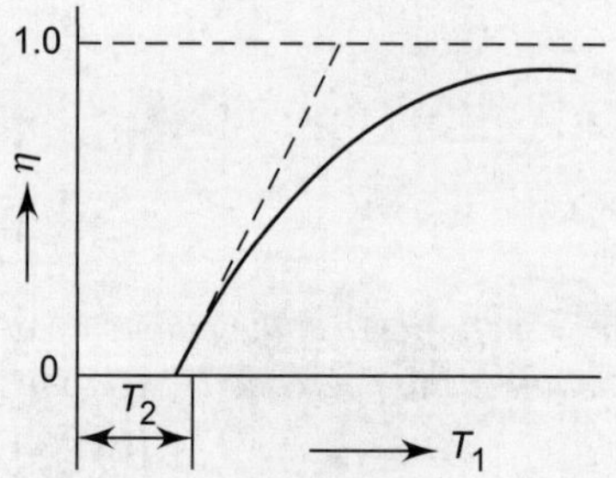

Fig. 6.34

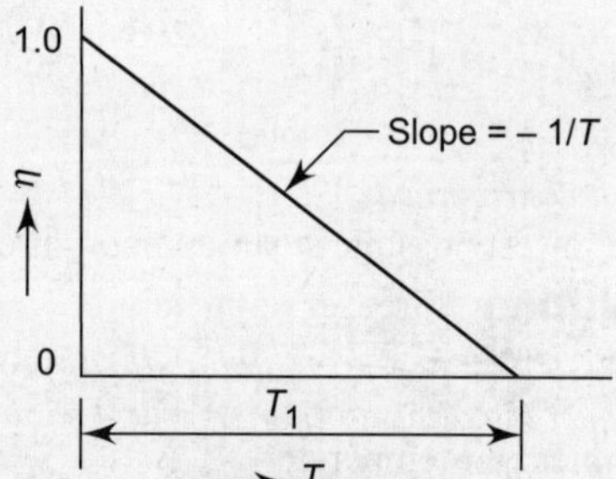

Fig. 6.35

Also $\left(\dfrac{\partial \eta}{\partial T_1}\right)_{T_1} = \dfrac{T_2}{T_1^2}$ and $\left(\dfrac{\partial \eta}{\partial T_2}\right)_{T_1} = -\dfrac{T_1}{T_1^2}$

Since $T_1 > T_2$, $\left(\dfrac{\partial \eta}{\partial T_2}\right)_{T_1} > \left(\dfrac{\partial \eta}{\partial T_1}\right)_{T_2}$

So, the more effective way to increase the efficiency is to decrease T_2.

Example 6.6 *It is proposed that solar energy be used to warm a large collector plate. This energy would, in turn, be transferred as heat to a fluid within a heat engine, and the engine would reject energy as heat to the atmosphere. Experiments indicate that about 1880 kJ/m² h of energy can be collected when the plate is operating at 90°C. Estimate the minimum collector area that would be required for a plant producing 1 kW of useful shaft power. The atmospheric temperature may be assumed to be 20°C.* E [LO 6.12]

Solution The maximum efficiency for the heat engine operating between the collector plate temperature and the atmospheric temperature (Fig. 6.40) as follows:

$$\eta_{max} = 1 - \frac{T_2}{T_1} = 1 - \frac{293}{363} = 0.192$$

The efficiency of any actual heat engine operating between these temperatures would be less than this efficiency.

$$\therefore \quad Q_{min} = \frac{W}{\eta_{max}} = \frac{1 \text{ kJ/s}}{0.192} = 5.21 \text{ kJ/s}$$

$$= 18{,}800 \text{ kJ/h}$$

∴ Minimum area required for the collector plate

$$= \frac{18{,}800}{1880} = \mathbf{10\ m^2}$$

Example 6.7 *A domestic food freezer maintains a temperature of – 15°C. The ambient air temperature is 30°C. If heat leaks into the freezer at the continuous rate of 1.75 kJ/s what is the least power necessary to pump this heat out continuously?* E [LO 6.12]

Solution Freezer temperature,

$$T_2 = -15 + 273 = 258 \text{ K}$$

Ambient air temperature,

$$T_1 = 30 + 273 = 303 \text{ K}$$

The refrigerator cycle removes heat from the freezer at the same rate at which heat leaks into it (Fig. 6.36). For minimum power requirement

$$\frac{Q_2}{T_2} = \frac{Q_1}{T_1}$$

$$\therefore \quad Q_1 = \frac{1.75}{2.8} \times 303 = 2.06 \text{ kJ/s}$$

$$\therefore \quad W = Q_1 - Q_2$$

$$= 2.06 - 1.75 = 0.31 \text{ kJ/s} = \mathbf{0.31 \text{ kW}}$$

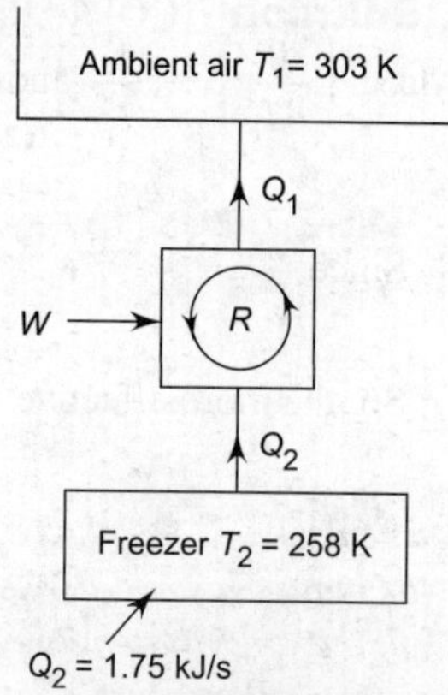

Fig. 6.36

Example 6.8 *A student was asked to do an experiment on a refrigerator test kit during an undergraduate laboratory class in an engineering college. He conducted the experiment for a period of 30 minute and measured that the refrigerated space is maintained at –5°C, while the temperature of the laboratory is 20°C. According to his experiment, it has found that during the experiment the refrigerator draws 1 kWh of electrical energy and 40000 kJ of heat is removed from the refrigerated space. Do you* think there is an error in his experiment?

E **[LO 6.12]**

Solution The work consumed by the refrigerator during the experiment is

$$W = 1 \text{ kWh} = 1 \text{ kJ/s} \times 3600 \text{ s} = 3600 \text{ kJ}$$

COP of the refrigerator is

$$[\text{COP}]_{\text{ref}} = \frac{Q_2}{W} = \frac{40000}{3600} = 11.11$$

The highest COP of the refrigerator is same as the COP of a reversible refrigerator operating between the same two temperatures and is found to be

$$[\text{COP}]_{\text{ref, rev}} = \frac{T_2}{T_1 - T_2} = \frac{268}{293 - 268} = 10.72$$

Since the actual COP of the refrigerator measured by the student is more than the maximum possible value, there is an error in the experiment.

Example 6.9 *A heat pump can use a maximum of 2 kW power for maintaining a house at 20°C during winter. The rate of heat transfer from the roof and walls is 3000 kJ/h per degree temperature difference between inside and outside. The coefficient of performance is 50% of a reversible heat pump. Determine the minimum outside temperature for which the heat pump can meet the heating requirements of the house.*

M **[LO 6.12]**

Solution COP of the heat pump is

$$[\text{COP}]_{HP} = \frac{\dot{Q}_1}{\dot{W}}$$

Where $\dot{W} = 2$ kW and

$$\dot{Q}_1 = 3000(T_1 - T_{amb}) \text{ kJ/h} = \frac{3000}{3600}(T_1 - T_{amb}) \text{ kJ/s} = \frac{5}{6}(T_1 - T_{amb}) \text{ kW}$$

Thus, $$[\text{COP}]_{HP} = \frac{\frac{5}{6}(T_1 - T_{amb})}{2} = \frac{5}{12}(T_1 - T_{amb})$$

COP of a reversible heat pump is

$$[\text{COP}]_{HP,rev} = \frac{T_1}{T_1 - T_{amb}}$$

From the given condition, we can write

$$[\text{COP}]_{HP} = 0.5[\text{COP}]_{HP,rev}$$

∴ $$\frac{5}{12}(T_1 - T_{amb}) = 0.5\frac{T_1}{T_1 - T_{amb}}$$

or $$(T_1 - T_{amb})^2 = 1.2T_1 = 1.2 \times 293 = 351.6$$

or $$293 - T_{amb} = 18.75$$

or $$T_{amb} = 274.25 \text{ K} = \mathbf{1.25°C}$$

Example 6.10 *An ideal gas cycle is represented by a rectangle on a p-V diagram. If p_1 and p_2 are the lower and higher pressures; and V_1 and V_2, the smaller and larger volumes, respectively, then (a) calculate the work done per cycle, (b) indicate which parts of the cycle involve heat flow into the gas. (c) Show that*

$$\eta = \frac{\gamma - 1}{\frac{\gamma p_2}{p_2 - p_1} + \frac{V_1}{V_2 - V_1}} \text{ if heat capacities are constant.}$$

M [LO 6.9]

Solution

(a) W = area of the cycle (Fig. 6.37)

$$= \mathbf{(p_2 - p_1)(V_2 - V_1)}$$

(b) Processes *ab* and *bc*

Heat absorbed by 1 mole of gas in one cycle,

$$Q = Q_{ab} + Q_{bc} = C_v(T_b - T_a) + C_p(T_c - T_b)$$

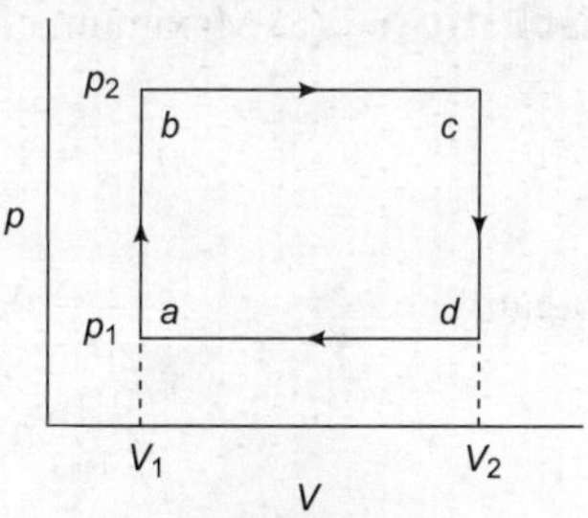

Fig. 6.37

$$\text{Now, } T_a = T_b \frac{p_1}{p_2} \text{ and } p_2 V_1 = R T_b$$

$$T_c = T_b \frac{V_2}{V_1}, \; T_b = \frac{p_2 V_1}{R}$$

$$Q = C_v T_b \left(1 - \frac{p_1}{p_2}\right) + C_p T_b \left(\frac{V_2}{V1} - 1\right)$$

$$= \frac{\mathbf{p_2 V_1}}{\mathbf{R}} \left(\mathbf{C_p} \frac{\mathbf{p_2 - p_1}}{\mathbf{p_2}} + \mathbf{C_p} \frac{\mathbf{V_2 - V_1}}{\mathbf{V_1}} \right)$$

$$\text{(c) } Q = \frac{W}{Q} = \frac{(p_2 - p_1)(V_2 - V_1)}{\dfrac{p_2 V_1}{R} \left[C_p \dfrac{p_2 - p_1}{p_2} + C_v \dfrac{V_2 - V_1}{V_1} \right]}$$

$$= \frac{R(p_2 - p_1)(V_2 - V_1)}{C_v V_1 (p_2 - p_1) + C_p p_2 (V_2 - V_1)}$$

$$= \frac{C_p - C_v}{C_v \dfrac{V_1}{V_2 - V_1} + C_p \dfrac{p_2}{p_2 - p_1}} = \frac{\boldsymbol{\gamma - 1}}{\dfrac{\mathbf{V_1}}{\mathbf{V_2 - V_1}} + \dfrac{\boldsymbol{\gamma} \mathbf{p_2}}{\mathbf{p_2 - p_1}}}$$

Example 6.11 *A reversible heat engine operates between two reservoirs at temperatures of 600 and 40°C. The engine drives a reversible refrigerator which operates between reservoirs at temperatures of 40°C and – 20°C. The heat transfer to the heat engine is 2000 kJ and the net work output of the combined engine refrigerator plant is 360 kJ.*

(a) Evaluate the heat transfer to the refrigerant and the net heat transfer to the reservoir at 40°C.

(b) Reconsider (a) given that the efficiency of the heat engine and the COP of the refrigerator are each 40% of their maximum possible values.

M **[LO 6.12]**

Solution (a) Maximum efficiency of the heat engine cycle (Fig. 6.38) is given by

$$\eta_{\max} = 1 - \frac{T_2}{T_1} = 1 - \frac{313}{873} = 1 - 0.358 = 0.642$$

Again $\dfrac{W_1}{Q_1} = 0.642$

$\therefore$ $W_1 = 0.642 \times 2000 = 1284 \text{ kJ}$

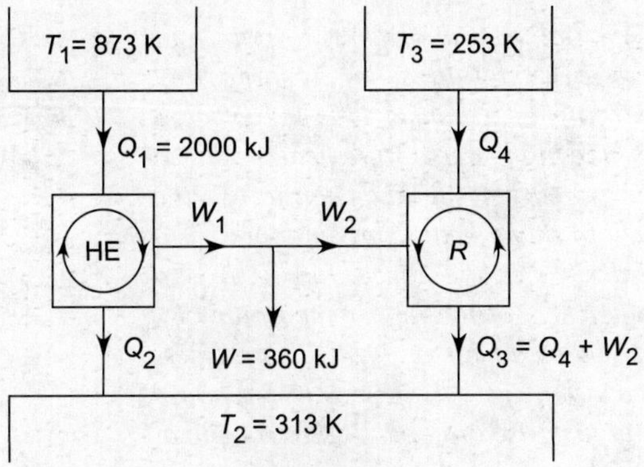

Fig. 6.38

Maximum COP of the refrigerator cycle

$$(\text{COP})_{\max} = \frac{T_3}{T_2 - T_3} = \frac{253}{313 - 253} = 4.22$$

Also $\quad \text{COP} = \dfrac{Q_4}{W_2} = 4.22$

Since $\quad W_1 - W_2 = W = 360 \text{ kJ}$

$\therefore \quad W_2 = W_1 - W = 1284 - 360 = 924 \text{ kJ}$

$\therefore \quad Q_4 = 4.22 \times 924 = 3899 \text{ kJ}$

$\therefore \quad Q_3 = Q_4 + W_2 = 924 + 3899 = 4823 \text{ kJ}$

$\quad Q_2 = Q_1 - W_1 = 2000 - 1284 = 716 \text{ kJ}$

Heat rejection to the 40°C reservoir

$$= Q_2 + Q_3 = 716 + 4823 = \mathbf{5539 \text{ kJ}}$$

(b) Efficiency of the actual heat engine cycle

$$\eta = 0.4\,\eta_{\max} = 0.4 \times 0.642$$

$\therefore \quad W_1 = 0.4 \times 0.642 \times 2000$

$\quad = 513.6 \text{ kJ}$

$\therefore \quad W_2 = 513.6 - 360 = 153.6 \text{ kJ}$

COP of the actual refrigerator cycle

$$\text{COP} = \frac{Q_4}{W_2} = 0.4 \times 4.22 = 1.69$$

Therefore,

$Q_4 = 153.6 \times 1.69 = \mathbf{259.6 \text{ kJ}}$

$Q_3 = 259.6 + 153.6 = \mathbf{413.2 \text{ kJ}}$

$Q_2 = Q_1 - W_1 = 2000 - 513.6 = \mathbf{1486.4 \text{ kJ}}$

Heat rejected to the 40°C reservoir

$$= Q_2 + Q_3 = 413.2 + 1486.4 = \mathbf{1899.6 \text{ kJ}}$$

Example 6.12 *Kelvin was the first to point out the thermodynamic wastefulness of burning fuel for the direct heating of a house. It is much more economical to use the high temperature heat produced by combustion in a heat engine and then to use the work so developed to pump heat from outdoors up to the temperature desired in the house. In Fig. 6.39 a boiler furnishes heat Q_1 at the high temperature T_1. This heat is absorbed by a heat engine, which extracts work W and rejects the waste heat Q_2 into the house at T_2. Work W is in turn used to operate a mechanical refrigerator or heat pump, which extracts Q_3 from outdoors at temperature T_3 and reject Q_2' (where $Q_2' = Q_3 + W$) into the house. As a result of this cycle of operations, a total quantity of heat equal to $Q_2 + Q_2'$ is liberated in the house, against Q_1 which would be provided directly by the ordinary combustion of the fuel. Thus the ratio $(Q_2 + Q_2')/Q_1$ represents the heat multiplication factor of this method. Determine this multiplication factor if $T_1 = 473$ K, $T_2 = 293$ K, and $T_3 = 273$ K.* M [LO 6.12]

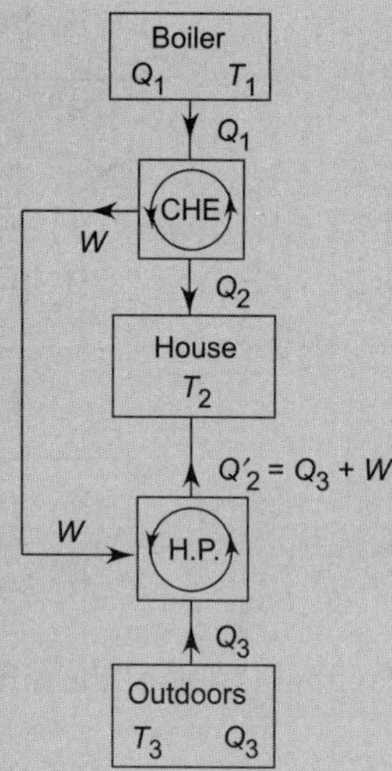

Fig. 6.39

Solution For the reversible heat engine (Fig. 6.39)

$$\frac{Q_2}{Q_1} = \frac{T_2}{T_1}$$

$\therefore$
$$Q_2 = Q_1\left(\frac{T_2}{T_1}\right)$$

Also
$$\eta = \frac{W}{Q_1} = \frac{T_1 - T_2}{T_1}$$

or
$$W = \frac{T_1 - T_2}{T_1} \cdot Q_1$$

For the reversible heat pump

$$\text{COP} = \frac{Q_2'}{W} = \frac{T_2}{T_2 - T_3}$$

$\therefore$
$$Q_2' = \frac{T_2}{T_2 - T_3} \cdot \frac{T_1 - T_2}{T_1} \cdot Q_1$$

$\therefore$ multiplication factor (M.F.)

$$= \frac{Q_2 + Q_2'}{Q_1} = \frac{Q_1 \dfrac{T_2}{T_1} + Q_1 \cdot \dfrac{T_2}{T_2 - T_3} \cdot \dfrac{T_1 - T_2}{T_1}}{Q_1}$$

or $$\text{M.F.} = \frac{T_2^2 - T_2T_3 + T_2T_1 - T_2^2}{T_1(T_2 - T_3)}$$

or $$\text{M.F.} = \frac{T_2(T_1 - T_3)}{T_1(T_2 - T_3)}$$

Here $T_1 = 473$ K, $T_2 = 293$ K and $T_3 = 273$ K

$\therefore$ $$\text{M.F.} = \frac{293(473-273)}{473(293-273)} = \frac{2930}{473} = \mathbf{6.3}$$

which means that every kg of coal burned would deliver the heat equivalent to over 6 kg. Of course, in an actual case, the efficiencies would be less than Carnot efficiencies, but even with a reduction of 50%, the possible savings would be quite significant.

Example 6.13 *A reversible heat engine in a satellite operates between a hot reservoir at T_1 and a radiating panel at T_2. Radiation from the panel is proportional to its area and to T_2^4. For a given work output and value of T_1 show that the area of the panel will be minimum when $\frac{T_2}{T_1} = 0.75$.*

Determine the minimum area of the panel for an output of 1 kW if the constant of proportionality is 5.67×10^{-8} W/m² K⁴ and T_1 is 1000 K. M [LO 6.12]

Solution For the heat engine (Fig. 6.40), the heat rejected Q_2 to the panel (at T_2) is equal to the energy emitted from the panel to the surroundings by radiation. If A is the area of the panel, $Q_2 \propto AT_2^4$, or $Q_2 = KAT_2^4$, where K is a constant.

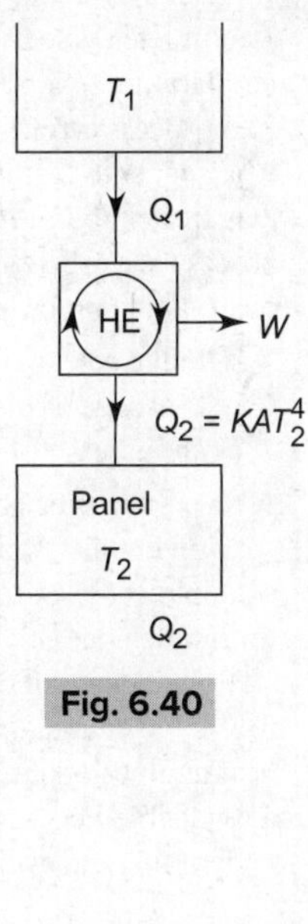

Fig. 6.40

Now, $$\eta = \frac{W}{Q_1} = \frac{T_1 - T_2}{T_1}$$

or $$\frac{W}{T_1 - T_2} = \frac{Q_1}{T_1} = \frac{Q_2}{T_2} = \frac{KAT_2^4}{T_2}$$

$$= KAT_2^3$$

$\therefore$ $$A = \frac{W}{KT_2^3(T_1 - T_2)} = \frac{W}{K(T_1T_2^3 - T_2^4)}$$

For a given W and T_1, A will be minimum when

$$\frac{dA}{dT_2} = -\frac{W}{K}(3T_1T_2^2 - 4T_2^3)\cdot(T_1T_2^3 - T_2^4)^{-2} = 0$$

Since $(T_1T_2^3 - T_2^4)^{-2} \neq 0$, $3T_1T_2^2 = 4T_2^3$

$\therefore$ $$\frac{T_2}{T_1} = 0.75 \text{ Proved.}$$

$$A_{\text{min}} = \frac{W}{K(0.75)^3T_1^3(T_1 - 0.75T_1)}$$

$$= \frac{W}{K\dfrac{27}{256}T_1^4} = \frac{256W}{27\,KT_1^4}$$

Here $W = 1$ kW, $K = 5.67 \times 10^{-8}$ W/m^2 K^4, and $T_1 = 1000$ K

$$\therefore \quad A_{\min} = \frac{256 \times 1\,\text{kW} \times \text{m}^2\text{K}^4}{27 \times 5.67 \times 10^{-8}\,\text{W} \times (1000)^4\,\text{K}^4}$$

$$= \frac{256 \times 10^3}{27 \times 5.67 \times 10^{-8} \times 10^{12}}\ \text{m}^2 = \mathbf{0.1672\ m^2}$$

SUMMARY

A heat engine cycle is a thermodynamic cycle in which there is a net flow of heat to the system and a net flow of work from the system. The system which executes a heat engine cycle is a heat engine. The efficiency of a heat engine, or of its cycle is

$$\eta = \frac{W_{\text{net}}}{Q_1}$$

where Q_1 is the heat transferred to the system in a cycle and W_{net} is the net work of the cycle. This efficiency is called thermal efficiency.

A heat pump or a refrigerator is a system to which there is a net flow of work and from which there is a net flow of heat in a cycle.

The second law is stated as follows (after Kelvin and Planck): It is impossible for a heat engine to produce net work in a complete cycle if it exchanges heat only with bodies at a single fixed temperature.

An equivalent statement (after Clausius) follows: It is impossible for a system working in a complete cycle to accomplish as its sole effect the transfer of heat from a body at a given temperature to a body at a higher temperature.

A process is reversible if, after the process has been carried out, it is possible by any means whatsoever to restore both the system and the entire surroundings to exactly the same states that were in before the process. A process that is not reversible is irreversible. A reversible process is a process that can be undone in such a way that no trace remains anywhere of the fact that the process had occurred. The causes of irreversibility of a process are:

1. Finite potential gradient causing the process like temperature, pressure, concentration, etc.
2. Presence of dissipative effects like friction in which macroscopic work dissipates into an increase of internal energy and then heat.

A reversible process would require infinite time.

A reversible cycle is a cycle composed entirely of reversible processes. The classical example is the Carnot cycle which consists of two reversible isothermal processes and two reversible adiabatic processes.

No heat engine operating between fixed temperature levels can be more efficient than a reversible engine operating between the same temperatures. The efficiency of all reversible heat engines operating between the same temperature levels is the same. The efficiency of a reversible heat engine is independent of the nature or amount of the working substance undergoing the cycle. The absolute thermodynamic scale is defined by the relation,

$$\frac{T_1}{T_2} = \frac{Q_1}{Q_2}$$

where Q_1 is the heat received from a source at T_1 and Q_2 is the heat rejected to a sink at T_2 by a reversible heat engine. The efficiency of a reversible heat engine receiving heat solely at T_1 and rejecting heat solely at T_2 is given by

$$\eta_{max} = \eta_{rev} = \frac{T_1 - T_2}{T_1} = 1 - \frac{T_2}{T_1}$$

The COPs of a Carnot refrigerator and a Carnot heat pump are

$$(COP_{max})_{Ref} = \frac{T_2}{T_1 - T_2}, \quad (COP_{max})_{HP} = \frac{T_1}{T_1 - T_2}$$

The absolute thermodynamic temperature scale or the Kelvin scale is equivalent to the ideal gas temperature scale, and the Kelvin temperature can be measured by a gas thermometer.

REVIEW QUESTIONS

(M) **6.1** What is the qualitative difference between heat and work? Why are heat and work not completely interchangeable forms of energy? **[LO 6.1]**

(E) **6.2** What is a cyclic heat engine? **[LO 6.3]**

(M) **6.3** Explain a heat engine cycle performed by a closed system. **[LO 6.3]**

(M) **6.4** Explain a heat engine cycle performed by a steady flow system. **[LO 6.3]**

(M) **6.5** Define the thermal efficiency of a heat engine cycle. Can this be 100%? **[LO 6.3]**

(E) **6.6** Draw a block diagram showing the four energy interactions of a cyclic heat engine. **[LO 6.3]**

(E) **6.7** What is a thermal energy reservoir? Explain the terms 'source' and 'sink'. **[LO 6.2]**

(E) **6.8** What is a mechanical energy reservoir? **[LO 6.2]**

(D) **6.9** Why can all processes in a TER or an MER be assumed to be quasi-static? **[LO 6.2]**

(E) **6.10** Give the Kelvin–Planck statement of the second law. **[LO 6.4]**

(M) **6.11** To produce net work in a thermodynamic cycle, a heat engine has to exchange heat with two thermal reservoirs. Explain. **[LO 6.4]**

(M) **6.12** What is a PMM2? Why is it impossible? **[LO 6.4]**

(E) **6.13** Give the Clausius' statement of the second law. **[LO 6.4]**

(M) **6.14** Explain the operation of a cyclic refrigerator plant with a block diagram. **[LO 6.5]**

(E) **6.15** Define the COP of a refrigerator. **[LO 6.5]**

(M) **6.16** What is a heat pump? How does it differ from a refrigerator? **[LO 6.5]**

(D) **6.17** Can you use the same plant as a heat pump in winter and as a refrigerator in summer? Explain. **[LO 6.5]**

(M) **6.18** Show that the COP of a heat pump is greater than the COP of a refrigerator by unity. **[LO 6.5]**

(M) **6.19** Why is direct heating thermodynamically wasteful? **[LO 6.5]**

(M) **6.20** How can a heat pump upgrade low grade waste heat? **[LO 6.5]**

(D) **6.21** Establish the equivalence of Kelvin–Planck and Clausius statements. **[LO 6.6]**

(D) **6.22** What is a reversible process? A reversible process should not leave any evidence to show that the process had ever occurred. Explain. **[LO 6.7]**

(D) **6.23** How is a reversible process only a limiting process, never to be attained in practice? **[LO 6.7]**

(D) **6.24** All spontaneous processes are irreversible. Explain. **[LO 6.8]**

(M) **6.25** What are the causes of irreversibility of a process? **[LO 6.7]**

(D) **6.26** Show that heat transfer through a finite temperature difference is irreversible. **[LO 6.7]**

(D) **6.27** Demonstrate, using the second law, that free expansion is irreversible. **[LO 6.7]**

(M) **6.28** What do you understand by dissipative effects? When is work said to be dissipated? **[LO 6.7]**

(E) **6.29** Explain perpetual motion of the third kind. **[LO 6.7]**

(D) **6.30** Demonstrate using the second law how friction makes a process irreversible. **[LO 6.7]**

(D) **6.31** When a rotating wheel is brought to rest by applying a brake, show that the molecular internal energy of the system (of the brake and the wheel) increases. **[LO 6.7]**

(M) **6.32** Show that the dissipation of stirring work to internal energy is irreversible. **[LO 6.7]**

(M) **6.33** Show by second law that the dissipation of electrical work into internal energy or heat is irreversible. **[LO 6.7]**

(M) **6.34** What is a Carnot cycle? What are the four processes which constitute the cycle? **[LO 6.9]**

(D) **6.35** Explain the Carnot heat engine cycle executed by (a) a stationary system, and (b) a steady flow system. **[LO 6.9]**

(M) **6.36** What is a reversed heat engine? **[LO 6.10]**

(D) **6.37** Show that the efficiency of a reversible engine operating between two given constant temperatures is the maximum. **[LO 6.10]**

(M) **6.38** Show that the efficiency of all reversible heat engines operating between the same temperature levels is the same. **[LO 6.11]**

(M) **6.39** Show that the efficiency of a reversible engine is independent of the nature or amount of the working substance going through the cycle. **[LO 6.11]**

(E) **6.40** How does the efficiency of a reversible cycle depend only on the two temperatures at which heat is transferred? **[LO 6.11]**

(M) **6.41** What is the absolute thermodynamic temperature scale? Why is it called absolute? **[LO 6.11]**

(M) **6.42** How is the absolute scale independent of the working substance? **[LO 6.11]**

(M) **6.43** How does Q play the role of thermometric property in the Kelvin scale? **[LO 6.11]**

(M) **6.44** Show that a definite zero point exists on the absolute temperature scale but that this point cannot be reached without a violation of the second law. **[LO 6.11]**

(E) **6.45** Give the Fowler–Guggenheim statement of the third law. **[LO 6.11]**

(D) **6.46** Is the third law an extension of the second law? Is it an independent law of nature? Explain. **[LO 6.11]**

(M) **6.47** How does the efficiency of a reversible engine vary as the source and sink temperatures are varied? When does the efficiency become 100%? **[LO 6.12]**

(E) **6.48** For a given T_2, show that the COP of a refrigerator increases as T_1 decreases. **[LO 6.12]**

(M) **6.49** Explain how the Kelvin temperature can be measured with a gas thermometer. **[LO 6.12]**

(D) **6.50** Establish the equality of ideal gas temperature and Kelvin temperature. **[LO 6.12]**

(M) **6.51** What do you understand by internal irreversibility and external irreversibility? **[LO 6.8]**

(M) **6.52** Explain mechanical, thermal and chemical irreversibilities. **[LO 6.8]**

(D) **6.53** A Carnot engine with a fuel burning device as source and a heat sink cannot be treated as a reversible plant. Explain. **[LO 6.8]**

PROBLEMS WITH HINTS

(E) **6.1** A reversible power cycle operates between a reservoir at temperature *T* and a lower temperature reservoir at 200 K. At steady state, the cycle develops 40 kW of power while rejecting 1000 kJ/min of energy by heat transfer to the cold reservoir. **[LO 6.11]**

Determine *T*.

Hint: $\dot{W}_{cycle} = \dot{Q}_1 - \dot{Q}_2$ where $\dot{W} = 40$ kW and $Q_2 = \frac{1000}{60}$ kW

Find $\dot{Q}_1$. For a reversible cycle, $\frac{\dot{Q}_1}{\dot{Q}_2} = \frac{T_1}{T_2}$ where

$$T_2 = 200 \text{ K, get } T_1 = T = \mathbf{951.9\ K}$$

(E) **6.2** At steady state, a refrigerator whose COP is 2.5 removes energy by heat transfer from a freezer cabinet at 0°C at the rate of 8000 kJ/h and discharges energy by heat transfer to the surroundings, which are at 20°C. Determine the power input to the refrigerator and compare with the power required by a reversible refrigerator cycle operating between reservoirs at these two reservoirs. **[LO 6.12]**

Hint: Find $\dot{W} = \frac{\dot{Q}_2}{\text{COP}}$ which is **3200 kJ/h**

$$\text{COP}_{rev} = \frac{T_2}{T_1 - T_2} \text{ would give 13.65.}$$

Find $\dot{W}_{min}$ = **586.1 kJ/h,** which is 5 ½ times less.

(E) **6.3** When the outside temperature is –10°C, a residential heat pump must provide 3.5×10^6 kJ per day to a dwelling to maintain its temperature at 20°C. (Fig. 6.41). If electricity cost ₹5.0 per kWh, determine the minimum theoretical operating cost for each day of operation. **[LO 6.12]**

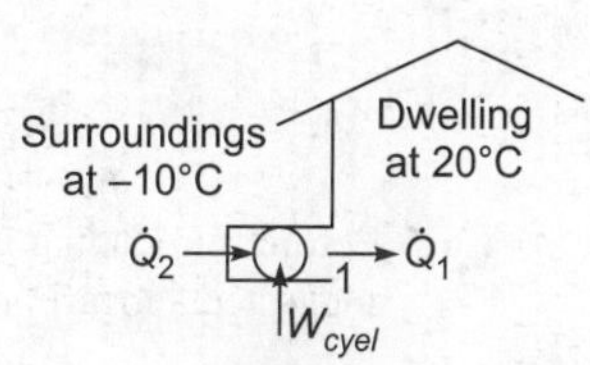

Fig. 6.41

Hint:

$$\text{COP}_{max} = \frac{T_1}{T_1 - T_2} = \frac{293}{293.263} = 9.77$$

$$\therefore \quad \dot{W}_{min} = \frac{\dot{Q}_1}{\text{COP}} = \frac{3.5 \times 10^6}{9.77} \frac{\text{kJ}}{\text{h}}$$

$$\therefore \text{ Operating cost per day} = \mathbf{\dot{W}_{min} \times 24\ h \times \frac{5}{3600}\ rupees.}$$

(E) **6.4** Air of mass 0.5 kg as an ideal gas executes a Carnot cycle having a thermal efficiency of 50%. The heat transfer to the air during the isothermal expansion is 40 kJ. At the beginning of isothermal expansion the pressure is 7 bar and the volume is 0.12 m^3. Determine (a) the maximum and minimum temperatures for the cycle, (b) the volume at the end of isothermal expansion, (c) the work and heat transfer for each of the four processes. **[LO 6.9, 6.12]**

Hint:

$$T_1 = \frac{P_1V_1}{RT_1} = \mathbf{585.4\ K.}$$

$$\frac{T_2}{T_1} = 1 - \eta = 0.5$$

$$\therefore\ T_2 = \mathbf{292.7\ K.}$$

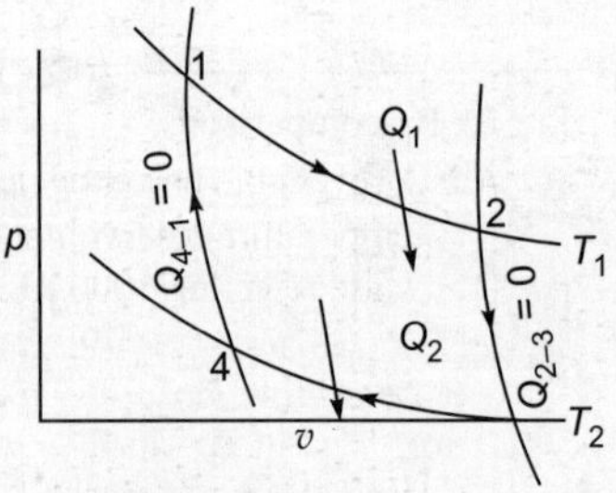

Fig. 6.42 Carnot cycle

Process 1–2

$$W_{1-2} = mRT_1 \ln \frac{V_2}{V_1}$$

$$\therefore\ \ln \frac{V_2}{V_1} = 0.4761$$

Find $V_2 = \mathbf{0.193\ m^3}$

Process 2–3

$$Q_{2-3} = 0,\ W_{23} = U_2 - U_3 = mc_u (T_2 - T_3)$$

$$= \mathbf{107.5\ kJ}$$

Process 3–4

$$\frac{Q_2}{Q_1} = \frac{T_2}{T_1} Q_2 = \frac{1}{2} Q_1 = 20 \text{ kJ}$$

$$= \mathbf{W_{3-4}}$$

Process 4–1

$$Q_{41} = 0,\ W_{41} = U_4 - U_1 = \mathbf{107.5\ kJ}$$

$$\eta_{\text{cycle}} = \frac{40 + 107.5 - 20 - 107.5}{40} = 0.5 \text{ (Proved).}$$

(M) **6.5** Before introducing the temperature scale, now known as the Kelvin scale, Kelvin suggested a logarithmic scale in which the function Q_1/Q_2 takes the form

$$\frac{Q_1}{Q_2} = \frac{\exp \theta_1}{\exp \theta_2}$$

where θ_1 and θ_2 denote the temperatures of the source and sink, respectively, on this scale.

(a) Show that the relation between the Kelvin temperature T and the temperature θ on the logarithmic scale is

$$\theta = \ln T + C.$$

where C is a constant.

(b) On the Kelvin scale, temperatures vary from O to $+\infty$. Determine the range of temperatures on the logarithmic scale.

(c) Obtain an expression for the thermal efficiency of any system undergoing a reversible power cycle operating between reservoirs at temperatures θ_1 and θ_2 on the logarithmic scale. **[LO 6.12]**

Hint: The two scales are

$$\frac{Q_2}{Q_1} = \frac{T_2}{T_1} \text{ Kelvin scale}$$

$$\frac{Q_2}{Q_1} = \frac{\exp \theta_2}{\exp \theta_1} \text{ logarithmic scale.}$$

By comparison,

$$\frac{T_2}{T_1} = \frac{\exp \theta_2}{\exp \theta_1} = \exp (\theta_2 - \theta_1)$$

$$\ln \frac{T_2}{T_1} = \theta_2 - \theta_1$$

or, $\ln T_2 - \ln T_1 = \theta_2 - \theta_1$

$\therefore$ $\theta = \ln T + C$ Proved.

(b) Temperature on the logarithmic scale varies from $-\infty$ to $+\infty$.

(c) $\dfrac{Q_2}{Q_1} = \dfrac{\exp \theta_2}{\exp \theta_1}$

$\therefore$ $\eta_{\text{cycle}} = 1 - \dfrac{\exp \theta_2}{\exp \theta_1}$

(E) **6.6** A heat engine (Fig. 6.43). Operating between two reservoirs at temperatures 600°C and 40°C drives a refrigerator operating between reservoirs at temperatures of 40°C and –15°C. The heat transfer to the heat engine is 2500 kJ and the net work output of the combined engine and refrigerator plant is 400 kJ. The efficiency of the heat engine and the COP of the refrigerator are each 40% of the maximum possible values. Estimate the heat transfer to the refrigerant and the net heat transfer to the reservoir at 40°C. **[LO 6.12]**

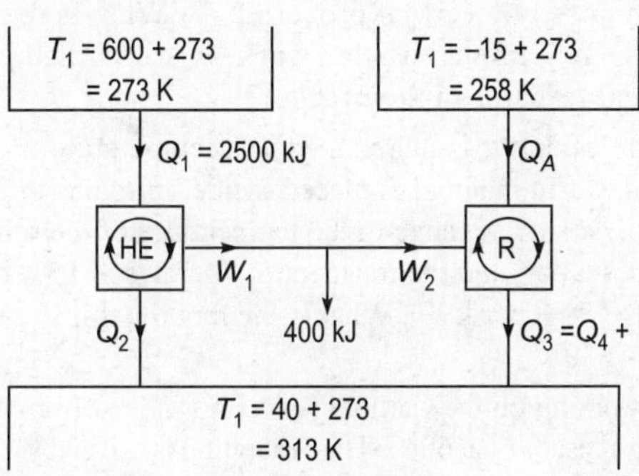

Fig. 6.43

Hint:

For the heat engine, $\eta_{\max} = 1 - \dfrac{T_2}{T_1} = 0.642$

$$\eta = 0.4\, \eta_{\max}, \; W_1 = \eta Q_1 = 642 \text{ kJ}$$

$$W_2 = 642 - 400 = 202 \text{ kJ}$$

$$\text{COP} = 0.4 \times \frac{258}{310 - 258} = 1.376 = \frac{Q_4}{W_2}$$

$$\therefore \quad Q_4 = 333 \text{ kJ}$$

$$\therefore \quad Q_3 = Q_4 + W_2, Q_2 = Q_1 - W_1$$

$$\therefore \quad Q_2 + Q_3 = \mathbf{2433 \ kJ}$$

PROBLEMS

(E) **6.1** An inventor claims to have developed an engine that takes in 105 MJ at a temperature of 400 K, rejects 42 MJ at a temperature of 200 K, and delivers 15 kWh of mechanical work. Would you advise investing money to put this engine in the market? **[LO 6.3]**

(E) **6.2** If a refrigerator is used for heating purposes in winter so that the atmosphere becomes the cold body and the room to be heated becomes the hot body, how much heat would be available for heating for each kW input to the driving motor? The COP of the refrigerator is 5, and the electromechanical efficiency of the motor is 90%. How does this compare with resistance heating? **[LO 6.5]**

(E) **6.3** Using an engine of 30% thermal efficiency to drive a refrigerator having a COP of 5, what is the heat input into the engine for each MJ removed from the cold body by the refrigerator?
If this system is used as a heat pump, how many MJ of heat would be available for heating for each MJ of heat input to the engine? **[LO 6.5]**

(M) **6.4** An electric storage battery which can exchange heat only with a constant temperature atmosphere goes through a complete cycle of two processes. In process 1–2, 2.8 kWh of electrical work flow into the battery while 732 kJ of heat flow out to the atmosphere. During process 2–1, 2.4 kWh of work flow out of the battery. (a) Find the heat transfer in process 2–1. (b) If the process 1–2 has occurred as above, does the first law or the second law limit the maximum possible work of process 2–1? What is the maximum possible work? (c) If the maximum possible work were obtained in process 2–1, what will be the heat transfer in the process? **[LO 6.7]**

(M) **6.5** A household refrigerator is maintained at a temperature of 2°C. Every time the door is opened, warm material is placed inside, introducing an average of 420 kJ, but making only a small change in the temperature of the refrigerator. The door is opened 20 times a day, and the refrigerator operates at 15% of the ideal COP. The cost of work is ₹2.50 per kWh. What is the monthly bill for this refrigerator? The atmosphere is at 30°C. **[LO 6.5]**

(M) **6.6** A heat pump working on the Carnot cycle takes in heat from a reservoir at 5°C and delivers heat to a reservoir at 60°C. The heat pump is driven by a reversible heat engine which takes in heat from a reservoir at 840°C and rejects heat to a reservoir at 60°C. The reversible heat engine also drives a machine that absorbs 30 kW. If the heat pump extracts 17 kJ/s from the 5°C reservoir, determine (a) the rate of heat supply from the 840°C source, and (b) the rate of heat rejection to the 60°C sink. **[LO 6.12]**

(E) **6.7** A refrigeration plant for a food store operates as a reversed Carnot heat engine cycle. The store is to be maintained at a temperature of – 5°C and the heat transfer from the store to the cycle is at the rate of 5 kW. If heat is transferred from the cycle to the atmosphere at a temperature of 25°C, calculate the power required to drive the plant. **[LO 6.12]**

(E) **6.8** A heat engine is used to drive a heat pump. The heat transfers from the heat engine and from the heat pump are used to heat the water circulating through the radiators of a building. The efficiency of the heat engine is 27% and the COP of the heat pump is 4. Evaluate the ratio of the heat transfer to the circulating water to the heat transfer to the heat engine. **[LO 6.3, 6.5]**

(E) **6.9** If 20 kJ are added to a Carnot cycle at a temperature of 100°C and 14.6 kJ are rejected at 0°C, determine the location of absolute zero on the Celsius scale. **[LO 6.9]**

(M) **6.10** Two reversible heat engines *A* and *B* are arranged in series, A rejecting heat directly to *B*. Engine *A* receives 200 kJ at a temperature of 421°C from a hot source, while engine *B* is in communication with a cold sink at a temperature of 4.4°C. If the work output of *A* is twice that of *B*, find (a) the intermediate temperature between *A* and *B*, (b) the efficiency of each engine, and (c) the heat rejected to the cold sink. **[LO 6.12]**

(M) **6.11** A heat engine operates between the maximum and minimum temperatures of 671°C and 60°C respectively, with an efficiency of 50% of the appropriate Carnot efficiency. It drives a heat pump which uses river water at 4.4°C to heat a block of flats in which the temperature is to be maintained at 21.1°C. Assuming that a temperature difference of 11.1°C exists between the working fluid and the river water, on the one hand, and the required room temperature on the other, and assuming the heat pump to operate on the reversed Carnot cycle, but with a COP of 50% of the ideal COP, find the heat input to the engine per unit heat output from the heat pump. Why is direct heating thermodynamically more wasteful? **[LO 6.12]**

(E) **6.12** An ice-making plant produces ice at atmospheric pressure and at 0°C from water. The mean temperature of the cooling water circulating through the condenser of the refrigerating machine is 18°C. Evaluate the minimum electrical work in kWh required to produce 1 tonne of ice. (The enthalpy of fusion of ice at atmospheric pressure is 333.5 kJ/kg). **[LO 6.12]**

(E) **6.13** A reversible engine works between three thermal reservoirs, *A*, *B* and *C*. The engine absorbs an equal amount of heat from the thermal reservoirs *A* and *B* kept at temperatures T_A and T_B respectively, and rejects heat to the thermal reservoir *C* kept at temperature T_C. The efficiency of the engine is α times the efficiency of the reversible engine, which works between the two reservoirs *A* and *C*. Prove that **[LO 6.12]**

$$\frac{T_A}{T_B} = (2\alpha - 1) + 2(1 - \alpha)\frac{T_A}{T_C}$$

(M) **6.14** A reversible engine operates between temperatures T_1 and T ($T_1 > T$). The energy rejected from this engine is received by a second reversible engine at the same temperature *T*. The second engine rejects energy at temperature T_2 ($T_2 < T_1$). Show that (a) temperature *T* is the arithmetic mean of temperatures T_1 and T_2 if the engines produce the same amount of work output, and (b) temperature *T* is the geometric mean of temperatures T_1 and T_2 if the engines have the same cycle efficiencies. **[LO 6.12]**

(M) **6.15** Two Carnot engines *A* and *B* are connected in series between two thermal reservoirs maintained at 1000 K and 100 K respectively. Engine *A* receives 1680 kJ of heat from the high-temperature reservoir and rejects heat to the Carnot engine *B*. Engine *B* takes in heat rejected by engine *A* and rejects heat to the low-temperature reservoir. If engines *A* and *B* have equal thermal efficiencies, determine (a) the heat rejected by engine *B*, (b) the temperature at which heat is rejected by engine, *A*, and (c) the work done during the process by engines, *A* and *B* respectively. If engines *A* and B deliver equal work, determine (d) the amount of heat taken in by engine B, and (e) the efficiencies of engines A and B. **[LO 6.12]**

(E) **6.16** A heat pump is to be used to heat a house in winter and then reversed to cool the house in summer. The interior temperature is to be maintained at 20°C. Heat transfer through the walls and roof is estimated to be 0.525 kJ/s per degree temperature difference between the inside and outside. (a) If the outside temperature in winter is 5°C, what is the minimum power required to drive the heat pump? (b) If the power output is the same as in part (a), what is the maximum outer temperature for which the inside can be maintained at 20°C? **[LO 6.5, 6.12]**

(M) **6.17** Consider an engine in outer space which operates on the Carnot cycle. The only way in which heat can be transferred from the engine is by radiation. The rate at which heat is radiated is proportional to the fourth power of the absolute temperature and to the area of the radiating surface. Show that for a given power output and a given T_1, the area of the radiator will be a minimum when

$$\frac{T_2}{T_1} = \frac{3}{4}$$

[LO 6.12]

(M) **6.18** It takes 10 kW to keep the interior of a certain house at 20°C when the outside temperature is 0°C. This heat flow is usually obtained directly by burning gas or oil. Calculate the power required if the 10 kW heat flow were supplied by operating a reversible engine with the house as the upper reservoir and the outside surroundings as the lower reservoir, so that the power were used only to perform work needed to operate the engine. **[LO 6.12]**

(M) **6.19** Prove that the COP of a reversible refrigerator operating between two given temperatures is the maximum. **[LO 6.12]**

(M) **6.20** A house is to be maintained at a temperature of 20°C by means of a heat pump pumping heat from the atmosphere. Heat losses through the walls of the house are estimated at 0.65 kW per unit of temperature difference between the inside of the house and the atmosphere. (a) If the atmospheric temperature is – 10°C, what is the minimum power required to drive the pump? (b) It is proposed to use the same heat pump to cool the house in summer. For the same room temperature, the same heat loss rate, and the same power input to the pump, what is the maximum permissible atmospheric temperature? **[LO 6.12]**

(M) **6.21** A solar-powered heat pump receives heat from a solar collector at T_h, rejects heat to the atmosphere at T_a, and pumps heat from a cold space at T_c. The three heat transfer rates are Q_h, Q_a, and Q_c respectively. Derive an expression for the minimum ratio Q_h/Q_c, in terms of the three temperatures.

If $T_h = 400$ K, $T_a = 300$ K, $T_c = 200$ K, $Q_c = 12$ kW, what is the minimum Q_h? If the collector captures 0.2 kW/m^2, what is the minimum collector area required? **[LO 6.12]**

(M) **6.22** A heat engine operating between two reservoirs at 1000 K and 300 K is used to drive a heat pump which extracts heat from the reservoir at 300 K at a rate twice that at which the engine rejects heat to it. If the efficiency of the engine is 40% of the maximum possible and the COP of the heat pump is 50% of the maximum possible, what is the temperature of the reservoir to which the heat pump rejects heat? What is the rate of heat rejection from the heat pump if the rate of heat supply to the engine is 50 kW? **[LO 6.12]**

(M) **6.23** A reversible power cycle is used to drive a reversible heat pump cycle. The power cycle takes in Q_1 heat units at T_1 and rejects Q_2 at T_2. The heat pump abstracts Q_4

from the sink at T_4 and discharges Q_3 at T_3. Develop an expression for the ratio Q_4/Q_1 in terms of the four temperatures. **[LO 6.12]**

(M) **6.24** Prove that the following propositions are logically equivalent: (a) A PMM2 is impossible, (b) A weight sliding at constant velocity down a frictional inclined plane executes an irreversible process. **[LO 6.7]**

(E) **6.25** A heat engine receives half of its heat supply at 1000 K and half at 500 K while rejecting heat to a sink at 300 K. What is the maximum thermal efficiency of the heat engine? **[LO 6.12]**

(E) **6.26** A heat pump provides 3×10^4 kJ/h to maintain a dwelling at 23°C on a day when the outside temperature is 0°C. The power input to the heat pump is 4 kW. Determine the COP of the heat pump and compare it with the COP of a reversible heat pump operating between the reservoirs at the same two temperatures. **[LO 6.5, 6.12]**

(E) **6.27** When the outside temperature is – 10°C, a residential heat pump must provide 3.5×10^6 kJ per day to a dwelling to maintain its temperature at 20°C. If the electricity costs ₹2.10 per kWh, find the minimum operating cost for each day of operation. **[LO 6.12]**

(D) **6.28** A reversible power cycle receives energy Q_H from a reservoir at temperature T_H and rejects Q_C to a reservoir at temperature T_C. The work developed by the power cycle is used to drive a reversible heat pump that removes Q'_C from a reservoir at T'_C and rejects energy Q'_H to a reservoir at temperature T'_H. (a) Develop an expression for the ratio Q'_H/Q_H in terms of the temperatures of the four reservoirs. (b) What must be the relationship of the temperatures T_H, T_C, T'_C and T'_H for $\dfrac{Q'_H}{Q_H}$ to exceed a value of unity? **[LO 6.12]**

ANSWERS TO PROBLEMS

6.1 No
6.2 5.4 kW
6.3 666.67 kJ
1.8 MJ
6.4 (a) – 708 kJ
(b) Second law, $W_{2-1} = 9348$ kJ
(c) $Q_{2-1} = 0$
6.5 ₹118.80
6.6 (a) 47.61 kW
(b) 34.61 kW
6.7 0.56 kW
6.8 1.81
6.9 – 270.37°C
6.10 143.4°C, 40% and 33.5%, 80 kJ
6.11 0.79 kJ/kJ heat input
6.12 6.11 kWh
6.15 (a) 168 kJ
(b) 316.2 K
(c) 1148.7, 363.3 kJ
(d) 924 kJ
(e) 45%, 81.8%
6.16 (a) 403 W
(b) 35.4°C
6.18 0.683 kW
6.20 2 kW, 50°C
6.21 26.25 kW, 131.25 m^2
6.22 326.5 K, 86 kW
6.23 $\dfrac{Q_4}{Q_1} = \dfrac{T_4(T_1 - T_2)}{T_1(T_3 - T_4)}$
6.25 55%
6.26 2.08, 12.87
6.27 ₹208.83
6.28 (a) $\dfrac{Q'_H}{Q_H} = \dfrac{T'_H(T_H - T_C)}{T_H(T'_H - T'_C)}$
(b) $\dfrac{T_C}{T'_C} < \dfrac{T_H}{T'_H}$

7

Entropy

Learning Objectives

LO 7.1 Recall three 'E's of thermodynamics
LO7.2 Discuss inequality of Clausius
LO7.3 Relate entropy concept and Clausius' theorem
LO7.4 Construct reversible process and cycles on temperature-entropy plot
LO7.5 Discuss entropy change in a process
LO7.6 Explain the principle of increase of entropy
LO7.7 Apply the principle of increase of entropy to quantify irreversibility
LO7.8 Contrast flow of heat and work related to entropy transfer
LO7.9 Demonstrate entropy increase of closed and open system processes
LO7.10 Combine principles of first and second law of thermodynamics
LO7.11 Explain reversible work for a steady flow system
LO7.12 Relate entropy with its direction of flow
LO7.13 Relate entropy with disorder
LO7.14 Explain absolute entropy and postulatory thermodynamics

7.1 INTRODUCTION

LO 7.1
Recall three 'E's of thermodynamics

The first law of thermodynamics was stated in terms of cycles first and it was shown that the cyclic integral of heat is equal to the cyclic integral of work. When the first law was applied for thermodynamic processes, the existence of a property, the internal energy, was found. Similarly, the second law was also first stated in terms of cycles executed by systems. When applied to processes, the second law also leads to the definition of a new property, known as entropy. If the first law is said to be the law of internal energy, then the second law may be stated to be the law of entropy. In fact, *thermodynamics is the study of three E's, namely, energy, equilibrium and entropy.*

7.2 THE INEQUALITY OF CLAUSIUS

LO 7.2
Discuss inequality of Clausius

Let us consider two heat engine cycles, one reversible (E_R) and the other irreversible (E_I), operating between the same two constant temperature reservoirs at temperatures T_1 and T_2 as shown in Fig. 7.1.

Our objective is to find the expression of cyclic integral of $đQ$ and $đQ/T$ both for reversible and irreversible heat engines, where T is the temperature of system boundary over which heat transfer takes place.

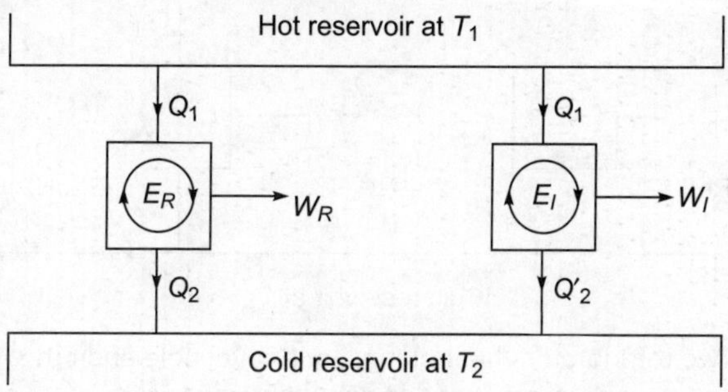

Fig. 7.1 Two heat engines, one reversible and other irreversible operating between same two reservoirs

From Kelvin-Planck statement of second law of thermodynamics, both for reversible and irreversible heat engines, we have

$$\oint đQ\big|_{HE} = Q_1 - Q_2 \text{ (or} = Q_1 - Q'_2) > 0 \tag{7.1}$$

For the reversible heat engine,

$$\oint \frac{đQ}{T}\bigg|_{\text{Rev, HE}} = \frac{Q_1}{T_1} - \frac{Q_2}{T_2} = 0 \tag{7.2}$$

Since the efficiency of the irreversible engine η_I is less than that of the reversible engine η_R working between the same two reservoirs, we have

$$\eta_I < \eta_R$$

or
$$\frac{W_I}{Q_1} < \frac{W_R}{Q_1}$$

For the supply of the same quantity of heat Q_1 to both the engines, we get

$$W_I < W_R$$

or
$$Q_1 - Q'_2 < Q_1 - Q_2$$

∴
$$Q'_2 > Q_2$$

For the irreversible heat engine

$$\oint \frac{đQ}{T}\bigg|_{\text{Irrev,HE}} = \frac{Q_1}{T_1} - \frac{Q_2'}{T_2} < 0 \tag{7.3}$$

So, in general, we have

$$\oint \frac{đQ}{T}\bigg|_{\text{HE}} \leq 0 \tag{7.4}$$

Next, we consider two refrigerator/heat pump cycles, one reversible (Ref/HP$_R$) and the other irreversible (Ref/HP$_I$), operating between the same two constant temperature reservoirs at temperatures T_1 and T_2 as shown in Fig. 7.2.

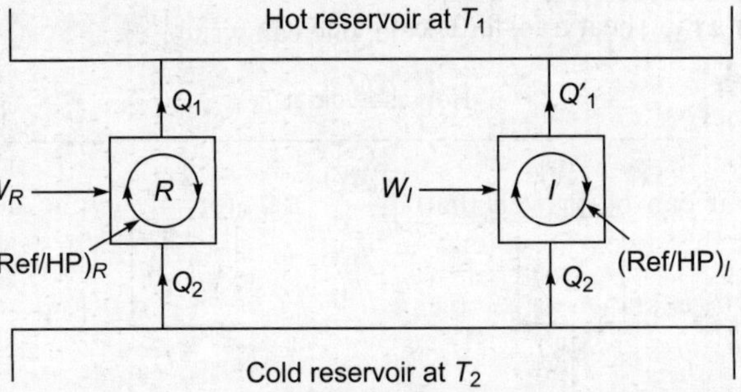

Fig. 7.2 Two refrigerators/heat pumps, one reversible and other irreversible operating between same two reservoirs

From Clausius statement of second law of thermodynamics, both for reversible and irreversible heat pumps/refrigerators, we have

$$\oint đQ\Big|_{\text{Ref/HP}} = Q_2 - Q_1 \text{ (or} = Q_2 - Q_1') < 0 \tag{7.5}$$

For the reversible heat pump/refrigerator,

$$\oint \frac{đQ}{T}\bigg|_{\text{Rev,Ref/HP}} = \frac{Q_2}{T_2} - \frac{Q_1}{T_1} = 0 \tag{7.6}$$

Since for the same Q_2 and working between same two reservoirs, the irreversible refrigerator/ heat pump requires more work input than that of reversible one, we have

$$W_I > W_R$$

or $$W_I + Q_2 > W_R + Q_2$$

or $$Q'_1 > Q_1$$

For the irreversible heat pump/refrigerator,

$$\oint \frac{đQ}{T}\bigg|_{\text{Irrev,Ref/HP}} = \frac{Q_2}{T_2} - \frac{Q_1'}{T_1} < 0 \tag{7.7}$$

So, in general, we have

$$\oint \frac{đQ}{T}\bigg|_{\text{Ref/HP}} \leq 0 \tag{7.8}$$

From the above discussion, important conclusions that can be drawn that although the sign of $\oint đQ$ differs, depending on whether it is heat engine or refrigerator/heat pump, $\oint \frac{đQ}{T} \le 0$ for all cyclic devices.

Using Eqs. (7.2) and (7.6), for any reversible cycle, one can write

$$\oint_R \frac{đQ}{T} = 0 \tag{7.9}$$

It implies that the cyclic integral of $đQ/T$ for a reversible cycle is equal to zero. This is known as *Clausius' theorem*. The letter R emphasizes the fact that the equation is valid only for a reversible cycle.

Using Eqs. (7.4) and (7.8), for any cycle, one can write

$$\oint \frac{đQ}{T} \le 0 \tag{7.10}$$

Similarly, it can be proved that $\oint \frac{đQ}{T} \le 0$ for any cyclic device operating between more than two reservoirs.

Equation (7.10) is known as the *inequality of Clausius. It provides the criterion of the reversibility of a cycle,*

If $\oint \frac{đQ}{T} = 0$, the cycle is reversible.

$\oint \frac{đQ}{T} < 0$, the cycle is irreversible and possible.

$\oint \frac{đQ}{T} > 0$, the cycle is impossible, since it violates the second law.

7.3 THE PROPERTY OF ENTROPY

LO 7.3
Relate entropy concept and Clausius' theorem

Let a system be taken from an initial equilibrium state i to a final equilibrium state f by following the reversible path R_1 (Fig. 7.3). The system is brought back from f to i by following another reversible path R_2. Then the two paths R_1 and R_2 together constitute a reversible cycle. From Clausius' theorem

$$\oint_{R_1R_2} \frac{đQ}{T} = 0$$

The above integral may be replaced as the sum of two integrals, one for path R_1 and the other for path R_2

$$\int_{\substack{i \\ R_1}}^{f} \frac{đQ}{T} + \int_{\substack{f \\ R_2}}^{i} \frac{đQ}{T} = 0$$

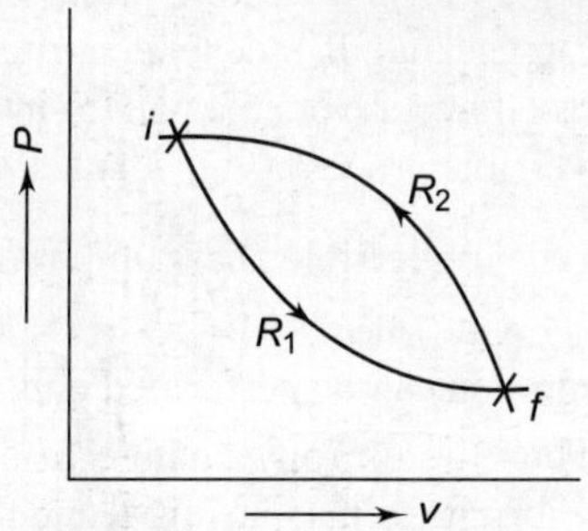

Fig. 7.3 Two reversible paths R_1 and R_2 between two equilibrium states i and f

$$\text{or} \quad \underset{R_1}{\int_i^f} \frac{đQ}{T} = -\underset{R_2}{\int_f^i} \frac{đQ}{T}$$

Since R_2 is a reversible path

$$\underset{R_1}{\int_i^f} \frac{đQ}{T} = \underset{R_2}{\int_i^f} \frac{đQ}{T}$$

Since R_1 and R_2 represent any two reversible paths, $\underset{R}{\int_i^f} \frac{đQ}{T}$ is independent of the reversible path connecting i and f. Therefore, there exists a property of a system whose value at the final state f minus its value at the initial state i is equal to $\underset{R}{\int_i^f} \frac{đQ}{T}$.

This property is called entropy, and is denoted by S. If S_i is the entropy at the initial state i, and S_f is the entropy at the final state f, then

$$\underset{R}{\int_i^f} \frac{đQ}{T} = S_f - S_i \tag{7.11}$$

Note that we are concerned with the changes in entropy during a process undergone by a system instead of its absolute value. Accordingly we have defined the change in entropy by Eq. (7.11)

When the two equilibrium states are infinitesimally near

$$\frac{đQ_R}{T} = dS \tag{7.12}$$

where dS is an exact *differential* because S is a point function and a property. The subscript R in $đQ$ indicates that heat $đQ$ is transferred *reversibly*.

The word 'entropy' was first used by Clausius, taken from the Greek word '*tropee*' meaning 'transformation'. It is an extensive property, and has the unit J/K. The specific entropy

$$s = \frac{S}{m} \text{ J/kg K}$$

If the system is taken from an initial equilibrium state i to a final equilibrium state f by an *irreversible path*, since entropy is a point or state function, and the entropy change is independent of the path followed, the non-reversible path is to be replaced by a reversible path to integrate for the evaluation of entropy change in the irreversible process (Fig. 7.4).

$$S_f - S_i = \int_i^f \frac{đQ_{\text{rev}}}{T} = (\Delta S)_{\text{irrev path}} \tag{7.13}$$

Integration can be performed only on a reversible path.

Thus the entropy change in any irreversible or real process connecting two equilibrium points, the initial and the final, can be estimated by replacing a reversible path between the same points.

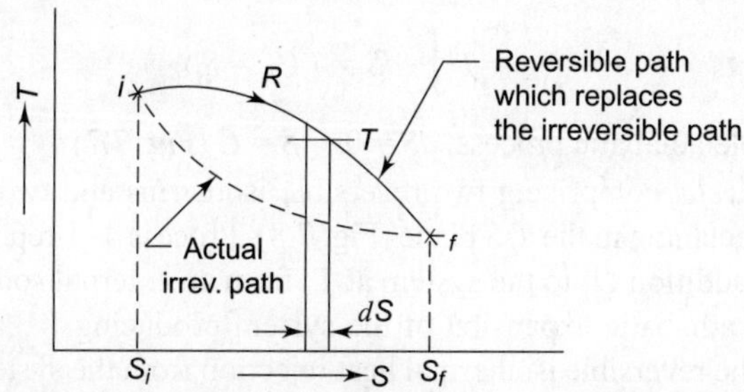

Fig. 7.4 Integration can be done only on a reversible path

7.4 TEMPERATURE-ENTROPY PLOT

LO 7.4 Construct reversible process and cycles on temperature-entropy plot

The infinitesimal change in entropy dS due to reversible heat transfer $đQ$ at temperature T is

$$dS = \frac{đQ_{\text{rev}}}{T}$$

If $đQ_{\text{rev}} = 0$, i.e. the process is reversible and adiabatic

$$dS = 0$$

and $$S = \text{constant}$$

A reversible adiabatic process is, therefore, an isentropic process.

Now

$$đQ_{\text{rev}} = TdS$$

or $$Q_{\text{rev}} = \int_i^f TdS$$

The system is taken from i to f reversibly (Fig. 7.5). The area under the curve $\int_i^f TdS$ is equal to the heat transferred in the process.

For reversible isothermal heat transfer (Fig. 7.6), T = constant.

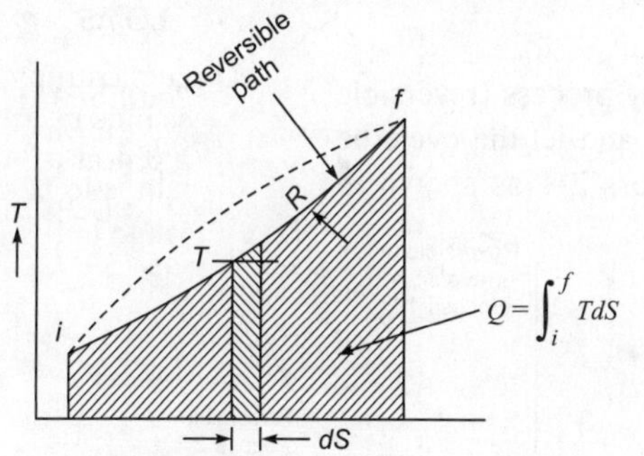

Fig. 7.5 Area under a reversible path on the T-s plot represents heat transfer

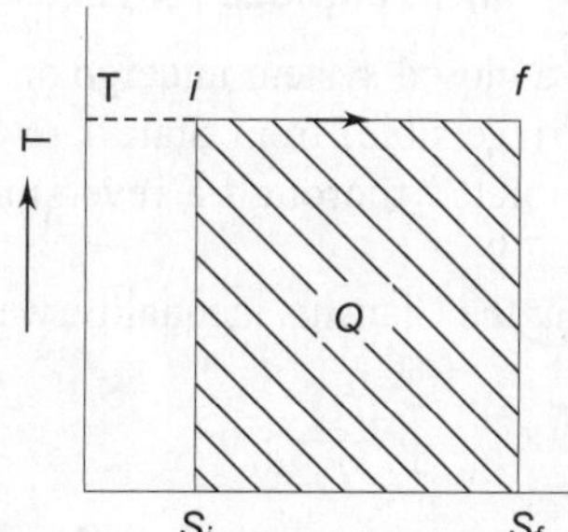

Fig. 7.6 Reversible isothermal heat transfer

$$\therefore \qquad Q_{rev} = T\int_i^f dS = T(S_f - S_i)$$

For a reversible adiabatic process, $dS = 0$, $S = C$ (Fig. 7.7).

The *Carnot cycle,* comprising two reversible isotherms and two reversible adiabatics, forms a rectangle in the *T-S* plane (Fig. 7.8). Process 4–1 represents reversible isothermal heat addition Q_1 to the system at T_1 from an external source, process 1–2 is the reversible adiabatic expansion of the system producing W_E amount of work, process 2 – 3 is the reversible isothermal heat rejection from the system to an external sink at T_2 and process 3– 4 represents reversible adiabatic compression of the system consuming W_c amount of work. Area 1234 represents the net work output per cycle and the area under 4 – 1 indicates the quantity of heat added to the system Q_1.

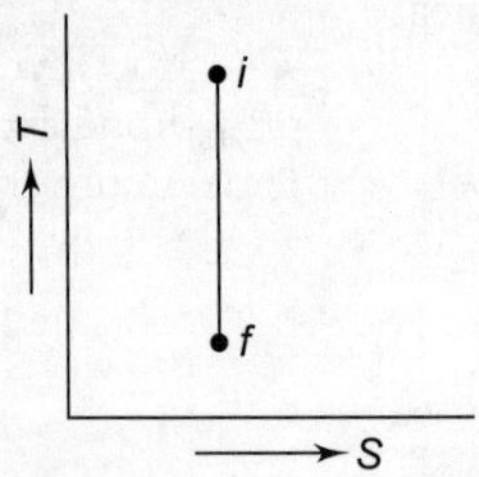

Fig. 7.7 Reversible adiabatic is isentropic

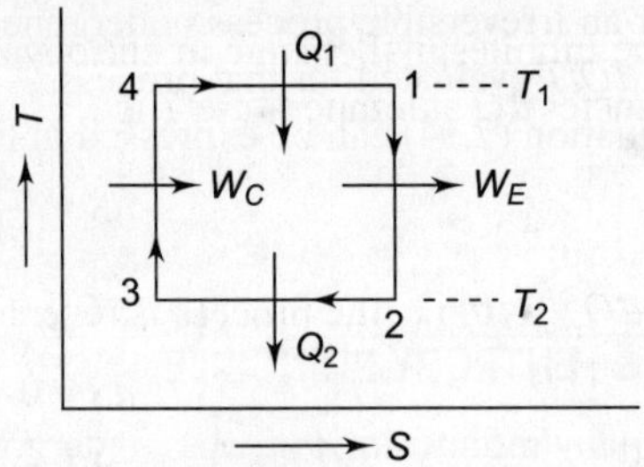

Fig. 7.8 Carnot cycle

$$\therefore \qquad \eta_{\text{Carnot}} = \frac{Q_1 - Q_2}{Q_1} = \frac{T_1(S_1 - S_4) - T_2(S_2 - S_3)}{T_1(S_1 - S_4)}$$

$$= \frac{T_1 - T_2}{T_1} = 1 - \frac{T_2}{T_1}$$

and $$W_{\text{net}} = Q_1 - Q_2 = (T_1 - T_2)(S_1 - S_4)$$

7.5 ENTROPY CHANGE IN AN IRREVERSIBLE PROCESS

LO 7.5 Discuss entropy change in a process

Let a closed system undergo an arbitrary process (reversible or irreversible) from State 1 to State 2, and let the cycle be completed thorough a reversible process 2-1 as shown in Fig. 7.9.

From the Clausius inequality, we have

$$\oint \frac{đQ}{T} \le 0$$

or $$\int_1^2 \left(\frac{đQ}{T}\right) + \int_2^1 \left(\frac{đQ}{T}\right)_{\text{rev}} \le 0$$

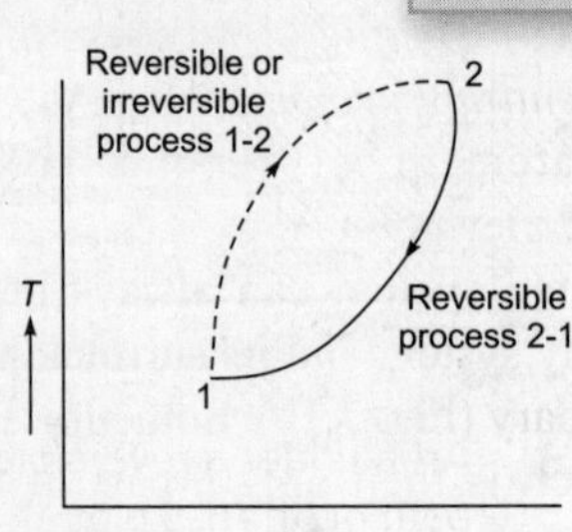

Fig. 7.9 Entropy change in an irreversible process

From the definition of entropy, we have $\int_2^1 \left(\frac{đQ}{T}\right)_{\text{rev}} = S_1 - S_2$. Therefore, the above equation reduces to

$$\int_1^2 \left(\frac{đQ}{T}\right) + S_1 - S_2 \le 0$$

or

$$S_2 - S_1 \ge \int_1^2 \left(\frac{đQ}{T}\right) \tag{7.14}$$

The equality sign holds good for a reversible process and the inequality sign for an irreversible process. It can be concluded from Eq. (7.14) that the change in entropy for an irreversible process undergone by a closed system is greater than the integral of $đQ/T$ evaluated for that process.

Equation (7.14) can be expressed in differential form as

$$dS \ge \frac{đQ}{T} \tag{7.15}$$

7.6 ENTROPY PRINCIPLE

> **LO 7.6**
> Explain the principle of increase of entropy

For any infinitesimal process undergone by a system, we have the following equation for the total mass

$$dS \ge \frac{đQ}{T}$$

For an isolated system which does not undergo any energy interaction with the surroundings, $đQ = 0$.

Therefore, for an isolated system

$$dS_{\text{iso}} \ge 0 \tag{7.16}$$

For a reversible process,

$$dS_{\text{iso}} = 0$$

or

$$S = \text{constant}$$

For an irreversible process

$$dS_{\text{iso}} > 0$$

It is thus proved that *the entropy of an isolated system can never decrease*. It always increases and remains constant only when the process is reversible. This is known as the *principle of increase of entropy*, or simply the *entropy principle*. It is the quantitative general statement of second law from the macroscopic viewpoint.

An isolated system can always be formed by including any system and its surroundings within a single boundary (Fig. 7.10). Sometimes the original system which is then only a part of the isolated system is called a 'subsystem'.

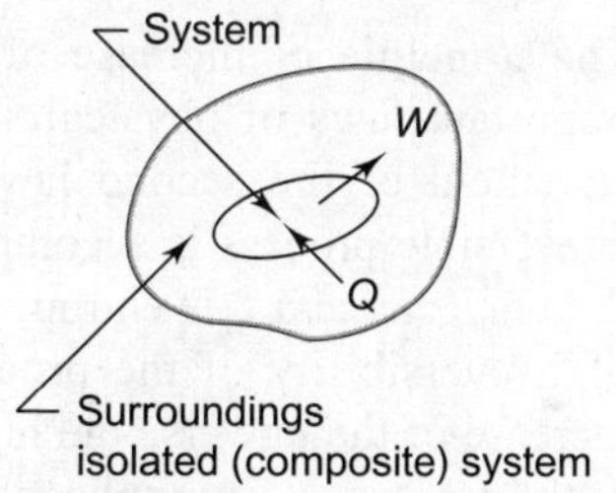

Fig. 7.10 Isolated system

The system and the surroundings together (the universe or the isolated system) include everything which is affected by the process. For all possible processes that a system in the given surroundings can undergo

$$dS_{\text{univ}} \geq 0$$

or
$$dS_{\text{sys}} + dS_{\text{surr}} \geq 0 \tag{7.17}$$

Entropy may decrease locally at some region within the isolated system, but it must be compensated by a greater increase of entropy somewhere within the system so that the net effect of an irreversible process is an entropy increase of the whole system. This implies that entropy of a closed system may decrease during a process without violation of the entropy principle provided that the increase in entropy of the universe during that process is more than that of decrease in entropy of the system. The entropy increase of an isolated system is a measure of the extent of irreversibility of the process undergone by the system.

Rudolf Clausius summarised the first and second laws of thermodynamics in the following words:

1. *Die Energie der Welt ist konstant.*
2. *Die Entropie der Welt strebt einem Maximum zu.*

[1. The energy of the world (universe) is constant.
2. The entropy of the world tends towards a maximum.]

The entropy of an isolated system always increases and becomes maximum at the state of equilibrium. If the entropy of an isolated system varies with some parameter x, then there is a certain value of x_e which maximises the entropy $\left(\text{when } \dfrac{dS}{dx} = 0\right)$ and represents the equilibrium state (Fig. 7.11). The system is then said to exist at the peak of the entropy hill, and $dS = 0$. *When the system is at equilibrium, any conceivable change in entropy would be zero.*

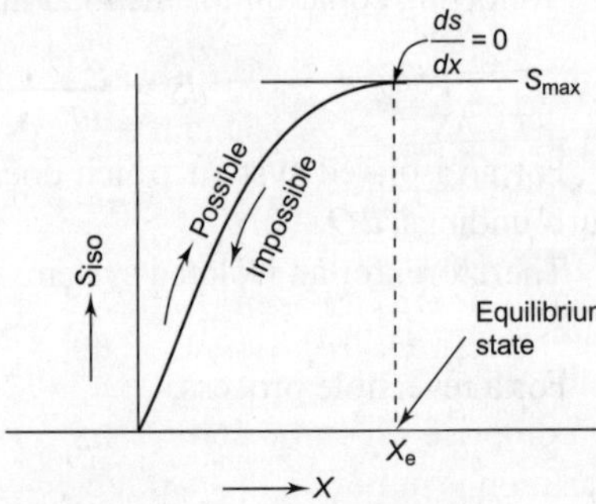

Fig. 7.11 Equilibrium state of an isolated system

7.7 APPLICATIONS OF ENTROPY PRINCIPLE

LO7.7 Apply the principle of increase of entropy to quantify irreversibility

The principle of increase of entropy is one of the most important laws of physical science. It is the quantitative statement of the second law of thermodynamics. Every irreversible process is accompanied by entropy increase of the universe, and this entropy increase quantifies the extent of irreversibility of the process. The higher the entropy increase of the universe the higher will be the irreversibility of the process. A few applications of the entropy principle are illustrated in the following sections.

7.7.1 Transfer of Heat through a Finite Temperature Difference

Let Q be the rate of heat transfer from reservoir A at T_1 to reservoir B at T_2, $T_1 > T_2$ (Fig. 7.12).

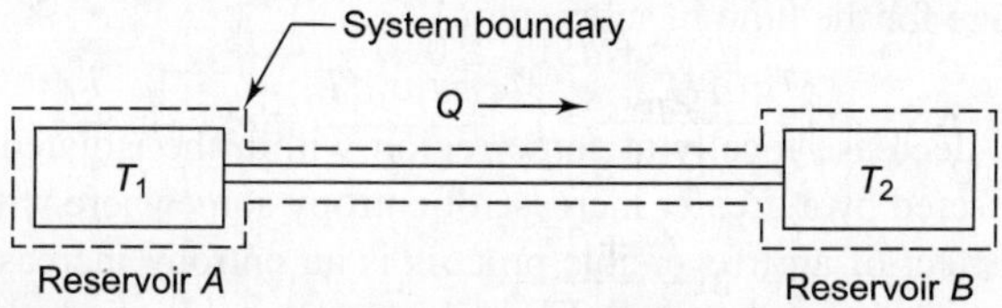

Fig. 7.12 Heat transfer through a finite temperature difference

For reservoir A, $\Delta S_A = -Q/T_1$. It is negative because heat Q flows out of the reservoir. For reservoir B, $\Delta S_B = +Q/T_2$. It is positive because heat flows into the reservoir. The rod connecting the reservoirs suffers no entropy change because once in the steady state, its coordinates do not change.

Therefore, for the isolated system comprising the reservoirs and the rod, and since entropy is an additive property

$$S = S_A + S_B$$

$$\Delta S_{\text{univ}} = \Delta S_A + \Delta S_B$$

or

$$\Delta S_{\text{univ}} = -\frac{Q}{T_1} + \frac{Q}{T_2} = Q \cdot \frac{T_1 - T_2}{T_2 T_2}$$

Since $T_1 > T_2$, ΔS_{univ} is positive, and the process is irreversible and possible. If $T_1 = T_2$, ΔS_{univ} is zero, and the process is reversible. If $T_1 < T_2$, ΔS_{univ} is negative and the process is impossible.

7.7.2 Mixing of Two Fluids

Subsystem 1 having a fluid of mass m_1, specific heat c_1 and temperature t_1 and subsystem 2 consisting of a fluid of mass m_2, specific heat c_2 and temperature t_2 comprise a composite system in an adiabatic enclosure (Fig. 7.13). When the partition is removed, the two fluids mix together, and at equilibrium let t_f be the final temperature, and $t_2 < t_f < t_1$. Since energy interaction is exclusively confined to the two fluids, the system being isolated

$$m_1 c_1 (t_1 - t_f) = m_2 c_2 (t_f - t_2)$$

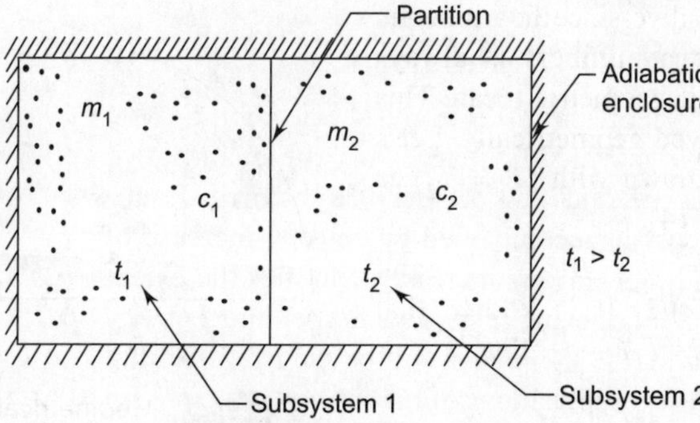

Fig. 7.13 Mixing of two fluids

$$\therefore \qquad t_f = \frac{m_1 c_1 t_1 + m_2 c_2 t_2}{m_1 c_1 + m_2 c_2}$$

Entropy change for the fluid in subsystem 1

$$\Delta S_1 = \int_{T_1}^{T_f} \frac{đQ_{\text{rev}}}{T} = \int_{T_i}^{T_f} \frac{m_1 c_1 dT}{T} = m_1 c_1 \ln \frac{T_f}{T_1}$$

$$= m_1 c_1 \ln \frac{t_f + 273}{t_1 + 273}$$

This will be negative, since $T_1 > T_f$

Entropy change for the fluid in subsystem 2

$$\Delta S_2 = \int_{T_2}^{T_f} \frac{m_2 c_2 dT}{T} = m_2 c_2 \ln \frac{T_f}{T_2} = m_2 c_2 \ln \frac{t_f + 273}{t_2 + 273}$$

This will be positive, since $T_2 < T_f$

$$\therefore \qquad \Delta S_{\text{univ}} = \Delta S_1 + \Delta S_2$$

$$= m_1 c_1 \ln \frac{T_f}{T_1} + m_2 c_2 \ln \frac{T_f}{T_2}$$

ΔS_{univ} will be positive definite, and the mixing process is irreversible.

Although the mixing process is irreversible, to evaluate the entropy change for the subsystems, the irreversible path was replaced by a reversible path on which the integration was performed.

If $m_1 = m_2 = m$ and $c_1 = c_2 = c$.

$$\Delta S_{\text{univ}} = mc \ln \frac{T_f^2}{T_1 \cdot T_2}$$

and

$$T_f = \frac{m_1 c_1 T_1 + m_2 c_2 T_2}{m_1 c_1 + m_2 c_2} = \frac{T_1 + T_2}{2}$$

$$\therefore \qquad \Delta S_{\text{univ}} = 2mc \ln \frac{(T_1 + T_2)/2}{\sqrt{T_1 \cdot T_2}}$$

This is always positive, since the arithmetic mean of any two numbers is always greater than their geometric mean. This can also be proved geometrically. Let a semi-circle be drawn with $(T_1 + T_2)$ as diameter (Fig. 7.14).

Here, $AB = T_1$, $BC = T_2$ and $OE = (T_1 + T_2)/2$. It is known that $(DB)^2 = AB \cdot BC = T_1 T_2$.

$$\therefore \qquad DB = \sqrt{T_1 T_2}$$

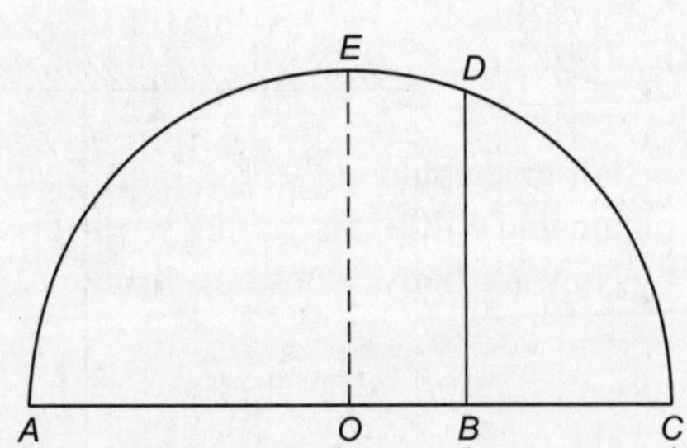

Fig. 7.14 Geometrical proof to show that *g.m.* < *a.m.*

Now, $OE > DB$

$$\frac{T_1 + T_2}{2} > \sqrt{T_1 T_2}$$

7.7.3 Maximum Work Obtainable from Two Finite Bodies at Temperatures T_1 and T_2

Let us consider two identical finite bodies of constant heat capacity at temperatures T_1 and T_2, respectively, T_1 being higher than T_2. If the two bodies are merely brought together into thermal contact, delivering no work, the final temperature T_f reached would be the maximum

$$T_f = \frac{T_1 + T_2}{2}$$

If a heat engine is operated between the two bodies acting as thermal energy reservoirs (Fig. 7.15), part of the heat withdrawn from body 1 is converted to work W by the heat engine, and the remainder is rejected to body 2. The lowest attainable final temperature T_f corresponds to the delivery of the largest possible amount of work, and is associated with a reversible process.

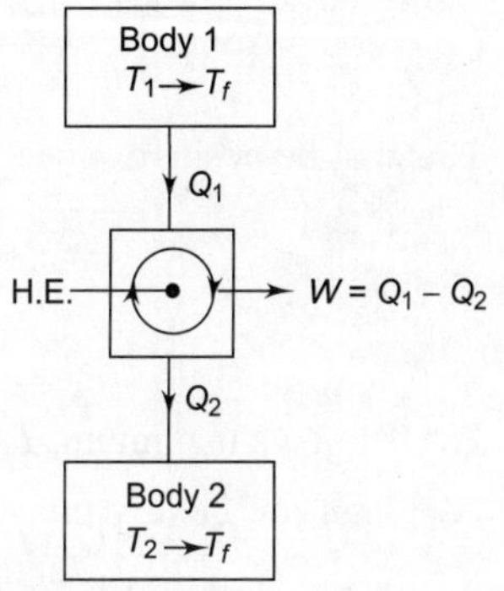

Fig. 7.15 Maximum work obtainable from two finite bodies

As work is delivered by the heat engine, the temperature of body 1 will be decreasing and that of body 2 will be increasing. When both the bodies attain the final temperature T_f, the heat engine will stop operating. Let the bodies remain at constant pressure and undergo no change of phase.

Total heat withdrawn from body 1

$$Q_1 = C_p (T_1 - T_f)$$

where C_p is the heat capacity of the two bodies at constant pressure

Total heat rejected to body 2

$$Q_2 = C_p (T_f - T_2)$$

$\therefore$ amount of total work delivered by the heat engine

$$W = Q_1 - Q_2$$

$$= C_p (T_1 + T_2 - 2T_f) \tag{7.18}$$

For given values of C_p, T_1 and T_2, the magnitude of work W depends on T_f. Work obtainable will be maximum when T_f is minimum.

Now, for body 1, entropy change ΔS_1 is given by

$$\Delta S_1 = \int_{T_1}^{T_f} C_p \frac{dT}{T} = C_p \ln \frac{T_f}{T_1}$$

For body 2, entropy change ΔS_2 would be

$$\Delta S_2 = \int_{T_2}^{T_f} C_p \frac{dT}{T} = C_p \ln \frac{T_f}{T_2}$$

Since the working fluid operating in the heat engine cycle does not undergo any entropy change, ΔS of the working fluid in heat engine $\oint dS = 0$. Applying the entropy principle

$$\Delta S_{\text{univ}} \geq 0$$

$$\therefore \quad C_p \ln \frac{T_f}{T_1} + C_p \ln \frac{T_f}{T_2} \geq 0$$

$$C_p \ln \frac{T_f^2}{T_1 T_2} \geq 0 \tag{7.19}$$

From Eq. (7.19), for T_f to be a minimum

$$C_p \ln \frac{T_f^2}{T_1 \cdot T_2} = 0$$

or

$$\ln \frac{T_f^2}{T_1 T_2} = 0 = \ln 1$$

$$\therefore \quad T_f = \sqrt{T_1 \cdot T_2} \tag{7.20}$$

For W to be a maximum, T_f will be $\sqrt{T_1 T_2}$. From Eq. (7.19)

$$W_{\max} = C_p\,(T_1 + T_2 - 2\sqrt{T_1 T_2}\,) = C_p(\sqrt{T_1} - \sqrt{T_2})^2$$

The final temperatures of the two bodies, initially at T_1 and T_2, can range from $(T_1 + T_2)/2$ with no delivery of work to $\sqrt{T_1 T_2}$ with maximum delivery of work.

7.7.4 Maximum Work Obtainable from a Finite Body and a TER

Let one of the bodies considered in the previous section be a thermal energy reservoir. The finite body has a thermal capacity C_p and is at temperature T and the TER is at temperature T_0, such that $T > T_0$. Let a heat engine operate between the two (Fig. 7.16). As heat is withdrawn from the body, its temperature decreases. The temperature of the TER would, however, remain unchanged at T_0. The engine would stop working when the temperature of the body reaches T_0. During that period, the amount of work delivered is W, and the heat rejected to the TER is $(Q - W)$. Then

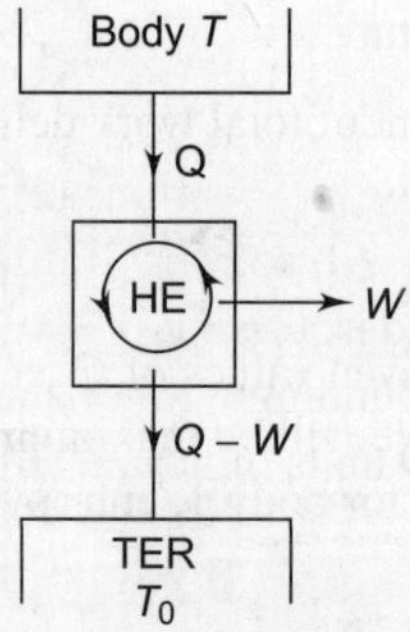

Fig. 7.16 Maximum work obtainable when one of the bodies is a TER

$$\Delta S_{\text{Body}} = \int_T^{T_0} C_p \frac{dT}{T} = C_p \ln \frac{T_0}{T}$$

$$\Delta S_{\text{HE}} = \oint dS = 0$$

$$Q = C_p\,(T - T_0)$$

$$\Delta S_{\text{TER}} = \frac{Q - W}{T_0}$$

$$\therefore \quad \Delta S_{\text{univ}} = C_p \ln \frac{T_0}{T} + \frac{Q - W}{T_0}$$

By the entropy principle,

$$\Delta S_{\text{univ}} \geq 0$$

$$C_p \ln \frac{T_0}{T} + \frac{Q - W}{T_0} \geq 0$$

or
$$C_p \ln \frac{T_0}{T} \geq \frac{W - Q}{T_0}$$

or
$$\frac{W - Q}{T_0} \leq C_p \ln \frac{T_0}{T}$$

or
$$W \leq Q + T_0 C_p \ln \frac{T_0}{T}$$

$$\therefore \quad W_{\max} = Q + T_0 C_p \ln \frac{T_0}{T}$$

or,
$$W_{\max} = C_p \left[(T - T_0) - T_0 \ln \frac{T}{T_0} \right] \tag{7.21}$$

It is called the *exergy* of the finite body at temperature T.

7.7.5 Processes Exhibiting External Mechanical Irreversibility

1. Isothermal dissipation of work Let us consider the isothermal dissipation of work through a system into the internal energy of a reservoir, as in the flow of an electric current I through a resistor in contact with a reservoir (Fig. 7.17). At steady state, the internal energy of the resistor and hence its temperature are constant. So, by first law

$$\dot{W} = \dot{Q}$$

The flow of current represents work transfer. At steady state, the work is dissipated isothermally into heat transfer to the surroundings. Since the surroundings absorb Q units of heat at temperature T,

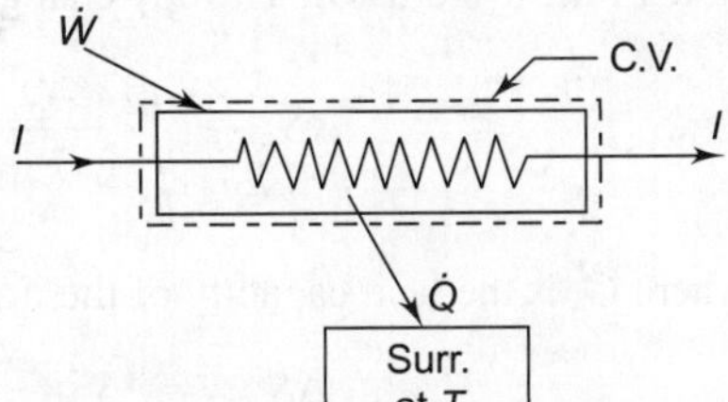

Fig. 7.17 External mechanical irreversibility

$$\dot{\Delta S}_{\text{surr}} = \frac{\dot{Q}}{T} = \frac{\dot{W}}{T}$$

At steady state, $\dot{\Delta S}_{\text{sys}} = 0$

$$\therefore \quad \dot{\Delta S}_{\text{univ}} = \dot{\Delta S}_{\text{sys}} + \dot{\Delta S}_{\text{surr}} = \frac{\dot{W}}{T} \tag{7.22}$$

The irreversible process is thus accompanied by an entropy increase of the universe.

2. Adiabatic dissipation of work Let W be the stirring work supplied to a viscous thermally insulated liquid, which is dissipated adiabatically into internal energy increase of the liquid, the temperature of which increases from T_i to T_f (Fig. 7.18). Since there is no flow of heat to or from the surroundings,

$$\Delta S_{\text{surr}} = 0$$

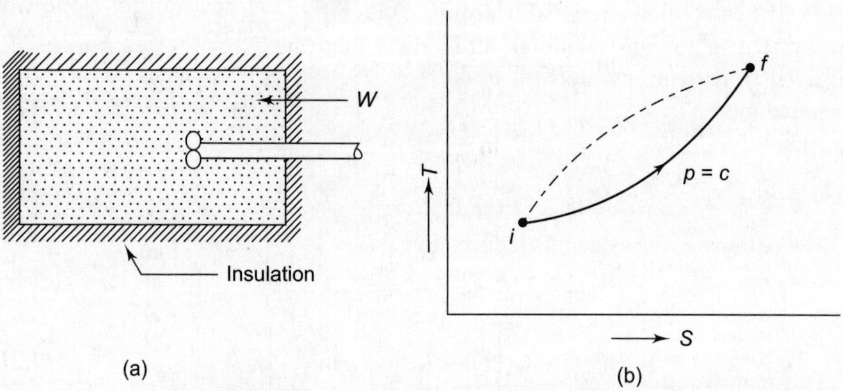

Fig. 7.18 Adiabatic dissipation of work

To calculate the entropy change of the system, the original irreversible path (dotted line) must be replaced by a reversible one between the same end states, *i* and *f*. Let us replace the irreversible performance of work by a reversible isobaric flow of heat from a series of reservoirs ranging from T_i to T_f to cause the same change in the state of the system. The entropy change of the system will be

$$\Delta S_{\text{sys}} = \int_{i \atop R}^{f} \frac{đQ}{T} = \int_{i \atop R}^{f} \frac{C_p dT}{T} = C_p \ln \frac{T_f}{T_i}$$

where C_p is the heat capacity of the liquid.

$$\Delta S_{\text{univ}} = \Delta S_{\text{sys}} + \Delta S_{\text{surr}} = C_p \ln \frac{T_f}{T_i} \tag{7.23}$$

which is positive.

7.8 ENTROPY TRANSFER WITH HEAT FLOW

LO 7.8 Contrast flow of heat and work related to entropy transfer

Since $dS = \frac{đQ_{\text{rev}}}{T}$, when heat is added to a system, $đQ$ is positive, and the entropy of the system increases. When heat is removed from the system, $đQ$ is negative, and the entropy of the system decreases.

Heat transferred to the system of fixed mass increases the internal energy of the system, as a result of which the molecules (of a gas) move with higher kinetic energy and collide more frequently, and so the disorder in the system increases. Heat is thus regarded as disorganised or disordered energy transfer which increases molecular chaos (*see Section 7.14*). If heat Q flows reversibly from the system to the surroundings at T_0 (Fig. 7.19), the entropy increase of the surroundings is

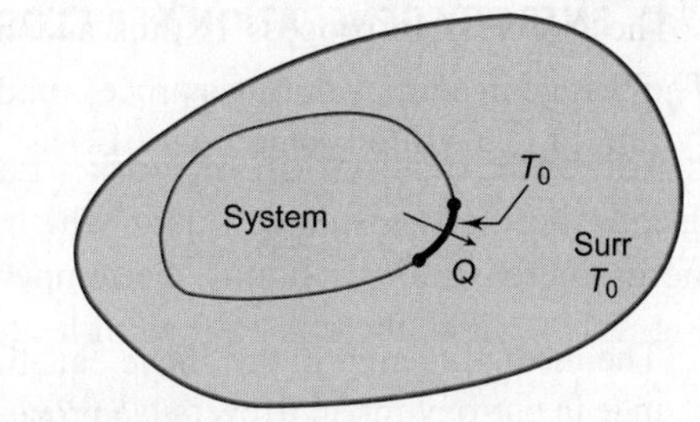

Fig. 7.19 Entropy transfer along with heat flow

$$\Delta S_{\text{surr}} = \frac{Q}{T_0}$$

The entropy of the system is reduced by

$$\Delta S_{\text{sys}} = -\frac{Q}{T_0}$$

The temperature of the boundary where heat transfer occurs is the constant temperature T_0. It may be said that the system has lost entropy to the surroundings. Alternatively, one may state that the surroundings have gained entropy from the system. Therefore, there is *entropy transfer* from the system to the surroundings along with heat flow. In other words, since the heat inflow increases the molecular disorder, there is flow of disorder along with heat.

On the other hand, *there is no entropy transfer associated with work.* In Fig. 7.20, the system delivers work to a flywheel where energy is stored in a fully recoverable form. The flywheel molecules are simply put into rotation around the axis in a perfectly organised manner, and there is no dissipation and hence no entropy increase of the flywheel. The same can be said about work transfer in the compression of a spring or in the raising of a weight by a certain height.

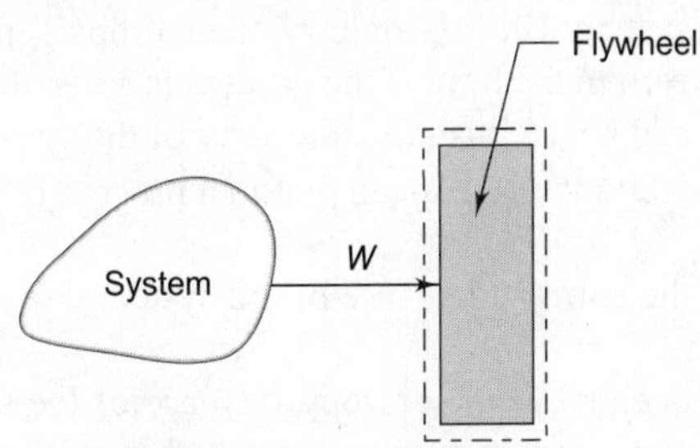

Fig. 7.20 No entropy transfer along with work transfer

There is thus no entropy transfer along with work. If work is dissipated adiabatically into internal energy increase of the system (Subsection 7.7.5), there is an entropy increase in the system, but there is as such no entropy transfer to it.

Thus, we can distinguish work and heat in this way: 'Work' is energy transfer without associated entropy transfer and 'Heat' is energy transfer with associated entropy transfer.

7.9 ENTROPY GENERATION IN A CLOSED SYSTEM

LO7.9 Demonstrate entropy increase of closed and open system processes

The change in entropy during a process undergone by a closed system can be written using Eq. (7.14) as

$$S_2 - S_1 \geq \int_1^2 \left(\frac{đQ}{T} \right)$$

The inequality sign in the above equation implies that the change in entropy for an irreversible process undergone by a closed system is greater than the integral of $đQ/T$ evaluated for that process and hence some entropy is generated during this process. Thus, the above equation can be written as

$$S_2 - S_1 = \int_1^2 \left(\frac{đQ}{T} \right) + S_{gen} \tag{7.24}$$

where S_{gen} is the entropy generation during the process due to internal irreversibilities present in the system. Thus, we can say the expression $dS - \frac{đQ}{T}$ represents the entropy generation due to internal irreversibilities.

It can be concluded from the above equation that the change of entropy during a process undergone by a closed system takes place due to two reasons:

1. By heat interaction in which there is entropy transfer.
2. By internal irreversibilities or dissipative effects in which work (or K.E.) is dissipated into internal energy and in this way internal energy of the system increases.

Note that the entropy transfer due to heat interaction for a system may be positive or negative depending on whether heat is transferred to the system or from the system. On the contrary, the entropy generation S_{gen} is always a positive quantity or zero in the limit if the process is reversible. The value of S_{gen} depends on the process and hence it is not a property of the system.

It may so happen that in a process (e.g., the expansion of a hot fluid in a turbine) the entropy decrease of the system due to heat loss to the surroundings $\left(-\int \frac{đQ}{T} \right)$ is equal to the entropy increase of the system due to internal irreversibilities such as friction, etc. (S_{gen}), in which case the entropy of the system before and after the process will remain the same $\left(\int dS = 0 \right)$. *Therefore, an isentropic process need not be adiabatic or reversible. But if the isentropic process is reversible, it must be adiabatic. Also, if the isentropic process is adiabatic, it cannot but be reversible. An adiabatic process need not be isentropic, since entropy can also increase due to friction etc. But if the process is adiabatic and reversible, it must be isentropic.*

For an infinitesimal reversible process by a closed system involving only pdV-work,

$$đQ_R = dU_R + pdV$$

If the process is irreversible,

$$đQ_I = dU_I + đW$$

Since U is a property,

$$dU_R = dU_I$$

$$\therefore \quad đQ_R - pdV = đQ_I - đW$$

or

$$\left(\frac{đQ}{T}\right)_R = \left(\frac{dQ}{T}\right)_I + \frac{pdV - đW}{T} \tag{7.25}$$

The difference ($pdV - đW$) indicates the work that is lost due to irreversibility, and is called the *lost work* $d(LW)$, which approaches zero as the process approaches reversibility as a limit. Equation (7.26) can be expressed in the form

$$dS = \frac{đQ}{T} + \frac{d(LW)}{T} \tag{7.26}$$

From the above equation it can be physically interpreted that the entropy generation S_{gen} is associated with the lost work due to internal irreversibility present in the system. The lost work is always positive and hence the entropy generation is a positive quantity.

In any process executed by a system, energy is always conserved, but entropy is produced internally. For any process between equilibrium states 1 and 2 (Fig. 7.21), the first law can be written as follows:

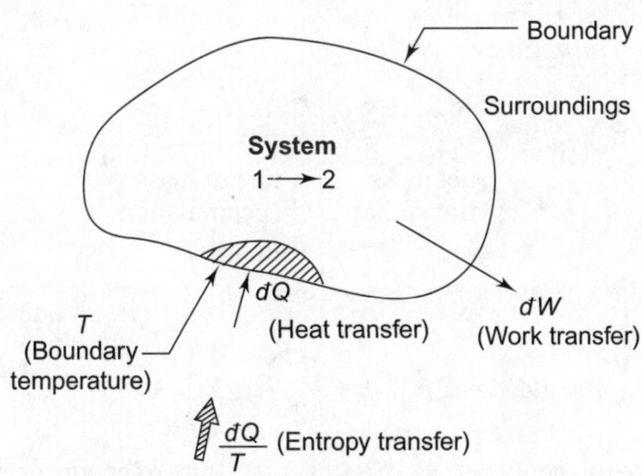

Fig. 7.21 Schematic of a closed system interacting with its surroundings

$$\int_1^2 đQ - \int_1^2 đW = E_2 - E_1$$

or Energy transfer Energy change

$$Q_{1-2} = E_2 - E_1 + W_{1-2}$$

By the second law,

$$S_2 - S_1 \geq \int_1^2 \frac{đQ}{T}$$

7.10 ENTROPY GENERATION IN AN OPEN SYSTEM

In an open system, there is transfer of three quantities: mass, energy and entropy. The control surface can have one or more openings for mass transfer (Fig. 7.22). It is rigid, and there is shaft work transfer across it.

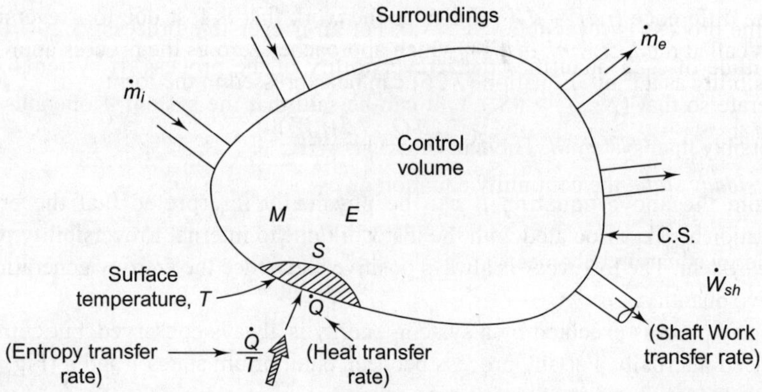

Fig. 7.22 Schematic of an open system and its interaction with surroundings

The *continuity equation* gives

$$\sum_i \dot{m}_1 - \sum_e \dot{m}_e = \frac{\partial M}{\partial \tau} \tag{7.27}$$

net mass transfer rate — rate of mass accumulation in the CV

The *energy equation* gives

$$\sum_i \dot{m}_1 \left(h + \frac{\mathbf{V}^2}{2} + gZ \right)_i - \sum_e \dot{m}_e \left(h + \frac{\mathbf{V}^2}{2} + gZ \right)_e + \dot{Q} - \dot{W}_{\text{sh}} = \frac{\partial E}{\partial \tau} \tag{7.28}$$

net rate of energy transfer — rate of energy accumulation in the CV

The *second law inequality* or the entropy principle gives

$$\sum_i \dot{m}_i S_i - \sum_e \dot{m}_e S_e + \frac{\dot{Q}}{T} \leq \frac{\partial S}{\partial \tau} \tag{7.29}$$

net rate of entropy transfer — rate of increase of entropy of the CV

Here $\dot{Q}$ represents the rate of heat transfer at the location of the boundary where the instantaneous temperature is T. The ratio $\dot{Q}/T$ accounts for the entropy transfer along with heat. The terms $\dot{m}_i S_i$ and $\dot{m}_e S_e$ account, respectively, for rates of entropy transfer into and out of the CV accompanying mass flow. The rate of entropy increase of the control volume exceeds, or is equal to, the net rate of entropy transfer into it. The difference is the entropy generated within the control volume due to irreversibility. Hence, the rate of entropy generation is given by

$$\dot{S}_{gen} = \frac{\partial S}{\partial \tau} - \sum_i \dot{m}_i S_i + \sum_e \dot{m}_e S_e - \frac{\dot{Q}}{T} \tag{7.30}$$

By the second law,

$$\dot{S}_{gen} \geq 0$$

If the process is reversible, $\dot{S}_{gen} = 0$. For an irreversible process, $\dot{S}_{gen} > 0$. The magnitude of $\dot{S}_{gen}$ quantifies the irreversibility of the process. If systems A and B operate so that $(\dot{S}_{gen})_A > (\dot{S}_{gen})_B$ it can be said that the system A operates more irreversibly than system B. The unit of $\dot{S}_{gen}$ is W/K.

At *steady state*, the continuity equation gives

$$\sum_i \dot{m}_i = \sum_e \dot{m}_e \tag{7.31}$$

the energy equation becomes

$$0 = \dot{Q} - \dot{W}_{sh} + \sum_i \dot{m}_i \left(h + \frac{\mathbf{V}^2}{2} + gZ \right)_i - \sum_e \dot{m}_e \left(h + \frac{\mathbf{V}^2}{2} + gZ \right)_e \tag{7.32}$$

and the entropy equation reduces to

$$0 = \frac{\dot{Q}}{T} + \sum_i \dot{m}_i S_i - \sum_e \dot{m}_e s_e + \dot{S}_{gen} \tag{7.33}$$

These equations often must be solved simultaneously, together with appropriate property relations.

Mass and energy are conserved quantities, but entropy is not generally conserved. The rate at which entropy is transferred out must exceed the rate at which entropy enters the CV, the difference being the rate of entropy generated within the CV owing to irreversibilities.

For one-inlet and one-exit control volumes, the entropy equation becomes

$$0 = \frac{\dot{Q}}{T} + \dot{m}(s_1 - s_2) + \dot{S}_{gen}$$

$$\therefore \quad s_2 - s_1 = \frac{1}{\dot{m}} \left(\frac{\dot{Q}}{T} \right) + \frac{\dot{S}_{gen}}{\dot{m}} \tag{7.34}$$

7.11 COMBINED CONSEQUENCES OF FIRST AND SECOND LAWS

> **LO 7.10**
> Combine principles of first and second law of thermodynamics

The heat transfer during a reversible process can be expressed using the second law as

$$đQ_{rev} = TdS$$

and by the first law, for a closed non-flow system involving only pdV work,

$$đQ_{\text{rev}} = dU + pdV$$

Combing the above two equations, we get

$$TdS = dU + pdV \tag{7.35}$$

Again the enthalpy

$$H = U + pV$$

or $$dH = dU + pdV + Vdp$$

or $$dH = TdS + Vdp$$

or $$TdS = dH - Vdp \tag{7.36}$$

Equations (7.35) and (7.36) are the thermodynamic equations relating the properties of the system and are used to estimate the entropy change of a system in a process between two states.

Let us now examine the following equations as obtained from the first and second laws:

1. $đQ = dE + đW$; This equation holds good for any process, reversible or irreversible, and for any system.
2. $đQ = dU + đW$; This equation holds good for any process undergone by a closed stationary system.
3. $đQ = dU + pdV$; This equation holds good for a reversible (quasi-static) process undergone by a closed system when only pdV-work is present.
4. $đQ = TdS$; This equation is true only for a reversible process.
5. $TdS = dU + pdV$; This equation holds good for any process reversible or irreversible, undergone by a closed system involving only pdV -work, since it is a relation among properties, which are independent of the path.
6. $TdS = dH - Vdp$; This equation also relates only the properties of the closed system. There is no path function term in the equation. Hence, the equation holds good for any process when only pdV-work is present.

The use of the term 'irreversible process' is not proper, since no irreversible path or process can be plotted on thermodynamic coordinates. It is more logical to state that 'the change of state is irreversible' rather than say 'it is an irreversible process'. A natural process which is inherently irreversible is indicated by a dotted line connecting the initial and final states, both of which are in equilibrium. The dotted line has no other meaning, since it can be drawn in any way. To determine the entropy change for a real process, a known reversible path is made to connect the two end states, and integration is performed on this path using either Eq. 5 or Eq. 6, as given above. Therefore, the entropy change of a system between two identifiable equilibrium states is the same whether the intervening process is reversible or the change of state is irreversible.

7.12 REVERSIBLE ADIABATIC WORK IN A STEADY FLOW SYSTEM

LO 7.11

Explain reversible work for a steady flow system

In the differential form, the steady flow energy equation per unit mass is given by Eq. (5.11),

$$đQ = dh + \mathbf{V}d\mathbf{V} + gdZ + đW_x$$

For a reversible process, $đQ = Tds$

$$\therefore \qquad Tds = dh + \mathbf{V}d\mathbf{V} + gdZ + đW_x \tag{7.37}$$

Using the property relation, Eq. (7.36), per unit mass

$$Tds = dh - vdp$$

In Eq. (7.37), we have

$$-v\delta p = \mathbf{V}d\mathbf{V} + gdZ + đW_x \tag{7.38}$$

On integration

$$-\int_1^2 vdp = \Delta\frac{\mathbf{V}^2}{2} + g\Delta Z + W_x \tag{7.39}$$

If the changes in K.E. and P.E. are neglected, Eq. (7.39) reduces to

$$W_x = -\int_1^2 vdp \tag{7.40}$$

If $đQ = 0$, implying $dS = 0$, the property relation gives

$$dh = vdp$$

or

$$h_2 - h_1 = \int_1^2 vdp \tag{7.41}$$

From Eqs. (7.40) and (7.41),

$$W_x = h_1 - h_2 = -\int_1^2 vdp \tag{7.42}$$

The integral $-\int_1^2 vdp$ represents an area on the p–v plane (Fig. 7.23). To make the integration, one must have a relation between p and v such as pv^n = constant.

$$\therefore \qquad W_{1-2} = h_1 - h_2 = -\int_1^2 vdp$$

$$= \text{area } 12\, ab\, 1$$

Equation (7.42) holds good for a steady flow work-producing machine like an engine or turbine as well as for a work-absorbing machine like a pump or a compressor, when the fluid undergoes reversible adiabatic expansion or compression.

Figure 7.23 shows some steady flow machines where the work done is $-\int_1^2 vdp$

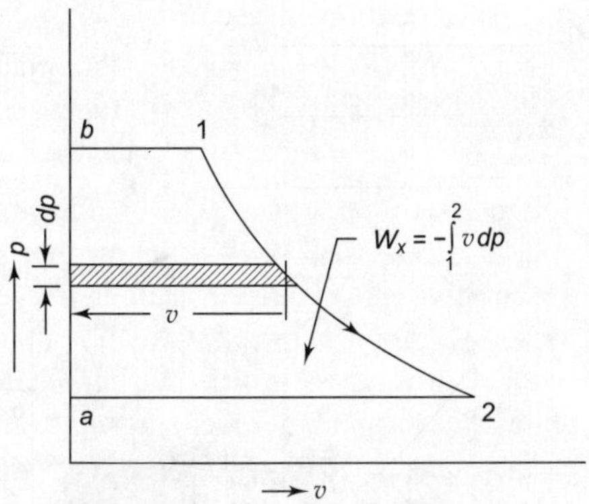

Fig. 7.23 Reversible steady flow work interaction

It may be noted that for a closed stationary system like a gas confined in a piston-cylinder machine (Fig. 7.24a), the reversible work done would be

$$W_{1-2} = \int_1^2 pdV = \text{Area } 12\, cd\, 1$$

The reversible work done by a steady flow system (Fig. 7.24b) would be

$$W_{1-2} = -\int_1^2 v\, dp = \text{Area } 12\, ab\, 1$$

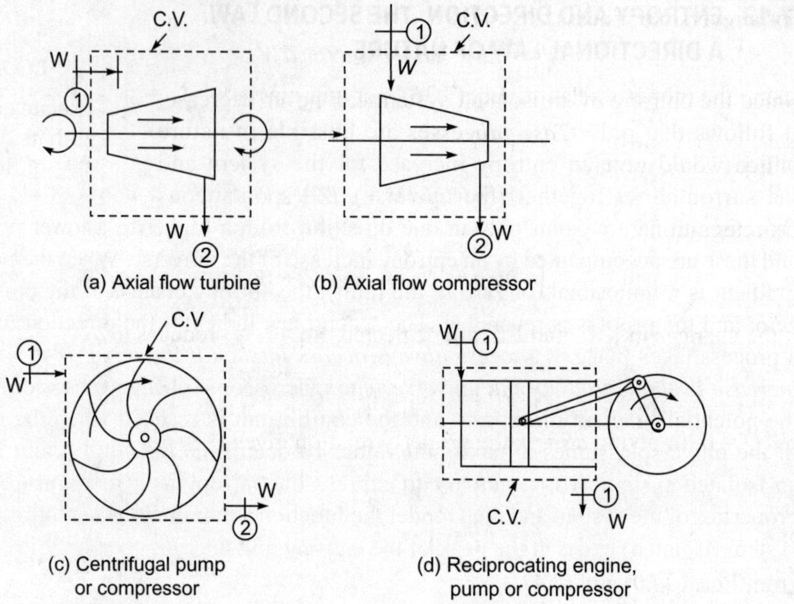

Fig. 7.24 Work done by steady flow machines, $-\int_1^2 vdp$

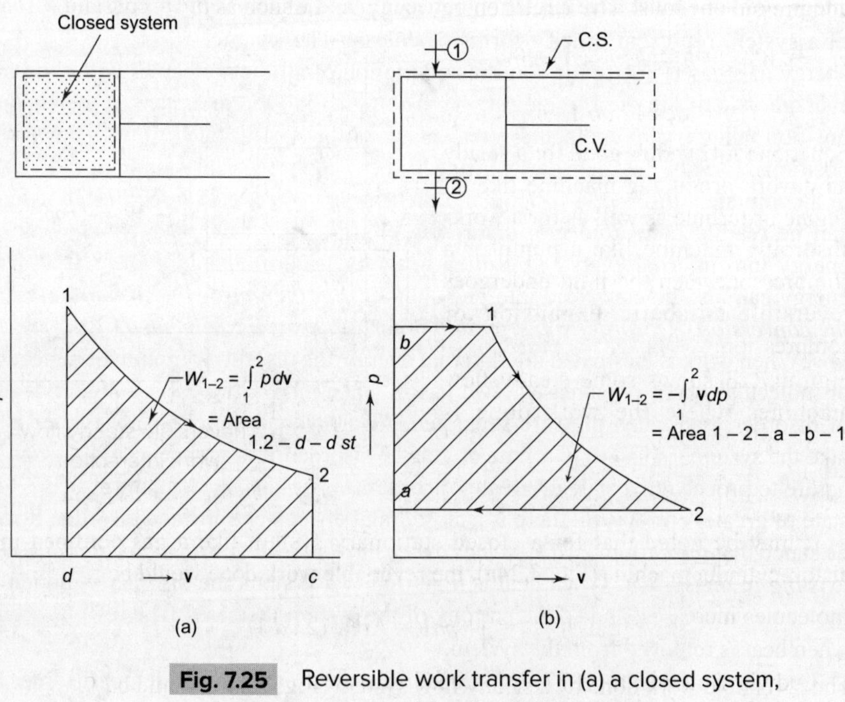

Fig. 7.25 Reversible work transfer in (a) a closed system, (b) a steady flow system

7.13 ENTROPY AND DIRECTION: THE SECOND LAW, A DIRECTIONAL LAW OF NATURE

LO 7.12

Relate entropy with its direction of flow

Since the entropy of an isolated system can never decrease, it follows that only those processes are possible in nature which would give an entropy increase for the system and the surroundings together (the universe). All spontaneous processes in nature occur only in one direction from a higher to a lower potential, and these are accompanied by an entropy increase of the universe. When the potential gradient is infinitesimal (or zero in the limit), the entropy change of the universe is zero, and the process is reversible. The second law indicates the direction in which a process takes place. *A process always occurs in such a direction as to cause an increase in the entropy of the universe*. The macroscopic change ceases only when the potential gradient disappears and the equilibrium is reached when the entropy of the universe assumes a maximum value. To determine the equilibrium state of an isolated system, it is necessary to express the entropy as a function of certain properties of the system and then render the function a maximum. At equilibrium, the system (isolated) exists at the peak of the entropy hill, and $dS = 0$ (Fig. 7.11).

7.14 ENTROPY AND DISORDER

LO 7.13

Relate entropy with disorder

Work is a macroscopic concept. Work involves order or the orderly motion of molecules, as in the expansion or compression of a gas. The kinetic energy and potential energy of a system represent orderly forms of energy. The kinetic energy of a gas is due to the coordinated motion of all the molecules with the same average velocity in the same direction. The potential energy is due to the vantage position taken by the molecules or displacements of molecules from their normal positions. Heat or thermal energy is due to the random thermal motion of molecules in a completely disorderly fashion and the average velocity is zero. Orderly energy can be readily converted into disorderly energy, e.g., mechanical and electrical energy into internal energy (and then heat) by friction and Joule effect. Orderly energy can also be converted into one another. *But there are natural limitations on the conversion of disorderly energy into orderly energy, as delineated by the second law*. When work is dissipated into internal energy, the disorderly motion of molecules or molecular chaos is increased. Two gases, when mixed, represent a higher degree of disorder than when they are separated. An irreversible process always tends to take the system (isolated) to a state of greater disorder. It is a tendency on the part of nature to proceed to a state of greater disorder. An isolated system always tends to a state of greater entropy. So there is a close link between entropy and disorder. It may be stated that *the entropy of a system is a measure of the degree of molecular disorder existing in the system*. When heat is imparted to a system, the disorderly motion of molecules increases, and so the entropy of the system increases. The reverse occurs when heat is removed from the system.

Ludwig Boltzmann (1877) introduced statistical concepts to define disorder by attaching to each state a *thermodynamic probability*, expressed by the quantity W, which is greater the more disordered the state is. The increase of entropy implies that the system proceeds by itself from one state to another with a higher thermodynamic probability (or disorder number). An irreversible process goes on until the *most probable state* (equilibrium state when W is maximum) corresponding to the maximum value of entropy is reached. Boltzmann assumed a functional relation between S and W. While entropy is additive, probability is multiplicative. If the two parts A and B of a system in equilibrium are considered, the entropy is the sum

$$S = S_A + S_B$$

and the thermodynamic probability is the product

$$W = W_A \cdot W_B$$

Again, $S = S(W)$, $S_A = S(W_A)$ and $S_B = S(W_B)$

$$\therefore \qquad S(W) = S(W_A W_B) = S(W_A) + S(W_B)$$

which is a well-known functional equation for the logarithm. Thus the famous relation is reached

$$S = K \ln W$$

where K is a constant, known as Boltzmann constant. This is engraved upon Boltzmann's tombstone in Vienna.

When $W = 1$, which represents the greatest order, $S = 0$. This occurs only at $T = 0\,K$. This state cannot be reached in a finite number of operations. This is the Nernst Simon statement of third law of thermodynamics. In the case of a gas, W increases due to an increase in volume V or temperature T. In the reversible adiabatic expansion of a gas, the increase in disorder due to an increase in volume is just compensated by the decrease in disorder due to a decrease in temperature, so that the disorder number or entropy remains constant.

7.15 CONCEPT OF ABSOLUTE ENTROPY

> **LO 7.14** Explain absolute entropy and postulatory thermodynamics

The entropy of a pure crystalline substance at absolute zero temperature is a state of perfect order ($W = 1$) and zero entropy. Since absolute zero temperature is not attainable in a finite number of operations (Section 6.16 in Chapter 6), the absolute zero entropy value is not attainable. It is the Nernst–Simon statement of third law of thermodynamics.

It is important to note that one is interested only in the amount by which the entropy of the system changes in going from an initial to final state, and not in the value of absolute entropy. In cases where it is necessary, a zero value of entropy of the system at an arbitrarily chosen standard state is assigned, and the entropy changes are calculated with reference to this standard state. At triple point of water, the entropy (also internal energy) is arbitrarily chosen to be zero.

7.16 POSTULATORY THERMODYNAMICS

The property 'entropy' plays the central role in thermodynamics. In the *classical approach*, as followed in this book, entropy is introduced via the concept of the heat engine. It follows the way in which the subject of thermodynamics developed historically, mainly through the contributions of Sadi Carnot, James Prescott Joule, William Thomson (Lord Kelvin), Rudolf Clausius, Max Planck and Josiah Willard Gibbs.

In the *postulatory approach*, as developed by HB Callen (see Reference), entropy is introduced at the beginning. The development of the subject has been based on four postulates. Postulate I defines the equilibrium state. Postulate II introduces the property 'entropy' which is rendered maximum at the final equilibrium state. Postulate III refers to the additive nature of entropy which is a monotonically increasing function of energy. Postulate IV mentions that the entropy of any system vanishes at the absolute zero of temperature. With the help of these postulates, the conditions of equilibrium under different constraints have been developed.

SOLVED EXAMPLES

Example 7.1 *Water flows through a turbine in which friction causes the water temperature to rise from 35°C to 37°C. If there is no heat transfer, how much does the entropy of the water change in passing through the turbine? (Water is incompressible and the process can be taken to be at constant volume.)*

E [LO 7.3, 7.5]

Solution The presence of friction makes the process irreversible and causes an entropy increase for the system. The flow process is indicated by the dotted line joining the equilibrium states 1 and 2 (Fig. 7.26). Since entropy is a state property and the entropy change depends only on the two end states and is independent of the path the system follows, to find the entropy change, the irreversible path has to be replaced by a reversible path, as shown in figure, because no integration can be made on a path other than a reversible path.

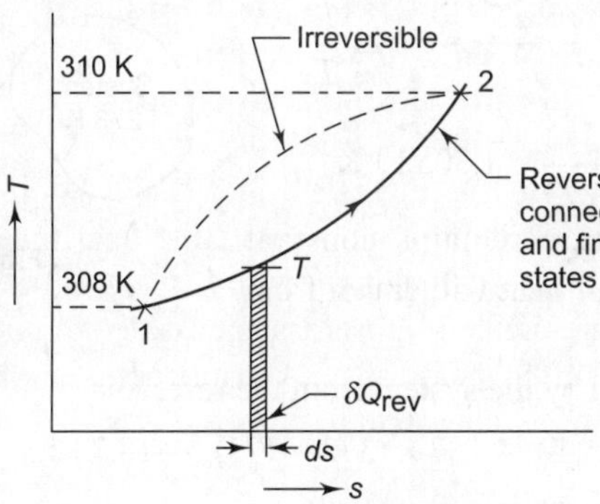

Fig. 7.26

$$T_2 = 37 + 273 = 310 \text{ K}$$
$$T_1 = 35 + 273 = 308 \text{ K}$$

We have

$$đQ_{rev} = TdS$$
$$dS = \frac{mc_v dT}{T}$$
$$S_2 - S_1 = mc_v \ln \frac{T_2}{T_1}$$
$$= 1 \times 4.187 \ln \frac{310}{308} = \mathbf{0.0243 \ kJ/K}$$

Example 7.2 *(a) One kg of water at 273 K is brought into contact with a heat reservoir at 373 K. When the water has reached 373 K, find the entropy change of the water, of the heat reservoir and of the universe. (b) If water is heated from 273 to 373 K by first bringing it in contact with a reservoir at 323 K and then with a reservoir at 373 K, what will the entropy change of the universe be? (c) Explain how water might be heated from 273 to 373 K with almost no change in the entropy of the universe.* M **[LO 7.3, 7.5]**

Solution (a) Water is being heated through a finite temperature difference (Fig. 7.27). The entropy of water would increase and that of the reservoir would decrease so that the net entropy change of the water (system) and the reservoir together would be positive definite. Water is being heated irreversibly, and to find the entropy change of water, we have to assume a reversible path between the end states that are at equilibrium.

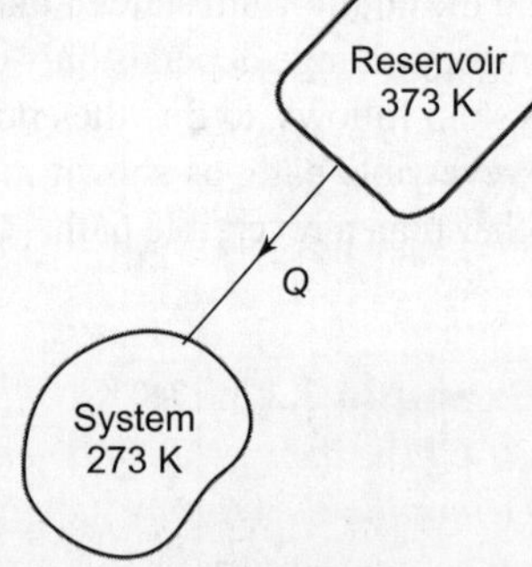

Fig. 7.27

$$(\Delta S)_{water} = \int \frac{đQ}{T} = \int_{T_1}^{T_1} \frac{mcdT}{T} = mc \ln \frac{T_2}{T_1}$$
$$= 1 \times 4.187 \ln \frac{373}{273} = 1.305 \text{ kJ/K}$$

The reservoir temperature remains constant irrespective of the amount of heat withdrawn from it.

Amount of heat absorbed by the system from the reservoir,

$$Q = 1 \times 4.187 \times (373 - 273) = 418.7 \text{ kJ}$$

∴ Entropy change of the reservoir

$$(\Delta S)_{res} = -\frac{Q}{T} = -\frac{418.7}{373} = -1.122 \text{ kJ/K}$$

∴ entropy change of the universe

$$(\Delta S)_{\text{univ}} = (\Delta S)_{\text{system}} + (\Delta S)_{\text{res}}$$
$$= 1.305 - 1.122 = \mathbf{0.183\ kJ/K}$$

(b) Water is being heated in two stages, first from 273 to 323 by bringing it in contact with a reservoir at 323 K, and then from 323 K to 373 K by bringing it in contact of a second reservoir at 373 K.

$$(\Delta S)_{\text{water}} = \int_{273\text{ K}}^{323\text{ K}} mc\frac{dT}{T} + \int_{323\text{ K}}^{373\text{ K}} mc\frac{dT}{T}$$

$$= 4.187\left(\ln\frac{323}{273} + \ln\frac{373}{323}\right) = (0.1673 + 0.1441)\ 4.187$$

$$= 1.305\text{ kJ/K}$$

$$(\Delta S)_{\text{res I}} = -\frac{1 \times 4.187 \times (323 - 273)}{323} = -\ 0.647\text{ kJ/K}$$

$$(\Delta S)_{\text{res II}} = \frac{1 \times 4.187 \times (373 - 323)}{373} = -\ 0.56\text{ kJ/K}$$

$$\therefore \quad (\Delta S)_{\text{univ}} = (\Delta S)_{\text{water}} + (\Delta S)_{\text{res I}} + (\Delta S)_{\text{res II}}$$
$$= 1.305 - 0.647 - 0.56 = \mathbf{0.098\ kJ/K}$$

(c) The entropy change of the universe would be less and less, if the water is heated in more and more stages, by bringing the water in contact successively with more and more heat reservoirs, each succeeding reservoir being at a higher temperature than the preceding one.

When water is heated in infinite steps, by bringing it in contact with an infinite number of reservoirs in succession, so that at any instant the temperature difference between the water and the reservoir in contact is infinitesimally small, then the net entropy change of the universe would be zero, and the water would be *reversibly* heated.

Example 7.3 *One kg of ice at – 5°C is exposed to the atmosphere which is at 20°C. The ice melts and comes into thermal equilibrium with the atmosphere. (a) Determine the entropy increase of the universe. (b) What is the minimum amount of work necessary to convert the water back into ice at – 5°C? c_p of ice is 2.093 kJ/kg K and the latent heat of fusion of ice is 333.3 kJ/kg.* **D [LO 7.7]**

Solution Heat absorbed by ice Q from the atmosphere (Fig. 7.28).

= Heat absorbed in solid phase + Latent heat + Heat absorbed in liquid phase

$$= 1 \times 2.093 \times [0 - (-5)] + 1 \times 333.3 + 1 \times 4.187 \times (20 - 0) = 427.5\text{ kJ}$$

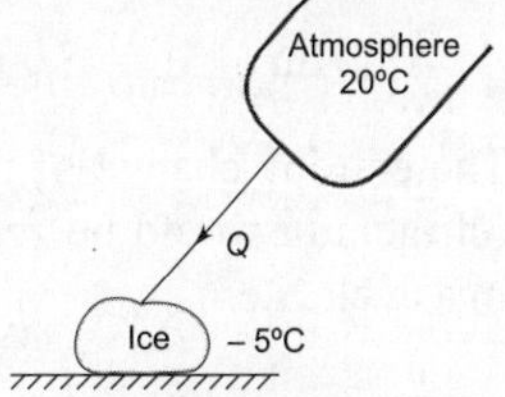

Fig. 7.28

Entropy change of the atmosphere

$$(\Delta S)_{atm} = -\frac{Q}{T} = -\frac{427.5}{293} = -1.46 \text{ kJ/K}$$

Entropy change of the system (ice) as it gets heated from – 5C to 0°C

$$(\Delta S_I)_{system} = \int_{268}^{293} mc_p \frac{dT}{T} = 1 \times 2.093 \ln \frac{273}{268} = 2.093 \times 0.0186$$
$$= 0.0389 \text{ kJ/K}$$

Entropy change of the system as ice melts at 0°C to become water at 0°C

$$(\Delta S_{II})_{system} = \frac{333.3}{273} = 1.22 \text{ kJ/K}$$

Entropy change of water as it gets heated from 0°C to 20°C

$$(\Delta S_{III})_{system} = \int_{273}^{293} mc_p \frac{dT}{T} = 1 \times 4.187 \ln \frac{293}{273} = 0.296 \text{ kJ/K}$$

Total entropy change of ice as it melts into water

$$(\Delta S)_{total} = \Delta S_I + \Delta S_{II} + \Delta S_{III}$$
$$= 0.0389 + 1.22 + 0.296 = 1.5549 \text{ kJ/K}$$

The entropy–temperature diagram for the system as ice at – 5°C converts to water at 20°C is shown in Fig. 7.29.

∴ entropy increase of the universe

$$(\Delta S)_{univ} = (\Delta S)_{system} + (\Delta S)_{atm}$$
$$= 1.5549 - 1.46 = \mathbf{0.0949 \text{ kJ/K}}$$

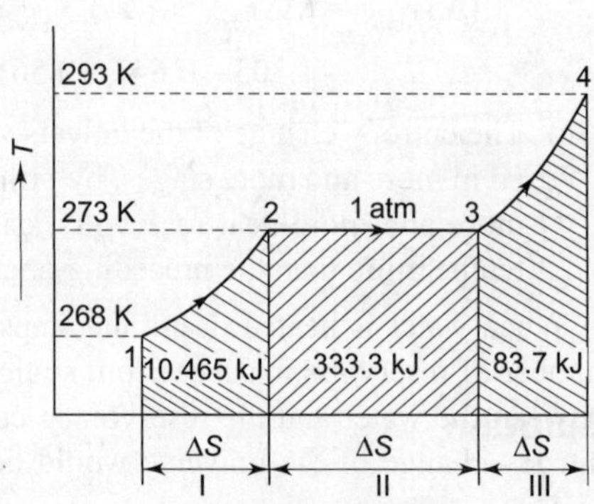

Fig. 7.29

(b) To convert 1 kg of water at 20°C to ice at – 5°C, 427.5 kJ of heat have to be removed from it, and the system has to be brought from state 4 to state 1 (Fig. 7.29). A refrigerator cycle, as shown in Fig. 7.30, is assumed to accomplish this.

The entropy change of the system would be the same, i.e. $S_4 - S_1$, with the only difference that its sign will be negative, because heat is removed from the system (Fig. 7.30).

$$(\Delta S)_{system} = S_1 - S_4 (\text{negative})$$

The entropy change of the working fluid in the refrigerator would be zero, since it is operating in a cycle, i.e.

$$(\Delta S)_{ref} = 0$$

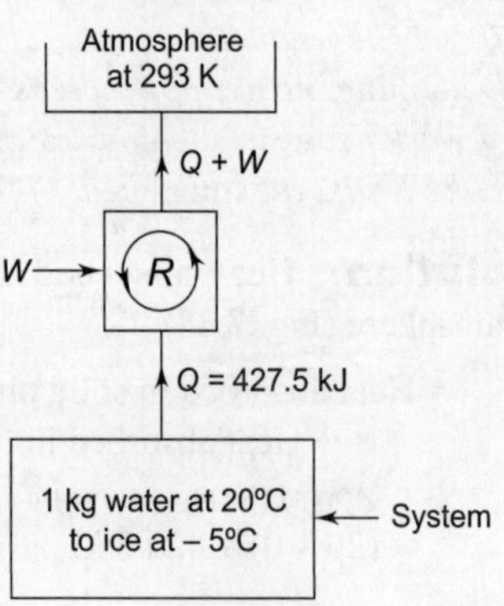

Fig. 7.30

The entropy change of the atmosphere (positive)

$$(\Delta S)_{\text{atm}} = \frac{Q+W}{T}$$

∴ Entropy change of the universe

$$(\Delta S)_{\text{univ}} = (\Delta S)_{\text{system}} + (\Delta S)_{\text{ref}} + (\Delta S)_{\text{atm}}$$

$$= (S_1 - S_4) + \frac{Q+W}{T}$$

By the principle of increase of entropy

$$(\Delta S)_{\text{univ or isolated system}} \geq 0$$

$$\therefore \quad \left[(S_1 - S_4) + \frac{Q+W}{T}\right] \geq 0$$

$$\therefore \quad \frac{Q+W}{T} \geq (S_4 - S_1)$$

$$W \geq T(S_4 - S_1) - Q$$

$$\therefore \quad W(\text{min}) = T(S_4 - S_1) - Q$$

Here

$$Q = 427.5 \text{ kJ}$$

$$T = 293 \text{ K}$$

$$S_4 - S_1 = 1.5549 \text{ kJ/K}$$

$$\therefore \quad W(\text{min}) = 293 \times 1.5549 - 427.5 = \mathbf{28.5 \text{ kJ}}$$

Example 7.4 *A body of constant heat capacity c_p and initial temperature T_1 is placed in contact with a heat reservoir at temperature T_2 and comes to thermal equilibrium with it. If $T_2 > T_1$, calculate the entropy change of the universe and show that this is always positive.* Ⓜ [LO 7.7]

Solution

$$\Delta S_{\text{body}} = mc_p \int_{T_1}^{T_2} \frac{dT}{T} = mc_p \ln \frac{T_2}{T_1}$$

$$\Delta S_{\text{reservoir}} = \frac{-mc_p (T_2 - T_1)}{T_2}$$

$$\Delta S_{\text{universe}} = mc_p \left[\ln \frac{T_2}{T_1} - \left(1 - \frac{T_1}{T_2}\right)\right]$$

Putting

$$1 - \frac{T_1}{T_2} = x, \quad \frac{T_2}{T_1} = \frac{1}{1-x}$$

Since $\frac{T_2}{T_1} > 1$, x is positive.

$$\frac{\Delta S_{\text{universe}}}{mc_p} = -\ln(1-x) - x$$

$$= \frac{x^2}{2} + \frac{x^2}{3} + \frac{x^4}{4} + \cdots$$

$\therefore \qquad \Delta S_{universe} > 0.$ Prored

Example 7.5 *In order to obtain the desired mechanical properties, heat treatment is done during manufacturing of metal products. For that purpose, 1 kg of molten steel, initially at 1200°C is quenched in a tank containing 80 kg of water, initially at 20°C. The specific heat of steel and water are 0.5 kJ/kgK and 4.18 kJ/kgK respectively. If there is negligible heat transfer between the tank and the ambient, determine the entropy generated within the tank due to quenching.*

Ⓜ [LO 7.7]

Solution Let T_f be the final temperature after attaining thermal equilibrium. Using the first law of thermodynamics, we have

$$\Delta U|_{steel} + \Delta U|_{water} = 0$$

or

$$m_{st} c_{p,st} (T_f - T_{i,st}) + m_w \, c_{p,w} (T_f - T_{i,w}) = 0$$

or

$$T_f = \frac{m_{st} c_{p,st} T_{i,st} + m_w c_{p,w} T_{i,w}}{m_{st} c_{p,st} + m_w c_{p,w}} = \frac{1 \times 0.5 \times 1473 + 80 \times 4.18 \times 293}{1 \times 0.5 + 80 \times 4.18} = 294.76 \text{ K}$$

Entropy change during this process is found to be

$$S_2 - S_1 = (\Delta S)_{st} + (\Delta S)_w = m_{st} c_{p,st} \ln \frac{T_f}{T_{i,st}} + m_w c_{p,w} \ln \frac{T_f}{T_{i,w}}$$

$$= 1 \times 0.5 \times \ln \frac{294.76}{1473} + 80 \times 4.18 \times \ln \frac{294.76}{293} = 1.198 \text{ kJ/K}$$

The entropy generated within the tank due to quenching can be found using the following equation

$$S_2 - S_1 = \cancel{\sum_j} \frac{\cancel{Q_j}^{=0}}{T_j} + S_{gen}$$

or

$$S_{gen} = \mathbf{1.198 \, kJ/K}$$

Example 7.6 *A reversible power cycle R and an irreversible power cycle I operate between the same two reservoirs. Each receives Q_H from the hot reservoir. The reversible cycle develops work W_R while the irreversible cycle develops work W_I. (Fig. 7.31).*

(a) Evaluate the rate of entropy generation σ for cycle I in terms of W_I, W_R and the temperature T_C of the cold reservoir.
(b) Demonstrate that σ must be positive.

Ⓜ [LO 7.2]

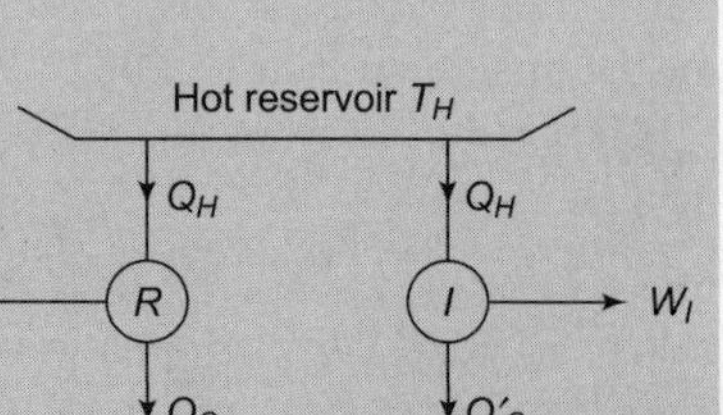

Fig. 7.31

Solution

(a) For cycle I,

$$\sigma = -\oint \frac{đQ}{T} = -\left[\frac{Q_H}{T_H} - \frac{Q_C'}{T_C}\right]$$

Energy balances for cycles R and I give

$$Q_H = W_R + Q_C$$

$$Q_H = W_I + Q_C'$$

$$\therefore \quad Q_C' = Q_C + W_R - W_I$$

Therefore,

$$\sigma = -\left[\left(\frac{Q_H}{T_H} - \frac{Q_C}{T_C}\right) - \left(\frac{W_R - W_I}{T_C}\right)\right]$$

Since R is reversible,

$$\frac{Q_H}{T_H} = \frac{Q_C}{T_C}$$

$$\therefore \quad \boldsymbol{\sigma = \frac{W_R - W_I}{T_C}}$$

Example 7.7 *Two identical bodies of constant heat capacity are at the same initial temperature T_i. A refrigerator operates between these two bodies until one body is cooled to temperature T_2. If the bodies remain at constant pressure and undergo no change of phase, show that the minimum amount of work needed to do this is* **M [LO 7.7]**

$$W(min) = C_p\left(\frac{T_i^2}{T_2} + T_2 - 2T_i\right)$$

Solution Both the finite bodies A and B are initially at the same temperature T_i. Body A is to be cooled to temperature T_2 by operating the refrigerator cycle, as shown in Fig. 7.32. Let T_2' be the final temperature of body B.

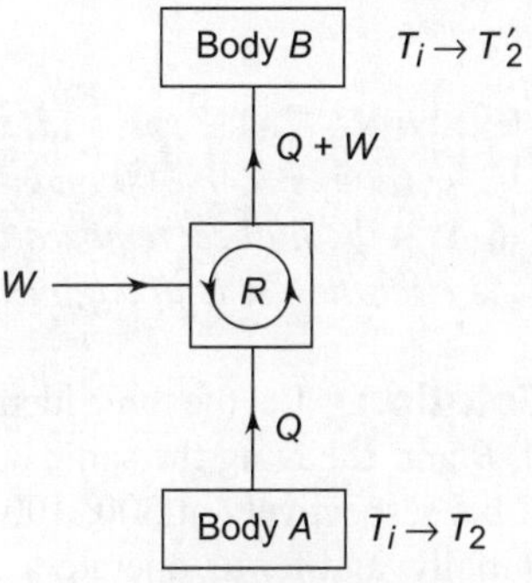

Fig. 7.32

Heat removed from body A to cool it from T_i to T_2

$$Q = C_p(T_i - T_2)$$

where C_p is the constant pressure heat capacity of the identical bodies A and B.

Heat discharged to body B

$$= Q + W = C_p(T_2' - T_i)$$

Work input, W

$$= C_p(T_2' - T_i) - C_p(T_i - T_2)$$

$$= C_p(T_2' + T_2 - 2T_i) \quad (7.43)$$

Now, the entropy change of body A

$$\Delta S_A = \int_{T_i}^{T_2} C_p \frac{dT}{T} = C_p \ln \frac{T_2}{T_i} \text{ (negative)}$$

The entropy change of body B

$$\Delta S_B = \int_{T_i}^{T_2'} C_p \frac{dT}{T} = C_p \ln \frac{T_2'}{T_i} \text{ (positive)}$$

Entropy change (cycle) of refrigerant = 0

∴ entropy change of the universe

$$(\Delta S)_{\text{univ}} = \Delta S_A + \Delta S_B$$

$$= C_p \ln \frac{T_2}{T_i} + C_p \ln \frac{T_2'}{T_i}$$

By the entropy principle

$$(\Delta S)_{\text{univ}} \geq 0$$

$$\left(C_p \ln \frac{T_2}{T_i} + C_p \ln \frac{T_2'}{T_i} \right) \geq 0$$

$$C_p \ln \frac{T_2 T_2'}{T_i^2} \geq 0 \tag{7.44}$$

In Eq. (7.43) with C_p, T_2 and T_i being given, W will be a minimum when T_2' is a minimum. From Eq. (7.44), the minimum value of T_2' would correspond to

$$C_p \ln \frac{T_2 T_2'}{T_i^2} = 0 = \ln 1$$

$$\therefore \quad T_2' = \frac{T_i^2}{T_2}$$

From Eq. (7.43)

$$W(\text{min}) = C_p \left(\frac{T_i^2}{T_2} + T_2 - 2T_i \right)$$

Example 7.8 *Three identical finite bodies of constant heat capacity are at temperatures 300, 300 and 100 K. If no work or heat is supplied from outside, what is the highest temperature to which any one of the bodies can be raised by the operation of heat engines or refrigerators?* M [LO 7.7]

Solution Let the three identical bodies A, B and C having the same heat capacity C be, respectively at 300, 100 and 300 K initially, and let us operate a heat engine and a refrigerator, as shown in Fig. 7.33. Let T_f be the final temperature of bodies A and B, and T_f' be the final temperature of body C. Now

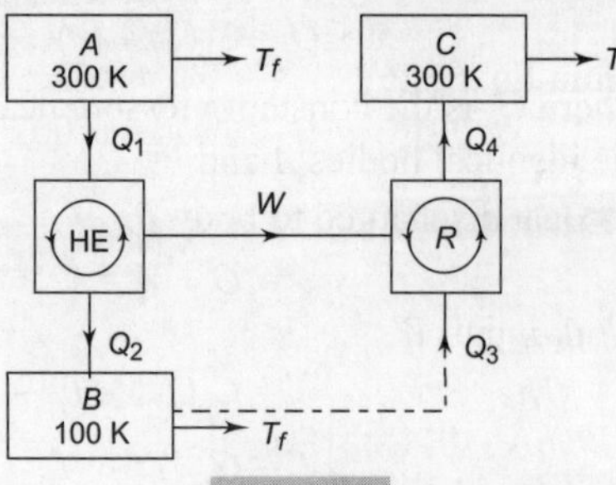

Fig. 7.33

$$(\Delta S)_A = C \ln \frac{T_f}{300}$$

$$(\Delta S)_B = C \ln \frac{T_f}{100}$$

$$(\Delta S)_C = C \ln \frac{T_f'}{300}$$

$$(\Delta S)_{\text{H.E.}} = 0$$

$$(\Delta S)_{\text{ref}} = 0$$

where C is the heat capacity of each of the three bodies.

Since $(\Delta S)_{\text{univ}} \geq 0$

$$\left(C \ln \frac{T_f}{300} + C \ln \frac{T_f}{100} + C \ln \frac{T_f'}{300} \right) \geq 0$$

$$C \ln \frac{T_f^2 T_f'}{9{,}000{,}000} \geq 0$$

For minimum value of T_f, $C \ln \dfrac{T_f^2 T_f'}{9 \times 10^6} = 0 = \ln 1$

$$\therefore \quad T_f^2 T_f' = 9{,}000{,}000 \tag{7.45}$$

Now

$$Q_1 = C\,(300 - T_f)$$

$$Q_2 = C\,(T_f - 100)$$

$$Q_4 = C\,(T_f' - 300)$$

Again

$$Q_1 = \text{Heat removed from body } A$$

$$= \text{Heat discharged to bodies } B \text{ and } C$$

$$= Q_2 + Q_4$$

$$\therefore \quad C\,(300 - T_f) = C\,(T_f - 100) + C\,(T_f' - 300)$$

$$\therefore \quad T_f' = 700 - 2T_f \tag{7.46}$$

T_f' will be the highest value when T_f is the minimum.

From Eqs. (7.45) and (7.46)

$$T_f^2\,(700 - 2T_f) = 9{,}000{,}000$$

$$\therefore \quad 2T_f^3 - 700\,T_f^2 + 9{,}000{,}000 = 0$$

or $T_f = 150$ K

From Eq. (7.46)

$$T_f' = (700 - 2 \times 150)\text{ K} = \mathbf{400\ K}$$

Example 7.9 *A system has a heat capacity at constant volume*

$$C_V = AT^2$$

where A = 0.042 J/K^3.

The system is originally at 200 K, and a thermal reservoir at 100 K is available. What is the maximum amount of work that can be recovered as the system is cooled down to the temperature of the reservoir? Ⓜ **[LO 7.7]**

Solution Heat removed from the system (Fig. 7.34)

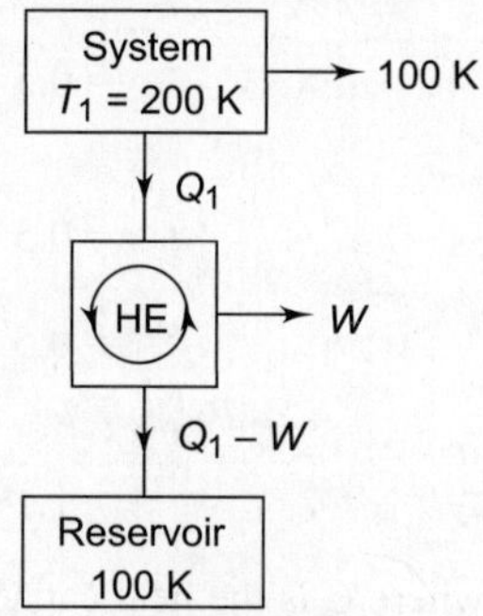

Fig. 7.34

$$Q_1 = \int_{T_1}^{T_2} C_V\, dT = \int_{T_1 = 200\text{ K}}^{T_2 = 100\text{ K}} 0.042\, T^2\, \delta T$$

$$= 0.042 \left[\frac{T^3}{3}\right]_{200\text{ K}}^{100\text{ K}}$$

$$= \frac{0.042}{3}\ \text{J/K}^3\,(100^3 - 200^3)\text{K}^3 = -98 \times 10^3\ \text{J}$$

$$(\Delta S)_{\text{system}} = \int_{200\text{K}}^{100\text{K}} C_V \frac{dT}{T} = \int_{200\text{K}}^{100\text{K}} 0.042\, T^2 \frac{dT}{T}$$

$$= \frac{0.042}{2}\ \text{J/K}^3\,[100^2 - 200^2]\text{K}^2 = -630\ \text{J/K}$$

$$(\Delta S)_{\text{res}} = \frac{Q_1 - W}{T_{\text{res}}} = \frac{98 \times 10^3 - W}{100}\ \text{J/K}$$

$$(\Delta S)_{\text{working fluid in H.E.}} = 0$$

$$\therefore \quad (\Delta S)_{\text{univ}} = (\Delta S)_{\text{system}} + (\Delta S)_{\text{res}}$$

$$= -630 + \frac{98 \times 10^3 - W}{100}$$

Since $(\Delta S)_{\text{univ}} \geq 0$

$$\therefore \quad -630 + \frac{98 \times 10^3 - W}{100} \geq 0$$

$$980 - \frac{W}{100} - 630 \geq 0$$

$$\frac{W}{100} \leq 350$$

$$W(\text{max}) = 35{,}000\ \text{J} = \mathbf{35\ kJ}$$

Example 7.10 *A fluid undergoes a reversible adiabatic compression from 0.5 MPa, 0.2 m^3 to 0.05 m^3 according to the law and $pv^{1.3}$ = constant. Determine the change in enthalpy, internal energy and entropy and the heat transfer and work transfer during the process.* M **[LO 7.11]**

Solution

$$TdS = dH - Vdp$$

For the reversible adiabatic process (Fig. 7.35)

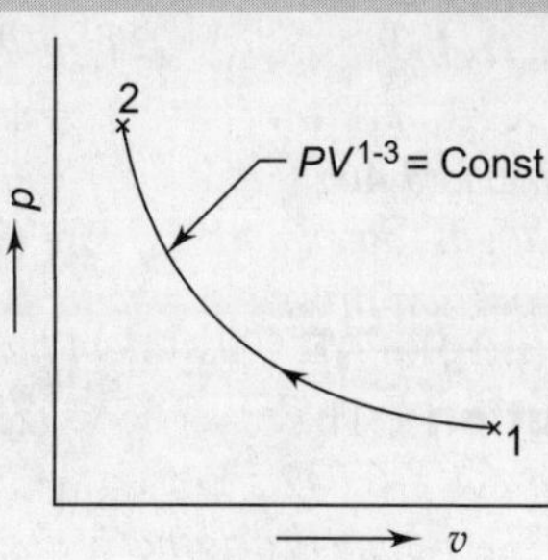

Fig. 7.35

$$dH = Vdp$$

$$p_1 = 0.5\ \text{MPa}, \quad V_1 = 0.2\ \text{m}^3$$

$$V_2 = 0.05\ \text{m}^3, \quad p_1 V_1^n = p_2 V_2^n$$

$$\therefore \quad p_2 = p_1 \left(\frac{V_1}{V_2}\right)^n$$

$$= 0.5 \times \left(\frac{0.20}{0.05}\right)^{1.3} \text{ MPa}$$

$$= 0.5 \times \left(\frac{0.20}{0.05}\right)^{1.3} \text{ MPa}$$

$$= 0.5 \times 6.061 \text{ MPa} = 3.0305 \text{ MPa}$$

$$p_1 V_1^n = pV^n$$

$$\therefore \quad V = \left(\frac{p_1 V_1^n}{p}\right)^{1/n}$$

$$\int_{H_1}^{H_2} dH = \int_{p_1}^{p_2} V dp$$

$$H_2 - H_1 = \int_{p_1}^{p_2} \left[\left(\frac{p_1 V_1^n}{p}\right)\right]^{1/n} dp$$

$$= (p_1 V_1^n)^{1/n} \left(\frac{p_1^{1-1/n} - p_1^{1-1/n}}{1 - 1/n}\right)$$

$$= \frac{n\left(p_2 V_2 - p_1 V_1\right)}{n - 1}$$

$$= \frac{1.3\left(3030.5 \times 0.05 - 500 \times 0.2\right)}{1.3 - 1} = 223.3 \text{ kJ}$$

$$H_2 - H_1 = (U_2 + p_2 V_2) - (U_1 + p_1 V_1)$$

$$= (U_2 - U_1) + (p_2 V_2 - p_1 V_1)$$

$$\therefore \quad U_2 - U_1 = (H_2 - H_1) - (p_2 V_2 - p_1 V_1)$$

$$= 223.3 - 51.53 = \mathbf{171.77 \text{ kJ}}$$

$$S_2 - S_1 = \mathbf{0}$$

$$Q_{1-2} = \mathbf{0}$$

$$Q_{1-2} = \boldsymbol{U_2 - U_1 + W_{1-2}}$$

$$\therefore \quad W_{1-2} = U_1 - U_2 = \mathbf{-171.77 \text{ kJ}}$$

Example 7.11 *Air flows steadily at a rate of 0.2 kg/s through an insulated converging duct having inlet and exit cross-sectional area of 1.3×10^{-3} m² and 4.3×10^{-4} m² respectively. Air enters the duct at a velocity of 130 m/s and pressure of 100 kPa and it leaves at 200 kPa with a velocity of 240 m/s. Assuming air as an ideal gas, determine the rate of entropy production in the duct. The average value of constant pressure specific heat is 1.005 kJ/kgK.* E **[LO 7.11]**

Solution The mass flow rate of air can be expressed as

$$w = \frac{A_1 V_1}{v_1} = \frac{A_2 V_2}{v_2}$$

where v_1 and v_2 can be found from the equation of state as

$$v_1 = \frac{RT_1}{p_1} \quad \text{and} \quad v_2 = \frac{RT_2}{p_2}$$

Thus, the temperature of air at inlet and exit can be found as

$$T_1 = \frac{A_1 V_1 p_1}{Rw} \quad \text{and} \quad T_2 = \frac{A_2 V_2 p_2}{Rw}$$

Substituting the respective values, we get

$$T_1 = \frac{1.3 \times 10^{-3} \times 130 \times 100}{0.287 \times 0.2} = 294.4 \text{ K}$$

$$T_2 = \frac{4.3 \times 10^{-4} \times 240 \times 200}{0.287 \times 0.2} = 359.58 \text{ K}$$

Entropy change during this process is found to be

$$\mathring{S}_2 - \mathring{S}_1 = wc_p \ln\frac{T_2}{T_1} - wR\ln\frac{p_2}{p_1}$$

$$= 0.2 \times 1.005 \times \ln\frac{359.58}{294.4} - 0.2 \times 0.287 \times \ln\frac{200}{100} = 1.198 \text{ kJ/K}$$

$$= 0.0402 - 0.0398 = 0.0004 \text{ kW/K}$$

The rate of entropy production in the duct can be found using the following equation

$$\mathring{S}_2 - \mathring{S}_1 = \sum_j \frac{\overset{=0}{\cancel{Q_j}}}{T_j} + \mathring{S}_{gen}$$

or $$S_{gen} = \mathbf{0.0004\ kW/K}$$

Example 7.12 *Two students A and B were asked to do experiments on an insulated steam turbine operating at steady state conditions. Both students did the experiments under same operating conditions. Steam at 2 MPa and 300°C enters the turbine with a velocity of 40 m/s and leaves the turbine as saturated liquid-vapour mixture at 0.3 MPa with a velocity of 100 m/s. The rate of steam flow through the turbine is 1 kg/s.*

The property values for steam are given as follows:

At 2 MPa and 300°C,	*h = 3024.2 kJ/kg,*	*s = 6.956 kJ/kgK*
At 0.3 MPa, when	*h = 2510 kJ/kg,*	*s = 6.4598 kJ/kgK*
At 0.3 MPa, when	*h = 2710 kJ/kg,*	*s = 6.964 kJ/kgK*

Student A reports that the turbine generates 510 kW of power, while student B reports that 310 kW of power is generated by the turbine. However, none of the student were confident about their observation. Evaluate the observations made by the individual student. D [LO 7.9]

Solution The entropy generated can be found using the following equation

$$S_2 - S_1 = \sum_j \frac{\overset{=0}{\cancel{Q_j}}}{T_j} + S_{gen}$$

Since entropy generation can not be a negative quantity, we can write

$$S_2 - S_1 \geq 0$$

The steady flow energy equation for the control volume as shown in Fig. 7.36 gives

$$w\left(h_1 + \frac{V_1^2}{2} + gz_1\right) + \overbrace{\frac{đQ}{d\tau}}^{=0} = w\left(h_2 + \frac{V_2^2}{2} + gz_2\right) + \frac{đW}{d\tau}$$

Substituting the respective values, we get

$$1\left(3024\times10^3 + \frac{40^2}{2}\right) = 1\left(h_2 + \frac{100^2}{2}\right) + \frac{đW}{d\tau}$$

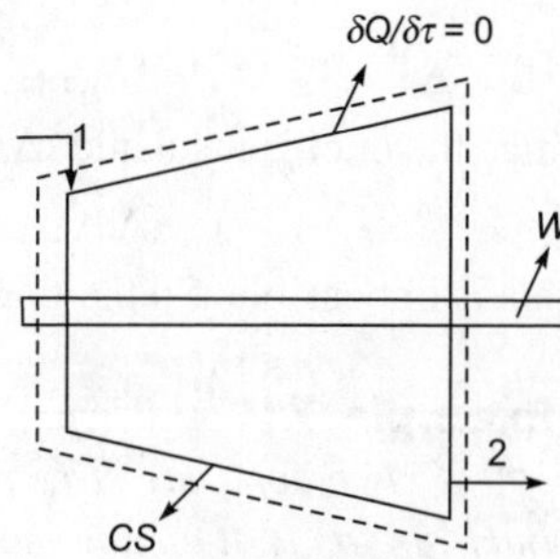

Fig. 7.36

From the results of student A, we have

$$1\left(3024.2\times10^3 + \frac{40^2}{2}\right) = 1\left(h_2 + \frac{100^2}{2}\right) + 510\times10^3$$

or $$h_2 = 2510 \text{ kJ/kg}$$

From the property values, we get $s_2 = 6.4598$ kJ/kgK.

Since $S_2 - S_1 < 0$, the observation made by student A cannot be correct.

From the results of student B, we have

$$1\left(3024.2\times10^3 + \frac{40^2}{2}\right) = 1\left(h_2 + \frac{100^2}{2}\right) + 310\times10^3$$

or $$h_2 = 2710 \text{ kJ/kgK.}$$

From the property values, we get $s_2 = 6.964$ kJ/kgK.

Since $S_2 - S_1 > 0$, the observation made by student B may be correct.

Example 7.13 *Air is flowing steadily in an insulated duct. The pressure and temperature measurements of the air at two stations A and B are given below. Establish the direction of the flow of air in the duct. Assume that for air, specific heat c_p is constant at 1.005 kJ/kg K, $h = c_p T$, and $\frac{v}{T} = \frac{0.287}{p}$, where p, v, and T are pressure (in kPa), volume (in m³/kg) and temperature (in K) respectively.*

	Station A	*Station B*
Pressure	130 kPa	100 kPa
Temperature	50°C	13°C

D [LO 7.10]

Solution From property relation

$$Tds = dh - vdp$$

$$ds = \frac{dh}{T} - v\frac{dp}{T}$$

For two states at A and B the entropy change of the system

$$\int_{s_A}^{s_B} ds = \int_{T_A}^{T_B} \frac{c_p dT}{T} - \int_{p_A}^{p_B} 0.287\frac{dp}{p}$$

$$\therefore \quad s_B - s_A = 1.005 \ln \frac{T_B}{T_A} - 0.287 \ln \frac{p_B}{p_A}$$

$$= 1.005 \ln \frac{273+13}{273+50} - 0.287 \ln \frac{100}{130}$$

$$= -0.1223 + 0.0753$$

$$= -0.047 \text{ kJ/kg K}$$

$$(\Delta s)_{\text{system}} = -0.047 \text{ kJ/kg K}$$

Since the duct is insulated $(\Delta s)_{\text{surr}} = 0$

$$\therefore \quad (\Delta s)_{\text{univ}} = \mathbf{-0.047 \text{ kJ/kg K}}$$

This is impossible. So the flow must be from B to A.

Example 7.14 *A hypothetical device is supplied with 2 kg/s of air at 4 bar, 300 K. Two separate streams of air leave the device, as shown in figure below. Each stream is at an ambient pressure of 1 bar, and the mass flow rate is the same for both streams. One of the exit streams is said to be at 330 K while the other is at 270 K. The ambient temperature is at 300 K. Determine whether such a device is possible.* M **[LO 7.10]**

Solution The entropy generation rate for the control volume (Fig. 7.37) is

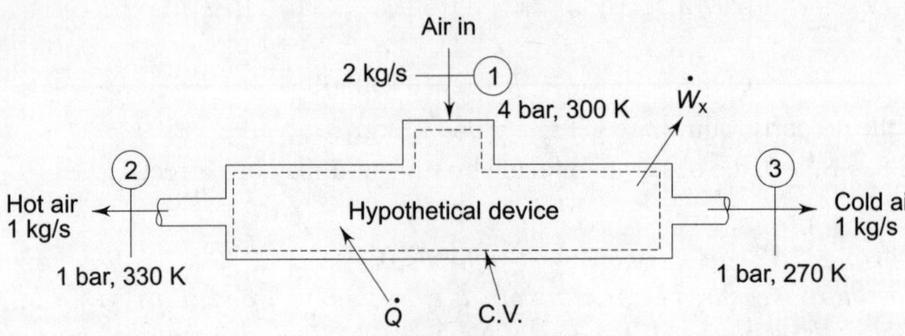

Fig. 7.37

$$\dot{S}_{\text{gen}} = \Sigma \dot{m}_e s_e - \Sigma \dot{m}_i s_i$$

$$= \dot{m}_2 s_2 + \dot{m}_3 s_3 - \dot{m}_1 s_1$$

$$= \dot{m}_2 s_2 + \dot{m}_3 s_3 - (\dot{m}_2 + \dot{m}_3) s_1$$

$$= \dot{m}_2 (s_2 - s_1) + \dot{m}_3 (s_3 - s_1)$$

Now,

$$s_2 - s_1 = c_p \ln \frac{T_2}{T_1} - R \ln \frac{p_2}{p_1}$$

$$= 1.005 \ln \frac{330}{300} - 0.287 \ln \frac{1}{4} = \mathbf{0.494 \text{ kJ/kg K}}$$

$$s_3 - s_1 = c_p \ln \frac{T_3}{T_1} - R \ln \frac{p_3}{p_1}$$

$$= 1.005 \ln \frac{270}{300} - 0.287 \ln \frac{1}{4} = 0.292 \text{ kJ/kgK}$$

$$\dot{S}_{gen} = 1 \times 0.494 + 1 \times 0.292 = \mathbf{0.786\ kW/K}$$

Since $\dot{S}_{gen} > 0$, the device is possible. Such devices actually exist and are called vortex tubes. Although they have low efficiencies, they are suitable for certain applications like rapid cooling of soldered parts, electronic component cooling, cooling of machining operations and so on. The vortex tube is essentially a passive device with no moving parts. It is relatively maintenance free and durable.

Example 7.15 *A room is maintained at 27°C while the surroundings are at 2°C. The temperatures of the inner and outer surfaces of the wall (k = 0.71 W/mK) are measured to be 21C and 6°C, respectively. Heat flows steadily through the wall 5 × 7 m in cross section and 0.32 m in thickness. Determine (a) the rate of heat transfer through the wall (b) the rate of entropy generation in the wall, and (c) the rate of total entropy generation with this heat transfer process.* Ⓜ **[LO 7.9]**

Solution

$$\dot{Q} = kA\frac{\Delta T}{L} = 0.71\ \frac{W}{mK} \times (5 \times 7)\ m^2 \times \frac{(21-6)K}{0.32m}$$
$$= \mathbf{1164.84\ W}$$

Taking the wall as the system, the entropy balance in rate form gives

$$\frac{dS_{wall}}{dt} = \dot{S}_{transfer} + \dot{S}_{gen.wall}$$

$$0 = \Sigma\frac{\dot{Q}}{T} + \dot{S}_{gen.wall}$$

$$0 = \frac{1164.84}{294} - \frac{1164.84}{279} + \dot{S}_{gen.wall}$$

Rate of entropy generation in the wall

$$\dot{S}_{gen.wall} = 4.175 - 3.962 = \mathbf{0.213\ W/K}$$

The entropy change of the wall is zero during this process, since the state and hence the entropy of the wall do not change anywhere in the wall.

To determine the rate of total entropy generation during this heat transfer process, we extend the system to include the regions on both sides of the wall,

$$\frac{dS_{total}}{dt} = \dot{S}_{transfer} + \dot{S}_{gen.total}$$

$$0 = \Sigma\frac{\dot{Q}}{T} + \dot{S}_{gen.total}$$

$$0 = \frac{1164.84}{300} - \frac{1164.84}{275} + \dot{S}_{gen.total}$$

$$\dot{S}_{gen.total} = 4.236 - 3.883 = \mathbf{0.353\ W/K}$$

SUMMARY

Two reversible adiabatic lines cannot intersect each other. So, a reversible adiabatic line is a constant property line. For a reversible cycle

$$\oint_R \frac{đQ}{T} = 0$$

The entropy of a system is a quantity that satisfies the relation

$$dS = \frac{\text{đ}Q_R}{R}$$

where the heat $\text{đ}Q$ is transferred reversibly.

$$Q_{\text{rev}} = \int T ds$$

If heat is supplied to a system, $\text{đ}Q$ is positive, then δs is positive, and entropy of the system increases. If heat is rejected from a system, $\text{đ}Q$ is negative, and dS is negative, and entropy of the system decreases. For a reversible adiabatic process, $\text{đ}Q$ is zero, and entropy remains constant, and the process is isentropic.

The inequality of Clausius states that the cyclic integral of $\frac{\text{đ}Q}{T}$ is negative for an irreversible or actual cycle and zero for a reversible cycle, or $\oint \frac{\text{đ}Q}{T} \le 0$. If it is positive, the cycle is impossible.

The entropy of an isolated system or universe always increases

$$dS_{\text{iso}} \ge 0$$

This is known as the principle of increase of entropy. The entropy of a system and its surroundings together always increase in any process, and remain constant only for a reversible process. The entropy increase of an isolated system is a measure of the extent of irreversibility. The entropy of a system can increase by external interaction (external irreversibility) and internal irreversibility (like friction)

$$dS = d_eS + d_iS = \frac{\text{đ}Q}{T} + d_iS$$

Since $\quad dS > \frac{\text{đ}Q}{T}, \quad \therefore \; d_iS \ge 0$

An isentropic process need not be reversible or adiabatic. If the isentropic process is reversible, it must be adiabatic. If the isentropic process is adiabatic, it cannot be reversible. An adiabatic process need not be isentropic, since entropy can also increase due to friction, etc.

By combining the first and second laws, the following property relations are obtained:

$$TdS = dU + pdV$$

$$TdS = dH - Vdp$$

All actual processes are irreversible. Between two equilibrium states a reversible path is assumed and integration $\int \frac{\text{đ}Q}{T}$ is performed on that path to find the entropy increase for the actual process.

REVIEW QUESTIONS

(M) **7.1** Show that through one point there can pass only one reversible adiabatic. **[LO 7.2]**

(D) **7.2** State and prove Clausius' theorem. **[LO 7.2]**

(D) **7.3** Show that entropy is a property of a system. **[LO 7.3]**

(D) **7.4** How is the entropy change of a reversible process estimated? Will it be different for an irreversible process between the same end states? **[LO 7.3]**

(E) **7.5** Why is the Carnot cycle on *T-s* plot a rectangle? **[LO 7.4]**

(D) **7.6** Establish the inequality of Clausius. **[LO 7.2]**

(M) **7.7** Give the criteria of reversibility, irreversibility and impossibility of a thermodynamic cycle. **[LO 7.2]**

(D) **7.8** What do you understand by the principle of increase of entropy? **[LO 7.6]**

(M) **7.9** When the system is at equilibrium, why would any conceivable change in entropy be zero? **[LO 7.6]**

(D) **7.10** Why is the entropy increase of an isolated system a measure of the extent of irreversibility of the process undergone by the system? **[LO 7.6]**

(M) **7.11** How did Rudolf Clausius summarise the first and second laws of thermodynamics? **[LO 7.6]**

(M) **7.12** Show that the transfer of heat through a finite temperature difference is irreversible. **[LO 7.7]**

(M) **7.13** Show that the adiabatic mixing of two fluids is irreversible. **[LO 7.7]**

(D) **7.14** What is the maximum work obtainable from two finite bodies at temperatures T_1 and T_2? **[LO 7.7]**

(D) **7.15** Determine the maximum work obtainable by using one finite body at temperature T and a thermal energy reservoir at temperature T_0, $T > T_0$. **[LO 7.7]**

(E) **7.16** What are the causes of entropy increase? **[LO 7.9]**

(M) **7.17** Why is an isentropic process not necessarily an adiabatic process? **[LO 7.9]**

(D) **7.18** What is the reversible adiabatic work for a steady flow system when K.E. and P.E. changes are negligibly small? How is it different from that for a closed stationary system? **[LO 7.11]**

(M) **7.19** Why are the equations **[LO 7.10]**

$$TdS = dU + pdV$$
$$TdS = dH - Vdp$$

valid for any process between two equilibrium end states?

(M) **7.20** Why is the second law called a directional law of nature? **[LO 7.13]**

(E) **7.21** How is entropy related to molecular disorder in a system? **[LO 7.13]**

(M) **7.22** Show that entropy varies logarithmically with the disorder number. **[LO 7.13]**

(M) **7.23** What do you understand by perfect order? **[LO 7.13]**

(E) **7.24** Give the Nernst–Simon statement of the third law of thermodynamics. **[LO 7.13]**

(E) **7.25** Why does entropy remain constant in a reversible adiabatic process? **[LO 7.13]**

(M) **7.26** What do you understand by the postulatory approach of thermodynamics? **[LO 7.13]**

(M) **7.27** What do you understand by 'lost work'? **[LO 7.9]**

(D) **7.28** The amount of entropy generation quantifies the intrinsic irreversibility of a process. Explain. **[LO 7.9]**

(M) **7.29** Show that S_{gen} is not a thermodynamic property. **[LO 7.9]**

(M) **7.30** Give the expression for the entropy generation rate for a control volume of a steady flow system. **[LO 7.9]**

(M) **7.31** What is the entropy generation in the isothermal dissipation of work? **[LO 7.7]**

(M) **7.32** What is the entropy generation in the adiabatic dissipation of work? **[LO 7.7]**

(D) **7.33** What do you understand by entropy transfer? Why is entropy transfer associated with heat transfer and not with work transfer? **[LO 7.8]**

PROBLEMS WITH HINTS

(E) **7.1** A quantity of air undergoes a thermodynamic cycle consisting of three processes. Process 1 – 2: Constant volume heating from P_1 = 0.1 MPa, T_1 = 15°C, V_1 = 0.02 m^3 to P_2 = 0.42 MPa. Process 2–3: Constant pressure cooling. Process 3–1: Isothermal heating to the initial state. Employing the ideal gas model with

$c_p = 1$ kJ/kgK, evaluate the change of entropy for each process. Sketch the cycle on *p-v* and *T-s* coordinates. **[LO 7.5]**

Hint:

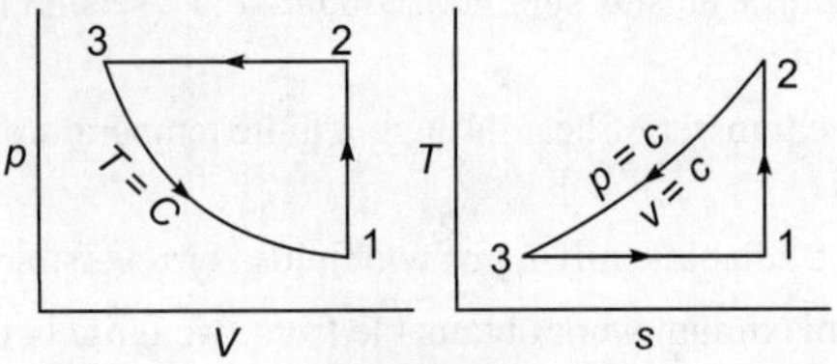

Fig. 7.38

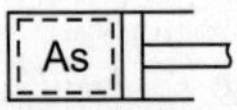

$$P_1 = 0.1 \text{ mPa}$$

$$T_1 = 15° + 273 = 288 \text{ K}$$

$$V_1 = 0.02 \text{ m}^3.$$

$$P_2 = 0.42 \text{ MPa}$$

$$m = \frac{P_1V_1}{RT_1} = \frac{100 \times 0.02}{0.287 \times 288} = 0.024 \text{ kg}$$

$$\frac{T_2}{T_1} = \frac{P_2}{P_1} \quad \text{Find } T_2 = 1210 \text{ K}$$

$$c_v = c_p - R = 1 - 0.287 = 0.713 \text{ kJ/kgK}$$

Process 1–2 $\quad S_2 - S_1 = mc_v \ln \frac{T_2}{T_1} = \mathbf{0.0246 \text{ kJ/K}}$

Process 2–3 $\quad S_3 - S_2 = mc_p \ln \frac{T_3}{T_2} = \mathbf{0.0344 \text{ kJ/K}}$

Process 3–1 $\quad S_1 - S_3 = -mR \ln \frac{P_1}{P_3} = \mathbf{0.0099.\ kJ/K}$

The cycle is shown on *p–v* and *T–s* coordinates in Fig. 7.38.

Ⓔ **7.2** Air initially occupying 1 m^3 at 1.5 bar and 20°C undergoes an internally reversible compression for which PV^n = constant to a final state where the pressure is 6 bar and the temperature is 120°C. Determine

(a) the value of *n*,

(b) the work and heat transfer and

(c) the change in entropy. Take $c_v = 0.718$ kJ/kgK **[LO 7.5]**

Hint:

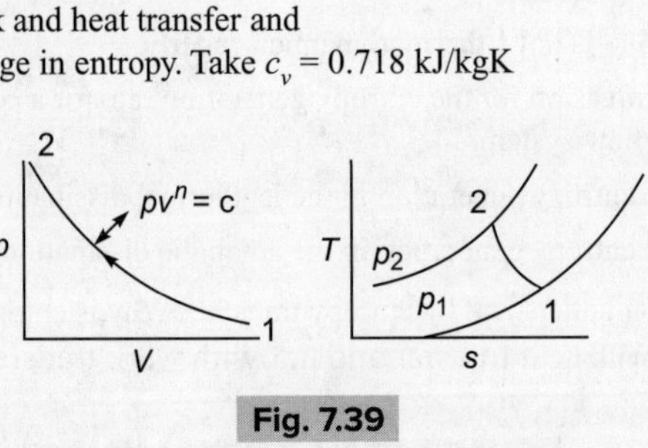

Fig. 7.39

(a) $\dfrac{n-1}{n} = \dfrac{\ln T_2/T_1}{\ln p_2/p_1} = \dfrac{\ln 393/293}{\ln 6/1}$ from which find $n = \mathbf{1.2687}$

(b) $m = \dfrac{p_1V_1}{RT_1} = 1.784$ kg

$$W = \frac{p_1V_1 - p_2V_2}{n-1} = \frac{mR(T_1 - T_2)}{n-1} = \mathbf{-190.54\ kJ}$$

$$Q = \Delta U + W = mc_v(T_2 - T_1) + W$$

$$= -61$$

(c) $S_2 - S_1 = \boldsymbol{m}\,\dfrac{\mathbf{n} - \gamma}{(\gamma - \mathbf{1})(\mathbf{n} - \mathbf{1})}\,\mathbf{R\ ln}\,\dfrac{\mathbf{T_2}}{\mathbf{T_2}}\,\dfrac{\mathbf{kJ}}{\mathbf{K}}$

(E) **7.3** The two cycles shown in Fig. 7.40 are composed of internally reversible processes. Determine an expression for the thermal efficiency of each cycle in terms of temperatures T_1 and T_3. **[LO 7.4]**

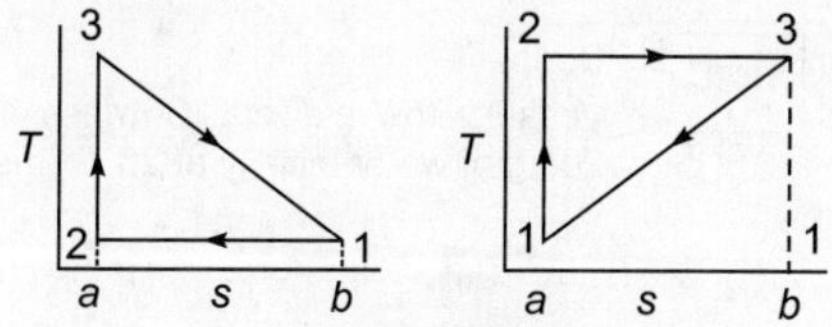

Fig. 7.40

Hint: For the cycle on the left

$$\eta = \frac{\text{Area } 1-2-3-1}{\text{Area } a-b-1-3-1} = \frac{\frac{1}{2}[(s_b - s_a)](T_3 - T_1)}{\frac{1}{2}[(T_3 + T_1)](s_b - s_a)}$$

$$= \frac{\mathbf{T_3 - T_1}}{\mathbf{T_3 + T_1}}$$

For the cycle on the right

$$\eta = \frac{\text{Area } 1-2-3-1}{\text{Area } a-b-3-2-a} = \frac{\mathbf{T_3 - T_1}}{\mathbf{2T_3}}$$

(M) **7.4** A gearbox operating at steady state receives 0.1 kW along the input shaft and delivers 0.095 kW along the output shaft. The outer surface of the gearbox is at 50°C. For the gear box, determine (a) the rate of heat transfer and (b) the rate of entropy generation. **[LO 7.9]**

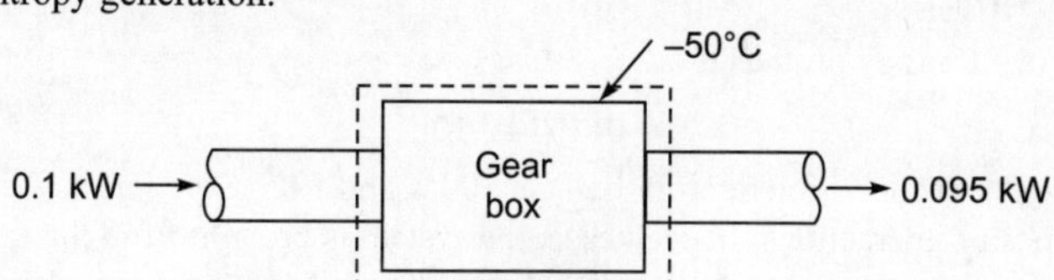

Fig. 7.41

Hint:

Entropy balance gives

$$\frac{ds}{dt} = \frac{d_e s}{dt} = \frac{d_i s}{dt} = 0$$

$$\therefore \qquad \sigma = \frac{d_i s}{dt} = \frac{d\dot{Q}}{T} \qquad \text{(i)}$$

Energy balance Fig. 7.40 gives

$$\frac{dE}{dt} = \dot{Q} - \dot{W} = 0$$

$$\therefore \qquad \dot{Q} = \dot{W} = 0.095 - 0.1$$

$$= \mathbf{-\ 0.005\ kW}$$

Substituting in Eq.(i) $\quad \sigma = \mathbf{1.59 \times 10^{-5} \frac{kW}{K}}$

(M) **7.5** An isolated system consists of a closed copper vessel of mass 0.1 kg immersed in 20 kg of liquid water initially at 20°C. The copper vessel contains 1 kg of liquid water, each initially at 5°C. The system is allowed to come to equilibrium. Determine (a) the final temperature, (b) the entropy change for the copper vessel and each of the liquid water masses and (c) the amount of entropy produced. Take c_p of copper = 0.89 and that of water = 4.18 kJ/kgK. **[LO 7.7]**

Hint:

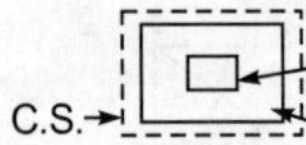

Fig. 7.42

(a) Since the system is isolated, the energy balance gives

$$\Delta U - Q - W = 0$$

$$\therefore \quad \Delta U_1 + \Delta U_2 + \Delta U_{copper} = 0$$

$$m_1 c_w (T_f - T_1) + m_2 c_w (T_f - T_2) + mc (T_f - T_1) = 0$$

where T_f = final temperature = **292.3 K**

(b) Entropy changes will be

$$\Delta S_1 = m_1 c_w \ln \frac{T_f}{T_1} = \mathbf{0.20967\ kJ/K}$$

$$\Delta S_2 = m_2 c_w \ln \frac{T_f}{T_2} = \mathbf{-0.19996\ kJ/K}$$

$$\Delta S_3 = \Delta S_c = mc \ln \frac{T_f}{T_1} = \mathbf{0.00996\ kJ/K}$$

(c) $\Delta S_{surr} = 0$

$$\therefore \sigma = \text{Entropy produced} = \Delta S_1 + \Delta S_2 + \Delta S_c$$

$$= \mathbf{0.01167\ kJ/K}$$

(M) **7.6** A system consisting of air initially at 300 K and 1 bar experiences the two different types of interactions. In each case, the system is brought from the initial state to a state where the temperature is 500 K, while volume remains constant.

For each case, determine the entropy produced and comment on the difference between the values.

(a) The temperature rise is brought about adiabatically by stirring the air with a paddle wheel.

(b) The temperature rise is brought about by heat transfer from a reservoir at 600 K while the temperature of the system where heat transfer occurs is also at 600 K.

[LO 7.9]

Hint:

Case 1: $\Delta S_{sys} = S_2 = S_1 = mc_v \ln \frac{T_2}{T_1} = 1 \times 0.718 \ln \frac{500}{300} \frac{kJ}{kg\,K}.$

$$\Delta S_{surr} = 0, \Delta S_{univ} = S_{gen} = \Delta S_{sys}.$$

Case 2: $Q = \Delta U = mc_v (T_2 - T_1) = 1 \times 0.718\ (500 - 300)$ kJ/kg

$$\Delta S_{suir} = \frac{Q}{600}\frac{\text{kJ}}{\text{kg K}},\ \Delta S_{sys} = \text{as given above.}$$

$$\therefore \quad S_{gen} = \mathbf{\Delta S_{univ}} = \mathbf{\Delta S_{sys}} - \frac{\mathbf{Q}}{\mathbf{600}}\ \mathbf{kJ/kgK}$$

(E) **7.7** A cylindrical copper rod of base area A and length L is insulated at its lateral surface. One end of the rod is in contact with a wall at temperature T_H. The other end is in contact with a wall at lower temperature T_c. At steady state, the rate of heat conduction from the hot wall is

$$\dot{Q}_H = \text{kA}\,\frac{(T_H - T_C)}{L}$$

where k is the thermal conductivity of the copper rod.

(a) For the rod as the system, find the rate of entropy generation in terms of A, L, T_H, T_C and k.

(b) If $T_H = 227°\text{C}$, $T_C = 27°\text{C}$, $k = 0.4$ kW/mK, $A = 0.1\ \text{m}^2$ and $L = 1$ m, calculate the rate heat transfer and the rate of entropy generation. **[LO 7.9]**

Hint:

$$\dot{Q}_H = +\text{kA}\,\frac{(T_H - T_C)}{L}$$
$$T_H = 500\ \text{K}$$
$$T_C = 300\ \text{K}$$
$$k = 0.4\ \text{kW/mK}$$
$$A = 0.1\ \text{m}^2,\ L = 1\ \text{m.}$$
$$\dot{Q}_H = Q_C$$

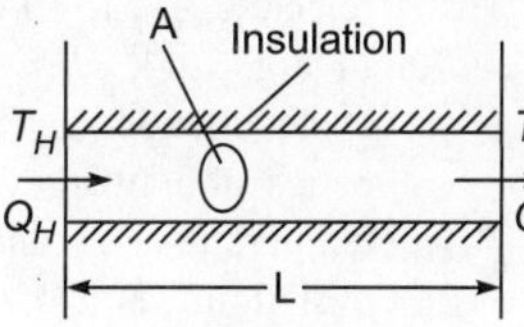

Fig. 7.43

Entropy balance gives

$$\frac{dS}{dt} = \overset{\circ}{S}_{gen} + \frac{Q_H}{T_H} - \frac{Q_C}{T_C} = 0$$

(a) $\therefore\ \dot{S}_{gen} = \frac{Q_C}{T_C} - \frac{Q_H}{T_H} = \dot{Q}_H\left(\frac{1}{T_C} - \frac{1}{T_H}\right) = \frac{kA(T_H - T_C)^2}{LT_HT_C}$ = +ve definite

(b) $\dot{S}_{gen} = \mathbf{5.14 \times 10^{-3}\ kW/K}$

PROBLEMS

(E) **7.1** On the basis of the first law fill in the blank spaces in the following table of imaginary heat engine cycles. On the basis of the second law classify each cycle as reversible irreversible, or impossible.

Cycle	*Temperature*		*Rate of Heat Flow*		*Rate of work*	*Efficiency*
	Source	*Sink*	*Supply*	*Rejection*	*Output*	
(a)	327°C	27°C	420 kJ/s	230 kJ/s	... kW	
(b)	1000°C	100°C	...kJ/min	4.2 MJ/min	...kW	65%
(c)	750 K	300 K	... kJ/s	... kJ/s	26 kW	60%
(d)	700 K	300 K	2500 kcal/h	... kcal/h	1 kW	—

[LO 7.2]

Ⓔ **7.2** The latent heat of fusion of water at 0°C is 335 kJ/kg. How much does the entropy of 1 kg of ice change as it melts into water in each of the following ways: (a) Heat is supplied reversibly to a mixture of ice and water at 0°C. (b) A mixture of ice and water at 0°C is stirred by a paddle wheel. **[LO 7.5]**

Ⓔ **7.3** Two kg of water at 80°C is mixed adiabatically with 3 kg of water at 30°C in a constant pressure process of 1 atmosphere. Find the increase in the entropy of the total mass of water due to the mixing process (c_p of water = 4.187 kJ/kg K). **[LO 7.5]**

Ⓜ **7.4** In a Carnot cycle, heat is supplied at 350°C and rejected at 27°C. The working fluid is water which, while receiving heat, evaporates from liquid at 350°C to steam at 350°C. The associated entropy change is 1.44 kJ/kg K. (a) If the cycle operates on a stationary mass of 1 kg of water, how much is the work done per cycle, and how much is the heat supplied? (b) If the cycle operates in steady flow with a power output of 20 kW, what is the steam flow rate? **[LO 7.2]**

Ⓜ **7.5** A heat engine receives reversibly 420 kJ/cycle of heat from a source at 327°C, and rejects heat reversibly to a sink at 27°C. There are no other heat transfers. For each of the three hypothetical amounts of heat rejected, in (a), (b) and (c) below, compute the cyclic integral of $đQ/T$. From these results show which case is irreversible, which reversible and which impossible: (a) 210 kJ/cycle rejected, (b) 105 kJ/cycle rejected, (c) 315 kJ/cycle rejected. **[LO 7.2]**

Ⓜ **7.6** In Fig. 7.44, *abcd* represents a Carnot cycle bounded by two reversible adiabatics and two reversible isotherms at temperatures T_1 and T_2 ($T_1 > T_2$). The oval figure is a reversible cycle, where heat is absorbed at temperature less than, or equal to, T_1, and rejected at temperatures greater than, or equal to, T_2. Prove that the efficiency of the oval cycle is less than that of the Carnot cycle. **[LO 7.2]**

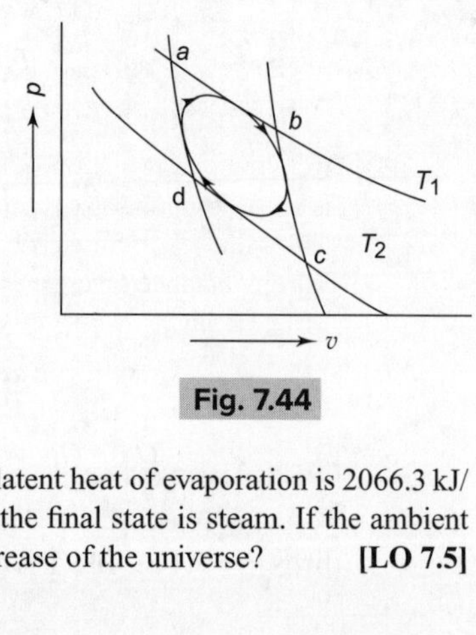

Fig. 7.44

Ⓔ **7.7** Water is heated at a constant pressure of 0.7 MPa. The boiling point is 164.97°C. The initial temperature of water is 0°C. The latent heat of evaporation is 2066.3 kJ/kg. Find the increase of entropy of water, if the final state is steam. If the ambient temperature is 30°C, what is the entropy increase of the universe? **[LO 7.5]**

Ⓜ **7.8** One kg of air initially at 0.7 MPa, 20°C changes to 0.35 MPa, 60°C by the three reversible non-flow processes, as shown in Fig. 7.45. Process 1-*a*-2 consists of a constant pressure expansion followed by a constant volume cooling, process 1-*b*-2 an isothermal expansion followed by a constant pressure expansion, and process 1-*c*-2 an adiabatic expansion followed by a constant volume heating. Determine the change of internal energy, enthalpy and entropy for each process, and find the work

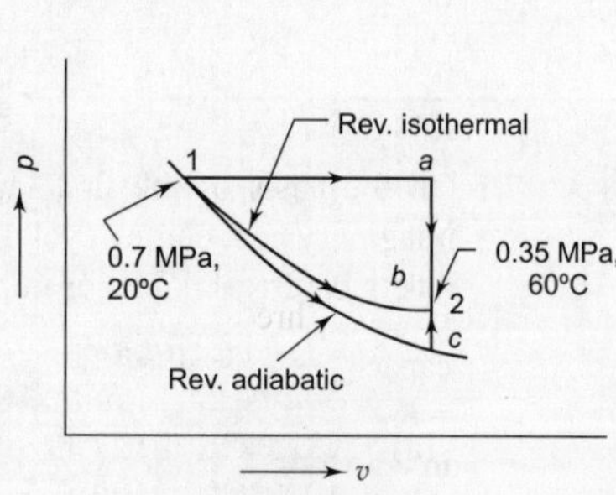

Fig. 7.45

transfer and heat transfer for each process. Take c_p = 1.005 and $c_v = 0.718$ kJ/kg K and assume the specific heat to be constant. Also assume for air $pv = 0.287$ T, where p is the pressure in kPa, v the specific volume in m^3/kg and T the temperature in K. **[LO 7.9]**

(M) **7.9** Ten grammes of water at 20°C is converted into ice at –10°C at constant atmospheric pressure. Assuming the specific heat of liquid water to remain constant at 4.2 J/gK and that of ice to be half of this value, and taking the latent heat of fusion of ice at 0°C to be 335 J/g, calculate the total entropy change of the system. **[LO 7.7]**

(M) **7.10** Calculate the entropy change of the universe as a result of the following processes:

(a) A copper block of 600 g mass and with C_p of 150 J/K at 100°C is placed in a lake at 8°C.

(b) The same block, at 8°C, is dropped from a height of 100 m into the lake.

(c) Two such blocks at 100 and 0°C are joined together. **[LO 7.10]**

(M) **7.11** A system maintained at constant volume is initially at temperature T_1, and a heat reservoir at the lower temperature T_0 is available. Show that the maximum work recoverable as the system is cooled to T_0 is

$$W = C_v \left[(T_1 - T_0) - T_0 \ln \frac{T_1}{T_0} \right]$$ **[LO 7.7]**

(E) **7.12** If the temperature of the atmosphere is 5°C on a winter day and if 1 kg of water at 90°C is available, how much work can be obtained. Take c_v of water as 4.186 kJ/kg K. **[LO 7.7]**

(M) **7.13** A body with the equation of state $U = CT$, where C is its heat capacity, is heated from temperature T_1 to T_2 by a series of reservoirs ranging from T_1 to T_2. The body is then brought back to its initial state by contact with a single reservoir at temperature T_1. Calculate the changes of entropy of the body and of the reservoirs. What is the total change in entropy of the whole system?

If the initial heating were accomplished merely by bringing the body into contact with a single reservoir at T_2, what would the various entropy changes be? **[LO 7.7]**

(M) **7.14** A body of finite mass is originally at temperature T_1 which is higher than that of a reservoir at temperature T_2. Suppose an engine operates in a cycle between the body and the reservoir until it lowers the temperature of the body from T_1 to T_2, thus extracting heat Q from the body. If the engine does work W, then it will reject heat Q-W to the reservoir at T_2. Applying the entropy principle, prove that the maximum work obtainable from the engine is

$$W\,(\text{max}) = Q - T_2\,(S_1 - S_2)$$

where $S_1 - S_2$ is the entropy decrease of the body.

If the body is maintained at constant volume having constant volume heat capacity $C_v = 8.4$ kJ/K which is independent of temperature, and if $T_1 = 373$ K and $T_2 = 303$ K, determine the maximum work obtainable. **[LO 7.7]**

(M) **7.15** Each of three identical bodies satisfies the equation $U = CT$, where C is the heat capacity of each of the bodies. Their initial temperatures are 200, 250 and 540 K. If C = 8.4 kJ/K, what is the maximum amount of work that can be extracted in a process in which these bodies are brought to a final common temperature? **[LO 7.7]**

(M) **7.16** In the temperature range between 0 and 100°C a particular system maintained at constant volume has a heat capacity.

$$C_V = A + 2BT$$

with $A = 0.014$ J/K and $B = 4.2 \times 10^{-4}$ J/K^2.

A heat reservoir at 0°C and a reversible work source are available. What is the maximum amount of work that can be transferred to the reversible work source as the system is cooled from 100°C to the temperature of the reservoir? **[LO 7.7]**

(M) **7.17** Each of the two bodies has a heat capacity at constant volume

$$C_V = A + 2BT$$

where $A = 8.4$ J/K and $B = 2.1 \times 10^{-2}$ J/K^2

If the bodies are initially at temperatures 200 and 400 K and if a reversible work source is available, what are the maximum and minimum final common temperatures to which the two bodies can be brought? What is the maximum amount of work that can be transferred to the reversible work source? **[LO 7.7]**

(M) **7.18** A reversible engine, as shown in Fig. 7.46 during a cycle of operations draws 5 MJ from the 400 K reservoir and does 840 kJ of work. Find the amount and direction of heat interaction with other reservoirs.

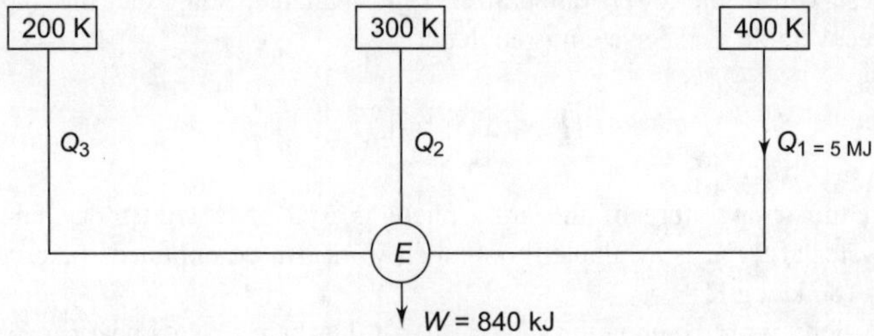

Fig. 7.46

[LO 7.2]

(D) **7.19** For a fluid for which pv/T is a constant quantity equal to R, show that the change in specific entropy between two states A and B is given by

$$s_B - s_A = \int_{T_A}^{T_B} \left(\frac{C_p}{T} \right) dT - R \ln \frac{p_B}{p_A}$$

A fluid for which R is a constant and equal to 0.287 kJ/kg K, flows steadily through an adiabatic machine, entering and leaving through two adiabatic pipes. In one of these pipes the pressure and temperature are 5 bar and 450 K and in the other pipe the pressure and temperature are 1 bar and 300 K, respectively. Determine which pressure and temperature refer to the inlet pipe.

For the given temperature range, c_p is given by

$$c_p = a \ln T + b$$

where T is the numerical value of the absolute temperature and $a = 0.026$ kJ/kg K, $b = 0.86$ kJ/kg K. Which one is the inlet pipe? **[LO 7.9]**

(M) **7.20** Two vessels, A and B, each of volume 3 m^3 may be connected by a tube of negligible volume. Vessel A contains air at 0.7 MPa, 95°C, while vessel B contains air at 0.35 MPa, 205°C. Find the change of entropy when A is connected to B by working from the first principles and assuming the mixing to be complete and adiabatic. For air take the relations as given in Example 7.8. **[LO 7.7]**

(M) **7.21** (a) An aluminium block ($c_p = 400$ J/kg K) with a mass of 5 kg is initially at 40°C in room air at 20°C. It is cooled reversibly by transferring heat to a completely reversible cyclic heat engine until the block reaches 20°C. The 20°C room air serves as a constant temperature sink for the engine. Compute (i) the change in entropy for the block, (ii) the change in entropy for the room air and (iii) the work done by the engine.

(b) If the aluminium block is allowed to cool by natural convection to room air, compute

(i) the change in entropy for the block, (ii) the change in entropy for the room air and

(iii) the net change in entropy for the universe. **[LO 7.7]**

(M) **7.22** Two bodies of equal heat capacities C and temperatures T_1 and T_2 form an adiabatically closed system. What will the final temperature be if one lets this system come to equilibrium (a) freely? (b) reversibly? (c) What is the maximum work that can be obtained from this system? **[LO 7.7]**

(M) **7.23** A resistor of 30 ohms is maintained at a constant temperature of 27°C while a current of 10 amperes is allowed to flow for 1 second. Determine the entropy change of the resistor and the universe.

If the resistor initially at 27°C is now insulated and the same current is passed for the same time, determine the entropy change of the resistor and the universe. The specific heat of the resistor is 0.9 kJ/kg K and the mass of the resistor is 10 g. **[LO 7.7]**

(E) **7.24** An adiabatic vessel contains 2 liter of water at 25°C. By paddle-wheel work transfer, the temperature of water is increased to 30°C. If the specific heat of water is assumed constant at 4.187 kJ/kg K, find the entropy change of the universe. **[LO 7.7]**

(E) **7.25** A copper rod is of length 1 m and diameter 0.01 m. One end of the rod is at 100°C, and the other at 0°C. The rod is perfectly insulated along its length and the thermal conductivity of copper is 380 W/mK. Calculate the rate of heat transfer along the rod and the rate of entropy production due to irreversibility of this heat transfer. **[LO 7.7]**

(M) **7.26** A body of constant heat capacity C_p and at a temperature T_i is put in contact with a reservoir at a higher temperature T_f. The pressure remains constant while the body comes to equilibrium with the reservoir. Show that the entropy change of the universe is equal to

$$C_p\left[\frac{T_i - T_f}{T_f} - \ln\left(1 + \frac{T_i - T_f}{T_f}\right)\right]$$

Prove that this entropy change is positive.

Given: $\ln(1 + x) = x - \frac{x^2}{2} + \frac{x^3}{3} - \frac{x^4}{4} \ldots -$

where $x < 1$. **[LO 7.7]**

(M) **7.27** An insulated 0.75 kg copper calorimeter can containing 0.2 kg water is in equilibrium at a temperature of 20°C. An experimenter now places 0.05 kg of ice at 0°C in the calorimeter and encloses the latter with a heat insulating shield. (a) When all the ice has melted and equilibrium has been reached, what will be the temperature of water and the can? The specific heat of copper is 0.418 kJ/kg K and the latent heat of fusion of ice is 333 kJ/kg. (b) Compute the entropy increase of the universe resulting from the process. (c) What will be the minimum work needed by a stirrer to bring back the temperature of water to 20°C? **[LO 7.7]**

(M) **7.28** Show that if two bodies of thermal capacities C_1 and C_2 at temperatures T_1 and T_2 are brought to the same temperature T by means of a reversible heat engine, then **[LO 7.2]**

$$\ln T = \frac{C_1 \ln T_1 + C_2 \ln T_2}{C_1 + C_2}$$

(E) **7.29** Two blocks of metal, each having a mass of 10 kg and a specific heat of 0.4 kJ/kg K, are at a temperature of 40°C. A reversible refrigerator receives heat from one block and rejects heat to the other. Calculate the work required to cause a temperature difference of 100°C between the two blocks. **[LO 7.2]**

(M) **7.30** A body of finite mass is originally at a temperature T_1, which is higher than that of a heat reservoir at a temperature T_2. An engine operates in infinitesimal cycles between the body and the reservoir until it lowers the temperature of the body from T_1 to T_2. In this process, there is a heat flow Q out of the body. Prove that the maximum work obtainable from the engine is $Q + T_2 (S_1 - S_2)$, where $S_1 - S_2$ is the decrease in entropy of the body. **[LO 7.7]**

(E) **7.31** A block of iron weighing 100 kg and having a temperature of 100°C is immersed in 50 kg of water at a temperature of 20°C. What will be the change of entropy of the combined system of iron and water? Specific heats of iron and water are 0.45 and 4.18 kJ/kg K, respectively. **[LO 7.7]**

(M) **7.32** 36 g of water at 30°C are converted into steam at 250°C at constant atmospheric pressure. The specific heat of water is assumed constant at 4.2 J/g K and the latent heat of vaporisation at 100°C is 2260 J/g. For water vapour, assume $pV = mRT$ where $R = 0.4619$ kJ/kg K and

$$\frac{c_p}{R} = a + bT + cT^2, \text{ where } a = 3.634,$$

$$b = 1.195 \times 10^{-3}\ \text{K}^{-1} \text{ and } c = 0.135 \times 10^{-6}\ \text{K}^{-2}$$

Calculate the entropy change of the system. **[LO 7.7]**

(E) **7.33** A 50 ohm resistor carrying a constant current of 1 A is kept at a constant temperature of 27°C by a stream of cooling water. In a time interval of 1 s, (a) what is the change in entropy of the resistor? (b) What is the change in entropy of the universe? **[LO 7.7]**

(M) **7.34** A lump of ice with a mass of 1.5 kg at an initial temperature of 260 K melts at the pressure of 1 bar as a result of heat transfer from the environment. After some time has elapsed the resulting water attains the temperature of the environment, 293 K. Calculate the entropy production associated with this process. The latent heat of fusion of ice is 333.4 kJ/kg, the specific heat of ice and water are 2.07 and 4.2 kJ/kg K, respectively, and ice melts at 273.15 K. **[LO 7.7]**

(M) **7.35** An ideal gas is compressed reversibly and adiabatically from state a to state b. It is then heated reversibly at constant volume to state c. After expanding reversibly and adiabatically to state d such that $T_b = T_d$, the gas is again reversibly heated at constant pressure to state e such that $T_e = T_e$. Heat is then rejected reversibly from the gas at constant volume till it returns to state a. Express T_a in terms of T_b and T_c. If $T_b = 555$ K and $T_c = 835$ K, estimate T_a. Take $\gamma = 1.4$. **[LO 7.2]**

(M) **7.36** Liquid water of mass 10 kg and temperature 20°C is mixed with 2 kg of ice at – 5°C till equilibrium is reached at 1 atm pressure. Find the entropy change of the system. *Given*: c_p of water = 4.18 kJ/kg K, c_p of ice = 2.09 kJ/kg K and latent heat of fusion of ice = 334 kJ/kg. **[LO 7.7]**

(E) **7.37** A thermally insulated 50-ohm resistor carries a current of 1 A for 1 s. The initial temperature of the resistor is 10°C. Its mass is 5 g and its specific heat is 0.85 J/g K. (a) What is the change in entropy of the resistor? (b) What is the change in entropy of the universe? **[LO 7.7]**

(M) **7.38** The value of c_p for a certain substance can be represented by $c_p = a + bT$. (a) Determine the heat absorbed and the increase in entropy of a mass m of the substance when its temperature is increased at constant pressure from T_1 to T_2. (b) Find the increase in the molal specific entropy of copper, when the temperature is increased at constant pressure from 500 to 1200 K. *Given for copper*: when $T = 500$ K, $c_p = 25.2 \times 10^3$ and when $T = 1200$ K, $c_p = 30.1 \times 10^3$ J/k mol K. **[LO 7.7]**

(M) **7.39** Assume that a particular body has the equation of state $U = CT$, where $C = 8.4$ J/K and assume that this equation of state is valid throughout the temperature range from 0.5 K to room temperature. How much work must be expended to cool this body from room temperature of 300 to 0.5 K, taking the ambient atmosphere as the hot reservoir? **[LO 7.7]**

(M) **7.40** An iron block of unknown mass at 85°C is dropped into an insulated tank that contains 0.1 m^3 of water at 20°C. At the same time a paddle wheel driven by a 200 W motor is activated to stir the water. Thermal equilibrium is established after 20 minutes when the final temperature is 24°C. Determine the mass of the iron block and the entropy generated during the process. **[LO 7.7]**

(M) **7.41** A piston cylinder device contains 1.2 kg of nitrogen gas at a 120 kPa and 27°C. The gas is now compressed slowly in a polytropic process during which $pV^{1.3}$ = constant. The process ends when the volume is reduced by one-half. Determine the entropy change of nitrogen during this process. **[LO 7.9]**

(M) **7.42** Air enters a compressor at ambient conditions of 96 kPa and 17°C with a low velocity and exits at 1 MPa, 327°C and 120 m/s. The compressor is cooled by the ambient air at 17°C at a rate of 1500 kJ/min. The power input to the compressor is 300 kW. Determine (a) the mass flow rate of air and (b) the rate of entropy generation. **[LO 7.9]**

(M) **7.43** A gearbox operating at steady state receives 0.1 kW along the input shaft and delivers 0.095 kW along the output shaft. The outer surface of the gearbox is at 50°C. For the gearbox, determine (a) the rate of heat transfer, (b) the rate at which entropy is produced. **[LO 7.9]**

(M) **7.44** At steady state, an electric motor develops power along its output shaft at the rate of 2 kW while drawing 20 amperes at 120 volts. The outer surface of the motor is at 50°C. For the motor, determine the rate of heat transfer and the rate of entropy generation. **[LO 7.9]**

(M) **7.45** Show that the minimum theoretical work input required by a refrigeration cycle to bring two finite bodies from the same initial temperature to the final temperatures of T_1 and T_2 ($T_2 < T_1$) is given by

$$W_{min} = mc\,[2(T_1T_2)^{1/2} - T_1 - T_2]$$ **[LO 7.7]**

(M) **7.46** A rigid tank contains an ideal gas at 40°C that is being stirred by a paddle wheel. The paddle wheel does 200 kJ of work on the ideal gas. It is observed that the temperature of the ideal gas remains constant during this process as a result of heat transfer between the system and the surroundings at 25°C. Determine (a) the entropy change of the ideal gas and (b) the total entropy generation. **[LO 7.9]**

(M) **7.47** A cylindrical rod of length L insulated on its lateral surface is initially in contact at one end with a wall at temperature T_1 and at the other end with a wall at a lower temperature T_2. The temperature within the rod initially varies linearly with position x according to

$$T_{(x)} = T_1 - \frac{T_1 - T_2}{L} x$$

The rod is insulated on its ends and eventually comes to a final equilibrium state where the temperature is T_f. Evaluate T_f in terms of T_1 and T_2, and show that the amount of entropy generated is:

$$S_{gen} = mc\left[1 + \ln T_{f+} + \frac{T_2}{T_1 - T_2}\ln T_2 - \frac{T_1}{T_1 - T_2}\ln T_1\right]$$

where c is the specific heat of the rod. **[LO 7.9]**

(M) **7.48** Air flowing through a horizontal, insulated duct was studied by students in a laboratory. One student group measured the pressure, temperature and velocity at a location in the duct as 0.95 bar, 67°C and 75 m/s. At another location, the respective values were found to be 0.8 bar, 22°C and 310 m/s. The group neglected to note the direction of flow, however. Using the known data, determine the direction. **[LO 7.9]**

(M) **7.49** Nitrogen gas at a 6 bar, 21°C enters an insulated control volume operating at steady state for which $W_{C.V.} = 0$. Half of the nitrogen exits the device at 1 bar, 82°C and the other half exits at 1 bar, – 40°C. The effects of K.E. and P.E. are negligible. Employing the ideal gas model, decide whether the device can operate as described. **[LO 7.9]**

(M) **7.50** An air compressor operates at steady state with air entering at p_1 = 1 bar, T_1 = 20°C and exiting at p_2 = 5 bar. Determine the work and heat transfer per unit mass passing through the device, if the air undergoes a polytropic process with n = 1.3. Neglect changes in K.E. and P.E. Use air as the ideal gas. **[LO 7.9]**

(E) **7.51** A cylinder fitted with a piston contains 2 kg steam at 500 kPa, 400°C. Find the entropy change and the work done when the steam expands to a final pressure of 200 kPa in each of the following ways: (a) adiabatically and reversibly and (b) adiabatically and irreversibly to an equilibrium temperature of 300°C. **[LO 7.7]**

ANSWERS TO PROBLEMS

7.1 (a) Irreversible
(b) Irreversible
(c) Reversible
(d) Impossible

7.2 1.2271 kJ/K

7.3 0.0576 kJ/K

7.4 (a) 465.12, 897.12 kJ/kg
(b) 0.043 kg/s

7.5 (a) Reversible
(b) Impossible
(c) Irreversible

7.7 6.6967 kJ/kg K

7.9 16.02 J/K

7.10 (a) 6.69 J/K
(b) 2.095 J/K
(c) 3.64 J/K

7.14 58.96 kJ

7.15 756 kJ

7.16 4.508 J

7.17 T_{min} = 292 K

7.18 Q_2 = + 4.98 MJ Q_3 = – 0.82 MJ

7.19 A is the inlet pipe,
$s_B - s_A$ = 0.0509 kJ/kg K, A is the inlet pipe.

7.20 0.959 kJ/K

7.21 (a) – 134 J/K, + 134 J/K, 740 J,

(b) – 134 J/K, + 136.5 J/K, 2.5 J/K

7.23 $(\Delta S)_{resistor} = 0$, $(\Delta S)_{univ} = 10$ J/K, $(\Delta S)_{univ} = 6.72$ J/K

7.24 0.139 kJ/K

7.25 2.985 W, 0.00293 W/K

7.27 (a) 4.68°C

(b) 0.00276 kJ/K

(c) 20.84 kJ

7.29 32 kJ

7.31 1.1328 kJ/K

7.32 277.8 J/K

7.33 (a) 0 (b) 0.167 J/K

7.34 0.1514 kJ/K

7.35 $T_a = \frac{T_b^{\gamma+1}}{T_c^{\gamma}}$, 313.29 K

7.36 190 J/K

7.37 (a) 0.173 J/K

(b) 0.173 J/K

7.38 (a) $m\left[a\left(T_2 - T_1\right) + \frac{b}{2}\left(T_2^2 - T_1^2\right)\right]$,

$m\left[a \ln \frac{T_2}{T_1} + b\left(T_2 - T_2\right)\right]$,

(b) 24.7 kJ/k mol K

7.40 52.2 kg, 1.285 kJ/K

7.41 – 0.0615 kJ/K.

7.42 (a) 0.851 kg/s

(b) 0.144 kW/K

7.43 (a) – 0.005 kW

(b) 1.54×10^{-5} kW/K

7.44 – 0.4 kW, 1.24×10^{-3} kW/K

7.46 (a) 0

(b) 0.671 kJ/K

7.47 $T_f = [T_1 + T_2]/2$

7.48 Flow is from right to left.

7.49 Yes, the device can operate as described.

7.50 – 164.2 kJ/kg, – 31 kJ/kg

7.51 (a) 0, 386.7 kJ

(b) 0.1976 kJ/K, 309.4 kJ

8

Exergy

Learning Objectives

LO 8.1 Define exergy and discuss its utility
LO 8.2 Discuss the dead state
LO 8.3 Discuss available and unavailable exergies
LO 8.4 Relate degradation of exergy with real heat flow
LO 8.5 Illustrate the exergy of a given mass of ideal gas at a specified state
LO 8.6 Relate second law with degradation of energy
LO 8.7 Compare reversible work with irreversible work of a closed system
LO 8.8 Demonstrate equality of reversible works for same end state points
LO 8.9 Evaluate the maximum work for different systems and relate Helmholtz and Gibbs functions with it
LO 8.10 Relate irreversibility with Gouy-Stodola theorem
LO 8.11 Demonstrate exergy balance for closed and steady flow systems
LO 8.12 Discuss second law efficiency and apply it to different engineering systems

LO 8.1

Define exergy and discuss its utility

The *exergy* of a system at a given state is the maximum work that can be extracted from it till it reaches the state of thermodynamic equilibrium with its surroundings. The concept of exergy is very important in all energy-producing, energy-consuming and energy-conveying systems. The first law provides an energy analysis of a thermodynamic system without making any discrimination of its quality. The second low emphasises that all forms of energy are not of the same quality and the quality of energy always degrades while conserving its quantity. Both first-law and second-law analyses, i.e. energy and exergy analyses require to be carried out for any energy conversion or energy conveying system, to make it more exergy efficient, which ultimately leads to energy saving or reduced energy consumption for a given task.

8.1 DEAD STATE

LO 8.2

Discuss the dead state

For any substance, if its pressure increases, the volume decreases. Therefore, in any thermodynamic plot in *p-v* coordinates, all processes must have $\frac{dp}{dv} < 0$, i.e. negative slope. No process having positive slope of dp/dv is possible in nature and cannot be drawn in *p-v* diagrams. On the other hand, for a substance, if its temperature increases the disorder of its molecules and hence its entropy increase. All processes that have $\frac{dT}{ds} > 0$, i.e. positive slope can only be drawn on *T-s* diagram. No process having negative slope of $\frac{dT}{ds}$ is possible in nature and cannot be drawn in *T-s* diagram.

Let p_o and T_o be the ambient pressure and temperature of the surroundings, respectively. If the state of a system departs from that of the surroundings, an opportunity exists for producing work (Fig. 8.1). However, as the system changes its state towards that of the surroundings, this opportunity diminishes, and it ceases to exist when the two are in equilibrium with each other. When the system is in equilibrium with the surroundings, it must be in pressure and temperature equilibrium with the surroundings, i.e. at p_o and T_o. It must also be in chemical equilibrium with the surroundings, i.e. there should not be any chemical reaction or mass transfer. The system must have zero velocity and minimum potential energy. This state of the system is known as the *dead state*, which is designated by affixing subscript '*o*' to the properties. It is shown by the point *O* in Fig. 8.1. Any change in the state of the system from the dead state is the measure of its exergy or maximum work that can be extracted from it. Further, the initial state of the system is from the dead state in terms of *p* and *t*, either above or below it, higher will be its exergy value (Fig. 8.1). *All spontaneous processes of a system terminate at the dead state, when* $S = S_{max}$. When a system is at the dead state, there cannot be any spontaneous departure of the system from it, when $S < S_{max}$, and the system must revert to the peak of he entropy hill when $S = S_{max}$ at equilibrium.

Whenever useful work is obtained during a process, in which a finite system undergoes a change of state, the process must terminate when the pressure and temperature of the system have become equal to the pressure and temperature of the surroundings, p_o and T_o, i.e. when the system has reached the dead state. An air engine operating with compressed air taken from a cylinder will continue to deliver work till the pressure of air in the cylinder becomes equal to that of the surroundings, p_o. A certain quantity of exhaust gases from an internal combination engine used as the high temperature source of the heat engine will deliver work till the temperature of the gas becomes equal to that of the surroundings, T_o.

The exergy of a given system is defined as the maximum work that is obtained in a process in which the system comes to equilibrium with its surroundings, i.e. it reaches the dead state. The exergy is thus a composite property depending on the state of both the system and the surroundings. If a system exists at the dead state, its exergy is zero, i.e. it is incapable of delivering any work. Further, it is away from the dead state, either above or below it, higher will be its exergy and more will be its capacity of delivering work.

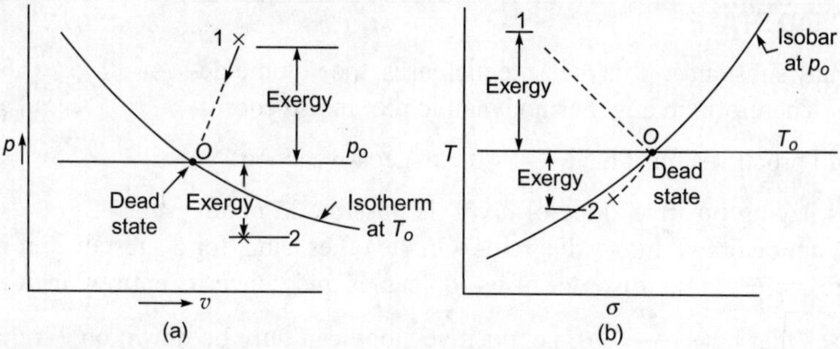

Fig. 8.1 Exergy of a system decreases as its state approaches the dead state at p_o, T_o, either above or below it

8.2 EXERGY OF HEAT INPUT IN A CYCLE

> **LO 8.3**
> Discuss available and unavailable exergies

The maximum work obtainable from a certain heat input in a cyclic heat engine is often called the available energy (A.E.) or the available part of the energy supplied. The minimum energy that has to be rejected by the second law is called the unavailable exergy (U.E.) or the unavailable part of the exergy supplied.

Let Q be the amount of heat supplied to a reversible engine at temperature T and the heat is rejected to the sink at the dead state T_o (Fig. 8.2). The maximum work obtained or exergy denoted by X is

$$X = W_{max} = \left(1 - \frac{T_o}{T}\right) Q = \text{AE}$$

$$= T(s_2 - s_1) - T_o(s_3 - s_4)$$

$$= (T - T_o)(s_3 - s_4)$$

$$= Q - T_o(s_3 - s_4)$$

$$= Q - \text{U.E} = Q - Q_o.$$

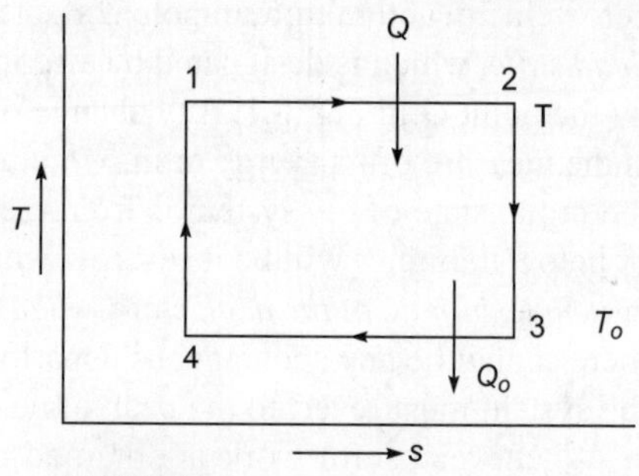

Fig. 8.2 Exergy of heat input

The exergy of heat input Q at temperature T is thus equal to $Q.\ \eta_{rev}$, i.e. $Q\left(1 - \frac{T_o}{T}\right)$.

For a given T_o, as T increases, the exergy increases. The exergy decreases as T decreases.

8.3 DECREASE IN EXERGY WHEN HEAT IS TRANSFERRED THROUGH A FINITE TEMPERATURE DIFFERENCE

> **LO 8.4**
> Relate degradation of exergy with real heat flow

Whenever heat is transferred through a finite temperature difference, there is a decrease in exergy output. Let us consider a reversible heat engine operating between T_1 and T_o (Fig. 8.3). Then

$$Q_1 = T_1\,\Delta s,\ Q_2 = T_o\,\Delta s \text{ and } W = \text{AE} = X = (T_1 - T_o)\,\Delta s$$

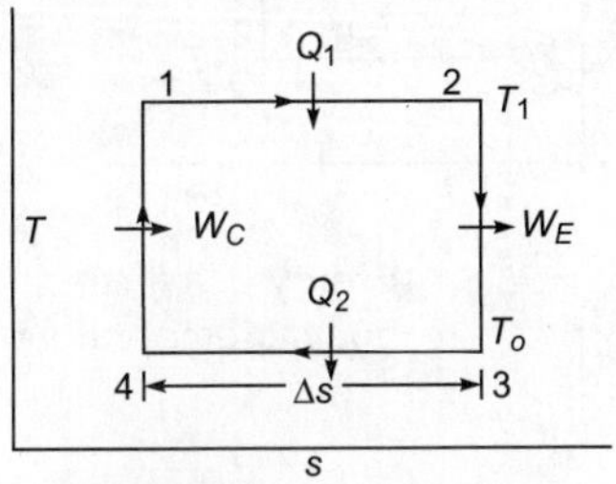

Fig. 8.3 Carnot cycle

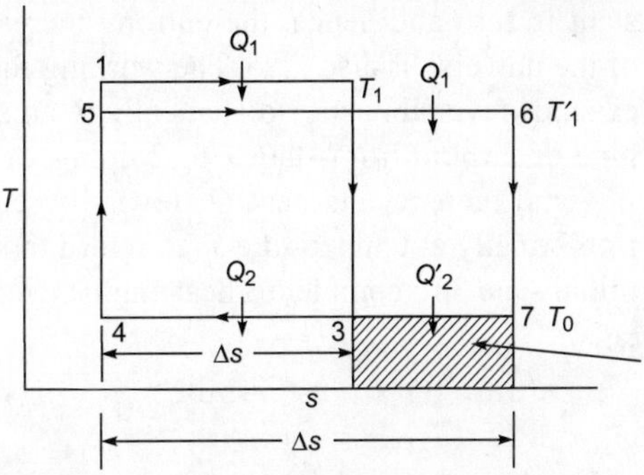

Fig. 8.4 Decrease in exergy due to heat transfer through a finite temperature difference

Let us now consider that heat Q_1 is transferred through a finite temperature difference from the reservoir or source at T_1 to the engine absorbing heat at T_1' lower than T_1 (Fig. 8.4). The exergy of Q_1 received by the engine at T_1' can be found by allowing the engine to operate reversibly in a cycle between T_1 and T_o, receiving Q_1 and rejecting Q_2.

Now, $Q_1 = T_1 \Delta s = T_1' \Delta s'$. Since $T_1 > T_1'$, $\Delta s' > \Delta s$

$$Q_2 = T_o\, \Delta s \text{ and } Q_2' = T_o\, \Delta s'$$

∴ $Q_2' > Q_2$

$$W' = Q_1 - Q_2' = T_1' \Delta s' - T_o\, \Delta s' = (T_1' - T_o)\, \Delta s'$$

and $W = (T_1 - T_o)\, \Delta s$

∴ $W' < W$

Exergy lost due to irreversible heat transfer through a finite temperature difference between the source and the working fluid during the heat addition process is given by

$$W - W' = Q_2' - Q_2 = T_o(\Delta s' - \Delta s) \tag{8.1}$$

Exergy destroyed in the process is thus the product of the lowest feasible temperature, i.e. the dead state, and the additional entropy change in the system while receiving heat irreversibly, compared to the case of reversible heat transfer from the same source.

The greater is the temperature difference $(T_1 - T_1')$, the greater is the heat rejection Q_2' and greater will be the exergy lost, or the unavailable part of the energy supplied Q_1. Exergy is said to be destroyed each time it flows through a finite temperature difference.

8.4 EXERGY OF A FINITE BODY AT TEMPERATURE *T*

> LO 8.5
>
> Illustrate the exergy of a given mass of ideal gas at a specified state

Let us consider a hot gas of mass m_g at temperature T when the environment temperature of the dead state is T_o (Fig. 8.5). Let the gas be cooled at constant pressure from state 1 at temperature T to state 3 at temperature T_o and the heat given up by the gas, Q_1, be utilised in heating up reversibly a working fluid of mass m_{wf} from state 3 to state 1 along the same path so that the temperature difference between the gas and the working fluid at any instant is zero and hence, the entropy increase of the universe is also zero. The working fluid expands reversibly and adiabatically in an engine or turbine from state 1 to 2 doing work W_E, and then rejects heat Q_2 reversibly and isothermally at temperature T_o to return to the initial state 3 to complete a heat engine cycle.

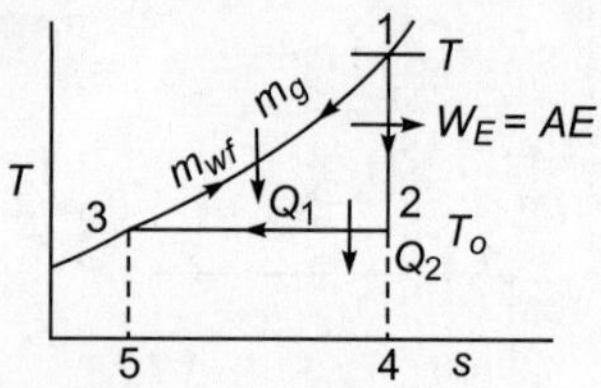

Fig. 8.5 Exergy of a finite bodyat temperature T

Now,

$$Q_1 = m_g c_{p_g} (T - T_o) = m_{wf} c_{p_{wf}} (T - T_o).$$

$$= \text{Area } 14531$$

$$\therefore \quad m_g c_{p_g} = m_{wf} c_{p_{wf}}$$

$$\therefore \quad \Delta S_{\text{gas}} = \int_T^{T_o} m_g c_{p_g} \frac{dT}{T} = m_g c_{p_g} \text{ in } \frac{T_o}{T} \text{ (negative)}$$

$$\Delta S_{wf} = \int_T^{T} m_{wf} c_{p_{wf}} \frac{dT}{T} = m_w c_{p_w} \text{ in } \frac{T}{T_o} \text{ (positive)}$$

$$\therefore \quad \Delta S_{\text{universe}} = \Delta S_{\text{gas}} + \Delta S_{wf} = 0$$

Also,

$$Q_2 = T_o \Delta S_{wf} = T_o m_{wf} c_{p_{wf}} \ln \frac{T}{T_o}$$

Exergy of the gas is thus

$$X = W_{\max} = Q_1 - Q_2$$

$$= m_g c_{p_g} (T - T_o) - T_o m_g c_{p_g} \ln \frac{T}{T_o}$$

$$= m_g c_{p_g} \left[T - T_o - T_o \ln \frac{T}{T_o} \right]$$

$$= \text{Area } 1231$$

Thus the exergy of any finite body of heat capacity C at temperature T is given by

$$X = C \left(T - T_o - T_o \ln \frac{T}{T_o} \right) \quad (8.2)$$

This is similar to Eq. (7.21) as derived from the entropy principle in Chapter 7.

8.5 QUALITY OF ENERGY

Let us assume that a hot gas is flowing through a pipeline (Fig. 8.6). Due to heat loss to the surroundings, the temperature of the gas decreases continuously from inlet at state 'a' to the exit at state 'b'. Although the process is irreversible, let us assume a reversible isobaric

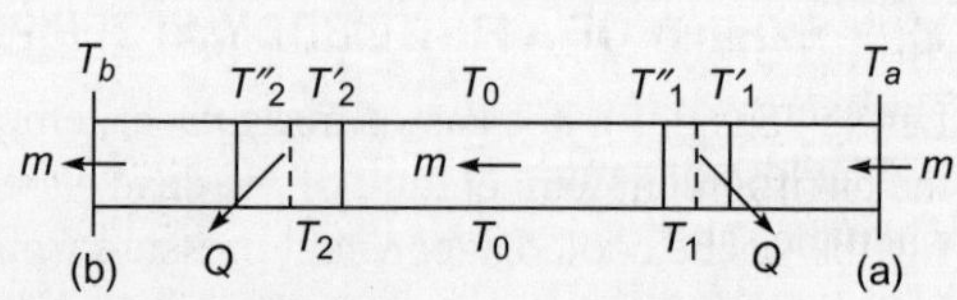

Fig. 8.6 Heat loss from a hot gas flowing through a pipeline

path between the inlet and exit states of the gas (Fig. 8.7). For an infinitesimal reversible process at constant pressure

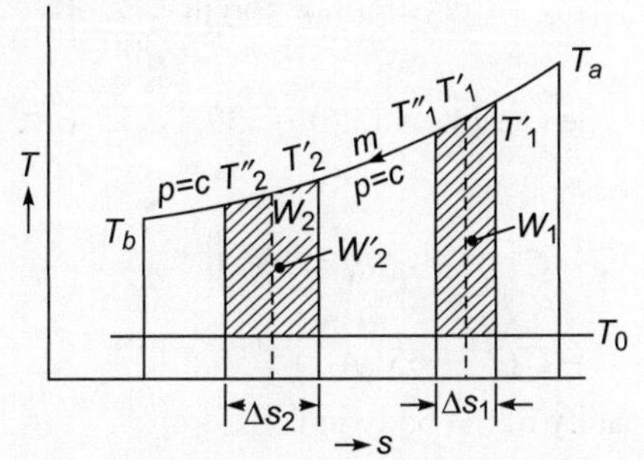

Fig. 8.7 Degradation of energy quality of gas as temperature of the gas decreases

$$dS = \frac{mc_p dt}{T}$$

$$\therefore \quad \frac{dT}{dS} = \frac{T}{mc_p} \tag{8.3}$$

where m is the mass of gas flowing and c_p is its specific heat. The slope $\frac{dT}{dS}$ depends on the gas temperature T. As T increases, the slope increases and if T decreases, slope decreases.

Let us assume that Q units of heat are lost to the surroundings at T_o as the temperature of the gas decreases from T_1' to T_1'', T_1 being the average of the two. Then

Heat loss, $$Q = mc_p (T_1' - T_1'') = T_1 \Delta S_1 \tag{8.4}$$

Exergy of this heat loss at temperature T_1 is

$$W_1 = Q - T_o \Delta S_1 \tag{8.5}$$

When the gas temperature has reached T_2 ($T_2 < T_1$), let us assume that the same heat loss Q occurs as the gas temperature decreases from T_2' to T_2'', T_2 being the average temperature of the gas.

Heat loss, $$Q = mc_p (T_2' - T_2'') = T_2 \Delta S_2 \tag{8.6}$$

Exergy of this heat loss at temperature T_2 is

$$W_2 = Q - T_o \Delta S_2 \tag{8.7}$$

From Eqs. (8.4) and (8.6),

Since $T_1 > T_2$, $\Delta S_1 < \Delta S_2$, therefore, from Eqs. (8.5) and (8.7),

$$W_1 > W_2 \tag{8.8}$$

The exergy of the gas at T_1 is more than that at T_2.

The loss of exergy is more when heat loss occurs at a higher temperature T_1 than when the same heat loss occurs at a lower temperature T_2. Therefore, a heat loss of 1 kJ at, say, 1000°C is more harmful than the same heat loss of 1 kJ at, say, 100°C. The exergy of a system at temperature T is given by Eq. (8.2),

$$X = C\left(T - T_o - T_o \ln \frac{T}{T_o}\right)$$

At $T = 1000 + 273 = 1273$ K and $T_o = 300$ K,

$$X_1 = C\left(1273 - 300 - 300 \ln \frac{1273}{300}\right)$$

$$= C(973 - 433.6) = 539.4\, C \text{ kJ}$$

At $T = 100 + 273 = 373$ K,

$$X_2 = C\left(373 - 300 - 300 \ln \frac{373}{300}\right)$$

$$= C(73 - 65.34) = 7.66\, C \text{ kJ}$$

where C is the heat capacity of the body in kJ/K.

Therefore, $X_1 > X_2$. Adequate insulation must be provided for high temperature fluids ($T >> T_o$) to reduce the precious heat loss. This may not be so important for low temperature fluids ($T \sim T_o$), since the loss of exergy from such fluids would be low. The cost of insulation may be more than the money saved by reducing the heat loss. Similarly, insulation must be provided adequately for very low temperature fluids ($T << T_o$) whose exergy is high to reduce heat gain from the surroundings and preserve its exergy.

We have seen that the exergy of a fluid at a higher temperature T_1 is more than that at a lower temperature T_2, and it decreases as the temperature decreases.

The second law thus affixes a quality to energy of a system at any state. The quality of energy of a gas at, say, 1000°C is superior to that at, say, 100°C. An awareness of this energy quality as of energy quantity is essential for the efficient use of our energy resources and for energy conservation. *The concept of exergy provides a useful measure of this energy quality.*

8.6 LAW OF DEGRADATION OF ENERGY

> **LO 8.6**
> Relate second law with degradation of energy

The exergy of a system decreases as its temperature or pressure decreases and approaches that of the dead state where its value is zero. When heat is transferred from a system, its temperature decreases and the quality of its energy deteriorates and its exergy decreases. The degradation of energy of the system is more when heat loss occurs at a higher temperature than that occurring at a lower temperature. Quality wise the energy loss may be the same, but quantity wise the losses are different. *While the first law states that energy is always conserved quantity wise, the second law emphasises that energy is always degraded quality wise.* When a gas is throttled adiabatically from high to low pressure, the enthalpy or energy of gas per unit mass remains the same, but there is a degradation of it energy or capacity of doing work. The same holds good for pressure drop due to friction of a fluid flowing through an insulated pipe. If the first law is the law of conservation of energy, the second law is called *the law of degradation of energy.*

Energy is always conserved, but its quality is always degraded.

8.7 MAXIMUM WORK IN A REVERSIBLE PROCESS

LO 8.7
Compare reversible work with irreversible work of a closed system

Let us consider a closed stationary system undergoing a reversible process R from state 1 to 2 interacting with the surroundings at p_o, To (Fig. 8.8). Then by the first law,

$$Q_R = U_2 - U_1 + W_R \tag{8.9}$$

If the process were irreversible, as represented by the dotted line I connecting the same equilibrium points 1 and 2,

$$Q_I = U_2 - U_1 + W_I \tag{8.10}$$

Therefore, from Eqs. (8.9) and (8.10),

$$Q_R - Q_I = W_R - W_I \tag{8.11}$$

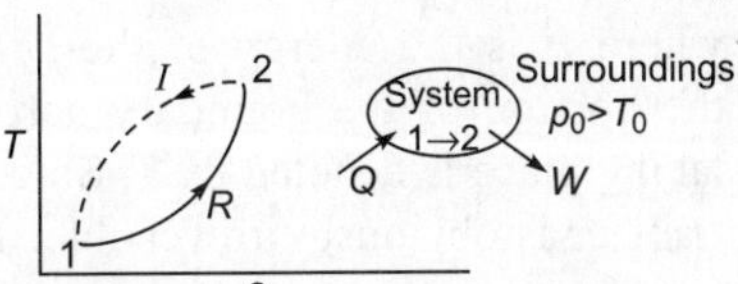

Fig. 8.8 Maximum work done by a closed system

Now,

$$\Delta S_{sys} = S_2 - S_1$$

and

$$\Delta S_{surr} = -\frac{Q}{T_o}$$

By second law,

$$\Delta S_{univ} \geqslant O$$

For a reversible process,

$$\Delta S_{univ} = S_2 - S_1 - \frac{Q_R}{T_o} = O$$

$$Q_R = T_o (S_2 - S_1) \tag{8.12}$$

For an irreversible process,

$$\Delta S_{univ} > O$$

or

$$S_2 - S_1 - \frac{Q_I}{T_o} > O$$

$\therefore$

$$Q_I < T_o (S_2 - S_1) \tag{8.13}$$

From Eqs. (8.12) and (8.13),

$$Q_R > Q_1$$

$\therefore$ From Eq. (8.11),

$$W_R > W_I \tag{8.14}$$

The work done by a closed system by interacting only with the surroundings at p_o, T_o in a reversible process is always more than that done by it in an irreversible process between the same end states.

8.8 WORK DONE IN ALL REVERSIBLE PROCESSES IS THE SAME

LO 8.8
Demonstrate equality of reversible works for same end state points

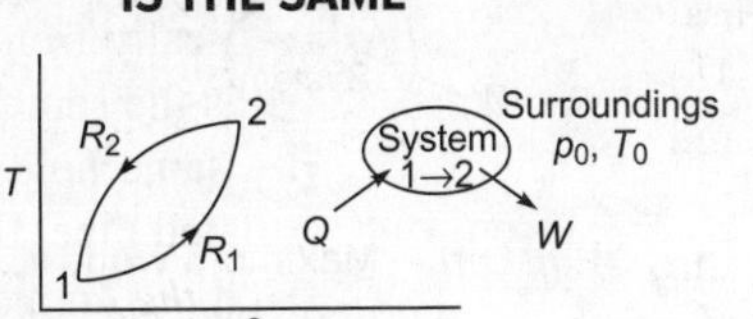

Fig. 8.9 Equal work done in all reversible process between the same end states

Let us assume two reversible processes R_1 and R_2 between the same end states 1 and 2 undergone by a closed system by exchanging energy only with the surroundings (Fig. 8.9). Let one of the processes be reversed.

Then the system would execute a reversible cycle 1-2-1 and produces net work represented by the area enclosed by exchanging energy with only one reservoir, i.e. the surroundings. This violates the Kelvin–Planck statement. Therefore, the *two reversible processes must coincide and produce equal amounts of work.*

8.9 EXERGY OF A CLOSED SYSTEM

LO 8.9 Evaluate the maximum work for different systems and relate Helmholtz and Gibbs functions with it

For a closed system having energy interactions Q and W and undergoing change of state from 1 to 2 (Fig. 8.10) and $T_o > T$ so that heat flows into the system, from the surroundings at T_o

$$Q = E_2 - E_1 + W$$

$$\therefore \quad W = E_1 - E_2 + Q \tag{8.15}$$

where E is the energy of the system. The temperature of the portion of the boundary across which heat transfer is taking place is denoted by T. Since $T_o > T$, the heat transfer is obviously irreversible. To make the heat transfer process reversible, let a Carnot engine be operated between T_o and T receiving heat Q_o from the surroundings, producing work W_c and rejecting heat Q at T, which enters the system. The maximum work delivered by the system will be

$$W_{\max} = W + W_C - p_o (V_2 - V_1) \tag{8.16}$$

where $p_o (V_2 - V_1)$ is the work that will be done by the system against the surroundings having a flexible boundary with the volume of the system increasing from V_1 to V_2 (which will be zero if the boundary is rigid). The work done by the Carnot engine,

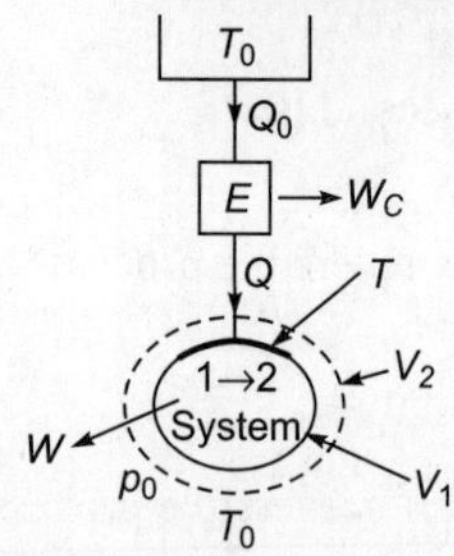

Fig. 8.10 Maximum work done by a closed system with heat flowing into the system ($T_o > T$)

$$W_C = Q_o - Q$$

For a reversible engine, $\dfrac{Q_o}{T_o} = \dfrac{Q}{T} = S_2 - S_1$

$$\therefore \quad W_C = T_o \frac{Q}{T} - Q, \tag{8.17}$$

From Eqs. (8.15), (8.16) and (8.17), neglecting K.E. and P.E. terms (when $E = U$),

$$W_{\max} = U_1 - U_2 + Q + T_o \frac{Q}{T} - Q - p_o (V_2 - V_1)$$

$$= U_1 - U_2 + T_o (S_2 - S_1) - p_o (V_2 - V_1)$$

$$= U_1 - U_2 + p_o (V_1 - V_2) - T_o (S_1 - S_2) = (W_u)_{\max} \tag{8.18}$$

This is the maximum useful work $(W_u)_{\max}$ delivered by the system.

If heat is rejected from the system across the boundary where the temperature is T and $T > T_o$, the first law gives (Fig. 8.11), neglecting K.E. and P.E. terms,

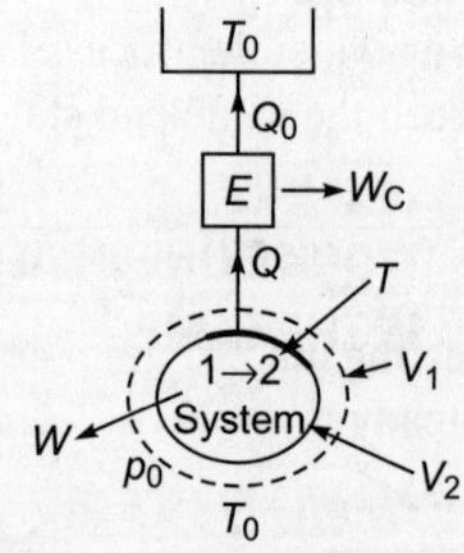

Fig. 8.11 Maximum work done by a closed system with heat rejection from the system ($T > T_o$)

$$-Q_1 = U_2 - U_1 + W$$

$$\therefore \qquad W = U_1 - U_2 - Q \tag{8.19}$$

The work done by the reversible engine operating between T and T_o.

$$W_C = Q - Q_o$$

By the second law, $\frac{Q_o}{T_o} = \frac{Q}{T}$ and $-\frac{Q}{T} = S_2 - S_1$ (8.20)

$$\begin{aligned}
\therefore \quad \dot{W}_{max} &= W + W_C - p_o(V_2 - V_1) \\
&= U_1 - U_2 - Q + Q - Q_o - p_o(V_2 - V_1) \\
&= U_1 - U_2 + T_o \frac{Q}{T} - p_o(V_1 - V_2) = U_1 - U_2 + T_o(S_2 - S_3) - p_o(V_2 - V_1) \\
&= U_1 - U_2 + p_o(V_1 - V_2) - T_o(S_1 - S_2)
\end{aligned}$$

This is the same as given by Eq. (8.18), which provides the maximum work done by the closed system, also called *availability A*, in the change of state from 1 to 2 by interacting only with the surroundings at p_o, T_o. This expression can also be written in the alternate from

$$W_{max} = A = (U_1 + p_o V_1 - T_o S_1) - (U_2 + p_o V_2 - T_o S_2) \tag{8.21a}$$

$$= \phi_1 - \phi_2 \tag{8.21b}$$

where ϕ_1 and ϕ_2 are exergy functions of a closed system at states 1 and 2, respectively. For unit mass of substance

$$\begin{aligned}
W_{max} &= a = (u_1 + p_o v_1 - T_o s_1) - (u_2 + p_o v_2 - T_o s_1) \\
&= (u_1 - u_2) + p_o(v_1 - v_2) - T_o(s_1 - s_2)
\end{aligned}$$

With reference to the dead state, the maximum work that can be delivered by the closed system at p, T is given by

$$\begin{aligned}
W_{max} &= A = (U - U_o) + p_o(V - V_o) - T_o(S - S_o) \\
&= (U + p_o V - T_o S) - (U_o + p_o V_o - T_o S_o) \\
&= \phi - \phi_o
\end{aligned} \tag{8.22}$$

where $\phi_o = U_o + p_o V_o - T_o S_o$ is the exergy function of the dead state. For unit mass of a substance,

$$\begin{aligned}
W_{max} &= a = (u + p_o v - T_o s) - (u_o + p_o v_o - T_o s_o) \\
&= \phi - \phi_o
\end{aligned}$$

8.10 EXERGY OF A STEADY FLOW SYSTEM

For an open or a steady flow system (Fig. 8.12), neglecting kinetic and potential energy changes, the steady flow energy equation (S.F.E.E.) gives

$$H_1 + Q = H_2 + W$$

or

$$W = H_1 - H_2 + Q \tag{8.23}$$

Now, with a reversible heat engine E operating between T_o and T, the work done will be

$$W_C = Q_o - Q$$

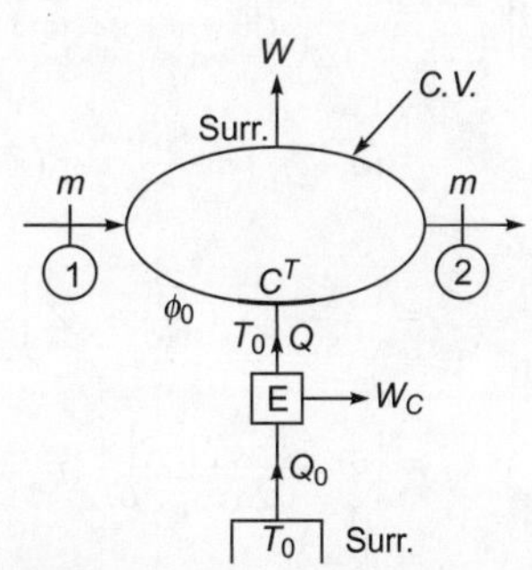

Fig. 8.12 Maximum work done by a steady flow system

Also, $\frac{Q}{T} = \frac{Q_o}{T_o}$ and $\frac{Q}{T} = S_2 - S_1$ (8.24)

∴ Maximum work obtained from the steady flow system would be from Eqs. (8.23) and (8.24), also called availability, A_f

$$\begin{aligned} W_{\text{Max}} &= A_f = W + W_c \\ &= H_1 - H_2 + Q + Q_o - Q \\ &= H_1 - H_2 + Q.\frac{T_o}{T} = H_1 - H_2 + T_o(S_2 - S_1) \\ &= H_1 - H_2 - T_o(S_1 - S_2) \\ &= (H_1 - T_oS_1) - (H_2 - T_oS_2) = B_1 - B_2 \end{aligned} \tag{8.25}$$

where $B = H - T_oS$ is called the *Keenan function*.

For unit mass of a substance,

$$\begin{aligned} W_{\max} &= a_f = (h_1 - h_2) - T_o(s_1 - s_2) \\ &= (h_1 - T_os_1) - (h_2 - T_os_2) \\ &= b_1 - b_2 \end{aligned} \tag{8.26}$$

If K.E. and P.E. terms are taken into consideration, Eqs. (8.25) and (8.26) become

$$\begin{aligned} W_{\max} &= A_f = \left(H_1 + \frac{mV_1^2}{2} + mgZ_1 - T_oS_1\right) - \left(H_2 + \frac{mV_2^2}{2} + mgZ_2 - T_oS_2\right) \\ &= \left(B_1 + \frac{mV_1^2}{2} + mgZ_1\right) - \left(B_2 + \frac{mV_2^2}{2} + mgZ_2\right) \\ &= \psi_1 - \psi_2 \end{aligned} \tag{8.27}$$

where $\psi = B + \frac{mV^2}{2} + mgZ$ is the *exergy function of a steady flow system*.

With reference to the dead state (Fig. 8.13), Eqs. (8.26) and (8.27) become

$$\begin{aligned} W_{\max} &= a_f = (h - T_os) - (h_o - T_os_o) \\ &= (h - h_o) - T_o(s - s_o) \\ &= b - b_o \end{aligned} \tag{8.28}$$

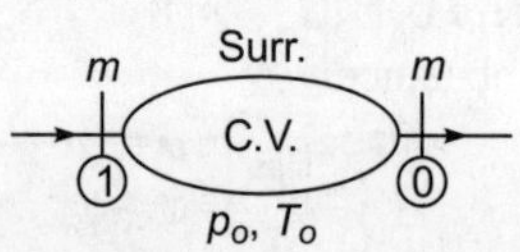

Fig. 8.13 Exergy of a steady flow system

$$\begin{aligned} W_{\max} &= a_f = \left(h + \frac{V^2}{2} + gZ\right) - (h_o + gZ_o) - T_o(s - s_o) \\ &= (h - h_o) + \frac{V^2}{2} + g(Z - Z_o) - T_o(s - s_o) \\ &= \left(h - T_oS + \frac{V^2}{2} + gZ\right) - (h_o - T_os_o + qZ_o) \\ &= \left(B + \frac{V^2}{2} + gZ\right) - (b_o + gZ_o) \\ &= \psi - \psi_o \end{aligned} \tag{8.29}$$

since the velocity of the system at the dead state is zero.

8.11 OBTAIN MAXIMUM WORK WHEN THE SYSTEM EXCHANGES HEAT WITH A THERMAL RESERVOIR IN ADDITION TO THE ATMOSPHERE

If the open system as discussed in Section 8.9 exchanges heat with a thermal reservoir at temperature T_R in addition to the atmosphere, the maximum work will be increased by $Q_R\left(1-\frac{T_o}{T_R}\right)$,

i.e. $$W_{max} = \left(H_1 - T_oS_1 + \frac{mV_1^2}{2} + mgZ_1\right) - \left(H_2 - T_oS_2 + \frac{mV_2^2}{2} + mgZ_2\right) + Q_R\left(1-\frac{T_o}{T_R}\right)$$

$$= \psi_1 - \psi_2 + Q_R\left(1-\frac{T_o}{T_R}\right) \tag{8.30}$$

For a closed system,

$$W_{max} = E_1 - E_2 + p_o(V_1 - V_2) - T_o(S_1 - S_2) + Q_R\left(1-\frac{T_o}{T_R}\right) \tag{8.31}$$

If K.E. and P.E. terms are neglected, for a steady flow system

$$W_{max} = H_1 - H_2 - T_o(S_1 - S_2) + Q_R\left(1-\frac{T_o}{T_R}\right)$$

and for a closed non-flow system,

$$W_{max} = U_1 - U_2 + p_o(V_1 - V_2) - T_o(S_1 - S_2) + Q_R\left(1-\frac{T_o}{T_R}\right) \tag{8.32}$$

8.12 EXERGY IN CHEMICAL REACTIONS

In many chemical reactions, the reactants are often in pressure and temperature equilibrium with the surroundings (before the reaction takes place) and so are the products after the reaction. An internal combustion engine can be cited as an example of such a process if we visualise the products being cooled to atmospheric temperature T_o before being discharged from the engine.

1. Let us consider a system which is in temperature equilibrium with the surroundings before and after the process. The maximum work obtainable during a change of state is given by Eq. (8.18),

$$W_{max} = E_1 - E_2 - T_o(S_1 - S_2)$$

$$= \left(U_1 + \frac{mV_1^2}{2} + mgZ_1\right) - \left(U_2 + \frac{mV_2^2}{2} + mgZ_2\right) - T_o(S_1 - S_2)$$

If K.E. and P.E. terms are neglected

$$W_{max} = U_1 - U_2 - T_o(S_1 - S_2)$$

Since the initial final temperature are the same as that of the surroundings,

$$T_1 = T_2 = T_o = T, \text{ say, then}$$

$$W_{max} = (U_1 - U_2)_T - T(S_1 - S_2)_T \tag{8.33}$$

Let a property called *Helmholtz functions F* be defined by the relation

$$F = U - T_5 \tag{8.34}$$

Thus for two equilibrium states 1 and 2 at the same temperature T,

$$(F_1 - F_2)_T = (U_1 - U_2)_T - T(S_1 - S_2)_T \tag{8.35}$$

From Eqs. (8.34) and (8.35)

$$(W_T)_{\max} = (F_1 - F_2)_T \tag{8.36}$$

Therefore, $$W_T \leqslant (F_1 - F_2)_T \tag{8.37}$$

The work done by a system in any process between two equilibrium states at the same temperature during which the system exchanges heat only when the environment is equal to or less than the decrease in the Helmholtz function of the system during the process. The maximum work is done when the process is reversible and the equality sign holds. If the process is irreversible, the work is less than the maximum.

2. Let us now consider a system which is in both pressure and temperature equilibrium with the surroundings before and after the process. When the volume of the system increases, same work is done by the system against the surroundings (pdV work) and this is not available for doing useful work. The exergy of the system, as defined by Eq. (8.22), neglecting the K.E. and P.E. terms, can be expressed in the form,

$$\begin{aligned} A = (W_u)_{\max} &= (U + p_o V - T_o S) - (U_o + p_o V - T_o S_o) \\ &= \phi - \phi_o \end{aligned}$$

The maximum work obtainable during a change of state is the decrease of its exergy given by

$$\begin{aligned} (W_u)_{\max} &= A_1 - A_2 = \phi_1 - \phi_2 \\ &= (U_1 - U_2) + p_o (V_1 - V_2) - T_o(S_1 - S_2) \end{aligned}$$

When the initial and final equilibrium states of the system are at the same pressure and temperature of the surroundings, say,

$$p_1 = p_2 = p_o = p \quad \text{and} \quad T_1 = T_2 = T_o = T, \text{ then}$$

$$(W_u)_{\max} = (U_1 - U_2)_{P,T} + p(V_1 - V_2)_{P,T} - T(S_1 - S_2)_{P,T} \tag{8.38}$$

The *Gibbs function F* is defined as

$$\begin{aligned} G &= H - TS \\ &= U + pV - TS \end{aligned} \tag{8.39}$$

Thus for two equilibrium states at the same pressure p and temperature T

$$(G_1 - G_2)_{P,T} = (U_1 - U_2)_{P,T} + P(V_1 - V_2)_{P,T} - T(S_1 - S_2)_{P,T} \tag{8.40}$$

From Eqs. (8.38) and (8.40),

$$(W_u)_{\max} = (G_1 - G_2)_{P,T}$$

$$\therefore \quad (W_u)_{P,T} \leqslant (G_1 - G_2)_{P,T} \tag{8.41}$$

The decrease in the Gibbs function of a system sets an upper limit to the work that can be performed, exclusive of the pdV work, in any process between two equilibrium points at the same pressure and temperature, provided the system exchanges heat only with the environment which is at the same temperature and pressure as the end states of the system. If the process is irreversible, the useful work is less than the maximum.

8.13 IRREVERSIBILITY AND GOUY–STODOLA THEOREM

LO 8.10

Relate irreversibility with Gouy-Stodola theorem

The actual work done by a system is always less than the idealised reversible work, and the difference between the two is called the *irreversibility* (I) of the process.

$$\therefore \quad I = W_{rev} - W = W_{max} - W \tag{8.42}$$

This is also often referred to as '*degradation*' or '*dissipation*'.

For a non-flow process between two equilibrium states, when the system exchanges heat only with the surroundings (Fig. 8.14), neglecting K.E. and P.E. terms, the first law gives from Eq. (8.42),

$$\begin{aligned} I &= [(U_1 - U_2) - T_o (S_1 - S_2)] - [(U_1 - U_2) + Q] \\ &= T_o (S_2 - S_1) - Q \\ &= T_o (\Delta S)_{system} + T_o (\Delta S)_{surr.} \\ &= T_o (\Delta S)_{univ} \end{aligned} \tag{8.43}$$

$$\therefore \quad I \geqslant O$$

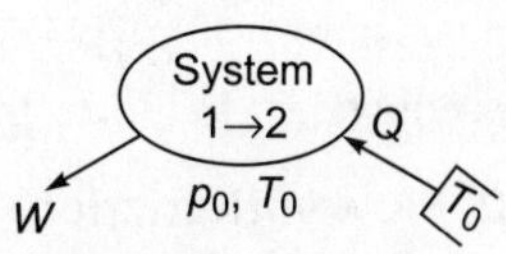

Fig. 8.14 Irreversibility in a process

Similarly, for the steady flow process

$$\begin{aligned} I &= W_{max} - W \\ &= \left(B_1 + \frac{mV_1^2}{2} + mgZ_1\right) - \left(B_2 + \frac{mV_2^2}{2} + mgZ_2\right) \\ &\quad - \left(H_1 + \frac{mV_1^2}{2} + mgZ_1\right) - \left(H_2 + \frac{mV_2^2}{2} + mgZ_2\right) \\ &= T_o(S_2 - S_1) - Q \\ &= T_o (\Delta S)_{sys} + T_o (\Delta S)_{surr} = T_o (\Delta S)_{univ.} \end{aligned} \tag{8.44}$$

The same expression for irreversibility holds good for both flow and non-flow processes. The quantity $T_o (\Delta S)_{univ}$ or $T_o S_{gen}$ represents the decrease in exergy of the system due to irreversibility.

This relation, Eq. (8.44), is called the *Gouy–Stodola theorem* which states that the rate of loss of exergy in a process is proportional to the rate of *entropy generation*. If (Eq. (8.44)) is written in the rate form

$$\dot{I} = \dot{W}_{lost} = T_o (\Delta \dot{S})_{univ} = T_o \dot{S}_{gen} \tag{8.45}$$

This is known as *Gouy–Stodola equation*. A thermodynamically efficient process would involve minimum exergy loss with minimum rate of entropy generation.

8.13.1 Application of Gouy–Stodola Equation

1. Heat transfer through a finite temperature difference If heat transfer $\dot{Q}$ occurs from the hot reservoir at T_1 to the cold reservoir at T_2 (Fig. 8.15), the rate of entropy generation

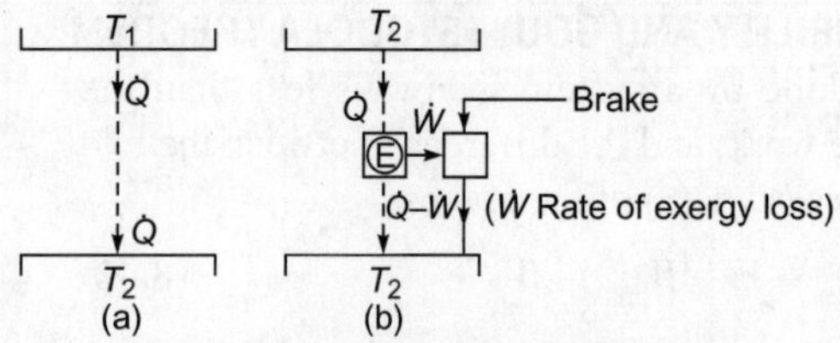

Fig. 8.15 Decrease of exergy by heat transfer through a finite temperature difference

$$\dot{S}_{gen} = \frac{\dot{Q}}{T_2} - \frac{\dot{Q}}{T_1} = \dot{Q}\frac{T_1 - T_2}{T_1 T_2}$$

and $$\dot{W}_{lost} = T_2 \dot{S}_{gen} = \dot{Q}\frac{T_1 - T_2}{T_1} = \dot{Q} . \eta_{rev} \tag{8.46}$$

If the heat transfer $\dot{Q}$ from T_1 to T_2 takes place through a reversible engine, work output $\dot{W}$ is dissipated in the brake, from which an equal amount of heat is rejected to the reservoir at T_2 (Fig. 8.15 b). Heat transfer through a finite temperature difference is equivalent to the destruction of its exergy.

2. Flow with friction Let us consider the steady and adiabatic flow of an *ideal gas* through the segment of a pipe (Fig. 8.16a), by the first law

$$h_1 = h_2$$

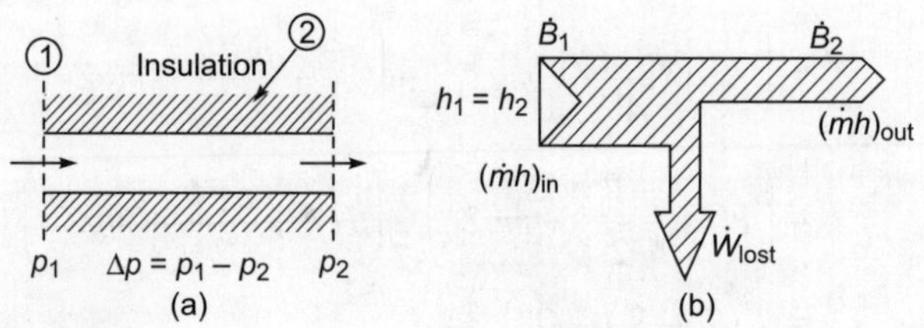

Fig. 8.16 Irreversibility in a duct flow due to fluid friction

and by the second law

$$Tds = dh - vdp$$

Since for an ideal gas, enthalpy is a function of temperature, which here remains constant, $dh = 0$, the entropy change

$$\int_1^2 ds = -\int_1^2 \frac{vdp}{T} = -\int_1^2 R\frac{dp}{p}$$

and the rate of entropy generation

$$\dot{S}_{gen} = \int_1^2 \dot{m}\, ds = -\int_1^2 \dot{m}\, R\frac{dp}{p} = \dot{m}\, R \ln\frac{p_1}{p_2}$$

$$= -\dot{m}\, R \ln\left(1 - \frac{\Delta p}{p_1}\right) = -\dot{m}R\left(-\frac{\Delta p}{p_1}\right)$$

$$= \dot{m}R\frac{\Delta p}{p_1} \tag{8.47}$$

where $\ln\left(1 - \frac{\Delta p}{p_1}\right) = -\frac{\Delta p}{p_1}$, since $\frac{\Delta p}{p_1} < 1$.

(We know: $\ln(1-x) = -x - \frac{x^3}{3} - \frac{x^5}{5} - \cdots = -x$, where $x < 1$, with higher terms being neglected)

$$\begin{aligned}\dot{W}_{\text{lost}} &= \dot{B}_1 - \dot{B}_2 \\ &= \dot{m}\,[(h_1 - T_o s_1) - (h_2 - T_o s_2)] \\ &= \dot{m} T_o (s_2 - s_1) \\ &= T_o \dot{S}_{\text{gen}} = \dot{m} R T_o \frac{\Delta p}{p} \qquad (8.48)\end{aligned}$$

The rate of exergy loss, $\dot{W}_{\text{lost}}$, is thus proportional to the pressure drop (Δp) and the mass flow rate ($\dot{m}$). It is shown on the right (Fig. 8.16b) by the *Grassmann diagram*, the width at any section being proportional to the *exergy* of the fluid (system). It is an adaptation of the *Sankey diagram* where the width at any section is proportional to the *energy* of the fluid in a pipeline.

3. Mixing of two fluids Two streams 1 and 2 of an incompressible fluid or an ideal gas mix adiabatically at constant pressure (Fig. 8.17).

Here, $\dot{m}_1 + \dot{m}_2 = \dot{m}_3 = \dot{m}$ (say)

Let $x = \dfrac{\dot{m}}{\dot{m}_1 + \dot{m}_2}$

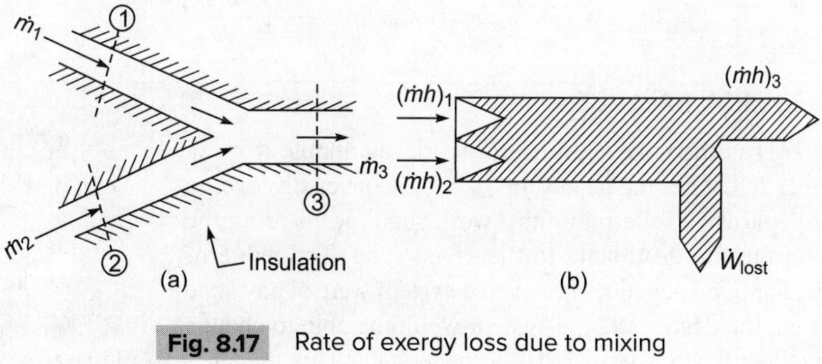

Fig. 8.17 Rate of exergy loss due to mixing

By the first law,

$$\dot{m}_1 h_1 + \dot{m}_2 h_2 = (\dot{m}_1 + \dot{m}_2)\, h_3$$

$$x h_1 + (1-x) h_2 = h_3$$

The preceding equation can be written in the following form since enthalpy is a function of temperature

$$x T_1 + (1-x) T_2 = T_3$$

or

$$\frac{T_3}{T_1} = x + (1-x)\tau \qquad (8.49)$$

where $\tau = T_2/T_1$

By the second law,

$$\dot{S}_{\text{gen}} = \dot{m}_3 s_3 - \dot{m}_1 s_1 - \dot{m}_2 s_2$$

or

$$\dot{S}_{\text{gen}} = \dot{m} s_3 - x \dot{m} s_1 - (1-x)\, \dot{m} s_2$$

$$\therefore \qquad \frac{\dot{S}_{gen}}{\dot{m}} = (s_3 - s_2) + x(s_2 - s_1).$$

$$= c_p \ln \frac{T_3}{T_2} + x\, c_p \ln \frac{T_2}{T_1}$$

$$\frac{\dot{S}_{gen}}{\dot{m} c_p} = \ln \frac{T_3}{T_2} + x \ln \frac{T_2}{T_1}$$

$$= \ln \left(\frac{T_3}{T_2}\right)\left(\frac{T_2}{T_1}\right)^x$$

or

$$N_s = \ln \frac{T_3}{T_1}\frac{T_1}{T_2}\left(\frac{T_2}{T_1}\right)^x$$

$$= \ln \frac{T_3/T_1}{(T_2/T_1)^{1-x}} = \frac{x + (1-x)z}{z^{1-x}} \qquad (8.50)$$

where N_S is a dimensionless quantity, called the *entropy generation number*, given by $\frac{\dot{S}_{gen}}{\dot{m} c_p}$. If $x = 1$ or $\tau = 1$, N_s becomes zero in each case. The rate of exergy loss due to mixing would be

$$\dot{W}_{lost} = \dot{I} = \dot{S}_{gen} = T_o\, \dot{m}\, c_p \ln \frac{x + (1-x)\tau}{\tau^{1-x}} \qquad (8.51)$$

8.14 EXERGY BALANCE

LO 8.11

Demonstrate exergy balance for closed and steady flow systems

The exergy is the maximum useful work obtainable from a system as it reaches the dead state (p_o, T_o). Conversely, exergy can be regarded as the minimum work required to bring the system from the dead state to the given state. The value of exergy cannot be negative. If a closed system were at any state other than the dead state, the system would be able to change its state spontaneously towards the dead state. This tendency would stop when the dead state is reached. No work is done to effect such a spontaneous change. Since any change in state of the closed system to the dead state can be accomplished with zero work, the *maximum work or exergy cannot be negative*.

While energy is always conserved, exergy is not generally conserved, but it is destroyed by irreversibilities. When a closed system is allowed to undergo a spontaneous change from the given state to the dead state, its exergy is completely destroyed without producing any useful work. The potential to develop work that exists originally at the given state is thus completely wasted in such a spontaneous process.

Therefore, at steady state

1. Energy in – Energy out = 0
2. Exergy in – Exergy out = Exergy destroyed

8.14.1 Exergy Balance for a Closed System

For a closed system, exergy transfer occurs through heat and work interactions (Fig. 8.18).

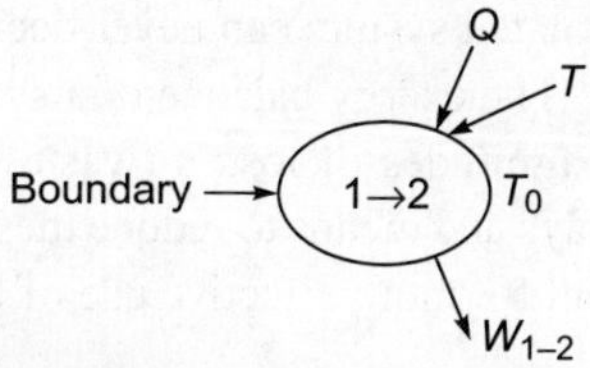

Fig. 8.18 Exergy balance for a closed system

First law: $$E_2 - E_1 = \int_1^2 \text{đ}Q - W_{1-2} \quad (8.52)$$

Second law: $$S_2 - S_1 - \int_1^2 \frac{\text{đ}Q}{T} = S_{\text{gen}}$$

or $$T_o(S_2 - S_1) - T_o \int_1^2 \frac{\text{đ}Q}{T} = T_o S_{\text{gen}} \quad (8.53)$$

Subtracting Eq. (8.53) from (Eq. 8.52),

$$E_2 - E_1 - T_o(S_2 - S_1) + T_o \int_1^2 \frac{\text{đ}Q}{T} = \int_1^2 \text{đ}Q - W_{1-2} - T_o S_{\text{gen}}$$

or $$E_2 - E_1 - T_o(S_2 - S_1) = \int_1^2 \text{đ}Q - W_{1-2} - T_o \int_1^2 \frac{\text{đ}Q}{T} - T_o S_{\text{gen}}$$

or $$E_2 - E_1 - T_o(S_2 - S_1) = \int_1^2 \left(1 - \frac{T_o}{T}\right) \delta Q - W_{1-2} - T_o S_{\text{gen}}$$

Since for a closed system, the change in exergy between the states 1 and 2 is given by

$$\phi_2 - \phi_1 = A_2 - A_1 = E_2 - E_1 + p_o(V_2 - V_1) - T_o(S_2 - S_1)$$

$$= p_o(V_2 - V_1) + \int_1^2 \left(1 - \frac{T_o}{T}\right) \delta Q - W_{1\text{-}2}\, T_o S_{\text{gen}}$$

$$A_2 - A_1 = \int_1^2 \left(1 - \frac{T_o}{T}\right) \delta Q - [W_{1-2} - p_o(V_2 - V_1)] - T_o S_{\text{gen}} \quad (8.54)$$

Change in exergy = Exergy transfer with heat – Exergy transfer with work + Exergy destruction

where A or ϕ is the availability or exergy.

8.14.2 Exergy Principle

For an isolated sysem, $Q = 0$ and $W = 0$, and there is no exergy transfer with heat or work. Therefore,

$$\phi_2 - \phi_1 = A_2 - A_1 = -T_o S_{\text{gen}} = -I \quad (8.55)$$

or in the rate form

$$\frac{d\phi}{d\tau} = \frac{dA}{d\tau} = -T_o \dot{S}_{\text{gen}} = -\dot{I} \quad (8.56)$$

Since $\dot{I} > 0$, the only processes allowed by the second law are those for which the exergy of the isolated system decreases, or $A_2 < A_1$ and $dA/d\tau < 0$. In other words, *the exergy of an isolated system can never increase, but always decreases.*

It is the *counterpart of the entropy principle* which states that the entropy of an isolated system can never decrease, but always increases.

The exergy balance of a system can be used to determine the locations, types and magnitudes of losses (waste) of the potential of energy resources (fuels) and find ways and means to reduce these losses so as to make the system more energy efficient and for more effective use of fuel.

8.14.3 Exergy Balance for a Steady Flow System

The steady flow energy equation (S.F.E.E.) between the inlet and exit sections of the control volume (c.v.) (Fig. 8.19), is given by

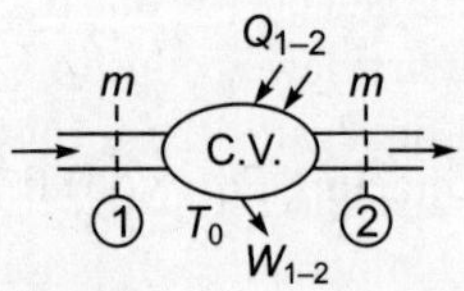

Fig. 8.19 Exergy balance for a steady flow system

First law: $$H_1 + \frac{mV_1^2}{2} + mgZ_1 + Q_{1-2}$$

$$= H_2 + \frac{mV_2^2}{2} + mgZ_2 + W_{1-2} \quad (8.57)$$

Second law:

$$S_1 - S_2 + \int_1^2 \frac{đQ}{T} = S_{\text{gen}}$$

or $$T_o(S_1 - S_2) + T_o \int_1^2 \frac{đQ}{T} = T_o S_{\text{gen}} = I \quad (8.58)$$

From Eqs. (8.57) and (8.58),

$$H_2 - H_1 - T_o(S_2 - S_1) + \frac{m(V_2^2 - V_1^2)}{2} + mg(Z_2 - Z_1)$$

$$= \int_1^2 \left(1 - \frac{T_o}{T}\right) đQ - W_{1-2} - I \quad (8.59)$$

or $$\psi_2 - \psi_1 = A_2 - A_1 = \int_1^2 \left(1 - \frac{T_o}{T}\right) đQ - W_{1-2} - I$$

$$= m\,(a_{f_2} - a_{f_1}) \quad (8.60)$$

For an isolated system,

$$\int_1^2 đQ = 0,\ W_{1-2} = 0,\ m = 0$$

$\therefore$ $$d\psi = dA = -I$$

In the rate form,

$$\sum_j \left(1 - \frac{T_o}{T}\right)\dot{Q}_j - \dot{W}_{1-2} + \dot{m}(a_{f_1} - a_{f_2}) - \dot{I}_{C.V.} = 0 \quad (8.61)$$

where $a_{f1} - a_{f2} = h_1 - h_2 - T_o\,(s_1 - s_2) + \dfrac{V_1^2 - V_2^2}{2} + g\,(Z_1 - Z_2)$, and

$\sum_j \left(1 - \frac{T_o}{T}\right)$ is the time rate of energy transfer along with heat Q_j occurring at the location on the boundary where the instantaneous temperature is T. For a single stream entering and leaving the c.v., the exergy balance gives

$$\left(1-\frac{T_o}{T}\right)\frac{\dot{Q}}{\dot{m}}+a_{f_1}-\frac{\dot{W}}{m}-a_{f_2}=\frac{\dot{I}}{\dot{m}} \tag{8.62}$$

8.15 SECOND LAW EFFICIENCY

> **LO 8.12**
> Discuss second law efficiency and apply it to different engineering systems

A common measure of energy use efficiency is the first law efficiency, η_I. The first law efficiency is defined as the ratio of the output energy of a device to the input energy of the device. The first law is concerned only with the quantities of energy, and disregards the forms in which the energy exists. It does not also discriminate between the energies available at different temperatures. It is the second law of thermodynamics which provides a means of assigning a quality index to energy. The concept of available energy or exergy provides a useful measure of energy quality (Section. 8.3).

With this concept, it is possible to analyse means of minimising the consumption of exergy to perform a given process, thereby ensuring the most efficient possible conversion of energy for the required task.

The second law efficiency, η_{II}, of a process is defined as the ratio of the minimum available energy (or exergy) which must be consumed to do a task divided by the actual amount of exergy consumed in performing the same task.

$$\eta_{II}=\frac{\text{Minimum exergy intake to perform the given task}}{\text{Actual exergy intake to perform the same task}}$$

or

$$\eta_{II}=\frac{A_{\min}}{A} \tag{8.63}$$

where A is the availability or exergy.

A power plant converts a fraction of exergy A or $W_{\max}$ to useful work W. For the desired output of W, $A_{\min} = W$ and $A = W_{\max}$. Here,

$$I = W_{\max} - W \quad \text{and} \quad \eta_{II}=\frac{W}{W_{\max}} \tag{8.64}$$

Now

$$\eta_I=\frac{W}{Q_1}=\frac{W}{W_{\max}}\cdot\frac{W_{\max}}{Q_1}$$

$$=\eta_{II}\cdot\eta_{\text{Carnot}} \tag{8.65}$$

$\therefore$

$$\eta_{II}=\frac{\eta_I}{\eta_{\text{Carnot}}} \tag{8.66}$$

Since $W_{\max}=Q_1\left(1-\frac{T_0}{T}\right)$, Eq. (8.66) can also be obtained directly as follows:

$$\eta_{II}=\frac{W}{Q_1\left(1-\frac{T_0}{T}\right)}=\frac{\eta_I}{\eta_{\text{Carnot}}}$$

If work is involved, $A_{min} = W$ (desired) and if heat is involved, $A_{min} = Q\left(1 - \frac{T_0}{T}\right)$.

If solar energy, Q_r is available at a reservoir storage temperature T_r and if quantity of heat Q_a is transferred by the solar collector at temperature T_a, then

$$\eta_I = \frac{Q_a}{Q_r}$$

and

$$\eta_{II} = \frac{\text{exergy output}}{\text{exergy input}} = \frac{Q_a\left(1 - \frac{T_0}{T_a}\right)}{Q_r\left(1 - \frac{T_0}{T_r}\right)} = \eta_I \frac{1 - \frac{T_0}{T_a}}{1 - \frac{T_0}{T_r}} \tag{8.67}$$

Table 8.1 shows the exergy values, and both η_I and η_{II} expressions for several common thermal tasks.

Table 8.1 $T_r > T_a > T_0 > T_c$

Task	*Energy input*	
	Input shaft work, W_i	*Q_r from reservoir at T_r*
Produce work, W_0	$A = W_i$ $A_{min} = W_0$ $\eta_I = \frac{W_0}{W_i}$ $\eta_{II} = \frac{A_{min}}{A}$ $\eta_{II} = \eta_I$ (electric motor)	$A = Q_r\left(1 - \frac{T_0}{T_r}\right)$ $A_{min} = W_0$ $\eta_I = \frac{W_0}{Q_r}$ $\eta_{II} = \eta_I \cdot \frac{1}{1 - \frac{T_0}{T_r}}$ (heat engine)

(Contd.)

Table 8.1 (Contd.)

Task	*Energy input*	
	Input shaft work, W_i	*Q_r from reservoir at T_r*
Add heat Q_a to reservoir at T_a	$A = W_i$ $A_{min} = Q_a\left(1-\frac{T_0}{T_a}\right)$ $*\eta_I = \frac{Q_a}{W_i}$ $*\eta_{II} = \eta_I\left(1-\frac{T_0}{T_a}\right)$ (heat pump)	$A = Q_r\left(1-\frac{T_0}{T_r}\right)$ $A_{min} = Q_a\left(1-\frac{T_0}{T_a}\right)$ $\eta_I = \frac{Q_a}{Q_r}$ $\eta_{II} = \eta_I \frac{1-\frac{T_0}{T_a}}{1-\frac{T_0}{T_r}}$ (solar water heater)
Extract heat Q_c from cold reservoir at T_c (below ambient)	$A = W_i$ $A_{min} = Q_c\left(\frac{T_0}{T_c}-1\right)$ $*\eta_I = \frac{Q_c}{W_i}$ $*\eta_{II} = \eta_I\left(\frac{T_0}{T_c}-1\right)$ (Refrigerator—electric motor driven)	$A = Q_r\left(1-\frac{T_0}{T_r}\right)$ $A_{min} = Q_c\left(\frac{T_0}{T_c}-1\right)$ $*\eta_I = \frac{Q_c}{Q_r}$ $*\eta_{II} = \eta_I\left(\frac{\frac{T_0}{T_c}-1}{1-\frac{T_0}{T_r}}\right)$ (Refrigerator—heat operated)

In the case of a heat pump, the task is to add heat Q_a to a reservoir to be maintained at temperature T_a and the input shaft work is W_i.

$$\text{COP} = \frac{Q_a}{W_i} = \eta_I, \text{ say}$$

$$(\text{COP})_{\max} = \frac{T_a}{T_a - T_0} = \frac{Q_a}{W_i} = \frac{Q_a}{A_{\min}}$$

* Strictly speaking, it is COP

$$\therefore \quad A_{\min} = Q_a\left(1-\frac{T_0}{T_a}\right)$$

$$\therefore \quad \eta_{II} = \frac{A_{\min}}{A} = \frac{Q_a\left(1-\frac{T_0}{T_a}\right)}{W_i}$$

$$\eta_{II} = \eta_I\left(1-\frac{T_0}{T_a}\right) \tag{8.68}$$

Similarly, expressions of η_I and η_{II} can be obtained for other thermal tasks.

8.15.1 Matching end use to Source

Combustion of a fuel releases the necessary energy for the tasks such as space heating, process steam generation and heating in industrial furnaces. When the products of combustion are at a temperature much greater than that required by a given task, the end use is not well matched to the source and results in inefficient utilisation of the fuel burned. To illustrate this, let us consider a closed system receiving a heat transfer $\dot{Q}_r$ at a source temperature T_r and delivering $\dot{Q}_a$ at a use temperature T_a (Fig. 8.20). Energy is lost to the surroundings by heat transfer at a rate $\dot{Q}_l$ across a portion of the surface at T_l. At a steady state, the energy and exergy rate balances become

$$\dot{Q}_r = \dot{Q}_a + \dot{Q}_l \tag{8.69}$$

$$Q_r\left(1-\frac{T_0}{T_r}\right) = Q_a\left(1-\frac{T_0}{T_a}\right) + \dot{Q}_l\left(1-\frac{T_0}{T_l}\right) + \dot{I} \tag{8.70}$$

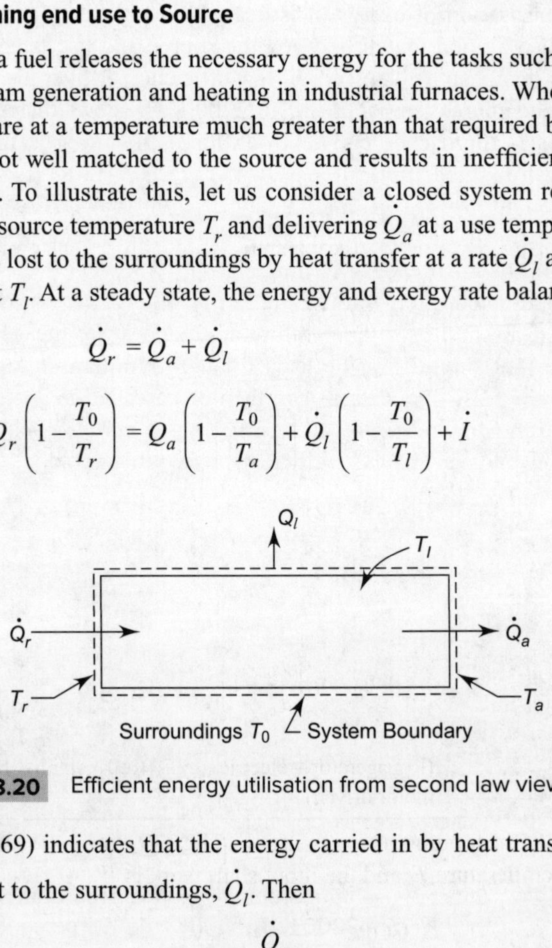

Fig. 8.20 Efficient energy utilisation from second law viewpoint

Equation (8.69) indicates that the energy carried in by heat transfer $\dot{Q}_r$ is either used, $\dot{Q}_a$, or lost to the surroundings, $\dot{Q}_l$. Then

$$\eta_I = \frac{\dot{Q}_a}{\dot{Q}_r} \tag{8.71}$$

The value of η_I can be increased by increasing insulation to reduce the loss. The limiting value, when $\dot{Q}_l = 0$, is $\eta_I = 1$ (100 percent).

Equation (8.70) shows that the exergy carried into the system accompanying the heat transfer $\dot{Q}_r$ is either transferred from the system accompanying the heat transfer $\dot{Q}_a$ and $\dot{Q}_l$ or destroyed by irreversibilities within the system, I.

Therefore,

$$\eta_{II} = \frac{\dot{Q}_a\left(1 - \dfrac{T_0}{T_a}\right)}{\dot{Q}_r\left(1 - \dfrac{T_0}{T_r}\right)} = \eta_I \frac{1 - \dfrac{T_0}{T_a}}{1 - \dfrac{T_0}{T_r}} \tag{8.72}$$

Both η_I and η_{II} indicate how effectively the input is converted into the product. The parameter η_I does this on energy basis, whereas η_{II} does it on an exergy basis.

For proper utilisation of exergy, it is desirable to make η_I as close to unity as practical and also a good match between the source and use temperatures, T_r and T_a. Figure 8.21 demonstrates the second law efficiency against the use temperature T_a for an assumed source temperature T_r = 2200 K. It shows that η_{II} tends to unity (100 percent) as T_a approaches T_r. The lower the T_a, the lower becomes the value of η_{II}. Efficiencies for three applications, viz., space heating at T_a = 320 K, process steam generation at T_a = 480 K, and heating in industrial furnaces at T_a = 700 K, are indicated in the figure. It suggests that fuel is used far more effectively in the high temperature use. An excessive temperature gap between T_r and T_a causes a low η_{II} and an inefficient energy utilisation. A fuel or any energy source is consumed efficiently when the first user temperature approaches the fuel temperature. This means that the fuel should first be used for high temperature applications. The heat rejected from these applications can then be cascaded to applications at lower temperatures, eventually to the task of, say, keeping a building warm. This is called *energy cascading* and ensures more efficient energy utilisation.

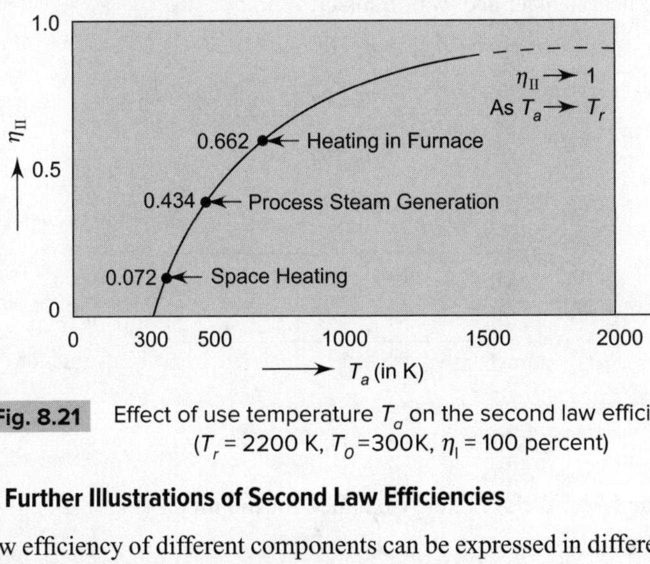

Fig. 8.21 Effect of use temperature T_a on the second law efficiency (T_r = 2200 K, T_0=300K, η_I = 100 percent)

8.15.2 Further Illustrations of Second Law Efficiencies

Second law efficiency of different components can be expressed in different forms. It is derived by using the exergy balance rate given as follows:

1. Turbines The steady state exergy balance (Fig. 8.22) gives

$$\frac{\dot{Q}}{\dot{m}}\left[1-\frac{T_0}{T_\sigma}\right]+a_{f_1}=\frac{\dot{W}}{\dot{m}}+a_{f_2}+\frac{\dot{I}}{\dot{m}}$$

where $a_{f_1} = h_1 - T_o s_1$ and $a_{f_2} = h_2 - T_o s_2$

If there is no heat loss,

$$a_{f_1}-a_{f_2}=\frac{\dot{W}}{\dot{m}}+\frac{\dot{I}}{\dot{m}} \qquad (8.73)$$

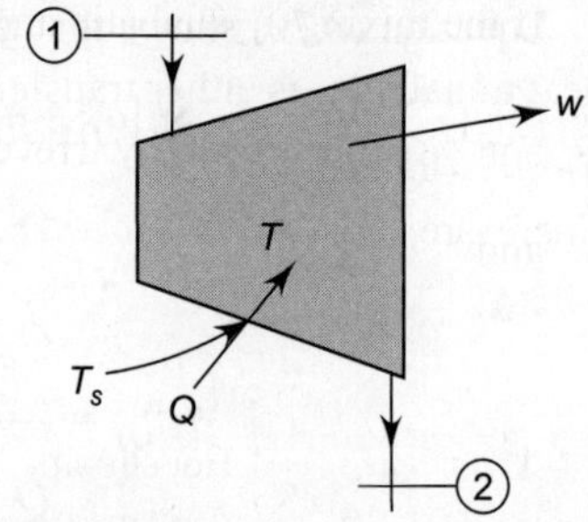

Fig. 8.22 Exergy balance of a turbine

The second law efficiency,
$$\eta_{II}=\frac{\dot{W}/\dot{m}}{a_{f_1}-a_{f_2}} \qquad (8.74)$$

2. Compressor and pump Similarly, for a compressor or a pump,

$$\frac{\dot{W}}{\dot{m}}=a_{f_2}-a_{f_1}+\frac{\dot{I}}{\dot{m}}$$

and
$$\eta_{II}=\frac{a_{f_2}-a_{f_2}}{-\dot{W}/\dot{m}} \qquad (8.75)$$

3. Heat exchanger Writing the exergy balance for the heat exchanger, (Fig. 8.23)

$$\sum\left[1-\frac{T_0}{T_j}\right]\dot{Q}_j-\dot{W}_{\text{C.V.}}+[\dot{m}_h a_{f_1}+\dot{m}_c a_{f_3}]-[\dot{m}_h a_{f_2}+\dot{m}_c a_{f_4}]-\dot{I}_{\text{C.V.}}=0 \text{ in}_c$$

If there is no heat transfer and work transfer,

$$\dot{m}_h[a_{f_1}-a_{f_2}]=\dot{m}_c[a_{f_4}-a_{f_3}]+\dot{I} \qquad (8.76)$$

$$\eta_{II}=\frac{\dot{m}_c[a_{f_4}-a_{f_3}]}{\dot{m}_h[a_{f_1}-a_{f_2}]} \qquad (8.77)$$

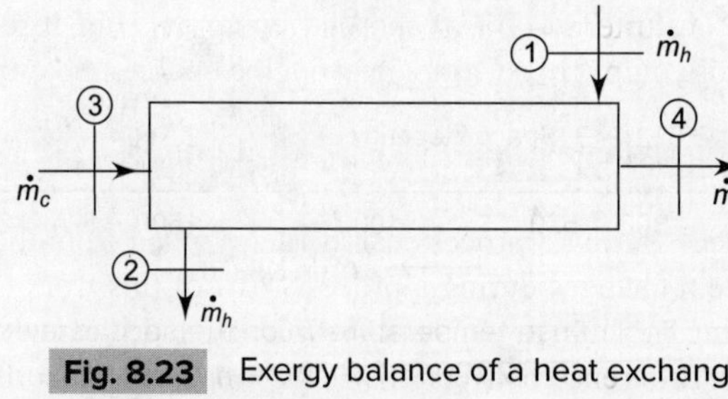

Fig. 8.23 Exergy balance of a heat exchanger

4. Mixing of two fluids Exergy balance for the mixer (Fig. 8.24) gives

$$\left[1-\frac{T_0}{T_\sigma}\right]\dot{Q}+\dot{m}_1 a_{f_1}+\dot{m}_2 a_{f_2}=\dot{W}_{\text{C.V.}}+\dot{m}_3 a_{f_3}+I_{\text{C.V.}}$$

If the mixing is adiabatic and since $\dot{W}_{C.V.} = 0$ and $\dot{m}_1 + \dot{m}_2 = \dot{m}_3$.

$$\dot{m}_1[a_{f1} - a_{f3}] = \dot{m}_2[a_{f3} - a_{f2}] + \dot{I} \tag{8.78}$$

and

$$\eta_{II} = \frac{\dot{m}_2\,[a_{f_3} - a_{f_2}]}{\dot{m}_1\,[a_{f_1} - a_{f_3}]} \tag{8.79}$$

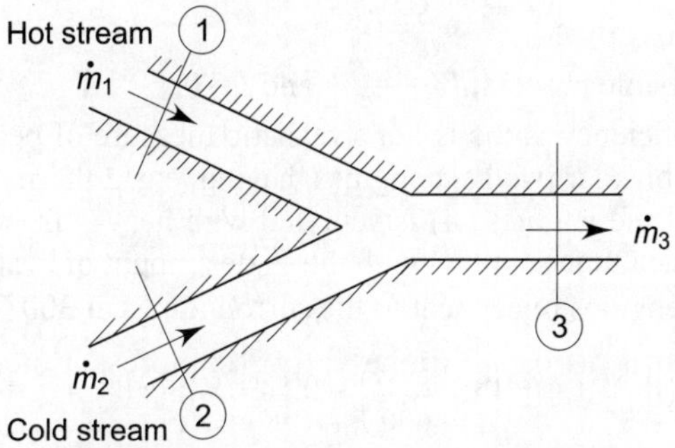

Fig. 8.24 Exergy loss due to mixing

8.16 COMMENTS ON EXERGY

The energy of the universe, like its mass, is constant. Yet at times, we are bombarded with speeches and articles on how to 'conserve' energy. As engineers, we know that energy is always conserved. What is not conserved is the exergy, i.e. the useful work potential of the energy. Once the exergy is wasted, it can never be recovered. When we use energy (electricity) to heat our homes, we are not destroying any energy, we are merely converting it to a less useful form, a form of less exergy value.

The maximum useful work potential of a system at the specified state is called exergy which is a composite property depending on the state of the system and the surroundings. A system which is in equilibrium with its surroundings is said to be at the dead state having zero exergy.

The mechanical forms of energy such as K.E. and P.E. are entirely available energy or exergy. The exergy (W) of thermal energy (Q) of reservoirs (TER) is equivalent to the work output of a Carnot heat engine operating between the reservoir at temperature T and environment at T_0, i.e. $W = Q\left[1 - \frac{T_0}{T}\right]$.

The actual work W during a process can be determined from the first law. If the volume of the system changes during a process, part of this work (W_{surr}) is used to push the surrounding medium at constant pressure p_0 and it cannot be used for any useful purpose. The difference between the actual work and the surrounding work is called *useful work*, W_u

$$W_u = W - W_{surr} = W - p_0(V_2 - V_1)$$

W_{surr} is zero for cyclic devices, for steady flow devices and for system with fixed boundaries (rigid walls).

The maximum amount of useful work that can be obtained from a system as it undergoes a process between two specified states is called *reversible work*, W_{rev}. If the final state of the system is the dead state, the reversible work and the exergy become identical.

The difference between the reversible work and useful work for a process is called *irreversibility*.

$$I = W_{rev} - W_u = T_0 S_{gen}$$

$$\dot{I} = T_0 \dot{S}_{gen}$$

For a total reversible process, $W_{rev} = W_u$ and $I = 0$.

The first law efficiency alone is not a realistic measure of performance for engineering devices. Consider two heat engines, having, e.g. a thermal efficiency of, say, 30 percent. One of the engines (A) is supplied with heat Q from a source at 600 K and the other engine (B) is supplied with the same amount of heat Q from a source at 1000 K. Both the engines reject heat to the surroundings at 300 K.

$$(W_A)_{rev} = Q\left(1 - \frac{300}{600}\right) = 0.5\,Q, \text{ while } (W_A)_{act} = 0.3Q$$

Similarly,

$$(W_B)_{rev} = Q\left(1 - \frac{300}{1000}\right) = 0.7\,Q, \text{ and } (W_B)_{act} = 0.3Q$$

At first glance, both engines seem to convert the same fraction of heat, that they receive, to work, thus performing equally well from the viewpoint of the first law. However, in the light of second law, the engine B has a greater work potential (0.7 Q) available to it and thus should do a lot better than engine A. Therefore, it can be said that engine B is performing poorly relative to engine A, even though both have the same thermal efficiency.

To overcome the deficiency of the first law efficiency, a second law efficiency hII can be defined as the ratio of actual thermal efficiency to the maximum possible thermal efficiency under the same conditions:

$$\eta_{II} = \frac{\eta_I}{\eta_{rev}}$$

So, for engine A, $\eta_{II} = 0.3/0.5 = 0.60$

and for engine B, $\eta_{II} = 0.3/0.7 = 0.43$

Therefore, the engine A is converting 60 percent of the available work potential (exergy) to useful work. This is only 43 percent for the engine B. Therefore,

$$\eta_{II} = \frac{\eta_{act}}{\eta_{rev}} = \frac{W_u}{W_{rev}} \text{ (for heat engines and other work producing devices)}$$

$$\eta_{II} = \frac{COP}{COP_{rev}} = \frac{W_{rev}}{W_u} \text{ (for refrigerators, heat pumps and other work absorbing}$$

devices).

The exergies of a closed system (ϕ) and a flowing fluid system (ψ) are given on unit mass basis:

$$\phi = (u - u_0) - T_0(s - s_0) + p_0(v - v_0) \text{ kJ/kg}$$

$$\psi = (h - h_0) - T_0(s - s_0) + \frac{V^2}{2} + gz \text{ kJ/kg}$$

8.16.1 Reversible Work Expressions

1. Cyclic devices

$$W_{rev} = \eta_{rev} Q_1 \text{ (Heat engines)}$$

$$-W_{rev} = \frac{Q_2}{(COP_{rev})_{Ref.}} \text{ (Refrigerators)}$$

$$-W_{rev} = \frac{Q_2}{(COP_{rev})_{HP}} \text{ (Heat pumps)}$$

2. Closed system

$$W_{rev} = U_1 - U_2 - T_0(S_1 - S_2) + p_0(V_1 - V_2)$$
$$= m(\phi_1 - \phi_2)$$

3. Steady flow system (single stream)

$$W_{rev} = m\left[\left(h_1 + \frac{V_1^2}{2} + gz_1 - T_0 s_1\right) - \left(h_2 + \frac{V_2^2}{2} + gz_2 - T_0 s_2\right)\right]$$
$$= m(\psi_1 - \psi_2)$$

When the system exchanges heat with another reservoir at temperature T_k other than the atmosphere,

$$\dot{W}_{rev} = \dot{m}(\psi_1 - \psi_2) + \dot{Q}_k\left(1 - \frac{T_0}{T_k}\right)$$

The first law efficiency is defined as the ratio of energy output and energy input, while their difference is the energy loss. Likewise, the second law efficiency is defined as the ratio of exergy output and exergy input and their difference is irreversibility. By reducing energy loss, first law efficiency can be improved. Similarly, by reducing irreversibilities, the second law efficiency can be enhanced.

SOLVED EXAMPLES

Example 8.1 *In a certain process, a vapour, while condensing at 420°C, transfers heat to water evaporating at 250°C. The resulting steam is used in a power cycle which rejects heat at 35°C. What is the fraction of the available energy in the heat transferred from the process vapour at 420°C that is lost due to the irreversible heat transfer at 250°C?* M [LO 8.4]

Solution *ABCD* (Fig. 8.25) would have been the power cycle, if there was no temperature difference between the vapour condensing and the water evaporating and the area under *CD* would have been the unavailable energy. *EFGD* is the power cycle when the vapour condenses at 420°C and the water evaporates at 250°C.

The unavailable energy becomes the area under DG. Therefore, the increase in unavailable energy due to irreversible heat transfer is represented by the area under CG.

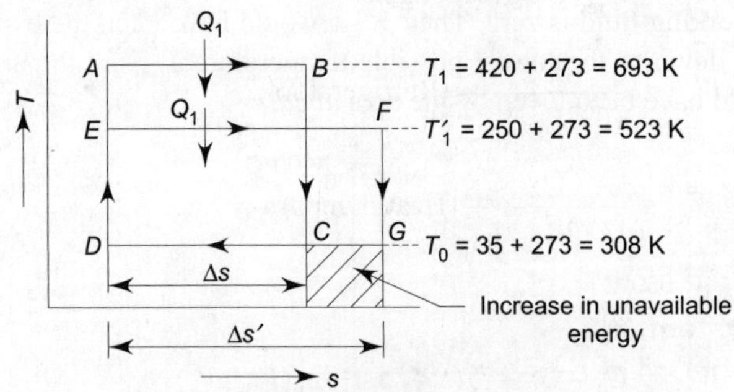

Fig. 8.25

Now $$Q_1 = T_1 \Delta S = T_1' \Delta S'$$

$$\frac{\Delta S'}{\Delta S} = \frac{T_1}{T_1'}$$

$$W = \text{Work done in cycle } ABCD$$
$$= (T_1 - T_0)\Delta S$$
$$W' = \text{Work done in cycle } EFGD$$
$$= (T_1' - T_0)\Delta S'$$

The fraction of energy that becomes unavailable due to irreversible heat transfer

$$= \frac{W - W'}{W} = \frac{T_0 (\Delta S' - \Delta S)}{(T_1 - T_0)\,\Delta S} = \frac{T_0\left(\dfrac{\Delta S'}{\Delta S} - 1\right)}{(T_1 - T_0)}$$

$$= \frac{T_0 (T_1 - T_1')}{T_1' (T_1 - T_0)} = \frac{308\,(693 - 523)}{523\,(693 - 308)} = \mathbf{0.26}$$

Example 8.2 *In a steam boiler, hot gases from a fire transfer heat to water which vapourises at constant temperature. In a certain case, the gases are cooled from 1100°C to 550°C while the water evaporates at 220°C. The specific heat of gases is 1.005 kJ/kg-K, and the latent heat of water at 220°C is 1858.5 kJ/kg. All the heat transferred from the gases goes to the water. How much does the total entropy of the combined system of gas and water increase as a result of the irreversible heat transfer? Obtain the result on the basis of 1 kg of water evaporated.*

If the temperature of the surroundings is 30°C, find the increase in unavailable energy due to irreversible heat transfer. **M [LO 8.4, 8.5]**

Solution Gas ($\dot{m}_g$) is cooled from state 1 to 2 (Fig. 8.26). For reversible heat transfer, the working fluid (w.f.) in the heat engine having the same c_p would have been heated along 2 – 1, so that at any instant, the temperature difference between gas and the working fluid is zero. Then 1 – *b* would have been the expansion of the working fluid down to the lowest possible temperature T_0, and the amount of heat rejection would have been given by the area *abcd*.

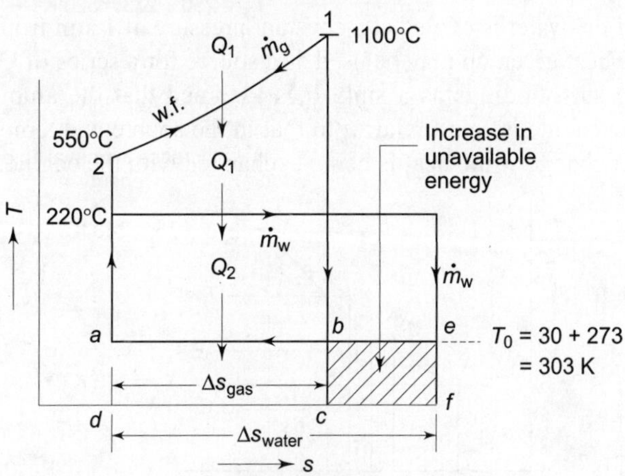

Fig. 8.26

When water evaporates at 220°C as the gas gets cooled from 1100°C to 550°C, the resulting power cycle has an unavailable energy represented by the area *aefd*. The increase in unavailable energy due to irreversible heat transfer is thus given by area *befc*.

Entropy increase of 1 kg water

$$(\Delta S)_{\text{water}} = \frac{\text{Latent heat absorbed}}{T} = \frac{1858.5}{(273+220)} = 3.77 \text{ kJ/kg-K}$$

$$\begin{aligned} Q_1 &= \text{Heat transferred from the gas} \\ &= \text{Heat absorbed by water during evaporation} \\ &= m_g c_{p_g} (1100 - 550) \\ &= 1 \times 1858.5 \text{ kJ} \end{aligned}$$

$$\therefore \quad m_g c_{p_g} = \frac{1858.5}{550} = 3.38 \text{ kJ/°C}$$

$$\Delta S_{\text{gas}} = \int_{T_{g1}}^{T_{g2}} \frac{đQ}{T} = \int_{T_{g1}}^{T_{g2}} m_g c_{pg} \frac{dT}{T}$$

$$= m_g c_{p_g} \ln \frac{T_g}{T_g} = 3.38 \ln \frac{823}{1373}$$

$$= -3.38 \times 0.51$$

$$= -1.725 \text{ kJ/K}$$

$$\therefore \quad \Delta S_{\text{total}} = (\Delta S)_{\text{water}} + (\Delta S)_{\text{gas}}$$

$$= 3.77 - 1.725 = \mathbf{2.045 \text{ kJ/K}}$$

Increase in unavailable energy

$$= T_0 (\Delta S)_{\text{total}} = 303 \times 2.045 = \mathbf{620\ kJ}$$

Example 8.3 *Calculate the available energy in 40 kg of water at 75°C with respect to the surroundings at 5°C, the pressure of water being 1 atm.* Ⓜ [LO 8.5]

Solution If the water is cooled at a constant pressure of 1 atm from 75°C to 5°C (Fig. 8.27), the heat given up may be used as a source for a series of Carnot engines each using the surroundings as a sink. It is assumed that the amount of energy received by any engine is small relative to that in the source and temperature of the source does not change while heat is being exchanged with the engine.

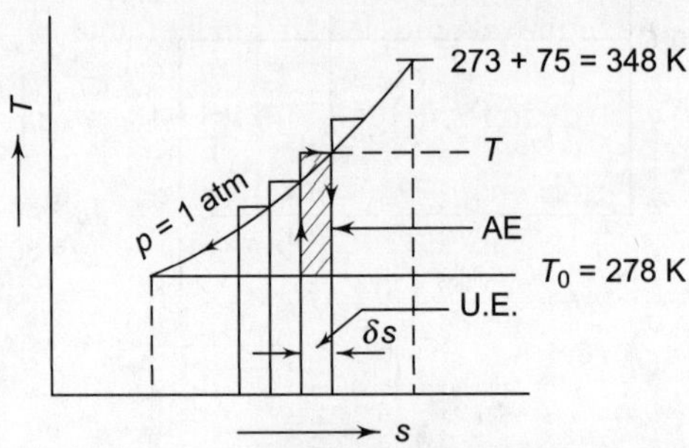

Fig. 8.27

Let us consider that the source has fallen to temperature T, at which level there operates a Carnot engine which takes in heat at this temperature and rejects heat at $T_0 = 278$ K. If ds is the entropy change of water, the work obtainable is

$$đW = -m(T - T_0)\,ds$$

where δs is negative.

$$đW = -40\,(T - T_0)\,\frac{c_p dT}{T}$$

$$\therefore \qquad = -40\,c_p\left(1 - \frac{T_0}{T}\right)dT$$

With a very great number of engines in the series, the total work (maximum) obtainable when the water is cooled from 348 to 278 K would be

$$W(\text{max}) = \text{AE} = -\lim \sum_{348}^{278} 40\,c_p\left(1 - \frac{T_0}{T}\right)dT$$

$$= \int_{278}^{348} 40\,c_p\left(1 - \frac{T_0}{T}\right)dT$$

$$= 40\,c_p\left[(348 - 278) - 278 \ln \frac{348}{278}\right]$$

$$= 40 \times 4.2\,(70 - 62) = \mathbf{1340\ kJ}$$

$$Q_1 = 40 \times 4.2\,(348 - 278) = \mathbf{11{,}760\ kJ}$$

$$\text{UE} = Q_1 - W(\text{max})$$

$$= 11{,}760 - 1340 = \mathbf{10{,}420\ kJ}$$

Example 8.4 *A rigid tank contains 100 m³ of air at 2 MPa and 20°C. If the environment conditions are 100 kPa and 20°C, determine how much work can be obtained from this compressed air.* Ⓜ [LO 8.3]

Solution The mass of air in the tank is found to be

$$m = \frac{p_1 V}{RT_1} = \frac{2000 \times 100}{0.287 \times 293} = 2378.38 \text{ kg}$$

The available energy of the compressed air can be found as

$$A_1 = m\left[\cancel{(u_1 - u_0)}^{=0} + p_0(v_1 - v_0) - T_0(s_1 - s_0)\right]$$

$$= m\left[p_0(v_1 - v_0) - T_0(s_1 - s_0)\right]$$

$$= m\left[p_0\left(\frac{RT_1}{p_1} - \frac{RT_0}{p_0}\right) - T_0\left(c_p \ln\frac{T_1}{T_0} - R\ln\frac{p_1}{p_0}\right)\right]$$

$$= m\left[RT_0\left(\frac{p_0}{p_1} - 1\right) - T_0\left(c_p \cancel{\ln\frac{T_1}{T_0}}^{=0} - R\ln\frac{p_1}{p_0}\right)\right]$$

$$= mRT_0\left[\frac{p_0}{p_1} - 1 + R\ln\frac{p_1}{p_0}\right]$$

$$= 2378.38 \times 0.287 \times 293\left[\frac{100}{2000} - 1 + R\ln\frac{2000}{100}\right]$$

$$= 409140.72 \text{ kJ} = \mathbf{409.14\ MJ}$$

Example 8.5 *Calculate the decrease in exergy when 25 kg of water at 95°C mix with 35 kg of water at 35°C, the pressure being taken as constant and the temperature of the surroundings being 15°C (c_p of water = 4.2 kJ/kg K).* Ⓜ [LO 8.3]

Solution The exergy of a system of mass m, specific heat c_p, and at temperature T, is given by

$$\text{X} = mc_p \int_{T_0}^{T}\left(1 - \frac{T_0}{T}\right) dT$$

∴ $$\text{X}_{25} = \text{Exergy of 25 kg of water at 95°C}$$

$$= 25 \times 4.2 \int_{273+15}^{273+95}\left(1 - \frac{288}{T}\right) dT$$

$$= 105\left[(368 - 288) - 288\ln\frac{368}{288}\right] = 987.49 \text{ kJ}$$

$$X_{35} = \text{Exergy of 35 kg of water at 35°C}$$

$$= 147\left[(308 - 288) - 288 \ln \frac{308}{288}\right] = 97.59 \text{ kJ}$$

Total exergy

$$(AE)_{total} = (AE)_{25} + (AE)_{35}$$
$$= 987.49 + 97.59$$
$$= 1085.08 \text{ kJ}$$

After mixing, if t is the final temperature

$$25 \times 4.2\,(95 - t) = 35 \times 4.2\,(t - 35)$$

$$\therefore \quad t = \frac{25 \times 95 + 35 \times 35}{25 + 35} = 60°C$$

Total mass after mixing = 25 + 35 = 60 kg

$$X_{60} = \text{Exergy of 60 kg of water at 60°C}$$

$$= 4.2 \times 60\left[(333 - 288) - 288 \ln \frac{333}{288}\right] = 803.27 \text{ kJ}$$

∴ decrease in exergy due to mixing

= Total available energy before mixing
– Total available energy after mixing
= 1085.08 – 803.27 = **281.81 kJ**

Example 8.6 *A cylinder fitted with piston contains 1.2 kg of water vapour at 200 kPa and 250°C. The water vapour is allowed to cool at constant pressure until the state becomes saturated liquid. Determine the work transfer, heat transfer and the exergy transfer accompanying work and heat. Properties of steam are as follows:*

Pressure (kPa)	*Temperature (°C)*	*Specific volume (m^3/kg)*	*Specific internal energy (kJ/kg)*	*Specific entropy (kJ/kg°C)*
200	Saturated liquid	0.001061	504.47	1.53
200	250	1.1988	2731.2	0.7.7085

Take p_0 = 100 kPa and T_0 = 20°C M **[LO 8.5]**

Solution The initial and final states are shown in the temperature-entropy diagram (Fig. 8.28).

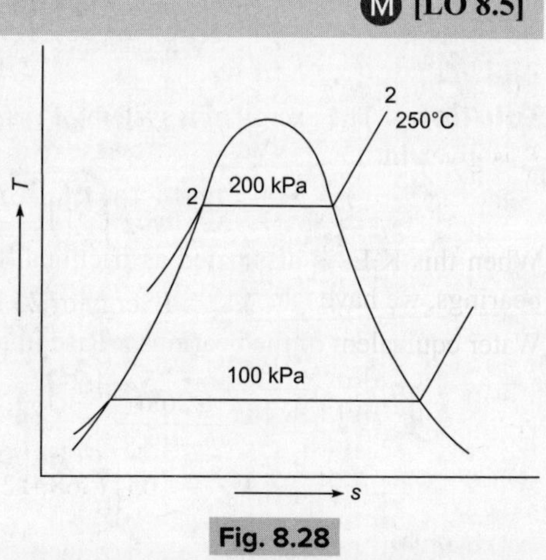

Fig. 8.28

The work transfer can be obtained as

$$W_{1-2} = \int_1^2 pdV = mp\,(v_2 - v_1)$$

$$W_{1-2} = 1.2 \times 200\,(0.001061 - 1.1988) = \mathbf{-287.46\ kJ}$$

By applying the first law of thermodynamics, one can find the heat transfer as

$$Q_{1-2} = U_2 - U_1 + W_{1-2} = m(u_2 - u_1) + W_{1-2}$$
$$= 1.2(504.47 - 2731.2) - 287.46 = \mathbf{-2959.54\ kJ}$$

The exergy transfer accompanying work transfer is

$$A_{w_{1-2}} = W_{1-2} - p_0 \Delta V = W_{1-2} - mp_0(v_2 - v_1)$$
$$= -287.46 - 1.2 \times 100(0.001061 - 1.1988) = \mathbf{-431.19\ kJ}$$

The exergy transfer accompanying heat transfer is

$$A_{Q_{1-2}} = \int_1^2 \left(1 - \frac{T_0}{T}\right) đQ = Q_{1-2} - T_0 \int_1^2 \frac{đQ}{T} = Q_{1-2} - T_0(S_2 - S_1)$$
$$= Q_{1-2} - mT_0(s_2 - s_1) = -2959.54 - 1.2 \times 293(1.53 - 7.7085) = \mathbf{-787.18\ kJ}$$

Example 8.7 *The moment of inertia of a flywheel is 0.54 kg · m² and it rotates at a speed 3000 RPM in a large heat insulated system, the temperature of which is 15°C. If the kinetic energy of the flywheel is dissipated as frictional heat at the shaft bearings which have a water equivalent of 2 kg, find the rise in the temperature of the bearings when the flywheel has come to rest. Calculate the greatest possible amount of this heat which may be returned to the flywheel as high-grade energy, showing how much of the original kinetic energy is now unavailable. What would be the final RPM of the flywheel, if it is set in motion with this available energy?* **D [LO 8.3]**

Solution Initial angular velocity of the flywheel

$$\omega_1 = \frac{2\pi N_1}{60} = \frac{2\pi \times 3000}{60} = 314.2 \text{ rad/s}$$

Initial available energy, or exergy of the flywheel

$$= (\text{K.E.})_{\text{initial}} = \frac{1}{2} I\omega_1^2$$

$$= 0.54 \text{ kg m}^2 \times (314.2)^2 \frac{\text{rad}^2}{s^2}$$

$$= 2.66 \times 10^4 \text{ Nm} = 26.6 \text{ kJ}$$

When this K.E. is dissipated as frictional heat, if Δt is the temperature rise of the bearings, we have

Water equivalent of the bearings × Rise in temperature = 26.6 kJ

$$\therefore \qquad \Delta t = \frac{26.6}{2 \times 4.187} = \mathbf{3.19°C}$$

∴ Final temperature of the bearings

$$t_f = 15 + 3.19 = 18.19°\text{C}$$

The maximum amount of energy which may be returned to the flywheel as high-grade energy is

$$\text{A.E.} = 2 \times 4.187 \int_{288}^{291.19} \left(1 - \frac{288}{T}\right) dT$$

$$= 2 \times 4.187 \left[(291.19 - 288) - 288 \ln \frac{291.19}{288} \right]$$

$$= \mathbf{0.1459\ kJ}$$

The amount of energy rendered unavailable is

$$\text{U.E.} = (\text{A.E.})_{\text{initial}} - (\text{A.E.})_{\text{returnable as high grade energy}}$$

$$= 26.6 - 0.1459$$

$$= \mathbf{26.4541\ kJ}$$

Since the amount of energy returnable to the flywheel is 0.146 kJ, if ω_2 is the final angular velocity, and the flywheel is set in motion with this energy

$$0.146 \times 10^3 = \frac{1}{2} \times 0.54\, \omega_2^2$$

$$\therefore \quad \omega_2^2 = \frac{146}{0.27} = 540.8$$

$$\therefore \quad \omega_2 = 23.246 \text{ rad/s}$$

If N_2 is the final RPM of the flywheel

$$\omega_2 = 23.246 = \frac{2\pi N_2}{60}$$

or

$$N_2 = \frac{23.246 \times 60}{2 \times \pi} = \mathbf{222\ RPM}$$

Example 8.8 *Two kg of air at 500 kPa, 80°C expands adiabatically in a closed system until its volume is doubled and its temperature becomes equal to that of the surroundings which is at 100 kPa, 5°C. For this process, determine (a) the maximum work, (b) the change in availability, and (c) the irreversibility. For air, take c_v = 0.718 kJ/kg K, $u = c_v T$ where c_v is contant, and $pV = mRT$ where p is pressure in kPa, V volume in m^3, m mass in kg, R a constant equal to 0.287 kJ/kg K, and T temperature in K.* Ⓜ **[LO 8.9]**

Solution From the property relation

$$TdS = dU + pdV$$

the entropy change of air between the initial and final states is

$$\int_1^2 dS = \int_1^2 \frac{mc_v dT}{T} + \int_1^2 \frac{mRdV}{V}$$

or
$$S_2 - S_1 = mc_v \ln \frac{T_2}{T_1} + mR \ln \frac{V_2}{V_1}$$

From Eq. (8.32),

$$W_{max} = (U_1 - U_2) - T_0 (S_1 - S_2)$$

$$= m\left[c_v (T_1 - T_2 + T_0 \left(c_v \ln \frac{T_2}{T_1} + R \ln \frac{V_2}{V_1}\right)\right]$$

$$= 2\left[0.718(80-5)+278\left(0.718\ln\frac{278}{353}+0.287\ln\frac{2}{1}\right)\right]$$

$$= 2\,[53.85 + 278\,(-\,0.172 + 0.199)]$$

$$= 2\,(53.85 + 7.51) = \mathbf{122.72\ kJ}$$

From Eq. (8.42), the change in availability

$$= \phi_1 - \phi_2$$

$$= (U_1 - U_2) - T_0(S_1 - S_2) + p_0(V_1 - V_2)$$

$$= W_{max} + p_0(V_1 - V_2)$$

$$= 122.72 + p_0(V_1 - 2V_1)$$

$$= 122.72 - 100 \times \frac{2 \times 0.287 \times 353}{500} = \mathbf{82.2\ kJ}$$

The irreversibility

$$I = W_{max,\ useful} - W_{act}$$

From the first law,

$$W_{act} = Q - \Delta U = -\,\Delta U = U_1 - U_2$$

$$\therefore \quad I = U_1 - U_2 - T_0(S_1 - S_2) - U_1 + U_2$$

$$= T_0(S_2 - S_1)$$

$$= T_0(\Delta S)_{system}$$

For adiabatic process, $(\Delta S)_{surr} = 0$

$$I = T_0\left[mc_v \ln \frac{T_2}{T_1} + mR \ln \frac{V_2}{V_1}\right]$$

$$= 278 \times 2\left[0.718 \ln \frac{278}{353} + 0.287 \ln 2\right]$$

$$= 278 \times 2(-\,0.172 + 0.199) = \mathbf{15.2\ kJ}$$

Example 8.9 *Air expands through a turbine from 500 kPa, 520°C to 100 kPa, 300°C. During expansion, 10 kJ/kg of heat is lost to the surroundings which is at 98 kPa, 20°C. Neglecting the K.E. and P.E. changes, determine per kg of air (a) the decrease in availability, (b) the maximum work and (c) the irreversibility. For air, take c_p = 1.005 kJ/kg K, $h = c_p T$ where c_p is constant, and the p, V and T relation as in Example 8.8.* Ⓜ **[LO 8.9, 8.10]**

Solution From the property relation

$$TdS = dH - Vdp$$

the entropy change of air in the expansion process is

$$\int_1^2 dS = \int_1^2 \frac{mc_p dT}{T} - \int_1^2 \frac{mRdp}{p}$$

or

$$S_2 - S_1 = mc_p \ln \frac{T_2}{T_1} - mR \ln \frac{p_2}{p_1}$$

For 1 kg of air,

$$s_2 - s_1 = c_p \ln \frac{T_2}{T_1} - R \ln \frac{p_2}{p_1}$$

The change in availability

$$\begin{aligned}\psi_1 - \psi_2 &= b_1 - b_2 \\ &= (h_1 - T_0 s_1) - (h_2 - T_0 s_2) \\ &= (h_1 - h_2) - T_0(s_1 - s_2) \\ &= c_p (T_1 - T_2) - T_0 \left(R \ln \frac{p_2}{p_1} - c_p \ln \frac{T_2}{T_1} \right) \\ &= 1.005 (520 - 300) - 293 \left(0.287 \ln \frac{1}{5} - 1.005 \ln \frac{573}{793} \right) \\ &= 1.005 \times 220 - 293 (0.3267 - 0.4619) \\ &= 221.1 + 39.6 = \mathbf{260.7 \ kJ/kg}\end{aligned}$$

The maximum work is

$$\begin{aligned}W_{max} &= \text{change in availability} = \psi_1 - \psi_2 \\ &= \mathbf{260.7 \ kJ/kg}\end{aligned}$$

From S.F.E.E.

$$\begin{aligned}Q + h_1 &= W + h_2 \\ W &= (h_1 - h_2) + Q \\ &= c_p (T_1 - T_2) + Q \\ &= 1.005 (520 - 300) - 10 = 211.1 \text{ kJ/kg}\end{aligned}$$

The irreversibility

$$\begin{aligned}I &= W_{max} - W \\ &= 260.7 - 211.1 = \mathbf{49.6 \ kJ/kg}\end{aligned}$$

Alternatively,

$$I = T_0\,(\Delta S_{\text{system}} + \Delta S_{\text{surr}})$$

$$= 293\left(1.005 \ln \frac{573}{793} - 0.287 \ln \frac{1}{5} + \frac{10}{293}\right)$$

$$= 293 \times 0.1352 + 10 = \mathbf{49.6\ kJ/kg}$$

Example 8.10 *An air preheater is used to cool the products of combustion from a furnace while heating the air to be used for combustion. The rate of flow of products is 12.5 kg/s and the products are cooled from 300 to 200°C, and for the products at this temperature, c_p = 1.09 kJ/kg K. The rate of air flow is 11.5 kg/s, the initial air temperature is 40°C and for the air c_p = 1.005 kJ/kg K. (a) Estimate the initial and final exergy of the products. (b) What is the irreversibility for the process? (c) If the heat transfer from the products occurs reversibly through heat engines, what is the final temperature of the air? What is the power developed by the heat engine? Take T_0 = 300 K and neglect pressure drop for both the fluids and heat transfer to the surroundings.* **D [LO 8.9, 8.10]**

Solution

(a)

$$\psi_1 = \text{initial exergy of the products}$$

$$= (h_1 - h_0) - T_0\,(s_1 - s_0)$$

$$= c_{pg}\,(T_{g1} - T_0) - T_0 c_{pg} \ln \frac{T_{g1}}{T_0}$$

$$= 1.09\,(573 - 300) - 300 \times 1.09 \ln \frac{573}{300}$$

$$= 297.57 - 211.6 = \mathbf{85.97\ kJ/kg}$$

$$\psi_2 = \text{final exergy of the products}$$

$$= (h_2 - h_0) - T_0(s_2 - s_0)$$

$$= 1.09\,(473 - 300) - 300 \times 1.09 \times \ln \frac{473}{300}$$

$$= 188.57 - 148.89 = \mathbf{39.68\ kJ/kg}$$

(b) Decrease in exergy of the products

$$= \psi_1 - \psi_2$$

$$= (h_1 - h_2) - T_0(s_1 - s_2)$$

$$= 1.09(573 - 473) - 300 \times 1.09 \ln \frac{573}{473}$$

$$= 109 - 62.72 = 46.28 \text{ kJ/kg}$$

By making an energy balance for the air preheater [Fig. 8.29 (a)].

$$\dot{m}_g c_{pg}\,(Tg_1 - Tg_2) = \dot{m}_a c_{pa}\,(T_{a2} - T_{a1})$$

$$12.5 \times 1.09\,(573 - 473) = 11.15 \times 1.005\,(T_{a2} - 313)$$

$$\therefore \qquad T_{a2} = \frac{12.5 \times 109}{11.5 \times 1.005} + 313 = 430.89 \text{ K}$$

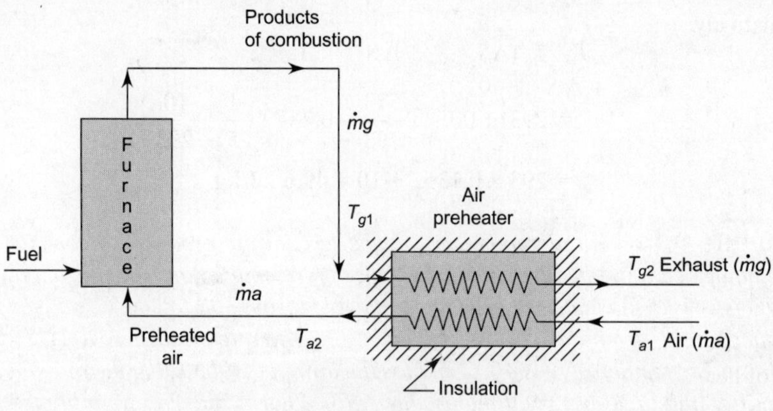

Fig. 8.29(a)

Increase in exergy for air

$$= \psi_2 - \psi_1$$

$$= (h_2 - h_1) - T_0(s_2 - s_1)$$

$$= c_{pa}\,(T_{a2} - T_{a1}) - T_0 c_{pa} \ln \frac{T_{a_2}}{T_{a_1}}$$

$$= 1.005 \times (430.89 - 313) - 300 \times 1.005 \ln \frac{430.89}{313}$$

$$= 118.48 - 96.37 = 22.11 \text{ kJ/kg}$$

$\therefore$ Irreversibility of the process

$$= 12.5 \times 46.28 - 11.5 \times 22.11$$

$$= 578.50 - 254.27 = \mathbf{324.23\ kW}$$

(c) Let us assume that the heat transfer from the products to air occurred through heat engines reversibly as shown in Fig. 8.29(b).

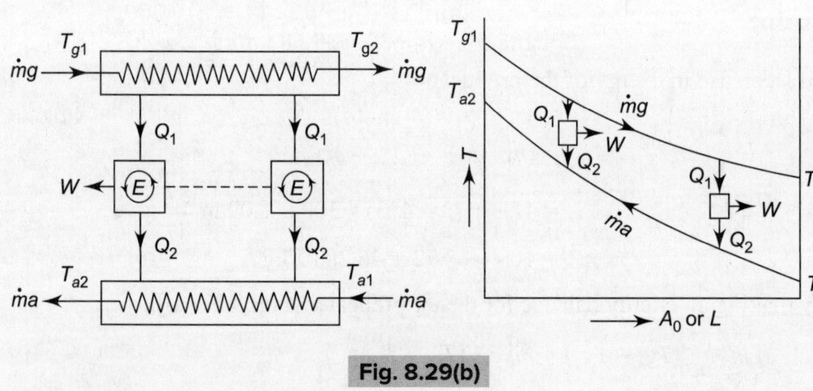

Fig. 8.29(b)

For reversible heat transfer,

$$\Delta \dot{S}_{\text{univ}} = 0$$

$$\Delta \dot{S}_{sys} + \Delta \dot{S}_{surr} = 0$$

$$\Delta \dot{S}_{gas} + \Delta \dot{S}_{air} = 0$$

$$\Delta \dot{S}_{gas} = -\Delta \dot{S}_{air}$$

$$\dot{m}_g c_{p_g} \ln \frac{T_{g2}}{T_{g1}} = \dot{m}_a c_{p_a} \ln \frac{T_{a_2}}{T_{a_1}}$$

$$12.5 \times 1.09 \ln \frac{473}{573} = -11.5 \times 1.005 \ln \frac{T_{a_2}}{313}$$

$$\therefore \quad T_{a2} = 392.41 \text{ K}$$

Rate of heat supply from the gas to the working fluid in the heat engine,

$$\dot{Q}_1 = \dot{m}_g c_{p_g} (T_{g_1} - T_{g_2})$$

$$= 12.5 \times 1.09\,(573 - 473) = 1362.50 \text{ kW}$$

Rate of heat rejection from the working fluid in the heat engine to the air,

$$\dot{Q}_2 = \dot{m}_a c_{p_a} (T_{a_2} - T_{a_1})$$

$$= 11.5 \times 1.005\,(392.41 - 313) = 917.78 \text{ kW}$$

Total power developed by the heat engine

$$\dot{W} = \dot{Q}_1 - \dot{Q}_2 = 1362.50 - 917.78 = \mathbf{444.72 \text{ kW}}$$

Example 8.11 *A gas is flowing through a pipe at the rate of 2 kg/s. Because of inadequate insulation, the gas temperature decreases from 800 to 790°C between two sections in the pipe. Neglecting pressure losses, calculate the irreversibility rate (or rate of energy degradation) due to this heat loss. Take T_0 = 300 K and a constant c_p = 1.1 kJ/kg K*

For the same temperature drop of 10°C when the gas cools from 80 to 70°C due to heat loss, what is the rate of energy degradation? Take the same values of T_0 and c_p. What is the inference you can draw from this example? M **[LO 8.10]**

Solution

$$\dot{S}_{gen} = \dot{S}_{sys} - \frac{\dot{Q}}{T_0}$$

$$= \dot{m}_{sys}(s_2 - s_1) - \frac{\dot{m}c_p(T_2 - T_1)}{T_0}$$

Irreversibility rate = Rate of energy degradation

= rate of exergy loss

$$\dot{I} = T_0 \dot{S}_{gen}$$

$$= \dot{m}T_0(s_2 - s_1) - \dot{m}(T_2 - T_1)$$

$$= \dot{m}c_p \left[(T_1 - T_2) - T_0 \ln \frac{T_1}{T_2} \right]$$

$$= 2 \times 1.1 \left[(1073 - 1063) - 300 \ln \frac{1073}{1063} \right] = \mathbf{15.818 \text{ kW}}$$

When the same heat loss occurs at lower temperature

$$\dot{I} = 2 \times 1.1 \left[(353 - 343) - 300 \ln \frac{353}{343} \right] = \mathbf{3.036\ kW}$$

It is thus seen that irreversibility rate or exergy destruction is more when the same heat loss occurs at higher temperature. Irreversibility rate decreases as the temperature of the gas decreases. Quantitatively, the heat loss may be the same, but qualitatively, it is different.

Example 8.12 *An ideal gas is flowing through an insulated pipe at the rate of 3 kg/s. There is a 10 percent pressure drop from inlet to exit of the pipe. What is the rate of exergy loss because of the pressure drop due to friction? Take $R = 0.287$ kJ/kg K and $T_0 = 300$ K.* E **[LO 8.10]**

Solution Rate of entropy generation from Eq. (8.47),

$$\dot{S}_{\text{gen}} = \dot{m}R \frac{\Delta p}{p_1}$$

$$= 3 \times 0.287 \frac{0.10 p_1}{p_1} = 0.0861 \text{ kW/K}$$

Rate of exergy loss

$$\dot{I} = T_0 \dot{S}_{\text{gen}}$$

$$= 300 \times 0.0861 = \mathbf{25.83\ kW}$$

Example 8.13 *Water at 90°C flowing at the rate of 2 kg/s mixes adiabatically with another stream of water at 30°C flowing at the rate of 1 kg/s. Estimate the entropy generation rate and the rate of exergy loss due to mixing. Take $T_0 = 300$ K.* M **[LO 8.10]**

Solution

$$\dot{m} = \dot{m}_1 + \dot{m}_2 = 2 + 1 = 3 \text{ kg/s}$$

Here

$$x = \frac{\dot{m}_1}{\dot{m}} = \frac{2}{3} = 0.67$$

$$\tau = \frac{T_2}{T_1} = \frac{303}{363} = 0.835$$

From Eq. (8.76),

$$\dot{S}_{\text{gen}} = \dot{m}c_p \ln \frac{x + \tau (1 - x)}{\tau^{1-x}}$$

$$= 3 \times 4.187 \ln \frac{0.67 \times 0.835 \times 0.33}{0.835^{0.33}}$$

$$= 12.561 \ln \frac{0.94555}{0.94223} = 0.0442 \text{ kW/K}$$

Rate of exergy loss due to mixing

$$\dot{I} = T_0 \dot{S}_{gen}$$

$$= 300 \times 0.0442 = \mathbf{13.26\ kW}$$

Alternatively,

Equilibrium temperature after mixing,

$$t = \frac{m_1 t_1 + m_2 t_2}{m_1 + m_2}$$

$$= \frac{2 \times 90 + 1 \times 30}{2 + 1} = 70°C$$

$$\Delta \dot{S}_{univ} = \dot{S}_{gen} = 2 \times 4.187 \ln \frac{343}{363} + 1 \times 4.187 \ln \frac{343}{303}$$

$$= 0.0447 \text{ kW/K}$$

$\therefore$ $$\dot{I} = 300 \times 0.0447 = \mathbf{13.41\ kW}$$

Example 8.14 *A well-insulated, thin walled, counter-flow heat exchanger (Fig. 8.30) is used to heat water (c_p = 4.18 kJ/kg°C) from 15°C to 60°C at a rate of 1.1 kg/s. The heating is done by hot oil (c_p = 4.18 kJ/kg°C) at 150°C at a rate of 1.5 kg/s. Determine the rate of heat transfer, the outlet temperature of oil and the rate of exergy destruction. Take T_0 = 20°C.*

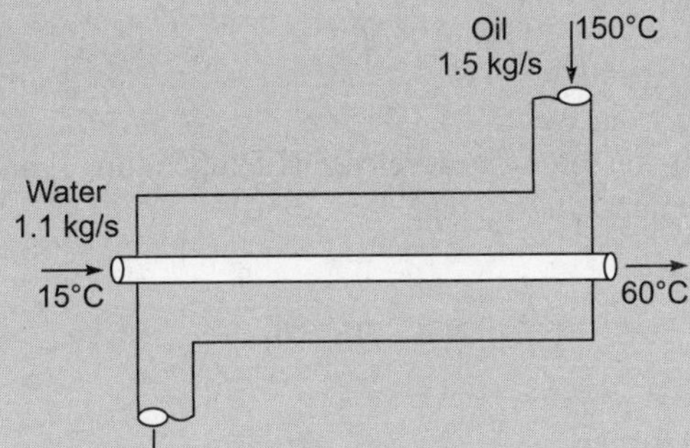

Fig. 8.30

M [LO 8.10]

Solution The rate of heat transfer to the cold water is found to be

$$\dot{Q} = \left[\dot{m}c_p(T_{out} - T_{in})\right]_{water} = 1.1 \times 4.18(60 - 15) = \mathbf{206.91\ kW}$$

The rise in temperature of cold water is accomplished by the heat loss of oil and hence the heat gain by the water is same as the heat loss by the oil. Thus

$$\dot{Q} = \left[\dot{m}c_p(T_{in} - T_{out})\right]_{oil} = 1.5 \times 2.3(150 - T_{out}) = 206.91$$

or $$T_{out} = \mathbf{90°C}$$

From the entropy balance equation for the heat exchanger under steady state condition, we obtain

$$\dot{m}_{\text{water}} c_{p,\text{water}} s_1 + \dot{m}_{\text{oil}} c_{p,\text{oil}} s_3 - \dot{m}_{\text{water}} c_{p,\text{water}} s_2 - \dot{m}_{\text{oil}} c_{p,\text{oil}} s_4 + \dot{S}_{\text{gen}} = 0$$

or

$$\dot{S}_{\text{gen}} = \dot{m}_{\text{water}} c_{p,\text{water}} (s_2 - s_1) + \dot{m}_{\text{oil}} c_{p,\text{oil}} (s_4 - s_3)$$

$$= \dot{m}_{\text{water}} c_{p,\text{water}} \ln \frac{T_2}{T_1} + \dot{m}_{\text{oil}} c_{p,\text{oil}} \ln \frac{T_4}{T_3}$$

$$= 1.1 \times 4.18 \ln \frac{60+273}{15+273} + 1.5 \times 2.3 \ln \frac{90+273}{150+273}$$

$$= 0.6675 - 0.5277 = 0.1398 \text{ kW/K}$$

Exergy destroyed can be found to be

$$\dot{A}_{\text{destroyed}} = T_0 \dot{S}_{\text{gen}} = 273 \times 0.1398 = \mathbf{40.96\ kW}$$

Example 8.15 *By burning a fuel, the rate of heat release is 500 kW at 2000 K. What would be the first law and the second law efficiencies if (a) energy is absorbed in a metallurgical furnace at the rate of 480 kW at 1000 K, (b) energy is absorbed at the rate of 450 kW for generation of steam at 500 K and (c) energy is absorbed in a chemical process at the rate of 300 kW at 320 K? Take T_0 = 300 K. (d) Had the energy absorption rate been equal to 450 kW in all these three cases, what would have been the second law efficiencies? What is the inference that you can draw from this example?* M **[LO 8.12]**

Solution If $\dot{Q}_r$ is the rate of heat release at temperature T_r and $\dot{Q}_a$ is the rate of heat absorption at temperature T_a, then

$$\eta_I = \frac{\dot{Q}_a}{\dot{Q}_r} \quad \text{and} \quad \eta_{II} = \eta_I \frac{1 - \frac{T_0}{T_a}}{1 - \frac{T_0}{T_r}}$$

(a) Metallurgical furnace

$$\eta_I = \frac{480}{500} \times 100 = 96 \text{ percent}$$

$$\eta_{II} = 0.96 \frac{1 - \frac{300}{1000}}{1 - \frac{300}{2000}} \times 100 = \mathbf{79\ percent}$$

(b) Steam generation

$$\eta_I = \frac{450}{500} \times 100 = 90 \text{ percent}$$

$$\eta_{II} = 0.90 \frac{1 - \frac{300}{500}}{1 - \frac{300}{2000}} \times 100 = \mathbf{42.3\ percent}$$

(c) Chemical process

$$\eta_I = \frac{300}{500} \times 100 = 60 \text{ percent}$$

$$\eta_{II} = 0.60 \frac{1 - \dfrac{300}{320}}{1 - \dfrac{300}{2000}} \times 100 = \textbf{4.41 percent}$$

(d) In all the three cases, η_I would remain the same, where

$$\eta_I = \frac{450}{500} \times 100 = 0.90$$

$$\eta_{II(a)} = 0.90 \times \frac{1 - \dfrac{300}{1000}}{1 - \dfrac{300}{2000}} \times 100 = \textbf{74.11 percent}$$

$$\eta_{II(b)} = 0.90 \times \frac{1 - \dfrac{300}{500}}{1 - \dfrac{300}{2000}} \times 100 = \textbf{42.3 percent}$$

$$\eta_{II(c)} = 0.90 \times \frac{1 - \dfrac{300}{320}}{1 - \dfrac{300}{2000}} \times 100 = \textbf{6.61 percent}$$

It is seen that as the energy loss ($\dot{Q}_r - \dot{Q}_a$) increases, the first law efficiency decreases. For the same heat loss, however, as the temperature difference between the source and the use temperature increases, the second law efficiency decreases, or in other words, the rate of exergy loss increases.

Example 8.16 *A system undergoes a power cycle while receiving energy Q_1 at temperature T_1 and discharging energy Q_2 at temperature T_2. There are no other heat transfers.*

(a) Show that the thermal efficiency of the cycle can be expressed as

$$\eta = 1 - \frac{T_2}{T_1} - \frac{T_2 I}{T_0 Q_1}$$

where T_0 is the ambient temperature and I is the irreversibility of the cycle.

(b) Obtain an expression for the maximum theoretical value for the thermal efficiency.

(c) Derive an expression for the irreversibility for which no net work is developed by the cycle. What conclusion do you derive from it? M [LO 8.3, 8.4]

Solution An availability balance for the cycle gives (Fig. 8.31)

$$(\Delta A)_{\text{cycle}} = 0 = \left(1 - \frac{T_0}{T_1}\right) Q_1 - \left(1 - \frac{T_0}{T_2}\right) Q_2 - W - I$$

since each property is restored to its initial state.

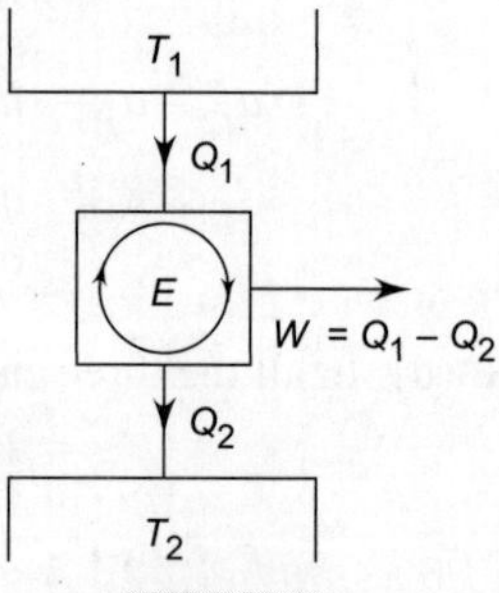

Fig. 8.31

Since $Q_2 = Q_1 - W$,

$$0 = \left(1 - \frac{T_0}{T_1}\right) Q_1 - \left(1 - \frac{T_0}{T_2}\right)(Q_1 - W) - W - I$$

$$W = \frac{T_2}{T_0}\left[\left(1 - \frac{T_0}{T_1}\right) - \left(1 - \frac{T_0}{T_2}\right)\right] Q_1 - \frac{T_2 I}{T_0}$$

$$\eta = \frac{W}{Q_1} = 1 - \frac{T_2}{T_1} - \frac{T_2 I}{T_0 Q_1}$$ Proved.

(b) When $I = 0$,

$$\eta_{\max} = 1 - \frac{T_2}{T_1}$$

(c) When $W = 0$

$$\eta = 0 = 1 - \frac{T_2}{T_1} - \frac{T_2 I}{T_0 Q_1}$$

$$I = T_0\left[\frac{1}{T_2} - \frac{1}{T_1}\right] Q_1 = T_0\left[\frac{Q_1}{T_2} - \frac{Q_1}{T_1}\right] = T_0 S_{\text{gen}}$$

The heat transfer Q_1 from T_1 to T_2 takes place through a reversible engine, and the entire work is dissipated in the brake, from which an equal amount of heat is rejected to the reservoir at T_2. Heat transfer through a finite temperature difference is thus equivalent to the destruction of its exergy. (*See Section 8.13.1.1*))

Example 8.17 *A compressor operating at steady state takes in 1 kg/s of air at 1 bar and 25°C and compresses it to 8 bar and 160°C. Heat transfer from the compressor to its surroundings occurs at a rate of 100 kW. (a) Determine the power input in kW. (b) Evaluate the second law efficiency for the compressor. Neglect K.E. and P.E. changes. Take T_0 = 25°C and P_0 = 1 bar.* Ⓜ **[LO 8.11]**

Solution SFEE for the compressor (Fig. 8.32) gives

$$\dot{W} = \dot{Q} + \dot{m}\,(h_1 - h_2)$$

$$= -100 + 1 \times 1.005\,(25 - 160) = \mathbf{-235.7\ kW}$$

Exergy balance for the compressor gives

$$\dot{m}\, a_{f_1} + \dot{Q}\left(1 - \frac{T_0}{T}\right) - \dot{W} - \dot{m}\, a_{f_2} = \dot{I}$$

$$-\dot{W} = \dot{m}\,(a_{f_2} - a_{f_1}) - \dot{Q}\left(1 - \frac{T_0}{T}\right) + \dot{I}$$

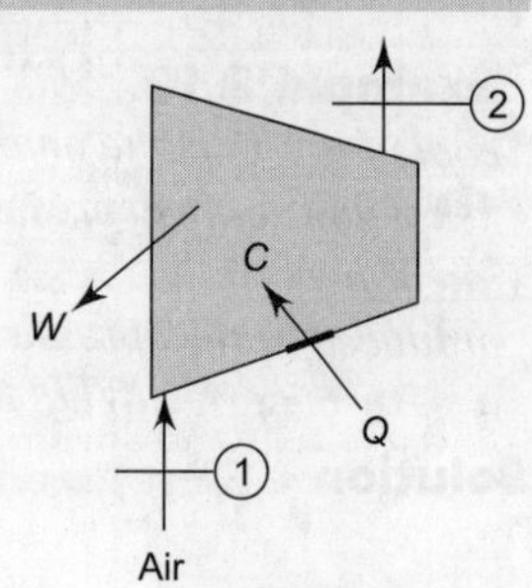

Fig. 8.32

$$\eta_{II} = \frac{\dot{m}(a_{f_2} - a_{f_1})}{\dot{W}}$$

$$a_{f2} - a_{f1} = h_2 - h_1 - T_0(s_2 - s_1)$$

$$= c_p(T_2 - T_1) - T_0\left(c_p \ln \frac{T_2}{T_1} - R \ln \frac{p_2}{p_1}\right)$$

$$= 1.005\,(160 - 25) - 298\left(1.005 \ln \frac{433}{298} - 0.287 \ln 8\right)$$

$$= 200.95 \text{ kJ/kg}$$

$$\eta_{II} = \frac{200.95}{235.7} = \mathbf{0.853} \text{ or, } \mathbf{85.3\%}$$

Example 8.18 *Determine the exergy of 1 m³ of complete vacuum.* E [LO 8.9]

Solution

$$\phi = U - U_0 + p_0(V - V_0) - T_0(S - S_0)$$

$$= H - H_0 - V(p - p_0) - T_0(S - S_0)$$

Since a vacuum has zero mass,

$$U = 0, H_0 = 0 \text{ and } S = 0$$

If the vacuum was reduced to the dead state

$$U_0 = 0, H_0 = S_0 = 0 \text{ and } V_0 = 0$$

The pressure p for the vacuum is zero.

But $p_0 = 1 \text{ bar} = 100 \text{ kPa and } V = 1 \text{ m}^3$

$$\phi = p_0 V = 100 \frac{\text{kN}}{\text{m}^2} \times 1 \text{ m}^3 = \mathbf{100\ kJ}$$

If an air motor operates between the atmosphere and the vacuum, this is the maximum useful work obtainable. Therefore, the vacuum has an exergy or work potential.

Example 8.19 *A mass of 1000 kg of fish initially at 1 bar, 300 K is to be cooled to – 20°C. The freezing point of fish is – 2.2°C, and the specific heats of fish below and above the freezing point are 1.7 and 3.2 kJ/kg K, respectively. The latent heat of fusion for the fish can be taken as 235 kJ/kg. Calculate the exergy produced in the chilling process. Take T_0 = 300 K and p_0 = 1 bar.* M [LO 8.9]

Solution Exergy produced $= H_2 - H_1 - T_0(S_2 - S_1)$

With reference to Fig. 8.33,

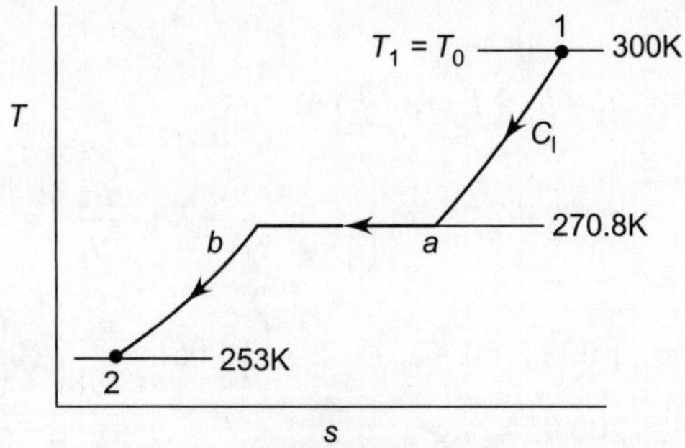

Fig. 8.33

$$H_1 - H_2 = 1000\,[1.7\,(270.8 - 253) + 235 + 3.2\,(300 - 270.8)]$$
$$= 1000\,[1.7 \times 17.8 + 235 + 3.2 \times 29.2]$$
$$= 1000\,[30.26 + 235 + 93.44] = 358.7 \text{ MJ}$$
$$H_2 - H_1 = -358.7 \text{ MJ}$$
$$S_1 - S_2 = 1000\left[1.7 \ln \frac{270.8}{253} + \frac{235}{270.8} + 3.2 \ln \frac{300}{270.8}\right]$$
$$= -1000\,[0.1156 + 0.8678 + 0.3277] = 1.311 \text{ MJ/K}$$
$$S_2 - S_1 = -1.311 \text{ MJ/K}$$
$$\text{Exergy produced} = -358.7 + 300 \times 1.311$$
$$= -358.7 + 393.3 = \mathbf{34.6\ MJ} \text{ or } \mathbf{9.54\ kWh}$$

Example 8.20 *A quantity of air initially at 1 bar, 300 K undergoes two types of interactions: (a) it is brought to a final temperature of 500 K adiabatically by paddle-wheel work transfer, (b) the same temperature rise is brought about by heat transfer from a thermal reservoir at 600 K. Take T_0 = 300 K, p_0 = 1 atm. Determine the irreversibility (in kJ/kg) in each case and comment on the results.*

E **[LO 8.10]**

Solution *Case (a):* As shown in the above figures (Fig. 8.34)

$$\Delta s_{\text{univ}} = s_{\text{gen}} = c_v \ln \frac{T_2}{T_1} = 0.718 \ln \frac{500}{300} = 0.367 \text{ kJ/kg K}$$

$$I = 300 \times 0.367 = \mathbf{110.1\ kJ/kg}$$

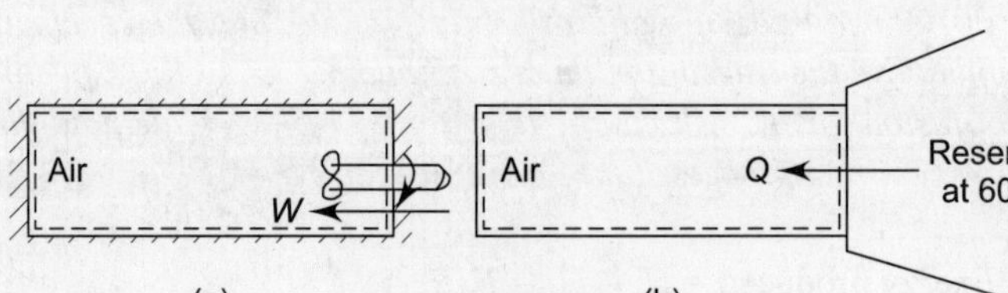

Fig. 8.34

Case (b):

$$Q = mc_v(T_2 - T_1)$$

$$= 1 \times 0.718\,(500 - 300) = \mathbf{143.6\ kJ/kg}$$

$$\Delta s_{\text{univ}} = s_2 - s_1 - \frac{Q_2}{T} = 0.367 - \frac{143.6}{600} = 0.1277 \text{ kJ/kg K}$$

$$I = 300 \times 0.1277 = \mathbf{38.31\ kJ/kg}$$

Comment:

The irreversibility in case (b) is less than in case (a).

$$I_a = T_0(s_2 - s_1),\ I_b = T_0(s_2 - s_1) - \frac{Q}{T}$$

$$I_a - I_b = \frac{Q}{T}$$

The irreversibility in case (b) is always less than in case (a) and the two values would approach each other only at high reservoir temperature, i.e

$$I_a \to I_b \text{ as } T \to \infty$$

Example 8.21 *Steam enters a turbine at 30 bar, 400°C (h = 3230 kJ/kg, s = 6.9212 kJ/g K) and with a velocity of 160 m/s. Steam leaves as saturated vapour at 100°C (h = 2676.1 kJ/kg, s = 7.3549 kJ/kg K) with a velocity of 100 m/s. At steady state, the turbine develops work at a rate of 540 kJ/ kg. Heat transfer between the turbine and its surroundings occurs at an average outer surface temperature of 500 K. Determine the irreversibility per unit mass. Give an exergy balance and estimate the second law efficiency of the turbine. Take* p_0 *= 1 atm,* T_0 *= 298 K and neglect PE effect.* Ⓜ **[LO 8.11, 8.12]**

Solution By exergy balance of the control volume (Fig. 8.35),

$$a_{f1} = W + Q\left(1 - \frac{T_0}{T_B}\right) + a_{f2} + I$$

where a_f is the exergy transfer per unit mass.

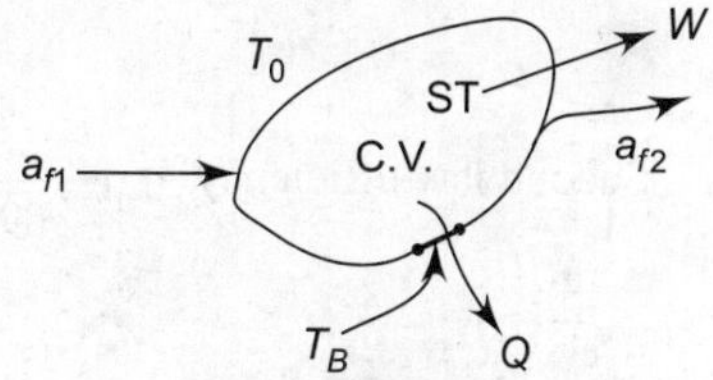

Fig. 8.35

$$I = a_{f1} - a_{f2} - W - Q\left(1 - \frac{T_0}{T_B}\right)$$

$$= (h_1 - h_2) - T_0(s_1 - s_2) + \frac{V_1^2 - V_2^2}{2} - W - Q\left(1 - \frac{T_0}{T_B}\right)$$

$$= (3230.9 - 2676.1) - 298(6.9212 - 7.3549) + \frac{160^2 - 100^2}{2}$$

$$\times 10^{-3} - 540 - Q\left(1 - \frac{298}{500}\right)$$

$$= 151.84 - Q(0.404) \qquad \text{(i)}$$

By S.F.E.E.,

$$h_1 + \frac{V_1^2}{2} = W + h_2 + Q + \frac{V_2^2}{2}$$

$$Q = (h_1 - h_2) + \frac{V_1^2 - V_2^2}{2} - W$$

$$= (3230.9 - 2676.1) + \frac{160^2 - 100^2}{2} \times 10^{-3} - 540 = 22.6 \text{ kJ/kg.}$$

From Eq. (i),

$$\text{I} = 151.84 - 22.6 \times 0.404 = \mathbf{142.71 \ kJ/kg}$$

Net exergy transferred to turbine

$$a_{f_1} - a_{f_2} = 691.84 \text{ kJ/kg}$$

$$\text{Work} = 540 \text{ kJ/kg}$$

$$\text{Exergy destroyed} = I = 142.71 \text{ kJ/kg}$$

Exergy transferred out accompanying heat transfer

$$= 22.6 \times 0.404 = 9.13 \text{ kJ/kg}$$

Exergy Balance

Exergy transferred	*Exergy utilized*
691.84 kJ/kg	Work = 540 kJ/kg (78 percent)
	Destroyed = 142.71 kJ/kg (20.6 percent)
	Transferred with heat = 9.13 kJ/kg (1.3%)
	691.84 kJ/kg

$$\text{Second law efficiency, } \eta_{II} = \frac{540}{691.84} = \mathbf{0.78 \text{ or } 78 \text{ percent}}$$

Example 8.22 *A furnace is heated by an electrical resistor. At steady state, electrical power is supplied to the resistor at a rate of 8.5 kW per metre length to maintain it at 1500 K when the furnace walls are at 500 K. Let $T_0 = 300\,K$ (a) For the resistor as the system, determine the rate of availability transfer accompanying heat and the irreversibility rate, (b) For the space between the resistor and the walls as the system, evaluate the irreversibility rate.* E [LO 8.3]

Solution *Case (a):* At steady state for the resistor (Fig. 8.36).

$$\dot{Q} = \Delta \dot{U} + \dot{W} = \dot{W} = 8.5 \text{ kW}$$

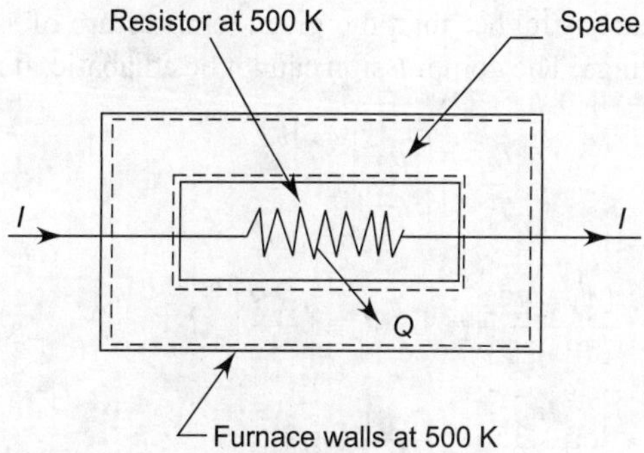

Fig. 8.36

Availability rate balance gives

$$\frac{dA}{dt} = \left(1 - \frac{T_0}{T}\right)\dot{Q} + \left(\dot{W} - p_0 \frac{dV}{dt}\right) - I = 0$$

I = Rate of irreversibility

$$= \left(1 - \frac{T_0}{T}\right)\dot{Q} + \dot{W} = \left(1 - \frac{300}{1500}\right)(-8.5) + 8.5 = \mathbf{1.7\ kW}$$

Rate of availability transfer with heat

$$= \left(1 - \frac{T_0}{T}\right)\dot{Q} = \left(1 - \frac{300}{1500}\right)(-8.5) = \mathbf{-6.8\ kW}$$

Case (b): Steady state,

$$\frac{dA}{dT} = \left(1 - \frac{T_0}{T}\right)\dot{Q} - \left(1 - \frac{T_0}{T}\right)\dot{Q} - \dot{W} - \dot{I} = 0$$

$$\dot{I} = \left(1 - \frac{300}{1500}\right)8.5 - \left(1 - \frac{300}{500}\right)8.5 = 6.8 - 3.4 = \mathbf{3.4\ kW}$$

Example 8.23 *Air enters a compressor at 1 bar, 30°C, which is also the state of the environment. It leaves at 3.5 bar, 141°C and 90 m/s. Neglecting inlet velocity and PE effect, determine (a) whether the compression is adiabatic or polytropic, (b) if not adiabatic, the polytropic index, (c) the isothermal efficiency, (d) the minimum work input and irreversibility and (d) the second law efficiency. Take c_p of air = 1.0035 kJ/kg K.* D **[LO 8.9, 8.12]**

Solution (a) After isentropic compression

$$\frac{T_{2s}}{T_1} = \left[\frac{p_2}{p_1}\right]^{(\gamma-1)/\gamma}$$

$$T_{2s} = 303\,(3.5)^{0.286} = 433.6\text{ K} = 160.6°\text{C}$$

Since this temperature is higher than the given temperature of 141°C, there is heat loss to the surroundings. The compression cannot be adiabatic. It must be polytropic.

(b)
$$\frac{T_2}{T_1} = \left[\frac{p_2}{p_1}\right]^{(n-1)/n}$$

$$\frac{141 + 273}{30 + 273} = 1.366 = \left[\frac{3.5}{1}\right]^{(n-1)/n}$$

$$\log 1.366 = \frac{n-1}{n} \log 3.5$$

$$1 - \frac{1}{n} = \frac{0.135}{0.544} = 0.248$$

$$n = 1.32978 = \mathbf{1.33}$$

(c) Actual work of compression

$$W_a = h_1 - h_2 - \frac{V_2^2}{2} = 1.0035\,(30 - 141) - \frac{90}{2} \times 10^{-3} = -\,115.7 \text{ kJ/kg}$$

Isothermal work

$$W_T = \int_1^2 v\,dp - \frac{V_2^2}{2} = -\,RT_1 \ln \frac{p_2}{p_1} - \frac{V_2^2}{2}$$

$$= -\,0.287 \times 303 \ln (3.5) - \frac{90^2}{2} \times 10^{-3} = -\,113 \text{ kJ/kg}$$

Isothermal efficiency:

$$\eta_T = \frac{W_T}{W_a} = \frac{113}{115.7} = \mathbf{0.977 \text{ or } 97.7 \text{ percent}}$$

(d) Decrease in availability or exergy

$$\psi_1 - \psi_2 = h_1 - h_2 - T_0(s_1 - s_2) + \frac{V_1^2 - V_2^2}{2}$$

$$= c_p(T_1 - T_2) - T_0\left[R \ln \frac{p_2}{p_1} - c_p \ln \frac{T_2}{T_1}\right] - \frac{V_2^2}{2}$$

$$= 1.0035\,(30 - 141)$$

$$-\,303\left[0.287 \ln 3.5 - 1.0035 \ln \frac{414}{303}\right] - \frac{90^2}{2000}$$

$$= -\,101.8 \text{ kJ/kg}$$

Minimum work input = **– 101.8 kJ/kg**

Irreversibility, $I = W_{\text{rev}} - W_a$

$$= -\,101.8 - (-\,115.7) = \mathbf{13.9 \text{ kJ/kg}}$$

(e) Second law efficiency

$$\eta_{II} = \frac{\text{Minimum work input}}{\text{Actual work input}} = \frac{101.8}{115.7} = \mathbf{0.88} \text{ or } \mathbf{88 \text{ percent}}$$

SUMMARY

The maximum work obtainable W_{max} from a certain heat input in a cyclic heat engine Q_1 is called the available energy (A.E.), $\left(1 - \frac{T_0}{T_1}\right) Q_1$ where T_0 is the temperature of the surroundings. The unavailable energy U.E. is the product of the lowest temperature of heat rejection and the change of entropy of the system during the process of supplying heat. The A.E. is also known as *exergy* and the U.E. as energy.

Whenever heat is transferred through a finite temperature difference there is a decrease in the availability of energy so transferred. Energy is said to be degraded each time it flows through a finite temperature difference. The available energy or exergy of a fluid at temperature T is given by

$$W_{max} = mc_p \left[(T - T_0) - T_0 \ln \frac{T}{T_0}\right]$$

where m is the mass of the fluid. The higher the temperature T, the more is the exergy. The exergy value provides a useful measure of the energy quality.

The first law states that energy is always conserved quantitywise, the second law emphasises that energy always degrades qualitywise.

The work done by a closed system in a reversible process between two states is always more than that done by any irreversible process between the same states. The reversible work in a steady flow process is given by

$$W_{max} = \left(h_1 - T_0 s_1 + \frac{V_1^2}{2} + gZ_1\right) - \left(h_2 - T_0 s_2 + \frac{V_2^2}{2} + gZ_2\right)$$

For a closed system it is

$$W_{max} = (u_1 - T_0 s_1) - (u_2 - T_0 s_2)$$

A certain portion of the work done by a closed system is spent in pushing the atmosphere. The useful work becomes

$$(W_u)_{max} = W_{max} - p_0 (V_2 - V_1)$$

When a system undergoes a process spontaneously due to a certain potential difference, the process must terminate when the system reaches the *dead state*, i.e. the pressure and temperature of the surroundings. The work done by a system between two equilibrium states at the same temperature as that of the surroundings is equal to or less than the change in Helmholtz function.

$$(W_T)_{max} \le (F_1 - F_2)_T$$

The decrease in the Gibbs function of a system sets an upper limit to the useful work done at the same pressure and temperature as those of the surroundings

$$(W_{u,\,max})_{p,\,T} = (G_1 - G_2)_{p,T}$$

The irreversibility of a process, I, is equal to

$$I = W_{max} - W$$

The Guoy–Stodola theorem states that the rate of loss of exergy in a process or irreversibility rate is given by

$$\dot{I} = T_0\, \Delta \dot{S} = T_0\, (\Delta \dot{S}_{yst} + \Delta \dot{S}_{surr}).$$

While energy is always conserved, exergy is always destroyed by irreversibilities.

Energy in – Energy out = 0

Exergy in – Exergy out = Exergy destroyed

The *second law efficiency* is given by

$$\eta_{II} = \frac{\text{Exergy output}}{\text{Exergy input}} = \frac{W}{W_{max}}$$

where $I = W_{max} - W = T_0\, S_{gen}.$

REVIEW QUESTIONS

(E) **8.1** What do you understand by high grade energy and low grade energy? **[LO 8.1]**

(M) **8.2** What is available energy and unavailable energy? **[LO 8.3]**

(D) **8.3** Who propounded the concept of availability? **[LO 8.9]**

(M) **8.4** What is the available energy referred to a cycle? **[LO 8.3]**

(M) **8.5** Show that there is a decrease in available energy when heat is transferred through a finite temperature difference. **[LO 8.4]**

(M) **8.6** Deduce the expression for exergy or available energy from a finite energy source at temperature T when the environmental temperature is T_0. **[LO 8.5]**

(M) **8.7** What do you understand by exergy and energy? **[LO 8.3]**

(M) **8.8** What is meant by quality of energy? **[LO 8.5]**

(E) **8.9** Why is exergy of a fluid at a higher temperature more than that at a lower temperature? **[LO 8.5]**

(M) **8.10** How does the exergy value provide a useful measure of the quality of energy? **[LO 8.5]**

(D) **8.11** Why is the second law called the law of degradation of energy? **[LO 8.6]**

(D) **8.12** Energy is always conserved, but its quality is always degraded. Explain. **[LO 8.6]**

(M) **8.13** Why is the work done by a closed system in a reversible process by interacting only with the surroundings the maximum? **[LO 8.7]**

(M) **8.14** Show that equal work is done in all reversible processes between the same end states of a system if it exchanges energy only with the surroundings. **[LO 8.8]**

(D) **8.15** Give the general expression for the maximum work of an open system which exchanges heat only with the surroundings. **[LO 8.9]**

(E) **8.16** What do you understand by Keenan function? **[LO 8.9]**

(M) **8.17** Give the expression for reversible work in a steady flow process under a given environment. **[LO 8.9]**

(M) **8.18** Give the expression for reversible work done by a closed system if it interacts only with the surroundings. **[LO 8.9]**

(D) **8.19** What do you understand by 'useful work'? Derive expressions for useful work for a closed system and a steady flow system which interact only with the surroundings. **[LO 8.9]**

(M) **8.20** What are the availability functions for a: (a) closed system, (b) steady flow system? **[LO 8.9]**

(E) **8.21** What do you understand by the dead state? **[LO 8.2]**

(E) **8.22** What is meant by availability? **[LO 8.9]**

(M) **8.23** Give expressions for availabilities of a closed system and a steady flow open system. **[LO 8.9]**

(M) **8.24** What are Helmholtz function and Gibbs function? **[LO 8.9]**

(M) **8.25** What is the availability in a chemical reaction if the temperature before and after the reaction is the same and equal to the temperature of the surroundings? **[LO 8.9]**

(M) **8.26** When is the availability of a chemical reaction equal to the decrease in the Gibbs function? **[LO 8.9]**

(M) **8.27** Derive the expression for irreversibility or exergy loss in a process executed by (a) a closed system, (b) a steady flow system, in a given environment. **[LO 8.9]**

(D) **8.28** State and explain the Gouy–Stodola theorem. **[LO 8.10]**

(D) **8.29** How is heat transfer through a finite temperature difference equivalent to the destruction of its exergy? **[LO 8.10]**

(D) **8.30** Considering the steady and adiabatic flow of an ideal gas through a pipe, show that the rate of decrease in availability or lost work is proportional to the pressure drop and the mass flow rate. **[LO 8.10]**

(M) **8.31** What do you understand by Grassman diagram? **[LO 8.10]**

(M) **8.32** What is entropy generation number? **[LO 8.10]**

(D) **8.33** Define the second law efficiency. How is it different from the first law efficiency in the case of a simple power plant? **[LO 8.11]**

(D) **8.34** Derive the second law efficiency for (a) a solar water heater and (b) a heat pump. **[LO 8.12]**

(M) **8.35** What is meant by energy cascading? How is it thermodynamically efficient? **[LO 8.12]**

(D) **8.36** Why is exergy always a positive value? Can it be negative? **[LO 8.11]**

(M) **8.37** Why and when is exergy completely destroyed? **[LO 8.11]**

(E) **8.38** Give the exergy balance for a closed system. **[LO 8.11]**

(M) **8.39** Explain the statement: The exergy of an isolated system can never increase. How is it related to the principle of increase of entropy? **[LO 8.11]**

(M) **8.40** Give the exergy balance of a steady flow system. **[LO 8.11]**

(D) **8.41** Derive expressions for the irreversibility and second law efficiency of a **[LO 8.12]**

(a) steam turbine (b) compressor
(c) heat exchanger and (d) mixer

(M) **8.42** What is the deficiency of the first law efficiency? How does the second law efficiency make up this deficiency? **[LO 8.12]**

(M) **8.43** How can you improve the first law efficiency and the second law efficiency? **[LO 8.12]**

PROBLEMS WITH HINTS

(E) **8.1** Water of mass 2 kg undergoes a process from an initial state where its $p = 1.985$ bar, $v = 891.9$ cm^3/g, $t = 1200°$C, $u = 2529.3$ kJ/kg and $s = 7.1296$ kJ/kgK,

$V = 30$ m/s and $Z = 6$ m to a final state where the water is at 10°C, $p = 0.01228$ bar, $v = 1.0004$ cm³/g, $u = 42$ kJ/kg, $s = 0.151$ kJ/kgK, $V = 25$ m/s and $Z = 3$ m. Determine (a) the exergy of water at the initial state, (b) the exergy at the final state and (c) the change in exergy. Take $T_o = 25$°C, $P_o = 1$ atm and $g = 9.8$ m/s². **[LO 8.9]**

Hint: The exergy at the initial state

$$A = m\left[(u - u_o) + p_o(v - v_o) - T_o(s - s_o) + \frac{V^2}{2} + gZ\right]$$

At $T_o = 25$°C and $p_o = 1$ atm, the water has

$$v_o = 1.0029 \text{ m}^3\text{/kg}, \; u_o = 104.88 \text{ kJ/kg and}$$
$$s_o = 0.3674 \text{ kJ/kgK}.$$

Substituting the values, you get

(a) $A_1 =$ **1000.1 kJ**
(b) $A_2 =$ **3.9 kJ**
(c) $A_1 - A_2 =$ **−996.2 kJ**

The minus sign indicates that the exergy decreases in this process.

(E) **8.2** An isolated system consists of two solid blocks. One block has a mass of 5 kg and is initially at 300°C. The other block has a mass of 10 kg and is initially at –50°C. The blocks are allowed to come into thermal equilibrium. Assuming the blocks to be incompressible with constant specific heats of 1.0 and 0.4 kJ/kgK, respectively, determine the final temperature and the irreversibility. Take $T_o = 27$°C. **[LO 8.10]**

Hint:

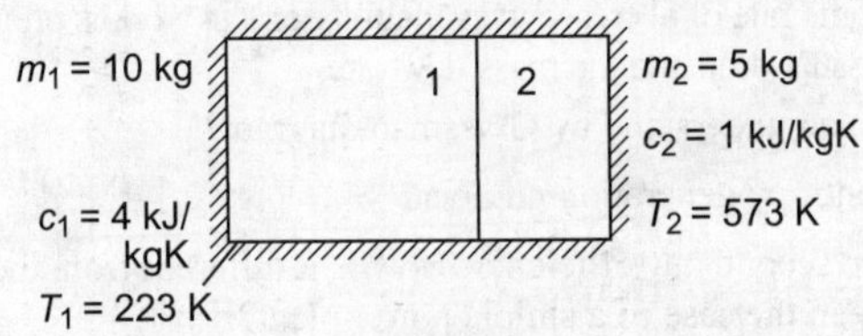

Fig. 8.37

By energy balance $m_1c_1(T_f - T_1) = m_2c_2(T_2 - T_f)$

$$T_f = \frac{m_1c_1T_1 + m_2c_2T_2}{m_1c_1 + m_2c_2} = \textbf{417.4 K.}$$

$$I = T_o S_{\text{gen}}$$

where $$S_{\text{gen}} = m_1c_1 \ln\frac{T_f}{T_1} + m_2c_2 \ln\frac{T_f}{T_2} = 0.92\,3 \text{ kJ/K}$$

∴ $$I = \textbf{276.99 kJ}$$

(E) **8.3** A thermal reservoir at 1000 K is separated from another thermal reservoir at 500 K by a cylindrical rod insulated on its lateral surfaces. At steady state, energy transfer by conduction takes place through the rod at a rate of 10 kW. Evaluate the irreversibility rate for the rod. Take $T_o = 300$ K. **[LO 8.10]**

Hint:

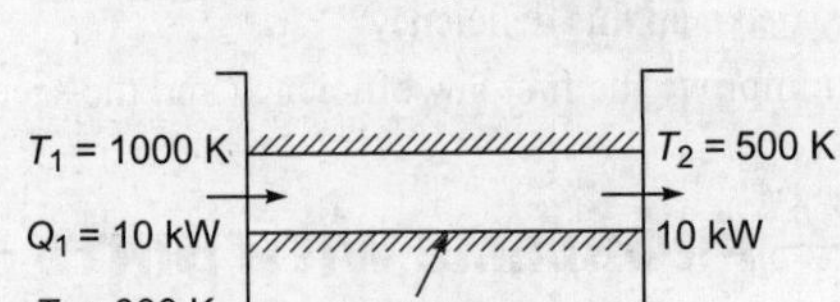

Fig. 8.38

At steady state,

$$\frac{dE}{dt} = \overset{\circ}{Q} - \overset{\circ}{W} = 0, \dot{W} = 0$$

∴ $\dot{Q} = 0$, i.e. rate of heat transfer in-rate of heat transfer out

Exergy balance gives

$$\frac{d\cancel{A}^{\circ 0}}{\cancel{dt}} = \left(1 - \frac{T_o}{T_1}\right)\dot{Q}_1 - \left(1 - \frac{T_o}{T_2}\right)\dot{Q}_1 - \left(\cancel{\dot{W}}^{=0} - p_o \frac{d\cancel{V}^{=0}}{\cancel{dt}}\right) - \dot{I}$$

∴ $$I = \overset{\circ}{Q}_1\left(\frac{T_o}{T_2} - \frac{T_o}{T_1}\right) = \mathbf{3\ kW}$$

(M) **8.4** An adiabatic turbine receives gas ($c_p = 1.00$ kJ/kgK, $\gamma = 1.3$) at 700 kPa and 1000°C and discharges it at 150 kPa and 665°C. Taking $T_o = 25$°C, determine the second law and isentropic efficiencies of the turbine. **[LO 8.11]**

Hint:

$c_p = 1.09$ kJ/kgK, $\gamma = \frac{c_p}{c_v} = 1.3$

∴ $$c_v = 0.838 \text{ and } R = 0.252 \text{ kJ/kgK.}$$

$$W_{rev} = h_1 - h_2 - T_o(s_1 - s_2)$$

$$= 1.09\,(1273 - 938) - 288\left(1.09 \ln \frac{1273}{938} - 0.252 \ln \frac{700}{150}\right)$$

$$= 381.6 \text{ kJ/kg}$$

$$W_{act} = h_1 - h_2$$

∴ $$n_\pi = \frac{W_{act}}{W_{rev}} = \mathbf{0.956 \text{ or } 95.6 \text{ percent}}$$

$$T_{2s} = T_1\left(\frac{p_2}{p_1}\right)^{\frac{v-1}{v}} = 802 \text{ K}$$

∴ $$\eta_{isen} = \frac{h_1 - h_2}{h_1 - h_{2s}} = \frac{T_1 - T_2}{T_1 - T_{2s}} = \mathbf{0.879 \text{ or } 87.9 \text{ percent}}$$

(E) **8.5** Air flows through an adiabatic compressor at 2 kg/s. The inlet conditions are 100 kPa and 310 K and the exit conditions are 700 kPa and 500 K. Taking $T_o = 298$ K, determine the change in exergy and the irreversibility **[LO 8.9]**

Hint:

$$m(\psi_2 - \psi_1) = m(h_2 - h_1) - mT(s_2 - s_1)$$

$$= 481.1 \text{ kW}$$

$$\dot{I} = mT(s_2 - s_1) = \mathbf{21.2\ kW}$$

(E) **8.6** An insulated rigid tank contains 0.9 kg of air at 150 kPa and 20°C. A paddle wheel inside the tank is now rotated by an external power source until the temperature in the tank rises to 55°C. If the surrounding air is at $T_o = 20$°C, determine the exergy destroyed and the reversible work for the process. Take c_v of air = 0.718 kJ/kgK. **[LO 8.9]**

Hint: Exergy destroyed $= T_o S_{gen} = T_o m c_v \ln \frac{T_2}{T_1}$

$$= \mathbf{21.4\ kJ}$$

Reversible work,

$$W_{rev} = E_2 - E_1 + p_o(V_2 - V_1) - T_o(S_2 - S_1)$$

$$= U_2 - U_1 - T_o(S_2 - S_1)$$

$$= m\,c_v(T_2 - T_1) - T_o(S_2 - S_1) = \mathbf{1.2\ kJ}$$

PROBLEMS

(E) **8.1** What is the maximum useful work which can be obtained when 100 kJ are abstracted from a heat reservoir at 675 K in an environment at 288 K? What is the loss of useful work if (a) a temperature drop of 50°C is introduced between the heat source and the heat engine, on the one hand and the heat engine and the heat sink, on the other, (b) the source temperature drops by 50°C and the sink temperature rises by 50°C during the heat transfer process according to the linear law $\frac{đQ}{dT} = \pm$ constant? **[LO 8.3]**

(E) **8.2** In a steam generator, water is evaporated at 260°C, while the combustion gas (c_p = 1.08 kJ/kg K) is cooled from 1300°C to 320°C. The surroundings are at 30°C. Determine the loss in exergy due to the above heat transfer per kg of water evaporated. (Latent heat of vaporisation of water at 260°C = 1662.5 kJ/kg.) **[LO 8.4]**

(E) **8.3** Exhaust gases leave an internal combustion engine at 800°C and 1 atm, after having done 1050 kJ of work per kg of gas in the engine (c_p of gas = 1.1 kJ/kg K). The temperature of the surroundings is 30°C. (a) How much exergy per kg of gas is lost by throwing away the exhaust gases? (b) What is the ratio of the lost exergy to the engine work? **[LO 8.4]**

(E) **8.4** A hot spring produces water at a temperature of 56°C. The water flows into a large lake, with a mean temperature of 14°C, at a rate of 0.1 m^3 of water per min. What is the rate of working of an ideal heat engine which uses all the available energy? **[LO 8.3]**

(E) **8.5** 0.2 kg of air at 300°C is heated reversibly at constant pressure to 2066 K. Find the available and unavailable energies of the heat added. Take T_0 = 30°C and c_p = 1.0047 kJ/kg K. **[LO 8.4]**

(M) **8.6** Eighty kg of water at 100°C are mixed with 50 kg of water at 60°C, while the temperature of the surroundings is 15°C. Determine the decrease in exergy due to mixing. **[LO 8.3]**

(M) **8.7** A lead storage battery used in an automobile is able to deliver 5.2 MJ of electrical energy. This energy is available for starting the car.

Let compressed air be considered for doing an equivalent amount of work in starting the car. The compressed air is to be stored at 7 MPa, 25°C. What is the volume of the tank that would be required to let the compressed air have an exergy of 5.2 MJ? For air, $pv = 0.287\ T$, where T is in K, p in kPa, and v in m^3/kg. **[LO 8.9]**

(E) **8.8** Ice is to be made from water supplied at 15°C by the process shown in Fig. 8.39. The final temperature of the ice is – 10°C, and the final temperature of the water that is used as cooling water in the condenser is 30°C. Determine the minimum work required to produce 1000 kg of ice.

Take c_p for water = 4.187 kJ/kg K, c_p for ice = 2.093 kJ/kg K, and latent heat of fusion of ice = 334 kJ/kg. **[LO 8.3]**

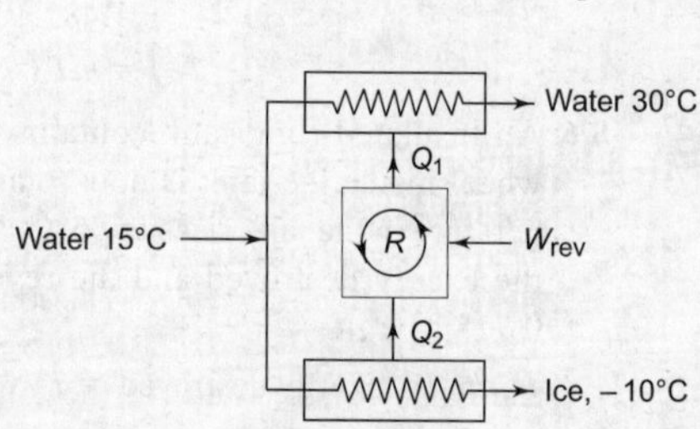

Fig. 8.39

(E) **8.9** A pressure vessel has a volume of 1 m^3 and contains air at 1.4 MPa, 175°C. The air is cooled to 25°C by heat transfer to the surroundings at 25°C. Calculate the exergy in the initial and final states and the irreversibility of this process. Take $p_0 = 100$ kPa. **[LO 8.9]**

(E) **8.10** Air flows through an adiabatic compressor at 2 kg/s. The inlet conditions are 1 bar and 310 K and the exit conditions are 7 bar and 560 K. Compute the net rate of exergy transfer and the irreversibility. Take $T_0 = 298$ K. **[LO 8.9]**

(E) **8.11** An adiabatic turbine receives a gas ($c_p = 1.09$ and $c_v = 0.838$ kJ/kg K) at 7 bar and 1000°C and discharges at 1.5 bar and 665°C. Determine the second law and isentropic efficiencies of the turbine. Take $T_0 = 298$ K. **[LO 8.12]**

(M) **8.12** Air enters an adiabatic compressor at atmospheric conditions of 1 bar, 15°C and leaves at 5.5 bar. The mass flow rate is 0.01 kg/s and the efficiency of the compressor is 75 percent. After leaving the compressor, the air is cooled to 40°C in an aftercooler. Calculate (a) the power required to drive the compressor, and (b) the rate of irreversibility for the overall process (compressor and cooler). **[LO 8.9]**

(D) **8.13** In a rotary compressor, air enters at 1.1 bar, 21°C where it is compressed adia batically to 6.6 bar, 250°C. Calculate the irreversibility and the entropy production for unit mass flow rate. The atmosphere is at 1.03 bar, 20°C. Neglect the K.E. changes. **[LO 8.11]**

(D) **8.14** In a steam boiler, the hot gases from a fire transfer heat to water which vaporises at a constant temperature of 242.6°C (3.5 MPa). The gases are cooled from 1100 to 430°C and have an average specific heat, $c_p = 1.046$ kJ/kg K over this temperature range. The latent heat of vapourisation of steam at 3.5 MPa is 1753.7 kJ/kg. If the steam generation rate is 12.6 kg/s and there is negligible heat loss from the boiler, calculate (a) the rate of heat transfer, (b) the rate of loss of exergy of the gas, (c) the rate of gain of exergy of the steam and (d) the rate of entropy generation. Take $T_0 = 21$°C. **[LO 8.10]**

(M) **8.15** An economiser, a gas-to-water finned tube heat exchanger, receives 67.5 kg/s of gas, $c_p = 1.0046$ kJ/kg K, and 51.1 kg/s of water, $c_p = 4.186$ kJ/kg K. The water rises in temperature from 402 to 469 K, where the gas falls in temperature from 682 K to 470 K. There are no changes of kinetic energy, and $p_0 = 1.03$ bar and $T_0 = 289$ K. Determine (a) rate of change of exergy of the water, (b) the rate of change of exergy of the gas and (c) the rate of entropy generation. **[LO 8.10]**

(M) **8.16** The exhaust gases from a gas turbine are used to heat water in an adiabatic counterflow heat exchanger. The gases are cooled from 260 to 120°C, while water enters at 65°C. The flow rates of the gas and water are 0.38 kg/s and 0.50 kg/s respectively. The constant pressure specific heats for the gas and water are 1.09 and 4.186 kJ/kg K respectively. Calculate the rate of exergy loss due to heat transfer. Take $T_0 = 35$°C. **[LO 8.10]**

(M) **8.17** The exhaust from a gas turbine at 1.12 bar, 800 K flows steadily into a heat exchanger which cools the gas to 700 K without significant pressure drop. The heat transfer from the gas heats an air flow at constant pressure, which enters the heat exchanger at 470 K. The mass flow rate of air is twice that of the gas and the surroundings are at 1.03 bar, 20°C. Determine (a) the decrease in exergy of the exhaust gases and (b) the total entropy production per kg of gas. (c) What arrangement would be necessary to make the heat transfer reversible and how much would this increase the power output of the plant per kg of turbine gas? Take c_p for exhaust gas as 1.08 and for air as 1.05 kJ/kg K. Neglect heat transfer to the surroundings and the changes in kinetic and potential energy. **[LO 8.9]**

(M) **8.18** An air preheater is used to heat up the air used for combustion by cooling the outgoing products of combustion from a furnace. The rate of flow of the products is 10 kg/s, and the products are cooled from 300°C to 200°C, and for the products at this temperature c_p = 1.09 kJ/kg K. The rate of air flow is 9 kg/s, the initial air temperature is 40°C, and for the air c_p = 1.005 kJ/kg K.

(a) What is the initial and final exergy of the products?

(b) What is the irreversibility for this process?

(c) If the heat transfer from the products were to take place reversibly through heat engines, what would be the final temperature of the air? What power would be developed by the heat engines? Take T_0 = 300 K. **[LO 8.9]**

(E) **8.19** A mass of 2 kg of air in a vessel expands from 3 bar, 70°C to 1 bar, 40°C, while receiving 1.2 kJ of heat from a reservoir at 120°C. The environment is at 0.98 bar, 27°C. Calculate the maximum work and the work done on the atmosphere. **[LO 8.9]**

(M) **8.20** Air enters the compressor of a gas turbine at 1 bar, 30°C and leaves the compressor at 4 bar. The compressor has an efficiency of 82 percent. Calculate per kg of air (a) the work of compression, (b) the reversible work of compression, and (c) the irreversibility. For air, use $\dfrac{T_{2s}}{T_1} = \left(\dfrac{p_2}{p_1}\right)^{\gamma - 1/\gamma}$

where T_{2s} is the temperature of air after isentropic compression and γ = 1.4. The compressor efficiency is defined as $(T_{2s} - T_1)/(T_2 - T_1)$, where T_2 is the actual temperature of air after compression. **[LO 8.11]**

(E) **8.21** A mass of 6.98 kg of air is in a vessel at 200 kPa, 27°C. Heat is transferred to the air from a reservoir at 727°C. until the temperature of air rises to 327°C. The environment is at 100 kPa, 17°C. Determine

(a) the initial and final exergy of air,

(b) the maximum useful work associated with the process. **[LO 8.9]**

(M) **8.22** Air enters a compressor in steady flow at 140 kPa, 17°C and 70 m/s and leaves it at 350 kPa, 127°C and 110 m/s. The environment is at 100 kPa, 7°C. Calculate per kg of air (a) the actual amount of work required, (b) the minimum work required, and (c) the irreversibility of the process. **[LO 8.9]**

(M) **8.23** Air expands in a turbine adiabatically from 500 kPa, 400 K and 150 m/s to 100 kPa, 300 K and 70 m/s. The environment is at 100 kPa, 17°C. Calculate per kg of air (a) the maximum work output, (b) the actual work output and (c) the irreversibility. **[LO 8.9]**

(E) **8.24** Calculate the specific exergy of air for a state at 2 bar, 393.15 K when the surroundings are at 1 bar, 293.15 K. Take c_p = 1 and R = 0.287 kJ/kg K. **[LO 8.3]**

(E) **8.25** Calculate the specific exergy of CO_2 (c_p = 0.8659 and R = 0.1889 kJ/kg K) for a state at 0.7 bar, 268.15 K and for the environment at 1.0 bar and 293.15 K. **[LO 8.3]**

(E) **8.26** A pipe carries a stream of brine with a mass flow rate of 5 kg/s. Because of poor thermal insulation the brine temperature increases from 250 K at the pipe inlet to 253 K at the exit. Neglecting pressure losses, calculate the irreversibility rate (or rate of energy degradation) associated with the heat leakage. Take T_0 = 293 K and c_p = 2.85 kJ/kg K. **[LO 8.10]**

(E) **8.27** In an adiabatic throttling process, energy per unit mass of enthalpy remains the same. However, there is a loss of exergy. An ideal gas flowing at the rate $\dot{m}$ is throttled from pressure p_1 to pressure p_2 when the environment is at tem perature T_0. What is the rate of exergy loss due to throttling? **[LO 8.10]**

(M) **8.28** Air at 5 bar and 20°C flows into an evacuated tank until the pressure in the tank is 5 bar. Assume that the process is adiabatic and the temperature of the surroundings is 20°C. (a) What is the final temperature of the air? (b) What is the reversible work produced between the initial and final states of the air? (c) What is the net entropy change of the air entering the tank? (d) Calculate the irreversibility of the process. **[LO 8.10]**

(M) **8.29** A Carnot cycle engine receives and rejects heat with a 20°C temperature differential between itself and the thermal energy reservoirs. The expansion and compression processes have a pressure ratio of 50. For 1 kg of air as the working substance, cycle temperature limits of 1000 and 300 K and T_0 = 280 K, determine the second law efficiency. **[LO 8.12]**

(M) **8.30** Energy is received by a solar collector at the rate of 300 kW from a source temperature of 2400 K. If 60 kW of this energy is lost to the surroundings at steady state and if the user temperature remains constant at 600 K, what are the first law and the second law efficiencies? Take T_0 = 300 K. **[LO 8.12]**

(M) **8.31** For flow of an ideal gas through an insulated pipeline, the pressure drops from 100 bar to 95 bar. If the gas flows at the rate of 1.5 kg/s and has c_p = 1.005 and c_v = 0.718 kJ/kg-K and if T_0 = 300 K, find the rate of entropy generation, and rate of loss of exergy. **[LO 8.10]**

(E) **8.32** The cylinder of an internal combustion engine contains gases at 2500°C, 58 bar. Expansion takes place through a volume ratio of 9 according to $pv^{1.38}$ = const. The surroundings are at 20°C, 1.1 bar. Determine the loss of exergy, the work transfer and the heat transfer per unit mass. Treat the gases as ideal having R = 0.26 kJ/kg-K and c_v = 0.82 kJ/kg-K. **[LO 8.10]**

(M) **8.33** In a counterflow heat exchanger, oil (c_p = 2.1 kJ/kg-K) is cooled from 440 to 320 K, while water (c_p = 4.2 kJ/g K) is heated from 290 K to temperature T. The respective mass flow rates of oil and water are 800 and 3200 kg/h. Neglecting pressure drop, K.E. and P.E. effects and heat loss, determine (a) the temperature T, (b) the rate of exergy destruction and (c) the second law efficiency. Take T_0 = 17°C and p_0 = 1 atm. **[LO 8.12]**

(E) **8.34** Oxygen enters a nozzle operating at steady state at 3.8 MPa, 387°C and 10 m/s. At the nozzle exit, the conditions are 150 kPa, 37°C and 750 m/s. Determine (a) the heat transfer per kg and (b) the irreversibility. Assume oxygen as an ideal gas, and take T_0 = 20°C, p_0 = 1 atm. **[LO 8.10]**

(M) **8.35** Argon gas expands adiabatically in a turbine from 2 MPa, 1000°C to 350 kPa. The mass flow rate is 0.5 kg/s and the turbine develops power at the rate of 120 kW. Determine (a) the temperature of argon at the turbine exit, (b) the irreversibility rate and (c) the second law efficiency. Neglect KE and PE effects and take T_0 = 20°C, p_0 = 1 atm. **[LO 8.9]**

(M) **8.36** In the boiler of a power plant, there are tubes through which water flows as it is brought from 0.8 MPa, 150°C (h = 632.6 kJ/kg, s = 1.8418 kJ/kg K) to 0.8 MPa, 250°C (h = 2950 kJ/kg, s = 7.0384 kJ/kg K). Combustion gases passing over the tubes cool from 1067 to 547°C. These gases may be considered as air (ideal gas) having

c_p = 1.005 kJ/kg K. Assuming steady state and neglecting any heat loss, and K.E. and P.E. effects, determine (a) the mass flow rate of combustion gases per kg of steam, (b) the loss of exergy per kg steam and (c) the second law efficiency. Take T_0 = 25°C, p_0 = 1 atm. **[LO 8.10]**

(M) **8.37** Air enters a hair dryer at 22°C, 1 bar with a velocity of 3.7 m/s and exits at 83°C, 1 bar with a velocity of 9.1 m/s through an area of 18.7 cm^2. Neglecting any heat loss and P.E. effect and taking T_0 = 22°C, (a) evaluate the power required in kW, and (b) derive and evaluate the second law efficiency. **[LO 8.12]**

(M) **8.38** An isolated system consists of two solid blocks. One block has a mass of 5 kg and is initially at 300°C. The other block has a mass of 10 kg and is initially at – 50°C. The blocks are allowed to come into thermal equilibrium. Assuming the blocks are incompressible with constant specific heats of 1 and 0.4 kJ/kg K, respectively, determine (a) the final temperature and (b) the irreversibility. Take T_0 = 300 K. **[LO 8.10]**

(M) **8.39** Air flows into a heat engine at ambient conditions 100 kPa, 300 K. Energy is supplied as 1200 kJ per kg air from a 1500 K source and in some part of the process, a heat loss of 300 kJ/kg air happens at 750 K. The air leaves the engine at 100 kPa, 800 K. Find the first and the second law efficiencies. **[LO 8.12]**

(M) **8.40** Consider two rigid containers each of volume 1 m^3 containing air at 100 kPa, 400 K. An internally reversible Carnot heat pump is then thermally connected between them so that it heats one up and cools the other down. In order to transfer heat at a reasonable rate, the temperature difference between the working fluid inside the heat pump and the air in the containers is set to 20°C. The process stops when the air in the coldest tank reaches 300 K. Find the final temperature of the air that is heated up, the work input to the heat pump, and the overall second law efficiency. **[LO 8.12]**

(M) **8.41** Argon enters an insulated turbine operating at steady state at 1000°C and 2 MPa and exhausts at 350 kPa. The mass flow rate is 0.5 kg/s and the turbine develops power at the rate of 120 kW. Determine (a) the temperature of the argon at the turbine exit, (b) the irreversibility of the turbine, (c) the second law efficiency. Neglect K.E. and P.E. effects. Take T_0 = 20°C, p_0 = 1 bar. **[LO 8.10, 8.12]**

(E) **8.42** A liquid (c_p = 6 kJ/kg K) is heated at an approximately constant pressure from 298 K to 90°C by passing it through tubes immersed in a furnace. The mass flow rate is 0.2 kg/s. Determine (a) the heating load in kW and (b) the exergy production rate in kW corresponding to the temperature rise of the fluid. **[LO 8.9]**

(M) **8.43** A flow of hot water at 80°C is used to heat cold water from 20 to 45°C in a heat exchanger. The cold water flows at the rate of 2 kg/s. When operated in parallel mode, the exit temperature of hot water stream cannot be less than 55°C, while in the counterflow mode, it can be as low as 30°C. Assuming the surroundings are at 300 K, compare the second law efficiencies for the two modes of operation. **[LO 8.12]**

(E) **8.44** Water at 90°C is flowing in a pipe. The pressure of the water is 3 bar, the mass flow rate is 10 kg/s, the velocity is 0.5 m/s and the elevation of the pipe is 200 m above the exit plane of the pipeline (ground level). Compute (a) the thermal exergy flux, (b) the pressure exergy flux, (c) the exergy flux from K.E., (d) the exergy flux from P.E. and (e) total exergy flux of the stream. **[LO 8.9]**

ANSWERS TO PROBLEMS

8.1 (a) 11.2 kJ
(b) 5.25 kJ
8.2 443.6 kJ
8.3 (a) 425.58 kJ
(b) 0.405
8.4 19.5 kW
8.5 211.9 and 78.1 kJ
8.6 236 kJ
8.7 0.228 m^3
8.8 33.37 MJ
8.9 135 kJ/kg, 114.6 kJ/kg, 222 kJ
8.10 481.1 kW and 21.2 kW
8.11 0.956, 0.879
8.12 (a) 2.42 kW
(b) 1 kW
8.13 19 kJ/kg, 0.064 kJ/kg K
8.14 (a) 22096 kW
(b) 15605.4 kW
(c) 9501.0 kW
(d) 20.76 kW/K
8.15 (a) 4802.2 kW
(b) 7079.8 kW
(c) 7.73 kW/K
8.16 12.5 kW
8.17 (a) 66 kJ/kg
(b) 0.0731 kJ/kg K
(c) 38.7 kJ/kg
8.18 (a) 85.97, 39.68 kJ/kg
(b) 256.5 kW
(c) 394.41 K, 353.65 kW
8.19 177 kJ, 112.5 kJ
8.20 (a) 180.5 kJ/kg
(b) 159.5 kJ/kg
(c) 21 kJ/kg
8.21 (a) 103.5, 621.9 kJ
(b) 582 kJ
8.22 (a) 114.4 kJ
(b) 97.3 kJ
(c) 17.1 kJ
8.23 (a) 159 kJ
(b) 109 kJ
(c) 50 kJ
8.24 72.31 kJ/kg
8.25 – 18.77 kJ/kg
8.26 7.05 kW
8.27 $\dot{I} = \dot{m} R T_0 \ln p_1/p_2$
8.28 (a) 410.2 K
(b) 98.9 kJ/kg
(c) 0.3376 kJ/kg K
(d) 98.9 kJ/kg
8.29 0.965
8.30 0.80, 0.457
8.31 0.0215 kW/K, 6.46 kW
8.32 1144 kJ/kg, 1074 kJ/kg, – 213 kJ/kg
8.33 (a) 305 K
(b) 41.4 MJ/h
(c) 10.9 percent
8.34 (a) – 37.06 kJ/kg
(b) 81.72 kJ/kg
8.35 (a) 538.1°C
(b) 18.78 kW
(c) 86.5 percent
8.36 (a) $m_g/m_w = 4.434$
(b) 802.29 kJ/kg steam
(c) 48.9 percent
8.37 (a) – 1.02 kW
(b) 9 percent
8.38 (a) 417.4 K
(b) 277 kJ
8.39 0.315, 0.672
8.40 550 K, 31.2 kJ, 0.816
8.41 (a) 812 K
(b) 18.87 kW
(c) 86.4 percent
8.42 (a) 78 kW
(b) 7.44 kW
8.44 (a) 260 kW
(b) 2.07 kW
(c) 1.25×10^3 kW
(d) 19.6 kW
(e) 282 kW

9

Properties of Pure Substances

Learning Objectives

LO 9.1 State pure substance

LO 9.2 Discuss *p-v* diagram of pure substance

LO 9.3 Discuss *p-T* diagram of pure substance

LO 9.4 Discuss *p-v-T* surface and its projections on *p-T* and *p-v* plane

LO 9.5 Discuss *T-s* diagram

LO 9.6 Construct Mollier diagram (*h-s*) and demonstrate its uses

LO 9.7 Explain dryness fraction

LO 9.8 Explain steam table

LO 9.9 Demonstrate property variations in *T-s* and *h-s* charts

LO 9.10 Discuss measurement of quality or dryness fraction of steam

What is a Pure Substance?

A pure substance is a substance of constant chemical composition throughout its mass. It is a one-component system. It may exist in one or more phases. Let us take water as the representative of a pure substance. We will study the behaviour of water in all the three phases in thermodynamic plots on $p–v$, $p–T$, $T–s$ and $h–s$ coordinates.

LO 9.1
State pure substance

9.1 p-v DIAGRAM FOR A PURE SUBSTANCE

LO 9.2 Discuss p-v diagram of pure substance

Assume a unit mass of ice (solid water) at –10°C and 1 atm contained in a cylinder and piston machine (Fig. 9.1). Let the ice be heated slowly so that its temperature is always uniform. The changes that occur in the mass of water would be traced as the temperature is increased while the pressure is held constant. Let the state changes of water be plotted on p–v coordinates. The distinct regimes of heating, as shown in Fig. 9.2, are:

1–2 The temperature of ice increases from –10 to 0°C. The volume of ice would increase, as would be the case for any solid upon heating. At state 2, i.e. 0°C, the ice would start melting.

2–3 Ice melts into water at a constant temperature of 0°C. At state 3, the melting process ends. *There is a decrease in volume which is a peculiarity of water.*

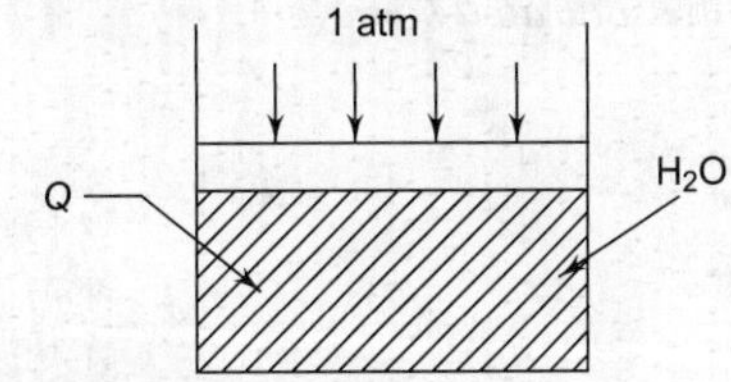

Fig. 9.1 Heating of H_2O at a constant pressure of 1 Atm

3–4 The temperature of water increases, upon heating, from 0 to 100°C. The volume of water increases because of thermal expansion.

4–5 The water starts boiling at state 4 and boiling ends at state 5. This phase change from liquid to vapour occurs at a constant temperature of 100°C (the pressure being constant at 1 atm). There is a large increase in volume.

5–6 The vapour is heated to, say, 250°C (state 6). The volume of vapour increases from v_5 to v_6.

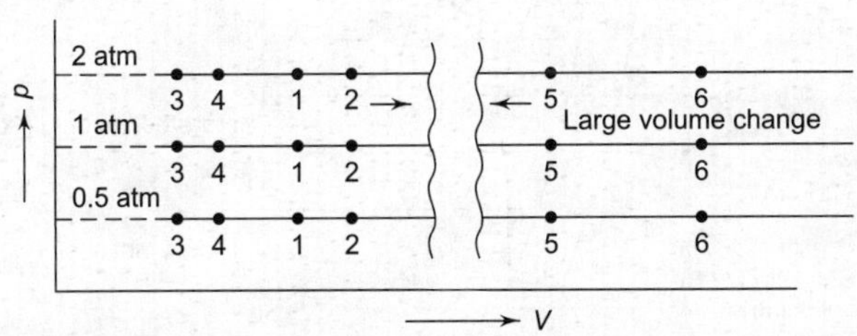

Fig. 9.2 Changes in the volume of water during heating at constant pressure

Water existed in the solid phase between 1 and 2, in the liquid phase between 3 and 4 and in the gas phase beyond 5. Between 2 and 3, the solid changed into the liquid phase by absorbing the latent heat of fusion and between 4 and 5, the liquid changed into the vapour phase by absorbing the latent heat of vaporisation, both at constant temperature and pressure.

The states 2, 3, 4 and 5 are known as *saturation states*. A saturation state is a state from which a change of phase may occur without a change of pressure or temperature. State 2 is a *saturated solid state* because a solid can change into liquid at constant pressure and temperature from state 2. States 3 and 4 are both saturated liquid states. In state 3, the liquid is saturated with respect to solidification, whereas in state 4, the liquid is saturated with respect to vaporisation. State 5 is a *saturated vapour state*, because from state 5, the vapour can condense into liquid without a change of pressure or temperature.

If the heating of ice at –10°C to steam at 250°C has done at a constant pressure of 2 atm, similar regimes of heating would have been obtained with similar saturation states 2, 3, 4 and 5, as shown in Fig. 9.2. All the state changes of the system can similarly be plotted on the *p-v* coordinates, when it is heated at different constant pressures. All the saturated solid states 2 at various pressures are joined by a line, as shown in Fig. 9.3.

Similarly, all the saturated liquid states 3 with respect to solidification, all the saturated liquid states 4 with respect to vaporisation and all the saturated vapour states 5, are joined together.

Figure 9.4 shows state changes of a pure substance other than water whose volume increases on melting.

The line passing through all the saturated solid states 2 (Figs. 9.3 and 9.4) is called the *saturated solid line*.

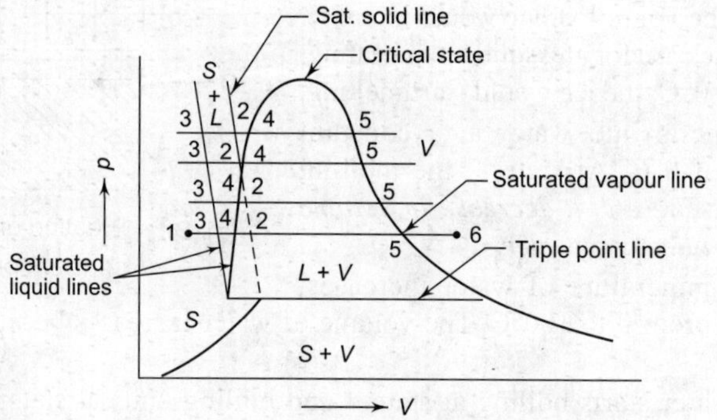

Fig. 9.3 *p-v* diagram of water whose volume decreases on melting

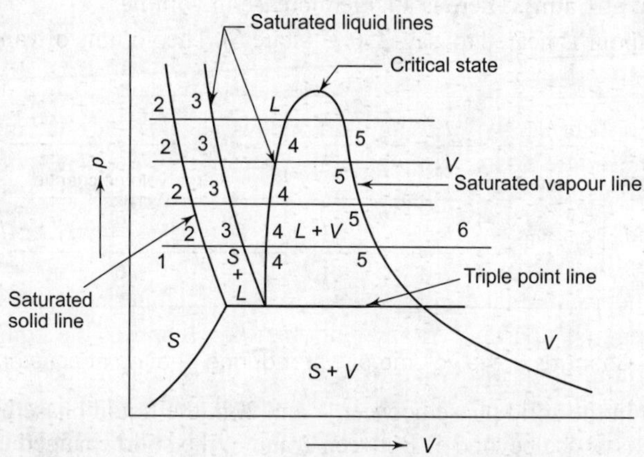

Fig. 9.4 *p-v* diagram of a pure substance other than water, whose volume increases on melting

The lines passing through all the saturated liquid states 3 and 4 with respect to solidification and vaporisation respectively are known as the *saturated liquid lines*, and the line passing through all the saturated vapour states 5, is the *saturated vapour line*. The saturated liquid line with respect to vaporisation and the saturated vapour line incline towards each other and form what is known as the *saturation* or *vapour dome*. The two lines meet at the *critical state*.

To the left of the saturated solid line is the *solid* (*S*) *region* (Fig. 9.4). Between the saturated solid line and saturated liquid line with respect to solidification there exists the *solid-liquid mixture* ($S + L$) region. Between the two saturated liquid lines is the *compressed liquid region*. The *liquid-vapour mixture region* ($L + V$) exists within the vapour dome between the saturated liquid and saturated vapour lines. To the right of the saturated vapour line is the *vapour region*. The *triple point* is a line on the *p-v* diagram, where all the three phases, solid, liquid and gas, exist in equilibrium. At a pressure below the triple point line, the substance cannot exist in the liquid phase, and the substance, when heated, transforms from solid to vapour (known as sublimation) by absorbing the latent heat of sublimation from the surroundings. The region below the triple point line is, therefore, the *solid-vapour* ($S + V$) *mixture region*. Table 9.1 gives the triple point data for a number of substances.

Table 9.1 *Triple-point data*

Substance	*Temperature, K*	*Pressure, mm Hg*
Acetylene, C_2H_2	192.4	962
Ammonia, NH_3	195.42	45.58
Argon, A	83.78	515.7
Carbon dioxide, CO_2	216.55	3885.1
Carbon monoxide, CO	68.14	115.14
Ethane, C_2H_6	89.88	0.006t
Enthylene, C_2H_4	104.00	0.9
Hydrogen, H_2	13.84	52.8
Methane, CH_4	90.67	87.7
Nitrogen, N_2	63.15	94.01
Oxygen, O_2	54.35	1.14
Water, H_2O	273.16	4.587

Liquid is, most often, the working fluid in power cycles, etc., and interest is often confined to the liquid-vapour regions only. So to locate the state points, the solid regions from Figs. 9.3 and 9.4 can be omitted. The *p-v* diagram then becomes as shown in Fig. 9.5. If the vapour at state *A* is compressed slowly and isothermally, the pressure will rise until there is saturated vapour at point *B*. If the compression is continued, condensation takes place, the pressure remaining constant so long as the temperature remains constant. At any point between *B* and *C*, the liquid and vapour are in equilibrium. Since a very large increase in pressure is needed to compress the liquid, line *CD* is almost vertical. *ABCD* is a typical isotherm of a pure substance on a *p-v* diagram. Some isotherms are shown in Fig. 9.5. As the temperature increases, the liquid-vapour transition, as represented by *BC*, decreases, and becomes zero at the critical point. Below the critical point only, there is a liquid-vapour transition zone, where a saturated liquid, on heating, absorbs the latent heat of vaporisation, and becomes saturated vapour at a constant pressure and temperature. Similarly, a saturated vapour, on cooling, releases the latent heat of condensation at constant pressure and temperature to become saturated liquid. Above the critical point, however, a liquid, upon heating, suddenly flashes into vapour, or a vapour, upon cooling, suddenly condenses into liquid. There is no distinct transition zone from liquid to vapour and vice versa. The isotherm passing through the critical point is called the critical isotherm, and the corresponding temperature is known as the critical temperature (t_c). The pressure and volume at the critical point are known as the critical pressure (p_c) and the critical volume (v_c), respectively. For water

$$p_c = 221.2 \text{ bar}$$
$$t_c = 374.15°\text{C}$$
$$v_c = 0.00317 \text{ m}^3/\text{kg}$$

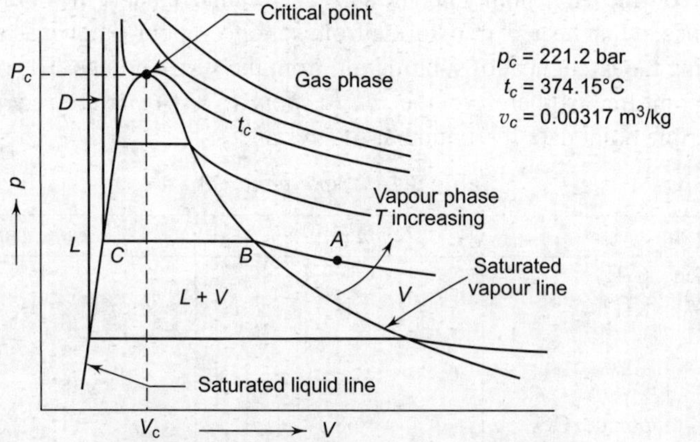

Fig. 9.5 Saturation curve on *p-v* diagram

The critical point data of certain substances are given in Appendix F. Above the critical point, the isotherms are continuous curves that at large volumes and low pressures approach equilateral hyperbolas.

When a liquid or solid is in equilibrium with its vapour at a given temperature, the vapour exerts a pressure that depends only on the temperature (Fig. 9.6). In general, the greater the temperature the higher is the *vapour pressure*. The temperature at which the vapour pressure is equal to 760 mm Hg is called the *normal boiling point (N.B.P)*.

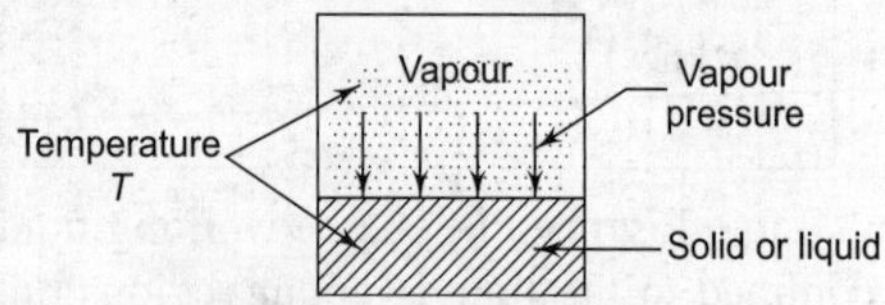

Fig. 9.6 Vapour pressure

Phase change occurs at constant pressure and temperature. A pure liquid at a given pressure will transform into vapour only at a particular temperature, known as *saturation temperature*, which is a function of pressure. Similarly, if the temperature is fixed, the liquid will boil (or condense) only at a particular pressure, called the *saturation pressure*, which is a function of temperature. In Fig. 9.7, if p_1 is the pressure, the corresponding saturation temperature is $(t_{sat})_1$, or if t_2 is the given temperature, the saturation pressure is $(p_{sat})_2$. As the pressure increases, the saturation temperature increases. Saturation states exist up to the critical point. At point *A*, the liquid starts boiling, and at point *B*, the boiling gets completed. At *A*, it is all liquid (saturated) and there is no vapour, while at *B*, it is all vapour (saturated) and there is no liquid. Vapour content progressively increases as the liquid changes its state from *A* towards *B*.

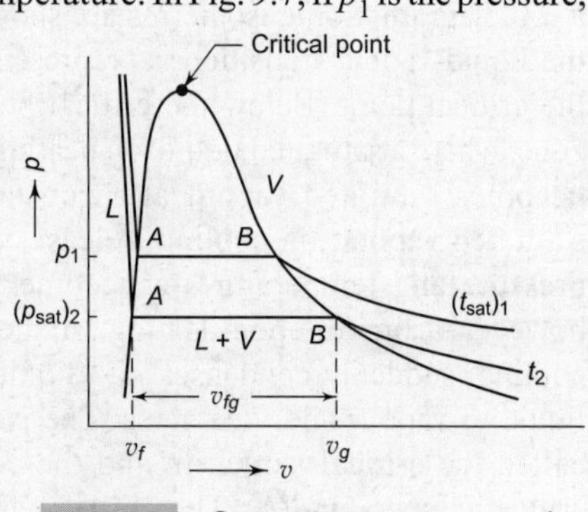

Fig. 9.7 Saturation pressure and temperature

If v_f is the specific volume of the saturated liquid at a given pressure, and v_g the specific volume of the saturated vapour, then $(v_g - v_f)$ or v_{fg} is the change in specific volume during phase transition (boiling or condensation) at the pressure. As pressure increases, v_{fg} decreases, and at the critical point v_{fg} becomes zero.

9.2 *p-T* DIAGRAM FOR A PURE SUBSTANCE

> **LO 9.3**
> Discuss *p-T* diagram of pure substance

The state changes of a pure substance, upon slow heating at different constant pressures, are shown on the *p-v* plane in Figs. 9.2, 9.3 and 9.4. If these state changes are plotted on *p-T* coordinates, the diagram, as shown in Fig. 9.8, will be obtained. If the heating of ice at – 10°C to steam at 250°C at the constant pressure of 1 atm is considered, 1–2 is the solid (ice) heating, 2–3 is the melting of ice at 0°C, 3–4 is the liquid heating, 4–5 is the vapourisation of water at 100°C and 5–6 is the heating in the vapour phase. The process will be reversed from state 6 to state 1 upon cooling. The curve passing through the 2, 3 points is called the *fusion curve*, and the curve passing through the 4, 5 points (which indicate the vapourisation or condensation at different temperatures and pressures) is called the *vapourisation curve*. If the vapour pressure of a solid is measured at different temperatures, and these are plotted, the *sublimation curve* will be obtained. The fusion curve, the vapourisation *curve* and the sublimation curve meet at the triple point.

The slopes of the sublimation and vapourisation curves for all substances are positive. The slope of the fusion curve for most substances is positive, but for water, it is negative. The temperature at which a liquid boils is very sensitive to pressure, as indicated by the vaporisation curve which gives the saturation temperatures at different pressures, but the temperature at which a solid melts is not such a strong function of pressure, as indicated by the small slope of the fusion curve.

The triple point of water is at 4.58 mm Hg and 273.16 K, whereas that of CO_2 is at 3885 mm Hg (about 5 atm) and 216.55 K. So when solid CO_2 ('dry ice') is exposed to 1 atm pressure, it gets transformed into vapour directly, absorbing the latent heat of sublimation from the surroundings which gets cooled or 'refrigerated'. So, dry ice is often used for low temperature cooling.

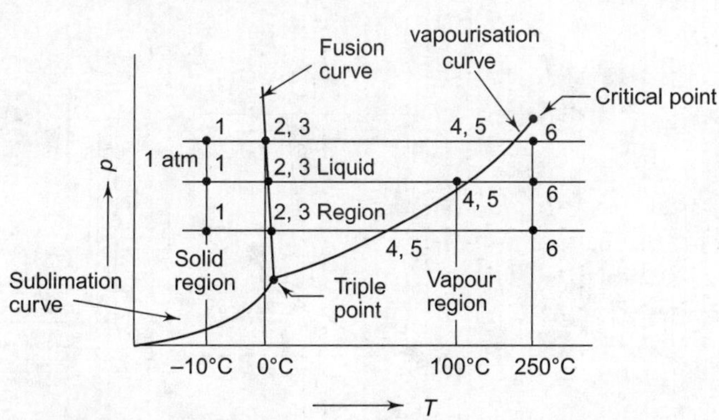

Fig. 9.8 Phase equilibrium diagram on *p-T* coordinates

9.3 *p-v-T* SURFACE

LO 9.4
Discuss *p-v-T* surface and its projections on *p-T* and *p-v* plane

The relationships between pressure, specific volume and temperature can be clearly understood with the aid of a three-dimensional *p-v-T* surface. Figure 9.9 illustrates a substance like water that expands upon freezing and Fig. 9.10 illustrates substances other than water which contract upon freezing. The projections on the *p-T* and *p-v* planes are also shown in these figures.

Any point on the *p-v-T* surface represents an equilibrium state of the substance. The triple point line when projected to the *p-T* plane becomes a point. The critical isotherm has a point of inflection at the critical point.

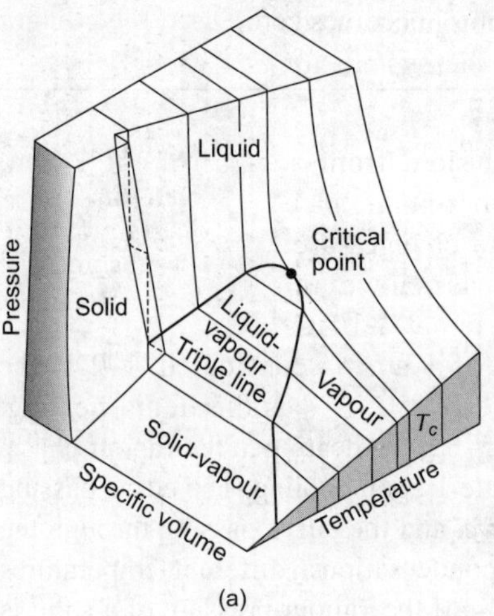

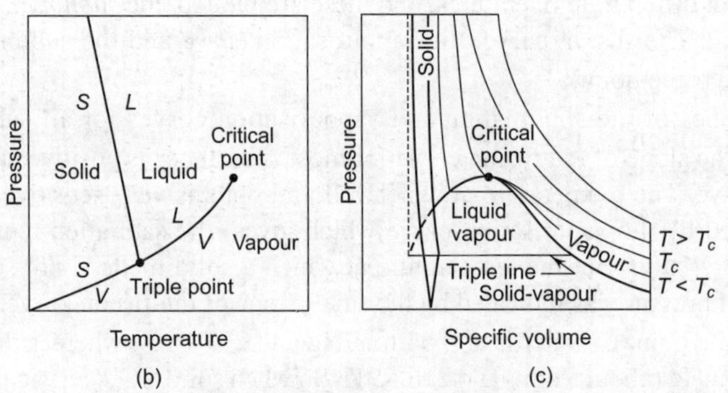

Fig. 9.9 *p-v-T* surface and projections for a substance that expands on freezing: (a) Three-dimensional view, (b) Phase diagram, (c) *p-v* diagram

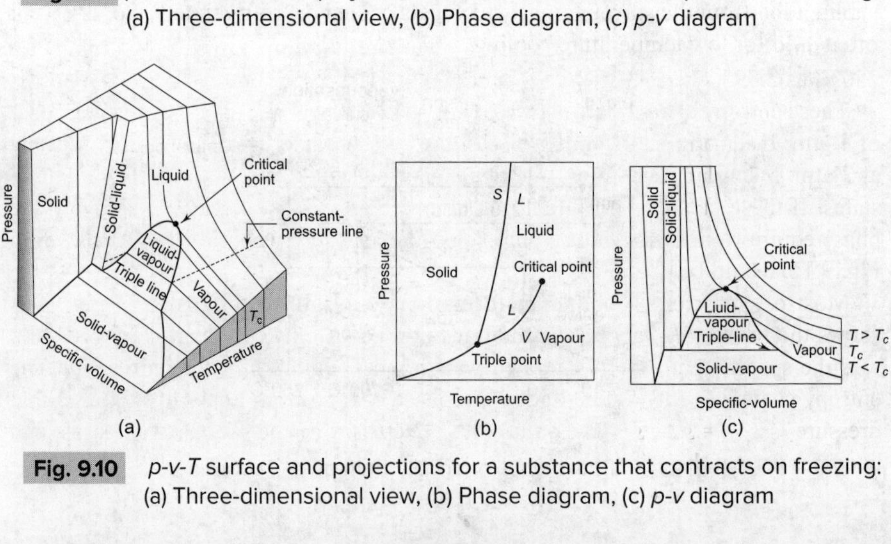

Fig. 9.10 *p-v-T* surface and projections for a substance that contracts on freezing: (a) Three-dimensional view, (b) Phase diagram, (c) *p-v* diagram

9.4 *T-s* DIAGRAM FOR A PURE SUBSTANCE

LO 9.5

Discuss *T-s* diagram

The heating of the system of 1 kg of ice at – 5°C to steam at 250°C is again considered, the pressure being maintained constant at 1 atm. The entropy increase of the system in different regimes of heating is given as follows:

1. The entropy increase of ice as it is heated from – 5 to 0°C at 1 atm. (c_{pice} = 2.093 kJ/kg K).

$$\Delta s_1 = s_2 - s_1 = \int \frac{đQ}{T} = mc_p \ln \frac{273}{268}$$

$$= 1 \times 2.093 \ln \frac{273}{268} = 0.0398 \text{ kJ/kg K}$$

2. The entropy increase of ice as it melts into water at 0°C (latent heat of fusion of ice = 334.96 kJ/kg)

$$\Delta s_2 = s_3 - s_2 = \frac{334.96}{273} = 1.23 \text{ kJ/kg K}$$

3. The entropy increase of water as it is heated from 0 to 100°C (c_{pwater} = 4.187 kJ/kg K)

$$\Delta s_3 = s_4 - s_3 = mc_p \ln \frac{T_3}{T_2} = 1 \times 4.187 \ln \frac{373}{273} = 1.305 \text{ kJ/kg K}$$

4. The entropy increase of water as it is vaporised at 100°C, absorbing the latent heat of vapourisation (2257 kJ/kg)

$$\Delta s_4 = s_5 - s_4 = \frac{2257}{273} = 6.05 \text{ kJ/kg K}$$

5. The entropy increase of vapour as it is heated from 100 to 250°C at 1 atm

$$\Delta s_5 = s_6 - s_5 = \int_{373}^{523} mc_p \frac{dT}{T} = 1 \times 2.093 \ln \frac{523}{373}$$

$$= 0.706 \text{ kJ/kg K}$$

assuming the average specific heat of steam in the temperature range of 100 to 250°C as 2.093 kJ/kg K.

These entropy changes are shown in Fig. 9.11. The curve 1-2-3-4-5-6 is the isobar of 1 atm. If, during the heating process, the pressure had been maintained constant at 2 atm, a similar curve would be obtained. The states 2, 3, 4 and 5 are saturation states. If these states for different pressures are joined, as in Figs. 9.3 and 9.4, the phase equilibrium diagram of a pure substance on the *T-s* coordinates, as shown in Fig. 9.12, would be obtained.

Most often, liquid-vapour transformations only are of interest, and Fig. 9.13 shows the liquid, the vapour and the transition zones only. At a particular pressure, s_f is the specific entropy of saturated water, and s_g is that of saturated vapour. The entropy change of the system during the phase change from liquid to vapour at that pressure is s_{fg} $(= s_g - s_f)$. The value of s_{fg} decreases as the pressure increases, and becomes zero at the critical point.

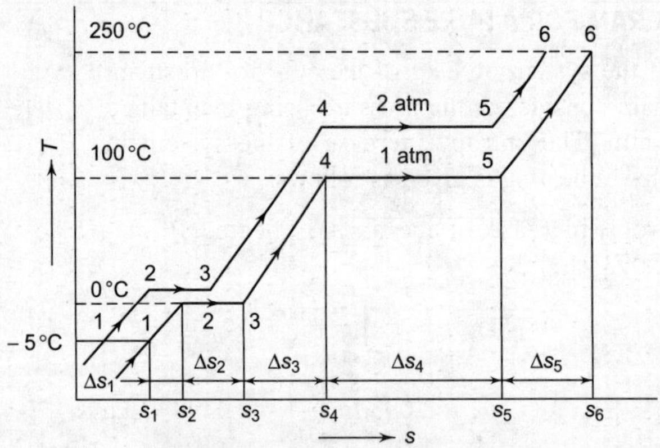

Fig. 9.11 Isobars on *T-s* plot

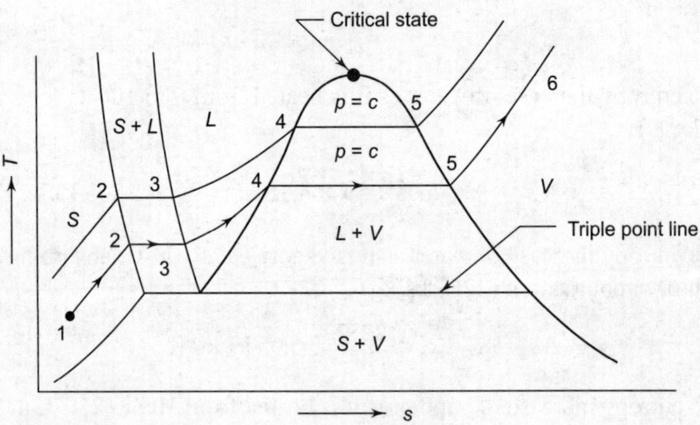

Fig. 9.12 Phase equilibrium diagram on *T-s* coordinates

9.5 *h-s* DIAGRAM OR MOLLIER DIAGRAM FOR A PURE SUBSTANCE

> **LO 9.6** Construct Mollier diagram (*h-s*) and demonstrate its uses

From the first and second laws of thermodynamics, the following property relation was obtained:

$$Tds = dh - vdp$$

or

$$\left(\frac{\partial h}{\partial s}\right)_p = T \tag{9.1}$$

This equation forms the basis of the *h-s* diagram of a pure substance, also called the Mollier diagram. The slope of an isobar on the *h-s* coordinates is equal to the absolute saturation temperature (t_{sat} + 273) at that pressure. If the temperature remains constant, the slope will remain constant. If the temperature increases, the slope of the isobar will increase.

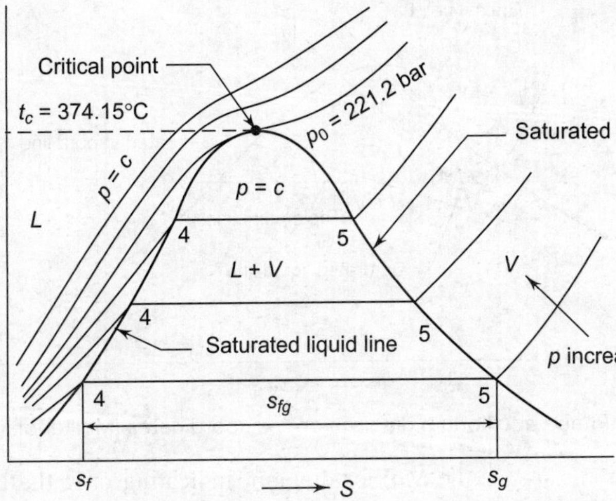

Fig. 9.13 Saturation (or vapour) dome for water

Consider the heating of a system of ice at – 5°C to steam at 250°C, the pressure being maintained constant at 1 atm. The slope of the isobar of 1 atm on the *h-s* coordinates (Fig. 9.14) first increases as the temperature of the ice increases from – 5 to 0°C (1–2). Its slope then remains constant as ice melts into water at the constant temperature of 0°C (2–3). The slope of the isobar again increases as the temperature of water rises from 0 to 100°C (3–4). The slope again remains constant as water vapourises into steam at the constant temperature of 100°C (4–5). Finally, the slope of the isobar continues to increase as the temperature of steam increases to 250°C (5–6) and beyond. Similarly, the isobars of different pressures can be drawn on the *h-s* diagram as shown in Figs. 9.14 and 9.15. States 2, 3, 4 and 5 are saturation states. Figure 9.15 shows the phase equilibrium diagram of a pure substance on the *h-s* coordinates, indicating the saturated solid line, saturated liquid lines and saturated vapour line, the various phases and the transition (mixture) zones.

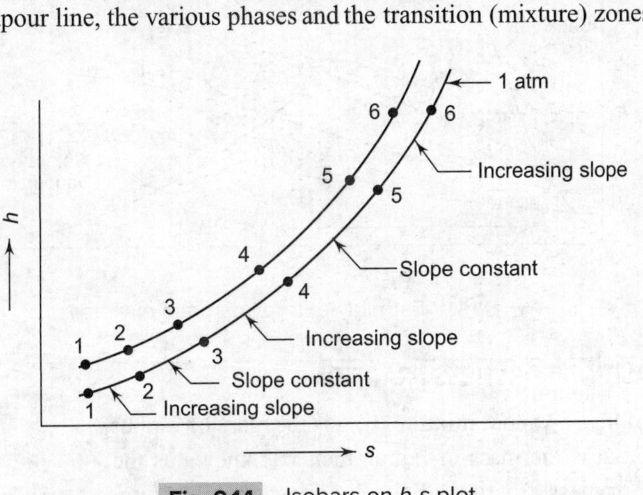

Fig. 9.14 Isobars on *h-s* plot

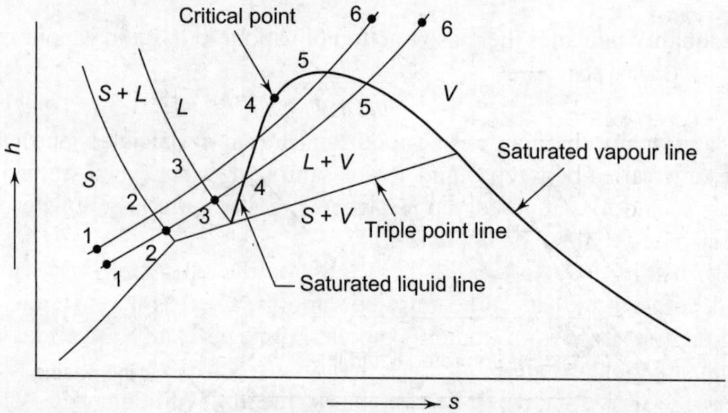

Fig. 9.15 Phase equilibrium diagram on *h-s* coordinates (Mollier diagram)

Figure 9.16 is the *h-s* or the Mollier diagram indicating only the liquid and vapour phases. As the pressure increases, the saturation temperature increases, and so the slope of the isobar also increases. Hence, the *constant pressure lines diverge from one another*, and the critical isobar is a tangent at the critical point, as show in figure. In the vapour region, the states of equal slopes at various pressures are joined by lines, as shown, which are the *constant temperature lines*. Although the slope of an isobar remains continuous beyond the saturated vapour line, the isotherm bends towards the right and its slope decreases asymptotically to zero, because in the ideal gas region it becomes horizontal and the constant enthalpy implies constant temperature. At a particular pressure, h_f is the specific enthalpy of saturated water, h_g is that of saturated vapour and h_{fg} $(= h_g - h_f)$ is the latent heat of vapourisation at that pressure. As the pressure increases, h_{fg} decreases, and at the critical pressure, h_{fg} becomes zero.

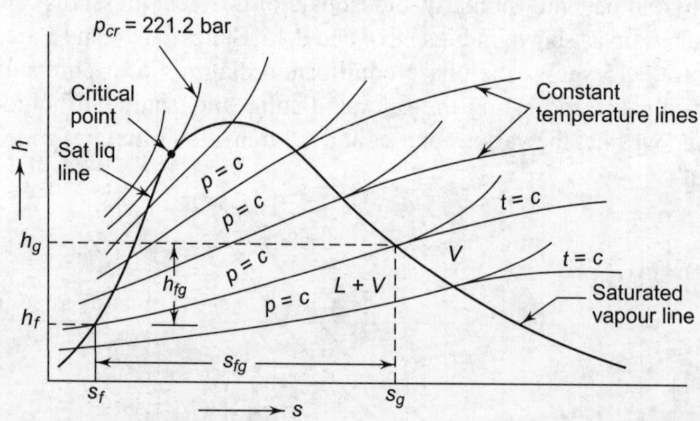

Fig. 9.16 Enthalpy-entropy diagram of water

9.6 QUALITY OR DRYNESS FRACTION

> **LO 9.7**
> Explain dryness fraction

If in 1 kg of liquid-vapour mixture, x kg is the mass of vapour and $(1 - x)$ kg is the mass of liquid, then x is known as the quality or dryness fraction of the liquid vapour mixture.

Therefore, quality indicates the mass fraction of vapour in a liquid vapour mixture or

$$x = \frac{m_v}{m_v + m_l}$$

where m_v and m_l are the masses of vapour and liquid, respectively in the mixture. The value of x varies between 0 and 1. For saturated water, when water just starts boiling, $x = 0$, and for saturated vapour, when vapourisation is complete, $x = 1$, for which the vapour is said to be *dry saturated*.

Points m in Fig. 9.17 (a), (b) and (c) indicate the saturated liquid states with $x = 0$, and points n indicate the saturated vapour states with $x = 1$, the lines mn indicating the transition from liquid to vapour. Points a, b and c at various pressures indicate the situations when the masses of vapour reached 25 per cent, 50 per cent, and 75 per cent of the total mass, i.e. at points a, the mass of liquid is 75 per cent and the mass of vapour is 25 per cent of the total mass, at points b, the mixture consists of 50 per cent liquid and 50 per cent vapour by mass and at points c, the mixture consists of 75 per cent vapour and 25 per cent liquid by mass. The lines passing through points a, b and c are the constant quality lines of 0.25, 0.50 and 0.75 respectively. Constant quality lines start from the critical point.

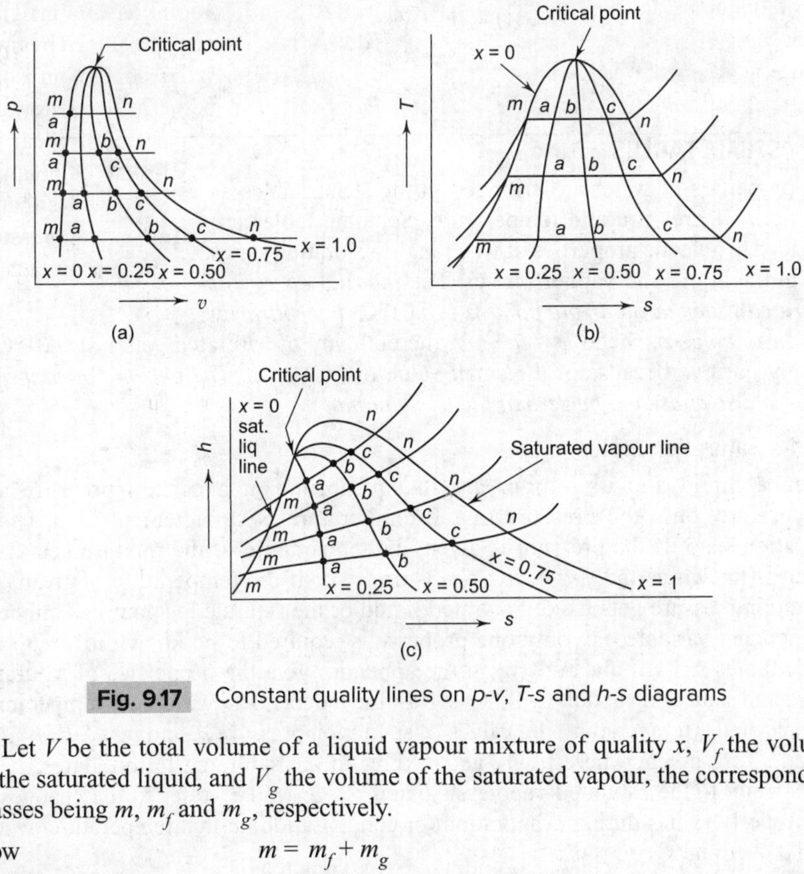

Fig. 9.17 Constant quality lines on *p-v*, *T-s* and *h-s* diagrams

Let V be the total volume of a liquid vapour mixture of quality x, V_f the volume of the saturated liquid, and V_g the volume of the saturated vapour, the corresponding masses being m, m_f and m_g, respectively.

Now $$m = m_f + m_g$$

and $$V = V_f + V_g$$

$$mv = m_f v_f + m_g v_g$$

$$= (m - m_g) v_f + m_g v_g$$

$$\therefore \qquad v = \left(1 - \frac{m_g}{m}\right) v_f + \frac{m_g}{m} v_g$$

$$v = (1 - x)\, v_f + x v_g \tag{9.2}$$

where $x = \frac{m_g}{m}$, v_f = specific volume of saturated liquid, v_g = specific volume of saturated vapour and v = specific volume of the mixture of quality x. Similarly,

$$h = (1 - x)\, h_f + x h_g \tag{9.3}$$

$$s = (1 - x)\, s_f + x s_g \tag{9.4}$$

$$u = (1 - x)\, u_f + x u_g \tag{9.5}$$

where s, h and u refer to the mixture of quality x, the suffix f and suffix g indicate the conditions of saturated liquid and saturated vapour, respectively. From Eq. (9.2)

$$\begin{aligned} v &= (1 - x)\, v_f + x v_g \\ &= v_f + x(v_g - v_f) \\ &= v_f + x\, v_{fg} \end{aligned}$$

or

$$v = v_f + x v_{fg} \tag{9.6}$$

Similarly,

$$h = h_f + x h_{fg} \tag{9.7}$$

$$s = s_f + x s_{fg} \tag{9.8}$$

$$u = u_f + x u_{fg} \tag{9.9}$$

9.7 STEAM TABLES

> **LO 9.8**
> Explain steam table

The properties of water are arranged in the steam tables as functions of pressure and temperature. Separate tables are provided to give the properties of water in the saturation states and in the liquid and vapour phases. The *internal energy of saturated water at the triple point* (t = 0.01°C) *is arbitrarily chosen to be zero*. Since $h = u + pv$, the enthalpy of saturated water at 0.01°C is slightly positive because of the small value of (pv) term. *The entropy of saturated water is also chosen to be zero at the triple point.*

9.7.1 Saturation States

When a liquid and its vapour are in equilibrium at a certain pressure and temperature, only the pressure or the temperature is sufficient to identify the saturation state. If the pressure is given, the temperature of the mixture gets fixed which is known as the *saturation temperature*, or if the temperature is given, the saturation pressure gets fixed. Saturated liquid or the saturated vapour has only one independent variable, i.e. only one property is required to be known to fix up the state. Tables A.1 (a) and A.1 (b) in the appendix give the properties of saturated liquid and saturated vapour. In Table A.1(a), the independent variable is temperature. At a particular temperature, the values of saturation pressure p, and v_f, v_g, h_f, h_{fg}, h_g, s_f and s_g are given, where v_f, h_f, and s_f refer to the saturated liquid states; v_g h_g and s_g refer to the saturated vapour state and v_{fg}, h_{fg} and s_{fg} refer to the changes in the property values during evaporation (or condensation) at that temperature, where $v_{fg} = v_g - v_f$ and $s_{fg} = s_g - s_f$.

In Table A.1(b), the independent variable is pressure. At a particular pressure, the values of saturation temperature t, and v_f, v_g, h_f, h_{fg}, h_g, s_f and s_g are given.

Depending upon whether the pressure or the temperature is given, either Table A.1(a) or Table A.1(b) can be conveniently used for computing the properties of saturation states.

If data are required for intermediate temperatures or pressures, *linear interpolation is normally accurate*. The reason for the two tables is to reduce the amount of interpolation required.

9.7.2 Liquid-Vapour Mixtures

Let us consider a mixture of saturated liquid water and water vapour in equilibrium at pressure p and temperature t. The composition of the mixture by mass will be given by its quality x, and its state will be within the vapour dome (Fig. 9.18). The properties of the mixture are as given in Section 9.6, i.e.

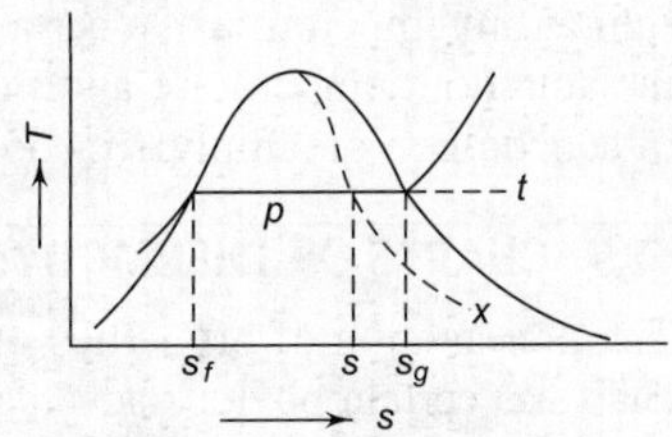

Fig. 9.18 Property in two-phase region

$$v = v_f + xv_{fg}$$
$$u = u_f + xu_{fg}$$
$$h = h_f + xh_{fg}$$
$$s = s_f + xs_{fg}$$

where v_f, v_{fg}, h_f, h_{fg}, s_f and s_{fg} are the saturation properties at the given pressure and temperature.

If p or t and the quality of the mixture are given, the properties of the mixture (v, u, h and s) can be evaluated from the above equations. Sometimes, instead of quality, one of the above properties, say, specific volume v, and pressure or temperature are given. In that case, the quality of the mixture x has to be calculated from the given v and p or t and then x being known, other properties are evaluated.

9.7.3 Superheated Vapour

When the temperature of the vapour is greater than the saturation temperature corresponding to the given pressure, the vapour is said to be *superheated* (state 1 in Fig. 9.19). The difference between the temperature of the superheated vapour and the saturation temperature at that pressure is called the *superheat* or the *degree of superheat*. As shown in Fig. 9.19, the difference $(t_1 - t_{sat})$ is the superheat.

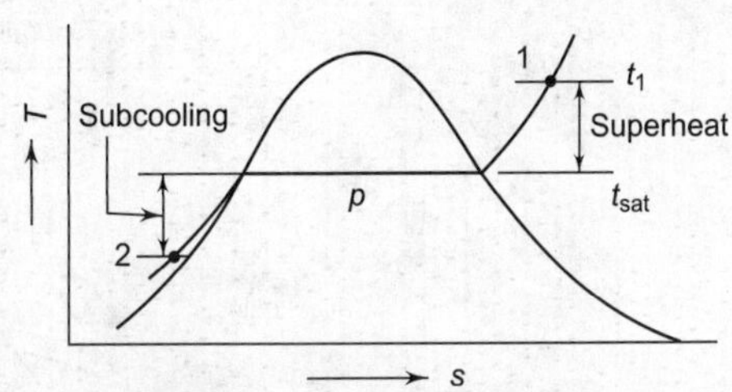

Fig. 9.19 Superheat and subcooling

In a superheated vapour at a given pressure, the temperature may have different values greater than the saturation temperature. Table A.2 in the appendix gives the values of the properties (volume, enthaly and entropy) of superheated vapour for each tabulated pair of values of pressure and temperature, both of which are now independent. Interpolation or extrapolation is to be used for pairs of values of pressure and temperature not given.

9.7.4 Compressed Liquid

When the temperature of a liquid is less than the saturation temperature at the given pressure, the liquid is called *compressed liquid* or *subcooled liquid* (state 2 in Fig. 9.19). The pressure and temperature of compressed liquid may vary independently, and a table of properties like the superheated vapour table could be arranged, to give the properties at any p and t. However, the *properties of liquids vary little with pressure.* Hence the properties are taken from the saturation tables at the temperature of the compressed liquid. When a liquid is cooled below its saturation temperature at a certain pressure it is said to be *subcooled*. The difference in saturation temperature and the actual liquid temperature is known as the degree of subcooling, or simply, subcooling (Fig. 9.19).

9.8 CHARTS OF THERMODYNAMIC PROPERTIES

> **LO 9.9**
> Demonstrate property variations in *T-s* and *h-s* charts

The presentation of properties of substances in the form of a chart has certain obvious advantages. The manner of variation of properties is clearly demonstrated in the chart and there is no problem of interpolation. However, the precision is not as much as in steam tables.

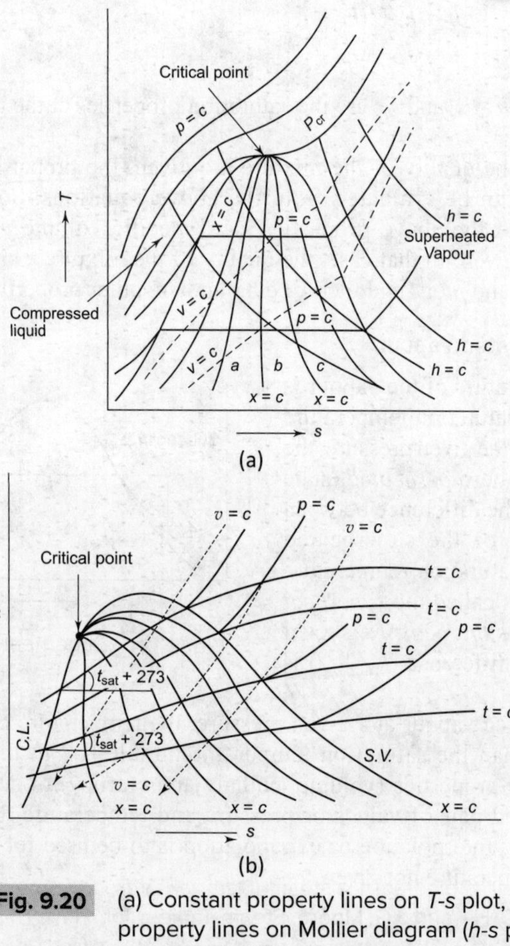

Fig. 9.20 (a) Constant property lines on *T-s* plot, and (b) Constant property lines on Mollier diagram (*h-s* plot)

The temperature-entropy plot and enthalpy-entropy plot (Figs. 9.20 and 9.21) are commonly used. The temperature-entropy plot shows the vapour dome and the lines of constant pressure, constant volume, constant enthalpy, constant quality and constant superheat. However, its scale is small and limited in use. The enthalpy-entropy plot or Mollier chart, has a larger scale to provide data suitable for many computations. It contains the same data as does the *T-s* chart. The Mollier chart for water is also given in Appendix E.I.

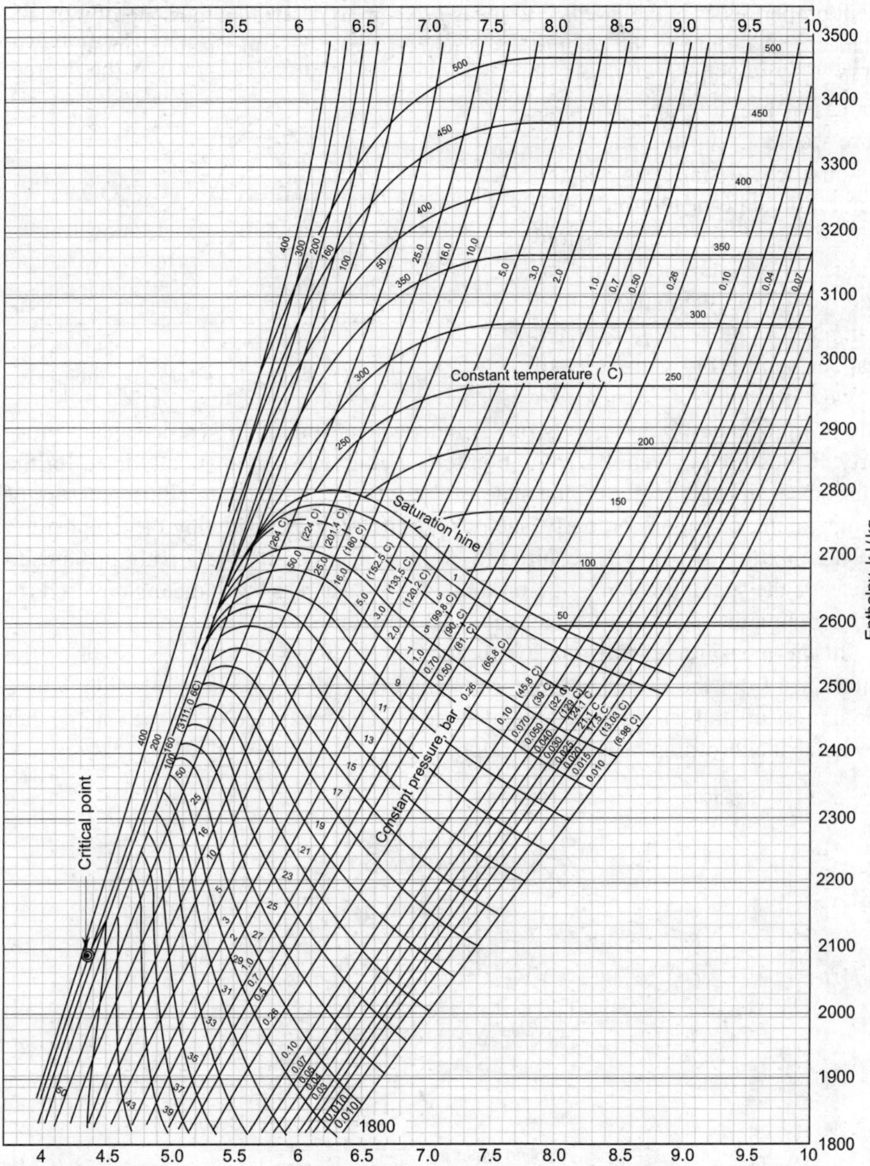

Fig. 9.21 Mollier diagram for steam (Data taken from Keenan, J.H., F.G. Keyes. P.C. Hill and J.G. Moore, Steam Tables, John Wiley, N.Y., 1969)

9.9 STEAM CALORIMETRY FOR MEASUREMENT OF STEAM QUALITY

LO 9.10 Discuss measurement of quality or dryness fraction of steam

The state of a pure substance gets fixed if two independent properties are given. A pure substance is thus said to have two degrees of freedom. Of all thermodynamic properties, it is easiest to measure the pressure and temperature of a substance. Therefore, whenever pressure and temperature are independent properties, it is the practice to measure them to determine the state of the substance. This is done in the compressed liquid region or the superheated vapour region (Fig. 9.22), where the measured values of pressure and temperature would fix up the state. But when the substance is in the saturation state or two-phase region (Fig. 9.22), the measured values of pressure and temperature could apply equally well to saturated liquid point *f*, saturated vapour point *g* or to mixtures of any quality points x_1, x_2 or x_3. Of the two properties, *p* and *t*, only one is independent; the other is a dependent property. If pressure is given, the saturation temperature gets automatically fixed for the substance. In order to fix up the state of the mixture, apart from either pressure or temperature, one more property, such as specific volume, enthalpy or composition of the mixture (quality) is required to be known. Since it is relatively difficult to measure the specific volume of a mixture, devices such as calorimeters are used for determining the quality or the enthalpy of the mixture.

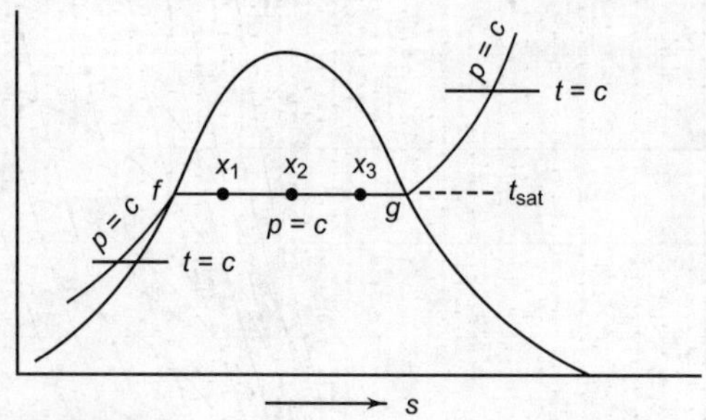

Fig. 9.22 Quality of liquid-vapour mixture

In the measurement of quality, the object is always to bring the state of the substance from the two-phase region to single-phase or superheated region, where both pressure and temperature are independent, and measured to fix the state, either by adiabatic throttling or electric heating.

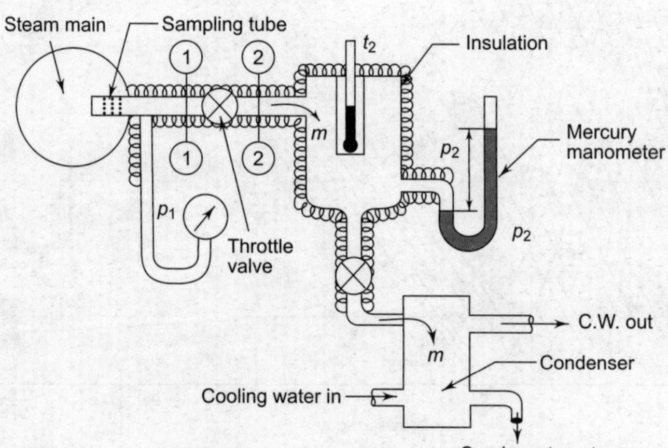

Fig. 9.23 Throttling calorimeter

In the *throttling calorimeter*, a sample of wet steam of mass m and at pressure p_1 is taken from the steam mainly through a perforated sampling tube (Fig. 9.23). Then it is throttled by the partially-opened valve (or orifice) to a pressure p_2, measured by mercury manometer, and temperature t_2, so that after throttling the steam is in the superheated region. The process is shown on the *T-s* and *h-s* diagrams in Fig. 9.24. The steady flow energy equation gives the enthalpy after throttling as equal to enthalpy before throttling. The initial and final equilibrium states 1 and 2 are joined by a dotted line since throttling is irreversible (adiabatic but not isentropic) and the intermediate states are non-equilibrium states not describable by thermodynamic coordinates. The initial state (wet) is given by p_1 and x_1, and the final state by p_2 and t_2 (superheated). Now

since
$$h_1 = h_2$$
$$h_{fp1} + x_1 h_{fgp1} = h_2$$
or
$$x_1 = \frac{h_2 - h_{fp1}}{h_{fgp1}}$$

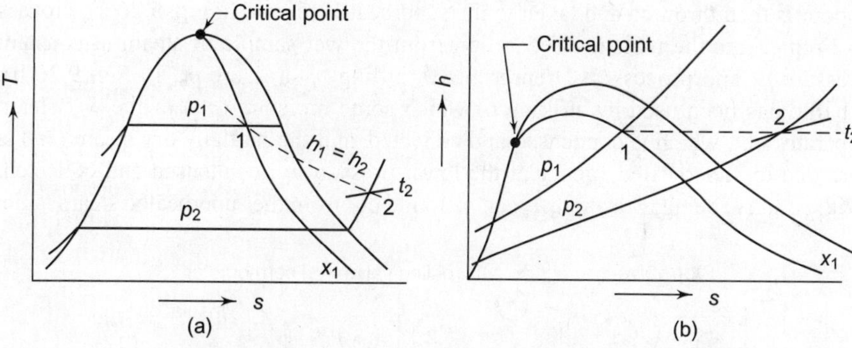

Fig. 9.24 Throttling process on (a) *T-s* and (b) *h-s* plot

With p_2 and t_2 being known, h_2 can be found out from the superheated steam table. The values of h_f and h_{fg} are taken from the saturated steam table corresponding to pressure p_1. Therefore, the quality of the wet steam x_1 can be calculated.

To be sure that steam after throttling is in the single-phase or superheated region, a minimum of 5°C superheat is desired. So, if the pressure after throttling is given and the minimum 5°C superheat is prescribed, then there is the minimum quality of steam (or the maximum moisture content) at the given pressure p_1 that can be measured by the throttling calorimeter. For example, if $p_2 = 1$ atm., then $t_2 = 105$°C and the state 2 after throttling gets fixed as shown in Fig. 9.25. From state 2, the constant enthalpy line intersects the constant pressure p_1 line at 1. Therefore, the quality x_1 is the minimum quality that can be measured simply by throttling. If the quality is, say, x'_1 less than x_1, then after throttling to $p_2 = 1$ atm, the superheat after throttling is less than 5°C. If the quality is x''_1, then throttling to 1 atm does not give any superheat at all.

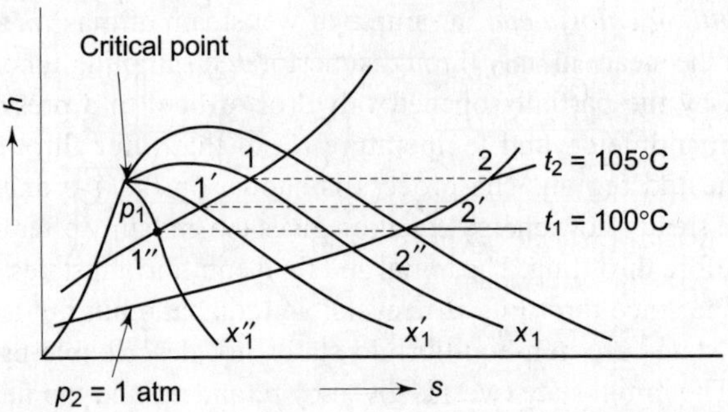

Fig. 9.25 Minimum quality that can be measured only by throttling

When the steam is very wet and the pressure after throttling is not low enough to take the steam to the superheated region, then a *combined separating and throttling calorimeter* is used for the measurement of quality. Steam from the main is first passed through a separator (Fig. 9.26), where some part of the moisture separates out due to the sudden change in direction and falls by gravity, and the partially dry vapour is then throttled and taken to the superheated region. In Fig. 9.26(a), process 1–2 represents the moisture separation from the wet sample of steam at constant pressure p_1 and process 2–3 represents throttling to pressure p_2. In Fig. 9.26(b), a baffle has been used by striking of which some moisture (m_1) of the wet steam separates out, which is condensed and collected, and the partially dry steam (m_2) is throttled to superheated vapour at the lower pressure p_2, condensed and collected. With p_2 and t_2 being measured, h_3 can be found out from the superheated steam table.

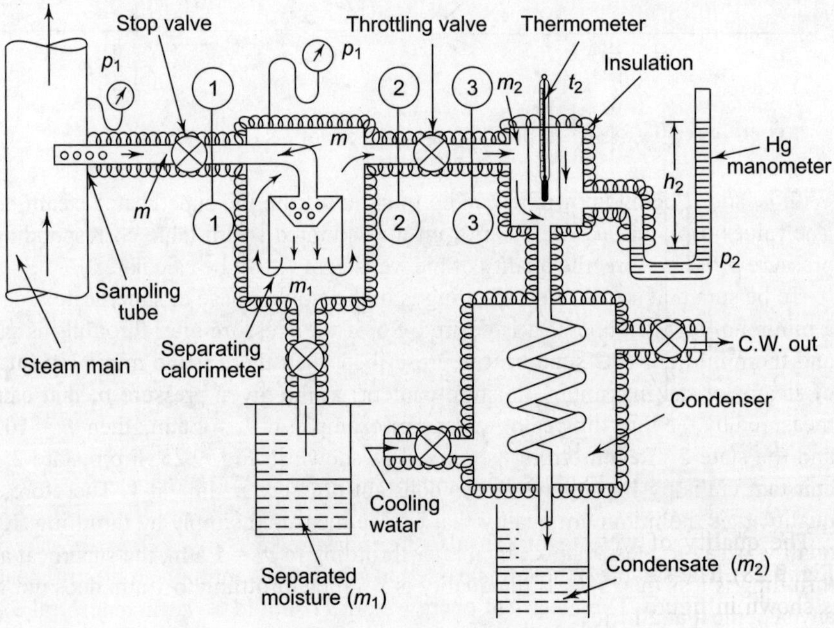

(a)

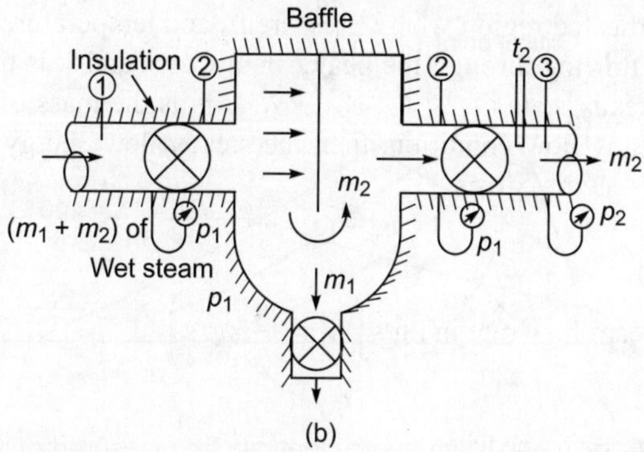

Fig. 9.26 Separating and throttling calorimeter

Now,

$$h_3 = h_2 = h_{fp1} + x_2 h_{fgp1}$$

Therefore, x_2, the quality of steam after partial moisture separation, can be evaluated (Fig. 9.27). If m kg of steam is taken through the sampling tube in t seconds, m_1 kg of it is separated, and m_2 kg is throttled and then condensed to water and collected, then $m = m_1 + m_2$, and at state 2, the mass of dry vapour will be $x_2 m_2$. Therefore, the quality of the sample of steam at state 1, x_1 is given by

$$x_1 = \frac{\text{Mass of dry vapour at state 1}}{\text{Mass of (liquid – vapour) mixture at state 1}}$$

$$= \frac{x_2 m_2}{m_1 + m_2}$$

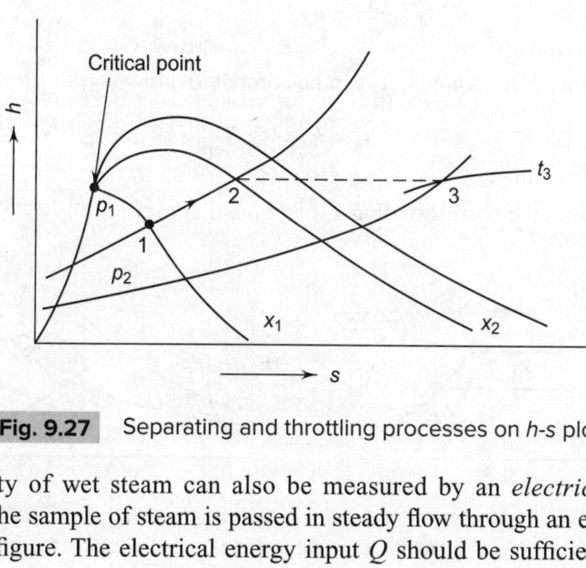

Fig. 9.27 Separating and throttling processes on *h-s* plot

The quality of wet steam can also be measured by an *electric calorimeter* (Fig. 9.28). The sample of steam is passed in steady flow through an electric heater, as shown in figure. The electrical energy input Q should be sufficient to take the steam to the superheated region where pressure p_2 and temperature t_2 are measured. If I is the current flowing through the heater in amperes and V is the voltage across the coil, then at steady state $Q = VI \times 10^{-3}$ kW. If m is the mass of steam taken in t seconds under steady flow condition, then the steady flow energy equation for the heater (as control volume) gives

$$w_1 h_1 + Q = w_1 h_2$$

where w_1 is the steam flow rate in kg/s $\left(w_1 = \frac{m}{t} \text{ kg/s} \right)$

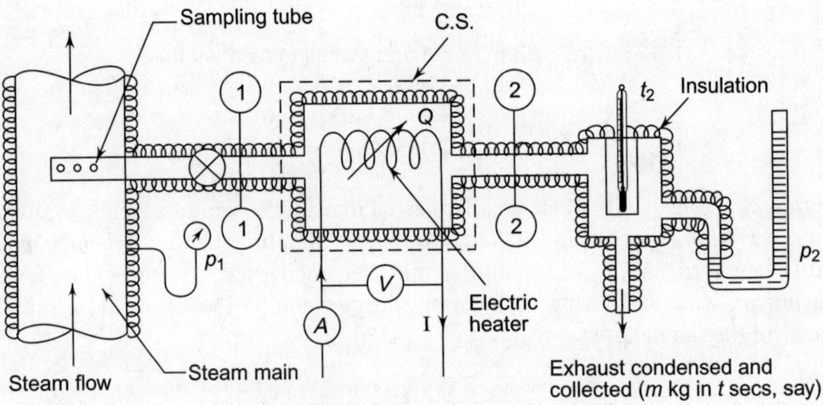

Fig. 9.28 Electrical calorimeter

$$\therefore \qquad h_1 + \frac{Q}{w_1} = h_2$$

With h_2, Q and w_1 being known, h_1 can be computed. Now

$$h_1 = h_{fp1} + x_1 h_{fgp1}$$

Hence x_1 can be evaluated.

Figure 9.29 shows the corresponding T-s and h-s diagrams.

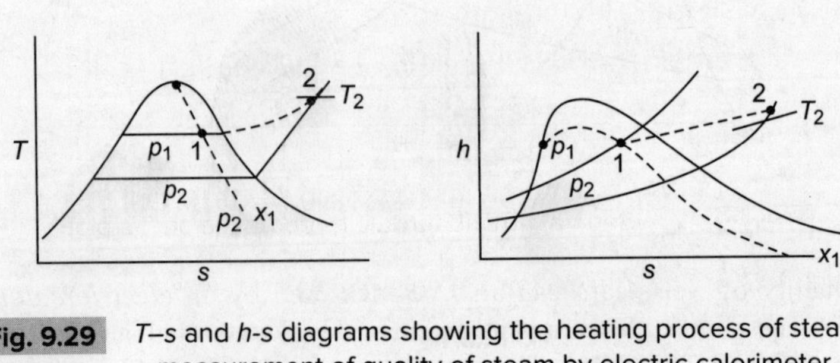

Fig. 9.29 T–s and h-s diagrams showing the heating process of steam in measurement of quality of steam by electric calorimeter

SOLVED EXAMPLES

Example 9.1 *Find the saturation temperature, the changes in specific volume and entropy during evaporation and the latent heat of vapourisation of steam at 1 MPa.* E [LO 9.7, 9.8]

Solution At 1 MPa, from Table A.1.2 in Appendix A

$$t_{sat} = 179.91°C$$
$$v_f = 0.001127 \text{ m}^3/\text{kg}$$
$$v_g = 0.19444 \text{ m}^3/\text{kg}$$
$$\therefore \quad v_{fg} = v_g - v_f = \mathbf{0.1933\ m^3/kg}$$
$$s_f = 2.1387 \text{ kJ/kg K}$$
$$s_g = 6.5865 \text{ kJ/kg K}$$
$$\therefore \quad s_{fg} = s_g - s_f = \mathbf{4.4478\ kJ/kg\ K}$$
$$h_{fg} = h_g - h_f = \mathbf{2015.3\ kJ/kg}$$

Example 9.2 *Saturated steam has an entropy of 6.76 kJ/kg K. What are its pressure, temperature, specific volume and enthalpy?* E [LO 9.8]

Solution In Table A.1.2 in Appendix A, when s_g = 6.76 kJ/kg K

$$p = \mathbf{0.6\ MPa}, \quad t = \mathbf{158.85°C}$$
$$v_g = \mathbf{0.3157\ m^3/kg}, \quad \text{and} \quad h_g = \mathbf{2756.8\ kJ/kg}$$

Example 9.3 *Find the enthalpy and entropy of steam when the pressure is 2 MPa and the specific volume is 0.09 m³/kg.* E [LO 9.7, 9.8]

Solution In Table A.1.2 in Appendix A, where p = 2 MPa, v_f = 0.001177 m³/kg and v_g = 0.09963 m³/kg. Since the given volume lies between v_f and v_g, the substance will be a mixture of liquid and vapour, and the state will be within the vapour dome. Also, in the two-phase region, the composition of the mixture or its quality has to be evaluated first. Now,

$$v = v_f + x\, v_{fg}$$
$$0.09 = 0.001177 + x\,(0.09963 - 0.001177)$$
$$\text{or} \quad x = 0.904 \text{ or } 90.4 \text{ per cent}$$
$$\text{At} \quad 2 \text{ MPa}, h_f = 908.79 \quad \text{and} \quad h_{fg} = 1890.7 \text{ kJ/kg}$$
$$s_f = 2.4474 \quad \text{and} \quad s_{fg} = 3.8935 \text{ kJ/kg K}$$
$$h = h_f + x\, h_{fg}$$
$$= 908.79 + 0.904 \times 1890.7 = \mathbf{2618.79\ kJ/kg}$$
$$s = s_f + x\, s_{fg}$$
$$= 2.4474 + 0.904 \times 3.8935$$
$$= \mathbf{5.9534\ kJ/kg\ K}$$

Example 9.4 *Find the enthalpy, entropy and volume of steam at 1.4 MPa, 380°C.* E [LO 9.8]

Solution At $p = 1.4$ MPa, in Table A.1.2, $t_{sat} = 195.07$°C. Therefore, the state of steam must be in the superheated region. In Table A.2, for properties of superheated steam,

at 1.4 MPa, 350°C $\quad v = 0.2003$ m³/kg

$h = 3149.5$ kJ/kg

$s = 7.1360$ kJ/kg K

and at 1.4 MPa, 400°C $\quad v = 0.2178$ m³/kg

$h = 3257.5$ kJ/kg

$s = 7.3026$ kJ/kg K

∴ By interpolation

at 1.4 MPa, 380°C $\quad v =$ **0.2108 m³/kg**

$h =$ **3214.3 kJ/kg**

$s =$ **7.2360 kJ/kg K**

Example 9.5 *A vessel of volume 0.04 m³ contains a mixture of saturated water and saturated steam at a temperature of 250°C. The mass of the liquid present is 9 kg. Find the pressure, the mass, the specific volume, the enthalpy the entropy and the internal energy.* E [LO 9.7, 9.8]

Solution From Table A.1.1 in Appendix A, at 250°C, $p_{sat} = 3.973$ MPa

$v_f = 0.0012512$ m³/kg, $\quad v_g = 0.05013$ m³/kg

$h_f = 1085.36$ kJ/kg, $\quad h_{fg} = 1716.2$ kJ/kg

$s_f = 2.7927$ kJ/kg K, $\quad s_{fg} = 3.2802$ kJ/kg K

Volume of liquid, $\quad V_f = m_f v_f$

$= 9 \times 0.0012512 = 0.01126$ m³

Volume of vapour, $\quad V_g = 0.04 - 0.01126 = 0.02874$ m³

∴ mass of vapour

$$m_g = \frac{V_g}{v_g} = \frac{0.02874}{0.05013} = 0.575 \text{ kg}$$

∴ total mass of mixture,

$m = m_f + m_g = 9 + 0.575 =$ **9.575 kg**

Quality of mixture,

$$x = \frac{m_g}{m_f + m_g} = \frac{0.575}{9.575} = 0.06$$

∴ $\quad v = v_f + x v_{fs}$

$= 0.0012512 + 0.06\,(0.05013 - 0.0012512)$

$=$ **0.00418 m³/kg**

$h = h_f + x h_{fg}$

$= 1085.36 + 0.06 \times 1716.2 =$ **1188.32 kJ/kg**

$s = s_f + x s_{fg}$

$= 2.7927 + 0.06 \times 3.2802 =$ **2.9895 kJ/kg K**

$$u = h - pv$$
$$= 1188.32 - 3.973 \times 10^3 \times 0.00418 = \mathbf{1171.72\ kJ/kg}$$

Also, at 250°C, $u_f = 1080.39$ and $u_{fg} = 1522.0$ kJ/kg

$\therefore$ $u = u_f + xu_{fg}$

$$= 1080.39 + 0.06 \times 1522 = \mathbf{1171.71\ kJ/kg}$$

Example 9.6 *A rigid tank of 1 m³ contains liquid-vapour mixture of water at 1.5 MPa and dryness fraction of 0.9. The mixture is allowed to cool until the pressure is reduced to 300 kPa. Determine the final mass of liquid, mass of vapour and the heat transfer during the process.* Ⓜ **[LO 9.7, 9.8]**

Solution From Table A.1.2 in the Appendix, we find at 1.5 MPa

$v_f = 0.001154\ \text{m}^3/\text{kg}$, $v_g = 0.13177\ \text{m}^3/\text{kg}$, $h_f = 844.87$ kJ/kg, $h_{fg} = 1947.3$ kJ/kg

The initial specific volume is found to be

$$v_1 = v_f + x(v_g - v_f)$$
$$v_1 = 0.001154 + 0.9(0.13177 - 0.001154) = 0.1187\ \text{m}^3/\text{kg}$$

The initial specific enthalpy is

$$h_1 = h_f + xh_{fg} = 844.87 + 0.9 \times 1947.3 = 2597.44\ \text{kJ/kg}$$

From Table A.1.2 in the Appendix, we find at 300 kPa

$v_f = 0.001073\ \text{m}^3/\text{kg}$, $v_g = 0.6028\ \text{m}^3/\text{kg}$, $h_f = 561.45$ kJ/kg, $h_{fg} = 2163.9$ kJ/kg

Since the vessel is rigid, the specific volume does not change. Hence the specific volume at a pressure of 300 kPa is also 0.1187 m³/kg. Since the value of specific volume is in between v_f and v_g at 300 kPa, the state of water will be saturated liquid-vapour mixture. The dryness fraction can be found as follows:

First, we can find the quality as follows:

$$v_1 = v_2 = v_{f,p=300\ \text{kPa}} + x(v_{g,p=300\ \text{kPa}} - v_{f,p=300\ \text{kPa}})$$

or $0.1187 = 0.001073 + x(0.6028 - 0.001073)$

or $x = 0.1955$

The total mass of water is

$$m = \frac{V}{v} = \frac{1}{0.1187} = 8.425\ \text{kg}$$

The mass of vapour in the mixture is

$$m_g = xm = 0.1955 \times 8.425 = \mathbf{1.647\ kg}$$

The mass of liquid is then

$$m_f = m - m_g = 8.425 - 1.647 = \mathbf{6.778\ kg}$$

The specific enthalpy of the mixture at final state is

$$h_2 = h_f + xh_{fg} = 561.45 + 0.1955 \times 2163.9 = 984.49\ \text{kJ/kg}$$

The total heat transfer is $Q_{1-2} = m(h_2 - h_1) = 1(984.49 - 2597.44) = \mathbf{-1612.95\ kJ}$ (Negative sign indicates that heat is transferred from the tank to the surrounding.)

Example 9.7 *Steam initially at 0.3 MPa, 250°C is cooled at constant volume. (a) At what temperature will the steam become saturated vapour? (b) What is the quality at 80°C? What is the heat transferred per kg of steam in cooling from 250°C to 80°C?* M [LO 9.7, 9.8]

Solution At 0.3 MPa, $t_{sat} = 133.55°C$

Since $t > t_{sat}$, the state would be in the superheated region (Fig. 9.30). From Table A.2 in Appendix A, for properties of superheated steam, at 0.3 MPa, 250°C

$$v_1 = 0.7964 \text{ m}^3/\text{kg}$$
$$h_1 = 2967.6 \text{ kJ/kg}$$
$$\therefore \quad v_1 = v_3 = v_2 = 0.7964 \text{ m}^3/\text{kg}$$

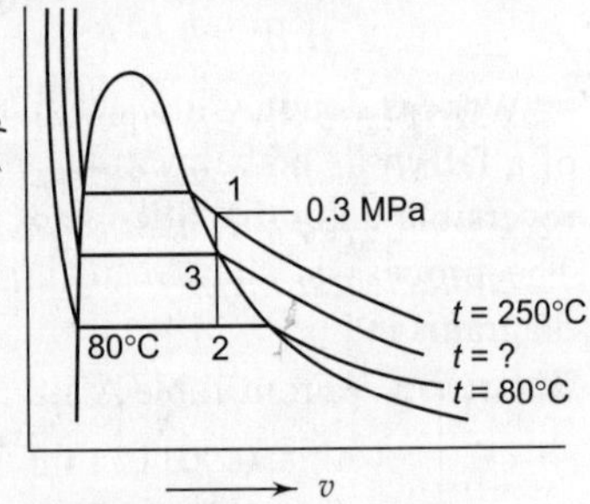

Fig. 9.30 Superheat and subcooling

In Table A.1 (Appendix A)

when $v_g = 0.8919$, $t_{sat} = 120°C$

when $v_g = 0.77076$, $t_{sat} = 125°C$

Therefore, when $v_g = 0.7964$, t_{sat}, by linear interpolation, would be 123.9°. Steam would become saturated vapour at $t = 123.9°C$

At 80°C, $v_f = 0.001029 \text{ m}^3/\text{kg}$, $v_g = 3.407 \text{ m}^3/\text{kg}$,

$h_f = 334.91 \text{ kJ/kg}$, $h_{fg} = 2308.8 \text{ kJ/kg}$, $p_{sat} = 47.39 \text{ kPa}$

$$v_1 = v_2 = 0.7964 \text{ m}^3/\text{kg} = v_{f80°C} + x_2 v_{fg80°C}$$
$$= 0.001029 + x_2 (3.407 - 0.001029)$$
$$\therefore \quad x_2 = \frac{0.79539}{3.40597} = 0.234$$
$$h_2 = 334.91 + 0.234 \times 2308.8 = 875.9 \text{ kJ/kg}$$
$$h_1 = 2967.6 \text{ kJ/kg}$$

From the first law of thermodynamics

$$đQ = du + pdv$$
$$\therefore \quad (đQ)_v = du$$
$$\text{or} \quad Q_{1-2} = u_2 - u_1 = (h_2 - p_2 v_2) - (h_1 - p_1 v_1)$$
$$= (h_2 - h_1) + v(p_1 - p_2)$$
$$= (875.9 - 2967.6) + 0.7964 (300 - 47.39)$$
$$= -2091.7 + 201.5 = \mathbf{-1890.2 \text{ kJ/kg}}$$

Example 9.8 *Steam initially at 1.5 MPa, 300°C expands reversibly and adiabatically in a steam turbine to 40°C. Determine the ideal work output of the turbine per kg of steam.* M [LO 9.7, 9.8]

Solution The steady flow energy equation for the control volume, as shown in Fig. 9.31, gives (other energy terms being neglected)

$$h_1 = h_2 + W$$

$$\therefore \quad W = h_1 - h_2$$

Work is done by the steam at the expense of a fall in its enthalpy value. The process is reversible and adiabatic, so it is isentropic. The process is shown on the *T–s* and *h–s* diagrams in Fig. 9.32.

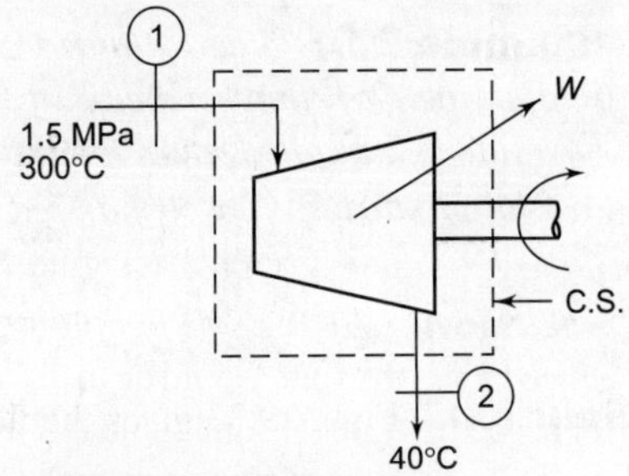

Fig. 9.31 Superheat and subcooling

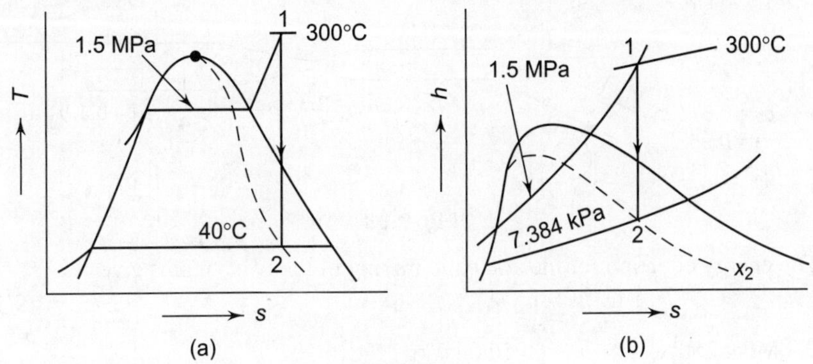

Fig. 9.32

From Table A.1(a) in Appendix A, at 40°C

$$p_{sat} = 7.384 \text{ kPa}, \qquad s_f = 0.5725, \quad s_{fg} = 7.6845 \text{ kJ/kg K}$$

$$h_f = 167.57 \quad \text{and} \quad h_{fg} = 2406.7 \text{ kJ/kg}$$

At p = 1.5 MPa, t = 300°C, from the tabulated properties of superheated steam (Table A.2 in Appendix A)

$$s_1 = 6.9189 \text{ kJ/kg K}$$

$$h_1 = 3037.6 \text{ kJ/kg}$$

Since $\quad s_1 = s_2$

$$6.9189 = s_f + x_2\, s_{fg40°C}$$

$$= 0.5725 + x_2 \times 7.6845$$

$$x_2 = \frac{6.3464}{7.6845} = 0.826 \text{ or } 82.6 \text{ per cent}$$

$$\therefore \quad h_2 = h_{f40°C} + x_2\, h_{fg40°C}$$

$$= 167.57 + 0.826 \times 2406.7 = 2152.57 \text{ kJ/kg}$$

$$\therefore \quad W = h_1 - h_2 = 3037.6 - 2152.57 = \mathbf{885.03 \text{ kJ/kg}}$$

Example 9.9 *Steam at 0.8 MPa, 250°C flowing at the rate of 1 kg/s passes into a pipe carrying wet steam at 0.8 MPa, 0.95 dry. After adiabatic mixing, the flow rate is 2.3 kg/s. Determine the condition of steam after mixing.*

The mixture is now expanded in a frictionless nozzle isentropically to a pressure of 0.4 MPa. Determine the velocity of the steam leaving the nozzle. Neglect the velocity of steam in the pipeline. **M [LO 9.7, 9.8]**

Solution Figure 9.33 gives the flow diagram.

$$w_2 = w_3 - w_1 = 2.3 - 1.0 = 1.3 \text{ kg/s}$$

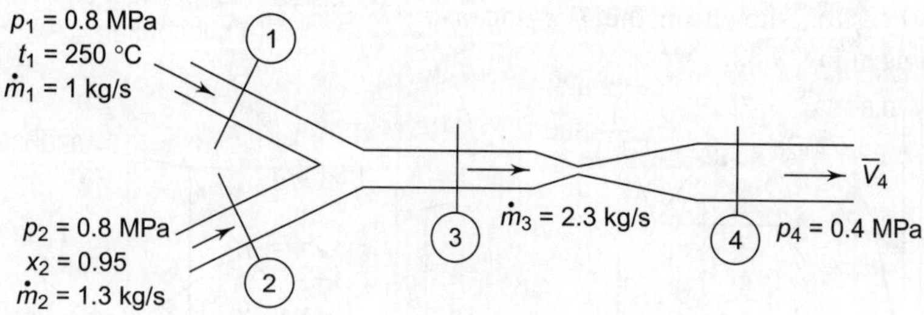

Fig. 9.33

The energy equation for the adiabatic mixing of the two streams gives

$$w_1 h_1 + w_2 h_2 = w_3 h_3 \tag{9.10}$$

At 0.8 MPa, 250°C, $h_1 = 2950.0$ kJ/kg

At 0.8 MPa, 0.95 dry

$$h_2 = h_f + 0.95\, h_{fg}$$
$$= 721.11 + 0.95 \times 2048.0 = 2666.71 \text{ kJ/kg}$$

∴ From Eq. (9.10)

$$1 \times 2950 + 1.3 \times 2666.71 = 2.3 \times h_3$$

∴ $h_3 = 2790$ kJ/kg

Since $(h_g)_{0.8\text{ MPa}} = 2769.1$ kJ/kg

and $h_3 > h_g$, the state must be in the superheated region. From the steam tables, when $p = 0.8$ MPa, $t = 200$°C

$$h = 2839.3 \text{ kJ/kg}$$

When $p = 0.8$ MPa, $t_{sat} = 170.43$°C

$$h_g = 2769.1 \text{ kJ/kg}$$

By linear interpolation

$$t_3 = 179\text{°C}$$

∴ Degree of superheat = 179 – 170.43 = 8.57°C

∴ Condition of steam after mixing = **0.8 MPa, 179°C**

The energy equation for the nozzle gives

$$h_3 = h_4 + \frac{V_4^2}{2}$$

Since V_3 = velocity of steam in the pipeline = 0

Steam expands is entropically in the nozzle to 0.4 MPa. By interpolation,

$$s_3 = 6.7087 \text{ kJ/kg K} = s_4$$

$$\therefore \quad 6.7087 = 1.7766 + x_4 \times 5.1193$$

$$x_4 = 0.964$$

$$\therefore \quad h_4 = 604.74 + 0.964 \times 2133.8 = 2660 \text{ kJ/kg}$$

$$V_4^2 \times 10^{-3} = 2(h_3 - h_4) = 2 \times 130 = 260$$

$$V_4 = \sqrt{26} \times 100 = \mathbf{509.9 \ m/s}$$

The processes are shown on the *h–s* and *T–s* diagrams in Fig. 9.34.

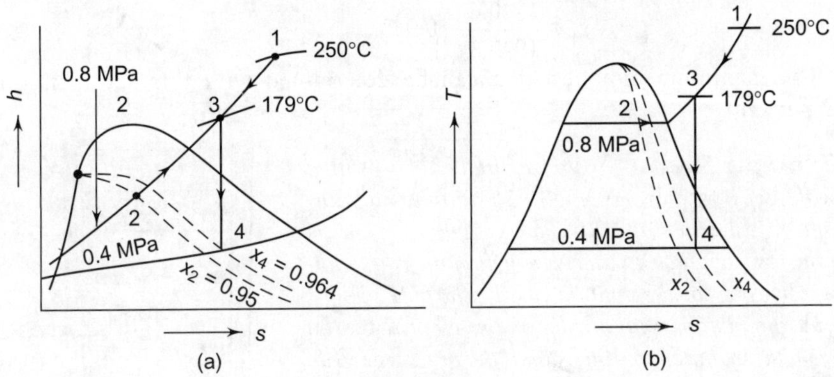

Fig. 9.34

Example 9.10 *Steam flows in a pipeline at 1.5 MPa. After expanding to 0.1 MPa in a throttling calorimeter, the temperature is found to be 120°C. Find the quality of steam in the pipeline. What is the maximum moisture at 1.5 MPa that can be determined with this set-up if at least 5°C of superheat is required after throttling for accurate readings?* **E [LO 9.10]**

Solution At state 2 (Fig. 9.35), when $p = 0.1$ MPa, $t = 120°C$ by interpolation

$$h_2 = 2716.2 \text{ kJ/kg}, \quad \text{and} \quad \text{at } p = 1.5 \text{ MPa}$$

$$h_f = 844.89 \quad \text{and} \quad h_{fg} = 1947.3 \text{ kJ/kg}$$

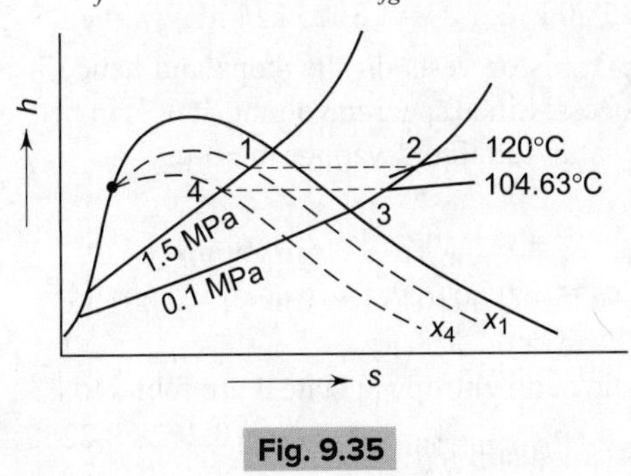

Fig. 9.35

Now, $h_1 = h_2$

or $h_{f1.5\text{MPa}} + x_1 h_{fg1.5\text{MPa}} = h_2$

$$844.89 + x_1 \times 1947.3 = 2716.2$$

$$x_1 = \frac{1871.3}{1947.3} = \mathbf{0.963}$$

When $p = 0.1$ MPa and $t = 99.63 + 5 = 104.63°C$

$$h_3 = 2685.5 \text{ kJ/kg}$$

Since $h_3 = h_4$

$$2685.5 = 844.89 + x_4 \times 1947.3$$

$$\therefore \quad x_4 = \frac{1840.6}{1947.3} = 0.948$$

The maximum moisture that can be determined with this set-up is only **5.2 per cent**.

Example 9.11 *A piston-cylinder device fitted with stops as shown in Fig. 9.36 initially contains one kg of water at 400 kPa and 300°C. Heat is rejected from the cylinder at constant pressure until water exists in the cylinder as a saturated vapour and the piston rests on the stops. Heat rejection from the cylinder continues until the pressure is 100 kPa. Find the heat transfer and the entropy change of the system.* Ⓜ **[LO 9.7, 9.8]**

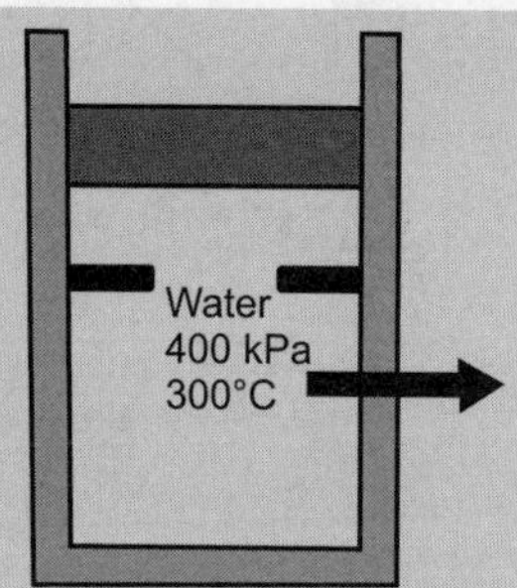

Fig. 9.36

Solution The saturation temperature at 400 kPa is $T_{sat} = 143.63°C$ and therefore initially water is in superheated state. From Table A.1.3 in the Appendix, we have

$$h_1 = 3066.7 \text{ kJ/kg}, \; s_1 = 7.5661 \text{ kJ/kgK}$$

Water at state 2 is saturated vapour at 400 kPa. From Table A.1.2 in the Appendix, we have

$$h_2 = 2738.5 \text{ kJ/kg}, \; s_2 = 6.8958 \text{ kJ/kgK}, \; v_2 = 0.4625 \text{ m}^3\text{/kg}$$

From Table A.1.2 in the Appendix, for $p = 100$ kPa

$$v_f = 0.001043 \text{ m}^3\text{/kg}, \; v_g = 1.6940 \text{ m}^3\text{/kg}, \; h_f = 417.4 \text{ kJ/kgK},$$
$$h_{fg} = 2258.0 \text{ kJ/kg}, \; s_f = 1.3025 \text{ kJ/kgK}, \; s_{fg} = 6.0568 \text{ kJ/kgK}$$

During process 2-3, piston rests on the stops and hence it is a constant volume (= 0.4625 m³/kg) process. Since specific volume at 3 is in between and at 100 kPa, the state at 3 will be saturated liquid-vapour mixture.

Therefore,

$$v_2 = v_3 = v_{f,\,p=100\text{ kPa}} + x(v_{g,\,p=100\text{ kPa}} - v_{f,\,p=100\text{ kPa}})$$

$$0.4625 = 0.001043 + x(1.6940 - 0.001043)$$

or $x = 0.2728$

The specific enthalpy and entropy at state 3 are found to be

$$h_3 = h_{f,\,p=100\text{ kPa}} + x h_{fg,\,p=100\text{ kPa}} = 417.44 + 0.2728 \times 2258 = 1033.42 \text{ kJ/kg}$$

$$s_3 = s_{f,p=100\text{ kPa}} + xs_{fg,p=100\text{ kPa}} = 1.3025 + 0.2728 \times 6.0568 = 2.9548 \text{ kJ/kgK}$$

The heat transfer during process 1-2 and 2-3 is

$$Q_{1-2} = m(h_2 - h_1) = 1(2738.5 - 3066.7) = -328.2 \text{ kJ}$$
$$Q_{2-3} = m(h_3 - h_2) = 1(1033.42 - 2738.5) = -1705.08 \text{ kJ}$$

The total heat transfer is

$$Q_{1-3} = Q_{1-2} + Q_{2-3} = -328.2 - 1705.08 = \mathbf{-2033.28 \text{ kJ}}$$

The entropy change of the system during process 1-2 and 2-3 is

$$S_2 - S_1 = m(s_2 - s_1) = 1(6.8958 - 7.5661) = -0.6703 \text{ kJ/K}$$
$$S_3 - S_2 = m(s_3 - s_2) = 1(2.9548 - 6.8958) = -3.941 \text{ kJ/K}$$

The total entropy change of the system is

$$S_3 - S_1 = -0.6703 - 3.941 = \mathbf{-4.6113 \text{ kJ/K}}$$

Example 9.12 *A piston-cylinder device contains 4 kg of saturated liquid-vapour mixture of water at 150 kPa. Initially, 3 kg of the water is in the vapour phase and the rest is in the liquid phase, and the piston is resting on a set of stops as shown in Fig. 9.37. Heat is now transferred to the water and as a result of which pressure of the water increases. When the pressure inside reaches 400 kPa, piston starts moving in a quasi-equilibrium manner and the heat transfer continues until the total volume increases by 10 percent. Determine the initial and final temperatures, the mass of liquid water when the piston starts moving, and the work done and the heat transfer during this process.*

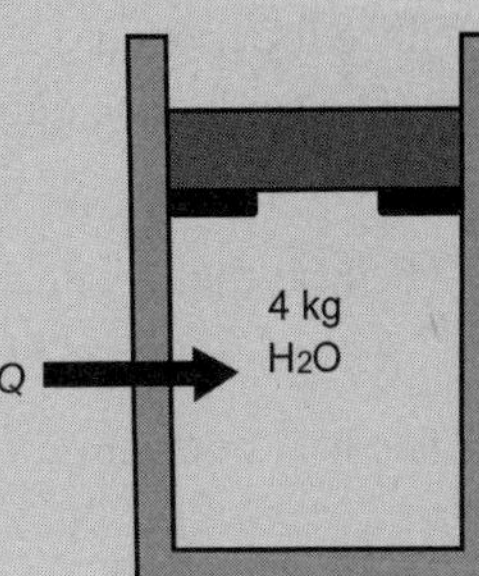

Fig. 9.37

D [LO 9.7, 9.8]

Solution The system initially is a saturated liquid-vapour mixture of water at 150 kPa pressure and thus the initial temperature is the saturation temperature at 150 kPa. From Table A.1.2 in the Appendix, we have

$$T_1 = \mathbf{111.37°C}$$
$$v_f = 0.001053 \text{ m}^3\text{/kg}, \quad v_g = 1.1593 \text{ m}^3\text{/kg}$$
$$h_f = 467.08 \text{ kJ/kg}, \quad h_g = 2693.5 \text{ kJ/kg}$$

The total initial volume of the mixture is then

$$V_1 = m_f v_f + m_g v_g = 1 \times 0.001053 + 3 \times 1.1593 = 3.479 \text{ m}^3$$

The total enthalpy of the mixture at initial state is

$$H_1 = m_f h_f + m_g h_g = 1 \times 467.08 + 3 \times 2693.5 = 8547.58 \text{ kJ}$$

The total volume at the final state is

$$V_3 = 1.1 V_1 = 1.1 \times 3.479 = 3.8269 \text{ m}^3$$

There are two different processes that occur from the initial state to final state, namely constant volume (=3.479 m^3) process 1-2 from a pressure of 150 to 400 kPa and constant pressure (=400 kPa) process 2-3 from a volume of 3.479 to 3.8269 m^3.

The specific volume at the final state is

$$v_3 = \frac{3.8268}{4} = 0.9567 \text{ m}^3\text{/kg}$$

At 400 kPa, $v_g = 0.4625$ m³/kg. Since $v_3 > v_g$, the state would be in the superheated region. When $v_3 = 0.9567$ m³/kg at 400 kPa, using linear interpolation, we get

$$T_3 = \mathbf{557.96°C},\ h_3 = 3610.97 \text{ kJ/kg}$$

When the piston first starts moving, $p_2 = 400$ kPa and $V_2 = V_1 = 3.479$ m³. The specific volume at this state is

$$v_2 = \frac{3.479}{4} = 0.8698 \text{ m}^3\text{/kg}$$

At 400 kPa, $v_g = 0.4625$ m³/kg. Since $v_2 > v_g$, the state would be in the superheated region and therefore **no liquid is left in the cylinder when the piston starts moving**.

When $v_2 = 0.9567$ m³/kg at 400 kPa, using linear interpolation, we get $h_2 =$ 3449.49 kJ/kg

Since the process 1-2 is a constant volume one, no work is done during this process. The work done during constant process 2-3 is found to be

$$W_{2-3} = p_2(V_3 - V_2) = 400(3.8269 - 3.479) = \mathbf{139.16\ kJ}$$

The heat transfer during process 1-2 is

$$Q_{1-2} = H_2 - H_1 = 4 \times 3449.49 - 8547.58 = \mathbf{5250.38\ kJ}$$

The heat transfer during process 2-3 is

$$Q_{2-3} = H_3 - H_2 = 4 \times (3610.97 - 3449.49) = \mathbf{645.92\ kJ}$$

Example 9.13 *The following data were obtained with a separating and throttling calorimeter:*

Pressure in pipeline	*1.5 MPa*
Condition after throttling	*0.1 MPa, 110°C*
During five minutes moisture collected in the separator	*0.150 litre at 70°C*
Steam condensed after throttling during 5 min	*3.24 kg*

Find the quality of steam in the pipeline E **[LO 9.10]**

Solution As shown in Fig. 9.38

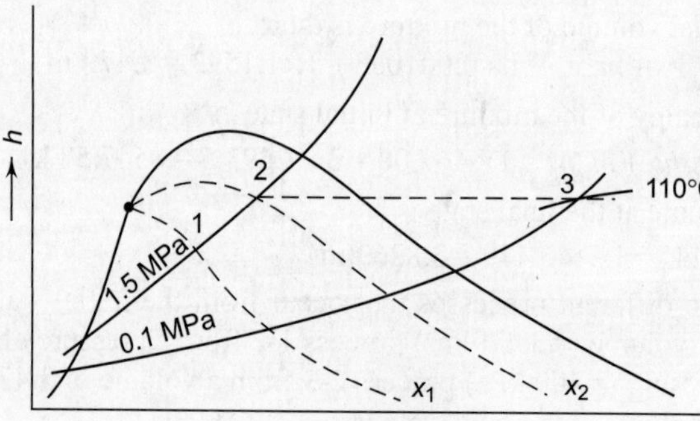

Fig. 9.38

at 0.1 MPa, 110°C

$$h_3 = 2696.2 \text{ kJ/kg}$$

Now $h_3 = h_2 = h_{f1.5\text{ MPa}} + x_2\, h_{fg1.5\text{ MPa}}$

or $2696.2 = 844.89 + x_2 \times 1947.3$

$$\therefore \quad x_2 = \frac{1851.31}{1947.3} = 0.955$$

If m_1 = mass of moisture collected in separator in 5 minutes and m_2 = mass of steam condensed after throttling in 5 minutes.

then $x_1 = \dfrac{x_2 m_2}{m_1 + m_2}$

At 70°C, $v_f = 0.001023 \text{ m}^3\text{/kg}$

$$m_1 = \frac{150 \times 10^{-6} \text{ m}^3}{1023 \times 10^{-6} \text{ m}^3\text{/kg}} = 0.1462 \text{ kg}$$

$$m_2 = 3.24 \text{ kg}$$

$$\therefore \quad x_1 = \frac{0.955 \times 3.24}{0.1462 + 3.24} = \frac{3.1}{3.3862} = \mathbf{0.915}$$

Example 9.14 *A steam boiler initially contains 5 m³ of steam and 5 m³ of water at 1 MPa. Steam is taken out at constant pressure until 4 m³ of water is left. What is the heat transferred during the process?* Ⓜ **[LO 9.7, 9.8]**

Solution At 1 MPa,

$$v_f = 0.001127, \quad \text{and} \quad v_g = 0.1944 \text{ m}^3\text{/kg}$$
$$h_g = 2778.1 \text{ kJ/kg}$$
$$u_f = 761.68, \quad u_g = 2583.6 \text{ kJ/kg}$$
$$u_{fg} = 1822 \text{ kJ/kg}$$

The initial mass of saturated water and steam in the boiler (Fig. 9.39).

$$\frac{V_f}{v_f} + \frac{V_g}{v_g} = \frac{5}{0.001127} + \frac{5}{0.1944} = (4.45 \times 10^3 + 25.70) \text{ kg}$$

where suffix f refers to saturated water and suffix g refers to saturated vapour. Final mass of saturated water and steam

$$= \frac{4}{0.001127} + \frac{6}{0.1944} = (3.55 \times 10^3 + 30.80) \text{ kg}$$

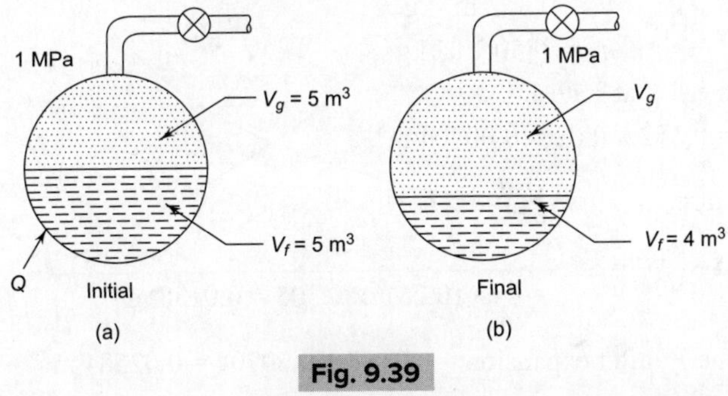

Fig. 9.39

∴ mass of steam taken out of the boiler (m_s)

$$= (4.45 \times 10^3 + 25.70) - (3.55 \times 10^3 + 30.80)$$

$$= 0.90 \times 10^3 - 5.1 = 894.9 \text{ kg}$$

Making an energy balance, we have: Initial energy stored in saturated water and steam + Heat transferred from the external source = Final energy stored in saturated water and steam + Energy leaving with the steam.

or $$U_1 + Q = U_f + m_s h_g$$

assuming that the steam taken out is dry ($x = 1$)

or $$4.45 \times 10^3 \times 761.68 + 25.70 \times 2583.6 + Q$$

$$= 3.55 \times 10^3 \times 761.68 + 30.8 \times 2583.6 + 894.9 \times 2778.1$$

or $$Q = 894.9 \times 2778.1 - (0.90 \times 10^3) \times 761.68 + 5.1 \times 2583.6$$

$$= 2425000 - 685500 + 13176$$

$$= 1752{,}676 \text{ kJ} = \mathbf{1752.676 \text{ MJ}}$$

Example 9.15 *A 280 mm diameter cylinder fitted with a frictionless leak-proof piston contains 0.02 kg of steam at a pressure of 0.6 MPa and a temperature of 200°C. As the piston moves slowly outwards through a distance of 305 mm, the steam undergoes a fully-resisted expansion during which the steam pressure p and the steam volume V are related by pV^n = constant, where n is a constant. The final pressure of the steam is 0.12 MPa. Determine (a) the value of n, (b) the work done by the steam, and (c) the magnitude and sign of heat transfer.* D **[LO 9.7, 9.8]**

Solution Since the path of expansion (Fig. 9.40) follows the equation:

$$pV^n = C$$

$$p_1 V_1^n = p_2 V_2^n$$

Taking logarithms and arranging the terms

$$n = \frac{\log \dfrac{p_1}{p_2}}{\log \dfrac{V_2}{V_1}}$$

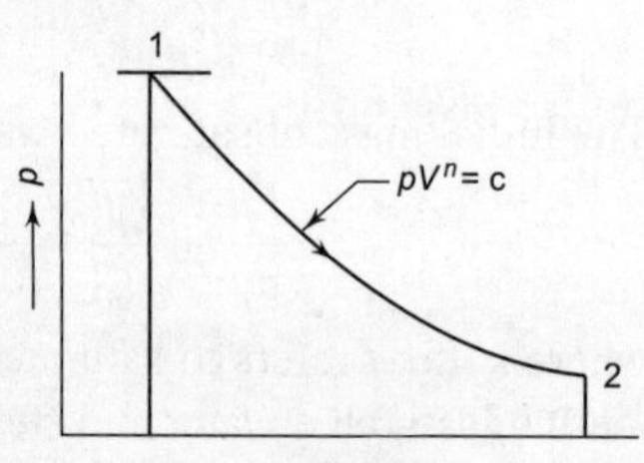

Fig. 9.40

Now, at 0.6 MPa, 200°C, from Tables A.2 in Appendix A

$$v_1 = 0.352 \text{ m}^3/\text{kg}$$

$$h_1 = 2850.1 \text{ kJ/kg}$$

∴ total volume, V_1, at state 1

$$= 0.352 \times 0.02 = 0.00704 \text{ m}^3$$

Displaced volume $$= \frac{\pi}{4} d^2 \cdot l$$

$$= \frac{\pi}{4} \times (0.28)^2 \times 0.305 = 0.0188 \text{ m}^3$$

∴ total volume V_2 after expansion = 0.0188 + 0.00704 = 0.02584 m³

$$n = \frac{\log \dfrac{0.6}{0.12}}{\log \dfrac{0.02584}{0.00704}} = \frac{\log 5}{\log 3.68} = \mathbf{1.24}$$

Work done by steam in the expansion process

$$W_{1-2} = \int_{V_1}^{V_2} p\,dV = \frac{p_1V_1 - p_2V_2}{n-1}$$

$$= \frac{6 \times 10^5 \text{ N/m}^2 \times 0.00704 \text{ m}^2 - 1.2 \times 10^5 \text{ N/m}^2 \times 0.02584 \text{ m}^3}{1.24 - 1}$$

$$= \frac{4224 - 3100.8}{0.24} \text{ N m}$$

$$= 4680 \text{ Nm} = \mathbf{4.68 \text{ kJ}}$$

Now $V_2 = 0.02584 \text{ m}^3$

$\therefore$ $v_2 = \dfrac{0.02584}{0.02} = 1.292 \text{ m}^3/\text{kg}$

Again

$$v_2 = v_{f0.12\text{MPa}} + x_2 v_{fg0.12\text{MPa}}$$

or $1.292 = 0.0010476 + x_2 \times 1.4271$

$\therefore$ $x_2 = \dfrac{1.291}{1.427} = 0.906$

At 0.12 MPa, $u_f = 439.3$ kJ/kg, $u_g = 2512.0$ kJ/kg

$\therefore$ $u_2 = 439.3 + 0.906\,(2512 - 439.3)$

$= 2314.3$ kJ/kg

Again $h_1 = 2850.1$ kJ/kg

$\therefore$ $u_1 = h_1 - p_1v_1 = 2850.1 - \dfrac{0.6 \times 10^6 \times 0.00704 \times 10^{-3}}{0.02}$

$= 2850.1 - 211.2 = 2638.9$ kJ/kg

By the first law

$$Q_{1-2} = U_2 - U_1 + W_{1-2}$$
$$= m\,(u_2 - u_1) + W_{1-2}$$
$$= 0.02\,(2314.3 - 2638.5) + 4.68$$
$$= -6.484 + 4.68 = \mathbf{-1.804 \text{ kJ}}$$

Example 9.16 *A large insulated vessel is divided into two chambers, one containing 5 kg of dry saturated steam at 0.2 MPa and the other 10 kg of steam, 0.8 quality at 0.5 MPa. If the partition between the chambers is removed and the steam is mixed thoroughly and allowed to settle, find the final pressure, steam quality and entropy change in the process.* Ⓜ **[LO 9.8, 9.9]**

Solution The vessel is divided into chambers, as shown in Fig. 9.41.

At 0.2 MPa, $v_g = v_1 = 0.8857 \text{ m}^3/\text{kg}$

$\therefore$ $V_1 = m_1 v_1 = 5 \times 0.8857 = 4.4285 \text{ m}^3$

At 0.5 MPa,

$$v_2 = v_f + x_2 v_{fg}$$
$$= 0.001093 + 0.8 \times 0.3749$$
$$= 0.30101 \text{ m}^3/\text{kg}$$

$\therefore$ $V_2 = m_2 v_2 = 10 \times 0.30101$
$$= 3.0101 \text{ m}^3$$

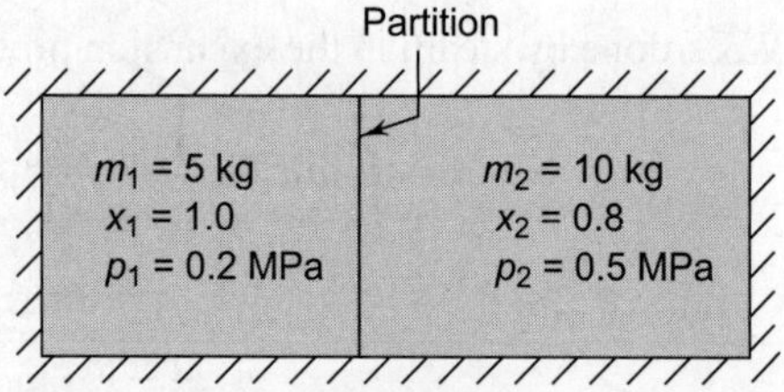

Fig. 9.41

$\therefore$ Total volume,

$$V_m = V_1 + V_2$$
$$= 7.4386 \text{ m}^3 \text{ (of mixture)}$$

Total mass of mixture, $m_m = m_1 + m_2$
$$= 5 + 10 = 15 \text{ kg}$$

$\therefore$ Specific volume of mixture

$$v_m = \frac{V_m}{m_m} = \frac{7.4386}{15} = 0.496 \text{ m}^3/\text{kg}$$

By energy balance

$$m_1 u_1 + m_2 u_2 = m_3 u_3$$

At 0.2 MPa, $h_g = h_1 = 2706.7 \text{ kJ/kg}$
$$u_1 = h_1 - p_1 v_1 \cong 2706.7 \text{ kJ/kg}$$

At 0.5 MPa, $h_2 = h_f + x_2 h_{fg}$
$$= 640.23 + 0.8 \times 2108.5 = 2327.03 \text{ kJ/kg}$$
$$u_2 = h_2 - p_2 v_2 \cong h_2 = 2327.03 \text{ kJ/kg}$$

$$h_3 = h_m = \frac{5 \times 2706.7 + 10 \times 2327.03}{15}$$

$\therefore$ $= 2453.6 \text{ kJ/kg} \simeq u_3$

Now for the mixture

$$h_3 = 2453.6 \text{ kJ/kg} = u_3$$
$$v_3 = 0.496 \text{ m}^3/\text{kg}$$

From the Mollier diagram, with the given values of h and v, point 3 after mixing is fixed (Fig. 9.42)

$$\mathbf{x_3 = 0.870 \quad s_3 = 6.29 \text{ kJ/kg K}}$$

$p_3 = 3.5$ bar Ans. $\quad s_4 = s_{g0.2\text{MPa}} = 7.1271 \text{ kJ/kg K}$

$$s_2 = s_{f0.5\text{MPa}} + 0.8 s_{fg\,0.5\text{MPa}}$$
$$= 1.8607 + 0.8 \times 4.9606 = 5.8292 \text{ kJ/kg K}$$

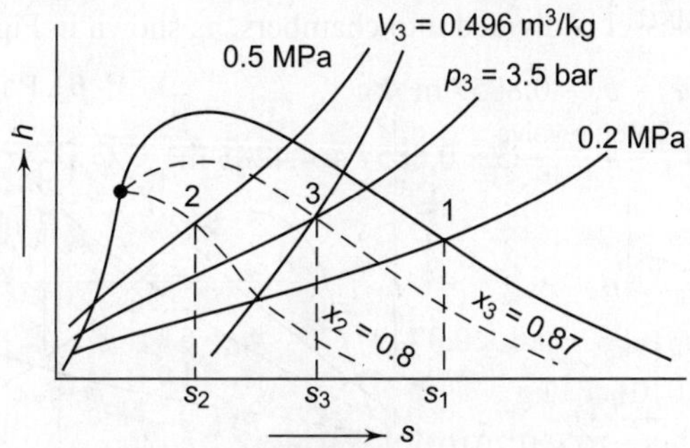

Fig. 9.42

∴ Entropy change during the process

$$= m_3 s_3 - (m_1 s_1 + m_2 s_2)$$

$$= 15 \times 6.298 - (5 \times 7.1271 + 10 \times 5.8292) = \mathbf{0.43\ kJ/K}$$

Example 9.17 *Steam generated at a pressure of 6 MPa and a temperature of 400°C is supplied to a turbine via a throttle valve which reduces the pressure to 5 MPa. Expansion in the turbine is adiabatic to a pressure of 0.2 MPa, the isentropic efficiency (actual enthalpy drop/isentropic enthalpy drop) being 82 per cent. The surroundings are at 0.1 MPa, 20°C. Determine the availability or exergy of steam before and after the throttle valve and at the turbine exhaust, and calculate the specific work output from the turbine. The K.E. and P.E. changes are negligible.* D **[LO 9.8, 9.9]**

Solution Steady flow availability or exergy ψ is given by

$$\Psi = (h - h_0) - T_0 (s - s_0) + \frac{1}{2} \mathbf{V}^2 + g(Z - Z_0)$$

where subscript 0 refers to the surroundings. Since the K.E. and P.E. changes are negligible

$$\Psi_1 = \text{Availability of steam before throttling}$$

$$= (h_1 - h_0) - T_0(s_1 - s_0)$$

At 6 MPa, 400°C (Fig. 9.43)

$$h_1 = 3177.2 \text{ kJ/kg}$$

$$s_1 = 6.5408 \text{ kJ/kg K}$$

At 20°C

$$h_0 = 83.96 \text{ kJ/kg}$$

$$s_0 = 0.2966 \text{ kJ/kg K}$$

∴ $$\Psi_1 = (3177.2 - 83.96) - 293 (6.5408 - 0.2966)$$

$$= 3093.24 - 1829.54 = 1263.7 \text{ kJ/kg} \quad \text{Ans.}$$

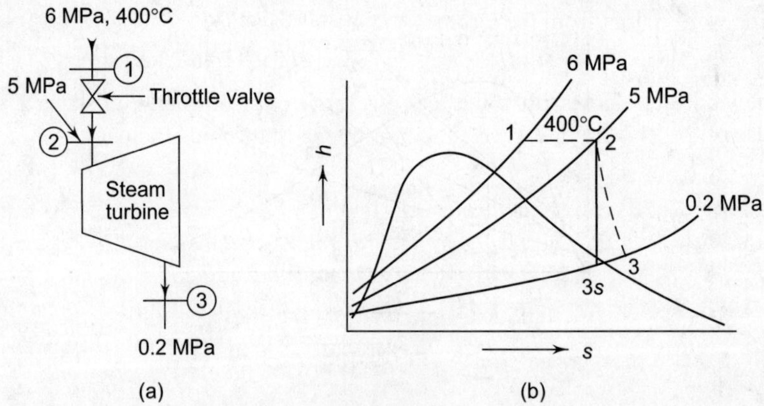

Fig. 9.43

Now $h_1 = h_2$, for throttling

At $h = 3177.2$ kJ/kg and $p = 5$ MPa, from the superheated steam table

$$\left.\begin{aligned} t_2 &= 390^\circ\text{C} \\ s_2 &= 6.63 \text{ kJ/kg K} \end{aligned}\right\} \text{by linear interpolation}$$

$\therefore$ ψ_2 = Availability or exergy of steam after throttling

$$= (h_2 - h_0) - T_0 (s_2 - s_0)$$

$$= (3177.2 - 83.96) - 293 (6.63 - 0.2966)$$

$$= 3093.24 - 1855.69 = 1237.55 \text{ kJ/kg} \quad \text{Ans.}$$

Decrease in availability or exergy due to throttling

$$= \psi_1 - \psi_2 = 1263.7 - 1237.55 = 26.15 \text{ kJ/kg}$$

Now

$$s_2 = s_{3s} = 6.63 = 1.5301 + x_{3s} (7.1271 - 1.5301)$$

$$\therefore \quad x_{3s} = \frac{5.10}{5.5970} = 0.9112$$

$$h_{3s} = 504.7 + 0.9112 \times 2201.9 = 2511.07 \text{ kJ/kg}$$

$$h_2 - h_{3s} = 3177.2 - 2511.07 = 666.13 \text{ kJ/kg}$$

$$\therefore \quad h_2 - h_3 = \eta_{is} (h_1 - h_{3s}) = 0.82 \times 666.13 = 546.12 \text{ kJ/kg}$$

$$\therefore \quad h_3 = 2631 \text{ kJ/kg} = 504.7 + x_2 \times 2201.7$$

$$\therefore \quad x_3 = \frac{2126.3}{2201.7} = 0.966$$

$$s_3 = 1.5301 + 0.966 \times 5.597 = 6.9368$$

$\therefore$ Ψ_3 = Exergy of steam at turbine exhaust

$$= (h_3 - h_0) - T_0 (s_3 - S_0)$$

$$= (2631 - 83.96) - 293 (6.9368 - 0.2966)$$

$$= 2547.04 - 1945.58 = 601.46 \text{ kJ/kg} \quad \text{Ans.}$$

Specific work output from the turbine

$$= h_2 - h_3 = 3177.2 - 2631 = \mathbf{546.2\ kJ/kg}$$

The work done is less than the loss of exergy of steam between states 2 and 3, because of the irreversibility accounted for by the isentropic efficiency.

Example 9.18 *A steam turbine receives 600 kg/h of steam at 25 bar, 350°C. At a certain stage of the turbine, steam at the rate of 150 kg/h is extracted at 3 bar, 200°C. The remaining steam leaves the turbine at 0.2 bar, 0.92 dry. During the expansion process, there is heat transfer from the turbine to the surroundings at the rate of 10 kJ/s. Evaluate per kg of steam entering the turbine (a) the exergy of steam entering and leaving the turbine, (b) the maximum work and (c) the irreversibility. The atmosphere is at 30°C.* D [LO 9.8]

Solution At 25 bar, 350°C

$$h_1 = 3125.87 \text{ kJ/kg}$$
$$s_1 = 6.8481 \text{ kJ/kg K}$$

At 30°C, $h_0 = 125.79$ kJ/kg

$$s_0 = s_{f30°C} = 0.4369 \text{ kJ/kg K}$$

At 3 bar, 200°C $h_2 = 2865.5$ kJ/kg

$$s_2 = 7.3115 \text{ kJ/kg K}$$

At 0.2 bar (0.92 dry) $h_f = 251.4$ kJ/kg

$$h_{fg} = -2358.3 \text{ kJ/kg}$$
$$s_f = 0.8320 \text{ kJ/kg K}$$
$$s_g = 7.9085 \text{ kJ/kg K}$$

$\therefore$ $h_3 = 251.4 + 0.92 \times 2358.3 = 2421.04$ kJ/kg

$$s_3 = 0.8320 + 0.92 \times 7.0765 = 7.3424 \text{ kJ/kg K}$$

The states of steam are shown in Fig. 9.44.

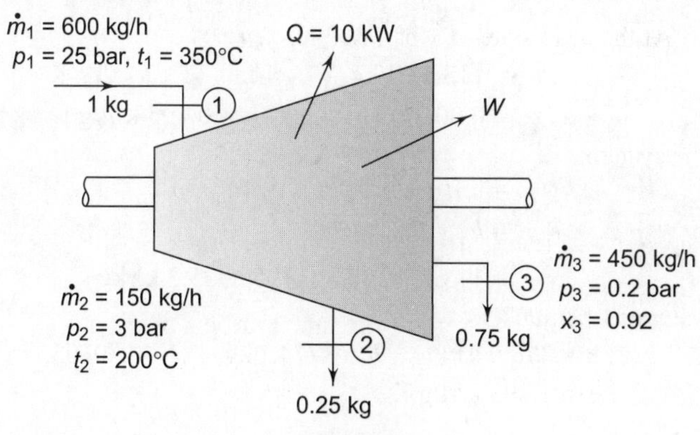

Fig. 9.44

(a) Exergy of steam entering the turbine

$$\Psi_1 = (h_1 - h_0) - T_0 (s_1 - s_0)$$
$$= (3125.87 - 125.79) - 303 (6.8481 - 0.4369)$$
$$= 3000.08 - 1942.60 = \mathbf{1057.48\ kJ/kg}$$

Exergy of steam leaving the turbine at state 2,

$$\Psi_2 = (h_2 - h_0) - T_0(s_2 - s_0)$$
$$= (2865.5 - 125.79) - 303 (7.3115 - 0.4369)$$
$$= 2739.71 - 2083.00 = 656.71 \text{ kJ/kg} \quad \text{Ans.}$$

Exergy of steam leaving the turbine at state 3,

$$\Psi_3 = (h_3 - h_0) - T_0 (s_3 - s_0) = (2421.04 - 125.79) - 303 (7.3524 - 0.4369)$$
$$= \mathbf{199.85\ kJ/kg}$$

(b) Maximum work per kg of steam entering the turbine

$$W_{rev} = \Psi_1 - \frac{m_2}{m_1}\Psi_2 - \frac{m_3}{m_1}\Psi_3 = 1057.48 - 0.25 \times 656.71 - 0.75 \times 199.85$$
$$= \mathbf{743.41\ kJ/kg}$$

(c) Irreversibility

$$I = T_0(w_2 s_2 + w_3 s_3 - w_1 s_1) - Q$$
$$= 303 (150 \times 7.3115 + 450 \times 7.3424 - 600 \times 6.8481) - (-10 \times 3600)$$
$$= 303 (1096.73 + 3304.08 - 4108.86) + 36000$$
$$= 124{,}460.85 \text{ kJ/h}$$
$$= 124.461 \text{ MJ/h} = \frac{124.461 \times 103}{600} = \mathbf{207.44\ kJ/kg}$$

Example 9.19 *Determine the exergy of (a) 3 kg of water at 1 bar and 90°C, (b) 0.2 kg of steam at 4 MPa, 500°C, (c) 0.4 kg of wet steam at 0.1 bar and 0.85 quality and (d) 3 kg of ice at 1 bar, 10°C. Assume a dead state of 1 bar and 300 K.* D [LO 9.8]

Solution At the dead state of 1 bar, 300 K,

$$u_0 = 113.1 \text{ kJ/kg}, \quad h_0 = 113.2 \text{ kJ/kg K}$$
$$v_0 = 0.001005 \text{ m}^3\text{/kg}, \quad s_0 = 0.395 \text{ kJ/kg}$$

Exergy of the system:

$$\phi = m[(u + p_0 v - T_0 s) - (u_0 + p_0 v_0 - T_0 s_0)]$$

Now, $u_0 + p_0 v_0 - T_0 s_0 = h_0 - T_0 s_0$

$$= 113.2 - 300 \times 0.395 = -5.3 \text{ kJ/kg}$$

(a) For water at 1 bar, 90°C

$$u = 376.9 \text{ kJ/kg}, \quad h = 377 \text{ kJ/kg}, \quad v = 0.001035 \text{ m}^3\text{/kg}$$
$$s = 1.193 \text{ kJ/kg K}$$

Since $p = p_0$,

$$u + p_0 v - T_0 s = u + pv - T_0 s = h - T_0 s$$
$$= 377 - 300 \times 1.193 = 19.1 \text{ kJ/kg}$$

Hence, $\phi = 3[19.1 - (-5.3)] = 3 \times 24.4 = \mathbf{73.2\ kJ}$

(b) At $p = 4$ MPa, $t = 500°C$

$$u = 3099.8, \quad h = 3446.3 \text{ kJ/kg}, \quad v = 0.08637 \text{ m}^3\text{/kg}$$

$$s = 7.090 \text{ kJ/kg K}$$

$$u + p_0 v - T_0 s = 30998\ 100 \times 0.08637 - 300 \times 7.090 = 981.4 \text{ kJ/kg}$$

$$\phi = 0.2\ [981.4 - (-5.3)] = \mathbf{197.34\ kJ}$$

(c) At 0.1 bar, 0.85 quality,

$$u = 192 + 0.85 \times 2245 = 2100.25 \text{ kJ/kg}$$

$$h = 192 + 0.85 \times 2392 = 2225.2 \text{ kJ/kg}$$

$$s = 0.649 + 0.85 \times 7.499 = 7.023 \text{ kJ/kg K}$$

$$v = 0.001010 + 0.85 \times 14.67 = 12.47 \text{ m}^3\text{/kg}$$

$$u + p_0 v - T_0 s = 2100.25 + 100 \times 12.47 - 300 \times 7.023 = 1240.4 \text{ kJ/kg}$$

$$\phi = 0.4\ [1240.4 - (-5.3)] = \mathbf{498.3\ kJ}$$

(d) Since $p = p_0$,

$$\phi = U - U_0 + p_0(V - V_0) - T_0(S - S_0)$$

$$= H - H_0 - V(p - p_0) - T_0(S - S_0)$$

$$= m[(h - h_0) - T_0(s - s_0)]$$

At 100 kPa, – 10°C

$$h = -354.1 \text{ kJ/kg and } s = -1.298 \text{ kJ/kg K}$$

$$\phi = 3\ [-354.1 - 113.2 - 300\ (-1.289 - 0.0395)] = \mathbf{81.2\ kJ}$$

Example 9.20 *A flow of hot water at 90°C is used to heat relatively cold water at 25°C to a temperature of 50°C in a heat exchanger. The cold water flows at the rate of 1 kg/s. When the heat exchanger is operated in the parallel mode, the exit temperature of the hot water stream must not be less than 60°C. In the counterflow operation, the exit temperature of hot water can be as low as 35°C. Compare the second law efficiency and the rate of exergy destruction in the two modes of operation. Take T_0 = 300 K.* D [LO 9.8]

Solution *Given:* $t_{h_1} = 90°C, \quad t_{c1} = 25°C, \quad t_{c_2} = 60°C,$

$$\dot{m}_c = 1 \text{ kg/s}, \quad T_0 = 300 \text{ K}$$

The two modes of operation of (a) parallel flow and (b) counterflow are shown in Fig. 9.45.

In parallel flow mode (a), $t_{h2} = 60°C$. Neglecting any heat loss,

$$\dot{m}_h c_h(t_{h_1} - t_{h_2}) = \dot{m}_c c_c(t_{c_2} - t_{c_1})$$

$$\dot{m}_h (90 - 60) = 1(50 - 25)$$

$$\dot{m}_h = 0.833 \text{ kg/s}$$

In counterflow mode, $t_{h_2} = 35°C$,

$$\dot{m}_h (90 - 35) = 1(50 - 25)$$

$$\dot{m}_h = \frac{25}{55} = 0.454 \text{ kg/s}$$

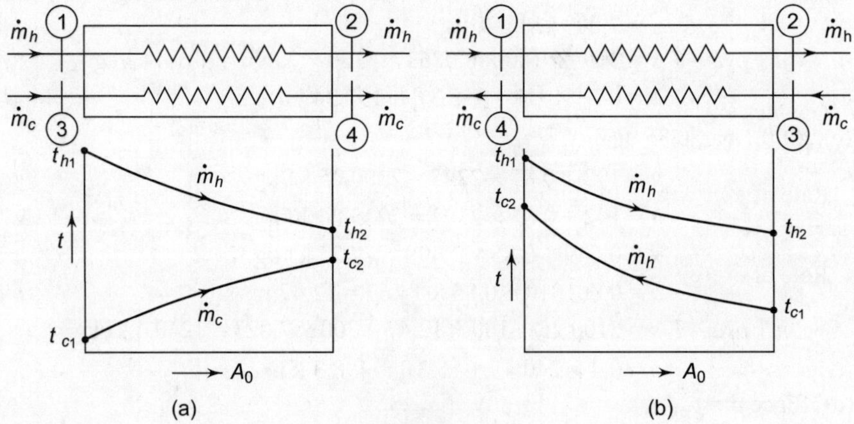

Fig. 9.45

Thus, the counterflow arrangement uses significantly less hot water.

Assuming that the hot water stream on exit from the heat exchanger is simply dumped into the drain, the energy flow rate of the hot water stream at entry is considered as the exergy input rate to the process.

$$a_{f_1} = \dot{m}_h[(h_1 - h_0) - T_0(s_1 - s_0)]$$

At 300 K or 27°C, $h_0 = 113.2$ kJ/kg and $s_0 = 0.395$ kJ/kg K

At 90°C, $h_1 = 376.92$ kJ/kg, $s_1 = 1.1925$ kJ/kgK

$$a_{f_1} = 0.833\,[(376.92 - 113.2) - 300\,(1.1925 - 0.395)]$$
$$= 0.833\,(263.72 - 239.25) = 20.38 \text{ kW}$$

Parallel flow:

At 60°C, $h_2 = 251.13$ kJ/kg, $s_2 = 0.8312$ kJ/kg K

At 25°C, $h_3 = 104.89$ kJ/kg, $s_3 = 0.3674$ kJ/kg K

At 50°C, $h_4 = 209.33$ kJ/kg, $s_4 = 0.7038$ kJ/kg K

Rate of exergy gain:

$$= \dot{m}_c[(h_4 - h_3) - T_0(s_4 - s_3)]$$
$$= 1\,[209.33 - 104.89) - 300(0.7038 - 0.3674)]$$
$$= 104.44 - 100.92 = 3.52 \text{ kW}$$

$$(\eta_{II})_p = \frac{3.52}{20.38} = 0.172 \text{ or } 17.2 \text{ per cent}$$

Rate of exergy loss by hot water:

$$= \dot{m}_h\,[(h_1 - h_2) - T_0\,(s_1 - s_2)]$$
$$= 0.833\,[(376.92 - 251.13) - 300\,(1.1925 - 0.8312)]$$
$$= 0.833\,(125.79 - 108.39) = 14.494 \text{ kW}$$

Rate of irreversibility or exergy destruction:

$$= 14.494 - 3.52 = 10.974 \text{ kW} \quad \text{Ans.}$$

If the hot water stream is not dumped to the drain,

$$\eta_{II,P} = \frac{3.52}{14.494} = 0.243 \quad \text{or} \quad \textbf{24.3 per cent}$$

Counterflow:

At 35°C, $h_2 = 146.68$ kJ/kg, $s_2 = 0.5053$ kJ/kg K

Rate of exergy gain of cold water $= \dot{m}_c\,[(h_4 - h_3) - T_0(s_4 - s_3)] = 3.52$ kW (same as in parallel flow)

Rate of exergy input (if existing hot water is dumped to the surroundings)

$$= 0.454\,(263.72 - 239.25) = 11.11 \text{ kW}$$

$$\eta_{II,C} = \frac{3.52}{11.11} = 0.3168 \text{ or } 31.68 \text{ per cent}$$

Rate of exergy loss of hot water:

$$= \dot{m}_h\,[(h_1 - h_2) - T_0(s_1 - s_2)]$$
$$= 0.454\,[(376.92 - 146.68) - 300\,(1.1925 - 0.5053)]$$
$$= 0.454\,(230.24 - 206.16) = 10.94 \text{ kW}$$

$$\eta_{II,C} = \frac{3.52}{10.94} = 0.3217 \quad \text{or} \quad \textbf{32.17 per cent}$$

Rate of irreversibility or exergy destruction:

$$= 10.94 - 3.52 = \mathbf{7.42\ kW}$$

The second law efficiency for the counterflow arrangement is significantly higher and the rate of irreversibility is substantially lower compared to the parallel flow arrangement.

Example 9.21 *A small geothermal well in a remote desert area produces 50 kg/h of saturated steam vapour at 150°C. The environment temperature is 45°C. This geothermal steam will be suitably used to produce cooling for homes at 23°C. The steam will emerge from this system as saturated liquid at 1 atm. Estimate the maximum cooling rate that could be provided by such a system.*

D **[LO 9.8]**

Solution The energy balance of the control volume as shown in Fig. 9.46 gives:

$$\dot{Q} + wh_1 = \dot{Q}_0 + wh_2$$

The entropy balance is

$$\dot{S}_{gen} = \left[\frac{\dot{Q}_0}{T_0} + ws_2\right] - \left[\frac{\dot{Q}}{T} + ws_1\right]$$

where T is the temperature maintained in the homes.

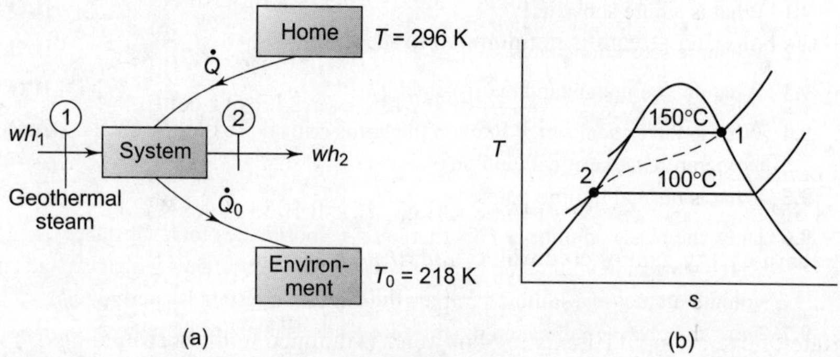

Fig. 9.46

Solving for $\dot{Q}$,

$$\dot{Q} = \frac{w[(h_1 - T_0 s_1) - (h_2 - T_0 s_2)] - T_0 \dot{S}_{gen}}{(T_0/T) - 1}$$

By second law, $\dot{S}_{gen} > 0$.

Therefore, for a given discharge state 2, the maximum $\dot{Q}$ would be

$$\dot{Q}_{max} = \frac{w(b_1 - b_2)}{(T_0/T) - 1}$$

State-1: $T_1 = 150°C = 423$ K, saturated vapour

$h_1 = 2746.4$ kJ/kg

$s_1 = 6.8387$ kJ/kg K

State-2: $T_2 - 100°C = 373$ K, saturated liquid

$h_2 = 419.0$ kJ/kg

$s_2 = 1.3071$ kJ/kg K

So since $T_0 = 318$ K,

$$b_1 = h_1 - T_0 s_1 = 2746.4 - 318 \times 6.8387 = 571.7 \text{ kJ/kg}$$

$$b_2 - h_2 - T_0 s_2 = 419.0 - 318 \times 1.3071 = 3.3 \text{ kJ/kg}$$

$$\dot{Q}_{max} = \frac{50 \times (571.7 - 3.3)}{(318/296) - 1} = 3.82 \times 10^5 \text{ kJ/h} = \mathbf{106\ kW}$$

SUMMARY

Water has been taken as the representative of a pure substance. Phase equilibrium diagrams of water on *p-v*, *p-T*, *T-s* and *h-s* coordinates have been described. Steam tables and Mollier chart are discussed. After defining the quality or dryness fraction of steam, different calorimetric methods have been elucidated.

REVIEW QUESTIONS

(E) **9.1** What is a pure substance? **[LO 9.1]**

(E) **9.2** What are saturation states? **[LO 9.2]**

(M) **9.3** What do you understand by triple point? **[LO 9.2]**

(D) **9.4** What is the critical state? Explain the terms critical pressure, critical temperature and critical volume of water? **[LO 9.2]**

(E) **9.5** What is normal boiling point? **[LO 9.2]**

(D) **9.6** Draw the phase equilibrium diagram on *p-v* coordinates for a substance which shrinks in volume on melting and then for a substance which expands in volume on melting. Indicate thereon the relevant constant property lines. **[LO 9.2]**

(D) **9.7** Draw the phase equilibrium diagram for a pure substance on *p-T* coordinates. Why does the fusion line for water have negative slope? **[LO 9.3]**

(D) **9.8** Draw the phase equilibrium diagram for a pure substance on *T-s* plot with relevant constant property lines. **[LO 9.5]**

(D) **9.9** Draw the phase equilibrium diagram for a pure substance on *h-s* plot with relevant constant property lines. **[LO 9.6]**

(M) **9.10** Why do the isobars on Mollier diagram diverge from one another? **[LO 9.6]**

(M) **9.11** Why do isotherms on Mollier diagram become horizontal in the superheated region at low pressures? **[LO 9.6]**

(E) **9.12** What do you understand by the degree of superheat and the degree of subcooling? **[LO 9.8]**

(D) **9.13** What is quality of steam? What are the different methods of measurement of quality? **[LO 9.7, 10]**

(D) **9.14** Why cannot a throttling calorimeter measure the quality if the steam is very wet? How is the quality measured then? **[LO 9.10]**

(M) **9.15** What is the principle of operation of an electrical calorimeter? **[LO 9.10]**

PROBLEMS WITH HINTS

(M) **9.1** As shown in Fig. 9.47, steam enters a desuperheater operating at steady state at 30 bar, 320°C where it is mixed with liquid water at 25 bar, 200°C to produce saturated vapour at 20 bar. Heat loss to the surroundings and K.E. and P.E. effects are neglected. Determine the mass flow rate of liquid water for a steam mass flow rate of 50, 000 kg/h. **[LO 9.8]**

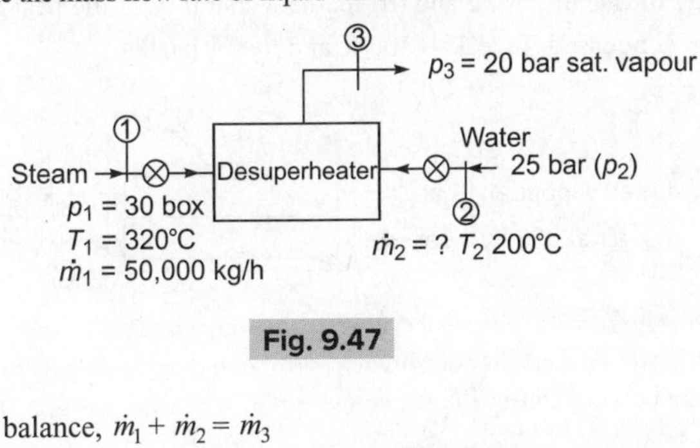

Fig. 9.47

Hint:

By mass balance, $\dot{m}_1 + \dot{m}_2 = \dot{m}_3$

$$0 = \dot{Q}_{cv} - \dot{W}_{cv} + \dot{m}_1\left(h_1 + \frac{V_1^2}{2} + gZ_1\right) + \dot{m}_2\left(h_2 + \frac{V_2^2}{2} + gZ_2\right) - (\dot{m}_1 + \dot{m}_2)(h_3 + V_3^2/2 + gZ_3)$$

Solving for $\dot{m}_2 = \dot{m}_1 \dfrac{h_3 - h_1}{h_2 - h_3}$

From steam tables, $h_2 = 852.8$ kJ/kg. $h_3 = 2799.5$ kJ/kg and $h_1 = 3043.4$ kJ/kg.

$\therefore \qquad \dot{m}_2 = \mathbf{6264\ kg/h}$

(M) **9.2** Steam at 7 bar, 355°C enters one open feed water heater operating at steady state. A separate stream of liquid water enters at 7 bar, 35°C. A single mixed stream exits at saturated liquid at 7 bar. Heat loss and K.E. and P.E. effects are neglected. Determine (a) the ratio of mass flow rates of the incoming streams, (b) the rate at which entropy is produced within the heater per kg steam at exit. **[LO 9.8]**

Hint:

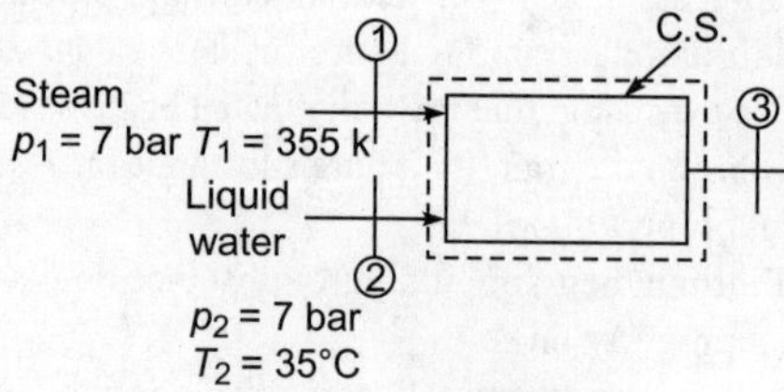

Fig. 9.48

By energy balance, $\dot{m}_1 h_1 + \dot{m}_2 h_2 = (\dot{m}_1 + \dot{m}_2) h_3$

$$\frac{\dot{m}_1}{\dot{m}_2} = \frac{h_3 - h_2}{h_1 - h_3} = \mathbf{0.222}$$

By entropy balance,

$$\dot{S}_{gen} = -\dot{m}_2 s_2 - \dot{m}_1 s_1 + (\dot{m}_1 + \dot{m}_1) s_3$$

$$\dot{S}_{gen} / \dot{m}_3 = s_3 - \frac{(\dot{m}_1/\dot{m}_2) s_1 + s_2}{1 + \dot{m}_1/\dot{m}_2} = \mathbf{0.2181\ \frac{kJ}{kg\,K}}$$

(M) **9.3** A counter flow heat exchange operating at a steady state has water entering as saturated vapour at 1 atm and exiting as saturated liquid at 1 bar. In a separate stream, air enters at 300 K, 1 bar and exits at 335 K with negligible change of pressure. Determine (a) the mass flow rate of the air required per kg of steam condensed, (b) the change in flow exergy of the steam per kg, (c) the change in flow exergy of the air per kg and (d) the irreversibility of the heat exchanger per kg of steam condensed. Take $T_o = 300$ K and $p_o = 1$ bar. **[LO 9.8]**

Hint:

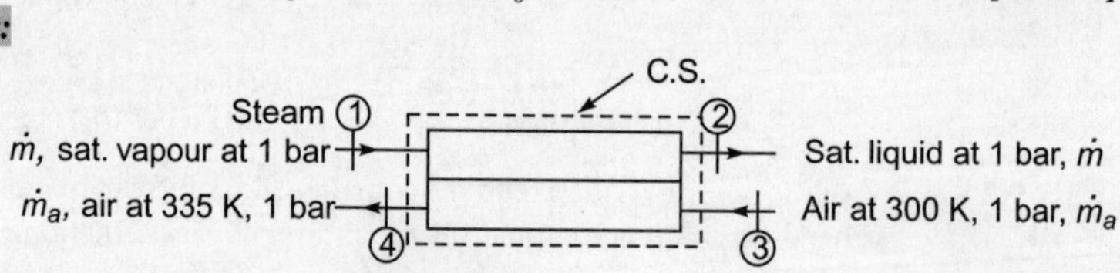

Fig. 9.49

By energy balance, $\dot{m}(h_1 - h_2) = \dot{m}_a (h_4 - h_3)$

$$\therefore \qquad \frac{\dot{m}_a}{m} = \mathbf{64.17\ \frac{kg\ air}{kg\ steam}}$$

(a) Change in flow exergy of steam

$$a_{f2} - a_{f1} = h_2 - h_1 - T_o(s_2 - s_1) = -\,\mathbf{440.96}\ \frac{\textbf{kg air}}{\textbf{kg steam}}$$

(b) Change in flow exergy for the air stream

$$a_{f4} - a_{f5} = h_4 - h_3 - T_o(s_4 - s_3) = \mathbf{1.929}\ \frac{\textbf{kJ}}{\textbf{kg air}}$$

(c) $\frac{\dot{m}_a}{\dot{m}}(a_{f4} - a_{f3}) = \mathbf{123.78}\ \frac{\textbf{kJ}}{\textbf{kg water}}$

By exergy balance

$$\dot{I}_{cv} = \dot{m}(a_{f1} - a_{f2}) + \dot{m}_a(a_{f3} - a_{f4})$$

$$\therefore \quad \frac{\dot{I}_{cv}}{\dot{m}} = (a_{f1} - a_{f2}) + \frac{\dot{m}_a}{\dot{m}}(a_{f3} - a_{f4}))$$

$$= \textbf{317.18 kJ/kg water}$$

PROBLEMS

(E) **9.1** Complete the following table of properties for 1 kg of water (liquid, vapour or mixture).

	p (bar)	*t* (°C)	*v* (m³/kg)	*x* (per cent)	Super-heat (°C)	*h* (kJ/kg)	*s* (kJ/kg K)
(a)	—	35	25.22	—	—	—	—
(b)	—	—	0.001044	—	—	419.04	—
(c)	—	212.42	—	90	—	—	—
(d)	1	—	—	—	—	—	6.104
(e)	10	320	—	—	—	—	—
(f)	5	—	0.4646	—	—	—	—
(g)	4	—	0.4400	—	—	—	—
(h)	—	500	—	—	—	3445.3	—
(i)	20	—	—	—	50	—	—
(j)	15	—	—	—	—	—	7.2690

[LO 9.8]

(E) **9.2** (a) A rigid vessel of volume 0.86 m³ contains 1 kg of steam at a pressure of 2 bar. Evaluate the specific volume, temperature, dryness fraction, internal energy, enthalpy, and entropy of steam.

(b) The steam is heated to raise its temperature to 150°C. Show the process on a sketch of the *p*–*v* diagram, and evaluate the pressure, increase in enthalpy, increase in internal energy, increase in entropy of steam and the heat transfer. Evaluate also the pressure at which the steam becomes dry saturated.

[LO 9.7, 9.8]

(E) **9.3** 10 kg of water at 45°C is heated at a constant pressure of 10 bar until it becomes superheated vapour at 300°C. Find the change in volume, enthalpy, internal energy and entropy. **[LO 9.7, 9.8]**

(E) **9.4** Water at 40°C is continuously sprayed into a pipeline carrying 5 tonnes of steam at 5 bar, 300°C per hour. At a section downstream, where the pressure is 3 bar, the quality is to be 95 per cent. Find the rate of water spray in kg/h. **[LO 9.8]**

(E) **9.5** A rigid vessel contains 1 kg of a mixture of saturated water and saturated steam at a pressure of 0.15 MPa. When the mixture is heated, the state passes through the critical point. Determine (a) the volume of the vessel, (b) the mass of liquid and of vapour in the vessel initially, (c) the temperature of the mixture when the pressure has risen to 3 MPa and (d) the heat transfer required to produce the final state (c). **[LO 9.7, 9.8]**

(E) **9.6** A rigid closed tank of volume 3 m^3 contains 5 kg of wet steam at a pressure of 200 kPa. The tank is heated until the steam becomes dry saturated. Determine the final pressure and the heat transfer to the tank. **[LO 9.8]**

(M) **9.7** Steam flows through a small turbine at the rate of 5000 kg/h entering at 15 bar, 300°C and leaving at 0.1 bar with 4 per cent moisture. The steam enters at 80 m/s at a point 2 m above the discharge and leaves at 40 m/s. Compute the shaft power assuming that the device is adiabatic but considering kinetic and potential energy changes. How much error would be made if these terms were neglected? Calculate the diameters of the inlet and discharge tubes. **[LO 9.8]**

(E) **9.8** A sample of steam from a boiler drum at 3 MPa is put through a throttling calorimeter in which the pressure and temperature are found to be 0.1 MPa and 120°C. Find the quality of the sample taken from the boiler. **[LO 9.10]**

(M) **9.9** It is desired to measure the quality of wet steam at 0.5 MPa. The quality of steam is expected to be not more than 0.9.
(a) Explain why a throttling calorimeter to atmospheric pressure will not serve the purpose.
(b) Will the use of a separating calorimeter, ahead of the throttling calorimeter, serve the purpose, if at best 5°C of superheat is desirable at the end of throttling? What is the minimum dryness fraction required at the exit of the separating calorimeter to satisfy this condition? **[LO 9.10]**

(M) **9.10** The following observations were recorded in an experiment with a combined separating and throttling calorimeter:
Pressure in the steam main—15 bar
Mass of water drained from the separator—0.55 kg
Mass of steam condensed after passing through the throttle valve—4.20 kg
Pressure and temperature after throttling—1 bar, 120°C
Evaluate the dryness fraction of the steam in the main, and state with reasons, whether the throttling calorimeter alone could have been used for this test. **[LO 9.10]**

(M) **9.11** Steam from an engine exhaust at 1.25 bar flows steadily through an electric calorimeter and comes out at 1 bar, 130°C. The calorimeter has two 1 kW heaters and the flow is measured to be 3.4 kg in 5 min. Find the quality in the engine exhaust. For the same mass flow and pressures, what is the maximum moisture that can be determined if the outlet temperature is at least 105°C? **[LO 9.10]**

(E) **9.12** Steam expands isentropically in a nozzle from 1 MPa, 250°C to 10 kPa. The steam flow rate is 1 kg/s. Find the velocity of steam at the exit from the nozzle, and the exit area of the nozzle. Neglect the velocity of steam at the inlet to the nozzle.

The exhaust steam from the nozzle flows into a condenser and flows out as saturated water. The cooling water enters the condenser at 25°C and leaves at 35°C. Determine the mass flow rate of cooling water. **[LO 9.8]**

(E) **9.13** A reversible polytropic process, begins with steam at $p_1 = 10$ bar, $t_1 = 200°C$, and ends with $p_2 = 1$ bar. The exponent n has the value 1.15. Find the final specific volume, the final temperature and the heat transferred per kg of fluid. **[LO 9.8]**

(M) **9.14** Two streams of steam, one at 2 MPa, 300°C and the other at 2 MPa, 400°C, mix in a steady flow adiabatic process. The rates of flow of the two streams are 3 kg/min and 2 kg/min, respectively. Evaluate the final temperature of the emerging stream, if there is no pressure drop due to the mixing process. What would be the rate of increase in the entropy of the universe? This stream with a negligible velocity now expands adiabatically in a nozzle to a pressure of 1 kPa. Determine the exit velocity of the stream and the exit area of the nozzle. **[LO 9.8]**

(M) **9.15** Boiler steam at 8 bar, 250°C, reaches the engine control valve through a pipeline at 7 bar, 200°C. It is throttled to 5 bar before expanding in the engine to 0.1 bar, 0.9 dry. Determine per kg of steam (a) the heat loss in the pipeline, (b) the temperature drop in passing through the throttle valve, (c) the work output of the engine, (d) the entropy change due to throttling and (e) the entropy change in passing through the engine. **[LO 9.8, 9.9]**

(M) **9.16** Tank *A* (Fig. 9.50) has a volume of 0.1 m^3 and contains steam at 200°C, 10 per cent liquid and 90 per cent vapour by volume, while tank *B* is evacuated. The valve is then opened, and the tanks eventually come to the same pressure, which is found to be 4 bar. During this process, heat is transferred such that the steam remains at 200°C. What is the volume of tank *B*? **[LO 9.7, 9.8]**

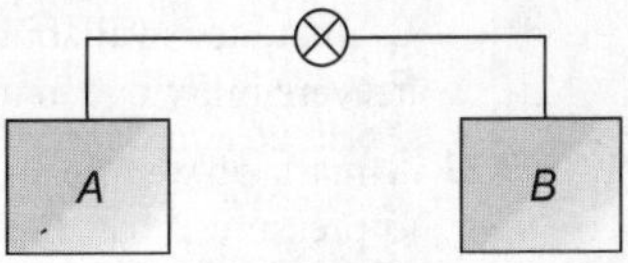

Fig. 9.50

(M) **9.17** Calculate the amount of heat which enters or leaves 1 kg of steam initially at 0.5 MPa and 250°C, when it undergoes the following processes:

(a) It is confined by a piston in a cylinder and is compressed to 1 MPa and 300°C as the piston does 200 kJ of work on the steam.

(b) It passes in steady flow through a device and leaves at 1 MPa and 300°C while, per kg of steam flowing through it, a shaft puts in 200 kJ of work. Changes in K.E. and P.E. are negligible.

(c) It flows into an evacuated rigid container from a large source which is maintained at the initial condition of the steam. Then 200 kJ of shaft work is transferred to the steam, so that its final condition is 1 MPa and 300°C. **[LO 9.7, 9.8]**

(M) **9.18** A sample of wet steam from a steam main flows steadily through a partially open valve into a pipeline in which is fitted an electric coil. The valve and the pipeline are well insulated. The steam mass flow rate is 0.008 kg/s while the coil takes 3.91 amperes at 230 volts. The main pressure is 4 bar, and the pressure and temperature of the steam downstream of the coil are 2 bar and 160°C, respectively. Steam velocities may be assumed to be negligible.

(a) Evaluate the quality of steam in the main.

(b) State, with reasons, whether an insulted throttling calorimeter could be used for this test. **[LO 9.7, 9.8]**

(M) **9.19** Two insulated tanks, *A* and *B*, are connected by a valve. Tank *A* has a volume of 0.70 m^3 and contains steam at 1.5 bar, 200°C. Tank *B* has a volume of 0.35 m^3 and contains steam at 6 bar with a quality of 90 per cent. The valve is then opened, and the two tanks come to a uniform state. If there is no heat transfer during the process, what is the final pressure? Compute the entropy change of the universe. **[LO 9.7, 9.8]**

(E) **9.20** A spherical aluminium vessel has an inside diameter of 0.3 m and a 0.62 cm thick wall. The vessel contains water at 25°C with a quality of 1 per cent. The vessel is then heated until the water inside is saturated vapour. Considering the vessel and water together as a system, calculate the heat transfer during this process. The density of aluminium is 2.7 g/cm^3 and its specific heat is 0.896 kJ/kg K. **[LO 9.8]**

(M) **9.21** Steam at 10 bar, 250°C flowing with negligible velocity at the rate of 3 kg/min mixes adiabatically with steam at 10 bar, 0.75 quality, flowing also with negligible velocity at the rate of 5 kg/min. The combined stream of steam is throttled to 5 bar and then expanded isentropically in a nozzle to 2 bar. Determine (a) the state of steam after mixing, (b) the state of steam after throttling, (c) the increase in entropy due to throttling, (d) the velocity of steam at the exit from the nozzle, and (e) the exit area of the nozzle. Neglect the K.E. of steam at the inlet to the nozzle. **[LO 9.7, 9.8]**

(M) **9.22** Steam of 65 bar, 400°C leaves the boiler to enter a steam turbine fitted with a throttle governor. At a reduced load, as the governor takes action, the pressure of steam is reduced to 59 bar by throttling before it is admitted to the turbine. Evaluate the availabilities of steam before and after the throttling process and the irreversibility due to it. **[LO 9.10]**

(E) **9.23** A mass of wet steam at temperature 165°C is expanded at constant quality 0.8 to pressure 3 bar. It is then heated at constant pressure to a degree of superheat of 66.5°C. Find the enthalpy and entropy changes during expansion and during heating. Draw the *T–s* and *h–s* diagrams. **[LO 9.9]**

(E) **9.24** Steam enters a turbine at a pressure of 100 bar and a temperature of 400°C. At the exit of the turbine, the pressure is 1 bar and the entropy is 0.6 J/g K greater than that at inlet. The process is adiabatic and changes in K.E. and P.E. may be neglected. Find the work done by the steam in J/g. What is the mass flow rate of steam required to produce a power output of 1 kW? **[LO 9.8]**

(M) **9.25** One kg of steam in a closed system undergoes a thermodynamic cycle composed of following reversible processes: (1–2) The steam initially at 10 bar, 40 per cent quality is heated at constant volume until the pressure rises to 35 bar; (2–3). It is then expanded isothermally to 10 bar; (3–1). It is finally cooled at constant pressure back to its initial state. Sketch the cycle on *T–s* coordinates, and calculate the work done, the heat transferred and the change of entropy for each of the three processes. What is the thermal efficiency of the cycle? **[LO 9.8, 9]**

(E) **9.26** Determine the exergy per unit mass for the steady flow of each of the following:
(a) steam at 1.5 MPa, 500°C,
(b) air at 1.5 MPa, 500°C,
(c) water at 4MPa, 300 K
(d) air at 4 MPa, 300 K and
(e) air at 1.5 MPa, 300 K.

[LO 9.8]

(E) **9.27** Steam expands isentropically in a nozzle from 1 MPa, 250°C to 10 kPa. The steam flow rate is 1 kg/s. Neglecting the K.E. of steam at inlet to the nozzle, find the velocity of steam at exit from the nozzle and the exit area of the nozzle. **[LO 9.8]**

(D) **9.28** Hot helium gas at 800°C is supplied to a steam generator and is cooled to 450°C while serving as a heat source for the generation of steam. Water enters the steam generator at 200 bar, 250°C and leaves as superheated steam at 200 bar, 500°C. The temperature of the surroundings is 27°C. For 1 kg helium, determine (a) the maximum work that could be produced by the heat removed from helium, (b) the mass of steam generated per kg of helium, (c) the actual work done in the steam cycle per kg of helium, (d) the net change for entropy of the universe and (e) the irreversibility. Take the average c_p for helium as 5.1926 kJ/kg K and the properties of water in inlet to the steam generator as those of saturated water at 250°C.

[LO 9.8]

ANSWERS TO PROBLEMS

9.2 **a.** 0.86 m^3/kg, 120.23°C, 0.97, 2468.54 k/kg, 2640.54 kJ/kg, 6.9592 kJ/kg K
b. 2.3 bar, 126 kJ/kg, 106.6 kJ/kg, 0.2598 kJ/kg K, 106.6 kJ/K)

9.3 2.569 m^3, 28627.5 kJ, 26047.6 kJ, 64.842 kJ/K

9.4 912.67 kg/h

9.5 (a) 0.003155 m^3
(b) 0.9982 kg, 0.0018 kg,
(c) 233.9°C
(d) 581.46 kJ/kg

9.6 304 kPa, 3346 kJ

9.7 765.6 kW, 0.44 per cent, 6.11 cm, 78.9 cm

9.8 0.951

9.9 0.97

9.10 0.85

9.11 0.944, 0.921

9.12 1224 m/s, 0.0101 m^2, 47.81 kg/s

9.14 340°C, 0.042 kJ/K min, 1530 m/s, 53.77 cm^2

9.15 a. 105.3 kJ/kg
b. 5°C
c. 499.35 kJ/kg
d. 0.1433 kJ/kg K
e. 0.3657 kJ/kg K

9.16 4.89 m^3

9.17 a. – 130 kJ
b. – 109 kJ
c. – 367 kJ

9.18 a. 0.97
b. not suitable

9.19 322.6 kPa, 0.1985 kJ/K

9.20 2682.82 kJ

9.21 a. 10 bar, 0.975 dry
b. 5 bar, 0.894 dry
c. 0.2669 kJ/kg K
d. 540 m/s
e. 1.864 cm^2

9.22 I = 21 kJ/kg

9.23 – 59 kJ/kg, 0.163 kJ/kgK during expansion and 676 kJ/kg, 1.588 kJ/kg K during heating

9.24 625 J/g, 1.6 kg/s

9.25 0, 1364 kJ, 2.781 kJ/K, 367.5 kJ, 404.6 kJ, 0.639 kJ/K, – 209.1 kJ, – 1611 kJ, – 3.419 kJ/K 8.93%

9.26 a. 1220 kJ/kg
b. 424 kJ/kg
c. 3.85 kJ/kg,
d. 316 kJ/kg
e. 232 kJ/kg

9.27 1223 m/s, 100 cm^2

9.28 a. 1202.4 kJ/kg He
b. 0.844 kg H_2O/kg He
c. 969.9 kJ/kg He
d. 0.775 kJ/(kg He-K)
e. 232.5 kJ/kg He)

10

Properties of Gases and Gas Mixtures

Learning Objectives

LO 10.1 State Avogadro's law

LO 10.2 State equation of state and discuss universal gas constant

LO 10.3 Differentiate between ideal and real gases

LO 10.4 Relate specific heats with internal energy and enthalpy of an ideal gas

LO 10.5 Demonstrate the change of entropy of an ideal gas with changes of p, v and T

LO 10.6 Explain property relations, work done by an ideal gas for a reversible adiabatic process

LO 10.7 Discuss work done and heat transfer for reversible isothermal process of an ideal gas

LO 10.8 Illustrate polytropic process, its entropy change, work and heat transfer

LO 10.9 Discuss equations of state for real gases

LO 10.10 Explain virial equations of state, compressibility factor and relate them

LO 10.11 Explain the law of corresponding states, the generalized compressibility chart

LO 10.12 Discuss other equations of state than ideal gas equation

LO 10.13 Discuss Dalton's law of partial pressure and Amagat's law of partial volume for a gas mixture

LO 10.14 Explain Gibb's theorem for properties of a gas mixture

10.1 AVOGADRO'S LAW

LO 10.1
State Avogadro's law

A mole of a substance has a mass numerically equal to the molecular weight of the substance.

One g mol of oxygen has a mass of 32 g, 1 kg mol of oxygen has a mass of 32 kg, 1 kg mol of nitrogen has a mass of 28 kg, and so on.

Avogadro's law states that the volume of a g mol of all gases at the pressure of 760 mm Hg and temperature of 0°C is the same, and is equal to 22.4 litres. Therefore, 1 g mol of a gas has a volume of 22.4×10^3 cm^3 and 1 kg mol of a gas has a volume of 22.4 m^3 at normal temperature and pressure (N.T.P.).

For a certain gas, if m is its mass in kg and μ is its molecular weight, then the number of kg moles of the gas, n, would be given by

$$n = \frac{m \text{ kg}}{\mu \dfrac{\text{kg}}{\text{kg mol}}} = \frac{m}{\mu} \text{ kg moles}$$

The molar volume, $\bar{v}$, is given by

$$\bar{v} = \frac{V}{n} \text{ m}^3/\text{kg mol}$$

where V is the total volume of the gas in m^3.

10.2 EQUATION OF STATE OF A GAS

LO 10.2
State equation of state and discuss universal gas constant

The functional relationship among the properties, pressure p, molar or specific volume v and temperature T, is known as an *equation of state,* which may be expressed in the form,

$$f(p, v, T) = 0$$

If two of these properties of a gas are known, the third can be evaluated from the equation of state.

It was discussed in Chapter 2 that gas is the best-behaved thermometric substance because of the fact that the ratio of pressure p of a gas at any temperature to pressure p_t of the same gas at the triple point, as both p and p_t approach zero, approaches a value independent of the nature of the gas. The ideal gas temperature T of the system at whose temperature the gas exerts pressure p (Section 2.5) is defined as

$$T = 273.16 \lim_{p_t \to 0} \frac{p}{p_t} \qquad \text{(Const. } V\text{)}$$

$$T = 273.16 \lim_{p \to 0} \frac{V}{V_t} \qquad \text{(Const. } p\text{)}$$

The relation between pv and p of a gas may be expressed by means of a power series of the form

$$pv = A(1 + B'p + C'p^2 + \ldots) \qquad (10.1)$$

where A, B', C', etc. depend on the temperature and nature of the gas.

A fundamental property of gases is that $\lim_{p \to 0} (pv)$ is independent of the nature of the gas and depends only on T. This is shown in Fig. 10.1, where the product pv is plotted against p for four different gases in the bulb (nitrogen, air, hydrogen and oxygen) at the boiling point of sulphur, at steam point and at the triple point of water. In each case, it is seen that as $p \to 0$, pv approaches the same value for all gases at the same temperature. From Eq. (10.1)

$$\lim_{p \to 0} pv = A$$

Therefore, the constant A is a function of temperature only and independent of the nature of the gas.

$$\lim \frac{p}{p_t} (\text{Const. } V) = \lim \frac{pV}{p_t V} = \frac{\lim pv}{\lim (pv)_t} = \frac{A}{A_t}$$

$$\lim \frac{V}{V_t} (\text{Const. } p) = \lim \frac{pV}{pV_t} = \frac{\lim pv}{\lim (pv)_t} = \frac{A}{A_t}$$

The ideal gas temperature T, is thus

$$T = 273.16 \frac{\lim (pv)}{\lim (pv)_t}$$

$$\therefore \qquad \lim (pv) = \left[\frac{\lim (pv)_t}{273.16} \right] T$$

The term within bracket is called the *universal gas constant* and is denoted by $\bar{R}$. Thus,

$$\bar{R} = \frac{\lim (pv)_t}{273.16} \tag{10.2}$$

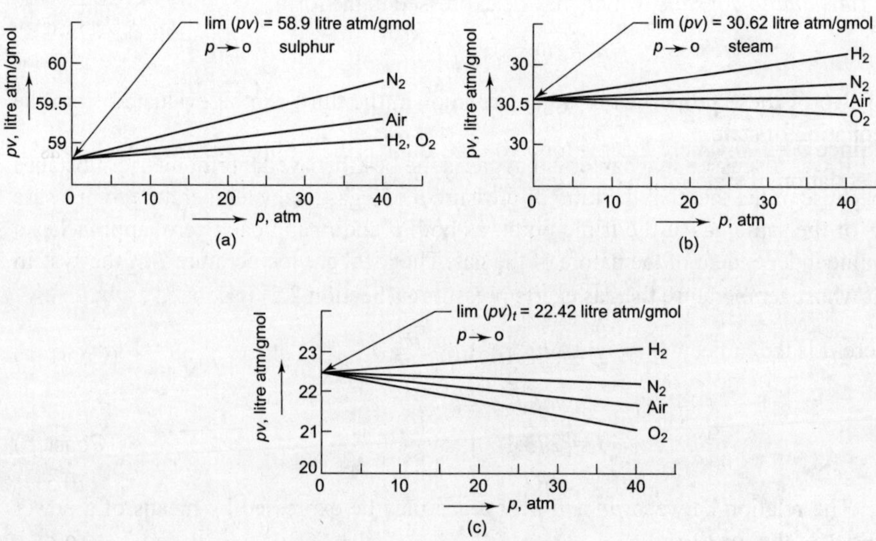

Fig. 10.1 For any gas $\lim_{p \to 0} (pV)_T$ is independent of the nature of the gas and depends only on T

The value obtained for $\lim_{p\to 0} (pv)_t$ is $22.4 \dfrac{\text{litre} - \text{atm}}{\text{g mol}}$

$$\therefore \qquad \bar{R} = \frac{22.4}{273.16} = 0.083 \frac{\text{litre} - \text{atm}}{\text{g mol K}}$$

The equation of state of a gas is thus

$$\lim_{p\to 0} p\bar{v} = \bar{R}T \tag{10.3}$$

where v is the molar volume.

10.3 IDEAL GAS

LO 10.3
Differentiate between ideal and real gases

A hypothetical gas which obeys the law $p\bar{v} = \bar{R}T$ at all pressures and temperatures is called an *ideal gas.*

Real gases do not conform to this equation of state with complete accuracy. As $p \to 0$, or $T \to \infty$, the real gas approaches the ideal gas behaviour. In the equation, $p\bar{v} = \bar{R}T$, as $T \to 0$, i.e. $t \to -273.15°C$, if $\bar{v}$ remains constant, $p \to 0$, or if p remains constant $\bar{v} \to 0$. Since negative volume or negative pressure is inconceivable, the lowest possible temperature is 0 K or – 273.15°C. T is, therefore, known as the *absolute temperature.*

There is no permanent or perfect gas. At atmospheric condition only, these gases exist in the gaseous state. They are subject to liquefication or solidification, as the temperature and pressure are sufficiently lowered.

From Avogadro's law, when p = 760 mm Hg = 1.013×10^5 N/m^2, T = 273.15 K, and $\bar{v}$ = 22.4 m^3/kg mol

$$\bar{R} = \frac{1.013 \times 10^5 \times 22.4}{273.15}$$

$$= 8314.3 \text{ Nm/kg mol K}$$

$$= 8.3143 \text{ kJ/kg mol K}$$

Since $\bar{v} = V/n$, where V is the total volume and n is the number of moles of the gas, the equation of state for an ideal gas may be written as

$$pV = n\bar{R}T \tag{10.4}$$

Also $$n = \frac{m}{\mu}$$

where μ is the molecular weight

$$\therefore \qquad pV = \frac{m\bar{R}T}{\mu}$$

or $$pV = mRT \tag{10.5}$$

where R = characteristic gas constant = $\dfrac{\bar{R}}{\mu}$ (10.6)

For oxygen, e.g.

$$R_{O_2} = \frac{8.3143}{32} = 0.262 \text{ kJ/kg K}$$

For air,

$$R_{air} = \frac{8.3143}{28.96} = 0.287 \text{ kJ/kg K}$$

There are 6.023×10^{23} molecules in a g mol of a substance. This is known as *Avogadro's number* (A).

$$\therefore \quad A = 6.023 \times 10^{26} \text{ molecules/kg mol}$$

In n kg moles of gas, the total number of molecules, N, are

$$N = nA$$

or

$$n = N/A$$

$$pV = N\frac{\bar{R}}{A}T$$

$$= NKT \tag{10.7}$$

where K = Boltzmann constant

$$= \frac{\bar{R}}{A} = \frac{8314.3}{6.023 \times 10^{26}} = 1.38 \times 10^{-23} \text{ J/molecule K}$$

Therefore, the equation of state of an ideal gas is given by

$$pV = mRT$$

$$= n\bar{R}T$$

$$= NKT$$

10.3.1 Specific Heats, Internal Energy and Enthalpy of an Ideal Gas

LO 10.4
Relate specific heats with internal energy and enthalpy of an ideal gas

An ideal gas not only satisfies the equation of state $pv = RT$, but its *specific heats are constant* also. For real gases, these vary appreciably with temperature, and little with pressure.

The properties of a substance are related by

$$Tds = du + pdv$$

or

$$ds = \frac{du}{T} + \frac{p}{T}dv \tag{10.8}$$

The internal energy u is assumed to be a function of T and v, i.e.

$$u = f(T, v)$$

or

$$du = \left(\frac{\partial u}{\partial T}\right)_v dT + \left(\frac{\partial u}{\partial v}\right)_T dv \tag{10.9}$$

From Eqs. (10.8) and (10.9)

$$ds = \frac{1}{T}\left(\frac{\partial u}{\partial T}\right)_v dT + \frac{1}{T}\left[\left(\frac{\partial u}{\partial v}\right)_T + p\right] dv \tag{10.10}$$

Again, let

$$s = f(T, v)$$

$$ds = \left(\frac{\partial s}{\partial T}\right)_v dT + \left(\frac{\partial s}{\partial v}\right)_T dv \tag{10.11}$$

Comparing Eqs. (10.10) and (10.11)

$$\left(\frac{\partial s}{\partial T}\right)_v = \frac{1}{T}\left(\frac{\partial u}{\partial T}\right)_v \tag{10.12}$$

$$\left(\frac{\partial s}{\partial v}\right)_T = \frac{1}{T}\left[\left(\frac{\partial u}{\partial v}\right)_T + p\right] \tag{10.13}$$

Differentiating Eq. (10.12) with respect to v when T is constant

$$\frac{\partial^2 s}{\partial T \partial v} = \frac{1}{T}\frac{\partial^2 u}{\partial T \partial v} \tag{10.14}$$

Differentiating Eq. (10.13) with respect to T when v is constant

$$\frac{\partial^2 s}{\partial v \partial T} = \frac{1}{T}\frac{\partial^2 u}{\partial v \partial T} + \frac{1}{T}\left(\frac{\partial p}{\partial T}\right)_v - \frac{1}{T^2}\left(\frac{\partial u}{\partial v}\right)_T - \frac{p}{T^2} \tag{10.15}$$

From Eqs. (10.14) and (10.15)

$$\frac{1}{T}\frac{\partial^2 u}{\partial T \partial v} = \frac{1}{T}\frac{\partial^2 u}{\partial v \partial T} + \frac{1}{T}\left(\frac{\partial p}{\partial T}\right)_v - \frac{1}{T^2}\left(\frac{\partial u}{\partial v}\right)_T - \frac{p}{T^2}$$

or

$$\left(\frac{\partial u}{\partial v}\right)_T + p = T\left(\frac{\partial p}{\partial T}\right)_v \tag{10.16}$$

For an ideal gas

$$pv = RT$$

$$v\left(\frac{\partial p}{\partial T}\right)_v = R$$

$$\left(\frac{\partial p}{\partial T}\right)_v = \frac{R}{v} = \frac{p}{T} \tag{10.17}$$

From Eqs. (10.16) and (10.17)

$$\left(\frac{\partial u}{\partial v}\right)_T = 0 \tag{10.18}$$

Therefore, u does not change when v changes at constant temperature. Similarly, if $u = f(T, p)$, it can be shown that $\left(\frac{\partial u}{\partial p}\right)_T = 0$. Therefore, u does not change with p either, when T remains constant.

u does not change unless T changes.

Then $$u = f(T) \tag{10.19}$$

only for an ideal gas. This is known as *Joule's law.*

If
$$u = f(T, v)$$
$$du = \left(\frac{\partial u}{\partial T}\right)_v dT + \left(\frac{\partial u}{\partial v}\right)_T dv$$

Since the last term is zero by Eq. (10.18), and by definition

$$c_v = \left(\frac{\partial u}{\partial T}\right)_v$$
$$du = c_v\, dT \tag{10.20}$$

The equation $du = c_v\, dT$ holds good for an ideal gas for any process, whereas for any other substance it is true for a constant volume process only.

Since c_v is constant for an ideal gas,

$$\Delta u = c_v\, \Delta T$$

The enthalpy of any substance is given by

$$h = u + pv$$

For an ideal gas

$$h = u + RT$$

Therefore,

$$h = f(T) \tag{10.21}$$

only for an ideal gas.

Now
$$dh = du + RdT$$

Since R is a constant

$$\begin{aligned}\Delta h &= \Delta u + R\Delta T\\ &= c_v\, \Delta T + R\Delta T\\ &= (c_v + R)\, \Delta T\end{aligned} \tag{10.22}$$

Since h is a function of T only, and by definition

$$c_p = \left(\frac{\partial h}{\partial T}\right)_p$$
$$dh = c_p\, dT \tag{10.23}$$

or
$$\Delta h = c_p\, \Delta T \tag{10.24}$$

From Eqs. (10.22) and (10.23)

$$c_p = c_v + R$$

or
$$c_p - c_v = R \tag{10.25}$$

The equation $dh = c_p\, dT$ holds good for an ideal gas, even when pressure changes, but for any other substance, this is true only for a constant pressure change.

The ratio of c_p / c_v is of importance in ideal gas computations and is designated by the symbol γ, i.e.

$$\frac{c_p}{c_v} = \gamma$$

or
$$c_p = \gamma c_v$$

From Eq. (10.25)

$$(\gamma - 1)\, c_v = R$$

$$\left.\begin{aligned} c_v &= \frac{R}{\gamma - 1} \\ \text{and} \qquad c_p &= \frac{\gamma R}{\gamma - 1} \end{aligned}\right\} \text{kJ/kg K} \tag{10.26}$$

If $R = \dfrac{\overline{R}}{\mu}$ is substituted in Eq. (10.26)

$$\left.\begin{aligned} \overline{c}_v &= \mu c_v = (c_v)_{\text{mol}} = \frac{\overline{R}}{\gamma - 1} \\ \text{and} \qquad \overline{c}_p &= \mu c_p = (c_p)_{\text{mol}} = \frac{\gamma \overline{R}}{\gamma - 1} \end{aligned}\right\} \text{kJ/(kg mol)(K)} \tag{10.27}$$

$\overline{c}_v$ and $\overline{c}_p$ are the *molar or molal specific heats* at constant volume and at constant pressure, respectively.

It can be shown by the classical kinetic theory of gases that the values of γ are 5/3 for monatomic gases and 7/5 for diatomic gases. When the gas molecule contains more than two atoms (i.e. for polyatomic gases) the value of γ may be taken approximately as 4/3. The minimum value of γ is thus 1 and the maximum is 1.67.

The value of γ thus depends only on the molecular structure of the gas, i.e. whether the gas is monatomic, diatomic or polyatomic having one, two or more atoms in a molecule. It may be noted that c_p and c_v of an ideal gas depend only on γ and R, i.e. the number of atoms in a molecule and the molecular weight of the gas. They are independent of temperature or pressure of the gas.

Further considerations An important ratio that we shall be using extensively is that of the specific heats,

$$\gamma = \frac{c_p}{c_v}$$

which has so far been defined on a mass basis. It can also be written on the mole basis. We denote the molal constant volume specific heat as

$$\overline{c}_v = \left(\frac{\partial \overline{u}}{\partial T}\right)_v \text{ kJ/kgmol-K}$$

and the molal constant pressure specific heat as

$$\overline{c}_p = \left(\frac{\partial \overline{h}}{\partial T}\right)_v \text{ kJ/kgmol-K}$$

The bar over any property denotes that it is on the mole basis; hence, both enthalpy $\overline{h}$ and internal energy $\overline{u}$ have units of kJ/kgmol. The property on the mass basis is multiplied by the molecular weight of the substance, μ, to obtain the property on the mole bases, e.g. $\overline{h} = \mu h$.

Variation of specific heat with temperature The specific heat of a real gas varies with temperature. The functional relationships denoting this variation are determined from experimental tests. Table 10.1 lists several formulas that predict the specific heat of a gas at a given temperature. Figure 10.2 shows a diagram of the specific heat variation with temperature for a few gases.

Table 10.1 Formulas for specific heat variation with temperature

Substance and temperature range	c_p *(kJ/kgK)*
Air, 280–1500 K	$0.9167 + 2.577 \times 10^{-4}T - 3.974 \times 10^{-8}T^2$
Ammonia, 300–100 K	$1.5194 + 1.936 \times 10^{-3}T - 1.789 \times 10^{-7}T^2$
Sulphur dioxide, 300–1000 K	$0.7848 + 0.7113 \times 10^{-4}T - 1.73 \times 10^{4}T^{-2}$
Hydrogen, 300–2200 K	$11.959 + 2.160 \times 10^{-3}T + 30.95T^{-1/2}$
Oxygen, 300–2750 K	$1.507 - 16.77T^{-1/2} + 111.1T^{-1}$
Nitrogen, 300–5000 K	$1.415 - 287.9T^{-1} + 5.35 \times 10^{4}T^{-2}$
Carbon monoxide, 300–5000 K	$1.415 - 273.3T^{-1} + 4.96 \times 10^{4}T^{-2}$
H_2O, 300–3000 K	$4.613 - 103.3T^{-1/2} + 967.5T^{-1}$
Carbon dioxide, 300–3500 K	$1.540 - 345.1T^{-1} + 4.13 \times 10^{4}T^{-2}$
Methane, 300–1500 K	$0.8832 + 4.71 \times 10^{-3}T - 1.123 \times 10^{-6}T^2$
Ethylene, 300–1500 K	$0.4039 + 4.35 \times 10^{-3}T - 1.35 \times 10^{-6}T^2$
Ethane, 300–1500 K	$0.306 + 5.34 \times 10^{-3}T - 1.53 \times 10^{-6}T^2$
n-butane (C_4H_{10}), 300–1500 K	$0.314 + 5.23 \times 10^{-3}T - 1.60 \times 10^{-6}T^2$
Propane, 300–1500 K	$0.214 + 5.48 \times 10^{-3}T - 1.67 \times 10^{-6}T^2$
Acetylene (C_2H_2), 280–1250 K	$1.921 + 7.06 \times 10^{-4}T - 3.73 \times 10^{4}T^{-2}$
Octane, 225–610 K	$0.290 + 3.97 \times 10^{-3}T$

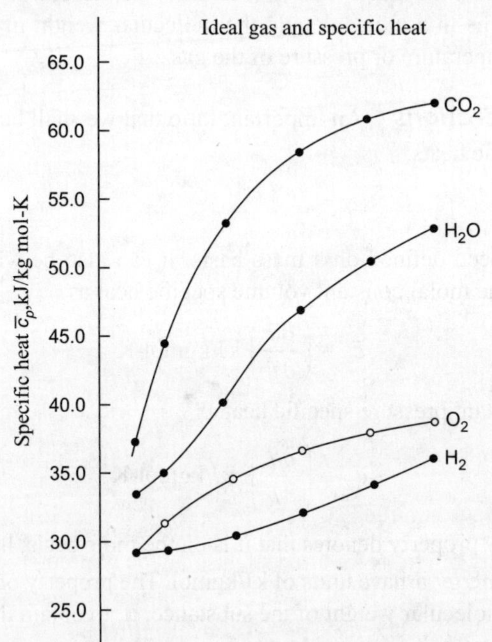

Fig. 10.2 Variation of c_p with temperature for four gases, platted from equations in Table 10.1

Gas tables Table A-12 in Appendix A provides the properties of air which account for the variation of specific heat with temperature for an ideal gas. Temperature is the independent variable and values of internal energy, enthalpy, etc. are given on the same line.

10.3.2 Entropy Change of an Ideal Gas

> **LO 10.5**
> Demonstrate the change of entropy of an ideal gas with changes of p, v and T

From the general property relations

$$Tds = du + pdv$$

$$Tds = dh - vdp$$

and for an ideal gas, $du = c_v\, dT$, $dh = c_p\, dT$ and $pv = RT$, the entropy change between any two states 1 and 2 may be computed as given below.

$$ds = \frac{du}{T} + \frac{p}{T} dv$$

$$= c_v \frac{dT}{T} + R \frac{dv}{v}$$

$$\therefore \quad s_2 - s_1 = c_v \ln \frac{T_2}{T_1} + R \ln \frac{v_2}{v_1} \tag{10.28}$$

Also

$$ds = \frac{dh}{T} - \frac{v}{T} dp$$

$$= c_p \frac{dT}{T} - R \frac{dp}{p}$$

or

$$s_2 - s_1 = c_p \ln \frac{T_2}{T_1} - R \ln \frac{p_2}{p_1} \tag{10.29}$$

Since $R = c_p - c_v$, Eq. (10.29) may be written as

$$s_2 - s_1 = c_p \ln \frac{T_2}{T_1} - c_p \ln \frac{p_2}{p_1} + c_v \ln \frac{p_2}{p_1}$$

or

$$s_2 - s_1 = c_p \ln \frac{v_2}{v_1} + c_v \ln \frac{p_2}{p_1} \tag{10.30}$$

Any one of the three Eqs. (10.28), (10.29) and (10.30), may be used for computing the entropy change between any two states of an ideal gas.

10.3.3 Reversible Adiabatic Process

> **LO 10.6**
> Explain property relations, work done by an ideal gas for a reversible adiabatic process

The general property relations for an ideal gas may be written as

$$Tds = du + pdv = c_v\, dT + pdv$$

and

$$Tds = dh - vdp = c_p\, dT - vdp$$

For a reversible adiabatic change, $ds = 0$

$$\therefore \quad c_v\, dT = -pdv \tag{10.31}$$

and $$c_p\,dT = vdp \tag{10.32}$$

By division

$$\frac{c_p}{c_v} = \gamma = -\frac{vdp}{pdv}$$

or $$\frac{dp}{p} + \gamma\frac{dv}{v} = 0$$

or $$d(\ln p) + \gamma d(\ln v) = d(\ln c)$$

where c is a constant.

$\therefore$ $$\ln p + \gamma \ln v = \ln c$$

$$pv^{\gamma} = c \tag{10.33}$$

Between any two states 1 and 2

$$p_1 v_1^{\gamma} = p_2 v_2^{\gamma}$$

or $$\frac{p_2}{p_1} = \left(\frac{v_1}{v_2}\right)^{\gamma}$$

For an ideal gas

$$pv = RT$$

From Eq. (10.33)

$$p = c \cdot v^{-\gamma}$$

$\therefore$ $$c \cdot v^{-\gamma} \cdot v = RT$$

$$c \cdot v^{1-\gamma} = RT$$

$$Tv^{\gamma-1} = \text{constant} \tag{10.34}$$

Between any two states 1 and 2, for a reversible adiabatic process in the case of an ideal gas

$$T_1 v_1^{\gamma-1} = T_2 v_2^{\gamma-1}$$

or $$\frac{T_2}{T_1} = \left(\frac{v_1}{v_2}\right)^{\gamma-1} \tag{10.35}$$

Similarly, substituting from Eq. (10.33)

$$v = \left(\frac{c}{p}\right)^{1/\gamma} \text{ in the Eq. } pv = RT$$

$$p \cdot \frac{C'}{p^{1/\gamma}} = RT$$

$$p^{1-(1/\gamma)} \cdot c' = RT$$

$\therefore$ $$Tp^{(1-\gamma)/\gamma} = \text{constant} \tag{10.36}$$

Between any two states 1 and 2

$$T_1 p_1^{(1-\gamma)/\gamma} = T_2 p_2^{(1-\gamma)/\gamma}$$

$$\therefore \qquad \frac{T_2}{T_1} = \left(\frac{p_2}{p_1}\right)^{(\gamma-1)/\gamma} \tag{10.37}$$

Equations (10.33), (10.34) and (10.36) give the relations among p, v and T in a reversible adiabatic process for an ideal gas.

The *internal energy change* of an ideal gas for a reversible adiabatic process is given by

$$Tds = du + pdv = 0$$

$$\text{or} \qquad \int_1^2 du = -\int_1^2 pdv = -\int_1^2 \frac{c}{v^\gamma}\, dv$$

$$\text{where} \qquad pv^\gamma = p_1 v_1^\gamma = p_2 v_2^\gamma = c$$

$$\therefore \qquad u_2 - u_1 = c\frac{v_2^{1-\gamma} - v_1^{1-\gamma}}{\gamma - 1} = \frac{p_2 v_2^\gamma \cdot v_2^{1-\gamma} - p_1 v_1^\gamma \cdot v_1^{1-\gamma}}{\gamma - 1}$$

$$= \frac{p_2 v_2 - p_1 v_1}{\gamma - 1}$$

$$= \frac{R(T_2 - T_1)}{\gamma - 1} = \frac{RT_1}{\gamma - 1}\left(\frac{T_2}{T_1} - 1\right)$$

$$= \frac{RT_1}{\gamma - 1}\left[\left(\frac{p_2}{p_1}\right)^{\gamma-1/\gamma} - 1\right] \tag{10.38}$$

The enthalpy change of an ideal gas for a reversible adiabatic process may be similarly derived.

$$Tds = dh - vdp = 0$$

$$\text{or} \qquad \int_1^2 dh = \int_1^2 vdp = \int_1^2 \frac{(c)^{1/\gamma}}{p^{1/\gamma}} vdp$$

$$\text{where} \qquad p_1 v_1^\gamma = p_2 v_2^\gamma = c$$

$$\therefore \qquad h_2 - h_1 = \frac{\gamma}{\gamma - 1} c^{1/\gamma}[p_2{}^{(\gamma-1)/\gamma} - p_1{}^{(\gamma-1)/\gamma}]$$

$$= \frac{\gamma}{\gamma - 1}(p_1 v_1^\gamma)^{1/\gamma} \cdot (p_1)^{(\gamma-1)/\gamma}\left[\left(\frac{p_2}{p_1}\right)^{(\gamma-1)/\gamma} - 1\right]$$

$$= \frac{\gamma p_1 v_1}{\gamma - 1}\left[\left(\frac{p_2}{p_1}\right)^{(\gamma-1)/\gamma} - 1\right]$$

$$= \frac{\gamma RT_1}{\gamma - 1}\left[\left(\frac{p_2}{p_1}\right)^{(\gamma-1)/\gamma} - 1\right] \tag{10.39}$$

The work done by an ideal gas in a reversible adiabatic process is given by

$$\delta Q = dU + \delta W = 0$$

or $$\delta W = -\, dU$$

i.e. work is done at the expense of the internal energy.

$$\therefore \quad W_{1-2} = U_1 - U_2 = m\,(u_1 - u_2)$$

$$= \frac{m(p_1 v_1 - p_2 v_2)}{\gamma - 1} = \frac{mR(T_1 - T_2)}{\gamma - 1} = \frac{mRT_1}{\gamma - 1}\left[1 - \left(\frac{p_2}{p_1}\right)^{(\gamma-1)/\gamma}\right] \tag{10.40}$$

where *m* is the mass of gas.

In a *steady flow process*, where both flow work and external work are involved, we have from SFEE,

$$W_x + \Delta\frac{V^2}{2} + g\Delta z = h_1 - h_2 = c_p\,(T_1 - T_2) = \frac{\gamma R(T_1 - T_2)}{\gamma - 1}$$

$$= \frac{\gamma(p_1 v_1 - p_2 v_2)}{\gamma - 1} = \frac{\gamma}{\gamma - 1} p_1 v_1 \left[1 - \left(\frac{p_2}{p_1}\right)^{\gamma-1/\gamma}\right] \tag{10.41}$$

If K.E. and P.E. changes are neglected,

$$W_x = \frac{\gamma}{\gamma - 1} p_1 v_1 \left[1 - \left(\frac{p_2}{p_1}\right)^{\gamma-1/\gamma}\right] \tag{10.42}$$

10.3.4 Reversible Isothermal Process

> **LO 10.7**
> Discuss work done and heat transfer for reversible isothermal process of an ideal gas

When an ideal gas of mass *m* undergoes a reversible isothermal process from state 1 to 2, the work done is given by

$$\int_1^2 \delta W = \int_{V_1}^{V_2} p dV$$

or $$W_{1-2} = \int_{V_1}^{V_2} \frac{mRT}{V}\, dV = mRT \ln \frac{V_2}{V_1}$$

$$= mRT \ln \frac{p_1}{p_2} \tag{10.43}$$

The heat transfer involved in the process

$$Q_{1-2} = U_2 - U_1 + W_{1-2}$$

$$= W_{1-2} = mRT \ln V_2/V_1 = T(S_2 - S_1) \tag{10.44}$$

10.3.5 Polytropic Process

> **LO 10.8**
> Illustrate polytropic process, its entropy change, work and heat transfer

An equation of the form pv^n = constant, where *n* is a constant can be used approximately to describe many processes which occur in practise. Such a process is called *polytropic process*. It is not adiabatic, but it can be reversible.

It may be noted that γ is a property of the gas, whereas n is not. The value of n depends upon the process. It is possible to find the value of n which more or less fits the experimental results. For two states on the process,

$$p_1 v_1^n = p_2 v_2^n \tag{10.45}$$

or

$$\left(\frac{v_2}{v_1}\right)^n = \frac{p_1}{p_2}$$

$$n = \frac{\log p_1 - \log p_2}{\log v_2 - \log v_1} \tag{10.46}$$

For known values of p_1, p_2, v_1 and v_2, n can be estimated from the above relation.

Two other relations of a polytropic process, corresponding to Eqs. (10.35) and (10.37) are:

$$\frac{T_2}{T_1} = \left(\frac{v_1}{v_2}\right)^{n-1} \tag{10.47}$$

$$\frac{T_2}{T_1} = \left(\frac{p_2}{p_1}\right)^{n-1/n} \tag{10.48}$$

1. Entropy change in a polytropic process In a reversible adiabatic process, the entropy remains constant. But in a reversible polytropic process, the entropy changes. Substituting Eqs. (10.45), (10.47) and (10.48) in Eqs. (10.28), (10.29) and (10.30) respectively, we have three expressions for entropy change given as follows:

$$s_2 - s_1 = c_v \ln \frac{T_2}{T_1} + R \ln \frac{v_2}{v_1}$$

$$= \frac{R}{\gamma - 1} \ln \frac{T_2}{T_1} + \frac{R}{n-1} \ln \frac{T_1}{T_2}$$

$$= \frac{n-\gamma}{(\gamma-1)(n-1)} R \ln \frac{T_2}{T_1} \tag{10.49}$$

Relations in terms of pressure and specific volume can be similarly derived. These are

$$s_2 - s_1 = \frac{n-\gamma}{n(\gamma-1)} R \ln \frac{p_2}{p_1} \tag{10.50}$$

and

$$s_2 - s_1 = -\frac{n-\gamma}{\gamma-1} R \ln \frac{v_2}{v_1} \tag{10.51}$$

It can be noted that when $n = \gamma$, the entropy change becomes zero. If $p_2 > p_1$, for $n \leq \gamma$, the entropy of the gas decreases, and for $n > \gamma$, the entropy of the gas increases.

The increase of entropy may result from reversible heat transfer to the system from the surroundings. Entropy decrease is also possible if the gas is cooled.

2. Heat and work in a polytropic process Using the first law to unit mass of an ideal gas,

$$Q - W = u_2 - u_1$$

$$= c_v (T_2 - T_1) = \frac{R(T_2 - T_1)}{\gamma - 1}$$

$$= \frac{p_2 v_2 - p_1 v_1}{\gamma - 1}$$

$$= \frac{p_1 v_1}{\gamma - 1}\left[\left(\frac{p_2}{p_1}\right)^{n-1/n} - 1\right] = \frac{p_1 v_1}{\gamma - 1}\left[\left(\frac{v_1}{v_2}\right)^{n-1} - 1\right] \tag{10.52}$$

For a steady flow system of unit mass of ideal gas, the SFEE Eq. (5.10) gives

$$Q - W_x - \Delta\left[\frac{\bar{V}_2}{2} + gz\right] = h_2 - h_1$$

$$= c_p (T_2 - T_1) = \frac{\gamma R(T_2 - T_1)}{\gamma - 1}$$

$$= \frac{\gamma}{\gamma - 1}(p_2 v_2 - p_1 v_1) \tag{10.53}$$

For a polytropic process,

$$Q - W_x - \Delta\left[\frac{\bar{V}_2}{2} + gz\right] = \frac{\gamma p_1 v_1}{\gamma - 1}\left[\left(\frac{p_2}{p_1}\right)^{n-1/n} - 1\right]$$

$$= \frac{\gamma p_1 v_1}{\gamma - 1}\left[\left(\frac{v_1}{v_2}\right)^{n-1} - 1\right] \tag{10.54}$$

Equations (10.52) and (10.54) can be used to determine heat and work quantities for a closed and a steady flow system, respectively.

3. Integral property relations in a polytropic process In a pv^n = constant process,

$$\int_1^2 p dv = \int_1^2 \frac{p_1 v_1^n}{v^n} dv = \frac{p_1 v_1}{n - 1}\left[1 - \left(\frac{v_1}{v_2}\right)^{n-1}\right]$$

$$= \frac{p_1 v_1}{n - 1}\left[1 - \left(\frac{p_2}{p_1}\right)^{n-1/n}\right] \tag{10.55}$$

Similarly,

$$-\int_1^2 vdp = \frac{np_1v_1}{n-1}\left[1-\left(\frac{v_1}{v_2}\right)^{n-1}\right]$$

$$= \frac{np_1v_1}{n-1}\left[1-\left(\frac{p_2}{p_1}\right)^{n-1/n}\right] \tag{10.56}$$

The integral of *Tds* is obtained from the property relation

$$Tds = du + pdv$$

$$\therefore \quad \int_1^2 Tds = \int_1^2 du + \int_1^2 pdv$$

$$= u_2 - u_1 + \int_1^2 pdv$$

Substituting from Eqs. (10.50) and (10.53)

$$\int_1^2 Tds = \frac{\gamma - n}{(\gamma-1)(n-1)} p_1v_1\left[1-\left(\frac{p_2}{p_1}\right)^{n-1/n}\right]$$

$$= \frac{\gamma - n}{(\gamma-1)(n-1)} p_1v_1\left[1-\left(\frac{v_1}{v_2}\right)^{n-1}\right]$$

$$= \frac{\gamma - n}{(\gamma-1)(n-1)} R(T_1 - T_2) \tag{10.57}$$

Since $R/(\gamma - 1) = c_v$, and putting $\Delta T = T_2 - T_1$, the reversible heat transfer

$$Q_R = \int_1^2 Tds = c_v \frac{\gamma - n}{1-n} \Delta T$$

$$= c_n \Delta T \tag{10.58}$$

where $c_n = c_v(\gamma - n)/(1-n)$ is called the *polytropic specific heat*. For $n > \gamma$, there will be positive heat transfer and gain in entropy. For $n < \gamma$, heat transfer will be negative and entropy of the gas would decrease.

Equation (10.58) can also be derived in the following way. For a polytropicprocess,

$$đQ = du + pdv$$

$$Q = c_v(T_2 - T_1) + \frac{R(T_1 - T_2)}{n-1} = c_v(T_2 - T_1) - \frac{c_v(\gamma-1)(T_2 - T_1)}{n-1}$$

$$= c_v(T_2 - T_1)\left[1 - \frac{\gamma - 1}{n-1}\right] = c_v\frac{n-\gamma}{n-1}(T_2 - T_1)$$

$$= c_n(T_2 - T_1)$$

Ordinarily, both heat and work are involved in a polytropic process. To evaluate the heat transfer during such a process, it is necessary to first evaluate the work via either $\int pdv$ or $-\int vdp$, depending on whether it is a closed or an open steady flow system. The application of the first law will then yield the heat transfer.

The polytropic processes for various values of n are shown in Fig. 10.3 on the p–v and T–s diagrams.

$$pv^n = C$$

On differentiation,

$$v^n\, dp + pnv^{n-1}\, dv = 0$$

$$\frac{dp}{dv} = -n\frac{p}{v} \tag{10.59}$$

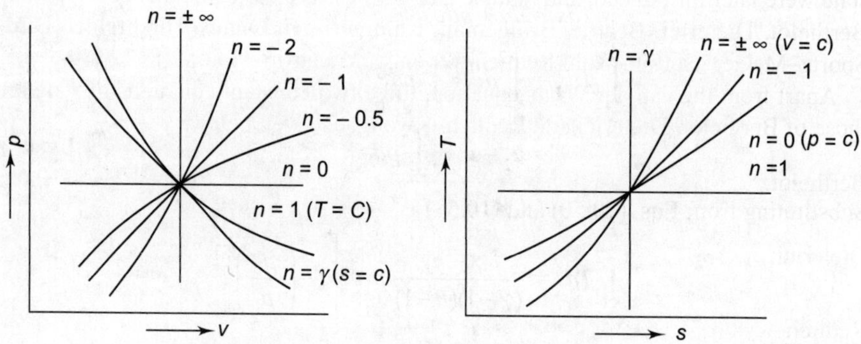

Fig. 10.3 Process in which pv^n = Constant

The slope of the curve increases in the negative direction with increase of n. The values of n for some familiar processes are given below.

Isobaric process ($p = c$), $n = 0$

Isothermal process ($T = c$), $n = 1$

Isentropic process ($s = c$), $n = \gamma$

Isometric or isochoric process ($v = c$), $n = \infty$.

10.4 EQUATIONS OF STATE: DIFFERENCE BETWEEN IDEAL AND REAL GASES

> **LO 10.9**
> Discuss equations of state for real gases

The ideal gas equation of state $p\bar{v} = \bar{R}T$ can be established from the postulates of the kinetic theory of gases developed by Clerk Maxwell, with two important assumptions that there is little or no attraction between the molecules of the gas and that the volume occupied by the molecules themselves is negligibly small compared to the volume of the gas. When pressure is very small or temperature is very large, the intermolecular attraction and the volume of the molecules compared to the total volume of the gas are not of much significance, and the real gas obeys very closely the ideal gas equation. But as pressure increases, the intermolecular forces of attraction and repulsion increase, and also the volume of the molecules becomes appreciable compared to the total gas volume. So then the real gases deviate considerably from the ideal gas equation. van der Waals, by applying the laws of mechanics to individual molecules, introduced two correction terms in the equation of ideal gas, and his equation is given below.

$$\left(p+\frac{a}{v^2}\right)(v-b)=RT \tag{10.60}$$

The coefficient a was introduced to account for the existence of mutual attraction between the molecules. The term a/v^2 is called the *force of cohesion*. The coefficient b was introduced to account for the volumes of the molecules, and is known as *co-volume.*

Real gases conform more closely with the van der Waals equation of state than the ideal gas equation of state, particularly at higher pressures. But it is not obeyed by a real gas in all ranges of pressures and temperatures. Many more equations of state were later introduced, and notable among these are the equations developed by Berthelot, Dieterici, Beattie–Bridgeman, Kammerlingh Onnes, Hirshfelder–Bird-Sportz–McGee–Sutton, Wohl Redlich–Kwong and Martin–Hou.

Apart from the van der Waals equation, three two-constant equations of state are those of Berthelot, Dieterici and Redlich–Kwong, as given below:

Berthelot:
$$p=\frac{RT}{v-b}-\frac{a}{Tv^2} \tag{10.61}$$

Dieterici:
$$p=\frac{RT}{v-b}\cdot e^{-a/RTv} \tag{10.62}$$

Redlich–Kwong:
$$p=\frac{RT}{v-b}-\frac{a}{T^{1/2}v(v+b)} \tag{10.63}$$

The constants, a and b are evaluated from the critical data, as shown for van der Waals equation in Section 10.7. The Berthelot and Dieterici equations of state, like the van der Waals equation, are of limited accuracy. But the Redlich–Kwong equation gives good results at high pressures and is fairly accurate for temperatures above the critical value.

Another two-constant equation which is again of limited accuracy is the Saha-Bose equation of state as given below.

$$p=-\frac{RT}{2b}\,e^{-a/RTv}\ln\left(\frac{v-2b}{v}\right) \tag{10.64}$$

It is, however, quite accurate for densities less than about 0.8 times the critical density.

One more widely used equation of state with good accuracy is the Beattie–Bridgeman equation:

$$p=\frac{RT(1-e)}{v^2}(v+B)-\frac{A}{v^2} \tag{10.65}$$

where

$$A=A_0\left(1-\frac{a}{v}\right),\ B=B_0\left(1-\frac{b}{v}\right),\ e=\frac{c}{vT^3}$$

There are five constants, A_0, B_0, a, b and c, which have to be determined experimentally for each gas. The Beattie–Bridgeman equation does not give satisfactory results in the critical point region.

The Benedict-Webb-Rubin Equation of State is another equation developed for real gases, featuring an increased number of experimentally determined constants. It is expressed as follows:

$$p=\rho RT+B_0RT-A_0-\frac{C_0}{T^2}\rho^2+(bRT-a)\rho^3+\alpha a p^6+\frac{c\rho^3}{T^2}\left(1+\gamma\rho^2\right)\exp\left(-\gamma\rho^2\right) \tag{10.66}$$

Where ρ represents the molar density, and all eight constants: $a, b, c, A, B, C, \alpha, \gamma$ are empirical, requiring experimental determination. Further improvements to this equation have been proposed, incorporating increased complexity in the calculation process.

All these equations mentioned above reduce to the ideal gas equation for large volumes and temperatures and for very small pressures.

10.5 VIRIAL EXPANSIONS

LO 10.10

Explain virial equations of state, compressibility factor and relate them

The relations between $p\bar{v}$ and p in the form of power series, as given in Eq. (10.1), may be expressed as

$$p\bar{v} = A(1 + B'p + C'p^2 + D'p^3 + ...)$$

For any gas, from Eq. (10.3)

$$\lim_{p \to 0} p\bar{v} = A = \bar{R}T$$

$$\therefore \quad \frac{p\bar{v}}{\bar{R}T} = 1 + B'p + C'p^2 + D'p^3 + ... \tag{10.67}$$

$$= 1 + \frac{B'\bar{R}T}{p} + C'\left(\frac{\bar{R}T}{v}\right)^2 + D'\left(\frac{\bar{R}T}{p}\right)^3 4$$

An alternative expression is

$$\frac{p\bar{v}}{\bar{R}T} = 1 + \frac{B}{\bar{v}} + \frac{C}{\bar{v}^2} + \frac{D}{\bar{v}^3} + \cdots \tag{10.68}$$

Both expressions in Eqs. (10.67) and (10.68) are known as *virial expansions* or virial equations of state, first introduced by the Dutch physicist, Kammerlingh Onnes, B', C', B, C, etc. are called *virial coefficients*. B' and B are called second virial coefficients, C' and C are called third virial coefficients and so on. For a given gas, these coefficients are the functions of temperature only.

The ratio $p\bar{v}/\bar{R}T$ is called the *compressibility factor*, Z. For an ideal gas $Z = 1$. The magnitude of Z for a certain gas at a particular pressure and temperature gives an indication of the extent of deviation of the gas from the ideal gas behaviour. The virial expansions become

$$Z = 1 + B'p + C'p^2 + D'p^3 + ... \tag{10.69}$$

and

$$Z = 1 + \frac{B}{\bar{v}} + \frac{C}{\bar{v}^2} + \frac{D}{\bar{v}^3} + \cdots \tag{10.70}$$

The relations between B', C' and B, C, ... can be derived as given below.

$$\frac{pv}{\bar{R}T} = 1 + B'p + C'p^2 + D'p^3 + ...$$

$$= 1 + B'\left[\frac{\bar{R}T}{\bar{v}}\left(1 + \frac{B}{\bar{v}} + \frac{C}{\bar{v}^2} + ...\right)\right]$$

$$+ C'\left[\left(\frac{\bar{R}T}{\bar{v}}\right)^2\left(1 + \frac{B}{\bar{v}} + \frac{C}{\bar{v}^2} ... +\right)^2\right] + ...$$

$$= 1 + \frac{B'\bar{R}T}{\bar{v}} + \frac{B'B\bar{R}T + C'(\bar{R}T)^2}{\bar{v}^2}$$

$$= + \frac{B'\overline{R}TC + C'(\overline{R}T)^2 + D'(\overline{R}T)^3}{\overline{v}^3} + ... \quad (10.71)$$

Comparing this equation with Eq. (10.82) and rearranging

$$B' = \frac{B}{\overline{R}T}, \; C' = \frac{C - B^2}{(\overline{R}T)^2},$$

$$D' = \frac{D - 3BC + 2B^3}{(\overline{R}T)^3}, \text{ and so on}$$

Therefore,

$$Z = \frac{p\overline{v}}{\overline{R}T} = 1 + B'p + C'p^2 + ...$$

$$= 1 + \frac{B}{\overline{R}T}p + \frac{C - B^2}{(\overline{R}T)^2}p^2 + ... \quad (10.72)$$

The terms $B/\overline{v}$, $C/\overline{v}^2$, etc. of the virial expansion arise on account of molecular interactions. If no such interactions exist (at very low pressures) $B = 0$, $C = 0$, etc. and $Z = 1$ and $p\overline{v} = \overline{R}T$.

10.6 LAW OF CORRESPONDING STATES

LO 10.11 Explain the law of corresponding states, the generalized compressibility chart

For a certain gas, the compressibility factor Z is a function of p and T [Eq. (10.72)], and so a plot can be made of lines of constant temperature on coordinates of p and Z (Fig. 10.4). From this plot, Z can be obtained for any value of p and T, and the volume can then be obtained from the equation $pv = ZRT$. The advantage of using Z instead of a direct plot of v is a smaller range of values in plotting.

For each substance, there is a compressibility factor chart. It would be very convenient if one chart could be used for all substances. the general shapes of the vapour dome and of the constant temperature lines on the p–v plane are similar for all substances, although the scales may be different. This physical similarity can be exploited by using dimensionless properties called *reduced properties*. The reduced pressure is the ratio of the existing pressure to the critical pressure of the substance, and similarly for reduced temperature and reduced volume. Then

$$p_r = \frac{p}{p_c}, \; T_r = \frac{T}{T_c}, \; v_r = \frac{v}{v_c}$$

where subscript r denotes the reduced property, and subscript c denotes the property at the critical state.

At the same pressure and temperature the specific or molal volumes of different gases are different. However, it is found from experimental data that at the same reduced pressure and reduced temperature, the reduced volumes of different gases are approximately the same. Therefore, for all substances

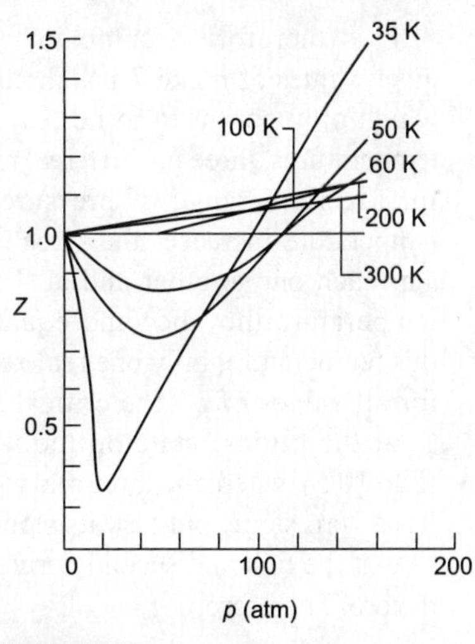

Fig. 10.4 Variation of the compressibility factor of hydrogen with pressure at constant temperature

$$v_r = f(p_r, T_r) \tag{10.73}$$

Now,

$$v_r = \frac{v}{v_c} = \frac{ZRTp_c}{Z_cRT_cp} = \frac{Z}{Z_c}\cdot\frac{T_r}{p_r} \tag{10.74}$$

where $Z_c = \frac{p_c v_c}{RT_c}$. This is called the critical compressibility factor. Therefore, from Eqs. (10.73) and (10.74),

$$Z = f(p_r, T_r, Z_c) \tag{10.75}$$

Experimental values of Z_c for most substances fall within a narrow range 0.20–0.30. Therefore, Z_c may be taken to be a constant and Eq. (10.75) reduces to

$$Z = f(p_r, T_r) \tag{10.76}$$

When T_r is plotted as a function of reduced pressure and Z, a single plot, known as *the generalised compressibility chart*, is found to be satisfactory for a great variety of substances. Although necessarily approximate, the plots are extremely useful in situations where detailed data on a particular gas are lacking but its critical properties are available.

The relation among the reduced properties, p_r, T_r and v_r, is known as the *law of corresponding states*. It can be derived from the various equations of state, such as those of van der Waals, Berthelot and Dieterici. For a van der Waals gas,

$$\left(p + \frac{a}{v^2}\right)(v - b) = RT$$

where a, b and R are the characteristic constants of the particular gas.

$$\therefore \qquad p = \frac{RT}{v-b} - \frac{a}{v^2}$$

or

$$pv^3 - (pb + RT)\, v^2 + av - ab = 0$$

It is therefore a cubic in v and for given values of p and T has three roots of which only one need to be real. For low temperatures, three positive real roots exist for a certain range of pressure. As the temperature increases the three real roots approach one another and at the critical temperature they become equal. Above this temperature, only one real root exists for all values of p. The critical isotherm T_c at the critical state on the p–v plane (Fig. 10.5), where the three real roots of the van der Waals equation coincide, not only has a zero slope, but also its slope changes at the critical state (point of inflection), so that the first and second derivatives of p with respect to v at $T = T_c$ are each equal to zero. Therefore,

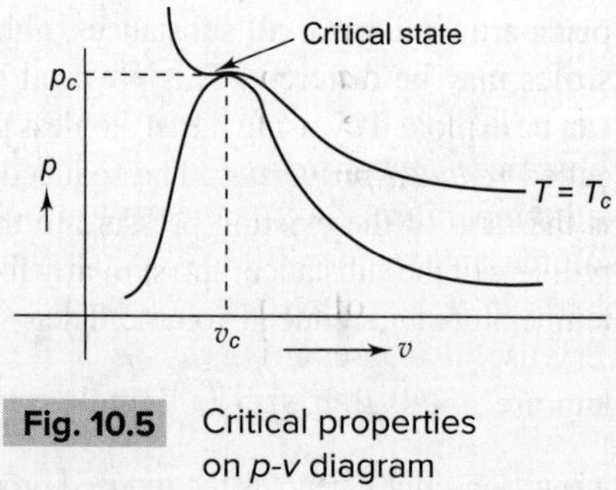

Fig. 10.5 Critical properties on p-v diagram

$$\left(\frac{\partial p}{\partial v}\right)_{T=T_c} = -\frac{RT_c}{(v_c - b)^2} + \frac{2a}{v_c^3} = 0 \tag{10.77}$$

$$\left(\frac{\partial^2 p}{\partial v^2}\right)_{T=T_c} = \frac{2 \cdot RT_c}{(v_c - b)^3} - \frac{6a}{v_c^4} = 0 \tag{10.78}$$

From these two equations, by rearranging and dividing, $b = \frac{1}{3} v_c$.
Substituting the value of b in Eq. (10.77)

$$R = \frac{8a}{9T_c v_c}$$

Substituting the values of b and R in Eq. (10.60)

$$\left(p_c + \frac{a}{v_c^2}\right)\left(\frac{2}{3}v_c\right) = \frac{8a}{9T_c v_c} \cdot T_c$$

$$a = 3p_c v_c^2$$

Therefore, the value of R becomes

$$R = \frac{8}{3}\frac{p_c v_c}{T_c}$$

The values of a, b and R have thus been expressed in terms of critical properties. Substituting these in the van der Waals equation of state

$$\left(p + \frac{3p_c v_c^2}{v^2}\right)\left(v - \frac{1}{3}v_c\right) = \frac{8}{3}\frac{p_c v_c}{T_c} \cdot T$$

or,

$$\left(\frac{p}{p_c} + \frac{3v_c^2}{v^2}\right)\left(\frac{v}{v_c} - \frac{1}{3}\right) = \frac{8}{3}\frac{T}{T_c}$$

Using the reduced parameters,

$$\therefore \quad \left(p_r + \frac{3}{v_r^2}\right)(3v_r - 1) = 8T_r \tag{10.79}$$

In the reduced equation of state (10.79), the individual coefficients a, b and R for a particular gas have disappeared. So this equation is an expression of the *law of corresponding states* because it reduces the properties of all gases to one formula. It is a 'law' to the extent that real gases obey van der Waals equation. Two different substances are considered to be in 'corresponding states', if their pressures, volumes and temperatures are of the same fractions (or multiples) of the critical pressure, volume and temperature of the two substances. The *generalised compressibility* charts in terms of reduced properties are shown in Fig. 10.6(a), (b) and (c). It is very useful in predicting the properties of are substances for which more precise data are not available. The value of Z at the critical state of a van der Waals gas is 0.375 $\left(\text{since } R = \frac{8}{3}\frac{p_c v_c}{T_c}\right)$. At very low pressures, Z approaches unity as a real gas approaches the ideal gas behaviour. Equation (10.79) can also be written in the following form:

$$\left(p_r v_r + \frac{3}{v_r}\right)(3v_r - 1) = 8T_r v_r$$

$$\therefore \quad p_r v_r = \frac{8T_r v_r}{3v_r - 1} - \frac{3}{v_r} \tag{10.80}$$

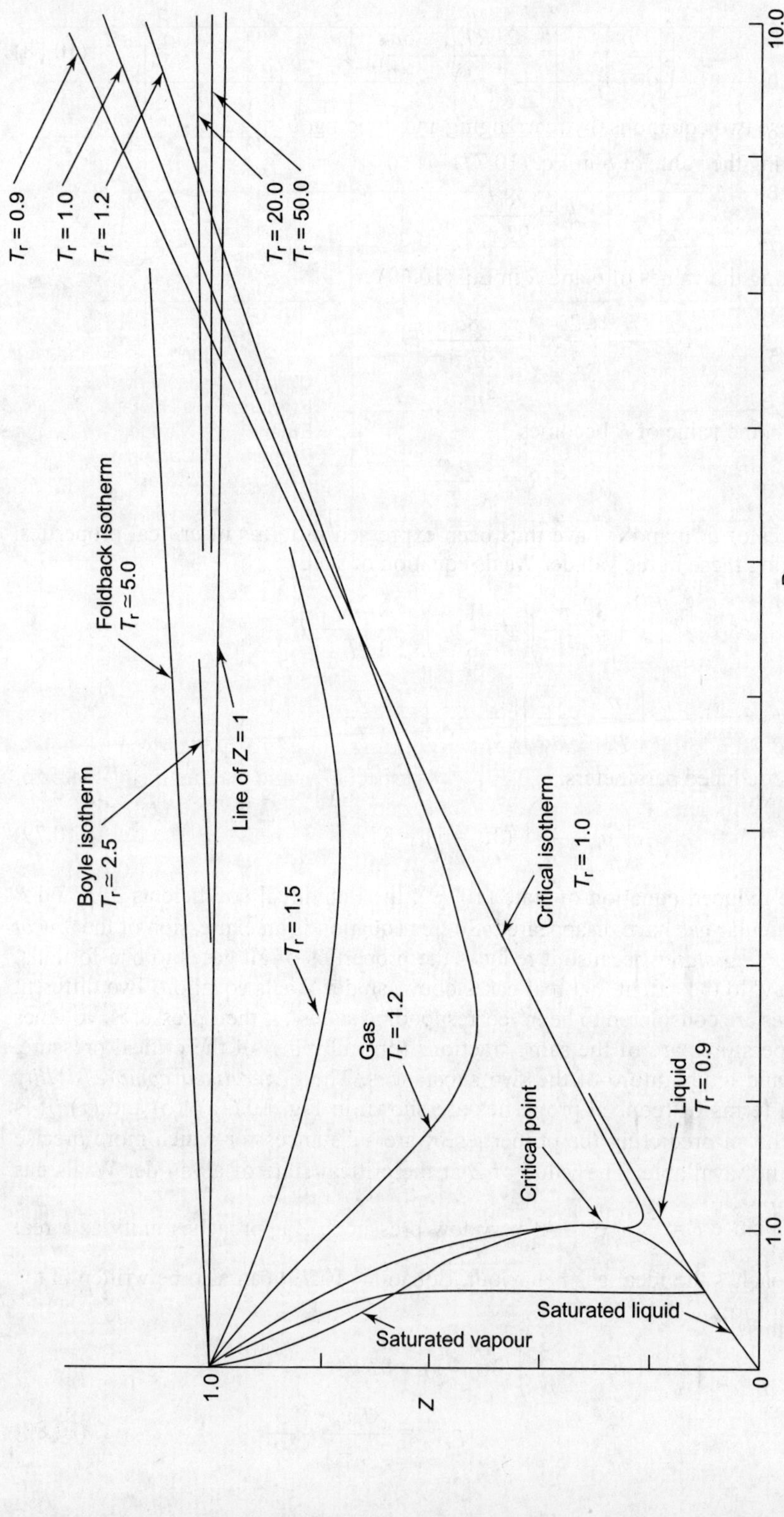

Fig. 10.6(a) Generalised compressibility chart

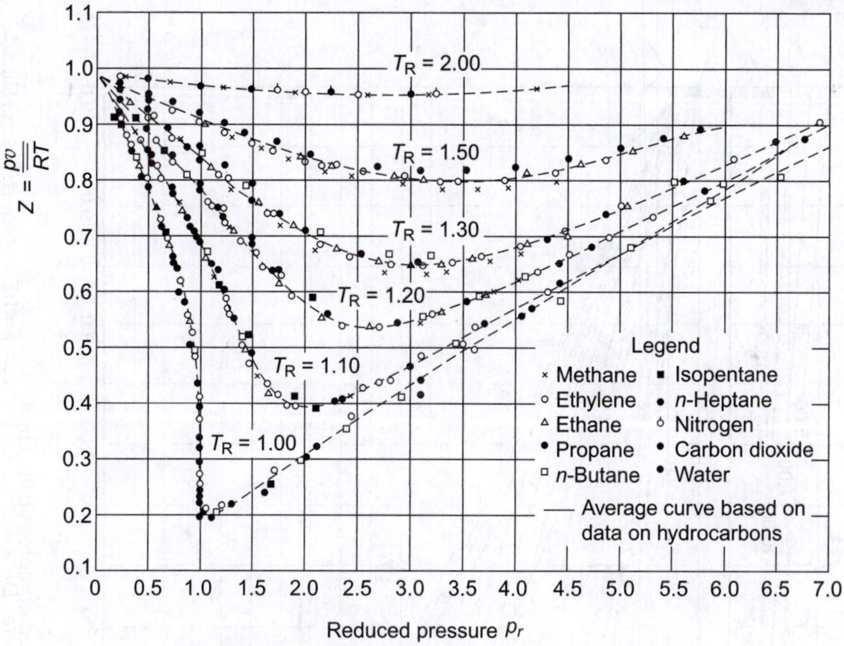

Fig. 10.6(b) Generalised compressibility chart

Figure 10.6 shows the law of corresponding states in reduced coordinates, $(p_r v_r)$ vs p_r. Differentiating Eq. (10.80) with respect to p_r and making it equal to zero, it is possible to determine the minima of the isotherms as shown below.

$$\frac{\partial}{\partial p_r}\left[\frac{8T_r v_r}{3v_r - 1} - \frac{3}{v_r}\right]_{T_r} = 0$$

or

$$\frac{\partial}{\partial v_r}\left[\frac{8T_r v_r}{3v_r - 1} - \frac{3}{v_r}\right]_{T_r}\left[\frac{\partial v_r}{\partial p_r}\right]_{T_r} = 0$$

$$\frac{\partial}{\partial v_r}\left[\frac{8T_r v_r}{3v_r - 1} - \frac{3}{v_r}\right]_{T_r} = 0$$

$\therefore$

$$\frac{8T_r}{(3v_r - 1)^2} = \frac{3}{v_r^2}$$

or

$$\frac{3(3v_r - 1)^2}{v_r^2} = 8T_r = \left[p_r + \frac{3}{v_r^2}\right](3v_r - 1)$$

Simplifying

$$(p_r v_r)^2 - 9(p_r v_r) + 6p_r = 0$$

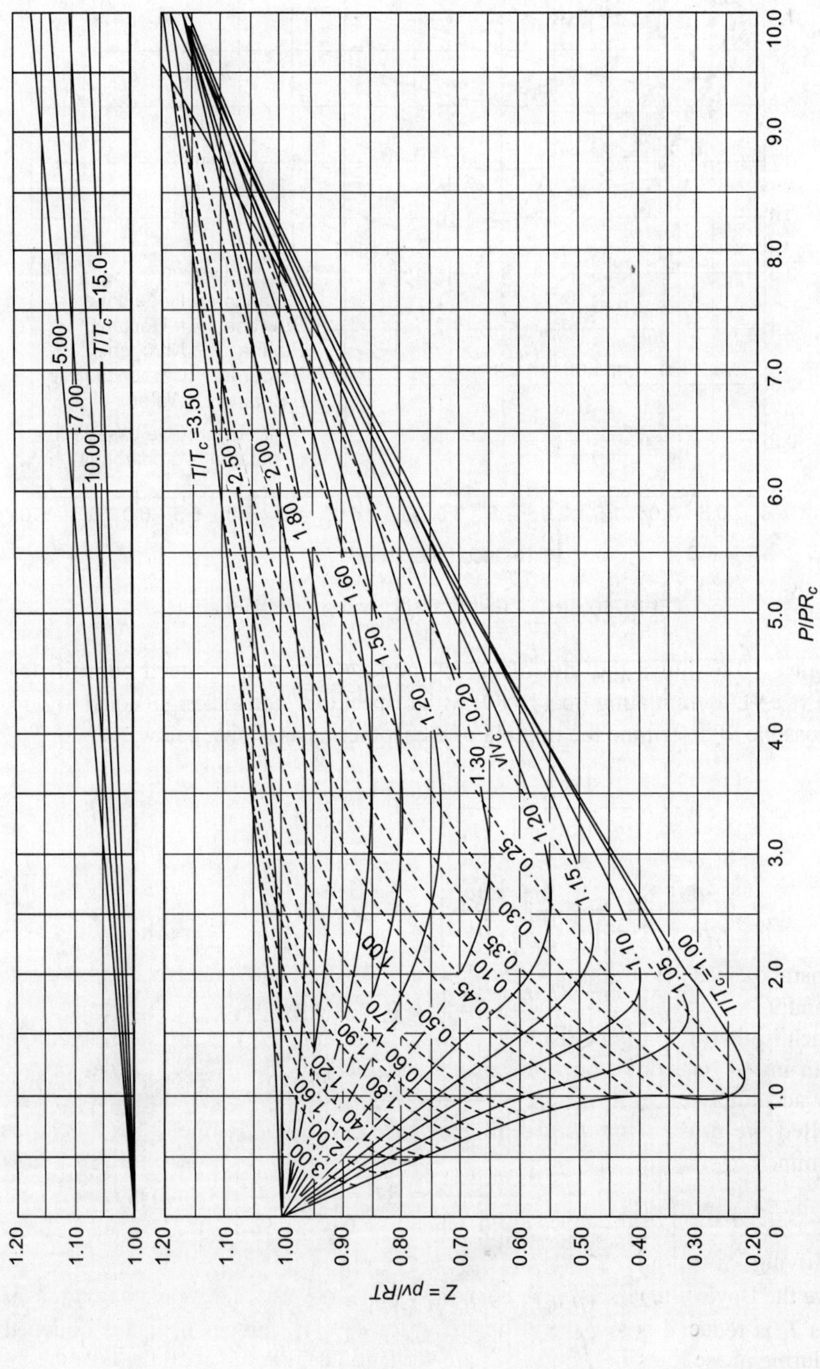

Fig. 10.6(c) Generalised compressibility chart–low pressure range. Data from L.C. Nelson and E.F. Obert, Generalised compressibility charts, Chem. En., vol. 61, p. 203, 1954. *Note:* $v/v_c = p_c\, v/RT_c$

This is the equation of a parabola passing through the minima of the isotherms (Fig. 10.7).

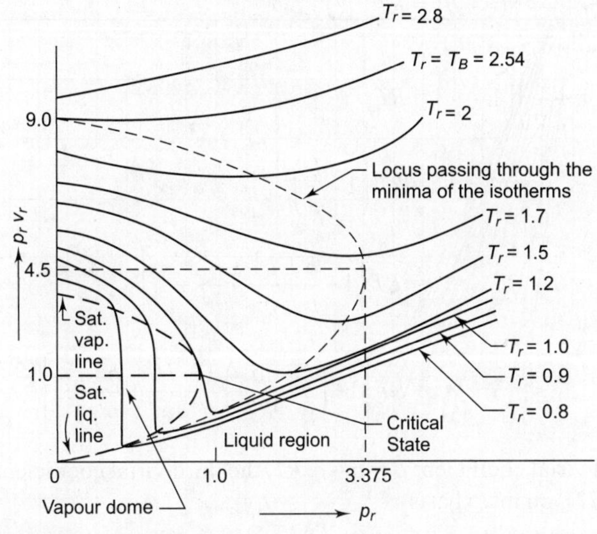

Fig. 10.7 Law of corresponding states in reduced coordinates

When
$$p_r = 0,$$
$$p_r v_r = 0, 9$$

Again

$$p_r = \frac{9(p_r v_r) - (p_r v_r)^2}{6}$$

$$\frac{dp_r}{d(p_r v_r)} = 9 - 2\,(p_r\, v_r) = 0$$

$\therefore$
$$p_r\, v_r = 4.5$$

$$p_r = \frac{9 \times 4.5 - (4.5)^2}{6} = 3.375$$

The parabola has the vertex at $p_r v_r = 4.5$ and $p_r = 3.375$, and it intersects the ordinate at 0 and 9.

Each isotherm up to that marked T_B has a minimum (Fig. 10.7). The T_B isotherm has an initial horizontal portion ($p_r v_r$ = constant) so that Boyle's law is obeyed fairly accurately up to moderate pressures. Hence, the corresponding temperature is called the *Boyle temperature* for that gas. The Boyle temperature T_B can be determined by making

$$\left[\frac{\partial (p_r v_r)}{\partial p_r}\right]_{T_r = T_B} = 0 \text{ when } p_r = 0$$

Above the Boyle temperature, the isotherms slope upwards and show no minima.

As T_r is reduced below the critical (i.e. for $T_r < 1$), the gas becomes liquefied, and during phase transition isotherms are vertical. The minima of all these isotherms lie in the liquid zone.

van der Waals equation of state can be expressed in the virial form given as follows.

$$\left(p+\frac{a}{v^2}\right)(v-b)=RT$$

$$\left(pv+\frac{a}{v}\right)\left(1-\frac{b}{v}\right)=RT$$

$$\therefore \quad pv+\frac{a}{v}=RT\left(1-\frac{b}{v}\right)^{-1}$$

$$=RT\left[1+\frac{b}{v}+\frac{b^2}{v^2}+\frac{b^3}{v^3}+\ldots\right] \quad \left(\text{where } \frac{b}{v}<1\right)$$

$$\therefore \quad pv=RT\left[1+\left(b-\frac{a}{RT}\right)\frac{1}{v}+\frac{b^2}{v^2}+\frac{b^3}{v^3}+\ldots\right] \tag{10.81}$$

$\therefore$ The second virial coefficient $B = b - a/RT$, the third virial coefficient $C = b^2$, etc. From Eq. (10.72), on mass basis

$$v=RT\left(1+\frac{B}{\overline{R}T}p+\frac{C-B^2}{\overline{R}T^2}p^2+\ldots\right)$$

To determine Boyle temperature, T_B

$$\left[\frac{\partial(pv)}{\partial p}\right]_{\substack{T=C\\p=0}}=0=\frac{B}{\overline{R}T}$$

$$\therefore \quad B=0$$

or

$$T_B=\frac{a}{bR}, \text{ because } B=b-\frac{a}{RT}$$

The point at which B is equal to zero gives the Boyle temperature. The second virial coefficient is the most important. Since $\left[\frac{\partial(pv)}{\partial p}\right]_{p=0} = B$, when B is known, the behaviour of the gas at moderate pressures is completely determined. The terms which contain higher powers (C/v^2, D/v^3, etc.) becomes significant only at very high pressures.

10.7 OTHER EQUATIONS OF STATE

van der Waals equation is one of the oldest equations of state introduced in 1899, where the constants a and b are related to the critical state properties as found earlier,

LO 10.12

Discuss other equations of state than ideal gas equation'

$$a=3p_c v_c^2=\frac{27}{64}\frac{R^2T_c^2}{p_c},\ b=\frac{1}{3}v_c=\frac{1}{8}\frac{RT_c}{p_c}$$

The Beattie–Bridgeman equation developed in 1928 is given in Eq. (10.65) which has five constants. It is essentially an empirical curve fit of data, and is reasonably accurate when values of specific volume are greater than v_c.

Benedict, Webb and Rubin extended the Beattie–Bridgeman equation of state to cover a broader range of states as given below.

$$p = \frac{\bar{R}T}{v} + \left(B\bar{R}T - A - \frac{c}{T^2} \right) \frac{1}{\bar{v}^2} + \frac{(b\bar{R}T - a)}{\bar{v}^3} + \frac{a}{\bar{v}^6}$$

$$+ \frac{c}{\bar{v}^3 T^2} \left(1 + \frac{\gamma}{\bar{v}^2} \right) \exp\left(-\frac{\gamma}{\bar{v}^2} \right)$$

It has eight constants and is quite successful in predicting the p–v–T behaviour of light hydrocarbons.

The Redlich–Kwong equation proposed in 1949 and given by Eq. (10.63) has the constants a and b in terms of critical properties as follows:

$$a = 0.4275 \frac{R^2 T_c^{2.5}}{p_c}, \quad b = 0.0867 \frac{RT_c}{p_c}$$

The values of the constants for the van der Waals, Redlich–Kwong and Benedict-Webb–Rubin equations of state are given in Table 10.2, while those for the Beattie-Bridgeman equation of state are given in Table 10.3. Apart from these, many other multi-constant equations of state have been proposed. With the advent of high speed computers, equations having 50 or more constants have been developed for representing the p–v–T behaviour of different substances.

Table 10.2 Constants for the van der Waals, Redlich–Kwong and Benedict–Webb–Rubin equations of state

1. van der Waals and Redlich–Kwong Constants for pressure in bars, specific volume in m3/kmol and temperature in K.

Substance	*van der Waals*		*Redlich–Kwong*	
	a	*b*	*a*	*b*
	$bar\left(\frac{m^3}{kmol}\right)^2$	$\frac{m^3}{kmol}$	$bar\left(\frac{m^3}{kmol}\right)^2 K^{1/2}$	$\frac{m^3}{kmol}$
Air	1.368	0.0367	15.989	0.02541
Butane (C_4H_{10})	13.86	0.1162	289.55	0.08060
Carbon dioxide (CO_2)	3.647	0.0428	64.43	0.02963
Carbon monoxide (CO)	1.474	0.0395	17.22	0.02737
Methane (CH_4)	2.293	0.0428	32.11	0.02965
Nitrogen (N_2)	1.366	0.0386	15.53	0.02677
Oxygen (O_2)	1.369	0.317	17.22	0.02197
Propane (C_3H_8)	9.349	0.0901	182.23	0.06242
Refrigerant 12	10.49	0.0971	208.59	0.06731
Sulfur dioxide (SO_2)	6.883	0.0569	144.80	0.03945
Water (H_2O)	5.531	0.0305	142.59	0.02111

Source: Calculated from critical data.

2. Benedict–Webb–Rubin Constants for pressure in bars, specific volume in m^3/kmol and temperature in K.

Substance a	A	b	B	c	C	α	γ
C_4H_{10} 1.9073	10.218	0.039998	0.12436	3.206×10^5	1.006×10^6	1.101×10^{-3}	0.0340
CO_2 0.1386	2.7737	0.007210	0.04991	1.512×10^4	1.404×10^5	8.47×10^{-5}	0.00539
CO 0.0371	1.3590	0.002632	0.05454	1.054×10^3	8.676×10^3	1.350×10^{-4}	0.0060
CH_4 0.0501	1.8796	0.003380	0.04260	2.579×10^3	2.287×10^4	1.244×10^{-4}	0.0060
N_2 0.0254	1.0676	0.002328	0.04074	7.381×10^2	8.166×10^1	1.272×10^{-4}	0.0053

Source: H.W. Copper, and J.C. Goldfrank, Hydrocarbon Processing, 45 (12); 141 (1967).

Table 10.3

(a) The Beattie–Bridgeman equation of state is

$$p = \frac{R_u T}{\bar{v}^2}\left(1 - \frac{c}{\bar{v}T^2}\right)(\bar{v} + B) - \frac{A}{\bar{v}}, \text{ where } A = A_0\left(1 - \frac{a}{\bar{v}}\right) \text{ and } B = B_0\left(1 - \frac{b}{\bar{v}}\right)$$

When p is in kPa, $\bar{v}$ is in m^3/k mol, T is in K and $R_u = 8.314$ kPa·m^3/(kmol · K), the five constants in the Beattie-Bridgeman equation are as follows:

Gas	A_0	a	B_0	b	c
Air	131.8441	0.01931	0.04611	– 0.001101	4.34×10^4
Argon, Ar	130.7802	0.02328	0.03931	0.0	5.99×10^4
Carbon dioxide, CO_2	507.2836	0.07132	0.10476	0.07235	6.60×10^5
Helium, He	2.1886	0.05984	0.01400	0.0	40
Hydrogen, H_2	20.0117	– 0.00506	0.02096	– 0.04359	504
Nitrogen, N_2	136.2315	0.02617	0.05046	– 0.00691	4.20×10^4
Oxygen, O_2	151.0857	0.02562	0.04624	0.004208	4.80×10^4

Source: Gordon J. Van Wylen and Richard E. Sonntag, Fundamentals of Classical Thermodynamics, English/SI Version, 3d Edition., Wiley New York, 1986, p. 46. Table 3.3.

10.8 PROPERTIES OF MIXTURES OF GASES—DALTON'S LAW OF PARTIAL PRESSURES

> **LO 10.13**
> Discuss Dalton's law of partial pressure and Amagat's law of partial volume for a gas mixture

Let us imagine a homogeneous mixture of inert ideal gases at a temperature T, a pressure p and a volume V. Let us suppose there are n_1 moles of gas A_1, n_2 moles of gas A_2, ... and up to n_c moles of gas A_c (Fig. 10.8). Since there is no chemical reaction, the mixture is in a state of equilibrium with the equation of state

$$pV = (n_1 + n_2 + \dots n_c)\, \bar{R}T$$

where $\bar{R} = 8.3143$ kJ/kg mol K

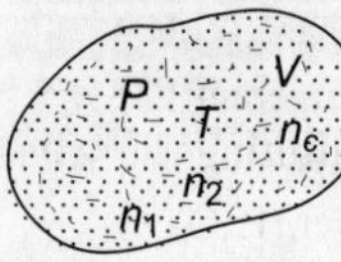

Fig. 10.8 Mixture of gases

$$\therefore \qquad p = \frac{n_1 \bar{R} T}{V} + \frac{n_2 \bar{R} T}{V} + \dots + \frac{n_c \bar{R} T}{V}$$

The expression $\frac{n_k \bar{R} T}{V}$ represents the pressure that the kth gas would exert if it occupied the volume V alone at temperature T. This is called the *partial pressure* of the kth gas and is denoted by p_k. Thus,

$$p_1 = \frac{n_1 \bar{R} T}{V}, p_2 = \frac{n_2 \bar{R} T}{V}, \dots, p_c = \frac{n_c \bar{R} T}{V}$$

and

$$p = p_1 + p_2 + \dots + p_c \qquad (10.82)$$

This is known as *Dalton's law of partial pressures* which states that the total pressure of a mixture of ideal gases is equal to the sum of the partial pressures.

Visualisation of Dalton's law of partial pressures is shown in Fig. 10.9. There are n_1 moles of gas A_1 in compartment 1, n_2 moles of gas A_2 in compartment 2 and so on, each at pressure p and temperature T. Compartments are separated from one another by partitions shown in (a). When the particulars are removed, the gases mix and diffuse into one another till equilibrium is reached as shown in (b). If each of these gases A_1, A_2, etc. Occupies the entire volume of the mixture at the same temperature, the pressure gauges will record the respective partial pressures shown in (c) and Dalton's states that $p = p_1 + p_2 + \dots + p_c$.

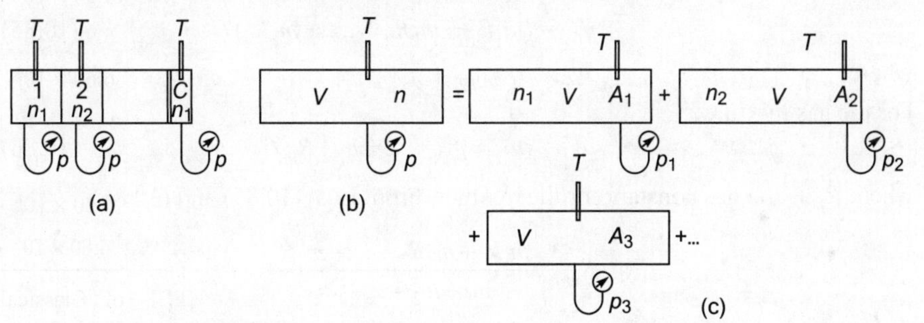

Fig. 10.9 Demonstration of two of partial pressures

Now

$$V = (n_1 + n_2 + \dots + n_c) \cdot \frac{\bar{R} T}{p}$$

$$= \Sigma n_K \cdot \frac{\bar{R} T}{p}$$

and the partial pressure of the kth gas is

$$p_k = \frac{n_k \bar{R} T}{V}$$

Substituting the value of V

$$p_K = \frac{n_K \cdot p}{\Sigma n_K} = \frac{n_K}{\Sigma n_K} \cdot p$$

Now

$$\Sigma n_K = n_1 + n_2 + ... + n_c$$

$$= \text{Total number of moles of gas}$$

The ratio $\frac{n_k}{\Sigma n_k}$ is called the *mole fraction of the* *k*th gas, and is denoted by x_K.

Thus

$$x_1 = \frac{n_1}{\Sigma n_K}, x_2 = \frac{n_2}{\Sigma n_K}, ..., x_c = \frac{n_c}{\Sigma n_K}$$

and

$$p_1 = x_1 p, p_2 = x_2 p, ..., p_c = x_c p$$

or

$$p_K = x_K \cdot p \tag{10.83}$$

Also

$$x_1 + x_2 + ... + x_c = 1 \tag{10.84}$$

In a mixture of gases, if not all but one mole fraction is determined, the last can be calculated from the above equation. Again, in terms of masses

$$p_1 V = m_1 R_1 T$$

$$p_2 V = m_2 R_2 T$$

$$.......................$$

$$p_c V = m_c R_c T$$

Adding, and using Dalton's law

$$pV = (m_1 R_1 + m_2 R_2 + ... + m_c R_c) T \tag{10.85}$$

where

$$p = p_1 + p_2 + ... + p_c$$

For the gas mixture

$$pV = (m_1 + m_2 + ... + m_c) R_m T \tag{10.86}$$

where R_m is the gas constant for the mixture. From Eqs. (10.85) and (10.86)

$$R_m = \frac{m_1 R_1 + m_2 R_2 + ... + m_c R_c}{m_1 + m_2 + ... + m_c} \tag{10.87}$$

The gas constant of the mixture is thus the weighted mean, on a mass basis, of the gas constants of the components.

The total mass of gas mixture m is

$$m = m_1 + ... + m_c$$

If μ denotes the *equivalent molecular* weight of the mixture having n total number of moles.

$$n\mu = n_1 \mu_1 + n_2 \mu_2 + ... + n_c \mu_c$$

∴

$$\mu = x_1 \mu_1 + x_2 \mu_2 + ... + x_c \mu_c$$

or

$$\mu = \Sigma x_k \mu_K \tag{10.88}$$

The atmospheric air is assumed to consist of 21 percent O_2, 78 per cent N_2 and 1 percent A by volume. Its molecular weight is

$$\mu_{air} = 0.78 \times 28 + 0.21 \times 32 + 0.01 \times 40$$

$$= 21.84 + 6.72 + 0.4 = 28.96$$

A quantity called the *partial volume* of a component of a mixture is the volume that the component alone would occupy at the pressure and temperature of the mixture. Designating the partial volumes by V_1, V_2, etc.

$$pV_1 = m_1R_1T, pV_2 = m_2R_2T, ..., pV_c = m_cR_cT$$

or
$$p(V_1 + V_2 + ... + V_c) = (m_1R_1 + m_2R_2 + ... + m_cR_c)T \tag{10.89}$$

From Eqs. (10.85), (10.86) and (10.87)

$$V = V_1 + V_2 + ... + V_c \tag{10.90}$$

The total volume is thus equal to the sum of the partial volumes.

This is known as *Amagat's law of additive volumes*. Figure 10.10 shows that each component in a gas mixture occupies the respective partial volume at the same pressure p and temperature T of the mixture and Amagat's law states that the total volume of the mixture is equal to the sum of the partial volumes.

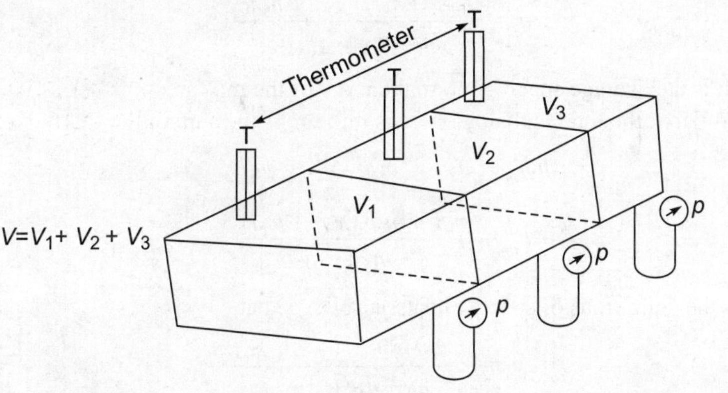

Fig. 10.10 Visualisation of Amagat's law of additive volumes

The specific *volume of the mixture*, v, is given by

$$v = \frac{V}{m} = \frac{V}{m_1 + m_2 + ... + m_c}$$

or
$$\frac{1}{v} = \frac{m_1 + m_2 + ... + m_c}{V}$$

$$= \frac{m_1}{V} + \frac{m_2}{V} + ... + \frac{m_c}{V}$$

or
$$\frac{1}{v} = \frac{1}{v_1} + \frac{1}{v_2} + ... + \frac{1}{v_c} \tag{10.91}$$

where v_1, v_2, ... denote specific volumes of the components, each component occupying the total volume.

Therefore, the density of the mixture

$$\rho = \rho_1 + \rho_2 + ... + \rho_c \tag{10.92}$$

10.9 INTERNAL ENERGY, ENTHALPY AND SPECIFIC HEATS OF GAS MIXTURES

LO 10.14
Explain Gibb's theorem for properties of a gas mixture'

When gases at equal pressures and temperatures are mixed adiabatically without work, as by inter-diffusion in a constant volume container, the first law requires that the internal energy of the gaseous system remains constant, and experiments show that the temperature remains constant. Hence, the internal energy of a mixture of gases is equal to the sum of the internal energies of the individual components, each taken at the temperature and volume of the mixture (i.e. sum of the 'partial' internal energies). This is also true for any of the thermodynamic properties like H, C_v, C_p, S, F and G, and is known as *Gibbs theorem*. Therefore, on a mass basis

$$mu_m = m_1u_1 + m_2u_2 + \dots + m_cu_c$$

$$u_m = \frac{m_1u_1 + m_2u_2 + \dots + m_cu_c}{m_1 + m_2 + \dots + m_c} \tag{10.93}$$

which is the average specific internal energy of the mixture.

Similarly, the total enthalpy of a gas mixture is the sum of the 'partial' enthalpies

$$mh_m = m_1h_1 + m_2h_2 + \dots + m_ch_c$$

and

$$h_m = \frac{m_1h_1 + m_2h_2 + \dots + m_ch_c}{m_1 + m_2 + \dots + m_c} \tag{10.94}$$

From the definitions of specific heats, it follows that

$$c_{vm} = \frac{m_1c_{v1} + m_2c_{v2} + \dots + m_cc_{vc}}{m_1 + m_2 + \dots + m_c} \tag{10.95}$$

and

$$c_{pm} = \frac{m_1c_{p1} + m_2c_{p2} + \dots + m_cc_{pc}}{m_1 + m_2 + \dots + m_c} \tag{10.96}$$

10.10 ENTROPY OF GAS MIXTURES

Gibbs theorem states that the total entropy of a mixture of gases is the sum of the partial entropies. The partial entropy of one of the gases of a mixture is the entropy that the gas would have if it occupied the whole volume alone at the same temperature. Let us imagine a number of inert ideal gases separated from one another by suitable partitions, all the gases being at the same temperature T and pressure p. The total entropy (initial)

$$S_i = n_1s_1 + n_2s_2 + \dots + n_cs_c$$

$$= \Sigma\, n_K s_K$$

From property relation

$$Tds = dh - vdp = c_p\, dT - vdp$$

$$\therefore \quad d\bar{s} = \bar{c}_p \frac{dT}{T} - \bar{R}\frac{dp}{p}$$

The entropy of 1 mole of the kth gas at T and p

$$\bar{s}_K = \int c_{p_K} \frac{dT}{T} - \bar{R} \ln p + \bar{s}_{0K}$$

where $\bar{s}_{0K}$ is the constant of integration.

$$\therefore \qquad S_i = \bar{R}\, \Sigma\, n_K \left(\frac{1}{\bar{R}} \int \bar{c}_{pK} \frac{dT}{T} + \frac{\bar{s}_{0K}}{\bar{R}} - \ln p \right)$$

Let

$$\sigma_K = \frac{1}{\bar{R}} \int \bar{c}_{pK} \frac{dT}{T} + \frac{\bar{s}_{0K}}{\bar{R}}$$

then

$$S_i = \bar{R}\, \Sigma\, n_k (\sigma_K - \ln p) \tag{10.97}$$

After the partitions are removed, the gases diffuse into one another at the same temperature and pressure, and by Gibbs theorem, the entropy of the mixture, S_f, is the sum of the partial entropies, with each gas exerting its respective partial pressure. Thus,

$$S_f = \bar{R} \Sigma\, n_K (\sigma_K - \ln p_K)$$

Since

$$p_K = x_K p$$

$$S_f = \bar{R} \Sigma\, n_K (\sigma_K - \ln x_K - \ln p) \tag{10.98}$$

A change in entropy due to the diffusion of any number of inert ideal gases is

$$S_f - S_i = -\bar{R} \Sigma\, n_K \ln x_K \tag{10.99}$$

or

$$S_f - S_i = -\bar{R}\, (n_1 \ln x_1 + n_2 \ln x_2 + \ldots + n_c \ln x_c)$$

Since the mole fractions are less than unity, $(S_f - S_i)$ is always positive, conforming to the second law.

Again

$$S_f - S_i = \left(n_1 \ln \frac{p_1}{p} + n_2 \ln \frac{p_2}{p} + \ldots + n_c \ln \frac{p_c}{p} \right) \tag{10.100}$$

which indicates that each gas undergoes in the diffusion process a *free expansion* from total pressure p to the respective partial pressure at constant temperature.

Similarly, on a mass basis, the entropy change due to diffusion

$$S_f - S_i = -\Sigma\, m_K R_K \ln \frac{p_K}{p}$$

$$= -\left(m_1 R_1 \ln \frac{p_1}{p} + m_2 R_2 \ln \frac{p_2}{p} + \ldots + m_c R_c \ln \frac{p_c}{p} \right)$$

Let us consider that there are n_a moles of the first component of a gas mixture in the first compartment and n_b moles of the second component in the second component, both are at the same pressure and temperature (Fig. 10.11). The partition separating the components is removed and the gases mix adiabatically. We are to determine the entropy change due to mixing of the gases.

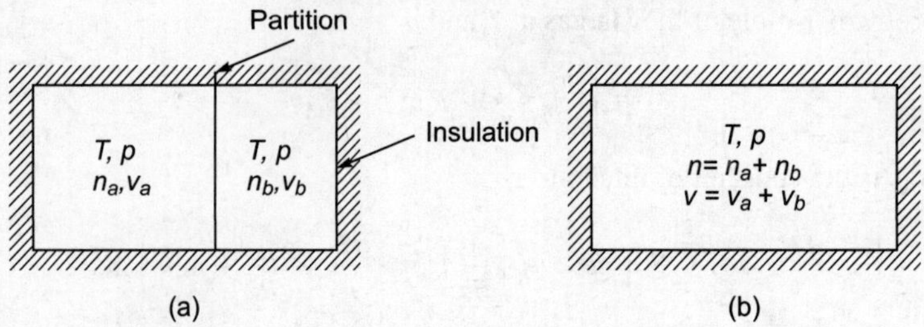

Fig. 10.11 Entropy change due to diffusion of gases

For constant temperature, the change of entropy is

$$S_2 - S_1 = -n\bar{R} \ln \frac{p_2}{p_1}$$

and for the component 'a', the final pressure is p_a, its partial pressure, and the initial pressure is p. Therefore,

$$(S_2 - S_1)_a = -n_a\bar{R} \ln \frac{p_a}{p} = -n_a \bar{R} \ln x_a$$

Similarly, for the second component 'b'

$$(S_2 - S_1)_b = -n_b\bar{R} \ln \frac{p_b}{p} = -n_b \bar{R} \ln x_b$$

where $\boldsymbol{x}$ is the mole fraction.

So, the entropy change of the mixture is the sum of the entropy changes of the components

$$S_2 - S_1 = -n_a \bar{R} \ln x_a - n_b \bar{R} \ln x_b$$

If the components a and b are the same gas, then the entropy change in mixing is zero. This is because we cannot distinguish between the gases to determine the partial pressures, and the final pressure therefore is also the partial pressure of the gas. Thus, for *distinguishable* components, the entropy change due to mixing of any number of components is

$$S_2 - S_1 = -\bar{R} \sum_i n_i \ln x_i$$

which is always positive.

10.11 GIBBS FUNCTION OF A MIXTURE OF INERT IDEAL GASES

From the equations

$$d\bar{h} = \bar{c}_p \, dT$$

$$d\bar{s} = \bar{c}_p \frac{dT}{T} - \bar{R}\frac{dp}{p}$$

the enthalpy and entropy of 1 mole of an ideal gas at temperature T and pressure p are

$$\bar{h} = \bar{h}_0 + \int c_p \, dT$$

$$\bar{s} = \int \bar{c}_p \frac{dT}{T} + \bar{s}_0 - \bar{R} \ln p$$

Therefore, the molar Gibbs function

$$\bar{g} = \bar{h} - Ts$$

$$= \bar{h}_0 + \int \bar{c}_p dT - T \int \bar{c}_p \frac{dT}{T} - T\bar{s}_0 + \bar{R}T \ln p$$

Now $\qquad \int d(uv) = \int u dv + \int v du = uv$

Let $\qquad u = \frac{1}{T}, v = \int c_p \, dT$

Then $\qquad \frac{1}{T} \int c_p \, dT = \int \frac{1}{T} c_p \, dT + \int c_p \, dT \left(-\frac{1}{T^2} \right) dT$

$$= \int \frac{1}{T} c_p \, dT - \int \frac{\int c_p dT}{T^2} dT$$

$$\int c_p \, dT - T \int c_p \frac{dT}{T} = -T \int \frac{\int c_p dT}{T^2} dT$$

Therefore,

$$\bar{g} = \bar{h}_0 - T \int \frac{\int \bar{c}_p dT}{T^2} dT - T_{\bar{s}_0} + \bar{R}T \ln p$$

$$= \bar{R}T \left(\frac{\bar{h}_0}{\bar{R}T} - \frac{1}{\bar{R}} \int \frac{\int \bar{c}_p dT}{T^2} dT - \frac{\bar{s}_0}{\bar{R}} + \ln p \right)$$

Let

$$\phi = \frac{\bar{h}_0}{\bar{R}T} - \frac{1}{\bar{R}} \int \frac{\int \bar{c}_p dT}{T^2} dT - \frac{\bar{s}_0}{\bar{R}} \tag{10.101}$$

Thus, $\qquad \bar{g} = \bar{R}T(\phi + \ln p) \qquad (10.102)$

where ϕ is a function of temperature only.

Let us consider a number of inert ideal gases separated from one another at the same T and p

$$G_i = \Sigma \, n_K g_K$$

$$= \bar{R}T \, \Sigma \, n_K(\phi_K + \ln p)$$

After the partitions are removed, the gases will diffuse, and the partial Gibbs function of a particular gas is the value of G, if that gas occupies the same volume at the same temperature exerting a partial pressure p_K. Thus,

$$G_f = \bar{R}T \, \Sigma \, n_K(\phi_K + \ln p_K)$$

$$= \bar{R}T \, \Sigma \, n_K (\phi_K + \ln p + \ln x_K)$$

Therefore,

$$G_f - G_i = \bar{R}T \, \Sigma \, n_K \ln x_K \qquad (10.103)$$

Since $x_K < 1$, $(G_f - G_i)$ is negative because G decreases due to diffusion. Gibbs function of a mixture of ideal gases at T and p is thus

$$G = \bar{R}T \, \Sigma \, n_K (\phi_K + \ln p + \ln x_K) \qquad (10.104)$$

SOLVED EXAMPLES

Example 10.1 *A fluid at 200 kPa and 300°C has a volume of 0.8 m³. In a frictionless process at constant volume the pressure changes to 100 kPa. Find the final temperature and the heat transferred (a) if fluid is air, and (b) if the fluid is steam.* E [LO 10.3]

Solution

(a) The fluid is air.

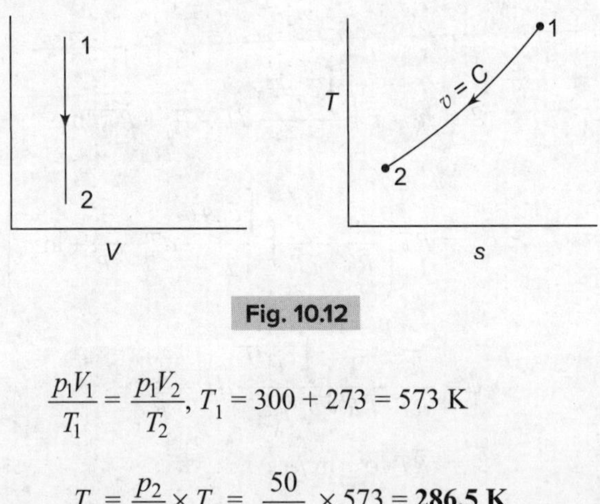

Fig. 10.12

$$\frac{p_1 V_1}{T_1} = \frac{p_1 V_2}{T_2}, \; T_1 = 300 + 273 = 573 \text{ K}$$

$$T_2 = \frac{p_2}{p_1} \times T_1 = \frac{50}{100} \times 573 = \mathbf{286.5 \text{ K}}$$

$$m = \frac{p_1 V_1}{R T_1} = \frac{100 \dfrac{\text{kN}}{\text{m}^2} \times 0.8 \text{ m}^3}{0.287 \dfrac{\text{kJ}}{\text{kg K}} \times 573 \text{ K}} = \mathbf{0.4865 \text{ kg}}$$

$$Q = mc_v (T_2 - T_1)$$

$$= 0.4865 \times 0.718 \, (286.5 - 573) = \mathbf{-100.11 \text{ kJ}}$$

(b) The fluid is steam (Fig. 10.13)

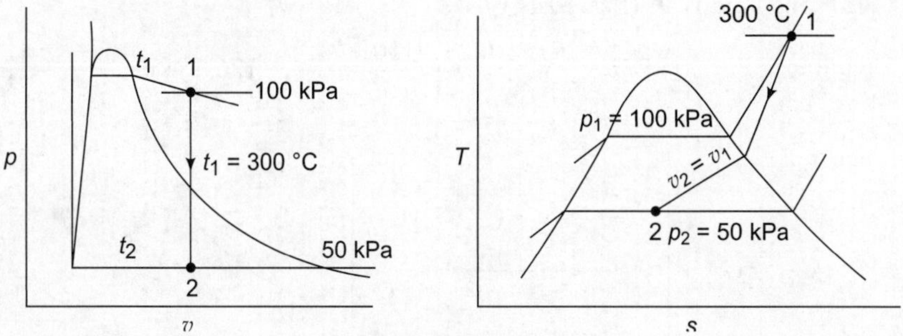

Fig. 10.13

$$p_2 = 50 \text{ kPa}$$

From Table A.1.1, $t_2 = 81.33°\text{C}, v_f = 0.00103 \text{ m}^3/\text{kg}$

$$v_g = 3.24 \text{ m}^2/\text{kg (state 2)}$$

At 100 kPa, 300°C, $v_1 = 2.6388 \text{ m}^3/\text{kg}$

$$u_1 = 2810.4 \text{ kJ/kg}$$

$$v_1 = v_2 = v_f + x_2 v_{fg}$$

$$2.6388 = 0.00103 + x_2 (3.24 - 0.00103)$$

$$= 0.00103 + x_2 \times 3.23897$$

$$x_2 = \frac{2.63777}{3.23897} = 0.8144$$

$\therefore$ $$u_2 = u_f + x_2 u_{fg} = 340.42 + 0.8144 \times 2143.4$$

$$= 2086 \text{ kJ/kg}$$

$$m = \frac{V}{v_1} = \frac{0.8}{2.6388} = 0.383 \text{ kg}$$

$\therefore$ $$Q = m (u_2 - u_1) = 0.383 (2086 - 2810.4)$$

$$= \mathbf{-277.44 \text{ kJ}}$$

$$T_2 = 81.33 + 273 = \mathbf{354.33 \text{ K}}$$

Example 10.2 *A fluid having a temperature of 150°C and a specific volume of 0.96 m³/kg at its initial state expands at constant pressure, without friction, until the volume is 1.55 m³/kg. Find, for 1 kg of fluid, the work, the heat transferred and the final temperature if (a) the fluid is air, and (b) the fluid is steam.*

E **[LO 10.3]**

Solution

(a) The fluid is air (Fig. 10.14)

$$p = \frac{RT}{v} = \frac{0.287\,(150 + 273)}{0.96} = 126.46 \text{ kPa}$$

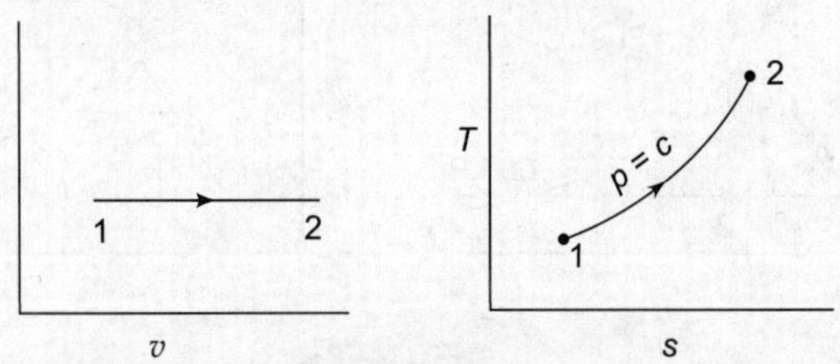

Fig. 10.14

$$W = p\,(v_2 - v_1) = 126.46\ \frac{\text{kN}}{\text{m}^2}\,(1.55 - 0.96)\ m^3$$
$$= \mathbf{74.61\ kJ}$$
$$T_2 = T_1\,\frac{V_2}{V_1} = 423 \times \frac{1.55}{0.96} = 682.97 \text{ K}$$
$$= \mathbf{409.97°C}$$
$$Q = C_p\,(T_2 - T_1) = 1.005\,(409.97 - 150)$$
$$= \mathbf{261.27\ kJ}.$$

(b) The fluid is steam (Fig. 10.15).

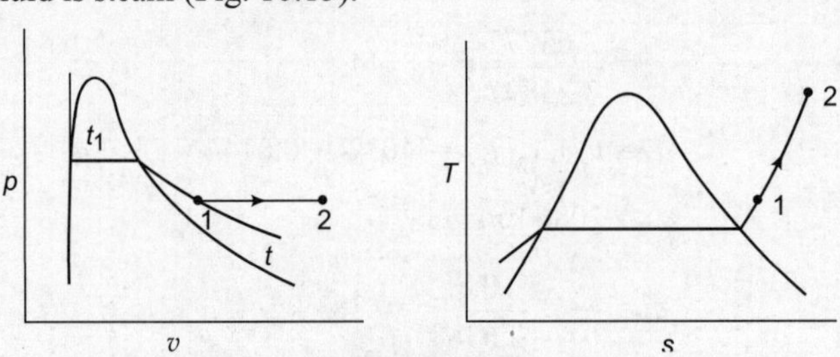

Fig. 10.15

At 150°C, $v_g = 0.3928$ m³/kg
Since $v_1 > v_g$, the state is superheated.
From superheated steam table,

$$p_1 = 200 \text{ kPa} = p_2$$

and $$h_1 = 2768.8 \text{ kJ/kg}$$
$$v_2 = 1.55 \text{ m}^3\text{/kg}$$

∴ $$t_2 = 400°\text{C},\ h_2 = 3276.5 \text{ kJ/kg}$$
$$Q = h_2 - h_1 = 3276.5 - 2768.8 = \mathbf{507.7\ kJ}$$
$$W = p\,(v_2 - v_1) = 200\,(1.55 - 0.96)$$
$$= \mathbf{118\ kJ}$$
$$t_2 = \mathbf{400°C}$$

Example 10.3 *A fluid at 250°C and 300 kPa is compressed reversibly and isothermally to 1/16th of its original volume. Find the final pressure, the workdone and the change of internal energy per kg of fluid (a) if the fluid in air, and (b) the fluid is steam.* M [LO 10.6]

Solution

(a) The fluid is air (Fig. 10.16)

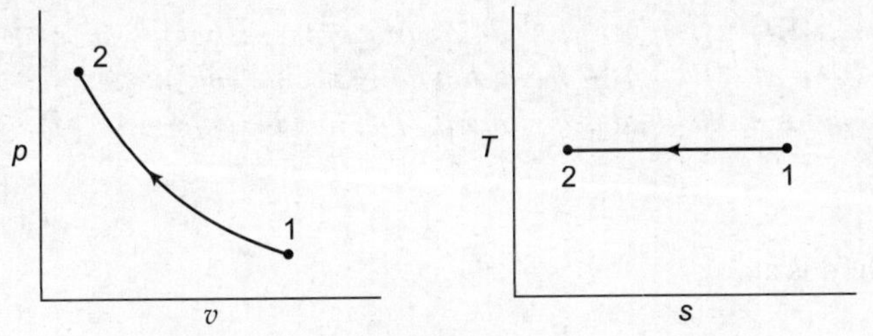

Fig. 10.16

$$\frac{p_1 v_1}{T_1} = \frac{p_2 v_2}{T_2} \quad \therefore\ p_2 = p_1 \frac{v_1}{v_2} = 300 \times 16 = \mathbf{4800\ kPa.}$$

$$W_{1-2} = p_1 v_1 \ln \frac{v_2}{v_1} = RT_1 \ln \frac{v_2}{v_1} = 0.287 \times 523 \ln \frac{1}{16}$$

$$= -2.773 \times 0.287 \times 523 = \mathbf{-416.17\ kJ}$$

$$\Delta u = u_2 - u_1 = \mathbf{0}$$

(b) The fluid is steam.

From the steam tables at 300 kPa, 250°C

$$v_1 = 0.7964\ \text{m}^3/\text{kg},\ u_1 = 2728.7\ \text{kJ/kg}$$

$$v_2 = \frac{1}{16} v_1 = 0.0498\ \text{m}^3/\text{kg},\ s_1 = 7.5165\ \text{kJ/kg K}$$

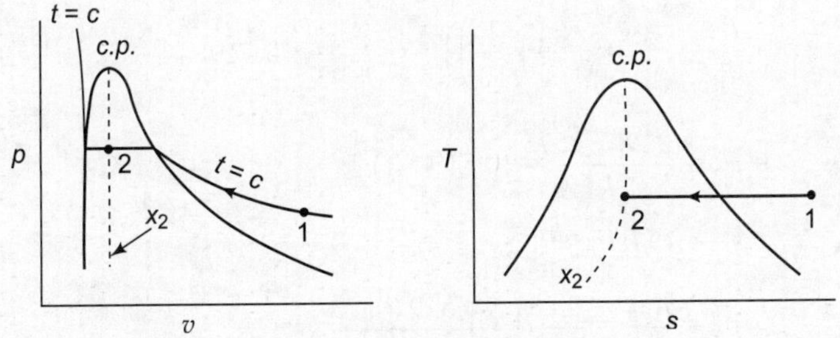

Fig. 10.17

$$v_2 = 0.0498 \text{ m}^3/\text{kg}$$
$$= (v_f + x_2\, v_{fg})_{250°\text{C}} = 0.00125 + x_2 \times (0.05013)$$
$$x_2 = 0.968,\ s_2 = 2.7927 + 0.968 \times 3.2802 = 5.968 \text{ kJ/kg K}$$
$$u_2 = 1080.37 + 0.968 \times 1522.0 = 2553.67 \text{ kJ/kg}$$
$$\Delta u = u_2 - u_1 = 2553.67 - 2728.7 = \mathbf{175\ kJ/kg.}$$
$$Q_{1\text{-}2} = T(s_2 - s_1) = (250 + 273)\,(5.968 - 7.5165)$$
$$= -\,523 \times 1.5485 = \mathbf{-\,809.87\ kJ/kg}$$

Example 10.4 *A reversible adiabatic process begins at p_1 = 10 bar, t_1 = 300°C and ends with p_2 = 1 bar. Find the specific volume and the work done per kg of fluid if (a) the fluid is air, and (b) the fluid is steam.* M **[LO 10.5]**

Solution

(a) The fluid is air.

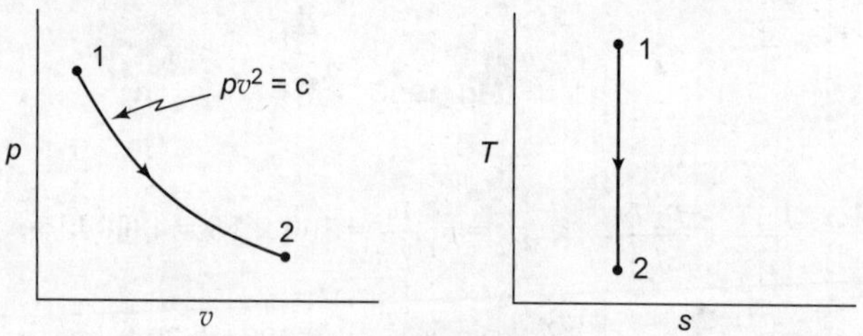

Fig. 10.18

$$p_1 v_1^{\gamma} = p_2 v_2^{\gamma}$$

$$W_{1\text{-}2} = \int_1^2 p\, dv = \frac{p_1 v_1 - p_2 v_2}{\gamma - 1} = \frac{R}{\gamma - 1}(T_1 - T_2)$$

$$= \frac{RT_1}{1-\gamma}\left(\frac{T_2}{T_1} - 1\right)$$

$$\frac{T_2}{T_1} = \left(\frac{p_2}{p_1}\right)^{\frac{\gamma-1}{\gamma}} = \left(\frac{1}{10}\right)^{\frac{1.4-1}{1.4}} = \frac{1}{1.93} = 0.519$$

$$\therefore \quad W_{1\text{-}2} = \frac{0.287 \times 573}{1 - 1.4}(0.519 - 1)$$

$$= \mathbf{197.75\ kJ/kg}$$

$$v_2 = \frac{RT_2}{p_2} = \frac{0.287 \times 0.519 \times 573}{100}$$
$$= \mathbf{0.854\ m^3/kg}$$

(b) The fluid is steam (Fig. 10.19)

$$u_1 = 2793.2 \text{ kJ/kg}, v_1 = 0.2579 \text{ m}^3\text{/kg}, s_1 = 7.1228 \frac{\text{kJ}}{\text{kg K}}$$

$$s_1 = s_2 = s_f + x_2 s_{fg}$$

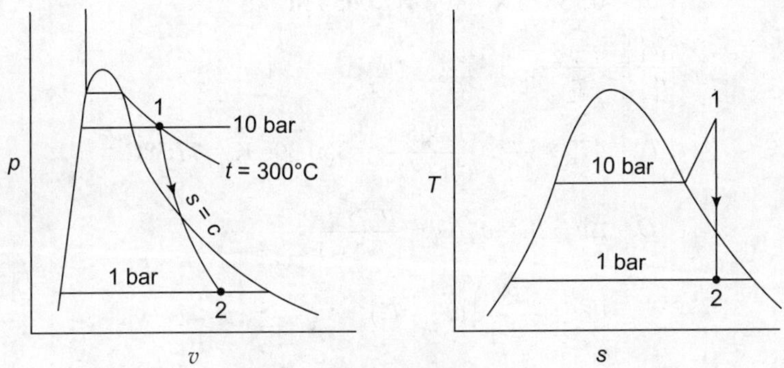

Fig. 10.19

$$7.1228 = 1.3025 + x_2 \times 6.0568$$

$$x_2 = \frac{5.8203}{6.0568} = 0.96$$

$$u_2 = 417.33 + 0.96 \times 2088.7 = 2422.5 \text{ kJ/g}$$

$$v_2 = 0.001043 + 0.96 \times 1.693 = \mathbf{1.626 \text{ m}^3\text{/kg}}$$

$$W_{1-2} = u_1 - u_2 = 2793.2 - 2422.5$$

$$= \mathbf{370.7 \text{ kJ/kg}}$$

Example 10.5 *A reversible polytropic process begins with a fluid at $p_1 = 10$ bar, $T_1 = 200°C$ and ends at $p_2 = 1$ bar. The exponent n has the value 1.15. Find the final specific volume, the final temperature and the heat transferred per kg of fluid if (a) the fluid is air, and (b) the fluid is steam.* M **[LO 10.8]**

Solution

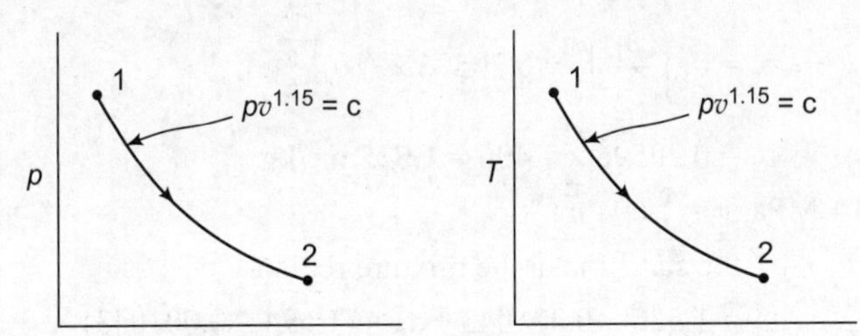

Fig. 10.20

(a) The fluid is air (Fig. 10.20).

$$v_2 = v_1 \left(\frac{p_1}{p_2}\right)^{\frac{1}{n}} = \frac{RT_1}{p_1}\left(\frac{p_1}{p_2}\right)^{\frac{1}{n}} = \frac{0.287 \times 473}{1000}(10)^{1/1.15}$$

$$v_2 = 0.13575 \times 7.406 = \mathbf{1.0054\ m^3/kg}$$

$$\frac{p_1 v_1}{T_1} = \frac{p_2 v_2}{T_2}$$

$$\therefore \quad T_2 = 473 \times \frac{1}{10} \times \frac{1.0054}{0.13575} = 350.306\ \text{K} = \mathbf{77.3°C}$$

$$Q = \Delta u + W = c_v\, \Delta T + \frac{p_2 v_2 - p_1 v_1}{1 - n}$$

$$= \left(c_v + \frac{R}{1-n}\right)(T_2 - T_1)$$

$$= \left(0.716 + \frac{0.287}{-0.15}\right)(350.306 - 473)$$

$$= (-1.195)(-123.306) = \mathbf{147.35\ kJ}$$

(b) The fluid is steam (Fig. 10.21)

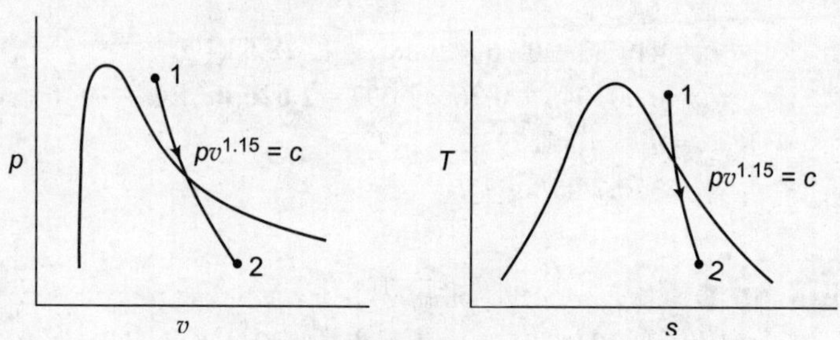

Fig. 10.21

At 10 bar, 200°C,

$$v_1 = 0.20596,\ u_1 = 2621.9$$

$$v_2 = v_1\left(\frac{p_1}{p_2}\right)^{\frac{1}{n}} = 0.20596 \times (10)^{\frac{1}{1.15}}$$

$$v_2 = 0.20596 \times 7.406 = \mathbf{1.525\ m^3/kg}$$

v_g at 0.1 MPa, $v_g = 1.694\ m^3/kg$

Since $v_2 < v_g$, the state 2 is in the mixture region.

$$v_2 = 1.525 = 0.001043 + x_2 \times (1.694 - 0.001043)$$

$$x_2 = 0.933$$

$$t_2 = (t_{sat})_{1\ \text{bar}} = \mathbf{99.62°C}$$

$$u_2 = 417.33 + 0.933 \times 2088.7 = 2366.1 \text{ kJ/kg}$$

$$W = \int_1^2 pdv = \frac{p_1 v_1 - p_2 v_2}{n-1} = \frac{1000 \times 0.13575 - 100 \times 1.525}{0.15}$$

$$= \frac{135.75 - 152.5}{0.15} = -111.67 \text{ kJ/kg}$$

$$Q = \Delta u + W = u_2 - u_1 + W$$

$$= (2366.1 - 2621.9) - 111.67$$

$$= -255.8 - 111.67 = \mathbf{-367.47 \text{ kJ/kg}}$$

Example 10.6 *An ideal monatomic gas occupies a volume of 10^{-3} m^3 at temperature 3 K and pressure 10^3 Pa. The internal energy of the gas is taken to be zero at this point. It undergoes the following cycle.*

The temperature is raised to 300 K at constant volume, the gas is expanded adiabatically till the temperature is 3 K, followed by isothermal compression to the original volume. Plot on a p-V diagram. Calculate (a) the work done and the heat transferred in each process and the internal energy at the end of each process, and (b) the thermal efficiency of the cycle. Ⓜ **[LO 10.3, 10.5, 10.7]**

Solution

$$pV = n\bar{R}T$$

At initial conditions

$$(10^3 \times 10^{-3}) \text{ J} = n\bar{R} \times 3$$

$$\therefore \quad n = \frac{1}{3\bar{R}} = \text{number of moles of the gas.}$$

(a) Process *ab*

Here $dV = 0, W_{ab} = 0$

$$Q_{ab} = n\bar{c}_v (T_b - T_a)$$

For a monatomic gas, $\bar{c}_v = \frac{3}{2}\bar{R}$

$$Q_{ab} = \frac{1}{3\bar{R}} \times \frac{3}{2}\bar{R} \times 297 = 148.5 \text{ J}$$

$$U_b - U_a = 148.5 \text{ J}$$

But $U_a = 0$ (given)

$$U_b = \mathbf{148.5 \text{ J}}$$

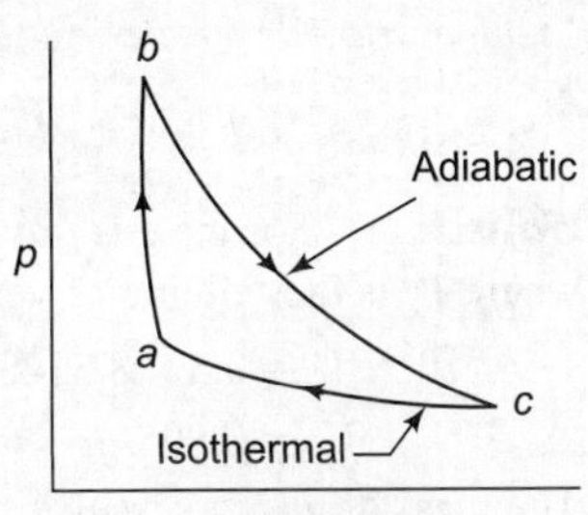

Fig. 10.22

Process *bc*

$$Q_{bc} = 0, W_{bc} = -\Delta U = -(U_C - U_b)$$

Since $T_C = 3$ K and U is a function of the temperature only and $U = 0$ at $T = 3$ K (given)

$$\therefore \quad U_C = 0 \quad \text{and} \quad W_{bc} = U_b = \mathbf{148.5 \text{ J}}$$

Process *ca*

$$W_{ca} = n\bar{R}T \ln \frac{V_a}{V_c} = - n\bar{R}T \ln \frac{V_c}{V_b}$$

For the adiabatic process *bc*,

$$TV^{\gamma-1} = C \quad \text{and} \quad \gamma = \frac{5}{3}$$

$$\left(\frac{V_c}{V_b}\right)^{2/3} = \frac{T_b}{T_c} = \frac{300}{3} = 100 = 10^2$$

$$\therefore \quad \frac{V_c}{V_b} = 10^3$$

$$W_{ca} = -\frac{1}{3\bar{R}} \cdot \bar{R} \cdot 3 \cdot \ln 10^3 = \mathbf{-\,6.9\ J}$$

$$Q_{ca} = -\,6.9 \text{ J} \quad \text{and} \quad U_a = U_C = 0$$

(b) $\eta = \dfrac{148.5 - 6.9}{148.5} = \mathbf{0.954}$

Example 10.7 *Two vessels, A and B, both containing nitrogen, are connected by a valve which is opened to allow the contents to mix and achieve an equilibrium temperature of 27°C. Before mixing, the following information is known about the gases in the two vessels.*

Vessel A	*Vessel B*
$p = 1.5$ *MPa*	$p = 0.6$ *MPa*
$t = 50°C$	$t = 20°C$
Contents = 0.5 kg mol	*Contents = 2.5 kg*

Calculate the final equilibrium pressure, and the amount of heat transferred to the surroundings. If the vessel had been perfectly insulated, calculate the final temperature and pressure which would have been reached. Take $\gamma = 1.4$

M **[LO 10.14]**

Solution For the gas in vessel A (Fig. 10.23)
where V_A is the volume of vessel *A*

$$1.5 \times 10^3 \times V_A = 0.5 \times 8.3143 \times 323$$

$$V_A = 0.895 \text{ m}^3$$

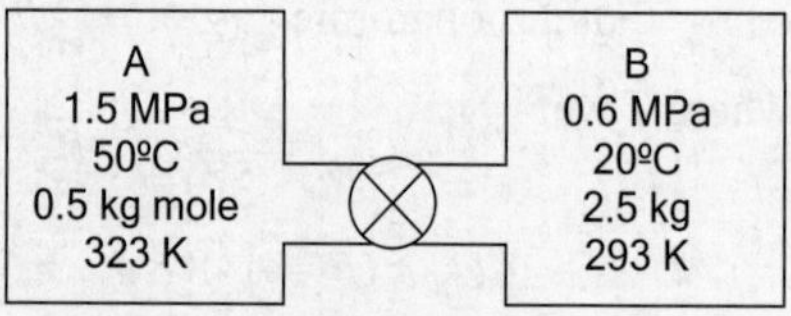

Fig. 10.23

The mass of gas in vessel *A*

$$m_A = n_A \mu_A$$

$$= 0.5 \text{ kg mol} \times 28 \text{ kg/kg mol} = 14 \text{ kg}$$

Characteristic gas constant *R* of nitrogen

$$R = \frac{8.3143}{28} = 0.297 \text{ kJ/kg K}$$

For the vessel *B*

$$p_B V_B = m_B R T_B$$

$$0.6 \times 10^3 \times V_B = 2.5 \times 0.297 \times 293$$

$\therefore \qquad V_B = 0.363 \text{ m}^3$

Total volume of A and B

$$V = V_A + V_B = 0.895 + 0.363 = 1.258 \text{ m}^3$$

Total mass of gas

$$m = m_A + m_B = 14 + 2.5 = 16.5 \text{ kg}$$

Final temperature after mixing

$$T = 27 + 273 = 300 \text{ K}$$

For the final condition after mixing

$$pV = mRT$$

where p is the final equilibrium pressure

$$\therefore \qquad p \times 1.258 = 16.5 \times 0.297 \times 300$$

$$\therefore \qquad p = \frac{16.5 \times 0.297 \times 300}{1.258} = 1168.6 \text{ kPa} = 1.168 \text{ MPa}$$

$$c_v = \frac{R}{\gamma - 1} = \frac{0.297}{0.4} = 0.743 \text{ kJ/kg K}$$

Since there is no work transfer, the amount of heat transfer

$$Q = \text{change of internal energy} = U_2 - U_1$$

Measuring the internal energy above the datum of absolute zero (at $T = 0$ K, $u = 0$ kJ/kg).

Initial internal energy U_1 (before mixing)

$$= m_A c_v T_A + m_B c_v T_B$$

$$= (14 \times 323 + 2.5 \times 293) \times 0.743 = 3904.1 \text{ kJ}$$

Final internal energy U_2 (after mixing)

$$= mc_v T$$

$$= 16.5 \times 0.743 \times 300 = 3677.9 \text{ kJ}$$

$$\therefore \qquad Q = 3677.9 - 3904.1 = \mathbf{-226.2 \text{ kJ}}$$

If the vessels were insulated

$$U_1 = U_2$$

$$m_A c_v T_A + m_B c_v T_B = mc_v T$$

where T would have been the final temperature.

$$\therefore \qquad T = \frac{m_A T_A + m_B T_B}{m}$$

$$= \frac{14 \times 323 + 2.5 \times 293}{16.5} = 318.5 \text{ K}$$

or $\qquad t = \mathbf{45.5°C}$

The final pressure

$$p = \frac{mRT}{V} = \frac{16.5 \times 0.297 \times 318.5}{1.258} = 1240.7 \text{ kPa} = \mathbf{1.24 \text{ MPa}}$$

Example 10.8 *A certain gas has $c_p = 1.968$ and $c_v = 1.507$ kJ/kg K. Find its molecular weight and the gas constant.*

A constant volume chamber of 0.3 m³ capacity contains 2 kg of this gas at 5°C. Heat is transferred to the gas until the temperature is 100°C. Find the work done, the heat transferred and the changes in internal energy, enthalpy and entropy.

E [LO 10.4]

Solution Gas constant,

$$R = c_p - c_v = 1.968 - 1.507 = \mathbf{0.461\ kJ/kg\ K}$$

Molecular weight,

$$\mu = \frac{\bar{R}}{R} = \frac{8.3143}{0.461} = \mathbf{18.04\ kg/kg\ mol}$$

At constant volume

$$Q_{1-2} = mc_v (t_2 - t_1)$$

$$= 2 \times 1.507\ (100 - 5) = \mathbf{286.33\ kJ}$$

$$W_{1-2} = \int_1^2 pdv = \mathbf{0}$$

Change in internal energy

$$U_2 - U_1 = Q_{1-2} = \mathbf{286.33\ kJ}$$

Change in enthalpy

$$H_2 - H_1 = mc_p (t_2 - t_1)$$

$$= 2 \times 1.968\ (100 - 5) = \mathbf{373.92\ kJ}$$

Change in entropy

$$S_2 - S_1 = mc_v \ln \frac{T_2}{T_1} = 2 \times 1.507 \ln \frac{373}{268} = \mathbf{0.921\ kJ/K}$$

Example 10.9 *(a) The specific heats of a gas are given by $c_p = a + kT$ and $c_v = b + kT$, where a, b and k are constants and T is in K. Show that for an isentropic expansion of this gas*

$$T^b v^{a-b} e^{kT} = constant$$

(b) *1.5 kg of this gas occupying a volume of 0.06 m³ at 5.6 MPa expands isentropically until the temperature is 240°C. If a = 0.946, b = 0.662 and k = 10^{-4}, calculate the work done in the expansion.*

M [LO 10.4]

Solution (a) $c_p - c_v = a + kT - b - kT$

$$= a - b = R$$

Now

$$ds = c_v \frac{dT}{T} + R \frac{dv}{v}$$

$$= (b + kT) \frac{dT}{T} + (a - b) \frac{dv}{v} = b \frac{dT}{T} + k\,dT + (a - b) \frac{dv}{v}$$

For an isentropic process

$$b \ln T + kT + (a - b) \ln v = \text{constant}$$

$$\therefore \quad T^b \cdot v^{a-b}\, e^{kT} = \text{constant} \qquad \text{(Q.E.D.)}$$

(b) $R = a - b = 0.946 - 0.662 = 0.284$ kJ/kg K

$$T_2 = 240 + 273 = 513 \text{ K}$$

$$T_1 = \frac{p_1 V_1}{mR} = \frac{5.6\times 10^3 \times 0.06}{1.5 \times 0.284} = 788.73 \text{ K} = 789 \text{ K}$$

$$TdS = dU + \delta W = 0$$

$$\therefore \quad W_{1-2} = -\int_{T_1}^{T_2} mc_v \, dT$$

$$= 1.5 \int_{513}^{789} (0.662 + 0.0001T)\, dT$$

$$= 1.5\,[0.662\,(789 - 513) + 10^{-4} \times 0.5\,\{(789)^2 - (513)^2\}]$$

$$= 1.5\,(182.71 + 19.97) = \mathbf{304\ kJ}$$

Example 10.10 *Show that for an ideal gas, the slope of the constant volume line on the T–s diagram is more than that of the constant pressure line.* E [LO 10.5]

Solution We have, for 1 kg of ideal gas

$$Tds = du + pdv$$

$$= c_v\, dT + pdv$$

$$\therefore \quad \left(\frac{\partial T}{\partial s}\right)_v = \frac{T}{c_v}$$

Also

$$Tds = dh - vdp$$

$$= c_p\, dT - vdp$$

$$\therefore \quad \left(\frac{\partial T}{\partial s}\right)_p = \frac{T}{c_p}$$

Since $c_p > c_v$, $\frac{T}{c_v} > \frac{T}{c_p}$

$$\therefore \quad \left(\frac{\partial T}{\partial s}\right)_v > \left(\frac{\partial T}{\partial s}\right)_p$$

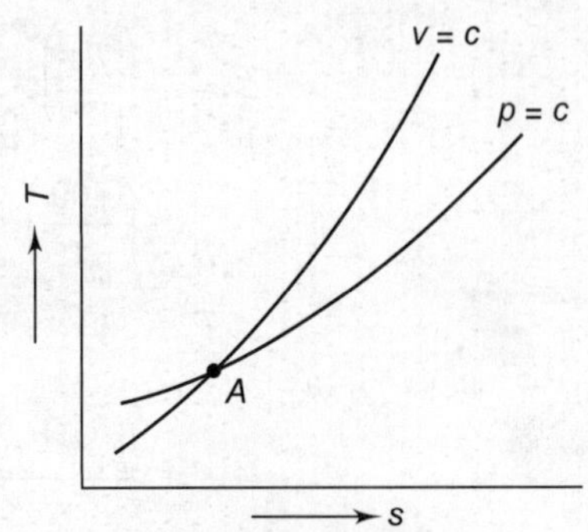

Fig. 10.24

This is shown in Fig. 10.24. The slope of the constant volume line passing through point A is steeper than that of the constant pressure line passing through the same point. (Q.E.D.)

Example 10.11 *0.5 kg of air is compressed reversibly and adiabatically from 80 kPa, 60°C to 0.4 MPa, and is then expanded at constant pressure to the original volume. Sketch these processes on the p–v and T–s planes. Compute the heat transfer and work transfer for the whole path.* E [LO 10.4]

Solution The processes have been shown on the p–v and T–s planes in Fig. 10.25.

At state 1

$$p_1 V_1 = mRT_1$$

$\therefore$ V_1 = volume of air at state 1

$$= \frac{mRT_1}{p_1} = \frac{1 \times 0.287 \times 333}{2 \times 80} = 0.597 \text{ m}^3$$

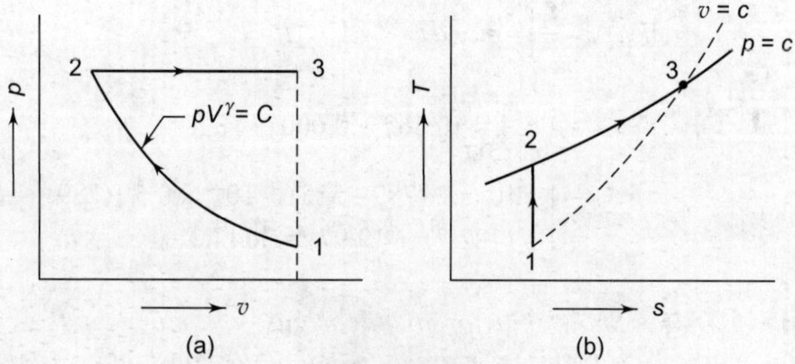

Fig. 10.25

Since the process 1–2 is reversible and adiabatic

$$\frac{T_2}{T_1} = \left(\frac{p_2}{p_1}\right)^{(\gamma-1)/\gamma}$$

$$\therefore \quad \frac{T_2}{T_1} = \left(\frac{400}{80}\right)^{(1.4-1)/1.4} = (5)^{2/7}$$

$$\therefore \quad T_2 = 333 \times (5)^{2/7} = 527 \text{ K}$$

For process 1–2, work done

$$W_{1-2} = \frac{p_1V_1 - p_2V_2}{\gamma - 1} = \frac{mR(T_1 - T_2)}{\gamma - 1}$$

$$= \frac{1/2 \times 0.287\,(333 - 527)}{0.4} = -\,69.6 \text{ kJ}$$

Again $\quad p_1 v_1^\gamma = p_2 v_2^\gamma$

$$\therefore \quad \left(\frac{v_2}{v_1}\right)^\gamma = \frac{p_1}{p_2} = \frac{80}{400} = \frac{1}{5}$$

$$\therefore \quad \frac{v_2}{v_1} = \left(\frac{1}{5}\right)^{1/1.4} = \frac{1}{3.162} = \frac{V_2}{V_1}$$

$$\therefore \quad V_2 = \frac{0.597}{3.162} = 0.189 \text{ m}^3$$

For process 2–3, work done

$$W_{2-3} = p_2 (V_1 - V_2) = 400\,(0.597 - 0.189) = 163.2 \text{ kJ}$$

∴ Total work transfer

$$W = W_{1-2} + W_{2-3}$$
$$= -69.6 + 163.2 = \mathbf{93.6\ kJ}$$

For states 2 and 3

$$\frac{p_2V_2}{T_2} = \frac{p_3V_3}{T_3}$$

∴
$$T_3 = T_2 \cdot \frac{V_3}{V_2} = 527 \times 3.162 = 1667\ \text{K}$$

Total heat transfer

$$Q = Q_{1-2} + Q_{2-3} = Q_{2-3} = mc_p\,(T_3 - T_2)$$
$$= 1/2 \times 1.005\,(1667 - 527) = \mathbf{527.85\ kJ}$$

Example 10.12 *A mass of air is initially at 260°C and 700 kPa, and occupies 0.028 m³. The air is expanded at constant pressure to 0.084 m³. A polytropic process with n = 1.50 is then carried out, followed by a constant temperature process which completes a cycle. All the processes are reversible. (a) Sketch the cycle in the p–v and T–s plane, (b) Find the heat received and heat rejected in the cycle and (c) Find the efficiency of the cycle.* Ⓜ **[LO 10.8]**

Solution (a) The cycle is sketched on the *p–v* and *T–s* planes in Fig. 10.26.

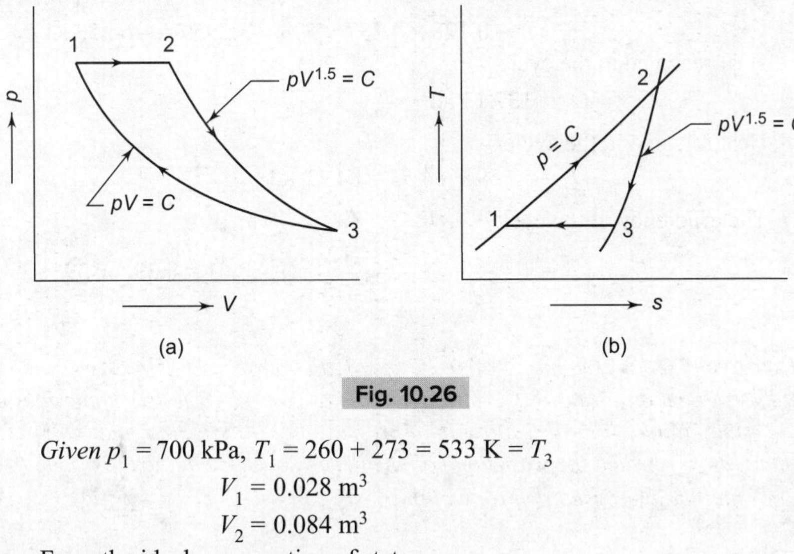

Fig. 10.26

Given $p_1 = 700$ kPa, $T_1 = 260 + 273 = 533\ \text{K} = T_3$

$$V_1 = 0.028\ \text{m}^3$$
$$V_2 = 0.084\ \text{m}^3$$

From the ideal gas equation of state

$$p_1V_1 = mRT_1$$

$$m = \frac{700 \times 0.028}{0.287 \times 533} = 0.128\ \text{kg}$$

Now
$$\frac{T_2}{T_1} = \frac{p_2V_1}{p_1V_1} = \frac{0.084}{0.028} = 3$$

$$\therefore \qquad T_2 = 3 \times 533 = 1599 \text{ K}$$

Again
$$\frac{p_2}{p_3} = \left(\frac{T_2}{T_3}\right)^{n/(n-1)} = \left(\frac{1.5/0.5}{533}\right)^{1.5/0.5} = (3)^3 = 27$$

Heat transfer in process 1–2

$$Q_{1-2} = mc_p (T_2 - T_1)$$
$$= 0.128 \times 1.005 (1599 - 533) = \mathbf{137.13\ kJ}$$

Heat transfer in process 2–3

$$Q_{2-3} = \Delta U + \int pdv$$
$$= m\, c_v (T_3 - T_2) + \frac{mR(T_2 - T_3)}{n-1}$$
$$= mc_v \frac{n-\gamma}{n-1} (T_3 - T_2) = 0.128 \times 0.718 \times \frac{1.5-1.4}{1.5-1} (533 - 1599)$$
$$= 0.128 \times 0.718 \times \frac{0.1}{0.5} (-1066) = \mathbf{-19.59\ kJ}$$

For process 3–1

$$đQ = dU + đW = đW$$
$$\therefore \qquad Q_{3-1} = W_{3-1} = \int_3^1 pdV = mRT_1 \ln \frac{V_1}{V_3}$$
$$= mRT_1 \ln \frac{p_3}{p_1} = 0.128 \times 0.287 \times 533 \ln \left(\frac{1}{27}\right)$$
$$= -0.128 \times 0.287 \times 533 \times 3.2959 = -64.53 \text{ kJ}$$

(b) Heat received in the cycle

$$Q_1 = \mathbf{137.13\ kJ}$$

Heat rejected in the cycle

$$Q_2 = 19.59 + 64.53 = \mathbf{84.12\ kJ}$$

(c) The efficiency of the cycle

$$\eta_{cycle} = 1 - \frac{Q_2}{Q_1} = 1 - \frac{84.12}{137.13} = 1 - 0.61 = \mathbf{0.39, or\ 39\%}$$

Example 10.13 *A mass of 0.25 kg of an ideal gas has a pressure of 300 kPa, a temperature of 80°C and a volume of 0.07 m³. The gas undergoes an irreversible adiabatic process to a final pressure of 300 kPa and final volume of 0.10 m³, during which the work done on the gas is 25 kJ. Evaluate the c_p and c_v of the gas and the increase in entropy of the gas.* **E [LO 10.4]**

Solution From

$$p_1V_1 = mRT_1$$
$$R = \frac{300 \times 0.07}{0.25 \times (273 + 80)} = 0.238 \text{ kJ/kg K}$$

Final temperature

$$T_2 = \frac{p_2V_2}{mR} = \frac{300 \times 0.1}{0.25 \times 0.238} = 505 \text{ K}$$

Now

$$Q = (U_2 - U_1) + W = mc_v(T_2 - T_1) + W$$

$$0 = 0.25\, c_v(505 - 353) - 25$$

$$\therefore \quad c_v = \frac{25}{0.25 \times 152} = \mathbf{0.658\ kJ/kg\ K}$$

Now

$$c_p - c_v = R$$

$$c_p = 0.658 + 0.238 = \mathbf{0.896\ kJ/kg\ K}$$

Entropy change

$$S_2 - S_1 = mc_v \ln \frac{p_2}{p_1} + mc_p \ln \frac{v_2}{v_1}$$

$$= mc_p \ln \frac{V_2}{V_1} = 0.25 \times 0.896 \ln \frac{0.10}{0.07}$$

$$= 0.224 \times 0.3569 = \mathbf{0.08\ kJ/kg\ K}$$

Example 10.14 *Two kg of N_2 at 100 kPa and 25°C is mixed with two kg of N_2 at 300 kPa and 25°C in a rigid insulated tank. Find the entropy generation due to such mixing.Also find the entropy generation when the initial conditions of both the gases are same.* D [LO 10.14]

Solution Let N_2 at 100 kPa and 25°C be denoted by symbol *A* and N_2 at 300 kPa and 25°C be denoted by symbol *B*. Since internal energy for an ideal gas is a function of temperature only and both the gases (N_2) initially at same temperature, we can write

$$u_{A1} = u_{B1}$$

Applying the first law of thermodynamics, we have

$$\cancelto{=0}{Q_{1-2}} = U_2 - U_1 + \cancelto{=0}{W_{1-2}}$$

or

$$m_2 u_2 - m_{A1} u_{A1} - m_{B1} u_{B1} = 0$$

or

$$u_2 = \frac{m_{A1} u_{A1} + m_{B1} u_{B1}}{m_2} = \frac{2u_{A1} + 2u_{A1}}{4} = u_{A1}$$

Therefore, final temperature of the gas will be same as the initial temperature. Characteristic gas constant *R* of nitrogen

$$R = \frac{8.3143}{28} = 0.297 \text{ kJ/kgK}$$

The total volume is the sum of the initial volume of the gas at different condition. Thus,

$$V_2 = V_A + V_B$$

or

$$\frac{m_2 R T_2}{p_2} = \frac{m_A R T_{A1}}{p_{A1}} + \frac{m_B R T_{B1}}{p_{B1}}$$

or

$$\frac{4}{p_2} = \frac{2}{p_{A1}} + \frac{2}{p_{B1}} = \frac{2}{100} + \frac{2}{300}$$

or

$$p_2 = 153.85 \text{ kPa}$$

Entropy change can be found using Eq.(10.29) as

$$S_2 - S_1 = -m_A R \ln \frac{p_2}{p_{B1}} - m_B R \ln \frac{p_2}{p_{B1}}$$

$$= -2 \times 0.297 \ln \frac{153.85}{100} - 2 \times 0.297 \ln \frac{153.85}{300} = 0.141 \text{ kJ/K}$$

The entropy generated can be found using the entropy balance equation as

$$S_2 - S_1 = \sum_j \frac{Q_j}{T_j}^{=0} + S_{gen}$$

or $$S_{\text{gen}} = \mathbf{0.141\ kJ / K}$$

When the initial conditions of both the gases are same, then the final pressure will be same as the initial one. The change in entropy and hence the entropy generation will be zero in that situation.

Example 10.15 *A rigid vessel of 1m^3 contains air at 150 kPa and ambient temperature of 25°C. An electric resistance heater placed in the vessel is turned on and kept on for 10 min, after which the temperature of air is measured using a temperature sensor to be 90°C. The total electrical energy (equivalent to mechanical work) supplied during the period is 10 kJ. Comment on whether the temperature sensor is working properly or not. Air can be treated as an ideal gas and the average value of constant volume specific heat within the given range of temperature is 0.718 kJ/kgK.* Ⓜ **[LO 10.5]**

Solution

The mass of the air in the vessel is

$$m = \frac{pV}{RT} = \frac{150 \times 1}{0.287 \times 298} = 1.754 \text{ kg}$$

The heat transfer during the process can be found by applying the first law of thermodynamics as

$$Q_{1-2} = U_2 - U_1 + W_{1-2} = mc_v(T_2 - T_1) + W_{1-2}$$

$$= 1.754 \times 0.718(90 - 25) - 10 = 71.859 \text{ kJ}$$

The change of specific entropy of air during the process can be obtained by using Eq. (10.28) as

$$s_2 - s_1 = c_v \ln \frac{T_2}{T_1} + R \ln \frac{v_2}{v_1}^{=0} = 0.718 \ln \frac{358}{298} = 0.1317 \text{ kJ/kgK}$$

The entropy generated can be found using the entropy balance equation as

$$S_2 - S_1 = \sum_j \frac{Q_j}{T_j} + S_{\text{gen}}$$

or $$S_{\text{gen}} = m(s_2 - s_1) - \frac{Q_{1-2}}{T_{\text{amb}}} = 1.754 \times 0.1317 - \frac{71.859}{298}$$

or $$S_{\text{gen}} = -0.01 \text{ kJ/K}$$

Since the entropy generation is negative, temperature sensor is not working properly.

Example 10.16 *One kg of air initially at 100 kPa and 300 K is contained in a rigid tank. In order to increase the temperature of air inside the tank to 600 K, two different methods have been adopted as follows: (a) The tank is insulated and one paddle-wheel is inserted into the tank. The paddle-wheel is turned by an external electric motor and the air inside the tank is stirred. (b) The air inside the tank is heated by means of transfer of heat from a source at temperature 600 K. In which method, the entropy generation will be more. Under which condition the entropy generation will be almost same for both the methods. Consider air as an ideal gas and the average constant volume specific heat over the range of temperature as* $c_v = 0.718$ kJ/kgK. M [LO 10.5]

Solution

The change of entropy of air during the change in temperature from 300 to 600 K is same for both the methods and the same can be obtained by using Eq. (10.28) as

$$S_2 - S_1 = mc_v \ln\frac{T_2}{T_1} + mR\ln\frac{v_2}{v_1} = 1\times 0.718\ln\frac{600}{300} = 0.498 \text{ kJ/K}$$

(a) The entropy generated can be found using the entropy balance equation as

$$S_2 - S_1 = \cancel{\sum_j \frac{Q_j}{T_j}}^{=0} + S_{gen}$$

or

$$S_{gen} = 0.498 \text{ kJ/K}$$

(b) The heat transfer required to raise the temperature can be found by applying the first law of thermodynamics as

$$Q_{1-2} = \cancel{W_{1-2}}^{=0} + U_2 - U_1 = mc_v(T_2 - T_1) = 1\times 0.718(600-300) = 215.4 \text{ kJ}$$

From the entropy balance equation, one can write

$$S_2 - S_1 = \sum_j \frac{Q_j}{T_j} + S_{gen}$$

or

$$S_{gen} = S_2 - S_1 - \frac{Q_{1-2}}{T_1} = 0.498 - \frac{215.4}{600} = 0.139 \text{ kJ/K}$$

Thus, the entropy generation will be more when the air inside the tank is stirred by the paddle-wheel. As the temperature of source increases in the second case, the entropy generation decreases, and hence for a very high source temperature (theoretically infinite) the entropy generation for both the cases will be same.

Example 10.17 *Atmospheric air is sucked in an internal combustion engine cylinder. The pressure, temperature and volume of air after suction are 1 bar, 75°C and 800 cm*^{3}*, respectively. The air is then reversibly compressed to 15 bar and 1/8th of its volume. Heat of combustion of fuel is then added to air at constant volume until its pressure is 50 bar. (a) Determine the index of compression process, (b) Calculate the change of entropy of air during each process and (c) Determine if there is any heat exchange between air and cylinder walls during compression. State the direction of heat flow.* M [LO 10.8]

Solution The processes are shown on p–V and T-s diagrams in Fig. 10.27.

(a) Index of compression process

$$n = \frac{\ln p_2/p_1}{\ln V_1/V_2} = \frac{\ln 15}{\ln 8} = \mathbf{1.3}$$

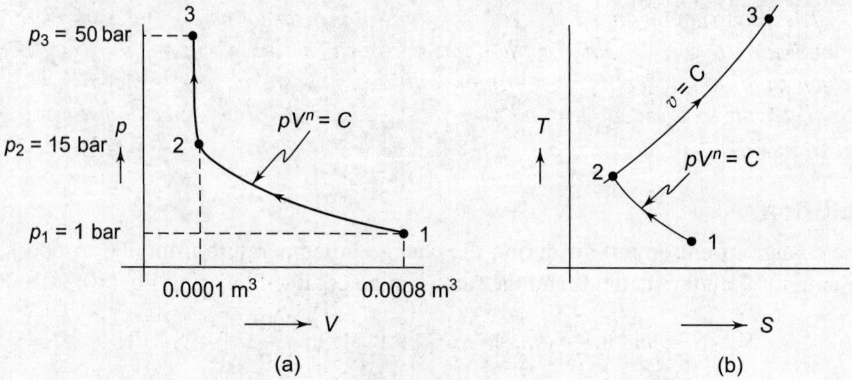

Fig. 10.27

(b) Mass of air in the cylinder

$$m = \frac{p_1V_1}{RT_1} = \frac{100 \times 800 \times 10^{-6}}{0.287 \times 348} = 0.0008 \text{ kg}$$

$$\frac{p_2V_2}{T_2} = \frac{p_1V_1}{T_1}$$

$$\therefore \quad T_2 = T_1 \frac{p_2V_2}{p_1V_1} = 348 \times 15 \times \frac{1}{8} = \mathbf{652.5\ K}$$

$$T_3 = T_2 \cdot \frac{p_3}{p_2} = 652.5 \times \frac{50}{15} = 2175 \text{ K}$$

$$S_2 - S_1 = m\left[c_2 \ln\frac{T_2}{T_1} + R\ln\frac{V_2}{V_1}\right]$$

$$= 0.0008\left[0.718 \ln\frac{652.5}{348} + 0.287 \ln\frac{1}{8}\right]$$

$$= 0.0008\,[0.4513 - 0.5968] = -\,0001164 \text{ kJ/K}$$

$$S_3 - S_2 = mc_v \ln\frac{T_3}{T_2} = 0.0008 \times 0.718 \ln\frac{2175}{652.5} = \mathbf{0.000692\ kJ/K}$$

(c) Heat transfer during polytropic compression process

$$Q = mc_v \frac{\gamma - n}{1 - n}(T_2 - T_1)$$

$$= 0.0008 \times 0.718 \frac{1.4 - 1.3}{1 - 1.3}(652.5 - 348) = -\,\mathbf{0.058\ kJ}$$

The direction of heat flow is from air to cylinder walls, which is thus negative.

Example 10.18 *At STP, 8.4 litres of oxygen and 14 litres of hydrogen mix with each other completely in an insulated chamber. Calculate the entropy change for the process assuming both gases behave as ideal gases.* E [LO 10.14]

Solution The entropy change during mixing of two gases can be found using Eq. (10.98) as

$$\Delta S = -\bar{R}(n_1 \ln x_1 + n_2 \ln x_2)$$

where $n_1 = \dfrac{8.4}{8.4+14} = 0.375$, $n_2 = \dfrac{14}{8.4+14} = 0.625$

The mole fraction of oxygen, $x_1 = \dfrac{n_1}{n_1+n_2} = 0.375$

The mole fraction of hydrogen, $x_2 = \dfrac{n_2}{n_1+n_2} = 0.625$

$$\therefore \quad \Delta S = -8.3143(0.375 \ln 0.375 + 0.625 \ln 0.625)$$
$$= \mathbf{5.49\ kJ/K}$$

Example 10.19 *A mixture of ideal gases consists of 3 kg of nitrogen and 5 kg of carbondioxide at a pressure of 300 kPa and a temperature of 20°C. Find (a) the mole fraction of each constituent, (b) the equivalent molecular weight of the mixture, (c) the equivalent gas constant of the mixture, (d) the partial pressures and the partial volumes, (e) the volume and density of the mixture and (f) the c_p and c_v of the mixture.*

If the mixture is heated at constant volume to 40°C, find the changes in internal energy, enthalpy and entropy of the mixture. Find the changes in internal energy, enthalpy and entropy of the mixture if the heating is done at constant pressure. Take γ for CO_2 and N_2 to be 1.286 and 1.4, respectively. D [LO 10.13, 10.14]

Solution (a) Since mole fraction

$$x_i = \frac{n_i}{\Sigma n_i}$$

$$x_{N_2} = \frac{\frac{3}{28}}{\frac{3}{28}+\frac{5}{44}} = \mathbf{0.485}$$

$$x_{CO_2} = \frac{\frac{5}{44}}{\frac{3}{28}+\frac{5}{44}} = \mathbf{0.515}$$

(b) Equivalent molecular weight of the mixture

$$M = x_1\,\mu_1 + x_2\,\mu_2$$
$$= 0.485 \times 28 + 0.515 \times 44 = \mathbf{36.25\ kg/kg\ mol}$$

(c) Total mass,

$$m = m_{N_2} + m_{CO_2} = 3 + 5 = 8\text{ kg}$$

Equivalent gas constant of the mixture

$$R = \frac{m_{N_2}R_{N_2} + m_{CO_2}R_{CO_2}}{m}$$

$$= \frac{3\times\frac{8.3143}{28} + 5\times\frac{8.3143}{44}}{8} = \frac{0.89+0.94}{8} = \mathbf{0.229\ kJ/kg\ K}$$

(d) $p_{N_2} = x_{N_2}\cdot p = 0.485 \times 300 = 145.5$ kPa

$p_{CO_2} = x_{CO2}\cdot p = 0.515 \times 300 = 154.5$ kPa

$$V_{N_2} = \frac{m_{N_2}R_{N_2}T}{p} = \frac{3\times\frac{8.3143}{28}\times 293}{300} = 0.87\ m^3$$

$$V_{CO_2} = \frac{m_{CO_2}R_{CO_2}T}{p} = \frac{5\times\frac{8.3143}{44}\times 293}{300} = \mathbf{0.923\ m^3}$$

(e) Total volume of the mixture

$$V = \frac{mRT}{p} = \frac{m_{N_2}R_{N_2}T}{p_{N_2}} = \frac{m_{CO_2}R_{CO_2}T}{p_{CO_2}}$$

$$\therefore \quad V = \frac{8\times 0.229\times 293}{300} = 1.79\ m^3$$

Density of the mixture

$$\rho = \rho_{N_2} + \rho_{CO_2} = \frac{m}{V} = \frac{8}{1.79} = \mathbf{4.46\ kg/m^3}$$

(f) $c_{pN_2} - c_{vN_2} = R_{N_2}$

$$\therefore \quad c_{vN_2} = \frac{R_{N_2}}{\gamma-1} = \frac{8.3143}{28\times(1.4-1)} - 0.742\ kJ/kg\ K$$

$$\therefore \quad c_{pN_2} = 1.4 \times 0.742 = 1.039\ kJ/kg\ K$$

For CO_2, $\gamma = 1.286$

$$\therefore \quad c_{vCO_2} = \frac{R_{CO_2}}{\gamma-1} = \frac{8.3143}{44\times 0.286} = 0.661\ kJ/kg\ K$$

$$c_{pCO_2} = 1.286 \times 0.661 = 0.85\ kJ/kg\ K$$

For the mixture

$$c_p = \frac{m_{N_2}c_{pN_2} + m_{CO_2}c_{pCO_2}}{m_{N_2} + m_{CO_2}}$$

$$= 3/8 \times 1.039 + 5/8 \times 0.85 = 0.92\ kJ/kg\ K$$

$$c_v = \frac{m_{N_2}c_{vN_2} + m_{CO_2}c_{vCO_2}}{m}$$

$$= 3/8 \times 0.742 + 5/8 \times 0.661 = \mathbf{0.69\ kJ/kg\ K}$$

If the mixture is heated at constant volume

$$U_2 - U_1 = mc_v(T_2 - T_1) = 8 \times 0.69 \times (40 - 20) = 110.4\ kJ$$

$$H_2 - H_1 = mc_p(T_2 - T_1) = 8 \times 0.92 \times 20 = 147.2\ kJ$$

$$S_2 - S_1 = mc_v \ln \frac{T_2}{T_1} + mR \ln \frac{V_2}{V_1}$$

$$= mc_v \ln \frac{T_2}{T_1} = 8 \times 0.69 \times \ln \frac{313}{293} = 0.368 \text{ kJ/kg K}$$

If the mixture is heated at constant pressure, ΔU and ΔH will remain the same. The change in entropy will be

$$S_2 - S_1 = mc_p \ln \frac{T_2}{T_1} - mR \ln \frac{p_2}{p_1}$$

$$= mc_p \ln \frac{T_2}{T_1} = 8 \times 0.92 \ln \frac{313}{293} = \mathbf{0.49 \text{ kJ/kg K}}$$

Example 10.20 *Find the increase in entropy when 2 kg of oxygen at 60°C is mixed with 6 kg of nitrogen at the same temperature. The initial pressure of each constituent is 103 kPa and is the same as that of the mixture.* E **[LO 10.14]**

$$x_{O_2} = \frac{p_{O_2}}{p} = \frac{\frac{2}{32}}{\frac{2}{32} + \frac{6}{28}} = 0.225$$

$$x_{N_2} = \frac{p_{N_2}}{p} = 0.775$$

Entropy increase due to diffusion

$$\Delta S = -m_{O_2} R_{O_2} \ln \frac{p_{O_2}}{p} - m_{N_2} R_{N_2} \ln \frac{p_{N_2}}{p}$$

$$= -2\left(\frac{8.3143}{32}\right) \ln 0.225 - 6\left(\frac{8.3143}{28}\right) \ln 0.775 = \mathbf{1.2314 \text{ kJ/kg K}}$$

Example 10.21 *The gas neon has a molecular weight of 20.183 and its critical temperature, pressure and volume are 44.5 K, 2.73 MPa and 0.0416 m³/kg mol, respectively. Reading from a compressibility chart for a reduced pressure of 2 and a reduced temperature of 1.3, the compressibility factor Z is 0.7. What are the corresponding specific volume, pressure, temperature and reduced volume?* E **[LO 10.11]**

Solution At $p_r = 2$ and $T_r = 1.3$ from chart (Fig. 10.28)

$$Z = 0.7$$

$$p = 2 \times 2.73 = \mathbf{5.46 \text{ MPa}}$$

$$\frac{T}{T_c} = 1.3$$

$$T = 1.3 \times 44.5 = \mathbf{57.85 \text{ K}}$$

$$p_v = ZRT$$

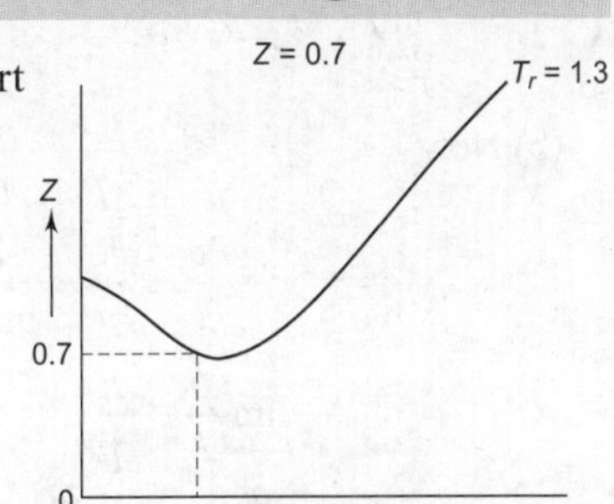

Fig. 10.28

$$\therefore \quad v = \frac{0.7 \times 8.3143 \times 57.85}{20.183 \times 5.46 \times 10^3} = 3.05 \times 10^{-3} \text{ m}^3\text{/kg}$$

$$\therefore \quad v_r = \frac{v}{v_c} = \frac{3.05 \times 10^{-3} \times 20.183}{4.16 \times 10^{-2}} = \mathbf{1.48}$$

Example 10.22 *For the Berthelot equation of state*

$$p = \frac{RT}{v-b} - \frac{a}{Tv^2}$$

show that (a) $\lim\limits_{\substack{p \to 0 \\ T \to \infty}} (RT - pv) = 0$

(b) $\lim\limits_{T \to \infty} \dfrac{v}{T} = \dfrac{R}{p}$,

(c) *Boyle temperature,* $T_B = \sqrt{\dfrac{a}{bR}}$,

(d) *Critical properties* $p_c = \dfrac{1}{12b}\sqrt{\dfrac{2aR}{3b}}$, $v_c = 3b$, $T_c = \sqrt{\dfrac{8a}{27bR}}$,

(e) *Law of corresponding states*

$$\left(p_r + \frac{3}{T_r \times v_r^2}\right)(3v_r - 1) = 8T_r$$

M [LO 10.9, 10.10]

Solution (a)

$$p = \frac{RT}{v-b} - \frac{a}{Tv^2}$$

$$\therefore \quad RT = \left(p + \frac{a}{Tv^2}\right)(v - b)$$

or

$$\frac{RT}{p} = v + \frac{a}{pvT} - b - \frac{ab}{pv^2T}$$

$$\therefore \quad RT - pv = \frac{a}{vT} - bp - \frac{ab}{v^2T}$$

$$\therefore \quad \lim_{\substack{p \to 0 \\ T \to \infty}} (RT - pv) = 0 \qquad \text{Proved (a)}$$

(b) Now

$$v = \frac{RT}{p} - \frac{a}{pvT} + b + \frac{ab}{pv^2T}$$

$$\therefore \quad \frac{v}{T} = \frac{R}{p} - \frac{a}{pvT^2} + \frac{b}{T} + \frac{ab}{pv^2T^2}$$

$$\therefore \quad \lim_{T \to \infty} \frac{v}{T} = \frac{R}{p} \qquad \text{Proved (b)}$$

(c) $pv = RT - \dfrac{a}{vT} + bp + \dfrac{ab}{v^2T}$

The last three terms of the equation are very small, except at very high pressures and small volume. Hence, substituting $v = RT/p$

$$pv = RT - \frac{ap}{RT^2} + bp + \frac{abp^2}{R^2T^3}$$

$$\therefore \quad \left[\frac{\partial (pv)}{\partial p}\right]_T = -\frac{a}{RT^2} + b + \frac{2abp}{R^2T^3} = 0$$

When $p = 0$, $T = T_B$, the Boyle temperature

$$\therefore \quad \frac{a}{RT_B^2} = b$$

or
$$T_B = \sqrt{\frac{a}{bR}}$$
Proved (c)

(d) $p = \dfrac{RT}{v-b} - \dfrac{a}{Tv^2}$

$$\left(\frac{\partial p}{\partial v}\right)_{T=T_c} = -\frac{RT_c}{(v_c - b)^2} + \frac{2a}{T_c \cdot v_c^3} = 0$$

$$\left(\frac{\partial^2 p}{\partial v^2}\right)_{T=T_c} = \frac{2RT_c}{(v_c - b)^3} - \frac{6a}{T_c \cdot v_c^4} = 0$$

$$\left(p_c + \frac{a}{T_c v_c^2}\right)(v_c - b) = RT_c$$

By solving the three equations, as was done in the case of van der Waals equation of state in Section 10.7

$$p_c = \frac{1}{12b}\sqrt{\frac{2aR}{3b}},\ v_c = 3b,\ \text{and}\ T_c = \sqrt{\frac{8a}{27bR}}$$
Proved (d)

(e) Solving the above three equations

$$a = \frac{8v_c^3 p_c^2}{R} = 3p_c \cdot v_c^2 \cdot T_c$$

$$b = \frac{v_c}{3},\ R = \frac{8}{3}\frac{p_v v_c}{T_c}\ \text{(so that } Z_c = 3/8)$$

Substituting in the equation

$$\left(p + \frac{a}{Tv^2}\right)(v - b) = RT$$

$$\left(p + \frac{3p_c v_c^2 T_c}{Tv^2}\right)\left(v - \frac{v_c}{3}\right) = \frac{8p_c v_c}{3T_c} \cdot T$$

$$\left(p_r + \frac{3}{T_r v_r^2}\right)(3v_r - 1) = 8T_r$$

This is the law of corresponding states. Proved (e)

SUMMARY

A mole of a substance has a mass numerically equal to the molecular weight of the substance. One kg-mol of a gas has a volume of 22.4 m^3 at a pressure of 760 mm Hg and 273.16°C. A hypothetical gas which obeys the law $p\bar{v} = \bar{R}T$ at all pressures and temperatures is called an *ideal gas*. As $p \to 0$ or $T \to \infty$, a real gas approaches the ideal gas behaviour. The *universal* gas constant $\bar{R}(= 8.3143$ kJ/kg-mol-K) divided by the molecular weight of the gas is called the *characteristic gas constant* (R).

The number of molecules per gmol of a substance is 6.023 × 10^{23} molecules and is called *Avogadro's number* (A). Boltzmann constant K is defined as $\frac{\bar{R}}{A}$ and is equals to 1.38 × 10^{-23} J/molecule-K. The ideal gas equation of state is expressed by $pV = \text{mRT} = n\bar{R}T = \text{NKT}$, where m = mass, n = number of moles and N = number of molecules of the gas.

The internal energy and enthalpy of an ideal gas are functions of temperature only. The specific heats of an ideal gas, c_p and c_v are constant for the gas depending on its molecular weight and the number of atoms in a molecule of the gas. The entropy change of a gas between two states can be evaluated by any of the three equations

(i) $s_2 - s_1 = c_p \ln \frac{v_2}{v_1} + c_v \ln \frac{p_2}{p_1}$

(ii) $s_2 - s_1 = c_p \ln \frac{T_2}{T_1} - \text{R} \ln \frac{p_2}{p_1}$

(iii) $s_2 - s_1 = c_v \ln \frac{T_2}{T_1} + \text{R} \ln \frac{v_2}{v_1}$

For a reversible adiabatic process of an ideal gas, the following relations hold good

(a) $pv^\gamma = c$, (b) $Tv^{\gamma - 1} = c$, (c) $T_p{}^{(1-\gamma)/\gamma} = c$, (d) $u_2 - u_1 = \frac{RT_1}{\gamma - 1}\left[\left(\frac{p_2}{p_1}\right)^{\frac{\gamma-1}{\gamma}} - 1\right]$,

(e) $h_2 - h_1 = \frac{\gamma RT_1}{\gamma - 1}\left[\left(\frac{p_2}{p_1}\right)^{\frac{\gamma-1}{\gamma}} - 1\right]$,

(f) $W_{1\text{-}2} = U_1 - U_2$. In a polytrophic process, $pv^n = c$, the entropy changes are given by

(i) $s_2 - s_1 = \frac{n-\gamma}{(\gamma-1)(n-1)} R \ln \frac{T_2}{T_1}$,

(ii) $s_2 - s_1 = \frac{n-\gamma}{n(\gamma-1)} R \ln \frac{p_2}{p_1}$ and

(iv) $s_2 - s_1 = -\frac{n-\gamma}{\gamma-1} R \ln \frac{v_2}{v_1}$.

The polytrophic specific heat is given by $c_n = c_v \frac{\gamma - n}{1-n}$ and $Q = c_n \Delta\text{T}$.
The van der Waals equation is given by

$$\left(p + \frac{a}{v^2}\right)(v - b) = RT$$

where $\frac{a}{v^2}$ is called the force of cohesion and b, the co-volume. The other two-constant equations of state are the Berthelot, Dietetics and Redlich–Kwong equations. The virial equations of state, first introduced by Kammerlingh Onnes, are given by

$$\frac{p\bar{v}}{RT} = Z = 1 + B'p + C'p^2 + D'p^3 + \ldots$$
$$= 1 + \frac{B}{v} + \frac{C}{v^2} + \frac{D}{v^3} + \ldots$$

where Z is the compressibility factor and $B, B', C, C'D \ldots$ are called virial coefficients. Each equation of state yields a *law of corresponding state* from which the *generalised compressibility chart* can be obtained.

Properties of gas mixtures, Dalton's law of partial pressures and Amagat's law of partial volumes are discussed. The entropy change due to diffusion of inert ideal gases is given by

$$S_f - S_i = -\bar{R}\left(n_1 \ln\frac{p_1}{p} + n_2 \ln\frac{p_2}{p} \ldots\right)$$
$$= -\bar{R}\sum n_k \ln x_k$$

REVIEW QUESTIONS

(E) **10.1** What is a mole? **[LO 10.1]**

(E) **10.2** What is Avogadro's law? **[LO 10.1]**

(E) **10.3** What is an equation of state? **[LO 10.2]**

(M) **10.4** What is the fundamental property of gases with respect to the product pv? **[LO 10.2]**

(E) **10.5** What is universal gas constant? **[LO 10.2]**

(E) **10.6** Define an ideal gas. **[LO 10.3]**

(E) **10.7** What is the characteristic gas constant? **[LO 10.3]**

(E) **10.8** What is Boltzmann constant? **[LO 10.3]**

(D) **10.9** Why do the specific heats of an ideal gas depend only on the atomic structure of the gas? **[LO 10.4]**

(M) **10.10** Show that for an ideal gas the internal energy depends only on its temperature. **[LO 10.4]**

(M) **10.11** Show that the enthalpy of an ideal gas is a function of temperature only. **[LO 10.4]**

(M) **10.12** Why is there no temperature change when an ideal gas is throttled? **[LO 10.4]**

(M) **10.13** Show that for an ideal gas, $c_p - c_v = R$. **[LO 10.4]**

(M) **10.14** Derive the equations used for computing the entropy change of an ideal gas. **[LO 10.5]**

(D) **10.15** Show that for a reversible adiabatic process executed by an ideal gas, the following relations hold good: (a) pv^{γ} = constant, (b) $Tv^{\gamma-1}$ = constant, and (c) $Tp^{(1-\gamma)/\gamma}$ = constant. **[LO 10.5]**

(M) **10.16** Express the changes in internal energy and enthalpy of an ideal gas in a reversible adiabatic process in terms of the pressure ratio. **[LO 10.5]**

(M) **10.17** Derive the expression of work transfer for an ideal gas in a reversible isothermal process. **[LO 10.7]**

(M) **10.18** What is a polytropic process? What are the relations among p, v and T of an ideal gas in a polytropic process? **[LO 10.8]**

(D) **10.19** Show that the entropy change between states 1 and 2 in a polytropic process, pv^n = constant, is given by the following relations: **[LO 10.8]**

(a) $s_2 - s_1 = \dfrac{n-\gamma}{(\gamma-1)(n-1)} R \ln \dfrac{T_2}{T_1}$

(b) $s_2 - s_1 = \dfrac{n-\gamma}{n(\gamma-1)} R \ln \dfrac{p_2}{p_1}$

(c) $s_2 - s_1 = -\dfrac{n-\gamma}{\gamma-1} R \ln \dfrac{v_2}{v_1}$

(M) **10.20** What are the expressions of work transfer for an ideal gas in a polytropic process, if the gas is: (a) a closed system and (b) a steady system? **[LO 10.8]**

(D) **10.21** Derive the expression of heat transfer for an ideal gas in a polytropic process. What is the polytropic specific heat? What would be the direction of heat transfer if (a) $n > \gamma$ and (b) $n < \gamma$? **[LO 10.8]**

(M) **10.22** Why is the external work supplied to a compressor equal to $-\int_{p_1}^{p_2} v\, dp$? **[LO 10.8]**

(M) **10.23** Why does isothermal compression need minimum work and adiabatic compression maximum work? **[LO 10.8]**

(E) **10.24** What is the isothermal efficiency of a compressor? **[LO 10.8]**

(D) **10.25** Write down the van der Waals equation of state. How does it differ from the ideal gas equation of state. What is force of cohesion? What is co-volume? **[LO 10.9]**

(E) **10.26** What are the two-constant equations of state? **[LO 10.9]**

(E) **10.27** Give the virial expansions for pv in terms of p and v. **[LO 10.10]**

(M) **10.28** What are virial coefficients? When do they become zero? **[LO 10.10]**

(E) **10.29** What is the compressibility factor? **[LO 10.10]**

(E) **10.30** What are reduced properties? **[LO 10.11]**

(M) **10.31** What is the generalised compressibility chart? **[LO 10.11]**

(E) **10.32** What is the law of corresponding states? **[LO 10.11]**

(M) **10.33** Express the van der Waals constants in terms of critical properties. **[LO 10.11]**

(D) **10.34** Draw the diagram representing the law of corresponding states in reduced coordinates indicating the isotherms and the liquid and vapour phases. **[LO 10.11]**

(M) **10.35** Define Boyle temperature? How is it computed? **[LO 10.11]**

(E) **10.36** State Dalton's law of partial pressures. **[LO 10.13]**

(E) **10.37** How is the partial pressure in a gas mixture related to the mole fraction? **[LO 10.13]**

(M) **10.38** How are the characteristic gas constant and the molecular weight of a gas mixture computed? **[LO 10.13]**

(E) **10.39** State and explain Amagat's law of partial volumes of a gas mixture. **[LO 10.13]**

(E) **10.40** What is Gibb's theorem? **[LO 10.14]**

(M) **10.41** Show that in a diffusion process a gas undergoes a free expansion from the total pressure to the relevant partial pressure. **[LO 10.14]**

(M) **10.42** Show that in a diffusion process at constant temperature the entropy increases and the Gibbs function decreases. **[LO 10.14]**

(M) **10.43** Express van der Waals equation of state in the virial form and find the Boyle temperature. **[LO 10.11]**

PROBLEMS WITH HINTS

(M) **10.1** Two tanks are connected by a valve. One tank contains 2 kg of CO_2 gas at 77°C and 0.2 bar. The other tank holds 8 kg of the same gas at 27°C and 1.2 bar. The valve is opened and the gases are allowed to mix while receiving energy by heat transfer from the surroundings. The final equilibrium temperature is 42°C. Determine the final equilibrium pressure and the heat transfer for the process. **[LO 10.14]**

Hint:

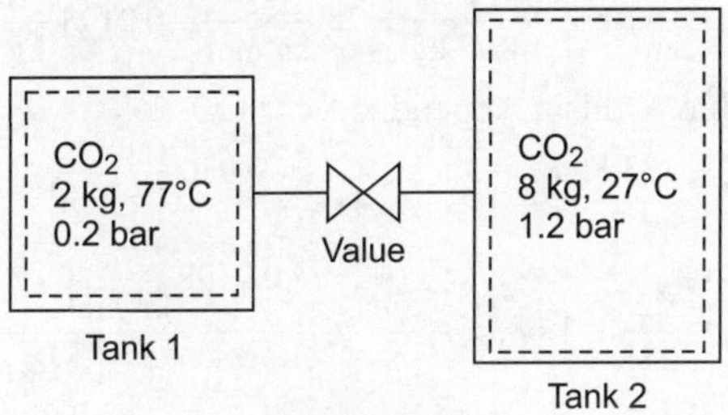

Fig. 10.29

$T_1 = 77 + 273 = 350$ K
$T_2 = 27 + 273 = 300$ K
$T_f = 42 + 273 = 315$ K

The final equilibrium pressure, $p_f = \dfrac{mRT_f}{V}$

$$\therefore \quad p_f = \frac{(m_{14} + m_2)RT_f}{\dfrac{m_1RT_1}{p_1} + \dfrac{m_2RT_2}{p_2}} = \frac{(m_{14} + m_2)T_f}{\dfrac{m_1T_1}{p_1} + \dfrac{m_2T_2}{p_2}} = \frac{(m_1 + m_2)RT_f}{V_1 + V_2}$$

Substituting the given values p_f = **1.05 bar**

$$Q = U_f - U_i = m_1 c_{v1}(T_f - T_i) + m_2 c_{v2}(T_f - T_i)$$

$$= 24 \times 0.745 \frac{\text{kJ}}{\text{kg K}} \times (315 - 350)\text{ K} + 8\text{ kg} \times 0.745 \frac{\text{kJ}}{\text{kg K}} \times (315 - 300)\text{K}$$

$$= \mathbf{-37.25\ kJ}$$

(E) **10.2** Consider a gas mixture of molecular weight 33, initially at 3 bar, 300 K and occupying a volume of 0.1 m^3. The gas undergoes an expansion to 0.2 m^3 during which the pressure – volume relation is $PV^{1.3}$ = Constant. Assuming $c_v = 0.6 + 2.5 \times 10^{-4}\ T$, where T is in K and c_v is in kJ/kgK, and neglecting K.E. and P.E. effects, determine (a) the mass of gas, (b) the final pressure, (c) the final temperature and (d) the work and heat transfer. **[LO 10.4]**

Hint:

$$m = \frac{pV_1}{RT_1} = \frac{3 \times 10^5 \times 10^{-3}}{\dfrac{8.3143}{33} \times 300} = \mathbf{0.397\ kg}$$

$$p_2 = p_1\left(\frac{V_1}{V_2}\right)^{1.3} = 1.218 \text{ bar}, \; T_2 = \frac{p_2V_2}{mR} = \mathbf{243.6\ K}$$

$$W_{1-2} = \frac{p_1V_1 - p_2V_2}{n-1} = 18.8 \text{ kJ}$$

$$Q = m\int_{T_1}^{T_2}(0.6 + 2.5\times10^{-4}T)\,dT + W = \mathbf{3.84\ kJ}$$

(M) **10.3** A cylindrical tank containing 4 kg of carbon monoxide gas at –50°C has an inner diameter of 0.2 m and a length of 1 m. Determine the pressure exerted by the gas using (a) the generalised compressibility chart, (b) the ideal gas equation of state and (c) the van der Walls equation of state, (d) the Redlich–Kwong equation of state. **[LO 10.9, 10.10]**

Hint:

$$V = \frac{\pi}{4}d^2L = 0.034 \text{ m}^3$$

$$\bar{v} = \frac{V}{m}\mu = \frac{0.0314}{4}\frac{\text{m}^3}{\text{kg}} \times 28 \frac{\text{kg}}{\text{kg mol}} = 0.2198 \frac{\text{m}^3}{\text{kg mol}}$$

(a) From Table A. 11 in Appendix A for CO,

$$T_C = 133 \text{ K}, P_C = 3.5 \text{ MPa}, v_c = 0.930 \frac{\text{m}^3}{\text{kg mol}}$$

$$\therefore \quad T_p = \frac{223}{133} = 1.68, \; p_r = \frac{p}{p_c}, \; v_r = \frac{0.2198\dfrac{\text{m}^3}{\text{kg mol}}\times 3.5\times100\dfrac{\text{kN}}{\text{m}^2}}{8.3143\dfrac{\text{kJ}}{\text{kg mol-K}}\times 133 \text{ K}}$$

$$= 0.696$$

Using generalised compressibility chart, (Fig. 10.5c), $Z = 0.9$

$$\therefore \quad P = \frac{Z\bar{R}T}{\bar{v}} = \frac{0.9\times 8.3143\dfrac{\text{kJ}}{\text{kg mol-K}}\times 223 \text{ K}}{0.2198\dfrac{\text{m}^3}{\text{kg mol}}}\frac{\text{kN}}{\text{m}^2} = \mathbf{75.9\ bar}$$

(b) For CO, the van der Waals constants a and b are given in Table 10.1,

$$a = 1.474 \text{ bar}\left(\frac{\text{m}^3}{\text{kg mol}}\right)^2 \text{ and } b = 0.0395 \frac{\text{m}^3}{\text{kg mol}}$$

$$\therefore \quad p = \frac{\bar{R}T}{\bar{v}-b} - \frac{a}{\bar{v}^2} = \mathbf{72.3\ bar} \text{ (on substitution).}$$

(c) The Redlich–Kwong constants a and b for CO are:

$$a = 17.22\left(\text{bar}\frac{\text{m}^2}{\text{kg mol K}}\right)\frac{K^{1/2}}{(\text{1k mol})^2} \text{ and } b = 0.02737 \frac{\text{m}^3}{\text{kg mol}}$$

In substitution,

$$p = \frac{\bar{R}T}{\bar{v}-b} - \frac{a}{\bar{v}(\bar{v}+b)T^{1/2}} = \mathbf{75.1\ bar}$$

(M) **10.4** A mixture of 60 % CH_4 and 40 % C_3H_8 by volume is compressed isentropically from 40°C, 0.4×10^6 N/m² to 1.2×16^6. Find the final temperature of the mixture, the work required per unit mass and the entropy change for each gas assuming ideal gas behaviour with constant specific heats. Take c_p's for CH_4 and C_3H_8 are 35.72 and 0.7456 kJ/kg, respectively. **[LO 10.14]**

Hint:

The specific heats of the mixture are

$$\bar{c}_p = 0.6 \times 35.72 + 0.4 \times 74.56 = 51.256 \text{ kJ/kgmol-K}$$

$$\bar{c}_v = \bar{c}_p - \bar{R} = 51.256 - 8.314 = 47.942 \text{ kJ/kgmol-K}$$

The molecular weight of the mixture

$$\mu_m = 0.60 \times 16 + 0.4 \times 44 = 27.2 \text{ kg/kgmol}$$

$$\gamma_m = \frac{\bar{c}_p}{\bar{c}_v} = 1.197$$

$$\frac{T_2}{T_1} = \left(\frac{p_2}{p_1}\right)^{\frac{\gamma-1}{\gamma}} = (3)^{1.193} \rightarrow T_2 = \mathbf{3140\ K}$$

$$W = 51.26\,(374 - 313) = 3127 \text{ kJ/kgmol} = \mathbf{114\ kJ/kg}$$

For methane,

$$s_2 - s_1 = \bar{c}_p \ln \frac{T_2}{T_1} - \bar{R} \ln \frac{p_2}{p_1} = 35.72 \ln \frac{374}{313} - 8.314 \ln 3$$

$$= \mathbf{-2.7686\ kJ/kgmol\text{-}K}$$

For propane,

$$s_2 - s_1 = 74.56 \ln \left(\frac{374}{313}\right) + 8.314 \ln 3$$

$$= +\,4.1529 \text{ kJ/kgmol-K} \quad \textbf{Ans.}$$

For the mixture,

$$(s_2 - s_1)_{\text{mix}} = 0.6 \times (-2.7686) + 0.4 \times 4.1569$$

$$= \mathbf{0.00\ kJ/kgmol\text{-}K\ which\ is\ zero.}$$

(E) **10.5** Air initially at 0.75 bar, 1000 K and occupying a volume of 0.12 m^3 undergoes two processes. The air is compressed isothermally until the volume is halved. Then it undergoes a constant pressure process until the volume is halved again. Assuming ideal gas behaviour,

(a) sketch the processes on *p-V* and *T-s* diagrams,
(b) determine the total work for the two processes and
(c) determine the total heat transfer in the two processes. **[LO 10.4]**

Hint:

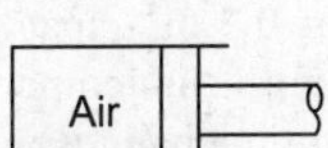

$p_1 = 0.75$ bar
$T_1 = T_2 = 1000$ K
$V_1 = 0.12$ m^3
$V_2 = \frac{1}{2} V_1 = 0.06$ m^3
$V_3 = \frac{1}{2} V_2 = 0.03$ m^3

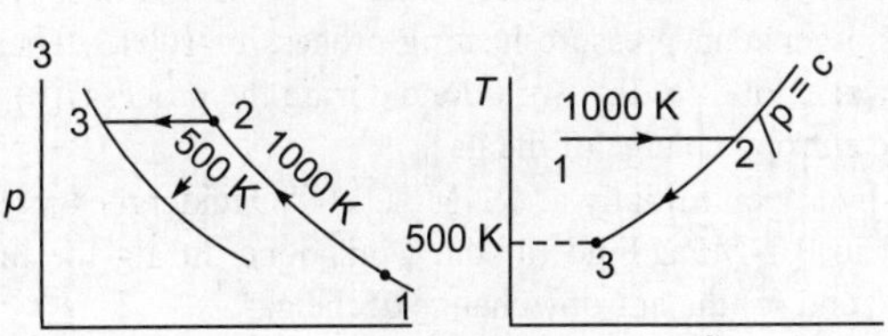

Fig. 10.30

(a) Process 1 – 2: $p_2 = p_1 \frac{V_1}{2} = 1.5$ bar

Process 2 – 3, $p_3 = p_2$, $T_3 = T_2 \frac{V_3}{V_2} = 500$ K

$$W_{1-3} = W_{12} + W_{2-3} = \int_{V_1}^{V_2} pdV + \int_{V_2}^{V_3} pdV = \text{mRT}_1 \ln \frac{V_2}{V_1} + p_2 (V_3 - V_2)$$

Now, $m = \dfrac{P_1V_1}{RT_1} = 0.03136$ kg

On substitution, $W_{13} = \mathbf{-10.74\ kJ}$

(b) $Q_{13} - W_{13} = mc_v (T_3 - T_1)$

$$Q_{13} = \mathbf{-23.27\ kJ}$$

PROBLEMS

(E) **10.1** What is the mass of air contained in a room 6 m × 9 m × 4 m if the pressure is 101.325 kPa and the temperature is 25°C? **[LO 10.3]**

(E) **10.2** The usual cooking gas (mostly methane) cylinder is about 25 cm in diameter and 80 cm in height. It is changed to 12 MPa at room temperature (27°C). (a) Assuming the ideal gas law, find the mass of gas filled in the cylinder. (b) Explain how the actual cylinder contains nearly 15 kg of gas. (c) If the cylinder to be protected against excessive pressure by means of a fusible plug, at what temperature should the plug melt to limit the maximum pressure to 15 MPa? **[LO 10.3]**

(E) **10.3** A certain gas has $c_p = 0.913$ and $c_v = 0.653$ kJ/kg K. Find the molecular weight and the gas constant R of the gas. **[LO 10.4]**

(E) **10.4** From an experimental determination, the specific heat ratio for acetylene (C_2H_2) is found to 1.26. Find the two specific heats. **[LO 10.4]**

(E) **10.5** Find the molal specific heats of monatomic, diatomic and polyatomic gases, if their specific heat ratios are, respectively, 5/3, 7/5 and 4/3. **[LO 10.4]**

(M) **10.6** A supply of natural gas is required on a site 800 m above storage level. The gas at – 150°C, 1.1 bar from storage is pumped steadily to a point on the site where its pressure is 1.2 bar, its temperature 15°C and its flow rate 1000 m³/hr. If the work transfer to the gas at the pump is 15 kW, find the heat transfer to the gas between the two points.Neglect the change in K.E. and assume that the gas has the properties of methane (CH_4) which may be treated as an ideal gas having $\gamma = 1.33$ ($g = 9.75$ m/s²). **[LO 10.5]**

(E) **10.7** A constant volume chamber of 0.3 m³ capacity contains 1 kg of air at 5°C. Heat is transferred to the air until the temperature is 100°C. Find the work done, the heat transferred and the changes in internal energy, enthalpy and entropy. **[LO 10.5]**

(E) **10.8** One kg of air in a closed system, initially at 5°C and occupying 0.3 m³ volume, undergoes a constant pressure heating process to 100°C. There is no work other than *pdv* work. Find (a) the work done during the process, (b) the heat transferred and (c) the entropy change of the gas. **[LO 10.5]**

(E) **10.9** 0.1 m³ of hydrogen initially at 1.2 MPa, 200°C undergoes a reversible isothermal expansion to 0.1 MPa. Find (a) the work done during the process, (b) the heat transferred and (c) the entropy change of the gas. **[LO 10.7]**

(E) **10.10** Air in a closed stationary system expands in a reversible adiabatic process from 0.5 MPa, 15°C to 0.2 MPa. Find the final temperature, and per kg of air, the change in enthalpy, the heat transferred and the work done. **[LO 10.5]**

(E) **10.11** If the above process occurs in an open steady flow system, find the final temperature, and per kg of air, the change in internal energy, the heat transferred and the shaft work. Neglect velocity and elevation changes. **[LO 10.5]**

(E) **10.12** The indicator diagram for a certain water-cooled cylinder and piston air compressor shows that during compression $pv^{1.3}$ = constant. The compression starts at 100 kPa, 25°C and ends at 600 kPa. If the process is reversible, how much heat is transferred per kg of air? (12) (0) **[LO 10.5]**

(E) **10.13** An ideal gas of molecular weight 30 and $\gamma = 1.3$ occupies a volume of 1.5 m^3 at 100 kPa and 77°C. The gas is compressed according to the law $pv^{1.25}$ = constant to a pressure of 3 MPa. Calculate the volume and temperature at the end of compression and heating, work done, heat transferred and the total change of entropy. **[LO 10.8]**

(E) **10.14** Calculate the change of entropy when 1 kg of air changes from a temperature of 330 K and a volume of 0.15 m^3 to a temperature of 550 K and a volume of 0.6 m^3. If the air expands according to the law, pv^n = constant, between the same end states, calculate the heat given to, or extracted from, the air during the expansion, and show that it is approximately equal to the change of entropy multiplied by the mean absolute temperature. **[LO 10.5, 10.8]**

(E) **10.15** 0.5 kg of air, initially at 25°C, is heated reversibly at constant pressure until the volume is doubled, and is then heated reversibly at constant volume until the pressure is doubled. For the total path, find the work transfer, the heat transfer and the change of entropy. **[LO 10.8]**

(E) **10.16** An ideal gas cycle of three processes uses Argon (Mol. wt. 40) as a working substance. Process 1–2 is a reversible adiabatic expansion from 0.014 m^3, 700 kPa, 280°C to 0.056 m^3. Process 2–3 is a reversible isothermal process. Process 3–1 is a constant pressure process in which heat transfer is zero. Sketch the cycle in the *p*-*v* and *T*-*s* planes, and find (a) the work transfer in process 1–2, (b) the work transfer in process 2–3 and (c) the net work of the cycle. Take $\gamma = 1.67$. **[LO 10.5, 10.7, 10.8]**

(E) **10.17** A gas occupies 0.024m^3 at 700 kPa and 95°C. It is expanded in the non-flow process according to the law $pv^{1.2}$ = constant to a pressure of 70 kPa after which it is heated at constant pressure back to its original temperature. Sketch the process on the *p*-*v* and *T*-*s* diagrams, and calculate for the whole process the work done, the heat transferred and the change of entropy. Take $c_p = 1.047$ and $c_v = 0.775$ kJ/kg K for the gas. **[LO 10.8]**

(E) **10.18** 0.5 kg of air at 600 kPa receives an addition of heat at constant volume so that its temperature rises from 110C to 650°C. It then expands in a cylinder polytropically to its original temperature and the index of expansion is 1.32. Finally, it is compressed isothermally to its original volume. Calculate (a) the change of entropy during each of the three stages and (b) the pressures at the end of constant volume heat addition and at the end of expansion. Sketch the processes on the *p*-*v* and *T*-*s* diagrams. **[LO 10.8]**

(E) **10.19** 0.5 kg of helium and 0.5 kg of nitrogen are mixed at 20°C and at a total pressure of 100 kPa. Find (a) the volume of the mixture, (b) the partial volumes of the components, (c) the partial pressures of the components, (d) the mole fractions of the components, (e) the specific heats c_p and c_v of the mixture and (f) the gas constant of the mixture. **[LO 10.13]**

(E) **10.20** A gaseous mixture consists of 1 kg of oxygen and 2 kg of nitrogen at a pressure of 150 kPa and a temperature of 20°C. Determine the changes in internal energy, enthalpy and entropy of the mixture when the mixture is heated to a temperature of 100°C (a) at constant volume and (b) at constant pressure. **[LO 10.13, 10.14]**

(M) **10.21** A closed rigid cylinder is divided by a diaphragm into two equal compartments, each of volume 0.1 m^3. Each compartment contains air at a temperature of 20°C. The pressure in one compartment is 2.5 MPa and in the other compartment it is 1 MPa. The diaphragm is ruptured so that the air in both the compartments mixes to bring the pressure to a uniform value throughout the cylinder which is insulated. Find the net change of entropy for the mixing process. **[LO 10.14]**

(M) **10.22** A vessel is divided into three compartments (a), (b) and (c) by two partitions. Part (a) contains oxygen and has a volume of 0.1 m^3, (b) has a volume of 0.2 m^3 and contains nitrogen, while (c) is 0.05 m^3 and holds CO_2. All three parts are at a pressure of 2 bar and a temperature of 13°C. When the partitions are removed and the gases mix, determine the change of entropy of each constituent, the final pressure in the vessel and the partial pressure of each gas. The vessel may be taken as being completely isolated from its surroundings. **[LO 10.14]**

(E) **10.23** A Carnot cycle uses 1 kg of air as the working fluid. The maximum and minimum temperatures of the cycle are 600 and 300 K. The maximum pressure of the cycle is 1 MPa and the volume of the gas doubles during the isothermal heating process. Show by the calculation of net work and heat supplied that the efficiency is the maximum possible for the given maximum and minimum temperatures. **[LO 10.7]**

(M) **10.24** An ideal gas cycle consists of three reversible processes in the following sequence: (a) constant volume pressure rise, (b) isentropic expansion to r times the initial volume and (c) constant pressure decrease in volume. Sketch the cycle on the p-v and T-s diagrams. Show that the efficiency of the cycle is

$$\eta_{\text{cycle}} = \frac{r^\gamma - 1 - \gamma(r-1)}{r^\gamma - 1}$$

Evaluate the cycle efficiency when $\gamma = \frac{4}{3}$ and $r = 8$. **[LO 10.8]**

(M) **10.25** Using the Dieterici equation of state

$$p = \frac{RT}{v-b} \cdot \exp\left(-\frac{a}{RTv}\right)$$

(a) Show that

$$p_c = \frac{a}{4e^2b^2},\ v_c = 2b,\ T_c = \frac{a}{4Rb}$$

(b) Expand in the form

$$pv = RT\left(1 + \frac{B}{v} + \frac{C}{v^2} + \ldots\right)$$

(c) Show that

$$T_B = \frac{a}{bR}$$

[LO 10.9]

(M) **10.26** The number of moles, the pressures and the temperatures of gases a, b and c are given below.

Gas	*m (kg mol)*	*p (kPa)*	*t (°C)*
N_2	1	350	100
CO	3	420	200
O_2	2	700	300

If the containers are connected, allowing the gases to mix freely, find (a) the pressure and temperature of the resulting mixture at equilibrium, and (b) the change of entropy of each constituent and that of the mixture. **[LO 10.14]**

(M) **10.27** Calculate the volume of 2.5 kg moles of steam at 236.4 atm. and 776.76 K with the help of compressibility factor versus reduced pressure graph. At this volume and given pressure, what would the temperature be in K, if steam behaved like a van der Waals gas? **[LO 10.11]**

The critical pressure, volume and temperature of steam are 218.2 atm., 57 cm^3/g mole, and 647.3 K, respectively.

(E) **10.28** Two vessels, *A* and *B*, each of volume 3 m^3 may be connected together by a tube of negligible volume. Vessel *A* contains air at 7 bar, 95°C while *B* contains air at 3.5 bar, 205°C. Find the change of entropy when *A* is connected to *B*. Assume the mixing to be complete and adiabatic. **[LO 10.14]**

(E) **10.29** An ideal gas at temperature T_1 is heated at constant pressure to T_2 and then expanded reversibly, according to the law pv^n = constant, until the temperature is once again T_1. What is the required value of *n*, if the changes of entropy during the separate processes are equal? **[LO 10.8]**

(M) **10.30** A certain mass of sulphur dioxide (SO_2) is contained in a vessel of 0.142 m^3 capacity, at a pressure and temperature of 23.1 bar and 18°C, respectively. A valve is opened momentarily and the pressure falls immediately to 6.9 bar. Sometimes later, the temperature is again 18°C and the pressure is observed to be 9.1 bar. Estimate the value of specific heat ratio. **[LO 10.14]**

(M) **10.31** A gaseous mixture contains 21% by volume of nitrogen, 50% by volume of hydrogen and 29% by volume of carbon dioxide. Calculate the molecular weight of the mixture, the characteristic gas constant *R* for the mixture and the value of the reversible adiabatic index γ. (At 10°C, the c_p values of nitrogen, hydrogen and carbon dioxide are 1.039, 14.235 and 0.828 kJ/kg K, respectively.)

A cylinder contains 0.085 m^3 of the mixture at 1 bar and 10°C. The gas undergoes a reversible non-flow process during which its volume is reduced to one-fifth of its original value. If the law of compression is $pv^{1.2}$ = constant, determine the work and heat transfer in magnitude and sense and the change in entropy. **[LO 10.13]**

(M) **10.32** Two moles of an ideal gas at temperature *T* and pressure *p* are contained in a compartment. In an adjacent compartment, there is one mole of an ideal gas at temperature 2*T* and pressure *p*. The gases mix adiabatically but do not react chemically when a partition separating the compartments is withdrawn. Show that the entropy increase due to the mixing process is given by

$$R\left(\ln\frac{27}{4}+\frac{\gamma}{\gamma-1}\ln\frac{32}{27}\right)$$

provided that the gases are different and that the ratio of specific heat γ is the same for both gases and remains constant.

What would the entropy change be if the mixing gases were of the same species? **[LO 10.14]**

(M) **10.33** n_1 moles of an ideal gas at pressure p_1 and temperature *T* are in one compartment of an insulated container. In an adjoining compartment, separated by a partition, are n_2 moles of an ideal gas at pressure p_2 and temperature *T*. When the partition is removed, calculate (a) the final pressure of the mixture, (b) the entropy change when the gases are identical and (c) the entropy change when the gases are different. Prove that the entropy change in (c) is the same as that produced by two independent free expansions. **[LO 10.14]**

(E) **10.34** Assume that 20 kg of steam are required at a pressure of 600 bar and a temperature of 750°C in order to conduct a particular experiment. A 140-litre heavy duty tank is available for storage.

Predict if this is an adequate storage capacity using

(a) the ideal gas theory,

(b) the compressibility factor chart,

(c) the van der Waals equation with a = 5.454 (litre)^2 atm/(g mol)^2, b = 0.03042 litres/gmol for steam,

(d) the Mollier chart and

(e) the steam tables.

Estimate the error in each. **[LO 10.3, 10.11]**

(E) **10.35** Estimate the pressure of 5 kg of CO_2 gas which occupies a volume of 0.70 m^3 at 75°C, using the Beattie–Bridgeman equation of state.

Compare this result with the value obtained using the generalised compressibility chart. Which is more accurate and why?

For CO_2 with units of atm, litres/g mol and K, A_0 = 5.0065, a = 0.07132, B_0 = 0.10476, b = 0.07235, $C \times 10^{-4}$ = 66.0. **[LO 10.9, 10.11]**

(M) **10.36** Measurements of pressure and temperature at various stages in an adiabatic air turbine show that the states of air lie on the line $pv^{1.25}$ = constant. If kinetic and gravitational potential energus are neglected, prove that the shaft work per kg as a function of pressure is given by the following relation:

$$W = 3.5\, p_1 v_1 \left[1 - \left(\frac{p_2}{p_1}\right)^{1/5}\right]$$

Take γ for air as 1.4. **[LO 10.5]**

(M) **10.37** A mass of an ideal gas exists initially at a pressure of 200 kPa, temperature 300 K and specific volume 0.5 m^3/kg. The value of γ is 1.4. (a) Determine the specific heats of the gas, (b) What is the change in entropy when the gas is expanded to pressure 100 kPa according to the law $pv^{1.3}$ = const? (c) What will be the entropy change if the path is $pv^{1.5}$ = const. (by the application of a cooling jacket during the process)? (d) What is the inference you can draw from this example? **[LO 10.5, 10.8]**

(E) **10.38** (a) A closed system of 2 kg of air initially at pressure 5 atm and temperature 227°C, expands reversibly to pressure 2 atm following the law $pv^{1.25}$ = const. Assuming air as an ideal gas, determine the work done and the heat transferred.

(b) If the system does the same expansion in a steady flow process,what is the work done by the system? **[LO 10.5]**

(E) **10.39** Air contained in a cylinder fitted with a piston is compressed reversibly according to the law $pv^{1.25}$ = const. The mass of air in the cylinder is 0.1 kg. The initial pressure is 100 kPa and the initial temperature 20°C. The final volume is 1/8 of the initial volume. Determine the work and the heat transfer. **[LO 10.5]**

(E) **10.40** Air is contained in a cylinder fitted with a frictionless piston. Initially, the cylinder contains 0.5 m^3 of air at 1.5 bar, 20°C. The air is then compressed reversibly according to the law pv^n = constant until the final pressure is 6 bar, at which point the temperature is 120°C. Determine (a) the polytropic index n, (b) the final volume of air, (c) the work done on the air and the heat transfer and (d) the net change in entropy. **[LO 10.5]**

(M) **10.41** The specific heat at constant pressure for air is given by

$$c_p = 0.9169 + 2.577 + 10^{-4}\,T - 3.974 \times 10^{-8}\,T^2 \text{ kJ/kg K}$$

Determine the change in internal energy and that in entropy of air when it undergoes a change of state from 1 atm and 298 K to a temperature of 2000 K at the same pressure. **[LO 10.5]**

(M) **10.42** A closed system allows nitrogen to expand reversibly from a volume of 0.25 to 0.75 m along the path $pv^{1.32}$ = const. The original pressure of the gas is 250 kPa and its initial temperature is 100°C. (a) Draw the *p-v* and *T-s* diagrams. (b) What are the final temperature and the final pressure of the gas? (c) How much work is done and how much heat is transferred? (d) What is the entropy change of nitrogen? **[LO 10.8]**

(M) **10.43** Methane has a specific heat at constant pressure given by c_p = 17.66 + 0.06188 *T* kJ/kg mol K when 1 kg of methane is heated at constant volume from 27 to 500°C. If the initial pressure of the gas is 1 atm, calculate the final pressure, the heat transfer, the work done and the change in entropy. **[LO 10.8]**

(E) **10.44** Air is compressed reversibly according to the law $pv^{1.25}$ = const. from an initial pressure of 1 bar and volume of 0.9 m³ to a final volume of 0.6 m³. Determine the final pressure and the change of entropy per kg of air. **[LO 10.5]**

(M) **10.45** In a heat engine cycle, air is isothermally compressed. Heat is then added at constant pressure, after which the air expands isentropically to its original state. Draw the cycle on *p-v* and *T-s* coordinates. Show that the cycle efficiency can be expressed in the following form:

$$\eta = 1 - \frac{(\gamma - 1)\ln r}{\gamma[r^{\gamma-1/\gamma} - 1]}$$

where *r* is the pressure ratio, p_2/p_1. Determine the pressure ratio and the cycle efficiency if the initial temperature is 27°C and the maximum temperature is 327°C. **[LO 10.7]**

(E) **10.46** What is the minimum amount of work required to separate 1 mole of air at 27°C and 1 atm pressure (assumed composed of 1/5 O_2 and 4/5 N_2) into oxygen and nitrogen each at 27°C and 1 atm pressure? **[LO 10.13]**

(M) **10.47** A closed adiabatic cylinder of volume 1 m³ is divided by a partition into two compartments 1 and 2. Compartment 1 has a volume of 0.6 m³ and contains methane at 0.4 MPa, 40°C, while compartment 2 has a volume of 0.4 m³ and contains propane at 0.4 MPa, 40°C. The partition is removed and the gases are allowed to mix. (a) When the equilibrium state is reached, find the entropy change of the universe. (b) What are the molecular weight and the specific heat ratio of the mixture?

The mixture is now compressed reversibly and adiabatically to 1.2 MPa. Compute (c) the final temperature of the mixture, (d) the work required per unit mass and (e) the specific entropy change for each gas. Take c_p of methane and propane as 35.72 and 74.56 kJ/kg mol K, respectively. **[LO 10.13, 10.14]**

(M) **10.48** An ideal gas cycle consists of the following reversible processes: (a) isentropic compression, (b) constant volume heat addition, (c) isentropic expansion, and (d) constant pressure heat rejection. Show that the efficiency of this cycle is given by

$$\eta = 1 - \frac{1}{r_k^{\gamma-1}}\left[\frac{\gamma(a^{1/\gamma} - 1)}{a - 1}\right]$$

where r_k is the compression ratio and *a* is the ratio of pressures after and before heat addition.

An engine operating on the above cycle with a compression ratio of 6 starts the compression with air at 1 bar, 300 K. If the ratio of pressures after and before heat addition is 2.5, calculate the efficiency and the m.e.p. of the cycle. Take $\gamma = 1.4$ and $c_v = 0.718$ kJ/kg K. **[LO 10.8]**

(E) **10.49** The relation between u, p and v for many gases is of the form $u = a + bpv$ where a and b are constants. Show that for a reversible adiabatic process pv^γ = constant, where $\gamma = (b + 1)/b$. **[LO 10.5]**

(E) **10.50** (a) Show that the slope of a reversible adiabatic process on p-v coordinates is

$$\frac{dp}{dv} = -\frac{1}{kv}\frac{c_p}{c_v} \text{ where } k = -\frac{1}{v}\left(\frac{\partial v}{\partial p}\right)_T$$

(b) Hence, show that for an ideal gas, pv^γ = constant in a reversible adiabatic process. **[LO 10.5]**

(E) **10.51** A certain gas obeys the Clausius equation of state $p(v - b) = RT$ and has its internal energy given by $u = c_v T$. Show that the equation for a reversible adiabatic process is $b\,p(v - b)^\gamma$ = constant, where $\gamma = c_p/c_v$. **[LO 10.5]**

(M) **10.52** (a) Two curves, one representing a reversible adiabatic process undergone by an ideal gas and the other an isothermal process by the same gas, intersect at the same point on the p-v diagram. Show that the ratio of the slope of the adiabatic curve to the slope of the isothermal curve is equal to γ. (b) Determine the ratio of work done during a reversible adiabatic process to the work done during an isothermal process for a gas having $\gamma = 1.6$. Both processes have a pressure ratio of 6. **[LO 10.5]**

(E) **10.53** Two containers p and q with rigid walls contain two different monatomic gases with masses m_p and m_q, gas constants R_p and R_q and initial temperatures T_p and T_q, respectively, are brought in contact with each other and allowed to exchange energy until equilibrium is achieved. Determine (a) the final temperature of the two gases, and (b) the change of entropy due to this energy exchange. **[LO 10.13, 10.14]**

(E) **10.54** The pressure of a certain gas (photon gas) is a function of temperature only and is related to the energy and volume by $p(T) = (1/3)\,(U/V)$. A system consisting of this gas confined by a cylinder and a piston undergoes a Carnot cycle between two pressures p_1 and p_2. (a) Find expressions for work and heat of reversible isothermal and adiabatic processes, (b) Plot the Carnot cycle on p-v and T-s diagrams, (c) Determine the efficiency of the cycle in terms of pressures and (d) What is the functional relation between pressure and temperature? **[LO 10.5, 10.7]**

(M) **10.55** The gravimetric analysis of dry air is approximately: oxygen = 23%, nitrogen = 77%. Calculate (a) the volumetric analysis, (b) the gas constant, (c) the molecular weight, (d) the respective partial pressures, (e) the specific volume at 1 atm, 15°C and (f) How much oxygen must be added to 2.3 kg air to produce a mixture which is 50% oxygen by volume? **[LO 10.13, 10.14]**

(M) **10.56** A vessel of volume $2V$ is divided into two equal compartments. These are filled with the same ideal gas, the temperature and pressure on one side of the partition being (p_1, T_1) and on the other (p_2, T_2). Show that if the gases on the two sides are allowed to mix slowly with no heat entering, the final pressure and temperature will be given by

$$p = \frac{p_1 + p_2}{2},\ T = \frac{T_1 T_2 (p_1 + p_2)}{p_1 T_2 + p_2 T_1}$$

Further, show that the entropy gain is

$$\Delta S = V\left[\left(\frac{c_p}{R}\right)\left\{\frac{p_1}{T_1}\ln\frac{T}{T_1}+\frac{p_2}{T_2}\ln\frac{T}{T_2}\right\}-\frac{p_1}{T_1}\ln\frac{p}{p_1}-\frac{p_2}{T_2}\ln\frac{p}{p_2}\right]$$ **[LO 10.13, 10.14]**

(M) **10.57** An ideal gas with a constant volume of c_p = 29.6 J/gmol-K is made to undergo a cycle consisting of the following reversible processes in a closed system:
Process 1–2: The gas expands adiabatically from 5 MPa, 550 K to 1 MPa;
Process 2–3: The gas is heated at constant volume until 550 K; Process 3–1: The gas is compressed isothermally back to its initial condition.
Calculate the work, the heat and the change of entropy of the gas for each of the three processes. Draw the *p-v* and *T-s* diagrams. **[LO 10.8]**

(E) **10.58** Air in a closed system expands reversibly and adiabatically from 3 MPa, 200°C to two times its initial volume, and then cools at constant volume until the pressure drops to 0.8 MPa. Calculate the work done and heat transferred per kg of air. Use c_p = 1.017 and c_v = 0.728 kJ/kgK. **[LO 10.5]**

(M) **10.59** A vessel is divided into three compartments (a), (b) and (c) by two partitions. Part (a) contains hydrogen and has a volume of 0.1 m^3, part (b) contains nitrogen and has a volume of 0.2 m^3 and part (c) contains carbon dioxide and has a volume of 0.05 m^3. All the three parts are at a pressure of 2 bar and a temperature of 13°C. The partitions are removed and the gases are allowed to mix. Determine (a) the molecular weight of the mixture, (b) the characteristics gas constant for the mixture, (c) the partial pressures of each gas, (d) the reversible adiabatic index γ and (e) the entropy change due to diffusion. The specific heats of hydrogen, nitrogen and carbon dioxide are 14.235, 1.039 and 0.828 kJ/kg K, respectively.
(f) The above gas mixture is then reversibly compressed to a pressure of 6 bar according to the law $pv^{1.2}$ = constant. Determine the work and heat interactions in magnitude and sense and (g) the change in entropy. **[LO 10.13, 10.14]**

(M) **10.60** Predict the pressure of nitrogen gas at T = 175 K and v = 0.00375 m^3/kg on the basis of (a) the ideal gas equation of state, (b) the van der Waals equation of state, (c) the Beattie-Bridgeman equation of state and (d) the Benedict–Webb–Rubin equation of state. Compare the values obtained with the experimentally determined value of 10,000 kPa. **[LO 10.11, 10.12]**

(E) **10.61** The pressure in an automobile tyre depends on the temperature of the air in the tyre. When the air temperature is 25°C, the pressure gauge reads 210 kPa. If the volume of the tyre is 0.025 m^3, determine the pressure rise in the tyre when the air temperature in the tyre rises to 50°C. Also find the amount of air that must be bled off to restore pressure to its original value at this temperature. Take atmospheric pressure as 100 kPa. **[LO 10.2]**

(M) **10.62** Two tanks are connected by a valve. One tank contains 2 kg of CO gas at 77°C and 0.7 bar. The other tank holds 8 kg of the same gas at 27°C and 1.2 bar. The valve is opened and the gases are allowed to mix while receiving energy by heat transfer from the surroundings. The final equilibrium temperature is 42°C. Using the ideal gas model, determine (a) the final equilibrium pressure, and (b) the heat transfer for the process. **[LO 10.13]**

ANSWERS TO PROBLEMS

10.1 256 kg

10.6 63.9 kW

10.22 0.0875, 0.0783, 0.0680 kJ/K; 2 bar; 0.5714, 1.1429, 0.2857 bar

10.24 $\eta = 0.378$

10.28 0.975 kJ/kg K

10.29 $\left(n = \dfrac{2\gamma}{\gamma + 1}\right)$

10.30 1.29

10.31 19.64 kg/kg mol, 0.423 kJ/kg K, 1.365, –16 kJ, –7.24 kJ, –0.31 kJ/kg K

10.37 (a) 1.166, 0.833 kJ/kg K
(b) 0.044 kJ/kg K
(c) – 0.039 kJ/kg K
(d) Entropy increases when $n < \gamma$ and decreases when $n > \gamma$)

10.38 (a) 193 kJ, 72 kJ
(b) 241 kJ

10.39 – 22.9 kJ, – 8.6 kJ

10.40 (a) 1,2685
(b) 0.1676 m^3
(c) – 95.3 kJ, – 31.5 kJ
(d) 0.0153 kJ/K

10.41 1470.4 kJ/kg, 2.1065 kJ/kg K

10.42 (b) 262.44 K, 58.63 kPa
(c) 57.89 kJ, 11.4 kJ
(d) 0.0362 kJ/K

10.43 2.577 atm, 1258.5 kJ/kg, 0, 2.3838 kJ/kg K

10.44 1.66 bar, – 0.0436 kJ/kg K

10.45 13.4, 32.4%

10.46 1250 J

10.47 (a) 0.8609 kJ/K
(b) 27.2, 1.193
(c) 100.9°C
(d) 396 kJ
(e) 0.255 kJ/kg K

10.48 0.579, 2.5322 bar

10.49 (a) 21% O_2, 79% N_2
(b) 0.288 kJ/kg K
(d) 21 kPa for O_2
(e) 0.84 m^3/kg
(f) 1.47 kg

10.57 $W_{1-2} = 4260$ J/gmol, $Q_{1-2} = 0$, D$s_{1-2} = 0$, $W_{2-3} = 0$, $Q_{2-3} = 4260$ J/gmol, D$s_{2-3} = 9.62$ J/g mol-K, $W_{3-1} = -$ 5290 J/mol $= Q_{3-1}$, D$s_{3-1} = -9.62$ J/gmol-K, $W_{net} - Q_{net} = -$ 1030 g/gmol, f dS = 0

10.58 82.7 kJ/kg, – 78.1 kJ/kg

10.59 (a) 22.8582
(b) 0.3637 kJ/kg K
(c) $p_{H2} = 0.5714$, $p_{N2} = 1.1428$, $p_{CO2} = 0.2858$ bar
(d) 1.384
(e) 0.3476 kJ/kgK
(f) – 70.455 kJ, – 33.772 kJ
(g) – 0.1063 kJ/K

10.60 (a) 13,860 kPa
(b) 9468 kPa
(c) 10,110 kPa
(d) 10,000 kPa

10.61 (a) 1.05 bar
(b) 37.25 kJ

11

Thermodynamic Relations, Equilibrium and Stability

Learning Objectives

LO 11.1 Relate exact differential relations with thermodynamic properties

LO 11.2 Formulate Maxwell's equations

LO 11.3 Construct thermodynamic property relations using *TdS* equations

LO 11.4 Relate heat capacity ratio with compression work

LO 11.5 Apply energy equation for useful property relations

LO 11.6 Explain Joule-Kelvin effect and inversion curve

LO 11.7 Formulate Clausius-Clapeyron equation

LO 11.8 Evaluate thermodynamic properties other than p, v, T from equation of state

LO 11.9 Construct property relations from equation of state

LO 11.10 Discuss properties of multi-component systems

LO 11.11 Formulate the conditions of thermodynamic equilibrium of a heterogeneous system

LO 11.12 Discuss the Gibbs Phase Rule for non reactive heterogeneous systems

LO 11.13 Formulate the conditions of four types of thermodynamic equilibriums and their stability

11.1 SOME MATHEMATICAL THEOREMS

LO 11.1 Relate exact differential relations with thermodynamic properties

Theorem 1 If a relation exists among the variables x, y and z, then z may be expressed as a function of x and y, or

$$dz = \left(\frac{\partial z}{\partial x}\right)_y dx + \left(\frac{\partial z}{\partial y}\right)_x dy$$

If
$$\left(\frac{\partial z}{\partial x}\right)_y = M, \quad \text{and} \quad \left(\frac{\partial z}{\partial y}\right)_x = N$$

then
$$dz = M\,dx + N\,dy,$$

where z, M and N are functions of x and y. Differentiating M partially with respect to y, and N with respect to x

$$\left(\frac{\partial M}{\partial y}\right)_x = \frac{\partial^2 z}{\partial x\,\partial y}$$

$$\left(\frac{\partial N}{\partial x}\right)_y = \frac{\partial^2 z}{\partial y\,\partial x}$$

$$\therefore \quad \left(\frac{\partial M}{\partial y}\right)_x = \left(\frac{\partial N}{\partial x}\right)_y \tag{11.1}$$

This is the *condition of exact* (*or perfect*) *differential*.

Theorem 2 If a quantity f is a function of x, y and z, and a relation exists among x, y and z, then f is a function of any two of x, y and z. Similarly, any one of x, y and z may be regarded to be a function of f and any one of x, y and z. Thus, if

$$x = x\,(f, y)$$

$$dx = \left(\frac{\partial x}{\partial f}\right)_y df + \left(\frac{\partial x}{\partial y}\right)_f dy$$

Similarly, if

$$y = y(f, z)$$

$$dy = \left(\frac{\partial y}{\partial f}\right)_z df + \left(\frac{\partial y}{\partial z}\right)_f dz$$

Substituting the expression of dy in the preceding equation

$$dx = \left(\frac{\partial x}{\partial f}\right)_y df + \left(\frac{\partial x}{\partial y}\right)_f \left[\left(\frac{\partial y}{\partial f}\right)_z df + \left(\frac{\partial y}{\partial z}\right)_f dz\right]$$

$$= \left[\left(\frac{\partial x}{\partial f}\right)_y + \left(\frac{\partial x}{\partial y}\right)_f \left(\frac{\partial y}{\partial f}\right)_z\right] df + \left(\frac{\partial x}{\partial y}\right)_f \left(\frac{\partial y}{\partial z}\right)_f dz$$

Again

$$dx = \left(\frac{\partial x}{\partial f}\right)_z df + \left(\frac{\partial x}{\partial z}\right)_f dz$$

$$\therefore \quad \left(\frac{\partial x}{\partial z}\right)_f = \left(\frac{\partial x}{\partial y}\right)_f \left(\frac{\partial y}{\partial z}\right)_f$$

$$\therefore \quad \left(\frac{\partial x}{\partial y}\right)_f \left(\frac{\partial y}{\partial z}\right)_f \left(\frac{\partial z}{\partial x}\right)_f = 1 \tag{11.2}$$

Theorem 3 Among the variables x, y and z, any one variable may be considered as a function of the other two. Thus,

$$x = x\,(y, z)$$

$$dx = \left(\frac{\partial x}{\partial y}\right)_z dy + \left(\frac{\partial x}{\partial z}\right)_y dz$$

Similarly,

$$dz = \left(\frac{\partial z}{\partial x}\right)_y dx + \left(\frac{\partial z}{\partial y}\right)_x dy$$

$$\therefore \quad dx = \left(\frac{\partial x}{\partial y}\right)_z dy + \left(\frac{\partial x}{\partial z}\right)_y \left[\left(\frac{\partial z}{\partial x}\right)_y dx + \left(\frac{\partial z}{\partial y}\right)_x dy\right]$$

$$= \left[\left(\frac{\partial x}{\partial y}\right)_z + \left(\frac{\partial x}{\partial z}\right)_y \left(\frac{\partial z}{\partial y}\right)_x\right] dy + \left(\frac{\partial x}{\partial z}\right)_y \left(\frac{\partial z}{\partial x}\right)_y dx$$

$$= \left[\left(\frac{\partial x}{\partial y}\right)_z + \left(\frac{\partial x}{\partial z}\right)_y \left(\frac{\partial z}{\partial y}\right)_x\right] dy + dx$$

$$\therefore \quad \left(\frac{\partial x}{\partial y}\right)_z + \left(\frac{\partial x}{\partial z}\right)_y \left(\frac{\partial z}{\partial y}\right)_x = 0$$

or

$$\left(\frac{\partial x}{\partial y}\right)_z \left(\frac{\partial z}{\partial x}\right)_y \left(\frac{\partial y}{\partial z}\right)_x = -1 \tag{11.3}$$

Among the thermodynamic variables p, V and T, the following relation holds good:

$$\left(\frac{\partial p}{\partial V}\right)_T \left(\frac{\partial V}{\partial T}\right)_p \left(\frac{\partial T}{\partial p}\right)_V = -1$$

11.2 MAXWELL'S EQUATIONS

> **LO 11.2**
> Formulate Maxwell's equations

A pure substance existing in a single phase has only two independent variables. Of the eight quantities p, V, T, S, U, H, F (Helmholtz function) and G (Gibbs function) *any one may be expressed as a function of any two others.*

For a pure substance undergoing an infinitesimal reversible process

1. $dU = TdS - pdV$
2. $dH = dU + pdV + Vdp = TdS + Vdp$
3. $dF = dU - TdS - SdT = -pdV - SdT$
4. $dG = dH - TdS - SdT = Vdp - SdT$

Since U, H, F and G are thermodynamic properties and exact differentials of the type

$$dz = M\,dx + N\,dy, \text{ then}$$

$$\left(\frac{\partial M}{\partial y}\right)_x = \left(\frac{\partial N}{\partial x}\right)_y$$

Applying this to the four equations

$$\left(\frac{\partial T}{\partial V}\right)_S = -\left(\frac{\partial p}{\partial S}\right)_V \tag{11.4}$$

$$\left(\frac{\partial T}{\partial p}\right)_S = \left(\frac{\partial V}{\partial S}\right)_p \tag{11.5}$$

$$\left(\frac{\partial p}{\partial T}\right)_V = \left(\frac{\partial S}{\partial V}\right)_T \tag{11.6}$$

$$\left(\frac{\partial V}{\partial T}\right)_p = -\left(\frac{\partial S}{\partial p}\right)_T \tag{11.7}$$

These four equations are known as *Maxwell's equations.*

11.3 *TdS* EQUATIONS

LO 11.3
Construct thermodynamic property relations using *TdS* equations

Let entropy S be imagined as a function of T and V. Then

$$dS = \left(\frac{\partial S}{\partial T}\right)_V dT + \left(\frac{\partial S}{\partial V}\right)_T dV$$

$$\therefore \quad TdS = T\left(\frac{\partial S}{\partial T}\right)_V dT + T\left(\frac{\partial S}{\partial V}\right)_T dV$$

Since $T\left(\frac{\partial S}{\partial T}\right)_V = C_v$, heat capacity at constant volume, and

$$\left(\frac{\partial S}{\partial V}\right)_T = \left(\frac{\partial p}{\partial T}\right)_V, \text{ Maxwell's third equation,}$$

$$TdS = C_v\,dT + T\left(\frac{\partial p}{\partial T}\right)_V dV \tag{11.8}$$

This is known as the *first TdS equation.*

If $S = S(T, p)$

$$dS = \left(\frac{\partial S}{\partial T}\right)_p dT + \left(\frac{\partial S}{\partial p}\right)_T dp$$

$$\therefore \quad TdS = T\left(\frac{\partial S}{\partial T}\right)_p dT + T\left(\frac{\partial S}{\partial p}\right)_T dp$$

Since $T\left(\frac{\partial S}{\partial T}\right)_p = C_p$, and $\left(\frac{\partial S}{\partial p}\right)_T = -\left(\frac{\partial V}{\partial T}\right)_p$

then

$$TdS = C_p\, dT - T\left(\frac{\partial V}{\partial T}\right)_p dp \tag{11.9}$$

This is known as the *second TdS equation.*

11.4 DIFFERENCE IN HEAT CAPACITIES

Equating the first and second *TdS* equations

$$TdS = C_p\, dT - T\left(\frac{\partial V}{\partial T}\right)_p dp = C_v\, dT + T\left(\frac{\partial p}{\partial T}\right)_V dV$$

$$(C_p - C_v)\, dT = T\left(\frac{\partial p}{\partial T}\right)_V dV + T\left(\frac{\partial V}{\partial T}\right)_p dp$$

$$\therefore \quad dT = \frac{T\left(\frac{\partial p}{\partial T}\right)_V}{C_p - C_v} dV + \frac{T\left(\frac{\partial V}{\partial T}\right)_p}{C_p - C_v} dp$$

Again $$dT = \left(\frac{\partial T}{\partial V}\right)_p dV + \left(\frac{\partial T}{\partial p}\right)_v dp$$

$$\therefore \quad \frac{T\left(\frac{\partial p}{\partial T}\right)_V}{C_p - C_v} = \left(\frac{\partial T}{\partial V}\right)_p \quad \text{and} \quad \frac{T\left(\frac{\partial V}{\partial T}\right)_p}{C_p - C_v} = \left(\frac{\partial T}{\partial p}\right)_V$$

Both these equations give

$$C_p - C_v = T\left(\frac{\partial p}{\partial T}\right)_v \left(\frac{\partial V}{\partial T}\right)_p$$

But $$\left(\frac{\partial p}{\partial T}\right)_v \left(\frac{\partial T}{\partial V}\right)_p \left(\frac{\partial V}{\partial p}\right)_T = -1$$

$$\therefore \quad C_p - C_v = -T\left(\frac{\partial V}{\partial T}\right)_T^2 \left(\frac{\partial p}{\partial V}\right)_T \tag{11.10}$$

This is a very important equation in thermodynamics. It indicates the following important facts:

1. Since $\left(\frac{\partial V}{\partial T}\right)_p^2$ is always positive, and $\left(\frac{\partial p}{\partial V}\right)_T$ for any substance is negative, $(C_p - C_v)$ is always positive. Therefore, C_p is always greater than C_v.
2. As $T \to 0\,K$, $C_p \to C_v$ or at absolute zero, $C_p = C_v$.
3. When $\left(\frac{\partial V}{\partial T}\right)_p = 0$ (e.g. for water at 4°C, when density is maximum, or specific volume minimum), $C_p = C_v$.
4. For an ideal gas, $pV = mRT$

$$\left(\frac{\partial V}{\partial T}\right)_p = \frac{mR}{p} = \frac{V}{T}$$

and

$$\left(\frac{\partial p}{\partial V}\right)_T = -\frac{mRT}{V^2}$$

∴

$$C_p - C_v = mR$$

or

$$c_p - c_v = R$$

Equation (11.10) may also be expressed in terms of *volume expansivity* (β), defined as

$$\beta = \frac{1}{V}\left(\frac{\partial V}{\partial T}\right)_p$$

and *isothermal compressibility* (k_T), defined as

$$k_T = -\frac{1}{V}\left(\frac{\partial V}{\partial p}\right)_T$$

$$C_p - C_v = \frac{TV\left[\frac{1}{V}\left(\frac{\partial V}{\partial T}\right)_p\right]^2}{-\frac{1}{V}\left(\frac{\partial V}{\partial p}\right)_T}$$

∴

$$C_p - C_v = \frac{TV\beta^2}{k_T} \tag{11.11}$$

11.5 RATIO OF HEAT CAPACITIES

LO 11.4
Relate heat capacity ratio with compression work

At constant S, the two TdS equations become

$$C_p\, dT_s = T\left(\frac{\partial V}{\partial T}\right)_p dp_s$$

$$C_v\, dT_s = -T\left(\frac{\partial p}{\partial T}\right)_V dV_s$$

$$\therefore \qquad \frac{C_p}{C_v} = -\left(\frac{\partial V}{\partial T}\right)_p \left(\frac{\partial T}{\partial p}\right)_V \left(\frac{\partial p}{\partial V}\right)_S = \frac{\left(\frac{\partial p}{\partial V}\right)_S}{\left(\frac{\partial p}{\partial V}\right)_T} = \gamma$$

Since $\gamma > 1$,

$$\left(\frac{\partial p}{\partial V}\right)_S > \left(\frac{\partial p}{\partial V}\right)_T$$

Therefore, the slope of an isentrope is greater than that of an isotherm on $p - v$ diagram (Fig. 11.1). For reversible and adiabatic compression, the work done is

$$W_s = h_{2s} - h_1 = \int_1^{2S} v dp$$

$$= \text{Area } 1 - 2S - 3 - 4 - 1$$

For reversible and isothermal compression, the work done would be

$$W_T = h_{2T} - h_1 = \int_1^{2T} v dp$$

$$= \text{Area } 1\text{–}2T - 3 - 4 - 1$$

$$\therefore \qquad W_T < W_S$$

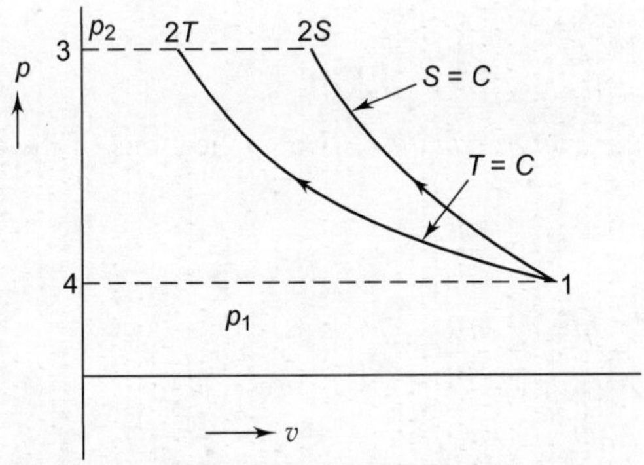

Fig. 11.1 Compression work in different reversible process

For polytropic compression with $1 < n < \gamma$, the work done will be between these two values. So, isothermal compression requires minimum work and isentropic compression requires the maximum work.

The adiabatic compressibility (k_s) is defined as

$$k_s = -\frac{1}{V}\left(\frac{\partial V}{\partial p}\right)_s$$

$$\therefore \qquad \frac{C_p}{C_v} = \frac{-\dfrac{1}{V}\left(\dfrac{\partial V}{\partial p}\right)_T}{-\dfrac{1}{V}\left(\dfrac{\partial V}{\partial p}\right)_s} = \gamma$$

or
$$\gamma = \frac{k}{k_s} \tag{11.12}$$

11.6 ENERGY EQUATION

LO 11.5 Apply energy equation for useful property relations

For a system undergoing an infinitesimal reversible process between two equilibrium states, the change of internal energy is

$$dU = TdS - pdV$$

Substituting the first *TdS* equation

$$dU = C_v\, dT + T\left(\frac{\partial p}{\partial T}\right)_V dV - pdV$$

$$= C_v\, dT + \left[T\left(\frac{\partial p}{\partial T}\right)_V - p\right] dV \tag{11.13}$$

If
$$U = U(T, V)$$

$$dU = \left(\frac{\partial U}{\partial T}\right)_V dT + \left(\frac{\partial U}{\partial V}\right)_T dV$$

$$\left(\frac{\partial U}{\partial V}\right)_T = T\left(\frac{\partial p}{\partial T}\right)_V - p \tag{11.14}$$

This is known as the *energy equation*. Two applications of the equation are given below:

1. For an ideal gas, $p = \dfrac{n\bar{R}T}{V}$

$$\therefore \qquad \left(\frac{\partial p}{\partial T}\right)_V = \frac{n\bar{R}}{V} = \frac{p}{T}$$

$$\therefore \qquad \left(\frac{\partial U}{\partial V}\right)_T = T\cdot\frac{p}{T} - p = 0$$

U does not change when *V* changes at $T = C$.

$$\left(\frac{\partial U}{\partial p}\right)_T \left(\frac{\partial p}{\partial V}\right)_T \left(\frac{\partial V}{\partial U}\right)_T = 1$$

$$\therefore \qquad \left(\frac{\partial U}{\partial p}\right)_T \left(\frac{\partial p}{\partial V}\right)_T = \left(\frac{\partial U}{\partial V}\right)_T = 0$$

Since $\left(\dfrac{\partial p}{\partial V}\right)_T \neq 0, \left(\dfrac{\partial U}{\partial p}\right)_T = 0$

U does not change either when p changes at $T = C$. So the internal energy of an ideal gas is a function of temperature only, as shown earlier in Chapter 10. Another important point to note is that in Eq. (11.13), for an ideal gas

$$pV = n\bar{R}T \text{ and } T\left(\frac{\partial p}{\partial T}\right)_v - p = 0$$

Therefore,

$$dU = C_v\, dT$$

holds good for an ideal gas in any process (even when the volume changes). But for any other substance

$$dU = C_v\, dT$$

is true only when the volume is constant and $dV = 0$.

Similarly,

$$dH = TdS + Vdp$$

and

$$TdS = C_p\, dT - T\left(\frac{\partial V}{\partial T}\right)_p dp$$

$$\therefore \quad dH = C_p\, dT + \left[V - T\left(\frac{\partial V}{\partial T}\right)_p\right] dp \tag{11.15}$$

$$\therefore \quad \left(\frac{\partial H}{\partial p}\right)_T = V - T\left(\frac{\partial V}{\partial T}\right)_p \tag{11.16}$$

As shown for internal energy, it can be similarly proved from Eq. (11.16) that the enthalpy of an ideal gas is not a function of either volume or pressure

$$\left[\text{i.e.} \left(\frac{\partial H}{\partial p}\right)_T = 0 \text{ and } \left(\frac{\partial H}{\partial V}\right)_T = 0\right]$$

but a function of temperature alone.

Since for an ideal gas, $pV = n\bar{R}T$

and

$$V - T\left(\frac{\partial V}{\partial T}\right)_p = 0$$

the relation $dH = C_p\, dT$ is true for any process (even when the pressure changes). However, for any other substance the relation $dH = C_p\, dT$ holds good only when the pressure remains constant or $dp = 0$.

2. Thermal radiation in equilibrium with the enclosing walls possesses an energy that depends only on the volume and temperature. The energy density (u), defined as the ratio of energy to volume, is a function of temperature only, or

$$u = \frac{U}{V} = f(T) \text{ only.}$$

The electromagnetic theory of radiation states that radiation is equivalent to a photon gas and it exerts a pressure, and that the pressure exerted by the black-body radiation in an enclosure is given by

$$p = \frac{u}{3}$$

Black-body radiation is thus specified by the pressure, volume and temperature of the radiation.

Since $$U = uV \quad \text{and} \quad p = \frac{u}{3}$$

$$\left(\frac{\partial U}{\partial V}\right)_T = u \text{ and } \left(\frac{\partial p}{\partial T}\right)_V = \frac{1}{3}\frac{du}{dT}$$

By substituting in the energy Eq. (11.13)

$$u = \frac{T}{3}\frac{du}{dT} - \frac{u}{3}$$

$$\therefore \quad \frac{du}{u} = 4\,\frac{dT}{T}$$

or $$\ln u = \ln T^4 + \ln b$$

or $$u = bT^4$$

where b is a constant. This is known as the *Stefan–Boltzmann Law.*

Since $$U = uV = VbT^4$$

$$\left(\frac{\partial U}{\partial T}\right)_V = C_v = 4VbT^3$$

and $$\left(\frac{\partial p}{\partial T}\right)_V = \frac{1}{3}\frac{du}{dT} = \frac{4}{3}bT^3$$

From the first *TdS* equation

$$TdS = C_v\,dT + T\left(\frac{\partial p}{\partial T}\right)_V dV$$

$$= 4VbT^3\,dT + \frac{4}{3}bT^4 \cdot dV$$

For a reversible isothermal change of volume, the heat to be supplied reversibly to keep temperature constant

$$Q = \frac{4}{3}\,bT^4\,\Delta V$$

For a reversible adiabatic change of volume

$$\frac{4}{3}\,bT^4\,dV = -\,4VbT^3\,dT$$

$$\therefore \quad \frac{dV}{V} = -3\,\frac{dT}{T}$$

or $$VT^3 = \text{const.}$$

If the temperature is one-half the original temperature, the volume of black-body radiation is to be increased adiabatically eight times its original volume so that the radiation remains in equilibrium with matter at that temperature.

11.7 JOULE–KELVIN EFFECT

LO 11.6 Explain Joule-Kelvin effect and inversion curve

A gas is made to undergo continuous throttling process by a valve, as shown in Fig. 11.2. The pressures and temperatures of the gas in the insulated pipe upstream and downstream of the valve are measured with suitable manometers and thermometers.

Let p_i and T_i be the arbitrarily chosen pressure and temperature before throttling and let them be kept constant. By operating the valve manually, the gas is throttled successively to different pressures and temperatures p_{f1}, T_{f1}; p_{f2}, T_{f2}; p_{f3}, T_{f3} and so on. These are then plotted on the *T-p* coordinates as shown in Fig. 11.3. All the points represent equilibrium states of some constant mass of gas, say, 1 kg, at which the gas has the *same enthalpy*.

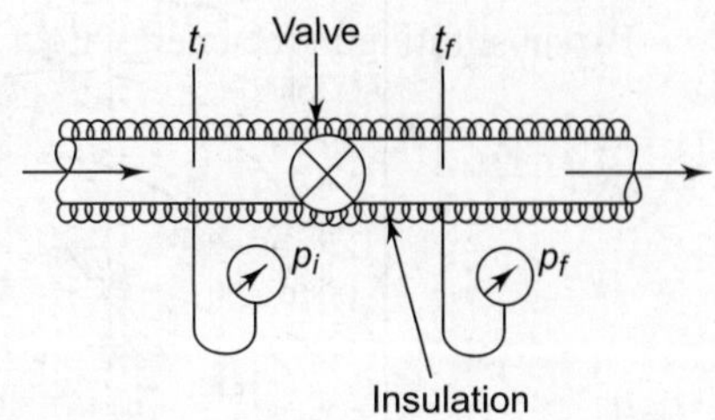

Fig. 11.2 Joule–Thomson expansion

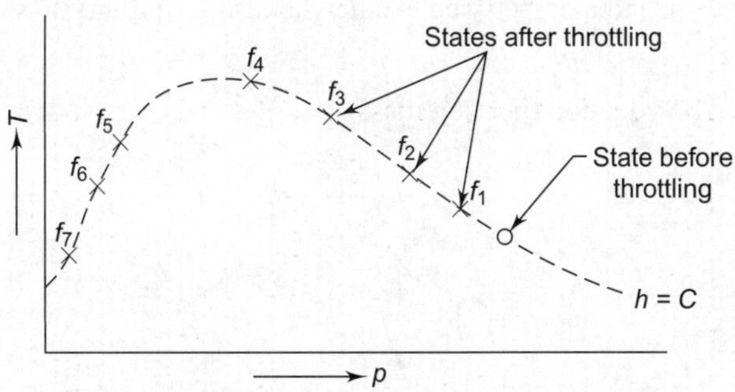

Fig. 11.3 Isenthalpic states of a gas

The curve passing through all these points is an isenthalpic curve or an *isenthalpe.* It is not the graph of a throttling process, but the graph through points of equal enthalpy.

The initial temperature and pressure of the gas (before throttling) are then set to new values, and by throttling to different states, a *family of isenthalpes* is obtained for the gas, as shown in Figs. 11.4 and 11.5. The curve passing through the maxima of these isenthalpes is called the *inversion curve*.

The numerical value of the slope of an isenthalpe on a *T-p* diagram at any point is called the *Joule–Kelvin coefficient* and is denoted by μ_J. Thus the locus of all points at which μ_J is zero is the inversion curve. The region inside the inversion curve where μ_J is positive is called the *cooling region* and the region outside where μ_J is negative is called the *heating region*. If a gas is throttled in the cooling region, it will get cooled. If it is throttled in the heating region, the gas will be heated. So,

$$\mu_J = \left(\frac{\partial T}{\partial p}\right)_h$$

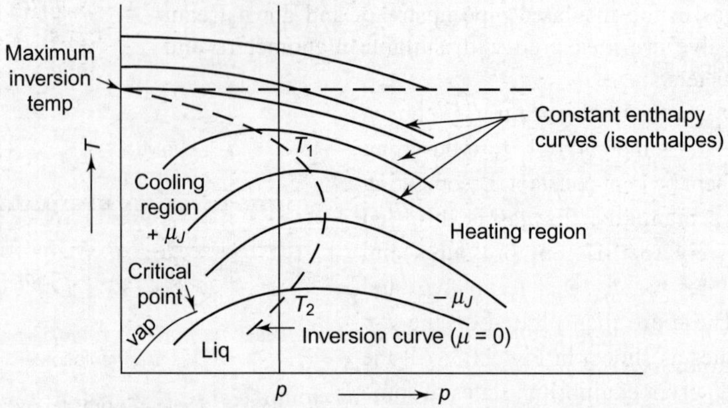

Fig. 11.4 Isenthalpic curves and the inversion curve

The difference in enthalpy between two neighbouring equilibrium states is

$$dh = Tds + vdp$$

and the second *TdS* equation (per unit mass)

$$Tds = c_p\, dT - T\left(\frac{\partial v}{\partial T}\right)_p dp$$

$$\therefore \qquad dh = c_p\, dT - \left[T\left(\frac{\partial v}{\partial T}\right)_p - v\right] dp$$

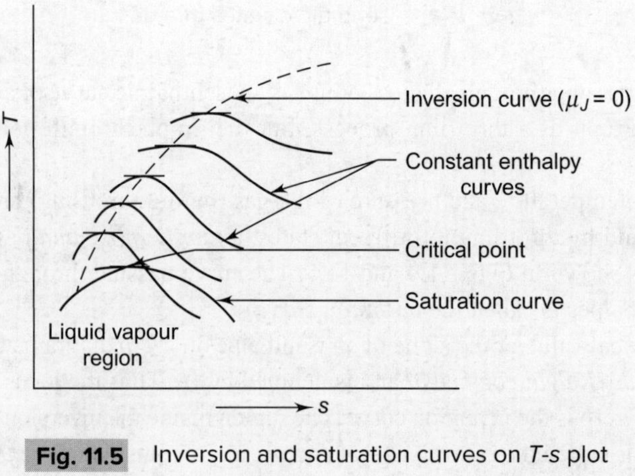

Fig. 11.5 Inversion and saturation curves on *T-s* plot

The second term in the above equation stands only for a real gas, because for an ideal gas, $dh = c_p\, dT$.

$$\mu_J = \left(\frac{\partial T}{\partial p}\right)_h = \frac{1}{c_p}\left[T\left(\frac{\partial v}{\partial T}\right)_p - v\right] \tag{11.17}$$

For an ideal gas

$$pv = RT$$

$$\therefore \quad \left(\frac{\partial v}{\partial T}\right)_p = \frac{R}{p} = \frac{v}{T}$$

$$\therefore \quad \mu_J = \frac{1}{c_p}\left(T \cdot \frac{v}{T} - v\right) = 0$$

There is no change in temperature when an ideal gas is made to undergo a Joule–Kelvin expansion (i.e. throttling).

For achieving the effect of cooling by Joule–Kelvin expansion, the initial temperature of the gas must be below the point where the inversion curve intersects the temperature axis, i.e. below the *maximum inversion temperature*. For nearly all substances, the maximum inversion temperature is above the normal ambient temperature, and hence cooling can be obtained by the Joule–Kelvin effect. In the case of hydrogen and helium, however, the gas is to be precooled in heat exchangers below the maximum inversion temperature before it is throttled. For liquefaction, the gas has to be cooled below the critical temperature.

Let the initial state of gas before throttling be at A (Fig. 11.6). The change in temperature may be positive, zero or negative, depending upon the final pressure after throttling. If the final pressure lies between A and B, there will be a rise in temperature or heating effect. If it is at C, there will be no change in temperature. If the final pressure is below p_C, there will be a cooling effect, and if the final pressure is p_D, the temperature drop will be $(T_A - T_D)$.

Maximum temperature drop will occur if the initial state lies on the inversion curve. In Fig. 11.6, it is $(T_B - T_D)$.

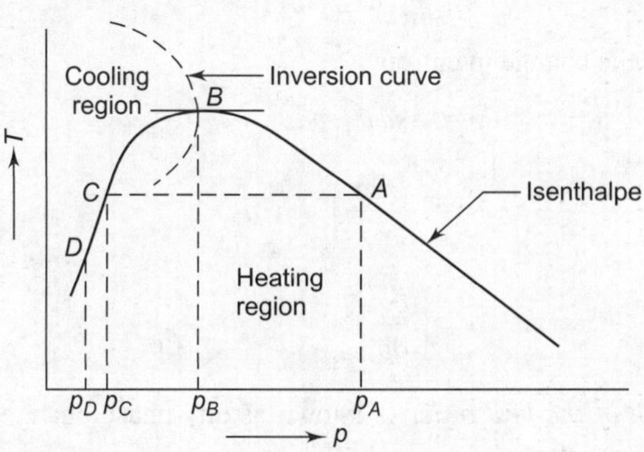

Fig. 11.6 Maximum cooling by Joule–Kelvin expansion

The volume expansivity is

$$\beta = \frac{1}{v}\left(\frac{\partial v}{\partial T}\right)_p$$

So the Joule–Kelvin coefficient μ_J is given by, from Eq. (11.17)

$$\mu_J = \frac{1}{c_p}\left[T\left(\frac{\partial v}{\partial T}\right)_p - v\right]$$

or

$$\mu_J = \frac{v}{c_p}[\beta T - 1]$$

For an ideal gas, $\beta = \frac{1}{T}$ and $\mu_J = 0$

There are two inversion temperatures for each pressure, e.g. T_1 and T_2 at pressure p (Fig. 11.4).

11.8 CLAUSIUS–CLAPEYRON EQUATION

LO 11.7 Formulate Clausius-Clapeyron equation

During phase transitions like melting, vaporisation and sublimation, the temperature and pressure remain constant, while the entropy and volume change. If x is the fraction of initial phase i which has been transformed into final phase f, then

$$s = (1 - x)\, s^{(i)} + x s^{(f)}$$

$$v = (1 - x)\, v^{(i)} + x v^{(f)}$$

where s and v are linear functions of x.

For reversible phase transition, the heat transferred per mole (or per kg) is the latent heat, given by

$$l = T\{s^{(f)} - s^{(i)}\}$$

which indicates the change in entropy.

Now

$$dg = -sdT + vdp$$

or

$$s = -\left(\frac{\partial g}{\partial T}\right)_p$$

and

$$v = \left(\frac{\partial g}{\partial p}\right)_T$$

A *phase change of the first order* is known as any phase change that satisfies the following requirements:

1. There are changes of entropy and volume.

2. The first-order derivatives of Gibbs function change discontinuously (Fig. 11.7a).

Let us consider the first-order phase transition of one mole of a substance from phase i to phase f. Using the first *TdS* equation

$$Tds = c_v\, dT + T\left(\frac{\partial p}{\partial T}\right)_V dv$$

for the phase transition which is reversible, isothermal and isobaric, and integrating over the whole change of phase, and since $\left(\frac{\partial p}{\partial T}\right)_V$ is independent of v

$$T\{s^{(f)} - s^{(i)}\} = T\frac{dp}{dT}\{v^{(f)} - v^{(i)}\}$$

$$\therefore \quad \frac{dp}{dT} = \frac{s^{(f)} - s^{(i)}}{v^{(f)} - v^{(i)}} = \frac{l}{T\{v^{(f)} - v^{(i)}\}} \qquad (11.18)$$

The above equation is known as the *Clausius–Clapeyron equation.*

The Clausius-Clapeyron equation can also be derived in another way. For a reversible process at constant T and p, the Gibbs function remains constant. Therefore, for the first-order phase change at T and p

$$g^{(i)} = g^{(f)}$$

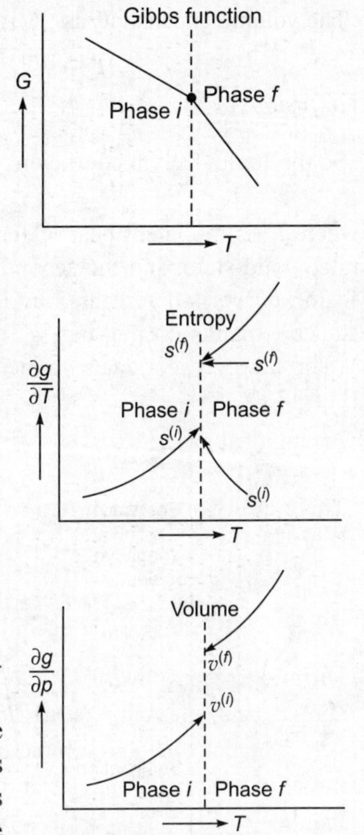

Fig. 11.7(a) First order phase transition

and for a phase change at $T + dT$ and $p + dp$ (Fig. 11.7b)

$$g^{(i)} + dg^{(i)} = g^{(f)} + dg^{(f)}$$

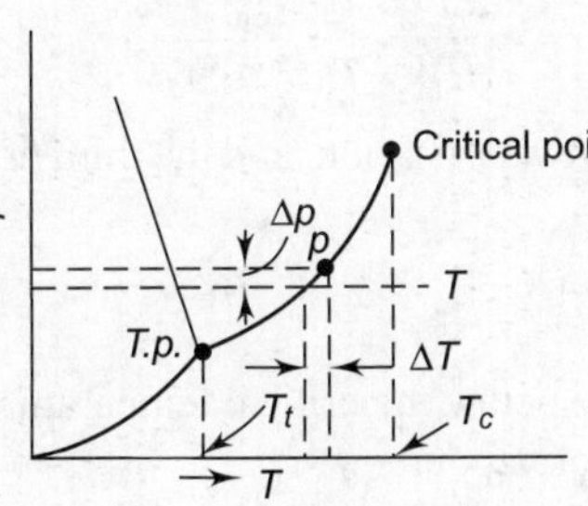

Fig. 11.7(b) Phase diagram in *p-T* coordinates where a solid shrinks on melting

Subtracting

$$dg^{(i)} = dg^{(f)}$$

or

$$-s^{(i)}\, dT + v^{(i)}\, dp = -s^{(f)}\, dT + v^{(f)}\, dp$$

$$\therefore \qquad \frac{dp}{dT} = \frac{s^{(f)} - s^{(i)}}{v^{(f)} - v^{(i)}} = \frac{l}{T\,(v^{(f)} - v^{(i)})}$$

For *fusion*

$$\frac{dp}{dT} = \frac{l_{fu}}{T\,(v'' - v')}$$

where l_{fu} is the latent heat of fusion, the first prime in v, i.e. v' indicates the saturated solid state, and the second prime the saturated liquid state. The slope of the fusion curve is determined by $(v'' - v')$, since l_{fu} and T are positive. If the substance expands on melting, $v'' > v'$, the slope is positive. This is the usual case. Water, however, contracts on melting and has the fusion curve with a negative slope (Fig. 11.8).

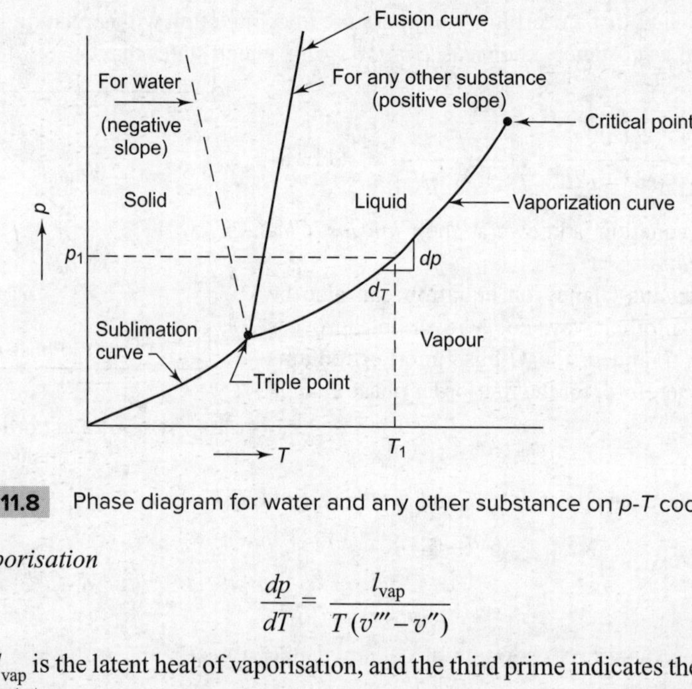

Fig. 11.8 Phase diagram for water and any other substance on *p-T* coordinates

For *vaporisation*

$$\frac{dp}{dT} = \frac{l_{\text{vap}}}{T\,(v''' - v'')}$$

where l_{vap} is the latent heat of vaporisation, and the third prime indicates the saturated vapour state.

$$\therefore \qquad l_{\text{vap}} = T\,\frac{dp}{dT}\,(v''' - v'')$$

At temperature considerably below the critical temperature, $v''' >> v''$ and using the ideal gas equation of state for vapour

$$v''' = \frac{RT}{p}$$

$$l_{\text{vap}} \cong T \cdot \frac{dp}{dT}\,\frac{RT}{p}$$

or

$$l_{\text{vap}} = \frac{RT^2}{p}\,\frac{dp}{dT} \qquad (11.19)$$

If the slope dp/dT at any state (e.g. point p_1, T_1 in Fig. 11.8) is known, the latent heat of vapourisation can be computed from the above equation.

The vapour pressure curve is of the form

$$\ln p = A + \frac{B}{T} + C \ln T + DT$$

where A, B, C and D are constants. By differentiating with respect to T

$$\frac{1}{p}\frac{dp}{dT} = -\frac{B}{T^2} + \frac{C}{T} + D \tag{11.20}$$

Equations (11.19) and (11.20) can be used to estimate the latent heat of vaporisation.

Clapeyron's equation can also be used to estimate approximately the vapour pressure of a liquid at any arbitrary temperature in conjunction with a relation for the latent heat of a substance, known as *Trouton's rule*, which states that

$$\frac{\bar{h}_{fg}}{T_B} \cong 88 \text{ kJ/kg mol K}$$

where h_{fg} is the latent heat of vapourisation in kJ/kg mol and T_B is the boiling point at 1.013 bar. On substituting this into Eq. (11.19)

$$\frac{dp}{dT} = \frac{88T_B}{\bar{R}T^2}p$$

or

$$\int_{101.325}^{p} \frac{dp}{p} = \frac{88T_B}{\bar{R}}\int_{T_B}^{T}\frac{dT}{T^2}$$

$$\ln \frac{p}{101.325} = -\frac{88T_B}{\bar{R}}\left(\frac{1}{T} - \frac{1}{T_B}\right)$$

$\therefore$

$$p = 101.325 \exp\left[\frac{88}{R}\left(1 - \frac{T_B}{T}\right)\right] \tag{11.21}$$

This gives the vapour pressure p in kPa at any temperature T.

For *sublimation*

$$\frac{dp}{dT} = \frac{l_{sub}}{T(v''' - v')}$$

where l_{sub} is the latent heat of sublimation.

Since $v''' >> v'$, and vapour pressure is low, $v''' = \dfrac{RT}{p}$

$$\frac{dp}{dT} = \frac{l_{sub}}{T \cdot \dfrac{RT}{p}}$$

or

$$l_{sub} = -2.303\, R\, \frac{d(\log p)}{d(1/T)}$$

the slope of log p *vs.* $1/T$ curve is negative, and if it is known, l_{sub} can be estimated. At the triple point (Fig. 9.12),

$$l_{sub} = l_{vap} + l_{fus} \tag{11.23}$$

$$\left(\frac{dp}{dT}\right)_{vap} = \frac{p_{tp} l_{vap}}{\bar{R} T_{tp}^2}$$

$$\left(\frac{dp}{dT}\right)_{sub} = \frac{p_{tp} l_{sub}}{\bar{R} T_{tp}^2}$$

Since $l_{sub} > l_{vap}$, at the triple point

$$\left(\frac{dp}{dT}\right)_{sub} > \left(\frac{dp}{dT}\right)_{vap}$$

Therefore, the slope of the sublimation curve at the triple point is greater than that of the vapourisation curve (Fig. 11.8).

11.9 EVALUATION OF THERMODYNAMIC PROPERTIES FROM AN EQUATION OF STATE

> **LO 11.8** Evaluate thermodynamic properties other than p, v, T from equation of state

Apart from calculating pressure, volume or temperature, an equation of state can also be used to evaluate other thermodynamic properties such as internal energy, enthalpy and entropy. The property relations to be used are:

$$du = c_v\, dT + \left[T\left(\frac{\partial p}{\partial T}\right)_v - p\right] dv \tag{11.24}$$

$$dh = c_p\, dT + \left[v - T\left(\frac{\partial v}{\partial T}\right)_p\right] dp \tag{11.25}$$

$$ds = \frac{1}{T}\left[c_v\, dT + T\left(\frac{\partial p}{\partial T}\right)_v dv\right]$$

$$= \frac{1}{T}\left[c_p\, dT - T\left(\frac{\partial v}{\partial T}\right)_p dp\right] \tag{11.26}$$

Integrations of the differential relations of the properties p, v and T in the above equations are carried out with the help of an equation of state. The changes in properties are independent of the path and depend only on the end states. Let us consider that the change in enthalpy per unit mass of a gas from a reference state 0 at p_0, T_0 having enthalpy, h_0 to some other state B at p, T with enthalpy h is to be calculated (Fig. 11.9). The reversible path $0B$ may be replaced for convenience by either path 0-*a*-*B* or path 0-*b*-*B*, both also being reversible.

Path 0-*a*-*B*:

From Eq. (11.25),

$$h_a - h_0 = \int_{T_0}^{T} c_p \, dT$$

$$h - h_a = \int_{p_0}^{p} \left[v - T\left(\frac{\partial v}{\partial T}\right)_p \right]_T dp$$

On addition

$$h - h_0 = \left[\int_{T_0}^{T} c_p dT \right]_{p_0} + \left\{ \int_{p_0}^{p} \left[v - T\left(\frac{\partial v}{\partial T}\right)_p \right] dp \right\}_T \tag{11.27}$$

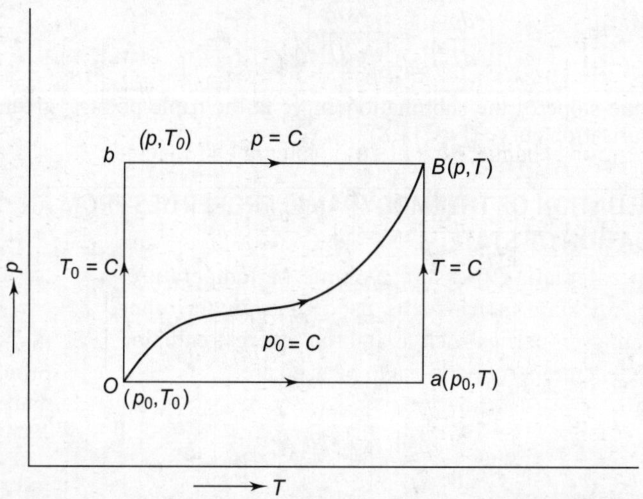

Fig. 11.9 Processes connecting states (p0, T0) and (p, T)

Similarly, for
Path 0-*b*-*B*:

$$h - h_0 = \left\{ \int_{p_0}^{p} \left[v - T\left(\frac{\partial v}{\partial T}\right)_p \right] dp \right\}_{T_0} + \left[\int_{T_0}^{T} c_p \, dT \right]_p \tag{11.28}$$

Equation (11.27) is preferred to Eq. (11.28) since c_p at lower pressure can be conveniently measured. Now,

$$\int_{p_0 v_a}^{pv} d\,(pv) = \int_{v_a}^{v} p\,dv + \int_{p_0}^{p} v\,dp$$

$$\int_{v_a}^{v} v\,dp = pv - p_0 v_a - \left[\int_{v_a}^{v} p dv \right]_T \tag{11.29}$$

Again,

$$\left[\frac{\partial v}{\partial T}\right]_p \left[\frac{\partial T}{\partial p}\right]_v \left[\frac{\partial p}{\partial v}\right]_T = -1$$

$$\therefore \qquad \left[\frac{\partial v}{\partial T}\right]_p = -\left[\frac{\partial p}{\partial T}\right]_v \left[\frac{\partial v}{\partial p}\right]_T$$

Substituting in Eq. (11.27),

$$h - h_0 = \left[\int_{T_0}^{T} c_p dT\right]_{p_0} + pv - p_0 v_a - \left[\int_{v_a}^{v} p dv\right]_T + \left\{\int_{p_0}^{p} T\left(\frac{\partial p}{\partial T}\right)_p \left(\frac{\partial v}{\partial p}\right)_T dp\right\}_T$$

$$= \left[\int_{T_0}^{T} c_p dT\right]_{p_0} + pv - p_0 v_a - \left\{\int_{v_a}^{v}\left[p - T\left(\frac{\partial p}{\partial T}\right)_v\right] dv\right\}_T \tag{11.30}$$

To find the entropy change, Eq. (11.26) is integrated to yield:

$$s - s_0 = \left[\int_{T_0}^{T} c_p \frac{dT}{T}\right]_{p_0} - \left[\int_{p_0}^{p}\left(\frac{\partial v}{\partial T}\right)_p dp\right]_T$$

$$= \left[\int_{T_0}^{T} c_p \frac{dT}{T}\right]_{p_0} - \left[-\int_{p_0}^{p}\left(\frac{\partial p}{\partial T}\right)_v \left(\frac{\partial v}{\partial p}\right)_T dp\right]_T$$

$$= \left[\int_{T_0}^{T} c_p \frac{dT}{T}\right]_{p_0} + \left[\int_{v_a}^{v}\left(\frac{\partial p}{\partial T}\right)_v dv\right]_T$$

11.10 GENERAL THERMODYNAMIC CONSIDERATIONS ON AN EQUATION OF STATE

LO 11.9 Construct property relations from equation of state

Certain general characteristics are common to all gases. These must be clearly observed in the developing and testing of an equation of state. It is edifying to discuss briefly some of the more important ones:

1. Any equation of state must reduce to the ideal gas equation as pressure approaches zero at any temperature. This is clearly seen in a generalised compressibility factor chart in which all isotherms converge to the point $z = 1$ at zero pressure. Therefore,

$$\lim_{p \to 0}\left[\frac{pv}{RT}\right] = 1 \text{ at any temperature}$$

Also, as seen from Fig. 10.6, the reduced isotherms approach the line $z = 1$ as the temperature approaches infinity, or

$$\lim_{T \to \infty}\left[\frac{pv}{RT}\right] = 1 \text{ at any pressure.}$$

2. The critical isotherm of an equation of state should have a point of inflection at the critical point on p–v coordinates, or

$$\left[\frac{\partial p}{\partial v}\right]_{T=T_c} = 0 \quad \text{and} \quad \left[\frac{\partial^2 p}{\partial v^2}\right]_{T=T_c} = 0$$

3. The isochores of an equation of state on a p–T diagram should be essentially straight, or

$$\left[\frac{\partial p}{\partial T}\right]_v = \text{constant}, \left[\frac{\partial^2 p}{\partial T^2}\right]_v = 0 \text{ as } p \to 0, \text{ or as } T \to \infty.$$

An equation of state can predict the slope of the critical isochores of a fluid. This slope is identical with the slope of the vapourisation curve at the critical point. From the Clapegron equation, $dp/dT = \Delta s/\Delta v$, the slope of the vapourisation curve at the critical point becomes

$$\frac{dp}{dT} = \left[\frac{\partial s}{\partial v}\right]_{T_c} = \left[\frac{\partial p}{\partial T}\right]_{v_c} \qquad \text{(by Maxwell's equation)}$$

Therefore, the vapour-pressure slope at the critical point, dp/dT, is equal to the slope of the critical isochore $(\partial p/\partial T)_{vc}$ (Fig. 11.10)

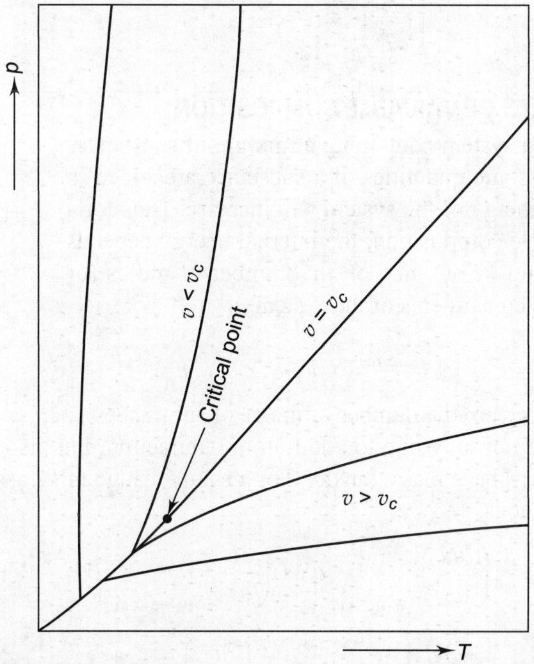

Fig. 11.10 Pressure-Temperature diagram with isochoric lines

4. The slopes of the isotherms of an equation of state on a Z–p compressibility factor chart as p approaches zero should be negative at lower temperatures and positive at higher temperatures. At the Boyle temperature, the slope is zero as p approaches zero, or

$$\lim_{p\to 0}\left[\frac{\partial z}{\partial p}\right]_T = 0 \text{ at } T = T_B$$

An equation of state should predict the Boyle temperature which is about 2.54 T_c for many gases.

An isotherm of maximum slope on the Z–p plot as p approaches zero, called the *foldback isotherm*, which is about $5T_c$ for many gases, should be predicted by an equation of state, for which

$$\lim_{p\to 0}\left[\frac{\partial^2 z}{\partial T\cdot\partial p}\right] = 0 \text{ at } T = T_F$$

where T_F is the foldback temperature (Fig. 10.5a). As temperature increases beyond T_F the slope of the isotherm decreases, but always remains positive.

5. An equation of state should predict the Joule–Thomson coefficient, which is

$$\mu_J = \frac{1}{c_p}\left[T\left(\frac{\partial v}{\partial T}\right)_p - v\right] = \frac{RT^2}{pc_p}\left(\frac{\partial z}{\partial T}\right)_p$$

For the inversion curve, $\mu_J = 0$,

or
$$\left(\frac{\partial z}{\partial T}\right)_p = 0$$

11.11 MIXTURES OF VARIABLE COMPOSITION

LO 11.10 Discuss properties of multi-component systems

Let us consider a system containing a mixture of substances 1, 2, 3, ..., K. If some quantities of a substance are added to the system, the energy of the system will increase. Thus for a system of variable composition, the internal energy depends not only on S and V, but also on the number of moles (or mass) of various constituents of the system.

Thus,

$$U = U(S, V, n_1, n_2, \ldots, n_K)$$

where $n_1, n_2, \ldots, n_K$ are the number of moles of substances 1, 2, ..., K. The composition may change not only due to addition or subtraction, but also due to chemical reaction and inter-phase mass transfer. For a small change in U, assuming the function to be continuous.

$$dU = \left(\frac{\partial U}{\partial S}\right)_{V, n_1, n_2, \ldots, n_K} dS + \left(\frac{\partial U}{\partial V}\right)_{S, n_1, n_2, \ldots, n_K} dV$$

$$+ \left(\frac{\partial U}{\partial n_1}\right)_{S, V, n_2, \ldots, n_K} dn_1 + \left(\frac{\partial U}{\partial n_2}\right)_{S, V, n_1, n_3, \ldots, n_K} dn_2$$

$$+ \ldots + \left(\frac{\partial U}{\partial n_K}\right)_{S, V, n_1, n_2, \ldots, n_{K-1}} dn_K$$

or $$dU = \left(\frac{\partial U}{\partial S}\right)_{V,n_i} dS + \left(\frac{\partial U}{\partial V}\right)_{S,n_i} dV + \sum_{i=1}^{K}\left(\frac{\partial U}{\partial n_i}\right)_{S,V,n_j} dn_i \qquad (11.31)$$

where subscript i indicates any substance and subscript j any other substance except the one whose number of moles is changing.

If the composition does not change

$$dU = TdS - pdV$$

$$\therefore \quad \left(\frac{\partial U}{\partial S}\right)_{V,n_i} = T, \quad \text{and} \quad \left(\frac{\partial U}{\partial V}\right)_{S,n_i} = -p$$

$$\therefore \quad dU = TdS - pdV + \sum_{i=1}^{K}\left(\frac{\partial U}{\partial n_i}\right)_{S,V,n_j} dn_i \qquad (11.32)$$

Molal *chemical potential*, μ, of component i is defined as

$$\mu_i = \left(\frac{\partial U}{\partial n_i}\right)_{S,V,n_j}$$

signifying the change in internal energy per unit mole of component i when S, V and the number of moles of all other components are constant.

$$\therefore \quad dU = TdS - pdV + \sum_{i=1}^{K} \mu_i \, dn_i$$

or $$TdS = dU + pdV - \sum_{i=1}^{K} \mu_i \, dn_i \qquad (11.33)$$

This is known as *Gibbs entropy equation*.

In a similar manner

$$G = G(p, T, n_1, n_2, \ldots, n_K)$$

or $$dG = \left(\frac{\partial G}{\partial p}\right)_{T,n_i} dp + \left(\frac{\partial G}{\partial T}\right)_{p,n_i} dT + \sum_{i=1}^{K}\left(\frac{\partial G}{\partial n_i}\right)_{T,p,n_j} dn_i$$

$$= Vdp - SdT + \sum_{i=1}^{K}\left(\frac{\partial G}{\partial n_i}\right)_{T,p,n_j} dn_i \qquad (11.34)$$

Since $$G = U + pV - TS$$

$$d(U + pV - TS) = VdP - SdT + \sum_{i=1}^{K}\left(\frac{\partial G}{\partial n_i}\right)_{T,p,n_j} dn_i$$

or $$dU + pdV + Vdp - TdS - SdT$$

$$= Vdp - SdT + \sum_{i=1}^{K}\left(\frac{\partial G}{\partial n_i}\right)_{T,p,n_j} dn_i$$

or $$dU = TdS - pdV + \sum_{i=1}^{K} \left(\frac{\partial G}{\partial n_i}\right)_{T,p,n_j} dn_i$$

Comparing this equation with Eq. (11.32)

$$\left(\frac{\partial U}{\partial n_i}\right)_{S,V,n_j} = \left(\frac{\partial G}{\partial n_i}\right)_{T,p,n_j} = \mu_i$$

$\therefore$ Eq. (11.34) becomes

$$dG = Vdp - SdT + \sum_{i=1}^{K} \mu_i \, dn_i$$

Similar equations can be obtained for changes in H and F. Thus,

$$dU = TdS - pdV + \sum_{i=1}^{K} \mu_i \, dn_i \tag{11.35a}$$

$$dG = Vdp - SdT + \sum_{i=1}^{K} \mu_i \, dn_i \tag{11.35b}$$

$$dH = TdS + Vdp + \sum_{i=1}^{K} \mu_i \, dn_i \tag{11.35c}$$

$$dF = -SdT - pdV + \sum_{i=1}^{K} \mu_i \, dn_i \tag{11.35d}$$

where

$$\mu_i = \left(\frac{\partial U}{\partial n_i}\right)_{S,V,n_j} = \left(\frac{\partial G}{\partial n_i}\right)_{T,p,n_j} = \left(\frac{\partial H}{\partial n_i}\right)_{S,p,n_j} = \left(\frac{\partial F}{\partial n_i}\right)_{T,V,n_j} \tag{11.36}$$

Chemical potential is an intensive property.

Let us consider a homogeneous phase of a multi-component system, for which

$$dU = TdS - pdV + \sum_{i=1}^{K} \mu_i \, dn_i$$

If the phase is enlarged in size, U, S and V will increase, whereas T, p and μ will remain the same. Thus,

$$\Delta U = T\Delta S - p\Delta V + \Sigma \mu_i \Delta n_i$$

Let the system be enlarged to K-times the original size. Then

$$\Delta U = KU - U = (K - 1)U$$
$$\Delta S = KS - S = (K - 1)S$$
$$\Delta V = (K - 1)V$$
$$\Delta n_i = (K - 1)\, n_i$$

Substituting

$$(K-1)\,U = T\,(K-1)\,S - p\,(K-1)\,V + \Sigma\mu_i\,(K-1)\,n_i$$

$$\therefore \quad U = TS - pV + \Sigma\,\mu_i n_i$$

$$\therefore \quad G_{T,p} = \Sigma\,\mu_i n_i \tag{11.37}$$

Let us now find a relationship if there is a simultaneous change in intensive property. Differentiating Eq. (11.37)

$$dG = \Sigma\,n_i\,d\mu_i + \Sigma\,\mu_i\,dn_i \tag{11.38}$$

at constant T and p, with only μ changing.

When T and p change

$$dG = -\,SdT + Vdp + \Sigma\mu_i\,dn_i \tag{11.39}$$

Combining Eq. (11.38) and (11.39)

$$-\,SdT + Vdp - \Sigma\,n_i\,d\,\mu_i = 0 \tag{11.40}$$

This is known as *Gibbs–Duhem equation*, which shows the necessary relationship for simultaneous changes in T, p and μ.

Now

$$G_{T\cdot p} = \Sigma\,\mu_i n_i = \mu_1 n_1 + \mu_2 n_2 \ldots + \mu_K n_K$$

For a phase consisting of only one constituent

$$G = \mu n$$

or

$$\mu = \frac{G}{n} = g$$

i.e. the chemical potential is the molar Gibbs function and is a function of T and p only.

For a single phase, multi-component system, μ_i is a function of T, p and the mole fraction x_i.

11.12 CONDITIONS OF EQUILIBRIUM OF A HETEROGENEOUS SYSTEM

> **LO 11.11**
>
> Formulate the conditions of thermodynamic equilibrium of a heterogeneous system

Let us consider a heterogeneous system of volume V, in which several homogeneous phases ($\phi = a, b, \ldots, r$) exist in equilibrium. Let us suppose that each phase consists of i (= 1, 2, ..., C) constituents and that the number of constituents in any phase is different from the others.

Within each phase, a change in internal energy is accompanied by a change in entropy, volume and composition, according to

$$dU_\phi = T_\phi\,dS_\phi - p_\phi\,dV_\phi + \sum_{i=1}^{C}(\mu_i\,dn_i)_\phi$$

A change in the internal energy of the entire system can, therefore, be expressed as follows:

$$\sum_{\phi=a}^{r} dU_\phi = \sum_{\phi=a}^{r} T_\phi\,dS_\phi - \sum_{\phi=a}^{r} p_\phi\,dV_\phi + \sum_{\phi=a}^{r}\sum_{i=1}^{C}(\mu_i\,dn_i)_\phi \tag{11.41}$$

Also, a change in the internal energy of the entire system involves changes in the internal energy of the constituent phases.

$$dU = dU_a + dU_b + \dots + dU_r = \sum_{\phi=a}^{r} dU_\phi$$

Likewise, changes in the volume, entropy, or chemical composition of the entire system result from contributions from each of the phases

$$dV = dV_a + dV_b + \dots + dV_r = \sum_{\phi=a}^{r} dV_\phi$$

$$dS = dS_a + dS_b + \dots + dS_r = \sum_{\phi=a}^{r} dS_\phi$$

$$dn = dn_a + dn_b + \dots + dn_r = \sum_{\phi=a}^{r} dn_\phi$$

In a closed system in equilibrium, the internal energy, volume, entropy and mass are constant.

$\therefore$

$$dU = dV = dS = dn = 0$$

or

$$\begin{aligned} dU_a &= -(dU_b + \dots + dU_r) = -\sum_j dU_j \\ dV_a &= -\sum_j dV_j \\ dS_a &= -\sum_j dS_j \\ dn_a &= -\sum_j dn_j \end{aligned} \tag{11.42}$$

where subscript j includes all phases except phase a.

Equation (11.41) can be written in terms of j independent variables and the dependent variable a (Eq. 11.42).

$$(T_a\, dS_a + \sum_j T_j\, dS_j) - (p_a\, dV_a + \sum_j p_j\, dV_j$$

$$+ \left[\sum_i (\mu_i\, dn_i)_a + \sum_j \sum_j (\mu_i\, dn_i)_j\right] = 0$$

Substituting from Eq. (11.34)

$$\left(-T_a \sum_j dS_j + \sum_j T_j\, dS_j\right) - \left(-p_a \sum_j dV_j + \sum_j p_j \cdot dV_j\right)$$

$$+ \left[-\sum_j \sum_t \mu_{ia}\, dn_{ij} + \sum_j \sum_i (\mu_i\, dn_i)_j\right] = 0$$

where subscript $i\,a$ refers to component i of phase a.

Rearranging and combining the coefficients of the independent variables, dS_j, dV_j and dn_j, give

$$\sum_j (T_j - T_a)dS_j - \sum_j (p_j - p_a)dV_j + \sum_j \sum_i (\mu_{ij} - \mu_{ia})\, dn_{ij} = 0$$

But since dS_j, dV_j, and dn_j are independent, their coefficients must each be equal to zero.

$$\therefore \qquad T_j = T_a, \quad p_j = p_a, \quad \mu_{ij} = \mu_{ia} \qquad (11.43)$$

These equations represent conditions that exist when the system is in thermal, mechanical and chemical equilibrium. The temperature and pressure of phase a must be equal to those of all other phases, and the chemical potential of the ith component in phase a must be equal to the chemical potential of the same component in all other phases.

11.13 GIBBS PHASE RULE

LO 11.12

Discuss the Gibbs Phase Rule for non reactive heterogeneous systems

Let us consider a heterogeneous system of C chemical constituents which do not combine chemically with one another. Let us suppose that there are ϕ phases, and every constituent is present in each phase. The constituents are denoted by subscripts and the phases by superscripts. The Gibbs function of the whole heterogeneous system is

$$G_{T,p} = \sum_{i=1}^{C} n_i^{(1)} \mu_i^{(1)} + \sum_{i=1}^{C} n_i^{(2)} \mu_i^{(2)} + \ldots + \sum_{i=1}^{C} n_i^{(\phi)} \mu_i^{(\phi)}$$

G is a function of T, p, and the n's of which there are $C\phi$ in number. Since there are no chemical reactions, the only way in which the n's may change is by the transport of the constituents from one phase to another. In this case, the total number of moles of each constituent will remain constant.

$$\begin{aligned} n_1^{(1)} + n_1^{(2)} + \ldots + n_1^{(\phi)} &= \text{constant} \\ n_2^{(1)} + n_2^{(2)} + \ldots + n_2^{(\phi)} &= \text{constant} \\ &\cdots\cdots \\ n_C^{(1)} + n_C^{(2)} + \ldots + n_C^{(\phi)} &= \text{constant} \end{aligned}$$

These are the *equations of constraint.*

At chemical equilibrium, G will be rendered a minimum at constant T and p, subject to these equations of constraint. At equilibrium, from Eq. (11.43).

$$\begin{aligned} \mu_{ij} &= \mu_{ia} \\ \mu_1^{(1)} &= \mu_1^{(2)} = \ldots = \mu_1^{(\phi)} \\ \mu_2^{(1)} &= \mu_2^{(2)} = \ldots = \mu_2^{(\phi)} \qquad (11.44) \\ &\cdots\cdots \\ \mu_C^{(1)} &= \mu_C^{(2)} = \ldots = \mu_C^{(\phi)} \end{aligned}$$

These are known as the *equations of phase equilibrium*. The equations of the phase equilibrium of one constituent are $(\phi - 1)$ in number. Therefore, for C constituents, there are $C(\phi - 1)$ such equations.

When equilibrium has been reached, there is no transport of matter from one phase to another. Therefore, in each phase, $\Sigma x = 1$. For ϕ phases, there are ϕ such equations available.

The state of the system at equilibrium is determined by the temperature, pressure and $C\phi$ mole fractions. Therefore,

Total number of variables $= C\phi + 2$.

Among these variables, there are $C(\phi - 1)$ equations of phase equilibrium and ϕ equations of $\Sigma x = 1$ type. Therefore,

Total number of equations $= C(\phi - 1) + \phi$

If the number of variables is equal to the number of equations, the system is *nonvariant*. If the number of variables exceeds the number of equations by one, then the system is called *monovariant* and is said to have a variance of 1.

The excess of variables over equations is called the *variance*, *f*. Thus,

$$f = (C\phi + 2) - [C(\phi - 1) + \phi]$$

or

$$f = C - \phi + 2 \tag{11.45}$$

This is known as the *Gibbs Phase Rule* for a nonreactive system. The variance '*f*' is also known as the *degree of freedom*.

For a pure substance existing in a single phase, $C = 1$, $\phi = 1$, and therefore, the variance is 2. There are two properties required to be known to fix up the state of the system at equilibrium.

If $C = 1$, $\phi = 2$, then $f = 1$, i.e. only one property is required to fix up the state of a single-component two-phase system.

If $C = 1$, $\phi = 3$, then $f = 0$. The state is thus unique for a substance and refers to the triple point where all the three phases exist in equilibrium.

11.14 TYPES OF EQUILIBRIUM

LO 11.13 Formulate the conditions of four types of thermodynamic equilibriums and their stability

The thermodynamic potential which controls equilibrium in a system depends on the particular *constraints* imposed on the system. Let δQ be the amount of heat transfer involved between the system and the reservoir in an infinitesimal irreversible process (Fig. 11.11). Let dS denote the entropy change of the system and dS_0 the entropy change of the reservoir. Then, from the entropy principle

$$dS_0 + dS > 0$$

Since

$$dS_0 = -\frac{đQ}{T}$$

$$-\frac{đQ}{T} + dS > 0$$

or

$$đQ - TdS < 0$$

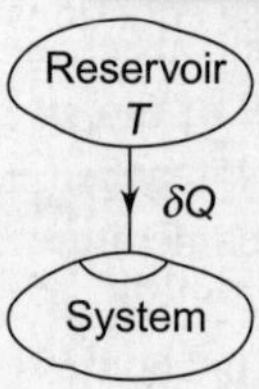

Fig. 11.11 Heat interaction between a system and its surroundings

During the infinitesimal process, the internal energy of the system changes by an amount dU, and an amount of work pdV is performed. So, by the first law

$$đQ = dU + pdV$$

Thus the inequality becomes

$$dU + pdV - TdS < 0 \tag{11.46}$$

If the constraints are constant U and V, then the Eq. (11.46) reduces to

$$dS > 0$$

The condition of constant U and V refers to an isolated system. Therefore, entropy is the critical parameter to determine the state of thermodynamic equilibrium of an isolated system. The entropy of an isolated system always increases and reaches a maximum value when equilibrium is reached.

If the constraints imposed on the system are constant T and V, the Eq. (11.40) reduces to

$$dU - d(TS) < 0$$

or $$d(U - TS) < 0$$

$\therefore$ $$dF < 0$$

which expresses that the Helmholtz function decreases, becoming a minimum at the final equilibrium state.

If the constraints are constant T and p, the Eq. (11.46) becomes

$$dU + d(pV) - d(TS) < 0$$

$$d(U + pV - TS) < 0$$

$\therefore$ $$dG < 0$$

The Gibbs function of a system at constant T and p decreases during an irreversible process, becoming a minimum at the final equilibrium state. For a system constrained in a process to constant T and p, G is the critical parameter to determine the state of equilibrium.

The thermodynamic potential and the corresponding constrained variables are shown below.

S	U	V
H		F
P	G	T

This trend of G, F or S establishes four types of equilibrium, namely (a) stable, (b) neutral, (c) unstable and (d) metastable.

A system is said to be in a state of *stable* equilibrium if, when the state is perturbed, the system returns to its original state. A system is not in equilibrium if there is a spontaneous change in the state. If there is a spontaneous change in the system, the entropy of the system increases and reaches a maximum when the equilibrium condition is reached (Fig. 11.12). Both A and B (Fig. 11.13) are assumed to be at the same temperature T. Let there be some spontaneous change; the temperature of A rises to $T + dT_1$, and that of B decreases to $T - dT_2$. For simplicity, let the heat capacities of the bodies be the same, so that $dT_1 = dT_2 = dT$. If Q is the heat interaction involved, then the entropy change

$$dS_A = \frac{đQ}{T+dT}, \quad dS_B = \frac{đQ}{T-dT}$$

$$\therefore \quad dS = dS_A + dS_B = đQ\left[\frac{1}{T+dT} - \frac{1}{T-dT}\right] = -\frac{2dT}{T^2}\,đQ$$

So there is a decrease in entropy for the isolated system of A and B together. It is thus clear that the variation in temperature dT cannot take place. The system, therefore, exists in a stable equilibrium condition. Perturbation of the state produces an absurd situation and the system must revert to the original stable state. It may be observed:

If for all the possible variations in state of the isolated system, there is a negative change in entropy, then the system is in stable equilibrium.

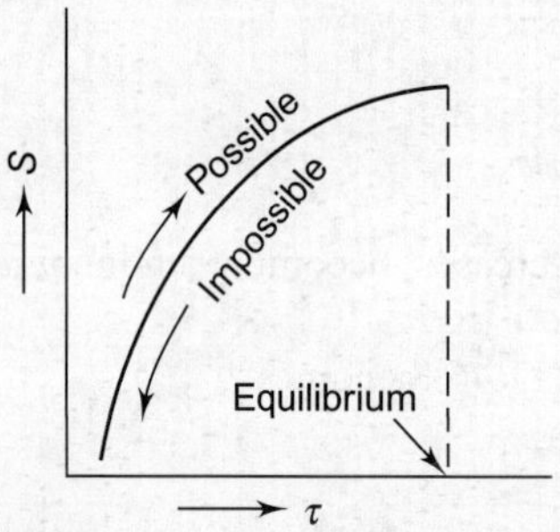

Fig. 11.12 Possible process for an isolated system

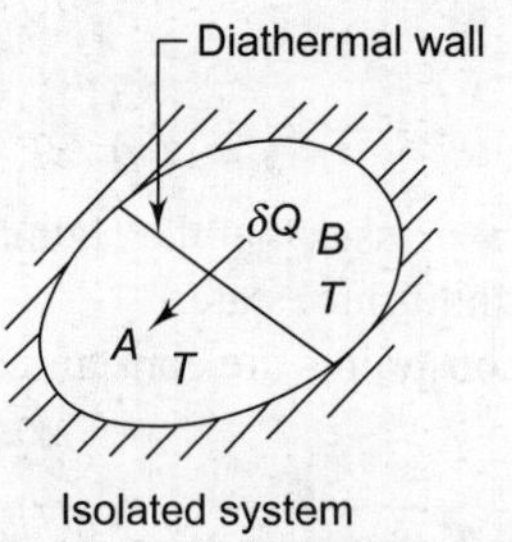

Fig. 11.13 Spontaneous changes in A and B due to heat interaction

$$\begin{aligned} &(dS)_{U,V} > 0 && \text{Spontaneous change} \\ &(dS)_{U,V} = 0 && \text{Equilibrium} \\ &(dS)_{U,V} < 0 && \text{Criterion of stability} \end{aligned} \tag{11.47}$$

Similarly,

$$\begin{aligned} &(dG)_{p,T} < 0,\ (dF)_{T,V} < 0 && \text{Spontaneous change} \\ &(dG)_{p,T} = 0,\ (dF)_{T,V} = 0 && \text{Equilibrium} \\ &(dG)_{p,T} > 0,\ (dF)_{T\cdot V} > 0 && \text{Criterion of stability} \end{aligned} \tag{11.48}$$

A system is in a state of stable equilibrium if, for any finite variation of the system at constant T and p, G increases, i.e. the stable equilibrium state corresponds to the minimum value of G.

A system is said to be in a state of *neutral equilibrium* when the thermodynamic criterion of equilibrium (G, F, S, U or H) remains at constant value for all possible variations of finite magnitude. If perturbed, the system does not revert to the original state.

For a system at constant T and p, the criterion of neutral equilibrium is

$$\delta G_{T,p} = 0$$

Similarly,

$$\delta F_{T,V} = 0,\ \delta H_{S,p} = 0,\ \delta U_{S,V} = 0,\ \delta S_{U,V} = 0$$

A system is in a state of *unstable equilibrium* when the thermodynamic criterion is neither an extremum nor a constant value for all possible variations in the system. If the system is in an unstable equilibrium, there will be a spontaneous change accompanied by

$$\delta G_{T,p} < 0,\ \delta F_{T,V} < 0,\ \delta U_{S,V} < 0,\ \delta H_{S,p} < 0,\ \delta S_{U,V} > 0$$

A system is in a state of *metastable equilibrium* if it is stable to small but not to large disturbances. A mixture of oxygen and hydrogen is in a metastable equilibrium. A little spark may start a chemical reaction. Such a mixture is not in its most stable state, even though in the absence of a spark it appears to be stable.

Figure 11.14 shows different types of equilibrium together with their mechanical analogies. S has been used as the criterion for equilibrium.

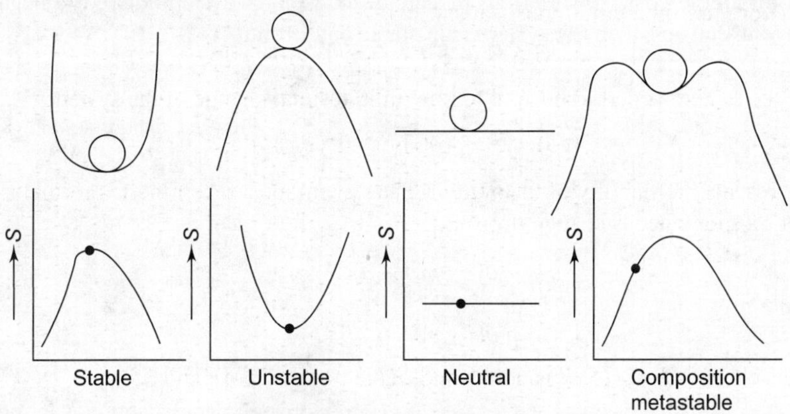

Fig. 11.14 Types of equilibrium

11.15 LOCAL EQUILIBRIUM CONDITIONS

Let an arbitrary division of an isolated system be considered, such that

$$S = S_1 + S_2,\ U = U_1 + U_2$$

Then for equilibrium, it must satisfy the condition

$$(\delta S)_{U,V} = 0$$

to first order in small displacements (otherwise δS could be made positive because of higher order terms). Now to the first order in a very small change

$$\delta_1 S = \left(\frac{\partial S}{\partial U_1}\right)_V \delta U_1 + \left(\frac{\partial S}{\partial U_2}\right)_V \delta U_2 + \left(\frac{\partial S}{\partial V_1}\right)_U \delta V_1 + \left(\frac{\partial S}{\partial V_2}\right)_U \delta V_2$$

Now $$TdS = dU + pdV$$

$$\therefore \quad \left(\frac{\partial S}{\partial U}\right)_V = \frac{1}{T};\ \left(\frac{\partial S}{\partial V}\right)_U = \frac{p}{T}$$

$$\therefore \quad \delta_1 S = \frac{1}{T_1}\delta U_1 + \frac{1}{T_2}\delta U_2 + \frac{p_1}{T_1}\delta V_1 + \frac{p_2}{T_2}\delta V_2$$

Again $\delta U_1 = -\delta U_2$ and $\delta V_1 = -\delta V_2$

$$\therefore \quad \delta_1 S = \left(\frac{1}{T_1} - \frac{1}{T_2}\right)\delta U_1 + \left(\frac{p_1}{T_1} - \frac{p_2}{T_2}\right)\delta V_1 + \text{Second-order terms}$$

When $\delta_1 S = 0$, at equilibrium

$$T_1 = T_2, p_1 = p_2$$

11.16 CONDITIONS OF STABILITY

At equilibrium, $S = S_{max}$, $F = F_{min}$, $G = G_{min}$ and $dS = 0$, $\delta F = 0$; $\delta G = 0$; these are necessary but not sufficient conditions for equilibrium. To prove that S is a maximum, and G or F a minimum, it must satisfy

$$\delta^2 S < 0, \delta^2 F > 0, \delta^2 G > 0$$

If the system is perturbed, and for any infinitesimal change of the system

$$(\delta S)_{U,V} < 0, (\delta G)_{p,T} > 0, (\delta F)_{T,V} > 0$$

it represents the stability of the system. The system must revert to the original state.

For a spontaneous change, from Eq. (11.46)

$$\delta U + p\delta V - T\delta S < 0$$

For stability

$$\delta U + p\delta V - T\delta S > 0$$

Let us choose $U = U(S, V)$ and expand δU in powers of δV and δS.

$$\delta U = \left(\frac{\partial U}{\partial S}\right)_V \delta S + \frac{1}{2}\left(\frac{\partial^2 U}{\partial S^2}\right)_V (\delta S)^2 + \left(\frac{\partial U}{\partial V}\right)_S \delta V$$

$$+ \frac{1}{2}\left(\frac{\partial^2 U}{\partial V^2}\right)_S (\delta V)^2 + \frac{\partial^2 U}{\partial V \partial S}\delta V \delta S + \ldots$$

$$= T\delta S - p\delta V + \frac{1}{2}\left(\frac{\delta^2 U}{\delta S^2}\right)_V (\delta S)^2$$

$$+ \frac{1}{2}\left(\frac{\partial^2 U}{\partial V^2}\right)_S (\delta V)^2 + \frac{\partial^2 U}{\partial V \partial S}\delta V \delta S + \ldots$$

The third, order and higher terms are neglected.

Since $\delta U + p\delta v - T\delta S > 0$, it must satisfy the conditions given below.

$$\left(\frac{\partial^2 U}{\partial S^2}\right)_V > 0, \left(\frac{\partial^2 U}{\partial V^2}\right)_S > 0, \frac{\partial^2 U}{\partial V \partial S} > 0$$

These inequalities indicate how the signs of some important physical quantities become restricted for a system to be stable.

Since $$\left(\frac{\partial U}{\partial S}\right)_V = T$$

$$\left(\frac{\partial^2 U}{\partial S^2}\right)_V = \left(\frac{\partial T}{\partial S}\right)_V = \frac{T}{C_v}$$

∴ $$\frac{T}{C_v} > 0$$

Since $T > 0$ K

∴ $$C_v > 0 \tag{11.49}$$

which is the condition of *thermal stability*.

Also $$\left(\frac{\partial U}{\partial V}\right)_S = -p$$

$$\left(\frac{\partial^2 U}{\partial V^2}\right)_S = -\left(\frac{\partial p}{\partial V}\right)_S$$

∴ $$\left(\frac{\partial p}{\partial V}\right)_S < 0 \tag{11.50}$$

i.e. the adiabatic bulk modulus must be negative.

Similarly, if $F = F(T, V)$, then by Taylor's expansion, and using appropriate substitution

$$\delta F = -S\delta T - p\delta V + \frac{1}{2}\left(\frac{\partial^2 F}{\partial V^2}\right)_T (\delta V)^2$$

$$+ \frac{1}{2}\left(\frac{\partial^2 F}{\partial T^2}\right)_S (\delta T)^2 + \frac{\partial^2 F}{\partial V\,\partial T}\,\delta V\,\delta T + \ldots$$

For stability

$$\delta F + S\delta T + p\delta V > 0$$

$$\left(\frac{\partial^2 F}{\partial V^2}\right)_T > 0$$

Again $$\left(\frac{\partial F}{\partial V}\right)_T = -p$$

∴ $$\left(\frac{\partial^2 F}{\partial V^2}\right)_T = -\left(\frac{\partial p}{\partial V}\right)_T$$

∴ $$\left(\frac{\partial p}{\partial V}\right)_T < 0 \tag{11.51}$$

which is known as the condition of *mechanical stability*. The isothermal bulk modulus must also be negative.

SOLVED EXAMPLES

Example 11.1

(a) *Derive the equation*

$$\left(\frac{\partial C_p}{\partial p}\right)_T = -T\left(\frac{\partial^2 V}{\partial T^2}\right)_p$$

(b) *Prove that C_p of an ideal gas is a function of T only.*

(c) *In the case of a gas obeying the equation of state*

$$\frac{pv}{RT} = 1 + B'p$$

where B′ is a function of T only, show that

$$C_p = \bar{R}Tp\frac{d^2}{dT^2}(B'T) + (C_p)_0$$

where $(C_p)_0$ is the value at very low pressures. M [LO 11.2]

Solution

(a)
$$C_p = T\left(\frac{\partial S}{\partial T}\right)_p$$

$$\left(\frac{\partial C_p}{\partial p}\right)_T = T\left(\frac{\partial^2 S}{\partial T \cdot \partial p}\right)$$

Now
$$\left(\frac{\partial S}{\partial p}\right)_T = -\left(\frac{\partial V}{\partial T}\right)_p, \text{ by Maxwell's relation}$$

$\therefore$
$$\frac{\partial^2 S}{\partial p \cdot \partial T} = -\left(\frac{\partial^2 V}{\partial T^2}\right)_p$$

$\therefore$
$$\left(\frac{\partial C_p}{\partial p}\right)_T = -T\left(\frac{\partial^2 V}{\partial T^2}\right)_p$$

Proved.

(b) For an ideal gas

$$V = \frac{n\bar{R}T}{p}$$

$$\left(\frac{\partial V}{\partial T}\right)_p = \frac{n\bar{R}}{p} \quad \text{and} \quad \left(\frac{\partial^2 V}{\partial T^2}\right)_p = 0$$

$\therefore$
$$\left(\frac{\partial \bar{C}_p}{\partial p}\right)_T = 0, \text{ i.e., } C_p \text{ is a function of } T \text{ alone.}$$

(c)
$$\frac{p\bar{v}}{\bar{R}T} = 1 + B'p$$

$\therefore$
$$B'p = \frac{p\bar{v}}{\bar{R}T} - 1$$

$$B'T = \frac{T}{p}\left(\frac{p\bar{v}}{\bar{R}T} - 1\right) = \left(\frac{\bar{v}}{\bar{R}} - \frac{T}{p}\right)$$

$$\left[\frac{\partial}{\partial T}(B'T)\right]_p = + \frac{1}{\bar{R}}\left(\frac{\partial \bar{v}}{\partial T}\right)_p - \frac{1}{p}$$

$$\left[\frac{\partial^2 (B'T)}{\partial T^2}\right]_p = \frac{1}{\bar{R}}\left(\frac{\partial^2 \bar{v}}{\partial T^2}\right)_p = -\frac{1}{\bar{R}T}\left(\frac{\partial \bar{C}_p}{\partial p}\right)_T$$

$\therefore$ On integration

$$\bar{C}_p = -\bar{R}Tp\frac{d^2}{dT^2}(B'T) + \bar{C}_{p0}$$

where $\bar{C}_{p0}$ (integration constant) is the value of $\bar{C}_p$ at very low values of pressure.

Example 11.2 *The Joule–Kelvin coefficient μ_J is a measure of the temperature change during a throttling process. A similar measure of the temperature change produced by an isentropic change of pressure is provided by the coefficient μ_s, where*

$$\mu_S = \left(\frac{\partial T}{\partial p}\right)_S$$

Prove that

$$\mu_S - \mu_J = \frac{V}{C_p}$$

E [LO 11.2, 11.6]

Solution The Joule–Kelvin coefficient, μ_j, is given by

$$\frac{T\left(\frac{\partial V}{\partial T}\right)_p - V}{C_p}$$

Since $C_p = T\left(\frac{\partial S}{\partial T}\right)_p$ and by Maxwell's relation

$$\left(\frac{\partial V}{\partial T}\right)_p = -\left(\frac{\partial S}{\partial p}\right)_T$$

$$\mu_J = \frac{-T\left(\frac{\partial S}{\partial p}\right)_T}{T\left(\frac{\partial S}{\partial T}\right)_p} - \frac{V}{C_p}$$

$$\therefore \qquad \mu_J = -\left(\frac{\partial S}{\partial p}\right)_T \left(\frac{\partial T}{\partial S}\right)_p - \frac{V}{C_p}$$

Since $\left(\frac{\partial S}{\partial p}\right)_T \left(\frac{\partial T}{\partial S}\right)_p \left(\frac{\partial p}{\partial T}\right)_s = -1$

$$\therefore \qquad \mu_J = +\mu_s - \frac{V}{C_p}$$

$$\therefore \qquad \mu_s - \mu_J = \frac{V}{C_p}$$

Proved.

Alternative method:

From the second *Tds* equation

$$TdS = C_p\, dT - T\left(\frac{\partial V}{\partial T}\right)_p dp$$

$$\left(\frac{\partial T}{\partial p}\right)_s = \mu_s = \frac{T}{C_p}\left(\frac{\partial V}{\partial T}\right)_p$$

Now $$\mu_J = \frac{1}{C_p}\left[T\left(\frac{\partial V}{\partial T}\right)_p - V\right]$$

$$\therefore \qquad \mu_s - \mu_J = \frac{V}{C_p}$$

Proved.

Example 11.3 *A vertical cylinder fitted with a piston contains 0.02 m³ of liquid water at 100 kPa and 25°C. The liquid water is compressed to 20 MPa at constant temperature due to movement of the piston. If the isothermal compressibility of liquid water at 25°C is* 0.46×10^{-9} m²/N, *determine the work required for such compression process. The density of water at 100 kPa and 25°C is 997 kg/m³.*

E **[LO 11.3, 11.4]**

Solution

The work done can be found as

$$W_{1-2} = \int_1^2 pdV = \int_1^2 p\left(\frac{\partial V}{\partial P}\right)_T dp = -\int_1^2 Vk_T\, pdp$$

For $V \approx$ constant and $k_T \approx$ constant, we can write

$$W_{1-2} = -Vk_T \frac{(p_2^2 - p_1^2)}{2}$$

Substituting the respective values, we have

$$W_{1-2} = -\frac{0.02 \times 0.46 \times 10^{-9}}{2}\left[(20 \times 10^6)^2 - (100 \times 10^3)^2\right] = \mathbf{1839.95\ J}$$

Example 11.4 *If the boiling point of benzene at 1 atm pressure is 353 K, estimate the approximate value of the vapour pressure of benzene at 303 K.*

E **[LO 11.7]**

Solution Using Clapeyron's equation and Trouton's rule, Eq. (11.21),

$$p = 101.325 \exp\left\{\frac{88}{\bar{R}}\left(1 - \frac{T_B}{T}\right)\right\}$$

$$= 101.325 \exp\left\{\frac{88}{8.3143}\left(1 - \frac{353}{303}\right)\right\} = \mathbf{17.7\ kPa}$$

Example 11.5 *The vapour pressure, in mm of mercury, of solid ammonia is given by*

$$ln\ p = 23.03 - \frac{3754}{T}$$

and that of liquid ammonia by

$$ln\ p = 19.49 - \frac{3063}{T}$$

(a) What is the temperature of the triple point? What is the pressure? (b) What are the latent heats of sublimation and vapourisation? (c) What is the latent heat of fusion at the triple point?

M **[LO 11.7]**

Solution At the triple point, the saturated solid and saturated liquid lines meet.

$$23.03 - \frac{3754}{T} = 19.49 - \frac{3063}{T}$$

$$\therefore \quad T = \mathbf{195.2\ K}$$

$$\ln p = 23.03 - \frac{3754}{195.2}$$

$$\ln p = 3.80$$

$$p = \mathbf{44.67\ mm\ Hg}$$

With the assumptions, $v''' >> v'$ and $v''' \sim \dfrac{\bar{R}T}{p}$, Clausius–Clapeyron equation reduces to

$$\frac{dp}{dT} = \frac{p}{\bar{R}T^2} \cdot l_{\text{sub}}$$

where l_{sub} is the latent heat of sublimation.
The vapour pressure of solid ammonia is given by

$$\ln p = 23.03 - \frac{3754}{T}$$

$$\therefore \quad \frac{1}{p} \cdot \frac{dp}{dT} = \frac{3754}{T^2}$$

$$\therefore \quad \frac{dp}{dT} = 3754 \frac{p}{T^2} = \frac{p}{\bar{R}T^2} \cdot l_{sub}$$

$$\therefore \quad l_{sub} = 3754 \times 8.3143 = \mathbf{31,200 \text{ kJ/kg mol}}$$

The vapour pressure of liquid ammonia is given by

$$\ln p = 19.49 - \frac{3063}{T}$$

$$\therefore \quad \frac{dp}{dT} = 3063 \frac{p}{T^2} = \frac{p}{\bar{R}T^2} l_{vap}$$

where l_{vap} is the latent heat of vapourisation.

$$\therefore \quad l_{vap} = 3063 \times 8.3143 = \mathbf{25,500 \text{ kJ/kg mol}}$$

At the triple point

$$l_{sub} = l_{vap} + l_{fu}$$

where l_{fu} is the latent heat of fusion.

$$\therefore \quad l_{fu} = l_{sub} - l_{vap}$$

$$= 31,200 - 25,500 = \mathbf{5,700 \text{ kJ/kg mol}}$$

Example 11.6 *Explain why the specific heat of a saturated vapour may be negative.* Ⓜ **[LO 11.3, 11.7]**

Solution As seen in Fig. 11.15, if heat is transferred along the saturation line, there is a decrease in temperature. The slope of the saturated vapour line is negative, i.e. when dS is positive, dT is negative. Therefore, the specific heat at constant saturation

$$C'''_{sat} = T\left(\frac{dS'''}{dT}\right)$$

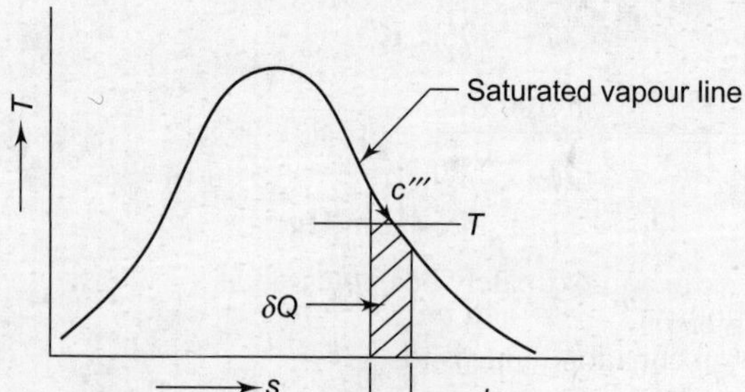

Fig. 11.15

is negative. From the second TdS equation

$$TdS = C_p dT - T\left(\frac{\partial V}{\partial T}\right)_p dp$$

$$T\frac{dS'''}{dT} = C_p - T\left(\frac{\partial V'''}{\partial T}\right)_p\left(\frac{dp}{dT}\right)_{sat}$$

$$= C_p - T\frac{n\bar{R}}{p}\cdot\frac{l_{vap}}{T\left(V''' - V''\right)} \quad \text{[using } pV''' = n\bar{R}T \text{ and Clapeyron's equation]}$$

$$C'''_{sat} = C_p - \frac{V'''}{T}\cdot\frac{l_{vap}}{V'''} \quad [\because V''' >> V'']$$

$$\therefore \quad C'''_{sat} = C_p - \frac{l_{vap}}{T}$$

Now the value of l_{vap}/T for common substances is about 83.74 J/kg mol K (*Trouton's rule*), where C_p is less than 41.87 J/g mol K. Therefore, C'''_{sat} can be negative. Proved.

Example 11.7 *A piston-cylinder device initially contains 1.5 kg saturated vapour of an unknown refrigerant at 200 kPa and –10°C. The state of the refrigerant is converted from saturated vapour to saturated liquid by the removal of 310 kJ of heat at constant pressure and temperature process. During this process, the volume of the system decreases by 0.15 m³. Determine s_{fg} of the refrigerant at –10°C. Also determine the boiling point temperature of the refrigerant at 240 kPa.*

Ⓜ **[LO 11.7]**

Solution

The latent heat of vapourisation is $h_{fg} = \frac{310}{1.5} = 206.67$ kJ/kg

From Clausius-Clapeyron equation, we have

$$\left.\frac{dp}{dT}\right|_{sat} = \frac{h_{fg}}{T_{sat}v_{fg}} \quad \text{and} \quad \left.\frac{dp}{dT}\right|_{sat} = \frac{s_{fg}}{v_{fg}}$$

Thus, $s_{fg} = \frac{h_{fg}}{T_{sat}} = \frac{206.67}{263} = \mathbf{0.7854\ kJ/\ kgK}$

Now, $\left.\frac{dp}{dT}\right|_{sat} = \frac{h_{fg}}{T_{sat}v_{fg}} = \frac{206.67}{263\times\frac{0.15}{1.5}} = 7.858$ kPa/K

Further, $\left.\frac{dp}{dT}\right|_{sat}$ can approximately be expressed as

$$\left.\frac{dp}{dT}\right|_{sat} \approx \left(\frac{p_2 - p_1}{T_2 - T_1}\right)_{sat}$$

or $T_2 = T_1 + \dfrac{(p_2 - p_1)_{sat}}{dp/dT|_{sat}} = 263 + \dfrac{240 - 200}{7.858} = 268.09 \text{ K} = \mathbf{-9.91°C}$

Example 11.8 *One unknown refrigerant flows steadily through the condenser of a refrigerator at a rate of 0.5 kg/s. It enters the condenser at 200 kPa and 40°C and exits as saturated liquid. It may be assumed that $v_g \gg v_f$ and the vapour approximately obeys the ideal gas law. The saturation pressure of the refrigerant varies with saturation temperature in the following manner: $\ln p_{sat} = 15.65 - 2720/T_{sat}$, where p_{sat} is in kPa and T_{sat} is in K. Determine the rate of heat transfer in the condenser. The molecular weight and constant pressure specific heat of the refrigerant are 102 and 0.8 kJ/kgK, respectively.* D [LO 11.7]

Solution The saturation temperature at 200 kPa is found to be

$$\ln (200) = 15.65 - 2720/T_{sat}$$

or $$T_{sat} = 262.76 \text{ K} = -10.24°\text{C}$$

Therefore the refrigerant enters the condenser as superheated vapour and the state is represented by 1 in the *T-s* diagram (Fig. 11.16).

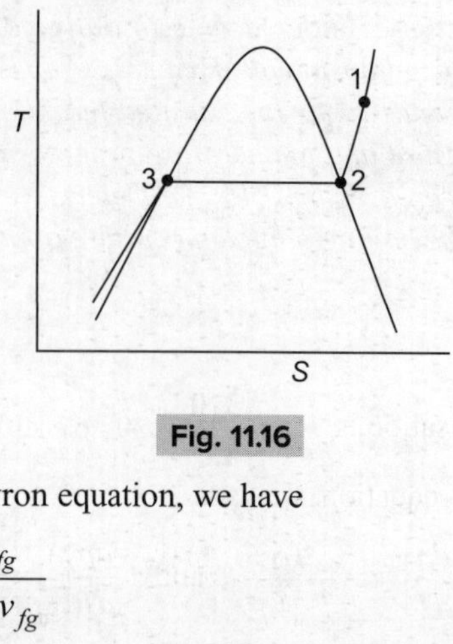

Fig. 11.16

From Clausius-Clapeyron equation, we have

$$\left.\frac{dp}{dT}\right|_{sat} = \frac{h_{fg}}{T_{sat} v_{fg}}$$

Now, $v_{fg} \approx v_g$ and $v_g = \dfrac{RT_{sat}}{p_{sat}}$

$$\therefore \quad \left.\frac{dp}{dT}\right|_{sat} = p_{sat} \frac{h_{fg}}{RT_{sat}^2}$$

Again $\ln p_{sat} = 15.65 - 2720/T_{sat}$

or $$\left.\frac{dp}{dT}\right|_{sat} = p_{sat} \frac{2720}{T_{sat}^2}$$

Thus, $$h_{fg} = 2720R = 2720 \times \frac{8.3143}{102} = 221.7 \text{ kJ/kg}$$

The rate of heat transfer in the condenser is

$$Q = w(h_3 - h_1) = w(h_3 - h_2) + w(h_2 - h_1) = -\, w h_{fg}\Big|_{T_2} + wc_p(T_2 - T_1)$$

$$= -0.5 \times 221.7 + 0.5 \times 0.8(-10.24 - 40) = \mathbf{-130.95\ kW}$$

Example 11.9 *(a) Establish the condition of equilibrium of a closed composite system consisting of two simple systems separated by a movable diathermal wall that is impervious to the flow of matter.*

(b) If the wall has rigid and diathermal, permeable to one type of material and impermeable to all others, state the condition of equilibrium of the composite system.

(c) Two particular systems have the following equations of state:

$$\frac{1}{T_1} = \frac{3}{2}\bar{R}\frac{N_1}{U_1},\ \frac{p_1}{T_1} = \frac{N_1}{V_1}\bar{R}$$

and

$$\frac{1}{T_2} = \frac{3}{2}\bar{R}\frac{N_2}{U_2},\ \frac{p_2}{T_2} = \bar{R}\frac{N_2}{V_2}$$

where $\bar{R}$ = 8.3143 kJ/kg mol K, and the subscripts indicate systems 1 and 2. The mole number of the first system is N_1 = 0.5, and that of the second is N_2 = 0.75. The two systems are contained in a closed adiabatic cylinder, separated by a movable diathermal piston. The initial temperatures are T_1 = 200 K and T_2 = 300 K and the total volume is 0.02 m^3. What is the energy and volume of each system in equilibrium? What is the pressure and temperature? D **[LO 11.10]**

Solution

For the composite system, as shown in Fig. 11.17

$$U_1 + U_2 = \text{constant}$$
$$V_1 = V_2 = \text{constant}$$

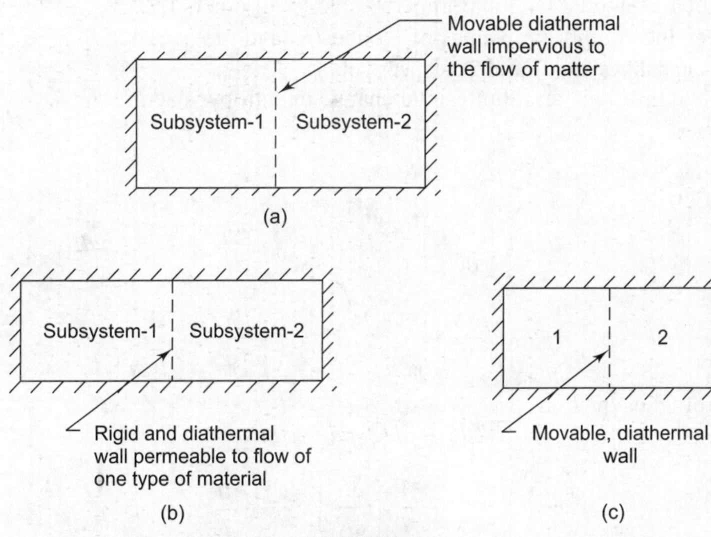

Fig. 11.17

The values of U_1, U_2, V_1 and V_2 would change in such a way as to maximise the value of entropy. Therefore, when the equilibrium condition is achieved

$$dS = 0$$

for the whole system. Since

$$S = S_1 + S_2$$
$$= S_1(U_1, V_1, ..., N_{k1} ...) + S_2 (U_2, V_2, ..., N_{k2} ...)$$

$$\therefore \quad dS = \left(\frac{\partial S_1}{\partial U_1}\right)_{V_1..., N_{k1}...} dU_1 + \left(\frac{\partial S_1}{\partial V_1}\right)_{U_1..., N_{k1}...,} dV_1$$
$$+ \left(\frac{\partial S_2}{\partial U_2}\right)_{V_2..., N_{k2}...,} dU_2 + \left(\frac{\partial S_2}{\partial V_2}\right)_{V_2..., N_{k2}...} dV_2$$
$$= \frac{1}{T_1} dU_1 + \frac{p_1}{T_1} dV_1 + \frac{1}{T_2} dU_2 + \frac{p_2}{T_2} dV_2$$

Since $dU_1 + dU_2 = 0$ and $dV_1 + dV_2 = 0$

$$\therefore \quad dS = \left(\frac{1}{T_1} - \frac{1}{T_2}\right) dU_1 + \left(\frac{p_1}{T_1} - \frac{p_2}{T_2}\right) dV_1 = 0$$

Since the expression must vanish for arbitrary and independent values of dU_1 and dV_1

$$\frac{1}{T_1} - \frac{1}{T_2} = 0 \text{ and } \frac{p_1}{T_1} - \frac{p_2}{T_2} = 0$$

or $\quad p_1 = p_2$ and $T_1 = T_2$

$\therefore$ These are the conditions of mechanical and thermal equilibrium.

(b) We will consider the equilibrium state of two simple subsystems (Fig. 11.17(b)) connected by a rigid and diathermal wall, permeable to one type of material (N_1) and impermeable to all others (N_2, N_3, ... N_r). We thus seek the equilibrium values of U_1 and U_2, and of N_{1-1} and N_{1-2} (i.e. material N_1 in subsystems 1 and 2, respectively.)

At equilibrium, an infinitesimal change in entropy is zero

$$dS = 0$$

Now $\quad dS = dS_1 + dS_2$

$$= \left(\frac{\partial S_1}{\partial U_1}\right)_{V_1, N_{1-1},...} dU_1 + \left(\frac{\partial S_1}{\partial N_{1-1}}\right)_{U_1, V_1, N_{1-2},...} dN_{1-1}$$
$$+ \left(\frac{\partial S_2}{\partial U_2}\right)_{V_2, N_{1-2}} dU_2 + \left(\frac{\partial S_2}{\partial N_{1-2}}\right)_{U_2, V_2, N_{2-2}} dN_{1-2}$$

From the equation

$$TdS = dU + pdV - \mu dN$$

$$\therefore \quad \left(\frac{\partial S}{\partial U}\right)_{V, N,...} = \frac{1}{T}, \left(\frac{\partial S}{\partial N}\right)_{U,V} = -\frac{\mu}{T}$$

and $$dN_{1-1} + dN_{1-2} = 0$$

$$dU_1 + dU_2 = 0$$

$$dS = \left(\frac{1}{T_1} - \frac{1}{T_2}\right) dU_1 - \left(\frac{\mu_{1-1}}{T_1} - \frac{\mu_{1-2}}{T_2}\right) dN_{1-1} = 0$$

As dS must vanish for arbitrary values of both dU_1 and dN_{1-1}

$$T_1 = T_2$$

$$\mu_{1-1} = \mu_{1-2}$$

which are the conditions of thermal and chemical equilibrium.

(c) $$N_1 = 0.5 \text{ g mol}, \quad N_2 = 0.75 \text{ g mol}$$

$$T_{1-1} = 200 \text{ K}, \quad T_{1-2} = 300 \text{ K}$$

$$V = V_1 + V_2 = 0.02 \text{ m}^3$$

$$U_1 + U_2 = \text{constant}$$

$$\Delta U_1 + \Delta U_2 = 0$$

Let T_f be the final temperature (Fig. 11.17(c))

$$(U_{f-1} - U_{i-1}) = -(U_{f-2} - U_{i-2})$$

$$\frac{3}{2}\bar{R}N_1 (T_f - T_{i-1}) = \frac{3}{2}\bar{R}N_2 (T_f - T_{i-2})$$

$$0.5 (T_f - 200) = -0.75 (T_f - 300)$$

$\therefore$ $$1.25\, T_f = 325$$

or $$T_f = \mathbf{260\ K}$$

$$U_{f-1} = \frac{3}{2}\bar{R}\, N_1 T_f = \frac{3}{2} \times 8.3143 \times 0.5 \times 10^{-3} \times 260 = 1.629 \text{ kJ}$$

$$U_{f-2} = \frac{3}{2} \times 8.3143 \times 0.75 \times 10^{-3} \times 260 = \mathbf{2430\ kJ}$$

$$V_{f-1} = \frac{\bar{R}N_1 T_{f-1}}{p_{f-1}}$$

At equilibrium

$$p_{f-1} = p_{f-2} = p_f$$

$$V_{f-2} = \frac{\bar{R}N_2 T_{f-2}}{p_{f-2}} \qquad T_{f-1} = T_{f-2} = T_f$$

$$V_{f-1} + V_{f-2} = \frac{\bar{R}T_f}{p_f} (N_1 + N_2) = 0.02 \text{ m}^3$$

$$\frac{8.3143 \times 260}{p_f} \times 1.25 \times 10^{-3} = 0.02 \text{ m}^3$$

$\therefore$ $$p_f = \frac{8.3143 \times 260 \times 1.25 \times 10^{-3}}{0.02} \text{ kN/m}^2$$

$$= 135 \text{ kN/m}^2 = 1.35 \text{ bar}$$

$$\therefore \qquad V_{f-1} = \frac{8.3143 \times 0.5 \times 10^{-3} \times 260}{135} = 0.008 \text{ m}^3$$

$$\therefore \qquad V_{f-2} = 0.02 - 0.008 = \mathbf{0.012\ m^3}$$

Example 11.10 *Show that for a van der Waals' gas*

(a) $\left(\frac{\partial c_v}{\partial v}\right)_T = 0$

(b) $(s_2 - s_1)_T = R \ln \frac{v_2 - b}{v_1 - b}$

(c) $T(v - b)^{R/c_v} = \text{constant, for an isentropic}$

(d) $c_p - c_v = \frac{R}{1 - 2a(v-b)^2/RTv^3}$

(e) $(h_2 - h_1)_T = (p_2 v_2 - p_1 v_1) + a\left(\frac{1}{v_1} - \frac{1}{v_2}\right)$ **[LO 11.3, 11.5]**

Solution (a) From the energy Eq. (11.13)

$$\left(\frac{\partial U}{\partial V}\right)_T = T\left(\frac{\partial p}{\partial T}\right)_V - p$$

$$\frac{\partial^2 U}{\partial V\, \partial T} = T\left(\frac{\partial^2 p}{\partial T^2}\right)_V + \left(\frac{\partial p}{\partial T}\right)_V - \left(\frac{\partial p}{\partial T}\right)_V$$

$$\frac{\partial^2 U}{\partial V\, \partial T} = T\left(\frac{\partial^2 p}{\partial T^2}\right)_V$$

$$C_v = \left(\frac{\partial U}{\partial T}\right)_V$$

$$\left(\frac{\partial C_v}{\partial V}\right)_T = \frac{\partial^2 U}{\partial T \cdot \partial V} = T\left(\frac{\partial^2 p}{\partial T^2}\right)_V = \left(\frac{\partial c_v}{\partial v}\right)_T$$

For a van der Waals' gas

$$\left(p + \frac{a}{v^2}\right)(v - b) = RT$$

$$p = \frac{RT}{v - b} - \frac{a}{v^2}$$

$\because \quad \left(\frac{\partial^2 p}{\partial T^2}\right)_V = 0$

$\therefore \quad \left(\frac{\partial c_v}{\partial v}\right)_T = 0$ Proved (a)

$\therefore$ c_v is independent of volume.

(b) From the first *Tds* Eq. (11.8)

$$Tds = c_v\, dT + T\left(\frac{\partial p}{\partial T}\right)_V dv$$

and energy Eq. (11.13), $\left(\frac{\partial U}{\partial V}\right)_T = T\left(\frac{\partial p}{\partial T}\right)_V - p$

$$ds = c_v \frac{dT}{T} + \frac{1}{T}\left[p + \left(\frac{\partial U}{\partial V}\right)_T\right] dv$$

For van der Waals' gas

$$\left(\frac{\partial U}{\partial V}\right)_T = \frac{a}{v^2}$$

$\therefore$
$$ds = c_v \frac{dT}{T} + \frac{1}{T}\left(p + \frac{a}{v^2}\right) dv$$

$$= c_v \frac{dT}{T} + \frac{R}{v-b}\, dv$$

$\therefore \quad (s_2 - s_1)_T = R \ln \frac{v_2 - b}{v_1 - b}$ Proved (b)

(c) At constant entropy

$$c_v \frac{dT}{T} + \frac{R}{v-b}\, dv = 0$$

or
$$\frac{dT}{T} + \frac{R}{c_v}\frac{dv}{v-b} = 0$$

by integration, $T(v-b)^{R/c_v} = \text{constant}$ Proved (c)

(d) $$c_p - c_v = T\left(\frac{\partial p}{\partial T}\right)_V \left(\frac{\partial v}{\partial T}\right)_p$$

$$= \left[\left(\frac{\partial U}{\partial V}\right)_T + p\right]\left(\frac{\partial v}{\partial T}\right)_p$$

$$= \left(\frac{a}{v^2} + p\right)\left(\frac{\partial v}{\partial T}\right)_p$$

$$= \left(\frac{RT}{v-b}\right)\left(\frac{\partial v}{\partial T}\right)_p$$

From the equation

$$\left(p + \frac{a}{v^2}\right)(v - b) = RT$$

$$\therefore \qquad (v-b)(-2av^{-3})\left(\frac{\partial v}{\partial T}\right)_p + \left(p + \frac{a}{v^2}\right)\left(\frac{\partial v}{\partial T}\right)_p = R$$

$$\left(\frac{\partial v}{\partial T}\right)_p = \frac{R/(v-b)}{\dfrac{RT}{(v-b)} - \dfrac{2a}{v^3}}$$

$$\therefore \qquad c_p - c_v = \frac{R}{1 - 2a(v-b)^2/RTv^3} \qquad \text{Proved (d)}$$

(e)

$$\left(\frac{\partial U}{\partial V}\right)_T = T\left(\frac{\partial p}{\partial T}\right)_V - p = \frac{a}{v^2}$$

$$du_T = \frac{a}{v^2}\, dv_T$$

$$\therefore \qquad (u_2 - u_1)_T = a\left(\frac{1}{v_1} - \frac{1}{v_2}\right)$$

$$\therefore \qquad (h_2 - h_1)_T = (p_2v_2 - p_1v_1) + a\left(\frac{1}{v_1} - \frac{1}{v_2}\right) \qquad \text{Proved (e)}$$

Example 11.11 *The virial equation of state of a gas is given by*

$$pv = RT(1 + B'p + C'p^2 + \cdots)$$

Show that

$$\lim_{p \to 0}\left[T\left(\frac{\partial v}{\partial T}\right)_p - v\right] = RT^2\frac{dB'}{dT}$$

Hence, prove that the inversion temperature of a van der Waals' gas is twice the Boyle temperature. **D [LO 11.6, 11.10]**

Solution

$$pv = RT(1 + B'p + C'p^2 + \cdots)$$

$$v = \frac{RT}{p} + RTB' + RTpC' + \cdots$$

$$\left(\frac{\partial v}{\partial T}\right)_p = \frac{R}{p} + RT\frac{dB'}{dT} + RB' + RTp\frac{dC'}{dT} + RpC' + \cdots$$

$$\therefore \quad T\left(\frac{\partial v}{\partial T}\right)_p = \frac{RT}{p} + RT^2\frac{dB'}{dT} + RTB' + RT^2p\frac{dC'}{dT} + RTpC' + \cdots$$

$$\therefore \quad T\left(\frac{\partial v}{\partial T}\right) - v = RT^2\frac{dB'}{dT} + RT^2p\frac{dC'}{dT} + \ldots$$

$$\mu_J = \frac{1}{c_p}\left[T\left(\frac{\partial v}{\partial T}\right)_p - v\right]$$

$$\therefore \quad = \frac{RT^2}{c_p}\left[\frac{dB'}{dT} + p\frac{dC'}{dT} + \ldots\right]$$

$$\lim_{p\to 0} \mu_J = \frac{RT^2}{c_p}\frac{dB'}{dT}$$

or $$\lim_{p\to 0}\left[T\left(\frac{dv}{dT}\right)_p - v\right] = RT^2\frac{dB'}{dT}$$ Proved.

For a van der Waals' gas, to find Boyle temperature T_B,

$$B = b - \frac{a}{RT} = 0$$

$$\therefore \quad T_B = \frac{a}{bR}$$

$$B' = \frac{B}{RT} = \frac{b}{RT} - \frac{a}{R^2T^2}$$

$$\frac{dB'}{dT} = -\frac{b}{RT^2} + \frac{2a}{R^2T^3}$$

$$\therefore \quad \lim_{p\to 0} \mu_J = \frac{RT^2}{c_p}\left(-\frac{b}{RT^2} + \frac{2a}{R^2T^3}\right) = 0$$

$$\therefore \quad \frac{b}{RT^2} = \frac{2a}{R^2T^3}$$

$$\therefore \quad T_i = \frac{2a}{bR}$$

$$\therefore \quad T_1 = 2T_B$$

or Inversion temperature = 2 × Boyle temperature Proved.

Example 11.12 *Over a certain range of pressures and temperatures the equation of a certain substance is given by the relation*

$$v = \frac{RT}{p} - \frac{C}{T^3}$$

where C is a constant. Derive an expression for (a) the change of enthalpy, and (b) the change of entropy, of this substance in an isothermal process. M **[LO 11.3, 11.5]**

Solution (a) From Eq. (11.15)

$$dh = c_p\, dT + \left[v - T\left(\frac{\partial v}{\partial T}\right)_p\right] dp$$

$$\therefore \qquad (h_2 - h_1)_T = \int_1^2 \left[v - T\left(\frac{\partial v}{\partial T}\right)_p\right] dp$$

Now,

$$v = \frac{RT}{p} - \frac{C}{T^3}$$

$$\left(\frac{\partial v}{\partial T}\right)_p = \frac{R}{p} + \frac{3C}{T^4}$$

$$\therefore \qquad T\left(\frac{\partial v}{\partial T}\right)_p = \frac{RT}{p} + \frac{3C}{T^3}$$

$$\therefore \qquad v - T\left(\frac{\partial v}{\partial T}\right)_p = \frac{RT}{p} - \frac{C}{T^3} - \frac{RT}{p} - \frac{3C}{T^3} = \frac{4C}{T^3}$$

on substitution,

$$(h_2 - h_1)_T = \int_{p_1}^{p_2} -\frac{4C}{T^3}\, dp$$

$$= \frac{4C}{T^3}(p_1 - p_2)_T$$

(b) Using second *Tds* equation

$$Tds = c_p\, dT - T\left(\frac{\partial v}{\partial T}\right)_p dp$$

$$\therefore \qquad ds_T = -\left(\frac{\partial v}{\partial T}\right)_p dp_T$$

$$= \left(\frac{R}{p} + \frac{3C}{T^4}\right) dp_T$$

$$(s_2 - s_1)_T = R \ln \frac{p_1}{p_2} + \frac{3C}{T^4}(p_1 - p_2)_T$$

SUMMARY

Deriving the Maxwell equations, the *TdS* equations

$$\text{(i)}\ TdS = C_v dT + T\left(\frac{\partial p}{\partial T}\right)_V dV$$

$$\text{(ii)}\ TdS = C_p dT - T\left(\frac{\partial V}{\partial T}\right)_p dp$$

are obtained. By subtracting Eq. (i) from Eq. (ii)

$$C_p - C_v = -T = -T\left(\frac{\partial V}{\partial T}\right)_p^2\left(\frac{\partial p}{\partial T}\right)_T \frac{TV\beta^2}{k_T}$$

are shown, where $\beta = \frac{1}{V}\left(\frac{\partial V}{\partial T}\right)_p$ is the volume expansively and $k_T = -\frac{1}{V}\left(\frac{\partial V}{\partial p}\right)_T$ is the isothermal compressibility. For an ideal gas, $pV = \text{mRT}$ and on substitution we obtain $\beta = \frac{1}{T}$, $k_T = \frac{1}{p}$ and $C_p - C_v = \text{mR}$ or $c_p - c_v = \text{R}$. Dividing Eqs. (i) and (ii), we get

$$\frac{c_p}{c_v} = \gamma = \frac{(\partial p/\partial V)_s}{(\partial p/\partial V)_T}$$

which shows $\left(\frac{\partial p}{\partial V}\right)_s > \left(\frac{\partial p}{\partial V}\right)_T$ and the slope of an isentrope is greater than that of an isotherm on *p-v* diagram. By using the first *TdS* relation, the following *energy equation* is obtained:

$$\left(\frac{\partial U}{\partial V}\right)_T = T\left(\frac{\partial p}{\partial T}\right)_V - p$$

and from the second *TdS* relation

$$\left(\frac{\partial H}{\partial V}\right)_T = V - T\left(\frac{\partial V}{\partial T}\right)_p$$

From these equations it is demonstrated that *U* and *H* for an ideal gas are functions of temperature only.
The second *TdS* equation yields the Joule–Thomson coefficient

$$\mu_J = \left(\frac{\partial T}{\partial p}\right)_h = \frac{1}{c_p}\left[T\left(\frac{\partial v}{\partial T}\right)_p - v\right]$$

which gives temperature drop or rise when a gas is throttled. For an ideal gas, however, there is no change in temperature when it is throttled. By using the first *TdS* equation, the Clausius–Clapeyron equation

$$\frac{dp}{dT} = \frac{l}{T(v^{(f)} - v^{(i)})}$$

is obtained where l is the latent heat and superscripts (f) and (i) in v indicate the final and initial phase, respectively. This equation helps us to evaluate the latent heats of fusion, vapourisation and sublimation and the vapour pressure approximately if the p, v, T data of the substance are available.

The conditions of equilibrium of a heterogeneous system have been derived

$$T_j = T_a, p_j = p_a, \mu_{ij} = \mu_{ia}$$

The temperature and pressure of phase 'a' are equal to those of all other phases and the chemical potential of the i-th component in phase 'a' is equal to the chemical potential of the same component in all other phases. Gibbs phase rule gives the variance for a non-reactive system

$$f = C - \phi + 2$$

where C = number of constituents, and ϕ = number of phases.

The criteria of spontaneous change, equilibrium and stability of a system have been discussed.

REVIEW QUESTIONS

(E) **11.1** What is the condition for exact differential? **[LO 11.1]**

(M) **11.2** Derive Maxwell's equations. **[LO 11.2]**

(D) **11.3** Write down the first and second *TdS* equations, and derive the expression for the difference in heat capacities, C_p and C_v. What does the expression signify? **[LO 11.3]**

(M) **11.4** Define volume expansivity and isothermal compressibility. **[LO 11.3]**

(D) **11.5** Show that the slope of an isentrope is greater than that of an isotherm on p–v plot. How is it meaningful for estimating the work of compression? **[LO 11.4]**

(D) **11.6** What is the energy equation? How does this equation leads to the derivation of the Stefan–Boltzman law of thermal radiation? **[LO 11.5]**

(M) **11.7** Show that the internal energy and enthalpy of an ideal gas are functions of temperature only. **[LO 11.5]**

(D) **11.8** Why are $dU = C_v\, dT$ and $dH = C_p\, dT$ true for an ideal gas in any process, whereas these are true for any other substance only at constant volume and at constant pressure, respectively? **[LO 11.5]**

(M) **11.9** Explain Joule–Kelvin effect. What is inversion temperature? **[LO 11.6]**

(M) **11.10** What is Joule–Thomson coefficient? Why is it zero for an ideal gas? **[LO 11.6]**

(M) **11.11** Why does the hydrogen gas need to be precooled before being throttled to get the cooling effect? **[LO 11.6]**

(E) **11.12** Why does the maximum temperature drop occur if the state before throttling lies on the inversion curve? **[LO 11.6]**

(M) **11.13** Why does the Gibbs function remain constant during phase transition? **[LO 11.7]**

(E) **11.14** What are the characteristics of the first-order phase transition? **[LO 11.7]**

(M) **11.15** Write down the representative equation for phase transition. Why does the fusion line for water have negative slope on the p–T diagram? **[LO 11.7]**

(M) **11.16** Why is the slope of the sublimation curve at the triple point on the p–T diagram greater than that of the vapourisation curve at the same point? **[LO 11.7]**

(M) **11.17** Explain how thermodynamic properties are evaluated from an equation of state. **[LO 11.8]**

(M) **11.18** Explain how enthalpy change and entropy change of a gas are estimated from an equation of state. **[LO 11.8]**

(E) **11.19** State the important thermodynamic criterion which an equation of state must satisfy. **[LO 11.9]**

(M) **11.20** Explain how Boyle temperature is yielded from **[LO 11.9]**

$$\lim_{p\to 0}\left(\frac{\partial z}{\partial p}\right)_T = 0$$

(E) **11.21** What is foldback temperature? **[LO 11.9]**

(E) **11.22** Show that for an inversion curve $\left(\frac{\partial z}{\partial T}\right)_p = 0.$ **[LO 11.9]**

(M) **11.23** Define chemical potential of a component in terms of U, H and G. **[LO 11.10]**

(E) **11.24** What is the use of the Gibbs entropy equation? **[LO 11.10]**

(E) **11.25** Explain the significance of the Gibbs–Duhem equation. **[LO 11.10]**

(E) **11.26** State the conditions of equilibrium of a heterogeneous system. **[LO 11.11]**

(E) **11.27** What do you understand by phase equilibrium? **[LO 11.12]**

(M) **11.28** Give the Gibbs phase rule for a non-reactive system. Why is the triple point of a substance nonvariant? **[LO 11.12]**

(M) **11.29** What are the four types of equilibrium? What is stable equilibrium? **[LO 11.13]**

(D) **11.30** State the conditions of spontaneous change, equilibrium and criterion of stability for (a) a system having constant U and V (i.e., isolated) and (b) a system having constant T and p. **[LO 11.13]**

(E) **11.31** What do you understand by neutral and unstable equilibrium? **[LO 11.13]**

(E) **11.32** What is metastable equilibrium? **[LO 11.13]**

(D) **11.33** Show that for a system to be stable, these conditions are satisfied: **[LO 11.13]**

(a) $C_V > 0$ (thermal stability)

(b) $\left(\frac{\partial p}{\partial V}\right)_T < 0$ (mechanical stability)

PROBLEMS WITH HINTS

(M) **11.1** For a gas which obeys the van der Waals equation of state, determine the relations between T and v and between p and v in a reversible adiabatic process, assuming c_v = constant. **[LO 11.6, 11.9]**

Hint: For a gas obeying the van der Waals equation, we have $s = s\,(T_1 v)$

$$ds = \left(\frac{\partial s}{\partial T}\right)_v dT + \left(\frac{\partial s}{\partial v}\right)_T dv$$

$$= \frac{c_v}{T}\,dT + \left(\frac{\partial p}{\partial T}\right)_v dv$$

$$= \frac{c_v}{T} dT + \frac{R}{v-b} dv$$

For a reversible adiabatic process, $ds = 0$

$c_v \ln T + R \ln (v - b) = \text{constant}$

$$\therefore \qquad \boldsymbol{T (v - l)^{R/c_v} = \textbf{constant}}$$

$$\left[\left(p + \frac{a}{v^2}\right)\frac{v-b}{R}\right](v-b)^{R/c_v} = \text{constant}$$

Hence, $$\left(p + \frac{a}{v^2}\right)(v-b)^{1+R/c_v} = \text{constant } \textbf{Ans.}$$

(D) **11.2** Derive the expression for the Joule–Thomson coefficient of a gas obeying the van der Waals equation of state and determine the equation of the inversion curve. **[LO 11.1, 11.6]**

Hint: Volume expansivity, $\beta = \frac{1}{v}\left(\frac{\partial v}{\partial T}\right)_p$

The van der Waals equation of state

$$p = \frac{RT}{v-b} - \frac{a}{v^2}$$

$$\left(\frac{\partial p}{\partial T}\right)_v = \frac{R}{v-b} \quad \text{and} \quad \left(\frac{\partial p}{\partial v}\right)_T = \frac{2a}{v^3} - \frac{RT}{(v-b)^2}$$

$$\left(\frac{\partial v}{\partial T}\right)_p \left(\frac{\partial T}{\partial p}\right)_v \left(\frac{\partial p}{\partial v}\right)_T = -1$$

$$\therefore \quad \left(\frac{\partial v}{\partial T}\right)_p = -\frac{\left(\frac{\partial p}{\partial T}\right)_v}{\left(\frac{\partial p}{\partial T}\right)_T} = -\frac{R(v-b)}{(2a/v^3) - RT/(v-b)^2} = \frac{Rv^3(v-b)}{RTv^3 - 2a(v-b)^3}$$

$$\therefore \quad \beta = \frac{1}{v}\left(\frac{\partial v}{\partial T}\right)_p = \frac{Rv^2(v-b)}{RTv^3 - 2a(v-b)^2}$$

Joule–Thomson coefficient, μ_J is given by

$$\mu_J = \left(\frac{\partial T}{\partial p}\right)_h = \frac{v}{c_p}[\beta T - 1] = \frac{v}{c_p}\left[\frac{RTv^2(v-b)}{RTv^3 - 2a(v-b)^2} - 1\right]$$

$$= \frac{v}{c_p}\left[\frac{2a(v-b)^2 - RTbv^2}{RTv^3 - 2a(v-b)^2}\right] \textbf{ Ans.}$$

The equation of the inversion curve is obtained by setting $\mu_J = 0$, which gives

$$T = \frac{2a}{Rb}\left(1 - \frac{b}{v}\right)^2$$

Substituting this equation into the van der Waals equation gives

$$p = \frac{a}{bv}\left(2 - \frac{3b}{v}\right)$$

The above two equations are the defining equations for the inversion curve in parametric form. The equation connecting the inversion temperature and the corresponding pressure can be obtained by eliminating v between the preceding two equations.

PROBLEMS

(E) **11.1** Derive the following equations

(a) $U = F - T\left(\frac{\partial F}{\partial T}\right)_V = -T^2\left(\frac{\partial F/T}{\partial T}\right)_V$

(b) $C_v = -T\left(\frac{\partial^2 F}{\partial T^2}\right)_v$

(c) $H = G - T\left(\frac{\partial G}{\partial T}\right)_p = -T^2\left(\frac{\partial G/T}{\partial T}\right)_p$

(d) $C_p = -T\left(\frac{\partial^2 G}{\partial T^2}\right)_p$ **[LO 11.1, 11.2]**

(M) **11.2** (a) Derive the equation

$$\left(\frac{\partial c_v}{\partial v}\right)_T = T\left(\frac{\partial^2 p}{\partial T^2}\right)_V$$

(b) Prove that c_v of an ideal gas is a function of T only.

(c) In the case of a gas obeying the equation of state

$$\frac{pv}{RT} = 1 + \frac{B''}{v}$$

where B'' is a function of T only, show that

$$c_v = -\frac{RT}{v}\frac{d^2}{dT^2}(B''\,T) + (c_v)_0$$

where $(c_v)_0$ is the value at very large volumes. **[LO 11.2]**

(M) **11.3** Derive the third TdS equation

$$TdS = C_v\left(\frac{\partial T}{\partial p}\right)_v dp + C_p\left(\frac{\partial T}{\partial V}\right)_p dV$$

and show that the three TdS equations may be written as

(a) $TdS = C_v\,dT + \frac{\beta T}{k_T}\,dV$

(b) $TdS = C_p\,dT - V\beta\,Tdp$

(c) $TdS = \frac{C_v}{\beta}\,k_T dp + \frac{C_p}{\beta V}\,dV$ **[LO 11.3]**

(M) **11.4** Derive the equations:

(a) $C_p = T\left(\frac{\partial V}{\partial T}\right)_p\left(\frac{\partial p}{\partial T}\right)_s$

(b) $\left(\dfrac{\partial p}{\partial T}\right)_s = \dfrac{C_p}{V\beta T}$

(c) $\dfrac{(\partial p/\partial T)_s}{(\partial p/\partial T)_v} = \dfrac{\gamma}{\gamma - 1}$ **[LO 11.2]**

M **11.5** Derive the equations:

(a) $C_v = -T\left(\dfrac{\partial p}{\partial T}\right)_v \left(\dfrac{\partial V}{dT}\right)_s$

(b) $\left(\dfrac{\partial V}{\partial T}\right)_s = -\dfrac{C_v k_T}{\beta T}$

(c) $\dfrac{(\partial V/\partial T)_s}{(\partial V/\partial T)_p} = \dfrac{1}{1-\gamma}$ **[LO 11.2]**

E **11.6** (a) Prove that the slope of a curve on a Mollier diagram representing a reversible isothermal process is equal to

$$T - \frac{1}{\beta}$$

(b) Prove that the slope of a curve on a Mollier diagram representing a reversible isochoric process is equal to

$$T + \frac{\gamma - 1}{\beta}$$ **[LO 11.2]**

M **11.7** (a) Show that

$$\mu_J c_p = T^2 \left(\frac{\partial V/T}{\partial T}\right)_p$$

For 1 mole of a gas, in the region of moderate pressures, the equation of sate may be written as

$$\frac{p\bar{v}}{\bar{R}T} = 1 + B'p + C'p^2$$

where B' and C' are functions of temperature only.

(b) Show that as $p \to 0$

$$\mu_J c_p \to \bar{R}T^2 \frac{dB'}{dT}$$

(c) Show that the equation of the inversion curve is

$$p = -\frac{dB'/dT}{dC'/dT}$$ **[LO 11.6]**

M **11.8** Prove the following functional relationship of the reduced properties for the inversion curve of a van der Waals gas

$$T_r = \frac{3(3v_r - 1)^2}{4v_r^2} \quad \text{and} \quad p_r = \frac{9(2v_r - 1)}{v_r^2}$$

Hence, show that

$$\frac{\text{Maximum inversion temperature}}{\text{Critical temperature}} = 6.75$$

and $$\frac{\text{Minimum inversion temperature}}{\text{Critical temperature}} = 0.75$$ **[LO 11.6]**

(M) **11.9** Estimate the maximum inversion temperature of hydrogen if it is assumed to obey the equation of state

$$pV = RT + B_1p + B_2p^2 + B_3p^3 + \ldots$$

For hydrogen, $B_1 \times 10^5 = a + 10^{-2}\,bT + 10^2\,c/T$

where $a = 166,\ b = -\,7.66,\ c = -\,172.33$ **[LO 11.6]**

(E) **11.10** The vapour pressure of mercury at 399 K and 401 K is found to be 0.988 and 1.084 mm of mercury, respectively. Calculate the latent heat of vapourisation of liquid mercury at 400 K. **[LO 11.7]**

(M) **11.11** In the vicinity of the triple point, the vapour pressure of liquid ammonia (in atmospheres) is represented by

$$\ln p = 15.16 - \frac{3063}{T}$$

This is the equation of the liquid-vapour boundary curve in a p–T diagram. Similarly, the vapour pressure of solid ammonia is

$$\ln p = 18.70 - \frac{3754}{T}$$

(a) What is the temperature and pressure at the triple point?
(b) What are the latent heats of sublimation and vapourisation?
(c) What is the latent heat of fusion at the triple point? **[LO 11.7]**

(E) **11.12** It is found that a certain liquid boils at a temperature of 95ºC at the top of a hill, whereas it boils at a temperature of 105ºC at the bottom. The latent heat is 4.187 kJ/g mole. What is the approximate height of the hill? **[LO 11.7]**

(M) **11.13** Show that for an ideal gas in a mixture of ideal gases

$$d\mu_k = \frac{\mu_k - h_k}{T}\,dT + v_k\,dp + RTd\ln x_k$$ **[LO 11.10]**

(E) **11.14** Compute μ_j for a gas whose equation of state is

$$p(v - b) = RT$$ **[LO 11.6]**

(E) **11.15** Show that

(a) $$\left(\frac{\partial \beta}{\partial p}\right)_T = -\left(\frac{\partial k_T}{\partial T}\right)_p$$

(b) $$\left(\frac{\partial u}{\partial p}\right)_T = -T\left(\frac{\partial v}{\partial T}\right)_p - p\left(\frac{\partial v}{\partial p}\right)_T$$ **[LO 11.3]**

(M) **11.16** Two particular systems have the following equations of state

$$\frac{1}{T^{(1)}} = \frac{3}{2}\,\bar{R}\,\frac{N^{(1)}}{U^{(1)}} \quad \text{and} \quad \frac{1}{T^{(2)}} = \frac{5}{2}\,\bar{R}\,\frac{N^{(2)}}{U^{(2)}}$$

where $\bar{R} = 8.3143$ kJ/kg mol K. The mole number of the first system is $N^{(1)} = 2$, and that of the second is $N^{(2)} = 3$. The two systems are separated by a diathermal wall, and the total energy in the composite system is 25.120 kJ. What is the internal energy of each system in equilibrium? **[LO 11.10, 1.11]**

(M) **11.17** Two systems with the equations of state given in Problem 11.16 are separated by a diathermal wall. The respective mole numbers are $N^{(1)} = 2$ and $N^{(2)} = 3$. The initial temperatures are $T^{(1)} = 250$ and $T^{(2)} = 350$ K. What are the values of $U^{(1)}$ and $U^{(2)}$ after equilibrium has been established? What is the equilibrium temperature? **[LO 11.11]**

(M) **11.18** Show that the change in latent heat L with temperature is given by the following relation:

$$\left(\frac{dL}{dT}\right) = (C_p''' - C_p'') + \frac{L}{T} - \frac{v'''\beta''' - v''\beta''}{v''' - v''} L$$ **[LO 11.7]**

(M) **11.19** Show that for a van der Waals gas, the Joule–Thomson coefficient is given by

$$\mu_j = \frac{v}{c_p}\left[\frac{2a(v-b)^2 - RTbv^2}{RTv^3 - 2a(v-b)^2}\right]$$ **[LO 11.9]**

(E) **11.20** At 273.15 K, the specific volumes of water and ice are 0.001 and 0.001091 m^3/kg, respectively, and the latent heat of fusion of ice is 334 kJ/kg. Determine the melting point increase due to increase of pressure by 1 atm (101.325 kPa) **[LO 11.7]**

(E) **11.21** Calculate the latent heat of vapourisation of steam formed by boiling water under a pressure of 101.325 kPa. At a pressure near this, a rise of temperature of 1 K causes an increase of vapour pressure of 3.62 kPa. **[LO 11.7]**

(M) **11.22** It is known that radiation exerts a pressure $p = 1/3\, u$, where u is the energy per unit volume.

(a) Show that

$$du = Tds + \frac{1}{V}\left(Ts - \frac{4}{3}u\right)dV$$

where s is the entropy per unit volume.

(b) Assuming u and s as functions of temperature only, show that

(i) $u = As^{4/3}$

(ii) $s = \frac{4}{3}aT^3$

(iii) $u = aT^4$

where A is the constant of integration and $a = 81/256\, A^3$.

(c) Show that the average time radiation remains in a spherical enclosure of radius r is given by

$$t = \frac{4r}{3c}$$

where c is the speed of radiation.

(d) If E_B is the energy emitted per unit area of spherical surface per unit time, show that

$$E_B = \sigma T^4$$

where $\sigma = ac/4$ and T is the temperature of the surface. **[LO 11.5]**

(E) **11.23** Show that the inversion temperature of a van der Waals gas is given by $T_i = 2a/bR$. **[LO 11.6]**

(M) **11.24** Show that **[LO 11.6]**

(a) $$\left(\frac{\partial u}{\partial v}\right)_T = T^2\left[\frac{\partial(p/T)}{\partial T}\right]_v$$

(b) $\left(\frac{\partial h}{\partial p}\right)_T = T^2\left[\frac{\partial (v/T)}{\partial T}\right]_v$

(M) **11.25** Show that for a van der Waals gas at low pressure, a Joule–Thomson expansion from pressure p_1 to p_2 produces a temperature change which can be found from the solution of

$$p_1 - p_2 = \frac{c_p}{b}(T_1 - T_2) + T_i \ln \frac{T_2 - T_i}{T_1 - T_i}$$

where T_i is the inversion temperature. **[LO 11.9]**

(D) **11.26** Using the Redlich–Kwong equation of state, develop expressions for the changes in entropy and internal energy of a gas in an isothermal process. **[LO 11.8]**

(E) **11.27** Find the change of entropy of a gas following Clausius equation of state at constant temperature:

$$p(v - b) = RT$$

[LO 11.8]

(D) **11.28** (a) Show that for a van der Waals gas

$$\beta = \frac{Rv^2(v-b)}{RTv^3 - 2a(v-b)^2}$$

$$k_T = \frac{v^2(v-b)^2}{RTv^3 - 2a(v-b)^2}$$

(b) What is the value of k_T/β expressed in its simplest form? **[LO 11.6]**

(c) What do the above relations become when $a = 0$, $b = 0$ (ideal gas)?

(M) **11.29** (a) Show that

(i) $\left(\frac{\partial u}{\partial p}\right)_v = \frac{k_T}{\beta} \cdot c_v$

(ii) $\left(\frac{\partial u}{\partial v}\right)_p = \frac{c_p}{v\beta} - p$

(b) Hence show that the slope of a reversible adiabatic process on p–v coordinates is

$$\frac{dp}{dv} = -\frac{\gamma}{k_T}$$

where k_T is the isothermal compressibility. **[LO 11.6]**

(M) **11.30** According to Berthelot, the temperature effect of the second virial coefficient is given by

$$B'(T) = \frac{b}{T} - \frac{a}{T^3}$$

where a and b are constants. Show that according to Berthelot,

$$T_{\text{inv}}/T_B = \sqrt{3}$$

[LO 11.6]

(M) **11.31** The following expressions for the equation of state and the specific heat c_p are obeyed by a certain gas:

$$v = \frac{RT}{p} + \alpha T^2 \quad \text{and} \quad c_p = A + BT + Cp$$

where α, A, B, C are constants. Obtain an expression for (a) the Joule–Thomson coefficient, and (b) the specific heat c_v. **[LO 11.9]**

(E) **11.32** Determine the maximum Joule–Thomson inversion temperature in terms of the critical temperature T_c predicted by the

(a) van der Waals' equations
(b) Redlich–Kwong equation
(c) Dieterici equation **[LO 11.9]**

(M) **11.33** From the virial form of the equation of state of a gas

$$v = \frac{RT}{p} + RTB'(T) + RTC'(T)p + \ldots$$

show that the Joule–Thomson coefficient is

$$\mu_J = \frac{RT^2}{c_p}\left[\frac{dB'}{dT} + \frac{dC'}{dT}p + \ldots\right]$$

(b) For a van der Waals gas

$$B'(T) = \frac{bRT - a}{R^2T^2}$$

Show that the limiting value of μ_J at low pressure is

$$\mu_J = \frac{1}{c_p}\left(\frac{2a}{RT} - b\right)$$ **[LO 11.9]**

(E) **11.34** Show that $k_T - k_s = \dfrac{Tv\beta^2}{c_p}$ **[LO 11.3]**

(M) **11.35** For a simple compressible system, show that

(a) $$\left[\frac{\partial u}{dv}\right]_T = T^2\left[\frac{\partial p/T}{\partial T}\right]_V$$

(b) $$\left[\frac{\partial h}{dp}\right]_T = -T^2\left[\frac{\partial v/T}{\partial T}\right]_p$$ **[LO 11.8]**

(M) **11.36** The liquid-vapour equilibrium curve for nitrogen over the range from the triple point to the normal boiling point may be expressed by the relation,

$$\log p = A - BT - \frac{C}{T}$$

where p is the vapour pressure in mm Hg, T is the temperature in K, and $A = 7.782$, $B = 0.006265$ and $C = 341.6$.

(a) Derive an expression for the enthalpy of vaporisation h_{fg} in terms of A, B, C, T and v_{fg}.

(b) Calculate h_{fg} for nitrogen at 71.9 K with $v_{fg} = 11.530$ cm³/gmol. **[LO 11.7]**

(M) **11.37** For a gas obeying the van der Waals equation of state, show that

(a) $$c_p - c_v = \frac{R}{1 - 2a(v-b)^2/RTv^3}$$

(b) $\left[\dfrac{\partial c_v}{dv}\right]_T = T\left[\dfrac{\partial^2 p}{\partial T^2}\right]_v = 0$ to prove that c_v is a function of temperature only

(c) $\left[\dfrac{\partial c_p}{dp}\right]_T = -T\left[\dfrac{\partial^2 v}{\partial T^2}\right]_p$

$$= R^2 T\left[\frac{2av^3 - 6abv^{-4}}{\left(p - av^{-2} + 2av^{-3}\right)^3}\right]$$

to prove that c_p for a van der Waals gas is not a function of temperature only.

(d) The relation between T and v is given by

$T(v - b)^{R/cv}$ = constant

(e) The relation between p and v is given by

$\left[p + \dfrac{a}{v^2}\right](v - b)^{1 + R/cv}$ = constant. **[LO 11.3]**

Ⓓ **11.38** Nitrogen at a pressure of 250 atm and a temperature of 400 K expands reversibly and adiabatically in a turbine to an exhaust pressure of 5 atm. The flow rate is 1 kg/s. Calculate the power output if nitrogen obeys the Redlich–Kwong equation of state. For nitrogen at 1 atm take.

$$c_p = 6.903 - 0.3753 \times 10^{-3}\, T + 1.930 \times 10^{-6}\, T^2 - 6.861 \times 10^{-9}\, T^3$$

where c_p is in cal/gmol-K and T is in K.

For nitrogen, $T_c = 126.2$ K,

$p_c = 33.5$ atm. **[LO 11.5, 11.8]**

Hints: See Art 11.9 and Ex. 11.9.

$h_1 - h_2 = (h_1 - h_4) + (h_4 - h_3) + (h_3 - h_2)$ and

$s_1 - s_2 = 0 = (s_1 - s_4) + (s_4 - s_3) + (s_3 - s_2)$

$a = 15.4 \times 10^6$ atm/K$^{1/2}$ cm^6/(gmol)^2, $b = 26.8$ cm^3/gmol

By trial and error, $v_1 = 143$ cm^3/gmol, $v_4 = 32{,}800$ cm^3/gmol

$T_2 = 124$ K, $h_1 - h_2 = 7.61$ kJ/gmol.

ANSWERS TO PROBLEMS

11.10 61,634.96 kJ/kg mol

11.11 195.2 K, 0.585 atm., 1498 kJ/kg, 1836 kJ/kg, 338 kJ/kg

11.12 394 m

11.14 $\mu_J = -b/c_p$

11.16 7.2 kJ, 17.92 kJ

11.17 8.02 kJ, 20.04 kJ, 321.4 K

11.20 0.00753 K

11.21 2257 kJ/kg

11.26 $(s_2 - s_1)_T = R\ln\dfrac{v_2 - b}{v_1 - b} + \dfrac{a}{2bT^{3/2}}\ln\left[\dfrac{v_2(v_1 + b)}{v_1(v_2 + b)}\right]$

$(u_2 - u_1)_T = \dfrac{3a}{2bT^{1/2}}\ln\left[\dfrac{v_2(v_1 + b)}{v_1(v_2 + b)}\right]$

11.27 $R\ln\dfrac{v_2 - b}{v_1 - b}$

11.31 (a) $m_J = \dfrac{\alpha T^2}{A + BT + Cp}$

(b) $c_v = A + BT + \dfrac{CRT}{v - \alpha T^2} - R\left(\dfrac{v + \alpha T^2}{v - \alpha T^2}\right)$

11.32 (a) 6.75 T_c, (b) 5.34 T_c (c) 8T_c

11.36 5,790 J/gmol

11.38 272 kW

12

Vapour Power Cycles

Learning Objectives

LO 12.1	Describe basic components of a steam power cycle
LO 12.2	Explain ideal Rankine cycle
LO 12.3	Identify different losses in actual steam power cycle
LO 12.4	Compare Carnot and Rankine cycles
LO 12.5	Explain the effects of mean temperature of heat addition on Rankine cycle performance
LO 12.6	Explain the reheat cycle
LO 12.7	Relate the ideal regenerative cycle with the Carnot cycle
LO 12.8	Discuss the practical regenerative cycle
LO 12.9	Apprise reheat-regenerative cycle as the cycle of the modern steam power plant
LO 12.10	Distinguish between open and closed feedwater heaters used in regenerative cycle
LO 12.11	Explain super critical and ultra supercritical steam power cycles
LO 12.12	Explain exergy analysis of vapour power cycles
LO 12.13	Identify salient features of an ideal working fluid for a vapour power cycle
LO 12.14	Explain binary (two-fluid) vapour power cycles
LO 12.15	Demonstrate overall efficiency of a coupled cycle
LO 12.16	Analyze cogeneration
LO 12.17	State different component efficiencies and the overall efficiency of a steam power plant

12.1 SIMPLE STEAM POWER CYCLE

LO 12.1 Describe basic components of a steam power cycle

A power cycle continuously converts heat (energy released by the burning of fuel) into work (shaft work), in which a working fluid repeatedly performs a succession of processes. The components of a simple vapour power plant are shown in Fig. 12.1(a). In the vapour power cycle the working fluid, which is water, undergoes a change of phase. Figure 12.1(b) gives the schematic of a simple steam power plant working on the vapour power cycle. Heat is transferred to water in the boiler from an external source (furnace, where fuel is continuously burnt) to raise steam, the high pressure, high temperature steam leaving the boiler expands in the turbine to produce shaft work, the steam leaving the turbine condenses into water in the condenser (where cooling water circulates), rejecting heat, and then the water is pumped back to the boiler. Figure 12.2 shows how a unit mass of the working fluid, sometimes in the liquid phase and sometimes in the vapour phase, undergoes various external heat and work interactions in executing a power cycle. Since the fluid is undergoing a cyclic process, there will be no net change in its internal energy over the cycle, and consequently the net energy transferred to the unit mass of the fluid as heat during the cycle must equal the net energy transfer as work from the fluid. Figure 12.3 shows the cyclic heat engine operating on the vapour power cycle, where the working substance, water, flows along the B-T-C-P (Boiler-Turbine-Condenser-Pump) path, interacting externally as shown, and converting net heat input to net work output continuously. By the first law,

$$\sum_{\text{Cycle}} Q_{\text{net}} = \sum_{\text{Cycle}} W_{\text{net}}$$

or

$$Q_1 - Q_2 = W_T - W_P$$

where Q_1 = heat transferred to the working fluid (kJ/kg)

Q_2 = heat rejected from the working fluid (kJ/kg)

W_T = work transferred from the working fluid (kJ/kg)

W_P = work transferred into the working fluid (kJ/kg)

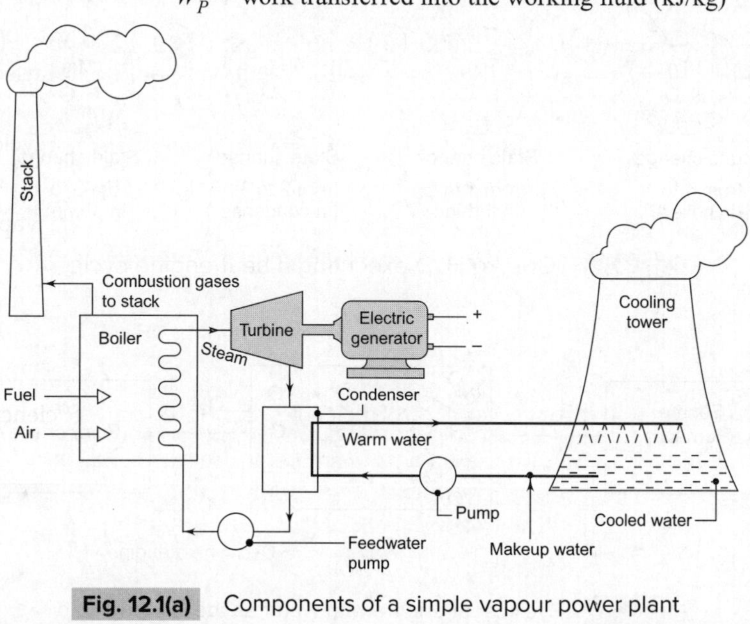

Fig. 12.1(a) Components of a simple vapour power plant

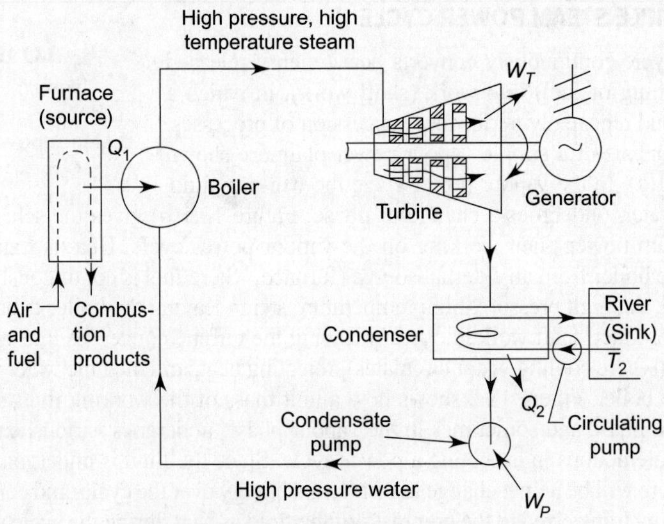

Fig. 12.1(b) Simple steam power plant

The efficiency of the vapour power cycle would be given by

$$\eta_{\text{cycle}} = \frac{W_{\text{net}}}{Q_1} = \frac{W_T - W_P}{Q_1} = \frac{Q_1 - Q_2}{Q_1}$$

$$= 1 - \frac{Q_2}{Q_1} \tag{12.1}$$

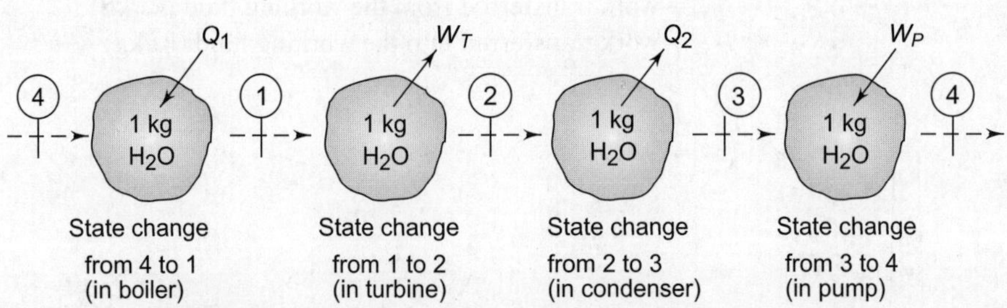

Fig. 12.2 One kg H_2O executing a heat engine cycle

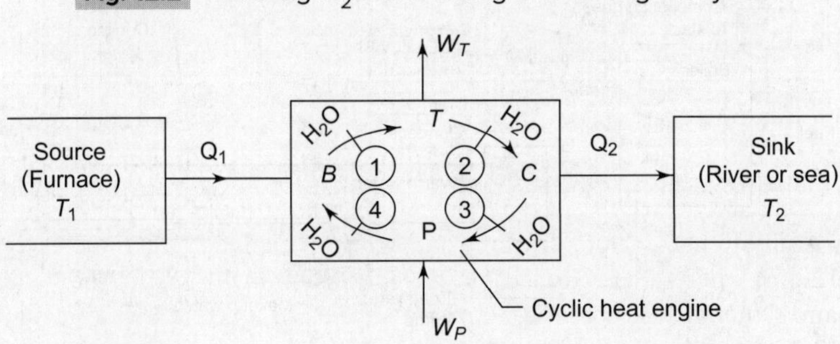

Fig. 12.3 Cyclic heat engine with water as the working fluid

12.2 RANKINE CYCLE

LO 12.2

Explain ideal Rankine cycle

For each process in the vapour power cycle, it is possible to assume a hypothetical or ideal process which represents the basic intended operation and involves no extraneous effects. For the steam boiler, this would be a reversible constant pressure heating process of water to form steam, for the turbine the ideal process would be a reversible adiabatic expansion of steam, for the condenser it would be a reversible constant pressure heat rejection as the steam condenses till it becomes saturated liquid, and for the pump, the ideal process would be the reversible adiabatic compression of this liquid ending at the initial pressure. When all these four processes are ideal, the cycle is an ideal cycle, called a *Rankine cycle*. This is a reversible cycle. Figure 12.4 shows the flow diagram of the Rankine cycle, and in Fig. 12.5, the cycle has been plotted on the *p-v*, *T-s* and *h-s* planes. The numbers on the plots correspond to the numbers on the flow diagram. For any given pressure, the steam approaching the turbine may be dry saturated (state 1), wet (state 1′) or superheated (state 1″), but the fluid approaching the pump is, in each case, saturated liquid (state 3). Steam expands reversibly and adiabatically in the turbine from state 1 to state 2 (or 1′ to 2′, or 1″ to 2″), the steam leaving the turbine condenses to water in the condenser reversibly at constant pressure from state 2 (or 2′, or 2″) to state 3, the water at state 3 is then pumped to the boiler at state 4 reversibly and adiabatically, and the water is heated in the boiler to form steam reversibly at constant pressure from state 4 to state 1 (or 1′ or 1″).

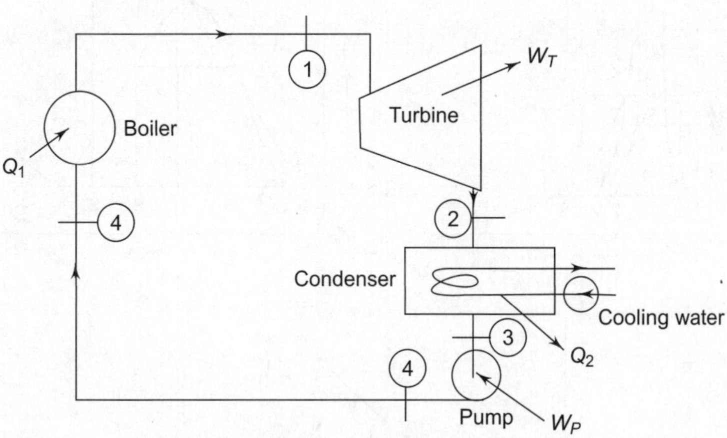

Fig. 12.4 A simple steam plant

For a simple nuclear power plant, the boiler is replaced by a nuclear reactor (Fig. 12.4 a) where heat released by nuclear fission is utilised in the generation of steam. Other features of the plant are similar to a conventional steam plant (Fig. 12.4).

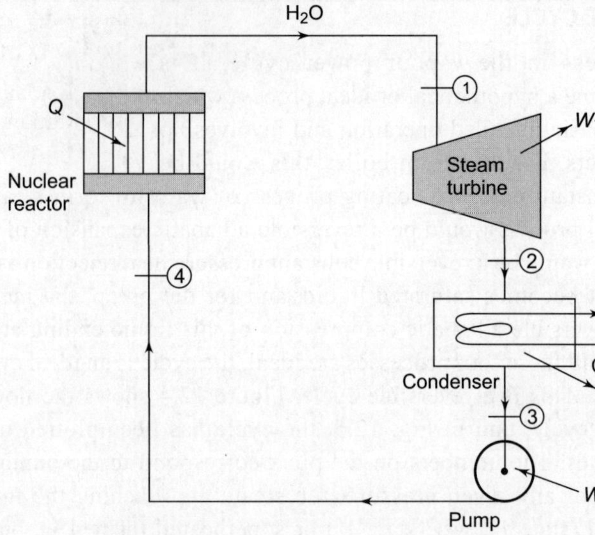

Fig. 12.4(a) A simple nuclear power plant

For purposes of analysis the Rankine cycle is assumed to be carried out in a steady flow operation. Applying the steady flow energy equation to each of the processes on the basis of unit mass of fluid, and neglecting changes in kinetic and potential energy, the work and heat quantities can be evaluated in terms of the properties of the fluid.

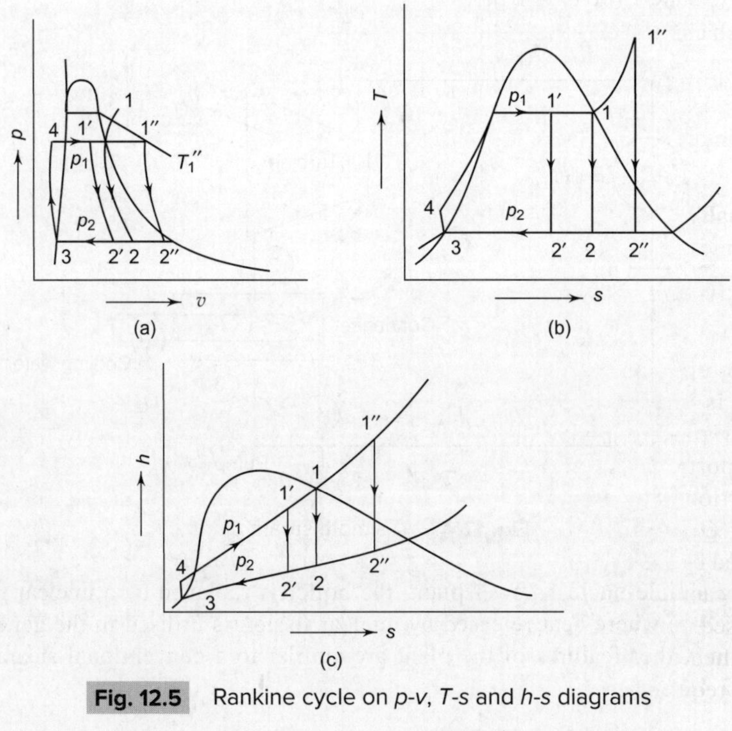

Fig. 12.5 Rankine cycle on *p-v*, *T-s* and *h-s* diagrams

For 1 kg fluid

The SFEE for the boiler (control volume) gives

$$h_4 + Q_1 = h_1$$

$$\therefore \quad Q_1 = h_1 - h_4 \tag{12.2}$$

The SFEE for the turbine as the control volume gives

$$h_1 = W_T + h_2$$

$$\therefore \quad W_T = h_1 - h_2 \tag{12.3}$$

Similarly, the SFEE for the condenser is

$$h_2 = Q_2 + h_3$$

$$\therefore \quad Q_2 = h_2 - h_3 \tag{12.4}$$

and the SFEE for the pump gives

$$h_3 + W_P = h_4$$

$$\therefore \quad W_P = h_4 - h_3 \tag{12.5}$$

The efficiency of the Rankine cycle is then given by

$$\eta = \frac{W_{net}}{Q_1} = \frac{W_T - W_P}{Q_1} = \frac{(h_1 - h_2) - (h_4 - h_3)}{h_1 - h_4} \tag{12.6}$$

The pump handles liquid water which is incompressible, i.e., its density or specific volume undergoes little change with an increase in pressure. For reversible adiabatic compression, by the use of the general property relation,

$$Tds = dh - vdp; \quad ds = 0$$

and $$dh = vdp$$

Since change in specific volume is negligible

$$\Delta h = v \, \Delta p$$

or $$h_4 - h_3 = v_3 \, (p_1 - p_2)$$

If v is in m^3/kg and p is in bar

$$h_4 - h_3 = v_3 \, (p_1 - p_2) \times 10^5 \text{ J/kg} \tag{12.7}$$

Usually, the pump work is quite small compared to the turbine work and is sometimes neglected. Then $h_4 = h_3$, and the cycle efficiency approximately becomes

$$\eta \cong \frac{h_1 - h_2}{h_1 - h_4}$$

The efficiency of the Rankine cycle is presented graphically in the *T-s* plot in Fig. 12.6. Thus Q_1 is proportional to area 1564, Q_2 is proportional to area 2563 and W_{net} $(= Q_1 - Q_2)$ is proportional to area 1234 enclosed by the cycle.

The capacity of a steam plant is often expressed in terms of *steam rate*, which is defined as the rate of steam flow (kg/h) required to produce unit shaft

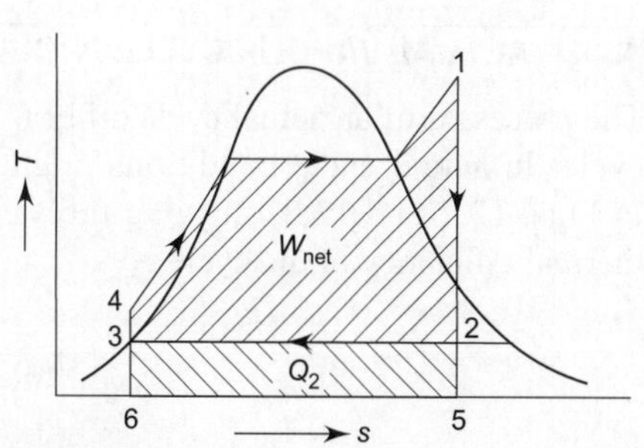

Fig. 12.6 Q_1, W_{net} and Q_2 are proportional to areas

output (1 kW). Therefore,

$$\text{Steam rate} = \frac{1}{W_T - W_P}\frac{\text{kg}}{\text{kJ}}\cdot\frac{1\,\text{kJ/s}}{1\,\text{kW}}$$

$$= \frac{1}{W_T - W_P}\frac{\text{kg}}{\text{kWs}} = \frac{3600}{W_T - W_P}\frac{\text{kJ}}{\text{kWh}} \tag{12.8}$$

The cycle efficiency is sometimes expressed alternatively as *heat rate* which is the rate of heat input (Q_1) required to produce unit work output (1 kW)

$$\text{Heat rate} = \frac{3600\,Q_1}{W_T - W_P} = \frac{3600}{\eta_{\text{cycle}}}\frac{\text{kJ}}{\text{kWh}} \tag{12.9}$$

From the equation $W_{\text{rev}} = -\int_3^4 v\,dp$, it is obvious that the reversible steady-flow work is closely associated with the specific volume of fluid flowing through the device. The larger the specific volume, the larger the reversible work produced or consumed by the steady-flow device. Therefore, every effort should be made to keep the specific volume of a fluid as small as possible during a compression process to minimise the work input and as large as possible, during an expansion process to maximise the work output.

In steam or gas power plants (Chapter 13), the pressure rise in the pump or compressor is equal to the pressure drop in the turbine if we neglect the pressure losses in various other components. In steam power plants, the pump handles liquid, which has a very small specific volume, and the turbine handles vapour, whose specific volume is many times larger. Therefore, the work output of the turbine is much larger than the work input to the pump. This is one of the reasons for the overwhelming popularity of steam power plants in electric power generation.

If we were to compress the steam exiting the turbine back to the turbine inlet pressure before cooling it first in the condenser in order to 'save' the heat rejected, we would have to supply all the work produced by the turbine back to the compressor. In reality, the required work input would be still greater than the work output of the turbine because of the irreversibilities or energy losses present in both processes (see Example 12.1).

12.3 ACTUAL VAPOUR CYCLE PROCESSES

LO 12.3

Identify different losses in actual steam power cycle

The processes of an actual cycle differ from those of ideal cycle. In actual cycle conditions might be as indicated in Figs. 12.7 and 12.8, showing the various losses. The thermal efficiency of the cycle is

$$\eta_{\text{th}} = \frac{W_{\text{net}}}{Q_1}$$

where the work and heat quantities are the measured values for the actual cycle, which are different from the corresponding quantities of the ideal cycle.

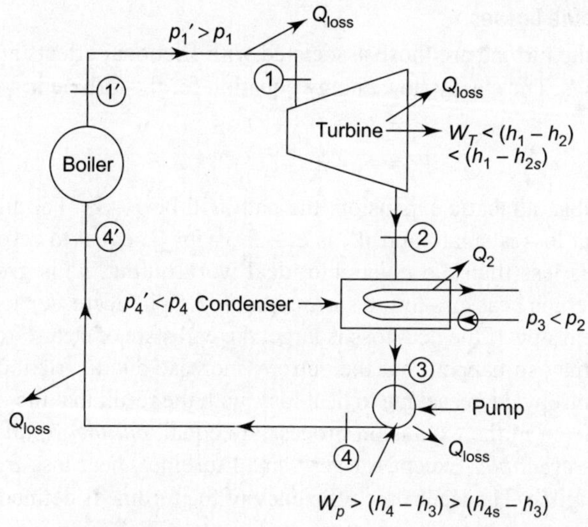

Fig. 12.7 Various losses in a steam plant

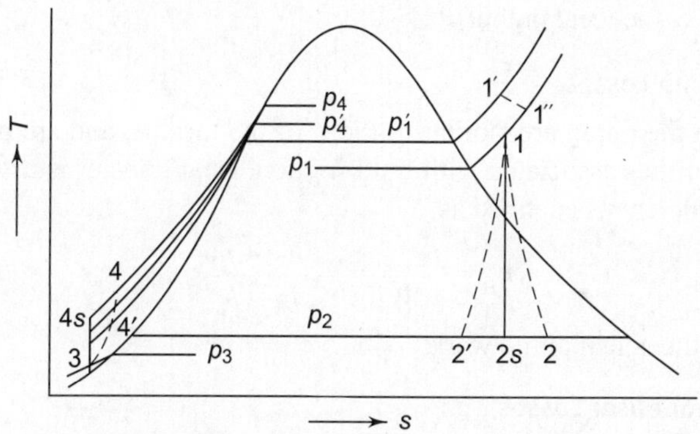

Fig. 12.8 Various losses on *T-s* plot

12.3.1 Piping Losses

Pressure drop due to friction and heat loss to the surroundings are the most important piping losses. States 1′ and 1 (Fig. 12.8) represent the states of the steam leaving the boiler and entering the turbine' respectively, 1′–1″ represents the frictional losses and 1″–1 shows the constant pressure heat loss to the surroundings. Both the pressure drop and heat transfer reduce the exergy or availability of steam entering the turbine.

A similar loss is the pressure drop in the boiler and also in the pipeline from the pump to the boiler. Due to this pressure drop, the water entering the boiler must be pumped to a much higher pressure than the desired steam pressure leaving the boiler, and this requires additional pump work.

12.3.2 Turbine Losses

The losses in the turbine are those associated with frictional effects and heat loss to the surroundings. The steady flow energy equation for the turbine in Fig. 12.7 gives

$$h_1 = h_2 + W_T + Q_{\text{loss}}$$

$$\therefore \quad W_T = h_1 - h_2 - Q_{\text{loss}} \tag{12.10}$$

For the reversible adiabatic expansion, the path will be 1 – 2*s*. For an ordinary real turbine the heat loss is small, and W_T is $h_1 - h_2$, with Q_2 equal to zero. Since actual turbine work is less than the reversible ideal work output, h_2 is greater than h_{2s}. However, if there is heat loss to the surroundings, h_2 will decrease, accompanied by a decrease in entropy. If the heat loss is large, the end state of steam from the turbine may be 2′. It may so happen that the entropy increase due to frictional effects just balances the entropy decrease due to heat loss, with the result that the initial and final entropies of steam in the expansion process are equal, *but the expansion is neither adiabatic nor reversible*. Except for very small turbines, heat loss from turbines is generally negligible. The isentropic efficiency of the turbine is defined as follows:

$$\eta_T = \frac{W_T}{h_1 - h_{2s}} = \frac{h_1 - h_2}{h_1 - h_{2s}} \tag{12.11}$$

where W_T is the actual turbine work, and $(h_1 - h_{2s})$ is the isentropic enthalpy drop in the turbine (i.e. the ideal output).

12.3.3 Pump Losses

The losses in the pump are similar to those of the turbine, and are primarily due to the irreversibilities associated with fluid friction. Heat transfer is usually negligible. The pump efficiency is defined as

$$\eta_P = \frac{h_{4s} - h_3}{W_P} = \frac{h_{4s} - h_3}{h_4 - h_3} \tag{12.12}$$

where W_P is the actual pump work.

12.3.4 Condenser Losses

The losses in the condenser are usually small. These include the loss of pressure and the cooling of condensate below the saturation temperature.

12.4 COMPARISON OF RANKINE AND CARNOT CYCLES

> **LO 12.4**
> Compare Carnot and Rankine cycles

Although the Carnot cycle has the maximum possible efficiency for the given limits of temperature, it is not suitable in steam power plants. Figure 12.9 shows the Rankine and Carnot cycles on the *T*–*s* diagram. The reversible adiabatic expansion in the turbine, the constant temperature heat rejection in the condenser and the reversible adiabatic compression in the pump, are similar characteristic features of both the Rankine and Carnot cycles. But whereas the heat addition process in the Rankine cycle is reversible and at constant pressure, in the Carnot cycle it is reversible and isothermal. In Figs. 12.9 (a) and (c), Q_2 is the same in both the cycles, but since Q_1 is more,

η_{Carnot} is greater than $\eta_{Rankine}$. The two Carnot cycles in Figs. 12.9(a) and (b) have the same thermal efficiency. Therefore, in Fig. 12.9 (b) also, $\eta_{Carnot} > \eta_{Rankine}$. But the Carnot cycle cannot be realised in practise because the pump work [in all the three cycles (a), (b) and (c)] is very large. Whereas in (a) and (c), it is impossible to add heat at infinite pressures and at constant temperature from state 4*c* to 1, in (b), it is difficult to control the quality at 3*c*, so that insentropic compression leads to saturated liquid state.

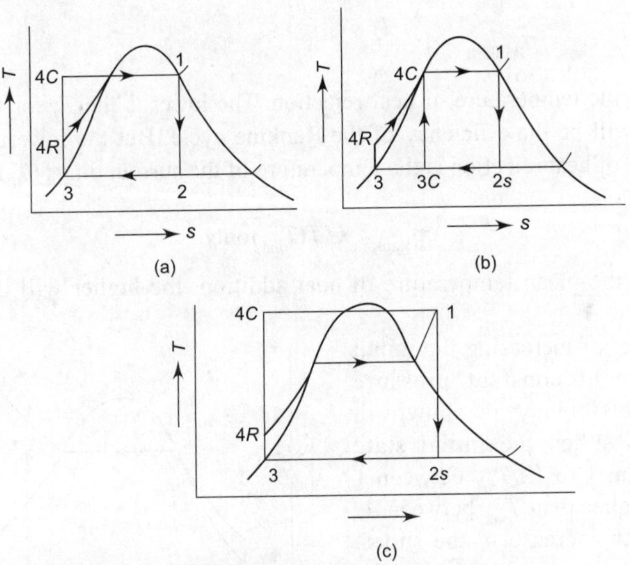

Fig. 12.9 Comparison of Carnot and Rankine cycles

12.5 MEAN TEMPERATURE OF HEAT ADDITION

> **LO 12.5**
> Explain the effects of mean temperature of heat addition on Rankine cycle performance

In the Rankine cycle, heat is added reversibly at a constant pressure, but at infinite temperatures. If T_{m1} is the mean temperature of heat addition, as shown in Fig. 12.10, so that the area under 4*s* and 1 is equal to the area under 5 and 6, then heat added

$$Q_1 = h_1 - h_{4s} = T_{m1}\,(s_1 - s_{4s})$$

$$\therefore \quad T_{m1} = \text{Mean temperature of heat addition}$$

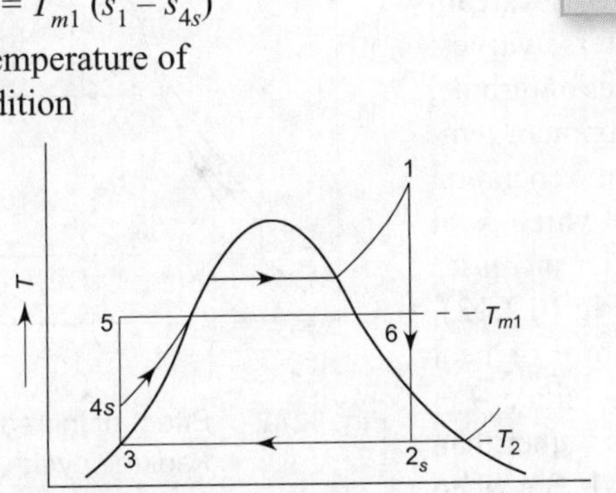

Fig. 12.10 Mean temperature of heat addition

$$= \frac{h_1 - h_{4s}}{s_1 - s_{4s}}$$

Since Q_2 = Heat rejected = $h_{2s} - h_3$

$$= T_2 (s_1 - s_4)$$

$$\eta_{\text{Rankine}} = 1 - \frac{Q_2}{Q_1} = 1 - \frac{T_2 (s_1 - s_{4s})}{T_{m1}(s_1 - s_{4s})}$$

$$\therefore \quad \eta_{\text{Rankine}} = 1 - \frac{T_2}{T_{m1}} \tag{12.13}$$

where T_2 is the temperature of heat rejection. The lower is the T_2 for a given T_{m1}, the higher will be the efficiency of the Rankine cycle. But the lowest practicable temperature of heat rejection is the temperature of the surroundings (T_0). This being fixed,

$$\eta_{\text{Rankine}} = f(T_{m1}) \text{ only} \tag{12.14}$$

The higher the mean temperature of heat addition, the higher will be the cycle efficiency.

The effect of increasing the initial temperature at constant pressure on cycle efficiency is shown in Fig. 12.11. When the initial state changes from 1 to 1′, T_{m1} between 1 and 1′ is higher than T_{m1} between 4s and 1. So an increase in the superheat at constant pressure increases the mean temperature of heat addition and hence the cycle efficiency.

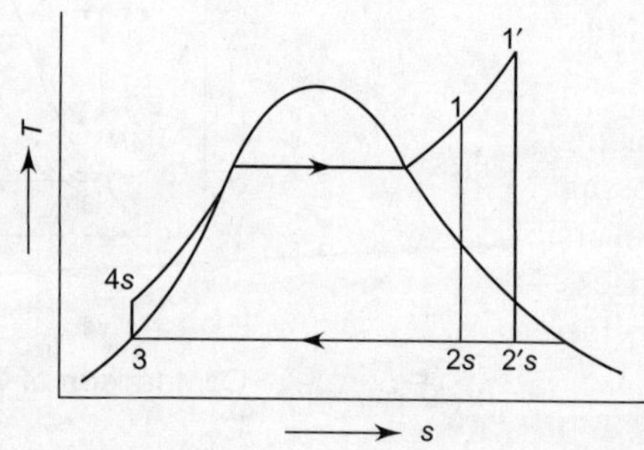

Fig. 12.11 Effect of superheat on mean temperature of heat addition

The maximum temperature of steam that can be used is *fixed from metallurgical considerations* (i.e. the materials used for the manufacture of the components which are subjected to high-pressure, high-temperature steam like the superheaters, valves, pipelines, inlet stages of turbine, etc.). When the maximum temperature is fixed, as the operating steam pressure at which heat is added in the boiler increases from p_1 to p_2 (Fig. 12.12), the mean temperature of heat addition increases, since T_{m1} between 7s and 5 is higher than that between 4s and 1. But when

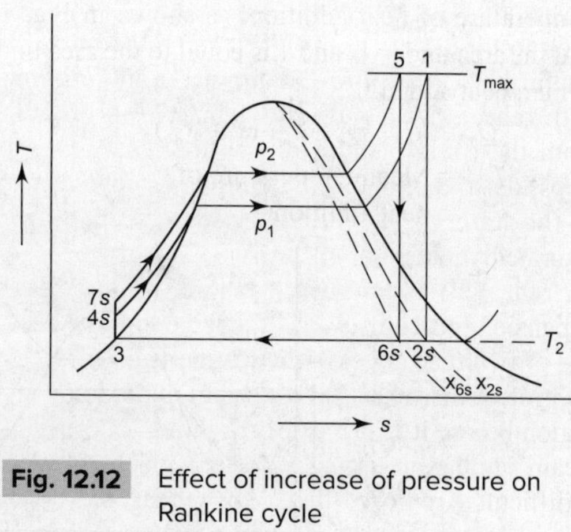

Fig. 12.12 Effect of increase of pressure on Rankine cycle

the turbine inlet pressure increases from p_1 to p_2, the ideal expansion line shifts to the left and the moisture content at the turbine exhaust increases (because $x_{6s} < x_{2s}$). If the moisture content of steam in the later stages of the turbine is higher, the entrained water particles along with the vapour coming out from the nozzles with high velocity strike the blades and erode their surfaces, as a result of which the longevity of the blades decreases. From a consideration of the erosion of blades in the later stages of the turbine, the maximum moisture content at the turbine exhaust is not allowed to exceed 15 per cent, or the quality to fall below 85 per cent. It is desirable that most of the turbine expansion should take place in the single phase or vapour region.

Therefore, with the maximum steam temperature at the turbine inlet, the minimum temperature of heat rejection, and the minimum quality of steam at the turbine exhaust being fixed, the maximum steam pressure at the turbine inlet also gets fixed (Fig. 12.13). The vertical line drawn from $2s$, fixed by T_2 and x_{2s}, intersects the T_{max} line, fixed by material, at 1, which gives the maximum steam pressure at the turbine inlet. The irreversibility in the expansion process has, however, not been considered.

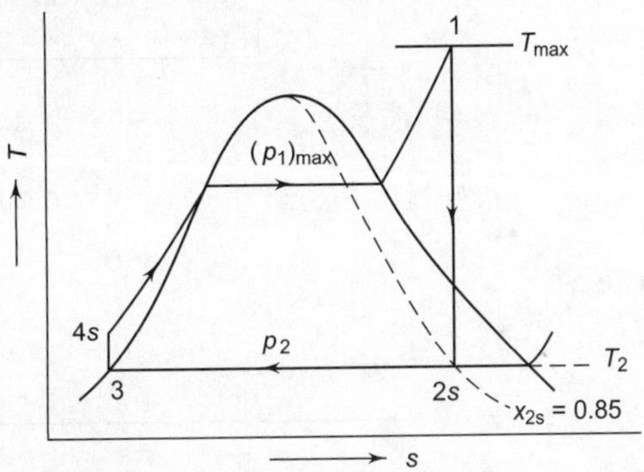

Fig. 12.13 Fixing of exhaust quality, maximum temperature and maximum pressure in Rankine cycle

12.6 REHEAT CYCLE

LO 12.6
Explain the reheat cycle

If a steam pressure higher than $(p_1)_{max}$ (Fig. 12.13) is used, in order to limit the quality to 0.85, at the turbine exhaust, reheat has to be adopted. In that case all the steam after partial expansion in the turbine is brought back to the boiler, reheated by combustion gases and then fed back to the turbine for further expansion. The flow, *T-s* and *h-s* diagrams for the ideal Rankine cycle with reheat are shown in Fig. 12.14. In the reheat cycle, the expansion of steam from the initial state 1 to the condenser pressure is carried out in two or more steps, depending upon the number of reheats used. In the first step, steam expands in the high pressure (HP) turbine from the initial state to approximately the saturated vapour line (process 1–$2s$ in Fig. 12.14). The steam is then resuperheated (or reheated) at constant pressure in the boiler (process $2s$–3) and the remaining expansion (process 3-$4s$) is carried out in the low pressure (LP) turbine. In the case of use of two reheats, steam is resuperheated twice at two different constant pressures. To protect the reheat tubes, steam is not allowed to expand deep into the two-phase region before it is taken for reheating, because in that case the moisture particles in steam while evaporating would leave behind solid deposits in the form of scale which is difficult to remove. Also, a low reheat pressure may bring down T_{m1} and hence, cycle efficiency. Again, a high reheat pressure increases the moisture content at turbine exhaust. Thus, the reheat pressure is optimised. The optimum reheat pressure for most of the modern power plants is about 0.2 to 0.25 of the initial steam pressure. For the cycle in Fig. 12.14, for 1 kg of steam

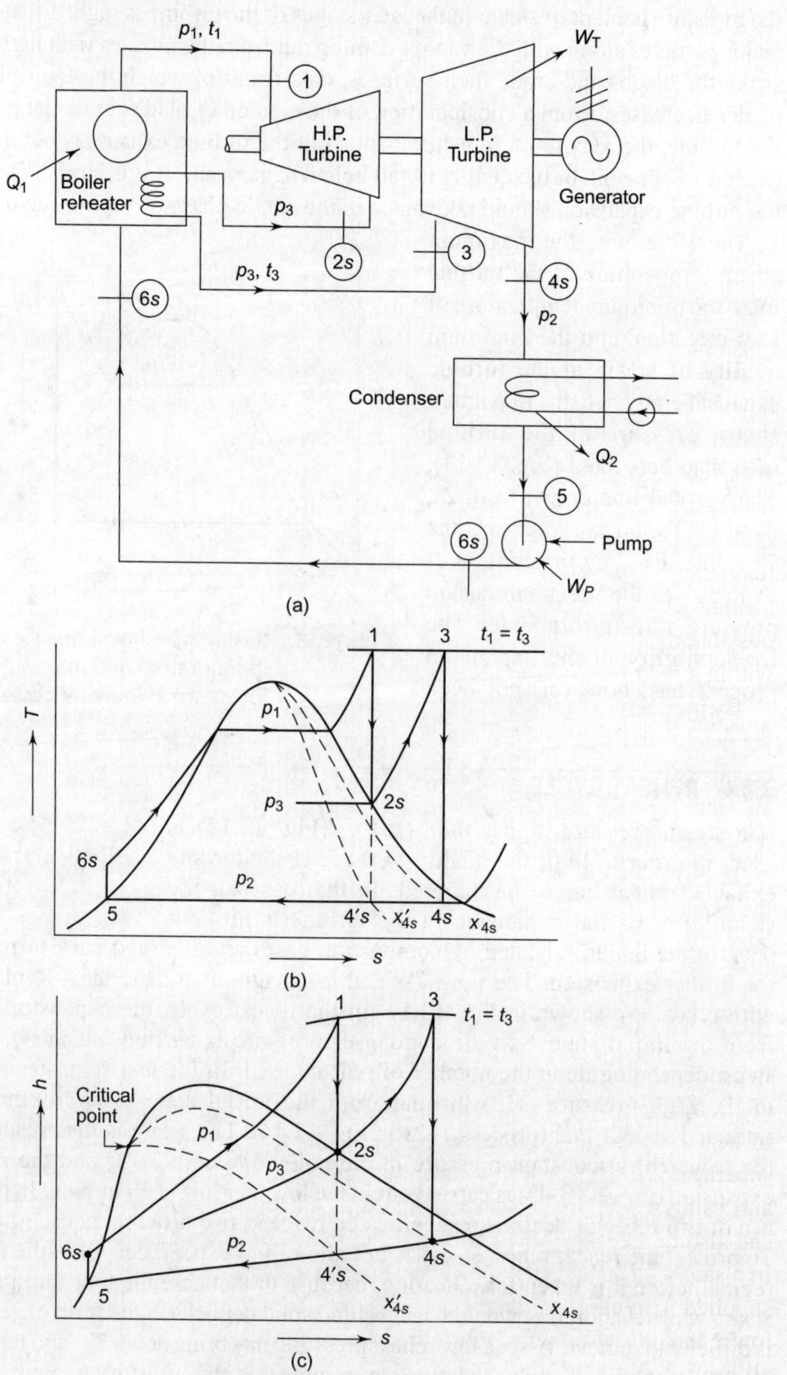

Fig. 12.14 Reheat cycle

$$Q_1 = h_1 - h_{6s} + h_3 - h_{2s}$$
$$Q_2 = h_{4s} - h_5$$
$$W_T = h_1 - h_{2s} + h_3 - h_{4s}$$
$$W_P = h_{6s} - h_5$$

$$\eta = \frac{W_T - W_P}{Q_1} = \frac{(h_1 - h_{2s} + h_3 - h_{4s}) - (h_{6s} - h_5)}{h_1 - h_{6s} + h_3 - h_{2s}} \quad (12.15)$$

$$\text{Steam rate} = \frac{3600}{(h_1 - h_{2s} + h_3 - h_{4s}) - (h_{6s} - h_5)} \text{ kg/kWh} \quad (12.16)$$

where enthalpy is in kJ/kg.

Since higher pressures are used in a reheat cycle, pump work may be appreciable.

Had the high pressure p_1 been used without reheat, the ideal Rankine cycle would have been $1 - 4's - 5 - 6s$. With the use of reheat, the area $2s - 3 - 4s - 4's$ has been added to the basic cycle. It is obvious that net work output of the plant increases with reheat, because $(h_3 - h_{4s})$ is greater than $(h_{2s} - h_{4's})$, and hence the steam rate decreases. Whether the cycle efficiency improves with reheat depends upon whether the mean temperature of heat addition in process $2s$–3 is higher than the mean temperature of heat addition in process $6s - 1$. In practise, the use of reheat only gives a small increase in cycle efficiency, but it increases the net work output by making possible the use of higher pressures, keeping the quality of steam at turbine exhaust within a permissible limit. The quality improves from x'_{4s} to x_{4s} by the use of reheat.

By increasing the number of reheats, still higher steam pressures could be used, but the mechanical stresses increase at a higher proportion than the increase in pressure, because of the prevailing high temperature. The cost and fabrication difficulties will also increase. In that way, the maximum steam pressure gets fixed, and more than two reheats have not yet been used so far.

In Fig. 12.14, only ideal processes have been considered. The irreversibilities in the expansion and compression processes have been considered in the example given later.

12.7 IDEAL REGENERATIVE CYCLE

> **LO 12.7**
> Relate the ideal regenerative cycle with the Carnot cycle

In order to increase the mean temperature of heat addition (T_{m1}), attention was so far confined to increase the amount of heat supplied at high temperatures, such as increasing superheat, using higher pressure and temperature of steam, and using reheat. The mean temperature of heat addition can also be increased by decreasing the amount of heat added at low temperatures. In a saturated steam Rankine cycle (Fig. 12.15), a considerable part of the total heat supplied is in the liquid phase when heating up water from 4 to 4′, at a temperature lower than T_1, the maximum temperature of the cycle. For maximum efficiency, all heat should be supplied at T_1, and feedwater should enter the boiler at state 4′.

This may be accomplished in what is known as an *ideal regenerative cycle*, the flow diagram of which is shown in Fig. 12.16 and the corresponding *T-s* diagram in Fig. 12.17.

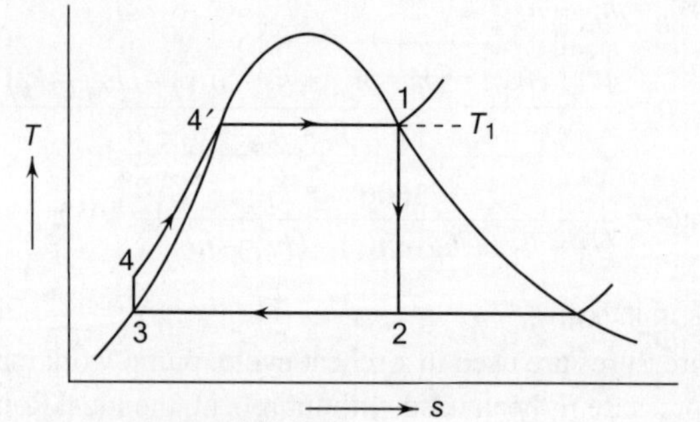

Fig. 12.15 Simple Rankine cycle

The unique feature of the ideal regenerative cycle is that the condensate, after leaving the pump circulates around the turbine casing, counterflow to the direction of vapour flow in the turbine (Fig. 12.16). Thus it is possible to transfer heat from the vapour as it flows through the turbine to the liquid flowing around the turbine. Let us assume that this is a reversible heat transfer, i.e. at each point the temperature of the vapour is only infinitesimally higher than the temperature of the liquid. The process 1–2′ (Fig. 12.17) thus represents reversible expansion of steam in the turbine with reversible heat rejection. For any small step in the process of heating the water,

$$\Delta T \text{ (water)} = -\Delta T \text{ (steam)}$$

and

$$\Delta s \text{ (water)} = -\Delta s \text{ (steam)}$$

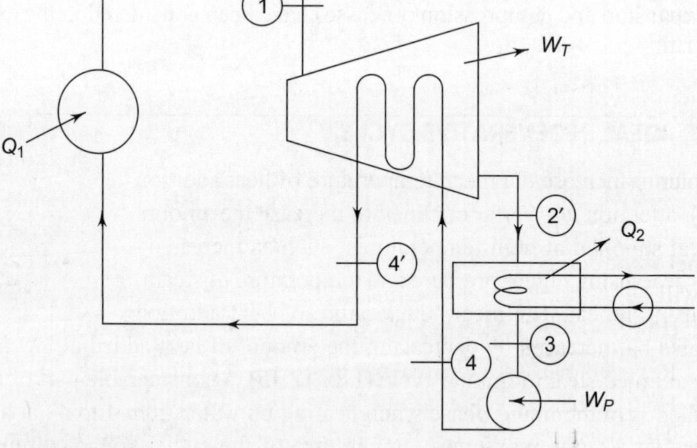

Fig. 12.16 Ideal regenerative cycle—Basic scheme

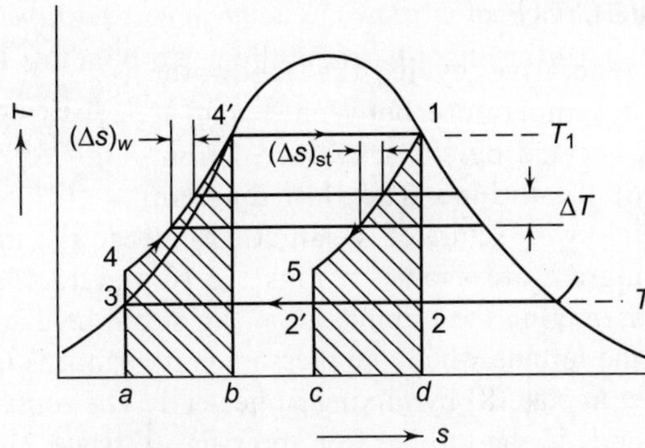

Fig. 12.17 Ideal regenerative cycle on *T-s* plot

Then the slopes of lines 1–2′ and 4′–3 (Fig. 12.17) will be identical at every temperature and the lines will be identical in contour. Areas 4–4′–b–*a*–4 and 2′–1–*d*–*c*-2′ are not only equal but congruous. Therefore, all the heat added from an external source (Q_1) is at the constant temperature T_1, and all the heat rejected (Q_2) is at the constant temperature T_2, both being reversible. Then

$$Q_1 = h_1 - h_{4'} = T_1(s_1 - s_{4'})$$

$$Q_2 = h_{2'} - h_3 = T_2(s_{2'} - s_3)$$

Since

$$s_{4'} - s_3 = s_1 - s_{2'}$$

or

$$s_1 - s_{4'} = s_{2'} - s_3$$

$\therefore$

$$\eta = 1 - \frac{Q_2}{Q_1} = 1 - \frac{T_2}{T_1}$$

The efficiency of the ideal regenerative cycle is thus equal to the Carnot cycle efficiency.

Writing the steady flow energy equation for the turbine

$$h_1 - W_T - h_{2'} + h_4 - h_{4'} = 0$$

or

$$W_T = (h_1 - h_{2'}) - (h_{4'} - h_4) \tag{12.17}$$

The pump work remains the same as in the Rankine cycle, i.e.

$$W_P = h_4 - h_3$$

The net work output of the ideal regenerative cycle is thus less, and hence its steam rate will be more, although it is more efficient, when compared with the Rankine cycle. However, the cycle is not practicable for the following reasons:

1. Reversible heat transfer cannot be obtained in finite time.
2. Heat exchanger in the turbine is mechanically impracticable.
3. The moisture content of the steam in the turbine will be high.

12.8 REGENERATIVE CYCLE

LO 12.8

Discuss the practical regenerative cycle

In the practical regenerative cycle, the feedwater enters the boiler at a temperature between 4 and 4′ (Fig. 12.17), and it is heated by steam extracted from intermediate stages of the turbine. The flow diagram of the regenerative cycle with saturated steam at the inlet to the turbine, and the corresponding *T-s* diagram are shown in Figs. 12.18 and 12.19, respectively. For every kg of steam entering the turbine, let m_1 kg steam be extracted from an intermediate stage of the turbine where the pressure is p_2, and it is used to heat up feedwater [$(1 - m_1)$ kg at state 8] by mixing in heater 1. The remaining $(1 - m_1)$ kg of steam then expands in the turbine from pressure p_2 (state 2) to pressure p_3 (state 3) when m_2 kg of steam is extracted for heating feed water in heater 2. So $(1 - m_1 - m_2)$ kg of steam then expands in the remaining stages of the turbine to pressure p_4, gets condensed into water in the condenser and then pumped to heater 2, where it mixes with m_2 kg of steam extracted at pressure p_3. Then $(1 - m_1)$ kg of water is pumped to heater 1 where it mixes with m_1 kg of steam extracted at pressure p_2. The resulting 1 kg of steam is then pumped to the boiler where heat from an external source is supplied. Heaters 1 and 2 thus operate at pressures p_2 and p_3, respectively. The amounts of steam m_1 and m_2 extracted from the turbine are such that at the exit from each of the heaters, the state is saturated liquid at the respective pressures. The heat and work transfer quantities of the cycle are as follows:

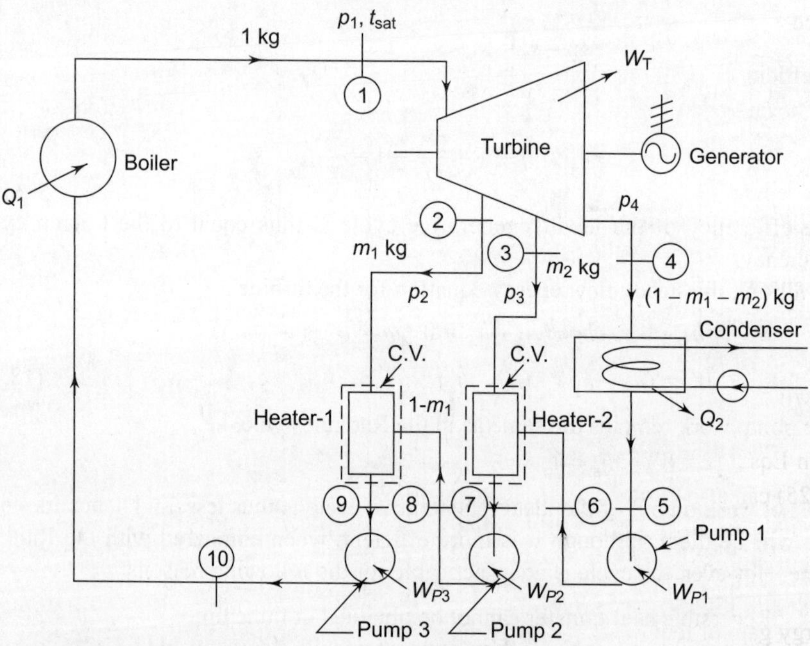

Fig. 12.18 Regenerative cycle flow diagram with two feedwater heaters

$$W_T = 1\,(h_1 - h_2) + (1 - m_1)\,(h_2 - h_3) + (1 - m_1 - m_2)\,(h_3 - h_4) \text{ kJ/kg} \qquad (12.18)$$

$$W_P = W_{P1} + W_{P2} + W_{P3}$$

$$= (1 - m_1 - m_2)\,(h_6 - h_5) + (1 - m_1)\,(h_8 - h_7) + 1\,(h_{10} - h_9) \text{ kJ/kg} \qquad (12.19)$$

$$Q_1 = 1\,(h_1 - h_{10}) \text{ kJ/kg} \qquad (12.20)$$

$$Q_2 = (1 - m_1 - m_2)\,(h_4 - h_5) \text{ kJ/kg} \qquad (12.21)$$

$$\text{Cycle efficiency, } \eta = \frac{Q_1 - Q_2}{Q_1} = \frac{W_T - W_P}{Q_1}$$

$$\text{Steam rate} = \frac{3600}{W_T - W_P} \text{ kg/kW h}$$

In the Rankine cycle operating at the given pressures, p_1 and p_4, the heat addition would have been from state 6 to state 1. By using two stages of regenerative feedwater heating, feedwater enters the boiler at state 10, instead of state 6, and heat addition is, therefore, from state 10 to state 1. Therefore,

$$(T_{m1})_{\text{with regeneration}} = \frac{h_1 - h_{10}}{s_1 - s_{10}} \qquad (12.22)$$

and

$$(T_{m1})_{\text{without regeneration}} = \frac{h_1 - h_6}{s_1 - s_6} \qquad (12.23)$$

Since

$$(T_{m1})_{\text{with regeneration}} > (T_{m1})_{\text{without regeneration}}$$

the efficiency of the regenerative cycle will be higher than that of the Rankine cycle.

The energy balance for heater 2 gives

$$m_1 h_2 + (1 - m_1)\,h_8 = 1 h_9$$

$$\therefore \quad m_1 = \frac{h_9 - h_8}{h_2 - h_8} \qquad (12.24)$$

The energy balance for heater 1 gives

$$m_2 h_3 + (1 - m_1 - m_2)\,h_6 = (1 - m_1)\,h_7$$

or

$$m_2 = (1 - m_1)\,\frac{h_7 - h_6}{h_3 - h_6} \qquad (12.25)$$

From Eqs. (12.24) and (12.25), m_1 and m_2 can be evaluated. Equations (12.24) and (12.25) can also be written alternatively as

$$(1 - m_1)\,(h_9 - h_8) = m_1\,(h_2 - h_9)$$

$$(1 - m_1 - m_2)\,(h_7 - h_6) = m_2\,(h_3 - h_7)$$

Energy gain of feedwater = Energy given off by vapour in condensation

Heaters have been assumed to be adequately insulated, and there is no heat gain from, or heat loss to, the surroundings.

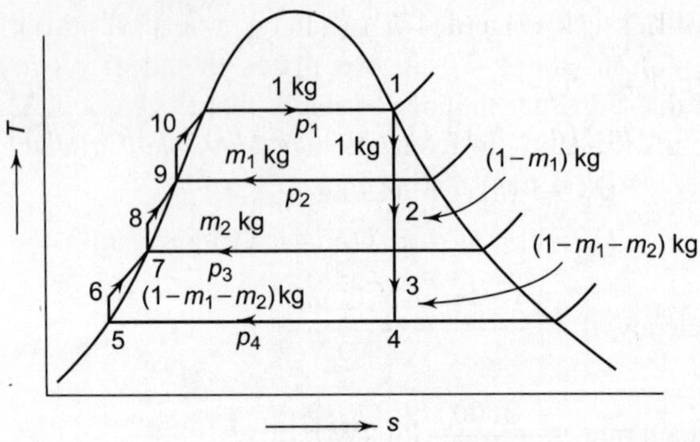

(a)

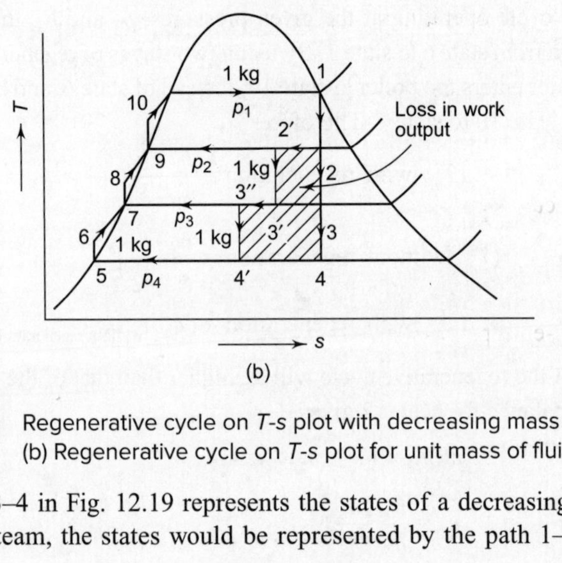

(b)

Fig. 12.19 Regenerative cycle on *T-s* plot with decreasing mass of fluid (b) Regenerative cycle on *T-s* plot for unit mass of fluid

Path 1–2–3–4 in Fig. 12.19 represents the states of a decreasing mass of fluid. For 1 kg of steam, the states would be represented by the path 1–2′–3′–4′. From Eq. (12.18),

$$W_T = (h_1 - h_2) + (1 - m_1)(h_2 - h_3) + (1 - m_1 - m_2)(h_3 - h_4)$$

$$= (h_1 - h_2) + (h_{2'} - h_{3'}) + (h_{3'} - h_{4'}) \tag{12.26}$$

where

$$(1 - m_1)(h_2 - h_3) = 1\,(h_{2'} - h_{3'}) \tag{12.27}$$

$$(1 - m_1 - m_2)(h_3 - h_4) = 1\,(h_{3''} - h_{4'}) \tag{12.28}$$

The cycle 1–2–2′–3′–3″–4′–5–6–7–8–9–10–1 represents 1 kg of working fluid. The heat released by steam condensing from 2 to 2′ is utilised in heating up the water from 8 to 9.

$$1(h_2 - h_{2'}) = (h_9 - h_8) \tag{12.29}$$

Similarly,

$$1(h_{3'} - h_{3''}) = 1(h_7 - h_6) \tag{12.30}$$

From Eqs. (12.26), (12.29) and (12.30)

$$W_T = (h_1 - h_{4'}) - (h_2 - h_{2'}) - (h_{3'} - h_{3''})$$

$$= (h_1 - h_{4'}) - (h_9 - h_8) - (h_7 - h_6) \tag{12.31}$$

The similarity of Eqs. (12.17) and (12.31) can be noticed. It is seen that the stepped cycle 1–2′–3′–4′–5–6–7–8–9–10 approximates the ideal regenerative cycle in Fig. 12.17, and that a greater number of stages would give a closer approximation (Fig. 12.20). Thus the heating of feedwater by steam 'bled' from the turbine, known as regeneration, *carnotises* the Rankine cycle.

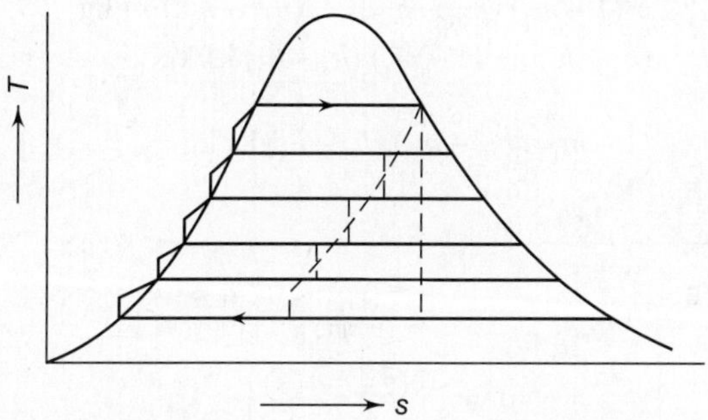

Fig. 12.20 Regenerative cycle with many stages of feedwater heating

The heat rejected Q_2 in the cycle decreases from $(h_4 - h_5)$ to $(h_{4'} - h_5)$. There is also loss in work output by the amount (Area under 2–2′ + Area under 3′–3″ – Area under 4–4′), as shown by the hatched area in Fig. 12.19 (b). So the steam rate increases by regeneration, i.e. more steam has to circulate per hour to produce unit shaft output.

The enthalpy-entropy diagram of a regenerative cycle is shown in Fig. 12.21.

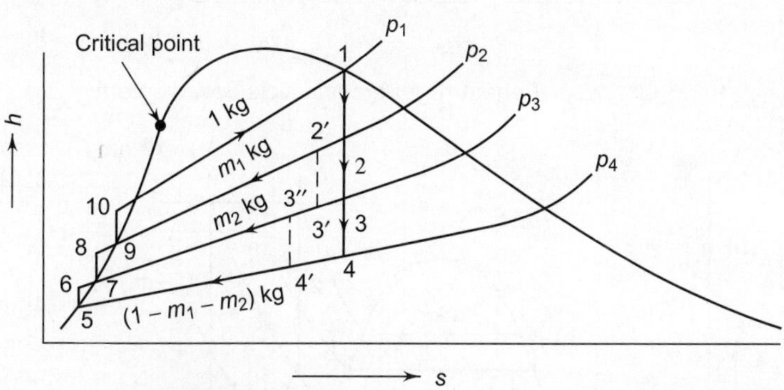

Fig. 12.21 Regenerative cycle on *h-s* diagram

12.9 REHEAT-REGENERATIVE CYCLE

> **LO 12.9**
> Apprise reheat-regenerative cycle as the cycle of the modern steam power plant

The reheating of steam is adopted when the vaporisation pressure is high. The effect of reheat alone on the thermal efficiency of the cycle is very small. Regeneration or the heating up of feedwater by steam extracted from the turbine has a marked effect on cycle efficiency. A modern steam power plant is equipped with both. Figures 12.22

and 12.23 give the flow and *T-s* diagrams of a steam plant with reheat and three stages of feedwater heating. Here

$$W_T = (h_1 - h_2) + (1 - m_1)(h_2 - h_3) + (1 - m_1)(h_4 - h_5) + (1 - m_1 - m_2)(h_5 - h_6) + (1 - m_1 - m_2 - m_3)(h_6 - h_7) \text{ kJ/kg}$$

$$W_P = (1 - m_1 - m_2 - m_3)(h_9 - h_8) + (1 - m_1 - m_2)(h_{11} - h_{10}) + (1 - m_1)(h_{13} - h_{12}) + 1(h_{15} - h_{14}) \text{ kJ/kg}$$

$$Q_1 = (h_1 - h_{15}) + (1 - m_1)(h_4 - h_3) \text{ kJ/kg}$$

and

$$Q_2 = (1 - m_1 - m_2 - m_3)(h_7 - h_8) \text{ kJ/kg}$$

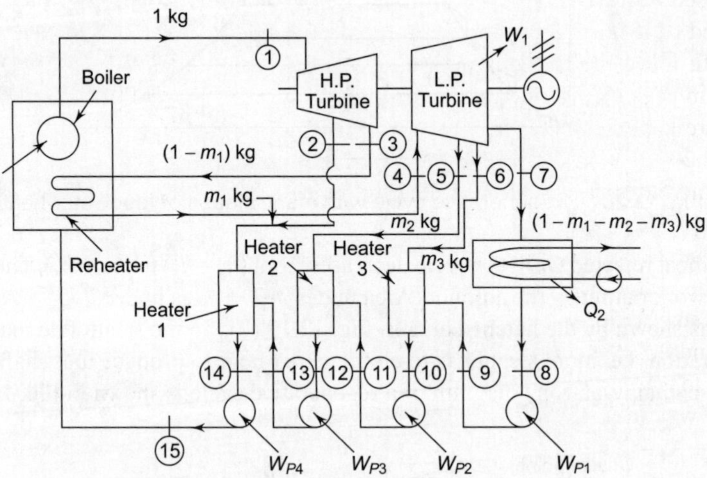

Fig. 12.22 Reheat-regenerative cycle flow diagram

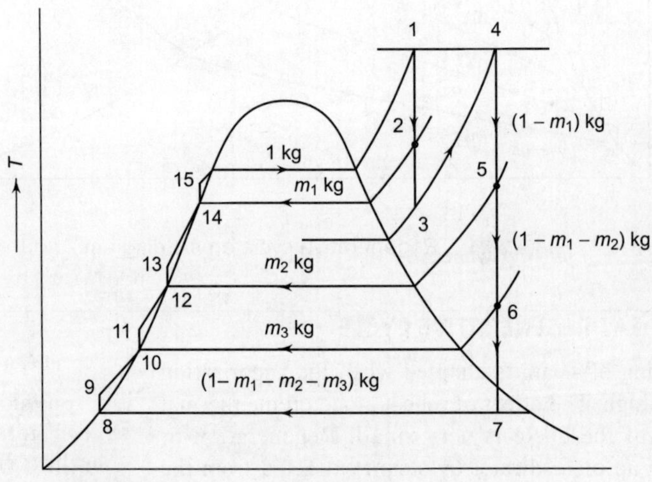

Fig. 12.23 *T-s* diagram of reheat-regenerative cycle

The energy balance of heaters 1, 2 and 3 give

$$m_1 h_2 + (1 - m_1)h_{13} = 1 \times h_{14}$$
$$m_2 h_5 + (1 - m_1 - m_2)h_{11} = (1 - m_1)h_{12}$$
$$m_3 h_6 + (1 - m_1 - m_2 - m_3)h_9 = (1 - m_1 - m_2)h_{10}$$

from which m_1, m_2 and m_3 can be evaluated.

12.10 FEEDWATER HEATERS

LO 12.10 Distinguish between open and closed feedwater heaters used in regenerative cycle

Feedwater heaters are of two types, viz. open heaters and closed heaters. In an open or contact-type heater, the extracted or bled steam is allowed to mix with feedwater, and both leave the heater at a common temperature, as shown in Figs. 12.18 and 12.22. In a closed heater, the fluids are kept separate, and not allowed to mix together (Fig. 12.24). The feedwater flows through the tubes in the heater and the extracted steam condenses on the outside of the tubes in the shell. The heat released by condensation is transferred to the feedwater through the walls of the tubes. The condensate (saturated water at the steam extraction pressure), sometimes called the heater-drip, then passes through a trap into the next lower pressure heater. This, to some extent, reduces the steam required by that heater. The trap passes only liquid and no vapour. The drip from the lowest pressure heater could similarly be trapped to the condenser, but this would be throwing away energy to the condenser cooling water. To avoid this waste, a drip pump feeds the drip directly into the feedwater stream.

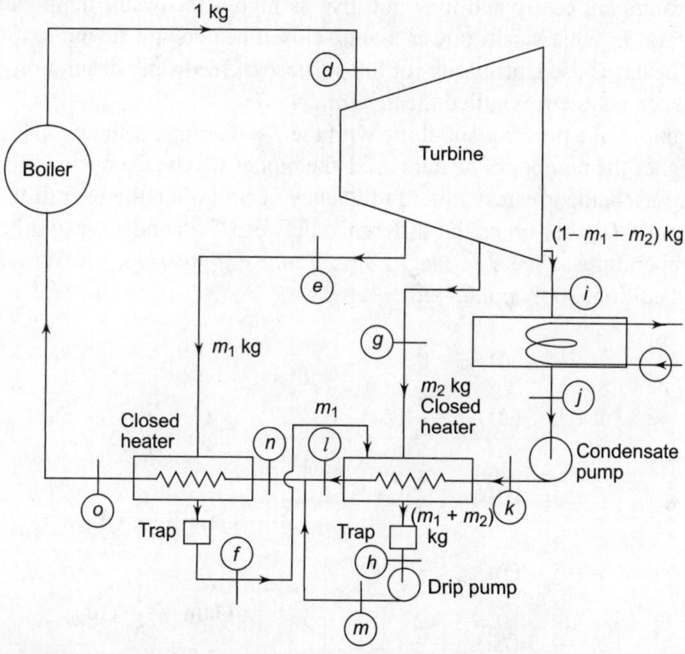

Fig. 12.24 Regenerative cycle flow diagram with closed feedwater heaters

Figure 12.25 shows the *T-s* plot corresponding to the flow diagram in Fig. 12.24. The temperature of the feedwater (at '*l*' or '*o*') leaving a particular heater is always less than the saturation temperature at the steam extraction pressure (at *e* or *g*), the difference being known as the *terminal temperature difference* of the heater.

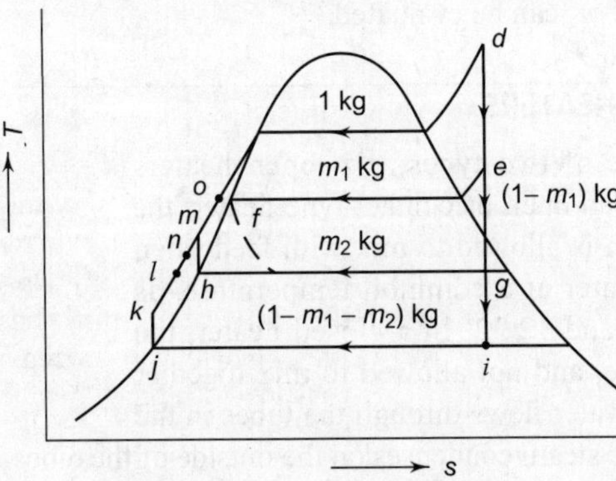

Fig. 12.25 *T-s* diagram of regenerative cycle with closed feedwater heaters

The advantages of the open heater are simplicity, lower cost and high heat transfer capacity. The disadvantage is the necessity of a pump at each heater to handle the large feedwater stream.

A closed heater system requires only a single pump for the main feedwater stream regardless of the number of heaters. The drip pump, if used, is relatively small. Closed heaters are costly and may not give as high a feedwater temperature as do open heaters. In most steam power plants, closed heaters are favoured, but at least one open heater is used, primarily for the purpose of feedwater deaeration. The open heater in such a system is called the *deaerator*.

An example of a power plant along with the *T-s* diagram is shown in Fig. 12.26.

The higher the number of heaters used, the higher will be the cycle efficiency. If *n* heaters are used, the greatest gain in efficiency occurs when the overall temperature rise is about $n/(n+1)$ times the difference between the condenser and boiler saturation temperatures. (See P.K. Nag, *Power Plant Engineering*, McGraw-Hill, New Delhi, 3rd edition, 2008, pp. 65–68).

$$n = 0 \quad (\Delta t)_{fw} = 0$$

$$n = 1, \quad (\Delta t)_{fw} = \frac{1}{2}(\Delta t)_0$$

$$\text{Gain} = \frac{1}{6}(\Delta t)_0$$

$$n = 2, \quad (\Delta t)_{fw} = \frac{2}{3}(\Delta t)_0$$

$$\text{Gain} = \frac{1}{12}(\Delta t)_0$$

$$n = 3, \quad (\Delta t)_{fw} = \frac{3}{4}(\Delta t)_0$$

$$\text{Gain} = \frac{1}{20}(\Delta t)_0$$

$$n = 4, \quad (\Delta t)_{fw} = \frac{4}{2}(\Delta t)_0$$

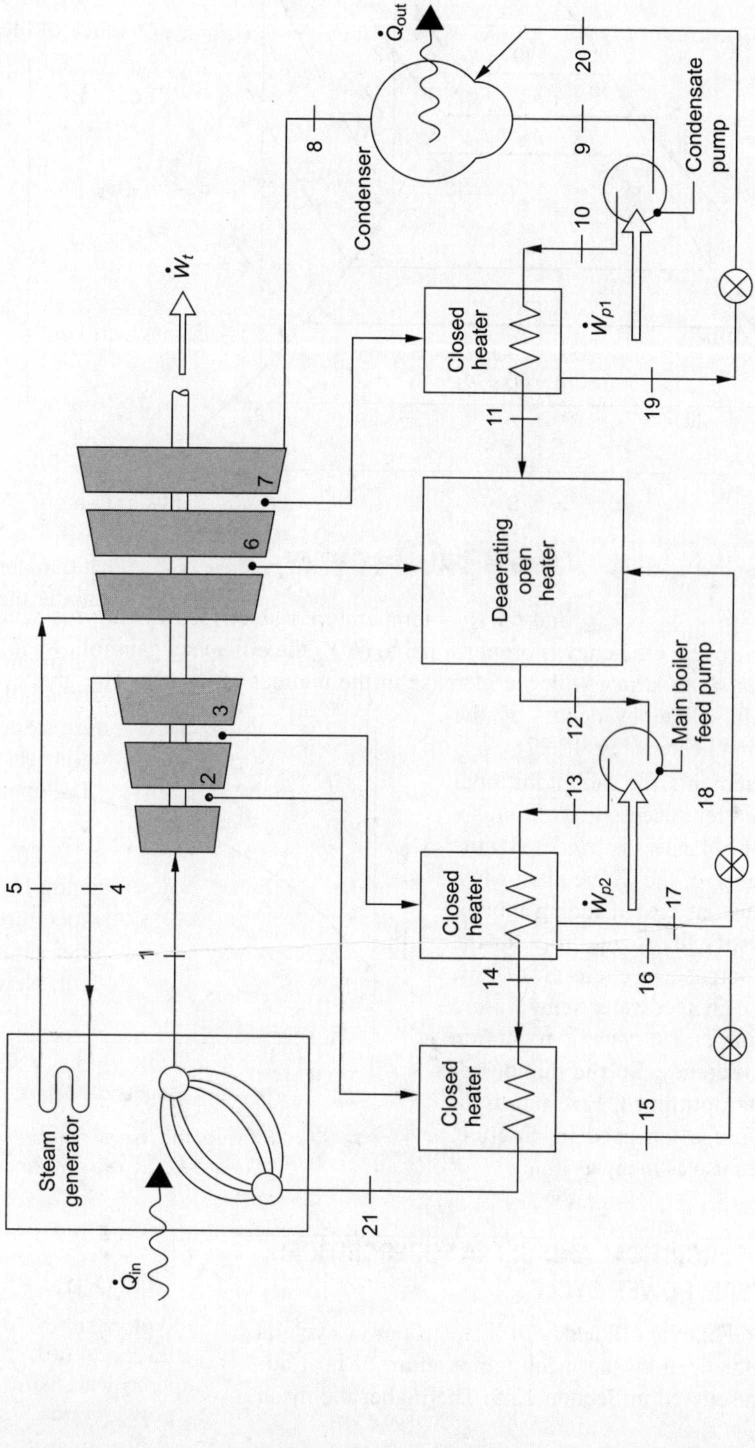

(a)

Fig. 12.26(a) Example of a power plant layout

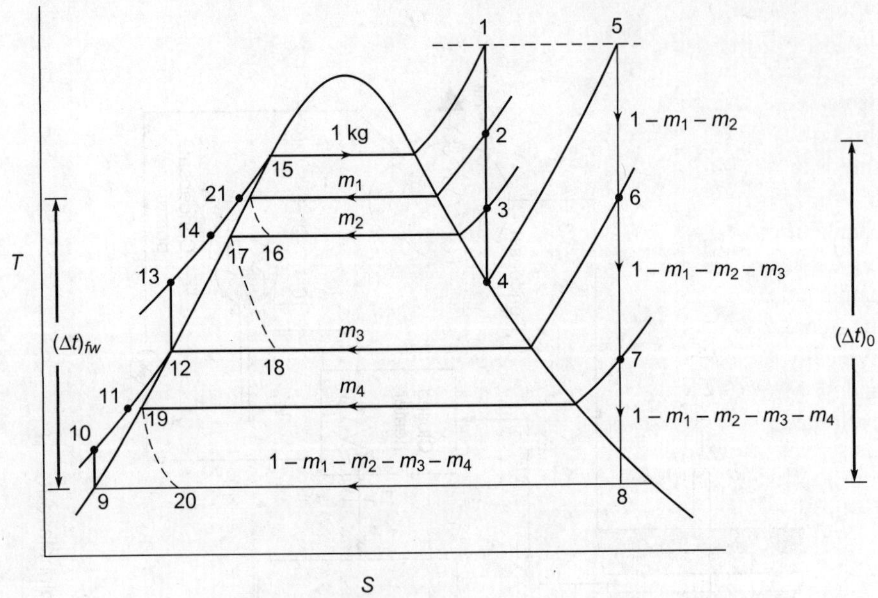

Fig. 12.26(b) *T-s* diagram

If $(\Delta t)_0 = t_{\text{boiler sat}} - t_{\text{cond}}$ and $(\Delta t)_{fw}$ = temperature rise of feedwaters, it is seen that since the cycle efficiency is proportional to $(\Delta t)_{fw}$ the efficiency gain follows *the law of diminishing return* with the increase in the number of heaters. The greatest increment in efficiency occurs by the use of the first heater. (Fig. 12.27)

The increments for each additional heater thereafter successively diminish. The number of heaters is fixed up by the energy balance of the whole plant when it is found that the cost of adding another does not justify the saving in Q_1 or the marginal increase in cycle efficiency. An increase in feedwater temperature may, in some cases, cause a reduction in boiler efficiency. So the number of heaters gets optimized. Five points of extraction are often used in practice. Some cycles use as many as nine.

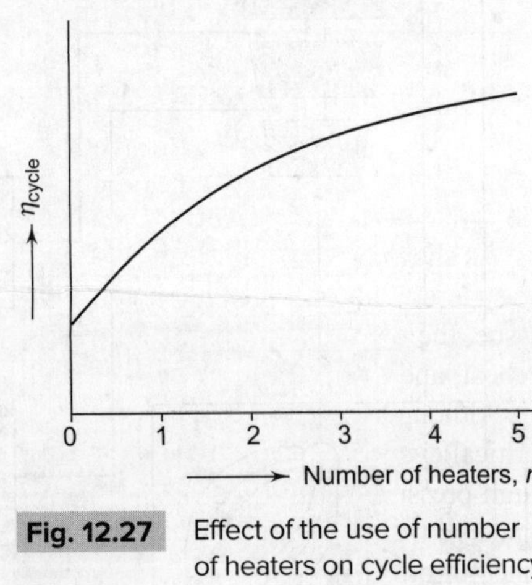

Fig. 12.27 Effect of the use of number of heaters on cycle efficiency

12.11 SUPERCRITICAL AND ULTRA SUPERCRITICAL STEAM POWER CYCLE

> **LO 12.11**
> Explain super critical and ultra supercritical steam power cycles

The thermodynamic efficiency of a steam power cycle is directly influenced by the mean temperature of heat addition, as discussed in Section 12.5. The higher the mean temperature of heat addition, the greater the efficiency of the cycle. Both increasing the pressure and temperature of steam generation serve to raise the mean temperature of heat addition. As pressure increases, the latent heat required for phase change decreases, ultimately becoming zero at the critical point. The isobar passing through this critical point is referred to as the critical isobar, and any isobar above this pressure is known as a supercritical isobar.

Above the critical pressure, steam generation occurs without any visible phase change. In supercritical steam generation, there is no defined saturated state, allowing the steam temperature to increase indefinitely. This offers significant thermodynamic potential to enhance the efficiency of a steam power cycle. Boilers that generate steam above the critical pressure of 221.2 bar are known as supercritical boilers. Typically, ultra-supercritical steam power cycles operate at pressures between 25 and 35 MPa, with steam temperatures exceeding 600 °C. The corresponding power cycles are referred to as supercritical steam power cycles, as illustrated in Fig. 12.28.

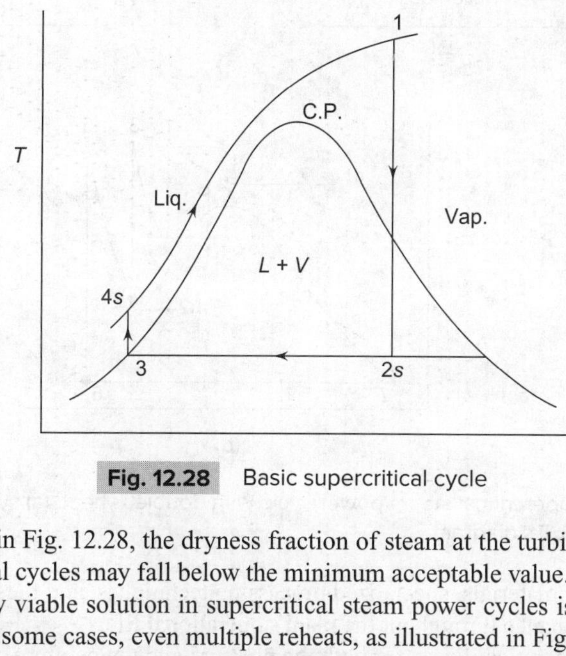

Fig. 12.28 Basic supercritical cycle

As shown in Fig. 12.28, the dryness fraction of steam at the turbine exit (i.e., 2s) in supercritical cycles may fall below the minimum acceptable value. To address this issue, the only viable solution in supercritical steam power cycles is to incorporate reheat, and in some cases, even multiple reheats, as illustrated in Fig. 12.29.

Although supercritical steam cycles consistently offer higher efficiency than subcritical cycles, they present several operational challenges due to the combination of high pressure and temperature of the steam. In supercritical boilers, water does not undergo the traditional evaporation process. As a result, these boilers are designed as once-through or drum-less systems, which eliminates the possibility of a conventional boiler blowdown to remove impurities from the feedwater. Consequently, supercritical boilers require feedwater of exceptionally high purity to prevent the accumulation of impurities in the boiler tubes, which in turn increases the cost of water treatment.

Furthermore, the materials used in supercritical boilers must meet higher standards to withstand the elevated temperatures and pressures, resulting in a higher initial capital cost compared to subcritical plants. This increase is due to the use of

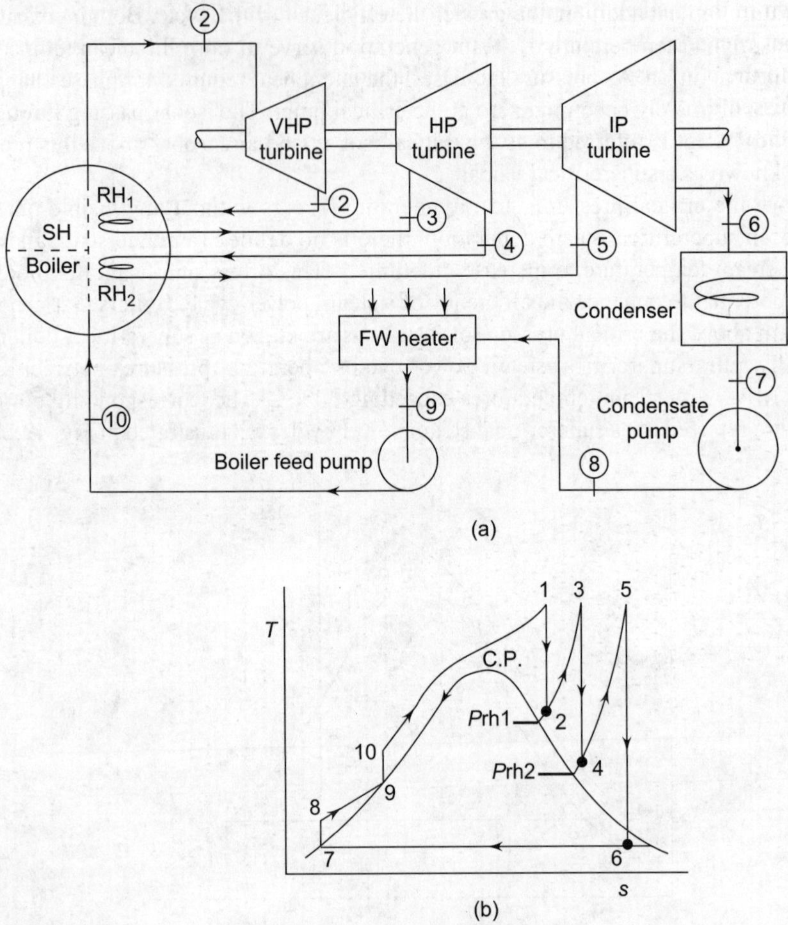

Fig. 12.29 Supercritical steam power cycle with double reheat: (a) system layout, (b) *T-s* diagram

more expensive materials, specialized processing techniques, and the need for more frequent maintenance throughout the plant's operational life.

While the theoretical limits for both the pressure and temperature of steam in supercritical cycles are virtually unrestricted, practical manufacturing and operational constraints impose limits. Current systems that operate at pressures up to 225 bar and temperatures up to 593°C are classified as supercritical cycles. Systems operating beyond these thresholds are considered ultra-supercritical.

12.12 EXERGY ANALYSIS OF VAPOUR POWER CYCLES

> **LO 12.12**
> Explain exergy analysis of vapour power cycles

Let the heating for steam generation in the boiler unit be provided by a stream of hot gases produced by burning of a fuel (Fig. 12.30(a)). The distribution of input energy is shown in the Sankey diagram (12.30(b)) which indicates that only about 30% of the input energy to the simple ideal plant is converted to shaft work and about 60% is lost to the condenser. The exergy analysis, however, gives a different distribution as discussed below.

Assuming that the hot gases are at atmospheric pressure, the exergy input is

$$a_{f1} = w_g\, c_{pg}\left[T_i - T_0 - T_0 \ln \frac{T_i}{T_0}\right]$$

$$= w_g\, c_{pg}\, T_0\left[\frac{T_i}{T_0} - 1 - \ln \frac{T_i}{T_0}\right]$$

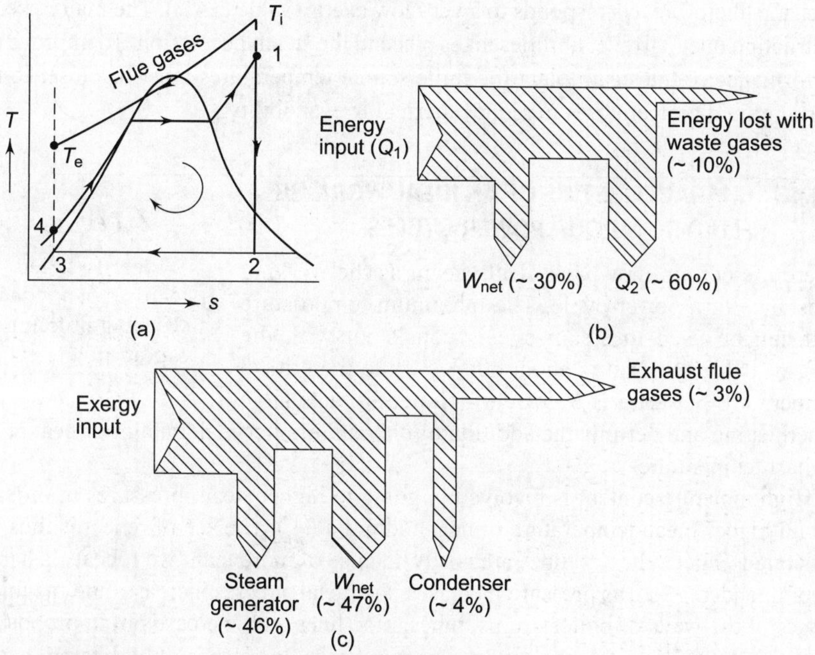

Fig. 12.30 (a) *T-s* diagram, (b) Sankey diagram, (c) Grassman diagram

Similarly, the exergy loss rate with the exhaust stream is

$$a_{f2} = w_g\, c_{pg}\, T_0\left[\frac{T_e}{T_0} - 1 - \ln \frac{T_e}{T_0}\right]$$

Net exergy input rate in the steam generation process

$$a_{i1} = a_{f1} - a_{f2}$$

The exergy utilisation rate in the steam generation is

$$a_{fu} = w_s\,[(h_1 - h_4)] - T_0(s_1 - s_4)]$$

Rate of exergy loss in the steam generator

$$I = a_{fi} - a_{fu}$$

The useful mechanical power output:

$$= W_{\text{net}} = w_s[(h_1 - h_2) - (h_4 - h_3)]$$

Exergy flow rate of the wet steam to the condenser:

$$a_{fc} = w_s\,[(h_2 - h_3) - T_0(s_2 - s_3)]$$

Second law efficiency, $\eta_{\text{II}} = \dfrac{W_{\text{net}}}{a_{f_1} - a_{f_2}}$

Exergy flow or Grassman diagram is shown in Fig. 12.30 (c). The energy disposition diagram (b) shows that the major energy loss (~60%) takes place in the condenser. This energy rejection, however, occurs at a temperature close to the ambient temperature, and, therefore, corresponds to a very low exergy value (~4%). The major exergy destruction due to irreversibilities takes place in the steam generation. To improve the performance of the steam plant the finite source temperatures must be closer to the working fluid temperatures to reduce thermal irreversibility.

12.13 CHARACTERISTICS OF AN IDEAL WORKING FLUID IN VAPOUR POWER CYCLES

LO 12.13

Identify salient features of an ideal working fluid for a vapour power cycle

There are certain drawbacks with steam as the working substance in a power cycle. The maximum temperature that can be used in steam cycles consistent with the best available material is about 600°C, while the critical temperature of steam is 375°C, which necessitates large superheating and permits the addition of only an infinitesimal amount of heat at the highest temperature.

High moisture content is involved in going to higher steam pressures in order to obtain higher mean temperature of heat addition (T_{m1}). The use of reheat is thus necessitated. Since reheater tubes are costly, the use of more than two reheats is hardly recommended. Also, as pressure increases, the metal stresses increase, and the thicknesses of the walls of boiler drums, tubes, pipe lines, etc. increase not in proportion to pressure increase, but much faster, because of the prevalence of high temperature.

It may be noted that high T_{m1} is only desired for high cycle efficiency. High pressures are only forced by the characteristics (weak) of steam.

If the lower limit is now considered, it is seen that at the heat rejection temperature of 40°C, the saturation pressure of steam is 0.075 bar, which is considerably lower than atmospheric pressure. The temperature of heat rejection can be still lowered by using some refrigerant as a coolant in the condenser. The corresponding vacuum will be still higher, and to maintain such low vacuum in the condenser is a big problem.

It is the low temperature of heat rejection that is of real interest. The necessity of a vacuum is a disagreeable characteristic of steam.

The saturated vapour line in the *T-s* diagram of steam is sufficiently inclined, so that when steam is expanded to lower pressures (for higher turbine output as well as cycle efficiency), it involves more moisture content, which is not desired from the consideration of the erosion of turbine blades in later stages.

The desirable characteristics of the working fluid in a vapour power cycle to obtain best thermal efficiency are given as follows:

1. The fluid should have a high critical temperature so that the saturation pressure at the maximum permissible temperature (metallurgical limit) is relatively low. It should have a large enthalpy of evaporation at that pressure.
2. The saturation pressure at the temperature of heat rejection should be above atmospheric pressure so as to avoid the necessity of maintaining vacuum in the condenser.
3. The specific heat of liquid should be small so that little heat transfer is required to raise the liquid to the boiling point.
4. The saturated vapour line of the *T-s* diagram should be steep, very close to the turbine expansion process so that excessive moisture does not appear during expansion.
5. The freezing point of the fluid should be below room temperature, so that it does not get solidified while flowing through the pipelines.
6. The fluid should be chemically stable and should not contaminate the materials of construction at any temperature.
7. The fluid should be nontoxic, noncorrosive, not excessively viscous and low in cost.

The characteristics of such an ideal fluid are approximated in the *T-s* diagram as shown in Fig. 12.31. Some superheat is desired to reduce piping losses and improve turbine efficiency. The thermal efficiency of the cycle is very close to the Carnot efficiency.

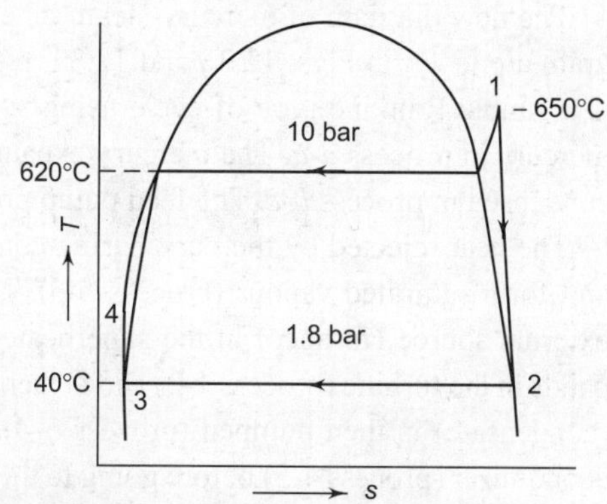

Fig. 12.31 *T-s* diagram of an ideal working fluid for a vapour power cycle

12.14 BINARY VAPOUR CYCLES

> **LO 12.14**
> Explain binary (two-fluid) vapour power cycles

No single fluid can meet all the requirements as mentioned above. Although in the overall evaluation, water is better than any other working fluid, however, in the high temperature range, there are a few better fluids, and notable among them are (a) diphenyl ether, $(C_6H_5)_2O$, (b) aluminium bromide, Al_2Br_6 and (c) mercury and other liquid metals like sodium or potassium. Among these, only mercury has actually been used in practise. Diphenyl ether could be considered, but it has not yet been used because, like most organic substances, it decomposes gradually at high temperatures. Aluminium bromide is a possibility and yet to be considered.

When $p = 12$ bar, the saturation temperature for water, aluminium bromide and mercury are 187°C, 482.5°C and 560°C, respectively. Mercury is thus a better fluid in the high temperature range, because at high temperature, its vapourisation pressure is relatively low. Its critical pressure and temperature are 1080 bar and 1460°C, respectively.

But in the low temperature range, mercury is unsuitable, because its saturation pressure becomes exceedingly low and it would be impractical to maintain such a high vacuum in the condenser. At 30°C, the saturation pressure of mercury is only 2.7×10^{-4} cm Hg. Its specific volume at such a low pressure is very large, and it would be difficult to accommodate such a large volume flow.

For this reason, mercury vapour leaving the mercury turbine is condensed at a higher temperature, and the heat released during the condensation of mercury is utilised in evaporating water to form steam to operate on a conventional turbine.

Thus in the binary (or two-fluid) cycle, two cycles with different working fluids are coupled in series, the heat rejected by one being utilised in the other.

The flow diagram of mercury-steam binary cycle and the corresponding *T-s* diagram are given in Figs. 12.32 and 12.33, respectively. The mercury cycle, *a-b-c-d*, is a simple Rankine type of cycle using saturated vapour. Heat is supplied to the mercury in process *d-a*. The mercury expands in a turbine (process *a-b*) and is then condensed in process *b-c*. The feed pump process, *c-d*, completes the cycle.

The heat rejected by mercury during condensation is transferred to boiled water and form saturated vapour (process 5-6). The saturated vapour is heated from the external source (furnace) in the superheater (process 6-1). Superheated steam expands in the turbine (process 1-2) and is then condensed (process 2-3). The feedwater (condensate) is then pumped (process 3-4), heated till it is saturated liquid in the economizer (process 4-5) before going to the mercury condenser-steam boiler, where the latent heat is absorbed. In an actual plant, the steam cycle is always a regenerative cycle, but for the sake of simplicity, this complication has been omitted.

Let m represent the flow rate of mercury in the mercury cycle per kg of steam circulating in the steam cycle. Then for 1 kg of steam,

$$Q_1 = m(h_a - h_d) + (h_1 - h_6) + (h_5 - h_4)$$

$$Q_2 = h_2 - h_3$$

$$W_T = m(h_a - h_b) + (h_1 - h_2)$$

$$W_P = m(h_d - h_c) + (h_4 - h_3)$$

$$\eta_{\text{cycle}} = \frac{Q_1 - Q_2}{Q_1} = \frac{W_T - W_P}{Q_1}$$

$$\text{and steam rate} = \frac{3600}{W_T - W_P} \text{ kg/kW h}$$

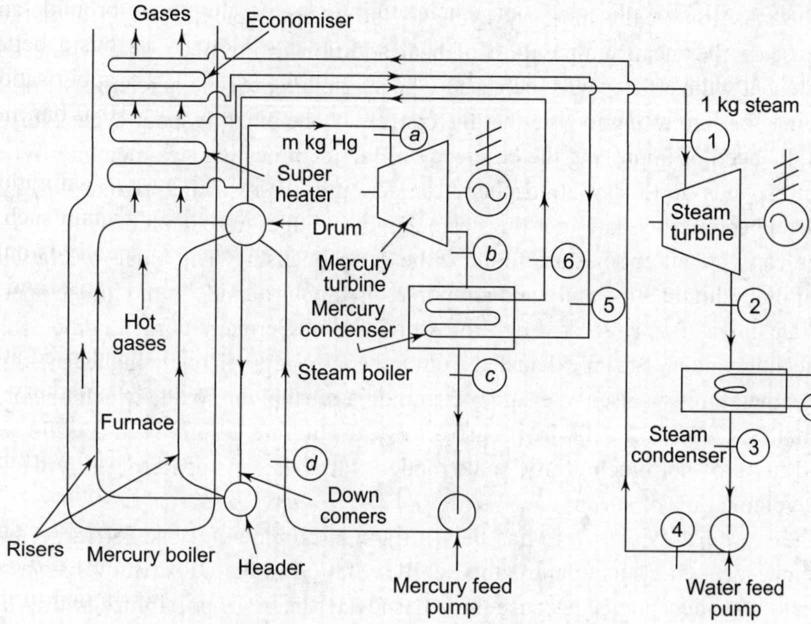

Fig. 12.32 Mercury-steam plant flow diagram

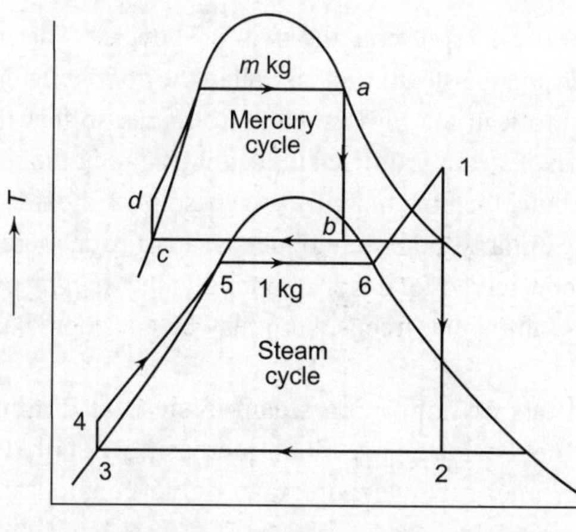

Fig. 12.33 Mercury-steam binary cycle

The energy balance of the mercury condenser-steam boiler gives

$$m(h_b - h_c) = h_6 - h_5$$

$\therefore$

$$m = \frac{h_6 - h_5}{h_b - h_c} \text{ kg Hg/kg } H_2O$$

To vapourise 1 kg of water, 7 to 8 kg of mercury must condense.

The addition of the mercury cycle to the steam cycle results in a marked increase in the mean temperature of heat addition to the plant as a whole and consequently the efficiency is increased. The maximum pressure is relatively low.

It may be interesting to note that the concept of the binary vapour cycle evolved from the need of improving the efficiency of the reciprocating steam engine. When steam expands up to, say, atmospheric temperature, the resultant volume flow rate of steam becomes too large for the steam engine cylinder to accommodate. So most of the early steam engines are found to be noncondensing. The binary cycle with steam in the high temperature and ammonia or sulphur dioxide in the low temperature range, was first suggested by Professor Josse of Germany in the middle of the nineteenth century. Steam exhausted from the engine at a relatively higher pressure and temperature was used to evaporate ammonia or sulphur dioxide which operated on another cycle. But with the progress in steam turbine design, such a cycle was found to be of not much utility, since modern turbines can cope efficiently with a large volume flow of steam.

The mercury-steam cycle has been in actual commercial use for more than three decades. One such plant is the Schiller Station in the USA. But it has never attained wide acceptance because there has always been the possibility of improving steam cycles by increasing pressure and temperature, and by using reheat and regeneration. Over and above, mercury is expensive, limited in supply and highly toxic.

The mercury-steam cycle represents the two-fluid cycles. The mercury cycle is called the *topping cycle* and the steam cycle is called the *bottoming cycle*. If a sulphur dioxide cycle is added to it in the low temperature range, so that the heat released during the condensation of steam is utilised in forming sulphur dioxide vapour which expands in another turbine, then the mercury-steam-sulphur dioxide cycle is a three-fluid or tertiary cycle. Similarly, other liquid metals, apart from mercury, like sodium or potassium, may be considered for a working fluid in the topping cycle. Apart from SO_2, other refrigerants (ammonia, freons, etc.) may be considered as working fluids for the bottoming cycle.

Since the possibilities of improving steam cycles are diminishing, and the incentives to reduce fuel cost are very much increasing, coupled cycles, like the mercury-steam cycle, may receive more favourable consideration in the near future.

12.15 THERMODYNAMICS OF COUPLED CYCLES

> **LO 12.15**
> Demonstrate overall efficiency of a coupled cycle

If two cycles are coupled in series where heat lost by one is absorbed by the other (Fig. 12.34), as in the mercury-steam binary cycle, let η_1 and η_2 be the efficiencies of the topping and bottom cycles, respectively, and η be the overall efficiency of the combined cycle.

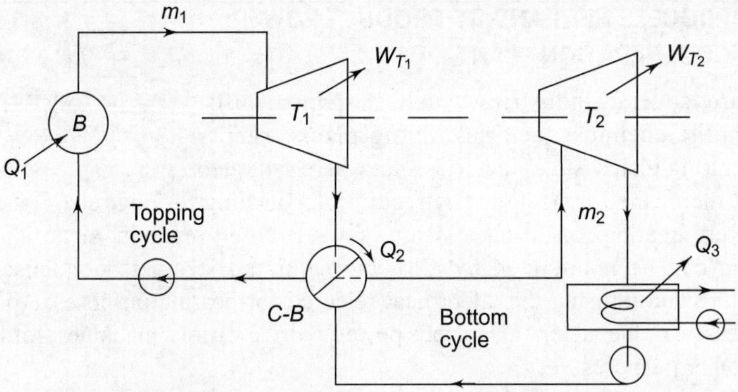

Fig. 12.34 Two vapour cycles coupled in series

$$\eta_1 = 1 - \frac{Q_2}{Q_1} \text{ and } \eta_2 = 1 - \frac{Q_3}{Q_2}$$

or $$Q_2 = Q_1(1 - \eta_1) \text{ and } Q_3 = Q_2(1 - \eta_2)$$

Now $$\eta = 1 - \frac{Q_3}{Q_1} = 1 - \frac{Q_2(1-\eta_2)}{Q_1}$$

$$= 1 - \frac{Q_1(1-\eta_1)(1-\eta_2)}{Q_1}$$

$$= 1 - (1 - \eta_1)(1 - \eta_2)$$

If there are *n* cycles coupled in series, the overall efficiency would be given by

$$\eta = 1 - \overset{\pi}{\underset{i=1}{\pi}} (1 - \eta_i)$$

i.e. $$\eta = 1 - (1 - \eta_1)(1 - \eta_2)(1 - \eta_3) \ldots (1 - \eta_n)$$

or $$1 - \eta = (1 - \eta_1)(1 - \eta_2)(1 - \eta_3) \ldots (1 - \eta_n)$$

∴ Total loss = Product of losses in all the cycles.

For two cycles coupled in series

$$\eta = 1 - (1 - \eta_1)(1 - \eta_2)$$

$$= 1 - (1 - \eta_1 - \eta_2 + \eta_1 \eta_2)$$

$$= \eta_1 + \eta_2 - \eta_1 \eta_2$$

or $$\eta = \eta_1 + \eta_2 - \eta_1 \eta_2$$

This shows that the overall efficiency of two cycles coupled in series equals the sum of the individual efficiencies minus their product.

By combining two cycles in series, even if individual efficiencies are low, it is possible to have a fairly high combined efficiency, which cannot be attained by a single cycle.

For example, if $\eta_1 = 0.50$ and $\eta_2 = 0.40$

$$\eta = 0.5 + 0.4 - 0.5 \times 0.4 = 0.70$$

It is almost impossible to achieve such a high efficiency in a single cycle.

12.16 PROCESS HEAT AND BY-PRODUCT POWER: COGENERATION PLANT

LO 12.16 Analyze cogeneration

There are several industries, such as paper mills, textile mills, chemical factories, dying plants, rubber manufacturing plants, sugar factories, etc. where saturated steam at the desired temperature is required for heating, drying, etc. For constant temperature heating (or drying), steam is a very good medium, since isothermal condition can be maintained by allowing saturated steam to condense at that temperature and utilising the latent heat released for heating purposes. Apart from the process heat, the factory also needs power to drive various machines, for lighting, and for other purposes.

Formerly it was the practice to generate steam for power purposes at a moderate pressure and to generate separately saturated steam for process work at a pressure which gave the desired heating temperature. Having two separate units for process heat and power is wasteful, *for of the total* heat supplied to the steam for power purposes, a greater part will normally be carried away by the cooling water in the condenser.

By modifying the initial steam pressure and exhaust pressure, it is possible to generate the required power and make available for process work the required quantity of exhaust steam at the desired temperature. In Fig. 12.35, the exhaust steam from the turbine is utilised for process heating, the process heater replacing the condenser of the ordinary Rankine cycle. The pressure at exhaust from the turbine is the saturation pressure corresponding to the temperature desired in the process heater. Such a turbine is called a *back pressure turbine*. A plant producing both power and process heat is sometimes known as a *cogeneration plant*. When the process steam is the basic need, and the power is produced incidentally as a by-product, the cycle is sometimes called a by-product power cycle. Figure 12.36 shows the *T-s* plot of such a cycle. If W_T is the turbine output in kW, Q_H the process heat required in kJ/h and w is the steam flow rate in kg/h

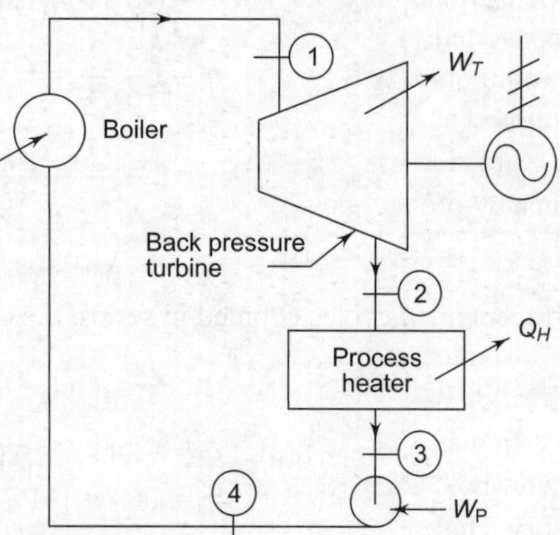

Fig. 12.35 Back pressure turbine

$$W_T \times 3600 = w(h_1 - h_2)$$

and

$$w(h_2 - h_3) = Q_H$$

$$\therefore \quad W_T \times 3600 = \frac{Q_H}{h_2 - h_3}(h_1 - h_2)$$

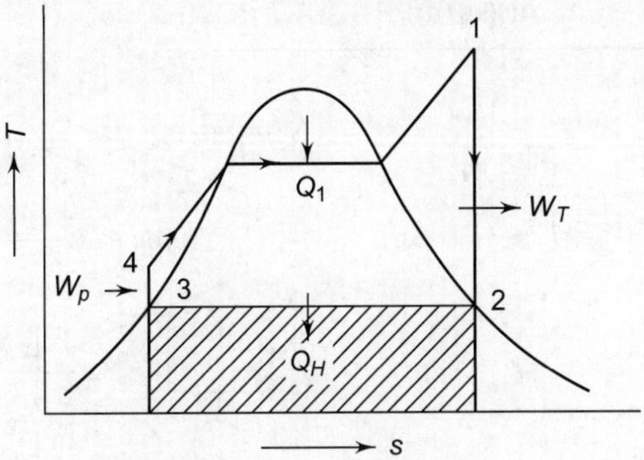

Fig. 12.36 By-product power cycle

or $$Q_H = \frac{W_T \times 3600 \times (h_2 - h_3)}{h_1 - h_2} \text{ kJ/h}$$

Of the total energy input Q_1 (as heat) to the by-product cycle, W_T part of it only is converted into shaft work (or electricity). The remaining energy $(Q_1 - W_T)$, which would otherwise have been a waste, as in the Rankine cycle (by the second law), is utilised as process heat.

Fraction of energy (Q_1) utilised in the form of work (W_T), and process heat (Q_H) in a by-product power cycle

$$= \frac{W_T + Q_H}{Q_1}$$

Condenser loss, which is the biggest loss in a steam plant, is here zero, and the fraction of energy utilised is very high.

In many cases, the power available from the back pressure turbine through which the whole of the heating steam flows is appreciably less than that required in the factory. This may be due to relatively high back pressure, or small heating requirement, or both. *Pass-out turbines* or extraction turbines are employed in these cases, where a certain quantity of steam is continuously extracted for heating purposes at the desired temperature and pressure. (Figs. 12.37 and 12.38).

$$Q_1 = w(h_1 - h_8) \text{ kJ/h}$$
$$Q_2 = (w - w_1)(h_3 - h_4) \text{ kJ/h}$$
$$Q_H = w_1(h_2 - h_6) \text{ kJ/h}$$
$$W_T = w(h_1 - h_2) + (w - w_1)(h_2 - h_3) \text{ kJ/h}$$
$$W_P = (w - w_1)(h_5 - h_4) + w_1(h_7 - h_6) \text{ kJ/h}$$
$$w_1 h_7 + (w - w_1)h_5 = w \times h_8$$

where w is the boiler capacity (kg/h) and w_1 is the steam flow rate required (kg/h) at the desired temperature for process heating.

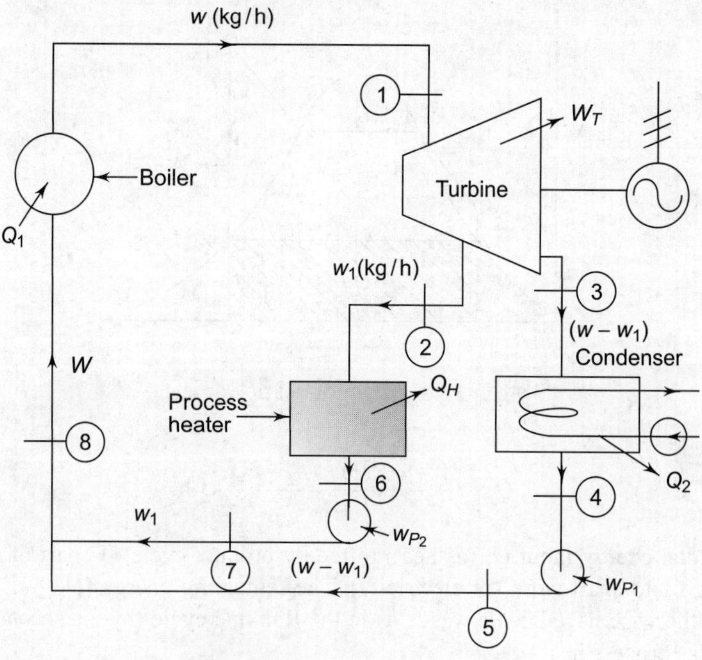

Fig. 12.37 Pass-out turbine

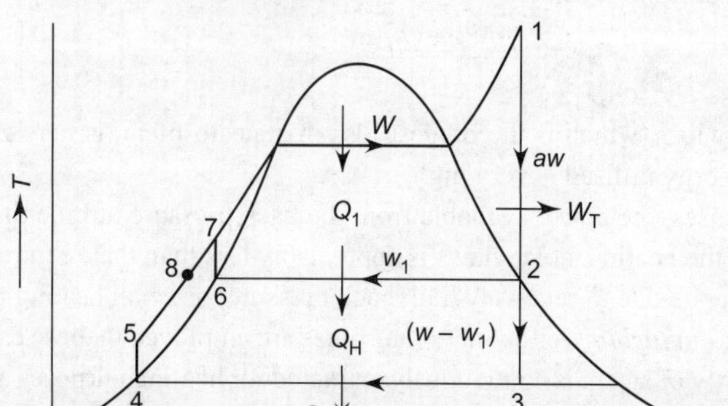

Fig. 12.38 T-s diagram of power and process heat plant

12.17 EFFICIENCIES IN STEAM POWER PLANT

LO 12.17 State different component efficiencies and the overall efficiency of a steam power plant

For the steady flow operation of a turbine, neglecting changes in K.E. and P.E. (Figs. 12.39 and 12.40).

Maximum or ideal work output per unit mass of steam

$$(W_T)_{\max} = (W_T)_{\text{ideal}} = h_1 - h_{2s}$$

= Reversible and adiabatic enthalpy drop in turbine

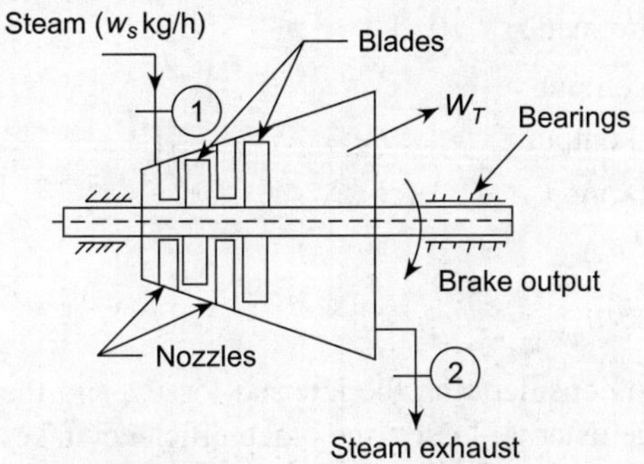

Fig. 12.39 Efficiencies in a steam turbine

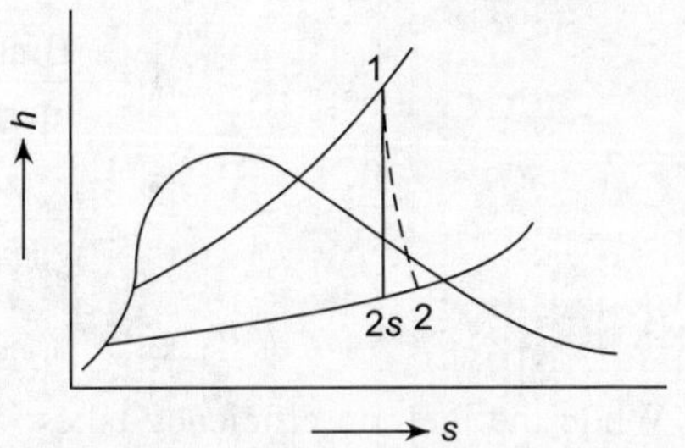

Fig. 12.40 Internal efficiency of a steam turbine

This work is, however, not obtainable, since no real process is reversible. The expansion process is accompanied by irreversibilities. The actual final state 2 can be defined, since the temperature, pressure and quality can be found by actual measurement. The actual path 1–2 is not known and its nature is immaterial, since the work output is here being expressed in terms of the change of a property, enthalpy. Accordingly, the work done by the turbine in irreversible adiabatic expansion from 1 to 2 is

$$(W_T)_{\text{actual}} = h_1 - h_2$$

This work is known as *internal work*, since only the irreversibilities within the flow passages of turbine are affecting the state of steam at the turbine exhaust.

∴ Internal output = Ideal output – Friction and other losses within the turbine casing

If w_s is the steam flow rate in kg/h

$$\text{Internal output} = w_s(h_1 - h_2) \text{ kJ/h}$$

$$\text{Ideal output} = w_s\,(h_1 - h_{2s}) \text{ kJ/h}$$

The *internal efficiency* of turbine is defined as

$$\eta_{\text{internal}} = \frac{\text{Internal output}}{\text{Ideal output}} = \frac{h_1 - h_2}{h_1 - h_{2s}}$$

Work output available at the shaft is less than the internal output because of the external losses in the bearings, etc.

∴ brake output or shaft output

= Internal output – External losses

= Ideal output – Internal and External losses

= (kW × 3600 kJ/h)

The *brake efficiency* of turbine is defined as

$$\eta_{\text{brake}} = \frac{\text{Brake output}}{\text{Ideal output}}$$

$$= \frac{\text{kW} \times 3600}{w_s\,(h_1 - h_{2s})}$$

The *mechanical efficiency* of turbine is defined as

$$\eta_{mech} = \frac{\text{Brake output}}{\text{Internal output}}$$

$$= \frac{\text{kW} \times 3600}{w_s(h_1 - h_2)}$$

$$\therefore \qquad \eta_{brake} = \eta_{internal} \times \eta_{mech}$$

While the internal efficiency takes into consideration the internal losses, and the mechanical efficiency considers only the external losses, the brake efficiency takes into account both the internal and external losses (with respect to turbine casing). The generator (or *alternator*) efficiency is defined as

$$\eta_{generator} = \frac{\text{Output at generator terminals}}{\text{Brake output of turbine}}$$

The *efficiency of the boiler* is defined as

$$\eta_{boiler} = \frac{\text{Energy utilised}}{\text{Energy supplied}} = \frac{w_s(h_1 - h_4)}{w_f \times \text{C.V.}}$$

where w_f is the fuel burning rate in the boiler (kg/h) and C.V. is the calorific value of the fuel (kJ/kg), i.e. the heat energy released by the complete combustion of unit mass of fuel.

The power plant is an *energy converter* from fuel to electricity (Fig. 12.41), and the *overall efficiency* of the plant is defined as

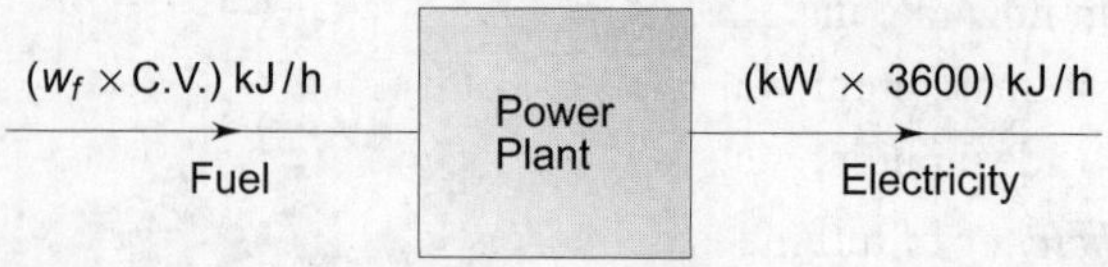

Fig. 12.41 Power plant–An energy converter from fuel to electricity

$$\eta_{overall} = \eta_{plant} = \frac{\text{kW} \times 3600}{w_f \times \text{C.V.}}$$

This may be expressed as follows:

$$\eta_{overall} = \frac{\text{kW} \times 3600}{w_f \times \text{C.V.}} = \frac{w_s(h_1 - h_4)}{w_f \times \text{C.V.}} \times \frac{w_s(h_1 - h_2)}{w_s(h_1 - h_4)}$$

$$\times \frac{\text{Brake output}}{w_s(h_1 - h_2)} \times \frac{\text{kW} \times 3600}{\text{Brake output}}$$

or $$\eta_{overall} = \eta_{boiler} \times \eta_{cycle} \times \eta_{turbine\,(mech)} \times \eta_{generator}$$

where pump work has been neglected in the expression for cycle efficiency.

SOLVED EXAMPLES

Example 12.1 *Determine the work required to compress steam isentropically from 1 bar to 10 bar, assuming that at the initial state the steam exists as (a) saturated liquid, and (b) saturated vapour. Neglect changes in kinetic and potential energies. What conclusion do you derive from this example?* E [LO 12.2]

Solution The compression processes are shown in Fig. 12.42.

(a) Steam is a saturated liquid initially, and its specific volume is

$$v_1 = (v_f)_{1\text{bar}} = 0.001043 \text{ m}^3/\text{kg}$$

Since liquid is incompressible, v_1 remains constant.

$$W_{\text{rev}} = \int_1^2 v\, \text{d}p = v_1(p_1 - p_2) = 0.001043\,(1 - 10) \times 10^2$$

$$= \mathbf{-\,0.9387\ kJ/kg.}$$

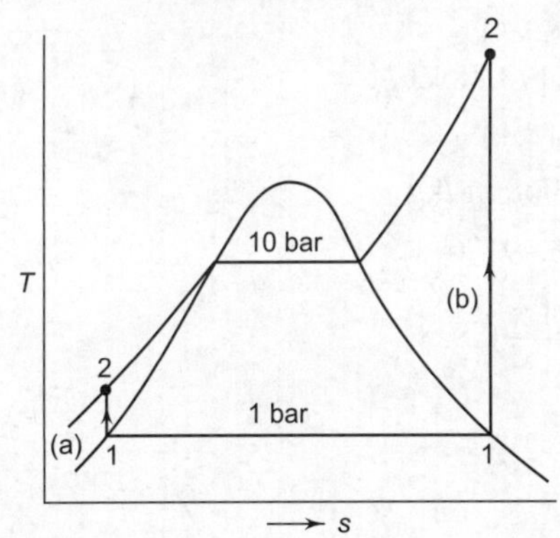

Fig. 12.42 Compression of steam isentropically

(b) Steam is a saturated vapour initially and remains a vapour during the entire compression process. Since the specific volume of a gas changes considerably during a compression process, we need to know how v varies with p to perform the integration $-\int v\, \text{d}p$. This relation is not readily available. But for an isentropic process, it is easily obtained from the property relation

$$T\text{d}s = \text{d}h - v\text{d}p = 0$$

or $$v\, \text{d}p = \text{d}h$$

$$W_{\text{rev}} = \int_1^2 v\, \text{d}p = -\int_1^2 \text{d}h = h_1 - h_2$$

From steam tables,

$$h_1 = (h_g)_{1\text{bar}} = 2675.5 \text{ kJ/kg}$$

$$s_1 = (s_g)_{1\text{bar}} = 7.3594 \text{ kJ/kg K} = s_2$$

For $p = 10$ bar $= 1$ MPa and $s = 7.3594$ kJ/kg K, by interpolation

$$h_2 = 3195.5 \text{ kJ/kg}$$

$$W_{\text{rev}} = 2675.5 - 3195.5 = \mathbf{-520 \text{ kJ/kg}}$$

It is thus observed that compressing steam in vapour form would require over 500 times more work than compressing it in liquid form for the same pressure rise.

Example 12.2 *Steam at 20 bar, 360°C is expanded in a steam turbine to 0.08 bar. It then enters a condenser, where it is condensed to saturated liquid water. The pump feeds back the water into the boiler. (a) Assuming ideal processes, find per kg of steams of the net work and the cycle efficiency. (b) If the turbine and the pump have each 80% efficiency, find the percentage reduction in the net work and cycle efficiency.* E **[LO 12.2]**

Solution The property values at different state points (Fig. 12.43) found from the steam tables are given below.

$$h_1 = 3159.3 \text{ kJ/kg} \qquad s_1 = 6.9917 \text{ kJ/kg K}$$

$$h_3 = h_{fp2} = 173.88 \text{ kJ/kg} \qquad s_3 = s_{fp2} = 0.5926 \text{ kJ/kg K}$$

$$h_{fgp2} = 2403.1 \text{ kJ/kg} \qquad s_{gp2} = 8.2287 \text{ kJ/kg K}$$

$$v_{fp2} = 0.001008 \text{ m}^3\text{/kg} \qquad \therefore \quad s_{fgp2} = 7.6361 \text{ kJ/kg K}$$

Now $\quad s_1 = s_{2s} = 6.9917 = s_{fp2} + x_{2s}\, s_{fgp2} = 0.5926 + x_2 \cdot 7.6361$

$$\therefore \quad x_{2s} = \frac{6.3991}{7.6361} = 0.838$$

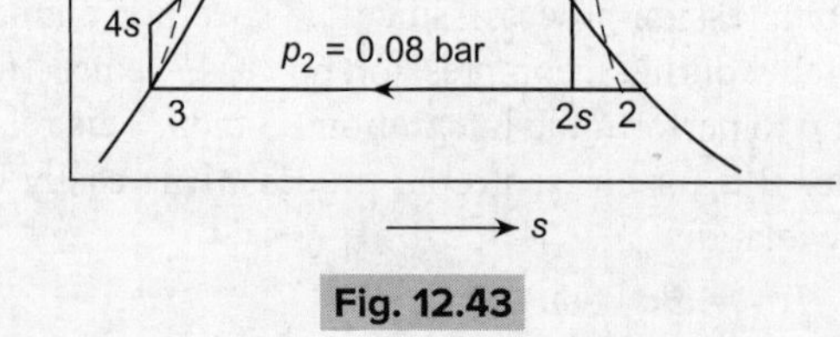

Fig. 12.43

$$\therefore \quad h_{2s} = h_{fp2} + x_{2s} h_{fgp2} = 173.88 + 0.838 \times 2403.1$$

$$= 2187.68 \text{ kJ/kg}$$

(a) $W_P = h_{4s} - h_3 = v_{fp2}\,(p_1 - p_2) = 0.001008 \dfrac{\text{m}^3}{\text{kg}} \times 19.92 \times 100 \dfrac{\text{kN}}{\text{m}^2}$

$$= 2.008 \text{ k/kg}$$

$$h_{4s} = 175.89 \text{ kJ/kg}$$

$$W_T = h_1 - h_{2s}$$

$$= 3159.3 - 2187.68 = 971.62 \text{ kJ/kg}$$

$$\therefore \quad W_{net} = W_T - W_P = \mathbf{969.61 \text{ kJ/kg}}$$

$$Q_1 = h_1 - h_{4s} = 3159.3 - 175.89$$

$$= 2983.41 \text{ kJ/kg}$$

$$\therefore \quad \eta_{cycle} = \frac{W_{net}}{Q_1} = \frac{969.61}{2983.41} = \mathbf{0.325, \text{ or } 32.5\%}$$

(b) If $\eta_P = 80\%$, and $\eta_T = 80\%$

$$W_P = \frac{2.008}{0.8} = 2.51 \text{ kJ/kg}$$

$$W_T = 0.8 \times 971.62 = 777.3 \text{ kJ/kg}$$

$$\therefore \quad W_{net} = W_T - W_P = 774.8 \text{ kJ/kg}$$

∴ % reduction in work output

$$= \frac{969.61 - 774.8}{969.61} \times 100 = \mathbf{20.1\%}$$

$$h_{4s} = 173.88 + 2.51 = 176.39 \text{ kJ/kg}$$

$$\therefore \quad Q_1 = 3159.3 - 176.39 = 2982.91 \text{ kJ/kg}$$

$$\therefore \quad \eta_{cycle} = \frac{774.8}{2982.91} = 0.2597, \text{ or } 25.97\%$$

∴ % reduction in cycle efficiency

$$= \frac{0.325 - 0.2597}{0.325} \times 100 = \mathbf{20.1\%}$$

Example 12.3 *A cyclic steam power plant is to be designed for a steam temperature at turbine inlet of 360°C and an exhaust pressure of 0.08 bar. After isentropic expansion of steam in the turbine, the moisture content at the turbine exhaust is not to exceed 15%. Determine the greatest allowable steam pressure at the turbine inlet, and calculate the Rankine cycle efficiency for these steam conditions. Estimate also the mean temperature of heat addition.* M **[LO 12.2]**

Solution As for state 2*s* (Fig. 12.44), the quality and pressure are known.

$$\therefore \quad s_{2s} = s_f + x_{2s}\, s_{fg} = 0.5926 + 0.85\,(8.2287 - 0.5926)$$

$$= 7.0833 \text{ kJ/kg K}$$

Since $s_1 = s_{2s}$

$$\therefore \quad s_1 = 7.0833 \text{ kJ/kg K}$$

At state 1, the temperature and entropy are thus known. At 360°C, $s_g = 5.0526$ kJ/kg K, which is less than s_1. So from the table of superheated steam, at $t_1 = 360°C$ and $s_1 = 7.0833$ kJ/kg K, the pressure is found to be 16.832 bar (by interpolation).

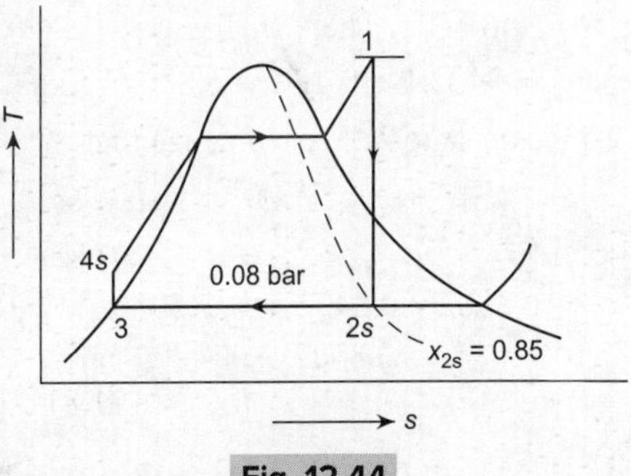

Fig. 12.44

∴ The greatest allowable steam pressure is

$$p_1 = \mathbf{16.832\ bar}$$
$$h_1 = 3165.54 \text{ kJ/kg}$$
$$h_{2s} = 173.88 + 0.85 \times 2403.1 = 2216.52 \text{ kJ/kg}$$
$$h_3 = 173.88 \text{ kJ/kg}$$
$$h_{4s} - h_3 = 0.001 \times (16.83 - 0.08) \times 100 = 1.675 \text{ kJ/kg}$$
$$\therefore \quad h_{4s} = 175.56 \text{ kJ/kg}$$
$$Q_1 = h_1 - h_{4s} = 3165.54 - 175.56$$
$$= 2990 \text{ kJ/kg}$$
$$W_T = h_1 - h_{2s} = 3165.54 - 2216.52 = 949 \text{ kJ/kg}$$
$$W_P = 1.675 \text{ kJ/kg}$$
$$\eta_{\text{cycle}} = \frac{W_{\text{net}}}{Q_1} = \frac{947.32}{2990} = \mathbf{0.3168 \text{ or } 31.68\%}$$

Mean temperature of heat addition

$$T_{m1} = \frac{h_1 - h_{4s}}{s_1 - s_{4s}} = \frac{2990}{7.0833 - 0.5926}$$
$$= 460.66 \text{ K} = \mathbf{187.51°C}$$

Example 12.4 *During on job training period, one trainee engineer was asked to design a steam power plant operating on Rankine cycle for which the maximum temperature of steam can not exceed 500°C. The condenser is to operate at a pressure of 10 kPa and the moisture content at the turbine exhaust is not to exceed 11%. If the turbine efficiency is 85%, determine the maximum allowable pressure at the turbine inlet.* M **[LO 12.2]**

Solution The property values at different state points (Fig. 12.45) found from the steam tables are given below.

$s_{fp2} = 0.6492$ kJ/kgK, $s_{fgp2} = 7.5010$ kJ/kgK,

$h_{fp2} = 191.81$ kJ/kg, $h_{fgp2} = 2392.8$ kJ/kg

$u_3 = v_{fp_2} = 0.00101 \text{ m}^3\text{/kg}, s_{gp2} = 6.3408$ kJ/kgK

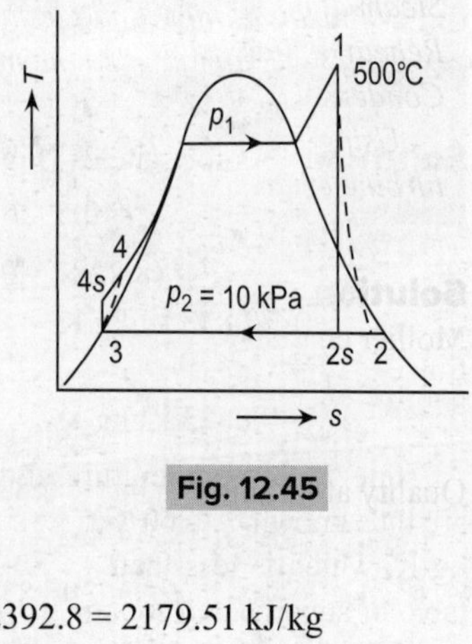

Fig. 12.45

Let us consider the turbine inlet pressure as

$p_1 = 6$ MPa. Then, we have

$h_1 = 3422.1$ kJ/kg, $s_1 = 6.8802$ kJ/kgK

To locate state 2s, we recognize that

$$s_{2s} = s_{fp2} + x_{2s} s_{fgp2}.$$
$$s_{2s} = s_{fp2} + x_{2s} s_{fgp2}$$
$$6.8802 = 0.6492 + x_{2s}\, 7.5010$$
$$\therefore \quad x_{2s} = 0.8307$$
$$\therefore \quad h_{2s} = h_{fp2} + x_{2s} h_{fgp2} = 191.81 + 0.8307 \times 2392.8 = 2179.51 \text{ kJ/kg}$$

From the efficiency of the turbine, we get

$$\eta_T = 0.85 = \frac{h_1 - h_2}{h_1 - h_{2s}} = \frac{3422.1 - h_2}{3422.1 - 2179.51}$$

or $\quad h_2 = 2365.9$ kJ/kg

The quality at the exit from the turbine can then found as

$$h_2 = 2365.9 = h_{fp2} + x_2 h_{fgp2} = 191.81 + x_2 \times 2392.8$$

$\therefore \quad x_4 = 0.9086$

Since the moisture content is less than the permissible limit, there is a scope for increasing the turbine inlet pressure.

Now, let us consider p_1 = 8 MPa. Then, we have

$$h_1 = 3398.3 \text{ kJ/kg}, s_1 = 6.7239 \text{ kJkgK}$$

To locate state 2s, we recognize that $s_1 = s_{2s}$ = 6.7239 kJ/kgK.

$$s_{2s} = s_{fp2} + x_{2s} s_{fgp2}$$

$$6.7239 = 0.6492 + x_{2s} 7.5010$$

$\therefore \quad x_{2s} = 0.8098$

$\therefore \quad h_{2s} = h_{fp2} + x_{2s} h_{fgp2} = 191.81 + 0.8098 \times 2392.8 = 2129.5$ kJ/kg

From the efficiency of the turbine, we get

$$\eta_T = 0.85 = \frac{h_1 - h_2}{h_1 - h_{2s}} = \frac{3398.3 - h_2}{3398.3 - 2129.5}$$

or $\quad h_2 = 2319.82$ kJ/kg

The quality at the exit from the turbine can then be found as

$$h_2 = 2319.82 = h_{fp2} + x_2 h_{fgp2} = 191.81 + x_2 \times 2392.8$$

$$x_4 \approx 0.89$$

Therefore, the maximum allowable pressure at the turbine inlet is 8 MPa.

Example 12.5 A *steam power station uses the following cycle:*
Steam at boiler outlet-150 bar, 550°C
Reheat at 40 bar to 550°C
Condenser at 0.1 bar.

Using the Mollier chart and assuming ideal processes, find the (a) quality at turbine exhaust, (b) cycle efficiency, and (c) steam rate. E [LO 12.6]

Solution The property values at different states (Fig. 12.46) are read from the Mollier chart.

$$h_1 = 3465, h_{2s} = 3065, h_3 = 3565,$$

$$h_{4s} = 2300 \text{ kJ/kg}, x_{4s} = 0.88, h_5 \text{ (steam table)} = 191.83 \text{ kJ/kg}$$

Quality at turbine exhaust = **0.88**

$$W_P = v\Delta p = 10^{-3} \times 150 \times 10^2 = 15 \text{ kJ/kg}$$

$$h_{6s} = 206.83 \text{ kJ/kg}$$

$$Q_1 = (h_1 - h_{6s}) + (h_3 - h_{2s})$$

$$= (3465 - 206.83) + (3565 - 3065) = 3758.17 \text{ kJ/kg}$$

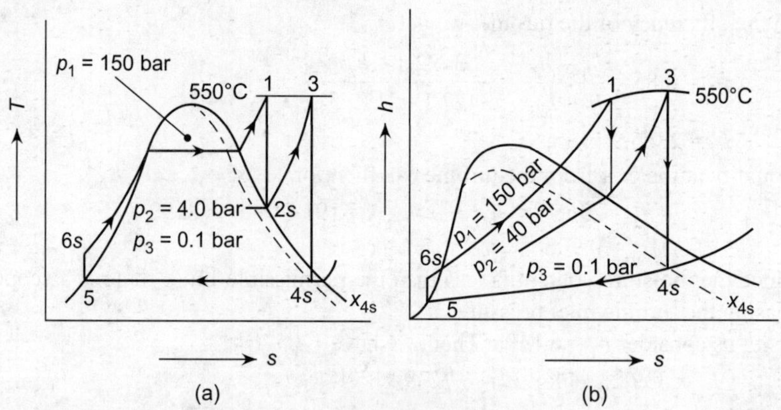

Fig. 12.46

$$W_T = (h_1 - h_{2s}) + (h_3 - h_{4s})$$
$$= (3465 - 3065) + (3565 - 2300) = 1665 \text{ kJ/kg}$$
$$W_{net} = W_T - W_P = 1665 - 15 = 1650 \text{ kJ/kg}$$
$$\eta_{cycle} = \frac{W_{net}}{Q_1} = \frac{1650}{3758.17} = \mathbf{0.4390, or\ 43.9\%}$$
$$\text{Steam rate} = \frac{3600}{1650} = \mathbf{2.18\ kg/kW\ h}$$

Example 12.6 *A coal-fired steam power plant operates on the reheat Rankine cycle. Steam enters the high-pressure (HP) turbine at 10 MPa, 450°C and leaves at 2 MPa. Steam is then reheated at constant pressure to 400°C before it expands in the low-pressure (LP) turbine to a pressure of 10 kPa. The efficiencies of the HP and LP turbines are 85 percent each and that of the pump is 90 percent. Steam leaves the condenser as a saturated liquid. If the mass flow rate of steam through the cycle is 10 kg/s, determine the quality of steam at the exit of low-pressure turbine, net power output from the cycle and the cycle efficiency.*

M **[LO 12.6]**

Solution The property values at different state points (Fig. 12.47) found from the steam tables are given below.

$h_1 = 3240.8$ kJ/kg , $s_1 = 6.4189$ kJ/kgK

$h_3 = 3247.6$ kJ/kg , $s_3 = 7.1270$ kJ/kgK

$s_5 = s_{fp5} = 0.6492.$ kJ/kgK, $s_{fgp5} = 7.5010$ kJ/kgK,

$v_5 = h_{fp5} = 191.81$ kJ/kg, $h_{fgp5} = 2392.8$ kJ/kg,

$v_5 = v_{fp5} = 0.00101$ m³/kg, $s_{gp2} = 6.3408$ kJ/kgK

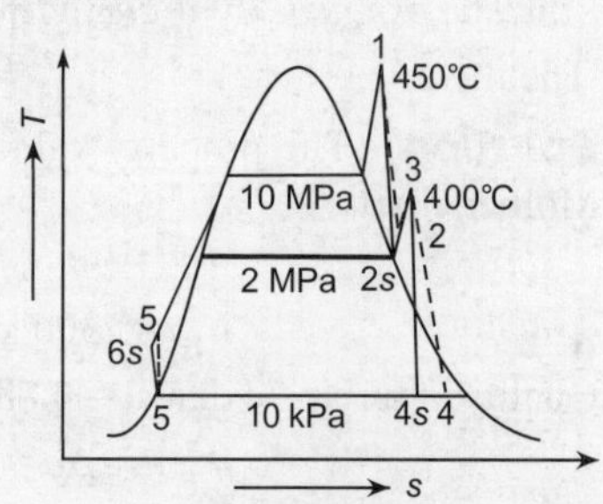

Fig. 12.47

To locate state 2s, we recognize that $s_1 = s_{2s} = 6.4189$ kJ/kgK. Since, $s_{2s} > s_{gp2}$, the state of steam at 2s is superheated vapour. Using linear interpolation, we find $T_{2s} = 225.82°C$ and $h_{2s} = 2837.99$ kJ/kg. From the efficiency of the high pressure turbine, we get

$$\eta_{HPT} = 0.85 = \frac{h_1 - h_2}{h_1 - h_{2s}} = \frac{3240.8 - h_2}{3240.8 - 2837.99}$$

or $\quad h_2 = 2898.41$ kJ/kg

To locate state 4s, we recognize that $s_3 = s_{4s} = 7.1270$ kJ/kgK.

$$s_{4s} = s_{fp4} + x_{4s} s_{fgp4}$$

$$7.1270 = 0.6492 + x_{4s}\, 7.5010$$

$\therefore \quad x_{4s} = 0.8636$

$\therefore \quad h_{4s} = h_{fp4} + x_{4s} h_{fgp4} = 191.81 + 0.8636 \times 2392.8 = 2258.23$ kJ/kg

From the efficiency of the low pressure turbine, we get

$$\eta_{LPT} = 0.85 = \frac{h_3 - h_4}{h_3 - h_{4s}} = \frac{3247.6 - h_4}{3247.6 - 2258.23}$$

or $\quad h_4 = 2406.64$ kJ/kg

The quality at the exit from the turbine can then be found to be

$$h_4 = 2406.64 = h_{fp4} + x_4 h_{fgp4} = 191.81 + x_4 \times 2392.8$$

$$x_4 = \mathbf{0.9256}$$

From the pump work, we get

$$h_{6s} = h_5 + v_{fp5}(p_6 - p_5) = 191.81 + 0.00101(10 \times 10^3 - 10) = 201.9 \text{ kJ/kg}$$

From the efficiency of the pump, we have

$$\eta_P = 0.90 = \frac{h_{6s} - h_5}{h_6 - h_5} = \frac{201.9 - 191.81}{h_6 - 191.81}$$

or $\quad h_6 = 203.02$ kJ/kg

Net work output of the cycle is $\dot{W}_{net} = w[(h_1 - h_2) + (h_3 - h_4) - (h_6 - h_5)]$

$$= 10[(3240.8 - 2898.41) + (3247.6 - 2406.64) - (203.02 - 191.81)]$$

$$= \mathbf{11721.4\ kW}$$

The total heat input is found as $Q_1 = w[(h_1 - h_6) + (h_3 - h_2)]$

$$= 10[(3240.8 - 203.02) + (3247.6 - 2898.41)] = 33869.7 \text{ kW}$$

The thermal efficiency of the cycle $\eta_{cycle} = \frac{\dot{W}_{net}}{\dot{Q}_1} = \frac{11721.4}{33869.7} = 0.3461$ or **34.61%**

Example 12.7 *In a vapour power plant, steam leaves the boiler at 8 MPa and 500°C. Steam while flowing through the steam pipe connecting the boiler end to the turbine inlet both heat loss and pressure drop and as a result the pressure and temperature of steam at the turbine inlet are 7.5 MPa and 450°C, respectively. The steam leaves the turbine at 10 kPa after expansion, while the liquid water leaves the condenser at 8 kPa, 30°C. The pump raises the pressure to 8.5 MPa. The efficiency of the turbine is 88% and that of the pump is 90%. If the mass flow rate of steam is 30 kg/s, determine the quality of steam at the exit of the turbine, net power output and the thermal efficiency of the plant.* D **[LO 12.2, 3]**

Solution The property values at different state points(Fig. 12.48) found from the steam tables are given below.

$$h_1 = 3398.3 \text{ kJ/kg}, s_1 = 6.7239 \text{ kJ/kgK}$$

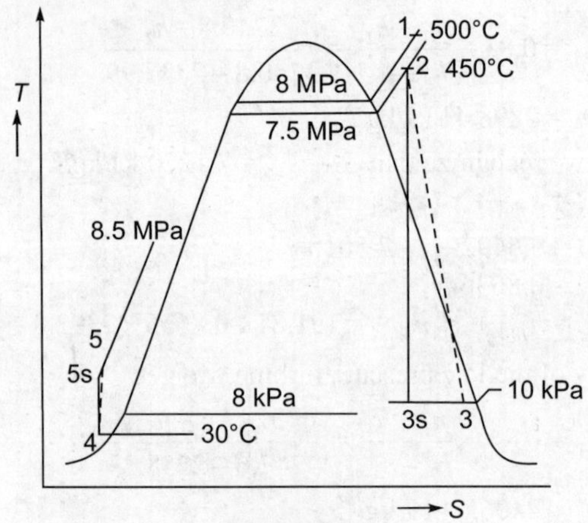

Fig. 12.48

Using linear interpolation, we further get

$$h_2 = 3279.5 \text{ kJ/kg}, s_2 = 6.5938 \text{ kJ/kgK}$$

To locate state 3s, we recognize that $s_2 = s_{35} = 6.5938$ kJ/kgK.

$$s_{3s} = s_{fp3} + x_{3s} s_{fgp3}$$

$$6.5938 = 0.6492 + x_{3s} 7.5010$$

$$\therefore \quad x_{3s} = 0.7925$$

$$h_{3s} = h_{fp3} + x_{3s} h_{fgp3} = 191.81 + 0.7925 \times 2392.8 = 2088.1 \text{ kJ/kg}$$

From the efficiency of the turbine, we get

$$\eta_T = 0.88 = \frac{h_2 - h_3}{h_2 - h_{3s}} = \frac{3279.5 - h_3}{3279.5 - 2088.1}$$

or $\quad h_3 = 2231.07$ kJ/kg

The quality at the exit from the turbine can then be found as

$$h_3 = 2231.07 = h_{fp3} + x_3 h_{fgp3} = 191.81 + x_3 \times 2392.8$$

$$\therefore \quad x_3 = \mathbf{0.8522}$$

$$h_4 = h_{fT4} = 125.77 \text{ kJ/kg}, v_4 = v_{fT4} = 0.001004 \text{ m}^3\text{/kg}$$

From the pump work, we get

$$h_{5s} = h_4 + v_{fT4}(p_5 - p_4) = 125.77 + 0.001004(8.5 \times 10^3 - 8) = 134.3 \text{ kJ/kg}$$

From the efficiency of the pump, we get

$$\eta_P = 0.90 = \frac{h_{5s} - h_4}{h_5 - h_4} = \frac{134.3 - 125.77}{h_5 - 125.77}$$

or $\quad h_5 = 135.25$ kJ/kg

Net work output of the cycle is $\dot{W}_{\text{net}} = w\left[(h_2 - h_3) - (h_5 - h_4)\right]$

$$= 30\left[(3279.5 - 2231.07) - (135.25 - 125.77)\right] = \mathbf{31168.5 \text{ kW}}$$

The heat input is found as $Q_1 = w(h_1 - h_5) = 30(3398.3 - 135.25) = 97891.5$ kW

The thermal efficiency of the cycle $\eta_{cycle} = \frac{\dot{W}_{net}}{\dot{Q}_1} = \frac{31168.5}{97891.5} = \mathbf{0.3184}$ or **31.84%**

Example 12.8 *In a single-heater regenerative cycle, the steam enters the turbine at 30 bar, 400°C and the exhaust pressure is 0.10 bar. The feedwater heater is a direct-contact type which operates at 5 bar. Find (a) the efficiency and the steam rate of the cycle, and (b) the increase in mean temperature of heat addition, efficiency and steam rate, as compared to the Rankine cycle (without regeneration). Neglect pump work.* M **[LO 12.8]**

Solution Figure 12.49 gives the flow, *T-s* and *h-s* diagrams. From the steam tables, the property values at various states have been obtained.

$$h_1 = 3230.9 \text{ kJ/kg}$$
$$s_1 = 6.9212 \text{ kJ/kg K} = s_2 = s_3$$
$$s_g \text{ at 5 bar} = 6.8213 \text{ kJ/kg K}$$

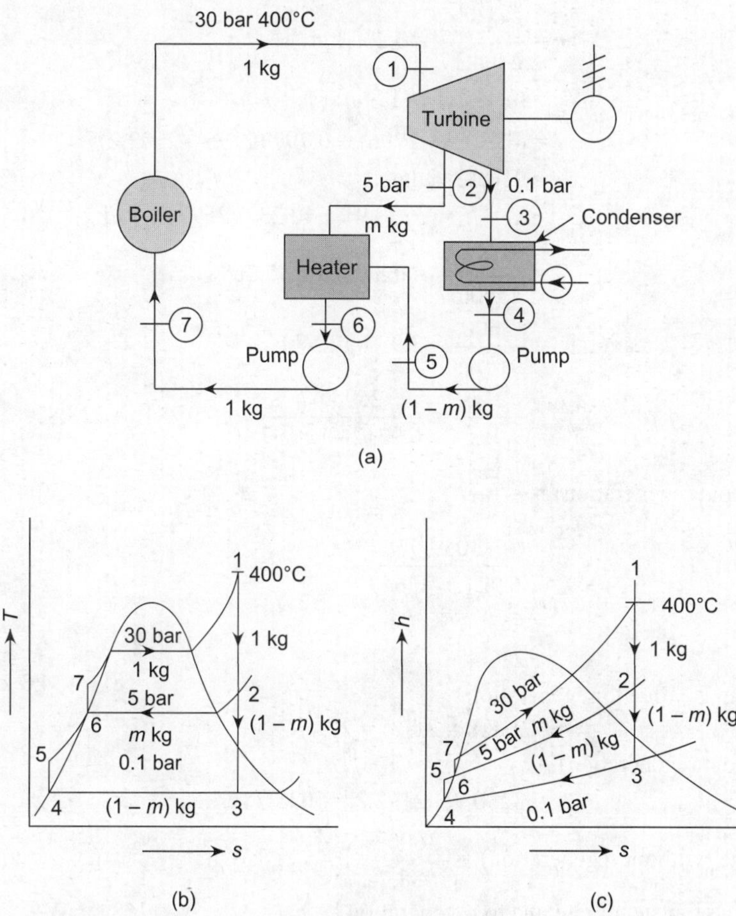

Fig. 12.49

Since $s_2 > s_g$, the state 2 must lie in the superheated region. From the table for superheated steam $t_2 = 172°\text{C}$, $h_2 = 2796$ kJ/kg

$$s_3 = 6.9212 = s_{f0.1\text{ bar}} + x_3 s_{fg0.1\text{ bar}}$$
$$= 0.6493 + x_3 7.5009$$
$$\therefore \quad x_3 = \frac{6.2719}{7.5009} = 0.836$$
$$\therefore \quad h_3 = 191.83 + 0.836 \times 2392.8 = 2192.2 \text{ kJ/kg}$$

Since pump work is neglected

$$h_4 = 191.83 \text{ kJ/kg} = h_5$$
$$h_6 = 640.23 \text{ kJ/kg} = h_7$$

Energy balance for the heater gives

$$m(h_2 - h_6) = (1 - m)(h_6 - h_5)$$
$$m(2796 - 640.23) = (1 - m)(640.23 - 191.83)$$
$$2155.77\, m = 548.4 - 548.4\, m$$
$$\therefore \quad m = \frac{548.4}{2704.17} = 0.203 \text{ kg}$$
$$\therefore \quad W_T = (h_1 - h_2) + (1 - m)(h_2 - h_3)$$
$$= (3230.9 - 2796) + 0.797\,(2796 - 2192.2)$$
$$= 916.13 \text{ kJ/kg}$$
$$Q_1 = h_1 - h_6 = 3230.9 - 640.23 = 2590.67 \text{ kJ/kg}$$
$$\eta_{\text{cycle}} = \frac{916.13}{2590.67} = \mathbf{0.3536, \text{ or } 35.36\%}$$
$$\text{Steam rate} = \frac{3600}{916.13} = \mathbf{3.93 \text{ kg/kW h}}$$
$$T_{m1} = \frac{h_1 - h_7}{s_1 - s_7} = \frac{2590.67}{6.9212 - 1.8607} = 511.95 \text{ K} = 238.8°\text{C}$$
$$T_{m1} \text{ (without regeneration)} = \frac{h_1 - h_4}{s_1 - s_4}$$
$$= \frac{3039.07}{6.9212 - 0.6493}$$
$$= 484.55 \text{ K}$$
$$= 211.4°\text{C}$$

Increase in T_{m1} due to regeneration

$$= 238.8 - 211.4 = \mathbf{27.4°C}$$
$$W_T \text{ (without regeneration)} = h_1 - h_3$$
$$= 3230.9 - 2192.2 = 1038.7 \text{ kJ/kg}$$
$$\text{Steam rate (without regeneration)} = \frac{3600}{1038.7} = 3.46 \text{ kg/kW h}$$

$\therefore$ Increase in steam rate due to regeneration

$$= 3.93 - 3.46 = \mathbf{0.476 \text{ kg/kW h}}$$

$$\eta_{cycle} \text{ (without regeneration)} = \frac{h_1 - h_3}{h_1 - h_4} = \frac{1038.7}{3039.07}$$
$$= 0.3418 \text{ or } 34.18\%$$

∴ Increase in cycle efficiency due to regeneration

$$= 35.36 - 34.18 = \mathbf{1.18\%}$$

Example 12.9 *In a steam power plant the condition of steam at inlet to the steam generator is 20 bar and 300°C and the condenser pressure is 0.1 bar. Two feedwater heaters operate at optimum temperature. Determine (a) the quality of steam at turbine exhaust, (b) net work per kg of steam, (c) cycle efficiency, and (d) the steam rate. Neglect pump work.* Ⓜ **[LO 12.8]**

Solution From Fig. 12.19 (a),

$$h_1 = 3023.5 \text{ kJ/kg}$$
$$s_1 = 6.7664 \text{ kJ/kg K} = s_2 = s_3 = s_4$$
$$t_{sat} \text{ at 20 bar} \cong 212°C$$
$$t_{sat} \text{ at 0.1 bar} \cong 46°C$$
$$\Delta t_{OA} = 212 - 46 = 166°C$$

∴ Temperature rise per heater $= \frac{166}{3} = 55°C$

∴ Temperature at which the first heater operates

$$= 212 - 55 = 157°C \cong 150°C \text{ (assumed)}$$

Temperature at which the second heater operates = 157 – 55 = 102°C ≅ 100°C (assumed)

At 0.1 bar,

$$h_f = 191.83,\ h_{fg} = 2392.8,\ s_f = 0.6493$$
$$s_g = 8.1502$$

At 100°C,

$$h_f = 419.04,\ h_{fg} = 2257.0,\ s_f = 1.3069,\ s_g = 7.3549$$

At 150°C,

$$h_f = 632.20,\ h_{fg} = 2114.3,\ s_f = 1.8418,\ s_g = 6.8379$$
$$6.7664 = 1.8418 + x_2 \times 4.9961$$

∴ $x_2 = 0.986$

∴ $h_2 = 632.2 + 0.986 \times 2114.3$

$$= 2716.9 \text{ kJ/kg}$$
$$6.7664 = 1.3069 + x_3 \times 6.0480$$
$$x_3 = 0.903$$

∴ $h_3 = 419.04 + 0.903 \times 2257.0$

or $h_3 = 2457.1$ kJ/kg

$$6.7664 = 0.6493 + x_4 \times 7.5010$$
$$x_4 = 0.816$$

∴ $h_4 = 191.83 + 0.816 \times 2392.8$

$$= 2144.3 \text{ kJ/kg}$$

Since pump work is neglected, $h_{10} = h_9$, $h_8 = h_7$, $h_6 = h_5$. By making an energy balance for the hp heater

$$(1 - m_1)(h_9 - h_8) = m_1(h_2 - h_9)$$

Rearranging

$$m_1 = \frac{h_9 - h_7}{h_2 - h_7} = \frac{213.16}{2297.86} = 0.093 \text{ kg}$$

By making an energy balance for the *lp* heater,

$$(1 - m_1 - m_2)(h_7 - h_6) = m_2 (h_3 - h_7)$$

$$(1 - 0.093 - m_2)(419.04 - 191.83) = m_2 (2457.1 - 419.04)$$

$$\therefore \quad m_2 = 0.091 \text{ kg}$$

$$\therefore \quad W_T = 1(h_1 - h_2) + (1 - m_1)(h_2 - h_3) + (1 - m_1 - m_2)(h_3 - h_4)$$

$$= (3023.5 - 2716.9) + (1 - 0.093)(2716.9 - 2457.1) + (1 - 0.093 - 0.091)(2457.1 - 2144.3) = \mathbf{797.48 \text{ kJ/kg}}$$

$$Q_1 = h_1 - h_9 = 3023.5 - 632.2 = 2391.3 \text{ kJ/kg}$$

$$\therefore \quad \eta_{\text{cycle}} = \frac{W_T - W_P}{Q_1} = \frac{797.48}{2391.3} = \mathbf{0.3334 \text{ or } 33.34\%}$$

$$\text{Steam rate} = \frac{3600}{W_{\text{net}}} = \frac{3600}{797.48} = \mathbf{4.51 \text{ kg/kW h}}$$

Example 12.10 *Dry saturated steam at 40 bar expands in a turbine isentropically to the condenser pressure of 0.075 bar. Hot gases available at 2000 K, and 1 atm pressure are used for steam generation and then exhausted at 450 K to the ambient atmosphere which is at 300 K and 1 atm. The heating rate provided by the gas stream is 100 MW. Assuming c_p of hot gases as 1.1 kJ/kg K, give an exergy balance of the plant and compare it with the energy balance, and find the second law efficiency.* D **[LO 12.11]**

Solution

$$\dot{Q}_1 = w_g c_{pg} (T_i - T_e) = 100 \text{ MW}$$

w_g = mass flow rate of hot gas

$$= \frac{100 \times 10^3}{1.1 \times (2000 - 450)} = 58.7 \text{ kg/s}$$

Exergy flow rate of inlet gas

$$a_{f1} = w_g c_{pg} T_0 \left[\frac{T_i}{T_0} - 1 - \ln \frac{T_i}{T_0} \right]$$

$$= 58.7 \times 1.1 \times 300 \left[\frac{2000}{300} - 1 - \ln \frac{2000}{300} \right]$$

$$= 73 \text{ MW}$$

Exergy flow rate of exhaust gas stream

$$a_{f2} = 58.7 \times 1.1 \times 300 \left[\frac{450}{300} - 1 - \ln \frac{450}{300} \right] = 1.83 \text{ MW}$$

The exergy loss rate is only about $\left[\frac{1.83}{73} \times 100 \right]$ or 2.5% of the initial exergy of the source gas.

The rate of exergy decrease of the gas stream,

$$a_{fi} = \text{Exergy input rate} = 73 - 1.83 = 71.17 \cong 71.2 \text{ MW}$$

The rate of exergy increase of steam = Exergy utilization rate

$$a_{fu} = w_s [h_1 - h_4 - T_0(s_1 - s_4)]$$

Now,

$$h_1 = (h_g)_{40 \text{ bar}} = 2801 \text{ kJ/kg}, h_3 = 169 \text{ kJ/kg}$$

$$s_3 = s_4 = 0.576 \text{ kJ/kgK}, h_4 = 172.8 \text{ kJ/kg}$$

$$s_1 = s_2 = 6.068 \text{ kJ/kgK}, h_2 = 1890.2 \text{ kJ/kg}$$

$$W_T = h_1 - h_2 = 2801 - 1890.2 = 910.8 \text{ kJ/kg}$$

$$W_P = h_4 - h_3 = 172.8 - 169 = 3.8 \text{ kJ/kg}$$

$$Q_1 = h_1 - h_4 = 2801 - 172.8 = 2628 \text{ kJ/kg}$$

$$Q_2 = h_2 - h_3 = 1890.2 - 169 = 1721 \text{ kJ/kg}$$

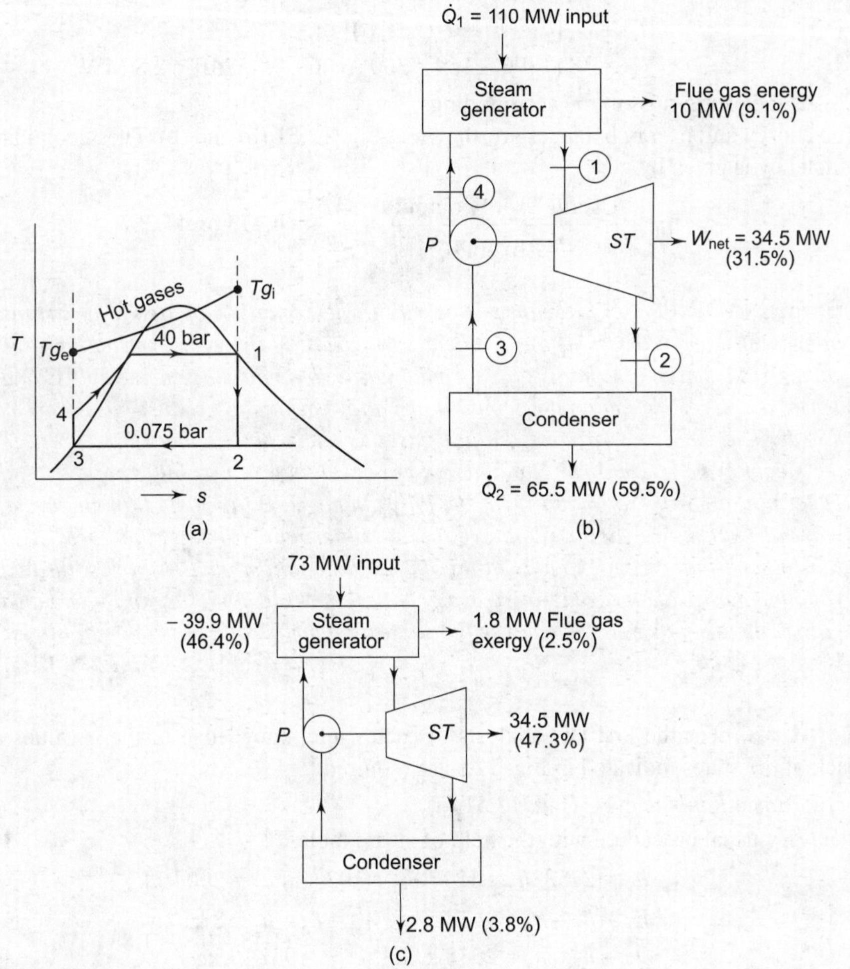

Fig. 12.50 *T-s* diagram, (b) Energy distribution diagram, (c) Exergy distribution diagram

$$W_{net} = W_T - W_P = Q_1 - Q_2 = 907 \text{ kJ/kg}$$
$$\dot{Q}_1 = w_s \times 2628 = 100 \times 10^3 \text{ kW}$$
$$w_s = 38 \text{ kg/s}$$
$$a_{fu} = 38\,[2801 - 172.8 - 300\,(6.068 - 0.576)]$$
$$= 37.3 \text{ MW}$$

Rate of exergy destruction in the steam generator

= Rate of exergy decrease of gases – Rate of exergy increase of steam.

$$\dot{I} = a_{fi} - a_{fu} = 71.2 - 37.3 = 33.9 \text{ MW}$$

Rate of useful mechanical power output

$$\dot{W}_{net} = 38 \times 907 = 34.5 \text{ MW}$$

Exergy flow rate of wet steam to the condenser

$$a_{fc} = w_s\,[h_2 - h_3 - T_0(s_2 - s_3)]$$
$$= 38\,[1890 - 169 - 300\,(6.068 - 0.576)] = 2.8 \text{ MW}$$

This is the exergy loss to the surroundings.

The energy and exergy balances are shown in Fig. 12.51 (b) and (c). The second law efficiency is given by

$$\eta_{II} = \frac{\text{Useful Exergy output}}{\text{Exergy input}} = \frac{34.5}{73} = \mathbf{0.473 \text{ or } 47.3\%}$$

Example 12.11 *In a steam power plant, the condition of steam at turbine inlet is 80 bar, 500°C and the condenser pressure is 0.1 bar. The heat source comprises a steam of exhaust gases from a gas turbine discharging at 560°C and 1 atm pressure. The minimum temperature allowed for the exhaust gas steam is 450 K. The mass flow rate of the hot gases is such that the heat input rate to the steam cycle is 100 MW. The ambient condition is given by 300 K and 1 atm. Determine η_I, work ratio and η_{II} of the following cycles: (a) Basic Rankine cycle, without superheat, (b) Rankine cycle with superheat, (c) Rankine cycle with reheat such that steam expands in the h.p. turbine until it exits as dry saturated vapour and (d) Ideal regenerative cycle, with the exit temperature of the exhaust gas steam taken as 320°C, because the saturation temperature of steam at 80 bar is close to 300°C.* **D [LO 12.2, 12.6, 12.7]**

Solution For the first law analysis of each cycle, knowledge of the h values at each of the states indicated in Fig. 12.51 is required.

(a) *Basic Rankine cycle* (Fig. 12.51 (a))

By usual procedure with the help of steam tables,

$$h_1 = 2758,\ h_2 = 1817,\ h_3 = 192 \text{ and } h_4 = 200 \text{ kJ/kg}$$
$$W_T = h_1 - h_2 = 941 \text{ kJ/kg},\ W_p = h_4 - h_3 = 8 \text{ kJ/kg}$$
$$Q_1 = h_1 - h_4 = 2558 \text{ kJ/kg},\ W_{net} = 933 \text{ kJ/kg}$$
$$\eta_I = \frac{W_{net}}{Q_1} = \frac{933}{2558} = 0.365 \text{ or } \mathbf{36.5\%}$$

$$\text{Work ratio} = \frac{W_T - W_P}{W_T} = \frac{933}{941} = \mathbf{0.991}$$

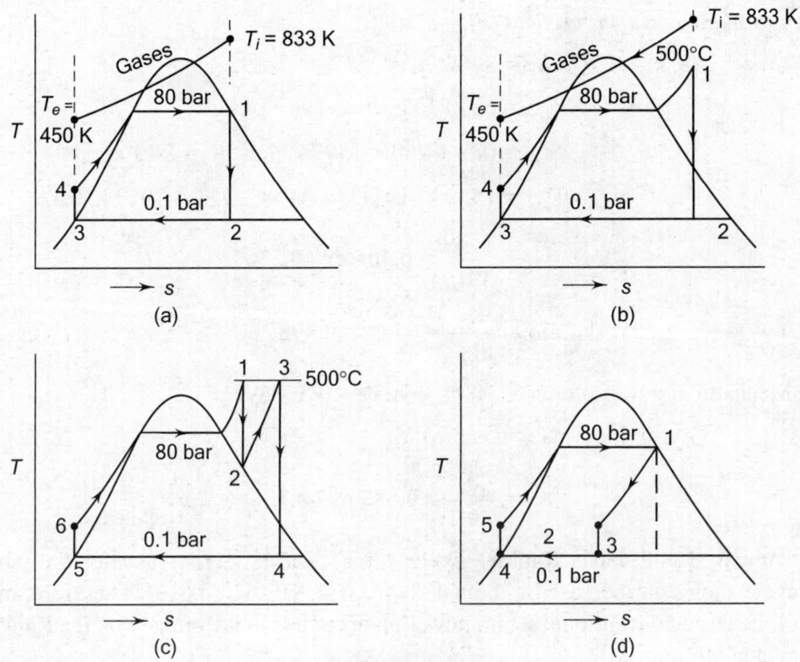

Fig. 12.51

$$\text{Power output} = \eta_I\, Q_1 = 0.365 \times 100 = 36.5 \text{ MW}$$

$$\text{Exergy input rate} = w_g\, c_{pg} \left[(T_i - T_0) - T_0 \ln \frac{T_i}{T_0} \right]$$

$$= \frac{100 \times 1000}{833 - 450} \left[(833 - 300) - 300 \ln \frac{833}{300} \right]$$

$$= 59.3 \text{ MW}$$

$$\eta_{II} = \frac{36.5}{59.3} = \mathbf{0.616 \text{ or } 61.6\%}$$

(b) *Rankine cycle with superheat* (Fig. 12.49 (b))

$$h_1 = 3398,\ h_2 = 2130,\ h_3 = 192 \text{ and } h_4 = 200 \text{ kJ/kg}$$

$$W_T = 1268 \text{ kJ/kg},\ W_P = 8 \text{ kJ/kg},\ Q_1 = 3198 \text{ kJ/kg}$$

$$\eta_I = \frac{1260}{3198} = \mathbf{0.394 \text{ or } 39.4\%}$$

$$\text{Work ratio} = \frac{1260}{1268} = \mathbf{0.994}$$

$$\text{Exergy input rate} = 59.3 \text{ MW},\ W_{net} = Q_1 \times \eta_I = 39.4 \text{ MW}$$

$$\eta_{II} = \frac{36.5}{59.3} = \mathbf{0.664 \text{ or } 66.4\%}$$

Improvements in both first law and second law efficiencies are achieved with superheating. The specific work output is also increased. Therefore, conventional vapour power plants are almost always operated with some superheat.

(c) *Rankine cycle with reheat* (Fig. 12.49 (c))

$h_1 = 3398$, $h_2 = 2761$, $h_3 = 3482$, $h_4 = 2522$, $h_5 = 192$ and $h_6 = 200$ kJ/kg

$$W_{T1} = 637 \text{ kJ/kg}, W_{T2} = 960 \text{ kJ/kg}$$

$$W_T = 637 + 960 = 1597 \text{ kJ/kg}, W_P = 8 \text{ kJ/kg}$$

$$W_{net} = 1589 \text{ kJ/kg}, Q_1 = 3198 + 721 = 3919 \text{ kJ/kg}$$

$$\eta_I = \frac{1589}{3919} = \mathbf{0.405 \text{ or } 40.5\%}$$

$$\text{Work ratio} = \frac{W_{net}}{W_T} = \frac{1589}{1597} = \mathbf{0.995}$$

$$\text{Mechanical power output} = 100 \times 0.405 = 40.5 \text{ MW}$$

$$\text{Exergy input rate} = 59.3 \text{ MW},$$

$$\eta_{II} = \frac{40.5}{59.3} = \mathbf{0.683 \text{ or } 68.3\%}$$

Compared with basic Rankine cycle, the second law efficiency for the reheat cycle shows an increase of about 11% [(0.683 – 0.616)/0.616]. Therefore, most of the large conventional steam power plants in use today operate on the Rankine cycle with reheat.

(d) *Rankine cycle with complete regeneration* (Fig. 12.49 (d))

$$t_{sat} \text{ at } 0.1 \text{ bar} = 45.8°\text{C} = 318.8 \text{ K and}$$

$$t_{sat} \text{ at } 80 \text{ bar} = 295°\text{C} = 568 \text{ K}$$

$$\eta_I = \eta_{Carnot} = 1 - \frac{T_3}{T_1} = 1 - \frac{318.8}{568.0} = \mathbf{0.439 \text{ or } 43.9\%}$$

$$Q_1 = h_1 - h_6 = 2758 - 1316 = 1442 \text{ kJ/kg}$$

$$W_{net} = Q_1 \times \eta_I = 1442 \times 0.439 = 633 \text{ kJ/kg}$$

$$W_P = 8 \text{ kJ/kg} \quad W_T = 641 \text{ kJ/kg}$$

$$\text{Work ratio} = \frac{633}{641} = \mathbf{0.988}$$

$$\text{Power output} = 0.439 \times 100 = 43.9 \text{ MW}$$

$$\text{Exergy input rate} = \frac{100 \times 1000}{833 - 593}\left[(833 - 300) - 300 \ln \frac{833}{300}\right]$$

$$= 94.583 \text{ MW} \cong 94.6 \text{ MW}$$

$$\eta_{II} = \frac{43.9}{94.6} = 0.464 \text{ or } \mathbf{46.4\%}$$

The second law efficiency is lower for regeneration because of the more substantial loss of exergy carried by the effluent gas steam at 593 K.

Example 12.12 *A certain chemical plant requires heat from process steam at 120°C at the rate of 5.83 MJ/s and power at the rate of 1000 kW from the generator terminals. Both the heat and power requirements are met by a back pressure turbine of 80% brake and 85% internal efficiency, which exhausts steam at 120°C dry saturated. All the latent heat released during condensation is utilised in the process heater. Find the pressure and temperature of steam at the inlet to the turbine. Assume 90% efficiency for the generator.* E [LO 12.16, 12.17]

Solution At 120°C, $h_{fg} = 2202.6$ kJ/kg $= h_2 - h_3$ (Fig. 12.52)

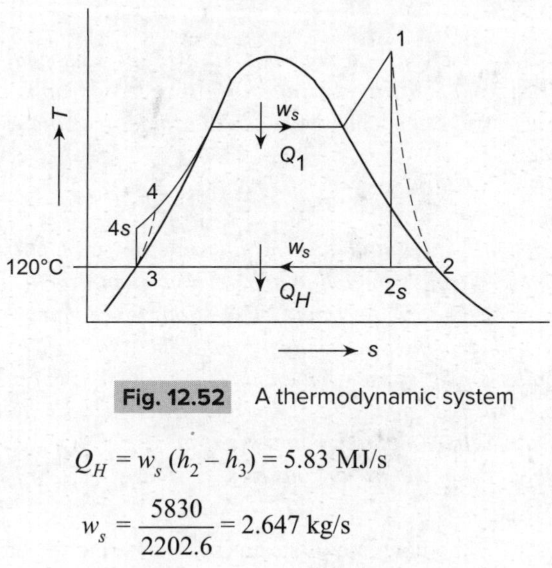

Fig. 12.52 A thermodynamic system

$$Q_H = w_s (h_2 - h_3) = 5.83 \text{ MJ/s}$$

$$\therefore \quad w_s = \frac{5830}{2202.6} = 2.647 \text{ kg/s}$$

$$W_{net} = \frac{1000}{0.9} \text{ kJ/s} = \text{Brake output}$$

Now
$$\eta_{brake} = \frac{\text{Brake output}}{\text{Ideal output}} = \frac{(1000)/0.9}{w_s(h_1 - h_{2s})} = 0.80$$

$$\therefore \quad h_1 - h_{2s} = \frac{1000}{0.9 \times 0.8 \times 2.647} = 524.7 \text{ kJ/kg}$$

Again
$$\eta_{internal} = \frac{h_1 - h_2}{h_1 - h_{2s}} = 0.85$$

$$\therefore \quad h_1 - h_2 = 0.85 \times 524.7 = 446 \text{ kJ/kg}$$

$$h_2 = h_g \text{ at } 120°\text{C} = 2706.3 \text{ kJ/kg}$$

$$\therefore \quad h_1 = 3152.3 \text{ kJ/kg}$$

$$h_{2s} = h_1 - 524.7 = 2627.6 \text{ kJ/kg}$$
$$= h_f + x_{2s} h_{fg}$$

$$= 503.71 + x_{2s} \times 2202.6$$

$$\therefore \quad x_{2s} = \frac{2123.89}{2202.6} = 0.964$$

$$\therefore \quad s_{2s} = s_f + x_{2s} \cdot s_{fg} = 1.5276 + 0.964 \times 5.6020$$

$$= 6.928 \text{ kJ/kg K}$$

$$= s_1$$

At state 1, $h_1 = 3152.3$ kJ/kg

$s_1 = 6.928$ kJ/kg K

From the Mollier chart

$p_1 = \mathbf{22.5\ bar}$

$t_1 = \mathbf{360°C}$

Example 12.13 *A certain factory has an average electrical load of 1500 kW and requires 3.5 MJ/s for heating purposes. It is proposed to install a single-extraction passout steam turbine to operate under the following conditions:*

Initial pressure 15 bar.

Initial temperature 300°C.

Condenser pressure 0.1 bar.

Steam is extracted between the two turbine sections at 3 bar, 0.96 dry, and is isobarically cooled without subcooling in heaters to supply the heating load. The internal efficiency of the turbine (in the LP Section) is 0.80 and the efficiency of the boiler is 0.85 when using oil of calorific value 44 MJ/kg.

If 10% of boiler steam is used for auxiliaries, calculate the oil consumption per day. Assume that the condensate from the heaters (at 3 bar) and that from the condenser (at 0.1 bar) mix freely in a separate vessel (hot well) before being pumped to the boiler. Neglect extraneous losses. D **[LO 12.16]**

Solution Let w_s be the flow rate of steam (kg/h) entering the turbine, and w the amount of steam extracted per hour for process heat (Fig. 12.53).

$$h_1 = 3037.3 \text{ kJ/kg}$$

$$h_2 = 561.47 + 0.96 \times 2163.8$$

$$= 2638.7 \text{ kJ/kg}$$

$$s_2 = 1.6718 + 0.96 \times 5.3201$$

$$= 6.7791 \text{ kJ/kg K}$$

$$= s_{3s}$$

$$s_{3s} = 6.7791 = 0.6493 + x_{3s} \times 7.5009$$

$$x_{3s} = \frac{6.1298}{7.5009} = 0.817$$

$$h_{3s} = 191.83 + 0.817 \times 2392.8 = 2146.75 \text{ kJ/kg}$$

$$h_2 - h_{3s} = 2638.7 - 2146.75 = 491.95 \text{ kJ/kg}$$

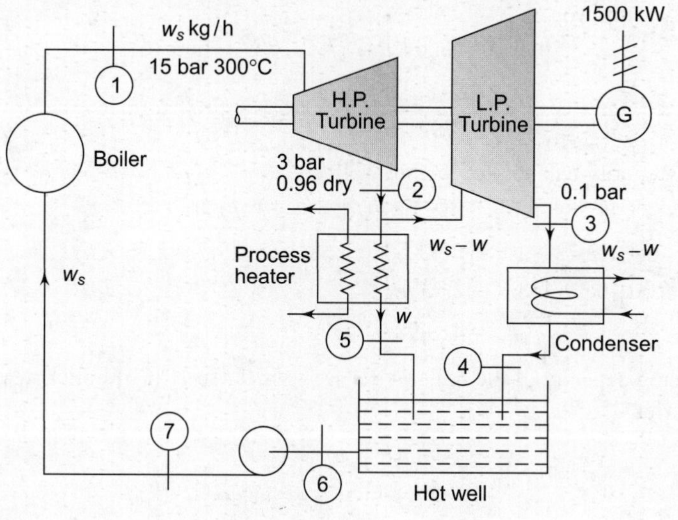

(a)

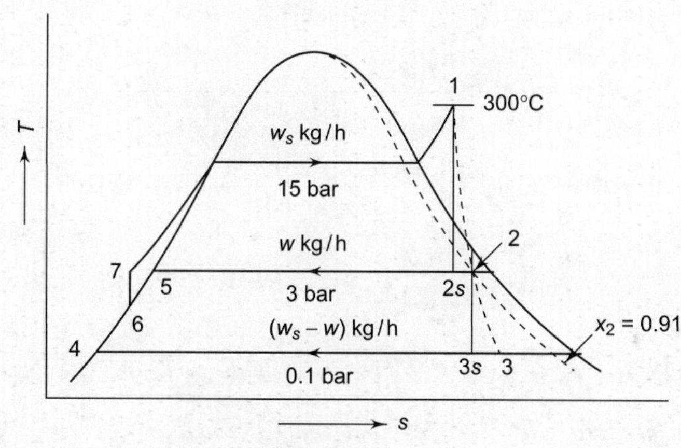

(b)

Fig. 12.53

$$h_2 - h_3 = 0.8 \times 491.95 = 393.56 \text{ kJ/kg}$$

$\therefore$ $$h_3 = 2638.7 - 393.56 = 2245.14 \text{ kJ/kg}$$

$$h_5 = 561.47 \text{ kJ/kg}, \quad h_4 = 191.83 \text{ kJ/kg}$$

$$Q_H = w(h_2 - h_5) = 2\,(2638.7 - 561.47) = 3.5 \text{ MJ/s}$$

$\therefore$ $$w = \frac{3500}{2077.2}\, 1.685 \text{ kg/s}$$

Now

$$\begin{aligned} W_T &= w_s(h_1 - h_2) + (w_s - w)\,(h_2 - h_3) \\ &= w_s(3037.3 - 2638.7) + (w_s - 1.685) \times 393.56 \\ &= w_s \times 398.6 + w_s \times 393.56 - 663.15 \\ &= 792.16\, w_s - 663.15 \end{aligned}$$

Neglecting pump work

$$W_T = 1500 \text{ kJ/s} = 792.16\, w_s - 663.15$$

$$\therefore \quad w_s = \frac{2163.15}{792.16} = 2.73 \text{ kg/s} = 9828 \text{ kg/h}$$

By making energy balance for the hot well

$$(w_s - w)h_4 + wh_5 = w_s h_6$$

$$(2.73 - 1.685)\ 191.83 + 1.685 \times 561.47 = 2.73 \times h_6$$

$$200.46 + 946.08 = 2.73\, h_6$$

$$\therefore \quad h_6 = 419.98 \text{ kJ/kg} \cong h_7$$

Steam raising capacity of the boiler = 1.1 w_s kg/h, since 10% of boiler steam is used for auxiliaries.

$$\therefore \quad \eta_{\text{boiler}} = \frac{1.1 w_s (h_1 - h_7)}{w_f \times \text{C.V.}}$$

where w_f = fuel burning rate (kg/h)
and CV = calorific value of fuel = 44 MJ/kg

$$\therefore \quad 0.85 = \frac{1.1 \times 9828 \times (3037.3 - 419.98)}{w_f \times 44000}$$

or

$$w_f = \frac{1.1 \times 9828 \times 2617.32}{0.85 \times 44000} = 756.56 \text{ kg/h}$$

$$= \frac{756.56 \times 24}{1000} = \mathbf{18.16\ tonnes/day}$$

Example 12.14 *A steam turbine gets its supply of steam at 70 bar and 450°C. After expanding to 25 bar in high pressure stages, it is reheated to 420°C at the constant pressure. Next, it is expanded in intermediate pressure stages to an appropriate minimum pressure such that part of the steam bled at this pressure heats the feedwater to a temperature of 180°C. The remaining steam expands from this pressure to a condenser pressure of 0.07 bar in the low pressure stage. The isentropic efficiency of the h.p. stage is 78.5%, while that of the intermediate and LP stages is 83% each. From the above data (a) determine the minimum pressure at which bleeding is necessary, and sketch a line diagram of the arrangement of the plant, (b) Sketch on the T-s diagram all the processes, (c) determine the quantity of steam bled per kg of flow at the turbine inlet and (d) calculate the cycle efficiency. Neglect pump work.* M **[LO 12.9]**

Solution Figure 12.54 gives the flow and *T-s* diagrams of the plant. It would be assumed that the feedwater heater is an open heater. Feedwater is heated to 180°C. So p_{sat} at 180°C ≅ 10 bar is the pressure at which the heater operates.

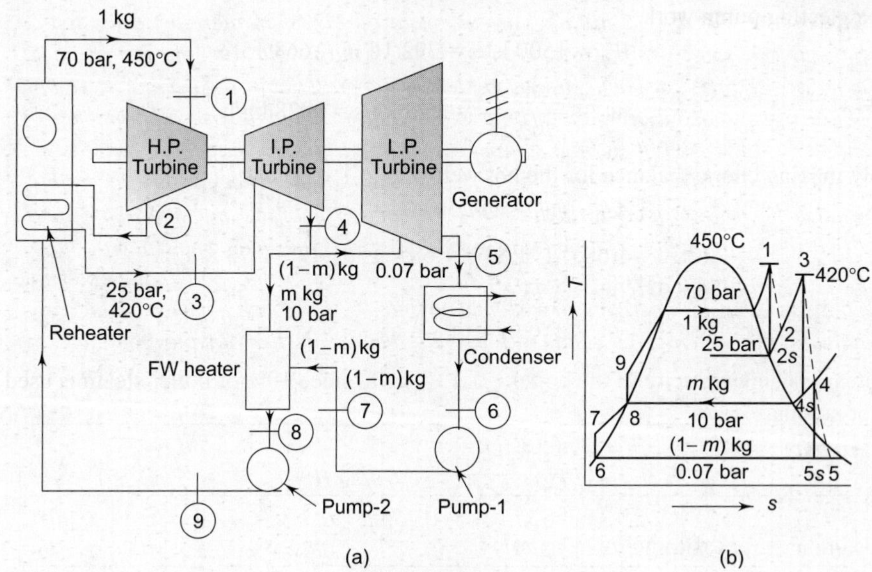

Fig. 12.54

Therefore, the pressure at which bleeding is necessary is **10 bar**. From the Mollier chart

$$h_1 = 3285,\ h_{2s} = 3010,\ h_3 = 3280,\ h_{4s} = 3030 \text{ kJ/kg}$$

$$h_3 - h_4 = 0.83\,(h_3 - h_{4s}) = 0.83 \times 250 = 207.5 \text{ kJ/kg}$$

$$h_4 = 3280 - 207.5 = 3072.5 \text{ kJ/kg}$$

$$h_{5s} = 2225 \text{ kJ/kg}$$

$$h_4 - h_5 = 0.83\,(h_4 - h_{5s}) = 0.83 \times 847.5 = 703.4 \text{ kJ/kg}$$

$$\therefore \quad h_5 = 3072.5 - 703.4 = 2369.1 \text{ kJ/kg}$$

$$h_6 = 162.7 \text{ kJ/kg}$$

$$h_8 = 762.81 \text{ kJ/kg}$$

$$h_1 - h_2 = 0.785\,(h_1 - h_{2s}) = 0.785 \times 275 = 215.9 \text{ kJ/kg}$$

$$h_2 = 3285 - 215.9 = 3069.1 \text{ kJ/kg}$$

Energy balance for the heater gives

$$m \times h_4 + (1 - m)h_7 = 1 \times h_8$$

$$m \times 3072.5 + (1 - m) \times 162.7 = 1 \times 762.81$$

$$m = \frac{600.11}{2909.8} = 0.206 \text{ kg/kg steam flow at turbine inlet.}$$

$$\therefore \quad \eta_{\text{cycle}} = \frac{(h_1 - h_2) + (h_3 - h_4) + (1 - m)(h_4 - h_5)}{(h_1 - h_8) + (h_3 - h_2)}$$

$$= \frac{215.9 + 207.5 + 0.794 \times 703.4}{2522.2 + 210.9}$$

$$= \frac{981.9}{2733.1} = \mathbf{0.3592 \text{ or } 35.92\%}$$

Example 12.15 *A binary-vapour cycle operates on mercury and steam. Saturated mercury vapour at 4.5 bar is supplied to the mercury turbine, from which it exhausts at 0.04 bar. The mercury condenser generates saturated steam at 15 bar which is expanded in a steam turbine to 0.04 bar. (a) Find the overall efficiency of the cycle. (b) If 50,000 kg/h of steam flows through the steam turbine, what is the flow through the mercury turbine? (c) Assuming that all processes are reversible, what is the useful work done in the binary vapour cycle for the specified steam flow? (d) If the steam leaving the mercury condenser is superheated to a temperature of 300°C in a superheater located in the mercury boiler, and if the internal efficiencies of the mercury and steam turbines are 0.85 and 0.87, respectively, calculate the overall efficiency of the cycle. The properties of saturated mercury are given below:* D [LO 12.14]

p (bar)	t(°C)	h_f (kJ/kg)	h_g	s_f (kJ/kg K)	s_g	v_f (m³/kg)	v_g
4.5	450	62.93	355.98	0.1352	0.5397	79.9×10^{-6}	0.068
0.04	216.9	29.98	329.85	0.0808	0.6925	76.5×10^{-6}	5.178

Solution The cycle is shown in Fig. 12.55.

For the mercury cycle, $h_a = 355.98$ kJ/kg

$$s_a = 0.5397 \text{ kJ/kg K} = s_b = s_f + x_b s_{fg} = 0.0808 + x_b (0.6925 - 0.0808)$$

$$\therefore \quad x_b = \frac{0.4589}{0.6117} = 0.75$$

$$h_b = 29.98 + 0.75 \times 299.87 = 254.88 \text{ kJ/kg}$$

$$(W_T)_m = h_a - h_b = 355.98 - 254.88 = 101.1 \text{ kJ/kg}$$

$$(W_P)_m = h_d - h_c = 76.5 \times 10^{-6} \times 4.46 \times 100 = 3.41 \times 10^{-2} \text{ kJ/kg}$$

$$\therefore \quad W_{net} = 101.1 \text{ kJ/kg}$$

$$Q_1 = h_a - h_d = 355.98 - 29.98 = 326 \text{ kJ/kg}$$

$$\therefore \quad \eta_m = \frac{W_{net}}{Q_1} = \frac{101.1}{326} = 0.31 \text{ or } 31\%$$

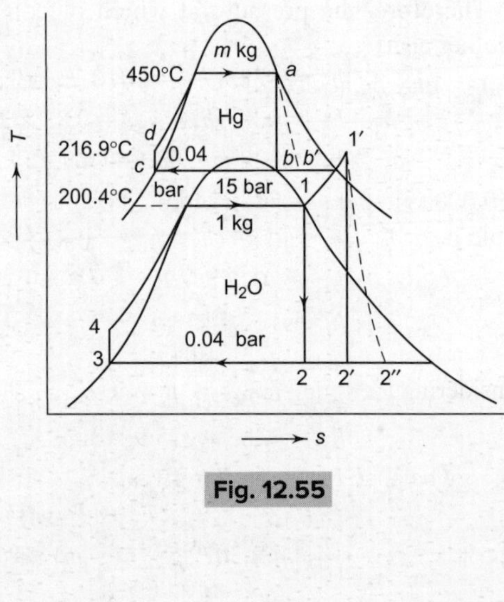

Fig. 12.55

For the steam cycle

$$h_1 = 2792.2 \text{ kJ/kg}$$

$$s_1 = 6.4448 \text{ kJ/kg K} = s_2 = s_f + x_2 s_{fg2} = 0.4226 + x_2 (8.4746 - 0.4226)$$

$$x_2 = \frac{6.0222}{8.0520} = 0.748$$

$$h_2 = 121.46 + 0.748 \times 2432.9 = 1941.27 \text{ kJ/kg}$$
$$(W_T)_{St} = h_1 - h_2 = 2792.2 - 1941.27 = 850.93 \text{ kJ/kg}$$
$$(W_P)_{St} = h_4 - h_3 = 0.001 \times 14.96 \times 100 = 1.496 \text{ kJ/kg} \cong 1.5 \text{ kJ/kg}$$
$$h_4 = 121.46 + 1.5 = 122.96 \text{ kJ/kg}$$
$$Q_1 = h_1 - h_4 = 2792.2 - 122.96 = 2669.24 \text{ kJ/kg}$$
$$(W_{net})_{St} = 850.93 - 1.5 = 849.43 \text{ kJ/kg}$$

$$\therefore \quad \eta_{st} = \frac{W_{net}}{Q_1} = \frac{849.43}{2669.24} = 0.318 \text{ or } 31.8\%$$

Overall efficiency of the binary cycle would be

$$\eta_{overall} = \eta_m + \eta_{St} - \eta_m \cdot \eta_{St}$$
$$= 0.31 + 0.318 - 0.31 \times 0.318$$
$$= \mathbf{0.5294 \text{ or } 52.94\%}$$

$\eta_{overall}$ can also be determined in the following way:

By writing the energy balance for a mercury condenser-steam boiler

$$m(h_b - h_c) = 1(h_1 - h_4)$$

where m is the amount of mercury circulating for 1 kg of steam in the bottom cycle.

$$\therefore \quad m = \frac{h_1 - h_4}{h_b - h_c} = \frac{2669.24}{254.88 - 29.98} = \frac{2669.24}{224.90} = 11.87 \text{ kg}$$
$$(Q_1)_{total} = m(h_a - h_d) = 11.87 \times 326 = 3869.6 \text{ kJ/kg}$$
$$(W_T)_{total} = m(h_a - h_b) + (h_1 - h_2)$$
$$= 11.87 \times 101.1 + 850.93 = 2051 \text{ kJ/kg}$$

$(W_P)_{total}$ may be neglected

$$\therefore \quad \eta_{overall} = \frac{W_T}{Q_1} = \frac{2051}{3869.6} = \mathbf{0.53 \text{ or } 53\%}$$

If 50,000 kg/h of steam flows through the steam turbine, the flow rate of mercury w_m would be

$$w_m = 50{,}000 \times 11.87 = \mathbf{59.35 \times 10^4 \text{ kg/h}}$$
$$(W_T)_{total} = 2051 \times 50{,}000 = 10255 \times 10^4 \text{ kJ/h}$$
$$= 0.2849 \times 10^5 \text{ kW} = \mathbf{28.49 \text{ MW}}$$

Considering the efficiencies of turbines

$$(W_T)_m = h_a - h'_b = 0.85 \times 101.1 = 84.95 \text{ kJ/kg}$$
$$\therefore \quad h'_b = 355.98 - 85.94 = 270.04 \text{ kJ/kg}$$
$$\therefore \quad m'(h_{b'} - h_{c'}) = (h_1 - h_4)$$
$$\therefore \quad m' = \frac{2669.24}{240.06} = 11.12 \text{ kg}$$
$$(Q_1)_{total} = m'(h_a - h_d) + 1(h'_1 - h_1)$$
$$= 11.12 \times 326 + (3037.3 - 2792.2)$$
$$= 3870.22 \text{ kJ/kg}$$
$$s'_1 = 6.9160 = 0.4226 + x'_2 (8.4746 - 0.4226)$$
$$x'_2 = \frac{6.4934}{8.0520} = 0.806$$
$$h'_2 = 121.46 = 0.806 \times 2432.9 = 2082.38 \text{ kJ/kg}$$
$$(W_T)_{st} = h'_1 - h''_2 = 0.87\,(3037.3 - 2082.38)$$
$$= 830.78 \text{ kJ/kg}$$

$$(W_T)_{\text{total}} = 11.12 \times 85.94 + 830.78 = 1786.43 \text{ kJ/kg}$$

Pump work is neglected.

$$\eta_{\text{overall}} = \frac{1786.43}{3870.22} = \mathbf{0.462 \text{ or } 46.2\%}$$

Example 12.16 *In a double reheat ultra-super critical steam power cycle, steam is supplied to the turbine at 30 MPa, 610⁰C. Two reheat pressures are 20 MPa and 10 MPa. Reheating at both pressures are also done up to 610⁰C. Assuming isentropic expansion in all turbines and condensation temperature being 55⁰C, find out the efficiency of the power cycle. Neglect pump work. Assume properties of steam at 30MPa, 610⁰C: h=3474 kJ/kg, s=6.2696kJ/kg K.* **D [LO 12.11]**

Solution

The T-s diagram of the double reheat ultra-supercritical steam power cycle is as shown below:

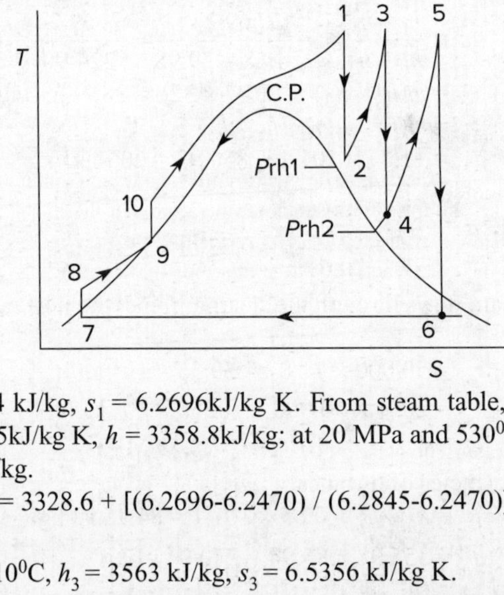

Here, h_1=3474 kJ/kg, s_1 = 6.2696kJ/kg K. From steam table, at 20 MPa and t = 540^{0}C: s = 6.2845kJ/kg K, h = 3358.8kJ/kg; at 20 MPa and 530^{0}C: s = 6.2470 kJ/kg K, h = 3328.6 kJ/kg.

As, $s_2 = s_1$, h_2 = 3328.6 + [(6.2696-6.2470) / (6.2845-6.2470)] × (3358.8-3328.6) = 3346.8 kJ/kg.

At 20 MPa, 610^{0}C, h_3 = 3563 kJ/kg, s_3 = 6.5356 kJ/kg K.

At 10 MPa, 480^{0}C, s = 6.5321 kJ/kg K ≈ s_3. So, h_4 = 3323 kJ/kg.

At 10 MPa, 610^{0}C, h_5 = 3647kJ/kg, s_5 = 6.9292 kJ/kg K

At 55^{0}C, s_g = 7.9912 kJ/kg K, s_f = 0.7679 kJ/kg K, h_g = 2601 kJ/kg, h_f = 230 kJ/kg = $h7$

As, $s_6 = s_5$, x_6 = (6.9292–0.7679)/(7.9912–0.7679) = 0.853

Hence, h_6 = 230 + 0.853 × (2601–230) = 2252.46 kJ/kg

So, neglecting pump work, efficiency:

$\eta = [(h_1 - h_2) + (h_3 - h_4) + (h_5 - h_6)]/[(h_1 - h_7) + (h_3 - h_2) + (h_5 - h_4)]$

= [(3474–3346.8) + (3563–3323) + (3647–2252.46)]/[(3474–230) + (3563–3346) + (3647–3323)] = 1761.74/3785

= **0.4655 or 46.55%**

SUMMARY

Rankine cycle is a phase-change cycle consisting of two reversible isobars (where heat is supplied and rejected) and two reversible adiabatics (where work is done by and upon the system) represents a simple steam power plant. The flow, $p-v$, $T-s$ and $h-s$ diagrams of the cycle are discussed.

The cycle efficiency, steam rate and heat rate are defined as

$$\eta_{\text{cycle}} = \frac{W_T - W_P}{Q_1} = \frac{Q_1 - Q_2}{Q_1} = \frac{(h_1 - h_4) - (h_2 - h_3)}{h_1 - h_4}$$

$$= 1 - \frac{T_2}{T_m}$$

$$\text{Steam rate} = \frac{3600}{W_{\text{net}}} \text{ kg/kWh}$$

$$\text{Heat rate} = 3600 \frac{Q_1}{W_{\text{net}}} \text{ kJ/kWh}$$

where T_{m1} is the mean temperature of heat addition $(h_1 - h_4)/(s_1 - s_4)$ and T_2 is the temperature of heat rejection. The minimum value of T_2 being fixed by the ambient condition, the cycle efficiency depends on T_{m1}. Higher the mean temperature of heat addition, higher is the cycle efficiency. The steam rate (SR) is the steam flow rate (kg/h) and the heat rate (HR) is the heat release rate in the furnace (kJ/h) to produce unit electrical output (kWh). The SR and the H.R. are the reciprocal of W_{net} and cycle efficiency, respectively.

T_{m1} during heat supply can be increased by going to higher temperature T_1, higher pressure, reheat and regeneration of steam. The maximum steam temperature is limited by the metallurgical point, i.e. the maximum steam temperature that the metals can withstand, the maximum steam pressure is restricted to a certain fixed value due to stress considerations and stresses subjected to the blades and the maximum moisture content that can be tolerated by the later stages of the turbine. The reheating of steam improves the turbine output significantly and not the cycle efficiency appreciably. The feedwater heating or regeneration by heating the feedwater with the steam extracted from the turbine significantly improves the cycle efficiency but reduces the turbine output. The closed heaters are always preferred in practise, but one open heater, feedwater, called deaerator, is used to get rid of the dissolved gases like CO_2 and O_2 in feedwater. The gain in cycle efficiency follows the law of diminishing return as the number of heaters increases, and the degree of regeneration is optimised.

Since the heat rejection temperature in the condenser is constrained by the local ambient temperature, one effective way to improve efficiency is by increasing the steam conditions at the turbine inlet. Following this principle, supercritical and ultra-supercritical power cycles achieve higher efficiencies compared to subcritical cycles.

The merits and demerits of multifluid coupled cycles have been discussed. The cogeneration plant meeting the dual demand of process heat and electricity is thermodynamically beneficial.

REVIEW QUESTIONS

(E) **12.1** What are the four basic components of a steam power plant? **[LO 12.1]**

(M) **12.2** What is the reversible cycle that represents the simple steam power plant? **[LO 12.2]** Draw the flow, p–v, T–s and h–s diagrams of this cycle.

(E) **12.3** What do you understand by steam rate and heat rate? What are their units? **[LO 12.2]**

(D) **12.4** Why is Carnot cycle not practicable for a steam power plant? **[LO 12.4]**

(D) **12.5** What do you understand by the mean temperature of heat addition? **[LO 12.5]**

(M) **12.6** For a given T_2, show how the Rankine cycle efficiency depends on the mean temperature of heat addition. **[LO 12.5]**

(E) **12.7** What is metallurgical limit? **[LO 12.5]**

(E) **12.8** Explain how the quality at turbine exhaust gets restricted. **[LO 12.5]**

(M) **12.9** How are the maximum temperature and maximum pressure in the Rankine cycle fixed? **[LO 12.5]**

(D) **12.10** When is reheating of steam recommended in a steam power plant? How does the reheat pressure get optimised? **[LO 12.6]**

(D) **12.11** What is the effect of reheat on (a) the specific output, (b) the cycle efficiency, (c) steam rate and (d) heat rate, of a steam power plant? **[LO 12.6]**

(D) **12.12** Give the flow and *T–s* diagrams of the ideal regenerative cycle. Why is the efficiency of this cycle equal to Carnot efficiency? Why is this cycle not practicable? **[LO 12.7]**

(D) **12.13** What is the effect of regeneration on the (a) specific output, (b) mean temperature of heat addition, (c) cycle efficiency, (d) steam rate and (e) heat rate of a steam power plant? **[LO 12.8]**

(D) **12.14** How does the regeneration of steam carnotise the Rankine cycle? **[LO 12.8]**

(M) **12.15** What are open and closed heaters? Mention their merits and demerits. **[LO 12.10]**

(E) **12.16** Why is one open feedwater heater used in a steam plant? What is it called? **[LO 12.10]**

(D) **12.17** How are the number of heaters and the degree of regeneration get optimised? **[LO 12.10]**

(D) **12.18** How does a supercritical steam power cycle differ from a subcritical power cycle, and what advantages does the former offer over the latter? **[LO 12.11]**

(M) **12.19** Draw the *T–s* diagram of an ideal working fluid in a vapour power cycle. **[LO 12.13]**

(M) **12.20** Discuss the desirable characteristics of a working fluid in a vapour power cycle. **[LO 12.13]**

(E) **12.21** Mention a few working fluids suitable in the high temperature range of a vapour power cycle. **[LO 12.14]**

(E) **12.22** What is a binary vapour cycle? **[LO 12.14]**

(E) **12.23** What are topping and bottoming cycles? **[LO 12.14]**

(M) **12.24** Show that the overall efficiency of two cycles coupled in series equals the sum of the individual cycle efficiencies minus their product. **[LO 12.15]**

(M) **12.25** What is a cogeneration plant? What are the thermodynamic advantages of such a plant? **[LO 12.16]**

(M) **12.26** What is a back pressure turbine? What are its applications? **[LO 12.16]**

(M) **12.27** What is the biggest loss in a steam plant? How can this loss be reduced? **[LO 12.16]**

(M) **12.28** What is a pass-out turbine? When is it used? **[LO 12.16]**

(M) **12.29** Define the following: (a) internal work, (b) internal efficiency, **[LO 12.17]**

(c) brake efficiency (d) mechanical efficiency and (e) boiler efficiency.

(E) **12.30** Express the overall efficiency of a steam plant as the product of boiler, turbine, generator and cycle efficiencies. **[LO 12.17]**

PROBLEMS WITH HINTS

(E) **12.1** A steam turbine with an internal efficiency of 90% receives steam at 7 MPa and 550°C and exhausts at 20 kPa. Determine the turbine work, exhaust enthalpy and exit quality of the steam. **[LO 12.2]**

Hint:

$$h_2 = ?, x_2 = ?\ W_T = h_1 - h_2 = ?$$

$$s_1 = s_{2s} = 6.9486$$

$$= s_f + x_{2s}\, s_{fg} = 0.8319 + x_{2s} \times 7.0766$$

$$x_{2s} = 0.864$$

$$h_{2s} = h_f + 0.864\, h_{fg}$$

$$= 2290 \text{ kJ/kg}$$

$$h_1 - h_2 = \eta_T (h_1 - h_{2s})$$

$$\therefore \quad h_2 = \mathbf{2414.1\ kJ/kg}$$

$$= h_f + x_2\, h_{fg}$$

$$\therefore \quad x_2 = \mathbf{0.917}$$

$$W_T = h_1 - h_2$$

$$= \mathbf{1116.81\ kJ/kg}$$

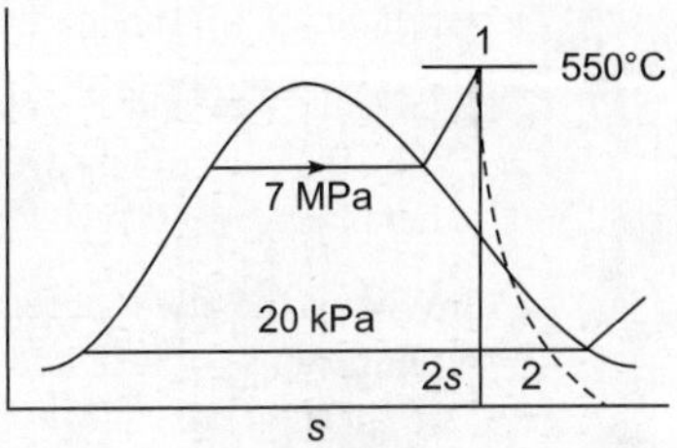

Fig. 12.56

(M) **12.2** Calculate the power plant efficiency and the net work of a steam power plant that has turbine inlet steam conditions of 6 MPa and 500°C, bleed steam to the feedwater heater occurring at 800 kPa and exhaust to the condenser at 15 kPa. The turbine and pump have internal efficiencies of 90 %. The flow rate is 63 kg/s **[LO 12.8]**

Hint:

$$h_1 = 3422.1 \text{ kJ/kg}$$

$$s_1 = 6.8802 \text{ kJ/kg K} = s_{2s} = s_{3s}$$

$$s_{3s} = 0.7548 + x_{3s} \times 7.2536$$

$$x_{3s} = 0.8442$$

$$h_{3s} = 225.91 \times 0.8442 \times 2373.1$$

$$= 2229.3 \text{ kJ/kg}$$

$$h_1 - h_3 = \eta_T (h_1 - h_{3s})$$

$$= 0.9 \times 1192.8$$

$$h_3 = 2348.6 \text{ kJ/kg}$$

$$h_6 = 721.1 \text{ kJ/kg}$$

$$h_7 = 727.0 \text{ kJ/kg}$$

$$h_4 = 225.9 \text{ kJ/kg}$$

$$h_5 = 226.7 \text{ kJ/kg}$$

$$s_{2s} = 6.8802 = 2.0461 + x_{2s}\, 4.6166\, x_{2s} > 1$$

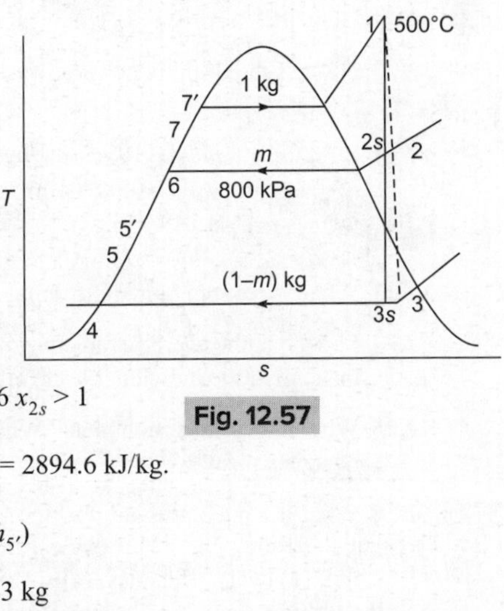

Fig. 12.57

From superheated steam table

$$\therefore \quad t_2 = 225°\text{C and } h_2 = 2894.6 \text{ kJ/kg.}$$

Now.,

$$m(h_2 - h_c) = (1 - m)(h_6 - h_{5'})$$

$$m = \frac{h_6 - h_5'}{h_2 - h_5} = 0.223 \text{ kg}$$

$$W_P = 6.35 + 0.88 = 7.23 \text{ kJ/kg}$$

$$W_T = h_1 - h_2 + (1 - m)(h_2 - h_3) = 1056.48$$
$$W_{net} = 1044.25 \text{ kJ/kg}, \; Q_1 = h_1 - h_{7'} = 2695.0 \text{ kJ/kg}$$
$$\eta_{cycle} = \mathbf{0.387 \text{ or } 38.7\%}$$
$$W_{net} = 1044.25 \times 63 = 6265.5 \text{ kW} = \mathbf{6.26 \text{ MW}}$$

(E) **12.3** A steam power plant operating on the ideal reheat Rankine cycle has steam condition at turbine inlet at 15 MPa, 600°C and is condensed at pressure of 10 kPa. If the moisture content at turbine is not to exceed 10.4%, determine (a) the pressure at which the steam should be reheated and (b) the thermal efficiency of the cycle. Assume the steam is reheated to the inlet temperature of the turbine. **[LO 12.6]**

Hint:

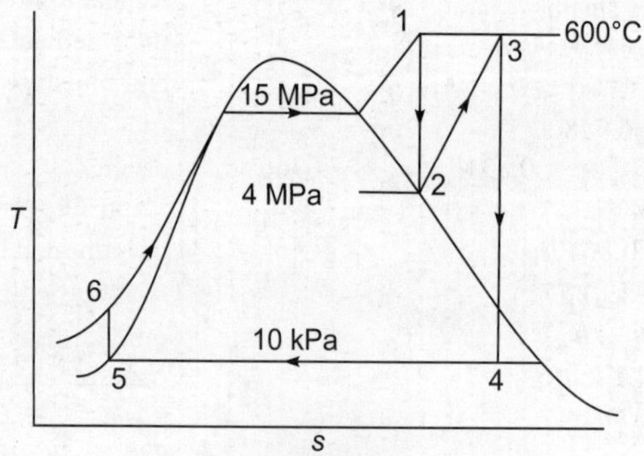

Fig. 12.58

State 1, $h_1 = 3583.1$ kJ/kg

$s_1 = 6.6796$ kJ/kgK

State 4, $P_4 = 10$ kPa

$x_4 = 0.896$

$s_4 = x_f + x_4 s_{fg}$

$= 7.3688$ kJ/kgK.

$h_4 = 191.81 + 0.896 \times 2392.1$

$= 2335.1$ kJ/kg

State 3 $\left.\begin{array}{l} T_3 = 600°\text{C} \\ s_3 = s_4 \end{array}\right\}$ $\therefore P_2 = \mathbf{4 \text{ MPa}}$

$h_3 = 3674.9$ kJ/kg

State 5

$h_5 = 191.81$ kJ/kg

$v_5 = 0.00101 \text{ m}^3/\text{kg}$

State 6 $W_p = v(p_1 - p_2) = 15.14$ kJ/kg

$h_6 = 206.95$ kJ/kg

$Q_1 = h_1 - h_6 + h_3 - h_2 = 3896.1$ kJ/kg

$Q_2 = h_4 - h_3 = 2143.3$ kJ/ks

$\therefore$ $h_{th} = 1 - \dfrac{Q_2}{Q_1} = \mathbf{0.45 \text{ or } 45\ \%}$

PROBLEMS

(E) **12.1** For the following steam cycles, find (a) W_T in kJ/kg (b) W_P in kJ/kg, (c) Q_1 in kJ/kg, (d) cycle efficiency, (e) steam rate in kg/kW h and (f) moisture at the end of the turbine process. Show the results in tabular form with your comments.

[LO 12.2, 12.6, 12.7, 12.8, 12.10]

Boiler Outlet	Condenser Pressure	Type of Cycle
10 bar, saturated	1 bar	Ideal Rankine Cycle
-do-	-do-	Neglect W_P
-do-	-do-	Assume 75% pump and turbine efficiency
-do-	0.1 bar	Ideal Rankine Cycle
10 bar, 300°C	-do-	-do-
150 bar, 600°C	-do-	-do-
-do-	-do-	Reheat to 600°C at maximum intermediate pressure to limit end moisture to 15%
-do-	-do-	-do- but with 85% turbine efficiency
10 bar, saturated	0.1 bar	Isentropic pump process ends on saturated liquid line
-do-	-do-	-do- but with 80% machine efficiencies
-do-	-do-	Ideal regenerative cycle
-do-	-do-	Single open heater at 110°C
-do-	-do-	Two open heaters at 90°C and 135°C
-do-	-do-	-do- but the heaters are closed heaters

(E) **12.2** A geothermal power plant utilises steam produced by natural means underground. Steam wells are drilled to tap this steam supply which is available at 4.5 bar and 175°C. The steam leaves the turbine at 100 mm Hg absolute pressure. The turbine isentropic efficiency is 0.75. Calculate the efficiency of the plant. If the unit produces 12.5 MW, what is the steam flow rate? **[LO 12.2, 12.17]**

(E) **12.3** A simple steam power cycle uses solar energy for the heat input. Water in the cycle enters the pump as a saturated liquid at 40°C, and is pumped to 2 bar. It then evaporates in the boiler at this pressure, and enters the turbine as saturated vapour. At the turbine exhaust the conditions are 40°C and 10% moisture. The flow rate is 150 kg/h. Determine (a) the turbine isentropic efficiency, (b) the net work output (c) the cycle efficiency and (d) the area of solar collector needed if the collectors pick up 0.58 kW/m^2. **[LO 12.2, 12.17]**

(E) **12.4** In a reheat cycle, the initial steam pressure and the maximum temperature are 150 bar and 550°C, respectively. If the condenser pressure is 0.1 bar and the moisture at the condenser inlet is 5%, and assuming ideal processes, determine (a) the reheat pressure, (b) the cycle efficiency and (c) the steam rate. **[LO 12.6]**

(M) **12.5** In a nuclear power-plant, heat is transferred in the reactor to liquid sodium. The liquid sodium is then pumped to a heat exchanger where heat is transferred to steam. The steam leaves this heat exchanger as saturated vapour at 55 bar, and is then superheated in an external gas-fired superheater to 650°C. The steam then enters the turbine, which has one extraction point at 4 bar, where steam flows to an open feedwater heater. The turbine efficiency is 75% and the condenser temperature is 40°C. Determine the heat transfer in the reactor and in the superheater to produce a power output of 80 MW. **[LO 12.8]**

(M) **12.6** In a reheat cycle, steam at 500°C expands in a HP turbine till it is saturated vapour. It is reheated at constant pressure to 400°C and then expands in a LP turbine to 40°C. If the maximum moisture content at the turbine exhaust is limited to 15%, find (a) the reheat pressure, (b) the pressure of steam at the inlet to the HP turbine, (c) the net specific work output, (d) the cycle efficiency and (e) the steam rate. Assume all ideal processes.

What would have been the quality, the work output and the cycle efficiency without the reheating of steam? Assume that the other conditions remain the same. **[LO 12.2, 12.6]**

(M) **12.7** A regenerative cycle operates with steam supplied at 30 bar and 300°C, and condenser pressure of 0.08 bar. The extraction points for two heaters (one closed and one open) are at 3.5 bar and 0.7 bar, respectively. Calculate the thermal efficiency of the plant, neglecting pump work. **[LO 12.8, 12.10]**

(M) **12.8** The net power output of the turbine in an ideal reheat-regenertive cycle is 100 MW. Steam enters the high-pressure (HP) turbine at 90 bar, 550°C. After expansion to 7 bar, some of the steam goes to an open heater and the balance is reheated to 400°C, after which it expands to 0.07 bar. (a) What is the steam flow rate to the HP turbine? (b) What is the total pump work? (c) Calculate the cycle efficiency. (d) If there is a 10°C rise in the temperature of the cooling water, what is the rate of flow of the cooling water in the condenser? (e) If the velocity of the steam flowing from the turbine to the condenser is limited to a maximum of 130 m/s, find the diameter of the connecting pipe. **[LO 12.9]**

(M) **12.9** A mercury cycle is superposed on the steam cycle operating between the boiler outlet condition of 40 bar, 400°C and the condenser temperature of 40°C. The heat released by mercury condensing at 0.2 bar is used to impart the latent heat of vapourisation to the water in the steam cycle. Mercury enters the mercury turbine as saturated vapour at 10 bar. Compute (a) kg of mercury circulated per kg of water and (b) the efficiency of the combined cycle.

The property values of saturated mercury are given below:

p (bar)	*t (°C)*	h_f	h_g	s_f	s_g	v_f	v_g
		(kJ/kg)		(kJ/kg K)		(m³/kg)	
10	515.5	72.23	363.0	0.1478	0.5167	80.9×10^{-6}	0.0333
0.2	277.3	38.35	336.55	0.0967	0.6385	77.4×10^{-6}	1.163

[LO 12.14]

(M) **12.10** In an electric generating station, using a binary vapour cycle with mercury in the upper cycle and steam in the lower, the ratio of mercury flow to steam flow is 10 : 1 on a mass basis. At an evaporation rate of 1,000,000 kg/h for the mercury, its specific enthalpy rises by 356 kJ/kg in passing through the boiler. Superheating the steam in the boiler furnace adds 586 kJ to the steam specific enthalpy. The mercury gives up 251.2 kJ/kg during condensation, and the steam gives up 2003 kJ/kg in its condenser. The overall boiler efficiency is 85%. The combined turbine mechanical and generator efficiencies are each 95% for the mercury and steam units. The steam auxiliaries require 5% of the energy generated by the units. Find the overall efficiency of the plant. **[LO 12.14, 12.17]**

(D) **12.11** A sodium-mercury-steam cycle operates between 1000°C and 40°C. Sodium rejects heat at 670°C to mercury. Mercury boils at 24.6 bar and rejects heat at 0.141 bar. Both the sodium and mercury cycles are saturated. Steam is formed at 30 bar and is superheated in the sodium boiler to 350°C. It rejects heat at 0.08 bar. Assume isentropic expansions, no heat losses and no regeneration and neglect pumping work. Find (a) the amounts of sodium and mercury used per kg of steam, (b) the heat added and rejected in the composite cycle per kg steam, (c) the total work done per kg steam, (d) the efficiency of the composite cycle, (e) the efficiency of the corresponding Carnot cycle and (f) the work, heat added and efficiency of a supercritical pressure steam (single fluid) cycle operating at 250 bar and between the same temperature limits.

For mercury, at 24.6 bar, $h_g = 366.78$ kJ/kg

$s_g = 0.48$ kJ/kg K and at 0.141 bar, $s_f = 0.09$

and $s_g = 0.64$ kJ/kg K, $h_f = 36.01$ and $h_g = 330.77$ kJ/kg

For sodium, at 1000°C, $h_g = 4982.53$ kJ/kg

At turbine exhaust, $h = 3914.85$ kJ/kg

At 670°C, $h_f = 745.29$ kJ/kg

For a supercritical steam cycle, the specific enthalpy and entropy at the turbine inlet may be computed by extrapolation from the steam tables. **[LO 12.14]**

(E) **12.12** A textile factory requires 10,000 kg/h of steam for process heating at 3 bar saturated and 1000 kW of power, for which a back pressure turbine of 70% internal efficiency is to be used. Find the steam condition required at the inlet to the turbine. **[LO 12.16, 12.17]**

(E) **12.13** A 10,000 kW steam turbine operates with steam at the inlet at 40 bar, 400°C and exhausts at 0.1 bar. Ten thousand kg/h of steam at 3 bar are to be extracted for process work. The turbine has 75% isentropic efficiency throughout. Find the boiler capacity required. **[LO 12.16, 12.17]**

(M) **12.14** A 50 MW steam plant built in 1935 operates with steam at the inlet at 60 bar, 450ºC and exhausts at 0.1 bar, with 80% turbine efficiency. It is proposed to scrap the old boiler and put in a new boiler and a topping turbine of efficiency 85% operating with inlet steam at 180 bar, 500°C. The exhaust from the topping turbine at 60 bar is reheated to 450°C and admitted to the old turbine. The flow rate is just sufficient to produce the rated output from the old turbine. Find the improvement in efficiency with the new set up. What is the additional power developed? **[LO 12.16, 12.17]**

(M) **12.15** A steam plant operates with an initial pressure at 20 bar and temperature 400°C, and exhausts to a heating system at 2 bar. The condensate from the heating system is returned to the boiler plant at 65°C, and the heating system utilises for its intended purpose 90% of the energy transferred from the steam it receives. The turbine efficiency is 70%. (a) What fraction of the energy supplied to the steam plant serves a useful purpose? (b) If two separate steam plants had been set up to produce the same useful energy, one to generate heating steam at 2 bar, and the other to generate power through a cycle working between 20 bar, 400°C and 0.07 bar, what fraction of the energy supplied would have served a useful purpose? **[LO 12.16, 12.17]**

(M) **12.16** In a nuclear power plant, saturated steam at 30 bar enters a HP turbine and expands isentropically to a pressure at which its quality is 0.841. At this pressure, the steam is passed through a moisture separator which removes all the liquid. Saturated vapour leaves the separator and is expanded isentropically to 0.04 bar in LP turbine, while the saturated liquid leaving the separator is returned via a feed pump to the boiler. The condensate leaving the condenser at 0.04 bar is also returned to the boiler via a second feed pump. Calculate the cycle efficiency and turbine outlet quality taking into account the feed pump term. Recalculate the same quantities for a cycle with the same boiler and condenser pressures but without moisture separation. **[LO 12.10]**

(M) **12.17** The net power output of an ideal regenerative-reheat steam cycle is 80 MW. Steam enters the HP turbine at 80 bar, 500°C and expands till it becomes saturated vapour. Some of the steam then goes to an open feedwater heater and the balance is reheated to 400°C, after which it expands in the LP turbine to 0.07 bar. Compute (a) the reheat pressure, (b) the steam flow rate to the HP turbine and (c) the cycle efficiency. Neglect pump work. **[LO 12.9]**

(D) **12.18** Figure 12.59 shows the arrangement of a steam plant in which steam is also required for an industrial heating process. The steam leaves boiler *B* at 30 bar, 320°C and expands in the HP turbine to 2 bar, the efficiency of the HP turbine being 75%. At this point, one half of the steam passes to the process heater *P* and the remainder enters separator *S* which removes all the moisture. The dry steam enters the LP turbine at 2 bar and expands to the condenser pressure 0.07 bar, the efficiency of the LP turbine being 70%. The drainage from the separator mixes with the condensate from the process heater and the combined flow enters the hotwell *H* at 80°C. Traps are provided at the exist from *P* and *S*. A pump extracts the condensate from condenser *C* and this enters the hotwell at 38°C. Neglecting the feed pump work and radiation loss, estimate the temperature of water leaving the hotwell which is at atmospheric pressure. Also calculate, as percentage of heat transferred in the boiler, (a) the heat transferred in the process heater, and (b) the work done in the turbines. **[LO 12.10]**

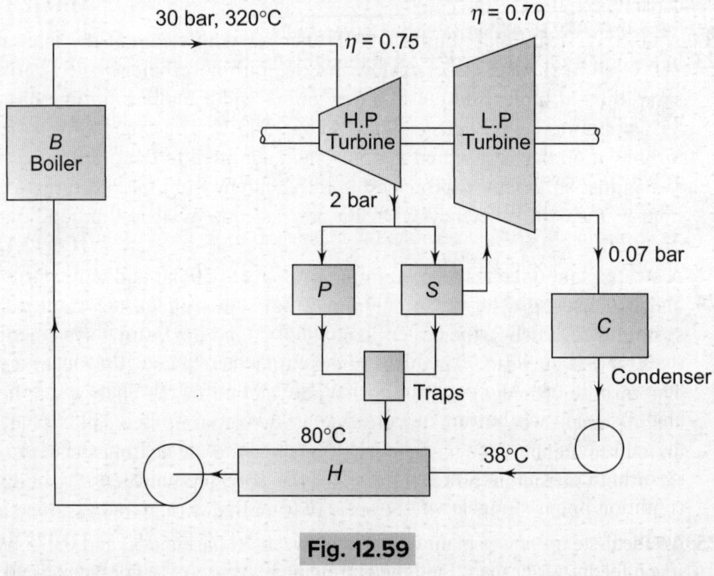

Fig. 12.59

(M) **12.19** In a combined power and process plant, the boiler generates 21,000 kg/h of steam at a pressure of 17 bar, and temperature 230°C. A part of the steam goes to a process heater which consumes 132.56 kW, the steam leaving the process heater 0.957 dry at 17 bar being throttled to 3.5 bar. The remaining steam flows through a HP turbine which exhausts at a pressure of 3.5 bar. The exhaust steam mixes with the process steam before entering the LP turbine which develops 1337.5 kW. At the exhaust the pressure is 0.3 bar, and the steam is 0.912 dry. Draw a line diagram of the plant and determine (a) the steam quality at the exhaust from the HP turbine, (b) the power developed by the HP turbine and (c) the isentropic efficiency of the HP turbine. **[LO 12.16, 12.17]**

(M) **12.20** In a cogeneration plant, the power load is 5.6 MW and the heating load is 1.163 MW. Steam is generated at 40 bar and 500°C and is expanded isentropically through a turbine to a condenser at 0.06 bar. The heating load is supplied by extracting steam from the turbine at 2 bar which condensed in the process heater to saturated liquid at 2 bar and then pumped back to the boiler. Compute (a) the steam generation capacity of the boiler in tonnes/h, (b) the heat input to the boiler in MW and (c) the heat rejected to the condenser in MW. **[LO 12.16]**

(E) **12.21** Steam is supplied to a pass-out turbine at 35 bar, 350°C and dry saturated process steam is required at 3.5 bar. The low pressure stage exhausts at 0.07 bar and the condition line may be assumed to be straight (the condition line is the locus passing through the states of steam leaving the various stages of the turbine). If the power required is 1 MW and the maximum process load is 1.4 kW, estimate the maximum steam flow through the high and low pressure stages. Assume that the steam just condenses in the process plant. **[LO 12.16]**

(M) **12.22** Geothermal energy from a natural geyser can be obtained as a continuous supply of steam 0.87 dry at 2 bar and at a flow rate of 2700 kg/h. This is utilised in a mixed-pressure cycle to augment the superheated exhaust from a high pressure turbine of 83% internal efficiency, which is supplied with 5500 kg/h of steam at 40 bar and 500°C. The mixing process is adiabatic and the mixture is expanded to a condenser pressure of 0.10 bar in a low pressure turbine of 78% internal efficiency. Determine the power output and the thermal efficiency of the plant. **[LO 12.16, 12.17]**

(D) **12.23** In a study for a space project, it is thought that the condensation of a working fluid might be possible at – 40°C. A binary cycle is proposed, using Refrigerant-12 as the low temperature fluid, and water as the high temperature fluid. Steam is generated at 80 bar, 500°C and expands in a turbine of 81% isentropic efficiency to 0.06 bar, at which pressure it is condensed by the generation of dry saturated refrigerant vapour at 30°C from saturated liquid at –40°C. The isentropic efficiency of the R-12 turbine is 83%. Determine the mass ratio of R-12 to water and the efficiency of the cycle. Neglect all losses. **[LO 12.14]**

(M) **12.24** Steam is generated at 70 bar, 500°C and expands in a turbine to 30 bar with an isentropic efficiency of 77%. At this condition it is mixed with twice its mass of steam at 30 bar, 400°C. The mixture then expands, with an isentropic efficiency of 80% to 0.06 bar. At a point in the expansion where the pressure is 5 bar, steam is bled for feedwater heating in a direct contact heater, which raises the feedwater to the saturation temperature of the bled steam. Calculate the mass of steam bled per kg of high pressure steam and the cycle efficiency. Assume that the LP expansion condition line is straight. **[LO 12.8]**

(M) **12.25** An ideal steam power plant operates between 70 bar, 550°C and 0.075 bar. It has seven feedwater heaters. Find the optimum pressure and temperature at which each of the heaters operate. **[LO 12.10]**

(E) **12.26** In a reheat cycle, steam at 550°C expands in an HP turbine till it is saturated vapour. It is reheated at constant pressure to 400°C and then expands in a LP turbine to 40°C. If the moisture content at turbine exhaust is given to be 14.67%, find (a) the reheat pressure, (b) the pressure of steam at inlet to the HP turbine,

(c) the net work output per kg and (d) the cycle efficiency. Assume all processes to be ideal.

[LO 12.6]

(M) **12.27** In a reheat steam cycle, the maximum steam temperature is limited to 500°C. The condenser pressure is 0.1 bar and the quality at turbine exhaust is 0.8778. Had there been no reheat, the exhaust quality would have been 0.7592. Assuming ideal processes, determine (a) reheat pressure, (b) the boiler pressure, (c) the cycle efficiency and (d) the steam rate. **[LO 12.6]**

(M) **12.28** Find out the efficiency of a supercritical steam power cycle with following parameters:
Turbine inlet condition: 23MPa, 600°C (h = 3508 kJ/kg, s = 6.4154 kJ/kg K); one reheat at 6MPa, 600°C and condensing steam at 50°C. Assuming all processes to be reversible and neglecting pump work, calculate the efficiency of the power cycle. **[LO 12.11]**

(M) **12.29** In a cogeneration plant, steam enters the HP stage of a two-stage turbine at 1 MPa, 200°C and leaves it at 0.3 MPa. At this point some of the steam is bled off and passed through a heat exchanger which it leaves as saturated liquid at 0.3 MPa. The remaining steam expands in the LP stage of the turbine to 40 kPa. The turbine is required to produce a total power of 1 MW and the heat exchanger to provide a heating rate of 500 kW. Calculate the required mass flow rate of steam into the HP stage of the turbine. Assume (a) steady condition throughout the plant, (b) velocity and gravity terms to be negligible and (c) both turbine stages are adiabatic with isentropic efficiencies of 0.80. **[LO 12.16]**

(M) **12.30** A steam power plant is designed to operate on the basic Rankine cycle. The heat input to the boiler is at the rate of 50 MW. The H_2O exits the condenser as saturated liquid and exits the boiler as saturated vapour. The pressure of steam at boiler exit is 120 bar and the condenser pressure is 0.04 bar. The heat input to the boiler is provided by a steady stream of hot gases initially at 2200 K and 1 atm. The hot gases exhaust at 600 K and 1 atm to the surroundings which are at 600 K and 1 atm. Taking the c_p of hot gases as 1.1 kJ/kgK, determine (a) the cycle efficiency, (b) the work output, (c) the power output (in MW), (d) the required mass flow rate of steam (in kg/h), (e) the specific steam consumption (in kg/k Wh), (f) the mass flow rate (in kg/h) of the stream of hot gases, (g) the exergy flux (in MW) of the inlet gases, (h) the exergy loss rate (in MW) with the exhaust gases, (i) the exergy consumption (in MW) in the steam generation process, (j) the exergy consumption (in MW) in the condensation process and (i) the second law efficiency. **[LO 12.2, 12.11]**

(M) **12.31** In a cogeneration plant the steam generator provides 10^6 kg/h of steam at 80 bar, 480°C, of which 4×10^5 kg/h is extracted between the first and second turbine stages at 10 bar and diverted to a process heating load. Condensate returns from the process heating load at 9.5 bar, 120°C and is mixed with liquid exiting the lower-pressure pump at 9.5 bar. The entire flow is then pumped to the steam generator pressure. Saturated liquid at 0.08 bar leaves the condenser. The turbine stages and the pumps operate with isentropic efficiencies of 86 and 80%, respectively. Determine (a) the heating load, in kJ/h, (b) the power developed by the turbine, in kW and (c) the rate of heat transfer in the steam generator, in kJ/h. **[LO 12.16, 12.17]**

(M) **12.32** A power plant operates on a regenerative vapour power cycle with one closed feedwater heater. Steam enters the first turbine stage at 120 bar, 520°C and expands to 10 bar, where some of the steam is extracted and diverted to a closed feedwater heater. Condensate exiting the feedwater heater as saturated liquid at

10 bar passes through a trap into the condenser. The feedwater exits the heater at 120 bar with a temperature of 170°C. The condenser pressure is 0.06 bar. For isentropic turbine and pump work, determine (a) the thermal efficiency, and (b) the net power developed for a mass flow rate of 10^6 kg/h into the first-stage turbine. **[LO 12.10]**

(M) **12.33** Reconsider the cycle above, and assume that each turbine has an isentropic efficiency of 82%, the pump efficiency remain 100%. **[LO 12.10, 12.17]**

ANSWERS TO PROBLEMS

12.2 42%, 5.91 kg/s
12.3 (c) 2.78%, (d) 18.2 m^2
12.4 13.5 bar, 43.6%, 2.05 kg/kW h
12.5 155.13 MW, 70.25 MW
12.6 (a) 20 bar,
(b) 150 bar
(c) 1528.36 kJ/kg
(d) 42.64%
(e) 2.356 kg/kWh, 0.75, 42.2%
12.7 36%
12.8 (a) 68.93 kg/s,
(b) 620.36 kW,
(c) 0.4523,
(d) 2890.66 kg/s,
(e) 3.184 m
12.9 (a) 3738.73 (b) 48.11%
12.10 46.86%
12.12 37.5 bar, 344°C
12.13 53.36 kg/s
12.15 91.2%, 55.3%
12.16 35.5%, 0.824; 35%; 0.716
12.17 6.5 bar, 58.4 kg/s, 43.7%
12.19 (a) 0.96,
(b) 1125 kW,
(c) 77%
12.20 (a) 19.07 t/h,
(b) 71.57 MW, and
(c) 9.607 MW
12.21 1.543 and 1.182 kg/s
12.22 1745 kW, 35%
12.23 10.86; 44.4%
12.24 0.53 kg; 31.9%
12.26 (a) 20 bar,
(b) 200 bar,
(c) 1604 kJ/kg,
(d) 43.8%
12.27 (a) 30 bar,
(b) 150 bar,
(c) 50.51%,
(d) 1.9412 kg/kWh
12.28 46%
12.29 2.457 kg/s
12.30 (f) 1.20 ¥ 10^5 kg/h,
(g) 41.5 MW,
(h) 3.14 W,
(i) 16.9 MW,
(j) 1.40 MW, (i) 48.3%
12.31 (a) 9.529 ¥ 10^8 kJ/h,
(b) 236,500 kW,
(c) 3.032 ¥ 10^9 kJ/h
12.32 (a) 43.4%, (b) 325 MW
12.33 (a) 36.8%, (b) 273.3 MW

13

Gas Power Cycles

Learning Objectives

LO 13.1 Illustrate and differentiate between Carnot, Stirling and Ericsson cycles
LO 13.2 Explain the mechanism of reciprocating engines
LO 13.3 Discuss the origin of air standard cycles
LO 13.4 Relate spark ignition engine with Otto cycle and illustrate the cycle
LO 13.5 Relate compression ignition engine with Diesel cycle and illustrate the cycle
LO 13.6 Discuss the significance of Dual cycle and illustrate it
LO 13.7 Compare Otto, Diesel and Dual Cycles
LO 13.8 Illustrate *p-V* and *T-s* diagrams of Lenoir and Atkinson cycles
LO 13.9 Relate gas turbine with Brayton cycle and illustrate the cycle
LO 13.10 Relate actual Brayton cycles with ideal Brayton cycles
LO 13.11 Explain principles of different types of aircraft propulsion systems
LO 13.12 Analyze Brayton-Rankine combined cycle

Here gas is the working fluid. It does not undergo any phase change. Engines operating on gas cycles may be either cyclic or non-cyclic. Hot air engines using air as the working fluid operate on a closed cycle. Internal combustion engines where the combustion of fuel takes place inside the engine cylinder are non-cyclic heat engines.

> **LO 13.1**
> Illustrate and differentiate between Carnot, Stirling and Ericsson cycles

13.1 CARNOT CYCLE (1824)

The Carnot cycle (Fig. 13.1) has been discussed in Chapters 6 and 7. It consists of two reversible isotherms and two reversible adiabatics. If an ideal gas is assumed as the working fluid. Then for 1 kg of gas,

$$Q_{1-2} = RT_1 \ln \frac{v_2}{v_1}; \quad W_{1-2} = RT_1 \ln \frac{v_2}{v_1}$$

$$Q_{2-3} = 0; \quad W_{2-3} = -c_v (T_3 - T_2)$$

$$Q_{3-4} = RT_2 \ln \frac{v_4}{v_3}; \quad W_{3-4} = RT_2 \ln \frac{v_4}{v_3}$$

$$Q_{4-1} = 0; \quad W_{4-1} = -c_v (T_1 - T_4)$$

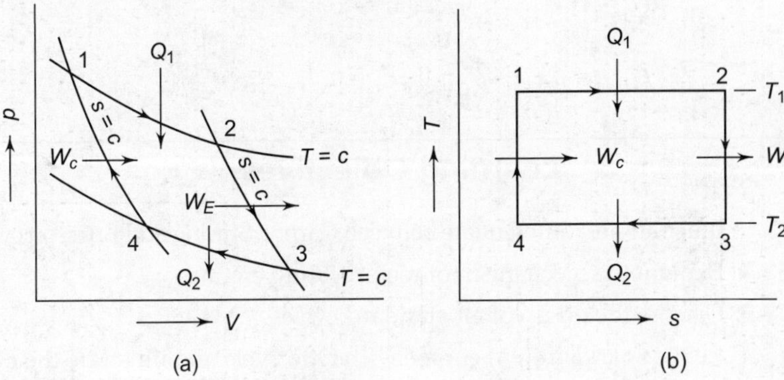

Fig. 13.1 Carnot cycle

$\therefore$
$$\sum_{\text{cycle}} đQ = \sum_{\text{cycle}} dQ$$

Now

$$\frac{v_2}{v_3} = \left(\frac{T_2}{T_1}\right)^{1/(\gamma-1)}$$

and

$$\frac{v_1}{v_4} = \left(\frac{T_2}{T_1}\right)^{1/(\gamma-1)}$$

$\therefore$
$$\frac{v_2}{v_3} = \frac{v_1}{v_4} \text{ or } \frac{v_2}{v_1} = \frac{v_3}{v_4}$$

Therefore

$$Q_1 = \text{Heat added} = RT_1 \ln \frac{v_2}{v_1}$$

$$W_{\text{net}} = Q_1 - Q_2 = R \ln \frac{v_2}{v_1} (T_1 - T_2)$$

$\therefore$
$$\eta_{\text{cycle}} = \frac{W_{\text{net}}}{Q_1} = \frac{T_1 - T_2}{T_1} \tag{13.1}$$

The large back work ($W_c = W_{4-1}$) is a big drawback for the Carnot gas cycle, as in the case of the Carnot vapour cycle.

13.2 STIRLING CYCLE (1827)

The Stirling cycle (Fig. 13.2) consists of two reversible isotherms and two reversible isochores. For 1 kg of ideal gas

$$Q_{1-2} = W_{1-2} = RT_1 \ln \frac{v_2}{v_1}$$

$$Q_{2-3} = -c_v(T_2 - T_1);\ W_{2-3} = 0$$

$$Q_{3-4} = W_{3-4} = -RT_2 \ln \frac{v_3}{v_4}$$

$$Q_{4-1} = c_v(T_1 - T_2);\ W_{4-1} = 0$$

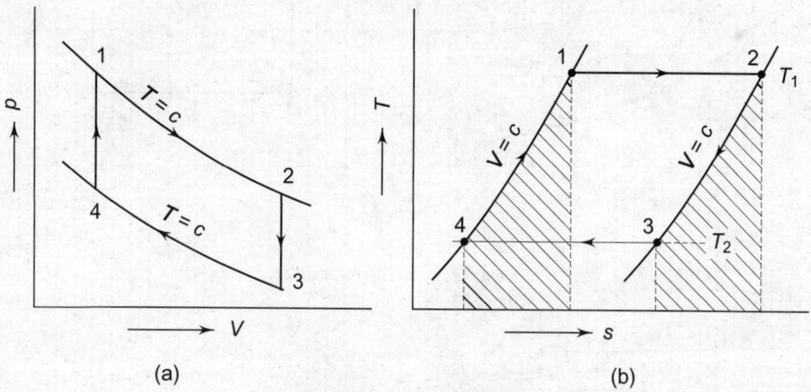

Fig. 13.2 Stirling cycle

Due to heat transfers at constant volume processes, the efficiency of the Stirling cycle is less than that of the Carnot cycle. However, if a regenerative arrangement is used such that $Q_{2-3} = Q_{4-1}$, i.e. the area under 2–3 is equal to the area under 4–1, then the cycle efficiency becomes

$$\eta = \frac{RT_1 \ln \frac{v_2}{v_1} - RT_2 \ln \frac{v_3}{v_4}}{RT_1 \ln \frac{v_2}{v_1}} = \frac{T_1 - T_2}{T_1} \tag{13.2}$$

So, the regenerative Stirling cycle has the same efficiency as the Carnot cycle.

13.3 ERICSSON CYCLE (1850)

The Ericsson cycle (Fig. 13.3) is made up of two reversible isotherms and two reversible isobars.

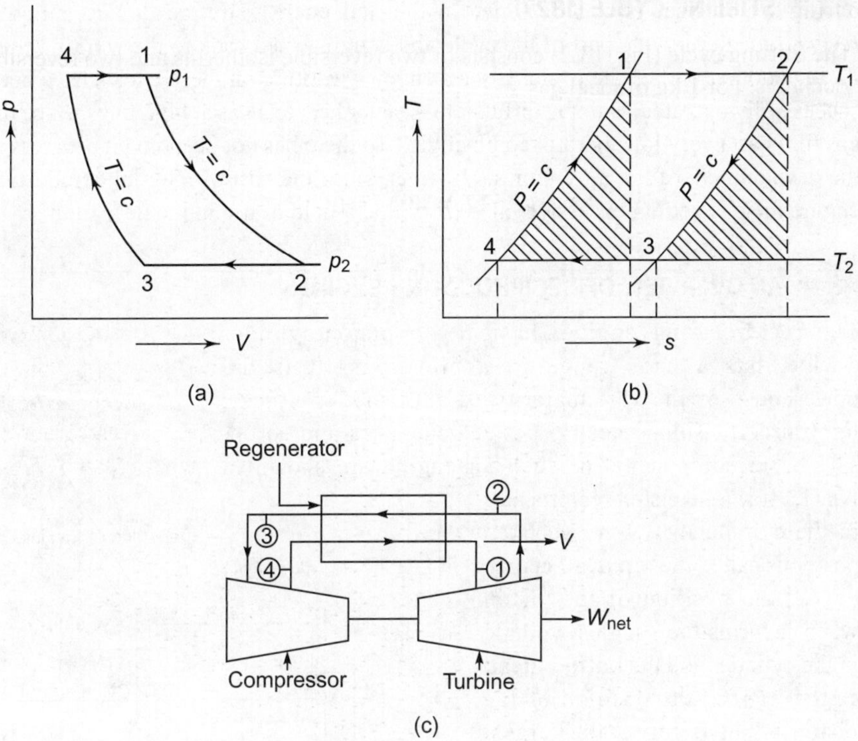

Fig. 13.3 Ericsson cycle

For 1 kg of ideal gas

$$Q_{1-2} = W_{1-2} = RT_1 \ln \frac{p_1}{p_2}$$

$$Q_{2-3} = c_p(T_2 - T_1);\ W_{2-3} = p_2(v_3 - v_2) = R(T_2 - T_1)$$

$$Q_{3-4} = W_{3-4} = -RT_2 \ln \frac{p_1}{p_2}$$

$$Q_{4-1} = c_p(T_1 - T_4);\ W_{4-1} = p_1(v_1 - v_4) = R(T_1 - T_2)$$

Since part of the heat is transferred at constant pressure and part at constant temperature, the efficiency of the Ericsson cycle is less than that of the Carnot cycle. But with ideal regeneration, $Q_{2-3} = Q_{4-1}$ so that all the heat is added from the external source at T_1 and all the heat is rejected to an external sink at T_2, the efficiency of the cycle becomes equal to the Carnot cycle efficiency, since

$$\eta = 1 - \frac{Q_2}{Q_1} = 1 - \frac{RT_2 \ln \frac{p_1}{p_2}}{RT_1 \ln \frac{p_1}{p_2}} = 1 - \frac{T_2}{T_1} \tag{13.3}$$

The regenerative, Stirling and Ericsson cycles have the same efficiency as the Carnot cycle, but much less back work. Both cycles utilise *regeneration*, a process during which heat is transferred to a thermal energy storage device called a *regenerator* during one part of the cycle and is transferred back to the working fluid during another part of the cycle. Hot air engines working on these cycles have been successfully operated. But it is difficult to transfer heat to a gas at high rates, since the gas film has a very low thermal conductivity. So there has not been much progress in the development of hot air engines. However, since the cost of internal combustion engine fuels is getting excessive, these may find a field of use in the near future.

13.4 AN OVERVIEW OF RECIPROCATING ENGINES

LO 13.2

Explain the mechanism of reciprocating engines

The reciprocating engine, basically a piston-cylinder device, has a wide range of applications. It is the powerhouse of the vast majority of automobiles, trucks, light aircraft, ships, electric power generators and so on. The basic components of such an engine are shown in Fig. 13.4(a). The piston reciprocates in the cylinder between two fixed positions called the **top dead centre (TDC)**-the position of the piston when it forms the smallest volume in the cylinder and the **bottom dead centre (BDC)**-the position of the piston when it forms the largest volume in the cylinder. The distance between the TDC and the BDC is the largest distance that the piston can travel in one direction, and it is called the **stroke** of the engine. The diameter of the piston is called the **bore**. The air or air-fuel mixture is drawn into the cylinder through the **intake valve** and the combustion products are expelled from the cylinder through the **exhaust valve**.

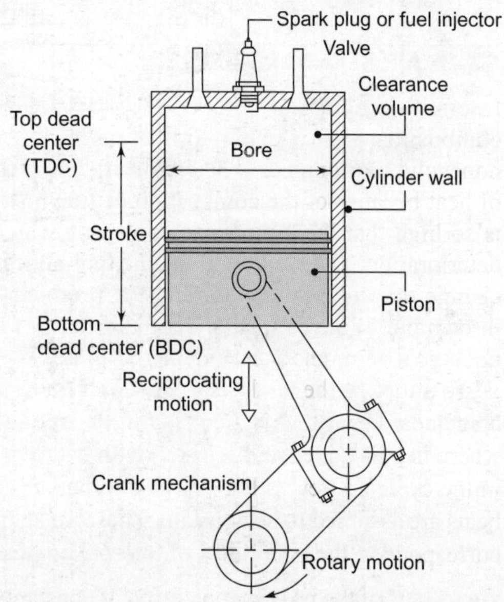

Fig. 13.4(a) Nomenclature for reciprocating piston–cylinder engine

The minimum volume formed in the cylinder when the piston is at TDC is called the **clearance volume**. The volume displaced by the piston as it moves between TDC and BDC is called the **displacement volume**. The ratio of the maximum volume formed in the cylinder to the minimum (clearance) volume is called the **compression ratio** r_k of the engine

$$r_k = V_{max}/V_{min} = V_{BDC}/V_{TDC}$$

Note that r_k is a **volume ratio** and should not be confused with the pressure ratio.

Another term frequently used in regard to reciprocating engines is the **mean effective pressure (mep)**, as mentioned earlier in Section 3.3 while studying indicator diagram. It is a fictitious pressure that, if it acted on the piston during the entire power stroke, would produce the same amount of net work as that produced during the actual cycle.

$$W_{net} = \text{m.e.p.} \times \text{piston area} \times \text{stroke} = \text{m.e.p.} \times \text{displacement volume}$$

or $$\text{m.e.p., } p_m = (W_{net})/[V_{max} - V_{min}]$$

The mean effective pressure can be used as a parameter to compare the performance of reciprocating engines of equal size. The engine with a larger value of m.e.p. will deliver more net work per cycle and thus will perform better.

Reciprocating engines are classified as **spark-ignition** (SI) **engines** or **compression-ignition** (CI) **engines**, depending on how the combustion process in the cylinder is initiated. In SI engines, also called petrol or gasoline engines, the combustion of the air-fuel mixture is initiated by a spark plug. In CI engines, also called diesel engines, the air-fuel mixture is self-ignited as a result of compressing the mixture above its self-ignition temperature.

13.5 AIR STANDARD CYCLES

> **LO 13.3**
> Discuss the origin of air standard cycles

Internal combustion engines (Fig. 13.4(a)) in which the combustion of fuel occurs in the engine cylinder itself are noncyclic heat engines. The temperature due to the evolution of heat because of the combustion of fuel inside the cylinder is so high that the cylinder is cooled by water circulation around it to avoid rapid deterioration. The working fluid, the fuel-air mixture, undergoes permanent chemical change due to combustion, and the products of combustion after doing work are thrown out of the engine, and a fresh charge is taken. So the working fluid does not undergo a complete thermodynamic cycle.

To simplify the analysis of IC engines, *air standard cycles* are conceived. In an air standard cycle, a certain mass of air operates in a complete thermodynamic cycle, where heat is added and rejected with external heat reservoirs, and all the processes in the cycle are reversible. Air is assumed to behave as an ideal gas, and its specific heats are assumed to be constant. These air standard cycles are so conceived that they correspond to the operations of internal combustion engines.

13.6 OTTO CYCLE (1876)

> **LO 13.4**
> Relate spark ignition engine with Otto cycle and illustrate the cycle

The Otto cycle is the air standard cycle of the SI engine. It is named after Nikolaus A. Otto, a German engineer, who first built a successful four-stroke SI engine in 1876 using the cycle proposed by a Frenchman Alphouse Beau de Rochas in 1862. In most spark-ignition engines, the piston executes four complete strokes within the cylinder, and the crankshaft completes two revolutions for each thermodynamic cycle. These engines are called **four-stroke** internal combustion engines. A schematic of each stroke as well as *p-v* diagram for an actual four-stroke SI engine is given in Fig. 13.4(b).

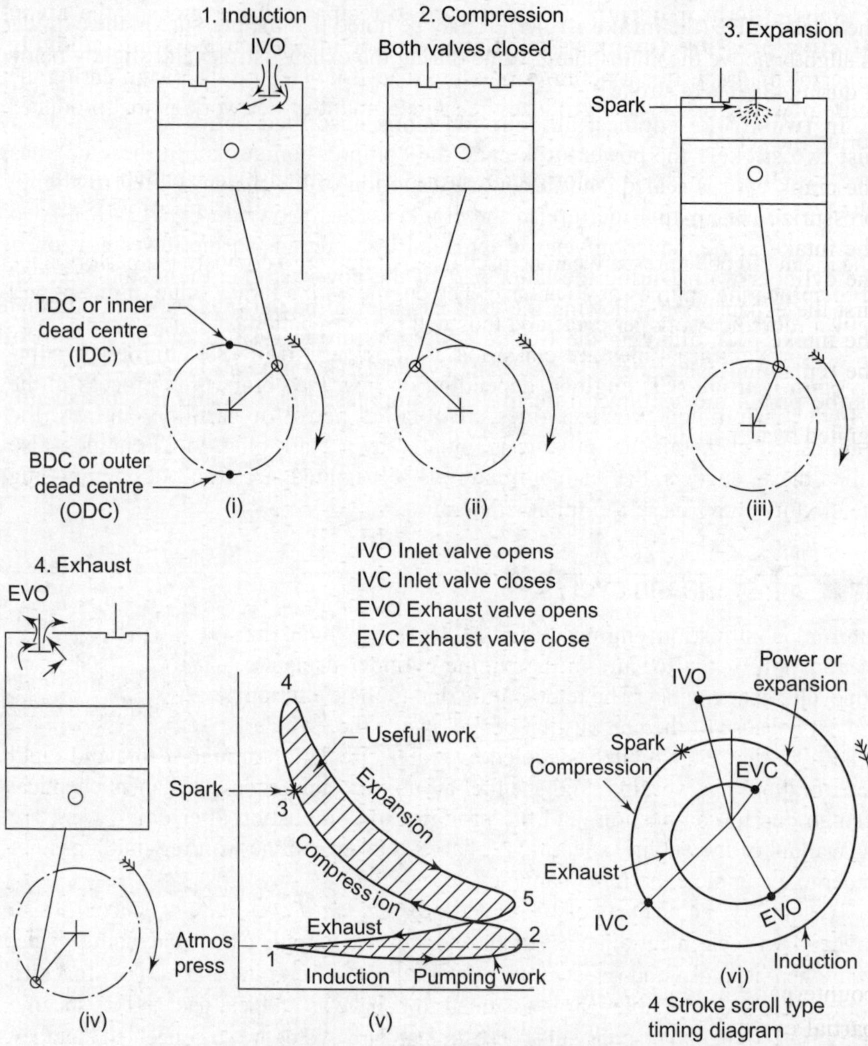

Fig. 13.4(b) The SI engine: Four-stroke cycle

Initially, the inlet valve opens (IVO) and fresh charge of fuel and air mixture is drawn into the cylinder. Then both the intake and exhaust valves are closed, and the piston is at its lowest position (BDC). During the **compression stroke**, the piston moves upwards compressing the air-fuel mixture. Shortly before the piston reaches its highest position (TDC), the spark plug fires and the mixture ignites, increasing the pressure and temperature of the system. The high pressure gases force the piston down, which in turn forces the crankshaft to rotate, producing a useful work output during the **expansion or power stroke**. At the end of this stroke, the piston is at its lowest position, and the cylinder is filled with the combustion products. Next the piston moves upwards again, purging the exhaust gases through the exhaust valve (the **exhaust stroke**), and down a second time, drawing in fresh air-fuel mixture through the intake valve (the **intake stroke**). It may be noted that the pressure in the cylinder is slightly above the atmospheric value during the exhaust stroke and slightly below it during the intake stroke.

In **two-stroke** engines, all four functions described above are executed in just two strokes: the power stroke and the compression stroke. In these engines, the crank case is sealed, and the outward motion of the piston is used to slightly pressurize the air-fuel mixture in the crankcase, as shown in Fig. 13.4(c). Also, the intake and exhaust valves are replaced by openings in the lower portion of the cylinder wall. During the latter part of the power stroke, the piston uncovers first the exhaust port, allowing the exhaust gases to be partially expelled and then the intake port, allowing the fresh air-fuel mixture to rush in and drive most of the remaining exhaust gases out of the cylinder. This mixture is then compressed as the piston moves upward during the compression stroke and is subsequently ignited by a spark plug.

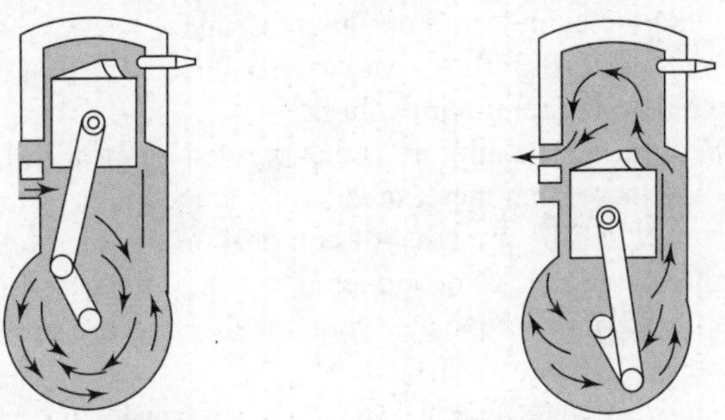

Cylinder: Compression and ignition
Crankcase: Intake

Cylinder: Intake and exhaust
Crankcase: Compression

Fig. 13.4(c) Operating principle of the two-stroke cycle SI engine

The two-stroke engines are generally less efficient than their four-stroke counterparts because of the incomplete expulsion of the exhaust gases and the partial expulsion of the fresh air-fuel mixture with the exhaust gases. However, they are relatively simple and inexpensive and they have high power-to-weight and power-to-volume ratios, which make them suitable for applications requiring small size and weight such as for motorcycles, scooters, chain saws and lawn mowers. Advances in several technologies, such as direct fuel injection, stratified charge combustion and electronic control, brought about a renewed interest in two-stroke engines that can offer high performance and fuel economy while satisfying the future stringent emission requirements.

The thermodynamic analysis of the actual four-stroke or two-stroke cycles is quite complex. To simplify the analysis; the air standard cycles (Otto cycle) are conceived. The execution of the Otto cycle in a piston-cylinder device together with p-v diagram is illustrated in Fig. 13.5(a) and described in the following paragraphs.

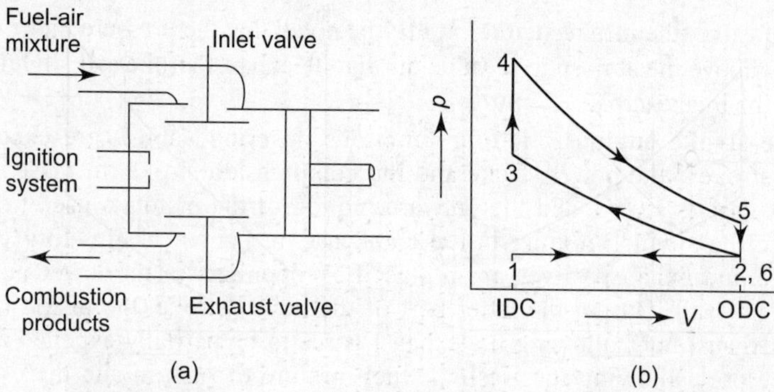

Fig. 13.5 (a) SI engine (Horizontal) (b) Indicator diagram

Process 1-2, Intake. The inlet valve is open, the piston moves to the right, admitting fuel-air mixture into the cylinder at constant pressure.
Process 2-3, Compression. Both the valves are closed, the piston compresses the combustible mixture to the minimum volume.
Process 3-4, Combustion. The mixture is then ignited by means of a spark, combustion takes place and there is an increase in temperature and pressure.
Process 4-5, Expansion. The products of combustion do work on the piston which moves to the right, and the pressure and temperature of the gases decrease.
Process 5-6, Blow-down. The exhaust valve opens, and the pressure drops to the initial pressure.
Process 6-1, Exhaust. With the exhaust valve open, the piston moves inwards to expel the combustion products from the cylinder at constant pressure.

The series of processes as described above constitute a *mechanical cycle*, and not a thermodynamic cycle. The cycle is completed in four strokes of the piston.

Figure 13.5(c) shows the air standard cycle (Otto cycle) corresponding to the above engine. It consists of two reversible adiabatics, one reversible isobar and one reversible isochore.

Air is compressed in process 1-2 reversibly and adiabatically. Heat is then added to air reversibly at constant volume in process 2-3. Work is done by air in expanding reversibly and adiabatically in process 3-4. Heat is then rejected by air reversibly at constant volume in process 4-1, and the system (air) comes back to its initial state. Heat transfer processes have been substituted for the combustion and blow-down processes of the engine. The intake and exhaust processes of the engine cancel each other.

Let m be the fixed mass of air undergoing the cycle of operations as described above.

$$\text{Heat supplied } Q_1 = Q_{2-3} = mc_v (T_3 - T_2)$$

$$\text{Heat rejected } Q_2 = Q_{4-1} = mc_v (T_4 - T_1)$$

$$\text{Efficiency} \quad \eta = 1 - \frac{Q_2}{Q_1} = 1 - \frac{mc_v(T_4 - T_1)}{mc_v(T_3 - T_2)}$$

$$= 1 - \frac{T_4 - T_1}{T_3 - T_2} \qquad (13.4)$$

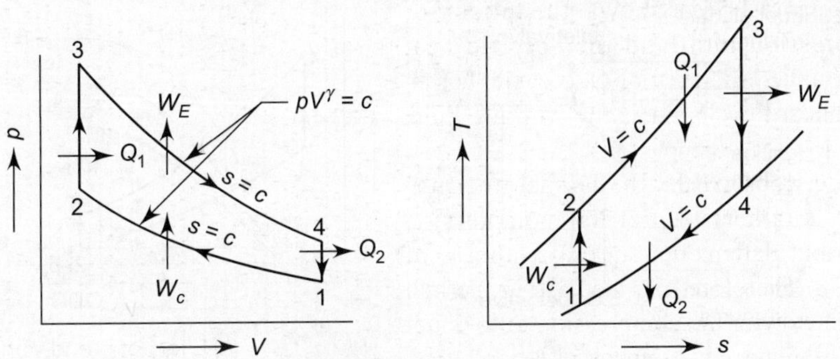

Fig. 13.5 (c) Otto cycle

Process 1-2, $$\frac{T_2}{T_1} = \left(\frac{v_1}{v_2}\right)^{\gamma-1}$$

Process 3-4, $$\frac{T_3}{T_4} = \left(\frac{v_4}{v_3}\right)^{\gamma-1} = \left(\frac{v_1}{v_2}\right)^{\gamma-1}$$

$\therefore$ $$\frac{T_2}{T_1} = \frac{T_3}{T_4}$$

or $$\frac{T_3}{T_2} = \frac{T_4}{T_1}$$

$$\frac{T_3}{T_2} - 1 = \frac{T_4}{T_1} - 1$$

$\therefore$ $$\frac{T_4 - T_1}{T_3 - T_2} = \frac{T_1}{T_2} = \left(\frac{v_2}{v_1}\right)^{\gamma-1}$$

$\therefore$ From Eq. (13.4), $$\eta = 1 - \left(\frac{v_2}{v_1}\right)^{\gamma-1}$$

or $$\boldsymbol{\eta_{\text{otto}} = 1 - \frac{1}{r_k^{\gamma-1}}} \quad (13.5)$$

where r_k is called the *compression ratio* and given by

$$r_k = \frac{\text{Volume at the beginning of compression}}{\text{Volume at the end of compression}}$$

$$= \frac{V_1}{V_2} = \frac{v_1}{v_2}$$

The efficiency of the air standard Otto cycle is thus a function of the compression ratio only. The higher the compression ratio, the higher the efficiency. It is independent of the temperature levels at which the cycle operates. The compression ratio cannot, however, be increased beyond a certain limit, because of a noisy and destructive combustion phenomenon, known as *detonation*. It also depends upon the fuel, the engine design and the operating conditions.

Figure 13.5(d) shows the effect of compression ratio and the specific heat ratio on the efficiency of Otto cycle. For an air standard cycle, air is the working fluid, $\gamma = 1.4$.

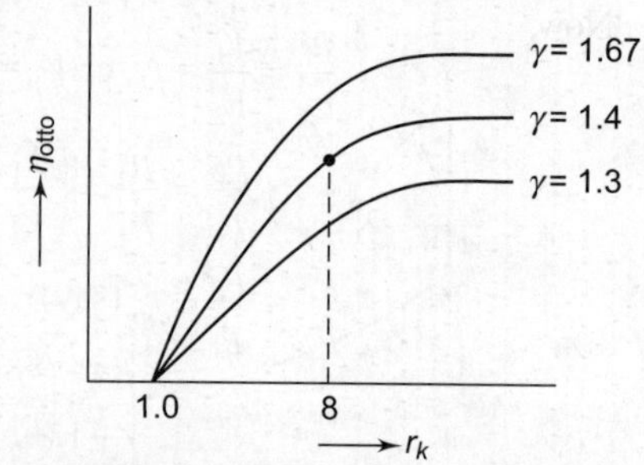

Fig. 13.5 (d) Effect of r_k and γ on Otto cycle efficiency

We can observe that the thermal efficiency curve is rather steep at low compression ratios but flattens out starting with a r_k of about 8. Therefore, the increase in thermal efficiency with the compression ratio is not that pronounced at high compression ratios. Also, when high compression ratios are used, the temperature of the air fuel mixture rises above the self-ignition temperature of the fuel when the mixture ignites without the spark, causing an early and rapid combustion of the fuel ahead of the flame front, followed by almost instantaneous burning of the remaining mixture. This premature ignition of the fuel, called **auto-ignition**, produces an audible noise, which is called **engine knock or detonation**. This auto-ignition hurts performance and can cause engine damage, thus setting the upper limit of the compression ratio that can be used in SI engines.

Improvement of the thermal efficiency of gasoline engines by utilising higher compression ratios (up to 12) without facing the auto-ignition problem has been made possible by using gasoline blends that have good antiknock characteristics, such as gasoline mixed with tetraethyl lead. Tetraethyl lead has been added to gasoline since the 1920s because it is the cheapest method of raising the **octane rating**, which is a measure of the engine knock resistance of a fuel. Leaded gasoline, however, has a very undesirable side effect as it forms compounds during the combustion process that are hazardous to health and pollute the environment. Most cars made since 1975 have been designed to use unleaded gasoline, and the compression ratios had to be lowered to avoid engine knock. As a result the thermal efficiency of car engines has somewhat decreased. However, owing to improvement in other areas like reduction in overall automobile weight, improved aerodynamic design, etc, today's cars have better fuel economy.

The second parameter affecting the thermal efficiency of the Otto cycle is the specific heat ratio γ. For a given r_k, the use of a monatomic gas such as argon or helium as the working fluid yields the highest thermal efficiency (Fig. 13.5d). The specific heat ratio γ and thus the thermal efficiency of the Otto cycle decrease as the molecules of the working fluid get larger. At room temperature, it is 1.4 for air, 1.3 for carbon dioxide and 1.2 for ethane. The working fluid in actual engines contains larger molecules such as CO_2, and γ decreases with temperature, because of which the thermal efficiencies of the actual engines are lower than those of the Otto cycle and vary from 25 to 30 per cent.

13.6.1 Work Output

The net work output for an Otto cycle (Fig. 13.5(c)) can be expressed

$$W_{\text{net}} = \frac{p_3V_3 - p_4V_4}{\gamma - 1} - \frac{p_2V_2 - p_1V_1}{\gamma - 1}$$

Now,
$$\frac{v_1}{v_2} = \frac{V_1}{V_2} = r_k, \text{ or } V_1 = V_2\, r_k = V_4$$

$$\frac{p_2}{p_1} = \frac{p_3}{p_4} = \left(\frac{V_1}{V_2}\right)^{\gamma} = r_k^{\gamma}$$

$$\frac{p_3}{p_2} = \frac{p_4}{p_1} = r_p \text{ (say)}$$

$$\therefore \quad W_{\text{net}} = \frac{p_1V_1}{\gamma - 1}\left(\frac{p_3V_3}{p_1V_1} - \frac{p_4V_4}{p_1V_1} - \frac{p_2V_2}{p_1V_1} + 1\right)$$

$$= \frac{p_1V_1}{\gamma - 1}\left(\frac{r_p r_k^{\gamma}}{r_k} - r_p - \frac{r_k^{\gamma}}{r_k} + 1\right)$$

$$= \frac{p_1V_1}{\gamma - 1}\left(r_p\, r_k^{\gamma-1} - r_p - r_k^{\gamma-1} + 1\right)$$

$$W_{\text{net}} = \frac{p_1V_1}{\gamma - 1}(r_p - 1)\,(r_k^{\gamma-1} - 1) \tag{13.6}$$

13.6.2 Mean Effective Pressure

The mean effective pressure (m.e.p) of the cycle is given by

$$p_m = \frac{\text{Network output}}{\text{Swept volume}}$$

where swept volume $= V_1 - V_2 = V_2\,(r_k - 1)$

$$\therefore \quad p_m = \frac{\dfrac{p_1V_1}{\gamma-1}(r_p - 1)(r_k^{\gamma-1} - 1)}{V_2\,(r_k - 1)}$$

$$= \frac{p_1 r_k (r_p - 1)(r_k^{\gamma-1} - 1)}{(\gamma - 1)(r_k - 1)} \tag{13.7}$$

Thus, it is seen that the net work output is directly proportional to the pressure ratio r_p. For given values of r_k and γ, p_m increases with r_p. For an Otto cycle, an increase in r_k leads to an increase in p_m, W_{net} and cycle efficiency.

13.7 DIESEL CYCLE (1892)

LO 13.5
Relate compression ignition engine with Diesel cycle and illustrate the cycle

The limitation on compression ratio in the SI engine can be overcome by compressing air alone, instead of the fuel-air mixture, and then injecting the fuel into the cylinder in spray form when combustion is desired. The CI engine, first proposed by Rudolph Diesel in the 1890s, is very similar to the SI engine, differing mainly in the method of initiating combustion. In SI engines, a mixture of air and fuel is compressed during compression stroke, and the compression ratios are limited by the onset of autoignition or engine knock. In CI engines, only air is compressed during the compression stroke. Therefore, diesel engines can operate at much higher compression ratios, typically between 12 and 24. The spark plug and carburettor (for mixing fuel and air) are replaced by a fuel injector in diesel engines. The temperature of air after compression must be high enough so that the fuel sprayed into the hot air burns spontaneously. The rate of burning can, to some extent, be controlled by the rate of injection of fuel. An engine operating in this way is called a *compression ignition (CI) engine*. The sequence of processes in the elementary operation of a CI engine, shown in Fig. 13.6, is given as follows.

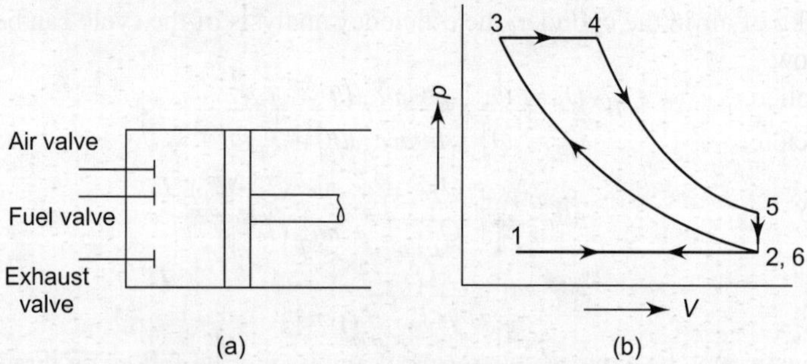

Fig. 13.6 (a) CI engine (b) Indicator diagram

Process 1-2, intake. The air valve is open. The piston moves out admitting air into the cylinder at constant pressure.
Process 2-3, Compression. The air is then compressed by the piston to the minimum volume with all the valves closed.
Process 3-4, Fuel injection and combustion. The fuel valve is open, fuel is sprayed into the hot air and combustion takes place at constant pressure.
Process 4-5, Expansion. The combustion products expand, doing work on the piston which moves out to the maximum volume.
Process 5-6, Blow-down. The exhaust valve opens, and the pressure drops to the initial pressure.
Pressure 6-1, Exhaust. With the exhaust valve open, the piston moves towards the cylinder cover driving away the combustion products from the cylinder at constant pressure.

The above processes constitute an engine cycle, which is completed in four strokes of the piston or two revolutions of the crank shaft.

Figure 13.7 shows the air standard cycle, called the *Diesel cycle*, corresponding to the CI engine, as described above. The cycle is composed of:

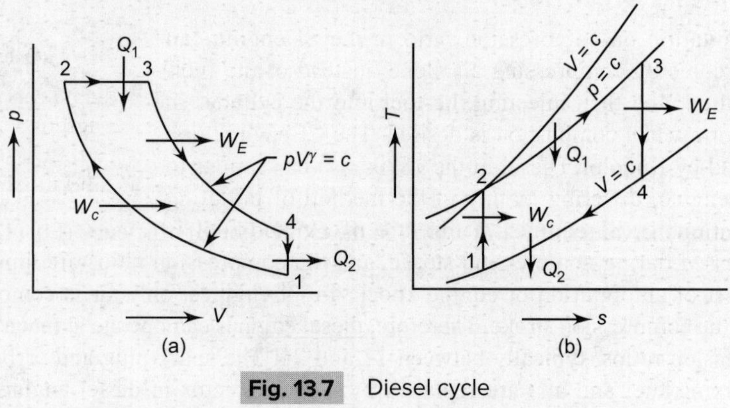

Fig. 13.7 Diesel cycle

Two reversible adiabatics, one reversible isobar, and one reversible isochore.

Air is compressed reversibly and adiabatically in process 1-2. Heat is then added to it from an external source reversibly at constant pressure in process 2-3. Air then expands reversibly and adiabatically in process 3-4. Heat is rejected reversibly at constant volume in process 4-1, and the cycle repeats itself.

For *m* kg of air in the cylinder, the efficiency analysis of the cycle can be made as given below.

Heat supplied, $Q_1 = Q_{2-3} = m\,c_p\,(T_3 - T_2)$

Heat rejected, $Q_2 = Q_{4-1} = m\,c_v\,(T_4 - T_1)$

Efficiency $\eta = 1 - \dfrac{Q_2}{Q_1} = 1 - \dfrac{mc_v(T_4 - T_1)}{mc_p(T_3 - T_2)}$

$$\therefore \quad \eta = 1 - \frac{T_4 - T_1}{\gamma\,(T_3 - T_2)} \tag{13.8}$$

The efficiency may be expressed in terms of any two of the following three ratios

Compression ratio, $r_k = \dfrac{V_1}{V_2} = \dfrac{v_1}{v_2}$

Expansion ratio, $r_e = \dfrac{V_4}{V_3} = \dfrac{v_4}{v_3}$

Cut-off ratio, $r_c = \dfrac{V_3}{V_2} = \dfrac{v_3}{v_2}$

It is seen that $r_k = r_e \cdot r_c$

Process 3-4

$$\frac{T_4}{T_3} = \left(\frac{v_3}{v_4}\right)^{\gamma-1} = \frac{1}{r_e^{\gamma-1}}$$

$$T_4 = T_3 \frac{r_c^{\gamma-1}}{r_k^{\gamma-1}}$$

Process 2-3

$$\frac{T_2}{T_3} = \frac{p_2 v_2}{p_3 v_3} = \frac{v_2}{v_3} = \frac{1}{r_c}$$

$$\therefore \quad T_2 = T_3 \cdot \frac{1}{r_c}$$

Process 1-2

$$\frac{T_1}{T_2} = \left(\frac{v_2}{v_1}\right)^{\gamma-1} = \frac{1}{r_k^{\gamma-1}}$$

$$\therefore \quad T_1 = T_2 \cdot \frac{1}{r_k^{\gamma-1}} = \frac{T_3}{r_c} \cdot \frac{1}{r_k^{\gamma-1}}$$

Substituting the values of T_1, T_2 and T_4 in the expression of efficiency (Eq. (13.8))

$$\eta = 1 - \frac{T_3 \cdot \dfrac{r_c^{\gamma-1}}{r_k^{\gamma-1}} - \dfrac{T_3}{r_c}\dfrac{1}{r_k^{\gamma-1}}}{\gamma\left(T_3 - T_3 \cdot \dfrac{1}{r_c}\right)}$$

$$\therefore \qquad \eta_{\text{Diesel}} = 1 - \frac{1}{\gamma} \cdot \frac{1}{r_k^{\gamma-1}} \cdot \frac{r_c^{\gamma} - 1}{r_c - 1} \tag{13.9}$$

As $r_c > 1$, $\frac{1}{\gamma}\left(\frac{r_c^{\gamma} - 1}{r_c - 1}\right)$ is also greater than unity. Therefore, the efficiency of the Diesel cycle is less than that of the Otto cycle for the same compression ratio.

13.8 LIMITED PRESSURE CYCLE, MIXED CYCLE OR DUAL CYCLE

The air standard Diesel cycle does not simulate exactly the pressure-volume variation in an actual compression ignition engine, where the fuel injection is started before the end of compression stroke. A closer approximation is the limited pressure cycle in which some part of heat is added to air at constant volume, and the remainder at constant pressure.

LO 13.6

Discuss the significance of Dual cycle and illustrate it

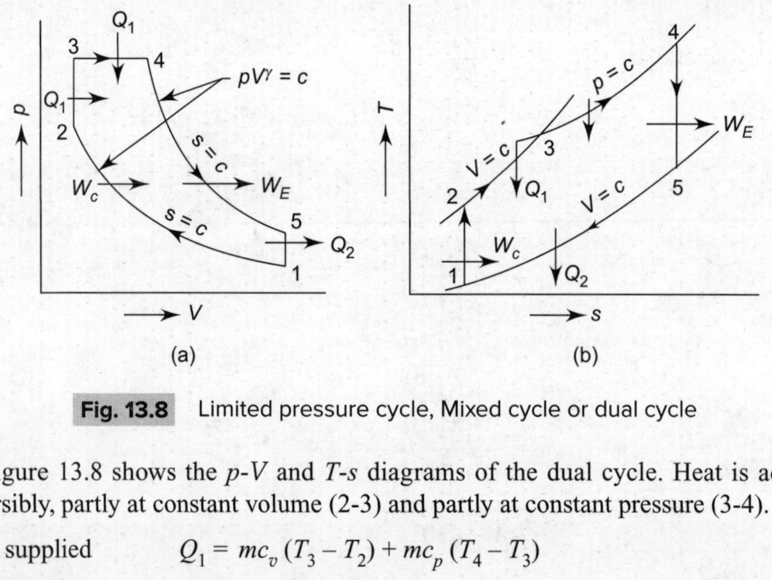

Fig. 13.8 Limited pressure cycle, Mixed cycle or dual cycle

Figure 13.8 shows the *p-V* and *T-s* diagrams of the dual cycle. Heat is added reversibly, partly at constant volume (2-3) and partly at constant pressure (3-4).

Heat supplied $\qquad Q_1 = mc_v\,(T_3 - T_2) + mc_p\,(T_4 - T_3)$

Heat rejected $\qquad Q_2 = mc_v\,(T_5 - T_1)$

Efficiency
$$\eta = 1 - \frac{Q_2}{Q_1}$$

$$= 1 - \frac{mc_v(T_5 - T_1)}{mc_v(T_3 - T_2) + mc_p(T_4 - T_3)}$$

$$= 1 - \frac{T_5 - T_1}{(T_3 - T_2) + \gamma(T_4 - T_3)} \tag{13.10}$$

The efficiency of the cycle can be expressed in terms of the following ratios:

Compression ratio, $r_k = \dfrac{V_1}{V_2}$

Expansion ratio, $r_e = \dfrac{V_5}{V_4}$

Cut-off ratio, $r_c = \dfrac{V_4}{V_3}$

Constant volume pressure ratio, $r_p = \dfrac{p_3}{p_2}$

It is seen, as before that

$$r_k = r_c \cdot r_e$$

or $$r_e = \frac{r_k}{r_c}$$

Process 3-4

$$r_c = \frac{V_4}{V_3} = \frac{T_4 p_3}{p_4 T_3} = \frac{T_4}{T_3}$$

$$T_3 = \frac{T_4}{r_c}$$

Process 2-3

$$\frac{p_2 V_2}{T_2} = \frac{p_3 V_3}{T_3}$$

$$T_2 = T_3 \frac{p_2}{p_3} = \frac{T_4}{r_p \cdot r_c}$$

Process 1-2

$$\frac{T_1}{T_2} = \left(\frac{v_2}{v_1}\right)^{\gamma-1} = \frac{1}{r_k^{\gamma-1}}$$

$\therefore$ $$T_1 = \frac{T_4}{r_p \cdot r_c \cdot r_k^{\gamma-1}}$$

Process 4-5

$$\frac{T_5}{T_4} = \left(\frac{v_4}{v_5}\right)^{\gamma-1} = \frac{1}{r_e^{\gamma-1}}$$

$\therefore$ $$T_5 = T_4 \cdot \frac{r_c^{\gamma-1}}{r_k^{\gamma-1}}$$

Substituting the values of T_1, T_2, T_3 and T_5 in the expression of efficiency (Eq. (13.8)).

$$\eta = 1 - \frac{T_4 \cdot \dfrac{r_c^{\gamma-1}}{r_k^{\gamma-1}} - \dfrac{T_4}{r_p \cdot r_c \cdot r_k^{\gamma-1}}}{\left(\dfrac{T_4}{r_c} - \dfrac{T_4}{r_p \cdot r_c}\right) + \gamma\left(T_4 - \dfrac{T_4}{r_c}\right)}$$

$$\therefore \qquad \eta_{\text{Dual}} = 1 - \frac{1}{r_k^{\gamma-1}} \frac{r_p \cdot r_c^{\gamma} - 1}{r_p - 1 + \gamma r_p (r_c - 1)} \qquad (13.11)$$

13.9 COMPARISON OF OTTO, DIESEL AND DUAL CYCLES

LO 13.7 Compare Otto, Diesel and Dual cycles

The three cycles can be compared on the basis of either the same compression ratio or the same maximum pressure and temperature.

Figure 13.9 shows the comparison of Otto, Diesel and Dual cycles for the same compression ratio and heat rejection. Here

1–2–6–5 —Otto cycle
1–2–7–5 —Diesel cycle
1–2–3–4–5 —Dual cycle

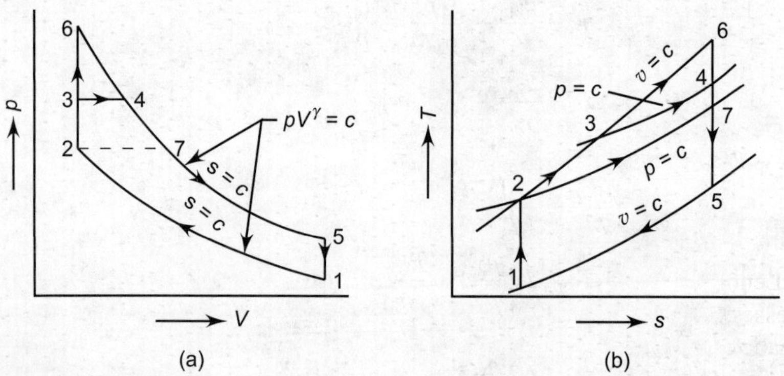

Fig. 13.9 Comparison of Otto, Diesel and Dual cycles for the same compression ratio

For the same Q_2, the higher the Q_1, the higher is the cycle efficiency. In the *T-s* diagram, the area under 2-6 represents Q_1 for the Otto cycle, the area under 2-7 represents Q_1 for the Diesel cycle and the area under 2-3-4 represents Q_1 for the Dual cycle. Therefore, for the same r_k and Q_2

$$\eta_{\text{otto}} > \eta_{\text{Dual}} > \eta_{\text{Diesel}}$$

Figure 13.10 shows a comparison of the three air standard cycles for the same maximum pressure and temperature (state 4), the heat rejection being also the same. Here

1–6–4–5 —Otto cycle
1–7–4–5 —Diesel cycle
1–2–3–4–5 —Dual cycle

Q_1 is represented by the area under 6-4 for the Otto cycle, by the area under 7-4 for the Diesel cycle and by the area under 2-3-4 for the Dual cycle in the *T-s* plot, Q_2 being the same.

$$\therefore \quad \eta_{\text{Diesel}} > \eta_{\text{dual}} > \eta_{\text{otto}}$$

This comparison is of greater significance, since the Diesel cycle would definitely have a higher compression ratio than the Otto cycle.

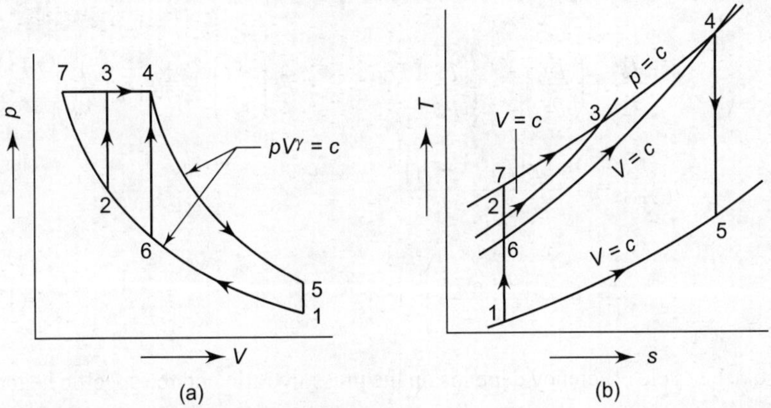

Fig. 13.10 Comparison of Otto, Diesel and dual cycles for the same maximum pressure and temperature

13.10 LENOIR CYCLE

> **LO 13.8**
> Illustrate *p-V* and *T-s* diagrams of Lenoir and Atkinson cycles

The Lenoir cycle (Fig. 13.11) consists of the following processes: consant volume heat addition (1–2); reversible adiabatic expansion (2–3) and constant pressure heat rejection (3–1). The Lenoir cycle is applicable to pulse jet engines.

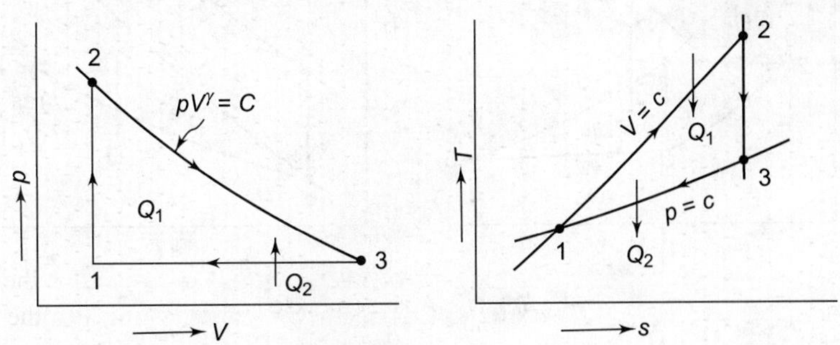

Fig. 13.11 Lenoir cycle

For 1 kg gas,

$$Q_1 = c_v (T_2 - T_1)$$

$$Q_2 = c_p (T_3 - T_1)$$

$$\eta = 1 - \frac{Q_2}{Q_1} = 1 - \frac{c_p(T_3 - T_1)}{c_v(T_2 - T_1)} = 1 - \frac{\gamma(T_3 - T_1)}{T_2 - T_1}$$

Using $$r_p = \frac{p_2}{p_1}, \frac{p_1 v_1}{T_1} = \frac{p_2 v_2}{T_2} \therefore T_2 = T_1 \frac{p_2}{p_1} = r_p \cdot T_1$$

$$\frac{T_3}{T_2} = \left(\frac{p_3}{p_2}\right)^{\frac{\gamma-1}{\gamma}} = \left(\frac{p_1}{p_2}\right)^{\frac{\gamma-1}{\gamma}} \therefore T_3 = T_2\left(\frac{1}{r_p}\right)^{\frac{\gamma-1}{\gamma}} = T_1 r_p \left(\frac{1}{r_p}\right)^{\frac{\gamma-1}{\gamma}}$$

$$\therefore \quad \eta_{\text{Lenoir}} = 1 - \frac{\gamma[T_1 r_p^{1/\gamma} - T_1]}{T_1 r_p - T_1}$$

$$= 1 - \gamma\left(\frac{r_p^{1/\gamma} - 1}{r_p - 1}\right) \tag{13.12}$$

Thus, the cycle efficiency depends on the pressure ratio and the specific heat ratio.

13.11 ATKINSON CYCLE

Atkinson cycle (Fig. 13.12) is an ideal cycle for an Otto engine exhausting to a gas turbine. In this cycle, the isentropic expansion (3-4) of an Otto cycle is allowed to further expand to the lowest cycle pressure (3-5) so as to increase the work output.

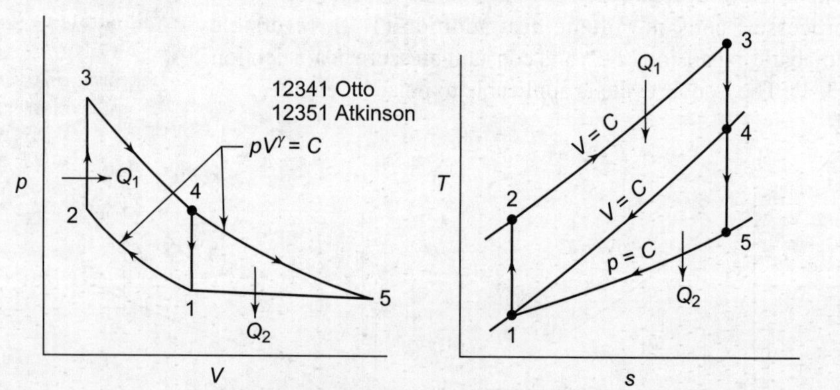

Fig. 13.12 Atkinson cycle

For 1 kg gas, $$Q_1 = c_v(T_3 - T_2)$$

$$Q_2 = c_p(T_5 - T_1)$$

$$\eta = 1 - \frac{c_p(T_5 - T_1)}{c_v(T_3 - T_2)} = 1 - \frac{\gamma(T_5 - T_1)}{T_3 - T_2})$$

Let r_k, compression ratio $= \dfrac{v_1}{v_2}$

r_e, expansion ratio $= \dfrac{v_5}{v_3}$

$$\frac{T_2}{T_1} = \frac{v_1}{v_2} \quad \therefore \quad T_2 = T_1 \, r_k^{\gamma-1}$$

$$\frac{T_3}{T_2} = \frac{p_3}{p_2} = \frac{p_3}{p_5} \cdot \frac{p_5}{p_2} = \frac{p_3}{p_5} \cdot \frac{p_1}{p_2}$$

$$\frac{p_3}{p_5} = \left(\frac{v_5}{v_3}\right)^{\gamma} = r_e^{\gamma}$$

$$\frac{p_1}{p_2} = \left(\frac{v_2}{v_1}\right)^{\gamma} = \frac{1}{r_k^{\gamma}}$$

$$\therefore \quad T_3 = T_2 \cdot r_e^{\gamma} \cdot \frac{1}{r_k^{\gamma}} = T_1 \, r_k^{\gamma-1} \cdot r_e^{\gamma} \cdot \frac{1}{r_k^{\gamma}} = T_1 \cdot \frac{r_e^{\gamma}}{r_k}$$

$$\frac{T_5}{T_3} = \left(\frac{v_3}{v_5}\right)^{\gamma-1} = \frac{1}{r_e^{\gamma-1}}$$

$$\therefore \quad T_5 = T_3 \cdot \frac{1}{r_e^{\gamma-1}} = T_1 \cdot \frac{r_e^{\gamma}}{r_k} \cdot \frac{1}{r_e^{\gamma-1}} = T_1 \frac{r_e}{r_k}$$

Substituting T_2, T_3 and T_5 in the expression of efficiency

$$\eta_{\text{Atkinson}} = 1 - \gamma \frac{T_1 \cdot \dfrac{r_e}{r_k} - T_1}{T_1 \dfrac{r_e^{\gamma}}{r_k} - T_1 r_k^{\gamma-1}}$$

$$= 1 - \gamma \frac{r_e - r_k}{r_e^{\gamma} - r_k^{\gamma}} \tag{13.13}$$

13.12 BRAYTON CYCLE

> **LO 13.9**
>
> Relate gas turbine with Brayton cycle and illustrate the cycle

A simple gas turbine power plant is shown in Fig. 13.13. Air is first compressed adiabatically in process *a-b*, it then enters the combustion chamber where fuel is injected and burned essentially at constant pressure in process *b-c*, and then the products of combustion expand in the turbine to the ambient pressure in process *c-d* and are thrown out to the surroundings. The cycle is open. The state diagram on the *p-V* coordinates is shown in Fig. 13.14. Open cycles are used in aircraft, automotive (buses and trucks) and industrial gas turbine installations.

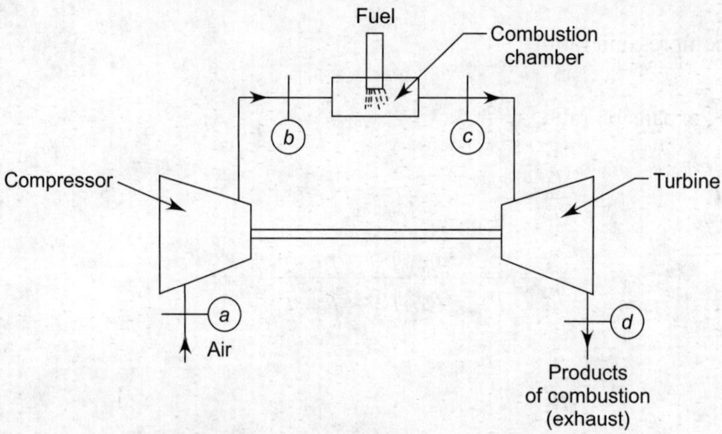

Fig. 13.13 A simple gas turbine plant

The Brayton cycle is the air standard cycle for the gas turbine power plant. Here air is first compressed reversibly and adiabatically, heat is added to it reversibly at constant pressure, air expands in the turbine reversibly and adiabatically and heat is then rejected from the air reversibly at constant pressure to bring it to the initial state. The Brayton cycle, therefore, consists of two reversible isobars and two reversible adiabatics.

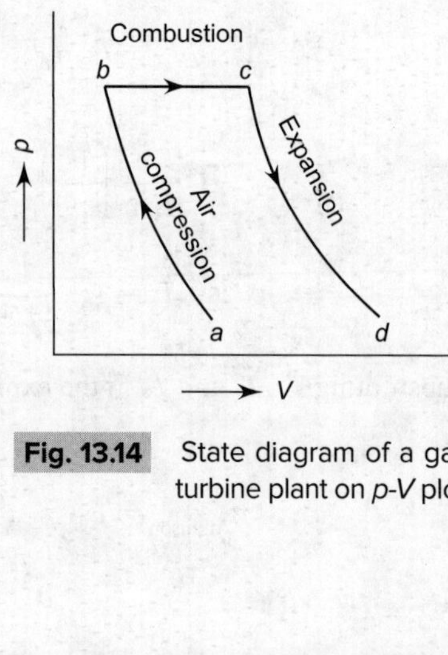

Fig. 13.14 State diagram of a gas turbine plant on *p-V* plot

The flow, *p-V* and *T-s* diagrams are shown in Fig. 13.15. For *m* kg of air

$$Q_1 = \text{heat supplied} = m\, c_p\, (T_3 - T_2)$$

$$Q_2 = \text{heat rejected} = m\, c_p\, (T_4 - T_1)$$

$$\therefore \text{ Cycle efficiency, } \eta = 1 - \frac{Q_2}{Q_1}$$

$$= 1 - \frac{T_4 - T_1}{T_3 - T_2} \tag{13.14}$$

Now

$$\frac{T_2}{T_1} = \left(\frac{p_2}{p_1}\right)^{(\gamma-1)/\gamma} = \frac{T_3}{T_4} \text{ (Since } p_2 = p_3, \text{ and } p_4 = p_1)$$

$$\therefore \qquad \frac{T_4}{T_1} - 1 = \frac{T_3}{T_2} - 1$$

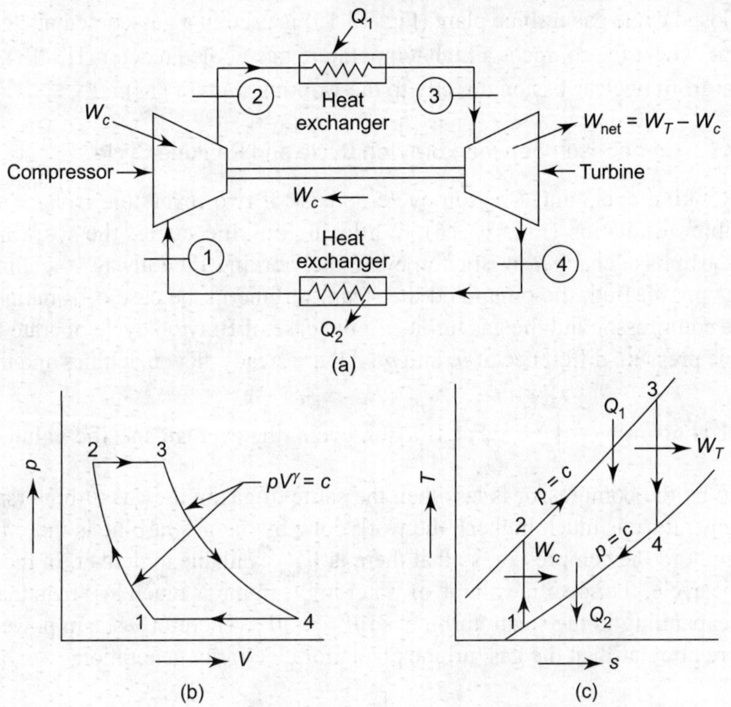

Fig. 13.15 Brayton cycle

or
$$\frac{T_4 - T_1}{T_3 - T_2} = \frac{T_1}{T_2} = \left(\frac{p_1}{p_2}\right)^{(\gamma-1)/\gamma} = \left(\frac{v_2}{v_1}\right)^{\gamma-1}$$

If r_k = compression ratio = v_1/v_2 the efficiency becomes (from Eq. (13.14))

$$\eta = 1 - \left(\frac{v_2}{v_1}\right)^{\gamma-1}$$

or
$$\eta_{\text{Brayton}} = 1 - \frac{1}{r_k^{\gamma-1}} \tag{13.15}$$

If r_p = pressure ratio = p_2/p_1 the efficiency may be expressed in the following form also

$$\eta = 1 - \left(\frac{p_1}{p_2}\right)^{(\gamma-1)/\gamma}$$

or
$$\boldsymbol{\eta_{\text{Brayton}} = 1 - \frac{1}{(r_p)^{(\gamma-1)/\gamma}}} \tag{13.16}$$

The efficiency of the Brayton cycle, therefore, depends upon either the compression ratio or the pressure ratio. For the same compression ratio, the Brayton cycle efficiency is equal to the Otto cycle efficiency.

A closed cycle gas turbine plant (Fig. 13.15) is used in a gas-cooled nuclear reactor plant, where the source is a high temperature gas-cooled reactor (HTGR) supplying heat from nuclear fission directly to the working fluid (a gas).

13.12.1 Comparison between Brayton Cycle and Rankine Cycle

Both Rankine cycle and Brayton cycle consist of two reversible isobars and two reversible adiabatics (Fig. 13.16). While in Rankine cycle, the working fluid undergoes phase change, in Brayton cycle the working fluid always remains in the gaseous phase. Both the pump and the steam turbine in the case of Rankine cycle, and the compressor and the gas turbine in the case of Brayton cycle operate through the same pressure difference of p_1 and p_2. All are steady-flow machines and the work transfer is given by $-\int_{p_1}^{p_2} v\, dp$. For Brayton cycle, the average specific volume of air handled by the compressor is less than the same of gas in the gas turbine (since the gas temperature is much higher), the work done by the gas turbine is more than the work input to the compressor, so that there is W_{net} available to deliver. In the case of Rankine cycle, the specific volume of water in the pump is much less than that of the steam expanding in the steam turbine, so $W_T >> W_P$. Therefore, steam power plants are more popular than the gas turbine plants for electricity generation.

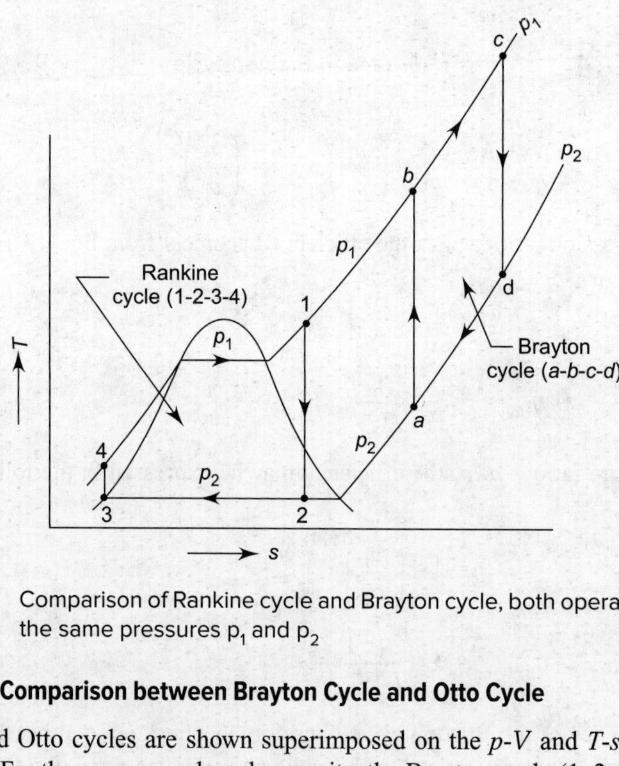

Fig. 13.16 Comparison of Rankine cycle and Brayton cycle, both operating between the same pressures p_1 and p_2

13.12.2 Comparison between Brayton Cycle and Otto Cycle

Brayton and Otto cycles are shown superimposed on the *p-V* and *T-s* diagrams in Fig. 13.17. For the same r_k and work capacity, the Brayton cycle (1–2–5–6) handles a larger range of volume and a smaller range of pressure and temperature than does the Otto cycle (1-2-3-4).

In the reciprocating engine field, the Brayton cycle is not suitable. A reciprocating engine cannot efficiently handle a large volume flow of low pressure gas, for which the engine size ($\pi/4\ D^2 L$) becomes large, and the friction losses also become more. So the Otto cycle is more suitable in the reciprocating engine field.

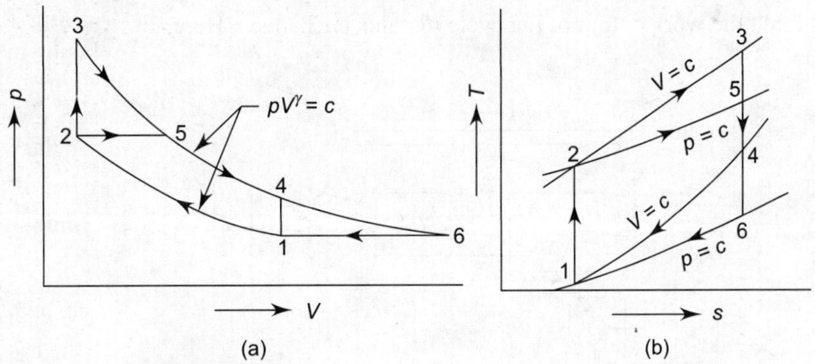

Fig. 13.17 Comparison of Otto and Brayton cycles

In turbine plants, however, the Brayton cycle is more suitable than the Otto cycle. An internal combustion engine is exposed to the highest temperature (after the combustion of fuel) only for a short while, and it gets time to become cool in the other processes of the cycle. On the other hand, a gas turbine plant, a steady flow device, is always exposed to the highest temperature used. So to protect material, the maximum temperature of gas that can be used in a gas turbine plant cannot be as high as in an internal combustion engine. Also, in the steady flow machinery, it is more difficult to carry out heat transfer at constant volume than at constant pressure. Moreover, a gas turbine can handle a large volume flow of gas quite efficiently. So we find that the Brayton cycle is the basic air standard cycle for all modern gas turbine plants.

13.12.3 Effect of Regeneration on Brayton Cycle Efficiency

The efficiency of the Brayton cycle can be increased by utilising part of the energy of the exhaust gas from the turbine in heating up the air leaving the compressor in a heat exchanger called a *regenerator*, thereby reducing the amount of heat supplied from an external source and also the amount of heat rejected. Such a cycle is illustrated in Fig. 13.18. The temperature of air leaving the turbine at 5 is higher than that of air leaving the compressor at 2. In the regenerator, the temperature of air leaving the compressor is raised by heat transfer from the turbine exhaust. The maximum temperature to which the cold air at 2 could be heated is the temperature of the hot air leaving the turbine at 5. This is possible only in an infinite heat exchanger. In the real case, the temperature at 3 is less than that at 5. The ratio of the actual temperature rise of air to the maximum possible rise is called the *effectiveness* of the regenerator. For this case,

$$\text{Effectiveness, } \in = \frac{t_3 - t_2}{t_5 - t_2}$$

When the regenerator is used in the idealised cycle (Fig. 13.18), the heat supplied and the heat rejected are each reduced by the same amount, Q_x. *The mean temperature of heat addition increases and the mean temperature of heat rejection decreases because of the use of the regenerator*. The efficiency is increased as a result, but the work output of the cycle remains unchanged. Here,

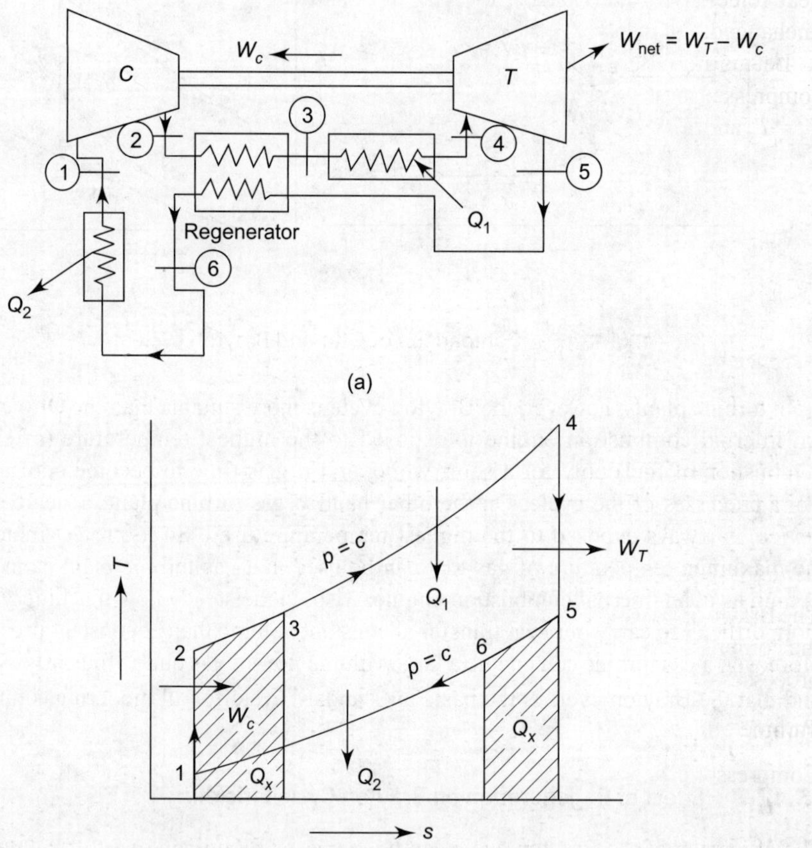

Fig. 13.18 Effect of regeneration on Brayton cycle

$$Q_1 = h_4 - h_3 = c_p\,(T_4 - T_3)$$

$$Q_2 = h_6 - h_1 = c_p\,(T_6 - T_1)$$

$$W_T = h_4 - h_5 = c_p\,(T_4 - T_5)$$

$$W_c = h_2 - h_1 = c_p\,(T_2 - T_1)$$

$$\eta = 1 - \frac{Q_2}{Q_1} = 1 - \frac{T_6 - T_1}{T_4 - T_3}$$

In practise, the regenerator is costly, heavy and bulky, and causes pressure losses which bring about a decrease in cycle efficiency. These factors have to be balanced against the gain in efficiency to decide whether it is worthwhile to use the regenerator.

Above a certain pressure ratio (p_2/p_1) the addition of a regenerator causes a loss in cycle efficiency when compared to the original Brayton cycle. In this situation, the compressor discharge temperature (T_2) is higher than the turbine exhaust gas temperature (T_5) (Fig. 13.18). The compressed air will thus be cooled in the regenerator and the exhaust gas will be heated. As a result both the heat supply and heat rejected are increased. However, the compressor and turbine works remain unchanged. So, the cycle efficiency (W_{net}/Q_1) decreases.

Let us now derive an expression for the ideal regenerative cycle when the compressed air is heated to the turbine exhaust temperature in the regenerator so that $T_3 = T_5$ and $T_2 = T_6$ (Fig. 13.17). Therefore,

$$\eta = 1 - \frac{Q_2}{Q_1} = 1 - \frac{T_6 - T_1}{T_4 - T_3} = 1 - \frac{T_1}{T_4}\left[\frac{(T_2/T_1) - 1}{1 - (T_5/T_4)}\right]$$

$$= 1 - \frac{T_1}{T_4} \cdot \frac{T_2}{T_1}\left[\frac{1 - (T_1/T_2)}{1 - (T_5/T_4)}\right]$$

Since $$\frac{T_2}{T_1} = \left(\frac{p_2}{p_1}\right)^{\gamma - 1/\gamma} = \frac{T_4}{T_5}$$

$$\eta = 1 - \frac{T_1}{T_4} \gamma_p^{\gamma - 1/\gamma} \tag{13.17}$$

For a fixed ratio of (T_1/T_4), the cycle efficiency drops with increasing pressure ratio.

13.12.4 Effect of Irreversibilities in Turbine and Compressor

The Brayton cycle is highly sensitive to the real machine efficiencies of the turbine and the compressor. Figure 13.19 shows the actual and ideal expansion and compression processes.

Turbine efficiency, $$\eta_T = \frac{h_3 - h_4}{h_3 - h_{4s}} = \frac{T_3 - T_4}{T_3 - T_{4s}}$$

Compressor efficiency,

$$\eta_C = \frac{h_{2s} - h_1}{h_2 - h_1} = \frac{T_{2s} - T_1}{T_2 - T_1}$$

$$W_{net} = W_T - W_C = (h_3 - h_4) - (h_2 - h_1)$$

$$Q_1 = h_3 - h_2 \text{ and } Q_2 = h_4 - h_1$$

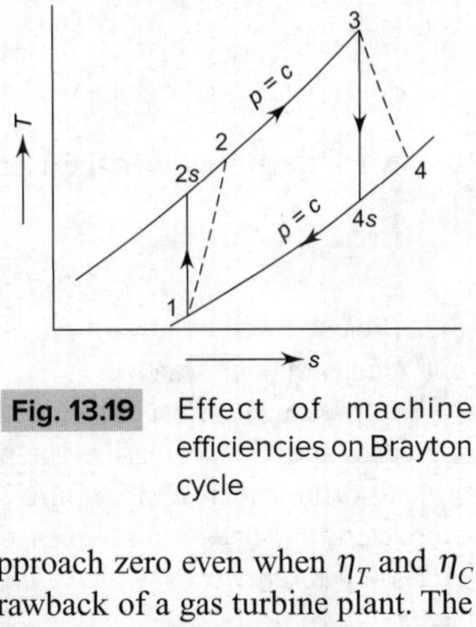

Fig. 13.19 Effect of machine efficiencies on Brayton cycle

The net output is reduced by the amount $[(h_4 - h_{4s}) + (h_2 - h_{2s})]$, and the heat supplied is reduced by the amount ($h_2 - h_{2s}$). The efficiency of the cycle will thus be less than that of the ideal cycle. As η_T and η_C decrease, η_{cycle} also decreases. The cycle efficiency may approach zero even when η_T and η_C are of the order of 60 to 70%. This is the main drawback of a gas turbine plant. The machines have to be highly efficient to obtain reasonable cycle efficiency.

13.12.5 Effect of Pressure Ratio on the Brayton Cycle

The efficiency of the Brayton cycle is a function of the pressure ratio as given by the equation

$$\eta = 1 - \frac{1}{(r_p)^{(\gamma-1)/\gamma}}$$

The more the pressure ratio, the more will be the efficiency.

Let T_1 = the lowest temperature of the cycle, which is the temperature of the surroundings (T_{min}), and T_3 = the maximum or the highest temperature of the cycle limited by the characteristics of the material available for burner and turbine construction (T_{max}).

Since the turbine, a steady-flow machine, is always exposed to the highest temperature gas, the maximum temperature of gas at the inlet to the turbine is limited to about 800°C by using a high air-fuel ratio. With turbine blade cooling, however, the maximum gas inlet temperature can be 1100°C or even higher.

Figure 13.20 shows the Brayton cycles operating between the same T_{max} and T_{min} at various pressure ratios. As the pressure ratio changes, the cycle shape also changes. For the cycle 1-2-3-4 of low pressure ratio r_p, since the average temperature of heat addition

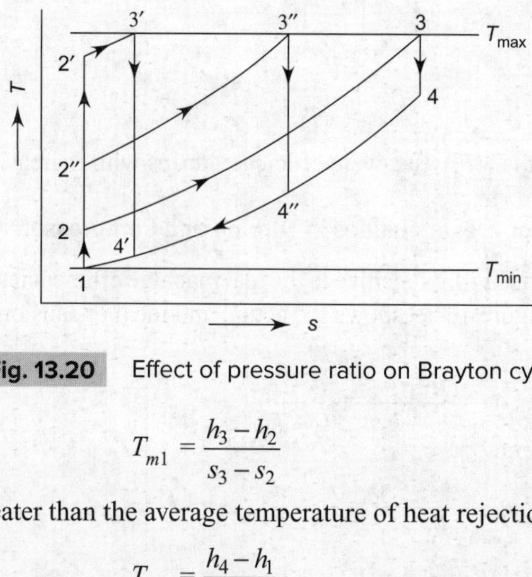

Fig. 13.20 Effect of pressure ratio on Brayton cycle

$$T_{m1} = \frac{h_3 - h_2}{s_3 - s_2}$$

is only a little greater than the average temperature of heat rejection

$$T_{m2} = \frac{h_4 - h_1}{s_4 - s_1}$$

the efficiency will be low. *At the lower limit of unity pressure ratio, both work output and efficiency will be zero.*

As the pressure ratio is increased, the efficiency steadily increases, because T_{m1} increases and T_{m2} decreases. The mean temperature of heat addition T_{m1} approaches T_{max} and the mean temperature of heat rejection T_{m2} approaches T_{min}, with the increase in r_p. *In the limit when the compression process ends at T_{max}, the Carnot efficiency is reached, r_p has the maximum value (r_{pmax}), but the work capacity again become zero.*

Figure 13.21 shows how the cycle efficiency varies with the pressure ratio, with r_p varying between two limiting values of 1 and $(r_p)_{max}$ when the Carnot efficiency is reached. When $r_p = (r_p)_{max}$, the cycle efficiency is given by

$$\eta = 1 - \frac{1}{((r_p)_{max}^{(\gamma-1)/\gamma}} = \eta_{Carnot}$$

$$= 1 - \frac{T_{min}}{T_{max}}$$

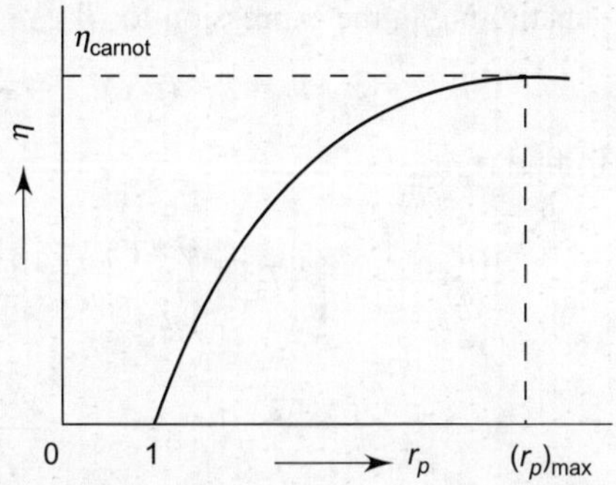

Fig. 13.21 Effect of pressure ratio on Brayton cycle efficiency

$$\therefore \quad ((r_p)_{max})^{(\gamma-1)/\gamma} = \frac{T_{max}}{T_{min}}$$

$$\therefore \qquad (r_p)_{max} = \left(\frac{T_{max}}{T_{min}}\right)^{\gamma/(\gamma-1)} \tag{13.18}$$

From Fig. 13.20, it is seen that the work capacity of the cycle, operating between T_{max} and T_{min}, is zero when $r_p = 1$ passes through a maximum, and then again becomes zero when the Carnot efficiency is reached. There is an optimum value of pressure ratio $(r_p)_{opt}$ at which work capacity becomes a maximum, as shown in Fig. 13.22.

For 1 kg,

$$W_{net} = c_p\,[(T_3 - T_4) - (T_2 - T_1)]$$

where $\qquad T_3 = T_{max}$ and $T_1 = T_{min}$

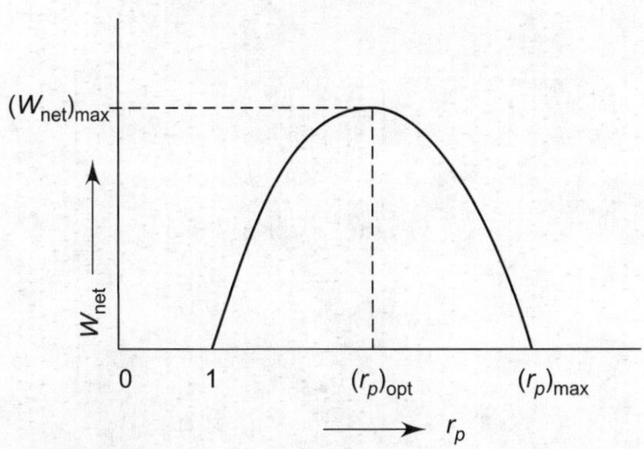

Fig. 13.22 Effect of pressure ratio on net output

Now $\qquad \dfrac{T_3}{T_4} = (r_p)^{(\gamma-1)/\gamma}$

$\therefore \qquad T_4 = T_3 \cdot r_p^{-(-\gamma-1)/\gamma}$

Similarly, $\qquad T_2 = T_1 \cdot r_p^{(\gamma-1)/\gamma}$

Substituting in the expression for W_{net}

$$W_{net} = c_p\,[T_3 - T_3 \cdot (r_p)^{-(\gamma-1)/\gamma} - T_1\, r_p^{(\gamma-1)/\gamma} + T_1] \tag{13.19}$$

To find $(r_p)_{opt}$

$$\frac{dW_{net}}{dr_p} = c_p\left[-T_3\left(-\frac{\gamma-1}{\gamma}\right)r_p^{(-1+(1/\gamma)-1)} - T_1\left(\frac{\gamma-1}{\gamma}\right)r_p^{(1-(1/\gamma)-1)}\right] = 0$$

$$\therefore \quad T_3\left(\frac{\gamma-1}{\gamma}\right)r_p^{(1/\gamma)-2} = T_1\left(\frac{\gamma-1}{\gamma}\right)\cdot r_p^{-1/\gamma}$$

$$\therefore \quad r_p^{-(1/\gamma)-(1/\gamma)+2} = \frac{T_3}{T_1}$$

$$(r_p)_{opt} = \left(\frac{T_3}{T_1}\right)^{\gamma/2(\gamma-1)}$$

$$\therefore \quad (r_p)_{opt} = \left(\frac{T_{max}}{T_{min}}\right)^{\gamma/2(\gamma-1)} \tag{13.20}$$

From Eqs. (13.18) and (13.20)

$$(r_p)_{opt} = \sqrt{(r_p)_{max}} \tag{13.21}$$

Substituting the values of $(r_p)_{opt}$ in Eq. (13.19)

$$W_{net} = (W_{net})_{max} = c_p\left[T_3 - T_3\left(\frac{T_1}{T_3}\right)\frac{\gamma}{2(\gamma-1)}\cdot\frac{\gamma-1}{\gamma} - T_1\left(\frac{T_3}{T_1}\right)\frac{\gamma}{2(\gamma-1)}\cdot\frac{\gamma-1}{\gamma} + T_1\right]$$

$$= c_p\,[T_3 - 2\sqrt{T_1T_3} + T_1]$$

or

$$(W_{net})_{max} = c_p\,(\sqrt{T_{max}} - \sqrt{T_{min}})^2 \tag{13.22}$$

$$\eta_{cycle} = 1 - \frac{1}{r_p^{\gamma-1/\gamma}} = 1 - \sqrt{\frac{T_{min}}{T_{max}}} \tag{13.23}$$

Considering the cycles 1-2′-3′-4′ and 1-2″-3″-4″ (Fig. 13.20), it is obvious that to obtain a reasonable work capacity, a certain reduction in efficiency must be accepted.

13.12.6 Effect of Intercooling and Reheating on Brayton Cycle

The efficiency of the Brayton cycle may often be increased by the use of staged compression with intercooling, or by using staged heat supply, called reheat.

Let the compression process be divided into two stages. Air, after being compressed in the first stage, is cooled to the initial temperature in a heat exchanger, called an intercooler, and then compressed further in the second stage (Fig. 13.23). 1-2′-5-6 is the ideal cycle without intercooling, having a single-stage compression, 1-2-3-4-5-6 is the cycle with intercooling, having a two-stage compression. The cycle 2-3-4-2′ is thus added to the basic cycle 1-2′-5-6. There is more work capacity, since the included area is more. There is more heat supply also. For the cycle 4-2′-2-3, T_{m1} is lower and T_{m2} higher (lower r_p) than those of the basic cycle 1-2′-5-6. So the efficiency of the cycle reduces by staging the compression and intercooling. But if a regenerator is used, the low temperature heat addition (4-2′) may be obtained by recovering energy from the exhaust gases from the turbine. So there may be a net gain in efficiency when intercooling is adopted in conjunction with a regenerator.

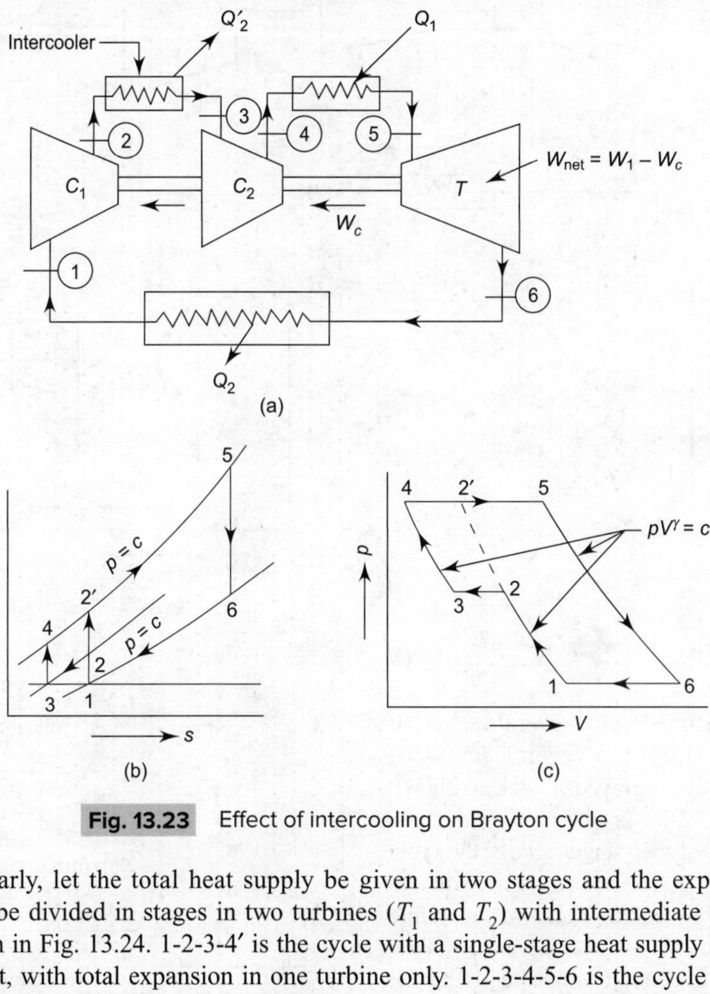

Fig. 13.23 Effect of intercooling on Brayton cycle

Similarly, let the total heat supply be given in two stages and the expansion process be divided in stages in two turbines (T_1 and T_2) with intermediate reheat, as shown in Fig. 13.24. 1-2-3-4′ is the cycle with a single-stage heat supply having no reheat, with total expansion in one turbine only. 1-2-3-4-5-6 is the cycle with a single-stage reheat, having the expansion divided into two stages. With the basic cycle, the cycle 4-5-6-4′ is added because of reheat. The work capacity increases, but the heat supply also increases.

In the cycle 4-5-6-4′, r_p is lower than in the basic cycle 1-2-3-4′, so its efficiency is lower. Therefore, the efficiency of the cycle decreases with the use of reheat. But T_6 is greater than T'_4. Therefore, if regeneration is employed, there is more energy that can be recovered from the turbine exhaust gases. So when regeneration is employed in conjunction with reheat, there may be a net gain in cycle efficiency.

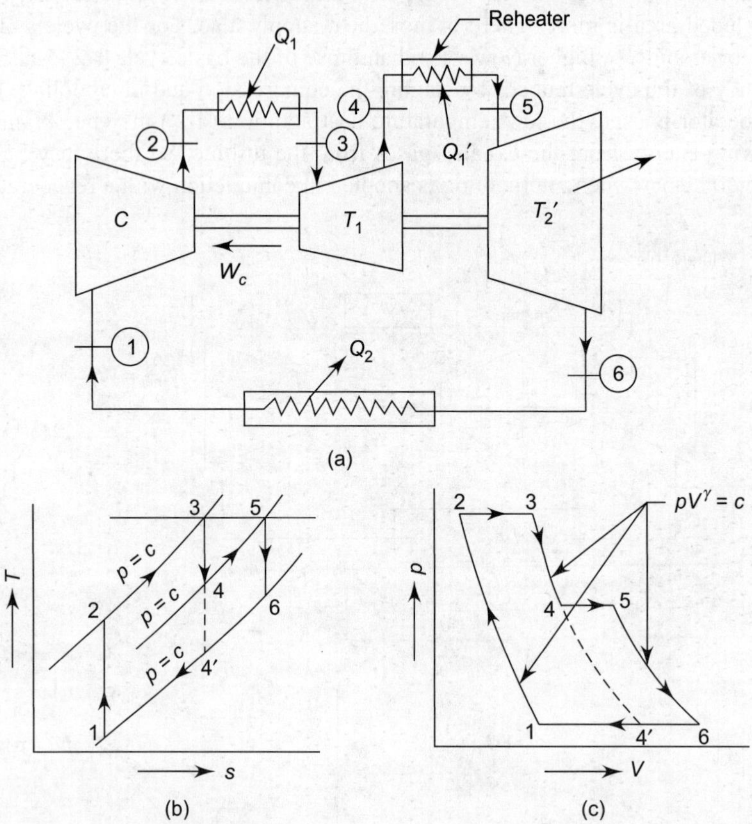

Fig. 13.24 Effect of reheat on Brayton cycle

If in one cycle, several stages of intercooling and several stages of reheat are employed, a cycle as shown in Fig. 13.25 is obtained. When the number of such stages is large the cycle reduces to the Ericsson cycle with two reversible isobars and two reversible isotherms. With ideal regeneration the cycle efficiency becomes equal to the Carnot efficiency.

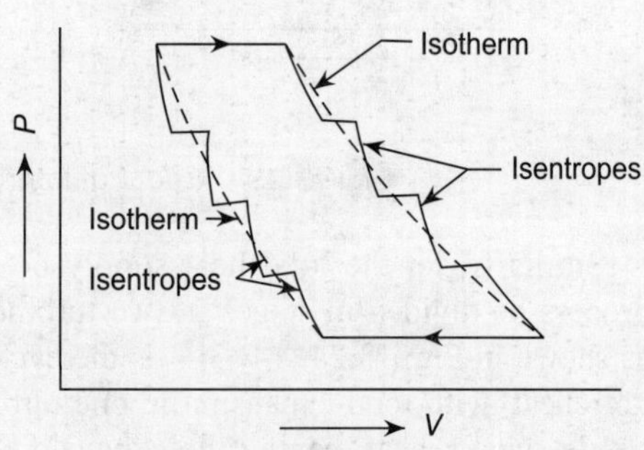

Fig. 13.25(a)

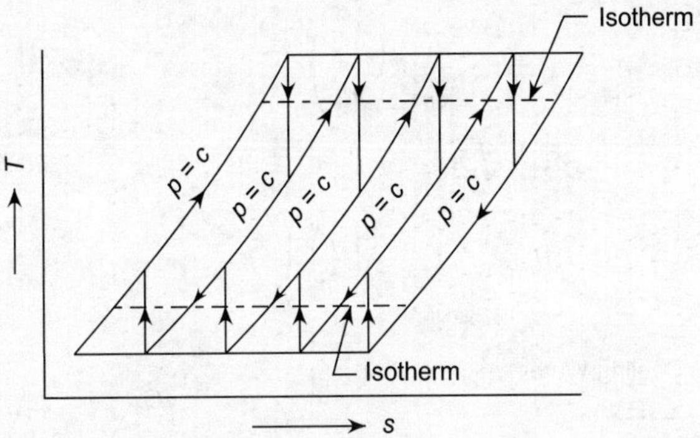

Fig. 13.25(b) Brayton cycle with many stages of intercooling and reheating approximates to Ericsson cycle

13.12.7 Ideal Regenerative Cycle with Intercooling and Reheat

Let us consider an ideal regenerative gas turbine cycle with two-stage compression and a single reheat. It assumes that both intercooling and reheating take place at the root mean square of the high and low pressure in the cycle, so that $p_3 = p_2 = p_7 = p_8 = \sqrt{p_1 p_4} = \sqrt{p_6 p_9}$ (Fig. 13.26). Also, the temperature after intercooling is equal to the compressor inlet temperature ($T_1 = T_3$) and the temperature after reheat is equal to the temperature entering the turbine initially ($T_6 = T_8$).

Here, $$Q_1 = c_p (T_6 - T_5) + c_p (T_8 - T_7)$$

Since $\dfrac{p_6}{p_7} = \dfrac{p_8}{p_9}$ and $T_6 = T_8$, it follows that $T_5 = T_7 = T_9$.

$\therefore$ $$Q_1 = 2\, c_p (T_6 - T_7)$$

Again, $$Q_2 = c_p (T_{10} - T_1) + c_p (T_2 - T_3)$$

but $$\frac{p_2}{p_1} = \frac{p_4}{p_3} \quad \text{and} \quad T_3 = T_1$$

so that $$T_2 = T_4 = T_{10}$$

$\therefore$ $$Q_2 = 2\, c_p (T_2 - T_1)$$

$$W_{net} = Q_1 - Q_2 = 2\, c_p [(T_6 - T_7) - (T_2 - T_1)]$$

$\therefore$ $$\eta_{cycle} = \frac{W_{net}}{Q_1} = 1 - \frac{T_2 - T_1}{T_6 - T_7}$$

$$= 1 - \frac{T_1}{T_6}\left[\frac{(T_2/T_1) - 1}{1 - (T_7/T_6)}\right]$$

$$= 1 - \frac{T_1}{T_6} \times \frac{T_2}{T_1}\left[\frac{1 - (T_1/T_2)}{1 - (T_7/T_6)}\right]$$

$$= 1 - \frac{T_1}{T_6} \times \left(\frac{p_2}{p_1}\right)^{\gamma-1/\gamma}$$

But $$p_2 = \sqrt{p_1 p_4},$$

$\therefore$ $$\eta_{\text{cycle}} = 1 - \frac{T_1}{T_6}\left(\frac{p_4}{p_1}\right)^{(\gamma-1)/2\gamma} \tag{13.24}$$

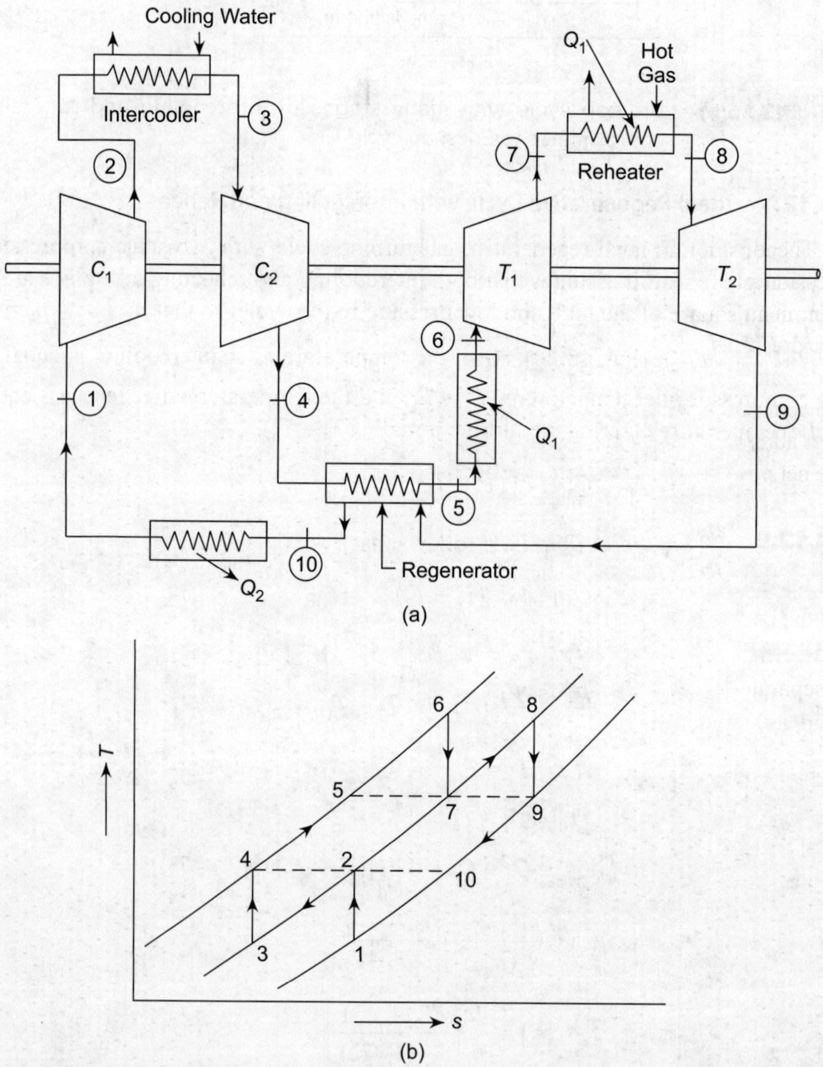

Fig. 13.26 Ideal regenerative gas turbine cycle with two-stage compression and reheat

13.12.8 Deviation in Actual Cycle from Ideal Brayton Cycle

LO 13.10

Relate actual Brayton cycles with ideal Brayton cycles

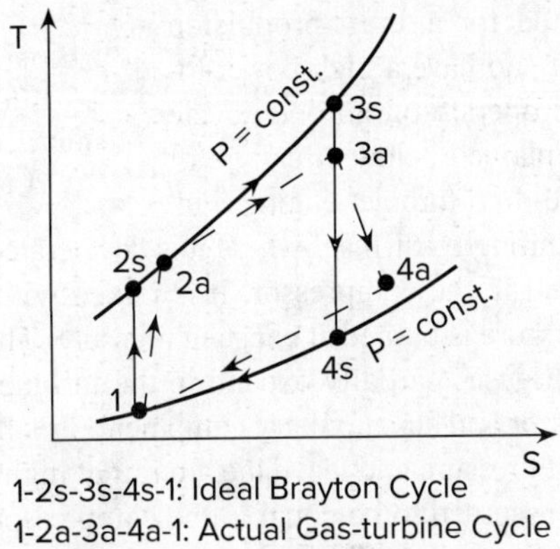

Fig. 13.27 Ideal Brayton cycle vs. actual gas turbine cycle

The ideal Brayton cycle assumes that all processes are reversible, with isentropic compression and expansion, and no pressure drops during heat addition and rejection. However, in a real cycle (depicted by the dotted lines), all processes are irreversible. Both the compression and expansion processes are non-isentropic, as discussed in Section 13.12.4 and again illustrated in Fig. 13.27. This deviation results in a reduced net power output in the actual cycle compared to the ideal cycle. Furthermore, pressure drops occur during heat addition (process 2a-3a) and heat rejection (process 4a-1), which further decrease the net power output relative to the ideal cycle.

13.12.9 Free-shaft Turbine

So far only a single shaft has been shown in flow diagrams, on which were mounted all the compressors and the turbines. Sometimes, for operating convenience and part-load efficiency, one turbine is used for driving the compressor only on one shaft, and a separate turbine is used on another shaft, known as free-shaft, for supplying the load, as shown in Fig. 13.28.

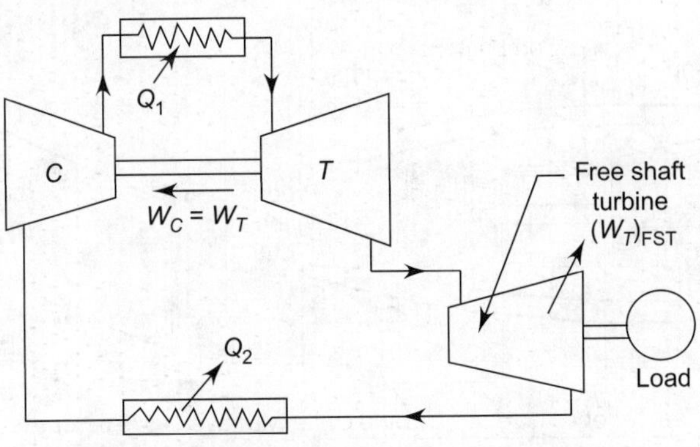

Fig. 13.28 Free shaft turbine

13.13 AIRCRAFT PROPULSION

LO 13.11 Explain principles of different types of aircraft propulsion systems

Gas turbines are particularly suited for aircraft propulsion because they are light and compact and have a high power-to-weight ratio. Aircraft gas turbines operate on an open cycle called '*jet propulsion cycle*', as explained below. It can be of turbojet, turbofan or turboprop type. In a turbojet engine (Fig. 13.29), high velocity air first flows through a diffuser where it is decelerated increasing its pressure. Air is then compressed in the compressor. It is mixed with fuel in the combustion chamber, where the mixture is burned at constant pressure. The high pressure, high temperature combustion gases partially expand in the turbine, producing enough power to drive the compressor and any auxiliary equipment. Finally, the gases expand in the nozzle to the ambient pressure and leave the aircraft at high velocity. In the ideal case, the turbine work is assumed to be equal to the compressor work. The processes in the diffuser, the compressor, the turbine and the nozzle are assumed to be reversible and adiabatic. Another view of a turbojet engine is given in Fig. 13.29(a).

The thrust developed in a turbojet engine is the unbalanced force caused by the difference in the momentum of the air entering the engine and the exhaust gases leaving the engine, so that

$$F = (\dot{m}\bar{V})_{exit} - (\dot{m}\bar{V})_{inlet} = \dot{m}(\bar{V}_{exit} - \bar{V}_{inlet}) \tag{13.25}$$

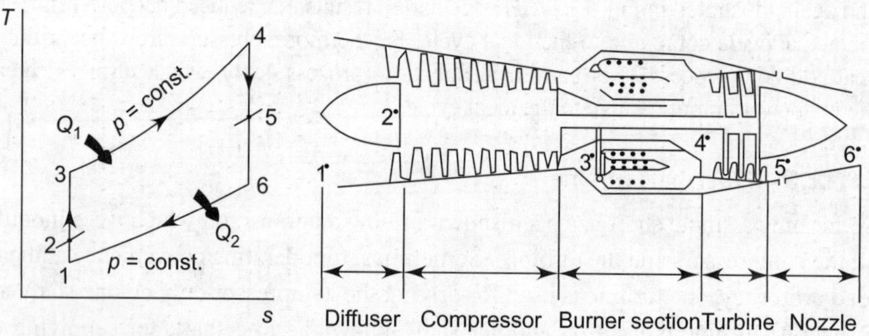

Fig. 13.29(a) Basic components of a turbojet engine and the *T-s* diagram of an ideal turbojet cycle

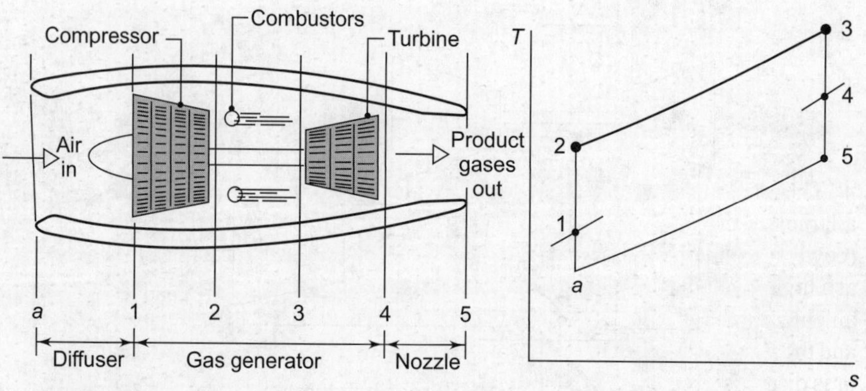

Fig. 13.29(b) Turbojet engine schematic and accompanying ideal *T-s* diagram

The pressures at inlet and exit of the engine are the ambient pressure. For an aircraft cruising in still air, $\bar{V}$ inlet is the aircraft velocity. The mass flow rates of the gases at the engine exit and the inlet are different, the difference being equal to the combustion rate of the fuel. But the air-fuel ratio used in jet propulsion engines is usually very high, making the difference very small. Thus $\dot{m}$ in Eq. (13.25) is taken as the mass flow rate of air through the engine. For an aircraft cruising at a steady speed, the thrust is used to overcome the fluid drag, and the net force acting on the body of the aircraft is zero. Commercial airplanes save fuel by flying at higher altitudes during long trips since the air at higher altitudes is of less density and exerts a smaller drag force on the aircraft.

The power developed from the thrust of the engine is called the *propulsive power*, W_p, given by

$$\dot{W}_p = F\bar{V}_{\text{aircraft}} = \dot{m}(\bar{V}_{\text{exit}} - \bar{V}_{\text{inlet}})\bar{V}_{\text{aircraft}} \tag{13.26}$$

The propulsive efficiency, η_p, is defined by

$$\eta_p = \frac{\text{Propulsive power}}{\text{Energy input rate}} = \frac{\dot{W}_p}{\dot{Q}_{\text{in}}} \tag{13.27}$$

It is a measure of how efficiently the energy released during combustion is converted to propulsive power.

Space and weight limitations prohibit the use of regenerators and intercoolers on aircraft engines. The counterpart of reheating is afterburning. The air-fuel ratio in a jet engine is so high that the turbine exhaust gases are sufficiently rich in oxygen to support the combustion of more fuel in an afterburner (Fig. 13.30). Such burning of fuel raises the temperature of the gas before it expands in the nozzle, increasing the KE change in the nozzle and consequently increasing the thrust. In the air-standard case, the combustion is replaced by constant pressure heat addition.

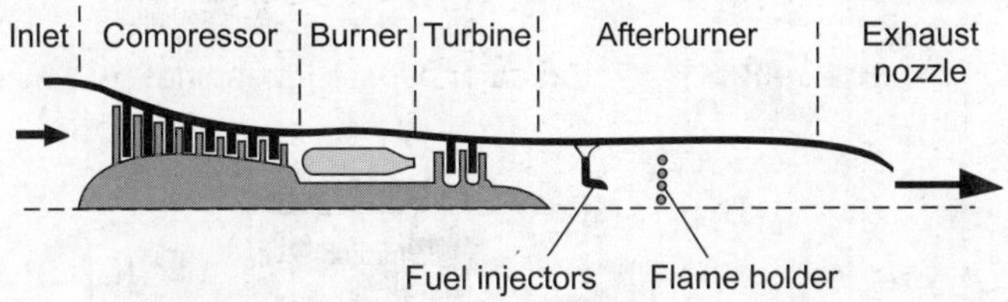

Fig. 13.30 Turbojet engine with afterburner

The most widely used engine in aircraft propulsion is the *turbofan engine* wherein a large fan driven by the turbine forces a considerable amount of air through a duct (cowl) surrounding the engine (Figs. 13.31 and 13.32). The fan exhaust leaves the duct at a higher velocity, enhancing the total thrust of the engine significantly. Some of the air entering the engine flows through the compressor, combustion chamber and turbine and the rest passes through the fan into a duct and is either mixed with the exhaust gases or is discharged separately. It improves the engine performance over a broad operating range. The ratio of the mass flow rates of the two streams is called the *bypass ratio.*

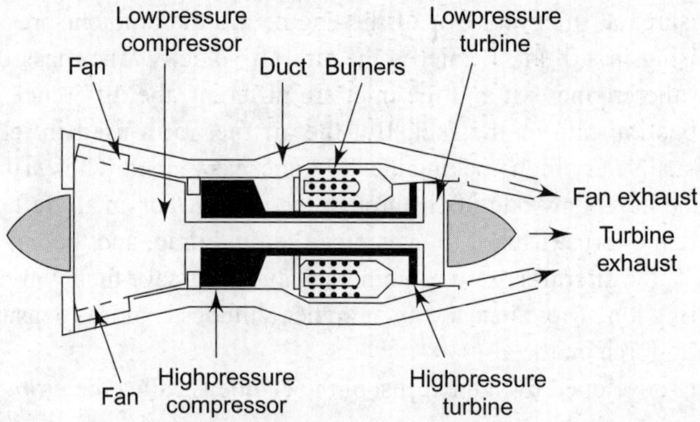

Fig. 13.31 Turbofan engine

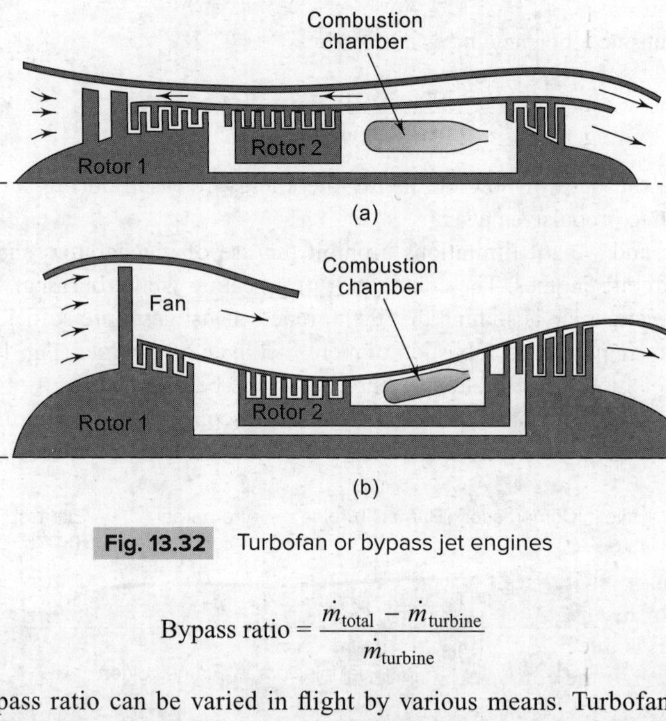

Fig. 13.32 Turbofan or bypass jet engines

$$\text{Bypass ratio} = \frac{\dot{m}_{\text{total}} - \dot{m}_{\text{turbine}}}{\dot{m}_{\text{turbine}}}$$

The bypass ratio can be varied in flight by various means. Turbofan engines deserve most of the credit for the success of jumbo jets, which weigh almost 400,000 kg and are capable of carrying 400 passengers for up to 10,000 km at speeds over 950 km/h with less fuel per passenger mile.

Increasing the bypass ratio of a turbofan engine increases thrust. If the cowl is removed from the fan the result is a *turboprop engine* (Fig. 13.33). Turbofan and turboprop engines differ mainly in their bypass ratio: 5 or 6 for turbofans and as high as 100 for turboprop. In general, propellers are more efficient than jet engines, but they are limited to low-speed and ow-altitude operation since their efficiency decreases at high speeds and altitudes.

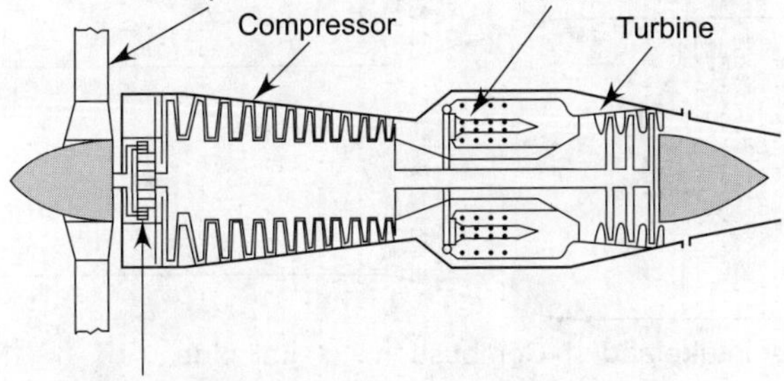

Fig. 13.33 Turboprop engine

A particularly simple type of engine known as a *ramjet* is shown in Fig. 13.34. This engine requires neither a compressor nor a turbine. A sufficient pressure rise is obtained by decelerating the high speed incoming air in the diffuser (ram effect) on being rammed against a barrier. For the ramjet to operate, the aircraft must already be in flight at a sufficiently great speed. The combustion products exiting the combustor are expanded through a nozzle to produce the thrust.

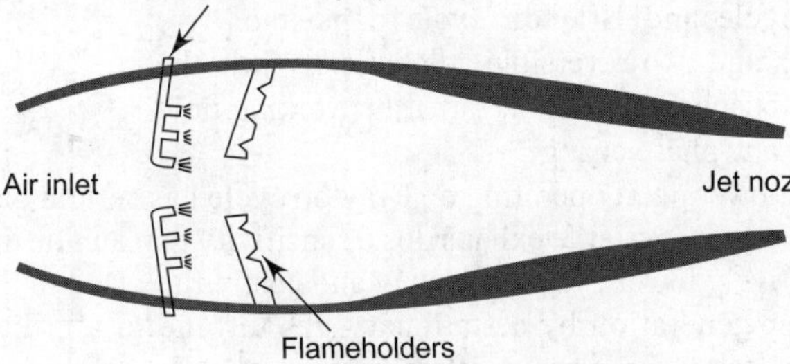

Fig. 13.34 Ramjet engine

The pulse jet engine is a simple engine that does not require the motion of the aircraft for its operation. As shown in Fig. 13.35, air enters the intake manifold where fuel is injected to mix with the air. The air-fuel mixture then passes through the valves into the combustion chamber, where combustion initiates with a spark. The exhaust gases exit through the nozzle at the other end of the tail pipe while the valves remain closed. The rapid expulsion of the combustion products through the nozzle generates thrust as a reaction force. Simultaneously, the fast exhaust creates an instant vacuum, drawing in a fresh air-fuel mixture through the valves for the next pulse combustion. Since the thrust in this engine occurs in pulses following each combustion cycle, it is referred to as a pulse jet engine. The only moving components in such engines are the valves, though in some variations, no valves are used at all, and the flow of air is controlled by aerodynamic principles. The pulse jet engine is a basic form of jet propulsion but has lower efficiency due to the low compression ratio of the air-fuel mixture.

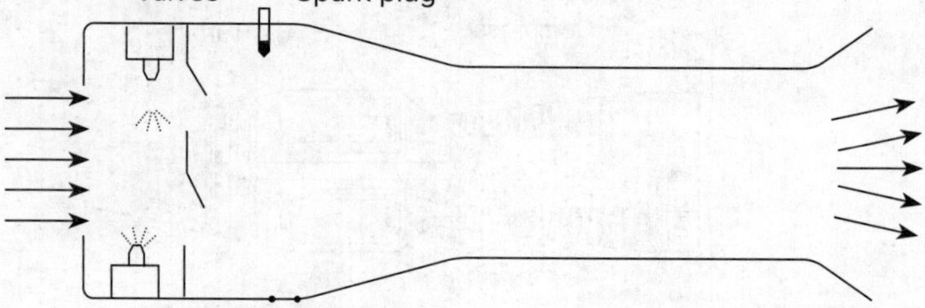

Fig. 13.35 Pulse jet engine

In each of the engines mentioned so far, combustion of the fuel is supported by air brought into the engines from the atmosphere. For very high altitude flights and space travel, where this is no longer possible, rockets may be employed. In a rocket, both fuel and an oxidiser (such as liquid oxygen) are carried on board of the craft. High pressure combustion gases are expanded in a nozzle. The gases leave the rocket at very high velocities, producing the thrust to propel the rocket.

13.14 BRAYTON–RANKINE COMBINED CYCLE

> **LO 13.12**
> Analyze Brayton-Rankine combined cycle

Both Rankine cycle and Brayton cycle consist of two reversible isobars and two reversible adiabatics. While the former is a phase change cycle, in the latter the working fluid does not undergo any phase change.

A gas turbine power plant operating on Brayton cycle has certain disadvantages like large compressor work, large exhaust loss, sensitivity to machine inefficiencies (η_T and η_C), relatively lower cycle efficiency and costly fuel. Due to these factors, the cost of power generation by a stationary gas turbine in a utility system is high. However, a gas turbine plant offers certain advantages also, such as less installation cost, less installation time, quick starting and stopping and fast response to load changes. So, a gas turbine plant is often used as a *peaking unit* for certain hours of the day, when the energy demand is high. To utilise the high temperature exhaust and to raise its plant efficiency, a gas turbine may be used in conjunction with a steam turbine plant to offer the gas turbine advantages of quick starting and stopping and permit flexible operation of the combined plant over a wide range of loads.

Let us consider two cyclic power plants coupled in series, the topping plant operating on Brayton cycle and the bottoming one operating on Rankine cycle (Fig. 13.36). Helium may be the working fluid in the topping plant and water in the bottoming plant. The overall efficiency of the combined plant is

$$\eta = \eta_1 + \eta_2 - \eta_1 \eta_2$$

where η_1 and η_2 are the efficiencies of the Brayton cycle, and Rankine cycle respectively.

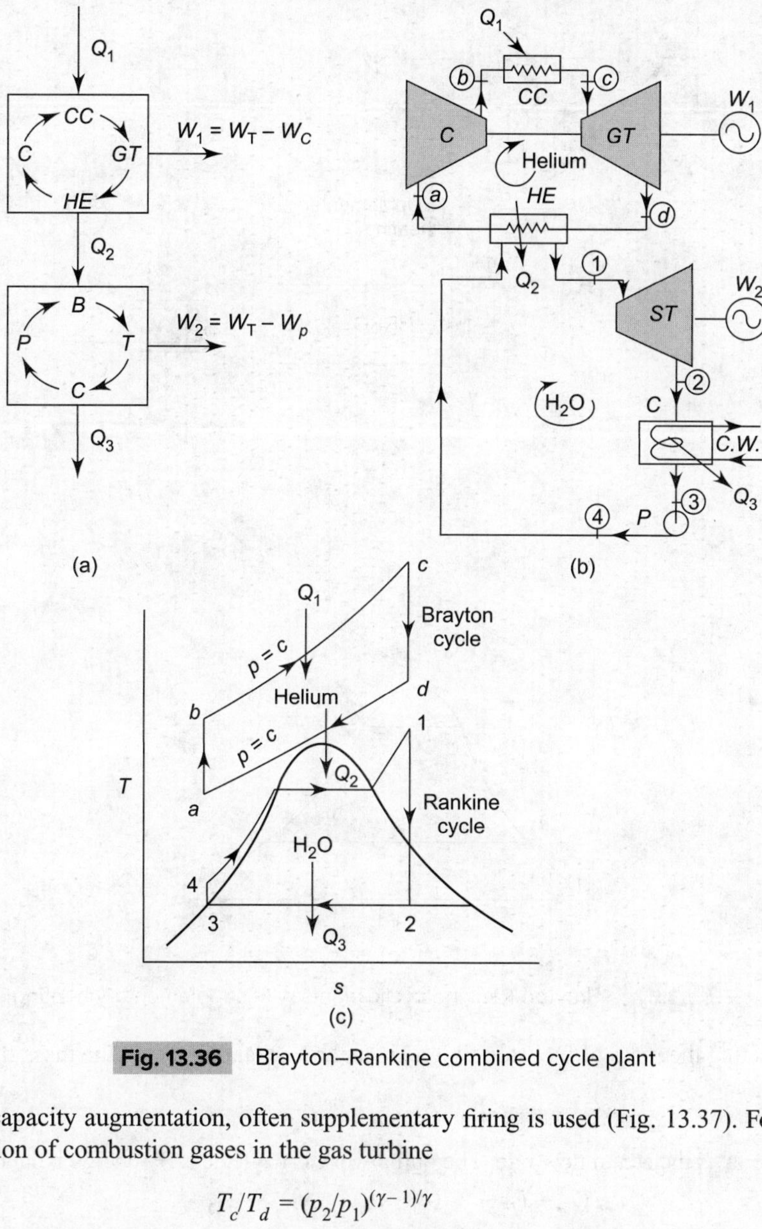

Fig. 13.36 Brayton–Rankine combined cycle plant

For capacity augmentation, often supplementary firing is used (Fig. 13.37). For expansion of combustion gases in the gas turbine

$$T_c/T_d = (p_2/p_1)^{(\gamma-1)/\gamma}$$

where $\gamma = 1.3$ (assumed)

For the compressor, $T_b/T_a = (p_2/p_1)^{(\gamma'-1)/\gamma'}$

where $\gamma' = 1.4$

$$W_{GT} = w_a\,[c_{pg}\,(T_c - T_d) - c_{pa}\,(T_b - T_a)]$$

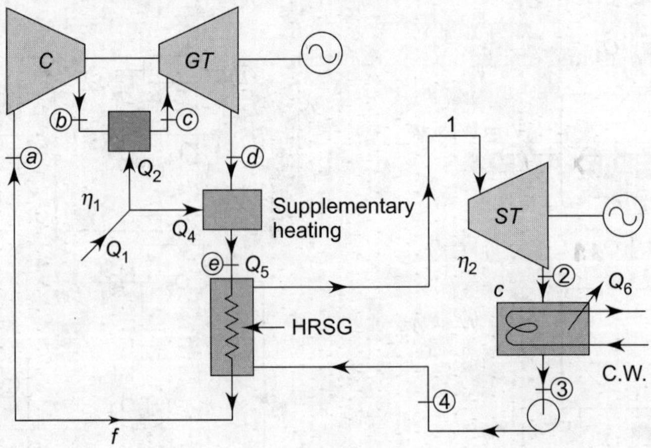

(a)

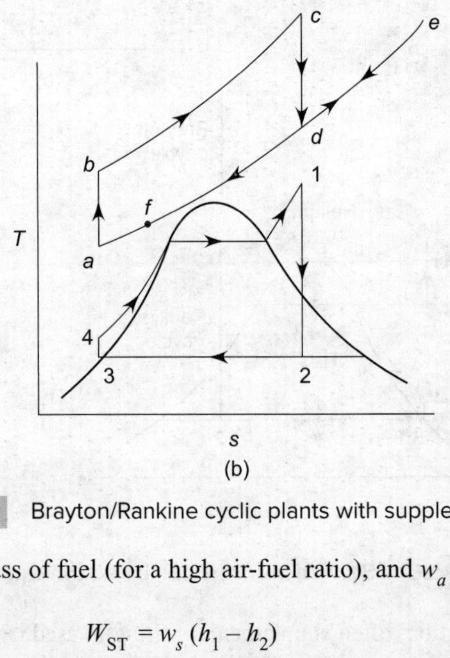

(b)

Fig. 13.37 Brayton/Rankine cyclic plants with supplementary heating

neglecting the mass of fuel (for a high air-fuel ratio), and w_a being the mass flow of air.

$$W_{ST} = w_s (h_1 - h_2)$$

where w_s is the steam flow rate. The pump work is neglected. By energy balance,

$$w_a c_{pg} (T_c - T_f) = w_s (h_1 - h_4)$$

Now,
$$Q_1 = w_a c_{pg} [(T_c - T_b) + (T_e - T_d)]$$

The overall efficiency of the plant is

$$\eta = \frac{W_{GT} + W_{ST}}{Q_1}$$

Again,
$$Q_1 = w_f \times \text{C.V.}$$

where w_f is the fuel burning rate.

High overall efficiency, low investment cost, less water requirement, large operating flexibility, phased installation and low environmental impact are some of the advantages of combined gas-steam cycles.

SOLVED EXAMPLES

Example 13.1 *An engine working on the Otto cycle is supplied with air at 0.1 MPa, 35°C. The compression ratio is 8. Heat supplied is 2100 kJ/kg. Calculate the maximum pressure and temperature of the cycle, the cycle efficiency and the mean effective pressure. (For air, $c_p = 1.005$, $c_v = 0.718$, and $R = 0.287$ kJ/kg K).*

E **[LO 13.4]**

Solution From Fig. 13.38

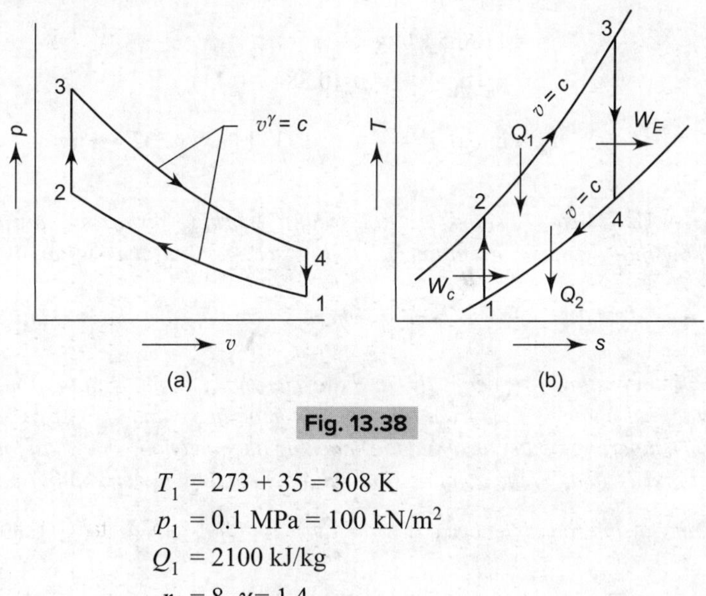

Fig. 13.38

$$T_1 = 273 + 35 = 308 \text{ K}$$

$$p_1 = 0.1 \text{ MPa} = 100 \text{ kN/m}^2$$

$$Q_1 = 2100 \text{ kJ/kg}$$

$$r_k = 8,\ \gamma = 1.4$$

$$\therefore \quad \eta_{\text{cycle}} = 1 - \frac{1}{r_k^{\gamma-1}} = 1 - \frac{1}{8^{0.4}} = 1 - \frac{1}{2.3}$$

$$= \mathbf{0.565 \text{ or } 56.5\%}$$

$$\frac{v_1}{v_2} = 8,\ v_1 = \frac{RT_1}{p_1} = \frac{0.287 \times 308}{100} = 0.844 \text{ m}^3\text{/kg}$$

$$\therefore \quad v_2 = \frac{0.884}{8} = 0.11 \text{ m}^3\text{/kg}$$

$$\frac{T_2}{T_1} = \left(\frac{v_1}{v_2}\right)^{\gamma-1} = (8)^{0.4} = 2.3$$

$$\therefore \quad T_2 = 2.3 \times 308 = 708.4 \text{ K}$$

$$Q_1 = c_v (T_3 - T_2) = 2100 \text{ kJ/kg}$$

$\therefore$ $$T_3 - 708.4 = \frac{2100}{0.718} = 2925 \text{ K}$$

or $$T_3 = T_{max} = \mathbf{3633\ K}$$

$$\frac{p_2}{p_1} = \left(\frac{v_1}{v_2}\right)^{\gamma} = (8)^{1.4} = 18.37$$

$\therefore$ $$p_2 = 1.873 \text{ MPa}$$

Again

$$\frac{p_3 v_3}{T_3} = \frac{p_2 v_2}{T_2}$$

$\therefore$ $$p_3 = p_{max} = 1.837 \times \frac{363}{708} = \mathbf{9.426\ MPa}$$

$$W_{net} = Q_1 \times \eta_{cycle} = 2100 \times 0.565$$

$$= 1186.5 \text{ kJ/kg}$$

$$= p_m(v_1 - v_2) = p_m(0.884 - 0.11)$$

$\therefore$ $$p_m = \text{m.e.p.} = \frac{1186.5}{0.774} = 1533 \text{ kPa} = \mathbf{1.533\ MPa}$$

Example 13.2 *(a) A spark-ignition engine working on air standard Otto cycle. Show that the net work output from the cycle per unit mass of air is given by* $w_{net} = c_v \left[T_3 - T_1 r_k^{\gamma-1} - \frac{T_3}{r_k^{\gamma-1}} + T_1 \right]$, *where T_3 and T_1 are the maximum and minimum temperatures, respectively, in the cycle and r is the compression ratio. (b) In practical engines, T_1 is fixed at ambient temperature and T_3 is fixed by the maximum temperature that can avoid the knocking phenomenon. Show that for fixed values of and T_1 and T_3, the compression ratio for maximum work output per unit mass of air flowing round the cycle is given by* $r_k = (T_1/T_3)^{\frac{1}{2(1-\gamma)}}$. **M [LO 13.4]**

Solution (a) The net shaft work output from the air-standard Otto cycle per unit mass of air is given by (see Fig. 13.39)

$$w_{net} = c_v[T_3 - T_2] - c_v[T_4 - T_1]$$

For the isentropic process compression process 1 – 2, we get

$$T_2 = T_1 \left(\frac{v_1}{v_2}\right)^{\gamma-1} = T_1 r_k^{\gamma-1}$$

Similarly, for the isentropic expansion process, 3–4, we have

$$T_4 = T_3 \left(\frac{v_3}{v_4}\right)^{\gamma-1} = \frac{T_3}{r_k^{\gamma-1}}$$

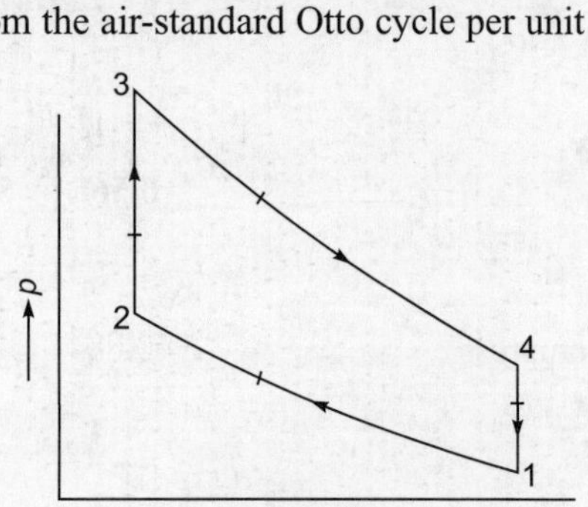

Fig. 13.39

Substituting the values of T_2 and T_4 and in the expression of w_{net}, we have

$$w_{net} = c_v\left[T_3 - T_1 r_k^{\gamma-1} - \frac{T_3}{r_k^{\gamma-1}} + T_1\right]$$

(b) For fixed values of T_1 and T_3, $w_{net,}$ will be maximum when $\frac{dw_{net}}{dr} = 0$. Thus

$$\frac{d}{dr}c_v\left[T_3 - T_1 r_k^{\gamma-1} - \frac{T_3}{r_k^{\gamma-1}} + T_1\right] = 0$$

or $$\frac{d}{dr}c_v\left[T_3 - T_1 r_k^{\gamma-1} - \frac{T_3}{r_k^{\gamma-1}} + T_1\right] = 0$$

or $$-T_1(\gamma-1)r_k^{\gamma-1-1} + (\gamma-1)T_3 r_k^{-(\gamma-1)-1} = 0$$

or $$T_3 r_k^{-\gamma} = T_1 r_k^{\gamma-2}$$

or $$r_k^{2(1-\gamma)} = \frac{T_1}{T_3}$$

Example 13.3 *In a spark ignition engine working on air-standard Otto cycle, the pressure in the cylinder at 20 and 80% of the compression stroke are measured to be 0.12 and 0.52 MPa, respectively. Determine the air-standard efficiency of the cycle.* E [LO 13.4]

Solution From Fig. 13.40, one can write

$$V_a = V_1 - 0.2V_s = V_2 + V_s - 0.2V_s = V_2 + 0.8V_s$$

$$V_b = V_1 - 0.8V_s = V_2 + V_s - 0.8V_s = V_2 + 0.2V_s$$

For the isentropic compression between a-b, we have

$$p_a V_a^{\gamma} = p_b V_b^{\gamma}$$

or $$\frac{V_a}{V_b} = \left(\frac{p_b}{p_a}\right)^{1/\gamma} = \left(\frac{0.52}{0.12}\right)^{1/1.4} = 2.85$$

or $$\frac{V_2 + 0.8V_s}{V_2 + 0.2V_s} = 2.85$$

or $$\frac{1 + 0.8\frac{V_s}{V_2}}{1 + 0.2\frac{V_s}{V_2}} = 2.85$$

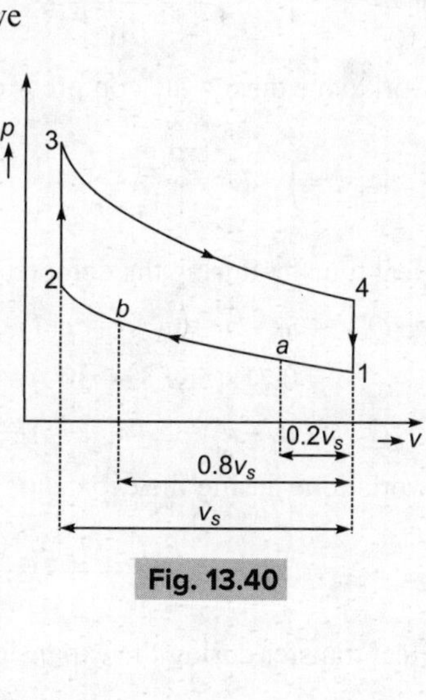

Fig. 13.40

Compression ratio $r_k = \frac{V_1}{V_2} = \frac{V_2 + V_s}{V_2} = 1 + \frac{V_s}{V_2}$

or $$\frac{V_s}{V_2} = r_k - 1$$

Thus, $$\frac{1 + 0.8(r_k - 1)}{1 + 0.2(r_k - 1)} = 2.85$$

or $$0.2 + 0.8r_k = 2.28 + 0.57r_k$$

or $$r_k = 9.04$$

The air-standard cycle efficiency is

$$\eta = 1 - \frac{1}{r_k^{\gamma-1}} = 1 - \frac{1}{(9.04)^{1.4-1}} = 0.5855 \text{ or } \mathbf{58.55\%}$$

Example 13.4 *The processes within the cylinder of a spark ignition engine are modelled as an air-standard Otto cycle. The state at the beginning of the compression is p_1 = 0.1 MPa and T_1 = 300 K. The compression ratio is 8 and the maximum temperature of the cycle is 1400 K. There is a modification in the cycle in which the isentropic compression and expansion processes are replaced with polytropic processes having polytropic index n = 1.3. Determine the net work output and the cycle efficiency due to such modification. (For air R = 0.287 kJ/kgK, c_p = 1.005 kJ/kgK and c_v = 0.718 kJ/kgK),* M **[LO 13.4]**

Solution From Fig. 13.41

$$\frac{T_2}{T_1} = \left(\frac{v_1}{v_2}\right)^{n-1}$$

or $$T_2 = 300(8)^{1.3-1} = 559.82 \text{ K}$$

$$\frac{T_3}{T_4} = \left(\frac{v_4}{v_3}\right)^{n-1} = r_k^{n-1}$$

or $$T_4 = \frac{T_3}{r_k^{n-1}} = \frac{1400}{(8)^{1.3-1}} = 750.24 \text{ K}$$

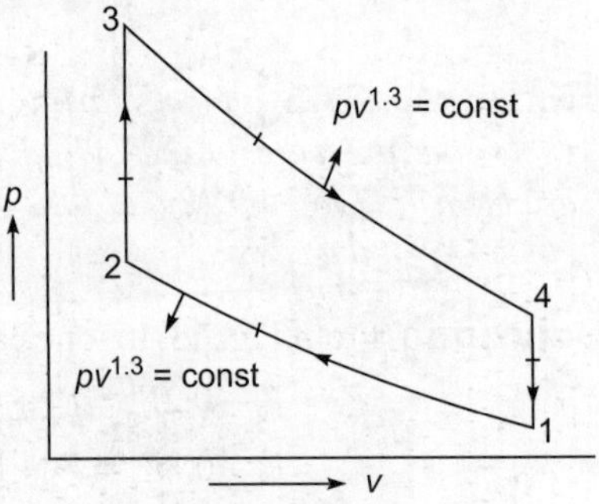

Fig. 13.41

Work done during the compression process 1-2

$$w_{1-2} = \int_1^2 pdv = \frac{R(T_1 - T_2)}{n-1} = \frac{0.287(300 - 559.82)}{1.3-1} = -248.56 \text{ kJ/kg}$$

Heat transfer during the compression process 1-2

$$Q_{1-2} = u_2 - u_1 + w_{1-2} = c_v(T_2 - T_1) + w_{1-2}$$
$$= 0.718(559.82 - 300) - 248.56 = -58.21 \text{ kJ/kg}$$

$$Q_{2-3} = u_3 - u_2 = c_v(T_3 - T_2) = 0.718(1400 - 559.82) = 603.25 \text{ kJ/kg}$$

Work done during the expansion process 3-4

$$w_{3-4} = \int_3^4 pdv = \frac{R(T_3 - T_4)}{n-1} = \frac{0.287(1400 - 750.24)}{1.3-1} = 621.6 \text{ kJ/kg}$$

Heat transfer during the expansion process 3-4

$$Q_{3-4} = u_4 - u_3 + w_{3-4} = c_v(T_4 - T_3) + w_{3-4} = 0.718(750.24 - 1400) + 621.66$$
$$= 155.07 \text{ kJ/kg}$$

Net work output from the cycle

$$w_{net} = w_{3-4} - w_{1-2} = 621.6 - 248.56 = \mathbf{373.04\ kJ/kg}$$

Total heat intput to the cycle

$$Q_{in} = Q_{2-3} + Q_{3-4} = 603.25 + 155.07 = 758.32\ \text{kJ/kg}$$

The air-standard cycle efficiency is

$$\eta = \frac{w_{net}}{Q_{in}} = \frac{373.04}{758.32} = 0.4919 \text{ or } \mathbf{49.19\%}$$

Example 13.5 *A Diesel engine has a compression ratio of 14 and cut-off takes place at 6% of the stroke. Find the air standard efficiency.* E [LO 13.5]

Solution From Fig. 13.42

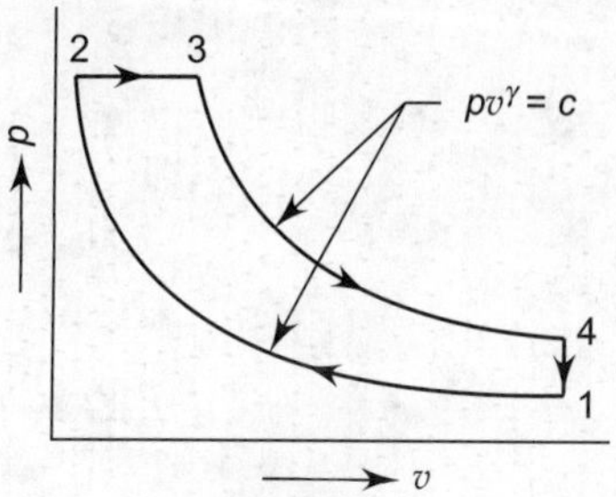

Fig. 13.42

$$r_k = \frac{v_1}{v_2} = 14$$

$$v_3 - v_2 = 0.06\,(v_1 - v_2)$$
$$= 0.06\,(14\,v_2 - v_2)$$
$$= 0.78\,v_2$$
$$v_3 = 1.78\,v_2$$

$\therefore$ Cut-off ratio, $r_c = \frac{v_3}{v_2} = 1.78$

$$\eta_{Diesel} = 1 - \frac{1}{\gamma}\cdot\frac{1}{r_k^{\gamma-1}}\cdot\frac{r_c^\gamma - 1}{r_c - 1}$$

$$= 1 - \frac{1}{1.4}\cdot\frac{1}{(14)^{0.4}}\cdot\frac{(1.78)^{1.4} - 1}{1.78 - 1}$$

$$= 1 - 0.248\cdot\frac{1.24}{0.78} = 0.605, \text{ i.e. } \mathbf{60.5\%}$$

Example 13.6 *In an air standard Diesel cycle, the compression ratio is 16, and at the beginning of isentropic compression, the temperature is 15°C and the pressure is 0.1 MPa. Heat is added until the temperature at the end of the constant pressure process is 1480°C. Calculate (a) the cut-off ratio, (b) the heat supplied per kg of air, (c) the cycle efficiency, and (d) the m.e.p.* M [LO 13.5]

Solution From Fig. 13.43

$$r_k = \frac{v_1}{v_2} = 16$$

$$T_1 = 273 + 15 = 288\ \text{K}$$
$$p_1 = 0.1\ \text{MPa} = 100\ \text{kN/m}^2$$
$$T_3 = 1480 + 273 = 1735\ \text{K}$$

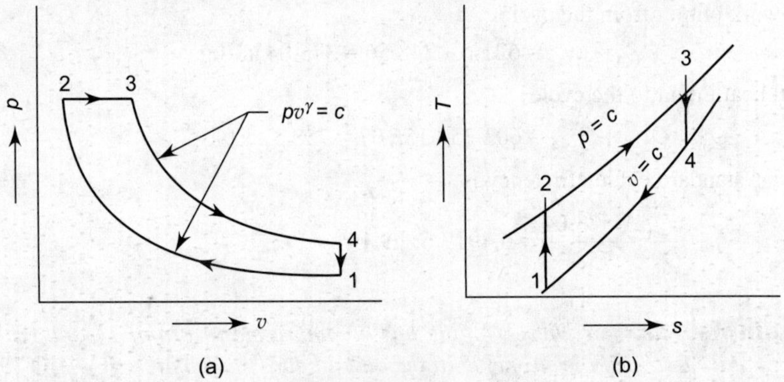

Fig. 13.43

$$\frac{T_2}{T_1} = \left(\frac{v_1}{v_2}\right)^{\gamma-1} = (16)^{0.4} = 3.03$$

$$\therefore \quad T_2 = 288 \times 3.03 = 873 \text{ K}$$

$$\frac{p_2 v_2}{T_2} = \frac{p_3 v_3}{T_3}$$

(a) Cut-off ratio, $$r_c = \frac{v_3}{v_2} = \frac{T_3}{T_2} = \frac{1753}{273} = \mathbf{2.01}$$

(b) Heat supplied, $$Q_1 = c_p (T_3 - T_2) = 1.005 (1753 - 873) = \mathbf{884.4 \text{ kJ/kg}}$$

$$\frac{T_3}{T_4} = \left(\frac{v_4}{v_3}\right)^{\gamma-1} = \left(\frac{v_1}{v_2} \times \frac{v_2}{v_3}\right)^{\gamma-1} = \left(\frac{16}{2.01}\right)^{0.4} = 2.29$$

$$\therefore \quad T_4 = \frac{1753}{2.29} = 766 \text{ K}$$

Heat rejected, $$Q_2 = c_v (T_4 - T_1) = 0.718 (766 - 288) = 343.2 \text{ kJ/kg}$$

(c) $$\text{Cycle efficiency} = 1 - \frac{Q_2}{Q_1} = 1 - \frac{343.2}{884.4} = \mathbf{0.612 \text{ or } 61.2\%}$$

It may also be estimated from the equation

$$\eta_{\text{cycle}} = 1 - \frac{1}{\gamma} \cdot \frac{1}{r_k^{\gamma-1}} \cdot \frac{r_c^{\gamma} - 1}{r_c - 1}$$

$$= 1 - \frac{1}{1.4}\frac{1}{(16)^{0.4}} \cdot \frac{(2.01)^{1.4} - 1}{2.01 - 1}$$

$$= 1 - \frac{1}{1.4} \cdot \frac{1}{3.03} \cdot 1.64 = \mathbf{0.612 \text{ or } 61.2\%}$$

$$W_{net} = Q_1 \times \eta_{cycle}$$

$$= 884.4 \times 0.612 = 541.3 \text{ kJ/kg}$$

$$v_1 = \frac{RT_1}{p_1} = \frac{0.287 \times 288}{100} = 0.052 \text{ m}^3\text{/kg}$$

$$v_2 = \frac{0.827}{16} = 0.052 \text{ m}^3\text{/kg}$$

$$\therefore \quad v_1 - v_2 = 0.827 - 0.052 = 0.775 \text{ m}^3\text{/kg}$$

(d) $$\text{m.e.p.} = \frac{W_{net}}{v_1 - v_2} = \frac{541.3}{0.775} = \mathbf{698.45 \text{ kPa}}$$

Example 13.7 *The processes within each cylinder of a four-stroke compression ignition engine having three cylinders are modelled as an air-standard Diesel cycle. The displacement volume is 1248 cc and the state of air at the beginning of compression is p_1 = 0.1 MPa, T_1 = 300 K and V_1 = 1326 cc. The cut-off takes place at 5% of the stroke. If the engine runs at 1800 rpm, determine the power output from the engine and the cycle efficiency.* M **[LO 13.5]**

Solution From Fig. 13.44

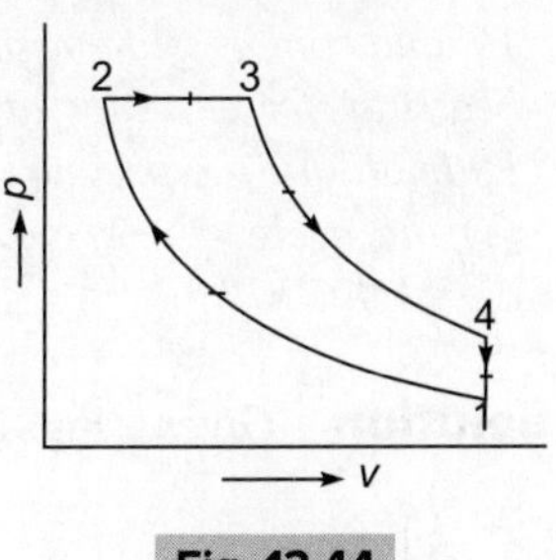

Fig. 13.44

$$V_1 = 1326 \text{ cc}, (V_1 - V_2) = 1248 \text{ cc}$$

$$\frac{V_1 - V_2}{V_1} = \frac{1248}{1326}$$

or $$r_k = \frac{V_1}{V_2} = 17$$

$$V_3 - V_2 = 0.05(V_1 - V_2)$$

or $$\frac{V_3}{V_2} - 1 = 0.05\left(\frac{V_1}{V_2} - 1\right)$$

or $$\frac{V_3}{V_2} = r_c = 1 + 0.05(17 - 1) = 1.8$$

Efficiency of the Diesel cycle is given by

$$\eta = 1 - \frac{1}{r_k^{\gamma-1}}\frac{1}{\gamma}\frac{r_c^{\gamma} - 1}{r_c - 1}$$

$$= 1 - \frac{1}{(17)^{1.4-1}}\frac{1}{1.4}\frac{(1.8)^{1.4} - 1}{1.8 - 1} = 0.6329 \text{ or } \mathbf{63.29\%}$$

From Fig. 13.44

$$\frac{T_2}{T_1} = \left(\frac{V_1}{V_2}\right)^{\gamma-1}$$

or $$T_2 = 300(17)^{1.4-1} = 931.75 \text{ K}$$

$$\frac{V_2}{T_2} = \frac{V_3}{T_3}$$

or $$V_3 = T_3 \frac{V_3}{V_2} = 931.75 \times 1.8 = 1677.15 \text{ K}$$

$$V_3 = T_3 \frac{V_3}{V_2} = 931.75 \times 1.8 = 1677.15 \text{ K}$$

$$Q_1 = c_p(T_3 - T_2) = 1.005(1677.15 - 931.75) = 749.127 \text{ kJ/kg}$$

Work done per kg of air is found to be

$$W_{\text{net}} = Q_1 \times \eta = 749.127 \times 0.6329 = 474.12 \text{ kJ/kg}$$

Power output = Work output × $\dot{m}_a$

$$\dot{m}_a = \frac{p_1 V_1}{RT_1} \times \frac{N}{2} = \frac{0.1 \times 10^3 \times 1326 \times 10^{-6}}{0.287 \times 300} \times \frac{1800}{2 \times 60} = 0.023 \text{ kg/s}$$

Power output = $474.12 \times 0.023 =$ **10.902 kW**

Example 13.8 *An air standard dual cycle has a compression ratio of 16, and compression begins at 1 bar, 50°C. The maximum pressure is 70 bar. The heat transferred to air at constant pressure is equal to that at constant volume. Estimate (a) the pressures and temperatures at the cardinal points of the cycle, (b) the cycle efficiency, and (c) the m.e.p. of the cycle, c_v = 0.718 kJ/kg K, c_p = 1.005 kJ/kg K.* M **[LO 13.6]**

Solution *Given:* (Fig. 13.45)

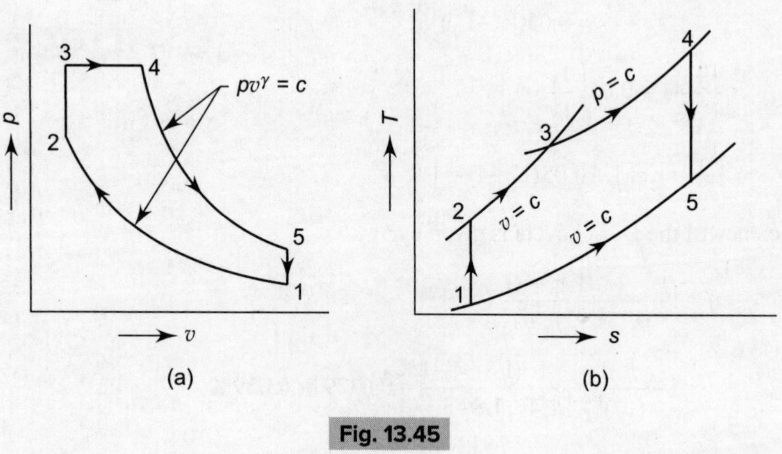

Fig. 13.45

$$T_1 = 273 + 50 = 323 \text{ K}$$

$$\frac{T_2}{T_1} = \left(\frac{v_1}{v_2}\right)^{\gamma-1} = (16)^{0.4}$$

$\therefore$
$$T_2 = 979 \text{ K}$$

$$p_2 = p_1 \left(\frac{v_1}{v_2}\right)^{\gamma} = 1.0 \times (16)^{1.4} = 48.5 \text{ bar}$$

$$T_3 = T_2 \cdot \frac{p_3}{p_2} = 979 \times \frac{70}{48.5} = 1413 \text{ K}$$

$$Q_{2-3} = c_v (T_3 - T_2) = 0.718\,(1413 - 979) = 312 \text{ kJ/kg}$$

Now

$$Q_{2-3} = Q_{3-4} = c_p (T_4 - T_3)$$

$\therefore$
$$T_4 = \frac{312}{1.005} + 1413 = 1723 \text{ K}$$

$$\frac{v_4}{v_3} = \frac{T_4}{T_3} = \frac{1723}{1413} = 1.22$$

$\therefore$
$$\frac{v_5}{v_4} = \frac{v_1}{v_2} \times \frac{v_3}{v_4} = \frac{16}{1.22} = 13.1$$

$\therefore$
$$T_5 = T_4 \left(\frac{v_4}{v_5}\right)^{\gamma-1} = 1723 \times \frac{1}{(13.1)^{0.4}} = 615 \text{ K}$$

$$p_5 = p_1 \left(\frac{T_5}{T_1}\right) = 1.0 \times \frac{615}{323} = 1.9 \text{ bar}$$

$$\eta_{\text{cycle}} = 1 - \frac{Q_2}{Q_1} = 1 - \frac{c_v(T_5 - T_1)}{c_v(T_3 - T_2) + c_p(T_4 - T_3)}$$

$$= 1 - \frac{0.718(615 - 323)}{312 + 312}$$

$$= 1 - \frac{0.718 \times 292}{624} = \mathbf{0.665 \text{ or } 66.5\%}$$

$$v_1 = \frac{RT_1}{p_1} = \frac{0.287 \text{ kJ/kg K} \times 323 \text{ K}}{10^2 \text{ kN/m}^2} = 0.927 \text{ m}^3\text{/kg}$$

$$v_1 - v_2 = v_1 - \frac{v_1}{16} = \frac{15}{16} v_1$$

$$W_{\text{net}} = Q_1 \times \eta_{\text{cycle}} = 0.665 \times 624 \text{ kJ/kg}$$

$\therefore$
$$\text{m.e.p.} = \frac{W_{\text{net}}}{v_1 - v_2} = \frac{0.665 \times 624 \text{ kJ/kg}}{\frac{15}{16} \times 0.927 \text{ m}^2\text{/kg}}$$

$$= 476 \text{ kN/m}^2 = \mathbf{4.76 \text{ bar}}$$

Example 13.9 *In a gas turbine plant, working on the Brayton cycle with a regenerator of 75% effectiveness, the air at the inlet to the compressor is at 0.1 MPa, 30°C, the pressure ratio is 6, and the maximum cycle temperature is 900°C. If the turbine and compressor have each an efficiency of 80%, find the percentage increase in the cycle efficiency due to regeneration.* M **[LO 13.9]**

Solution *Given:* (Fig. 13.46)

$$p_1 = 0.1 \text{ MPa}$$
$$T_1 = 303 \text{ K}$$
$$T_3 = 1173 \text{ K}$$
$$r_p = 6, \eta_T = \eta_C = 0.8$$

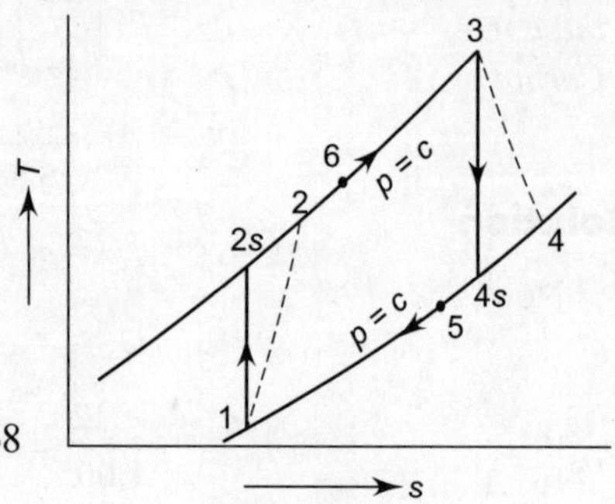

Fig. 13.46

Without a regenerator

$$\frac{T_{2s}}{T_1} = \left(\frac{p_2}{p_1}\right)^{(\gamma-1)/\gamma} = \frac{T_3}{T_{4s}} = (6)^{0.4/1.4} = 1.668$$

$$T_{2s} = 303 \times 1.668 = 505 \text{ K}$$

$$T_{4s} = \frac{1173}{1.668} = 705 \text{ K}$$

$$T_2 - T_1 = \frac{T_{2s} - T_1}{\eta_C} = \frac{505 - 303}{0.8} = 252 \text{ K}$$

$$T_3 - T_4 = \eta_T (T_3 - T_{4s}) = 0.8\,(1173 - 705) = 375 \text{ K}$$

$$W_T = h_3 - h_4 = c_p (T_3 - T_4) = 1.005 \times 375 = 376.88 \text{ kJ/kg}$$

$$W_C = h_2 - h_1 = c_p (T_2 - T_1) = 1.005 \times 252 = 253.26 \text{ kJ/kg}$$

$$T_2 = 252 + 303 = 555 \text{ K}$$

$$Q_1 = h_3 - h_2 = c_p (T_3 - T_2) = 1.005\,(1173 - 555) = 621.09 \text{ kJ/kg}$$

$$\therefore \quad \eta = \frac{W_T - W_C}{Q_1} = \frac{376.88 - 253.26}{621.09} = 0.199 \text{ or } 19.9\%$$

With regenerator

$$T_4 = T_3 - 375 = 1173 - 375 = 798 \text{ K}$$

$$\text{Regenerator effectiveness} = \frac{T_6 - T_2}{T_4 - T_2} = 0.75$$

$$\therefore \quad T_6 - 555 = 0.75\,(798 - 555)$$

or $\quad T_6 = 737.3 \text{ K}$

$$\therefore \quad Q_1 = h_3 - h_6 = c_p (T_3 - T_6)$$
$$= 1.005\,(1173 - 737.3) = 437.88 \text{ kJ/kg}$$

W_{net} remains the same.

$$\therefore \quad \eta = \frac{W_{net}}{Q_1} = \frac{123.62}{437.9} = \mathbf{0.2837 \text{ or } 28.37\%}$$

∴ Percentage increase due to regeneration

$$= \frac{0.2837 - 0.199}{0.199} = \mathbf{0.4256, \text{ or } 42.56\%}$$

Example 13.10 *A gas turbine plant operates on the Brayton cycle between T_{min} = 300 K and T_{max} = 1073 K. Find the maximum work done per kg of air, and the corresponding cycle efficiency. How does this efficiency compare with the Carnot cycle efficiency operating between the same two temperatures?*

E [LO 13.9]

Solution

$$(W_{net})_{max} = c_p\,(\sqrt{T_{max}} - \sqrt{T_{min}})^2$$

$$= 1.005\,(\sqrt{1073} - \sqrt{300})^2$$

$$= 1.005\,(15.43)^2 = \mathbf{239.28\ kJ/kg}$$

$$\eta_{cycle} = 1 - \frac{1}{(r_p)^{(\gamma-1)/\gamma}} = 1 - \sqrt{\frac{T_{min}}{T_{max}}}$$

$$= 1 - \sqrt{\frac{300}{1073}} = \mathbf{0.47 \text{ or } 47\%}$$

$$\eta_{Carnot} = 1 - \frac{T_{min}}{T_{max}} = 1 - \frac{300}{1073} = 0.721 \text{ or } 72.1\%$$

$$\frac{\eta_{Brayton}}{\eta_{Carnot}} = \frac{0.47}{0.721} = \mathbf{0.652}$$

Example 13.11 *In an ideal Brayton cycle, air from the atmosphere at 1 atm, 300 K is compressed to 6 atm and the maximum cycle temperature is limited to 1100 K by using a large air-fuel ratio. If the heat supply is 100 MW, find (a) the thermal efficiency of the cycle, (b) work ratio, (c) power output, and (d) exergy flow rate of the exhaust gas leaving the turbine.*

M [LO 13.9]

Solution The cycle efficiency,

$$\eta_{cycle} = 1 - \frac{1}{r_p^{(\gamma-1)/\gamma}} = 1 - \frac{1}{6^{0.4/1.4}}$$

$$= \mathbf{0.401 \text{ or } 40.1\%}$$

$$T_2/T_1 = (r_p)^{(\gamma-1)\gamma} = 1.67$$

$$T_2 = 501\text{ K}$$

$$T_3/T_4 = 1.67,\ T_4 = 1100/1.67 = 658.7\text{ K}$$

$$W_C = 1.005\,(501 - 300) = 202\text{ kJ/kg}$$

$$W_T = 1.005\,(1100 - 658.7) = 443.5\text{ kJ/kg}$$

$$\text{Work ratio} = \frac{W_T - W_C}{W_T} = \frac{241.5}{443.5} = \mathbf{0.545}$$

$$\text{Power output} = 100 \times 0.401 = 40.1 \text{ MW}$$

$$Q_1 = \dot{m}c_p (T_3 - T_2) = 100{,}000 \text{ kW}$$

$$\dot{m} = \mathbf{166.1 \text{ kg/s}}$$

Exergy flow rate of the exhaust gas stream

$$= \dot{m}c_p T_0 \left(\frac{T_4}{T_0} - 1 - \ln \frac{T_4}{T_0} \right)$$

$$= 166.1 \times 1.005 \times 300 \left(\frac{658}{300} - 1 - \ln \frac{658.7}{300} \right)$$

$$= \mathbf{20.53 \text{ MW}}$$

Example 13.12 *The following refer to a stationary gas turbine:*

Compressor inlet temperature = 311 K
Compressor pressure ratio = 8
Combustion chamber pressure drop = 5% of inlet pressure
Turbine inlet temperature = 1367 K
Turbine exit and compressor inlet pressures are atmospheric.

There exists a facility to take air from the compressor exit for use in cooling the turbine. Find the percentage of air that may be taken from the compressor for this purpose so that the overall cycle efficiency drops by 5% from that of the case of no usage of compressed air for cooling of turbine. For simplicity, assume the following: (a) Take properties of gas through the turbine as those of air, (b) Addition of cooling air to the turbine and addition of fuel to the combustion chamber do not affect the turbine power and (c) Compressor and turbine efficiencies are 0.87 and 0.90, respectively. D [LO 13.9]

Solution *Given:* $\eta_C = 0.87$, $\eta_T = 0.9$, $T_1 = 311$ K, $p_2/p_1 = 8$,
$p_3 = 0.95p_2$, $T_3 = 1367$ K, $p_4 = p_1 = 1$ atm, $\gamma = 1.4$

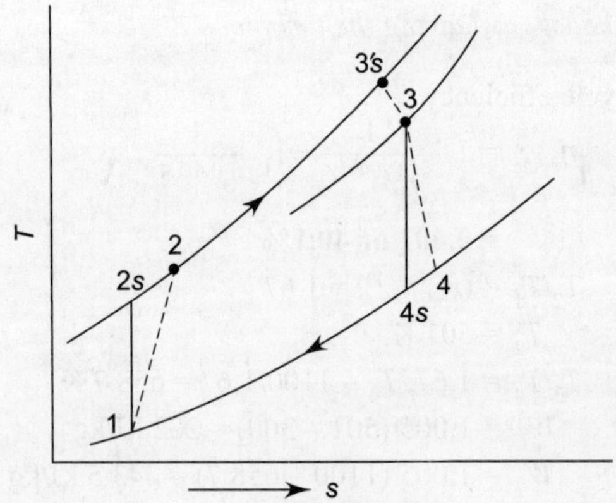

Fig. 13.47

Case 1: No cooling

$$W_C = \frac{\dot{m}_a c_{p_a}(T_{2s} - T_1)}{\eta_c}$$

$$W_T = \dot{m}_g\, c_{pg}\,(T_3 - T_{4s})\,\eta_T$$

$$Q_1 = \dot{m} c_{p_c}(T_3 - T_2)$$

Here,

$$\dot{m}_a = \dot{m}_g = \dot{m}_c = \dot{m}$$

$$c_{pa} = c_{pg} = c_{pc} = c_p$$

$$(T_{2s}/T_1 = p_{2s}/p_1)^{(\gamma-1)/\gamma} = 8^{0.4/1.4} = 1.181$$

$$T_{2s} = 563.3 \text{ K}$$

$$\frac{T_{2s} - T_1}{T_2 - T_1} = 0.87,\ T_2 = 601 \text{ K}$$

$$T_3/T_{4s} = (p_3/p_{4s})^{(\gamma-1)/\gamma} = \left(\frac{0.95p_2}{p_1}\right)^{0.4/1.4} = 1.785$$

$$T_{4s} = 765.83 \text{ K}$$

$$W_C = 290\, \dot{m}c_p,\ W_T = 541.06\, \dot{m}c_p \text{ and } Q_1 = 766\, \dot{m}c_p$$

$$\eta_{cycle} = \frac{541.06 - 290}{766} = 0.328$$

Case 2: With cooling

$$\eta_{cycle} = 0.328 - 0.05 = 0.278$$

Since the extraction of compressed air for turbine cooling does not contribute to turbine work or burner fuel flow, it can be treated as an increment x added to the compressor mass flow.

$$\frac{541.06 - 290(1 + x)}{766} = 0.278$$

$$\therefore \quad x = 0.13$$

$$\% \text{ of compressor delivery air flow} = \frac{0.13}{1.13} \times 100 = \mathbf{11.6\%}$$

Example 13.13 *In a gas turbine plant the ratio of T_{max}/T_{min} is fixed. Two arrangements of components are to be investigated: (a) single-stage compression followed by expansion in two turbines of equal pressure ratios with reheat to the maximum cycle temperature, and (b) compression in two compressors of equal pressure ratios, with intercooling to the minimum cycle temperature, followed by single-stage expansion. If η_C and η_T are the compressor and turbine efficiencies, respectively, show that the optimum specific output is obtained at the same overall pressure ratio for each arrangement.*

If η_C is 0.85 and η_T is 0.9, and T_{max}/T_{min} is 3.5, determine the above pressure ratio for optimum specific output and show that with arrangement (a) the optimum output exceeds that of arrangement (b) by about 11%. D [LO 13.9]

Solution

(a) With reference to Fig. 13.48(a),

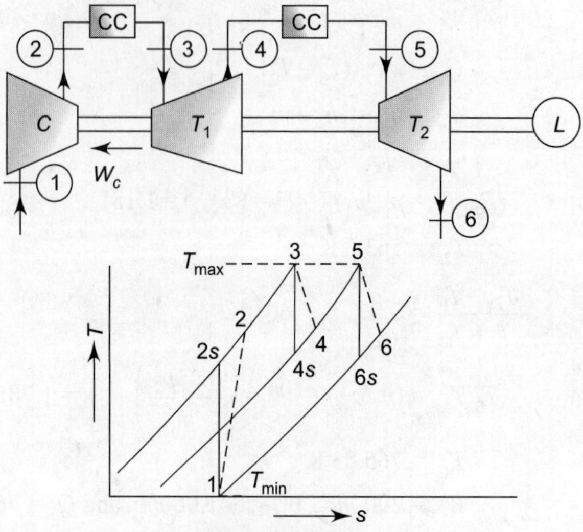

Fig. 13.48(a)

$$T_1 = T_{\min},\ T_3 = T_5 = T_{\max},\ \frac{p_2}{p_4} = \frac{p_4}{p_1}$$

$\therefore$ $$p_4 = \sqrt{p_1 p_2}$$

$$\frac{p_{2s}}{p_1} = r, \text{ pressure ratio}$$

$\therefore$ $$p_{2s} = p_2 = r p_1$$

$\therefore$ $$p_4 = \sqrt{r} \cdot p_1$$

$$\frac{T_{2s}}{T_1} = \left(\frac{p_2}{p_1}\right)^{(\gamma-1)/\gamma} = r^x$$

where $$x = \frac{\gamma - 1}{\gamma}$$

$\therefore$ $$T_{2s} = T_{\min}\, r^x$$

$$(\Delta T_s)_{\text{comp}} = T_{2s} - T_1 = T_{\min}\, r^x - T_{\min} = T_{\min}\,(r^x - 1)$$

$\therefore$ $$(\Delta T)_{\text{comp}} = \frac{T_{\min}\,(r^x - 1)}{\eta_C}$$

$$\frac{T_3}{T_{4s}} = \left(\frac{p_3}{p_4}\right)^{(\gamma-1)/\gamma} = \left(\frac{r p_1}{\sqrt{r} \cdot p_1}\right)^x r^{x/2}$$

$$\therefore \quad T_{4s} = T_3\, r^{-x/2} = T_{\max}\, r^{-x/2}$$

$$(\Delta T_s)_{\text{turb}} = T_3 - T_{4s} = T_{\max} - T_{\max} \cdot r^{-x/2}$$

$$= T_{\max}\,(1 - r^{-x/2})$$

$$\therefore \quad (\Delta T)_{\text{turb 1}} = \eta_T\, T_{\max}\,(1 - r^{-x/2}) = (\Delta T)_{\text{turb 2}}$$

$$\therefore \quad W_{\text{net}} = c_p\left[2\eta_T T_{\max}\left(1 - r^{-x/2}\right) - \frac{T_{\min}}{\eta_C}\left(r^x - 1\right)\right]$$

$$\frac{dW_{\text{net}}}{dr} = c_p\left[2\eta_T T_{\max}\frac{x}{2}\cdot r^{-x/2-1} - \frac{T_{\min}}{\eta_C}\, x \cdot r^{x-1}\right] = 0$$

On simplification

$$r^{3x/2} = \eta_T\, \eta_C\, \frac{T_{\max}}{T_{\min}}$$

$$\therefore \quad r_{\text{opt}} = \left(\eta_T \eta_C\, \frac{T_{\max}}{T_{\min}}\right)^{2\gamma/3(\gamma-1)}$$

(b) With reference to Fig. 13.48(b)

$$\frac{T_{2s}}{T_{\min}} = \left(\frac{p_2}{p_1}\right)^{(\gamma-1)/\gamma} = \left(\sqrt{r}\right)^x = r^{x/2}$$

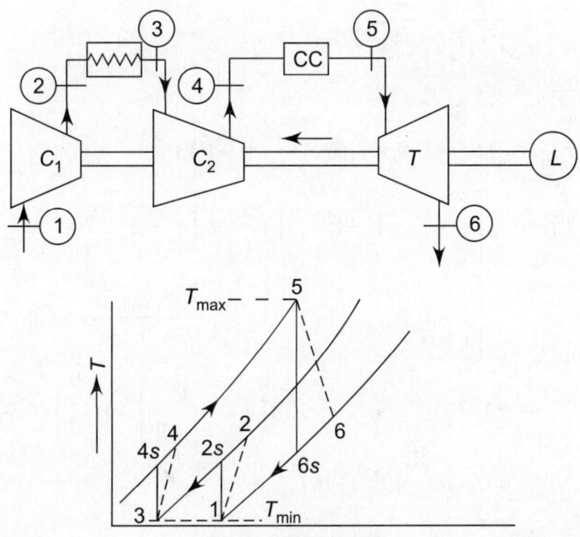

Fig. 13.48(b)

$$(\Delta T_s)_{\text{comp 1}} = T_{2s} - T_1 = T_{\min}\,(r^{x/2} - 1)$$

$$(\Delta T)_{\text{comp1}} = \frac{T_{\min(r^{x/2}-1)}}{\eta_C} = (\Delta T)_{\text{comp2}}$$

$$\frac{T_{\max}}{T_{6s}} = \left(\frac{p_5}{p_6}\right)^{(\gamma-1)/\gamma} = r^x$$

$$\therefore \quad T_{6s} = T_{\max} \cdot \gamma^{-x}$$

$$(\Delta T_s)_{\text{turb}} = T_{\max} - T_{6s} = T_{\max}\,(1 - r^{-x})$$

$$(\Delta T)_{\text{turb}} = \eta_T\, T_{\max}\,(1 - r^{-x})$$

$$W_{\text{net}} = c_p \left[\eta_T T_{\max}\left(1 - r^{-x}\right) - \frac{2T_{\min}\left(r^{x/2} - 1\right)}{\eta_C}\right]$$

$$\frac{dW_{\text{net}}}{dr} = c_p \left[\eta_T T_{\max}\; x \cdot r^{-(x+1)} - \frac{2T_{\min}}{\eta_C}\frac{x}{2} \cdot r^{(x/2)-1}\right] = 0$$

On simplification

$$r_{\text{opt}} = \left(\eta_T \eta_C \frac{T_{\max}}{T_{\min}}\right)^{2\gamma/3(\gamma-1)}$$

This is same as in (a).

If $\quad \eta_C = 0.85,\ \eta_T = 0.9$

$$\frac{1}{x} = \frac{\gamma}{\gamma - 1} = \frac{1.4}{0.4} = \frac{7}{2}$$

$$\frac{T_{\max}}{T_{\min}} = 3.5$$

$$\therefore \quad r_{\text{opt}} = (0.85 \times 0.9 \times 3.5)^{2/3 \times 7/2} = \mathbf{9.933}$$

$$W_{\text{net}}\,(a) = c_p \left[2\eta_T T_{\max}\left(1 - r^{-x/2}\right) - \frac{T_{\min}}{\eta_C}\left(r^x - 1\right)\right]$$

$$= c_p \left[2 \times 0.9 T_{\max}\left(1 - 9.933^{-0.143}\right) - \frac{T_{\min}}{0.85}\left(9.933^{0.286} - 1\right)\right]$$

$$= c_p \cdot T_{\min}\left[2 \times 0.9 \times 3.5\left(1 - \frac{1}{1.388}\right) - 1.178\,(0.928)\right]$$

$$= 0.670\; c_p \cdot T_{\min}$$

$$W_{\text{net}}\,(b) = c_p\, T_{\min}\left[0.9 \times 3.5\left(1 - 9.933^{-0.286}\right) - \frac{2}{0.85}\left(9.933^{0.143} - 1\right)\right]$$

$$= c_p\, T_{\min}\,(1.518 - 0.914)$$

$$= 0.604\; c_p\, T_{\min}$$

$$\frac{W_{net}(a) - W_{net}(b)}{W_{net}(a)} \times 100$$

$$= \frac{0.670 - 0.604}{0.670} \times 100 = 10.9\%$$

Example 13.14 *A turbojet aircraft flies with a velocity of 300 m/s at an altitude where the air is at 0.35 bar and – 40°C. The compressor has a pressure ratio of 10, and the temperature of the gases at the turbine inlet is 1100°C. Air enters the compressor at a rate of 50 kg/s. Estimate (a) the temperature and pressure of the gases at the turbine exit, (b) the velocity of gases at the nozzle exit and (c) the propulsive efficiency of the cycle.* M **[LO 13.11]**

Solution (a) For isentropic flow of air in the diffusor (Fig. 13.49)

$$\overset{=0}{\cancel{Q}_{1-2}} - \overset{=0}{\cancel{W}_{1-2}} = h_2 - h_1 + \overset{=0}{\frac{\cancel{V_2^2} - V_1^2}{2}}$$

$$0 = c_p (T_2 - T_1) - \frac{V_1^2}{2}$$

$$T_2 = T_1 + \frac{V_1^2}{2c_p} = 233 + \frac{300^2}{2 \times 1.005} \times 10^{-3}$$

$$= 277.78 \text{ K}$$

$$p_2 = p_1 (T_2/T_1)^{\gamma/(\gamma-1)}$$

$$= 35 \frac{\text{kN}}{\text{m}^2} \left(\frac{277.78}{233}\right)^{1.4/0.4} = 64.76 \text{ kPa}$$

$$p_3 = r_p p_2 = 10 \times 64.76 = 647.6 \text{ kPa}$$

$$T_3 = \left(\frac{p_3}{p_2}\right)^{(\gamma-1)/\gamma} T_2 = 277.78\ (10)^{0.4/1.4}$$

$$= 536.66 \text{ K}$$

$$W_C = W_T$$

$$h_3 - h_2 = h_4 - h_5$$

or $$T_3 - T_2 = T_4 - T_5$$

$$T_5 = T_4 - T_3 + T_2 = 1373 - 536.66 + 277.78$$

$$= \mathbf{1114.12 \text{ K}}$$

$$p_5 = p_4 \left(\frac{T_5}{T_4}\right)^{\gamma/(\gamma-1)} = 647.6 \left(\frac{1114.12}{1373}\right)^{3.5}$$

$$= \mathbf{311.69 \text{ K}}$$

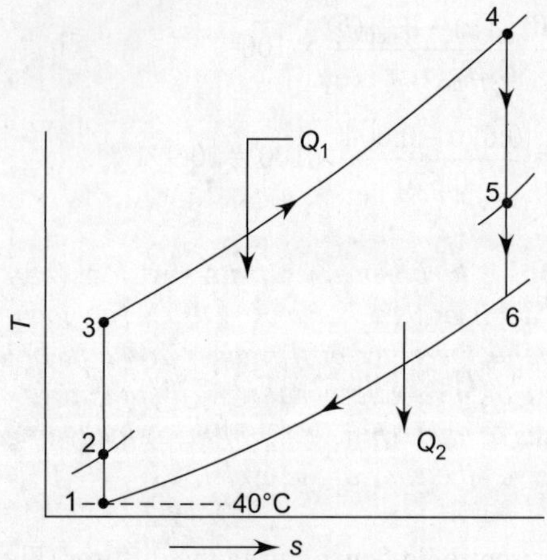

Fig. 13.49

(b) For isentropic expansion of gases in the nozzle,

$$T_6 = T_5 \left(\frac{p_6}{p_5}\right)^{(\gamma-1)/\gamma} = 1114.12 \left(\frac{35}{311.69}\right)^{0.286}$$

$$= 596.12 \text{ K}$$

Neglecting the KE of gas at nozzle inlet,

$$V_6 = [2c_p (T_5 - T_6) \times 1000]^{1/2}$$

$$= [2 \times 1.005 (1114.12 - 596.12) \times 1000]^{1/2}$$

$$= \mathbf{1020.4 \ m/s}$$

(c) The propulsive efficiency of a turbojet engine is the ratio of the propulsive power developed $\dot{W}_P$ to the total heat transfer to the fluid

$$\dot{W}_P = w[V_{exit} - V_{inlet}] V_{aircraft}$$

$$= 50 [1020.4 - 300] \times 300 \frac{\text{kg}}{\text{s}} \times \frac{\text{m}^2}{\text{s}^2}$$

$$= 10.806 \text{ MW}$$

$$Q_1 = w (h_4 - h_3) = 50 \times 1.005 (1373 - 536.66)$$

$$= 42.026 \text{ MW}$$

$$\eta_P = \frac{10.806}{42.026} = \mathbf{0.257 \text{ or } 25.7\%}$$

Example 13.15 *In a combined GT—ST plant, the exhaust gas from the GT is the supply of gas to the steam generator at which a further supply of fuel is burned in the gas. The pressure ratio for the GT is 8, the inlet air temperature is 15°C and the maximum cycle temperature is 800°C.*

Combustion in the steam generator raises the gas temperature to 800°C and the gas leaves the generator at 100°C. The steam condition at supply is at 60 bar, 600°C and the condenser pressure is 0.05 bar. Calculate the flow rates of air and steam required for a total power output of 190 MW and the overall efficiency of the combined plant. What would be the air-fuel ratio? Assume ideal processes. Take $c_{pg} = 1.11$ and $c_{pa} = 1.005$ kJ/kg K, $\gamma_g = 1.33$, $\gamma_a = 1.4$, CV of fuel = 43.3 MJ/kg. Neglect the effect of fuel flow on the total mass flow of gas expanding in the gas turbine. Ⓜ **[LO 13.12]**

Solution With reference to Fig. 13.37.

$$T_b = T_a (p_b/p_a)^{(\gamma-1)/\gamma} = 288 \times 8^{0.4/1.4} = 522 \text{ K}$$

$$T_d = \frac{T_C}{r_p(\gamma-1)/\gamma} = \frac{1073}{8^{0.33 1.33}} = \frac{1073}{1.682} = 638 \text{ K}$$

$$W_{GT} = c_{pg}(T_c - T_d) - c_{pa}(T_b - T_a)$$
$$= 1.11\,(1073 - 638) - 1.005\,(522 - 288)$$
$$= 249 \text{ kJ/kg}$$

$$Q_1 = c_{pg}(T_c - T_b) = 1.11\,(1073 - 522)$$
$$= 612 \text{ kJ/kg}$$

$$Q'_1 = c_{pg}(T_e - T_d) = 1.11(1073 - 638)$$
$$= 483 \text{ kJ/kg}$$

$$h_1 = 3775,\ h_2 = 2183,\ h_3 = 138 = h_4, \text{ all in kJ/kg}$$

$$(Q_1)_{St} = h_1 - h_3 = 3775 - 138 = 3637 \text{ kJ/kg}$$

$$Q''_{fe} = 1.11\,(800 - 100) = 777 \text{ kJ/kg}$$

By energy balance of the steam generator,

$$w_a \times 777 = w_s \times 3637$$

$$w_a/w_s = 4.68$$

$$W_{ST} = h_1 - h_2 = 3775 - 2183 = 1592 \text{ kJ/kg}$$

$$w_a \times 249 + w_s \times 1592 = 190 \times 10^3 \text{ kW}$$

$$w_s\,(4.68 \times 249 + 1592) = w_s \times 2.757 \times 10^3 = 190 \times 10^3$$

$$w_s = \mathbf{68.9\ kg/s} \text{ and } w_a = \mathbf{322.5\ kg/s}$$

Now, $$w_a(612 + 483) = w_f \times 43{,}300$$

$$w_a/w_f = \text{A/F ratio} = \frac{43{,}300}{1095} = \mathbf{39.5}$$

$$\text{Fuel energy input} = \frac{322.5}{39.5} \times 43{,}300 = 353525 \text{ kW}$$
$$= 353.5 \text{ MW}$$

$$\eta_{0A} = \frac{190}{353.5} = \mathbf{0.537 \text{ or } 53.7\%}$$

SUMMARY

Stirling cycle consisting of two reversible isochores (constant volume lines) and two reversible isotherms, and Ericsson cycle consisting of two reversible isobars and two reversible isotherms, were used in engines with air as the working fluid. However, they lost importance after the advent of I.C. engines. The regenerative Stirling and Ericsson cycles have efficiencies equal to the Carnot cycle operating between the same two temperatures.

Air standard cycles were conceived to simplify the analysis of I.C Engines where the working fluid is assumed to behave as an ideal gas with constant specific heats. Otto cycle, consisting of two reversible isochores and two reversible adiabatics represent the speak ignition engine, the efficiency of which depends only on the compression ratio. The Diesel cycle, consisting of two reversible adiabatics, one reversible isobar in which heat is supplied and one reversible isochore in which heat is rejected is the air standard cycle of the Diesel engine. In a limited pressure or dual cycle, heat is supplied partially it constant volume and partially at constant pressure. For the same compression ratio and heat rejection

$$\eta_{\text{Otto}} > \eta_{\text{Dual}} > \eta_{\text{Diesel}}.$$

But for the same maximum pressure and temperature and heat rejection,

$$\eta_{\text{Diesel}} > \eta_{\text{Dual}} > \eta_{\text{Otto}}$$

Brayton cycle consisting of two reversible isobars and two reversible adiabatics is the air standard cycle for a gas turbine power plant. Its efficiency depends only on the pressure ratio and for the same compression ratio is equal to that of Otto cycle. Although Brayton and Rankine cycles consist of similar processes, Brayton is the single phase (air) cycle whereas Rankine is a two phase (water) cycle. The effect of regeneration where the turbine exhaust is utilized in heating the compressed air before it goes to the combustion chamber improves the cycle efficiency. The optimum pressure ratio r_p at which net work done in the cycle is given by

$$(r_p)_{\text{opt}} = \left(\frac{T_{\max}}{T_{\min}}\right)^{\frac{\gamma}{2(\gamma-1)}}$$

and the maximum possible value of r_p which is equal to Carnot efficiency is given by

$$(r_p)_{\max} = \left(\frac{T_{\max}}{T_{\min}}\right)^{\frac{\gamma}{\gamma-1}} = (r_{p,\text{opt}})^2$$

The corresponding values of W_{net} and η_{cycle} are

$$(W_{\text{net}})_{\max} = c_p\left(\sqrt{T_{\max}} - \sqrt{T_{\min}}\right)^2$$

and
$$\eta_{\text{cycle}} = 1 - \sqrt{\frac{T_{\min}}{T_{\max}}}$$

By dividing the compression process with intercooling and the heat supply with reheating, an attempt may be made to improve the cycle efficiency. However, it is hardly used in aircraft because of a large increase in its weight. A simple cycle is used, even without any regeneration.

Aircraft gas turbines operate on an open jet propulsion cycle. In a turbojet engine, a ram or diffuser, compressor, combustor, turbine and nozzle make up the engine, whereas in a turboprop engine, a propeller is used to augment the thrust. A ramjet engine however, requires neither a compressor nor a turbine. A pulse jet engine can operate independent of the movement of the aircraft.

However, in actual gas turbine cycles, there will be non-isentropic compression and expansion as well as pressure drops during heat addition and rejection. A pulse jet engine can operate independent of the movement of the aircraft.

REVIEW QUESTIONS

(M) **13.1** What are cyclic and noncyclic heat engines? Give examples. **[LO 13.1]**

(E) **13.2** What are the four processes which constitute the Stirling cycle? Show that the regenerative Stirling cycle has the same efficiency as the Carnot cycle. **[LO 13.1]**

(D) **13.3** State the four processes that constitute the Ericsson cycle. Show that the regenerative Ericsson cycle has the same efficiency as the Carnot cycle. **[LO 13.1]**

(M) **13.4** Mention the merits and demerits of the Stirling and Ericsson cycles. **[LO 13.1]**

(M) **13.5** What is an air standard cycle? Why are such cycles conceived? **[LO 13.3]**

(M) **13.6** What is a spark ignition engine? What is the air standard cycle of such an engine? What are its four processes? **[LO 13.4]**

(D) **13.7** Show that the efficiency of the Otto cycle depends only on the compression ratio. **[LO 13.4]**

(E) **13.8** How is the compression ratio of an SI engine fixed? **[LO 13.4]**

(M) **13.9** What is a compression ignition engine? Why is the compression ratio of such an engine more than that of an SI engine? **[LO 13.5]**

(E) **13.10** State the four processes of the Diesel cycle. **[LO 13.5]**

(M) **13.11** Explain the mixed or dual cycle. **[LO 13.6]**

(D) **13.12** For the same compression ratio and heat rejection, which cycle is most efficient: Otto, Diesel or Dual? Explain with *p–v* and *T–s* diagrams. **[LO 13.7]**

(D) **13.13** With the help of *p–v* and *T–s* diagrams, show that for the same maximum pressure and temperature of the cycle and the same heat rejection, **[LO 13.7]**

$$\eta_{\text{Diesel}} > \eta_{\text{Dual}} > \eta_{\text{Otto}}$$

(M) **13.14** What are the three basic components of a gas turbine plant? What is the air standard cycle of such a plant? What are the processes it consists of? **[LO 13.9]**

(M) **13.15** Show that the efficiency of the Brayton cycle depends only on the pressure ratio. **[LO 13.9]**

(E) **13.16** What is the application of the closed cycle gas turbine plant? **[LO 13.9]**

(M) **13.17** How does Brayton cycle compare with Rankine cycle? **[LO 13.9]**

(D) **13.18** Discuss the merits and demerits of Brayton and Otto cycles applied to reciprocating and rotating plants. **[LO 13.9]**

(D) **13.19** What is the effect of regeneration on Brayton cycle efficiency? Define the effectiveness of a regenerator. **[LO 13.9]**

(M) **13.20** What is the effect of irreversibilities in turbine and compressor on Brayton cycle efficiency? **[LO 13.9]**

(M) **13.21** Explain the effect of pressure ratio on the net output and efficiency of a Brayton cycle. **[LO 13.9]**

(D) **13.22** Drive the expression of optimum pressure ratio for maximum net work output in an ideal Brayton cycle. What is the corresponding cycle efficiency? **[LO 13.9]**

(D) **13.23** Explain the effects of (a) intercooling and (b) reheating on Brayton cycle. **[LO 13.9]**

(M) **13.24** How actual gas turbine cycle deviates from the ideal Brayton cycle? **[LO 13.10]**

(E) **13.25** What is a free shaft turbine? **[LO 13.10]**

(M) **13.26** With the help of flow and T–s diagrams explain the air standard cycle for a jet propulsion plant. **[LO 13.11]**

(M) **13.27** Define propulsive power and propulsive efficiency. **[LO 13.11]**

(M) **13.28** Why are regenerators and intercoolers not used in aircraft engines? What is after burning? Why is it used? **[LO 13.11]**

(D) **13.29** Explain the working of a turbofan engine with the help of a neat sketch. Define 'bypass ratio'. How does it influence the engine thrust? **[LO 13.11]**

(E) **13.30** How does a turboprop engine differ from a turbofan engine? **[LO 13.11]**

(M) **13.31** What is a ramjet? How is the thrust produced here? **[LO 13.11]**

(D) **13.32** What is a pulse jet engine, and what are its advantages and disadvantages compared to a ramjet engine? **[LO 13.11]**

(M) **13.33** What is a rocket? How is it propelled? **[LO 13.11]**

(M) **13.34** Explain the advantages and disadvantages of a gas turbine plant for a utility system. **[LO 13.12]**

(E) **13.35** What are the advantages of a combined gas turbine-steam turbine power plant? **[LO 13.12]**

(D) **13.36** With the help of flow and T–s diagrams explain the operation of a combined GT–ST plant. Why is supplementary firing often used? **[LO 13.12]**

PROBLEMS WITH HINTS

(E) **13.1** An engine operating on the air standard Otto cycle. The conditions at the start of the compression are 27°C and 100 kPa. The heat added is 1840 kJ/kg. The compression ratio is 8. Determine the temperature and pressure at each point in the cycle, the thermal efficiency and the mean effective pressure. **[LO 13.4]**

Hint:

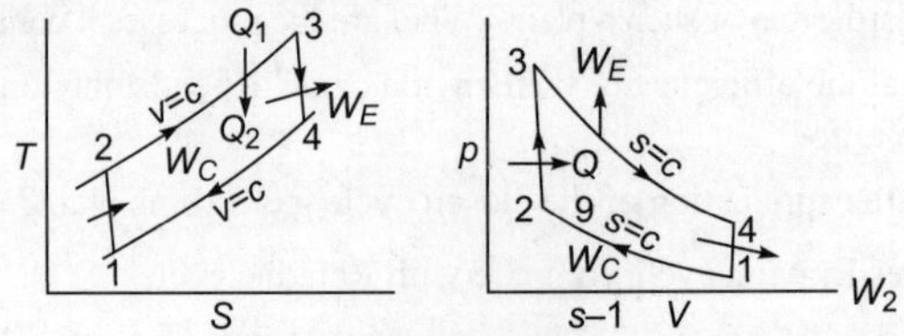

Fig. 13.50

$T_1 = 300$ K, $p_1 = 100$ kPa $\qquad \dfrac{T_2}{T_1} = \left(\dfrac{V_2}{V_1}\right) \Rightarrow T_2 = 300 \times 8^{0.4} = \mathbf{680.2\ K}$

$$\frac{p_2}{p_1} = \left(\frac{V_1}{V_2}\right)^8 \qquad \therefore p_2 = 100\,(8)^{1.4} = \mathbf{1837.9\ kPa.}$$

$$Q_1 = u_3 - u_2 = c_v(T_3 - T_2) = 1840 \text{ kJ/kg}$$

$$\therefore \qquad p_3 = p_2 \frac{T_3}{T_2} = \mathbf{8676.1\ kPa}$$

$$\frac{T_4}{T_3} = \left(\frac{V_3}{V_4}\right)^{8-1} \qquad T_4 = \mathbf{1416.2\ K}$$

$$p_4 = p_3\left(\frac{1}{r_k}\right)^{1.4} = \mathbf{472.1\ kPa}$$

$$\eta_{th} = 1 - \frac{1}{r_k^{8-1}} = \mathbf{0.565\ or\ 56.5\ \%}$$

$$W_{net} = Q_1 \times \eta_{th} = \mathbf{1039.6\ kJ/kg}$$

$$v_1 = \frac{RT_1}{P_1} = 0.861\ m^3/kg,\ v_2 = 0.1076\ m^3/kg$$

$$\therefore \quad p_m = \frac{W_{net}}{v_1 - v_2} = \mathbf{1379.9\ kPa}$$

(E) **13.2** An ideal Diesel cycle with air as the working fluid has a compression ratio of 18 and a cut-off ratio of 2. At the beginning of compression, the air is at 100 kPa, 27°C and 1917 cm³. Determine (a) the pressure and temperature of air at each point, (b) the net work and the thermal efficiency and (c) the mean effective pressure. **[LO 13.5]**

Hint:

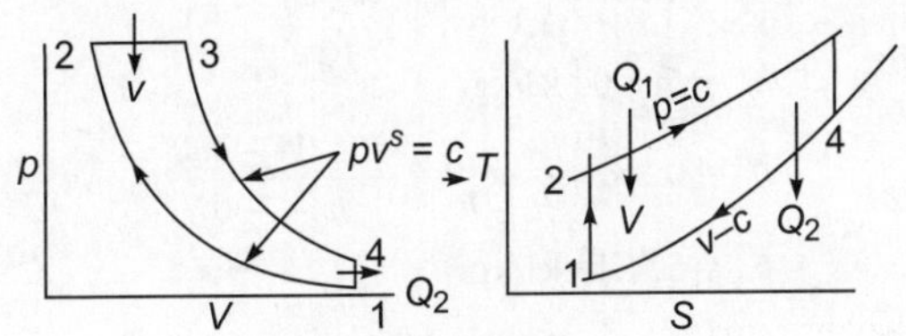

Fig. 13.51

$$V_2 = \frac{V_1}{r_k} = 106.5\ cm^3, \qquad V_3 = r_c V_2 = 213\ cm^3$$

$$V_4 = V_1 = 1917\ cm^3; \qquad T_2 = T_1\, r^{1.4-1}{}_K = \mathbf{953\ K}$$

$$p_2 = p_1\left(\frac{V_1}{V_2}\right)^8 = 5720\ kPa,\ p_3 = p_2 = \mathbf{5720\ kPa}$$

$$T_4 = T_3\left(\frac{V_3}{V_4}\right)^{8-1} = \mathbf{794\ K},\ \text{where } T_3 = T_2 \cdot \frac{V_3}{V_2} = \mathbf{1906\ K}$$

$$p_4 = p_3\left(\frac{V_3}{V_4}\right)^{1.4} = \mathbf{264\ kPa} \qquad m = \frac{P_1V_1}{RT_1} = 0.00223\ kg.$$

$$Q_1 = m\, c_p(T_3 - T_2) = 2.136\ kJ,\ Q_2 = m\, c_v\,(T_4 - T_1) = 0.786\ kJ$$

$$W_{net} = 1.35\ kJ.$$

$$\eta_{th} = W_{net}/Q_1 = 0.632\ or\ \mathbf{63.2\ \%}$$

$$m.e.p. = \frac{W_{net}}{V_1 - V_2} = \frac{1.35\ kJ}{(1917 - 106.5)\times 10^{-6}\ m^3} = \mathbf{746\ kPa}$$

(M) **13.3** A gas turbine unit receives air at 100 kPa and 300 K and compresses it adiabatically to 620 kPa with $\eta_c = 0.88$. The fuel has a heating value of 44,186 kJ/kg, and the fuel-air ratio is 0.017 kg fuel/kg air. The turbine internal efficiency is 90 %. Calculate the turbine work, compressor work and thermal efficiency. **[LO 13.9]**

Hint:

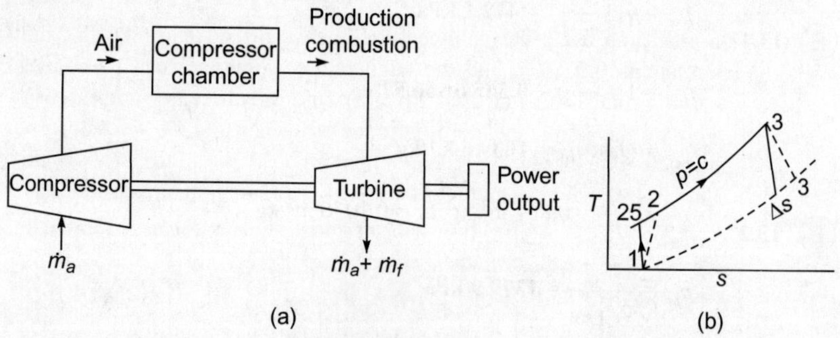

Fig. 13.52

From Table A.12 in Appendix A,

$$h_1 = 300.19 \text{ kJ/kg}, pr_1 = 1.386$$

Process 1-2

$$pr_2 = 1.3860\left(\frac{620}{100}\right) = 8.593$$

$$\therefore \quad h_2 = 506.04 \text{ kJ/kg}$$

$$\eta_c = 0.88 = \frac{h_{25} - h_1}{h_2 - h_1} = \frac{506.04 - 300}{h_2 - 300}$$

$$\therefore \quad h_2 = 534.11 \text{ kJ/kg}$$

Process 2-3 Energy balance gives

$$h_2 + \frac{F}{A} \times C.V. - \left(1 + \frac{F}{A}\right) h_3 \text{ or } CV = \text{calorific/heating value}$$

$$534.11 + 0.017 \times 44,186 = 1.017 \times h_3$$

$$\therefore \quad h_3 = 1263.8 \text{ kJ/kg}$$

$$\therefore \quad T_3 = 1166 \text{ K}, Pr_3 = 236.0$$

Process 3-4

$$Pr_4 = 236 \times \frac{100}{620} = 38.06$$

$$h_4 = 711 \text{ kJ/kg}$$

$$\eta_t = 0.90 = \frac{h_3 - h_4}{h_3 - h_{45}} \rightarrow h_4 = 820.6 \text{ kJ/kg}$$

By energy balance of each component yields

$$W_T = \left(1 + \frac{F}{A}\right)(h_3 - h_4) = 1.017\,(1263.8 - 820.6)$$

$$= \mathbf{450.7\ kJ/kg}$$

$$W_C = -(h_2 - h_1) = -(534.1 - 300.19) = \mathbf{-233.92\ kJ/kg}$$

$$W_{\text{net}} = W_T + W_C = 216.78 \text{ kJ/kg}$$

$$Q_1 = \frac{F}{A} \times C.V = 751.2 \text{ kJ/kg}$$

$$\eta_{\text{th}} = W_{\text{net}}/Q_1 = 0.288 \text{ or } \mathbf{28.8\ \%}$$

PROBLEMS

(E) **13.1** In a Stirling cycle the volume varies between 0.03 and 0.06 m^3, the maximum pressure is 0.2 MPa, and the temperature varies between 540 and 270°C. The working fluid is air (an ideal gas). (a) Find the efficiency and the work done per cycle for the simple cycle. (b) Find the efficiency and the work done per cycle for the cycle with an ideal regenerator, and compare with the Carnot cycle having the same isothermal heat supply process and the same temperature range. **[LO 13.1]**

(E) **13.2** An Ericsson cycle operating with an ideal regenerator works between 1100 and 288 K. The pressure at the beginning of isothermal compression is 1.013 bar. Determine (a) the compressor and turbine work per kg of air, and (b) the cycle efficiency. **[LO 13.1]**

(E) **13.3** Plot the efficiency of the air standard Otto cycle as a function of the compression ratio for compression ratios from 4 to 16. **[LO 13.4]**

(E) **13.4** Find the air standard efficiencies for Otto cycles with a compression ratio of 6 using ideal gases having specific heat ratios 1.3, 1.4 and 1.67. What are the advantages and disadvantages of using helium as the working fluid? **[LO 13.4]**

(E) **13.5** An engine equipped with a cylinder having a bore of 15 cm and a stroke of 45 cm operates on an Otto cycle. If the clearance volume is 2000 cm^3, compute the air standard efficiency. **[LO 13.4]**

(E) **13.6** In an air standard Otto cycle the compression ratio is 7, and compression begins at 35°C, 0.1 MPa. The maximum temperature of the cycle is 1100°C. Find (a) the temperature and pressure at the cardinal points of the cycle, (b) the heat supplied per kg of air, (c) the work done per kg of air, (d) the cycle efficiency and (e) the m.e.p. of the cycle. **[LO 13.4]**

(E) **13.7** An engine working on the Otto cycle has an air standard cycle efficiency of 56% and rejects 544 kJ/kg of air. The pressure and temperature of air at the beginning of compression are 0.1 MPa and 60°C, respectively. Compute (a) the compression ratio of the engine, (b) the work done per kg of air, (c) the pressure and temperature at the end of compression and (d) the maximum pressure in the cycle. **[LO 13.4]**

(E) **13.8** For an air standard Diesel cycle with a compression ratio of 15 plot the efficiency as a function of the cut-off ratio for cut-off ratios from 1 to 4. Compare with the results of Problem 13.3. **[LO 13.5]**

(E) **13.9** In an air standard Diesel cycle, the compression ratio is 15. Compression begins at 0.1 MPa, 40°C. The heat added is 1.675 MJ/kg. Find (a) the maximum temperature of the cycle, (b) the work done per kg of air, (c) the cycle efficiency, (d) the temperature at the end of the isentropic expansion, (e) the cut-off ratio, (f) the maximum pressure of the cycle and (g) the m.e.p. of the cycle. **[LO 13.5]**

(M) **13.10** Two engines are to operate on Otto and Diesel cycles with the following data: Maximum temperature 1400 K, exhaust temperature 700 K. State of air at the beginning of compression 0.1 MPa, 300 K.
Estimate the compression ratios, the maximum pressures, efficiencies and rate of work outputs (for 1 kg/min of air) of the respective cycles. **[LO 13.4, 13.5]**

(M) **13.11** An air standard limited pressure cycle has a compression ratio of 15 and compression begins at 0.1 MPa, 40°C. The maximum pressure is limited to 6 MPa and the heat added is 1.675 MJ/kg. Compute (a) the heat supplied at constant volume per kg of air, (b) the heat supplied at constant pressure per kg of air, (c) the work done per kg of air, (d) the cycle efficiency, (e) the temperature at the end of the constant volume heating process, (f) the cut-off ratio and (g) the m.e.p. of the cycle. **[LO 13.6]**

Ⓜ **13.12** In an ideal cycle for an internal combustion engine the pressure and temperature at the beginning of adiabatic compression are respectively 0.11 MPa and 115°C, the compression ratio being 16. At the end of compression heat is added to the workign fluid, first, at constant volume, and then at constant pressure reversibly. The working fluid is then expanded adiabatically and reversibly to the original volume.

If the working fluid is air and the maximum pressure and temperature are respectively 6 MPa and 2000°C, determine, per kg of air (a) the pressure, temperature, volume, and entropy of the air at the five cardinal points of the cycle (take s_1 as the entropy of air at the beginning of compression), and (b) the work output and efficiency of the cycle. **[LO 13.6]**

Ⓜ **13.13** Show that the air standard efficiency for a cycle comprising two constant pressure processes and two isothermal processes (all reversible) is given by

$$\eta = \frac{(T_1 - T_2)\ln(r_p)^{(\gamma-1)\gamma}}{T_1\left[1 + \ln(r_p)^{(\gamma-1)/\gamma} - T_2\right]}$$

where T_1 and T_2 are the maximum and minimum temperatures of the cycle, and r_p is the pressure ratio. **[LO 13.9]**

Ⓜ **13.14** Obtain an expression for the specific work done by an engine working on the Otto cycle in terms of the maximum and minimum temperatures of the cycle, the compression ratio r_k and constants of the working fluid (assumed to be an ideal gas).

Hence show that the compression ratio for maximum specific work output is given by

$$r_k = \left(\frac{T_{\min}}{T_{\max}}\right)^{1/2(1-\gamma)}$$ **[LO 13.4]**

Ⓜ **13.15** A dual combustion cycle operates with a volumetric compression ratio $r_k = 12$, and with a cut-off ratio 1.615. The maximum pressure is given by $p_{\max} = 54p_1$, where p_1 is the pressure before compression. Assuming indices of compression and expansion of 1.35, show that the m.e.p. of the cycle

$$p_m = 10\,p_1$$

Hence evaluate (a) temperatures at cardinal points with $T_1 = 335$ K, and (b) cycle efficiency. **[LO 13.6]**

Ⓜ **13.16** Recalculate (a) the temperatures at the cardinal points, (b) the m.e.p., and (c) the cycle efficiency when the cycle of Problem 13.15 is a Diesel cycle with the same compression ratio and a cut-off ratio such as to give an expansion curve coincident with the lower part of that of the dual cycle of Problem 13.15. **[LO 13.5]**

Ⓜ **13.17** In an air standard Brayton cycle the compression ratio is 7 and the maximum temperature of the cycle is 800ºC. The compression begins at 0.1 MPa, 35ºC. Compare the maximum specific volume and the maximum pressure with the Otto cycle of Problem 13.6. Find (a) the heat supplied per kg of air, (b) the net work done per kg of air, (c) the cycle efficiency and (d) the temperature at the end of the expansion process. **[LO 13.4, 13.9]**

Ⓔ **13.18** A gas turbine plant operates on the Brayton cycle between the temperatures 27 and 800ºC. (a) Find the pressure ratio at which the cycle efficiency approaches the Carnot cycle efficiency, (b) find the pressure ratio at which the work done per kg of air is maximum and (c) compare the efficiency at this pressure ratio with the Carnot efficiency for the given temperatures. **[LO 13.9]**

(E) **13.19** In a gas turbine plant working on the Brayton cycle the air at the inlet is at 27ºC, 0.1 MPa. The pressure ratio is 6.25 and the maximum temperature is 800ºC. The turbine and compressor efficiencies are each 80%. Find (a) the compressor work per kg of air, (b) the turbine work per kg of air, (c) the heat supplied per kg of air, (d) the cycle efficiency, and (e) the turbine exhaust temperature. **[LO 13.9]**

(E) **13.20** Solve Problem 13.19 if a regenerator of 75% effectiveness is added to the plant. **[LO 13.9]**

(M) **13.21** Solve Problem 13.19 if the compression is divided into two stages, each of pressure ratio 2.5 and efficiency 80%, with intercooling to 27ºC. **[LO 13.9]**

(M) **13.22** Solve Problem 13.21 if a regenerator of 75% effectiveness is added to the plant. **[LO 13.9]**

(E) **13.23** Solve Problem 13.19 if a reheat cycle is used. The turbine expansion is divided into two stages, each of pressure ratio 2.5 and efficiency 80%, with reheat to 800ºC. **[LO 13.9]**

(E) **13.24** Solve Problem 13.23 if a regenerator of 75% effectiveness is added to the plant. **[LO 13.9]**

(E) **13.25** Solve Problem 13.25 if the staged compression of Problem 13.21 is used in the plant. **[LO 13.9]**

(M) **13.26** Find the inlet condition for the free-shaft turbine if a two-shaft arrangement is used in Problem 13.19. **[LO 13.9]**

(M) **13.27** A simple gas turbine plant operating on the Brayton cycle has air inlet temperature 27ºC, pressure ratio 9, and maximum cycle temperature 727ºC. What will be the improvement in cycle efficiency and output if the turbine process is divided into two stages each of pressure ratio 3, with intermediate reheating to 727ºC? **[LO 13.9]**

(M) **13.28** Obtain an expression for the specific work output of a gas turbine unit in terms of pressure ratio, isentropic efficiencies of the compressor and turbine, and the maximum and minimum temperatures, T_3 and T_1. Hence show that the pressure ratio r_p for maximum power is given by

$$r_p = \left(\eta_T \eta_C \frac{T_3}{T_1}\right)^{\gamma/2(\gamma-1)}$$

If $T_3 = 1073$ K, $T_1 = 300$ K, $\eta_C = 0.8$, $\eta_T = 0.8$ and $\gamma = 1.4$ compute the optimum value of pressure ratio, the maximum net work output per kg of air, and corresponding cycle efficiency. **[LO 13.9]**

(M) **13.29** A gas turbine plant draws in air at 1.013 bar, 10ºC and has a pressure ratio of 5.5. The maximum temperature in the cycle is limited to 750ºC. Compression is conducted in an uncooled rotary compressor having an isentropic efficiency of 82%, and expansion takes place in a turbine with an isentropic efficiency of 85%. A heat exchanger with an efficiency of 70% is fitted between the compressor outlet and combustion chamber. For an air flow of 40 kg/s, find (a) the overall cycle efficiency, (b) the turbine output and (c) the air-fuel ratio if the calorific value of the fuel used is 45.22 MJ/kg. **[LO 13.9]**

(M) **13.30** A gas turbine for use as an automotive engine is shown in Fig. 13.53. In the first turbine, the gas expands to just a low enough pressure p_5, for the turbine to drive the compressor. The gas is then expanded through a second turbine connected to the drive wheels. Consider air as the working fluid, and assume that all processes are ideal. Determine (a) pressure p_5, (b) the net work per kg and mass flow rate, (c) temperature T_3 and cycle thermal efficiency and (d) the *T-s* diagram for the cycle. **[LO 13.9]**

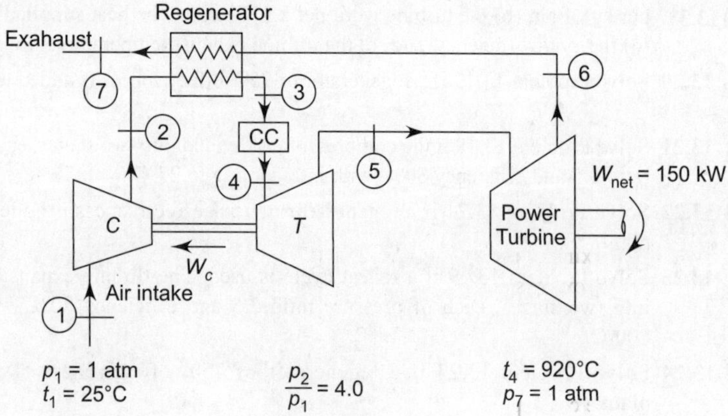

Fig. 13.53

(M) **13.31** Repeat Problem 13.30 assuming that the compressor has an efficiency of 80%, both the turbines have efficiencies of 85%, and the regenerator has an efficiency of 72%. **[LO 13.9]**

(M) **13.32** An ideal air cycle consists of isentropic compression, constant volume heat transfer, isothermal expansion to the original pressure, and constant pressure heat transfer to the original temperature. Deduce an expression for the cycle efficiency in terms of volumetric compression ratio r_k, and isothermal expansion ratio, r_e. In such a cycle, the pressure and temperature at the start of compression are 1 bar and 40ºC, the compression ratio is 8 and the maximum pressure is 100 bar. Determine the cycle efficiency and the m.e.p. **[LO 13.8]**

(E) **13.33** For a gas turbine jet propulsion unit, shown in Fig. 13.28, the pressure and temperature entering the compressor are 1 atm and 15ºC, respectively. The pressure ratio across the compressor is 6 to 1 and the temperature at the turbine inlet is 1000ºC. On leaving the turbine the air enters the nozzle and expands to 1 atm. Determine the pressure at the nozzle inlet and the velocity of the air leaving the nozzle. **[LO 13.10]**

(E) **13.34** Repeat Problem 13.33, assuming that the efficiency of the compressor and turbine are both 85%, and that the nozzle efficiency is 95%. **[LO 13.10]**

(M) **13.35** Develop expressions for work output per kg and the efficiency of an ideal Brayton cycle with regeneration, assuming maximum possible regeneration. For fixed maximum and minimum temperatures, how do the efficiency and work outputs vary with the pressure ratio? What is the optimum pressure ratio? **[LO 13.9]**

(M) **13.36** For an air standard Otto cycle with fixed intake and maximum temperatures, T_1 and T_3, find the compression ratio that renders the net work per cycle a maximum. Derive the expression for cycle efficiency at this compression ratio.

If the air intake temperature, T_1, is 300 K and the maximum cycle temperature, T_3, is 1200 K, compute the compression ratio for maximum net work, maximum work output per kg in a cycle and the corresponding cycle efficiency.

Find the changes in work output and cycle efficiency when the compression ratio is increased from this optimum value to 8. Take $c_v = 0.718$ kJ/kg K. **[LO 13.4]**

(M) **13.37** Show that the mean effective pressure, p_m, for the Otto cycle is given by

$$p_m = \frac{\left(p_3 - p_1 r_k^{\gamma}\right)\left(1 - \dfrac{1}{r_k^{\gamma-1}}\right)}{(\gamma - 1)(r_k - 1)}$$

where $p_3 = p_{max}$, $p_1 = p_{min}$ and r_k is the compression ratio. **[LO 13.4]**

(M) **13.38** A gas turbine plant operates on the Brayton cycle using an optimum pressure ratio for maximum net work output and a regenerator of 100% effectiveness. Derive expressions for net work output per kg of air and corresponding efficiency of the cycle in terms of the maximum and the minimum temperatures.

If the maximum and minimum temperatures are 800 and 30ºC, respectively, compute the optimum value of pressure ratio, the maximum net work output per kg and the corresponding cycle efficiency. **[LO 13.9]**

(M) **13.39** A gas turbine unit is to provide peaking power for an electrical utility with a net power output of 10 MW. The pressure ratio across the compressor is 7, the efficiency of the compressor 80%, and the efficiency of the turbine is 92%. In order to conserve fuel, a regenerator with an effectiveness of 85% is used. The maximum temperature of the cycle is 1200 K. The air at compressor inlet is at 20ºC, 1.1 bar. Assume the working fluid to be air which behaves as an ideal gas with $c_p = 1.005$ kJ/kg K and $\gamma = 1.4$. Neglect pressure drops in the combustion chamber and regenerator. Determine the required air flow and the fuel flow rates for a fuel heating value of 42 MJ/kg, and the power plant efficiency. **[LO 13.9]**

(D) **13.40** Show that for the Stirling cycle with all the processes occurring reversibly but where the heat rejected is not used for regenerative heating, the efficiency is given by

$$\eta = 1 - \frac{\left(\dfrac{T_1}{T_2} - 1\right) + (\gamma - 1)\ln r}{\left(\dfrac{T_1}{T_2} - 1\right) + (\gamma - 1)\dfrac{T_1}{T_2}\ln r}$$

where r is the compression ratio and T_1/T_2 the maximum to minimum temperature ratio. **[LO 13.1]**

Determine the efficiency of this cycle using hydrogen (R = 4.307 kJ/kg K, c_p = 14.50 kJ/kg K) with a pressure and temperature prior to isothermal compression of 1 bar and 300 K respectively, a maximum pressure of 2.55 MPa and heat supplied during the constant volume heating of 9300 kJ/kg. If the heat rejected during the constant volume cooling can be utilised to provide the constant volume heating, what will be the cycle efficiency? Without altering the temperature ratio, can the efficiency be further improved in the cycle?

(M) **13.41** Helium is used as the working fluid in an ideal Brayton cycle. Gas enters the compressor at 27ºC and 20 bar and is discharged at 60 bar. The gas is heated to

1000°C before entering the turbine. The cooler returns the hot turbine exhaust to the temperature of the compressor inlet. Determine (a) the temperatures at the end of compression and expansion, (b) the heat supplied, the heat rejected and the net work per kg of He and (c) the cycle efficiency and the heat rate. Take $c_p = 5.1926$ kJ/kg K. **[LO 13.9]**

(M) **13.42** In an air standard cycle for a gas turbine jet propulsion unit, the pressure and temperature entering the compressor are 100 kPa and 290 K, respectively. The pressure ratio across the compressor is 6 to 1 and the temperature at the turbine inlet is 1400 K. On leaving the turbine the air enters the nozzle and expands to 100 kPa. Assuming that the efficiency of the compressor and turbine are both 85% and that the nozzle efficiency is 95%, determine the pressure at the nozzle inlet and the velocity of the air leaving the nozzle. **[LO 13.10]**

(M) **13.43** A stationary gas turbine power plant operates on the Brayton cycle and delivers 20 MW to an electric generator. The maximum temperature is 1200 K and the minimum temperature is 290 K. The minimum pressure is 95 kPa and the maximum pressure is 380 kPa. If the isentropic efficiencies of the turbine and compressor are 0.85 and 0.80 respectively, find (a) the mass flow rate of air to the compressor, (b) the volume flow rate of air to the compressor, (c) the fraction of the turbine work output needed to drive the compressor and (d) the cycle efficiency. If a regenerator of 75% effectiveness is added to the plant, what would be the changes in the cycle efficiency and the net work output? **[LO 13.9]**

(E) **13.44** Air enters the compressor of a gas turbine operating on Brayton cycle at 1 bar, 27°C. The pressure ratio in the cycle is 6. Calculate the maximum temperature in the cycle and the cycle efficiency. Assume $W_T = 2.5\ W_C$ and $\gamma = 1.4$. **[LO 13.9]**

(M) **13.45** In an ideal jet propulsion cycle, air enters the compressor at 1 atm, 15°C. The pressure of air leaving the compressor is 5 atm and the maximum temperature is 870°C. The air expands in the turbine to such a pressure that the turbine work is equal to the compressor work. On leaving the turbine, the air expands reversibly and adiabatically in a nozzle to 1 atm. Determine the velocity of air leaving the nozzle. **[LO 13.10]**

(M) **13.46** In a gas turbin the compressor is driven by the h.p. turbine. The exhaust from the HP turbine goes to a free-shaft LP turbine which runs the load. The air flow rate is 20 kg/s, and the minimum and maximum temperatures are, respectively 300 and 1000 K. The compressor pressure ratio is 4. Calculate the pressure ratio of the LP turbine and the temperature of the exhaust gases from the unit. The compressor and turbine are isentropic. Take c_p of air and exhaust gases as 1 kJ/kg K and $\gamma = 1.4$. **[LO 13.9]**

(M) **13.47** A regenerative gas turbine with intercooling and reheat operates at steady state. Air enters the compressor at 100 kPa, 300 K with a mass flow rate of 5.807 kg/s. The pressure ratio across the two-stage compressor as well as the turbine is 10. The intercooler and reheater each operate at 300 kPa. At the inlets to the turbine stages, the temperature is 1400 K. The temperature at inlet to the second compressor stage is 300 K. The efficiency of each compressor and turbine stage is 80%. The regenerator effectiveness is 80%. Determine (a) the thermal efficiency, (b) the back work ratio, W_C/W_T and (c) the net power developed. **[LO 13.9]**

(M) **13.48** In a regenerative gas turbine power plant air enters the compressor at 1 bar, 27°C and is compressed to 4 bar. The isentropic efficiency of the compressor is 80% and the regenerator effectiveness is 90%. All of the power developed by the HP turbine is used to drive the compressor and the LP turbine provides the net power output of 97 kW. Each turbine has an isentropic efficiency of 87% and the temperature at inlet to the HP turbine is 1200 K. Determine (a) the mass flow rate of air into the compressor, (b) the thermal efficiency and (c) the temperature of the air at the exit of the regenerator. **[LO 13.9]**

ANSWERS TO PROBLEMS

13.1 (a) 27.7%, 53.7 kJ/kg,
(b) 32.2%

13.2 (a) W_T = 465 kJ/kg, W_C = 121.8 kJ/kg
(b) 0.738

13.5 47.4%

13.10 Otto—r_k = 5.656, p_{max} = 2.64 MPa, W = 2872 kJ/kg, η = 50% Diesel—r_k = 7.456, p_{max} = 1.665 MPa, W = 446.45 kJ/kg, η = 60.8%

13.11 (a) 235 kJ/kg,
(b) 1440 kJ/kg,
(c) 1014 kJ/kg,
(d) 60.5%,
(e) 1252 K,
(f) 2.144
(g) 1.21 MPa)

13.15 (a) T_2 = 805 K, p_2 = 29.2 p_1, T_3 = 1490 K,
T_4 = 2410 K, T_5 = 1200 K,
(b) η = 0.67)

13.16 (a) T_2 = 805 K, T_3 = 1970 K, T_4 = 1142 K
(b) 6.82 p_1,
(c) η = 0.513)

13.19 (a) 259.4 kJ/kg,
(b) 351.68 kJ/kg,
(c) 569.43 kJ/kg,
(d) 16.2%,
(e) 723 K)

13.27 – 18.3%, 30.6%

13.28 4.263, 100 kJ/kg, 17.2%

13.29 (a) 30.4%,
(b) 4272 kW,
(c) 115

13.32 51.5%, 3.45 bar

13.36 5.65, 215 kJ/kg, 50%, ΔW_{net} = – 9 kJ/kg, $\Delta\eta$ = + 6.4%

13.38 $(W_{net})_{max} = c_p\left(\sqrt{T_{max}} - \sqrt{T_{min}}\right)^2$
$(\eta_{cycle})_{max} = 1 - \sqrt{\dfrac{T_{min}}{T_{max}}}$, $(r_p)_{opt}$ = 9.14,
$(W_{net})_{max}$ = 236.79 kJ/kg; η_{cycle} = 0.469

13.41 (a) 465.5, 820.2 K,
(b) 4192.5, 2701.2, 1491.3 kJ/kg,
(c) 0.3557, 10,121 kJ/kWh

13.42 285 kPa, 760 m/s

13.43 (a) 126.37 kg/s,
(b) 110.71 m^3/s,
(c) 0.528,
(d) 0.2146, $\Delta\eta$ = 0.148 ΔW_{net} = 0

13.44 1251.4 K, 40%

13.45 710.3 m/s

13.46 2.3, 673 K

13.47 (a) 0.443,
(b) 0.454,
(c) 2046 kW

13.48 (a) 0.562 kg/s,
(b) 0.432,
(c) 523.2 K

14

Refrigeration Cycles

Learning Objectives

LO 14.1 State traditional refrigeration using ice/dry ice
LO 14.2 Relate reversed heat engine with refrigerator
LO 14.3 Explain vapour compression refrigeration systems
LO 14.4 Discuss refrigerants including green refrigerants
LO 14.5 Differentiate between vapour compression and vapour absorption systems
LO 14.6 Discuss heat pump systems
LO 14.7 Explain gas cycle refrigeration cycles for aviation cooling
LO 14.8 Discuss Linde-Hampson and Claude systems of air liquefaction
LO 14.9 Explain dry ice production process

14.1 REFRIGERATION BY NON-CYCLIC PROCESSES

LO 14.1
State traditional refrigeration using ice/dry ice

Refrigeration is the cooling of a system below the temperature of its surroundings.

The melting of ice or snow was one of the earliest methods of refrigeration and is still employed. Ice melts at 0°C. So when ice is placed in a given space warmer than 0°C, heat flows into the ice and the space is cooled or refrigerated. The latent heat of fusion of ice is supplied from the surroundings, and the ice changes its state from solid to liquid.

Another medium of refrigeration is solid carbon dioxide or dry ice. At atmospheric pressure, CO_2 cannot exist in a liquid state, and consequently, when solid CO_2 is exposed to atmosphere, it sublimates, i.e., it goes directly from solid to vapour, by absorbing the latent heat of sublimation (620 kJ/kg at 1 atm, –78.5°C) from the surroundings (Fig. 14.1). Thus dry ice is suitable for low temperature refrigeration.

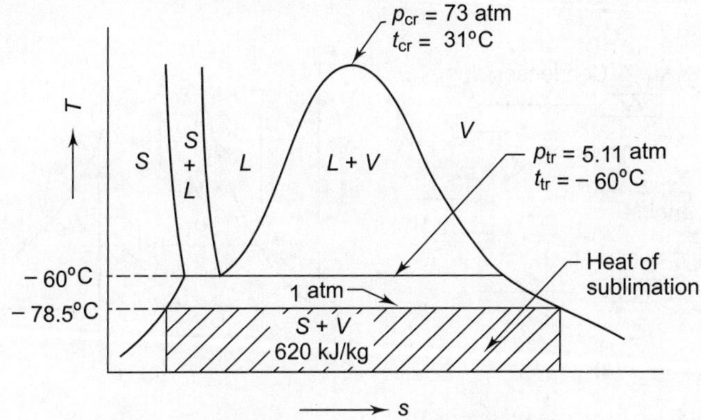

Fig. 14.1 T-s diagram of CO_2

In these two examples, it is observed that the refrigeration effect has been accomplished by non-cyclic processes. Of greater importance, however, are the methods in which the cooling substance is not consumed and discarded, but used again and again in a thermodynamic cycle.

14.2 REVERSED HEAT ENGINE CYCLE

LO 14.2
Relate reversed heat engine with refrigerator

A reversed heat engine cycle, as explained in Section 6.12, is visualised as an engine operating in the reverse way, i.e., receiving heat from a low temperature region, discharging heat to a high temperature region and receiving a net inflow of work (Fig. 14.2). Under such conditions the cycle is called a *heat pump cycle* or a *refrigeration cycle* (see Section 6.6). For a heat pump

$$(\text{COP})_{\text{H.P.}} = \frac{Q_1}{W} = \frac{Q_1}{Q_1 - Q_2}$$

and for a refrigerator

$$(\text{COP})_{\text{ref}} = \frac{Q_2}{W} = \frac{Q_2}{Q_1 - Q_2}$$

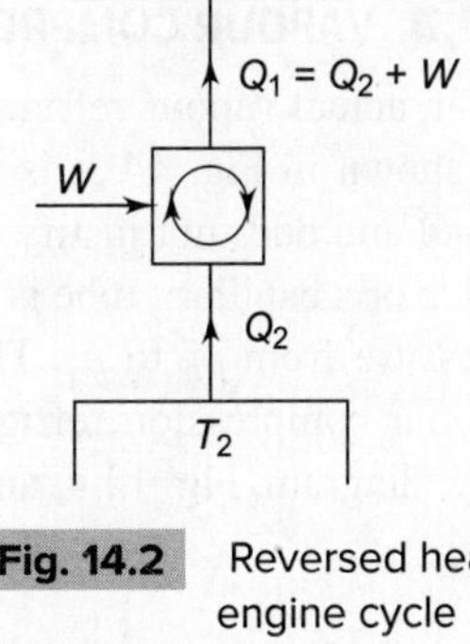

Fig. 14.2 Reversed heat engine cycle

The working fluid in a refrigeration cycle is called a *refrigerant*. In a reversed Carnot cycle (Fig. 14.3), the refrigerant is first compressed reversibly and adiabatically in process 1-2 where the work input per kg of refrigerant is W_c, then it is condensed reversibly in process 2-3 where the heat rejection is Q_1, the refrigerant then expands reversibly and adiabatically in process 3-4 where the work output is W_E, and finally it absorbs heat Q_2 reversibly by evaporation from the surroundings in process 4-1.

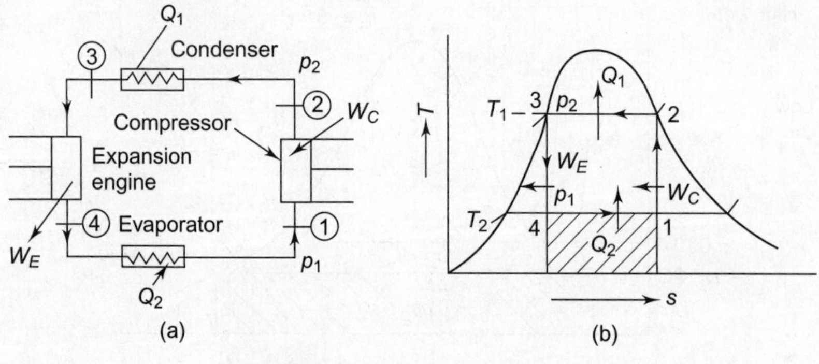

Fig. 14.3 Reversed Carnot cycle

Here $$Q_1 = T_1 (s_2 - s_3),\ Q_2 = T_2 (s_1 - s_4)$$

and $$W_{net} = W_C - W_E = Q_1 - Q_2 = (T_1 - T_2)(s_1 - s_4)$$

where T_1 is the temperature of heat rejection and T_2 the temperature of heat absorption.

$$(\text{COP}_{ref})_{rev} = \frac{Q_2}{W_{net}} = \frac{T_2}{T_1 - T_2} \tag{14.1a}$$

and $$(\text{COP}_{H.P.})_{rev} = \frac{Q_1}{W_{net}} = \frac{T_1}{T_1 - T_2} \tag{14.1b}$$

As shown in Section. 6.17, these are the maximum values for any refrigerator or heat pump operating between T_1 and T_2. It is important to note that for the same T_2 or T_1, the COP increases with the decrease in the temperature difference $(T_1 - T_2)$, i.e. the closer the temperatures T_1 and T_2, the higher the COP.

14.3 VAPOUR COMPRESSION REFRIGERATION CYCLE

> **LO 14.3**
> Explain vapour compression refrigeration systems

In an actual vapour refrigeration cycle, an expansion engine, as shown in Fig. 14.3, is not used, since power recovery is small and does not justify the cost of the engine. A throttling valve or a capillary tube is used for expansion in reducing the pressure from p_2 to p_1. The basic operations involved in a vapour compression refrigeration plant are illustrated in the flow diagram, Fig. 14.4, and the property diagrams, Fig. 14.5.

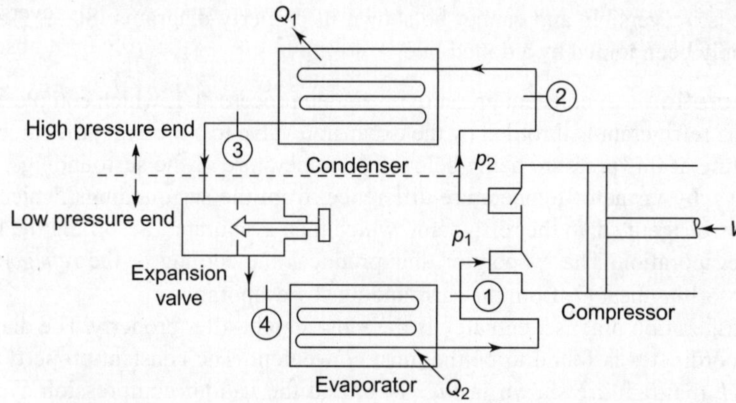

Fig. 14.4 Vapour compression refrigeration plant-flow diagram

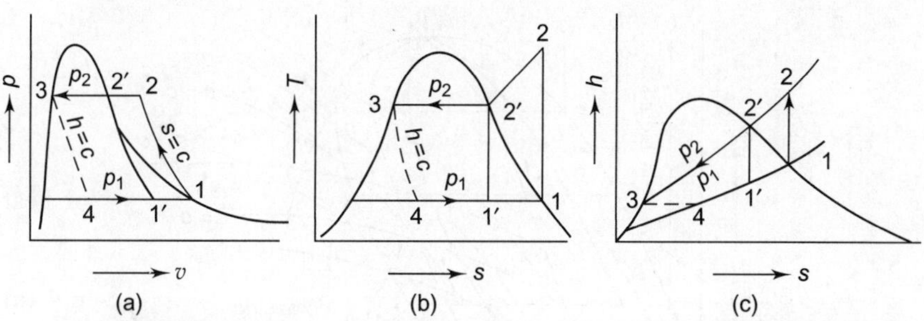

Fig. 14.5 Vapour compression refrigeration cycle-property diagrams

The operations represented are as follows for an idealised plant:

1. Compression A reversible adiabatic process 1–2 or 1′–2′ either starting with saturated vapour (state 1), called *dry compression*, or starting with wet vapour (state 1′), called *wet compression*. Dry compression (1-2) is always preferred to wet compression (1′-2′), because with wet compression there is a danger of the liquid refrigerant being trapped in the head of the cylinder by the rising piston which may damage the valves or the cylinder head, and the droplets of liquid refrigerant may wash away the lubricating oil from the walls of the cylinder, thus accelerating wear.

2. Cooling and condensing A reversible constant pressure process, 2-3, first desuperheated and then condensed, ending with saturated liquid. Heat Q_1 is transferred out.

3. Expansion An adiabatic throttling process 3-4, for which enthalpy remains unchanged. States 3 and 4 are equilibrium points. Process 3-4 is adiabatic (then only $h_3 = h_4$ by SFEE), but not isentropic.

$$Tds = dh - vdp,$$

or

$$s_4 - s_3 = -\int_{p_1}^{p_2} \frac{vdp}{T}$$

Hence it is irreversible and cannot be shown in property diagrams. States 3 and 4 have simply been joined by a dotted line.

4. Evaporation A constant pressure reversible process, 4-1, which completes the cycle. The refrigerant is throttled by the expansion valve to a pressure, the saturation temperature at this pressure being below the temperature of the surroundings. Heat then flows, by virtue of temperature difference, from the surroundings, which gets cooled or refrigerated, to the refrigerant, which then evaporates, absorbing the latent heat of evaporation. The evaporator thus produces the cooling or the *refrigerating effect*, absorbing heat Q_2 from the surroundings by evaporation.

In refrigeration practise, enthalpy is the most sought-after property. The diagram in *p-h* coordinates is found to be the most convenient. The constant property lines in the *p-h* diagram are shown in Fig. 14.6, and the vapour compression cycle in Fig. 14.7.

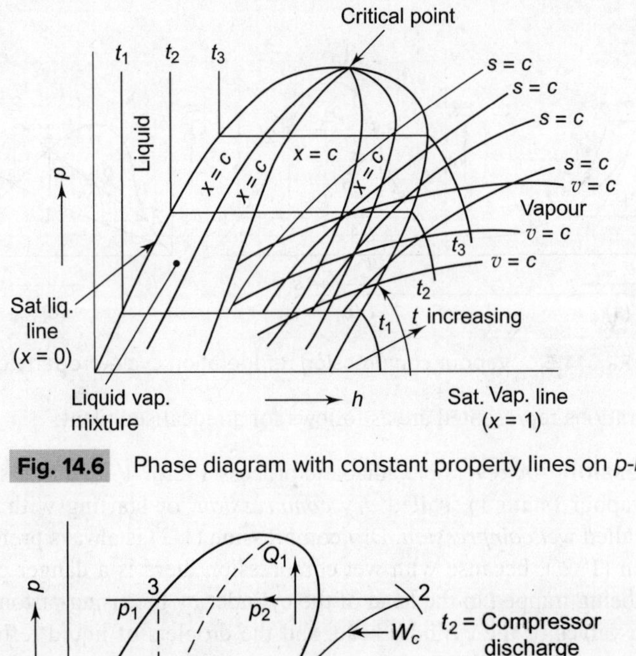

Fig. 14.6 Phase diagram with constant property lines on *p-h* plot

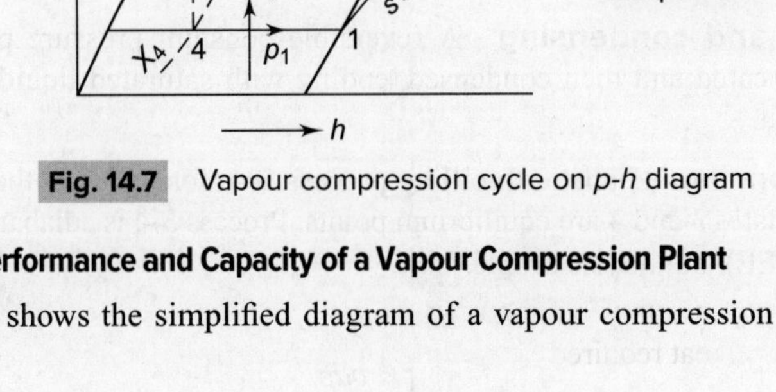

Fig. 14.7 Vapour compression cycle on *p-h* diagram

14.3.1 Performance and Capacity of a Vapour Compression Plant

Figure 14.8 shows the simplified diagram of a vapour compression refrigeration plant.

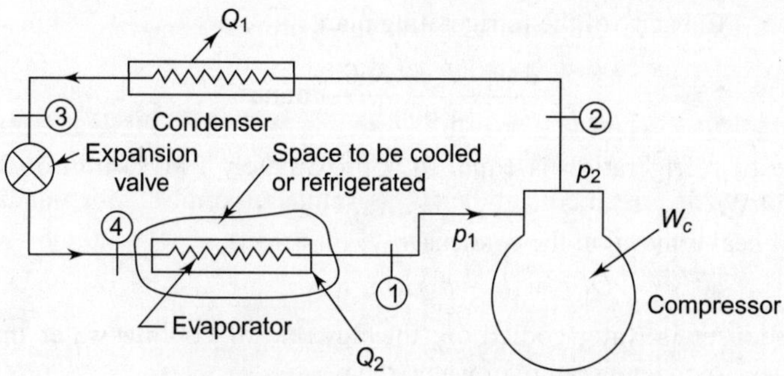

Fig. 14.8 Vapour compression plant

When steady state has been reached, for 1 kg flow of refrigerant through the cycle, the steady flow energy equations (neglecting K.E. and P.E. changes) may be written for each of the components in the cycle as given below.

Compressor $h_1 + W_c = h_2$

$\therefore$ $W_c = (h_2 - h_1)$ kJ/kg

Condenser $h_2 = Q_1 + h_3$

$\therefore$ $Q_1 = (h_2 - h_3)$ kJ/kg

Expansion valve $h_3 = h_4$

or $(h_f)_{p2} = (h_f)_{p1} + x_4\,(h_{fg})_{p2}$

$\therefore$ $$x_4 = \frac{(h_f)_{p_2} - (h_f)_{p_1}}{(h_{fg})_{p_1}}$$

This is the quality of the refrigerant at the inlet to the evaporator (mass fraction of vapour in liquid-vapour mixture), often termed as *flash gas fraction*.

Evaporator

$$h_4 + Q_2 = h_1$$

$\therefore$ $Q_2 = (h_1 - h_4)$ kJ/kg

This is known as the *refrigerating effect*, the amount of heat removed from the surroundings per unit mass flow of refrigerant.

If the *p-h* chart for a particular refrigerant is available with the given parameters, it is possible to obtain from the chart the values of enthalpy at all the cardinal points of the cycle. Then for the cycle

$$\text{COP} = \frac{Q_2}{W_c} = \frac{h_1 - h_4}{h_2 - h_1} \tag{14.2}$$

If w is the mass flow of refrigerant in kg/s, then the rate of heat removal from the surroundings

$$= w(h_1 - h_4)\text{ kJ/s} = w(h_1 - h_4) \times 3600\text{ kJ/h}$$

One tonne of refrigeration is defined as the rate of heat removal from the surroundings equivalent to the heat required for melting 1 tonne of ice in one day. If the latent heat of fusion of ice is taken as 336 kJ/kg, then 1 tonne is equivalent to heat removal at the rate of (1000 × 336)/24 or 14,000 kJ/h

∴ Capacity of the refrigerating plant

$$= \frac{w(h_1 - h_4) \times 3600}{14{,}000} \text{ tonnes}$$

One tonne of refrigeration is equal to approximately 211 kJ/min or 3.5 kW or 1 TR = 3.5 kW.

The rate of heat removal in the condenser

$$Q_1 = w(h_2 - h_3) \text{ kJ/s}$$

If the condenser is water-cooled, $\dot{m}_c$ the flowrate of cooling water in kg/s and $(t_{c2} - t_{c1})$ the rise in temperature of water, then

$$Q_1 = w(h_2 - h_3) = \dot{m}_c c_c (t_{c2} - t_{c1}) \text{ kJ/s}$$

provided the heat transfer is confined only between the refrigerant and water, and there is no heat interaction with the surroundings.

The rate of work input to the compressor

$$W_c = w(h_2 - h_1) \text{ kJ/s}$$

14.3.2 Actual Vapour Compression Cycle

In order to ascertain that there is no droplet of liquid refrigerant being carried over into the compressor, some superheating of vapour is recommended after the evaporator.

A small degree of subcooling of the liquid refrigerant after the condenser is also used to reduce the mass of vapour formed during expansion, so that too many vapour bubbles do not impede the flow of liquid refrigerant through the expansion valve.

Both the superheating of vapour at the evaporator outlet and the subcooling of liquid at the condenser outlet contribute to an increase in the refrigerating effect, as shown in Fig. 14.9. The compressor discharge temperature, however, increases, due to superheat, from t_2' to t_2, and the load on the condenser also increases.

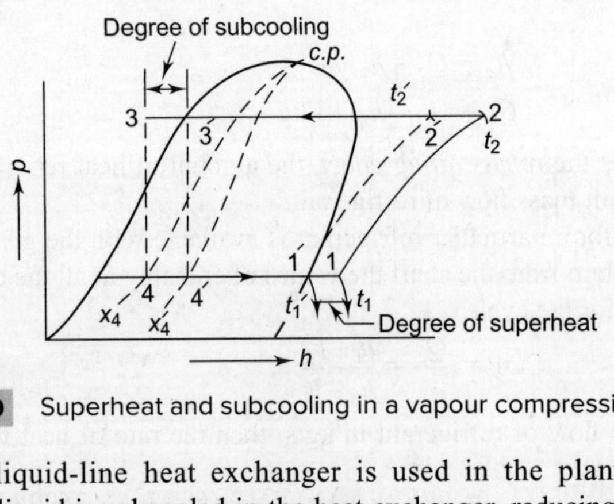

Fig. 14.9 Superheat and subcooling in a vapour compression cycle

Sometimes, a liquid-line heat exchanger is used in the plant, as shown in Fig. 14.10. The liquid is subcooled in the heat exchanger, reducing the load on the condenser and improving the COP. For 1 kg flow

$$Q_2 = h_6 - h_5,\ Q_1 = h_2 - h_3$$

$$W_c = h_2 - h_1 \text{ and } h_1 - h_6 = h_3 - h_4$$

14.3.3 Components in a Vapour Compression Plant

1. Condenser It must desuperheat and then condense the compressed refrigerant. Condensers may be either air-cooled or water-cooled. An air-cooled condenser is used in small self-contained units. Water-cooled condensers are used in larger installations.

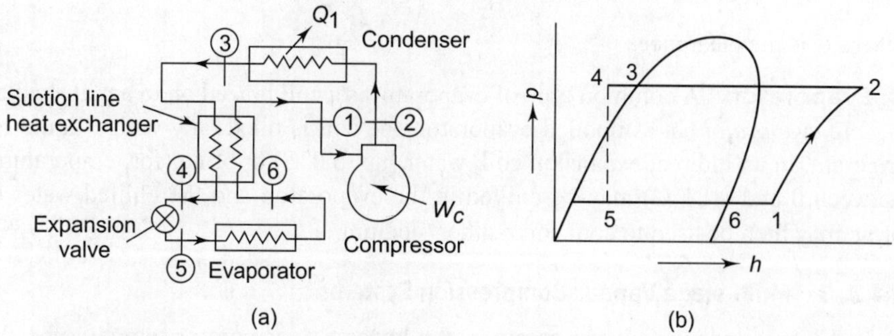

Fig. 14.10 Vapour compression cycle with a suction-line heat exchanger

2. Expansion device It reduces the pressure of the refrigerant, and also regulates the flow of the refrigerant to the evaporator. Two widely used types of expansion devices are: capillary tubes and throttle valves (thermostatic expansion valves). Capillary tubes are used only for small units. Once the size and length are fixed, the evaporator pressure, etc., get fixed. No modification in operating conditions is possible. Throttle valves are used in larger units. These regulate the flow of the refrigerant according to the load on the evaporator.

3. Compressor Compressors may be of three types: (i) reciprocating, (ii) rotary, and (iii) centrifugal. When the volume flow rate of the refrigerant is large, centrifugal compressors are used. Rotary compressors are used for small units. Reciprocating compressors are used in plants up to 100 tonnes capacity. For plants of higher capacities, centrifugal compressors are employed.

In reciprocating compressors, which may be single-cylinder or multi-cylinder ones, because of clearance, leakage past the piston and valves, and throttling effects at the suction and discharge valves, the actual volume of gas drawn into the cylinder is less than the volume displaced by the piston. This is accounted for in the term *volumetric efficiency*, which is defined as

$$\eta_{\text{vol}} = \frac{\text{Actual volume of gas drawn at evaporator pressure and temperature}}{\text{Piston displacement}}$$

∴ Volume of gas handled by the compressor

$$= w \cdot v_1 (\text{m}^3/\text{s}) = \left(\frac{\pi}{4} D^2 L \frac{N}{60} n \right) \times \eta_{\text{vol}}$$

where w is the refrigerant flow rate,

v_1 is the specific volume of the refrigerant at the compressor inlet,

D and *L* are the diameter and stroke of the compressor,
n is the number of cylinders in the compressor and
N is the r.p.m.

The clearance volumetric efficiency is given by Eq. (18.13)

$$\eta_{\text{vol}} = 1 + C - C\left(\frac{p_2}{p_1}\right)^{1/n}$$

where *C* is the clearance.

4. Evaporator A common type of evaporator is a coil brazed on to a plate, called a *plate evaporator*. In a 'flooded evaporator' the coil is filled only with a liquid refrigerant. In an indirect expansion coil, water (up to 0°C) or brine (for temperatures between 0 and – 21°C) may be chilled in the evaporator, and the chilled water or brine may then be used to cool some other medium.

14.3.4 Multi-stage Vapour Compression Systems

For a given condensation temperature, the lower the evaporator temperature, the higher becomes the compressor pressure ratio. For a reciprocating compressor, a high pressure ratio across a single stage means low volumetric efficiency. Also, with dry compression the high pressure ratio results in high compressor discharge temperature which may damage the refrigerant. To reduce the work of compression and improve the COP, multi-stage compression with intercooling may be adopted. Since the intercooler temperature may be below the temperature of available cooling water used for the condenser, the refrigerant itself may be used as the intercooling medium. Figure 14.11 shows a two-stage compression system with a direct contact heat exchanger.

As shown in Section. 18.5.1, for minimum work, the intercooler pressure p_i is the geometric mean of the evaporator and condenser pressures, p_1 and p_2, or

$$p_i = \sqrt{p_1 \cdot p_2}$$

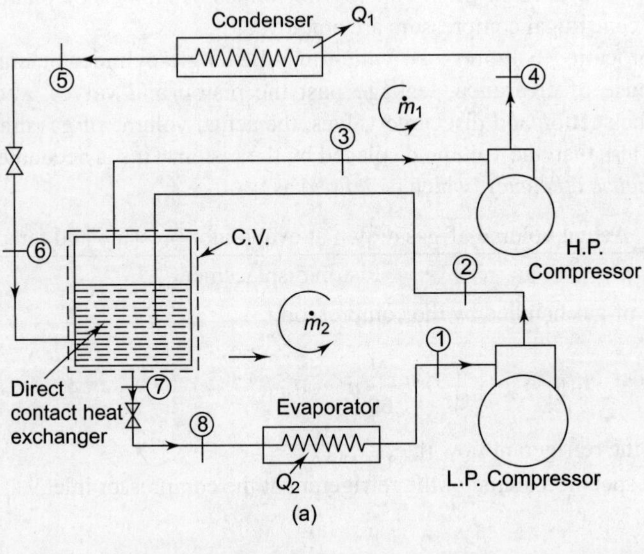

(a)

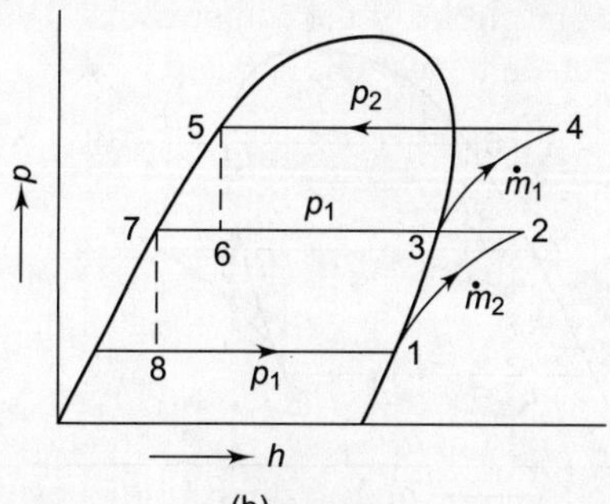

(b)

Fig. 14.11 Two-stage vapour compression system

By making an energy balance of the direct contact heat exchanger,

$$\dot{m}_2 h_2 + \dot{m}_1 h_6 = \dot{m}_2 h_7 + \dot{m}_1 h_3$$

$$\therefore \qquad \frac{\dot{m}_1}{\dot{m}_2} = \frac{h_2 - h_7}{h_3 - h_6}$$

The desired refrigerating effect determines the flow rate in the low pressure loop, $\dot{m}_2$, as given below

$$\dot{m}_2(h_1 - h_8) = \frac{14000}{3600} \times P$$

where P is the capacity, in tonnes of refrigeration.

$$\therefore \qquad \dot{m}_2 = \frac{3.89P}{h_1 - h_8} \text{ kg/s}$$

Figure 14.12 shows a two-stage vapour compression system with a flash chamber intercooler, where the vapour from the flash chamber (state 9) mixes with the vapour from the LP compressor (state 2) to form vapour at state 3, which enters the HP compressor.

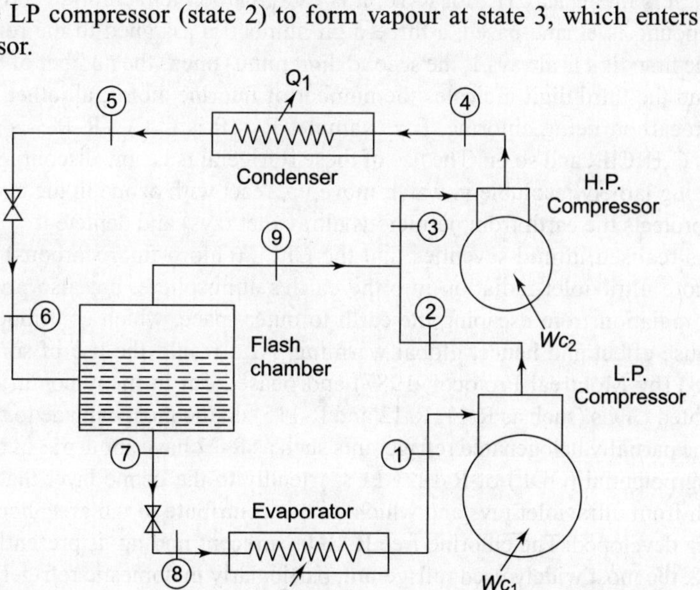

(a)

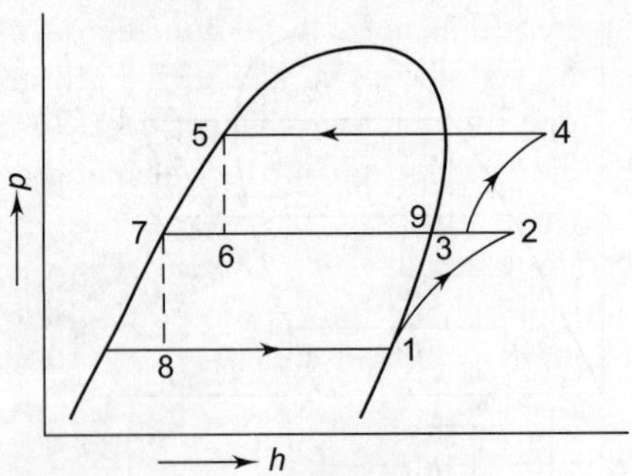

(b)

Fig. 14.12 Two-stage vapour compression system with a flash intercooler

14.3.5 Refrigerants

> **LO 14.4**
> Discuss refrigerants including green refrigerants

The most widely used refrigerants till recently were a group of halogenated hydrocarbons, marketed under the various proprietary names of freon, genetron, arcton, isotron, frigen, etc. These are either methane-based or ethane-based, where the hydrogen atoms are replaced by chlorine or fluorine atoms. Methane-based compounds are denoted by a number of two digits, where the first digit minus one is the number of hydrogen atoms and the second digit indicates the number of fluorine atoms, while the other atoms are chlorine. For Refrigerant-12 (R–12), e.g. the number of hydrogen atoms is zero, the number of fluorine atoms is two and hence the other two atoms must be chlorine. Therefore, the compound is CCl_2F_2, dichloro-difluoro methane. Similarly, for R-22, it is $CHClF_2$, monochloro-difluoro methane; for R-50, it is methane, CH_4; for R-10, it is CCl_4, carbon tetrachloride and so on. If the compound is ethane-based, a three-digit number is assigned to the refrigerant, where the first digit is always 1, the second digit minus one is the number of hydrogen and the third digit indicates the number of fluorine atoms, all other atoms in the hydrocarbon being chlorine. For example, R-110 is C_2Cl_6, R-113 is $C_2Cl_3F_3$, R-142 is $C_2H_3ClF_2$ and so on. The use of these refrigerants is now discouraged, since these, being largely insoluble in water, move up, react with ozone in the ozone layer (which protects the earth from pernicious ultraviolet rays) and deplete it.

It was realised in mid-seventies that the CFCs (chloro-fluorocarbons) not only allow more ultraviolet radiation into the earth's atmosphere, but also prevent the infrared radiation from escaping the earth to outer space, which contributes to the greenhouse effect and hence, global warming. As a result, the use of some CFCs is banned (by Montreal Protocol, 1987) and phased out in many countries. Fully halogenated CFCs (such as R-11, R-12 and R-115) do the most damage to the ozone layer. The partially halogenated refrigerants such as R-22 have about 5% of the ozone depleting potential (ODP) of R-12. CFCs, friendly to the ozone layer that protects the earth from ultraviolet rays and which do not contribute to the greenhouse effect are being developed. The chlorine free R-134a, a recent finding, is presently replacing R-12, the most widely used refrigerant, particularly in domestic refrigerators and freezers and automotive air conditioners.

Two important parameters that need to be considered in the selection of a refrigerant are the temperatures of the two media (the refrigerated space and the environment), with which the refrigerant exchanges heat. To have reasonable heat transfer rate, a temperature difference of 5 to 10°C should be maintained between the refrigerant and the medium. If a space is to be maintained at – 10°C, e.g. the refrigerant should evaporate at about –20°C (Fig. 14.13), the saturation pressure at which should be above atmospheric pressure to prevent any air leakage into the system. Again, the temperature of the refrigerant in the condenser should be above the cooling medium by about 10°C, as shown in the figure, the saturation pressure at which must be below the critical pressure of the refrigerant. If a single refrigerant cannot meet the temperature requirements (–20 to 50°C range), two cycles with two different refrigerants can be used in series (Fig. 14.14). Such a coupled cycle makes a *cascade refrigeration system*.

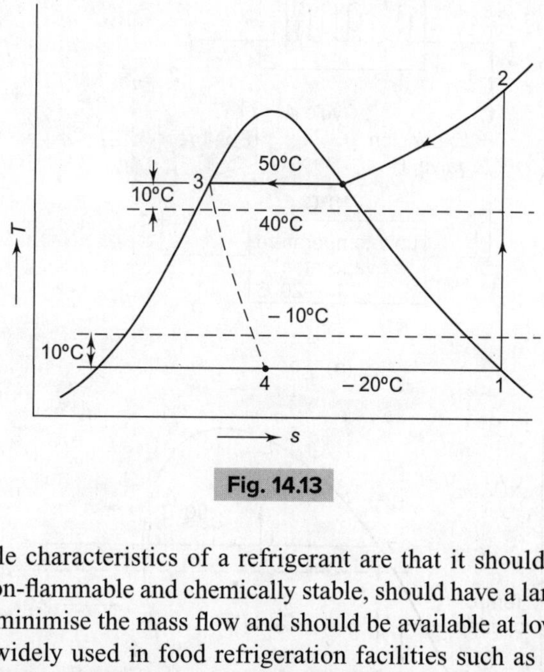

Fig. 14.13

Other desirable characteristics of a refrigerant are that it should be non-toxic, non-corrosive, non-flammable and chemically stable, should have a large enthalpy of vapourisation to minimise the mass flow and should be available at low cost.

Ammonia is widely used in food refrigeration facilities such as the cooling of fresh fruits, vegetables, meat and fish, refrigeration of beverages and dairy products such as beer, wine, milk and cheese, freezing of ice cream and ice production, low temperature refrigeration in the pharmaceutical and other process industries. The advantages of ammonia are its low cost, higher COPs and thus lower energy costs, greater detectability in the event of a leak, no effect on the ozone layer and more favourable thermodynamic and transport properties and thus higher heat transfer coefficients requiring smaller and lower cost heat exchangers. The major drawback of ammonia is its toxicity which makes it unsuitable for domestic use.

Other fluids used as refrigerants are sulphur dioxide, methyl chloride, ethyl chloride, hydrocarbons like propane, butane, ethane, ethylene, etc., carbon dioxide, air and water.

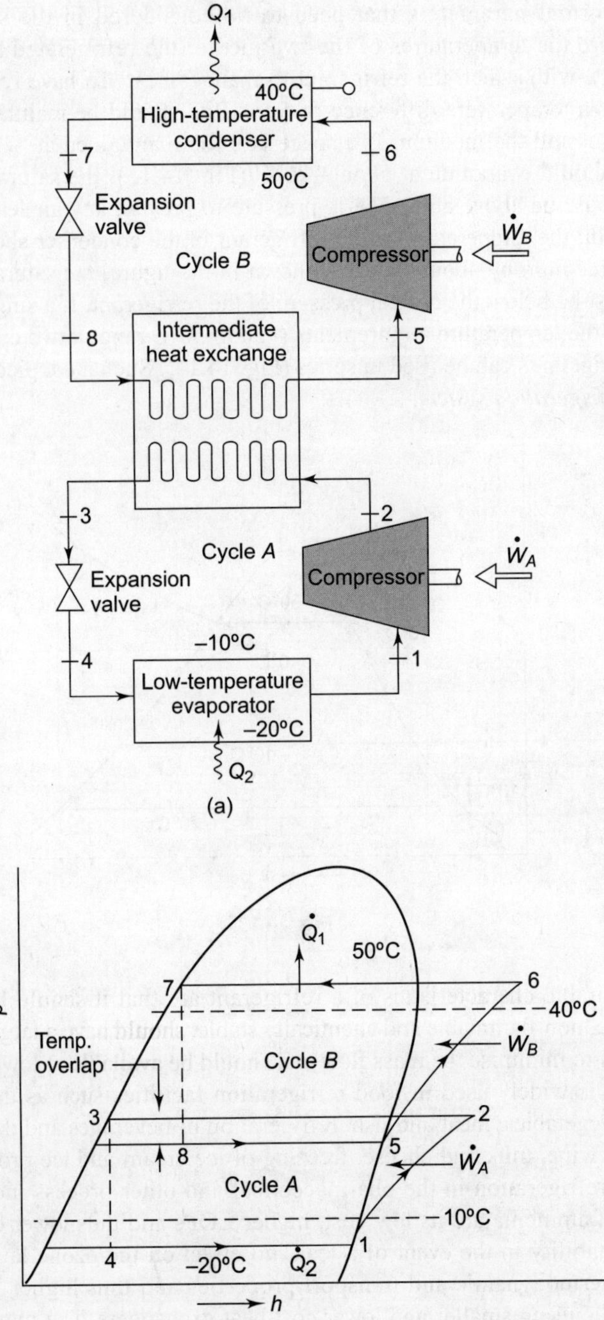

Fig. 14.14 Cascade refrigeration cycle

The selection of refrigerants is currently undergoing a global transition. The primary function of refrigerants is to absorb heat at low temperatures (in the evaporator) and release it at higher temperatures (in the condenser). The first generation of refrigerants was chosen mainly based on these basic functional requirements. The second generation prioritized safety and stability during operation, while the third generation focused on minimizing ozone depletion potential (ODP). The fourth generation has also aimed to reduce global warming potential (GWP).

The Montreal Protocol, an international treaty adopted in 1987 to phase out ozone-depleting refrigerants, has undergone several amendments, including the Kigali Amendment in 2016. As a result, substances such as chlorofluorocarbons (CFCs), hydrochlorofluorocarbons (HCFCs), and hydrofluorocarbons (HFCs) have been largely phased out. This initiative is widely regarded as one of the most successful international efforts to restore the ozone layer.

The search for "green refrigerants" continues, with an emphasis on solutions that are safe, non-toxic, and have both low or zero ODP and GWP, all while being cost-effective. Emerging candidates include various hydrofluoroolefins (HFOs) and their blends (e.g., R-1234yf, R-1234ze, R-1233zd, R-513A, R-1336mzz, R-514A, R-452B, R-454B) as well as natural refrigerants such as hydrocarbons, carbon dioxide, ammonia, and water.

14.4 ABSORPTION REFRIGERATION CYCLE

LO 14.5 Differentiate between vapour compression and vapour absorption systems

The absorption refrigeration system is a *heat operated unit* which uses a refrigerant that is *alternately absorbed and liberated from the absorbent*. In the basic absorption system, the compressor in the vapour compression cycle is *replaced by an absorber-generator assembly* involving less mechanical work. Figure 14.15 gives the basic absorption refrigeration cycle, in which *ammonia is the refrigerant and water is the absorbent.* This is known as the *aqua-ammonia absorption* system.

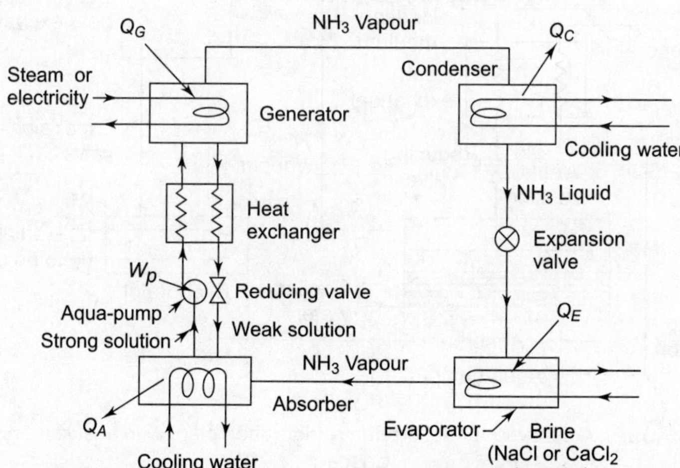

Fig. 14.15 Vapour absorption refrigeration plant-flow diagram

Ammonia vapour is vigorously absorbed in water. So when low-pressure ammonia vapour from the evaporator comes in contact in the absorber with the weak solution (the concentration of ammonia in water is low) coming from the generator, it is readily absorbed, releasing the latent heat of condensation. The temperature of the solution tends to rise, while the absorber is cooled by the circulating water, absorbing the *heat of solution* (Q_A) and maintaining a constant temperature. Strong solution, rich in ammonia, is pumped to the generator where heat (Q_G) is supplied from an external source (steam, electricity, gas flame, etc.). Since the boiling point of ammonia is less than that of water, the ammonia vapour is given off from the aqua-ammonia solution at high pressure, and the weak solution returns to the absorber through a pressure reducing valve. The heat exchanger preheats the strong solution and precools the weak solution, reducing both Q_G and Q_A, the heat to be supplied in the generator and the heat to be removed in the absorber respectively. The ammonia vapour then condenses in the condenser, is throttled by the expansion valve, and then evaporates, absorbing the heat of evaporation from the surroundings or the brine to be chilled.

In driving the ammonia vapour out of the solution in the generator, it is impossible to avoid evaporating some of the water. This water vapour going to the condenser along with the ammonia vapour, after condensation, may get frozen to ice and block the expansion valve. So an *analyser-rectifier combination* (Fig. 14.16) is used to eliminate water vapour from the ammonia vapour going into the condenser.

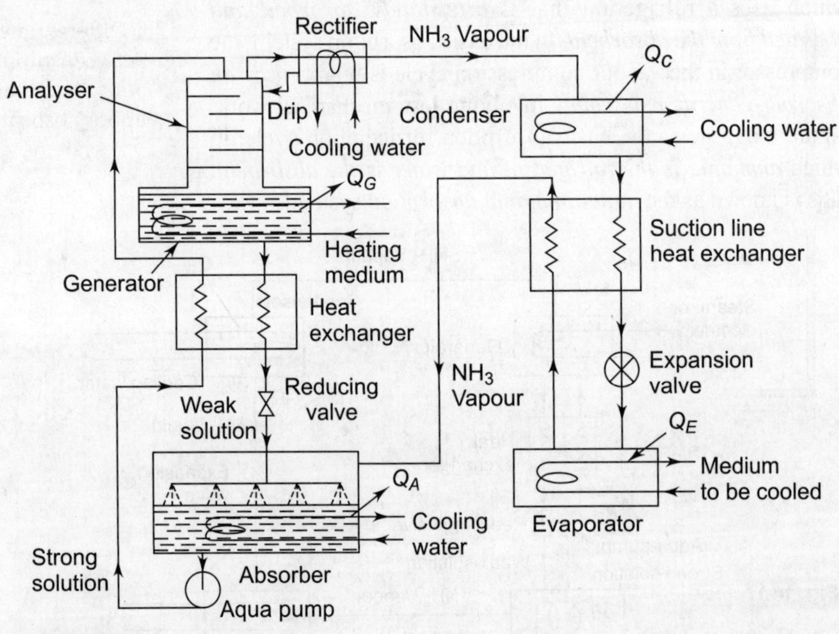

Fig. 14.16 Actual vapour absorption refrigeration plant with analyzer and rectifier

The analyser is a direct-contact heat exchanger consisting of a series of trays mounted above the generator. The strong solution from the absorber flows downwards over the trays to cool the outgoing vapours. Since the saturation temperature of water is higher than that of ammonia at a given pressure, it is the water vapour which condenses first. As the vapour passes upwards through the analyser, it is cooled and enriched by ammonia, and the liquid is heated. Thus the vapour going to the condenser is lower in temperature and richer in ammonia, and the heat input to the generator is decreased.

The final reduction in the percentage of water vapour in the ammonia going to the condenser occurs in the rectifier which is a water-cooled heat exchanger which condenses water vapour and returns it to the generator through the drip line, as shown in Fig. 14.16. The use of a suction-line heat exchanger is to reduce Q_A and increase Q_E, thus achieving a double benefit. In the absorber, the weak solution is sprayed to expose a larger surface area so as to accelerate the rate of absorption of ammonia vapour.

Figure 14.17 demonstrates the use of solar energy as a heat source in the aqua-ammonia absorption refrigeration system.

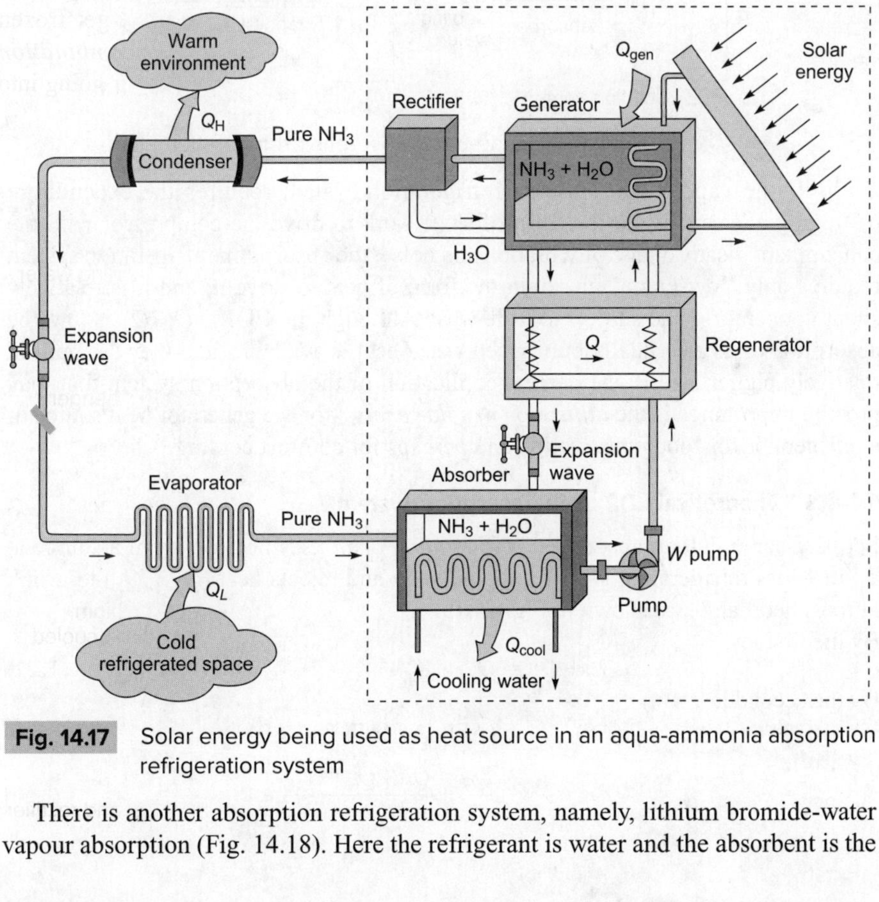

Fig. 14.17 Solar energy being used as heat source in an aqua-ammonia absorption refrigeration system

There is another absorption refrigeration system, namely, lithium bromide-water vapour absorption (Fig. 14.18). Here the refrigerant is water and the absorbent is the solution of lithium bromide salt in water. Since water cannot be cooled below 0°C, it can be used as a refrigerant in air conditioning units. Lithium bromide solution has a strong affinity for water vapour because of its very low vapour pressure. It absorbs water vapour as fast as it is released in the evaporator.

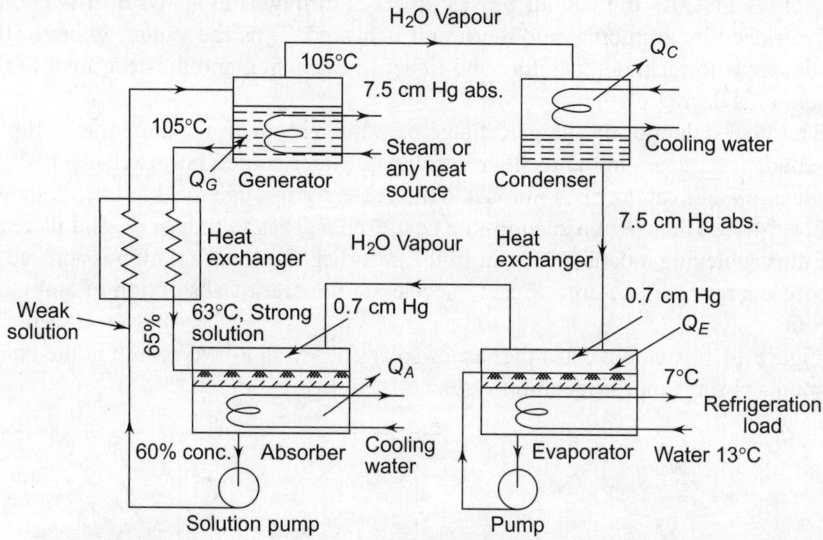

Fig. 14.18 Lithium bromide-water absorption refrigeration plant

While the vapour compression refrigeration system requires the expenditure of 'high-grade' energy in the form of shaft work to drive the compressor with the concomitant disadvantage of vibration and noise, the absorption refrigeration system requires only 'low-grade' energy in the form of heat to drive it, and it is relatively silent in operation and subject to little wear. Although the COP = Q_E/Q_G is low, the absorption units are usually built when waste heat is available, and they are built in relatively bigger sizes. One current application of the absorption system that may grow in importance is the *utilisation of solar energy* for the generator heat source of a refrigerator for food preservation and perhaps for comfort cooling.

14.4.1 Theoretical COP of an Absorption System

Let us assume that an absorption refrigeration plant uses heat Q_G from a source at T_1, provides refrigeration Q_E for a region at T_R and rejects heat $(Q_A + Q_C)$ to a sink (atmosphere) at T_2, as shown in Fig. 14.19.

By the first low

$$Q_E + Q_G = Q_C + Q_A \tag{14.3}$$

By the second law

$$(\Delta S)_{\text{source}} + (\Delta S)_{\text{sink}} + (\Delta S)_{\text{region}} \geq 0$$

$$\therefore \quad -\frac{Q_G}{T_1} + \frac{Q_E + Q_G}{T_2} - \frac{Q_E}{T_R} \geq 0$$

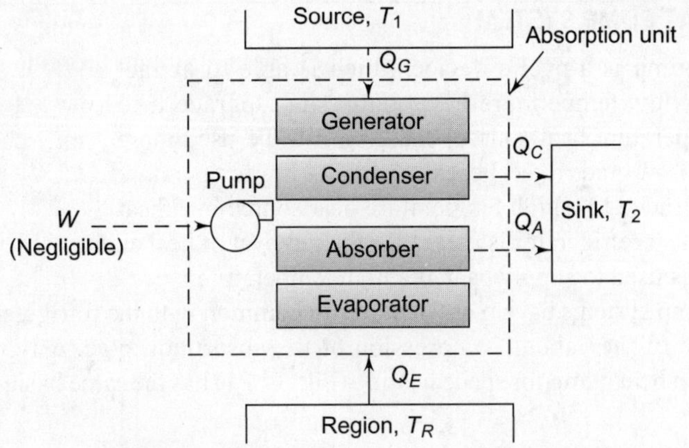

Fig. 14.19 Energy fluxes in vapour absorption plant

From Eq. (14.3)

$$-\frac{Q_G}{T_1}+\frac{Q_E+Q_G}{T_2}-\frac{Q_E}{T_R}\geq 0$$

Or $$\frac{T_1-T_2}{T_1T_2}Q_G+\frac{T_R-T_2}{T_2T_R}Q_E\geq 0$$

Or $$\frac{T_2-T_R}{T_2T_R}Q_E\leq\frac{T_1-T_2}{T_1T_2}Q_G$$

Or $$\frac{Q_E}{Q_G}\leq\frac{(T_1-T_2)T_R}{(T_2-T_R)T_1}$$

or $$\text{COP}\leq\frac{(T_1-T_2)T_R}{(T_2-T_R)T_1}$$

$$\therefore\quad (\text{COP})_{max}=\frac{(T_1-T_2)T_R}{(T_2-T_R)T_1}\qquad(14.4)$$

or $$(\text{COP})_{max}=\frac{T_R}{T_2-T_R}\times\frac{T_1-T_2}{T_1}$$

Therefore, the maximum possible COP is the product of the ideal COP of a refrigerator working between T_R and T_2, and the ideal thermal efficiency of an engine working between T_1 and T_2. The cyclic heat engine and the refrigerator together, as shown in Fig. 14.20, is equivalent to the absorption cycle.

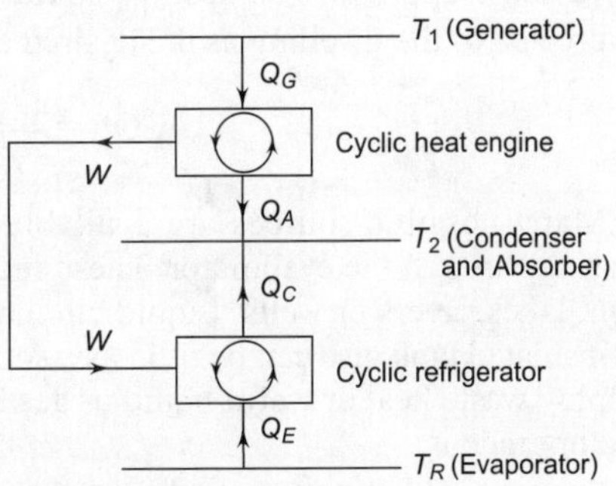

Fig. 14.20 An absorption cycle as equivalent to a cyclic heat engine and a cyclic refrigerator

14.5 HEAT PUMP SYSTEM

LO 14.6
Discuss heat pump systems

The heat pump is a cyclic device which is able to extract energy at a low temperature heat source and upgrade it to a high temperature heat source, enabling it to be used more effectively. Low grade reject heat available at a low temperature may be upgraded to a high temperature heat source by a heat pump. While a refrigerator is meant for the removal of heat and to achieve cooling, a heat pump is used to supply heat at a high temperature.

Heat pump systems have many features in common with the refrigeration systems and may be of the vapour compression or the absorption type. A typical vapour-compression heat pump for space heating (Fig. 14.21) has the same basic components

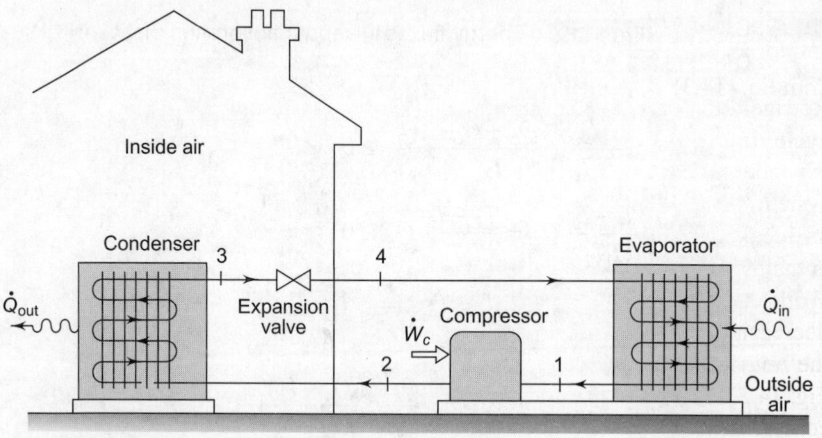

Fig. 14.21 Vapour-compression heat pump system for space heating

as the vapour-compression refrigeration system: compressor, condenser, expansion valve and evaporator. The heat $\dot{Q}_{in}$ comes from the surroundings (cold air), and $\dot{Q}_{out}$ is directed to the dwelling as the desired effect, with the COP given by

$$\text{COP} = \frac{\dot{Q}_{out}}{W_C} = \frac{h_2 - h_3}{h_2 - h_1}$$

Many possible sources are available for heat transfer ($\dot{Q}_{in}$) to the refrigerant passing through the evaporator. These include the outside air, the ground and water from lakes, rivers or wells. Liquid circulated through a solar collector and stored in an insulated tank also can be used as a source for a heat pump. Industrial heat pumps employ waste heat or warm liquid or gas streams as the low temperature heat source and upgrade it.

An air-air heat pump can be used for year-round air conditioning, to achieve heating during winter and cooling during summer, by using a reversing valve with the evaporator and condenser executing opposite duties. A heat pump in industry will be very effective if both condenser and evaporator are utilised for heating and cooling, respectively, e.g. for cooling a process stream and heating another one (Fig. 14.22).

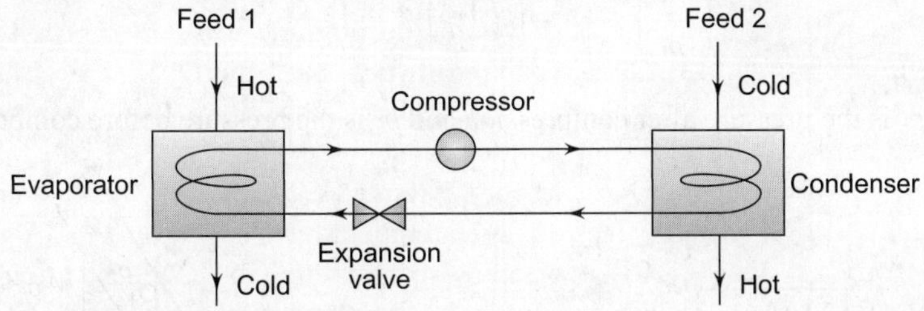

Fig. 14.22 Heating and cooling of two process streams

14.6 GAS CYCLE REFRIGERATION: REVERSE BRAYTON OR BELL-COLEMAN CYCLE

> **LO 14.7**
> Explain gas cycle refrigeration cycles for aviation cooling

Refrigeration can also be accomplished by means of a gas cycle. In the gas cycle, also called Bell-Coleman cycle, an expander replaces the throttle valve of a vapour compression system, because the drop in temperature by throttling a real gas is very small. For an ideal gas, enthalpy is a function of temperature only, and since in throttling enthalpy remains unchanged, there would not be any change in temperature also. Work output obtained from the expander is used as an aid in compression, thus decreasing the net work input. The ideal gas-refrigeration cycle is the same as the *reversed Brayton cycle*. The flow and *T-s* diagrams of the cycle are shown in Fig. 14.23. *Since there is no* phase change, the condenser and evaporator in a vapour compression system are here called the cooler and refrigerator, respectively. At 1, the air is isentropically compressed from p_1 to p_2(1 – 2), after which it is cooled at constant pressure p_2 to state 3 and then expanded isentropically to cool to state 4. The COP of the refrigeration cycle, assuming the gas to be ideal, is given by

$$\text{COP} = \frac{Q_2}{W_{\text{net}}} = \frac{h_1 - h_4}{(h_2 - h_1) - (h_3 - h_4)}$$

$$= \frac{T_1 - T_4}{(T_2 - T_1) - (T_3 - T_4)} = \frac{T_1 - T_4}{T_1\left(\dfrac{T_2}{T_1} - 1\right) - T_4\left(\dfrac{T_3}{T_4} - 1\right)}$$

For isentropic compression and expansion

$$\frac{T_2}{T_1} = \left(\frac{p_2}{p_1}\right)^{(\gamma-1)/\gamma} = \frac{T_3}{T_4}$$

$$\therefore \quad \text{COP} = \frac{T_1 - T_4}{(T_1 - T_4)\left(\dfrac{T_3}{T_4} - 1\right)} = \frac{T_4}{T_3 - T_4}$$

Also $$\text{COP} = \frac{1}{\left(\frac{p_2}{p_1}\right)^{(\gamma-1)/\gamma} - 1} = \frac{1}{(r_p)^{\frac{\gamma-1}{\gamma}} - 1} \tag{14.5}$$

where p_2 is the pressure after compression and p_1 is the pressure before compression.

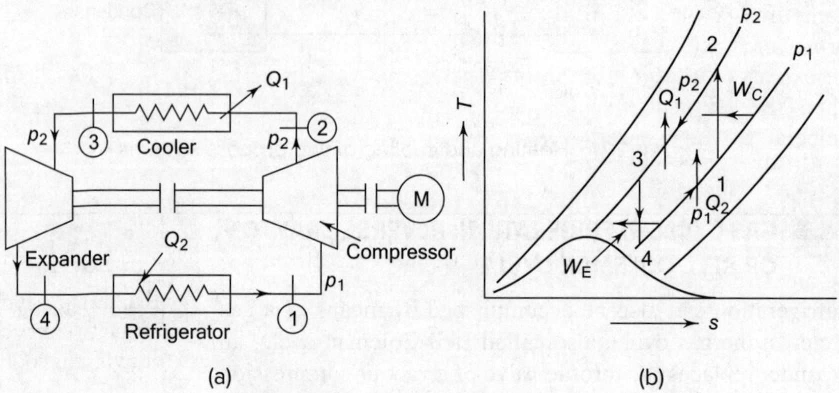

Fig. 14.23 Gas refrigeration cycle

The COP of a gas-cycle refrigeration system is low. The power requirement per unit capacity is high. Its prominent application is in aircrafts and missiles, where the vapour compression refrigeration system becomes heavy and bulky. Figure 14.24 shows the open cycle aircraft cabin cooling.

The compressed air is available and is a small percentage of the amount handled by the compressor of a turbojet or a supercharged aircraft engine. Large amounts of cool ambient air are available for cooling the compressed air. In addition to cooling, the replacement of stale air in the cabin is possible. At high altitudes, the pressurisation of cabin air is also possible. Because of these considerations, air cycle refrigeration is favoured in aircrafts.

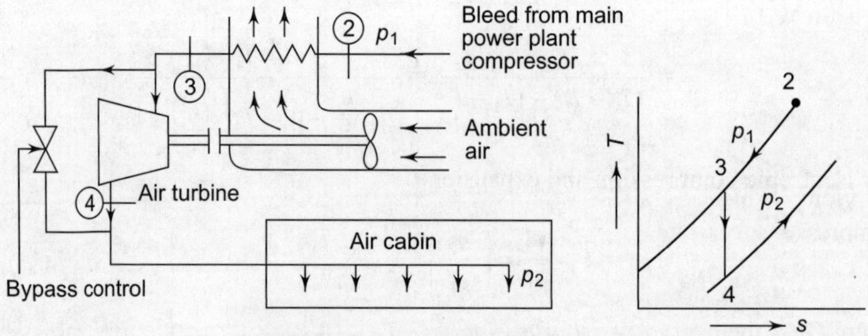

Fig. 14.24 Open cycle aircraft cabin cooling

14.7 LIQUEFACTION OF GASES

LO 14.8

Discuss Linde-Hampson and Claude systems of air liquefaction

An important application of gas refrigeration processes is the liquefaction of gases. A gas may be cooled either by making it expand isentropically in an expander, thus performing work (sometimes known as *external work method*), or making the gas undergo Joule–Kelvin expansion through a throttle valve (sometimes called the *internal work method*). While the former method always brings about a temperature decrease, the expansion through the throttle valve may yield a temperature decrease only when the temperature before throttling is below the maximum inversion temperature (see Section 11.7).

14.7.1 Linde–Hampson System for Liquefaction of Air

In this system, Joule–Kelvin effect is utilised for cooling, and ultimately, liquefying the air. The schematic and the *T-s* diagrams are shown in Fig. 14.25. Ideally, the compression would be isothermal as shown on the *T-s* diagram. A two-stage compressor with intercooling and aftercooling is shown. The *yield*, *Y*, of the system is defined as the ratio of the mass of liquid produced to the mass of gas compressed. The energy required per unit mass of liquid produced is known as the *specific work consumption*, *W*.

The theoretical yield, assuming perfect insulation, can be determined by a mass and energy balance for the control volume (Joule–Kelvin refrigeration system) as shown in Fig. 14.24. Let $\dot{m}_f$ be the rate at which liquid air is produced (the same given to the system as make-up), and $\dot{m}$ the rate at which air is compressed and then expanded. Then the yield is

$$Y = \frac{\dot{m}_f}{\dot{m}}$$

and the energy balance gives

$$\dot{m}h_2 - \dot{m}_f h_5 - (\dot{m} - \dot{m}_f)h_7 = 0$$

$$\therefore \quad \dot{m}(h_2 - h_7) - \dot{m}_f(h_5 - h_7) = 0$$

$$Y = \frac{\dot{m}_f}{\dot{m}} = \frac{h_2 - h_7}{h_5 - h_7}$$

$$\therefore \quad Y = \frac{h_7 - h_2}{h_7 - h_5} \tag{14.6}$$

No yield is thus possible unless h_7 is greater than h_2. The energy balance for the compressor gives

$$\dot{m}h_1 + W_c = \dot{m}h_2 + Q_R$$

where Q_R is the heat loss to the surroundings from the compressor

$$\therefore \quad \frac{W_c}{\dot{m}} = T_1(s_1 - s_2) - (h_1 - h_2)$$

This is the *minimum work requirement.*
Specific work consumption, W

$$= \frac{W_c}{\dot{m}} \times \frac{\dot{m}}{\dot{m}_f} = \frac{W_c}{\dot{m}} \frac{1}{Y} = \frac{h_7 - h_5}{h_7 - h_2} [T_1(s_1 - s_2) - (h_1 - h_2)]$$

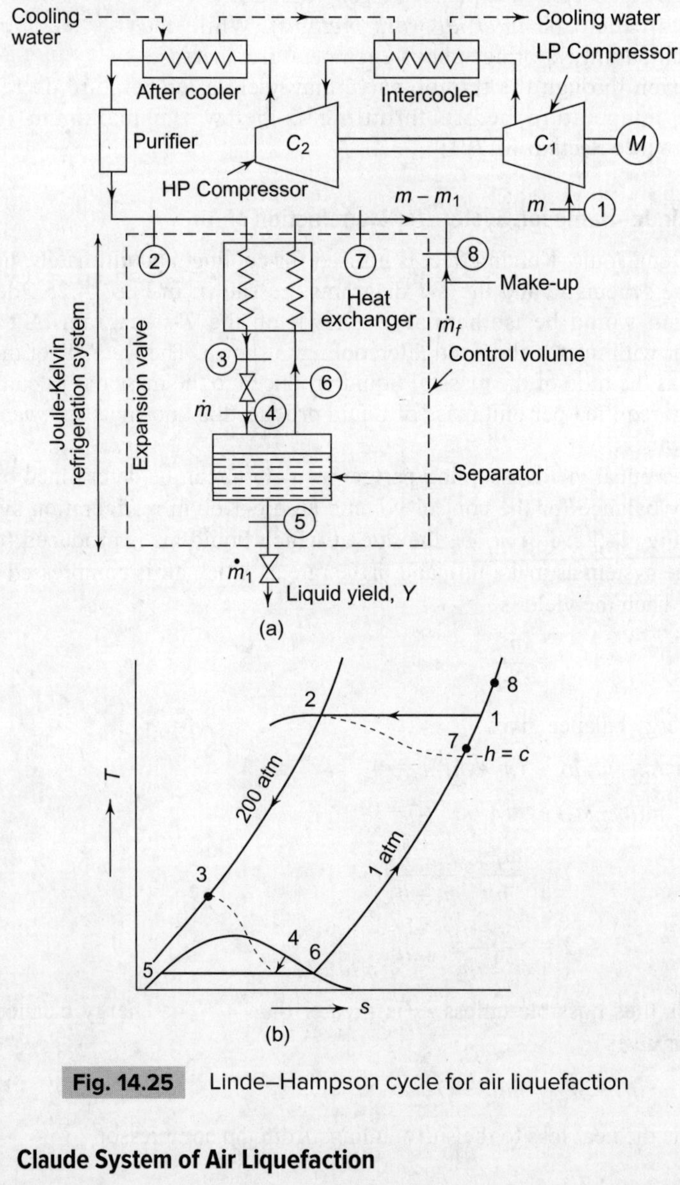

Fig. 14.25 Linde–Hampson cycle for air liquefaction

14.7.2 Claude System of Air Liquefaction

In the Claude system, energy is removed from the gas stream by allowing it to do some work in an expander. The flow and *T-s* diagrams are given in Fig. 14.26.

The gas is first compressed to pressures of about 40 atm and then passed through the first heat exchanger. Approximatly 80% of the gas is then diverted from the main stream, expanded through an expander and reunited with the return stream below the second heat exchanger. The stream to be liquefied continues through the second and third heat exchangers, and is finally expanded through an expansion valve to the liquid receiver. The cold vapour from the liquid receiver is returned through the heat exchangers to cool the incoming gas.

The yield and the specific work consumption may be computed by making the mass and energy balance as in the Linde–Hampson system.

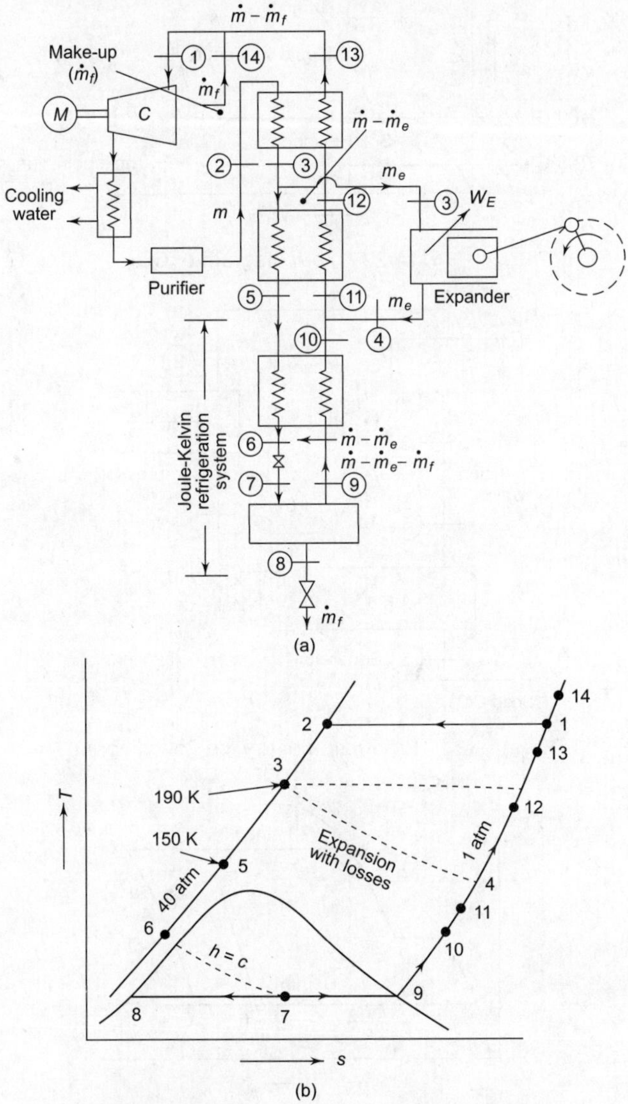

Fig. 14.26 Claude cycle for air liquefaction

14.8 PRODUCTION OF SOLID ICE

LO 14.9
Explain dry ice production process

Dry ice is used for low temperature refrigeration, such as to preserve ice cream and other perishables. The property diagram of CO_2 on the *p-h* coordinates is given in Fig. 14.27. The schematic diagram of producing solid CO_2 and the corresponding *p-h* diagram are shown in Fig. 14.28 and Fig. 14.29, respectively.

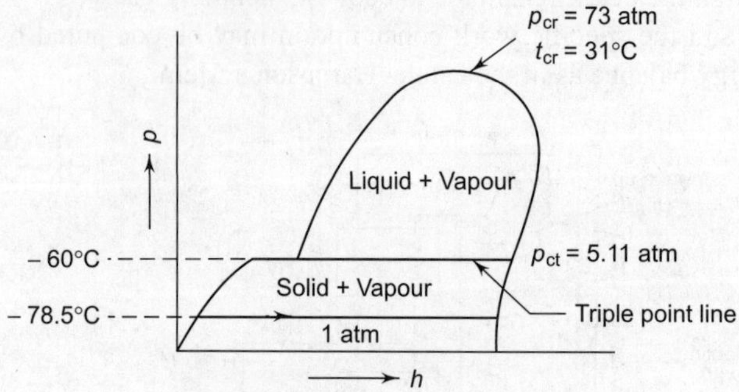

Fig. 14.27 *p-h* diagram of CO_2

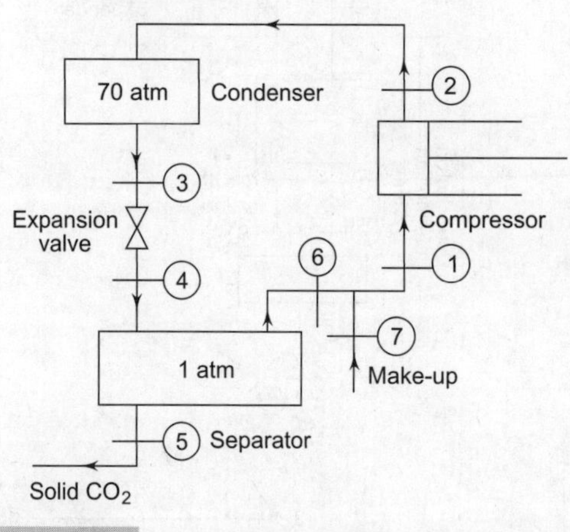

Fig. 14.28 Production of dry ice-flow diagram

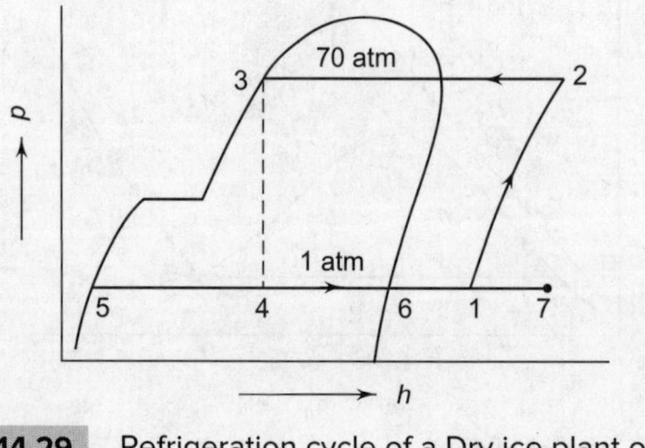

Fig. 14.29 Refrigeration cycle of a Dry ice plant on *p-h* plot

SOLVED EXAMPLES

Example 14.1 *A cold storage is to be maintained at – 5°C while the surroundings are at 35°C. The heat leakage from the surroundings into the cold storage is estimated to be 29 kW. The actual COP of the refrigeration plant used is one-third of an ideal plant working between the same temperatures. Find the power required (in kW) to drive the plant.* E [LO 14.2]

Solution $\text{COP (Ideal)} = \frac{T_2}{T_1 - T_2} = \frac{268}{308 - 268} = 6.7$

$\therefore$ $\text{Actual COP} = 1/3 \times 6.7 = 2.23 = \frac{Q_2}{W}$

$\therefore$ Power required to drive the plant (Fig. 14.30)

$$W = \frac{Q_2}{2.23} = \frac{29}{2.23} = \mathbf{13\ kW}$$

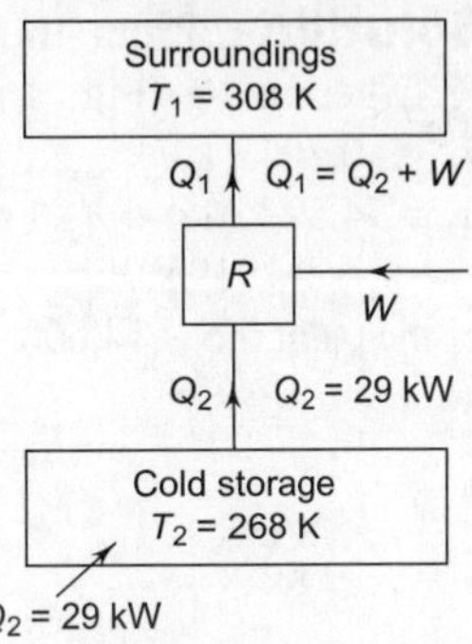

Fig. 14.30

Example 14.2 *A refrigerator uses R-134a as the working fluid and operates on an ideal vapour compression cycle between 0.14 and 0.8 MPa. If the mass flow rate of the refrigerant is 0.06 kg/s, determine (a) the rate of heat removal from the refrigerated space, (b) the power input to the compressor, (c) the heat rejection rate in the condenser and (d) the COP.* E [LO 14.3]

Solution From the R-134*a* tables, the enthalpies at the four states (Fig. 14.31) are:

$h_1 = 236.04$ kJ/kg

$s_1 = 0.9322$ kJ/kg K $= s_2$

For $p_2 = 0.8$ MPa, $s_2 = 0.9322$ kJ/kgK,

$h_2 = 272.05$ kJ/kg, $h_3 = h_4 = 93.42$ kJ/kg

$Q_2 = 0.06\,(236.04 - 93.42) = \mathbf{8.56\ kW}$

$W_c = 0.06\,(272.05 - 236.04) = \mathbf{2.16\ kW}$

$Q_1 = 0.06\,(272.05 - 93.42) = \mathbf{10.72\ kW}$

$$\text{COP} = \frac{Q_2}{W_c} = \frac{8.56}{2.16} = \mathbf{3.963}$$

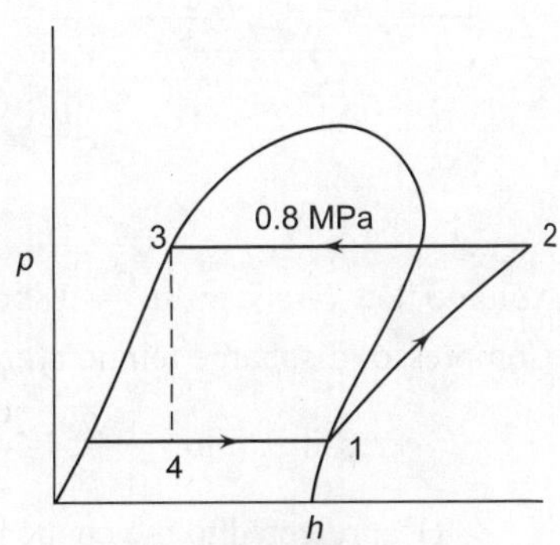

Fig. 14.31

Example 14.3 *A simple R-12 plant is to develop 5 tonnes of refrigeration. The condenser and evaporator temperatures are to be 40 and –10°C, respectively. Determine (a) the refrigerant flow rate in kg/s, (b) the volume flow rate handled by the compressor in m³/s, (c) the compressor discharge temperature, (d) the pressure ratio, (e) the heat rejected to the condenser in kW, (f) the flash gas percentage after throttling, (g) the COP and (h) the power required to drive the compressor.*

How does this COP compare with that of a Carnot refrigerator operating between 40 and –10°C? M [LO 14.3]

Solution From the table of the thermodynamic properties of a saturated Refrigerant-12 (Fig. 14.32), $p_1 = (p_{sat})_{-10°C} = 2.1912$ bar, $h_1 = 183.19$ kJ/kg, $s_1 = 0.7019$ kJ/kg K, $v_1 = 0.077$ m³/kg, $p_2 = (psat)_{40°C} = 9.6066$ bar, and $h_3 = 74.59$ kJ/kg $= h_4$. From the superheated table of R-12, when $p_2 = 9.6066$ bar, and $s_2 = s_1 = 0.7019$ kJ/kg K, by interpolation, $t_2 = 48$°C, and $h_2 = 209.41$ kJ/kg. The capacity of the plant $= 5 \times 14{,}000 = 70{,}000$ kJ/h. If w is the refrigerant flow rate in kg/s

$$w(h_1 - h_4) = \frac{70{,}000}{3600} = 19.44 \text{ kW}$$

$$\therefore \quad w = \frac{19.44}{183.19 - 74.59} = \mathbf{0.18\ kg/s}$$

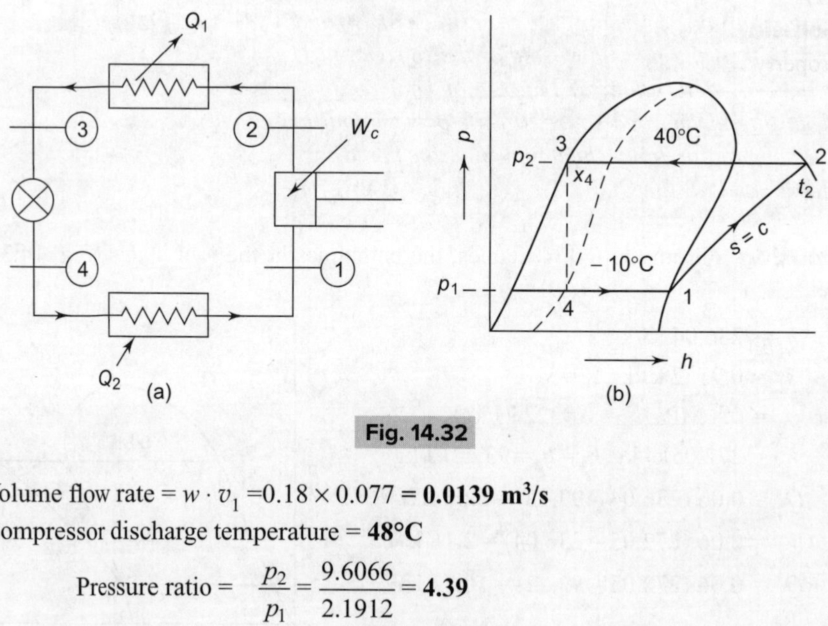

Fig. 14.32

Volume flow rate $= w \cdot v_1 = 0.18 \times 0.077 = \mathbf{0.0139\ m^3/s}$

Compressor discharge temperature = **48°C**

$$\text{Pressure ratio} = \frac{p_2}{p_1} = \frac{9.6066}{2.1912} = \mathbf{4.39}$$

$$\text{Heat rejected to the condenser} = w(h_2 - h_3)$$

$$= 0.18\,(209.41 - 74.59) = \mathbf{24.27\ kW}$$

$$h_4 = h_f + x_4 h_{fg} = 26.87 + x_4 \times 156.31 = 74.59$$

$$\therefore \quad x_4 = \frac{47.72}{156.31} = 0.305$$

$\therefore$ Flash gas percentage = **30.5%**

$$\text{COP} = \frac{h_1 - h_4}{h_2 - h_1} = \frac{183.19 - 74.59}{209.41 - 183.19}$$

$$= \frac{108.60}{26.22} = \mathbf{4.14}$$

Power required to drive the compressor

$$= w(h_2 - h_1) = 0.18 \times 26.22 = \mathbf{4.72\ kW}$$

$$\text{COP(Reversible)} = \frac{T_2}{T_1 - T_2} = \frac{263}{50} = 5.26$$

$$\therefore \quad \frac{\text{COP (Vap.Comp.cycle)}}{\text{COP (Carnot cycle)}} = \frac{4.14}{5.26} = \mathbf{0.787}$$

Example 14.4 *A refrigerator working on vapour compression refrigeration cycle used R-134a as the working fluid. The mass flow rate of the refrigerant is 0.08 kg/s. The refrigerant enters the compressor at 100 kPa and and leaves at 0.9 MPa and 60°C. If the refrigerant leaves the condenser at –20°C, determine (a) the efficiency of the compressor, (b) the quality of refrigerant at the evaporator inlet, (c) the rate of heat removal from the refrigerated space and (d) the COP of the refrigerator.* M [LO 14.3]

Solution The property values at different state points (Fig. 14.33) found from the property tables of Refrigerant R-134a are given below.

$$h_1 = 387.215 \text{ kJ/kg}, \; s_1 = 1.76655 \text{ kJ/kgK}, \; h_2 = 445.223 \text{ kJ/kg}$$
$$h_{f,100\text{kPa}} = 165.802 \text{ kJ/kg}, \; h_{fg,100\text{kPa}} = 216.36 \text{ kJ/kg}$$
$$h_3 = h_{f,30°\text{C}} = 241.79 \text{ kJ/kg}$$

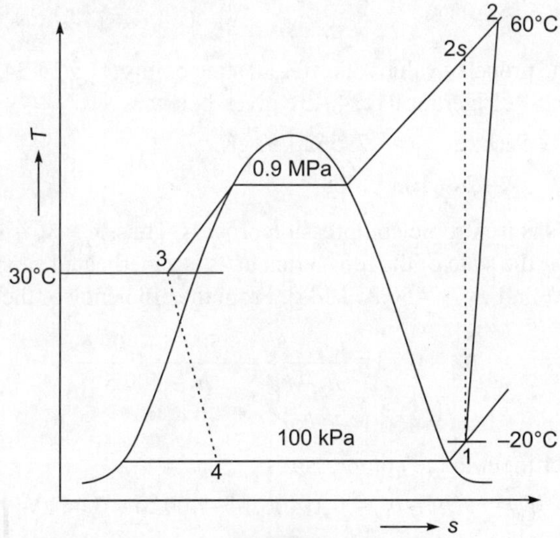

Fig. 14.33

(a) The process 1-2s is isentropic compression process. Thus, $s_1 = s_2 = 1.76655$ kJ/kgK. Since, $s_{2s} > s_{gp2}$, the state of the refrigerant at 2s is a superheated vapour. Using linear interpolation, we find $h_{2s} = 453.478$ kJ/kg.
The efficiency of the compressor is found to be

$$\eta_C = \frac{h_{2s} - h_1}{h_2 - h_1} = \frac{434.478 - 287.215}{445.223 - 387.215} = 0.8148 \text{ or } \mathbf{81.48\%}$$

(b) The quality of refrigerant at the evaporator inlet (x_4) can be found from the consideration that during throttling process (3-4), enthalpy remains constant. Thus,

$$h_3 = h_4 = h_{f,p4} + h_4 s_{fg,p4}$$

$$241.79 = 165.802 + x_{4s} 216.36$$

$$\therefore \quad x_{4s} = \mathbf{0.3512}$$

(c) The rate of heat removal from the refrigerated space is

$$Q_L = w(h_1 - h_4) = 0.08(387.215 - 241.79) = \mathbf{11.634\ kW}$$

(d) The COP of the refrigerator is

$$\text{COP} = \frac{Q_L}{W_{in}} = \frac{w(h_1 - h_4)}{w(h_2 - h_1)} = \frac{387.215 - 241.79}{445.223 - 387.215} = \mathbf{2.51}$$

Example 14.5 *Refrigerant R134 acirculates through a vapour compression refrigeration system at a rate of 0.1 kg/s. The refrigerant enters the compressor at 0.2 MPa and 10°C and leaves at 1.2 MPa. The efficiency of the compressor is 73%. The pressure drops as the refrigerant flows through the condenser and evaporator are negligibly small. If the refrigerant leaves the condenser at 35°C, determine (a) the power input to the compressor, (b) the refrigerating capacity and (c) the coefficient of performance.* M **[LO 14.3]**

Solution The property values at different state points (Fig. 14.34) found from the property tables of Refrigerant R-134a are given below.

$$h_1 = 409.5 \text{ kJ/kg}, \; s_1 = 1.79562 \text{ kJ/kgK}$$

$$h_3 = h_{f,35°C} = 249.103 \text{ kJ/kg}$$

The process 1-2s is isentropic compression process. Thus, $s_1 = s_2 = 1.79562$ kJ/kgK. Since, $s_{2s} > s_{gp2}$, the state of the refrigerant at 2s is superheated vapour. Using linear interpolation, we find $h_{2s} = 450.85$ kJ/kg. From the efficiency of the compressor, we get

$$\eta_C = 0.73 = \frac{h_{2s} - h_1}{h_2 - h_1} = \frac{450.85 - 409.5}{h_2 - 409.5}$$

or $$h_2 = 466.14 \text{ kJ/kg}$$

(a) The power input to the compressoris

$$W_{in} = w(h_2 - h_1) = 0.1(466.14 - 409.5) = \mathbf{5.66\ kW}$$

(b) The refrigerating capacity is

$$Q_L = w(h_1 - h_4) = 0.1(409.5 - 249.103) = \mathbf{16.04\ kW}$$

(c) The coefficient of performance is

$$\text{COP} = \frac{Q_L}{W_{in}} = \frac{w(h_1 - h_4)}{w(h_2 - h_1)} = \frac{409.5 - 249.103}{466.14 - 409.5} = \mathbf{2.83}$$

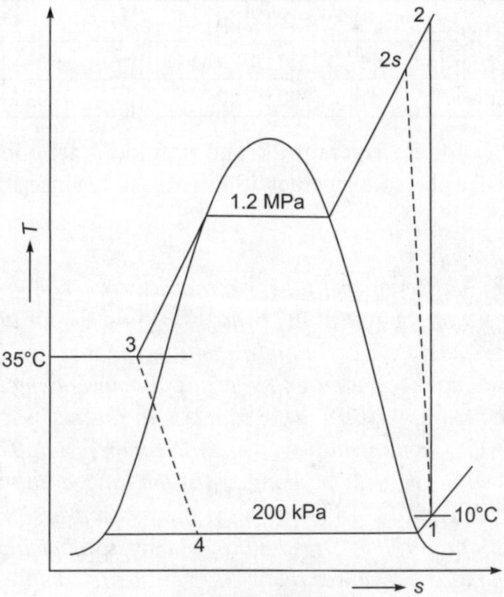

Fig. 14.34

Example 14.6 *A vapour compression refrigeration system maintains the temperature of the refrigerated space as –10°C while rejecting heat to the surrounding air at 25°C (Fig. 14.35). If the allowable maximum and minimum pressures of the refrigerants are 700 and 150 kPa, respectively, which of the followingrfrigerants can be used as the working fluid: Refrigerant 22, Refrigerant 134a or ammonia.*

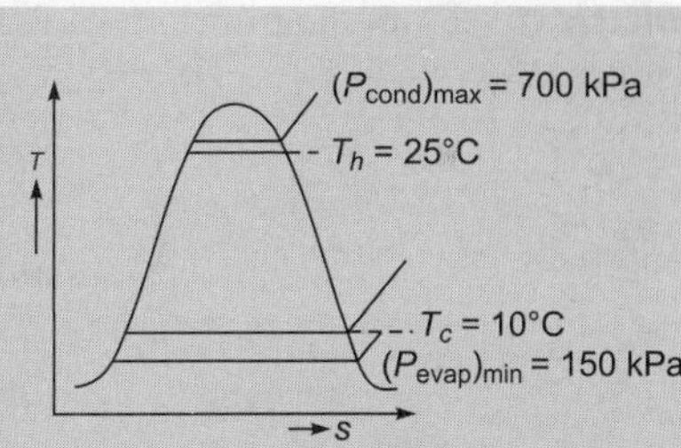

Fig. 14.35

M [LO 14.3]

Solution When the working fluid passes through the evaporator, there is a heat transfer from the refrigerated space to the working fluid. For that, it should be

$$(p_{\text{evap}})_{\min} \le p_{\text{evap}} \le p_{\text{sat}}(T_c)$$

or $$150 \text{ kPa} \le p_{\text{sat}}(T_c)$$

Similarly, when the working fluid passes through the condenser, heat transfer takes place from the working fluid to the surrounding. For that, we must have

$$(p_{\text{cond}})_{\max} \ge p_{\text{cond}} \ge p_{\text{sat}}(T_h)$$

or $$700 \text{ kPa} \ge p_{\text{sat}}(T_h)$$

For the three refrigerants, the saturation pressure corresponding to the temperatures of refrigerated space and surrounding are tabulated below:

Working fluid	p_{sat} (–10°C)	p_{sat} (25°C)
Refrigerant 22	354.3 kPa	1043.9 kPa
Refrigerant 134a	201.7 kPa	666.3 kPa
Ammonia	290.95 kPa	1003 kPa

Since p_{sat} (25°C) for Refrigerant 22 and ammonia are too high, these two refrigerants are not feasible. Refrigerant 134a is feasible since it satisfies both the criteria.

Example 14.7 *A Refrigerant-12 vapour compression plant producing 10 tonnes of refrigeration operates with condensing and evaporating temperatures of 35 and –10°C, respectively. A suction line heat exchanger is used to subcool the saturated liquid leaving the condenser. Saturated vapour leaving the evaporator is superheated in the suction line heat exchanger to the extent that a discharge temperature of 60°C is obtained after isentropic compression. Determine (a) the subcooling achieved in the heat exchanger, (b) the refrigerant flow rate in kg/s, (c) the cylinder dimensions of the two-cylinder compressor, if the speed is 900 rpm, stroke-to-bore ratio is 1.1, and the volumetric efficiency is 80%, (d) the COP of the plant and (e) the power required to drive the compressor in kW.*

M **[LO 14.3]**

Solution From the p-h chart of R-12, the property values at the states, as shown in Fig. 14.36.

$$h_3 = 882,\ h_2 = 1034$$

$$h_6 = 998,\ h_1 = 1008 \text{ kJ/kg}$$

$$v_1 = 0.084 \text{ m}^3\text{/kg}$$

$$h_3 - h_4 = h_1 - h_6$$

$$882 - h_4 = 1008 - 998 = 10$$

$$\therefore \quad h_4 = 872 \text{ kJ/kg}$$

$$\therefore \quad t_4 = 25°\text{C}$$

So 10°C subcooling is achieved in the heat exchanger. Refrigeration effect $= h_6 - h_5 = 998 - 872 =$ **126 kJ/kg**

$$\therefore \quad \text{Refrigerant flow rate } \frac{10 \times 14000}{126} = \mathbf{1110\ kg/h = 0.31\ kg/s}$$

$$\text{Volume flow rate} = w \cdot v_1 = 1110 \times 0.084 = 93 \text{ m}^3\text{/h}$$

$$\text{Compressor displacement} = \frac{93}{0.8} = 116 \text{ m}^3\text{/h} = 1.94 \text{ m}^3\text{/min}$$

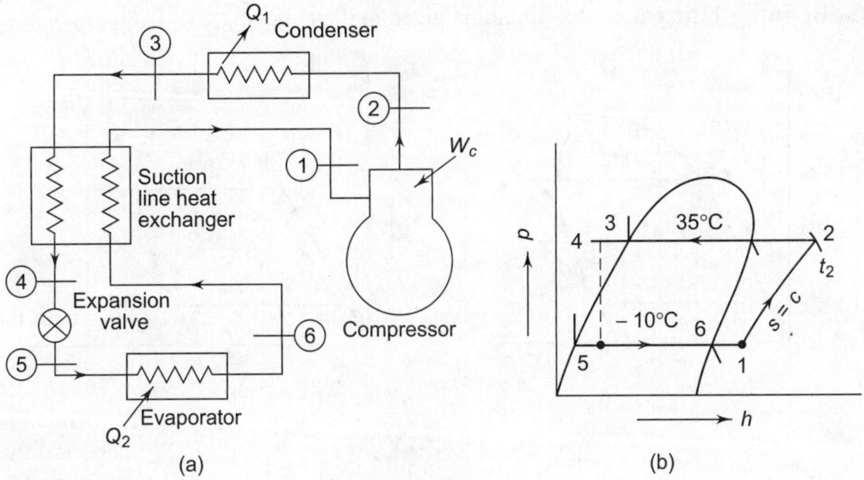

Fig. 14.36

This is equal to $\frac{\pi}{4} D^2$ LNn

where D = diameter,
L = stroke,
N = rpm and
n = number of cylinders of the compressor.

$$\frac{\pi}{4} D^2 \times 1.1D \times 900 \times 2 = 1.94 \text{ m}^3/\text{min}$$

$$D^3 = 1250 \text{ cm}^3$$

or $$D = \mathbf{10.8\ cm} \quad \text{and} \quad L = \mathbf{11.88\ cm}$$

$$\text{COP} = \frac{h_6 - h_5}{h_2 - h_1} = \frac{126}{1034 - 1008} = \mathbf{4.85}$$

Power required to drive the compressor

$$= w(h_2 - h_1) = \frac{1110 \times 26}{3600} = \mathbf{8.02\ kW}$$

Example 14.8 *A two-stage vapour compression refrigeration system with a direct contact heat exchanger (flash chamber) operates with ammonia as the refrigerant. The evaporator and condenser temperatures are –30 and 40°C, respectively. If the capacity of the plant is 30 tonnes of refrigeration, estimate the total work of compression and the COP. Had the compression been done in a single stage, what would have been the percentage increase in the work of compression? What is the percentage increase in the COP owing to the staging of the compression process?* Ⓜ [LO 14.3]

Solution Using tables for ammonia given in the appendix,

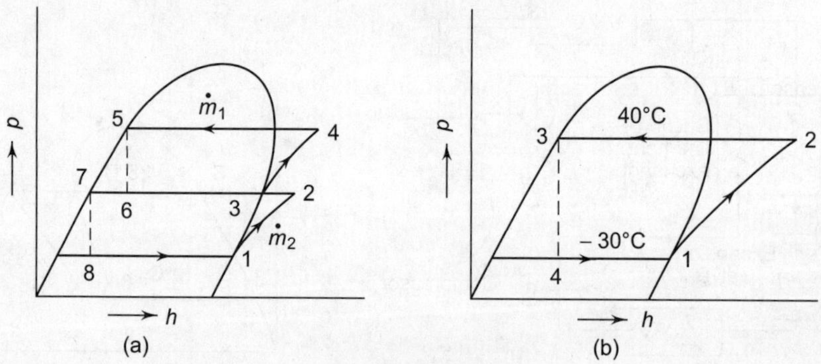

Fig. 14.37

at 40°C, $p_2 = 1554.3$ kPA

at –30°C, $p_1 = 119.5$ kPa

$\therefore$ $p_i = \sqrt{1554.3 \times 119.5} = 431$ kPa

$\therefore$ $h_1 = 1404.6 \quad s_2 = s_1 \quad h_2 = 1574.3$

$h_3 = 1443.5 \quad s_4 = s_3 \quad h_4 = 1628.1$

$h_5 = 371.7 = h_6,\ h_7 = 181.5$ hg

$$\therefore \quad \dot{m}_2 = \frac{3.89 \times 30}{1404.6 - 181.5} = \frac{116.7}{1223.1} = 0.0954 \text{ kg/s}$$

$$\dot{m}_1 = \dot{m}_2 \frac{h_2 - h_7}{h_3 - h_6} = 0.0954 \times \frac{1392.8}{1071.8} = 0.124 \text{ kg/s}$$

$$\dot{W}_c = \dot{m}_c(h_2 - h_1) + \dot{m}_1(h_4 - h_3)$$

$$= 0.0954 \times 169.7 + 0.124 \times 184.6 = 16.19 + 22.89$$

$$= 39.08 \text{ kW}$$

$$\text{COP} = \frac{30 \times 3.89}{39.08} = 2.986$$

Single stage

$h_1 = 1404.6 \qquad h_2 = 1805.1$

$h_3 = 371.7 = h_4$

$$\dot{m} = \frac{30 \times 3.89}{1404.6 - 371.7} = \frac{116.7}{1032.9} = 0.113 \text{ kg/s}$$

$$W_c = \dot{m}(h_2 - h_1) = 0.113 \times 400.5 = 45.26 \text{ kW}$$

$$\therefore \quad \text{COP} = \frac{116.7}{45.26} = 2.578$$

Increase in work of compression (for single stage)

$$= \frac{45.26 - 39.08}{39.08} \times 100 = \mathbf{15.81\%}$$

Increase in COP for two-stage compression

$$= \frac{2.986 - 2.578}{2.578} \times 100 = \mathbf{15.82\%}$$

Example 14.9 *In an aqua-ammonia absorption refrigerator system, heat is supplied to the generator by condensing steam at 0.2 MPa, 90% quality. The temperature to be maintained in the refrigerator is –10°C, and the ambient temperature is 30°C. Estimate the maximum COP of the refrigerator.*

If the actual COP is 40% of the maximum COP and the refrigeration load is 20 tonnes, what will the required steam flow rate be? Ⓜ **[LO 14.5]**

Solution At 0.2 MPa, from the steam table (Fig. 14.38)

$$t_{sat} = 120.2°C, h_{fg} = 2201.9 \text{ kJ/kg}$$

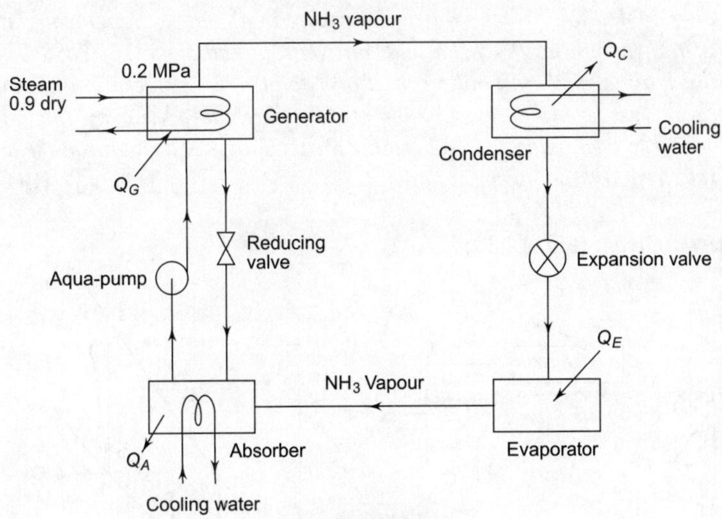

Fig. 14.38

The maximum COP of the absorption refrigeration system is given by Eq. (14.4)

$$(\text{COP})_{max} = \frac{(T_1 - T_2)T_R}{(T_2 - T_R)T_1}$$

where T_1 = generator temperature

$= 120.2 + 273 = 393.2$ K

T_2 = condenser and absorber temperature

$= 30 + 273 = 303$ K

$$T_R = \text{evaporator temperature}$$
$$= -10 + 273 = 263 \text{ K}$$

$$\therefore \qquad (\text{COP})_{\max} = \frac{(393.2 - 303) \times 263}{(303 - 263) \times 393.2} = \frac{90.2 \times 263}{40 \times 393.2} = 1.5$$

$$\therefore \qquad \text{Actual COP} = 1.5 \times 0.4 = 0.60$$

Since
$$\text{COP} = \frac{Q_E}{Q_G}$$

$$Q_G = \frac{Q_E}{\text{COP}} = \frac{20 \times 14000}{0.60 \times 3600} = 129.6 \text{ kW}$$

Heat transferred by 1 kg of steam on condensation

$$= (h_f + xh_{fg}) - h_f = 0.9 \times 2201.9 = 1981.71 \text{ kJ/kg}$$

∴ Steam flow rate required

$$= \frac{129.6}{1981.71} = \mathbf{0.0654 \text{ kg/s}}$$

Example 14.10 *In an aircraft cooling system, air enters the compressor at 0.1 MPa, 4°C, and is compressed to 0.3 MPa with an isentropic efficiency of 72%. After being cooled to 55°C at constant pressure in a heat exchanger the air then expands in a turbine to 0.1 MPa with an isentopic efficiency of 78%. The low temperature air absorbs a cooling load of 3 tonnes of refrigeration at constant pressure before re-entering the compressor which is driven by the turbine. Assuming air to be an ideal gas, determine the COP of the refrigerator, the driving power required and the air mass flow rate.* Ⓜ **[LO 14.7]**

Solution *Given*: (Fig. 14.39)

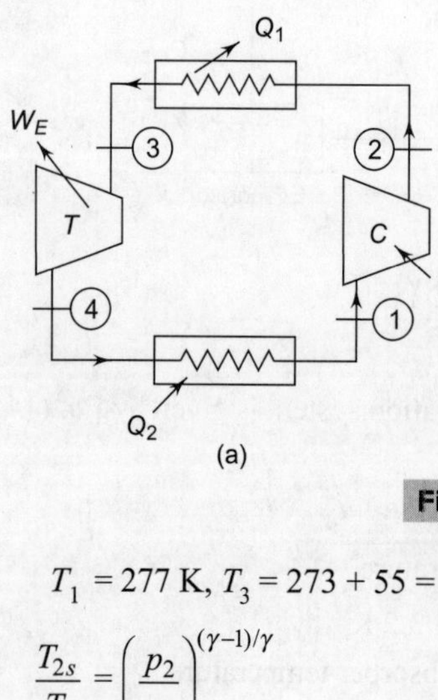

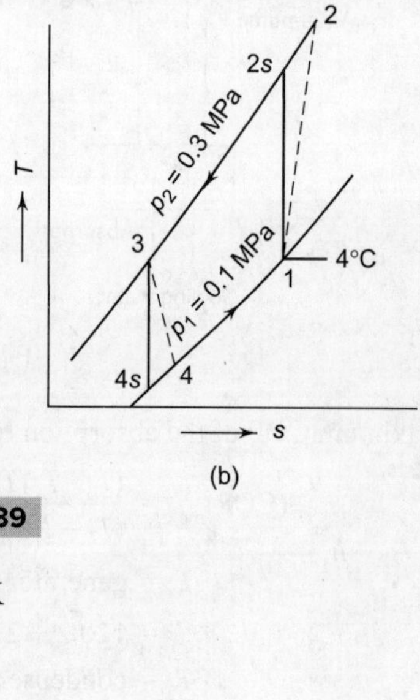

Fig. 14.39

$$T_1 = 277 \text{ K}, T_3 = 273 + 55 = 328 \text{ K}$$

$$\frac{T_{2s}}{T_1} = \left(\frac{p_2}{p_1}\right)^{(\gamma-1)/\gamma}$$

$$\therefore \quad T_{2s} = 277(3)^{0.4/1.4} = 379 \text{ K}$$

$$T_{2s} - T_1 = 102 \text{ K}$$

$$\therefore \quad T_2 - T_1 = \frac{102}{0.72} = 141.8 \text{ K}$$

$$\frac{T_{4s}}{T_3} = \left(\frac{p_2}{p_1}\right)^{(\gamma-1)/\gamma}$$

$$\therefore \quad T_{4s} = 328\,(3)^{0.4/1.4} = \frac{328}{1.368} = 240 \text{ K}$$

$$T_3 - T_{4s} = 88 \text{ K}$$

$$\therefore \quad T_3 - T_4 = 0.78 \times 88 = 68.6 \text{ K}$$

$$\therefore \quad T_4 = 259.4 \text{ K}$$

Refrigerating effect = $c_p(T_1 - T_4) = 17.6\, c_p$ kJ/kg

Net work input = $c_p[(T_2 - T_1) - (T_3 - T_4)]$

$$= c_p[141.8 - 68.6] = 73.2\, c_p \text{ kJ/kg}$$

$$\text{COP} = \frac{17.6 c_p}{73.2 c_p} = \mathbf{0.24}$$

Driving power required

$$= \frac{3 \times 14000}{0.24 \times 3600} = \mathbf{48.6 \text{ kW}}$$

Mass flow rate of air

$$= \frac{3 \times 14000}{1.005 \times 17.6} = 2374.5 \text{ kg/h} = \mathbf{0.66 \text{ kg/s}}$$

Example 14.11 *A vapour-compression heat pump system uses R-12 as the working fluid. The refrigerant enters the compressor at 2.4 bar, 0°C with a volumetric flow rate of 0.6 m³/min. Compression is adiabatic to 9 bar, 60°C and the saturated liquid exits the condenser at 9 bar. Determine (a) the power input to the compressor, (b) the heating capacity of the system, (c) the coefficient of performance and (d) the isentropic compressor efficiency.*

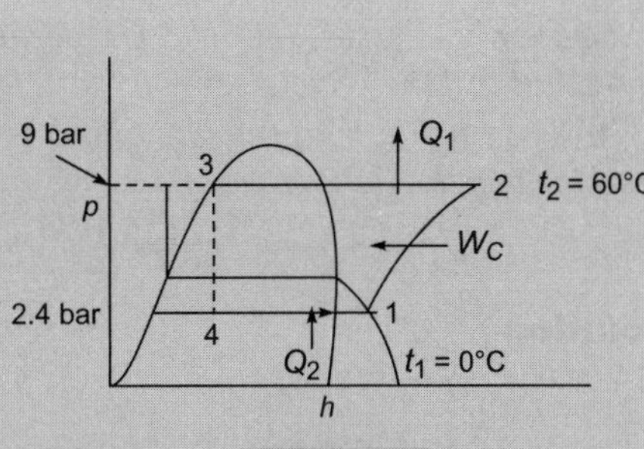

Fig. 14.40

M [LO 14.6]

Solution

At $p_1 = 2.4$ bar, $T_1 = 0$°C

$h_1 = 188.99$ kJ/kg

$s_1 = 0.7177$ kJ/kg K

$v_1 = 0.0703$ m³/kg

At $p_2 = 9$ bar, $T_2 = 60$°C

$h_2 = 219.37$ kJ/kg

$$h_2s = 213.27 \text{ kJ/kg}$$

At $p_2 = 9$ bar, sat. liquid

$$h_3 = 71.93 \text{ kJ/kg} = h_4$$

$$\dot{m} = \frac{A_1V_1}{v_1} = \frac{0.6}{60}\frac{\text{m}^3}{\text{s}} \times \frac{1}{0.0703}\text{kg m}^3$$

$$= 0.1422 \text{ kg/s}$$

(a) Power input $\dot{W}_c = \dot{m}(h_2 - h_1)$

$$= 0.1422 \,(219.37 - 188.99)$$

$$= \mathbf{4.32 \text{ kW}}$$

(b) Heating capacity, $\dot{Q}_1 = \dot{m}\,(h_2 - h_3)$

$$= 0.1422 \,(219.37 - 71.93)$$

$$= 20.97 \text{ kW} = \mathbf{5.963 \text{ tonnes}}$$

(c) $\text{COP} = \frac{\dot{Q}_1}{\dot{W}_c} = \frac{20.97}{4.32} = \mathbf{4.854}$

(d) $\eta_{\text{is}} = \frac{h_{2s} - h_1}{h_2 - h_1} = \frac{213.27 - 188.99}{219.37 - 188.99} = 0.799$ or **79.9%**

Example 14.12 *In an air-refrigeration system working on reversed Brayton cycle, the temperature of air at entrance to compressor (pressure ratio = 4, efficiency = 0.8) is 275 K and the inlet pressure is 1 bar. The pressure loss in the cooler is 0.1 bar and in the cold chamber it is 0.08 bar. The temperature of air at turbine (efficiency = 0.85) inlet is 310 K. Estimate the pressure ratio for the turbine and the COP of the cycle.*

M [LO 14.7]

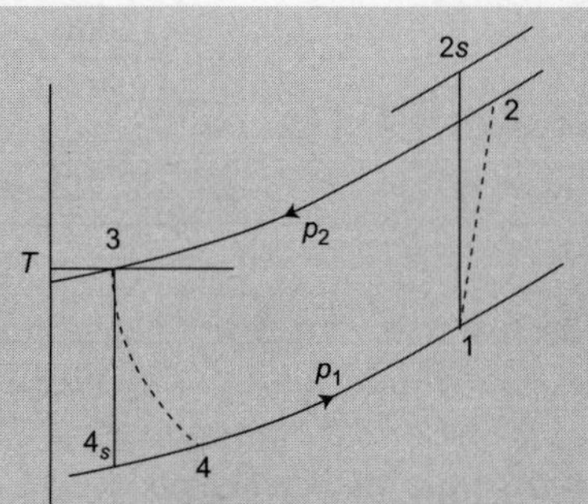

Fig. 14.41

Solution

$$T_1 = 275 \text{ K},$$

$$T_3 = 310 \text{ K}$$

$$\frac{p_2}{p_1} = 4, p_1 = 1 \text{ bar}$$

$$T_{2s} = T_1\,(p_2/p_1)^{\frac{\gamma-1}{v}}$$

$$= 275 \times 40.286$$

$$= 408.81 \text{ K}$$

$$T_2 = T_1 + (T_{2s} - T_1)\,\eta_c$$

$$= 442.26 \text{ K}$$

$$p_3 = p_2 - \text{pressure loss in the cooler} = 4 - 0.1 = 3.9 \text{ bar}$$

$$p_4 = p_1 + \text{pressure loss in the cold chamber} = 1 + 0.08$$

$$= 1.08 \text{ bar}$$

$\therefore$ Pressure ratio for the turbine $= \dfrac{3.9}{1.08} = \mathbf{3.611}$

$$\frac{T_3}{T_{4s}} = (p_3/p_{4s})^{0.286} = (3.611)^{0.286} = 1.4437$$

$$T_{4s} = 214.72 \text{ K}$$

$$T_4 = T_3 - (T_3 - T_{4s})\eta_t = 310 - (310 - 214.72) \times 0.85$$

$$= 229.015 \text{ K}$$

$$\therefore \quad \text{COP} = \frac{T_1 - T_4}{(T_2 - T_3) - (T_1 - T_4)} = \frac{275 - 229.015}{(442.26 - 310) - (275 - 229.015)}$$

$$= \mathbf{0.533}$$

SUMMARY

In the reversed Carnot cycle, the working fluid, the refrigerant, flows through compressor, condenser, expander and evaporator in a refrigerator or heat pump. In the actual vapour compression refrigeration cycle, the expander is replaced by a throttle value or capillary tube. Dry compression is always preferred. Subcooling of condensate and superheating of vapour after the evaporator improve the COP. The capacity of a refrigeration plant is expressed in tonnes. One tone of refrigeration, TR, is approximately equal to the rate of heat removal of 3.5 kW. To reduce the work of compression and improve the COP, multi-stage compression with intercooling may be adopted.

To counteract the effect of global warming and greenhouse effect, the CFCs (Chlorofluoro carbons) are not recommended for use in domestic refrigerators, and the chlorine-free refrigerant R-134a is used. When the evaporator temperature is too low and the condenser temperature is too high, two cycles with two different refrigerants are often used in series and such a coupled cycle makes a cascade refrigeration cycle.

The ongoing pursuit of improved refrigerants has led to the development of modern 'green refrigerant' solutions, designed to be safe, non-toxic, and have minimal or zero ozone depletion and global warming potential.

The absorption refrigeration system does not require a compressor and uses a refrigerant that is alternately absorbed and liberated from the absorbent. The compressor is replaced by an absorber-generator assembly. In aqua-ammonia absorption system, ammonia is the refrigerant and water is the absorbent, whereas in lithium bromide absorption system, water is the refrigerant and lithium bromide is the absorbent. The COP of absorption systems is usually low, but they are made in big sizes, particularly when waste heat is available in plenty.

A heat pump can be used for year-round air conditioning. In industry, a heat pump can be effectively utilised where both heating and cooling of process streams are required. In gas (or air) cycle refrigeration cycle, an expander is used instead of throttle value for expansion from the heater to the cooler. It is used in aircrafts to maintain the proper pressure and temperature inside the cabin for the comfort of passengers.

REVIEW QUESTIONS

(E) **14.1** What is refrigeration? How is (a) ice and (b) dry ice used for the purpose of refrigeration? **[LO 14.1]**

(M) **14.2** Explain the vapour compression cycle with the help of flow, *T-s* and *p-h* diagrams. Can this cycle be reversible? If not, why? **[LO 14.3]**

(E) **14.3** What do you understand by dry and wet compression? Which is preferred and why? **[LO 14.3]**

(E) **14.4** What is refrigerating effect? **[LO 14.3]**

(E) **14.5** What is a tonne of refrigeration? **[LO 14.3]**

(M) **14.6** Explan the effect of superheat and subcooling on the vapour compression cycle. **[LO 14.3]**

(M) **14.7** How does the actual vapour compression cycle differ from the ideal one? **[LO 14.3]**

(M) **14.8** What is a suction line heat exchanger? When is it used? **[LO 14.5]**

(E) **14.9** What are the expansion devices used in a vapour compression plant? When are they used? **[LO 14.3]**

(M) **14.10** What are the different types of compressors used in vapour compression plants and what are their applications? **[LO 14.3]**

(E) **14.11** What is a multi-stage vapour compression plant? When is it used? **[LO 14.3]**

(E) **14.12** What is a flash chamber? What is its advantage? **[LO 14.3]**

(E) **14.13** What are the most widely used refrigerants? **[LO 14.3]**

(M) **14.14** How are refrigerants numbered? **[LO 14.3]**

(E) **14.15** Why is the use of halogenated hydrocarbons as refrigerants now discouraged? **[LO 14.3]**

(E) **14.16** What are the effects of CFCs on the environment? How do they affect the ozone layer? **[LO 14.3]**

(M) **14.17** What is ODP? Why has R-22 less ODP than R-12? **[LO 14.3]**

(M) **14.18** What are the parameters to be considered in the selection of a refrigerant? **[LO 14.3]**

(E) **14.19** What are 'green refrigerants'? Give examples of a few such refrigerants. **[LO 14.4]**

(M) **14.20** What do you understand by cascade refrigeration system? Explain it with the help of flow and *T-s* diagrams. **[LO 14.3]**

(M) **14.21** Evaluate ammonia as a refrigerant. **[LO 14.3]**

(M) **14.22** What is an absorption refrigeration cycle? How does it differ from a vapour compression cycle? **[LO 14.5]**

(E) **14.23** How is the refrigerant liberated from the absorbent in an aqua-ammonia system? **[LO 14.5]**

(M) **14.24** What are the functions of the analyser and the rectifier? **[LO 14.5]**

(M) **14.25** What is the advantage of using a suction line heat exchanger in an absorption refrigeration system? **[LO 14.5]**

(E) **14.26** In a lithium bromide-water absorption system, which is the refrigerant? What is its limitation? **[LO 14.5]**

(M) **14.27** Derive the expression for the maximum COP of an absorption refrigeration system. **[LO 14.5]**

(E) **14.28** How does a heat pump upgrade low grade reject heat? **[LO 14.6]**

(M) **14.29** How can a heat pump be used for (a) space heating (b) year-round airconditioning? **[LO 14.6]**

(M) **14.30** How is a reversed Brayton cycle used for refrigeration? **[LO 14.7]**

(E) **14.31** Why is the COP of a gas cycle refrigeration system low? **[LO 14.7]**

(E) **14.32** Why is gas cycle refrigeration preferred in an aircraft? **[LO 14.7]**

(D) **14.33** What is the principle of the Linde–Hampson system for liquefaction of air? **[LO 14.8]**

(D) **14.34** Derive the expressions of liquid yield and the minimum work requirement in a Linde–Hampson system. **[LO 14.8]**

(M) **14.35** How does Claude cycle differ from a Linde–Hampson cycle in the context of the liquefaction of air? **[LO 14.8]**

(M) **14.36** With the help of flow and *p-h* diagrams, explain how dry ice is produced. **[LO 14.9]**

PROBLEMS WITH HINTS

(M) **14.1** In a vapour compression cycle with R-12 as the working fluid, the cold region is at 20°C and the hot region at 40°C. Saturated vapour enters the compressor at 20°C and saturated liquid leaves the condenser at 40°C. The mass flow rate of the refrigerant is 0.008 kg/s. Determine (a) the compressor power in kW, (b) the refrigeration capacity in tour, (c) the COP and (d) the COP of a Carnot refrigerator operating between 20 and 40°C. **[LO 14.3]**

Hint:

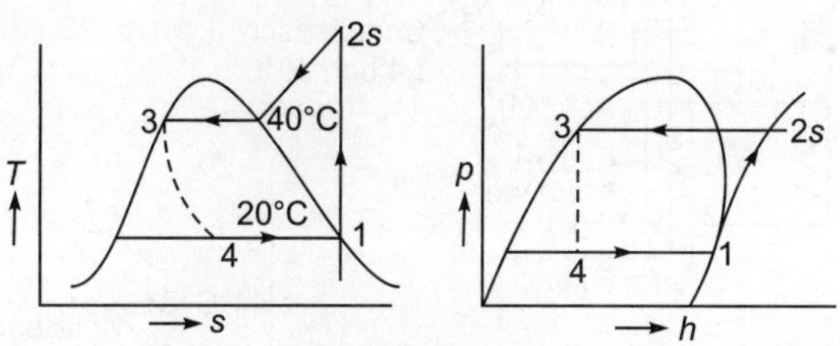

Fig. 14.42

Using Table A-2 for properties of R-12,

$$h_1 = 95.78 \text{ kJ/kg}, s_1 = 0.6884 \text{ kJ/kgK}$$

$$P_{sat} \text{ at } 40°C = 9.6065 \text{ bar}$$

$$h_{25} = 205.1 \text{ kJ/kg}$$

$$h_3 = 74.59 \text{ kJ/kg} = h_4$$

(a) $\dot{W} = \dot{m}(h_{25} - h_1) = 0.008\ (205.1 - 195.78)$

$= \mathbf{0.075\ kW}$

(b) $\dot{Q} = \dot{m}(h_1 - h_4) = 0.008\ (195.78 - 74.59) \times \dfrac{60}{211}$

$= \mathbf{0.276\ ton}$

(c) $COP = \dfrac{h_1 - h_4}{h_2 - h_1} = \mathbf{13.0}$

(d) COP Carnot $= \dfrac{T_c}{T_H - T_C} = \mathbf{14.65}$

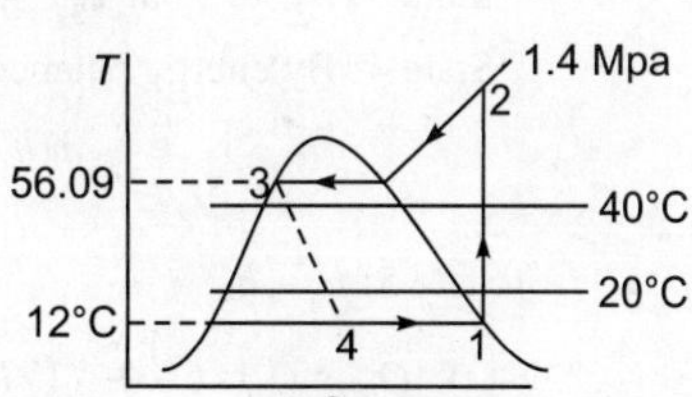

Fig. 14.43

(M) **14.2** Allowing temperature differences between the refrigerant and the warm and cold regions, saturated vapour enters the compressor at 12°C and saturated liquid leaves the condenser at 1.4 MPa. Determine for this modified cycle (a) the compressor power, (b) the refrigeration capacity and (c) the COP. **[LO 14.3]**

Hint:

From Table A.2 of Appendix A,

$$h_1 = 192.56 \text{ kJ/kg}$$
$$s_1 = 0.6913 \text{ kJ/kgK}$$
$$P_2 = 1.4 \text{ MPa, Interpolating, } h_2 = 212.71 \text{ kJ/kg}, h_3 = 91.46 \text{ kJ/kg}$$

(a) $\dot{W}_c$ = **0.161 kW**, (b) R. Capacity = **0.23 T**
(c) COP = **5.02**

(M) **14.3** An ideal vapour compression refrigeration cycle is modified to include a counter low heat exchanger (Fig. 14.44). Refrigerant 12 leaves the evaporates as saturated vapour at 1.4 bar and is heated at constant pressure to 20°C before entering the compressor. Following isentropic compression to 12 bar, the refrigerant passes through the condenser exiting at 44°C, 12 bar. The liquid then passes through the heat exchanger, entering the expansion valve at 12 bar. If the mass flow of refrigerant is 6 kg/min, determine (a) the refrigeration capacity, (b) the compressor power input and (c) the COP. **[LO 14.3]**

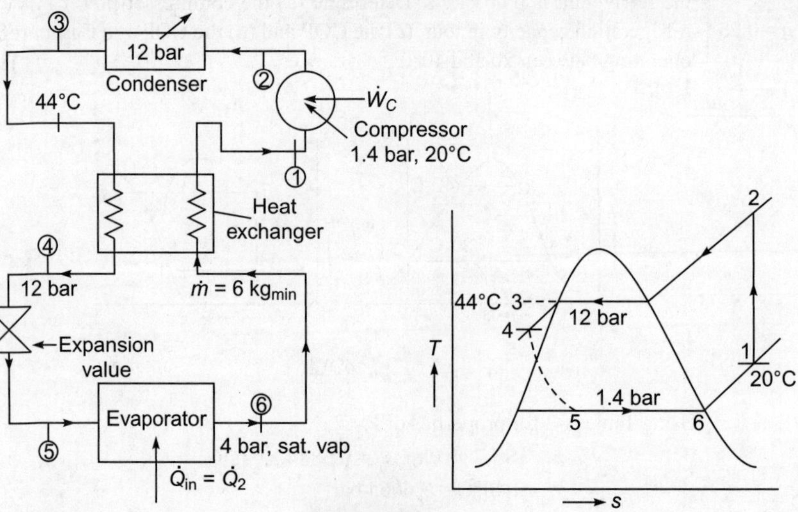

Fig. 14.44

Hint:

State 6: 1.4 bar, sat *vap*.

$$h_6 = 177.87 \text{ kJ/kg}$$

State 1: $P_1 = 1.4$ bar, $T_1 = 20°\text{C} \rightarrow h_1 = 203.23$ kJ/kg, $s_1 = 0.5035$ kJ/kgK.

State 2: $P_2 = 12$ bar, $s_2 = s_1 \rightarrow h_2 = 249.46$ kJ/kg

State 3: $P_3 = 12$ bar $T_3 = 44°\text{C}$, comp. liquid, $h_3 = (h_f)_{44°\text{C}} = 78.68 \dfrac{\text{kJ}}{\text{kg}}$

State 4: By energy balance of heat exchanger,

$$\dot{m}(h_1 - h_2) = \dot{m}(h_3 - h_4)$$

$$\therefore \quad h_4 = 53.32 \text{ kJ/kg}$$

State 5: $h_4 = h_5$

(a) $\dot{Q}_{\text{in}} = \dot{Q}_2 = 6 \dfrac{\text{kg}}{\text{min}} (177.87 - 53.32) \dfrac{\text{kJ}}{\text{kg}} \times \dfrac{1 \text{ ton}}{211 \dfrac{\text{kg}}{\text{min}}}$

= **3.54 Tonnes**

(b) $\dot{W}_c = \dot{m}(h_2 - h_1) = \mathbf{4.623\ kW}$

(c) COP = **2.69**

Ⓜ **14.4** A vapour compressor heat pump with a heating capacity of 500 kJ/min is driven by a power cycle with the thermal efficiency of 25%. For the heat pump, R-12 is compressed from saturated vapour at –10°C to the condenser pressure of 10 bar. The isotropic efficiency of the compressor is 80%. Liquid enters the expansion valve at 9.6 bar, 30°C. For the power cycle 80% of the heat rejected is transferred to the heated space.

(a) Determine the power input to the compressor. (b) Evaluate the ratio of the total rate that heat is delivered to the heated space to the rate of heat input to the power cycle. Discuss **[LO 14.3, 14.6]**

Hint:

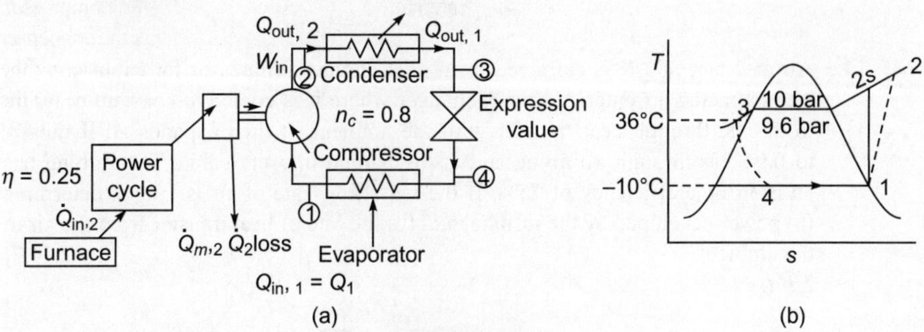

Fig. 14.45

State 1: $T_1 = -10°C$, saturated vapour

$$h_1 = 183.19 \text{ kJ/kg}$$

$$s_1 = 0.7019 \text{ kJ/kgK}$$

State 2: For isentropic compression, $P_2 = 10$ bar.

$$s_2 = s_1 \rightarrow h_{25} = 210.1 \text{ kJ/kg}$$

$$\eta_c = \frac{h_{25} - h_1}{h_2 - h_1} = 0.8 \rightarrow h_2 = 216.83 \text{ kJ/kg}$$

State 3: $P_3 = 9.6$ bar, $T_3 = 36°C$ comp. liq., $h_3 = (h_f)_{36°C} = 70.55$ kJ/kg

State 4: $h_4 = h_3 = 70.55$ kJ/kg

(a) $\dot{m} = \dfrac{\dot{Q}_{out}}{h_2 - h_3} = 0.0570$ kg/s

$$\dot{W}_c = \dot{m}(h_2 - h_1) = \mathbf{1.917\ kW}$$

(b) For the power cycle, $\eta = 0.25$ $\dot{W} = \mathbf{1.917\ kW}$

(c) $Q_{in}, 2 = \dfrac{1.917}{0.25} = 7.668$ kW

$$\dot{Q}_{rej} \text{ from power cycle} = 7.668 - 1.917 = 5.751 \text{ kW.}$$

$$Q_{out}, 2 = 0.8\ Q_{rej} = 4.601 \text{ kW}$$

$$\therefore \quad \frac{Q_{out,1} + Q_{out,2}}{Q_{in,2}} = \frac{500/60 + 4.601}{7.668} = \mathbf{1.687}$$

Discuss: The engine-driven heat pump delivers more energy to the heated space than could be obtained by burning the fuel directly.

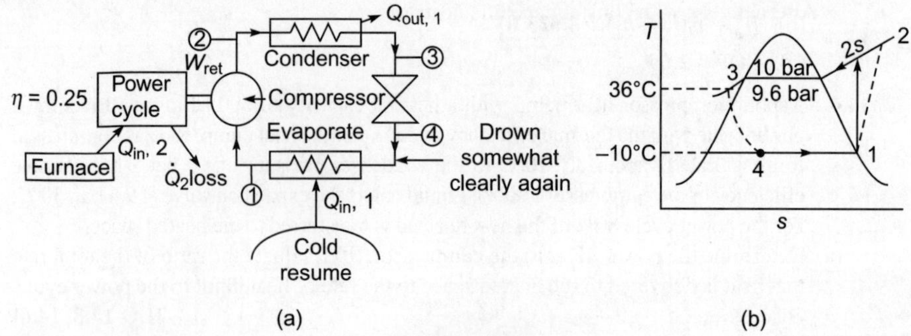

Fig. 14.46

14.5 Air at 2 bar, 380 K is extracted from a jet engine compressor for cabin cooling. The extracted air enters a heat exchanger where it is cooled at constant pressure to 320 K through heat transfer with the ambient. It then expands adiabatically to 0.95 bar through a turbine and is discharged into the cabin. The turbine has an isentropic efficiency of 75%. If the mass flow rate of air is 1 kg/s, determine (a) power developed by the turbine, and (b) the rate of heat transfer from the air to the ambient. **[LO 14.7]**

Hint:

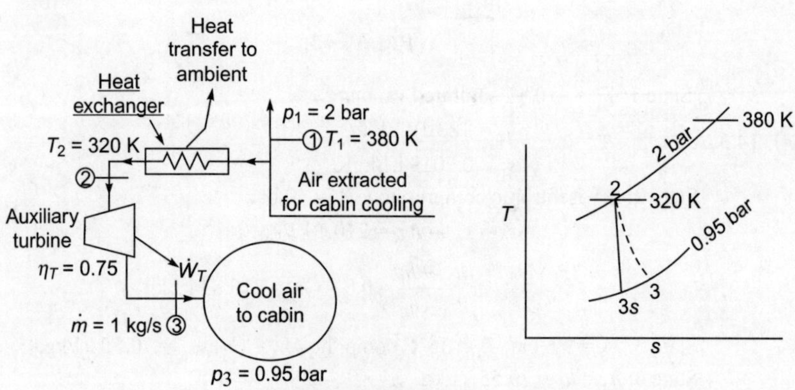

Fig. 14.47

(a) From Table A.2 of Appendix A

$$h_2 = 320.29 \text{ kJ/kg}, Pr_2 = 1.7375$$

$$s_2 = s_{35} \rightarrow pr_{35} = Pr_2\left(\frac{P_3}{P_2}\right) = 0.82531$$

$$h_{35} = 258.67 \text{ kJ/kg, (a) } \dot{W}_T = \dot{m}\eta_T(h_2 - h_{35})$$

$$\text{or, } \dot{W}_T = 1 \times 0.75 \times (320.29 - 258.67) = \mathbf{46.2 \text{ kW}}$$

(b) From Table A.2 of Appendix A

$$h_1 = 380.77 \text{ kJ/s} \quad \textbf{Ans.}$$

$$\dot{Q}_{cv} = m(h_2 - h_1) = 1(320.29 - 380.77) = \mathbf{60.48 \text{ kW}}$$

Alternatively, with constant specific heats of air

$$\frac{T_2}{T_{35}} = \left(\frac{P_2}{P_3}\right)^{\frac{\gamma-1}{\gamma}} = \frac{320}{T_{35}} = \left(\frac{2}{0.95}\right)^{0.286} \rightarrow T_{35} = 259.63 \text{ K.}$$

$$T_2 - T_3 = \eta_T (T_2 - T_{35})$$

$$\therefore \quad T_3 = 320 - 0.75\,(320 - 259.63) = 274.73 \text{ K}$$

$$\therefore \quad \dot{W}_I = 1.005\,(320 - 274.73) = \mathbf{45.53\ kW}$$

$$\dot{Q}_{cv} = \dot{m}(h_1 - h_2) = 1 \times 1.005 \times (380 - 320) = \mathbf{60.3\ kW}$$

PROBLEMS

(E) **14.1** A refrigerator using R-134*a* operates on an ideal vapour compression cycle between 0.12 and 0.7 MPa. The mass flow of refrigerant is 0.05 kg/s. Determine (a) the rate of heat removal from the refrigerated space, (b) the power input to the compressor, (c) the heat rejection to the environment and (d) the COP. **[LO 14.3]**

(M) **14.2** A Refrigerant-12 vapour compression cycle has a refrigeration load of 3 tonnes. The evaporator and condenser temperatures are –20°C and 40°C, respectively. Find (a) the refrigerant flow rate in kg/s, (b) the volume flow rate handled by the compressor in m^3/s, (c) the work input to the compressor in kW, (d) the heat rejected in the condenser in kW and (e) the isentropic discharge temperature.

If there is 5°C of superheating of vapour before it enters the compressor, and 5°C deg. subcooling of liquid before it flows through the expansion valve, determine the above quantities. **[LO 14.3]**

(M) **14.3** A 5 tonne R-12 plant maintains a cold store at –15°C. The refrigerant flow rate is 0.133 kg/s. The vapour leaves the evaporator with 5°C superheat. Cooling water is available in plenty at 25°C. A suction line heat exchanger subcools the refrigerant before throttling. Find (a) the compressor discharge temperature, (b) the COP, (c) the amount of subcooling in °C and (d) the cylinder dimensions of the compressor, if the speed is 900 rpm, stroke-to-bore ratio is 1.2 and volumetric efficiency is 95%.

Allow approximately 5°C temperature difference in the evaporator and condenser. **[LO 14.3]**

(M) **14.4** A vapour compression refrigeration system uses R-12 and operates between pressure limits of 0.745 and 0.15 MPa. The vapour entering the compressor has a temperature of –10°C and the liquid leaving the condenser is at 28°C. A refrigerating load of 2 kW is required. Determine the COP and the swept volume of the compressor if it has a volumetric efficiency of 76% and runs at 600 rpm. **[LO 14.3]**

(M) **14.5** A food-freezing system requires 20 tonnes of refrigeration at an evaporator temperature of – 35°C and a condenser temperature of 25°C. The refrigerant, R-12, is subcooled 4°C before entering the expansion valve, and the vapour is superheated 5°C before leaving the evaporator. A six-cylinder single-acting compressor with stroke equal to bore is to be used, operating at 1500 rpm. Determine (a) the refrigerating effect, (b) the refrigerant flow rate, (c) the theoretical piston displacement per sec, (d) the theoretical power required in kW, (e) the COP, (f) the heat removed in the condenser and (g) the bore and stroke of the compressor. **[LO 14.3]**

(M) **14.6** An R-12 vapour compression refrigeration system is operating at a condenser pressure of 9.6 bar and an evaporator pressure of 2.19 bar. Its refrigeration capacity is 15 tonnes. The values of enthalpy at the inlet and outlet of the evaporator are 64.6 and 195.7 kJ/kg, respectively. The specific volume at inlet to the reciprocating compressor is 0.082 m3/kg. The index of compression for the compressor is 1.13. Determine: (a) the power input in kW required for the compressor, and (b) the COP. Take 1 tonne of refrigeration as equivalent to heat removal at the rate of 3.517 kW. **[LO 14.3]**

(E) **14.7** A refrigeration plant produces 0.139 kg/s of the ice at –5°C from water at 30°C. If the power required to drive the plant is 22 kW, determine the capacity of the ice plant in tonnes and the actual COP (c_p of ice = 2.1 kJ/kg K). **[LO 14.3]**

(M) **14.8** An air conditioning unit using R-12 as a refrigerant is mounted in the window of a room. During steady operation 1.5 kW of heat is transferred from the air in the room to the evaporator coils of R-12. If this air is at 22°C and the temperature of R-12 in the evaporator is 15°C, determine (a) the refrigerant flow rate, and (b) the minimum power required to drive the compressor if the outside air is at 43°C and the temperature of the refrigerant during condensation is 50°C. **[LO 14.3]**

(E) **14.9** In a solar energy operated aqua-ammonia absorption refrigeration system, water is cooled at the rate of 10 kg/s from 38 to 13°C. If the incident solar energy is 640 W/m^2 and the COP of the system is 0.32, estimate the area of the solar collector needed. **[LO 14.5]**

(E) **14.10** A gas refrigerating system using air as a refrigerant is to work between –12 and 27°C using an ideal reversed Brayton cycle of pressure ratio 5 and minimum pressure 1 atm, and to maintain a load of 10 tonnes. Find (a) the COP, (b) the air flow rate in kg/s, (c) the volume flow rate entering the compressor in m^3/s and (d) the maximum and minimum temperatures of the cycle. **[LO 14.7]**

(E) **14.11** An open cycle (Brayton) aircraft cabin cooler expands air at 27°C through a turbine which is 30% efficient from 2 to 1 atm. The cabin temperature is not to exceed 24°C. Estimate the mass flow rate of air required (kg/s) for each tonne of cooling. **[LO 14.7]**

(E) **14.12** Determine the ideal COP of an absorption refrigerating system in which the heating, cooling and refrigeration take place at 197°C, 17°C and –3°C, respectively. **[LO 14.5]**

(M) **14.13** A heat pump is to use an R-12 cycle to operate between outdoor air at –1°C and air in a domestic heating system at 40°C. The temperature difference in the evaporator and the condenser is 8°C. The compressor efficiency is 80%, and the compression begins with saturated vapour. The expansion begins with saturated liquid. The combined efficiency of the motor and belt drive is 75%. If the required heat supply to the warm air is 43.6 kW, what will be the electrical load in kW? **[LO 14.3, 14.6]**

(E) **14.14** An ideal (Carnot) refrigeration system operates between the temperature limits of – 30 and 25°C. Find the ideal COP and the power required from an external source to absorb 3.89 kW at high temperature. **[LO 14.2]**

(M) **14.15** A heat pump that operates on the ideal vapour compression cycle with R-134*a* is used to heat a house and maintain it at 20°C, using underground water at 10°C as the heat source. The house is losing heat at a rate of 75 MJ/h. The evaporator and condenser pressures are 320 and 800 kPa, respectively. Determine the power input to the heat pump and the electric power saved by using a heat pump instead of a resistance heater. **[LO 14.3, 14.6]**

(D) **14.16** An ammonia-absorption system has an evaporator temperature of –12°C and a condenser temperature of 50°C. The generator temperature is 150°C. In this cycle, 0.42 kJ is transferred to the ammonia in the evaporator for each kJ transferred to the ammonia solution in the generator from the high temperature source.

It is desired to compare the performance of this cycle with the performance of a similar vapour compression cycle. For this, it is assumed that a reservoir is available at 150°C, and that heat is transferred from this reservoir to a reversible engine which rejects heat to the surroundings at 25°C. This work is then used to drive an ideal vapour compression system with ammonia as the refrigerant. Compare the amount of refrigeration that can be achieved per kJ from the high temperature source in this case with the 0.42 kJ that can be achieved in the absorption system. **[LO 14.3, 14.5]**

(M) **14.17** An R-12 plant is to cool milk from 30 to 1°C involving a refrigeration capacity of 10 tonnes. Cooling water for the condenser is available at 25 and 5°C rise in its temperature is allowable. Determine the suitable condensing and evaporating temperatures, providing a minimum of 5°C differential, and calculate the theoretical power required in kW and the cooling water requirement in kg/s. Also, find the percentage of flash gas at the end of the throttling. Assume a 2°C subcooling in the liquid refrigerant leaving the condenser. **[LO 14.7]**

(E) **14.18** The following data pertain to an air cycle refrigeration system for an aircraft.

Capacity 5 tonnes

Cabin air inlet temperature 15°C and outlet temperature 25°C

Pressure ratio across the compressor 5

The aircraft is flying at 0.278 km/s where the ambient conditions are 0°C and 80 kPa. Find the COP and the cooling effectiveness of the heat exchanger. The cabin is at 0.1 MPa, and the cooling turbine powers the circulating fans. **[LO 14.7]**

(E) **14.19** A water cooler supplies chilled water at 7°C when water is supplied to it at 27°C at a rate of 0.7 litres/min., while the power consumed amounts to 200 watts. Compare the COP of this refrigeration plant with that of the ideal refrigeration cycle for a similar situation. **[LO 14.2]**

(M) **14.20** A refrigerating plant of 8 tonnes capacity has an evaporation temperature of –8°C and condenser temperature of 30°C. The refrigerant, R-12, is subcooled 5°C before entering the expansion valve and the vapour is superheated 6°C before leaving the evaporator coil. The compression of the refrigerant is isentropic. If there is a suction pressure drop of 20 kPa through the valves, and discharge pressure drop of 10 kPa through the valves, determine the COP of the plant, theoretical piston displacement per second and the heat removal rate in the condenser. **[LO 14.3]**

(M) **14.21** An ultra-low-temperature freezer system employs a coupling of two vapour compression cycles of R-12 and R-13, as shown in Fig. 14.48. The states and properties of both cycles are shown on the *T-s* plot. Determine the ratio of the circulation rates of the two refrigerants, w_1/w_2 and the overall COP. How does this COP compare with the Carnot COP operating between 42 and –70°C? **[LO 14.3]**

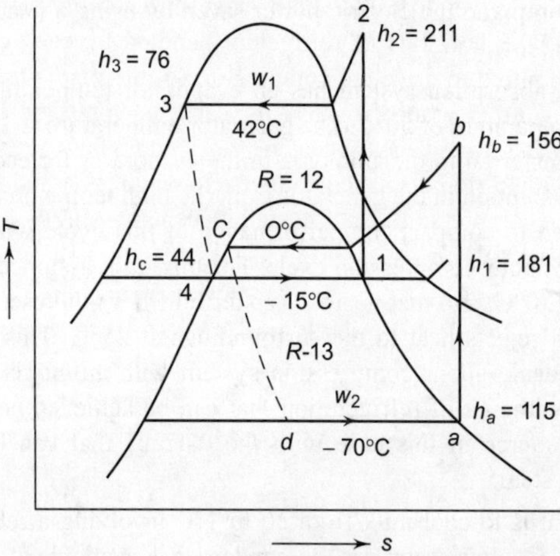

Fig. 14.48

(M) **14.22** Derive an expression for the COP of an ideal gas refrigeration cycle with a regenerative heat exchanger. Express the result in terms of the minimum gas temperature during heat rejection (T_h), maximum gas temperature during heat absorption (T_1) and pressure ratio for the cycle (p_2/p_1). **[LO 14.7]**

(M) **14.23** Large quantities of electrical power can be transmitted with relatively little loss when the transmission cable is cooled to a superconducting temperature. A regenerated gas refrigeration cycle operating with helium is used to maintain an electrical cable at 15 K. If the pressure ratio is 10 and heat is rejected directly to the atmosphere at 300 K, determine the COP and the performance ratio with respect to the Carnot cycle. **[LO 14.7]**

(M) **14.24** A 100 tonne refrigerating plant using R-12 has a condensing temp. of 35°C and an evaporating temperature of 5°C. Calculate the power requirement of the compressor in kW, the volume flow rate of compressor and the compressor displacement volume if the volumetric efficiency is 75% and the mechanical efficiency is 80%.

If a liquid suction heat exchanger is installed in the above plant which subcools the condensed refrigerant to 30°C, what would be the refrigeration capacity of the plant and the power required by the compressor? **[LO 14.3]**

(M) **14.25** A heat pump installation is proposed for a home heating unit with an output rated at 30 kW. The evaporator temperature is 10°C and the condenser pressure is 0.5 bar. Using an ideal vapour compression cycle, estimate the power required to drive the compressor if steam/water mixture is used as the working fluid, the COP and the mass flow rate of the fluid. Assume saturated vapour at compressor inlet and saturated liquid at condenser outlet. **[LO 14.3, 14.6]**

(M) **14.26** A 100 tonne low temperature R-12 system is to operate on a two-stage vapour compression refrigeration cycle with a flash chamber, with the refrigerant evaporating at – 40°C, an intermediate pressure of 2.1912 bar, and condensation at 30°C. Saturated vapour enters both the compressors and saturated liquid enters each expansion valve. Consider both stages of compression to be isentropic. Determine (a) the flow rate of refrigerant handled by each compressor, (b) the total power required to drive the compressor, (c) the piston displacement of each compressor, if the clearance is 2.5% for each machine and (d) the COP of the system. (e) What would have been the refrigerant flow rate, the total work of compression, the piston displacement in each compressor and the compressor and the COP, if the compression had occurred in a single stage? **[LO 14.3]**

(M) **14.27** A vapour compression plant uses R–134a as refrigerant. The evaporator temperature is – 10°C. The condenser pressure is 7.675 bar and there is no subcooling of the condensate. The flash chamber pressure is 4.139 bar and the dry saturated vapour bled off from the flash chamber is mixed with the refrigerant from the LP compressor before entering the HP compressor. The liquid from the flash chamber is throttled before entering the evaporator. Assuming isentropic compression, calculate (a) COP of the plant and (b) mass flow of refrigerant in the evaporator when the power input to the plant is 100 kW. **[LO 14.3]**

(M) **14.28** The working fluid in a heat pump installation is ammonia. The ammonia after evaporation to a dry saturated state at 2°C is compressed to a pressure of 12.38 bar at which it is cooled and condensed to a saturated liquid state. It then passes through a throttle valve and returns to the evaporator. Calculate the COP assuming that the isentropic efficiency of the compressor is 0.85. **[LO 14.3, 14.6]**

ANSWERS TO PROBLEMS

14.1 (a) 7.35 kW,
(b) 1.85 kW,
(c) 9.20 kW,
(d) 3.97

14.3 (a) 66°C,
(b) 4.1
(c) 125°C,
(d) 104.5 mm, 125 mm

14.4 4.15, 243 cm^3

14.6 (a) 11.57 kW,
(b) 4.56

14.12 5.16

14.15 2.27 kW, 18.56 kW

14.22 $COP = \dfrac{T_1}{T_h r_p^{(\gamma-1)/\gamma} - T_1}$

14.23 0.02, 0.38

14.25 8.0 kW, 3.77, 0.001012 kg/s

14.26 (a) 2.464, 3.387 kg/s,
(b) 123 kW,
(c) 0.6274, 0.314 m^3/s,
(d) 2.86,
(e) 3.349 kg/s, 144.54 kW, 1.0236 m^3/s, 2.433

14.27 (a) 5.93,
(b) 3.38 kg/s

14.28 7.88

15

Psychrometrics

Learning Objectives

LO 15.1 Describe psychrometrics and its importance

LO 15.2 Discuss different terms related to dry air mixed with water vapour

LO 15.3 Illustrate psychrometric chart

LO 15.4 Demonstrate psychrometric processes on a psychrometric chart

> **LO 15.1**
> Describe psychrometrics and its importance

The properties of the mixtures of ideal gases were presented in Chapter 10. The name 'psychrometrics' is given to the study of the properties of air-water vapour mixtures. Atmospheric air is considered to be a mixture of dry air and water vapour. The control of moisture (or water vapour) content in the atmosphere is essential for the satisfactory operation of many processes involving *hygroscopic* materials like paper and textiles, and is important in comfort air conditioning.

15.1 PROPERTIES OF ATMOSPHERIC AIR

> **LO 15.2**
> Discuss different terms related to dry air mixed with water vapour

Dry air is a mechanical mixture of the gases: oxygen, nitrogen, carbon dioxide, hydrogen, argon, neon, krypton, helium, ozone and xenon. However, oxygen and nitrogen make up the major part of the combination. Dry air is considered to consist of 21% oxygen and 79% nitrogen by volume, and 23% oxygen and 77% nitrogen by mass. Completely dry air does not exist in nature. Water vapour in varying amounts is diffused through it. If p_a and p_w are the partial pressures of dry air and water vapour, respectively, then by Dalton's law of partial pressures

$$p_a + p_w = p$$

where p is the atmospheric pressure.

$\therefore$ Mole-fraction of dry air, x_a

$$= \frac{p_a}{p} = p_a \qquad (\because p = 1 \text{ atm.})$$

and mole reaction of water vapour, x_w

$$= \frac{p_w}{p} = p_w$$

Since p_w is very small, the saturation temperature of water vapour at p_w is less than atmospheric temperature, t_{atm} (Fig. 15.1). So the water vapour in air exists in the *superheated state*, and air is said to be *unsaturated*.

Relative humidity (RH ϕ) is defined as the ratio of partial pressure of water vapour, p_w, in a mixture to the saturation pressure, p_s, of pure water, at the same temperature of the mixture (Fig. 15.1).

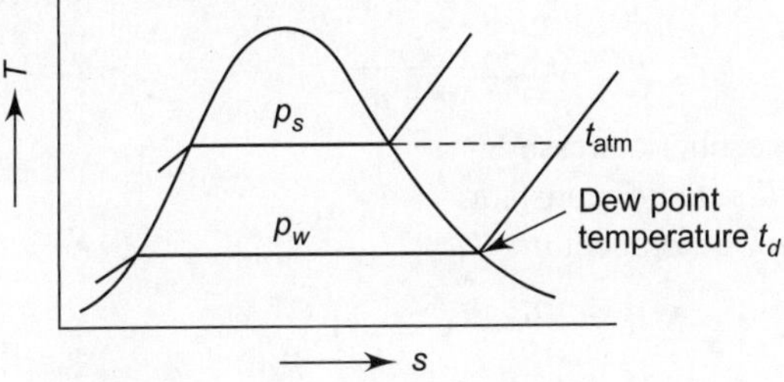

Fig. 15.1 States of water vapour in mixture

$$\therefore \qquad \text{RH } (\phi) = \frac{p_w}{p_s}$$

If water is injected into unsaturated air in a container, water will evaporate, which will increase the moisture content of the air, and p_w will increase. This will continue till air becomes saturated at that temperature, and there will be no more evaporation of water. For saturated air, the relative humidity is 100%. Assuming water vapour as an ideal gas

$$p_w V = m_w R_{H_2O} T = n_w \bar{R} T$$

and

$$p_s V = m_s R_{H_2O} T = n_s \bar{R} T$$

where V is the volume and T the temperature of air, the subscripts w and s indicating the unsaturated and saturated states of air, respectively.

$$\phi = \frac{p_w}{p_s} = \frac{m_w}{m_s}$$

$$= \frac{\text{Mass of water vapour in a given volume of air at temperature } T}{\text{Mass of water vapour when the same volume of air is saturated at temperature } T}$$

$$= \frac{n_w}{n_s} = \frac{x_w}{x_s}$$

Specific humidity or *humidity ratio*, W, is defined as the mass of water vapour (or moisture) per unit mass of dry air in a mixture of air and water vapour.

If G = mass of dry air

m = mass of water vapour

$$W = \frac{m}{G}$$

Specific humidity is the maximum when air is saturated at temperature T, or

$$W_{max} = W_s = \frac{m_s}{G}$$

If dry air and water vapour behave as ideal gases

$$p_w V = m R_w T$$

$$p_a V = G R_a T$$

$$\therefore \quad W = \frac{m}{G} = \frac{p_w}{p_a} \cdot \frac{R_a}{R_w} = \frac{p_w}{p - p_w} \frac{8.3143/28.96}{8.3143/18}$$

$$\therefore \quad W = 0.622 \frac{p_w}{p - p_w} \tag{15.1}$$

where p is the atmospheric pressure.

If p_w is constant, W remains constant.

If air is saturated at temperature T

$$W = W_s = 0.622 \frac{p_s}{p - p_s}$$

where p_s is the saturation pressure of water vapour at temperature T.

The degree of saturation, μ, is the ratio of the actual specific humidity and the saturated specific humidity, both at the same temperature T.

$$\therefore \quad \mu = \frac{W}{W_s} = \frac{0.622 \dfrac{p_w}{p - p_w}}{0.622 \dfrac{p_s}{p - p_s}} = \frac{p_w}{p_s} \cdot \frac{p - p_s}{p - p_w}$$

If $\phi = \dfrac{p_w}{p_s} = 0, p_w = 0, x_w = 0, W = 0$, i.e. for dry air,

$$\mu = 0$$

If $\phi = 100\%, p_w = p_s, W = W_s, \mu = 1$

Therefore, μ varies between 0 and 1.

If a mixture of air and superheated (or unsaturated) water vapour is cooled at constant pressure, the partial pressure of each constituent remains constant until the water vapour reaches the saturated state. Further cooling causes condensation. The temperature at which water vapour starts condensing is called the *dew point temperature*, t_{dp}, of the mixture (Fig. 15.1). It is equal to the saturation temperature at the partial pressure, p_w, of the water vapour in the mixture.

Dry bulb temperature (DBT) is the temperature recorded by the thermometer with a dry bulb.

Wet bulb temperature (WBT) is the temperature recorded by a thermometer when the bulb is enveloped by a cotton wick saturated with water. As the air stream flows past it, some water evaporates, taking the latent heat from the water-soaked wick, thus decreasing its temperature. Energy is then transferred to the wick from the air. When equilibrium condition is reached, there is a balance between energy removed from the water film by evaporation and energy supplied to the wick by heat transfer, and the temperature recorded is the wet bulb temperature.

A *psychrometer* is an instrument which measures both the dry bulb and the wet bulb temperatures of air. Figure 15.2 shows a continuous psychrometer with a fan for drawing air over the thermometer bulbs. A sling psychrometer has two thermometers mounted on a frame with a handle. The handle is rotated so that there is good air motion. The wet bulb temperature is the *lowest temperature* recorded by the moistened bulb.

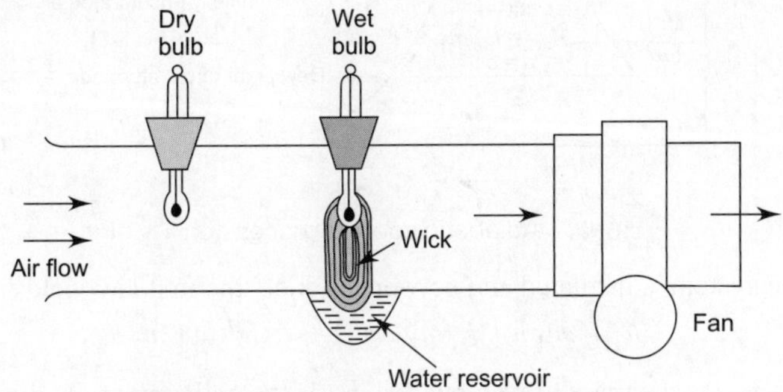

Fig. 15.2 Dry and wet bulb temperatures

At any DBT, the greater the depression (difference) of the WBT reading below the DBT, the smaller is the amount of water vapour held in the mixture.

When unsaturated air flows over a long sheet of water (Fig. 15.3) in an insulated chamber, the water evaporates, and the specific humidity of the air increases. Both the air and water are cooled as evaporation takes places. The process continues until the energy transferred from the air to the water is equal to the energy required to vapourise the water. When this point is reached, thermal equilibrium exists with respect to the water, air and water vapour, and consequently the air is saturated. The equilibrium temperature is called the *adiabatic saturation temperature* or the *thermodynamic wet bulb temperature*. The make-up water is introduced at this temperature to make the water level constant.

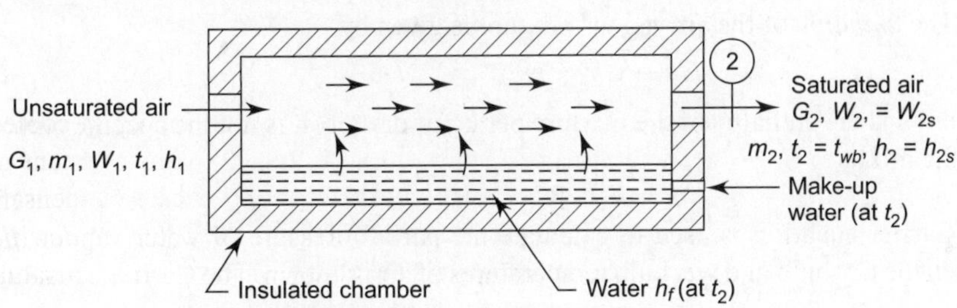

Fig. 15.3 Adiabatic saturation process

The 'adiabatic' cooling process is shown in Fig. 15.4 for the vapour in the air-vapour mixture. Although the total pressure of the mixture is constant, the partial pressure of the vapour increases, and in the saturated state corresponds to the adiabatic saturation temperature. The vapour is initially at the DBT t_{db1} and is cooled adiabatically to the DBT t_{db2} which is equal to the adiabatic saturation temperature t_{wb2}. The adiabatic saturation temperature and the wet bulb temperature are taken to be equal for all practical purposes. The WBT lies between the DBT and DPT.

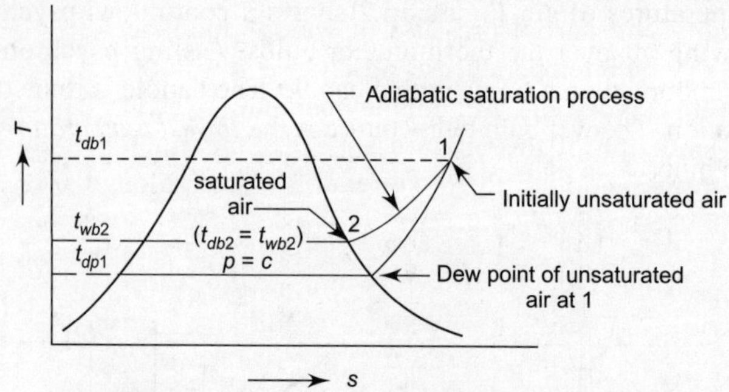

Fig. 15.4 Adiabatic cooling process on T-s plot

Since the system is insulated and no work is done, the first law yields

$$Gh_{a1} + m_1 h_{w1} + (m_2 - m_1) h_{f2} = Gh_{a2} + m_2 h_{w2}$$

where $(m_2 - m_1)$ is the mass of water added, h_{f2} is the enthalpy of the liquid water at $t_2 (= t_{wb2})$, h_a is the specific enthalpy of dry air and h_w is the specific enthalpy of water vapour in air. Dividing by G, and since $h_{g2} = h_{g2}$

$$h_{a1} + W_1 h_{w1} + (W_2 - W_1) h_{f2} = h_{a2} + W_2 h_{g2} \tag{15.2}$$

Solving for W_1

$$W_1 = \frac{(h_{a2} - h_{a1}) + W_2 (h_{g2} - h_{f2})}{h_{w1} - h_{f2}}$$

$$= \frac{C_{pa}(T_2 - T_1) + W_2 \cdot h_{fg2}}{h_{w1} - h_{f2}} \tag{15.3}$$

where $W_2 = \dfrac{m_2}{G} = \dfrac{m_s}{G} = 0.622 \dfrac{p_s}{p - p_s}$

The *enthalpy* of the air-vapour mixture is given by

$$Gh = Gh_a + mh_w$$

where h is the enthalpy of the mixture per kg of dry air (it is not the specific enthalpy of the mixture)

$$\therefore \quad h = h_a + Wh_w \tag{15.4}$$

Carrier equation is used to calculate the partial pressure of water vapour in air when the dry bulb and wet bulb temperatures of air is known. The Carrier equation is:

$$p_w = p'_w - \frac{(p - p'_w)(t_{db} - t_{wb}) \times 1.8}{2800 - 1.3(1.8t_{db} + 32)}$$

Where, p_w is the partial pressure of water vapour in air, p_w' is the saturation pressure of water at wet bulb temperature, p is the pressure of the air and water vapour mixture at

t_{db}. t_{db} and t_{wb} are the dry bulb and wet bulb temperatures respectively. In this equation, temperatures are in °C and all pressures are in any suitable but consistent unit.

15.2 PSYCHROMETRIC CHART

LO 15.3 Illustrate psychrometric chart

The psychrometric chart (Fig. 15.5) is a graphical plot with specific humidity and partial pressure of water vapour as ordinates, and dry bulb temperature as abscissa. The volume of the mixture (m^3/kg dry air), wet bulb temperature, relative humidity and enthalpy of the mixture appear as parameters. Any two of these properties fix the condition of the mixture. The chart is plotted for one barometric pressure, usually 760 mm Hg.

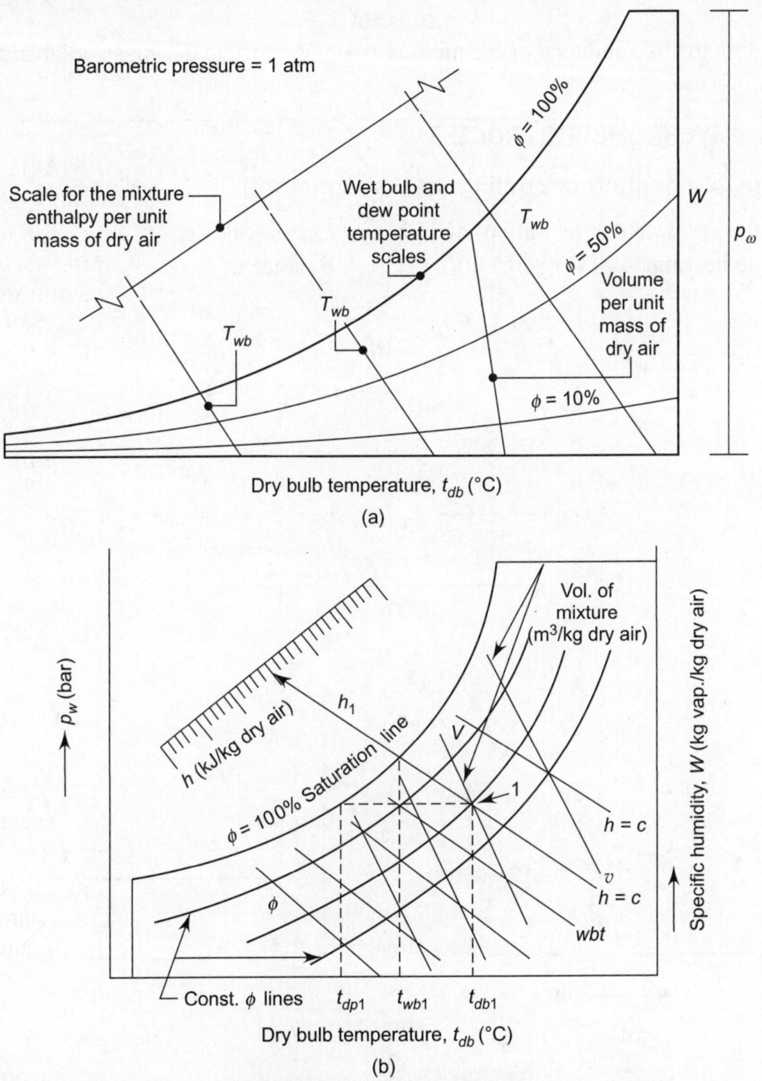

Fig. 15.5 Psychrometric chart

The constant WBT line represents the adiabatic saturation process. It also coincides with the constant enthalpy line. To show this, let us consider the energy balance for the adiabatic saturation process (Eq. (15.2)).

$$h_{a1} + W_1 h_{w1} + (W_2 - W_1)\, h_{f2} = ha_2 + W_2 h_{w2}$$

Since $h_a + Wh_w = h$ kJ/kg dry air (Eq. 15.4)

$$h_1 - W_1 h_{f2} = h_2 - W_2 h_{f2}$$

where subscript 2 refers to the saturation state, and subscript 1 denotes any state along the adiabatic saturation path. Therefore,

$$h - Wh_{f2} = \text{constant}$$

Since Wh_{f2} is small compared to h (of the order of 1 or 2%)

$$h = \text{constant}$$

indicating that the enthalpy of the mixture remains constant during an adiabatic saturation process.

15.3 PSYCHROMETRIC PROCESS

LO 15.4 Demonstrate psychrometric processes on a psychrometric chart

1. Sensible heating or cooling (at W= Constant)

Only the dry bulb temperature of air changes. Let us consider sensible heating of air [Figs. 15.6(a), (b), (c)]. Balance of

Dry air $G_1 = G_2$

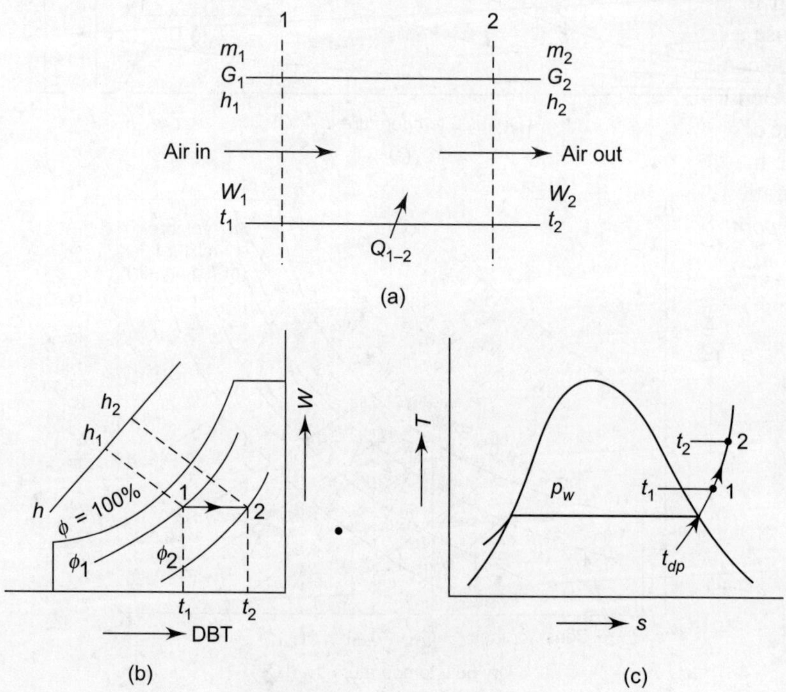

Fig. 15.6 Sensible heating

Moisture $m_1 = m_2$

$$G_1 W_1 = G_2 W_2$$

Energy $G_1 h_1 + Q_{1-2} = G_2 h_2$

$$Q_{1-2} = G(h_2 - h_1)$$

or

$$Q_{1-2} = G\,[h_{a2} + W_2\,h_{w2} - (h_{a1} + W_1\,h_{w1})]$$

$$= G\,[1.005\,(t_2 - t_1) + W_2\,[h_{g2} + 1.88\,(t_2 - t_{dp2})] - W_1\,[h_{g1} + 1.88\,(t_1 - t_{dp1})]\text{ kJ}$$

where c_p of water vapour in a superheated state has been assumed to be 1.88 kJ/kg °C.

Since

$$W_1 = W_2,\ h_{g2} = h_{g1}$$

$$t_{dp2} = t_{dp1}$$

$$Q_{1-2} = G\,[1.005\,(t_2 - t_1) + 1.88\,W\,(t_2 - t_1)]$$

$$= G\,(1.005 + 1.88\,W)\,(t_2 - t_1)$$

Similar equations can be obtained for sensible cooling at constant humidity ratio.

2. Cooling and dehumidification When the humidity ratio of air decreases, air is said to be dehumidified, and when it increases, air is humidified. Air may be cooled and dehumidified (i) by placing the evaporator coil across the air flow (Fig. 15.7b), (ii) by circulating chilled water (or brine) in a tube placed across the air flow (Fig. 15.7c), or (iii) by spraying chilled water to air in the form of fine mist (Fig. 15.7d) to expose a large surface area. The temperature of the cooling surface or the spray water must be below the dew point at state 1 (Fig. 15.7a). If the cooling surface or the spray shower is of large magnitude, the air may come out at the saturation state, 2s, known as the *apparatus dew point* (ADP) (Fig. 15.8).

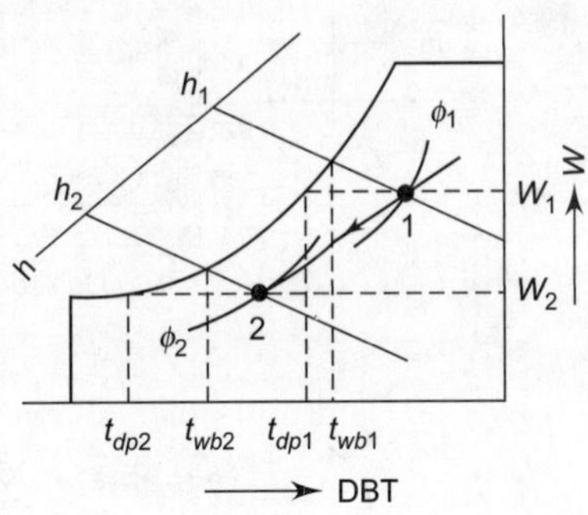

(a)

$$G_1 = G_2 = G$$

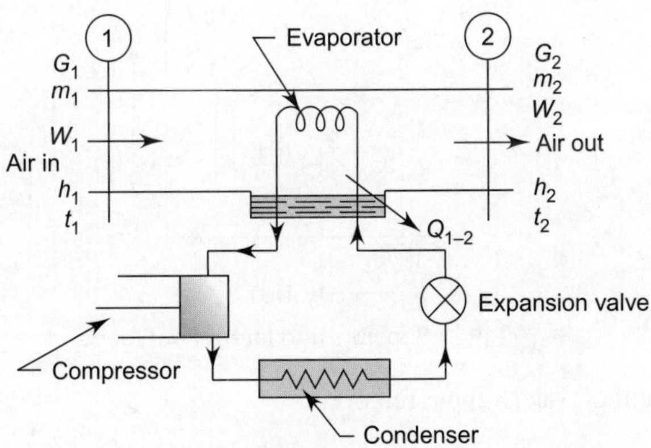

(b)

Fig. 15.7

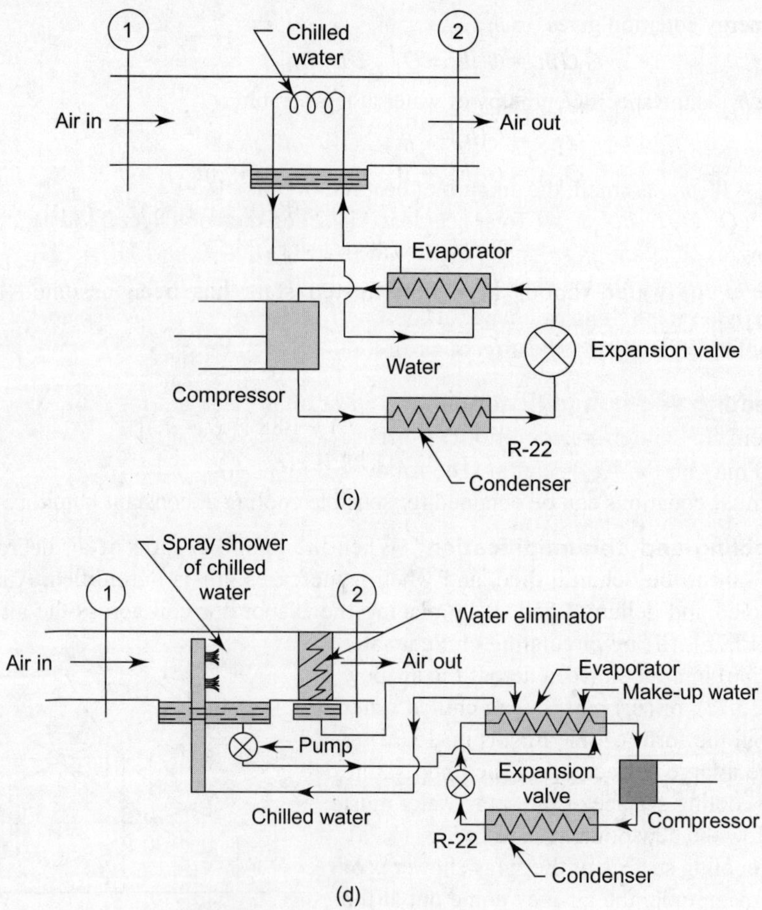

Fig. 15.7 Cooling and dehumidification

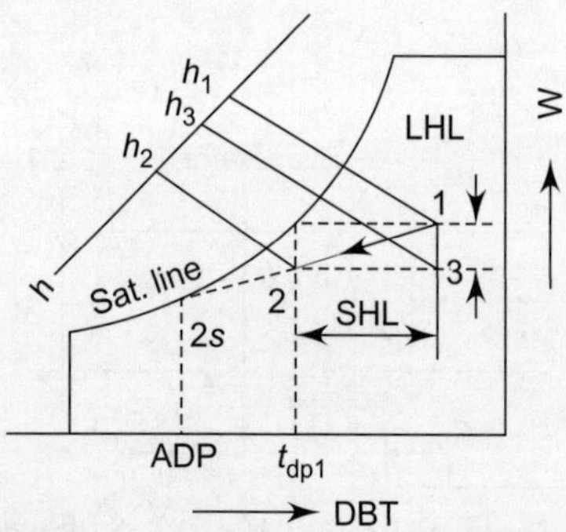

Fig. 15.8 Sensible and latent heat loads

If L is the amount of water vapour removed

$$m_1 = m_2 + L$$

or

$$L = G\,(W_1 - W_2)$$

The energy equation gives

$$G_1h_1 = G_2h_2 + Q_{1-2} + L \cdot h_{f2}$$

where h_{f2} is the specific enthalpy of water at temperature t_2.

$$\therefore \quad Q_{1-2} = G[(h_1 - h_2) - (W_1 - W_2)h_{f2}]$$

If $(W_1 - W_2)h_{f2}$ is small, the amount of heat removed is

$$Q_{1-2} = G(h_1 - h_2) = \text{Total heat load (THL) on the cooling coil (kJ/h).}$$

Now, as shown in Fig. 15.8, $h_1 - h_2 = (h_1 - h_3) + (h_3 - h_2) = \text{LHL} + \text{SHL} = \text{THL}$ where $h_1 - h_3$ = enthalpy change at constant DBT, known as *latent heat load* (*LHL*), and $(h_3 - h_2)$ is the enthalpy change at constant W, known as *sensible heat load* (*SHL*).

Cooling and dehumidification of air is common in summer air conditioning.

3. Heating and humidification The addition of heat and moisture to air is a problem for winter air conditioning. Figure 15.9 shows this process. The water added may be liquid or vapour. The following equations apply:

$$G_1 = G_2 = G$$

$$m_1 + L = m_2$$

$$\therefore \quad L = m_2 - m_1 = G(W_2 - W_1)$$

$$G_1h_1 + L \cdot h_f + Q_{1-2} = G_2h_2$$

$$Q_{1-2} = G(h_2 - h_1) - G(W_2 - W_1)h_f$$

$$= G[(h_2 - h_1) - (W_2 - W_1)h_f]$$

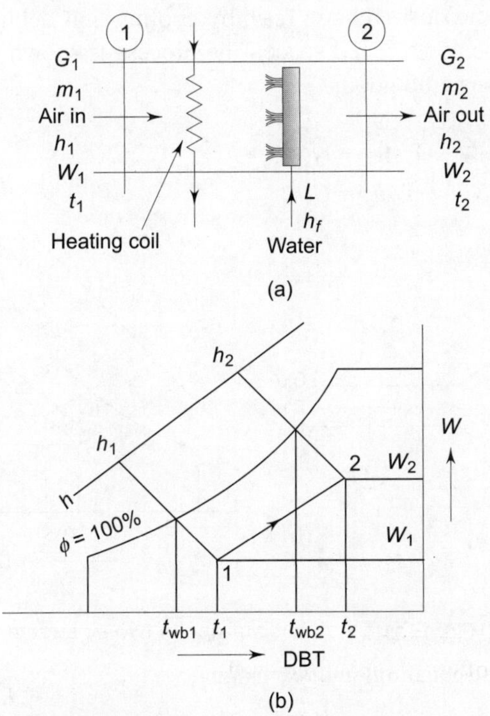

Fig. 15.9 Heating and humidification

Figure 15.10 shows first dehumidification and then heating.

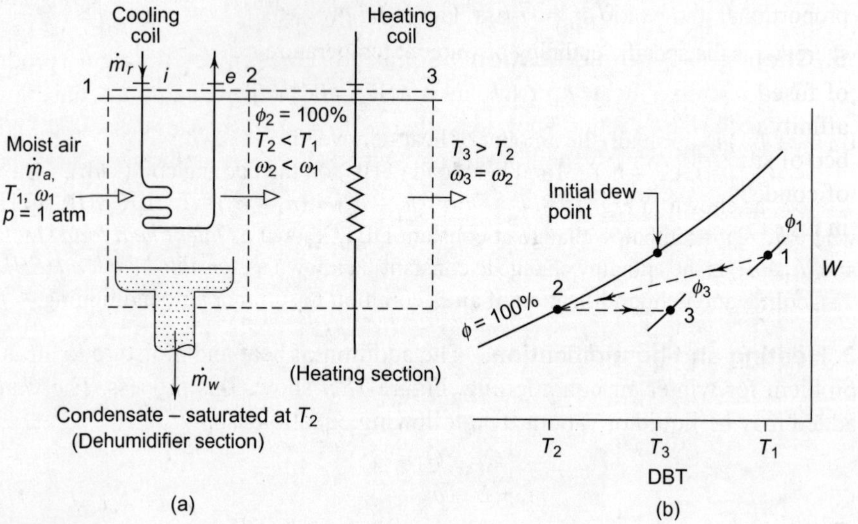

Fig. 15.10 Dehumidification with heating: (a) Equipment schematic, (b) Psychrometric chart representation

4. Adiabatic mixing of two streams This is a common problem in air conditioning, where ventilation air and some room air are mixed prior to processing it to the desired state (say, by cooling and dehumidification), and supplying it to the conditioned space. The process is shown in Fig. 15.11. The following equations hold good:

$$G_1 + G_2 = G_3$$
$$G_1W_1 + G_2W_2 = G_3W_3$$
$$G_1h_1 + G_2h_2 = G_3h_3$$

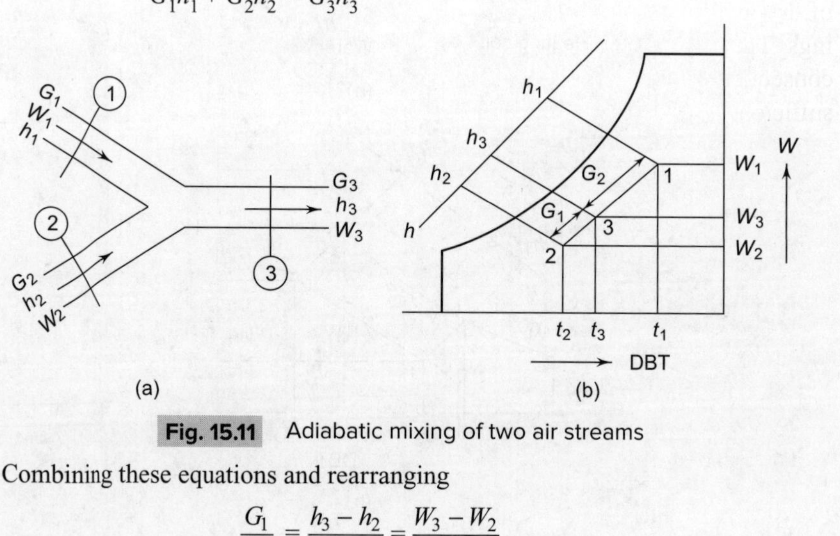

Fig. 15.11 Adiabatic mixing of two air streams

Combining these equations and rearranging

$$\frac{G_1}{G_2} = \frac{h_3 - h_2}{h_1 - h_3} = \frac{W_3 - W_2}{W_1 - W_3}$$

The points 1, 2 and 3 fall in a straight line, and the division of the line is inversely proportional to the ratio of the mass flow rates.

5. Chemical dehumidification Some substances like silica gel (product of fused sodium silicate and sulphuric acid), and activated alumina have great affinity with water vapour. They are called *adsorbents*. When air passes through a bed of silica gel, water vapour molecules get adsorbed on its surface. Latent heat of condensation is released. So the DBT of air increases. The process is shown in Fig. 15.12.

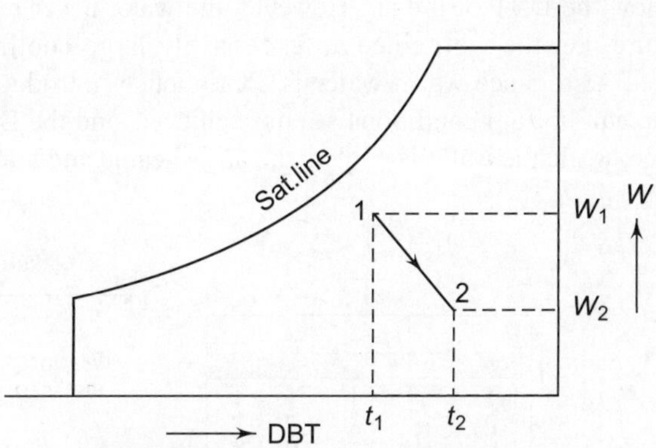

Fig. 15.12 Chemical dehumidification

6. Adiabatic evaporative cooling A large quantity of water is constantly circulated through a spray chamber. The air-vapour mixture is passed through the spray and, in doing so, evaporates some of the circulating water. The air may leave at a certain humidity ratio or in a saturated state (Fig. 15.13). The increase in specific humidity is equal to the quantity of water evaporated per unit mass of dry air. No heat transfer takes place between the chamber and the surroundings. Therefore, the energy required for evaporation is supplied by the air, and consequently, the DBT is lowered. After the process has been in operation for a sufficient length of time, the circulating water approaches the WBT of air.

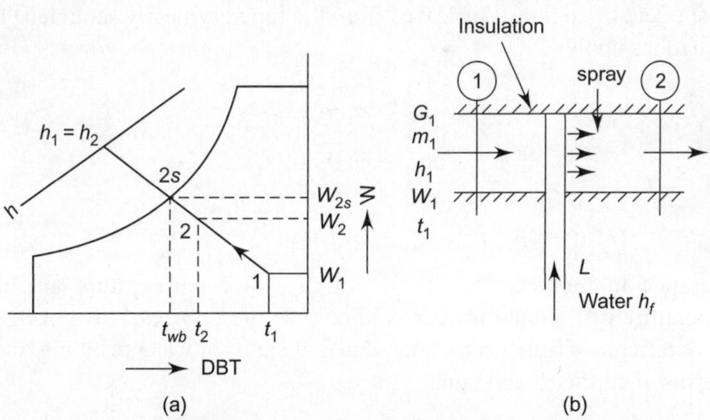

Fig. 15.13 Adiabatic evaporative cooling

$$G_1 = G_2 = G$$
$$G_1 W_1 + L = G_2 W_2$$
$$\therefore \quad L = G(W_2 - W_1)$$
$$G_1 h_1 + L h_f = G_2 \cdot h_2$$
$$\therefore \quad G(h_1 - h_2) + G(W_2 - W_1) h_f = 0$$
$$\therefore \quad h_1 - W_1 h_f = h_2 - W_2 h_f$$

The *cooling tower* utilises the phenomenon of evaporative cooling to cool warm water below the DBT of the air. However, the water never reaches the minimum temperature, i.e. the wbt, since an excessively large cooling tower would then be required. Also, since warm water is continuously introduced to the tower (Fig. 15.14), the equilibrium conditions are not achieved, and the DBT of the air is increased. Hence, while the water is cooled, the air is heated and humidified.

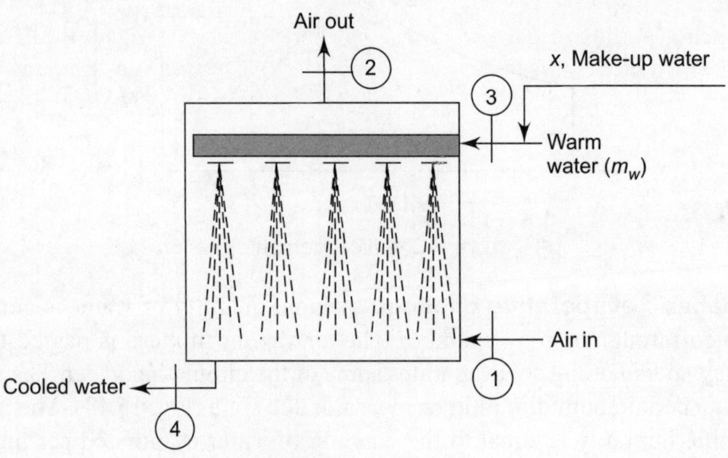

Fig. 15.14 Cooling tower

The warm water is introduced at the top of the tower in the form of spray to expose a large surface area for evaporation to take place. The more the water evaporates, the more is the effect of cooling. Air leaves from the top very nearly saturated. The following equations apply:

$$G_1 = G_2 = G$$
$$G_1 W_1 + m_{w3} = G_2 W_2 + m_{w4}$$
$$\therefore \quad m_{w3} - m_{w4} = G(W_2 - W_1)$$
$$G_1 h_1 + m_{w3} h_{w3} = G_2 h_2 + m_{w4} h_{w4}$$
$$\therefore \quad G(h_1 - h_2) + m_{w3} h_{w3} = m_{w4} h_{w4}$$

The difference in temperature of the cooled-water temperature and the wet bulb temperature of the entering air is known as the *approach*. The *range* is the temperature difference between the inlet and exit states of water. Cooling towers are rated in terms of approach and range.

If x is the make-up water supplied, then (Fig. 15.14)

$$x = G(W_2 - W_1)$$

and $$m_{w3} = m_{w4} = m_w$$

By energy balance,

$$G_1 h_1 + m_{w3}\, h_{w2} + x h_{w3} = G_2 h_2 + m_{w3}\, h_{w4}$$

$$\therefore \quad m_w (h_{w3} - h_{w4}) = G(h_2 - h_1) - G(W_2 - W_1) h_w$$

$$\therefore \quad h_{w3} - h_{w4} = \frac{G}{m_w} [(h_2 - h_1) - (W_2 - W_1) h_w] \tag{15.5}$$

$$\therefore \quad \text{Range} = t_{w3} - t_{w4} = \frac{G}{m_w c_{pw}} [(h_2 - h_1) - (W_2 - W_1) h_w] \tag{15.6}$$

where c_{pw} is the specific heat of water and h_w is the enthalpy of make-up water.

$$\text{Approach} = t_{w3} - t_{wb1} \tag{15.7}$$

The active portion of the tower in which energy exchange occurs is filled with a packing which breaks up the flow and exposes large surface area of water in contact with the air.

The schematic layout of a cooling tower is shown in Fig. 15.15.

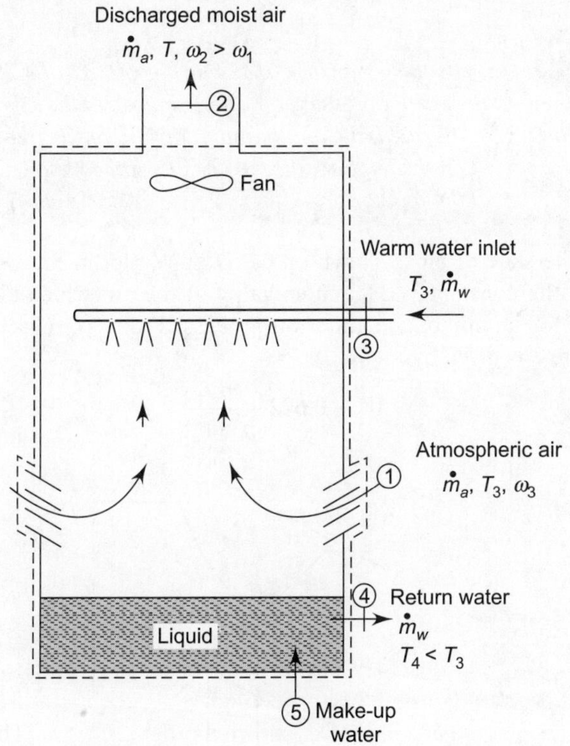

Fig. 15.15 Schematic of a cooling tower

SOLVED EXAMPLES

Example 15.1 *An air sample at atmospheric pressure (1.01325 bar) has dry bulb and wet bulb temperatures 43°C and 29°C, respectively. Calculate (a) the partial pressure of water vapour, (b) specific humidity and (c) relative humidity.* E [LO 15.2]

Solution From steam table, saturation pressure of water vapour at 29°C is 4.013 kPa

Hence, from Carrier equation: $p_{w=} 4.013 - \dfrac{(101.325 - 4.013)(43-29)(1.8)}{2800 - 1.3(1.8 \times 43 + 32)}$ = **3.09 kPa**

From steam table at 43°C, saturation pressure of water vapour is 8.65 kPa
Hence, specific humidity,

$$W = 0.622 \times \frac{3.09}{(101.325 - 3.09)} = \mathbf{0.01956\ kg\ vap/dry\ air}$$

Relative humidity,

$$\phi = \frac{3.09}{8.65} \times 100 = \mathbf{35.72\%}$$

Example 15.2 *Atmospheric air at 1.0132 bar has a DBT of 32°C and a WBT of 26°C. Compute (a) the partial pressure of water vapour, (b) the specific humidity, (c) the dew point temperature, (d) the relative humidity, (e) the degree of saturation, (f) the density of the air in the mixture, (g) the density of the vapour in the mixture and (h) the enthalpy of the mixture.* D [LO 15.2]

Solution The state of air is shown on the DBT-W plot in Fig. 15.16. The path 1–2 represents the constant WBT and enthalpy of the air, which also holds good approximately for an adiabatic saturation process. From Eq. (15.1), the specific humidity at state 2 is given by

$$W_2 = 0.622 \frac{p_s}{p - p_s}$$

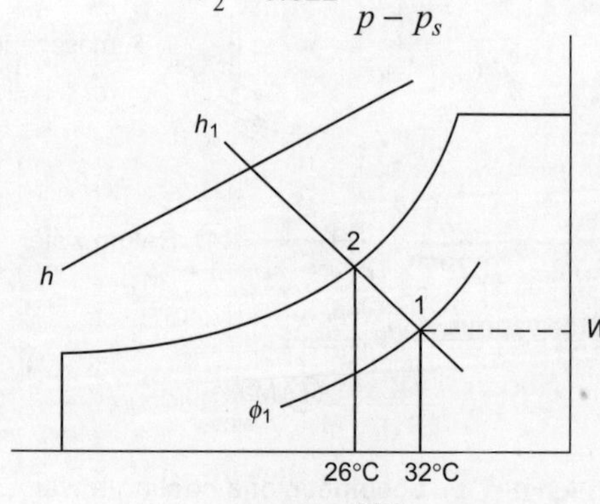

Fig. 15.16

The saturation pressure p_s at the wbt of 26°C is 0.03363 bar.

$$\therefore \quad W_2 = 0.622 \frac{0.03363}{1.0132 - 0.03363} = 0.021148 \text{ kg vap./kg dry air}$$

From Eq. (15.3), for adiabatic saturation

$$W_1 = \frac{c_{pa}(T_2 - T_1) + W_2 \cdot h_{fg2}}{h_{w1} - h_{f2}}$$

From the steam tables, at 26°C

$$h_{fg2} = 2439.9 \text{ kJ/kg}, \; h_{f2} = 109.1 \text{ kJ/kg}$$

At 32°C, $h_{w1} = h_g = 2559.9$ kJ/kg

(a)
$$W_1 = \frac{1.005(25 - 32) + 0.021148\,(2439.9)}{2559.9 - 109.1}$$

$$= \mathbf{0.0186 \text{ kg vap./kg dry air}}$$

(b)
$$W_1 = 0.622 \frac{p_w}{p - p_w} = 0.0186$$

$$\frac{p - p_w}{p_w} = \frac{0.622}{0.0186} = 33.44$$

$$p_w = \mathbf{0.03 \text{ bar}}$$

(c) Saturation temperature at 0.03 bar, DPT = **24.1°C**

(d) Relative humidity, $\phi = \dfrac{p_w}{p_{sat}}$

At 32°C, $p_{sat} = 0.048$ bar

$$\therefore \quad \phi = \frac{0.03}{0.048} = \mathbf{0.625} \text{ or } \mathbf{62.5\%}$$

(e) Degree of saturation $\mu = \dfrac{W}{W_s} = \dfrac{p_w}{p_s}\dfrac{p - p_s}{p - p_w} = \dfrac{0.03\,(1.0132 - 0.048)}{0.048\,(1.0132 - 0.03)}$

$$= \mathbf{0.614}$$

(f) Partial pressure of dry air

$$p_a = p - p_w = 1.0132 - 0.03 = 0.9832 \text{ bar}$$

∴ Density of dry air

$$\rho_a = \frac{p_a}{R_a T_{db}} = \frac{0.9832 \times 100}{0.287 \times (273 + 32)} = \mathbf{1.12 \text{ kg/m}^3 \text{ dry air}}$$

(g) Density of water vapour

$$\rho_w = 0.0186 \frac{\text{kg vap.}}{\text{kg dry air}} \times 1.12 \frac{\text{kg dry air}}{\text{m}^3 \text{ dry air}}$$

$$= \mathbf{0.021 \text{ kg vap/m}^3 \text{ dry air}}$$

(h) Enthalpy of the mixture

$$h = h_a + Wh_w = c_p t_a + W[h_g + 1.88\,(t_{db} - t_{dp})]$$
$$= 1.005 \times 32 + 0.0186\,[2559.9 + 1.88\,(32 - 24.1)]$$
$$= \mathbf{80.55\ kJ/kg}$$

Example 15.3 *When the temperature of outside air at 35°C and 100 kPa is decreased to 27°C, condensation of the moisture in the air takes place. Find the partial pressure of the dry air, the relative humidity and the humidity ratio.*

M **[LO 15.2]**

Solution The saturation pressure corresponding to 35°C is $p_{sat} = 5.628$ kPa. Using linear interpolation, the partial pressure of the water vapour at 27°C is $p_w = 3.6$ kPa.

The partial pressure of the dry air is

$$p_a = p - p_w = 100 - 3.6 = \mathbf{96.4\ kPa}$$

The relative humidity is

$$\phi = \frac{p_w}{p_{sat}} = \frac{3.6}{5.628} = \mathbf{0.6397\ or\ 63.97\%}$$

The humidity ratio is found to be

$$W = 0.622\frac{p_w}{p - p_w} = 0.622\frac{3.6}{100 - 3.6} = \mathbf{0.023228\ kg\ vap./kg\ dry\ air}$$

Example 15.4 *One of the major concern in the cold weather condition is the condensation of the moisture present in the air. Usually the temperature inside a house is not uniform and the condensation frequently takes place at the region of lower air temperature especially on the inner surfaces of the windows. If the house contains air at 15°C and relative humidity of 75%, what minimum temperature should be maintained to avoid the condensation?*

E **[LO 15.2]**

Solution The condensation takes place when the air temperature reaches its dew point temperature.

The partial pressure of water vapour at 15°C is found to be

$$p_w = \phi\, p_{sat}\big|_{15°C} = 0.75 \times 1.7051 = 1.2788\ \text{kPa}$$

The dew point temperature is then

$$T_{dp} = T_{sat}\big|_{p_w} = \mathbf{10.35°C}$$

Example 15.5 *A winter air conditioner operating at steady state takes outside air at 12°C, 100 kPa and relative humidity of 50% at a rate of 50 m³/min. The air is first heated to 20°C by passing the air through the heating section and then humidified to achieve a higher level of comfort by introducing water vapour. If the temperature inside the room is to be maintained at 24°C with a relative humidity of 65%, determine the rate of heat transfer to the air and the rate of steam added to the air.*

M **[LO 15.2, 15.4]**

Solution The process of heating and humidification is shown in Fig. 15.17.

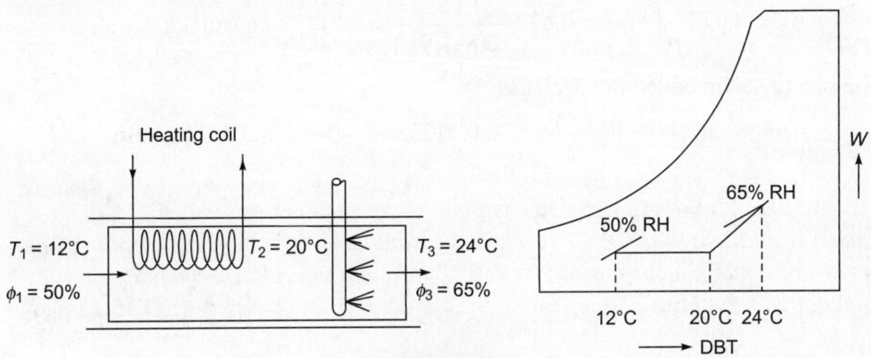

Fig. 15.17

The saturation pressure corresponding to 12°C is $p_{sat} = 1.4186$ kPa.
The partial pressure of the water vapour at 12°C is

$$p_{w1} = \phi\, p_{sat}\big|_{12°C} = 0.50 \times 1.4186 = 0.7093 \text{ kPa}.$$

The partial pressure of the dry air is

$$p_{a1} = p - p_{w1} = 100 - 0.7093 = 99.2907 \text{ kPa}$$

The specific volume of dry air is

$$v_1 = \frac{R_a T_1}{p_{a1}} = \frac{0.287 \times 285}{99.2907} = 100 - 0.7093 = 0.8238 \text{ m}^3/\text{kg}$$

The mass flow rate of dry air is $\dot{m}_a = \frac{\forall}{v_1} = \frac{50}{0.8238} = 60.694$ kg/min

The humidity ratio at state 1 is found to be

$$W_1 = 0.622 \frac{p_{w1}}{p - p_{w1}} = 0.622 \frac{0.7093}{100 - 0.7093} = 0.00444 \text{ kg vap./kg dry air}$$

The specific enthalpy at state 1 is found to be

$$h_1 = h_{a1} + W_1 h_{w1} = c_{pa} T_1 + W_1 h_{g1}$$

$$= 1.005 \times 12 + 0.00444 \times 2523.38 = 23.26 \text{ kJ/kg dry air}$$

There is no change in specific humidity during heating and hence the specific enthalpy at state 2 is

$$h_2 = h_{a2} + W_2 h_{w2} = c_{pa} T_2 + W_2 h_{g2}$$

$$= 1.005 \times 20 + 0.00444 \times 2538.1 = 31.369 \text{ kJ/kg dry air}$$

The rate of heat added to the air is

$$Q = \dot{m}_a (h_2 - h_1) = 60.694(31.369 - 23.26) = \mathbf{492.17 \text{ kJ/min}}$$

The saturation pressure corresponding to 24°Cis $p_{sat} = 3.003$ kPa.
The partial pressure of the water vapour at 65°C is

$$p_{w3} = \phi\, p_{sat}\big|_{12°C} = 0.65 \times 3.003 = 1.952 \text{ kPa}.$$

The humidity ratio at state 3 is found to be

$$W_3 = 0.622\frac{p_{w3}}{p - p_{w3}} = 0.622\frac{1.952}{100 - 1.952} = 0.01238 \text{ kg vap./kg dry air}$$

The rate of steam added to the air is

$$\dot{m}_w = \dot{m}_a (W_2 - W_1) = 60.694(0.01238 - 0.00444) = \mathbf{0.482\ kg/min}$$

Example 15.6 *An air-water vapour mixture enters an adiabatic saturator at 30°C and leaves at 20°C, which is the adiabatic saturation temperature. The pressure remains constant at 100 kPa. Determine the relative humidity and the humidity ratio of the inlet mixture.* **E [LO 15.2, 15.4]**

Solution The specific humidity at the exit

$$W_2 = 0.622\frac{p_s}{p - p_s}$$

$$= 0.622\left(\frac{2.339}{100 - 2.339}\right) = 0.0149\ \frac{\text{kg vap.}}{\text{kg dry air}}$$

The specific humidity at the inlet (Eq. (15.3))

$$W_1 = \frac{C_{pa}(T_2 - T_1) + W_2 h_{fg2}}{h_{w1} - h_{f2}}$$

$$= \frac{1.005(20 - 30) + 0.0149 \times 2454.1}{2556.3 - 83.96}$$

$$= 0.0107 \text{ kg vap./kg dry air}$$

$$W_1 = 0.622\left(\frac{p_{w1}}{100 - p_{w1}}\right) = 0.0107$$

$$p_{w1} = 1.691 \text{ kPa}$$

$$\therefore \quad \phi_1 = \frac{p_{w1}}{p_{s1}} = \frac{1.691}{4.246} = \mathbf{0.398 \text{ or } 39.8\%}$$

Example 15.7 *Saturated air at 2°C is required to be supplied to a room where the temperature must be held at 20°C with a relative humidity of 50%. The air is heated and then water at 10°C is sprayed in to give the required humidity. Determine the temperature to which the air must be heated and the mass of spray water required per m³ of air at room conditions. Assume that the total pressure is constant at 1.013 bar and neglect the fan power.* **M [LO 15.2, 15.4]**

Solution The process is shown in Fig. 15.18. From the steam tables, at 20°C, $p_{\text{sat}} = 2.339$ kPa

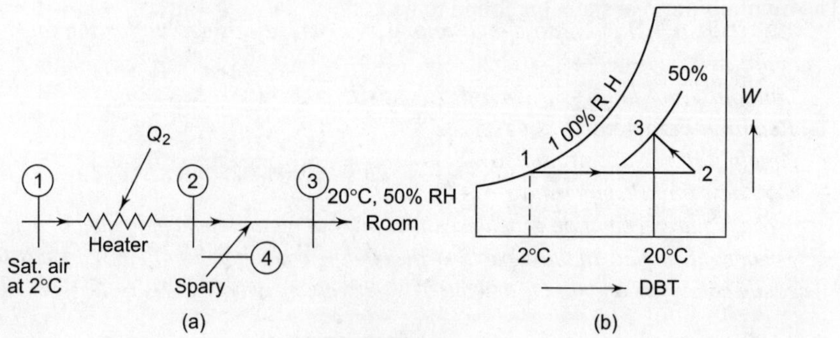

Fig. 15.18

$$\phi_3 = \frac{p_{w3}}{(p_{\text{sat}})_{t3}} = \frac{p_{w3}}{2.339} = 0.50$$

$\therefore$ $$p_{w3} = 1.17 \text{ kPa}$$

$\therefore$ $$p_{a3} = 101.3 - 1.17 = 100.13 \text{ kPa}$$

$$W_3 = 0.622 \frac{p_{w3}}{p_{a3}} = 0.622 \times \frac{1.17}{100.13} = 0.00727$$

$$\phi_1 = \frac{p_{w1}}{(p_{\text{sat}})_{2°\text{C}}} = 1.00$$

At 2ºC, $p_{\text{sat}} = 0.7156$ kPa

$\therefore$ $$p_{w1} = 0.7156 \text{ kPa}$$

$$p_{a1} = 101.3 - 0.7156 = 100.5844 \text{ kPa}$$

$$W_1 = 0.622 \frac{0.7156}{100.5844} = 0.00442$$

$$W_3 - W_1 = 0.00727 - 0.00442 = 0.00285 \text{ kg vap./kg dry air}$$

$$v_{a3} = \frac{R_a T_3}{p_{a3}} = \frac{0.287 \times 293}{100.13} = 0.84 \text{ m}^3\text{/kg dry air}$$

$\therefore$ $$\text{Spray water} = 0.00285 \frac{\text{kg vap.}}{\text{kg dry air}} \times \frac{\text{kg dry air}}{0.84 \text{ m}^3}$$

$$= \mathbf{0.00339 \text{ kg moisture/m}^3}$$

$$G_2 h_2 + m_{w4} h_4 = G_3 h_3$$

$\therefore$ $$h_2 + (W_3 - W_2) h_4 = h_3$$

$$h_{a2} + W_2 h_{w2} + (W_3 - W_2) h_4 = h_{a3} + W_3 h_{w3}$$

$\therefore$ $$c_p (t_3 - t_2) + W_3 h_{w3} - W_2 h_{w2} - (W_3 - W_2) h_4 = 0$$

From the steam tables, at $p_w = 1.17$ kPa

$$h_g = 2518 \text{ kJ/kg and } t_{\text{sat}} = 9.65\text{ºC}$$

$$1.005\,(20 - t_2) + 0.00727\,[2518 + 1.884\,(20 - 9.65)]$$
$$- 0.00442\,[2518 + 1.884\,(t_2 - 9.65)]$$
$$- 0.00285 \times 10 = 0,$$

$\therefore$ $$t_2 = \mathbf{27.2ºC}$$

Example 15.8 *An air conditioning system is designed under the following conditions:*

Outdoor conditions—30°C DBT, 75% R.H.

Required indoor conditions—22°C DBT, 70% R.H.

Amount of free air circulated—3.33 m^3*/s*

Coil dew point temperature—14°C

The required condition is achieved first by cooling and dehumidification, and then by heating. Estimate (a) the capacity of the cooling coil in tonnes, (b) the capacity of the heating coil in kW and (c) the amount of water vapour removed in kg/s. M **[LO 15.4]**

Solution The processes are shown in Fig. 15.19. The property values, taken from the psychrometric chart are:

$$h_1 = 82,\ h_2 = 52,\ h_3 = 47,\ h_4 = 40 \text{ kJ/kg dry air}$$

$$W_1 = 0.020,\ W_2 = W_3 = 0.0115 \text{ kg vap./kg dry air}$$

$$v_1 = 0.887 \text{ m}^3\text{/kg dry air}$$

$$G = \frac{3.33}{0.887} = 3.754 \text{ kg dry air/sec}$$

$\therefore$ Cooling coil capacity $= G(h_1 - h_3) = 3.754\ (82 - 47)$ kJ/s

$$= \frac{3.754 \times 35 \times 3600}{14,000} = \mathbf{33.79\ tonnes}$$

Capacity of the heating coil $= G(h_2 - h_3) = 3.754\ (52 - 47)$ kJ/s

$$= 3.754 \times 5 = \mathbf{18.77\ kW}$$

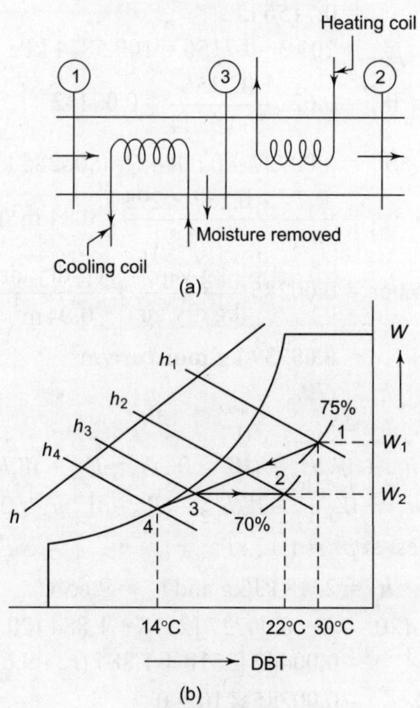

Fig. 15.19

Rate of water vapour removed = $G(W_1 - W_3)$
$= 3.754 \times (0.0200 - 0.0115)$
$= \mathbf{0.0319\ kg/s}$

Example 15.9 *Air at 20°C, 40% RH is mixed adiabatically with air at 40°C, 40% RH in the ratio of 1 kg of the former with 2 kg of the latter (on dry basis). Find the final condition of air.* E [LO 15.4]

Solution Figure 15.20 shows the mixing process of two air streams. The equations

$$G_1 + G_2 = G_3$$
$$G_1W_1 + G_2W_2 = G_3W_3$$
$$G_1h_1 + G_2h_2 = G_3h_3$$

result in

$$\frac{W_2 - W_3}{W_3 - W_1} = \frac{h_2 - h_3}{h_3 - h_1} = \frac{G_1}{G_2}$$

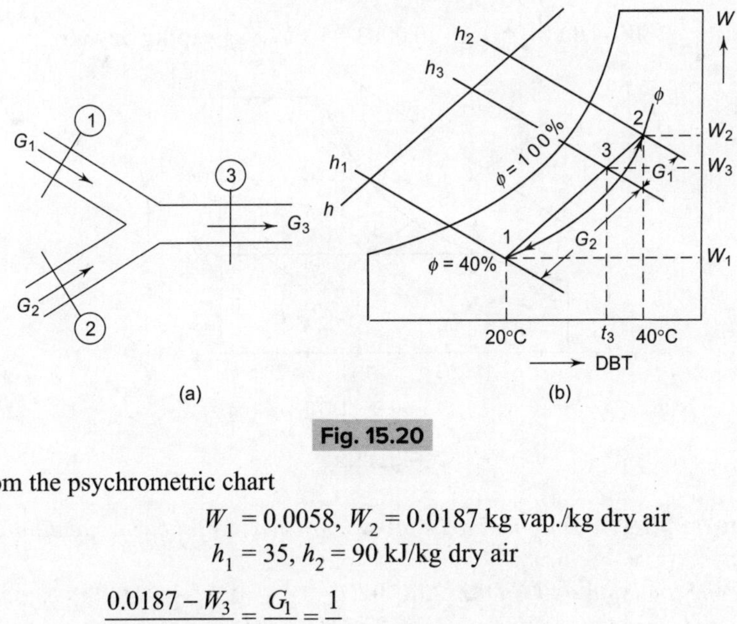

Fig. 15.20

From the psychrometric chart

$$W_1 = 0.0058,\ W_2 = 0.0187 \text{ kg vap./kg dry air}$$
$$h_1 = 35,\ h_2 = 90 \text{ kJ/kg dry air}$$

$$\therefore \quad \frac{0.0187 - W_3}{W_3 - 0.0058} = \frac{G_1}{G_2} = \frac{1}{2}$$

$$\therefore \quad W_3 = \frac{2}{3} \times 0.187 + \frac{1}{3} \times 0.0058$$
$$= 0.0144 \text{ kg vap./kg dry air}$$

Again

$$\frac{h_2 - h_3}{h_3 - h_1} = \frac{G_1}{G_2} = \frac{1}{2}$$

$$\therefore \quad h_3 = \frac{2}{3}h_2 + \frac{1}{3}h_1 = \frac{2}{3} \times 90 + \frac{1}{3} \times 35$$
$$= 71.67 \text{ kJ/kg dry air}$$

∴ Final condition of air is given by

$$W_3 = \mathbf{0.0144\ kg\ vap./kg\ dry\ air}$$
$$h_3 = \mathbf{71.67\ kJ/kg\ dry\ air}$$

Example 15.10 *Saturated air at 21°C is passed through a drier so that its final relative humidity is 20%. The drier uses silica gel adsorbent. The air is then passed through a cooler until its final temperature is 21°C without a change in specific humidity. Find out the temperature of air at the end of the drying process, the heat rejected in kJ/kg dry air during the cooling process, the relative humidity at the end of the cooling process, the dew point temperature at the end of the drying process and the moisture removed during the drying process in kg vap./kg dry air.* E **[LO 15.4]**

Solution From the psychrometric chart (Fig. 15.21)

$$T_2 = \mathbf{38.5°C}$$
$$h_1 - h_3 = 60.5 - 42.0 = \mathbf{18.5\ kJ/kg\ dry\ air}$$
$$\phi_3 = \mathbf{53\%},\ t_4 = \mathbf{11.2°C}$$
$$W_1 - W_2 = 0.0153 - 0.0083 = \mathbf{0.0070\ kg\ vap/kg\ dry\ air}$$

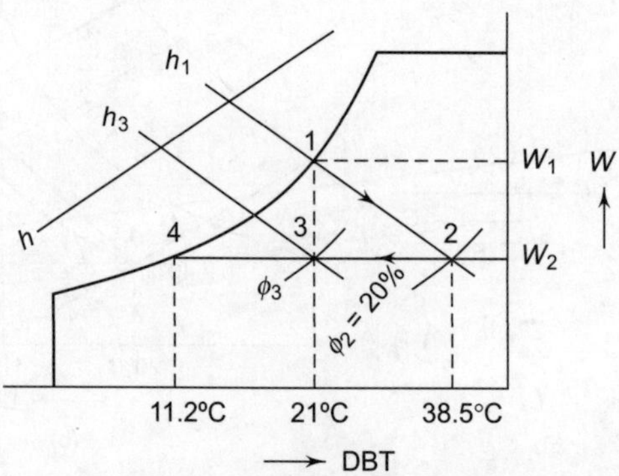

Fig. 15.21

Example 15.11 *For a hall to be air-conditioned, the following conditions are given:*

Outdoor condition—40°C DBT, 20°C WBT
Required comfort condition—20°C DBT, 60% RH
Seating capacity of hall—1500
Amount of outdoor air supplied—0.3 m³/min per person

If the required condition is achieved first by adiabatic humidification and then by cooling, estimate (a) the capacity of the cooling coil in tonnes and (b) the capacity of the humidifier in kg/h. E **[LO 15.4]**

Solution From the psychrometric chart (Fig. 15.22)

$$h_1 = h_2 = 57.0,\ h_3 = 42.0\ \text{kJ/kg dry air}$$
$$W_1 = 0.0065,\ W_2 = W_3 = 0.0088\ \text{kg vap./kg dry air}$$
$$t_2 = 34.5°\text{C},\ v_1 = 0.896\ \text{m}^3/\text{kg dry air}$$

Amount of dry air supplied

$$G = \frac{1500 \times 0.3}{0.896} = 502 \text{ kg/min}$$

$\therefore$ Capacity of the cooling coil

$$= G(h_2 - h_3) = 502\ (57 - 42) \text{ kJ/min}$$

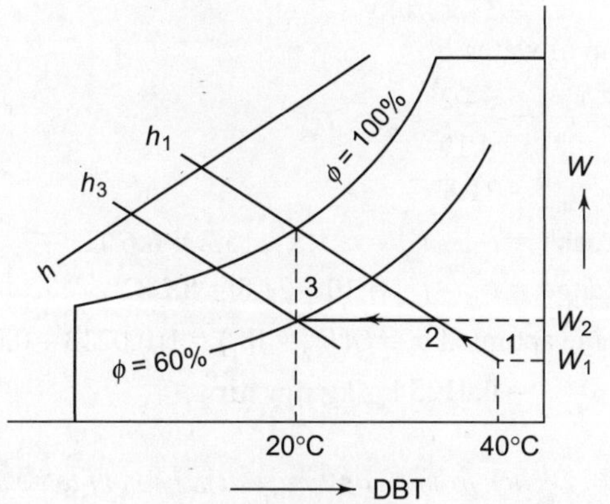

Fig. 15.22

$$= \frac{502 \times 15 \times 60}{14,000} = \mathbf{32.27 \text{ tonnes}}$$

Capacity of the humidifier

$$= G\,(W_2 - W_1) = 502\ (0.0088 - 0.0065) \text{ kg/min}$$
$$= 502 \times 23 \times 10^{-4} \times 60 = \mathbf{69.3 \text{ kg/h}}$$

Example 15.12 *Water at 30°C flows into a cooling tower at the rate of 1.15 kg per kg of air. Air enters the tower at a DBT of 20°C and a relative humidity of 60% and leaves it at a DBT of 28°C and 90% relative humidity. Make-up water is supplied at 20°C. Determine the temperature of water leaving the tower, the fraction of water evaporated and approach and range of the cooling tower.*

M [LO 15.4]

Solution Properties of air entering and leaving the tower (Fig. 15.13) are:

$t_{wb1} = 15.2°C$
$t_{wb2} = 26.7°C$
$h_1 = 43$ kJ/kg dry air
$h_2 = 83.5$ kJ/kg dry air
$W_1 = 0.0088$ kg water vapour/kg dry air
$W_2 = 0.0213$ kg water vapour/kg dry air

Enthalpies of the water entering the tower and the make-up water are:

$h_{w3} = 125.8$ kJ/kg $\qquad h_w = 84$ kJ/kg

From the energy balance Eq. (15.5),

$$h_{w3} - h_{w4} = \frac{G}{m_w}[(h_2 - h_1) - (W_2 - W_1)\,h_w]$$

$$= \frac{1}{1.15}[(83.5 - 43) - (0.0213 - 0.0088)\,84]$$

$$= 34.2 \text{ kJ/kg}$$

Temperature drop of water

$$t_{w3} - t_{w4} = \frac{34.2}{4.19} = 30 - t_{w4}$$

$$\therefore \quad t_{w4} = \mathbf{21.8°C}$$

$$\therefore \quad \text{Approach} = t_{w4} - t_{wb1} = 21.8 - 15.2 = \mathbf{6.6°C}$$

$$\text{Range} = t_{w3} - t_{w4} = 30 - 21.8 = \mathbf{8.2°C}$$

Fraction of water evaporated, $x = G(W_2 - W_1) = 1(0.0213 - 0.0088)$

$$= \mathbf{0.0125 \text{ kg/kg dry air}}$$

Example 15.13 *Water from a cooling system is itself to be cooled in the cooling tower at a rate of 2.78 kg/s. The water enters the tower at 65°C and leaves a collecting tank at the base at 30°C. Air flows through the tower, entering the base at 15°C, 0.1 MPa, 55% RH and leaving the top at 35°C, 0.1 MPa, saturated. Make-up water enters the collecting tank at 14°C. Determine the air flow rate into the tower in m³/s and the make-up water flow rate in kg/s.* **D [LO 15.4]**

Solution Figure 15.23 shows the flow diagram of the cooling tower. From the steam tables,

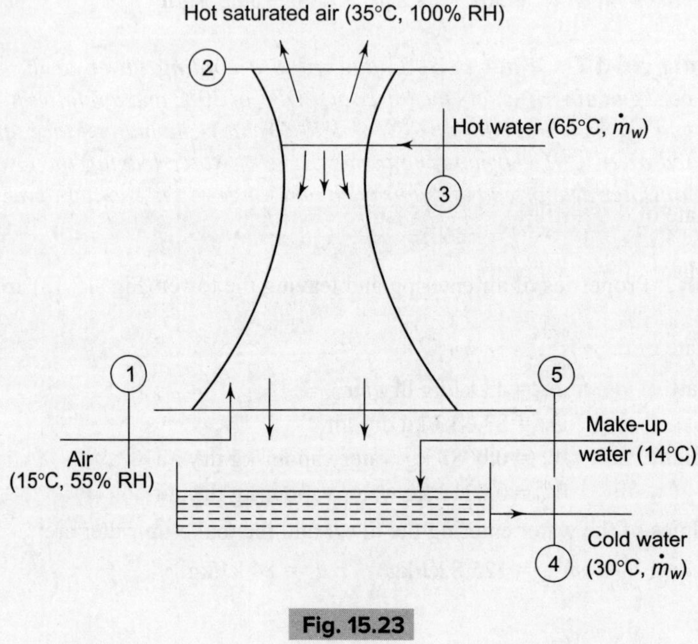

Fig. 15.23

$$\text{at } 15°\text{C}, p_{sat} = 0.01705 \text{ bar}, h_g = 2528.9 \text{ kJ/kg}$$
$$\text{at } 35°\text{C}, p_{sat} = 0.05628 \text{ bar}, h_g = 2565.3 \text{ kJ/kg}$$

$$\phi_1 = \frac{p_w}{(p_{sat})_{15°C}} = 0.55$$

$$\therefore \quad p_{w1} = 0.55 \times 0.01705 = 0.938 \times 10^{-2} \text{ bar}$$

$$\phi_2 = \frac{p_w}{(p_{sat})_{35°C}} = 1.00$$

$$\therefore \quad p_{w2} = 0.05628 \text{ bar}$$

$$W_1 = 0.622 \frac{p_w}{p - p_w} = 0.622 \times \frac{0.938 \times 10^{-2}}{1.00 - 0.00938}$$

$$= 0.00589 \text{ kg vap./kg dry air}$$

$$W_2 = 0.622 \times \frac{0.05628}{1.00 - 0.05628} = 0.0371 \text{ kg vap./dry air}$$

$$\therefore \quad \text{Make-up water} = W_2 - W_1 = 0.0371 - 0.00589$$
$$= 0.03121 \text{ kg vap./kg dry air}$$

Energy balance gives

$$H_2 + H_4 - H_1 - H_3 - H_5 = 0$$

For 1 kg of dry air

$$c_{pa}(t_2 - t_1) + W_2 h_2 - W_1 h_1 + \dot{m}_w (h_4 - h_3) - (W_2 - W_1) h_5 = 0$$

$$\therefore \quad 1.005(35 - 15) + 0.0371 \times 2565.3 - 0.00589 \times 2528.9$$
$$+ \dot{m}_w(-35)\, 4.187 - 0.03121 \times 4.187 \times 14 = 0$$

or $\quad 146.55\, \dot{m}_w = 98.54$

$$\therefore \quad \dot{m}_w = 0.672 \text{ kg water/kg dry air}$$

Since water flow rate is 2.78 kg/s

$$\therefore \quad \text{Rate of dry air flow} = \frac{2.78}{0.672} = 4.137 \text{ kg/s}$$

$\therefore$ Make-up water flow rate

$$= 0.03121 \times 4.137 = \mathbf{0.129 \text{ kg/s}}$$

Rate of dry air flow = 4.137 kg/s

Rate of wet air flow = $4.137(1 + W_1)$

$$= 4.137 \times 1.00589 = 4.16 \text{ kg/s}$$

$\therefore$ Volume flow rate of air

$$= \frac{\dot{m}_a RT}{p} = \frac{4.16 \times 0.287 \times 288}{100} = \mathbf{3.438 \text{ m}^3/\text{s}}$$

SUMMARY

Psychometrics is the study of the properties of atmospheric air considered as a mixture of dry air and water vapour. By Dalton's law, the atmospheric pressure 'p' is the sum of the partial pressure of dry air, p_a and water vapour p_w, or $p = p_a + p_w$. Since p_w is very small, water vapour in the air exists in the superheated state and is said to be unsaturated. *Relative humidity* (ϕ) is the ratio of p_w and p_s, the saturation pressure at the temperature of the mixture. The *specific humidity* or *humidity ratio* (W) is the mass of water vapour per kg of dry air (G), which is given by

$$W = 0.622 \frac{p_w}{p - p_w}$$

If air is saturated at temperature T, $W = W_{max} = W_s = 0.622 \dfrac{p_s}{p - p_s}$

The *degree of saturation* (μ) is defined as W/W_s and varies between 0 and 1.

Dry bulb temperature (DBT) is measured by a thermometer with a dry bulb. The *wet bulb temperature* (WBT) is the temperature recorded by a thermometer when the bulb is enveloped by a cotton wick saturated with water. When unsaturated air flows over a long sheet of water enclosed in an adiabatic container, water evaporates till the air becomes saturated, when the *adiabatic saturation temperature* is reached, which is close to WBT. The saturation temperature at the partial pressure of water vapour in air is called the *dew point temperature* (DPT). When the specific humidity of air is increased, say by water spray, air is said to be humidified. If W is decreased air is dehumidified.

Carrier equation may be used to calculate the partial pressure of water vapour in air when the dry bulb and wet bulb temperatures of air are known.

The psychrometric chart is a graphical plot with W and p_w as ordinates and DBT as abscissa. The volume of the mixture (m^3/kg dry air), WBT, ϕ and enthalpy of the mixture appear as parameters. The psychometric processes are usually (a) Sensible heating or cooling (W = constant), (b) cooling and dehumidification (in summer), (c) heating and humidification (in winter), (d) adiabatic mixing of two air streams, (e) chemical dehumidification and (f) adiabatic evaporative cooling. The *cooling tower* utilises evaporative cooling to cool warm water by air, with water never reaching the minimum temperature; the WBT.

REVIEW QUESTIONS

(E) **15.1** What is psychrometrics? **[LO 15.1]**

(E) **15.2** What are hygroscopic materials? **[LO 15.1]**

(E) **15.3** What do you understand by saturated air and unsaturated air? **[LO 15.2]**

(E) **15.4** What is relative humidity? How is it defined as the ratio of two mole fractions? **[LO 15.2]**

(E) **15.5** What is specific humidity? When does it become maximum? **[LO 15.2]**

(E) **15.6** What is degree of saturation? What are its limiting values? **[LO 15.2]**

(E) **15.7** What do you understand by dry bulb and wet bulb temperatures? **[LO 15.2]**

(E) **15.8** Define dew point temperature. **[LO 15.2]**

(E) **15.9** When do the DBT, WBT and DPT become equal? **[LO 15.2]**

(E) **15.10** What is a psychrometer? **[LO 15.2]**

(M) **15.11** What is an adiabatic saturation process? **[LO 15.2]**

(M) **15.12** What is thermodynamic wet bulb temperature? **[LO 15.2]**

(M) **15.13** What is the enthalpy of an air-vapour mixture? **[LO 15.2]**

(E) **15.14** What is Carrier's Equation? Why is it important in psychrometry? **[LO 15.2]**

(M) **15.15** Why does the enthalpy of an air-vapour mixture remains constant during an adiabatic saturation process? **[LO 15.3]**

(E) **15.16** What is sensible heating or cooling? **[LO 15.4]**

(M) **15.17** Explain the process of cooling and dehumidification. **[LO 15.4]**

(D) **15.18** What is apparatus dew point? **[LO 15.4]**

(D) **15.19** What do you understand by sensible heat load and latent heat load? **[LO 15.4]**

(M) **15.20** Explain the process of heating and humidification. **[LO 15.4]**

(E) **15.21** What is chemical dehumidification? What is an adsorbent? **[LO 15.4]**

(E) **15.22** Explain what do you understand by evaporative cooling. **[LO 15.4]**

(D) **15.23** What is a cooling tower? How is it specified? Where is it used? **[LO 15.4]**

(D) **15.24** Define the approach and range of a cooling tower. **[LO 15.4]**

PROBLEMS WITH HINTS

(M) **15.1** Air at 30°C and 50 % relative humidity enters a dehumidifier operating at steady state with a volumetric flow rate of 280 m^3/min. The air passes over a cooling coil and water vapour condenses. Condensate exits the dehumidifier saturated at 10°C. Saturated air exits in a separate stream at the same temperature. There is no significant heat transfer to the surroundings and pressure remains constant at 1.013 bar. Determine (a) the mass flow rate of dry air, (b) the rate at which water vapour is condensed in kg/kg dry air and (c) the required refrigeration capacity. **[LO 15.1, 15.2]**

Hint:

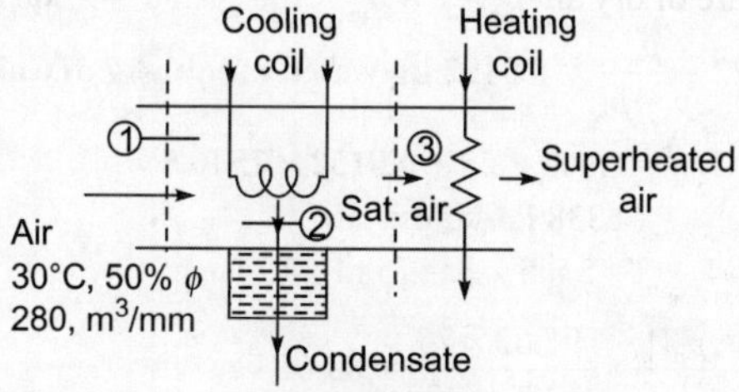

Fig. 15.24

At steady state, the mass of dry air entering = mass of dry air leaving

$$\text{(a)}\ \dot{m}_{a_1} = \frac{280\dfrac{\text{m}^3}{\text{min}} \times 99.177\dfrac{\text{kN}}{\text{m}^2}}{0.287\dfrac{\text{kN m}}{\text{kg K}} \times 303\ \text{K}}$$

$$= \mathbf{319\ kg/min} = G_1$$

$$\dot{m}_{a_1} = \frac{(AV)_1}{RT_1/p_{a_1}} = G_1$$

p_s at 30°C from steam tables = 0.04246 bar

$$p_w = \phi p_s = 0.5 \times 0.04246 = 0.02123 \text{ bar}$$
$$p_{a1} = 1.013 - 0.02123 = 0.99177 \text{ bar}$$

$$(30°\text{C})\ W_1 = 0.622 \frac{p_w}{p - p_w} = 0.622 \times \frac{0.02123}{0.99177} = 0.0133 \frac{\text{kg water vapour}}{\text{kg dry air}}$$

$$(10°\text{C})\ W_2 = 0.622 \frac{0.01228}{1.013 - 0.01228} = 0.0057 \frac{\text{kg water vapour}}{\text{kg dry air}}$$

(b) $\dfrac{\dot{m}_w}{\dot{m}_a} = 0.0133 - 0.0057 =$ **0.0076 kg vap./kg dry air**

(c) By energy balance, where $G = \dot{m}_a$

$$Q_{cv} = Q[(h_{a_2} - h_{a_1}) - W_1 hg_1 + W_2 hg_2 + (W_1 - W_2) h_{f_2}]$$
$$= G[(h_{a_2} + w_2 hg_2) - (h_{a_1} + w_1 hg_1) + (w_1 - w_2) h_{f_2}]$$
$$= 319.35[(283.1 - 303.2) - 0.0133 \times 1256.3 + 0.0076 \times 2519.8 + 0.0057 \times 42.01]$$
$$= -11.084 \text{ kJ/min} = -\frac{11084}{211} = \mathbf{52.5\ tons}$$

where enthalpies of water vapour at 1 and 2 are evaluated as saturated vapour values corresponding to T_1 and T_2, respectively, and h_{f_2} is sat. liquid at 10°C. For air, h_{a_1} and h_{a_2} are obtained from Table A_2 at 30 and 10°C, respectively.

(E) **15.2** A room having dimensions of 5 m × 5 m × 3 m contains air at 25°C and 100 kPa at a relative humidity of 75%. Determine (a) the partial pressure of dry air, (b) the specific humidity, (c) the enthalpy per unit mass of dry air and (d) the masses of dry air and water vapour in the room. **[LO 15.2]**

Hint:

For water at 25°C (from steam tables), $p_{sat} = p_s = 3.169$ kPa and $h_g = 2546.5$ kJ/kg

Now relative humidity $\phi = \dfrac{p_w}{p_s} = 0.75 \rightarrow p_w = 0.75 \times 3.169 = 2.38$ kPa

Partial pressure of dry air, $p_a = p - p_w = 100 - 2.38 =$ **97.68 kPa**

(b) $W = 0.622 \dfrac{p_w}{p - p_w} =$ **0.0152 kg water vapour/kg dry air**

(c) $h = h_a + Wh_w = \mathbf{1.005 \times 25 + 0.0152 \times 2546.5}$
$= \mathbf{6338\ kJ/kg\ dry\ air}$

(a) $V_a = V_w = V_{room} = 5 \times 5 \times 3 = 75 \text{ m}^3$

$$\therefore \quad m_a = \frac{p_a V_a}{R_w T} = \frac{97.62 \times 75}{0.287 \times 298} = 85.61 \text{ kg}$$

$$\dot{m}_w = \frac{p_w V_w}{R_w T} = \frac{2.38 \times 75}{(8.3143/18) \times 298} = \mathbf{1.30\ kg}$$

Also, $m_w = Wm_a = 0.0152 \times 85.61 =$ **1.30 kg**

(M) **15.3** The air in a room 6 m × 4 m × 4 m is at 101 kPa, 38°C and relative humidity 70%. Determine the specific humidity and the dew point of the air. The air is cooled at constant pressure until the temperature is 10°C. Find the amount of water vapour that has condensed. **[LO 15.2]**

Hint:

Volume V of the room $= 6 \times 4 \times 4 = 96 \text{ m}^3$

At 38°C, $p_{sat} = 6.632 \text{ kPa} = p_s$

$$\therefore \qquad p_w = \phi\, p_s = 4.642 \text{ kPa}$$

The dew point is T_{sat} at p_w = **31.5°C**

$$p_a = p - p_w = 101 - 4.6 = 96.4 \text{ kPa}$$

$$\therefore \qquad W = 0.622 \frac{p_w}{p - p_w} = \mathbf{0.0299 \text{ kg water vapour/kg dry air}}$$

At 10°C, the mixture is saturated because 10°C is less than the dew point

$$pw_2 = ps_2 = 1.2276 \text{ kPa.}$$

$$pa_2 = 101 - 1.22 = 99.78 \text{ kPa}$$

$$W_2 = 0.622 \frac{p_{w_2}}{p - p_{w_2}} = 0.00765 \text{ kg w.v./kg d.a.}$$

$\therefore$ Mass of water condensed = $m_a (W_1 - W_2)$

where m_a = mass of dry air = $\dfrac{p_a V}{R_a T} = \dfrac{96.4 \times 96}{0.287 \times 311} = 103.7$ kg

$\therefore$ Mass of condensate = $103.7 \times (0.0299 - 0.00765)$

= **2.31 kg**

(E) **15.4** Moist air at 32°C and 60% relative humidity enters the refrigeration coils of a dehumidifier (Fig. 15.25) with a flow rate of 1.5 kg air/s. The air leaves as saturated air at 15°C. Calculate the condensate removed and the refrigeration capacity. **[LO 15.4]**

Hins:

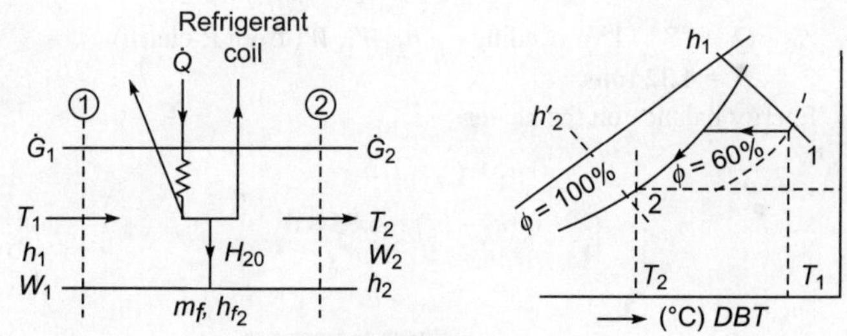

Fig. 15.25

By energy balance,

$$Gh_1 = \dot{m}_f h_{f2} + Gh_2 + \dot{Q}$$

$$h_{f2} = 62.99 \text{ kJ/kg (steam tables at 15°C)}$$

Now, $\dot{m}_f = G(W_1 - W_2) = 1.5\,(0.018 - 0.0106)$

= **0.0111 kg water/s**

$$\dot{Q} = G(h_1 - h_2) - \dot{m}_f h_{f2}$$

$$= 1.5\,(78.2 - 42.0) - 0.0111 \times 62.99$$

$$= 53.6 \text{ kW}$$

$\therefore$ Refrigeration capacity = $\dfrac{53.6}{3.5}$ = **15.2 tons**

(E) **15.5** Moist air enters a humidifier-heater unit. At 26°C and 80% relative humidity

(Fig. 15.26). It is to leave at 26°C and 50% relative humidity. For an air flow rate of 0.47 m^3/s, find the refrigeration capacity and the heating required. **[LO 15.4]**

Hint:

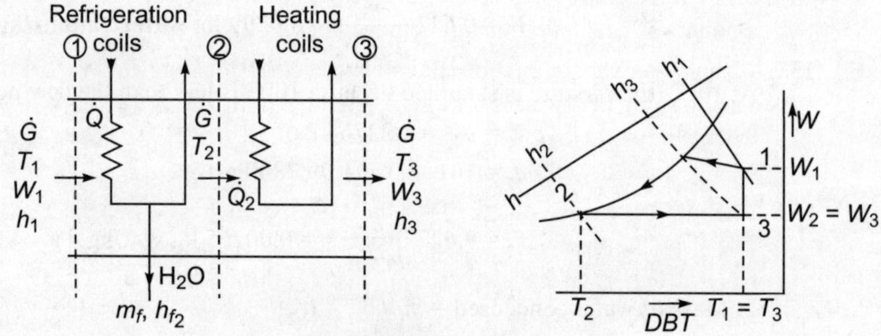

Fig. 15.26

Energy balance of the dehumidifier (1 – 2) gives

$$\dot{G}h_1 = m_f h_{f_2} + \dot{G}h_3 + \dot{Q}_1$$

$$\dot{G}\, v_{a_1} = \dot{V}_1 = 0.47 \text{ m}^3/\text{s},\ v_{a_1} = 0.87 \frac{\text{m}^3}{\text{kg d.a.}} \text{ (from P. chart)}$$

Find $\dot{G} = 0.54$ kg/s, $\dot{Q}_1 = \dot{G}(h_1 - h_2) - \dot{m}_f h_{f_2}$

$$\dot{m}_f = \dot{G}(W_1 - W_2),\ h_{f_2} = h_f \text{ at } T_2$$

$\therefore$ $\dot{Q}_1 = 15.17$ kW, (finding h_1, h_2, W_1, W_2 from P. chart.)

= **4.32 tons**

Energy balance on the heater

$$\dot{G}h_2 + \dot{Q}_2 = \dot{G}h_3$$

$$\therefore \quad \dot{Q}_2 = \dot{G}(h_2 - h_3) = \mathbf{6.53\ kW}$$

PROBLEMS

(E) **15.1** The air passing through an adiabatic saturator has an inlet and exit temperatures of 42ºC and 30ºC respectively. At the entry, what are the partial pressure of water vapour, humidity ratio and relative humidity of the air? **[LO 15.2]**

(E) **15.2** An air-water vapour mixture at 0.1 MPa, 30ºC, 80% RH has a volume of 50 m^3. Calculate the specific humidity, dew point, WBT, mass of dry air and mass of water vapour. **[LO 15.2]**

If the mixture is cooled at constant pressure to 5ºC, calculate the amount of water vapour condensed.

(E) **15.3** A sling psychrometer reads 40ºC DBT and 36ºC WBT. Find the humidity ratio, relative humidity, dew point temperature, specific volume and enthalpy of air. **[LO 15.2]**

(E) **15.4** Calculate the amount of heat removed per kg of dry air if the initial condition of air is 35ºC, 70% RH, and the final condition is 25ºC, 60% RH. **[LO 15.2]**

(E) **15.5** Two streams of air 25ºC, 50% RH and 25ºC, 60% RH are mixed adiabatically to obtain 0.3 kg/s of dry air at 30°C. Calculate the amounts of air drawn from both the streams and the humidity ratio of the mixed air. **[LO 15.4]**

(E) **15.6** Air at 40°C DBT and 27°C WBT is to be cooled and dehumidified by passing it over a refrigerant-filled coil to give a final condition of 15°C and 90% RH. Find the amounts of heat and moisture removed per kg of dry air. **[LO 15.4]**

(E) **15.7** An air-water vapour mixture enters a heater-humidifier unit at 5°C, 100 kPa, 50% RH. The flow rate of dry air is 0.1 kg/s. Liquid water at 10°C is sprayed into the mixture at the rate of 0.002 kg/s. The mixture leaves the unit at 30°C, 100 kPa. Calculate (a) the relative humidity at the outlet and (b) the rate of heat transfer to the unit. **[LO 15.4]**

(M) **15.8** A laboratory has a volume of 470 m^3, and is to be maintained at 20°C, 52.5% RH. The air in the room is to be completely changed once every hour and is drawn from the atmosphere at 1.05 bar, 32°C, 86% RH, by a fan absorbing 0.45 kW. This air passes through a cooler which reduces its temperature and causes condensation, the condensate being drained off at 8°C. The resulting saturated air is heated to room condition. The total pressure is constant throughout. Determine (a) the temperature of the air leaving the cooler, (b) the rate of condensation, (c) the heat transfer in the cooler and (d) the heat transfer in the heater. **[LO 15.4]**

(M) **15.9** In an air conditioning system, air is to be cooled and dehumidified by means of a cooling coil. The data are as follows:

Initial condition of the air at inlet to the cooling coil: DBT = 25°C, partial pressure of water vapour = 0.019 bar, absolute total pressure = 1.02 bar

Final condition of air at exit of the cooling coil:

DBT = 15°C, RH = 90%, absolute total pressure = 1.02 bar.

Other data are as follows:

Characteristic gas constant for air = 287 J/kg K

Characteristic gas constant for water vapour = 461.5 J/kg K

Saturation pressure for water at 15°C = 0.017 bar

Enthalpy of dry air = 1.005 t kJ/kg

Enthalpy of water vapour = (2500 +1.88 t) kJ/kg where t is in °C

Determine (a) the moisture removed from air per kg of dry air, (b) the heat removed by the cooling coil per kg of dry air. **[LO 15.4]**

(M) **15.10** Air at 30°C, 80% RH is cooled by spraying water at 12°C. This causes saturation, followed by condensation, the mixing being assumed to take place adiabatically and the condensate being drained off at 16.7°C. The resulting saturated mixture is then heated to produce the required conditions of 60% RH at 25°C. The total pressure is constant at 101 kPa. Determine the mass of water supplied to the sprays to provide 10 m^3/h of conditioned air. What is the heater power required? **[LO 15.4]**

(M) **15.11** An air-conditioned room requires 30 m^3/min of air at 1.013 bar, 20°C, 52.5% RH. The steady flow conditioner takes in air at 1.013 bar, 77% RH, which it cools to adjust the moisture content and reheats to room temperature. Find the temperature to which the air is cooled and the thermal loading on both the cooler and heater. Assume that a fan before the cooler absorbs 0.5 kW, and that the condensate is discharged at the temperature to which the air is cooled. **[LO 15.4]**

(M) **15.12** An industrial process requires an atmosphere having a RH of 88.4% at 22°C, and involves a flow rate of 2000 m^3/h. The external conditions are at 44.4% RH, 15°C. The air intake is heated and then humidified by water spray at 20°C. Determine the mass flow rate of spray water and the power required for heating, if the pressure throughout is 1 bar. **[LO 15.4]**

(E) **15.13** Cooling water enters a cooling tower at a rate of 1000 kg/h and 70°C. Water is pumped from the base of the tower at 24°C and some make-up water is added afterwards. Air enters the tower at 15°C, 50% RH, 1.013 bar, and is drawn from the tower saturated at 34°C, 1 bar. Calculate the flow rate of the dry air in kg/h and the make-up water required per hour. **[LO 15.4]**

(M) **15.14** A grain dryer consists of a vertical cylindrical hopper through which hot air is blown. The air enters the base at 1.38 bar, 65°C, 50% RH. At the top, saturated air is discharged into the atmosphere at 1.035 bar, 60°C.

Estimate the moisture picked up by 1 kg of dry air, and the total enthalpy change between the entering and leaving streams expressed per unit mass of dry air. **[LO 15.4]**

(D) **15.15** Air enters a counterflow cooling tower at a rate of 100 m^3/s at 30°C DBT and 40% relative humidity. Air leaves at the top of the tower at 32°C and 90% relative humidity. Water enters the tower at 35°C and the water flow rate is 1.2 times the mass flow rate of air. Make-up water is supplied at 20°C. What are the range and approach of the tower? At what rate is heat absorbed from the load by the stream of water on its way back to the top of the tower? What percentage of the water flow rate must be supplied as make-up water to replace the water evaporated into the air stream? **[LO 15.4]**

(M) **15.16** An auditorium has to be air-conditioned (first by cooling and dehumidifying and then heating) for summer when the outdoor conditions are: dbt 35°C, RH 70%; indoor conditions: DBT 20°C, RH 60%, cooling coil dew point temperature 10°C. Amount of free air to be circulated is 300 m^3/min. Estimate (a) the capacity of the cooling coil and its bypass factor, (b) capacity of the heating coil and its surface to temperature when the bypass factor is 0.25 and (c) the mass of water vapour removed. **[LO 15.4]**

(M) **15.17** Water is drawn from a river at 17°C and has to be heated to 64°C. Calculate the advantage of using the heat pump plant described below over direct heating of the water. Assume that a heat source above 450°C is available.

(a) *R–12 heat pump:* Compressor driven by steam turbine; condenser temperature = 60°C and evaporator temperature = – 5°C; R–12 compressor as saturated vapour and enters throttle valve as saturated liquid.

(b) *Steam turbine plant:* Steam pressure and temperature leaving the boiler at 30 bar and 450°C, respectively; condenser temperature = 76°C.

Take 100% efficiency for turbine and compressor and neglect all heat losses. The water drawn from the river passes through the refrigerant condenser and then through steam condenser. **[LO 15.2, 15.4]**

ANSWERS TO PROBLEMS

15.1 3.467 kPa, 0.02204 kg vap/dry air, 42.28%

15.8 (a) 10°C,
(b) 10.35 kg/h,
(c) 11.33 kW,
(d) 1.63 kW

15.9 (a) 0.0023 kg/kg d.a.
(b) 16.1 kJ/kg d.a.

15.10 2224 kg/h, 2.75 kW

15.11 10°C, 25 kW, 6.04 kW

15.12 23.4 kg/h, 20.5 kW

15.13 2088 kg/h, 62.9 kg/h

15.14 0.0864 kJ/kg air, 220 kJ/kg air

15.15 Range = 8.7°C, Approach = 6.3°C, Q = 5005 kW, % make-up = 1.39%

15.16 (a) 100.8 tonnes, 5.9%
(b) 44.1 kW, 22.67°C
(c) 5.226 kg/min

15.17 1.92

16

Reactive Systems

Learning Objectives

LO 16.1 State thermodynamic equilibrium of chemically reacting gas mixture

LO 16.2 Discuss degree of reaction

LO 16.3 Explain reaction equilibrium

LO 16.4 Demonstrate heat capacity of reacting gases

LO 16.5 Formulate the law of mass action

LO 16.6 Formulate the equation of heat of reaction

LO 16.7 Explain Saha ionization equation

LO 16.8 Illustrate Gibbs free energy change

LO 16.9 Illustrate Fugacity and Activity

LO 16.10 Describe fuels and combustion

LO 16.11 Discuss enthalpy of formation

LO 16.12 Illustrate first law for reactive systems

LO 16.13 Interpret adiabatic flame temperature

LO 16.14 Explain internal energy and enthalpy of combustion or heat values

LO 16.15 Discuss the third law of thermodynamics and absolute entropy

LO 16.16 Relate second law analysis of reactive systems with Gibbs function of formation

LO 16.17 Interpret chemical exergy

LO 16.18 Express second law efficiency of a reactive system

LO 16.19 Discuss thermodynamics of combustion in internal combustion engines

LO 16.20 Discuss thermodynamics of combustion in boilers

LO 16.21 Describe fuel cells

LO 16.1

State thermodynamic equilibrium of chemically reacting gas mixture

In this chapter, we shall study the thermodynamics of mixtures that may be undergoing chemical reaction. With every chemical reaction is associated a chemical equation which is obtained by balancing the atoms of each of the atomic species involved in the reaction. The initial constituents which start the reaction are called the *reactants*, and the final constituents which are formed by chemical reaction with the rearrangement of the atoms and electrons are called the *products*. The reaction between the reactants, hydrogen and oxygen, to form the product water can be expressed as

$$H_2 + \frac{1}{2} O_2 \rightleftharpoons H_2O \tag{16.1}$$

The equation indicates that one mole of hydrogen and half a mole of oxygen combine to form one mole of water. The reaction can also proceed in the reverse direction. The coefficients 1, $\frac{1}{2}$, 1 in the chemical Eq. (16.1) are called *stoichiometric coefficients.*

16.1 DEGREE OF REACTION

LO 16.2

Discuss degree of reaction

Let us suppose that we have a mixture of four substances, A_1, A_2, A_3 and A_4, capable of undergoing a reaction of the type

$$\nu_1 A_1 + \nu_2 A_2 \rightleftharpoons \nu_3 A_3 + \nu_4 A_4$$

where the ν's are the stoichiometric coefficients.

Starting with arbitrary amounts of both initial and final constituents, let us imagine that the reaction proceeds completely to the right with the disappearance of at least one of the initial constituents, say, A_1. Then the original number of moles of the initial constituents is given in the form

$$n_1 \text{ (original)} = n_0 \nu_1$$
$$n_2 \text{ (original)} = n_0 \nu_2 + N_2$$

where n_0 is an arbitrary positive number, and N_2 is the residue (or excess) of A_2, i.e. the number of moles of A_2 which cannot combine. If the reaction is assumed to proceed completely to the left with the disappearance of the final constituent A_3, then

$$n_3 \text{ (original)} = n'_0 \nu_3$$
$$n_4 \text{ (original)} = n'_0 \nu_4 + N_4$$

where n'_0 is an arbitrary positive number and N_4 is the excess number of moles of A_4 left after the reaction is complete from right to left.

For a reaction that has occurred completely to the left, there is a maximum amount possible of each initial constituent, and a minimum amount possible of each final constituent, so that

$$n_1\text{(max)} = \underset{\text{(Original number of moles of } A_1)}{n_0 \nu_1} + \underset{\text{(Number of moles of } A_1 \text{ formed by chemical reaction)}}{n'_0 \nu_1}$$

$$(n'_0 \nu_3 A_3 + n'_0 \nu_4 A_4 \rightarrow n'_0 \nu_1 A_1 + n'_0 \nu_2 A_2)$$

$$= (n_0 + n'_0)\, \nu_1$$

n_2 (max) $= (n_0\ \nu_2 + N_2)$ + $n'_0\ \nu_2$

(Original number of moles of A_2) (Number of moles of A_2 formed by chemical reaction)

$= (n_0 + n'_0)\ \nu_2 + N_2$

n_3 (min) = 0 (The constituent A_3 completely disappears by reaction)

n_4 (min) = N_4 (The excess number of moles of A_4 that are left after the reaction is complete to the left)

Similarly, if the reaction is imagined to proceed completely to the right, there is a minimum amount of each initial constituent and a maximum amount of each final constituent, so that

$$n_1\ (\text{min}) = 0$$

$$n_2\ (\text{min}) = N_2$$

$$n_3\ (\text{max}) = n'_0\ \nu_3 + n_0\ \nu_3$$

(Original amount) (Amount formed by chemical reaction)

$$(n_0\ \nu_1\ A_1 + n_0\ \nu_2\ A_2 \rightarrow n_0\ \nu_3\ A_3 + n_0\ \nu_4\ A_4)$$

$$= (n_0 + n'_0)\nu_3$$

$$n_4\ (\text{max}) = (n_0 + n'_0)\ \nu_4 + N_4$$

Let us suppose that the reaction proceeds partially either to the right or to the left to the extent that there are n_1 moles of A_1, n_2 moles of A_2, n_3 moles of A_3 and n_4 moles of A_4. The *degree (or advancement) of reaction* ε is defined in terms of any one of the initial constituents, say, A_1, as the fraction

$$\varepsilon = \frac{n_1(\text{max}) - n_1}{n_1(\text{max}) - n_1(\text{min})}$$

It is seen that when $n_1 = n_1$ (max), $\varepsilon = 0$, the reaction will start from left to right. When $n_1 = n_1$ (min), $\varepsilon = 1$, reaction is complete from left to right.
The degree of reaction can thus be written in the form

$$\varepsilon = \frac{(n_0 + n'_0)\nu_1 - n_1}{(n_0 + n'_0)\nu_1}$$

Therefore,

$$n_1 = (n_0 + n'_0)\ \nu_1 - (n_0 + n'_0)\ \nu_1\ \varepsilon$$

$= n$ (at start) $- n$ (consumed)

= Number of moles of A_1 at start – number of moles of A_1 consumed in the reaction

$$= (n_0 + n'_0)\ \nu_1(1 - \varepsilon)$$

$n_2 = n$ (at start) $- n$ (consumed)

$$= (n_0 + n'_0)\ \nu_2 + N_2 - (n_0 + n'_0)\ \nu_2\ \varepsilon$$

$$= (n_0 + n'_0)\ \nu_2\ (1 - \varepsilon) + N_2$$

$n_3 = n$ (at start) $+ n$ (formed)

$$= 0 + (n_0 + n'_0)\ \nu_3\ \varepsilon$$

$$= (n_0 + n'_0)\ \nu_3\ \varepsilon$$

$$n_4 = n \text{ (at start)} + n \text{ (formed)}$$
$$= N_4 + (n_0 + n'_0)\, \nu_4\, \varepsilon$$
$$= (n_0 + n'_0)\, \nu_4\, \varepsilon + N_4 \tag{16.2}$$

The number of moles of the constituents change during a chemical reaction, not independently but restricted by the above relations. These equations are the *equations of constraint*. The n's are functions of ε only. In a homogeneous system, in a given reaction, the mole fraction x's are also functions of ε only, as illustrated below.

Let us take the reaction

$$H_2 + \frac{1}{2} O_2 \rightleftharpoons H_2O$$

in which n_0 moles of hydrogen combine with $n_0/2$ moles of oxygen to form n_0 moles of water. The n's and x's as functions of ε are shown in Table 16.1 given below.

Table 16.1

A	ν	n	x
$A_1 = H_2$	$\nu_1 = 1$	$n_1 = n_0 (1 - \varepsilon)$	$x_1 = \frac{n_1}{\Sigma n} = \frac{2(1-\varepsilon)}{3-\varepsilon}$
$A_2 = O_2$	$\nu_2 = \frac{1}{2}$	$n_2 = \frac{n_0}{2}(1 - \varepsilon)$	$x_2 = \frac{1-\varepsilon}{3-\varepsilon}$
$A_3 = H_2O$	$\nu_3 = 1$	$n_3 = n_0\, \varepsilon$	$x_3 = \frac{2\varepsilon}{3-\varepsilon}$
		$\Sigma n = \frac{n_0}{2}(3 - \varepsilon)$	

If the reaction is imagined to advance to an infinitesimal extent, the degree of reaction changes from ε to $\varepsilon + d\varepsilon$, and the various n's will change by the amounts

$$dn_1 = -(n_0 + n'_0)\, \nu_1\, d\varepsilon$$
$$dn_2 = -(n_0 + n'_0)\, \nu_2\, d\varepsilon$$
$$dn_3 = (n_0 + n'_0)\, \nu_3\, d\varepsilon$$
$$dn_4 = (n_0 + n'_0)\, \nu_4\, d\varepsilon$$

or

$$\frac{dn_1}{-\nu_1} = \frac{dn_2}{-\nu_2} = \frac{dn_3}{\nu_3} = \frac{dn_4}{\nu_4} = (n_0 + n'_0)\, d\varepsilon$$

which shows that the dn's are proportional to the ν's.

16.2 REACTION EQUILIBRIUM

LO 16.3 Explain reaction equilibrium

Let us consider a homogeneous phase having arbitrary amounts of the constituents, A_1, A_2, A_3 and A_4, capable of undergoing the reaction

$$\nu_1 A_1 + \nu_2 A_2 \rightleftharpoons \nu_3 A_3 + \nu_4 A_4$$

The phase is at uniform temperature T and pressure p. The Gibbs function of the mixture is

$$G = \mu_1 n_1 + \mu_2 n_2 + \mu_3 n_3 + \mu_4 n_4$$

where the n's are the number of moles of the constituents at any moment, and the μ's are the chemical potentials.

Let us imagine that the reaction is allowed to take place at constant T and p. The degree of reaction changes by an infinitesimal amount from ε to $\varepsilon + d\varepsilon$. The change in the Gibbs function is

$$\begin{aligned} dG_{T,p} &= \Sigma\, \mu_k\, dn_k \\ &= \mu_1\, dn_1 + \mu_2\, dn_2 + \mu_3\, dn_3 + \mu_4\, dn_4 \end{aligned}$$

The equations of constraint in differential form are

$$dn_1 = -(n_0 + n'_0)\, \nu_1\, d\varepsilon, \qquad dn_3 = (n_0 + n'_0)\, \nu_3\, d\varepsilon$$

$$dn_2 = -(n_0 + n'_0)\, \nu_2\, d\varepsilon, \qquad dn_4 = (n_0 + n'_0)\, \nu_4\, d\varepsilon$$

On substitution

$$dG_{T,p} = (n_0 + n'_0)(-\nu_1\mu_1 - \nu_2\mu_2 + \nu_3\mu_3 + \nu_4\mu_4)\, d\varepsilon \qquad (16.3)$$

When the reaction proceeds *spontaneously* to the right, $d\varepsilon$ is positive, and since $dG_{T,p} < 0$

$$(\nu_1\mu_1 + \nu_2\mu_2) > (\nu_3\mu_3 + \nu_4\mu_4)$$

When the reaction proceeds *spontaneously* to the left, $d\varepsilon$ is negative

$$(\nu_1\mu_1 + \nu_2\mu_2) < (\nu_3\mu_3 + \nu_4\mu_4)$$

At equilibrium, the Gibbs function will be minimum, and $dG_{T,p} = 0$

or
$$\nu_1\mu_1 + \nu_2\mu_2 = \nu_3\mu_3 + \nu_4\mu_4 \qquad (16.4)$$

which is called the *equation of reaction equilibrium*.

Therefore, it is the value of $\Sigma \nu_k\, \mu_k$ which causes or forces the spontaneous reaction and is called the '*chemical affinity*'.

16.3 HEAT CAPACITY OF REACTING GASES IN EQUILIBRIUM

LO 16.4

Demonstrate heat capacity of reacting gases

For a reaction of four ideal gases, such as

$$\nu_1 A_1 + \nu_2 A_2 \rightleftharpoons \nu_3 A_3 + \nu_4 A_4$$

the enthalpy of mixture at equilibrium is $H = \Sigma\, n_k\, h_k$

where
$$n_1 = (n_0 + n'_0)\, \nu_1(1 - \varepsilon_e),\; n_2 = (n_0 + n'_0)\, \nu_2$$
$(1 - \varepsilon_e) + N_2$

$$n_3 = (n_0 + n'_0)\, \nu_3\, \varepsilon_e \text{ and } n_4 = (n_0 + n'_0)\, \nu_4\, \varepsilon_e + N_4$$

Let us suppose that an infinitesimal change in temperature takes place at constant pressure in such a way that equilibrium is maintained.
ε_e will change to the value $\varepsilon_e + d\varepsilon_e$, and the enthalpy will change by the amount

$$dH_P = \Sigma\, n_k\, dh_k + \Sigma\, h_k\, dn_k$$

where
$$dh_k = c_{pk}\, dT,\; dn_k = \pm\, (n_0 + n'_0)\, \nu_k\, d\varepsilon_e$$

Therefore,

$$dH_p = \Sigma n_k c_{pk} dT + (n_0 + n'_0)(v_3 h_3 + v_4 h_4 - v_1 h_1 - v_2 h_2)\, d\varepsilon_e$$

The heat capacity of the reacting gas mixture is

$$C_p = \left(\frac{\partial H}{\partial T}\right)_p = \Sigma n_k c_{pk} + (n_0 + n'_0)\,\Delta H \left(\frac{\partial \varepsilon_e}{\partial T}\right)_p$$

Using Eq. (16.12)

$$C_p = \Sigma n_k c_{pk} + (n_0 + n'_0) \frac{(\Delta H)^2}{\bar{R}T^2 \dfrac{d}{d\varepsilon} \ln \left[\dfrac{x_3^{v3} \cdot x_4^{v4}}{x_1^{v1} \cdot x_2^{v2}}\right]_{\varepsilon=\varepsilon_e}} \tag{16.5}$$

16.4 LAW OF MASS ACTION

LO 16.5 Formulate the law of mass action

For a homogeneous phase chemical reaction at constant temperature and pressure, when the constituents are *ideal gases*, the chemical potentials are given by the expression of the type

$$\mu_k = \bar{R}T\ (\phi_k + \ln p + \ln x_k)$$

where the ϕ's are functions of temperature only (Section 10.11).

Substituting in the equation of reaction equilibrium (16.4)

$$v_1 (\phi_1 + \ln p + \ln x_1) + v_2 (\phi_2 + \ln p + \ln x_2)$$
$$= v_3 (\phi_3 + \ln p + \ln x_3) + v_4 (\phi_4 + \ln p + \ln x_4)$$

On rearranging

$$v_3 \ln x_3 + v_4 \ln x_4 - v_1 \ln x_1 - v_2 \ln x_2 + (v_3 + v_4 - v_1 - v_2) \ln p$$
$$= -(v_3\phi_3 + v_4\phi_4 - v_1\phi_1 - v_2\phi_2)$$

$$\therefore \quad \ln \frac{x_3^{v_3} \cdot x_4^{v_4}}{x_1^{v_1} \cdot x_2^{v_2}} p^{v_3 + v_4 - v_1 - v_2}$$
$$= -(v_3\phi_3 + v_4\phi_4 - v_1\phi_1 - v_2\phi_2)$$

Denoting

$$\ln K = -(v_3\phi_3 + v_4\phi_4 - v_1\phi_1 - v_2\phi_2)$$

where K, known as the *equilibrium constant*, is a function of temperature only

$$\left[\frac{x_3^{v_3}\, x_4^{v_4}}{x_1^{v_1}\, x_2^{v_2}}\right]_{\varepsilon=\varepsilon_e} p^{v_3 + v_4 - v_1 - v_2} = K \tag{16.6}$$

This equation is called the *law of mass action*. K has the dimension of pressure raised to the $(v_3 + v_4 - v_1 - v_2)$th power. Here the x's are the values of mole fractions at equilibrium when the degree of reaction is ε_e. To make K dimentionless, K can be defined as

$$K = (x_3^{\gamma_3}\, x_4^{\gamma_4}) \cdot \left(\frac{p}{p_0}\right)^{(v_3 + v_4 - v_1 - v_2)}$$

where both p and p_0 are in atm, $p_0 = 1$ atm.

The law of mass action can also be written in this form

$$\frac{p_3^{\nu_3} \cdot p_4^{\nu_4}}{p_1^{\nu_1} \cdot p_2^{\nu_2}} = K$$

where the p's are the partial pressures.

16.5 HEAT OF REACTION

LO 16.6 Formulate the equation of heat of reaction

The equilibrium constant K is defined by the expression

$$\ln K = -(\nu_3\phi_3 + \nu_4\phi_4 - \nu_1\phi_1 - \nu_2\phi_2)$$

Differentiating $\ln K$ with respect to T

$$\frac{d}{dT}\ln K = -\left(\nu_3\frac{d\phi_3}{dT} + \nu_4\frac{d\phi_4}{dT} - \nu_1\frac{d\phi_1}{dT} - \nu_2\frac{d\phi_2}{dT}\right)$$

Now, from Eq. (10.100) in Chapter 10

$$\phi = \frac{h_0}{\bar{R}T} - \frac{1}{\bar{R}}\int\frac{\int c_p dT}{T^2}\cdot dT - \frac{s_0}{\bar{R}}$$

Therefore,

$$\frac{d\phi}{dT} = -\frac{h_0}{\bar{R}T^2} - \frac{\int c_p dT}{\bar{R}T^2}$$

$$= -\frac{1}{\bar{R}T^2}\left(h_0 + \int c_p dT\right) = -\frac{h}{\bar{R}T^2}$$

Therefore,

$$\frac{d}{dT}\ln K = \frac{1}{\bar{R}T^2}(\nu_3h_3 + \nu_4h_4 - \nu_1h_1 - \nu_2h_2)$$

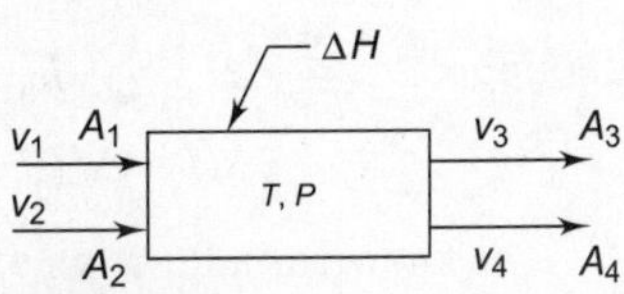

Fig. 16.1 Heat of reaction

where the h's refer to the same temperature T and the same pressure p. If ν_1 moles of A_1 and ν_2 moles of A_2 combine to form ν_3 moles of A_3 and ν_4 moles of A_4 at constant temperature and pressure, the heat transferred would be, as shown in Fig. 16.1, equal to the final enthalpy $(\nu_3h_3 + \nu_4h_4)$ minus the initial enthalpy $(\nu_1h_1 + \nu_2h_2)$. This is known as the *heat of reaction*, and is denoted by ΔH.

$$\Delta H = \nu_3h_3 + \nu_4h_4 - \nu_1h_1 - \nu_2h_2 \tag{16.7}$$

Therefore, $$\frac{d}{dT}\ln K = \frac{\Delta H}{\bar{R}T^2} \tag{16.8}$$

This is known as the *van't Hoff equation*. The equation can be used to calculate the heat of reaction at any desired temperature or within a certain temperature range.

By rearranging. Eq. (16.8)

$$\frac{\frac{d\ln K}{dT}}{T^2} = \frac{\Delta H}{\bar{R}}$$

$$\frac{d\ln K}{d(1/T)} = -\frac{\Delta H}{\bar{R}}$$

Therefore,

$$\Delta H = -2.303\,\bar{R}\frac{d\log K}{d\left(\frac{1}{T}\right)} = -19.148\frac{d\log K}{d\left(\frac{1}{T}\right)}\ \text{kJ/kg mol}$$

If K_1 and K_2 are the equilibrium constants evaluated at temperatures T_1 and T_2, respectively.

$$\Delta H = -19.148\frac{\log K_1 - \log K_2}{\frac{1}{T_1} - \frac{1}{T_2}}$$

or

$$\Delta H = 19.148\frac{T_1T_2}{T_1 - T_2}\log\frac{K_1}{K_2}$$

If ΔH is positive, the reaction is said to be *endothermic*. If ΔH is negative, the reaction is *exothermic*.

16.6 TEMPERATURE DEPENDENCE OF THE HEAT OF REACTION

$$\Delta H = \nu_3 h_3 + \nu_4 h_4 - \nu_1 h_1 - \nu_2 h_2$$

$$h = h_0 + \int c_p dT$$

Therefore,

$$\Delta H = \nu_3 h_{03} + \nu_4 h_{04} - \nu_1 h_{01} - \nu_2 h_{02} + \int(\nu_3 c_{p3} + \nu_4 c_{p4} - \nu_1 c_{p1} - \nu_2 c_{p2})dT$$

Denoting

$$\Delta H_0 = \nu_3 h_{03} + \nu_4 h_{04} - \nu_1 h_{01} - \nu_2 h_{02}$$

$$\Delta H = \Delta H_0 + \int(\nu_3 c_{p3} + \nu_4 c_{p4} - \nu_1 c_{p1} - \nu_2 c_{p2})dT$$

If c_p is known as a function of temperature and if at any temperature ΔH is known, than at any other temperature, ΔH can be determined for a certain chemical reaction from the above relation.

16.7 TEMPERATURE DEPENDENCE OF THE EQUILIBRIUM CONSTANT

$$\ln K = -(\nu_3\phi_3 + \nu_4\phi_4 - \nu_1\phi_1 - \nu_2\phi_2)$$

where

$$\phi = \frac{h_0}{\bar{R}T} - \frac{1}{\bar{R}}\int\frac{\int c_p dT}{T^2}dT - \frac{s_0}{\bar{R}}$$

On substitution

$$\ln K = \frac{1}{\bar{R}T}(\nu_3 h_{03} + \nu_4 h_{04} - \nu_1 h_{01} - \nu_2 h_{02})$$

$$+\frac{1}{\bar{R}}\int\frac{\int(\nu_3c_{p3}+\nu_4c_{p4}-\nu_1c_{p1}-\nu_2c_{p2})dT}{T^2}\cdot dT$$

$$+\frac{1}{\bar{R}}(\nu_3\, s_{03}+\nu_4\, s_{04}-\nu_1\, s_{01}-\nu_2\, s_{02})$$

If
$$\Delta H_0 = \nu_3 h_{03} + \nu_4 h_{04} - \nu_1 h_{01} - \nu_2 h_{02}$$
$$\Delta S_0 = \nu_3 s_{03} + \nu_4 s_{04} - \nu_1 s_{01} - \nu_2 s_{02}$$

$$\ln K = -\frac{\Delta H_0}{\bar{R}T}+\frac{1}{\bar{R}}\int\frac{\int(\nu_3c_{p3}+\nu_4c_{p4}-\nu_1c_{p1}-\nu_2c_{p2})dT}{T^2}dT+\frac{\Delta S_0}{\bar{R}} \quad (16.9)$$

This equation is sometimes known as the *Nernst's equation.*

16.8 THERMAL IONISATION OF A MONATOMIC GAS

LO 16.7
Explain Saha ionization equation

One interesting application of Nernst's equation was made by Dr. M.N. Saha to the thermal ionisation of a monatomic gas. If a monatomic gas is heated to a high enough temperature (5000 K and above), some ionisation occurs, with the electrons in the outermost orbit being shed off, and the atoms, ions and electrons may be regarded as a mixture of three ideal monatomic gases, undergoing the reaction

$$A \rightleftharpoons A^+ + e$$

Starting with n_0 moles of atoms, it is shown in Table 16.2.

Table 16.2

A	ν	n	x
$A_1 = A$	$\nu_1 = 1$	$n_1 = n_0(1-\varepsilon_e)$	$x_1 = \frac{1-\varepsilon_e}{1+\varepsilon_e}$
$A_3 = A^+$	$\nu_3 = 1$	$n_3 = n_0\,\varepsilon_e$	$x_3 = \frac{\varepsilon_e}{1+\varepsilon_e}$
$A_4 = e$	$\nu_4 = 1$	$n_4 = n_0\,\varepsilon_e$ $\Sigma n = n_0(1+\varepsilon_e)$	$x_4 = \frac{\varepsilon_e}{1+\varepsilon_e}$

$$\ln K = \ln\left\{\frac{x_3^{\nu_3}\cdot x_4^{\nu_4}}{x_1^{\nu_1}\cdot x_2^{\nu_2}}\right\}_{\varepsilon_e}\cdot p^{\nu_3+\nu_4-\nu_1-\nu_2}$$

$$\ln K = \ln\frac{\frac{\varepsilon_e}{1+\varepsilon_e}\cdot\frac{\varepsilon_e}{1+\varepsilon_e}}{\frac{1-\varepsilon_e}{1+\varepsilon_e}}\cdot p^{\nu_3+\nu_4-\nu_1} = \ln\frac{\varepsilon_e^2}{1-\varepsilon_e^2}\cdot p$$

Since the three gases are monatomic, $c_p = \frac{5}{2}\bar{R}$ which, on being substituted in the Nernst's equation, gives

$$\ln \frac{\varepsilon_e^2}{1-\varepsilon_e^2} \cdot p = -\frac{\Delta H_0}{\bar{R}T} + \frac{5}{2} \ln T + \ln B \tag{16.10}$$

where

$$\frac{\Delta S_0}{\bar{R}} = \ln B$$

$$\ln \frac{\varepsilon_e^2}{1-\varepsilon_e^2} \cdot \frac{p}{T^{5/2} \cdot B} = -\frac{\Delta H_0}{\bar{R}T}$$

$$\therefore \quad \frac{\varepsilon_e^2}{1-\varepsilon_e^2} = Be^{-\Delta H_0/\bar{R}T} \cdot \frac{T^{5/2}}{p} \tag{16.11}$$

where ε_e is the equilibrium value of the degree of ionisation. This is known as the *Saha's equation*. For a particular gas the degree of ionisation increases with an increase in temperature and a decrease in pressure. Here, ΔH_0 is the amount of energy necessary to ionise 1 g mol of atoms and is called the '*ionisation potential*'.

16.9 DISPLACEMENT OF EQUILIBRIUM DUE TO A CHANGE IN TEMPERATURE OR PRESSURE

The degree of reaction at equilibrium ε_e changes with temperature and also with pressure. From the law of mass action

$$\ln K = \ln \left[\frac{x_3^{\nu 3} \cdot x_4^{\nu 4}}{x_1^{\nu 1} \cdot x_2^{\nu 2}} \right]_{\varepsilon=\varepsilon_e} + (\nu_3 + \nu_4 - \nu_1 - \nu_2) \ln p$$

where ln K is a function of temperature only and the first term on the right-hand side is a function of ε_e only. Therefore,

$$\left(\frac{\partial \varepsilon_e}{\partial T}\right)_p = \left(\frac{\partial \varepsilon_e}{\partial \ln K}\right)_p \left(\frac{\partial \ln K}{\partial T}\right)_p$$

$$= \frac{\left(\frac{\partial \ln K}{\partial T}\right)_p}{\left(\frac{\partial \ln k}{\partial \varepsilon_e}\right)_p} = \frac{\Delta H}{\bar{R}T^2 \dfrac{d}{d\varepsilon_e} \ln \left[\dfrac{x_3^{\nu 3} \cdot x_4^{\nu 4}}{x_1^{\nu 1} \cdot x_2^{\nu 2}} \right]_{\varepsilon=\varepsilon_e}} \tag{16.12}$$

It can be shown (Example 16.4) that

$$\frac{d}{d\varepsilon} \ln \frac{x_3^{\nu 3} \cdot x_4^{\nu 4}}{x_1^{\nu 1} \cdot x_2^{\nu 2}} = \frac{n_0}{\Sigma n_k} \cdot \frac{(\nu_1 + \nu_2)(\nu_3 + \nu_4)}{\varepsilon (1-\varepsilon)}$$

which is always positive.

Therefore, for endothermic reaction, when ΔH is positive, $\left(\dfrac{\partial \varepsilon_e}{\partial T}\right)_p$ is positive, and for exothermic reaction, when ΔH is negative, $\left(\dfrac{\partial \varepsilon_e}{\partial T}\right)_p$ is negative.

Again,
$$\left(\frac{\partial \varepsilon_e}{\partial p}\right)_T = -\left(\frac{\partial \varepsilon_e}{\partial \ln K}\right)_p \left(\frac{\partial \ln K}{\partial p}\right)_{\varepsilon_e}$$

$$= \frac{-\left(\dfrac{\partial \ln K}{\partial p}\right)_{\varepsilon_e}}{\left(\dfrac{\partial \ln K}{\partial \varepsilon_e}\right)_p}$$

Using the law of mass action

$$\left(\frac{\partial \varepsilon_e}{\partial p}\right)_T = -\frac{\nu_3 + \nu_4 - \nu_1 - \nu_2}{p \dfrac{d}{d\varepsilon_e} \ln \left[\dfrac{x_3^{\nu 3} \cdot x_4^{\nu 4}}{x_1^{\nu 1} \cdot x_2^{\nu 2}}\right]_{\varepsilon=\varepsilon_e}} \tag{16.13}$$

If $(\nu_3 + \nu_4) > (\nu_1 + \nu_2)$, i.e. the number of moles increase or the volume increases due to reaction, $\left(\dfrac{\partial \varepsilon_e}{\partial p}\right)_T$ is negative. If $(\nu_3 + \nu_4) < (\nu_1 + \nu_2)$, i.e. volume decreases in an isothermal reaction, $\left(\dfrac{\partial \varepsilon_e}{\partial p}\right)_T$ is positive.

16.10 GIBBS FUNCTION CHANGE

> **LO 16.8**
> Illustrate Gibbs free energy change

Molar Gibbs function of an ideal gas at temperature T and pressure p is equal to (Section 10.11)

$$g = \bar{R}T\,(\phi + \ln p)$$

For the reaction of the type

$$\nu_1 A_1 + \nu_2 A_2 \rightleftharpoons \nu_3 A_3 + \nu_4 A_4$$

the Gibbs function change of the reaction ΔG is defined by the expression

$$\Delta G = \nu_3\, g_3 + \nu_4\, g_4 - \nu_1\, g_1 - \nu_2\, g_2 \tag{16.14}$$

where the g's refer to the gases completely separated at T, p. ΔG is also known as the *free energy change*. Substituting the values of the g's

$$\Delta G = \bar{R}T\,(\nu_3\,\phi_3 + \nu_4\,\phi_4 - \nu_1\,\phi_1 - \nu_2\,\phi_2) + \bar{R}T \ln p^{\nu_3 + \nu_4 - \nu_1 - \nu_2}$$

But
$$\ln K = -(\nu_3\,\phi_3 + \nu_4\,\phi_4 - \nu_1\,\phi_1 - \nu_2\,\phi_2)$$

$\therefore$
$$\Delta G = -\bar{R}T \ln K + \bar{R}T \ln p^{\nu_3 + \nu_4 - \nu_1 - \nu_2}$$

If p is expressed in atmospheres and ΔG calculated from each gas is at a pressure of 1 atm., the second term on the right drops out. Under these conditions ΔG is known as the *standard Gibbs function change* and is denoted by ΔG°

$$\Delta G^\circ = -\bar{R}T \ln K \tag{16.15}$$

This is an important equation which relates the standard Gibbs function change with temperature and the equilibrium constant. From this the equilibrium constant can be calculated from changes in the standard Gibbs function or vice versa.

For dissociation of water vapour, $H_2O \rightleftharpoons H_2 + \frac{1}{2} O_2$, the value of $\ln K_{298}$ is found to be –93.7. Therefore,

$$\Delta G^\circ_{298} = -8.3143 \times 298 \times (-93.7)$$
$$= 232{,}157 \text{ J/gmol.}$$

Substituting ln K from Nernst's equation

$$\Delta G^\circ = \Delta H_0 - T \frac{(v_3 c_{p3} + v_4 c_{p4} - v_1 c_{p1} - v_2 c_{p2}) dT}{T^2} dT - T \Delta S_0$$

from which also ΔG° may be calculated directly. Values of ΔH_0, ΔS_0 and ΔG°_{298} for fundamental ideal gas reactions are given in Table 16.3.

Both ΔS_0 and ΔG°_{298} may be added and subtracted in the same manner as ΔH_0. For example

$$H_2O \rightleftharpoons H_2 + \frac{1}{2} O_2 \qquad \Delta G^\circ_{298} = 232{,}545 \text{ J/gmol}$$

$$CO_2 \rightleftharpoons CO + \frac{1}{2} O_2 \qquad \Delta G^\circ_{298} = 258{,}940 \text{ J/gmol}$$

$$CO + H_2O \rightleftharpoons CO_2 + H_2 \quad \Delta G^\circ_{298} = -26{,}395 \text{ J/gmol}$$

$$\ln K_{298} = \frac{\Delta G^\circ_{298}}{\bar{R}T} = \frac{26{,}395}{8.3143 \times 298} = 10.653$$

$$K_{298} = 42{,}330$$

From this, the value of the degree of reaction at equilibrium ε_e may be calculated. From Eq. (16.3), if $n_0 = 1$ and $n'_0 = 0$

Table 16.3 Fundamental ideal gas reactions

Reaction	ΔH_0, *J/gmol*	ΔS_0 *J/gmol-K*	ΔG°_{298} *J/gmol*
$H_2 \rightleftharpoons 2H$	427,380	4.90	404,335
$HCl \rightleftharpoons \frac{1}{2} H_2 + \frac{1}{2} Cl_2$	91,760	– 22.25	95,110
$HBr \rightleftharpoons \frac{1}{2} H_2 + \frac{1}{2} Br_2$	50,280	– 24.05	54,050

Reaction	ΔH_0, *J/gmol*	ΔS_0 *J/gmol-K*	ΔG°_{298} *J/gmol*
$HI \rightleftharpoons \frac{1}{2}H_2 + \frac{1}{2}I_2$	5,320	– 21.00	8,380
$NO \rightleftharpoons \frac{1}{2}N_2 + \frac{1}{2}O_2$	– 90,500	– 10.48	– 87,570
$H_2O \rightleftharpoons H_2 + \frac{1}{2}O_2$	239,250	14.70	232,545
$H_2S \rightleftharpoons H_2 + \frac{1}{2}S_2$	80,450	6.90	71,230
$CO_2 \rightleftharpoons CO + \frac{1}{2}O_2$	279,890	18.69	258,940
$NO_2 \rightleftharpoons NO + \frac{1}{2}O_2$	59,500	11.44	37,290
$SO_2 \rightleftharpoons \frac{1}{2}S_2 + O_2$	349,025	3.77	329,330
$NH_3 \rightleftharpoons \frac{1}{2}N_2 + \frac{3}{2}H_2$	39,800	40.27	16,380
$SO_3 \rightleftharpoons SO_2 + \frac{1}{2}O_2$	94,690	89.50	67,880
$N_2O_4 \rightleftharpoons 2NO_2$	56,980	174.30	5,030

$$\left(\frac{\partial G}{\partial \varepsilon}\right)_{T,p} = (\nu_3\mu_3 + \nu_4\mu_4 - \nu_1\mu_1 - \nu_2\mu_2)$$

Since $\quad \mu_k = \bar{R}T\,(\phi_k + \ln p + \ln x_k)$

and $\quad g_k = \bar{R}T\,(\phi_k + \ln p)$

so $\quad \mu_k = g_k\,\bar{R}T \ln x_k$

Therefore,

$$\left(\frac{\partial G}{\partial \varepsilon}\right)_{T,p} = \nu_3 g_3 + \nu_4 g_4 - \nu_1 g_1 - \nu_2 g_2 + \bar{R}T \ln \frac{x_3^{\nu_3}\, x_4^{\nu_4}}{x_1^{\nu_1}\, x_2^{\nu_2}}$$

$$= \Delta G + \bar{R}T \ln \frac{x_3^{\nu_3}\, x_4^{\nu_4}}{x_1^{\nu_1}\, x_2^{\nu_2}}$$

At $\quad \varepsilon = 0,\ x_3 = 0,\ x_4 = 0$

$$\left(\frac{\partial G}{\partial \varepsilon}\right)_{T,p} = -\infty$$

and at $$\varepsilon = 1, x_1 = 0, x_2 = 0$$

$$\left(\frac{\partial G}{\partial \varepsilon}\right)_{T,p} = +\infty$$

At $$\varepsilon = \frac{1}{2}, x_1 = \frac{\nu_1}{\Sigma\nu}, x_2 = \frac{\nu_2}{\Sigma\nu}$$

$$x_3 = \frac{\nu_3}{\Sigma\nu}, x_4 = \frac{\nu_4}{\Sigma\nu}$$

where $$\Sigma\nu = \nu_3 + \nu_4 + \nu_1 + \nu_2$$

$$\left(\frac{\partial G}{\partial \varepsilon}\right)_{\substack{T,p\\ \varepsilon=\frac{1}{2}}} = \Delta G + \bar{R}T \ln \frac{\left(\frac{\nu_3}{\Sigma\nu}\right)^{\nu_3}\left(\frac{\nu_4}{\Sigma\nu}\right)^{\nu_4}}{\left(\frac{\nu_1}{\Sigma\nu}\right)^{\nu_1}\left(\frac{\nu_2}{\Sigma\nu}\right)^{\nu_2}}$$

If $$p = 1 \text{ atm., } T = 298 \text{ K}$$

$$\left(\frac{\partial G}{\partial \varepsilon}\right)_{\substack{p=1\,\text{atm}\\ T=298\,\text{K}\\ \varepsilon=\frac{1}{2}}} = \Delta G^\circ_{298} \tag{16.16}$$

because the magnitude of the second term on the right-hand side of the equation is very small compared to ΔG°. The slope $\left(\frac{\partial G}{\partial \varepsilon}\right)_{T,p}$ at $\varepsilon = \frac{1}{2}$ is called the 'affinity' of the reaction, and it is equal to ΔG° at the standard reference state. The magnitude of the slope at $\varepsilon = 1/2$ (Fig. 16.2) indicates the direction in which the reaction will proceed. For water vapour reaction, ΔG°_{298} is a large positive number, which indicates that the equilibrium point is far to the left of $\varepsilon = 1/2$, and therefore, ε_e is very small. Again for the reaction NO $\rightleftharpoons \frac{1}{2} N_2 + \frac{1}{2} O_2$, ΔG°_{298} is a large negative value, which shows that the equilibrium point is far to the right of $\varepsilon = 1/2$, and so ε_e is close to unity.

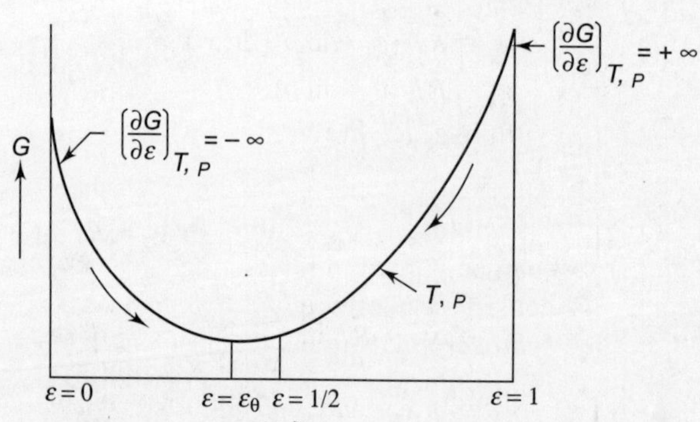

Fig. 16.2 Plot of G against ε at constants T and p

16.11 FUGACITY AND ACTIVITY

LO 16.9

Illustrate Fugacity and Activity

The differential of the Gibbs function of an ideal gas undergoing an isothermal process is

$$dG = Vdp = \frac{n\bar{R}T}{p}\,dp = n\bar{R}Td\,(\ln p)$$

Analogously, the differential of the Gibbs function for a real gas is

$$dG = n\bar{R}Td\,(\ln f) \tag{16.17}$$

where f is called the *fugacity*, first used by Lewis. The value of fugacity approaches the value of pressure as the latter tends to zero, i.e. when ideal gas conditions apply. Therefore,

$$\lim_{p\to 0} \frac{f}{p} = 1$$

This equation represents the Gibbs-Dalton Law. It states that, in a mixture of real gases, the fugacity of each gas component approaches its partial pressure as the pressure tends to zero. This can also be interpreted as real gases behaving like ideal gases when the pressure approaches zero.

For an ideal gas $f = p$. Fugacity has the same dimension as pressure. Integrating Eq. (16.17)

$$G - G^0 = n\bar{R}T \ln \frac{f}{f^0}$$

where G^0 and f^0 refer to the reference state when $p_0 = 1$ atm. The ratio f/f^0 is called the *activity*.

Therefore,

$$G - G^0 = n\bar{R}T \ln a \tag{16.18}$$

For ideal gases, the equilibrium constant is given by

$$K = \frac{p_3^{\nu 3} \cdot p_4^{\nu 4}}{p_1^{\nu 1} \cdot p_2^{\nu 2}}$$

For real gases

$$K_{\text{real}} = \frac{f_3^{\nu 3} \cdot f_4^{\nu 4}}{f_1^{\nu 1} \cdot f_2^{\nu 2}}$$

Similarly, it can be shown that

$$\Delta G^0 = -\,n\bar{R}T \ln K_{\text{real}} \text{ and } \frac{d \ln K_{\text{real}}}{dT} = \frac{\Delta H^0}{\bar{R}T^2}$$

16.12 INTRODUCTION TO FUELS AND COMBUSTION

LO 16.10

Describe fuels and combustion

A fuel is any material that can be burned to release thermal energy. All major fuels consist primarily of carbon and hydrogen. They are called *hydrocarbon fuels* and are denoted by the general formula C_nH_n. Hydrocarbon fuels exist in all phases like coal (s), gasoline (l) and natural gas (g).

Coal is the most abundant fuel and it contains, besides carbon, varying amounts of hydrogen, oxygen, nitrogen, sulphur, moisture and ash. Its composition varies widely from one geographical area to another. Most liquid hydrocarbon fuels are a mixture of many hydrocarbons and obtained from crude oil by distillation like gasoline, kerosene, diesel oil, lubricating oils, etc. The most volatile fuels obtained during distillation vapourise first, forming what we know as gasoline. The less volatile fuels obtained during distillation are kerosene, light diesel oil, heavy diesel oil and fuel oil.

Liquid hydrocarbon fuels are mixtures of many different hydrocarbons. However, for convenience, they are considered to be a single hydrocarbon, like *octane*, C_8H_{18} for gasoline petrol, and *dodecane*, $C_{12}H_{26}$ for diesel. Other common hydrocarbon fuels are methyl alcohol or *methanol*, CH_3OH and ethyl alcohol or ethanol, C_2H_5OH (largely obtained from starch in sugar factories).

Natural gas produced for gas wells or oil wells rich in natural gas, is composed of methane and small amounts of ethane, propane, CO_2, H_2O vapour, hydrogen, H_2S and nitrogen. Vehicles are run by compressed natural gas (CNG) which is a natural gas compressed to 150 to 250 atm or liquefied natural gas (LNG) which is mainly propane stored in liquid phase at –162°C. Liquefied petroleum gas (LPG) is a byproduct of natural gas processing or cude oil refining. Vehicles are the major source of air pollutants such as nitric noxides, carbon monoxide and hydrocarbons, as well as the greenhouse gas CO_2. Hence, transportation industry is gradually switching to *alternative fuels* such as gasohol (a blend of gasoline and alcohol), LNG, LPG and hydrogen. Electric and hybrid cars are also being increasingly used.

Combustion is a chemical reaction between a fuel and oxygen which proceeds at a fast rate with the release of energy in the form of heat. In the combustion of methane, e.g.

$$\underset{\text{Reactants}}{CH_4 + 2O_2} \rightarrow \underset{\text{Products}}{CO_2 + 2H_2O}$$

One mole of methane reacts with 2 moles of oxygen to form 1 mole of carbon dioxide and 2 moles of water. The water may be in the liquid or vapour phase depending on the temperature and pressure of the products of combustion. Only the initial and final products are being considered without any concern for the intermediate products that usually occur in a reaction.

Atmospheric air contains 21% oxygen, 78% nitrogen and 1% argon by volume. In combustion calculations, however, the argon is usually neglected, and air is assumed to consist of 21% oxygen and 79% nitrogen by volume (or molar basis). On a mass basis, air contains 23% oxygen and 77% nitrogen.

For each mole of oxygen taking part in a combustion reaction, there are 79.0/21.0 = 3.76 moles of nitrogen. So for the combustion of methane, the reaction can be written as

$$CH_4 + 2O_2 + 2(3.76)\, N_2 \rightarrow CO_2 + 2H_2O + 7.52\, N_2$$

The minimum amount of air which provides sufficient oxygen for the complete combustion of all the elements like carbon, hydrogen, etc. which may oxidise is called the *theoretical or stoichiometric air*. There is no oxygen in the products when complete combustion (oxidation) is achieved with this theoretical air. In practise, however, more air than this theoretical amount is required to be supplied for complete combustion. Actual air supplied is usually expressed in terms of percent theoretical air; 150% theoretical air means that 1.5 times the theoretical air is supplied. Thus, with 150% theoretical air, the methane combustion reaction can be written as

$$CH_4 + 2(1.5)\, O_2 + 2(3.76)\,(1.5)\, N_2$$
$$\rightarrow CO_2 + 2H_2O + O_2 + 11.28\, N_2$$

Another way of expressing the actual air quantity supplied is in terms of excess air. Thus 150% theoretical air means 50% excess air. *Equivalence ratio* is often used in this regard. It is the ratio of actual fuel-air ratio to the stoichiometric fuel-air ratio.

With less than theoretical air supply, combustion will remain incomplete with some CO present in the products. Even with excess air supply also, there may be a small amount of CO present, depending on mixing and turbulence during combustion, e.g. with 115% theoretical air

$$CH_4 + 2(1.15)\, O_2 + 2(1.15)\, 3.76\, N_2$$
$$\rightarrow 0.95\, CO_2 + 0.05\, CO + 2H_2O + 0.325\, O_2 + 8.65\, N_2$$

By analysing the product of combustion, the actual amount of air supplied in a combustion process can be computed. Such analysis is often given on the 'dry' basis, i.e. the fractional analysis of all the components, except water vapour. Following the principle of the conservation of mass of each of the elements, it is possible to make a carbon balance, hydrogen balance, oxygen balance and nitrogen balance from the combustion reaction equation, from which the actual air-fuel ratio can be determined. It has been illustrated in Example 16.7 later.

16.12.1 Stoichiometric Calculation and Exhaust Gas Analysis

Combustion is an exothermic reaction in which the combustible elements of a fuel react with oxygen from the air. The amount of oxygen (in kg) required for the complete combustion of 1 kg of fuel can be calculated using the mass balance of the balanced chemical equations for the reactions between the combustible elements and oxygen. These calculations are initially performed on a mole basis and can then be converted using known data for the molar mass of each combustible element. Only the reactions corresponding to the complete combustion of each element are considered in this calculation. The oxygen required for the complete combustion of 1 kg of fuel, with known compositions, is determined. This value represents the theoretical minimum oxygen required for complete combustion of 1 kg of fuel.

Since the mass percentage of oxygen in the air is also known, we can calculate the theoretically minimum air required for the complete combustion of 1 kg of fuel. This is referred to as the stoichiometric air. The ratio of the mass of this air to the mass of fuel is known as the stoichiometric air-fuel ratio.

The combustible elements in a fuel, particularly for common hydrocarbon fossil fuels, are carbon (C) and hydrogen (H). Sulfur (S) may also be present as a combustible element in some fuels. In addition to these combustible elements, fuels may contain non-combustible components such as moisture (M) and nitrogen (N), which do not participate in the combustion process. Ash (A) is another common incombustible component, especially in solid fuels like coal. Some oxygen (O) may also be present in the fuel, which can partially reduce the amount of oxygen required from the air during combustion.

For the purpose of calculating the stoichiometric air-fuel ratio, we assume mass fractions of each of these elements in the fuel, where the total of all these fractions equals 1 kg of fuel. The process of calculating the stoichiometric air-fuel ratio based on the known composition (ultimate analysis) of such a fuel, including all these possible components, is demonstrated below:

The ultimate analysis of the fuel is given by

$$C + H + O + N + S + M + A = 1.0$$

Oxygen needed for the oxidation processes can be calculated as follows:

C	+	O_2	=	CO_2	
12 kg		32 kg		44 kg	
1 kg		2.67 kg		3.67 kg	
C kg		2.67C kg		3.67C kg	
$2H_2$	+	O_2	=	$2H_2O$	
4 kg		32 kg		36 kg	
1 kg		8 kg		9 kg	(16.19)
H kg		8 H kg		9H kg	
S	+	O_2	=	SO_2	
32 kg		32 kg		64 kg	
1 kg		1 kg		2 kg	
S kg		S kg		2S kg	

Oxygen required for complete combustion of 1 kg fuel is

$$W_{O2} = 2.67C + 8H + S - O$$

where O is the oxygen in the fuel.

Air contains 23.2% oxygen by mass. Therefore, theoretically air required for complete combustion of 1 kg fuel is

$$W_T = \frac{W_{O_2}}{0.232} = \frac{2.67}{0.232}C + \frac{8}{0.232}\left(H - \frac{0}{8}\right) + \frac{1}{0.232}S$$

or, $$W_T = 11.5C + 34.5\left(H - \frac{0}{8}\right) + 4.3S \tag{16.20}$$

where C, H, O and S are the mass fractions of carbon, hydrogen, oxygen and sulphur in the fuel as given by the ultimate analysis.

As stated earlier, complete combustion of fuel cannot be achieved if only the theoretical or stoichiometric air is supplied. Excess air is always needed for complete combustion. It is expressed as a percentage or by the use of a dilution coefficient. The per cent excess air supplied is

$$\% \text{ excess air} = \frac{W_A - W_T}{W_T} \times 100 \tag{16.21}$$

where W_A is the actual amount of air supplied for complete combustion of 1 kg fuel. The dilution coefficient, d, is given by,

$$d = \frac{W_A}{W_T} \tag{16.22}$$

The percentage of excess air varies between 15 and 30% for most large utility boilers.

The products of combustion can be determined using the methodology outlined above, provided the composition of the fuel and the percentage of excess air are known. Conversely, by analyzing the composition of the combustion products and knowing the composition of the fuel, the percentage of excess air can be calculated, assuming complete combustion occurs.

Since all combustion products (except ash) are in gaseous form, volumetric analysis is necessary for exhaust gas analysis. Additionally, when incomplete combustion of carbon to carbon monoxide (CO) occurs, special considerations must be taken into account during the exhaust gas analysis.

The actual amount of air supplied per kg fuel can be ascertained from the measured volumetric composition of combustion products. There are several methods to perform these measurements: (a) Orsat analyser, (b) Haldane apparatus, (c) Infra-red gas analyser, and (d) Gas chromatograph. Of these, Orsat gas analyser is very simple, portable and most frequently used. It measures the volume or mole fractions of CO_2, CO and O_2 in the dry flue gas (dfg).

Samples of flue gas are first collected by suction in sampling bottles. An Orsat analyser contains three pipettes containing chemical solutions (Fig. 16.3). The reagents normally used are a KOH solution to absorb the CO_2 gas, pyrogallol solution to absorb the O_2 gas, and a cuprous chloride mixture ($CuCl_2$) to absorb the CO gas. By opening the stop valve and lowering the levelling bottle which contains water, 100 cm^3 of flue gas sample is collected over water in the graduated measuring burette. The valve of pipette 1 is opened and the levelling bottle is raised, so that all the gas is passed into the pipette 1. The CO_2 content in the flue gas sample is absorbed in the KOH solution. The levelling bottle is lowered and the volume of gas is measured, say, it is 89 cm^3. That means 11% of the dry gas is CO_2 by volume. In the same manner, volumetric percentages of O_2 and CO in the flue gas are measured by opening successively the pipettes 2 and 3. The remaining unabsorbed gas is nitrogen. Since the sample is collected over water, any water vapour in the flue gas would have condensed during the collection process. The SO_2 gas will react with water in the container. So, the flue gas analysis thus measured is on the dry basis:

$$CO_2 + CO + O_2 + N_2 = 100\%, \quad \text{by volume} \tag{16.23}$$

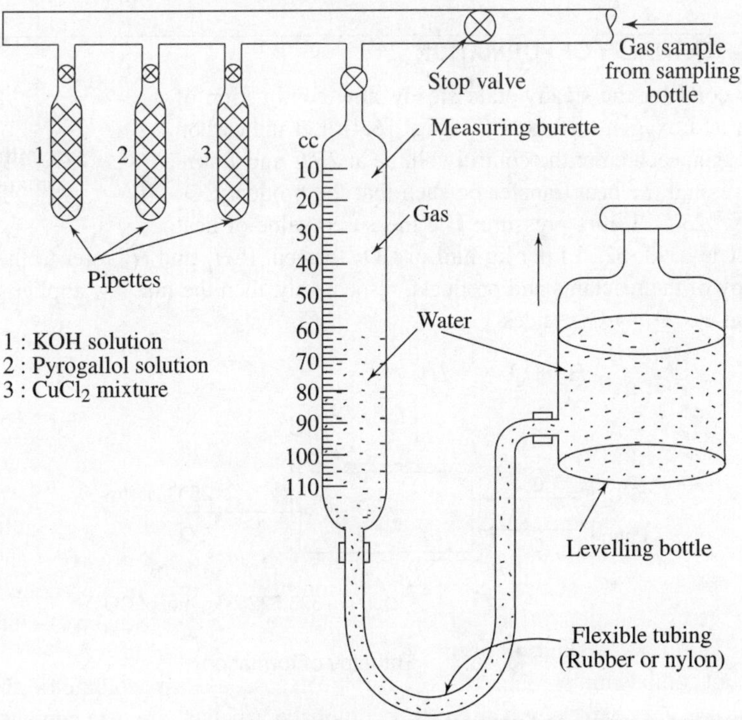

Fig. 16.3 Orsat analyser

Let us assume that the dry flue gas (dfg) analysis of a gas sample is: 12% CO_2, 3% CO, 5% O_2 and 80% N_2 by volume. Therefore, 1 mole of dfg contains 0.12

mol of CO_2, 0.03 mol of CO, 0.05 mole of O_2 and 0.80 mole of N_2. The molecular weight of the dfg then becomes equal to (0.12 × 44 + 0.03 × 28 + 0.05 × 32 + 0.80 × 28), or 30.12 kg/kg mol. Thus, in general, the molecular weight of dfg is given by

$$M_{dfg} = 44CO_2 + 28CO + 32O_2 + 28N_2 \tag{16.24}$$

where CO_2, CO, O_2 and N_2 are the volumetric percentages in dfg as measured, Therefore,

$$\% \, CO_2 \text{ by mass} = \frac{44CO_2}{44CO_2 + 28CO + 32O_2 + 28N_2} \times 100$$

$$\% \, O_2 \text{ by mass} = \frac{32O_2}{44CO_2 + 28CO + 32O_2 + 28N_2} \times 100 \tag{16.25}$$

$$\% \, N_2 \text{ by mass} = \frac{28N_2}{44CO_2 + 28CO + 32O_2 + 28N_2} \times 100$$

$$\% \, CO \text{ by mass} = \frac{28CO}{44CO_2 + 28CO + 32O_2 + 28N_2} \times 100$$

This is the gravimetric composition of the dry flue gas. From Eq. (16.25), it is seen that 44 kg of CO_2 gas has 12 kg carbon in it. So, ($44CO_2$) kg of CO_2 has ($12CO_2$) kg of carbon. Similarly, from Eq. (16.24), (28CO) kg of CO has (12CO) kg of carbon. So, 1 kg mol of dfg has $12(CO_2 + CO)$ kg of carbon.

16.13 ENTHALPY OF FORMATION

LO 16.11 Discuss enthalpy of formation

Let us consider the steady state steady flow combustion of carbon and oxygen to form CO_2 (Fig. 16.4). Let the carbon and oxygen each enter the control volume at 25°C and 1 atm. pressure, and the heat transfer be such that the product CO_2 leaves at 25°C, 1 atm. pressure. The measured value of heat transfer is –393, 522 kJ per kg mol of CO_2 formed. If H_R and H_P refer to the total enthalpy of the reactants and products, respectively, then the first law applied to the reaction $C + O_2 \rightarrow CO_2$ gives

$$H_R + Q_{C.V.} = H_P$$

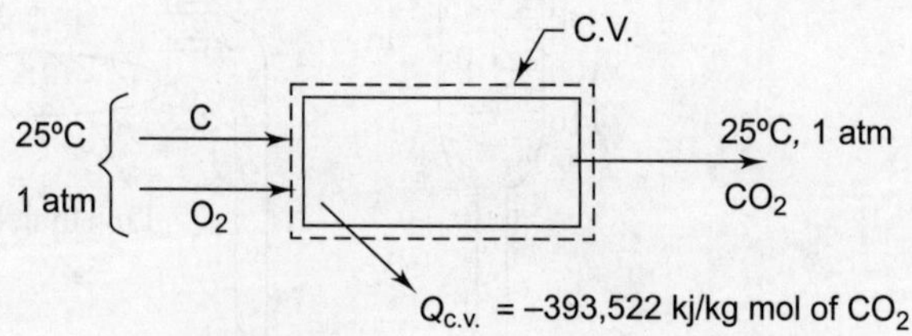

Fig. 16.4 Enthalpy of formation

For all the reactants and products in a reaction, the equation may be written as

$$\sum_R n_i \bar{h}_i + Q_{c.v} = \sum_P n_e \bar{h}_e$$

where R and P refer to the reactants and products, respectively.

The enthalpy of all the elements at the standard reference state of 25°C, 1 atm. is assigned the value of zero. In the carbon-oxygen reaction, $H_R = 0$. So the energy equation gives

$$Q_{C.V} = H_p = -393{,}522 \text{ kJ/kg mol}$$

This is what is known as the *enthalpy of formation of* CO_2 at 25°C, 1 atm., and designated by the symbol $\bar{h}_f^0$. So

$$\left(\bar{h}_f^0\right)_{CO_2} = -393{,}522 \text{ kJ/kg mol}$$

In most cases, however, the reactants and products are not at 25°C, 1 atm. Therefore, the change of enthalpy (in the case of constant pressure process or SSSF, steady state, steady flow, process) between 25°C, 1 atm. and the given state must be known. Thus the enthalpy at any temperature and pressure, $\bar{h}_{T,p}$ is

$$\bar{h}_{T,p} = \left(\bar{h}_f^0\right)_{298\text{ K, 1 atm.}} + \left(\Delta\bar{h}\right)_{298\text{ K, 1 atm} \rightarrow T,\, p}$$

For convenience, the subscripts are usually dropped, and

$$\bar{h}_{T,p} = \bar{h}_f^0 + \Delta\bar{h}$$

where $\Delta\bar{h}$ represents the difference in enthalpy between any given state and the enthalpy at 298.15 K, 1 atm.

Table 16.4 gives the values of enthalpy of formation of a number of substances in kJ/kg mol.

Table C in the appendix gives the values of $\Delta\bar{h} = \Delta\bar{h}^0 - \bar{h}_{298}^0$ in kJ/kg mol for various substances at different temperatures.

16.14 FIRST LAW FOR REACTIVE SYSTEMS

LO 16.12 Illustrate first law for reactive systems

For the SSSF process as shown in Fig. 16.5 the first law gives

$$H_R + Q_{C.V} = H_P + W_{C.V.}$$

or

$$\sum_R n_i\bar{h}_i + Q_{C.V.} = \sum_P n_e\bar{h}_e + W_{C.V.}$$

$$\sum_R n_i\left[\bar{h}_f^0 + \Delta\bar{h}\right]_i + Q_{C.V} = \sum_P n_e\left[\bar{h}_f^0 + \Delta\bar{h}\right]_e + W_{C.V.} \tag{16.26}$$

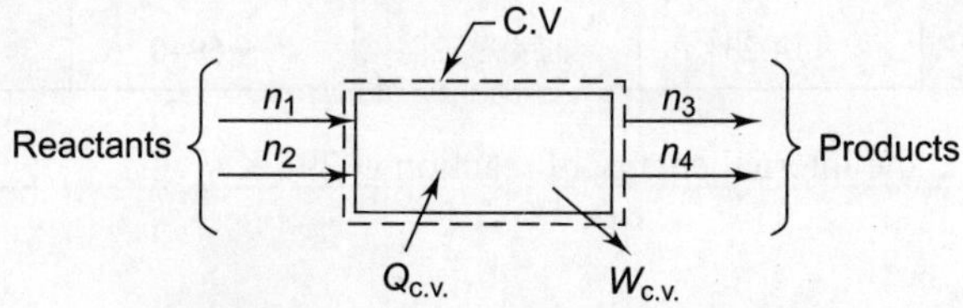

Fig. 16.5 First law for a reactive system

When the states of reactants and products are not in the standard reference state (298K, 1 atm), then, as shown in Fig. 16.5,

$$Q_{C.V} = H_P - H_R = (H_P - H_{P0}) + (H_{P0} - H_{R0}) + (H_{R0} - H_R)$$

$$= \sum_P n_e\left(\bar{h}_p - \bar{h}_{p0}\right) + \Delta\bar{h}^0_{RP} - \sum_R n_i\left(\bar{h}_R - \bar{h}_{R0}\right) \tag{16.27}$$

where $\Delta\bar{h}^0_{PR}$ is the enthalpy of reaction at the standard temperature (298 K). The variation of enthalpy with pressure is not significant.

For a constant volume reaction,

$$Q_{C.V} = U_P - U_R = (U_P - U_{P0}) + (U_{P0} - U_{R0}) + (U_{R0} - U_R)$$

$$= \sum_P n_e\left(\bar{u}_p - \bar{u}_{p0}\right) + \Delta\bar{u}^0_{RP} - \sum_R n_i\left(\bar{u}_R - \bar{u}_{R0}\right) \tag{16.28}$$

Table 16.4 Enthalpy of formation, Gibbs function of formation, and absolute entropy of various substances at 25°C, 1 atm. pressure

Substance	*Molecular Weight, M*	$\bar{h}^0_f$ *kJ/kg mol*	$\bar{g}^0_f$ *kJ/kg mol*	s^0 *kJ/kg mol K*
$CO(g)$	28.011	– 110529	– 137150	197.653
$CO_2(g)$	44.001	– 393522	– 394374	213.795
$H_2O(g)$	18.015	– 241827	– 228583	188.833
$H_2O(l)$	18.015	– 285838	– 237178	70.049
$CH_4(g)$	16.043	– 74873	– 50751	186.256
$C_2H_2(g)$	26.038	+ 226731	+ 209234	200.958
$C_2H_4(g)$	28.054	+ 52283	+ 68207	219.548
$C_2H_6(g)$	30.070	– 84667	– 32777	229.602
$C_3H_8(g)$	44.097	– 103847	– 23316	270.019
$C_4H_{10}(g)$	58.124	– 126148	– 16914	310.227
$C_8H_{18}(g)$	114.23	– 208447	+ 16859	466.835
$C_8H_{18}(l)$	114.23	– 249952	+ 6940	360.896

where $\Delta\bar{u}^0_{RP}$ is the internal energy of reaction at 298 K.

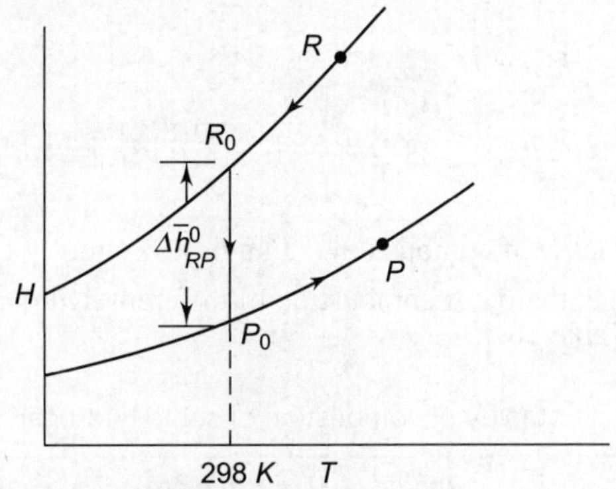

Fig. 16.6 Enthalpy of reactants and products varying with temperature

16.15 ADIABATIC FLAME TEMPERATURE

If a combustion process occurs adiabatically in the absence of work transfer or changes in K.E. and P.E. then the energy equation becomes

$$H_R = H_P$$

LO 16.13
Interpret adiabatic flame temperature

or

$$\sum_R n_i \bar{h}_i = \sum_R n_e \bar{h}_e$$

or

$$\sum_R n_i \left[\bar{h}_f^0 + \Delta\bar{h}\right]_i = \sum_P n_e \left[\bar{h}_f^0 + \Delta\bar{h}\right]_e \quad (16.29)$$

For such a process, the temperature of the products is called the *adiabatic flame temperature* which is the maximum temperature achieved for the given reactants. The adiabatic flame temperature can be controlled by the amount of excess air supplied; it is maximum with a stoichiometric mixture. Since the maximum permissible temperature in a gas turbine is fixed from metallurgical considerations, close control of the temperature of the products is achieved by controlling the excess air.

For a given reaction the adiabatic flame temperature is computed by trial and error method. The energy of the reactants H_R being known, a suitable temperature is chosen for the products so that the energy of the products at that temperature becomes equal to the energy of the reactants.

16.16 ENTHALPY AND INTERNAL ENERGY OF COMBUSTION HEATING VALUE

LO 16.14
Explain internal energy and enthalpy of combustion or heat values

The *enthalpy of combustion* is defined as the difference between the enthalpy of the products and the enthalpy of reactants when complete combustion occurs at a given temperature and pressure.

Therefore,

$$\bar{h}_{RP} = H_P - H_R$$

or

$$\bar{h}_{RP} = \sum_P n_e[\bar{h}_f^0 + \Delta\bar{h}]_e - \sum_R n_i[\bar{h}_f^0 + \Delta\bar{h}]_i \tag{16.30}$$

where $\bar{h}_{RP}$ is the enthalpy of combustion (kJ/kg or kJ/kg mol) of the fuel.

The values of the enthalpy of combustion of different hydrocarbon fuels at 25°C, 1 atm. are given in Table 16.5.

Table 16.5 Enthalpy of combustion of some hydrocarbons at 25°C

		Liquid H_2O in Products (Negative of Higher Heating Value)		*Vapour H_2O in Products (Negative of Lower Heating Value)*	
Hydrocarbon	*Formula*	*Liquid Hydro-carbon kJ/kg fuel*	*Gaseous Hydro-carbon kJ/kg fuel*	*Liquid Hydro-carbon kJ/kg fue*	*Gaseous Hydro-carbon kJ/kg fue*
Paraffin Family					
Methane	CH_4		– 55496		– 50010
Ethane	C_2H_6		– 51875		– 47484
Propane	C_3H_8	– 49975	– 50345	– 45983	– 46353
Butane	C_4H_{10}	– 49130	– 49500	– 45344	– 45714
Pentane	C_5H_{12}	– 48643	– 49011	– 44983	– 45351
Hexane	C_5H_{14}	– 48308	– 48676	– 44733	– 45101
Heptane	C_7H_{16}	– 48071	– 48436	– 44557	– 44922
Octane	C_8H_{18}	– 47893	– 48256	– 44425	– 44788
Decane	$C_{10}H_{22}$	– 47641	– 48000	– 44239	– 44598
Dodecane	$C_{12}H_{26}$	– 47470	– 47828	– 44100	– 44467
Olefin Family					
Ethene	C_2H_4		– 50296		– 47158
Propene	C_3H_6		– 48917		– 45780
Butene	C_4H_8		– 48453		– 45316
Pentene	C_5H_{10}		– 48134		– 44996
Hexene	C_6H_{12}		– 47937		– 44800
Heptene	C_7H_{14}		– 47800		– 44662
Octene	C_8H_{16}		– 47693		– 44556
Nonene	C_9H_{18}		– 47612		– 44475
Decene	$C_{10}H_{20}$		– 47547		– 44410
Alkylbenzene Family					

Hydrocarbon	Formula	Liquid H_2O in Products (Negative of Higher Heating Value)		Vapour H_2O in Products (Negative of Lower Heating Value)	
		Liquid Hydro-carbon kJ/kg fuel	Gaseous Hydro-carbon kJ/kg fuel	Liquid Hydro-carbon kJ/kg fue	Gaseous Hydro-carbon kJ/kg fue
Benzene	C_6H_6	– 41831	– 42266	– 40141	– 40576
Methylbenzene	C_7H_8	– 42473	– 42847	– 40527	– 40937
Ethylbenzene	C_8H_{10}	– 42997	– 43395	– 40924	– 41322
Propylbenzene	C_9H_{12}	– 43416	– 43800	– 41219	– 41603
Butylbenzene	$C_{10}H_{14}$	– 43748	– 44123	– 41453	– 41828

The *internal energy of combustion*, u_{RP}, is defined in a similar way.

$$\bar{u}_{RP} = U_P - U_R$$

$$= \sum_P n_e[\bar{h}_f^0 + \Delta\bar{h} - p\bar{v}]_e - \sum_R n_i[\bar{h}_f^0 + \Delta\bar{h} - p\bar{v}]_i$$

If all the gaseous constituents are considered ideal gases and the volume of liquid and solid considered is assumed to be negligible compared to gaseous volume

$$\bar{u}_{RP} = \bar{h}_{RP} - \bar{R}T\,(\eta_{\text{gaseous products}} - \eta_{\text{gaseous reactants}}) \tag{16.31}$$

In the case of a constant pressure or steady flow process, the negative of the enthalpy of combustion is frequently called the *heating value at constant pressure*, which represents the heat transferred from the chamber during combustion at constant pressure.

Similarly, the negative of the internal energy of combustion is sometimes designated as the *heating value at constant volume* in the case of combustion, because it represents the amount of heat transfer in the constant volume process.

The *higher heating value* (HHV) or higher calorific value (HCV) is the heat transferred when H_2O in the products is in the liquid phase. The *lower heating value* (LHV) or lower calorific value (LCV) is the heat transferred in the reaction when H_2O in the products is in the vapour phase.

Therefore,

$$\text{LHV} = \text{HHV} - m_{H_2O} \cdot h_{fg}$$

where m_{H_2O} is the mass of water formed in the reaction.

16.17 ABSOLUTE ENTROPY AND THE THIRD LAW OF THERMODYNAMICS

LO 16.15

Discuss the third law of thermodynamics and absolute entropy

So far only the first law aspects of chemical reactions have been discussed. The second law analysis of chemical reactions needs a base for the entropy of various substances. The entropy of substances at the absolute zero of temperature, called absolute entropy, is dealt with the third law of thermodynamics formulated in the early twentieth century primarily by W.H. Nernst (1864–1941) and Max Planck (1858–1947). The third law states that the entropy of a perfect crystal is zero at the absolute zero of temperature and it represents the maximum degree of order. A substance not having a perfect crystalline structure and possessing a degree of randomness such as a solid solution or a glassy solid, has a finite value of entropy at absolute zero. The third law provides an absolute base from which the entropy of each substance can be measured. The entropy relative to this base is referred to as the absolute entropy. Table 16.4 gives the absolute entropy of various substances at the standard state 25°C, 1 atm. For any other state

$$\bar{S}_{T,p} = \bar{S}_T^0 + (\Delta\,\bar{s})_{T,\,1\text{ atm.},\,\to T,p}$$

where $\bar{s}_T^0$ refers to the absolute entropy at 1 atm. and temperature T, and $(\Delta\,\bar{s})_{T,\,1\text{ atm}\to T,p}$ refers to the change of entropy for an isothermal change of pressure from 1 atm. to pressure p (Fig. 16.6). Table C in the appendix gives the values of $\bar{s}^0$ for various substances at 1 atm. and at different temperatures. Assuming ideal gas behaviour $(\Delta\,\bar{s})_{T,\,1\text{ atm}\to T,p}$ can be determined (Fig. 16.7)

$$\bar{s}_2 - \bar{s}_1 = -\bar{R}\ln\frac{p_2}{p_1} \tag{16.31a}$$

or

$$(\Delta\,\bar{s})_{T,\,1\text{ atm}\to T,p} = -\bar{R}\ln p$$

where p is in atm.

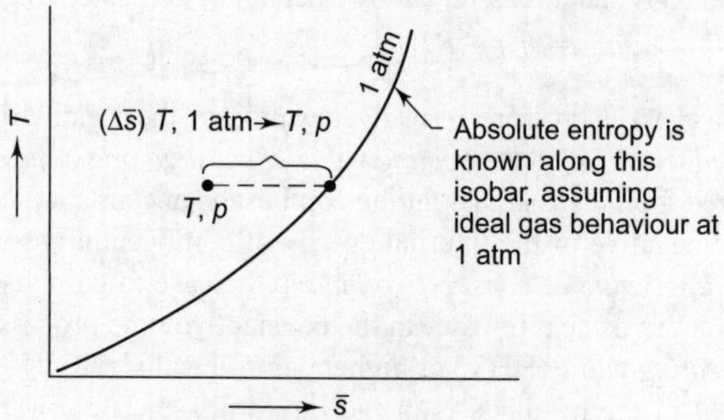

Fig. 16.7 Absolute entropy

16.18 SECOND LAW ANALYSIS OF REACTIVE SYSTEMS

LO 16.16
Relate second law analysis of reactive systems with Gibbs function of formation

The reversible work for a steady state steady flow process, in the absence of changes in K.E. and P.E. is given by

$$W_{\text{rev}} = \Sigma n_i(h_i - T_0\,s_i) - \Sigma n_e(h_e - T_0 s_e)$$

For an SSSF process involving a chemical reaction

$$W_{\text{rev}} = \sum_R n_i[\bar{h}_f^0 + \Delta\bar{h} - T_0\bar{s}]_i - \sum_P n_e[\bar{h}_f^0 + \Delta\bar{h} - T_0\bar{s}]_e \tag{16.32}$$

The irreversibility for such a process is

$$I = \sum_R n_e T_0 \bar{s}_e - \sum_R n_i T_0 \bar{s}_i - Q_{\text{C.V.}}$$

The availability, ψ, in the absence of K.E. and P.E. changes, for an SSSF process is

$$\psi = (h - T_0 s) - (h_0 - T_0 s_0)$$

When an S.S.S.F. chemical reaction takes place in such a way that both the reactants and products are in temperature equilibrium with the surroundings, the reversible work is given by

$$W_{\text{rev}} = \sum_R n_i \bar{g}_i - \sum_P n_e \bar{g}_e \tag{16.33}$$

where the $\bar{g}$'s refer to the Gibbs function. The *Gibbs function of formation*, $\bar{g}_f^0$, is defined similar to enthalpy of formation, $\bar{h}_f^0$. The Gibbs function of each of the elements at 25°C and 1 atm. pressure is assumed to be zero, and the Gibbs function of each substance is found relative to this base. Table 16.4 gives $\bar{g}_f^0$ for some substances at 25°C, 1 atm.

16.19 CHEMICAL EXERGY

LO 16.17
Interpret chemical exergy

In Chapter 8, it was stated that when a system is at the dead state, it is in thermal and mechanical equilibrium with the environment, and the value of its exergy is zero. To state it more precisely, the *thermomechanical* contribution to exergy is zero. However, the contents of a system at the dead state may undergo chemical reaction with environmental components and produce additional work. We will here study a combined system formed by an environment and a system having a certain amount of fuel at T_0, p_0. The work obtainable by allowing the fuel to react with oxygen from the environment to produce the environmental components of CO_2 and H_2O is evaluated. The *chemical exergy* is thus defined as the maximum theoretical work that could be developed by the combined system. Thus for a given system at a specific state

Total exergy = Thermomechanical exergy + Chemical exergy

Let us consider a hydrocarbon fuel (C_aH_b) at T_0, p_0 reacting with oxygen from the environment (Fig. 16.8) which is assumed to be consisting of an ideal gas mixture at T_0, p_0. The oxygen that reacts with the fuel is at a partial pressure of x_{02}, p_0, where x_{02} is

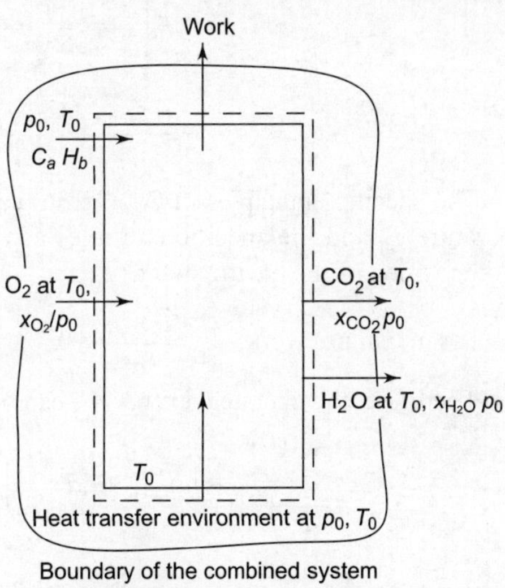

Fig. 16.8 Fuel exergy concept

the mole fraction of oxygen in the environment. The fuel and oxygen react completely to produce CO_2 and H_2O, which exit in separate streams at T_0 and respective partial pressures of $x_{CO2} \cdot p_0$ and $x_{H2O} \cdot p_0$. The reaction is given by

$$C_a H_b + \left(a + \frac{b}{4}\right) O_2 \rightarrow a\,CO_2 + \frac{b}{2} H_2O$$

At steady state, the energy balance gives

$$\dot{Q}_{CV} + \dot{n}H_R = \dot{W}_{CV} + \dot{n}H_P$$

or,

$$\frac{\dot{W}_{CV}}{\dot{n}} = \frac{\dot{Q}_{CV}}{\dot{n}} + H_R - H_P$$

$$= \frac{\dot{Q}_{CV}}{n} + \left(\bar{h}_f^0 + \Delta\bar{h}\right) C_a H_b + \left(a + \frac{b}{4}\right)\bar{h}_{O_2} - a\bar{h}_{CO_2} - \frac{b}{2}\bar{h}_{H_2O} \tag{16.34}$$

where $\dot{n}$ is the rate of fuel flow in moles, and KE and PE effects are neglected.

An entropy balance for the control volume gives

$$0 = \frac{\dot{Q}_{CV}/\dot{n}}{T_0} + \bar{s}_{C_aH_b} + \left(a + \frac{b}{4}\right)\bar{s}_{O_2} - a\bar{s}_{CO_2} - \frac{b}{2}\bar{s}_{H_2O} + \frac{\dot{s}_{gen}}{\dot{n}} \tag{16.35}$$

Eliminating $\dot{Q}_{CV}$ between Eqs. (16.34) and (16.35),

$$\frac{\dot{W}_{CV}}{\dot{n}} = \left[\bar{h}_{C_aH_b} + \left(a + \frac{b}{4}\right)\bar{h}_{O_2} - a\bar{h}_{CO_2} - \frac{b}{2}\bar{h}_{H_2O}\right] - T_0\left[\bar{s}_{C_aH_b} + \left(a + \frac{b}{4}\right)\bar{s}_{O_2} - a\bar{s}_{CO_2} - \frac{b}{2}\bar{s}_{H_2O}\right] - \frac{T_0\dot{s}_{gen}}{\dot{n}} \tag{16.36}$$

The specific enthalpies in Eq. (16.36) can be determined knowing only the temperature T_0 and the specific entropies can be determined knowing T_0, p_0 and the composition of the environment.

For maximum work, $\dot{I} = \frac{T_0\dot{s}_{gen}}{n} = 0$

Therefore, the chemical exergy a_{ch} can be expressed as

$$\bar{a}_{ch} = \left[\bar{h}_{C_aH_b} + \left(a + \frac{b}{4}\right)\bar{h}_{O_2} - a\bar{h}_{CO_2} - \frac{b}{2}\bar{h}_{H_2O}\right] - T_0\left[\bar{s}_{C_aH_b} + \left(a + \frac{b}{4}\right)\bar{s}_{O_2} - a\bar{s}_{CO_2} - \frac{b}{2}\bar{s}_{H_2O}\right] \tag{16.37}$$

The specific entropies for O_2, CO_2 and H_2O are written from (Eq. 16.32 (a)),

$$\bar{s}_i\,(T_0, x_1\,p_0) = s_i\,(T_0, p_0) - \bar{R}\ln x_i \tag{16.38}$$

where the first term on the right is the absolute entropy at T_0 and p_0, and x_i is the mole fraction of component i in the environment. Therefore, Eq. (16.37) becomes

$$\bar{a}_{ch} = \left[\bar{h}_{C_aH_b} + \left(a+\frac{b}{4}\right)\bar{h}_{O_2} - a\bar{h}_{CO_2} - \frac{b}{2}\bar{h}_{H_2O}\right] (\text{at } T_0, p_0)$$

$$- T_0\left[\bar{s}_{C_aH_b} + \left(a+\frac{b}{4}\right)\bar{s}_{O_2} - a\bar{s}_{CO_2} - \frac{b}{2}\bar{s}_{H_2O}\right] (\text{at } T_0, p_0)$$

$$+ RT_0 \ln \frac{(x_{O_2})^{a+b/4}}{\left(x_{CO_2}\right)^a \left(x_{H_2O}\right)^{b/2}} \tag{16.39}$$

In terms of Gibbs functions of respective substances,

$$\bar{a}_{ch} = \left[\bar{g}_{C_aH_b} + \left(a+\frac{b}{4}\right)\bar{g}_{O_2} - a\bar{g}_{CO_2} - \frac{b}{2}\bar{g}_{H_2O}\right] (\text{at } T_0, p_0)$$

$$+ \bar{R}\,T_0 \ln \frac{\left(x_{O_2}\right)^{a+b/4}}{\left(x_{CO_2}\right)^a \left(x_{H_2O}\right)^{b/2}} \tag{16.40}$$

where $\bar{g}\,(T_0, p_0) = \bar{g}_f^0 + \Delta\bar{g}_{T_0 . p_0 \to T_{ref}\,P_{ref}}$

For the special case when T_0 and p_0 are the same as T_{ref} and $p_{ref.}$ $\Delta\bar{g}$ will be zero. The chemical exergy of pure CO at T_0, p_0 where the reaction is given by

$$CO + \frac{1}{2}O_2 \rightarrow CO_2$$

$$a(\bar{a}_{ch})_{CO} = \left[\bar{g}_{CO} + \frac{1}{2}\bar{g}_{O_2} - \bar{g}_{CO_2}\right] (\text{at } T_0, p_0)$$

$$+ \bar{R}T_0 \ln \frac{\left(x_{O_2}\right)^{1/2}}{x_{CO_2}}$$

Water is present as a vapour within the environment, but normally is a liquid at T_0, p_0. The chemical exergy of liquid water is

$$H_2O(l) \rightarrow H_2O(g)$$

$$\left(\bar{a}_{ch}\right)_{H_2O(l)} = \left[\bar{g}_{H_2O(l)} - \bar{g}_{H_2O(g)}\right] (\text{at } T_0, p_0)$$

$$+ \bar{R}\,T_0 \ln \frac{1}{x_{H_2O(g)}}$$

The specific exergy of system is

$$a = a_{\text{thermo-mech}} + a_{\text{chem}}$$

$$= (u - u_0) + p_0(v - v_0) - T_0(s - s_0) + \frac{V^2}{2} + gz + a_{\text{ch}} \quad (16.41)$$

and the specific flow exergy is given by

$$a = (h - h_0) - T_0(s - s_0) + \frac{V^2}{2} + gz + a_{\text{ch}} \quad (16.42)$$

Between two states of a system of constant composition, a_{ch} cancels, leaving just the thermomechanical contribution. For example, to find the chemical exergy of liquid octane, the reaction is

$$C_8H_{18}(1) + 12.5O_2 \rightarrow 8CO_2 + 9H_2O(g)$$

Assume the composition of the environment on molar basis be
N_2 75.67%, O_2 20.35%, H_2O 3.12%, CO_2 0.03%, others 0.83%. Then,

$$\bar{a}_{\text{ch}} = \left[\bar{g}_{C_8H_{18}(1)} + 12.5\,\bar{g}_{O_2} - 8\bar{g}_{CO_2} - 9\bar{g}_{H_2O(g)}\right] (\text{at } T_0, p_0)$$

$$+ \bar{R}\,T_0 \ln \frac{\left(x_{O_2}\right)^{12.5}}{\left(x_{CO_2}\right)^8 \left(x_{H_2O}\right)^9}$$

Using the values given in Table 16.4

$$\left(\bar{a}_{\text{ch}}\right)_{C_8H_{18}(1)} = 6610 + 12.5(0) - 8(-394,380) - 9(-228,590)$$

$$+ 8.3143\,(298.15) \ln \frac{(0.2035)^{12.5}}{(0.0003)^8\,(0.0312)^9}$$

$$= 5{,}407{,}843 \text{ kJ/kg mol} = 47{,}346 \text{ kJ/kg}$$

16.20 SECOND LAW EFFICIENCY OF A REACTIVE SYSTEM

LO 16.18 Express second law efficiency of a reactive system

For a fuel at T_0, p_0, the chemical exergy is the maximum theoretical work that could be obtained through reaction with environmental substances. However, due to various irreversibilities like friction and heat loss, the actual work obtained is only a fraction of this maximum theoretical work. The second law efficiency may thus be defined as the ratio of

$$\eta_{\text{II}} = \frac{\text{Actual work done}}{\text{Maximum theoretical work}} = \frac{W_{\text{CV}}}{m_{\text{fuel}} \times a_{\text{ch}}} \quad (16.43)$$

The associated irreversibilities and the consequent exergy losses require to be reduced to enhance the second law efficiency, which in turn, reduces the fuel consumption and also increases the cost. The trade off between the fuel savings and the additional costs must be carefully weighed.

16.21 THERMODYNAMICS OF COMBUSTION IN INTERNAL COMBUSTION ENGINES

LO 16.19
Discuss thermodynamics of combustion in internal combustion engines

In internal combustion (IC) engines, fuel combustion occurs inside the cylinder, where power is generated. The combustion process, however, can vary significantly depending on the type of fuel used. The two primary fuels in IC engines are petrol and diesel, although gaseous fuels are increasingly being adopted. The combustion process in gas engines is generally similar to that of petrol engines. Additionally, alcohol is now commonly mixed with petrol to help reduce emissions. Various other additives are also blended with base fuels to minimize pollutant emissions from liquid fossil fuels. Despite these modifications, the basic combustion process remains largely unchanged for the base fuel.

In petrol engines, ignition occurs through a spark, which is why they are called Spark Ignition (SI) engines. In contrast, diesel engines initiate ignition through compression, a process known as Compression Ignition (CI).

16.21.1 Combustion in SI engines

In Spark Ignition (SI) engines, the fuel is mixed with air before being inducted into the cylinder for combustion. The device responsible for mixing the fuel - whether liquid or gaseous - with air in SI engines is called the carburetor. The configuration of the air-fuel induction system is illustrated in Fig. 16.9.

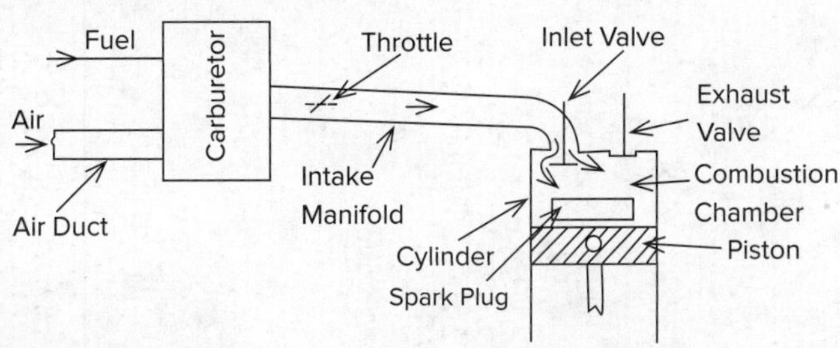

Fig. 16.9 Arrangement of air-fuel induction into the cylinder of a SI engine

The carburetor is responsible for controlling the air-fuel (A/F) ratio during engine operation. This ratio can vary on either side of the stoichiometric A/F ratio. When the air is less than the stoichiometric amount for a given fuel, the mixture is called a "rich mixture." Conversely, when the air exceeds the stoichiometric amount, the mixture is referred to as a "lean mixture." On both sides of the stoichiometric ratio, the air-fuel mixture will burn within a specific range, as shown in Fig. 16.10. The carburetor must be designed to supply a homogeneous air-fuel mixture that falls within this range during engine operation.

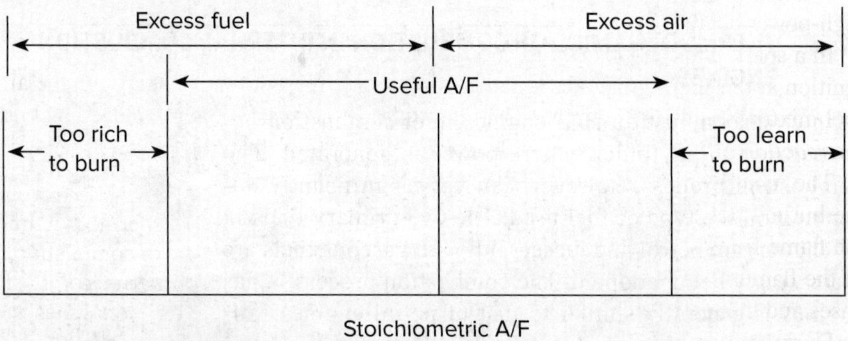

Fig. 16.10 Useful A/F for combustion in SI engine

The "best power mixture" corresponds to the air-fuel ratio that produces maximum power output, while the "best economy mixture" corresponds to the ratio that results in the minimum specific fuel consumption, as shown in Fig. 16.11.

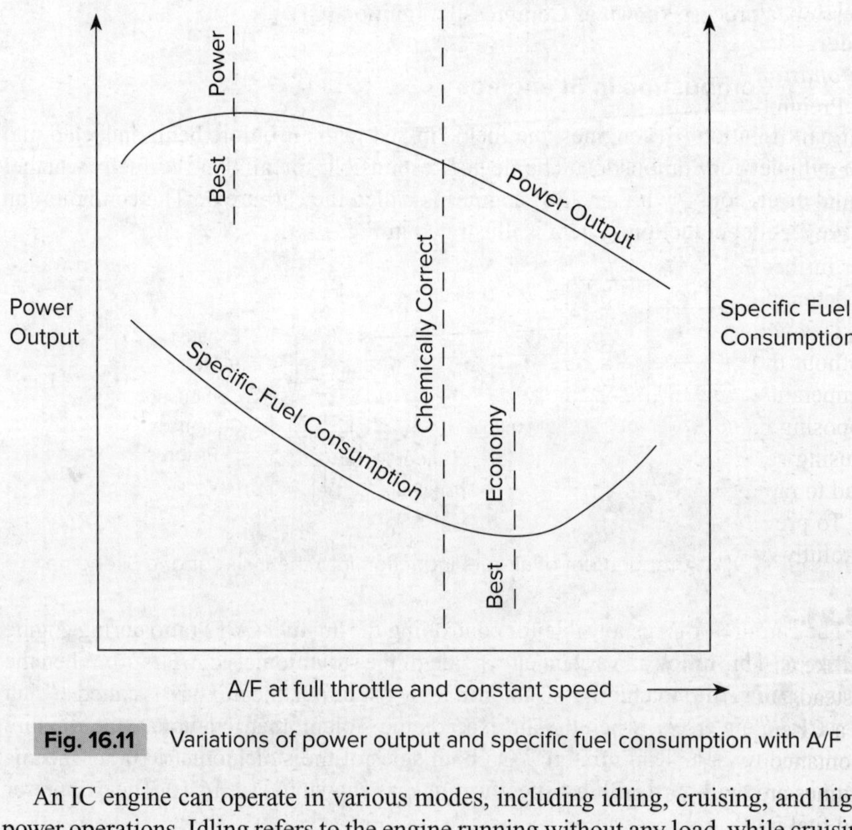

Fig. 16.11 Variations of power output and specific fuel consumption with A/F

An IC engine can operate in various modes, including idling, cruising, and high-power operations. Idling refers to the engine running without any load, while cruising involves the engine operating at a constant, nominal speed. High-power operations occur when the load on the engine increases. Although the best economy air-fuel ratio is preferred during cruising, a richer mixture is necessary for both idling and high-power operations.

In a spark ignition (SI) engine, normal combustion begins at the spark plug. The ignition at the spark generates a flame front, which then propagates through the air-fuel mixture inside the cylinder. The rate of flame propagation is influenced by both the *reaction rate* and the *transportation rate*.

The *reaction rate* refers to the speed at which the chemical reaction spreads from the burned portion of the air-fuel mixture to the remaining unburned fuel, driving the flame front forward. The *transportation rate* describes the additional movement of the flame front, which occurs due to the pressure difference between the burning gases and the unburned mixture within the cylinder walls.

During normal combustion, the flame front moves smoothly through the air-fuel mixture at an average *flame speed*. This flame speed is affected by several factors, with turbulence and the air-fuel (A/F) ratio being the two most significant. Other influencing factors include the temperature and pressure of the incoming charge, humidity, spark timing, compression ratio, and the amount of residual gas left after each combustion cycle.

Under certain conditions, *abnormal combustion* can occur within the cylinder. The two primary forms of abnormal combustion are *preignition* and *detonation*.

Preignition occurs when an overheated engine component - such as a valve, spark plug, or even a carbon particle - ignites the air-fuel mixture before the spark is delivered. Although combustion begins due to preignition, the regular spark will still ignite the mixture again, leading to multiple ignition events in a cycle. This can result in erratic combustion. Preignition may also create localized *hotspots*, which can trigger further instances of preignition, potentially damaging the engine or even leading to detonation.

Detonation refers to the autoignition of the final portion of the air-fuel mixture without the presence of a spark. This typically occurs under certain pressure and temperature conditions, where the mixture ignites spontaneously. In detonation, two opposing flame fronts propagate toward each other, generating a pressure shock and causing severe, destructive vibrations within the engine. Prolonged detonation can lead to rapid engine damage or even complete failure.

To prevent detonation, the compression ratio in a spark ignition engine must be carefully controlled.

16.21.2 Combustion in CI engines

Unlike SI engines, CI engines do not rely on a spark plug to initiate combustion. Instead, at the end of the compression stroke, the temperature of the compressed air in a CI engine reaches the ignition temperature of the fuel. This causes the fuel to spontaneously ignite due to the heat generated by compression, which is why these engines are termed "compression ignition" engines. The general arrangement of air and fuel induction in a CI engine is shown in Fig. 16.12 on the next page.

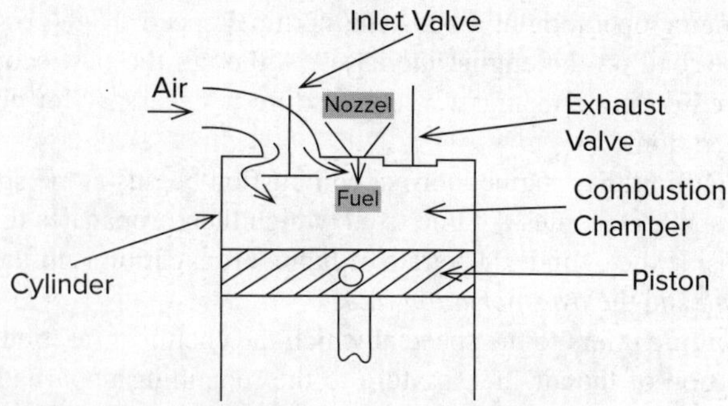

Fig. 16.12 General arrangement of air and fuel induction in a CI engine

CI engines also differ from SI engines in that they do not use a carburetor. Instead, fuel is directly injected into the compressed air via nozzles, rather than being mixed with air before entering the cylinder. This fundamental difference creates unique challenges in the combustion process of CI engines compared to SI engines.

In CI engines, the air-fuel ratio (A/F) is primarily controlled by the amount of fuel injected into the compressed air, while the air inside the cylinder remains nearly constant. Unlike the homogeneous air-fuel mixture in SI engines, the fuel in a CI engine is injected directly into the compressed air, resulting in a non-homogeneous mixture. Proper fuel distribution is essential, and the degree of air-fuel mixing depends on the dispersion of the injected fuel and the turbulence within the cylinder. The air-fuel ratio within the cylinder can vary significantly, creating both rich and lean mixtures. Even when the overall A/F ratio is stoichiometric or lean, certain areas may experience rich mixtures, which can impact combustion quality.

Because the air temperature is already above the ignition temperature when the fuel is injected, combustion in a CI engine is not initiated by a single spark but by multiple spontaneous ignition points throughout the cylinder. In this sense, it is as though the engine has multiple "spark plugs" distributed across the air. Unlike SI engines, which have a defined flame front, combustion in CI engines occurs simultaneously at various locations where the air-fuel ratio is conducive to ignition.

The power output in a CI engine increases with richer mixtures, but thermal efficiency tends to decrease as the mixture becomes richer. If the fuel quantity exceeds a certain threshold, visible black smoke may be emitted, which is primarily unburnt carbon (soot) being expelled.

Abnormal combustion in CI engines often results from ignition delay or ignition lag. Since fuel injection cannot occur instantaneously, the injection process begins before compression reaches its maximum and may continue into the expansion stroke. This delay can cause the fuel to accumulate before ignition, leading to a rapid pressure rise when combustion finally occurs. This phenomenon is actually the result of an accumulated fuel charge igniting all at once in CI engines, causing a rapid pressure buildup. While detonation in SI engines happens at the end of the combustion process, it occurs at the beginning of combustion in CI engines.

With longer ignition delays, the fuel accumulation can be significant, leading to extreme pressure differentials and violent gas vibrations, resulting in audible knocking. This knocking condition can severely reduce engine life and must be avoided.

To minimize ignition delay and reduce knock, factors such as a higher compression ratio, increased inlet air and coolant temperatures, and enhanced turbulence can help improve the combustion process in CI engines.

It is well-established that the power output of an engine increases with the amount of fuel burned, which in turn requires more air to be inducted into the cylinder. One way to achieve this is by increasing the cylinder volume, which allows for greater air induction. However, this results in a larger and bulkier engine. In aviation applications, where compactness and weight are crucial, this approach is less acceptable. Additionally, at higher altitudes, the air density decreases, leading to a reduction in power output. A more efficient solution to increase power without significantly altering the engine's size is to enhance the density of the inducted air by increasing its pressure. This process is known as **supercharging**, and the device responsible for supplying air at higher pressure is called a **supercharger**. The engine provides the necessary power to drive the supercharger, effectively delivering more air (and thus more power) through the supercharging process.

16.22 THERMODYNAMICS OF COMBUSTION IN BOILER

LO 16.20

Discuss thermodynamics of combustion in boilers

The primary fuel used in boilers is coal, whose combustion characteristics differ significantly from those of liquid or gaseous fuels. When coal is introduced into the combustion chamber, it is first heated to release volatile matter, which ignites readily. This initial combustion is typically spontaneous and largely complete.

Following this, the remaining solid portion of the coal, known as char, begins to burn. Unlike the volatiles, complete combustion of char depends heavily on the availability of oxygen and the extent to which the char particles come into contact with it. Incomplete combustion is often the result of unburned char particles and is a major source of inefficiency in coal-fired systems. This different process of combustion of coal needs proximate analysis in addition to the ultimate analysis as discussed above. According to proximate analysis:

$$FC + VM + M + A = 100\% \text{ by mass} \tag{16.44}$$

Where FC is fixed carbon, VM is volatile matter, M is moisture and A is ash (incombustible part of coal).

Proximate analysis provides insight into the behavior of coal when it is heated. To determine the moisture content, a 1-gram sample of coal is heated at approximately 105°C for one hour. The loss in weight during this process represents the moisture content of the coal.

To determine the volatile matter in coal, a 1-gram sample is placed in a covered platinum crucible and heated to 950°C, where it is maintained for approximately 7 minutes. During this process, a loss in weight occurs due to the release of moisture and volatile matter. Since the moisture content is determined separately through a prior test, the remaining weight loss corresponds to the volatile matter. This fraction primarily consists of hydrogen and hydrogen–carbon compounds that are driven off simply by heating the coal.

To determine the ash content of coal, a 1-gram sample is placed in an uncovered crucible and heated to approximately 720°C until complete combustion occurs. The process continues until a constant weight is achieved, indicating that only ash remains. Repeated weighing of the sample ensures that all combustible material has been burned off, confirming complete combustion.

Fixed carbon is calculated as the difference between 100% and the combined percentages of moisture, ash, and volatile matter in coal. However, this value does not represent all the carbon originally present in the coal. A portion of the carbon may exist in the form of hydrocarbons, which can be lost during the determination of volatile matter, as they are distilled off. Additionally, the fixed carbon fraction may also contain small amounts of other elements such as sulfur, nitrogen, and oxygen.

In order to burn a fuel completely, four basic conditions must be fulfilled:

1. Supply *enough* air for complete combustion of fuel.
2. Secure enough *turbulence* for thorough mixing of fuel and air.
3. Maintain a furnace *temperature* high enough to ignite the incoming fuel air mixture.
4. Provide a furnace volume large enough to allow *time* for combustion to be completed.

In addition to providing an adequate air supply, three critical factors—time, temperature, and turbulence—must be carefully considered when designing a furnace. These are often summarized by the acronym **MATT**:

- **M** stands for *mixture*, representing turbulence, which ensures thorough interaction between fuel and oxygen molecules.
- **A** stands for the *air–fuel ratio*, which must be properly balanced to support efficient combustion.
- **T** represents *temperature*, which must be high enough to sustain combustion.
- The second **T** stands for *time*, referring to the residence time needed for combustion to complete.

Since perfect mixing of fuel and air is practically unattainable, excess air is typically supplied to promote complete combustion. However, the better the mixing—or turbulence—the less excess air is required, improving overall combustion efficiency.

The coal feeding system in boilers plays a critical role in determining the efficiency and effectiveness of the combustion process. Based on the method of coal feeding, furnaces can be broadly classified into two main types:

1. Grate-fired furnaces
2. Chamber-type or flame furnaces

Grate-fired furnaces are well-suited for burning solid fuels such as coal, wood chips, bagasse, and urban waste. In these systems, fuel is burned on a grate in a fixed bed, as illustrated in Figure 16.13. The combustion of volatile matter, fine carbon particles, and carbon monoxide primarily occurs in the space above the fuel bed.

In contrast, chamber-type or flame furnaces (depicted in Figure 16.14) are designed for firing pulverized coal, fuel oil, or gas. In these furnaces, a mixture of fuel and air is injected directly into the combustion chamber, where the fuel ignites and burns while flowing through the furnace space.

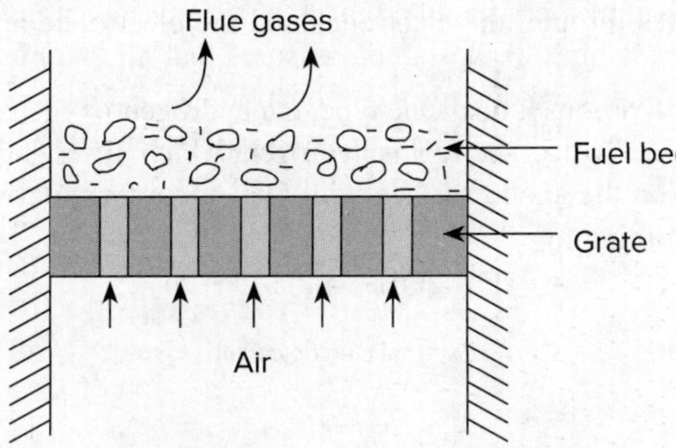

Fig. 16.13 Grate fired furnace

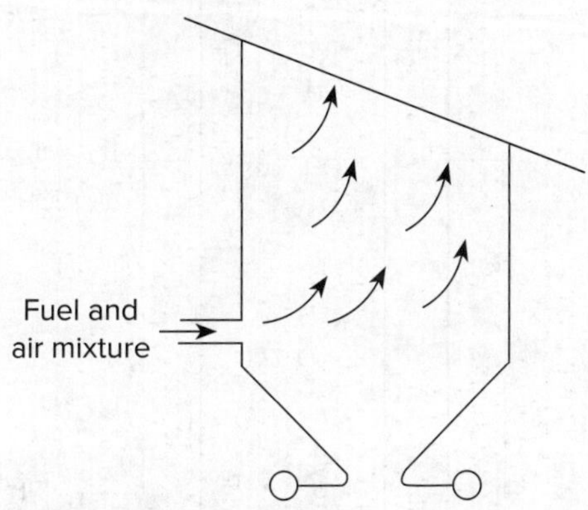

Fig. 16.14 Chamber type or flame furnace

Grate-fired boilers are capable of handling larger fuel sizes, whereas flame furnaces require pulverized fuel (PF) for combustion. In large power station boilers, PF firing is essential because combustion efficiency is critical due to the high fuel consumption rate, which significantly impacts the overall operating cost of the plant. Conversely, grate-fired boilers are often preferred for smaller-scale applications or when the fuel type cannot be pulverized effectively. These systems offer more flexibility in fuel handling, making them suitable for a wider range of solid fuels.

16.23 FUEL CELLS

> **LO 16.21**
> Describe fuel cells

Fuel cells are receiving prominent attention as clean, efficient and affordable alternative energy technologies for the future. A *fuel cell* is a direct energy conversion device in which fuel and an oxidiser undergo a controlled chemical reaction, producing products and supplying electrical current directly to an external circuit. The fuel and oxidiser react in stages on two separate electrodes, a positive electrode called *cathode*, and a negative electrode called *anode*. The two electrodes are separated by an electrolyte. The rates of reaction are limited by the time it takes for diffusion of chemical species through the electrodes and the electrolyte and by the reaction kinetics.

Fuel cells are of various kinds. One type, the hydrogen-oxygen fuel cell, is shown schematically in Fig. 16.15. The fuel is hydrogen, which diffuses through the porous anode and reacts on the anode surface with OH^- ions, forming water and yielding free electrons according to

$$H_2 + 2OH^- \rightarrow 2H_2O + 2e^-$$

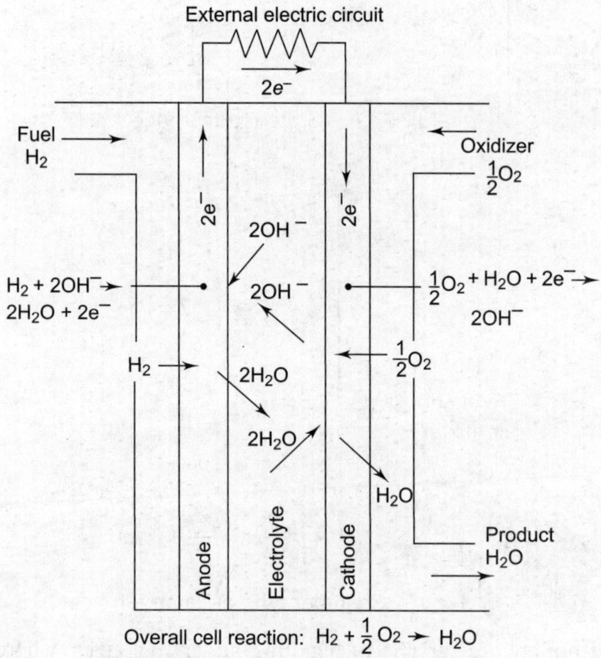

Fig. 16.15 Hydrogen-oxygen fuel cell

The electrons flow into the external circuit, and the water passes on to the electrolyte. On the cathode surface, oxygen fed into the cell combines with water from the electrolyte and electrons from the external circuit to produce OH^- ions and water according to

$$\frac{1}{2}O_2 + 2H_2O + 2e^- \rightarrow 2(OH)^- + H_2O$$

The electrolyte separating the electrodes transports the OH^- ions, completing, the circuit, and the water (products) is removed from the cell. The overall reaction is

$$H_2 + \frac{1}{2}O_2 \rightarrow H_2O$$

which is the same as the equation for the highly exothermic combustion reaction. However, in a fuel cell, only a relatively small amount of heat transfer between the cell and its surroundings takes place, and the temperature rise is also relatively much smaller.

Energy is removed from the fuel cell as electrical energy, whereas energy is removed from a combustion reaction as heat or as heat and work together. Because the fuel cell operates almost isothermally and continuously, the extent of its conversion of chemical energy to electrical energy is not limited by the second law of thermodynamics.

In a fuel cell, there is a continuous supply of the reactants. The overall reaction, as stated, is divided into two reactions that occur on separate electrodes. The fuel and the oxidiser do not come directly into contact with each other, because direct contact would generally involve a non-isothermal (exothermic) reaction as in a normal combustion process.

One reaction, occuring on the surface of one electrode, ionises the fuel and sends released electrons into an external electric circuit. On the surface of the other electrode, a reaction occurs that accepts electrons from the external circuit and when combined with the oxidiser creates ions. The ions from each reaction are combined in the electrolyte to complete the overall reaction. The electrolyte between the electrodes is necessary to transport ions, and it is not electrically conductive, thus, not allowing the flow of electrons through it.

The maximum work obtainable in a fuel cell is given by Eq. (16.33),

$$W_{max} = -\Delta G = \sum_R n_i \bar{g}_i - \sum_P n_e \bar{g}_e$$

where $\bar{g} = g_f^0 + \Delta \bar{g}$

Also, from Eq. (16.2)

$$W_{max} = W_{rev} = \sum_R n_i \left[\bar{h}_f^0 + \Delta \bar{h} - T_0 \bar{s}\right]_i - \sum_P n_e \left[\bar{h}_f^0 + \Delta \bar{h} - T_0 \bar{s}\right]_e$$

$$= \Delta H - T_0 \left[\sum_R n_i \bar{s}_i - \sum_P n_e \bar{s}_e\right]$$

The fuel cell efficiency is defined as

$$\varepsilon = \Delta G / \Delta H \tag{16.45}$$

which can be less than, equal to or greater than unity. The work done in a fuel cell can be written as

$$W_{max} = \text{moles of reactant} \times \frac{\text{molecules}}{\text{kg mol}} \times \frac{\text{electrons}}{\text{molecule}} \times \frac{\text{charge}}{\text{electron}} \times \text{voltage}$$

$$= n \times (6.023 \times 10^{26}) j\, (1.6022 \times 10^{19}) V = F n j V,$$

where F is Faraday constant (96.487×10^6 C/kmol) of electrons or 96,487 kJ/V-kmol of electrons, n is the number of moles of reactant having a valence j and V is the cell terminal voltage. The ideal cell terminal voltage is given by

$$V_i = \frac{\Delta G}{Fnj} \tag{16.46}$$

SOLVED EXAMPLES

Example 16.1 *Starting with n_0 moles of NH_3 which dissociates according to the equation*

$$NH_3 \rightleftharpoons \frac{1}{2}N_2 + \frac{3}{2}H_2$$

show that at equilibrium

$$K = \frac{\sqrt{27}}{4}\frac{\varepsilon_e^2}{1-\varepsilon_e^2}\cdot p$$

E [LO 16.2]

Solution

$$NH_3 \rightleftharpoons \frac{1}{2}N_2 + \frac{3}{2}H_2$$

Comparing the equation with

$$\nu_1 A_1 + \nu_2 A_2 \rightleftharpoons \nu_3 A_3 + \nu_4 A_4$$

we have

$$A_1 = NH_3,\ \nu_1 = 1,\ n_1 = n_0\,(1-\varepsilon),\ x_1 = \frac{1-\varepsilon}{1+\varepsilon}$$

$$A_3 = N_2,\ \nu_3 = \frac{1}{2},\ n_3 = \frac{n_0\varepsilon}{2},\ x_3 = \frac{\varepsilon}{2(1+\varepsilon)}$$

$$A_4 = H_2,\ \nu_4 = \frac{3}{2},\ n_4 = \frac{3n_0\varepsilon}{2},\ x_4 = \frac{3\varepsilon}{2(1+\varepsilon)}$$

$$\Sigma n = n_0(1+\varepsilon)$$

Substituting in the law of mass action

$$K = \left[\frac{x_3^{\nu_3}\cdot x_4^{\nu_4}}{x_1^{\nu_1}\cdot x_2^{\nu_2}}\right]_{\varepsilon=\varepsilon_e} p^{\nu_3+\nu_4-\nu_1-\nu_2}$$

$$= \frac{\left[\dfrac{\varepsilon_e}{2(1+\varepsilon_e)}\right]^{1/2}\left[\dfrac{3\varepsilon_e}{2(1+\varepsilon_e)}\right]^{3/2}}{\left(\dfrac{1-\varepsilon_e}{1+\varepsilon_e}\right)}\cdot p^{\frac{3}{2}+\frac{1}{2}-1} = \frac{\sqrt{27}\varepsilon^2}{4(1+\varepsilon_e)^2}\cdot\frac{1+\varepsilon_e}{1-\varepsilon_e}\cdot p$$

$$= \frac{\sqrt{27}}{4}\cdot\frac{\varepsilon_e^2}{1-\varepsilon_e^2}\cdot p$$

Proved.

Example 16.2 *At 35°C and 1 atm. the degree of dissociation of N_2O_4 at equilibrium is 0.27. (a) Calculate K, (b) Calculate ε_e at the same temperature when the pressure is 100 mm Hg and (c) The equilibrium constant for the dissociation of N_2O_4 has the values 0.664 and 0.141 at temperatures 318 and 298 K, respectively. Calculate the average heat of reaction within this temperature range.*

M [LO 16.1, 16.2]

Solution Nitrogen tetroxide dissociates according to the equation

$$N_2O_4 \rightleftharpoons 2NO_2$$

Comparing this equation with

$$\nu_1 A_1 + \nu_2 A_2 \rightleftharpoons \nu_3 A_3 + \nu_4 A_4$$

$$A_1 = N_2O_4,\ \nu_1 = 1,\ n_1 = n_0(1-\varepsilon),\ x_1 = \frac{1-\varepsilon}{1+\varepsilon}$$

$$A_3 = NO_2,\ \nu_3 = 2,\ n_3 = 2n_0\,\varepsilon,\ x_3 = \frac{2\varepsilon}{1+\varepsilon}$$

$$\Sigma n = n_0(1+\varepsilon)$$

$$\nu_2 = 0,\ \nu_4 = 0$$

Substituting in the law of mass action

$$K = \left[\frac{x_3^{\nu_3}\cdot x_4^{\nu_4}}{x_1^{\nu_1}\cdot x_2^{\nu_2}}\right]_{\varepsilon_e} \cdot p^{\nu_3+\nu_4-\nu_1-\nu_2}$$

$$= \frac{\left(\dfrac{2\varepsilon_e}{1+\varepsilon_e}\right)^2}{\dfrac{1-\varepsilon_e}{1+\varepsilon_e}} \cdot p^{2+0-1+0} = \frac{4\varepsilon_e^2 p}{1-\varepsilon_e^2}$$

(a) When $\varepsilon_e = 0.27$ and $p = 1$ atm.

$$K = \frac{4(0.27)^2\cdot 1}{1-(0.27)^2} = \mathbf{0.3145\ atm.}$$

(b) Taking the value of K to be the same at the same temperature

$$0.3145\text{ atm.} = \frac{4\varepsilon_e^2}{1-\varepsilon_e^2}\cdot\frac{100}{760}$$

$$\therefore\quad \varepsilon_e = \mathbf{0.612}$$

(c) The heat of reaction is given by

$$\Delta H = -2.30\,\bar{R}\frac{d\log K}{d\dfrac{1}{T}}$$

$$= 2.30\,\bar{R}\frac{T_1T_2}{T_1-T_2}\log\frac{K_1}{K_2}$$

$$= 2.30 \times 8.3143 \times \frac{318\times 298}{318-298}\log\frac{0.664}{0.141}$$

$$= 90700 \log 4.7 = \mathbf{61{,}000\ kJ/kg\ mol}$$

Example 16.3 *For the chemical reaction*

$$CO_2 + H_2 \rightleftharpoons CO + H_2O$$

the equilibrium value of the degree of reaction at 1200 K is 0.56. Determine the equilibrium constant and the Gibbs function change. E **[LO 16.1, 16.2]**

Solution

$$CO_2 + H_2 \rightleftharpoons CO + H_2O$$

Here, $\nu_1 = \nu_2 = \nu_3 = \nu_4 = 1$

At equilibrium, $n_1 = 1 - \varepsilon, n_2 = 1 - \varepsilon,$

$n_3 = \varepsilon$ and $n_4 = \varepsilon$

$\therefore$ $\Sigma n = 2$

Therefore, $x_1 = \dfrac{1-\varepsilon}{2}, \quad x_2 = \dfrac{1-\varepsilon}{2}, \quad x_3 = \dfrac{\varepsilon}{2}, \quad x_4 = \dfrac{\varepsilon}{2}$

Substituting $\varepsilon = 0.56$,

$$x_1 = 0.22, \quad x_2 = 0.22, \quad x_3 = 0.28, \quad x_4 = 0.28$$

$$K = \frac{x_3^{\nu_3}\, x_4^{\nu_4}}{x_1^{\nu_1}\, x_2^{\nu_2}} \times p^{\nu_3 + \nu_4 - \nu_1 - \nu_2}$$

$$= \frac{0.28 \times 0.28}{0.22 \times 0.22} = 1.62$$

$$\Delta G = -\bar{R}T \ln K + \bar{R}T \ln p^{(\nu_3 + \nu_4 - \nu_1 - \nu_2)}$$

$$= \bar{R}T \ln K = -8.3143 \times 1200 \ln 1.62 = \mathbf{-4813.2\ J/gmol}$$

Example 16.4 *Prove that for a mixture of reacting ideal gases,*

$$\frac{d}{d\varepsilon} ln \frac{x_3^{\nu_3}\, x_4^{\nu_4}}{x_1^{\nu_1}\, x_2^{\nu_2}} = \frac{(n_0 + n_0')}{\Sigma n_k} \frac{(V_1 + V_2)(V_3 + V_4)}{\varepsilon(1-\varepsilon)}$$

which is always positive. D [LO 16.2]

Solution From the law of mass action, the equilibrium constant is given by

$$K = \frac{x_3^{\nu_3}\, x_4^{\nu_4}}{x_1^{\nu_1}\, x_2^{\nu_2}} p^{\nu_3 + \nu_4 - \nu_1 - \nu_2}$$

$$= \frac{[n_3/\Sigma n]^{\nu_3}\,[n_4/\Sigma n]^{\nu_4}}{[n_1/\Sigma n]^{\nu_1}\,[n_2/\Sigma n]^{\nu_2}} p^{\nu_3 + \nu_4 - \nu_1 - \nu_2}$$

$$= \frac{n_3^{\nu_3}\, n_4^{\nu_4}}{n_1^{\nu_1}\, n_2^{\nu_2}} [p/\Sigma n]^{\Delta \nu}$$

where $\Sigma n = n_1 + n_2 + n_3 + n_4$ and $\Delta \nu = \nu_3 + \nu_4 - \nu_1 - \nu_2$.

By logarithmic differentiation,

$$\frac{dK}{K} - \Delta\nu \frac{dp}{p} + \Delta\nu \frac{d\Sigma n}{\Sigma n} = \nu_3 \frac{dn_3}{n_3} + \nu_4 \frac{dn_4}{n_4} - \nu_1 \frac{dn_1}{n_1} - \nu_2 \frac{dn_2}{n_2} \quad \text{(i)}$$

Now, $n_1 = (n_0 + n_0')\, \nu_1\, (1 - \varepsilon)$

$n_2 = (n_0 + n_0')\, \nu_2\, (1 - \varepsilon)$

$$n_3 = (n_0 + n'_0)\, \nu_3\, \varepsilon$$

$$n_4 = (n_0 + n'_0)\, \nu_4\, \varepsilon$$

$$\Sigma n = (n_0 + n'_0)\, (\nu_1 + \nu_2 + \varepsilon \Delta \nu) \qquad \text{(ii)}$$

Again, $$-\frac{dn_1}{\nu_1} = -\frac{dn_2}{\nu_2} = \frac{dn_3}{\nu_3} = \frac{dn_4}{\nu_4} = (n_0 + n'_0)\, d\varepsilon$$

$$d\,\Sigma n = dn_1 = dn_2 + dn_3 + dn_4 = (n_0 + n'_0)\, \Delta\nu\, d\varepsilon \qquad \text{(iii)}$$

From Eqs. (i), (ii) and (iii),

$$\frac{dK}{K} = -\Delta\nu \frac{dp}{p} + \Delta\nu\, (n_0 + n'_0)\, \Delta\nu\, d\varepsilon$$

$$= \nu_3 \frac{(n_0 + n'_0)\nu_3 d\varepsilon}{n_3} + \nu_4 \frac{(n_0 + n'_0)\nu_4 d\varepsilon}{n_4}$$

$$- \nu_1 \frac{(n_0 + n'_0)\nu_1 d\varepsilon}{n_1} - \nu_2 \frac{-(n_0 + n'_0)\nu_2 d\varepsilon}{n_2}$$

$$\frac{dK}{K} - \Delta\nu \frac{dp}{p} = (n_0 + n'_0)\left[\frac{\nu_3^2}{n_3} + \frac{\nu_4^2}{n_4} + \frac{\nu_1^2}{n_1} + \frac{\nu_2^2}{n_2} - \frac{(\Delta\nu)^2}{\Sigma n}\right] d\varepsilon$$

$$= \frac{n_0 + n'_0}{\Sigma n_K}\left[\frac{\nu_1^2}{x_1} + \frac{\nu_2^2}{x_2} + \frac{\nu_3^2}{x_3} + \frac{\nu_4^2}{x_4} - (\Delta\nu)^2\right] d\varepsilon$$

$$\frac{d(\ln K - \Delta\nu \ln p)}{d\varepsilon} = \frac{n_0 + n'_0}{\Sigma n_K} \frac{1}{\Psi}$$

where $$\frac{1}{\Psi} = \frac{\nu_1^2}{x_1} + \frac{\nu_2^2}{x_2} + \frac{\nu_3^2}{x_3} + \frac{\nu_4^2}{x_4} - (\Delta\nu)^2$$

L.H.S. $$= \frac{d}{d\varepsilon} \ln K p^{-\Delta\nu} = \frac{d}{d\varepsilon} \ln \frac{x_3^{\nu_3} x_4^{\nu_4}}{x_1^{\nu_1} x_2^{\nu_2}} p^{\Delta\nu} p^{-\Delta\nu}$$

$$\frac{d}{d\varepsilon} \ln \frac{x_3^{\nu_3} x_4^{\nu_4}}{x_1^{\nu_1} x_2^{\nu_2}} = \frac{n_0 + n'_0}{\Sigma n_K} \frac{1}{\Psi} \qquad \text{(iv)}$$

where x's are the equilibrium values.

Now,

$$\frac{1}{\Psi} = \frac{\nu_1^2}{x_1} + \frac{\nu_2^2}{x_2} + \frac{\nu_3^2}{x_3} + \frac{\nu_4^2}{x_4} - (\Delta\nu)^2$$

$$= \left[\frac{\nu_1^2}{n_1} + \frac{\nu_2^2}{n_2} + \frac{\nu_3^2}{n_3} + \frac{\nu_4^2}{n_4}\right] \Sigma n - (\Delta\nu)^2$$

$$= \left[\frac{\nu_1^2}{\nu_1(1-\varepsilon)} + \frac{\nu_2^2}{\nu_2(1-\varepsilon)} + \frac{\nu_3^2}{\nu_3\varepsilon} + \frac{\nu_4^2}{\nu_4\varepsilon}\right]$$

$$= \frac{\Sigma n}{n_0 + n_0'} (\Delta v)^2$$

$$= \left[\frac{v_1\varepsilon + v_2\varepsilon + v_3 - v_3\varepsilon + v_4 - v_4\varepsilon}{\varepsilon(1-\varepsilon)}\right]$$

$$(v_1 + v_2 + \varepsilon\Delta v) - (\Delta v)^2$$

$$= \frac{(v_3 + v_4) - \varepsilon\Delta v}{\varepsilon(1-\varepsilon)} [(v_1 + v_2) + \varepsilon\Delta v] - (\Delta v)^2$$

$$= \frac{(v_1 + v_2)(v_3 + v_4) - \varepsilon\Delta v(v_1 + v_2) + (v_3 + v_4) - \varepsilon^2(\Delta v)^2}{\varepsilon(1-\varepsilon)} - (\Delta v)^2$$

$$= \frac{(v_1 + v_2)(v_3 + v_4) + \varepsilon(\Delta v)^2 - \varepsilon^2(\Delta v)^2}{\varepsilon(1-\varepsilon)} - (\Delta v)^2$$

$$= \frac{(v_1 + v_2)(v_3 + v_4)}{\varepsilon(1-\varepsilon)} + \frac{\varepsilon(1-\varepsilon)(\Delta v)^2}{\varepsilon(1-\varepsilon)} - (\Delta v)^2$$

$$= \frac{(v_1 + v_2)(v_3 + v_4)}{\varepsilon(1-\varepsilon)}$$

Example 16.5 *For the dissociation of nitrogen tetraoxide according to the equation*

$$N_2O_4 \rightleftharpoons 2NO_2$$

Show that the degree of dissociation at equilibrium is $\varepsilon = \frac{V_e}{V_0} - 1$

where V_0 *= initial volume and* V_e *volume at equilibrium. At 50°C and 0.124 atm, there is a 77.7% increase in volume when equilibrium is reached. Find the value of the equilibrium constant.* M **[LO 16.2]**

Solution

$$N_2O_4 \rightleftharpoons 2NO_2$$

Starting with n_0 moles of N_2O_4 at temperature T and pressure p, the initial volume V_0 is

$$V_0 = n_0 \frac{\bar{R}T}{p}$$

If V_e denotes the volume at equilibrium, the temperature and pressure remaining the same, then

$$V_e = [n_0(1 - \varepsilon_e) + 2n_0\varepsilon_e] \frac{\bar{R}T}{p}.$$

where ε_e is the value of the degree of dissociation at equilibrium. This can be written as

$$V_e = (1 + \varepsilon_e)V_0$$

or, $$\varepsilon_e = \frac{V_e}{V_0} - 1$$ Proved.

Given $V_e/V_0 = 1.777,\ \varepsilon_e = 0.777$

Now,

$$n_1 = n_0\ \nu_1(1-\varepsilon_e) = n_0(1-\varepsilon_e)$$

$$n_3 = n_0\ \nu_3,\ \varepsilon_e = n_0 \cdot 2\varepsilon_e$$

$$\Sigma n = n_0(1+\varepsilon_e)$$

$$x_1 = \frac{1-\varepsilon_e}{1+\varepsilon_e},\ x_3 = \frac{2\varepsilon_e}{1+\varepsilon_e}$$

$$K = \left[\frac{x_3^{\nu_3}\cdot x_4^{\nu_4}}{x_1^{\nu_1}\cdot x_2^{\nu_2}}\right]_{\varepsilon=\varepsilon_a} p^{\nu_3+\nu_4-\nu_1-\nu_2}$$

$$= \frac{\left[\dfrac{2\varepsilon_e}{1+\varepsilon_e}\right]^2}{\left[\dfrac{1-\varepsilon_e}{1+\varepsilon_e}\right]} p^{2-1} = \frac{4\varepsilon_e^2}{1-\varepsilon_e^2} p$$

$$= \frac{4\times(0.777)^2}{1-(0.777)^2}\,0.124 = \frac{0.2994}{0.3963} = \mathbf{0.755\ atm}$$

Example 16.6 *Consider the equilibrium mixture of H_2O vapour, H_2 and O_2 caused by the dissociation of 1 gmol of H_2O at 1 atm and 1900 K. If ΔH = 250, 560 J/gmol, $\varepsilon_e = 3.2 \times 10^{-3}$, estimate $C_p - \Sigma n_k C_{Pk}$* Ⓜ **[LO 16.2]**

Solution $H_2O \rightleftharpoons H_2 + \frac{1}{2} O_2$

$$\nu_1 = 1,\ \nu_2 = 0,\ \nu_3 = 1,\ \nu_4 = \frac{1}{2}$$

$$n_1 = n_0\ \nu_1\ (1-\varepsilon_e) = n_0(1-\varepsilon_e),$$

$$n_3 = n_0\ \nu_3\ \varepsilon_e = n_0\ \varepsilon_e,\ n_4 = n_0\ \nu_4\ \varepsilon_e = n_0\frac{\varepsilon_e}{2}$$

$$\Sigma n = n_0\left(1+\frac{\varepsilon_e}{2}\right)$$

Now,

$$\frac{d}{d\varepsilon_e}\ln\frac{x_3^{\nu_3}\,x_4^{\nu_4}}{x_1^{\nu_2}\,x_2^{\nu_2}} = \frac{n_0}{\Sigma n}\frac{(\nu_1+\nu_2)(\nu_3+\nu_4)}{\varepsilon_e(1-\varepsilon_e)}$$

$$C_P - \Sigma n_k\,c_{Pk} = \frac{(\Delta H)^2(1+\varepsilon_e/2)\,\varepsilon_e(1+\varepsilon_e)}{\overline{R}T^2(\nu_1+\nu_2)(\nu_3+\nu_4)}$$

$$= \frac{(250{,}560)^2\times 3.2\times10^{-3}}{8.3143\times(1900)^2\times\dfrac{3}{2}}$$

$$= \mathbf{4.462\ J/gmol\ K}$$

Example 16.7 *The products of combustion of an unknown hydrocarbon $C_x H_y$ have the following composition as measured by an Orsat apparatus:*

$$CO_2\ 8.0\%,\ CO\ 0.9\%,\ O_2\ 8.8\% \text{ and } N_2\ 82.3\%$$

Determine: (a) the composition of the fuel, (b) the air-fuel ratio and (c) the percentage excess air used. **E [LO 16.10]**

Solution Let a mole of oxygen be supplied per mole of fuel. The chemical reaction can be written as follows.

$$C_xH_y + aO_2 + 3.76\,a\,N_2$$
$$\rightarrow 8\,CO_2 + 0.9\,CO + 8.8\,O_2 + 82.3\,N_2 + \frac{y}{2}H_2O$$

Oxygen balance gives

$$2a = 16 + 0.9 + 17.6 + \frac{y}{2} \tag{i}$$

Nitrogen balance gives

$$3.76a = 82.3$$
$$\therefore \quad a = 21.89$$

By substituting the value of a in Eq. (i),

$$y = 18.5$$

Carbon balance gives $x = 8 + 0.9 = 8.9$

Therefore, the chemical formula of the fuel is $C_{8.9}H_{18.5}$. The composition of the fuel is

$$\%\text{ carbon} = \frac{8.9 \times 12}{8.9 \times 12 + 18.5 \times 1} \times 100 = \mathbf{85.23\%}$$

$\therefore$ % hydrogen = **14.77%**

$$\text{(b) Air fuel ratio} = \frac{32a + 3.76a \times 28}{12x + y}$$
$$= \frac{21.89(32 + 3.76 \times 28)}{12 \times 8.9 + 18.5} = \frac{3005}{125.3} = \mathbf{24}$$

(c) % excess air used

$$= \frac{8.8 \times 32}{21.89 \times 32 - 8.8 \times 32} \times 100 = \mathbf{67.22\%}$$

Example 16.8 *Ethylene (C_2H_4) is burned with 180% theoretical air during a combustion process. Assuming that the combustion is complete and the pressure of the combustion products is 100 kPa, determine (a) the air-fuel ratio and (b) the dew point temperature of the products.* **E [LO 16.10]**

Solution The combustion equation with 180% theoretical air can be written as

$$C_2H_4 + 1.8a_{th}O_2 + 1.8a_{th}(3.76)N_2 \rightarrow 2CO_2 + 2H_2O + 0.8a_{th}O_2 + 1.8a_{th}(3.76)N_2$$

where a_{th} is the stoichiometric coefficient for air.

Oxygen balance gives

$$1.8a_{th} = 2 + 1 + 0.8a_{th}$$

or $\quad a_{th} = 3$

The combustion equation with 180% theoretical air is given by

$$C_2H_4 + 5.4O_2 + 5.4(3.76)N_2 \rightarrow 2CO_2 + 2H_2O + 2.4O_2 + 20.304N_2$$

(a) The air-fuel ratio is found to be

$$\text{AF} = \frac{m_{air}}{m_{fuel}} = \frac{5.4\times 4.76\times 28.96}{2\times 12+4\times 1} = \textbf{26.59 kg of air / kg of fuel}$$

(b) The partial pressure of the water vapour in the product gases is

$$p_v = \left(\frac{n_{\text{H}_2\text{O, vap}}}{n_{total,\text{ Prod}}}\right) p_{\text{Prod}} = \left(\frac{2\text{ kmol}}{26.704\text{ kmol}}\right) 100\text{ kPa} = 7.49\text{ kPa}$$

The dew point temperature is the saturation temperature of water vapour corresponding to the partial pressure in the combustion product gases. Thus,

$$T_{dp} = T_{sat}\big|_{7.49\text{ kPa}} = \textbf{40.29°C}$$

Example 16.9 *The composition of a natural gas is found from the volumetric analysis as follows: 68%* CH_4*, 7%* H_2*, 17%*N_2*, 3%*O_2 *and 5%*CO_2*. This gas is burned with 120% theoretical air during a combustion process. Assuming that the combustion is complete, determine the air-fuel ratio. If the product gases are cooled to 20°C, 101 kPa, determine the fraction of water vapour that will condense.* Ⓜ **[LO 16.10]**

Solution The combustion equation can be written with 1 kmol of gas as

$$0.68\text{CH}_4 + 0.07\text{H}_2 + 0.17\text{N}_2 + 0.03\text{O}_2 + 0.05\text{CO}_2 + 1.2a_{th}\text{O}_2 + 1.2a_{th}(3.76)\text{N}_2$$
$$\rightarrow x\text{CO}_2 + y\text{H}_2\text{O} + 0.2a_{th}\text{O}_2 + z\text{N}_2$$

Carbon balance gives

$$0.68 + 0.05 = x$$

or $x = 0.73$

Hydrogen balance gives

$$0.68\times 4 + 0.07\times 2 = 2y$$

or $y = 1.43$

Oxygen balance gives

$$0.03 + 0.05 + a_{th} = x + \frac{y}{2}$$

or $a_{th} = 1.365$

Nitrogen balance gives

$$0.17 + 1.2\times 3.76a_{th} = z$$

or $z = 6.329$

Thus,

$$0.68\text{CH}_4 + 0.07\text{H}_2 + 0.17\text{N}_2 + 0.03\text{O}_2 + 0.05\text{CO}_2 + 1.638\text{O}_2 + 1.638(3.76)\text{N}_2$$
$$\rightarrow 0.73\text{CO}_2 + 1.43\text{H}_2\text{O} + 0.273\text{O}_2 + 6.329\text{N}_2$$

The air-fuel ratio is found to be

$$\text{AF} = \frac{m_{\text{air}}}{m_{\text{fuel}}} = \frac{1.638\times 4.76\times 28.96}{0.68\times 16 + 0.07\times 2 + 0.17\times 28 + 0.03\times 32 + 0.05\times 44}$$
$$= \textbf{11.92 kg of air/kg}\text{ of fuel}$$

Let us assume that the dew-point temperature of the combustion products is more than 20°C. Let n_w kmol of H_2O condenses when the temperature of the combustion products decreases to 20°C. Total number of kmol of the products in the gas phase is 8.762 and after condensation it becomes $8.762 - n_w$. After condensation, total number of kmol of H_2O is $1.43 - n_w$. Using the relation between mole fraction with partial pressure, we have

$$\frac{1.43 - n_w}{8.762 - n_w} = \frac{p_w}{P_{total}}$$

where $p_w = p_{sat}\big|_{20°C} = 2.339$ kPa

$$\therefore \quad \frac{1.43 - n_w}{8.762 - n_w} = \frac{2.339}{101}$$

or $$n_w = 1.256$$

The fraction of water vapour that will condense is $1.256/1.43 = 0.8783$ or **87.83%**.

Example 16.10 *Determine the heat transfer per kg mol of fuel for the following reaction*

$$CH_4 + 2O_2 \rightarrow CO_2 + 2H_2O\ (l)$$

The reactants and products are each at a total pressure of 100 kPa and 25°C.

E **[LO 16.12]**

Solution By the first law

$$Q_{C.V.} + \sum_P n_i \bar{h}_i = \sum_P n_e \bar{h}_e$$

From Table C in the appendix

$$\sum_R n_i \bar{h}_i = (\bar{h}_f)_{CH_4} = -74{,}874 \text{ kJ}$$

$$\sum_P n_e \bar{h}_e = (\bar{h}_f^0)_{CO_2} + 2(\bar{h}_f^0)_{H_2O(l)}$$

$$= -393{,}522 + 2(-285{,}838) = -965{,}198 \text{ kJ}$$

$$\therefore \quad Q_{C.V.} = -965{,}198 - (-74{,}873) = \mathbf{-890{,}325\ kJ}$$

Example 16.11 *A gasoline engine delivers 150 kW. The fuel used is C_8H_{18} (l) and it enters the engine at 25°C. 150% theoretical air is used and it enters at 45°C. The products of combustion leave the engine at 750 K, and the heat transfer from the engine is 205 kW. Determine the fuel consumption per hour, if complete combustion is achieved.*

M **[LO 16.12]**

Solution The stoichiometric equation (Fig. 16.16) gives

$$C_8H_{18}\ (l) + 12.5\ O_2 + 3.76 \times 12.5\ N_2$$
$$\rightarrow 8CO_2 + 9H_2O + 3.76 \times 12.5\ N_2$$

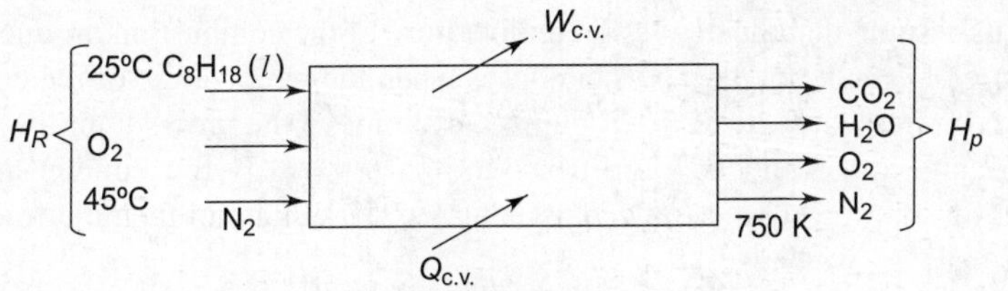

Fig. 16.16

With 150% theoretical air

$$C_8H_{18} + 18.7O_2 + 70\,N_2$$

$$\rightarrow 8CO_2 + 9H_2O + 6.25O_2 + 70N_2$$

$$H_R = \sum_R n_i[\bar{h}_f^0 + (\bar{h}_T^0 - \bar{h}_{298})]_i$$

$$= 1\,(\bar{h}_f)_{C_8H_{18}(1)} + 18.7\,(\bar{h}_f^0 + (\bar{h}_T - \bar{h}_{298})]_{O_2}$$

$$+ 70\,(\bar{h}_f^0 + (\bar{h}_T - \bar{h}_{298})]_{N_2}$$

$$= (-249{,}952) + 18.7 \times 560 + 70 \times 540$$

$$= -201{,}652 \text{ kJ/kg mol fuel}$$

$$H_P = 8[-393{,}522 + 20288] + 9\,[-241{,}827 + 16087]$$

$$+ 6.25\,[14171] + 70\,[13491]$$

$$= 8\,(-373{,}234) + 9\,[-225{,}740] + 6.25\,(14171) + 70\,(13491]$$

$$= -3{,}884{,}622 \text{ kJ/kg mol fuel}$$

Energy output from the engine, $W_{C.V.} = 150$ kW

$$Q_{C.V.} = -205 \text{ kW}$$

Let n kgmol of fuel be consumed per second.
By the first law

$$n\,(H_R - H_P) = W_{C.V.} - Q_{C.V.}$$

$$n[-201{,}652 + 3{,}884{,}662] = 150 - (-205) = 355 \text{ kW}$$

$$\therefore \quad n = \frac{355 \times 3600}{3683010} \text{ kg mol/h} = 0.346 \text{ kg mol/h}$$

$\therefore$ Fuel consumption rate = 0.346 × 114 = **39.4 kg/h**

Example 16.12 *Determine the adiabatic flame temperature when liquid octane at 25°C is burned with 300% theoretical air at 25°C in a steady flow process.*

E [LO 16.13]

Solution The combustion equation with 300% theoretical air is given by

$$C_8H_{18}(1) + 3(12.5)O_2 + 3(12.5)\,(3.76)\,N_2 \rightarrow$$

$$\rightarrow 8CO_2 + 9H_2O + 25O_2 + 141N_2$$

By the first law $H_R = H_P$

$$\sum_R n_i \left[h_{f0} + \Delta \bar{h}\right]_i = \sum_P n_e [\bar{h}_{f0} + \Delta \bar{h}]_e$$

Now $H_R = (\bar{h}_{f0})_{C_8H_{18}(1)} = -249{,}952$ kJ/kg mol fuel

$$H_P = \sum_P n_e[\bar{h}_{f0} + \Delta \bar{h}]_e = 8\,[-393{,}522 + \Delta \bar{h}_{CO_2}]$$

$$+ 9\,[-241{,}827 + \Delta \bar{h}_{H_2O}] + 25\ \Delta \bar{h}_{O_2} + 141\ \Delta \bar{h}_{N_2}$$

The exit temperature, which is the adiabatic flame temperature, is to be computed by trial and error satisfying the above equation.

Let $T_e = 1000$ K, then

$$H_P = 8(-393{,}522 + 33{,}405) + 9(-241{,}827 + 25{,}978) + 25(22.707) + 141(21{,}460) = -1226{,}577 \text{ kJ/kg mol}$$

If $T_e = 1200$ K

$$H_P = 8(-393{,}522 + 44{,}484) + 9(-241{,}827 + 34{,}476) + 25(29{,}765) + 141(28{,}108) = +46{,}537 \text{ kJ/kg mol}$$

If $T_e = 1100$ K,

$$H_P = 8(-393{,}522 + 38{,}894) + 9(-241{,}827 + 30{,}167) + 25(26{,}217) + 141(24{,}757) = -595{,}964 \text{ kJ/kg mol}$$

Therefore, by interpolation, the adiabatic flame temperature is **1182 K.**

Example 16.13 *Propane (C_3H_8) at 25°C is burned steadily with 30% excess air at 25°C, 1 atm and 70% relative humidity in a combustion chamber. It is found that 65000 kJ of heat per kmol of propane is lost from the combustion chamber to the surroundings. Assuming that the combustion is complete, calculate the exit temperature of the product gases.* Ⓜ **[LO 16.12, 16.13]**

Solution The combustion equation with 180% theoretical air can be written as

$$C_3H_8 + 1.3a_{th}O_2 + 1.3a_{th}(3.76)N_2 \rightarrow 3CO_2 + 4H_2O + 0.3a_{th}O_2 + 1.3a_{th}(3.76)N_2$$

where a_{th} is the stoichiometric coefficient for air.

Oxygen balance gives

$$1.3a_{th} = 3 + 2 + 0.3a_{th}$$

or

$$a_{th} = 5$$

The combustion equation with 180% theoretical air is given by

$$C_3H_8 + 6.5O_2 + 24.44N_2 \rightarrow 3CO_2 + 4H_2O + 1.5O_2 + 24.44N_2$$

The partial pressure of the water vapour present in the incoming air is

$$p_{v,in} = \phi\, p_{sat}\big|_{25°C} = 0.70 \times 3.169 = 2.218 \text{ kPa}$$

Let $n_{H_2O,\text{ in}}$ be the number of moles of moisture that accompanies with incoming dry air. Then

$$n_{H_2O,\text{ in}} = \left(\frac{p_{v,\text{in}}}{p_{\text{total}}}\right) n_{\text{total}} = \left(\frac{2.218}{101}\right)(31.94 + n_{H_2O,\text{in}})$$

or $$n_{H_2O,\text{ in}} = 0.72$$

Finally, the combustion equation is given by

$$C_3H_8 + 6.5O_2 + 24.44N_2 + 0.72H_2O \rightarrow 3CO_2 + 4.72H_2O + 1.5O_2 + 24.44N_2$$

By the first law, we have

$$\sum_R n_i\left(\bar{h}_f^0 + \bar{h} - \bar{h}^o\right)_i - \sum_P n_e\left(\bar{h}_f^0 + \bar{h} - \bar{h}^o\right)_e = -Q_{CV}$$

$$-103850 + 0.72(-241820) - 3\left(-393522 + \bar{h}_{CO_2} - 9364\right) - 4.72\left(-241820 + \bar{h}_{H_2O} - 9904\right)$$
$$-1.5\left(0 + \bar{h}_{O_2} - 8682\right) - 24.44\left(0 + \bar{h}_{N_2} - 8669\right) = -(-65000)$$

or $$3\bar{h}_{CO_2} + 4.72\bar{h}_{H_2O} + 1.5\bar{h}_{O_2} + 24.44\bar{h}_{N_2} = 2278728.24$$

The exit temperature of the product gases is obtained from a trial and error solution.

At 2000 K

$$3\bar{h}_{CO_2} + 4.72\bar{h}_{H_2O} + 1.5\bar{h}_{O_2} + 24.44\bar{h}_{N_2} = 3 \times 100804 + 4.72 \times 82593$$
$$+1.5 \times 67881 + 24.44 \times 64810 = 2376732 \text{ (Higher than 2278728.24)}$$

At 1900 K

$$3\bar{h}_{CO_2} + 4.72\bar{h}_{H_2O} + 1.5\bar{h}_{O_2} + 24.44\bar{h}_{N_2} = 3 \times 94793 + 4.72 \times 77517$$
$$+1.5 \times 64116 + 24.44 \times 61220 = 2241425.64 \text{ (Higher than 2278728.24)}$$

At 1920 K

$$3\bar{h}_{CO_2} + 4.72\bar{h}_{H_2O} + 1.5\bar{h}_{O_2} + 24.44\bar{h}_{N_2} = 3 \times 95995 + 4.72 \times 78527$$
$$+1.5 \times 64868 + 24.44 \times 61936 = 2268411.56$$

Therefore, the exit temperature is **1920 K**.

Example 16.14 *(a) Propane (g) at 25°C and 100 kPa is burned with 400% theoretical air at 25°C and 100 kPa. Assume that the reaction occurs reversibly at 25°C, that the oxygen and nitrogen are separated before the reaction takes place (each at 100 kPa, 25°C), that the constituents in the products are separated and that each is at 25°C, 100 kPa. Determine the reversible work for this process. (b) If the above reaction occurs adiabatically, and each constituent in the products is at 100 kPa pressure and at the adiabatic flame temperature, compute (i) the increase in entropy during combustion, (ii) the irreversibility of the process and (iii) the availability of the products of combustion.* D **[LO 16.12, 16.13]**

Solution The combustion equation (Fig. 16.17) is

$$C_3H_8(g) + 5(4)O_2 + 5(4)(3.76)\,N_2 \rightarrow$$

$$3CO_2 + 4H_2O(g) + 15O_2 + 75.2\ N_2$$

(a) $$W_{rev} = \sum_R n_i \bar{g}_i - \sum_P n_e \bar{g}_e$$

From Table 16.4

$$W_{rev} = (\bar{g}_{f0})_{C_3H_8(g)} - 3(\bar{g}_{f0})_{CO_2} - 4(\bar{g}_{f0})_{H_2O(g)}$$
$$= -23,316 - 3(-394,374) - 4(-228,583)$$
$$= \mathbf{2,074,128\ kJ/kg\ mol}$$
$$= \frac{2,074,128}{44.097} = \mathbf{47,035.6\ kJ/kg}$$

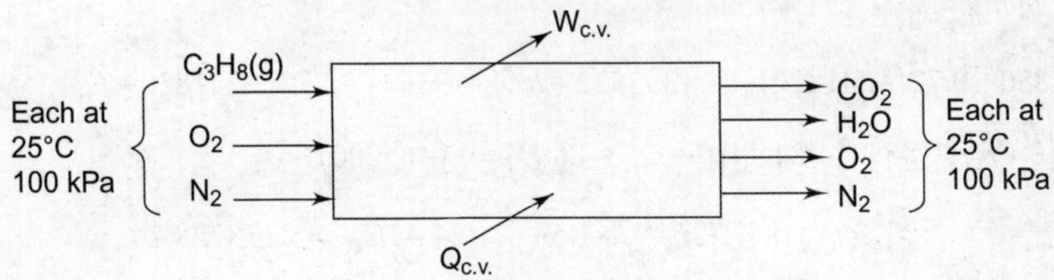

Fig. 16.17

(b) $H_R = H_P$

$$(\bar{h}_{f0})_{C_3H_8(g)} = 3(\bar{h}_{f0} + \Delta\bar{h})_{CO_2} + 4(\bar{h}_{f0} + \Delta\bar{h})_{H_2O(g)} + 15\Delta\bar{h}_{O2} + 75.2\Delta\bar{h}_{N2}$$

From Table 16.4

$$-103,847 = 3(-393,522 + \Delta\bar{h})_{CO2} + 4(-241,827 + \Delta\bar{h})_{H2O}(g) + 15\ \Delta\bar{h}_{O_2} + 75.2\ \Delta\bar{h}_{N_2}$$

Using Table C in the appendix, and by trial and error, the adiabatic flame temperature is found to be 980 K.

The entropy of the reactants

$$S_R = \sum_R (n_i \bar{s}_i^0)_{298} = (\bar{s}^0_{C_3H_8(g)} + 20\bar{s}^0_{O_2} + 75.2\bar{s}^0_{N_2})_{298}$$
$$= 270.019 + 20\ (205.142) + 75.2(191.611)$$
$$= 18,782.01\ \text{kJ/kg mol K}$$

The entropy of the products

$$S_P = \sum_P (n_e \bar{s}_e^0)_{980} = (3\bar{s}^0_{CO_2} + 4\bar{s}^0_{H_2O(g)} + 15\bar{s}^0_{O_2} + 75.2\bar{s}^0_{N_2})_{980}$$
$$= 3(268.194) + 4(231.849) + 15(242.855) + 75.2(227.485)$$
$$= 22,481.68\ \text{kJ/kg mol K}$$

∴ The increase in entropy during combustion

$$S_P - S_R = \mathbf{3,699.67\ kJ/kg\ mol\ K}$$

The irreversibility of the process

$$I = T_0 \left[\sum_P n_e \bar{s}_e - \sum_R n_i \bar{s}_i \right]$$

$$= 298 \times 3699.67$$
$$= 1{,}102{,}501.66 \text{ kJ/kg mol}$$
$$= \frac{1{,}102{,}501.66}{44.097} = \mathbf{25{,}001 \text{ kJ/kg}}$$

The availability of combustion products

$$\psi = W_{\text{rev}} - \text{I}$$
$$= 47{,}035.6 - 25{,}001 = \mathbf{22{,}034.6 \text{ kJ/kg}}$$

Example 16.15 *Determine the chemical exergy of (a) carbon, (b) hydrogen, (c) methane (d) carbon monoxide, (e) liquid methanol, (f) nitrogen, (g) oxygen, (h) carbon dioxide, and (i) water in kJ/kg, in respect of the environment in which the gas phase obeys the ideal gas model.*
Environment:
T_0 = 298.15 K, p_0 = 1 atm, N_2 75.67%, O_2 20.35%, $H_2O(g)$ 3.12%, CO_2 0.03% and others 0.83%, by volume. M **[LO 16.17]**

Solution (a) Carbon: $C + O_2 \rightarrow CO_2$

$$\bar{a}_{\text{ch}} = \left[\bar{g}_c + \bar{g}_{O_2} - \bar{g}_{CO_2}\right]_{T_0, p_0} + \bar{R}T_0 \ln \frac{x_{O_2}}{x_{CO_2}}$$

$$= 0 + 0 - (-394{,}380) + 8.3143 \times 298.15 \ln \frac{0.2035}{0.0003}$$

$$= 410{,}541 \text{ kJ/k mol} = \mathbf{34{,}212 \text{ kJ/kg}}$$

(b) Hydrogen: $H_2 + \frac{1}{2} O_2 \rightarrow H_2O$

$$\bar{a}_{\text{ch}} = \left[\bar{g}_{H_2} + \frac{1}{2}\bar{g}_{O_2} - \bar{g}_{H_2O}\right]_{p_0, T_0} + \bar{R}T_0 \ln \frac{(x_{O_2})^{1/2}}{x_{H_2O(g)}}$$

$$= 0 + 0 - (-228590) + 8.3143 \times 298.15 \ln \frac{(0.2035)^{1/2}}{0.0312}$$

$$= 235{,}212 \text{ kJ/kg mol} = \mathbf{116{,}557 \text{ kJ/kg}}$$

(c) Methane: $CH_4 + 2O_2 \rightarrow CO_2 + 2H_2O$

$$\bar{a}_{\text{ch}} = \left[\bar{g}_{CH_4} + 2\bar{g}_{O_2} - \bar{g}_{CO_2} - 2\bar{g}_{H_2O(g)}\right]$$

$$+ \bar{R}T_0 \ln \frac{x_{O_2}^2}{(x_{CO_2})(x_{H_2O})}$$

$$= -50{,}790 + 0 - (-394{,}380) - 2(-228{,}590)$$

$$+ 8.3143 \times 298.15 \ln \frac{(0.2035)^2}{(0.0003)(0.0312)^2}$$

$$= 830{,}174 \text{ kJ/kg mol} = \mathbf{51{,}756 \text{ kJ/kg}}$$

(d) Carbon monoxide: $CO + \frac{1}{2} O_2 \rightarrow CO_2$

$$\bar{a}_{cH} = \left[\bar{g}_{CO} + \frac{1}{2} \bar{g}_{O_2} - \bar{g}_{CO_2} \right] - \bar{R}T_0 \ln \frac{x_{O_2}^{1/2}}{x_{CO_2}}$$

$$= -137{,}150 + 0 - (-394{,}380)$$

$$+ 8.3143 \times 298.15 \ln \frac{(0.2035)^{1/2}}{0.0003}$$

$$= 275{,}364 \text{ kJ/kg mol} = \mathbf{9831\ kJ/kg}$$

(e) Liquid methanol: $CH_3OH(l) + 1.5O_2 \rightarrow CO_2 + 2H_2O$

$$\bar{a}_{ch} = \left[\bar{g}_{CH_3OH(l)} + 1.5\bar{g}_{O_2} - \bar{g}_{CO_2} - 2\bar{g}_{H_2O(g)} \right]$$

$$+ \bar{R}T_0 \ln \frac{x_{O_2}^{1.5}}{\left(x_{CO_2}\right)\left(x_{H_2O}\right)^2}$$

$$= -166{,}240 + 0 - (-394{,}380) - 2(-228{,}590)$$

$$+ 8.3143 \times 298.15 \ln \frac{(0.2035)^{1.5}}{(0.0003)(0.0312)^2} = 716{,}647 \text{ kJ/kg mol}$$

$$= \mathbf{22{,}360\ kJ/kg}$$

(f) Nitrogen: $N_2 \rightarrow N_2$

$$\bar{a}_{ch} = \bar{R}T_0 \ln \frac{1}{x_{N_2}} = 8.3143 \times 298.15 \ln \frac{1}{0.7567}$$

$$= 691.1 \text{ kJ/k mol} = \mathbf{24.7\ kJ/kg}$$

(g) Oxygen: $O_2 \rightarrow O_2$

$$\bar{a}_{ch} = \bar{R}T_0 \ln \frac{1}{x_{O_2}} = 8.3143 \times 298.15 \ln \frac{1}{0.2035}$$

$$= 3947 \text{ kJ/kmol} = \mathbf{123.3\ kJ/kg}$$

(h) Carbon dioxide:

$$\bar{a}_{ch} = \bar{R}T_0 \ln \frac{1}{x_{CO_2}} = 8.3143 \times 298.15 \ln \frac{1}{0.0003}$$

$$= 20{,}108 \text{ kJ/kg mol} = \mathbf{456.9\ kJ/kg}$$

(i) Water: $H_2O(l) \rightarrow H_2O(g)$

$$\bar{a}_{ch} = \bar{g}_{H_2O(l)} - \bar{g}_{H_2O(g)} + \bar{R}T_0 \ln \frac{1}{x_{H_2O}}$$

$$= -237{,}180 - (-228{,}590) + 8.3143$$

$$\times 298.15 \ln \frac{1}{0.0312}$$

$$= 4.9 \text{ kJ/kg mol} = \mathbf{0.272\ kJ/kg}$$

Example 16.16 *Liquid octane (C_8H_{18}) at 25°C, 1 atm and a mass flow rate of 0.57 kg/h enters a small internal combustion engine operating at steady state. The fuel burns with air entering the engine in a separate stream at 25°C, 1 atm. Combustion products exit at 670 K, 1 atm with a dry molar analysis of 11.4% CO_2, 2.9% CO, 1.6%O_2 and 84.1% N_2. If the engine develops power at the rate of 1 kW, determine (a) the rate of heat transfer from the engine in kW, and (b) the second law efficiency of the engine. Use the environment as given below and neglect K.E. and P.E. effects.*

Environment: 1 atm, 298.15 K and N_2 75.67%, O_2 20.25%, $H_2O(g)$ 3.12%, CO_2 0.03% and others 0.83%, by volume.

Ⓜ **[LO 16.10, 16.12, 16.16]**

Solution

$C_8H_{18}(1) + a(O_2 + 3.76N_2) \rightarrow b\,[0.114\,CO_2 + 0.029\,CO + 0.016O_2 + 8.41\,N_2] + CH_2O$

By carbon balance,

$$b(0.114 + 0.029) = 8 \quad b = 55.9$$

By hydrogen balance, $18 = 2C \quad C = 9$

By oxygen balance,

$$a = b \times 0.114 + \frac{b}{2} \times 0.029 + b \times 0.016 + \frac{C}{2}$$

$$= b(0.114 + 0.0145 + 0.016) + \frac{9}{2} = 12.58$$

The combustion equation becomes,

$C_8H_{18}(1) + 12.58\,(O_2 + 3.76\,N_2) \rightarrow 55.9\,[0.114\,CO_2 + 0.029\,CO + 0.016\,O_2 + 0.841\,N_2] + 9\,H_2O$

(a) By first law, $Q + H_R = W + H_P$

$$\frac{Q_{CV}}{\dot{n}_{fuel}} = \frac{\dot{W}_{CV}}{\dot{n}_{fuel}} - (-249{,}910) + 55.9\,[0.114\,(-393{,}520 + 25284) + 0.029$$

$$(-110{,}530 + 11{,}089) + 0.016\,(1151.5) + 0.841\,(11016) + 9\,(-241{,}820 + 13064)$$

$$= \frac{\dot{W}_{CV}}{\dot{n}_{fuel}} - 3{,}845{,}872 \text{ kJ/kg mol } C_8H_{18}\,(1)$$

$$\dot{Q}_{CV} = \dot{W}_{CV} - \dot{n}_{fuel} \times 3{,}845{,}872 \text{ kW}$$

$$= 1 - 3{,}845{,}872\,\frac{\text{kJ}}{\text{kg mol}} \times \frac{0.57\text{ kg/h}}{3600\text{ s/h}} \times \frac{1\text{ kg mol}}{114.22\text{ kg}}$$

$$= 1 - 5.33 = \mathbf{-\,4.33\ kW}$$

(b) Chemical exergy of C_8H_{18} was found to be 5,407,843 kJ/kg mol (Section 16.19).

$$\eta_{II} = \frac{\dot{W}_{CV}}{\dot{n}_{fuel} \times \bar{a}_{ch}}$$

$$= \frac{1\,\text{kW}}{5407843 \dfrac{\text{kJ}}{\text{kg mol}} \times \dfrac{0.57\,\text{kg}}{3600\,\text{s}} \times \dfrac{1\,\text{kg mol}}{114.22\,\text{kg}}}$$

$$= \frac{1}{7.496} = \mathbf{0.133 \text{ or } 13.3\%}.$$

SUMMARY

In the reversible chemical reaction,

$$\nu_1 A_1 + \nu_2 A_2 \rightleftharpoons \nu_3 A_3 + \nu_4 A_4$$

where A_1 and A_2 are the reactants and A_3 and A_4 are the products for reaction occurring from left to right, and ν's are the stoichiometric coefficients. The initial number of moles of A_1, A_2 are $n = n_0\gamma_1$ and $n_2 = n_0\gamma_2 + N_2$, where n_0 is an arbitrary positive number and N_2 is the excess number of moles of A_2 which cannot combine. When the reaction is completed, the final constituents are $N_3 = n_o'\gamma_3$ and $N_4 = n_o'\gamma_4 + N_4$ where n_o' is also an arbitrary positive number. When the reaction is completed to the left,

$$n_1(\max) = (n_o + n_o')\gamma_1,\ n_2(\max) = (n_o + n_o')\gamma_2 + N_2$$

and $n_3(\min) = 0$ and $n_4(\min) = N_4$.
Similarly, when the reaction is completed to the right

$$n_1(\min) = 0,\ n_2(\min) = N_2,\ n_3(\max) = (n_o + n_o')\gamma_3$$

and $n_4(\max) = (n_o + n_o')\gamma_3 + N_4$.
When the reaction is completed partially either to the left or to the right, the numbers of moles of the elements are

$$n_1 = (n_o + n_o')\gamma_3(1 - \varepsilon),\ n_2 = (n_o + n_o')\gamma_2(1 - \varepsilon) + N_2,$$

$$n_3 = (n_o + n_o')\gamma_3\,\varepsilon \text{ and } n_4 = (n_o + n_o')\gamma_4\,\varepsilon + N_4$$

where ε is the *degree or advancement of reaction*, which is defined by $\varepsilon = \dfrac{n(\max) - n_1}{n_1(\max) - n_1(\min)}$.

When the reaction advances a little, ε changes from ε to $\varepsilon + d\varepsilon$ and the various n's change by amounts

$dn_1 = -(n_o + n_o')\gamma_1\, d\varepsilon,\ dn_2 = dn_2 = -(n_o + n_o')\gamma_2\, d\varepsilon,\ dn_3 = (n_o + n_o')\gamma_3\, d\varepsilon$ and $dn_A = (n_o + n_o')\gamma_A\, d\varepsilon$. The change in Gibbs function becomes

$$dG_{T,p} = \mu_1 dn_1 + \mu_2 dn_2 + \mu_3 dn_3 + \mu_4 dn_4$$

At equilibrium, $dG_{T,p} = 0$, and *the equation of reaction equilibrium* is obtained as given below

$$\nu_1\mu_1 + \nu_2\mu_2 = \nu_3\mu_3 + \nu_4\mu_4$$

Substituting for ideal gases,

$$\mu_k = \bar{R}T(\phi_k + \ln p + \ln x_k)$$

where ϕ's are functions of temperature only, the *law of mass action*

$$\left(\frac{x_3^{\nu_3} x_4^{\nu_4}}{x_1^{\nu_1} x_2^{\nu_2}}\right) p^{\nu_3 + \nu_4 - \nu_1 - \nu_2} = K$$

where, K being called the *equilibrium constant* given by

$$\ln K = -(\gamma_3\phi_3 + \gamma_4\phi_4 - \gamma_1\phi_1 - \gamma_2\phi_2)$$

Since
$$\phi = \frac{h_0}{\bar{R}T} - \frac{1}{\bar{R}}\int \frac{\int c_p dT}{T^2} dT - \frac{s_o}{R}$$

$$\frac{d}{dT}\ln K = \frac{1}{\bar{R}T^2}(\gamma_3 h_3 + \gamma_4 h_4 - \gamma_1 h_1 - \gamma_2 h_2)$$

$$= \frac{\Delta H}{\bar{R}T^2}$$

This is known as *van't Hoff equation*, where the *heat of reaction* $\Delta H = \gamma_3 h_3 + \gamma_4 h_4 - \gamma_1 h_1 - \gamma_2 h_2$.

If ΔH is positive, the reaction is *endothermic* and if ΔH is negative, the reaction is exothermic.

The Nernst's equation is given by

$$\ln K = \frac{\Delta H}{\bar{R}T^2} + \frac{1}{\bar{R}}\int \frac{\int (\nu_3 c_{p_3} + \nu_4 c_{p_4} - \nu_1 c_{p_1} - \nu_2 c_{p_2})dT}{T^2} dT + \frac{\Delta S_o}{\bar{R}}$$

From which Saha obtained the equation of *degree of thermal ionisation* of a monatomic gas

$$\frac{\varepsilon_e^2}{1-\varepsilon_e^2} = B \cdot e^{-\frac{\Delta H_0}{\bar{R}T}} \frac{T^{5/2}}{p}.$$

The standard Gibbs function change is

$$\Delta G^\circ = -\bar{R}T \ln K$$

For a real gas,

$$G - G^\circ = n\bar{R}T \ln \frac{f}{f^0} = n\,\bar{R} \ln a$$

where 'f' is called the fugacity such that $\lim_{p\to 0} \frac{f}{f^0} = 1$ and 'a' is called the activity.

The heat capacity of reaching gases in equilibrium is

$$C_p = \sum n_k c_{p_k} + (n_0 + n_0')\frac{(\Delta H)^2}{\bar{R}T^2 \frac{d}{d\varepsilon}\ln\left(\frac{x_3^{\nu_3} x_4^{\nu_4}}{x_1^{\nu_1} x_2^{\nu_2}}\right)_{\varepsilon=\varepsilon_e}}$$

Combustion is the reaction of a fuel and oxygen with the release of heat. For all the reactants and products, the equation may be written as

$$\sum_R n_i \bar{h}_i + Q_{cv} = \sum_P n_e \bar{h}_e$$

The enthalpy of all elements at the standard reference state of 25°C is zero. If the reactants are elements, $H_R = 0$ and $Q_{C.V.} = H_p$ is called the *enthalpy of formation*. The first law for reactive systems is thus

$$H_R + Q_{C.V.} = H_P + W_{C.V.}$$

If the reaction occurs adiabatically in the absence of work, then the energy equation becomes

$$\sum_R n_i[\bar{h}_f^0 + \Delta\bar{h}] = \sum_P n_e[\bar{h}_f^0 + \Delta\bar{h}]$$

The temperature of the products is called the *adiabatic flame temperature* which is the maximum temperature achieved for the given reactants.

The *enthalpy of combustion* is given by

$$\bar{h}_{RP} = H_p - H_R$$

which is also called the *heating value at constant pressure*. The *internal energy of combustion* is given by

$$\bar{u}_{RP} = U_p - U_R$$

$$= \sum_P n_e[\bar{h}_f^0 + \Delta\bar{h} - p\bar{v}]_e - \sum_R n_i[\bar{h}_f^0 + \Delta\bar{h} - p\bar{v}]_i$$

which is also called the *heating value at constant volume*.

The *higher heating value* (HHV) is the heat released when the H_2O in the products is liquid. If the H_2O in the products is in vapour form, the *lower heating value* (LHV) results so that

$$LHV = HHV - m_{H_2O} h_{fg}$$

where m_{H_2O} is the mass of water formed in the reaction.

The maximum work obtained in a steady flow system

$$W_{rev} = W_{max} = \sum n_i(\bar{h}_i + T_o\bar{s}_i] - \sum n_e(\bar{h}_e + T_o\bar{s}_i]$$

$$= \sum_i n_i\bar{g}_i - \sum_e n_e\bar{g}_e$$

where g refers to the Gibbs function. The irreversibility of the reaction

$$I = \sum_P n_e T_o \bar{s}_e - \sum_P n_i T_o \bar{s}_i - Q_{iv}.$$

The *chemical energy* of a fuel, a_{ch}, for the reaction

$$C_aH_b + \left(a + \frac{b}{4}\right)O_2 \rightarrow aCO_2 + \frac{b}{2}H_2O$$

is given by

$$a_{ch} = \left[\bar{h}_{C_aH_b} + \left(a + \frac{b}{4}\right)\bar{h}_{O_2} - a\bar{h}_{CO_2} - \frac{b}{2}\bar{h}_{H_2O}\right]$$

which is the maximum work. Obtainable second law of efficiency of a reactive system is given by

$$\eta = \frac{W_{CV}}{m_{fuel} \times a_{ch}}$$

where W_{cv} is the actual work done.

In real-life applications, combustion processes vary significantly across different devices, depending on factors such as the type of fuel used, the fuel feeding system, the mixing of air and fuel, the time available for combustion, and other operational parameters. Two major devices that heavily rely on fuel combustion for their performance are internal combustion (IC) engines and boilers.

IC engines primarily use liquid fuels like petrol or diesel, or gaseous fuels. The method of air-fuel mixing and ignition varies between fuels: petrol and gases are typically pre-mixed with air and ignited using a spark (spark ignition or SI engines), whereas diesel is injected directly into compressed air and ignites due to high temperature (compression ignition or CI engines). Abnormal combustion can occur in both types of engines - detonation in petrol engines and knock in diesel engines - both of which can significantly reduce engine life and must be carefully controlled. Supercharging is a technique used to boost engine power by increasing the amount of air introduced into the cylinder.

In contrast, boilers typically use solid fuels such as coal. The fuel feeding and combustion processes in boilers differ substantially from those in IC engines. Solid fuels can be supplied in various forms - lump, crushed, or pulverized - depending on operational needs and design considerations. Since solid fuels release volatile compounds and gases upon heating before actual combustion begins, a proximate analysis (which evaluates moisture, volatile matter, ash, and fixed carbon) is essential in addition to an ultimate analysis (which assesses elemental composition) for proper evaluation. The selection of appropriate fuel firing and combustion methods in boilers is strongly influenced by the combined results of these two analyses.

REVIEW QUESTIONS

(E) **16.1** What are stoichiometric coefficients? **[LO 16.10]**

(E) **16.2** What is the degree of reaction? What are its limiting values? **[LO 16.1]**

(E) **16.3** When does a chemical reaction reach equilibrium? **[LO 16.2]**

(E) **16.4** What do you understand by the equilibrium constant of a chemical reaction? **[LO 16.2]**

(E) **16.5** What is the law of mass action? **[LO 16.2]**

(E) **16.6** What is the heat of reaction? When is it positive and when negative? **[LO 16.6]**

(E) **16.7** What is the van't Hoff equation? **[LO 16.6]**

(M) **16.8** Give Nernst's equation. What does it signify? **[LO 16.6]**

(E) **16.9** What is thermal ionisation? **[LO 16.7]**

(M) **16.10** What is the significance of Saha's equation? **[LO 16.7]**

(E) **16.11** What do you understand by the standard Gibbs function change? **[LO 16.8]**

(E) **16.12** What is chemical affinity? **[LO 16.8]**

(M) **16.13** Define fugacity and activity. **[LO 16.9]**

(M) **16.14** How can the heat capacity of reacting gases in equilibrium be estimated? **[LO 16.4]**

(E) **16.15** What is stoichiometric air? **[LO 16.10]**

(E) **16.16** What is excess air? Why is it required? **[LO 16.10]**

(D) **16.17** What is Orsat analyzer? Describe briefly. **[LO 16.10]**

(E) **16.18** What do you understand by dry flue gas analysis? **[LO 16.10]**

(M) **16.19** What do you understand by the enthalpy of formation? **[LO 16.11]**

(D) **16.20** Define adiabatic flame temperature. How is it estimated? **[LO 16.13]**

(D) **16.21** What is enthalpy of combustion? What is internal energy of combustion? **[LO 16.14]**

(E) **16.22** What do you understand by higher heating value and lower heating value of a fuel? **[LO 16.14]**

(E) **16.23** State Planck's formulation of the third law of thermodynamics. **[LO 16.15]**

(M) **16.24** What is the Gibbs function of formation? **[LO 16.16]**

(D) **16.25** Explain thermomechanical exergy and chemical exergy. **[LO 16.17]**

(D) **16.26** How is second law efficiency of a reactive system defined? **[LO 16.18]**

(D) **16.27** Explain the principle of operation of a hydrogen-oxygen fuel cell. What is the maximum work obtainable in the cell? What is ideal terminal voltage? **[LO 16.21]**

(E) **16.28** Why petrol engines are also called SI engines? **[LO 16.19]**

(M) **16.29** What is the useful air fuel ratio of a SI engine? Explain it. **[LO 16.19]**

(D) **16.30** What is detonation? Why it happens? **[LO 16.19]**

(E) **16.31** What is the fuel of a CI engine? Why is it called so? **[LO 16.19]**

(D) **16.32** What is knock? Why it happens? **[LO 16.19]**

(M) **16.33** What is the proximate analysis of coal? Why is it required in addition to ultimate analysis or elemental analysis? **[LO 16.20]**

(M) **16.34** Briefly describe grate fired and chamber type furnaces of a boiler. **[LO 16.20]**

PROBLEMS WITH HINTS

(M) **16.1** One kgmol of H_2O dissociates to form an equilibrium mixture of H_2O, H_2 and O_2. Determine the equilibrium composition for a pressure of 1 atm if the temperature is 2000 K. Repeat the problem for the pressure of 10 atm. **[LO 16.1, 16.2]**

Hint: By conservation of mass principle, let 1 kgmol of H_2O dissociate to form an equilibrium mixture of $(1 - \alpha)$ kmol of H_2O, β kmol of H_2 and γ kgmol of O_2, i.e.,

$$1\,H_2O \rightarrow (1 - \alpha)\,H_2O + \beta H_2 + \gamma O_2$$

By mass balance,

$$H:\ 2 = 2\,(1 - \alpha) + 2\beta \rightarrow \beta = \alpha$$

$$O:\ 1 = (1 - \alpha) + 2\beta \rightarrow \gamma = \alpha/2$$

Thus,

$$1\,H_2O \rightarrow (1 - \alpha)\,H_2O + \alpha H_2 + \frac{\alpha}{2} O_2$$

where $(1 - \alpha)$ is the amount of H_2O in kgmol present in the mixture. The amount of mixture in kgmol is

$$n = (1 - \alpha) + \alpha + \frac{\alpha}{2} = (2 - \alpha)/2$$

At equilibrium, $H_2O = H_2 + \dfrac{\alpha}{2} O_2$

$$K = \left(\frac{x_3^{\gamma_3} x_4^{\gamma_4}}{x_1^{\gamma_1} x_2^{\gamma_2}}\right) p^{\gamma_3+\gamma_4-\gamma_2} = \left[\text{where } x_3 = \frac{n_3}{n}, x_4 = \frac{n_4}{n} \ldots\right]$$

$$= \frac{\alpha\left(\dfrac{\alpha}{2}\right)^{1/2}}{(1-\alpha)}\left[\frac{p}{(2+\alpha)/2}\right]^{1+\frac{1}{2}-1}$$

$$= \frac{\alpha}{1-\alpha}\left(\frac{\alpha}{2+\alpha}\right)^{1/2} p^{1/2}$$

From Table A.8 of Appendix A, at 300 K

$$\log_{10} K = 1.343 \rightarrow K = 0.04539$$

For $p = 1$ atm,

$$0.04539 = \frac{\alpha}{1-\alpha}\left(\frac{2}{2+\alpha}\right)^{1/2}$$

Solve it to obtain $\alpha = 0.148$

Thus, the equilibrium mixture is

$$0.852\ H_2O,\ 0.148\ H_2 \text{ and } 0.074\ O_2.$$

or **85.2% H_2O, 14.8% H_2 and 7.4% O_2 by volume.**

At $p = 10$ atm,

$$0.04539 = \left(\frac{\alpha}{1-\alpha}\right)\left(\frac{\alpha}{2+\alpha}\right)^{1/2} 10^{1/2}$$

which gives $\alpha = 0.0716$

The equilibrium mixture is

0.9284 H_2O, 0.0716 H_2 and 0.0358 O_2

(M) **16.2** An equimolar mixture of CO and O_2 reacts to form an equilibrium mixture of CO_2, CO and O_2 at 3000 K, 5 atm. Determine the composition of the equilibrium mixture. Will lowering the pressure while keeping the temperature fixed increase or decrease the amount of CO_2 present? Explain. **[LO 16.2]**

Hint: $CO + O_2 \rightarrow xCO_2 + \alpha CO + \beta O_2$

Balancing:

$$C: 1 = x + \alpha \quad \therefore \alpha = 1 - x$$

$$O: 3 = 2x + \alpha + 2\beta = 2x + 1 - x + 2\beta$$

$$\therefore \quad \beta = \frac{2-x}{2}$$

Thus,

$$CO + O_2 \rightarrow x\ CO_2 + (1-x)CO + \left(1 - \frac{x}{2}\right)O_2$$

Amount of mixture is $n = x + 1 - x + 1 - 1 - \frac{x}{2} = \frac{4-x}{2}$

At equilibrium, $CO_2 = CO + \frac{1}{2}O_2$

$$\therefore \quad K = \frac{(1-x)\left(\frac{2-x}{2}\right)^{\frac{1}{2}}}{x}\left[\frac{p}{(4-x)/2}\right]^{1+\frac{1}{2}-1}$$

$$= \left(\frac{1-x}{x}\right)\left(\frac{2-x}{4-x}\right)^{1/2} (p)^{1/2} \qquad \text{(i)}$$

From Table A.8 of Appendix A, at 3000 K,

$$\log_{10} K = -0.485 \rightarrow K = 0.32724$$

For $p = 5$ atm, from Eq. (i) obtain $x = 0.807$

The equilibrium mixture is

0.807 CO_2, 0.193 CO and 0.5965 O_2

Substituting $p = 4$ atm in Eq. (i) and solving, obtain $x = 0.79$

Accordingly, less CO_2 appears as pressure decreases.

(M) **16.3** An equilibrium mixture at 2000 K, 1 atm consists of *Cs*, *Cs*+ and e^-. Based on 1 kgmol of *Cs* percent initially, determine the percent ionisation of cesium. At 2000 K, the ionisation equilibrium constant for $Cs \rightleftharpoons Cs^+ + e^-$ is $K = 15.63$. **[LO 16.3]**

Hint: $Cs \rightarrow (1 - Z)Cs + ZCs^+ + Ze^-$

Amount of mixture, $n = 1 - Z + Z + Z = 1 + Z$

At equilibrium, $Cs \rightleftharpoons Cs^+ + e^-$

$$K = \frac{(Z)(Z)}{1-Z}\left(\frac{P}{1+Z}\right)^{1+-1} = \frac{Z^2}{1-Z^2} = 15.6$$

which gives $Z = 0.969$

The amount of Cs in the mixture is 0.031 kgmol.

∴ Per cent of ionised Cs is **96.9%**

Ⓜ **16.4** Steam at 5 bar, 240°C leaks from a line in a vapour power plant. Evaluate the flow exergy of the steam relative to an environment at 25°C, 1 atm in which the mole fraction of water vapour is 0.0303. **[LO 16.17]**

Hint: Specific flow exergy is

$$a_f = (h - h_o) - T_o(s - s_o) + a^{ch}$$
$$= (2939.9 - 104.9) - 298\,(7.2307 - 0.3674)$$
$$= 789.7 \text{ kJ/kg} + \bar{a}^{ch} + a^{ch}$$

where h_o and s_o are taken as the saturated liquid values at 298 K. The chemical exergy is

$$a^{ch} = \frac{1}{M}\left[(\bar{g}_{H_2O(l)} - \bar{g}_{H_2O(g)})\right]_{T_0, p_0} + \bar{R}T_0 \ln \frac{1}{x^e_{H_2O}}$$
$$= \frac{1}{18}[-237,180 - (-228590)] + 8.3143 \times 298 \ln \frac{1}{0.0303}$$
$$= 4.1 \text{ kJ/kg}$$

$$\therefore \quad a_f = \mathbf{792.8\ kJ/kg}$$

Ⓜ **16.5** Calculate the enthalpy of combustion of gaseous propane at 500 K. The average c_p of C_3H_8 between 25°C and 500 K is 2.1 kJ/kmol-K. **[LO 16.11]**

Hint: Combustion equation is

$$C_3H_8\,(g) + 5O_2 \rightarrow 3\,CO_2 + 4\,H_2O\,(g)$$

Enthalpy of combustion is

$$(\bar{h}_{RP})_T = (\bar{h}_{RP})_T \text{ ne } (\bar{h}_f^0 + \Delta\bar{h})_e - \sum_R n_i(\bar{h}_f^0 + \Delta\bar{h})_i$$

$$H_R = [\bar{h}_f^0 + \bar{c}_p\Delta T]_{C_3J_8(g)} + n_{O_2}(\Delta\bar{h})_{O_2} = -54738 \text{ kJ}$$

From Table 16.4, find h_f for C_3H_8 (g) and take $\bar{c}_p = 2.1 \times 44.097$ kJ/kgmol-K and take $(\Delta\bar{h})_{O_2}$ from Table A.9 of Appendix A. This gives $H_R = -54.738$.

$$H_P = (\bar{h}_f^0 + \Delta\bar{h})_{CO_2} + (\bar{h}_f^0 + \Delta\bar{h})_{H_2O}$$
$$= -2095395 \text{ kJ}$$

$$\therefore \quad \bar{h}_{RP} = \frac{H_P - H_R}{M} = \mathbf{-46277\ kJ/kg}$$

Ⓜ **16.6** Ethane (g) at 25°C, 100 kPa is burned with 400% theoretical air at 25°C, 100 kPa. Assume that the reaction takes place reversibly at 25°C, 100 kPa and that the products leave at 25°C, 100 kPa. Assume also that oxygen and nitrogen are separated before the reaction takes place, each at 25°C, 100 kPa. Determine the reversible work for this process. **[LO 16.10]**

Hint: The combustion equation can be written as

$$C_2H_4\,(g) + 3(4)\,O_2 + 3(4)\,(3.76)\,N_2 \rightarrow 2\,CO_2 + 2\,H_2O\,(g) + 9O_2 + 45.1\,N_2$$

$$W_{rev} = \sum_R n_i \bar{g}_i - \sum_P n_e \bar{g}_e$$

From Table 16.4, using the values $\bar{g}_f^0$, obtain

$$W_{rev} = \mathbf{46{,}851\ kJ/kg}$$

(M) **16.7** If the reaction in the above problem occurs adiabatically and the adiabatic flame temperature is 1000 K, determine (a) the increase in entropy during combustion, (b) the irreversibility of the process and (c) the exergy of the products of combustion. **[LO 16.13, 16.16, 16.17]**

Hint:

$S_p - S_R = \sum_R n_e \bar{s} = (2\bar{s}^0_{CO_2} + 2\bar{s}^0_{H_2O(g)} + 9\bar{s}_{O_2} + 45.1\bar{s}^0_{N_2})_{1000} - (\bar{s}^0_{C_2H_4(g)} + 12\bar{s}_0 +$
$45.1\ \bar{s}^0_{N_2})_{298}$

$$= 13518.277 - 11322.705$$

$$= \mathbf{2195.752\ kJ/kgmol\text{-}K}$$

$$I = \frac{1}{M} T_o (S_P - S_R) = 23,\ \mathbf{394\ kJ/kg\ fuel.}$$

$\therefore$ Exergy of combustion products = 46, 850 – 23394

$$= \mathbf{23,\ 517\ kJ/kg}$$

PROBLEMS

(E) **16.1** Starting with n_0 moles of NO, which dissociates according to the equation

$$NO \rightleftharpoons \frac{1}{2} N_2 + \frac{1}{2} O_2$$

show that at equilibrium $K = \frac{1}{2} \cdot \frac{\varepsilon_e}{1-\varepsilon_e}$ **[LO 16.2]**

(M) **16.2** A mixture of $n_0\ \nu_1$ moles of A_1 and $n_0\ \nu_2$ moles of A_2 at temprature T and pressure p occupies a volume V_0. When the reaction

$$\nu_1 A_1 + \nu_2 A_2 \rightleftharpoons \nu_3 A_3 + \nu_4 A_4$$

has come to equilibrium at the same T and p, the volume is V_e. Show that

$$\varepsilon_e = \frac{V_e - V_0}{V_0} \cdot \frac{\nu_1 + \nu_2}{\nu_3 + \nu_4 - \nu_1 - \nu_2}$$ **[LO 16.2]**

(M) **16.3** The equilibrium constant of the reaction **[LO 16.2]**

$SO_3 \rightleftharpoons SO_2 + O_2$ has the following values

T	800 K	900 K	1000 K	1105 K
K	0.0319	0.153	0.540	1.59

Determine the average heat of dissociation graphically. **[LO 16.2]**

(M) **16.4** At high temperature the potassium atom is ionised according to the equation $K \rightleftharpoons K^+ + e^-$. The values of the equilibrium constant at 3000 and 3500 K are

8.33×10^{-6} and 1.33×10^{-4}, respectively. Compute the average heat of reaction in the given temperature range. **[LO 16.2]**

(M) **16.5** (a) Show that $\Delta G = \Delta H + T\left(\frac{\partial \Delta G}{\partial T}\right)_p$

(b) Show that $\Delta G = -\bar{R}T \ln \frac{x_3^{\nu_3} \cdot x_4^{\nu_4}}{x_1^{\nu_1} \cdot x_2^{\nu_2}}$ **[LO 16.3]**

where the x's are equilibrium values.

(M) **16.6** When 1 kg mol of HI dissociates according to the reaction

$$HI \rightleftharpoons \frac{1}{2} H_2 + \frac{1}{2} I_2$$

at $T = 675$ K, $K = 0.0174$ and $\Delta H = 5910$ kJ/kg mol.

Calculate $(\partial \varepsilon / \partial T)_p$ at this temperature. **[LO 16.3]**

(M) **16.7** (a) Prove that for a mixture of reacting ideal gases

$$\frac{d}{d\varepsilon} \ln \frac{x_3^{\nu_3} \cdot x_4^{\nu_4}}{x_1^{\nu_1} \cdot x_2^{\nu_2}} = \frac{n_0 + n_0'}{\Sigma n_K} \cdot \frac{1}{\psi}$$

where $$\frac{1}{\psi} = \frac{\nu_1^2}{x_1} + \frac{\nu_2^2}{x_2} + \frac{\nu_3^2}{x_3} + \frac{\nu_4^2}{x_4} - (\Delta\, \nu)^2$$

$$\Delta \nu = \nu_3 + \nu_4 - \nu_1 - \nu_2$$

(b) If we start with $n_0\, \nu_1$ moles of A_1, $n_0\, \nu_2$ moles of A_2, and $n_0\, A_3$ or A_4, show that

$$\psi = \frac{\varepsilon(1-\varepsilon)}{(\nu_1 + \nu_2)(\nu_3 + \nu_4)}$$ **[LO 16.3]**

(M) **16.8** Prove that, for a mixture of reacting ideal gases in equilibrium

(a) $$\left(\frac{\partial V}{\partial p}\right)_T = -\frac{V}{p} - \frac{(n_0 + n_0')\bar{R}T(\Delta \nu)^2}{p^2 \cdot \frac{d}{d\varepsilon_e} \ln \frac{x_3^{\nu_3} \cdot x_4^{\nu_4}}{x_1^{\nu_1} \cdot x_2^{\nu_2}}}$$

(b) $$\left(\frac{\partial V}{\partial T}\right)_p = \frac{V}{T} + \frac{(n_0 + n_0')\Delta \nu \Delta H}{pT \frac{d}{d\varepsilon_e} \ln \frac{x_3^{\nu_3} \cdot x_4^{\nu_4}}{x_1^{\nu_1} \cdot x_2^{\nu_2}}}$$

(c) $$\left(\frac{\partial p}{\partial T}\right)_{\varepsilon_e} = \frac{p \Delta H}{\bar{R}T^2 \Delta \nu}$$ **[LO 16.3]**

(M) **16.9** Oxygen dissociates according to the relation $O_2 \rightarrow 2O$. Show that the equilibrium constant is given by

$$K = \frac{4\varepsilon_e^2 p}{1 - \varepsilon_e^2}$$

where ε_e is the degree of dissociation. Obtain an expression for the rate of variation of ε_e with temperature, $\left(\frac{\partial \varepsilon_e}{\partial T}\right)_p$, in terms of ε_e, $\frac{dK}{dT}$ and the total pressure p. At a temperature of 3800 K, $K = 1$ atm and $\dfrac{\Delta H}{\bar{R}} = 59$ K, calculate the value of ε and $\left(\dfrac{\partial \varepsilon_e}{\partial T}\right)_p$ at a pressure of 1 atm. **[LO 16.2, 16.3]**

(M) **16.10** (a) Starting with n_0 moles of water vapour which dissociate according to the equation $H_2O \rightleftharpoons H_2 + \frac{1}{2}O_2$, show that at equilibrium

$$K = \frac{\varepsilon_e^{3/2}}{(2+\varepsilon_e)^{1/2}(1-\varepsilon_e)} \cdot p^{1/2}$$

(b) At an average temperature of 1900 K, the slope of the graph of log K against $1/T$ for the dissociation of water vapour is found to be – 13,000. Determine the heat of dissociation. Is it exothermic or endothermic? **[LO 16.6]**

(E) **16.11** The volumetric composition of the 'dry' products of combustion of an unknown hydrocarbon fuel, $C_x H_y$, gives: CO_2 12.1%, O_2 3.8%, CO 0.9% and N_2 83.2%. Detrmine (a) the chemical formula of the fuel, (b) the air fuel ratio and (c) the percentage of excess air used. **[LO 16.10]**

(E) **16.12** Liquid octane is burned with air in a combustion test, and the dry volumetric analysis of the products reveals the following composition: CO_2 11%, O_2 3.6%, CO 1.5% and N_2 83.9%. Compute the actual air-fuel ratio used during the test and the percentage excess air. **[LO 16.10]**

(E) **16.13** Propane (C_3H_8) is reacted with air in such a ratio that an analysis of the products of combustion gives CO_2 11.5%, O_2 2.7% and CO 0.7%. What is the percent theoretical air used during the test? **[LO 16.10]**

(M) **16.14** Carbon monoxide and 300% theoretical air enter a steady flow combustor at 400K and a low pressure. The energy released by the reaction heats the product gases to 1400 K. If the combustion is complete, estimate the heat gained or lost through the walls of the combustor. **[LO 16.10, 16.12]**

(M) **16.15** Liquid octane enters the combustion chamber of a gas turbine at 25°C and air enters from the compressor at 227°C. It is determined that 98% of the carbon in the fuel burns to form CO_2 and the remaining 2% burns to form CO. The temperature of the products is limited to 827°C. Estimate the air-fuel ratio used and the percentage excess air. **[LO 16.10]**

(M) **16.16** A mixture of methane and oxygen, in the proper ratio for complete combustion and at 25°C and 1 atm, reacts in a constant volume calorimeter bomb. Heat is transferred until the products of combustion are at 400 K. Determine the heat transfer per mole of methane. **[LO 16.12]**

(D) **16.17** Liquid hydrazine (N_2H_4) and oxygen gas, both at 25°C, 0.1 MPa are fed to a rocket combustion chamber in the ratio of 0.5 kg O_2/kg N_2H_4. The heat transfer from the chamber to the surroundings is estimated to be 100 kJ/kg N_2H_4. Determine the temperature of the products, assuming only H_2O, H_2 and N_2 to be present. The enthalpy of the formation of N_2H_4 (l) is + 50,417 kJ/kg mol. **[LO 16.12]**
If saturated liquid oxygen at 90 K is used instead of 25°C oxygen gas in the combustion process, what will the temperature of the products be?

(M) **16.18** Liquid ethanol (C_2H_5OH) is burned with 150% theoretical oxygen in a steady flow process. The reactants enter the combustion chamber at 25°C, and the products are cooled and leave at 65°C, 0.1 MPa. Calculate the heat transfer per kg mol of ethanol. The enthalpy of formation of C_2H_5OH (1) is – 277,634 kJ/kg mol. **[LO 16.10, 16.12]**

Ⓜ **16.19** A small gas turbine uses C_8H_{18} (1) for fuel and 400% theoretical air. The air and fuel enter at 25°C and the combustion products leave at 900 K. If the speicific fuel consumption is 0.25 kg/s per MW output, determine the heat transfer from the engine per kg mol of fuel, assuming complete combustion. **[LO 16.10, 16.12]**

Ⓜ **16.20** Hydrogen peroxide (H_2O_2) enters a gas generator at the rate of 0.1 kg/s, and is decomposed to steam and oxygen. The resulting mixture is expanded through a turbine to atmospheric pressure, as shown in Fig. 16.18. Determine the power output of the turbine and heat transfer rate in the gas generator. The enthalpy of formation of H_2O_2(l) is –187,583 kJ/kg mol. **[LO 16.12]**

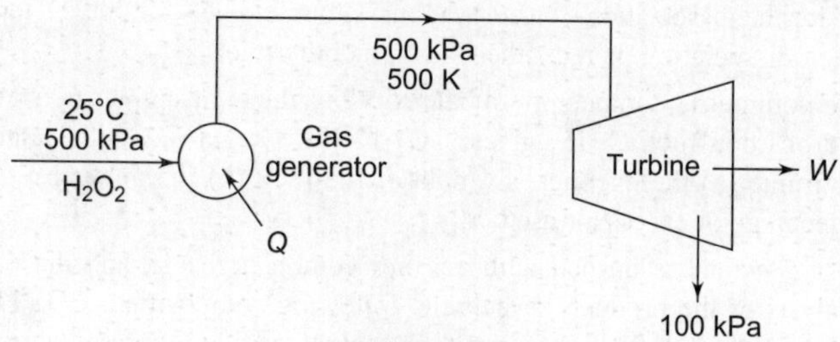

Fig. 16.18

Ⓜ **16.21** An internal combustion engine burns liquid octane and uses 150% theoretical air. The air and fuel enter at 25°C, and the products leave the engine exhaust ports at 900 K. In the engine 80% of the carbon burns to CO_2 and the remainder burns to CO. The heat transfer from this engine is just equal to the work done by the engine. Determine (a) the power output of the engine if the engine burns 0.006 kg/s of fuel, and (b) the composition and the dew point of the products of combustion. **[LO 16.10, 16.11]**

Ⓜ **16.22** Gaseous butane at 25°C is mixed with air at 400 K and burned with 400% theoretical air. Determine the adiabatic flame temperature. **[LO 16.13]**

Ⓜ **16.23** A mixture of butane and 150% theoretical air enters a combustion chamber of 25°C, 150 kPa, and the products of combustion leave at 1000 K, 150 kPa. Determine the heat transfer from the combustion chamber and the irreversibility for the process. **[LO 16.10, 16.12, 16.18]**

Ⓜ **16.24** The following data are taken from the test of a gas turbine:

Fuel—C_4H_{10} (g) 25°C, 0.1 MPa

Air—300% theoretical air at 25°C, 0.1 MPa

Velocity of inlet air—70 m/s

Velocity of products at exit—700 m/s

Temperature and pressure of products—900 K, 0.1 MPa

Assuming complete combustion, determine (a) the net heat transfer per kg mol of fuel, (b) the net increase of entropy per kg mol of fuel and (c) the irreversibility for the process. **[LO 16.10, 16.12, 16.18]**

Ⓜ **16.25** Calculate the equilibrium composition if argon gas is heated in an arc to 10,000 K, 1 kPa, assuming the plasma to consist of A, A^+, e^-. The equilibrium constant for the reaction

$$A \rightleftharpoons A^+ + e^-$$ **[LO 16.3]**

at this temperature is 0.00042.

Ⓜ **16.26** Methane is to be burned with oxygen in an adiabatic steady-flow process. Both enter the combustion chambar at 298 K, and the products leave at 70 kPa. What percent of excess O_2 should be used if the flame temperature is to be 2800 K? Assume that some of the CO_2 formed dissociates to CO and O_2, such that the products leaving the chamber consist of CO_2, CO, H_2O and O_2 at equilibrium. **[LO 16.12, 16.13]**

Ⓜ **16.27** (a) Octane burns with the stoichiometric amount of air. Determine the air-fuel ratio and the partial pressure of CO_2 in the dry products of combustion when the total pressure is 1 atm. (b) If 25% excess air is used in burning octane and combustion is complete determine the dry volumetric analysis of the products of combustion and the dew point temperature of the products. **[LO 16.10, 16.14]**

Ⓜ **16.28** Methane gas at 25°C, 1 atm and a volumetric flow rate of 27 m^3/h enters a furnace operating at steady state. The methane burns completely with 140% theoretical air, entering at 127°C at 1 atm. Products of combustion leave at 427°C, 1 atm. Determine (a) the volumetric flow rate of air in m^3/h, (b) the rate of heat transfer from the furnace in kW. Neglect K.E. and P.E. changes. **[LO 16.10, 16.12]**

Ⓜ **16.29** Octane gas at 25°C, 1 atm enters a jet engine and burns completely with 300% theoretical air at 25°C, 1 atm. Products of combustion leave at 990K, 1 atm. If the fuel and air enter with relatively low air velocities, determine the velocity of the combustion products at exit. Neglect PE effects and heat transfer between the engine and surroundings. **[LO 16.10, 16.12]**

Ⓜ **16.30** Liquid octane (C_8H_{18}) at 25°C, 1 atm enters the combustion chamber of a simple open gas turbine power plant and burns completely with 400% theoretical air entering the compressor at 25°C, 1 atm. Products of combustion exit the turbine at 627°C, 1 atm. If the rate of heat transfer from the gas turbine is estimated as 15% of the net power developed, determine the net power developed in kJ per kg mol of fuel. Kinetic and potential energy effects are negligible. **[LO 16.10, 16.12]**

Ⓜ **16.31** Gaseous ethanol (C_6H_5OH) at 25°C, 1 atm enters a reactor operating at steady state and burns completely with 130% theoretical air entering at 25°C, 1 atm. Products of combustion exit at 127°C, 1 atm. If the rate of heat transfer from the reactor is 900 kW, determine the mass flow rate of the fuel, in kg/s, neglecting the K.E. and P.E. effects. **[LO 16.10, 16.11, 16.12]**

Ⓜ **16.32** Carbon monoxide at 25°C, 1 atm enters an insulated reactor operating at steady state and reacts completely with the theoretical amount of air entering in a separate stream at 25°C. 1 atm. The products exit as mixture at 1 atm. Determine in kJ/kg mol of carbon (a) the availability entering with CO, (b) the availability exiting with the products, (c) the irreversibility rate and (d) the second law efficiency. Assume the environment as given in Example 16.13. **[LO 16.10, 16.18]**

Ⓜ **16.33** Liquid octane (C_8H_{18}) at 25°C, 1 atm enters an adiabatic reactor operated at steady state and burns completely with air entering at 227°C, 1 atm. If the combustion products exit at 1127°C, determine the percent excess air used, neglecting K.E. and P.E. effects. **[LO 16.12]**

ANSWERS TO PROBLEMS

16.4 490641 J/g · mol

16.10 248,885 kJ/kg mol, endothermic

16.11 (a) C_{13} H_{23}, (b) 17.03, (c) 18%

16.12 17.267, 14.19%

16.13 111%

16.14 – 17,360 kJ/kg mol

16.15 66,338%

16.16 – 794414 kJ/kg mol

16.17 2855 K

16.19 – 48.830 kJ/kg mol.

16.20 38.66 kW, –83.3 kW

16.27 (a) 15.11, 0.1455 atm, (b) CO_2, 11.45% O_2 4.47%, N_2 84.08; 47.356°C)

16.28 (a) 483.03 m^3/h, (b) – 202.5 kW

16.29 641.84 m/s

16.30 439,750 kJ/kg mol

16.31 0.034 2 kg/s

16.32 (a) 275,364 kJ/k mol CO,
(b) 226,105 kJ/k mol CO,
(c) 49,259 kJ/mol CO, (d) 82.1%)

16.33 168.5%

17

Compressible Fluid Flow

Learning Objectives

LO 17.1 State compressible fluid and basic laws applicable for its flow
LO 17.2 Express the velocity of sound in both ideal and non-ideal fluids
LO 17.3 Recognize stagnation state to relate stagnation and static properties
LO 17.4 Differentiate between nozzle and diffuser
LO 17.5 Explain critical pressure ratio for an isentropic compressible fluid flow through a duct
LO 17.6 Explain normal shock due to localized irreversibility in a compressible flow
LO 17.7 Compare the choking conditions for adiabatic and diabatic flows

LO 17.1
State compressible fluid and basic laws applicable for its flow

A fluid is defined as a substance which continuously deforms under the action of shearing forces. Liquids and gases are termed as fluids. A fluid is said to be *incompressible* if its density (or specific volume) does not change (or changes very little) with a change in pressure (or temperature or velocity). Liquids are incompressible. A fluid is said to be *compressible* if its density changes with a change in pressure or temperature or velocity. Gases are compressible. The effect of compressibility must be considered in flow problems of gases. Thermodynamics is an essential tool in studying compressible flows, because of which Theodore von Karman suggested the name 'Aerothermodynamics' for the subject which studies the dynamics of compressible fluids.

The basic principles in compressible flow are:

1. Conservation of mass (continuity equation)
2. Newton's second law of motion (momentum principle)

3. Conservation of energy (first law of thermodynamics)
4. Second law of thermodynamics (entropy principle)
5. Equation of state

For the first two principles, the student is advised to consult a book on fluid mechanics and the last three principles have been discussed in the earlier chapters of this book.

17.1 VELOCITY OF PRESSURE PULSE IN A FLUID

LO 17.2 Express the velocity of sound in both ideal and non-ideal fluids

Let us consider an infinitesimal pressure wave initiated by a slight movement of a piston to the right (Fig. 17.1) in a pipe of uniform cross section. The pressure wave front propagates steadily with a velocity c, which is known as the velocity of sound, sonic velocity or acoustic velocity. The fluid near the piston will have a slightly increased pressure and will be slightly more dense, than the fluid away from the piston.

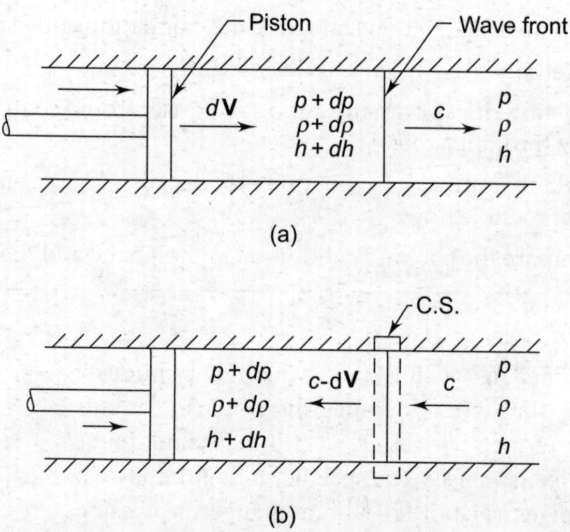

Fig. 17.1 Diagram illustrating sonic velocity: (a) Stationary observer (b) Observer travelling with the wave front

To simplify the analysis, let the observer be assumed to travel with the wave front to the right with the velocity c. Fluid flows steadily from right to left and as it passes through the wave front, the velocity is reduced from c to $c\text{-}d\mathbf{V}$. At the same time, the pressure rises from p to $p + dp$ and the density from ρ to $\rho + d\rho$.

The continuity equation for the control volume gives

$$\rho A c = (\rho + d\rho)\, A(c - d\mathbf{V})$$

$$\rho c = \rho c - \rho\, dV + c d\rho - d\rho \cdot d\mathbf{V}$$

Neglecting the product $d\rho \cdot d\mathbf{V}$, both being very small

$$\rho\, d\mathbf{V} = c d\rho \tag{17.1}$$

The momentum equation for the control volume gives

$$[p - (p + dp)]\, A = w\, [(c - d\mathbf{V}) - c]$$

$$-\, dp\, A = \rho A c\, (c - d\mathbf{V} - c)$$

$$dp = \rho\, c d\mathbf{V} \tag{17.2}$$

From Eqs. (17.1) and (17.2)

$$\frac{dp}{c} = c d\rho$$

$$\therefore \qquad c = \sqrt{\frac{dp}{d\rho}}$$

Since the variations in pressure and temperature are negligibly small and the change of state is so fast as to be essentially adiabatic, and in the absence of any internal friction or viscosity, the process is reversible and isentropic. Hence, the sonic velocity is given by

$$c = \sqrt{\left(\frac{\partial p}{\partial \rho}\right)_s} \tag{17.3}$$

No fluid is truly incompressible, although liquids show little change in density. The velocity of sound in common liquids is of the order of 1650 m/s.

17.1.1 Velocity of Sound in an Ideal Gas

For an ideal gas, in an isentropic process

$$pv^{\gamma} = \text{constant}$$

or

$$\frac{p}{\rho^{\gamma}} = \text{constant}$$

By logarithmic differentiation (i.e. first taking logarithm and then differentiating)

$$\frac{dp}{p} - \gamma \frac{d\rho}{\rho} = 0$$

$$\therefore \qquad \frac{dp}{d\rho} = \gamma \frac{p}{\rho}$$

Since

$$c^2 = \frac{dp}{d\rho} \text{ and } p = \rho\, RT$$

$$c^2 = \gamma RT$$

or

$$c = \sqrt{\gamma RT} \tag{17.4}$$

where R = characteristic gas constant = $\dfrac{\text{Universal gas constant}}{\text{Molecular weight}}$

The lower the molecular weight of the fluid and higher the value of γ, the higher is the sonic velocity at the same temperature. c is a thermodynamic property of the fluid.

17.1.2 Mach Number

The Mach number, M, is defined as the ratio of the actual velocity $\mathbf{V}$ to the sonic velocity c.

$$M = \frac{\mathbf{V}}{c}$$

When $M > 1$, the flow is supersonic, when $M < 1$, the flow is subsonic, and when $M = 1$, the flow is sonic.

17.2 STAGNATION PROPERTIES

LO 17.3
Recognize stagnation state to relate stagnation and static properties

The isentropic *stagnation state* is defined as the state a fluid in motion would reach if it was brought to rest isentropically in a steady flow, adiabatic, zero work output device. This is a reference state in a compressible fluid flow and is commonly designated with the subscript zero. The stagnation enthalpy h_0 (Fig. 17.2) is related to the enthalpy and velocity of the moving fluid by

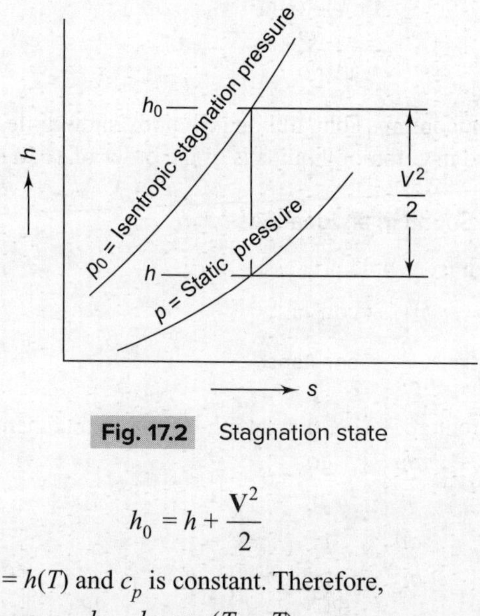

Fig. 17.2 Stagnation state

$$h_0 = h + \frac{\mathbf{V}^2}{2} \tag{17.5}$$

For an ideal gas, $h = h(T)$ and c_p is constant. Therefore,

$$h_0 - h = c_p (T_0 - T) \tag{17.6}$$

From Eqs. (17.5) and (17.6)

$$c_p (T_0 - T) = \frac{\mathbf{V}^2}{2}$$

$$\frac{T_0}{T} = 1 + \frac{\mathbf{V}^2}{2c_pT}$$

The properties without any subscript denote static properties.

Since $$c_p = \frac{\gamma R}{\gamma - 1}$$

$$\frac{T_0}{T} = 1 + \frac{\mathbf{V}^2(\gamma - 1)}{2\gamma RT}$$

Using Eq. (17.4) and the Mach number

$$\frac{T_0}{T} = 1 + \frac{\gamma - 1}{2} M^2 \tag{17.7}$$

The stagnation pressure p_0 is related to the Mach number and static pressure in the case of an ideal gas by the following equation

$$\frac{p_0}{p} = \left(\frac{T_0}{T}\right)^{\gamma/(\gamma-1)} = \left(1 + \frac{\gamma - 1}{2} M^2\right)^{\gamma/(\gamma-1)} \tag{17.8}$$

Similarly,

$$\frac{\rho_0}{\rho} = \left(1 + \frac{\gamma - 1}{2} M^2\right)^{1/(\gamma-1)} \tag{17.8a}$$

17.3 ONE-DIMENSIONAL STEADY ISENTROPIC FLOW

LO 17.4
Differentiate between nozzle and diffuser

From a one-dimensional point of view, the three most common factors which tend to produce continuous changes in the state of a flowing stream are

1. Changes in cross-sectional area
2. Wall friction
3. Energy effects, such as external heat exchange, combustion, etc.

A study will first be made of the effects of area change in the absence of friction and energy effects. The process, which has been called isentropic flow, might aptly be termed as simple area change.

A nozzle is any duct which increases the kinetic energy of a fluid at the expense of its pressure. A diffuser is a passage through which a fluid loses kinetic energy and gains pressure. The same duct or passage may be either a nozzle or diffuser depending upon the end conditions across it. A nozzle or diffuser with both a converging and diverging section is shown in Fig. 17.3. The minimum section is known as the *throat*. For the control volume shown in Fig. 17.3, since stagnation enthalpy and stagnation temperature do not change in adiabatic flow.

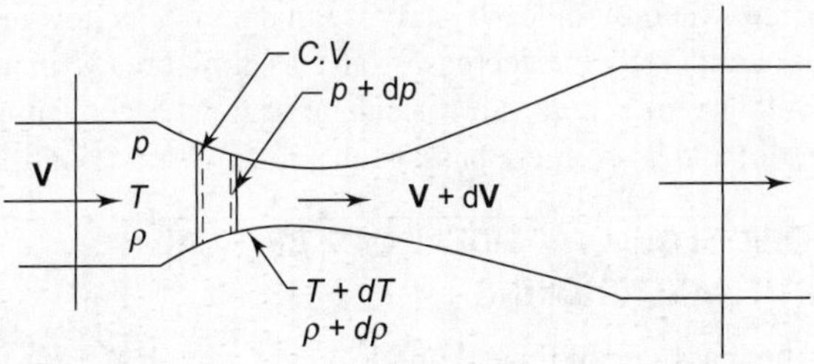

Fig. 17.3 Reversible adiabatic flow through a nozzle

$$h_0 = h + \frac{\mathbf{V}^2}{2}$$

$$dh = -\mathbf{V}\, d\mathbf{V} \tag{17.9}$$

From the property relation

$$Tds = dh - v\, dp$$

For isentropic flow

$$dh = \frac{dp}{\rho} \tag{17.10}$$

From Eqs. (17.9) and (17.10)

$$dp = -\rho \mathbf{V}\, d\mathbf{V} \tag{17.11}$$

or

$$\frac{dp}{d\mathbf{V}} < 0 \tag{17.12}$$

As pressure decreases, velocity increases, and vice versa.

The continuity equation gives

$$w = \rho A \mathbf{V}$$

By logarithmic differentiation

$$\frac{d\rho}{\rho} + \frac{dA}{A} + \frac{d\mathbf{V}}{\mathbf{V}} = 0$$

$$\therefore \quad \frac{dA}{A} = \frac{d\mathbf{V}}{\mathbf{V}} - \frac{d\rho}{\rho}$$

Substituting from Eq. (17.11)

$$\frac{dA}{A} = \frac{dp}{\rho \mathbf{V}^2} - \frac{d\rho}{\rho} = \frac{dp}{\rho \mathbf{V}^2}\left[1 - \mathbf{V}^2 \frac{d\rho}{dp}\right]$$

or

$$\frac{dA}{A} = \frac{dp}{\rho \mathbf{V}^2}(1 - M^2) \tag{17.13}$$

Also

$$\frac{dA}{A} = (M^2 - 1)\frac{d\mathbf{V}}{\mathbf{V}} \tag{17.14}$$

When $M < 1$, i.e. the inlet velocity is subsonic, as flow area A decreases, the pressure decreases and velocity increases, and when flow area A increases, pressure increases and velocity decreases. So for subsonic flow, a convergent passage becomes a nozzle (Fig. 17.4a) and a divergent passage becomes a diffuser (Fig. 17.4b).

When $M > 1$, i.e. when the inlet velocity is supersonic, as flow area A decreases, pressure increases and velocity decreases, and as flow area A increases, pressure decreases and velocity increases. So for subsonic flow, a convergent passage is a diffuser (Fig. 17.4c) and a divergent passage is a nozzle (Fig. 17.4d).

17.4 ISENTROPIC FLOW OF AN IDEAL GAS: CRITICAL PROPERTIES AND CHOKING

LO 17.5

Explain critical pressure ratio for an isentropic compressible fluid flow through a duct

Let us consider the mass rate of flow of an ideal gas through a nozzle. The flow is isentropic

$$w = \rho A \mathbf{V}$$

or

$$\frac{w}{A} = \frac{p}{RT} cM = \frac{p}{RT}\sqrt{\gamma RT}\, M$$

$$= \frac{p}{p_0} p_0 \sqrt{\frac{T_0}{T}} \sqrt{\frac{1}{T_0}} \sqrt{\frac{\gamma}{R}} M$$

$$= \left(\frac{T_0}{T}\right)^{-\gamma/\gamma-1} \left(\frac{T_0}{T}\right)^{1/2} \frac{p_0}{\sqrt{T_0}} \sqrt{\frac{\gamma}{R}} M$$

$$= \sqrt{\frac{\gamma}{R}} \frac{p_0 M}{\sqrt{T_0}} \frac{1}{\left(1+\frac{\gamma-1}{2}M^2\right)^{(\gamma+1)/2(\gamma-1)}} \quad (17.15)$$

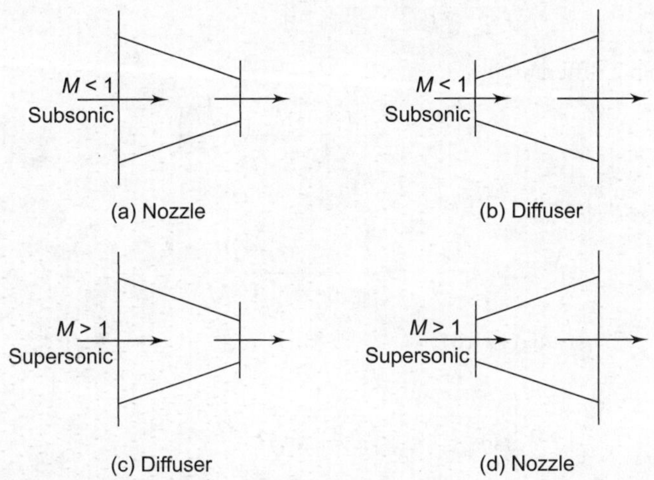

Fig. 17.4 Effect of area change in subsonic and supersonic flow at inlet to duct

Since p_0, T_0, γ and R are constant, the discharge per unit area w/A is a function of M only. There is a particular value of M when w/A is a maximum. Differentiating Eq. (17.15) with respect to M and equating it to zero,

$$\frac{d(w/A)}{dM} = \sqrt{\frac{\gamma}{R}} \frac{p_0}{\sqrt{T_0}} \frac{1}{\left(1+\frac{\gamma-1}{2}M^2\right)^{(\gamma+1)/2(\gamma-1)}}$$

$$+\sqrt{\frac{\gamma}{R}} \frac{p_0 M}{\sqrt{T_0}} \left[-\frac{\gamma+1}{2(\gamma-1)}\right]\left(1+\frac{\gamma-1}{2}M^2\right)^{-\frac{\gamma+1}{2(\gamma-1)}-1}$$

$$\left(\frac{\gamma-1}{2}2M\right) = 0$$

or

$$1 - \frac{M^2(\gamma+1)}{2\left(1+\frac{\gamma-1}{2}M^2\right)} = 0$$

$$M^2(\gamma+1) = 2 + (\gamma-1)M^2$$

$$M^2 = 1$$

or $$M = 1$$

So, the discharge w/A is maximum when $M = 1$.

Since $\mathbf{V} = cM = \sqrt{\gamma RT}\, M$, by logarithmic differentiation

$$\therefore \quad \frac{d\mathbf{V}}{\mathbf{V}} = \frac{dM}{M} + \frac{1}{2}\frac{dT}{T} \tag{17.16}$$

and $$\frac{T}{T_0} = \left(1 + \frac{\gamma-1}{2}M^2\right)^{-1}$$

by logarithmic differentiation

$$\therefore \quad \frac{dT}{T} = -\frac{(\gamma-1)M^2}{1+\dfrac{\gamma-1}{2}M^2}\frac{dM}{M} \tag{17.17}$$

From Eqs. (17.16) and (17.17)

$$\therefore \quad \frac{d\mathbf{V}}{\mathbf{V}} = \frac{1}{1+\dfrac{\gamma-1}{2}M^2}\frac{dM}{M} \tag{17.18}$$

From Eqs. (17.14) and (17.18)

$$\frac{dA}{A}\left(\frac{1}{M^2-1}\right) = \frac{1}{1+\dfrac{\gamma-1}{2}M^2}\frac{dM}{M}$$

$$\therefore \quad \frac{dA}{A} = \frac{(M^2-1)dM}{M\left(1+\dfrac{\gamma-1}{2}M^2\right)} \tag{17.19}$$

By substituting $M = 1$ in any one of the Eqs. (17.13), (17.14) or (17.19), $dA = 0$ or A = constant. So $M = 1$ *occurs only at the throat and nowhere else, and this happens only when the discharge is the maximum.*

If the convergent-divergent duct acts as a nozzle, in the divergent part also, the pressure will fall continuously to yield a continuous rise in velocity. The velocity of the gas is subsonic before the throat, becomes sonic at the throat and then becomes supersonic till its exit in isentropic flow, provided the exhaust pressure is low enough. The reverse situation prevails when the inlet velocity is supersonic. The whole duct then becomes a diffuser. The transition from subsonic flow to supersonic flow and vice versa can occur only when the compressible fluid flows through a throat and the exit pressure is maintained at the appropriate value.

When $M = 1$, the discharge is maximum and the nozzle is said to be choked. The properties at the throat are then termed as critical properties and these are designated by a superscript asterisk (*). Substituting $M = 1$ in Eq. (17.7).

$$\frac{T_0}{T^*} = 1 + \frac{\gamma-1}{2}M^2 = 1 + \frac{\gamma-1}{2} = \frac{\gamma+1}{2}$$

$$\therefore \qquad \frac{T^*}{T_0} = \frac{2}{\gamma+1} \tag{17.20}$$

The *critical pressure ratio* p^*/p_0 is then given by

$$\frac{p^*}{p_0} = \left(\frac{2}{\gamma+1}\right)^{\gamma/\gamma-1} \tag{17.20a}$$

For diatomic gases, like air, $\gamma = 1.4$

$$\frac{p^*}{p_0} = \left(\frac{2}{2.4}\right)^{1.4/0.4} = 0.528$$

The critical pressure ratio for air is 0.528.

For superheated steam, $\gamma = 1.3$ and p^*/p_0 is 0.546.

For air, $$\frac{T^*}{T_0} = 0.833$$

and $$\frac{\rho^*}{\rho_0} = \left(\frac{2}{\gamma+1}\right)^{1/(\gamma-1)} = 0.634$$

By substituting $M = 1$ in Eq. (17.15)

$$\frac{w}{A^*} = \sqrt{\frac{\gamma}{R}}\frac{p_0}{\sqrt{T_0}}\frac{1}{\left(\frac{\gamma+1}{2}\right)^{(\gamma+1)/2(\gamma-1)}} \tag{17.21}$$

Dividing Eq. (17.21) by Eq. (17.15)

$$\frac{A}{A^*} = \left[\left(\frac{2}{\gamma+1}\right)\left(1+\frac{\gamma-1}{2}M^2\right)\right]^{(\gamma+1)/2(\gamma-1)}\frac{1}{M} \tag{17.22}$$

The area ratio A/A^* is the ratio of the area at the point where the Mach number is M to the throat area A^*. Figure 17.5 shows a plot of A/A^* vs. M, which shows that a subsonic nozzle is converging and a supersonic nozzle is diverging.

17.4.1 Dimensionless Velocity, M*

Since the Mach number M is not proportional to the velocity alone and it tends towards infinity at high speed, one more dimensionless parameter M^* is often used, which is defined as

$$M^* = \frac{\mathbf{V}}{c^*} = \frac{\mathbf{V}}{\mathbf{V}^*} \tag{17.23}$$

where $c^* = \sqrt{\gamma R T^*} = \mathbf{V}^*$

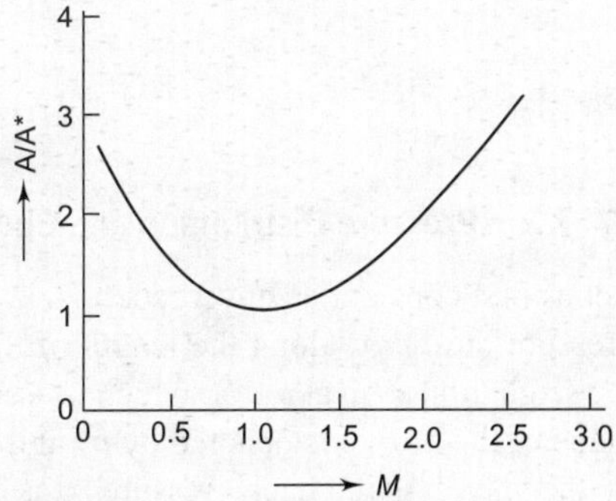

Fig. 17.5 Area ratio as a function of mach number in an isentropic nozzle

$$\therefore \qquad M^{*2} = \frac{\mathbf{V}^2}{c^{*2}} = \frac{\mathbf{V}^2}{c^2}\frac{c^2}{c^{*2}}$$

$$= M^2 \frac{c^2}{c^{*2}}$$

For the adiabatic flow of an ideal gas

$$\frac{\mathbf{V}^2}{2} + c_p T = \text{constant} = h_0 = c_p T_0$$

$$\frac{\mathbf{V}^2}{2} + \frac{\gamma RT}{\gamma - 1} = \frac{\gamma RT_0}{\gamma - 1}$$

$$\frac{\mathbf{V}^2}{2} + \frac{c^2}{\gamma - 1} = \frac{c_0^2}{\gamma - 1}$$

Since $$\frac{c_0}{c^*} = \sqrt{\frac{\gamma RT_0}{\gamma RT^*}} = \sqrt{\frac{T_0}{T^*}} = \sqrt{\left(\frac{\gamma + 1}{2}\right)}$$

$$\therefore \qquad \frac{\mathbf{V}^2}{2} + \frac{c^2}{\gamma - 1} = \frac{\gamma + 1}{2} c^{*2} \frac{1}{\gamma - 1}$$

$$\frac{\mathbf{V}^2}{c^{*2}} + \frac{2}{\gamma - 1}\frac{c^2}{c^{*2}} = \frac{\gamma + 1}{\gamma - 1}$$

$$M^{*2} + \frac{2}{\gamma - 1}\frac{M^{*2}}{M^2} = \frac{\gamma + 1}{\gamma - 1}$$

On simplification

$$M^{*2} = \frac{\frac{\gamma + 1}{2} M^2}{1 + \frac{\gamma - 1}{2} M^2} \tag{17.24}$$

When $M < 1, \quad M^* < 1$

When $M > 1, \quad M^* > 1 \qquad M = 0, \quad M^* = 0$

$M = 1, \quad M^* = 1 \qquad M = \infty, \quad M^* = \sqrt{\frac{\gamma + 1}{\gamma - 1}}$

17.4.2 Pressure Distribution and Choking in a Nozzle

Let us first consider a convergent nozzle as shown in Fig. 17.6, which also shows the pressure ratio p/p_0 along the length of the nozzle. The inlet condition of the fluid is the stagnation state at p_0, T_0, which is assumed to be constant. The pressure at the exit plane of the nozzle is denoted by p_E and the back pressure is p_B which can be varied by the valve. As the back pressure p_B is decreased, the mass flow rate w and the exit plane pressure p_E/p_0 vary, as shown in Fig. 17.7.

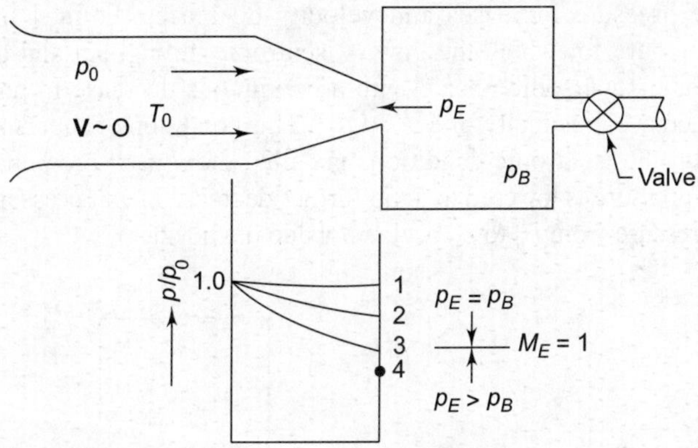

Fig. 17.6 Compressible flow through a converging nozzle

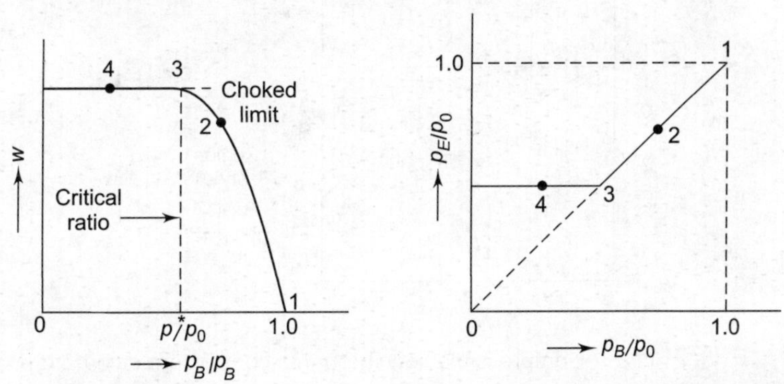

Fig. 17.7 Mass flow rate and exit pressure as a function of back pressure in a converging nozzle

When $p_B/p_0 = 1$, there is no flow, and $p_E/p_0 = 1$, as designated by point 1. If the back pressure p_B is now decreased to a value as designated by point 2, such that p_B/p_0 is greater than the critical pressure ratio, the mass flow rate has a certain value and $p_E = p_B$. The exit Mach number M_E is less than 1. Next the back pressure p_B is lowered to the critical pressure, denoted by point 3. The exit Mach number M_E is now, unity, and $p_E = p_B$. When p_B is decreased below the critical pressure, indicated by point 4, there is no increase in the mass flow rate, and p_E remains constant at a value equal to critical pressure, and $M_E = 1$. The drop in pressure from p_E to p_B occurs outside the nozzle exit. This is the choking limit which means that for given stagnation conditions the nozzle is passing the maximum possible mass flow.

Let us next consider a convergent-divergent nozzle, as shown in Fig. 17.8. Point 1 designates the condition when $p_B = p_0$, and there is no flow. When p_B is lowered to the pressure denoted by point 2, so that p_B/p_0 is less than 1 but greater than the critical pressure ratio, the velocity increases in the convergent section, but $M < 1$ at the throat. The divergent section acts as a subsonic diffuser in which the pressure increases and velocity decreases. Point 3 indicates the back pressure at which $M = 1$ at the throat, but the diverging section acts as a subsonic diffuser in which the pressure increases and velocity decreases. Point 4 indicates one other back pressure for which the flow is isentropic throughout and the diverging section acts as a supersonic nozzle with a continuous decrease in pressure and a continuous increase in velocity, and $p_{E4} = p_{B4}$. This condition of supersonic flow past the throat with the isentropic conditions is called the *design pressure ratio of the nozzle*. If the pressure is lowered to 5, no further decrease in exit pressure occurs and the drop in pressure from p_E to p_B occurs outside the nozzle.

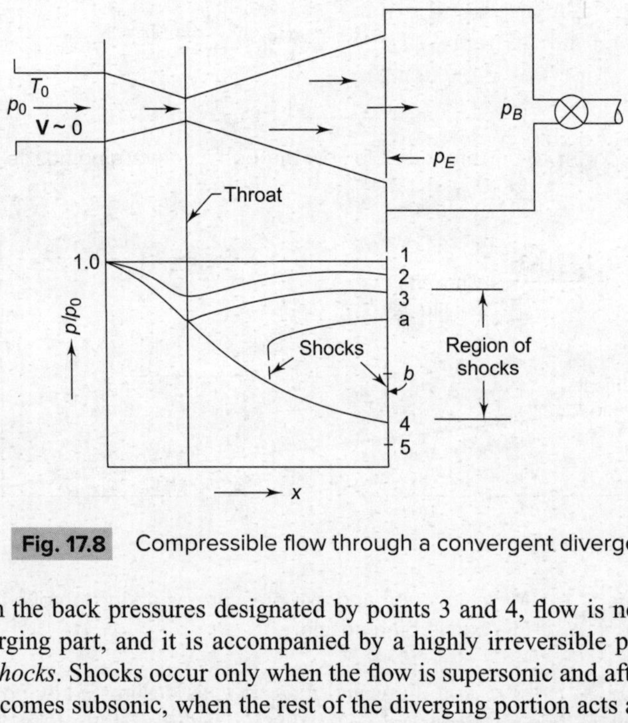

Fig. 17.8 Compressible flow through a convergent divergent nozzle

Between the back pressures designated by points 3 and 4, flow is not isentropic in the diverging part, and it is accompanied by a highly irreversible phenomenon, known as *shocks*. Shocks occur only when the flow is supersonic and after the shock the flow becomes subsonic, when the rest of the diverging portion acts as a diffuser. Properties vary discontinuously across the shock. When the back pressure is as indicated by point b (Fig. 17.8), the flow throughout the nozzle is isentropic, with pressure continuously decreasing and velocity increasing, but a shock appears just at the exit of the nozzle. When the back pressure is increased from b to a, the shock moves upstream, as indicated. When the back pressure is further increased, the shock moves further upstream and disappears at the nozzle throat where the back pressure corresponds to 3. Since flow throughout is subsonic, no shock is possible.

17.4.3 Gas Tables for Isentropic Flow

The values of M^*, A/A^*, p/p_0, ρ/ρ_0 and T/T_0 computed for an ideal gas having $\gamma = 1.4$ for various values of Mach number M from the Eqs. (17.24), (17.22), (17.8) and (17.7) respectively, are given in Table D.1 in Appendix D. These may be used with advantage for computations of problems of isentropic flow.

17.5 NORMAL SHOCKS

> **LO 17.6**
> Explain normal shock due to localized irreversibility in a compressible flow

Shock waves are highly localised irreversibilities in the flow. Within the distance of mean free path of a molecule, the flow passes from a supersonic to a subsonic state, the velocity decreases abruptly, and the pressure rises sharply. Figure 17.9 shows a control surface that includes such a normal shock. Normal shocks may be treated as shock waves perpendicular to the flow. The fluid is assumed to be in thermodynamic equilibrium upstream and downstream of the shock, the properties of which are designated by the subscripts x and y, respectively. For the control surface Continuity equation

$$\frac{w}{A} = \rho_x \mathbf{V}_x = \rho_y \mathbf{V}_y = G \qquad (17.25)$$

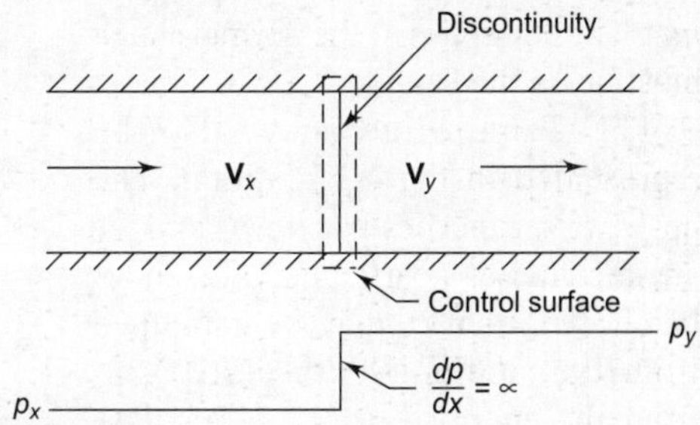

Fig. 17.9 One-dimensional normal shock

where G is the mass velocity (kg/m²s).
Momentum equation

$$p_x - p_y = \frac{w}{A}(\mathbf{V}_y - \mathbf{V}_x)$$
$$= \rho_y \mathbf{V}_y^2 - \rho_x \mathbf{V}_x^2$$

$$\therefore \quad p_x + \rho_x \mathbf{V}_x^2 = p_y + \rho_y \mathbf{V}_y^2 \qquad (17.26)$$

$$\text{or} \quad F_x = F_y \qquad (17.26a)$$

where $F = pA + \rho A \mathbf{V}^2$ is known as the *impulse function*.
Energy equation

$$h_{0x} = h_{0y} = h_0 = h_x + \frac{\mathbf{V}_x^2}{2} = h_y + \frac{\mathbf{V}_y^2}{2} = h_{0x} + h_{0y} = h_0 \qquad (17.27)$$

where h_0 is the stagnation enthalpy on both sides of the shock. Second law states

$$s_y - s_x \geq 0 \qquad (17.28)$$

The equation of state of the fluid may be written implicitly in the form

$$h = h(s, \rho) \qquad (17.29a)$$

$$dh = \left(\frac{\partial h}{\partial s}\right)_\rho ds + \left(\frac{\partial h}{\partial \rho}\right)_s d\rho$$

$$\text{or} \quad s = s(p, \rho) \qquad (17.29b)$$

Combining the continuity and energy equations and dropping the subscripts x and y

$$h_0 = h + \frac{G^2}{2\rho^2} = \text{constant} \qquad (17.30)$$

or
$$h = h_0 - \frac{G^2}{2\rho^2}$$

Given the values of G and h_0, this equation relates h and ρ (Fig. 17.10). The line representing the locus of points with the same mass velocity and stagnation enthalpy is called a *Fanno line*. The end states of the normal shock must lie on the Fanno line.

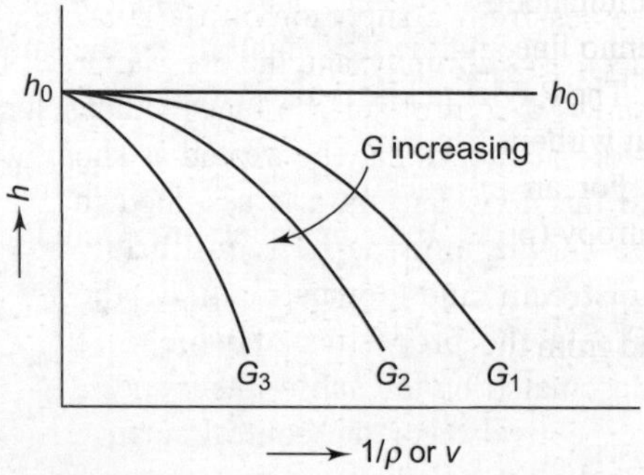

Fig. 17.10 Fanno line on *h*-1/ρ coordinate

The Fanno line may also be represented in the *h-s* diagram. The upstream properties h_x, ρ_x, p_x and $\mathbf{V}_x$ are known. Let a particular value of $\mathbf{V}_y$ be chosen, then ρ_y may be computed from the continuity Eq. (17.25), h_y from the energy Eq. (17.27) and s_y from Eq. (17.29a). By repeating the calculation for various values of $\mathbf{V}_y$ the Fanno line may easily be constructed (Fig. 17.11). Since the momentum equation has not been introduced, the Fanno line represents states with the same mass velocity and stagnation enthalpy, but not the same value of the impulse function.

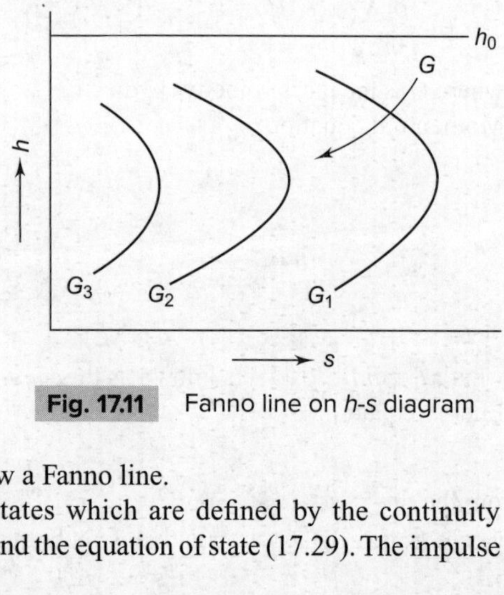

Fig. 17.11 Fanno line on *h-s* diagram

Adiabatic flow in a constant area duct with friction, in a one-dimensional model, has both constant G and constant h_0, and hence must follow a Fanno line.

Let us next consider the locus of states which are defined by the continuity Eq. (17.25), the momentum Eq. (17.26) and the equation of state (17.29). The impulse function in this case becomes

$$F = pA + \rho A\mathbf{V}^2$$

or the *impulse pressure I* is given by

$$= \frac{F}{A} = p + \rho\mathbf{V}^2 = p + \frac{G^2}{\rho} \quad (17.31)$$

Given the values for I and G, the equation relates p and ρ. The line representing the locus of states with the same impulse pressure and mass velocity is called the *Rayleigh line*. The end states of the normal shock must lie on the Rayleigh line, since $I_x = I_y$, and $G_x = G_y$.

The Rayleigh line may also be drawn on the *h-s* plot. The properties upstream of the shock are all known. The downstream properties are to be known. Let a particular value of $\mathbf{V}_y$ be chosen. Then ρ_y may be computed from the continuity Eq. (17.25) and p_y from the momentum Eq. (17.26), and s_y from Eq. (17.29b) may be found.

By repeating the calculations for various values of $\mathbf{V}_y$, the locus of possible states reachable from, say, state x may be plotted, and this is the Rayleigh line (Fig. 17.12).

Since the normal shock must satisfy Eqs. (17.25), (17.26), (17.27) and (17.29) simultaneously, the end states x and y of the shock must lie at the intersections of the Fanno line and the Rayleigh line for the same G (Fig. 17.12).

The Rayleigh line is also a model for flow in a constant area duct with heat transfer, but without friction.

For an infinitesimal process in the neighbourhood of the point of maximum entropy (point a) on the Fanno line, from the energy equation

$$dh + \mathbf{V}\, d\mathbf{V} = 0 \tag{17.32}$$

and from the continuity equation

$$\rho\, d\mathbf{V} + \mathbf{V} d\rho = 0 \tag{17.33}$$

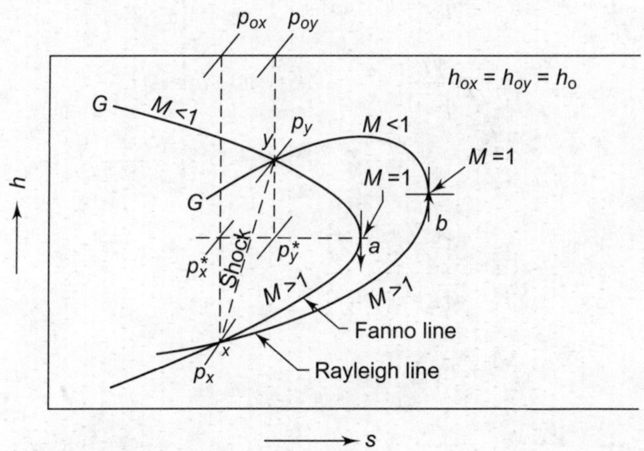

Fig. 17.12 End states of a normal shock on *h-s* diagram

From the thermodynamic relation

$$Tds = dh - vdp$$

or

$$dh = \frac{dp}{\rho} \tag{17.34}$$

By combining Eqs. (17.32), (17.33) and (17.34)

$$\frac{dp}{\rho} + \mathbf{V}\left(-\frac{\mathbf{V}d\rho}{\rho}\right) = 0$$

$$\frac{dp}{d\rho} = \mathbf{V}^2$$

or $\mathbf{V} = \sqrt{\left(\dfrac{\partial p}{\partial \rho}\right)_s}$, since the flow is isentropic.

This is the local sound velocity.

So the Mach number is unity at point *a*. Similarly, it can be shown that at point *b* on the Rayleigh line, $M = 1$. It may also be shown that the upper branches of the Fanno and Rayleigh lines represent subsonic speeds ($M < 1$) and the lower branches represent supersonic speeds ($M > 1$).

The normal shock always involves a change from supersonic to subsonic speed with a consequent pressure rise, and never the reverse. By the second law, entropy always increases during irreversible adiabatic change.

17.5.1 Normal Shock in an Ideal Gas

The energy equation for an ideal gas across the shock becomes

$$c_p T_x + \frac{\mathbf{V}_x^2}{2} = c_p T_y + \frac{\mathbf{V}_y^2}{2} = c_p T_0$$

Now $\quad h_{0x} = h_{0y} = h_0$, and $T_{0x} = T_{0y} = T_0$

Substituting $\quad c_p = \frac{\gamma R}{\gamma - 1}$, $c_x = \sqrt{\gamma R T_x}$ and $c_y = \sqrt{\gamma R T_y}$

$$\frac{T_0}{T_x} = 1 + \frac{\gamma - 1}{2} M_x^2, \text{ and } \frac{T_0}{T_y} = 1 + \frac{\gamma - 1}{2} M_y^2$$

$$\therefore \quad \frac{T_y}{T_x} = \frac{1 + \frac{\gamma - 1}{2} M_x^2}{1 + \frac{\gamma - 1}{2} M_y^2} \tag{17.35}$$

Again $\quad \rho_x \mathbf{V}_x = \rho_y \mathbf{V}_y$

$$\therefore \quad \frac{p_x}{RT_x} \mathbf{V}_x = \frac{p_y}{RT_y} \mathbf{V}_y$$

or
$$\frac{T_y}{T_x} = \frac{p_y}{p_x} \frac{\mathbf{V}_y}{\mathbf{V}_x} = \frac{p_y}{p_x} \frac{M_y c_y}{M_x c_x} = \frac{p_y}{p_x} \frac{M_y}{M_x} \sqrt{\frac{T_y}{T_x}}$$

or
$$\frac{T_y}{T_x} = \left(\frac{p_y}{p_x}\right)^2 \left(\frac{M_y}{M_x}\right)^2 \tag{17.36}$$

From Eqs. (17.35) and (17.36)

$$\frac{p_y}{p_x} = \frac{M_x}{M_y} \sqrt{\frac{1 + \frac{\gamma - 1}{2} M_x^2}{1 + \frac{\gamma - 1}{2} M_y^2}} \tag{17.37}$$

Also

$$p_x + \rho_x \mathbf{V}_x^2 = p_y + \rho_y \mathbf{V}_y^2$$

or
$$p_x + \frac{\gamma p_x}{\gamma R T_x} \mathbf{V}_x^2 = p_y + \frac{\gamma p_y}{R T_y \gamma} \mathbf{V}_y^2$$

$$p_x(1 + \gamma M_x^2) = p_y(1 + \gamma M_y^2)$$

$$\therefore \quad \frac{p_y}{p_x} = \frac{1+\gamma M_x^2}{1+\gamma M_y^2} \tag{17.38}$$

From Eqs. (17.37) and (17.38), upon rearrangement

$$M_y^2 = \frac{M_x^2 + \dfrac{2}{\gamma-1}}{\dfrac{2\gamma}{\gamma-1}M_x^2 - 1} \tag{17.39}$$

Then from Eqs. (17.37), (17.38) and (17.39)

$$\frac{p_y}{p_x} = \frac{2\gamma}{\gamma+1}M_x^2 - \frac{\gamma-1}{\gamma+1} \tag{17.40}$$

and from Eqs. (17.35) and (17.39)

$$\frac{T_y}{T_x} = \frac{\left(1+\dfrac{\gamma-1}{2}M_x^2\right)\left(\dfrac{2\gamma}{\gamma-1}M_x^2 - 1\right)}{\dfrac{(\gamma+1)^2}{2(\gamma-1)}M_x^2} \tag{17.41}$$

Then

$$\frac{\rho_y}{\rho_x} = \frac{p_y}{T_y}\frac{T_x}{p_x}$$

$$= \frac{\left(\dfrac{2\gamma}{\gamma+1}M_x^2 - \dfrac{\gamma-1}{\gamma+1}\right)\left(\dfrac{(\gamma+1)^2}{2(\gamma-1)}M_x^2\right)}{\left(1+\dfrac{\gamma-1}{2}M_x^2\right)\left(\dfrac{2\gamma}{\gamma-1}M_x^2 - 1\right)} \tag{17.42}$$

The ratio of the stagnation pressure is a measure of the irreversibility of the shock process. Now

$$\frac{p_{oy}}{p_{ox}} = \frac{p_{oy}}{p_y}\frac{p_y}{p_x}\frac{p_x}{p_{ox}}$$

and

$$\frac{p_{oy}}{p_y} = \left(1+\frac{\gamma-1}{2}M_y^2\right)^{\gamma/(\gamma-1)}$$

$$\frac{p_{ox}}{p_x} = \left(1+\frac{\gamma-1}{2}M_x^2\right)^{\gamma/(\gamma-1)}$$

$$\therefore \quad \frac{p_{oy}}{p_{ox}} = \left\{\frac{\dfrac{\gamma+1}{2}M_x^2}{1+\dfrac{\gamma-1}{2}M_x^2}\right\}^{\gamma(\gamma-1)} \Bigg/ \left[\frac{2\gamma}{\gamma+1}M_x^2 - \frac{\gamma-1}{\gamma+1}\right]^{1/(\gamma-1)} \tag{17.43}$$

$$\frac{p_{oy}}{p_x} = \frac{p_{oy}}{p_y} \cdot \frac{p_y}{p_x}$$

$$= \left(1 + \frac{\gamma - 1}{2} M_y^2\right)^{\gamma/(\gamma-1)} \left[\frac{2\gamma}{\gamma + 1} M_x^2 - \frac{\gamma - 1}{\gamma + 1}\right] \quad (17.44)$$

For different values of M_x, and for $\gamma = 1.4$, the values of M_y, p_y/p_x, T_y/T_x, ρ_y/ρ_x, p_{oy}/p_{ox} and p_{oy}/p_x computed from Eqs. (17.39), (17.40), (17.42), (17.43) and (17.44), respectively, are given in Table D.2 in Appendix D.

To evaluate the entropy change across the shock, for an ideal gas

$$ds = c_p \frac{dT}{T} - R \frac{dp}{p}$$

$$\therefore \quad s_y - s_x = c_p \ln \frac{T_y}{T_x} - R \ln \frac{p_y}{p_x}$$

$$= c_p \left[\ln \frac{T_y}{T_x} - \ln \left(\frac{p_y}{p_x}\right)^{(\gamma-1)/\gamma}\right]$$

since $$R = \frac{c_p(\gamma - 1)}{\gamma}$$

$$\therefore \quad s_y - s_x = c_p \ln \frac{T_y/T_x}{(p_y/p_x)^{(\gamma-1)/\gamma}}$$

$$= c_p \ln \frac{T_{oy}/T_{ox}}{(p_{oy}/p_{ox})^{(\gamma-1)/\gamma}} = -R \ln \frac{p_{oy}}{p_{ox}} \quad (17.45)$$

The strength of a shock wave, P, is defined as the ratio of the pressure increase to the initial pressure, i.e.

$$P = \frac{p_y - p_x}{p_x} = \frac{p_y}{p_x} - 1$$

Substituting from Eq. (17.40)

$$P = \frac{2\gamma}{\gamma + 1} M_x^2 - \frac{\gamma - 1}{\gamma + 1} - 1 = \frac{2\gamma}{\gamma + 1} (M_x^2 - 1) \quad (17.46)$$

17.6 ADIABATIC FLOW WITH FRICTION AND DIABATIC FLOW WITHOUT FRICTION

LO 17.7 Compare the choking conditions for adiabatic and diabatic flows

It was stated that the Fanno line representing the states of constant mass velocity and constant stagnation enthalpy also holds for adiabatic flow in a constant area duct with friction. For adiabatic flow the entropy must increase in the flow direction. Hence, a Fanno process must follow its Fanno line to the right, as shown in

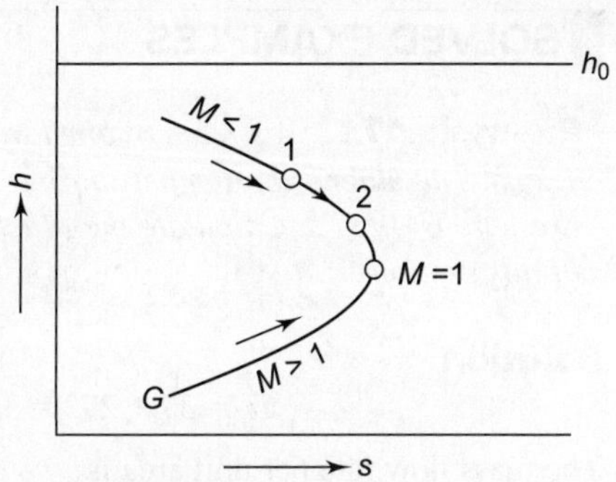

Fig. 17.13 A Fanno line on *h-s* plot

Fig. 17.13. Since friction will tend to move the state of the fluid to the right on the Fanno line, the Mach number of subsonic flows increases in the downstream section (Fig. 17.13), and in supersonic flows friction acts to decrease the Mach number. Hence, friction tends to drive the flow to the sonic point.

Let us consider a short duct with a given h_0 and G, i.e. a given Fanno line, with a given subsonic exit Mach number represented by point 1 in Fig. 17.13. If some more length is added to the duct, the new exit Mach number will be increased due to friction, as represented by, say, point 2. The length of the duct may be further increased till the exit Mach number is unity. Any further increase in duct length is not possible without incurring a reduction in the mass flow rate. Hence subsonic flows can be choked by friction. There is a maximum flow rate that can be passed by a pipe with given stagnation conditions. Choking also occurs in supersonic flow with friction, usually in a very short length. It is thus difficult to use such flows in applications.

Diabatic flows, i.e. flows with heating or cooling, in a constant area duct, in the absence of friction, can be treated by the Rayleigh process (Fig. 17.14). The process is reversible, and the direction of entropy change is determined by the sign of the heat transfer. Heating a compressible flow has the same effect as friction, and the Mach number goes towards unity. Therefore, there is a maximum heat input for a given flow rate which can be passed by the duct, which is then choked. Although the cooling of the fluid increases the flow stagnation pressure with a decrease in entropy, a nonmechanical pump is not feasible by cooling a compressible flow, because of the predominating effect of friction.

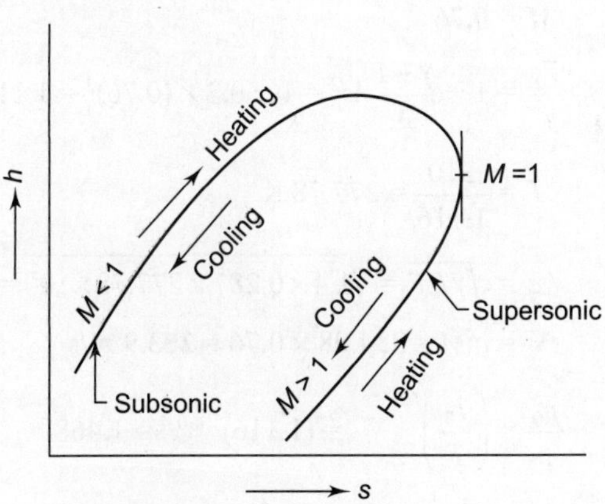

Fig. 17.14 A Rayleigh line on *h-s* plot

SOLVED EXAMPLES

Example 17.1 *A stream of air flows in a duct of 100 mm diameter at a rate of 1 kg/s. The stagnation temperature is 37°C. At one section of the duct the static pressure is 40 kPa. Calculate the Mach number, velocity and stagnation pressure at this section.* M [LO 17.2, 17.3]

Solution

$$T_0 = 37 + 273 = 310 \text{ K}, p = 40 \text{ kPa}, \gamma = 1.4$$

The mass flow rate per unit area is

$$\frac{w}{A} = \rho \mathbf{V} = \frac{p}{RT}\sqrt{\gamma RT} \cdot M$$

$$= \sqrt{\frac{\gamma}{R}} \cdot \frac{pM}{\sqrt{T_0}}\sqrt{\frac{T_0}{T}} = \sqrt{\frac{\gamma}{R}} \cdot \frac{pM}{\sqrt{T_0}}\left(1+\frac{\gamma-1}{2}M^2\right)^{1/2}$$

$$\frac{1}{\frac{\pi}{4}(0.1)^2} \text{ kg/m}^2\text{s} = \sqrt{\frac{1.4}{0.287 \text{ kJ/kg K}}} \frac{40 \text{ kN/m}^2 \times M}{\sqrt{310 \text{ K}}} (1 + 0.2 M^2)^{1/2}$$

$$127.39 = \sqrt{\frac{1.4}{0.287 \times 10^3 \times 310}}\, 40 \times 10^3 M (1 + 0.2 M^2)^{1/2}$$

$$\therefore \quad M(1 + 0.2 M^2)^{1/2} = 0.803$$

$$M^2 (1 + 0.2 M^2) = 0.645$$

$$M^4 + 5M^2 - 3.225 = 0$$

$$M^2 = \frac{-5 \pm \sqrt{25 + 12.9}}{2} = \frac{-5 + 6.16}{2} = 0.58$$

$$\therefore \quad \mathbf{M = 0.76}$$

$$\frac{T_0}{T} = 1 + \frac{\gamma - 1}{2} M^2 = 1 + 0.2 \times (0.76)^2 = 1.116$$

$$\therefore \quad T = \frac{310}{1.116} = 277.78 \text{ K}$$

$$c = \sqrt{\gamma RT} = \sqrt{1.4 \times 0.287 \times 277.78 \times 10^3} = \mathbf{334.08 \text{ m/s}}$$

$$\therefore \quad \mathbf{V} = c\,M = 334.08 \times 0.76 = \mathbf{253.9 \text{ m/s}}$$

$$\frac{p_0}{p} = \left(\frac{T_0}{T}\right)^{\gamma/(\gamma-1)} = (1.116)^{1.4/0.4} = 1.468$$

$$\therefore \quad p_0 = 40 \times 1.468 = \mathbf{58.72 \text{ kPa}}$$

Example 17.2 *Air flows from a large tank through a converging nozzle and discharges to the atmosphere (p = 101 kPa). The flow is steady and isentropic. Find the range of tank pressures for which the flow at the nozzle exit will be sonic. The exit area of the nozzle is 0.0014 m². If the tank pressure and temperature are 800 kPa (abs) and 700 K, respectively, calculate the mass flow rate through the nozzle.*

Ⓜ **[LO 17.2, 17.3]**

Solution The nozzle will be chocked (i.e. $M = 1$) for

$$\frac{p^*}{p_0} \le \left(\frac{2}{\gamma+1}\right)^{\frac{\gamma}{\gamma-1}} = \left(\frac{2}{1.4+1}\right)^{\frac{1.4}{1.4-1}} = 0.528$$

The nozzle discharges air to the atmosphere and hence $p_b = 101$ kPa

Thus, $$\frac{p_b}{p_0} \le 0.528$$

or $$p_0 \ge \frac{p_b}{0.528} = \frac{101}{0.528} = \mathbf{191.29\ kPa}$$

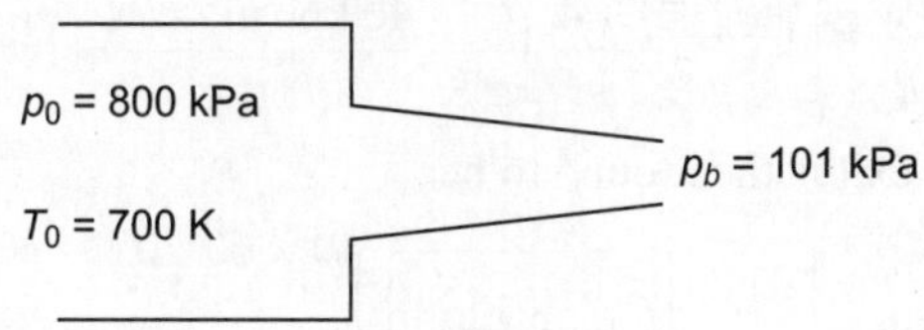

Fig. 17.15

For $p_0 = 800$ kPa, $T_0 = 700$ K and $M = 1$ (Fig. 17.15), we have

$$\frac{T_0}{T} = 1 + \frac{\gamma-1}{2}M^2$$

or $$T = \frac{T_0}{1+\frac{\gamma-1}{2}M^2} = \frac{700}{1+\frac{1.4-1}{2}1^2} = 583.33 \text{ K}$$

$$\frac{p_0}{p} = \left(1+\frac{\gamma-1}{2}M^2\right)^{\frac{\gamma}{\gamma-1}}$$

or $$p = \frac{p_0}{\left(1+\frac{\gamma-1}{2}M^2\right)^{\frac{\gamma}{\gamma-1}}} = \frac{800}{\left(1+\frac{1.4-1}{2}1^2\right)^{\frac{1.4}{1.4-1}}} = 422.63 \text{ kPa}$$

$$c = \sqrt{\gamma RT} = \sqrt{1.4 \times 287 \times 583.33} = 484.13 \text{ m/s}$$

$$V = Mc = 484.13 \text{ m/s}$$

$$\rho = \frac{422.63}{0.287 \times 583.33} = 2.52 \text{ kg/m}^3$$

The mass flow rate of air is

$$w = \rho AV = 2.52 \times 0.0014 \times 484.13 = \mathbf{1.708\ kg/s}$$

Example 17.3 *A converging-diverging nozzle is attached to a large tank of air (Fig.17.16) where the pressure is 1.20 MPa (abs) and temperature is 140°C.*

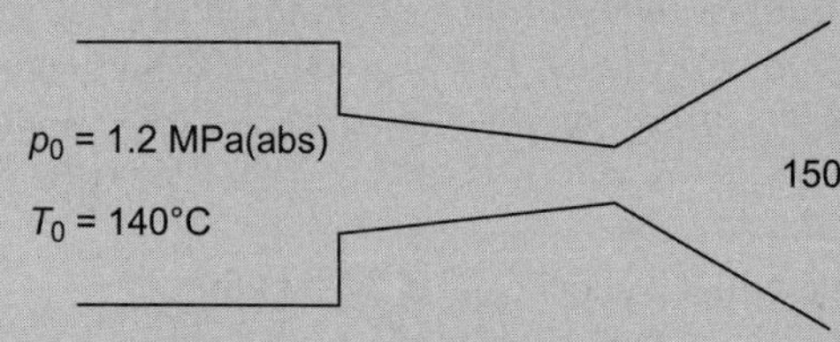

Fig. 17.16

The pressure at the exit of the nozzle is 150 kPa(abs). If the air flows isentropically with a mass flow rate of 2.2 kg/s, determine the throat and exit areas of the nozzle.

Ⓜ **[LO 17.2, 17.3]**

Solution We know that

$$\frac{p_0}{p} = \left[1 + \frac{\gamma - 1}{2} M^2\right]^{\frac{\gamma}{\gamma-1}}$$

Mach number at the exit is then found to be

$$M_1 = \left[\frac{2}{\gamma - 1}\left\{\left(\frac{p_0}{p_1}\right)^{\frac{\gamma-1}{\gamma}} - 1\right\}\right]^{1/2}$$

$$= \left[\frac{2}{1.4 - 1}\left\{\left(\frac{1.2}{0.15}\right)^{\frac{1.4-1}{1.4}} - 1\right\}\right]^{1/2} = 2.01$$

Temperature of air at the exit is found as follows

$$\frac{T_0}{T_1} = 1 + \frac{\gamma - 1}{2} M_1^2$$

$$T_1 = \frac{T_0}{1 + \frac{\gamma - 1}{2} M_1^2} = \frac{413}{1 + \frac{1.4 - 1}{2}(2)^2} = 229.44 \text{ K}$$

Velocity of air at exit is

$$V_1 = M_1 c_1 = M_1\sqrt{\gamma R T_1} = 2.01\sqrt{1.4 \times 287 \times 229.44} = 610.29 \text{ m/s}$$

The density of air at exit is given by

$$\rho_1 = \frac{p_1}{RT_1} = \frac{150}{0.287 \times 229.44} = 2.278 \text{ kg/m}^3$$

Since $M_1 > 1$, the nozzle must be chocked and $M^* = 1.0$.

Pressure and temperature at throat are found to be

$$p^* = \frac{p_0}{\left[1+\frac{\gamma-1}{2}M^{*2}\right]^{\frac{\gamma}{\gamma-1}}} = \frac{1.2\times10^3}{\left[1+\frac{1.4-1}{2}(1)^2\right]^{\frac{1.4}{1.4-1}}} = 633.94 \text{ kPa}$$

$$T^* = \frac{T_0}{1+\frac{\gamma-1}{2}M^{*2}} = \frac{413}{1+\frac{1.4-1}{2}(1)^2} = 344.17 \text{ K}$$

$$\rho^* = \frac{p^*}{RT^*} = \frac{633.94}{0.287\times344.17} = 6.42 \text{ kg/m}^3$$

Velocity of air at throat is

$$V^* = M^*c^* = M^*\sqrt{\gamma RT^*} = 1.0\sqrt{1.4\times287\times344.17} = 371.87 \text{ m/s}$$

Cross-sectional area at throat is

$$A^* = \frac{w}{\rho^* V^*} = \frac{2.2}{6.42\times371.87} = \mathbf{0.00092\ m^2}$$

Cross-sectional area at exit is

$$A_1 = \frac{w}{\rho_1 V_1} = \frac{2.2}{2.278\times610.29} = \mathbf{0.00158\ m^2}$$

Example 17.4 *A conical air diffuser has an intake area of 0.11 m² and an exit area of 0.44 m². Air enters the diffuser with a static pressure of 0.18 MPa, static temperature of 37°C and velocity of 267 m/s. Calculate the mass flow rate of air through the diffuser, the Mach number, static temperature and static pressure of air leaving the diffuser and the net thrust acting upon the diffuser, assuming that its outer surfaces are wetted by atmospheric pressure at 0.1 MPa.*

M **[LO 17.5, 17.6]**

Solution The conical diffuser is shown in Fig. 17.17. The mass flow rate of air through it

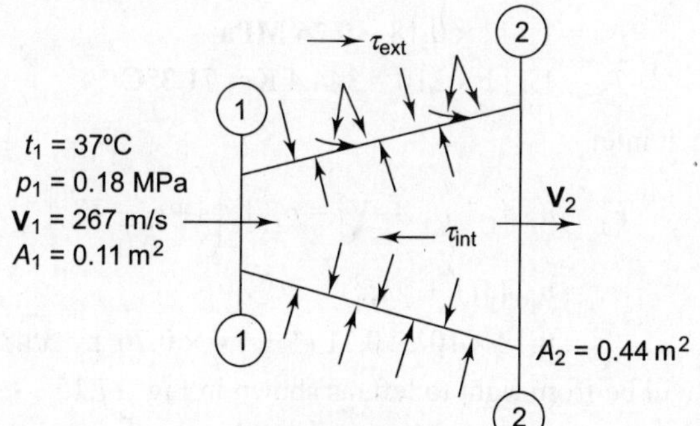

Fig. 17.17

$$\omega = \rho A \mathbf{V}$$

$$= \frac{p_1}{RT_1} A_1 \mathbf{V}_1 = \frac{0.18 \times 10^3}{0.287 \times 310} \times 0.11 \times 267 = 59.42 \text{ kg/s}$$

$$c_1 = \sqrt{\gamma R T_1}$$

$$= 20.045\sqrt{T_1} = 20.045\sqrt{310} = 352 \text{ m/s}$$

$$\therefore \quad M_1 = \frac{\mathbf{V}_1}{c_1} = \frac{267}{352} = 0.76$$

From the gas tables for the isentropic flow of air ($\gamma = 1.4$), given in the appendix, when $M_1 = 0.76$

$$\frac{A_1}{A^*} = 1.0570, \frac{p_1}{p_{01}} = 0.68207, \frac{T_1}{T_{01}} = 0.89644$$

$$\frac{F_1}{F^*} = 1.0284$$

Now
$$\frac{A_2}{A_1} = \frac{0.44 \text{ m}^2}{0.11 \text{ m}^2} = 4 = \frac{A_2/A^*}{A_1/A^*}$$

$$\therefore \quad \frac{A_2}{A^*} = 4 \times 1.0570 = 4.228$$

Again, from the isentropic flow tables, when $\frac{A_2}{A^*} = 4.228$

$$M_2 \cong 0.135, \frac{p_2}{p_{02}} = 0.987, \frac{T_2}{T_{02}} = 0.996, \frac{F_2}{F^*} = 3.46$$

$$\frac{p_2}{p_1} = \frac{p_2/p_{02}}{p_1/p_{01}} = \frac{0.987}{0.682} = 1.447$$

$$\frac{T_2}{T_1} = \frac{T_2/T_{02}}{T_1/T_{01}} = \frac{0.996}{0.89644} = 1.111$$

$$\frac{F_2}{F_1} = \frac{F_2/F^*}{F_1/F^*} = \frac{3.46}{1.0284} = 3.364$$

$$\therefore \quad p_2 = 1.447 \times 0.18 = \mathbf{0.26 \text{ MPa}}$$

$$T_2 = 1.111 \times 310 = 344.4 \text{ K} = \mathbf{71.3°C}$$

Impulse function at inlet

$$F_1 = p_1 A_1 + \rho_1 A_1 \mathbf{V}_1^2 = p_1 A_1 \left(1 + \frac{1}{RT_1}\mathbf{V}_1^2\right)$$

$$= p_1 A_1 (1 + \gamma M_1^2)$$

$$= 0.18 \times 10^3 \times 0.11 (1 + 1.4 \times 0.76^2) = 35.82 \text{ kN}$$

Internal thrust τ will be from right to left, as shown in Fig. 17.15.

$$\tau_{\text{int}} = F_2 - F_1 = 3.364 F_1 - F_1$$

$$= 2.364 \times 35.82 = 84.68 \text{ kN}$$

External thrust is from left to right

$$\tau_{ext} = p_0(A_2 - A_1)$$
$$= 0.1 \times 10^3\ (0.44 - 0.11) = 33 \text{ kN}$$

Net thrust $= \tau_{int} - \tau_{ext}$

$$= 84.68 - 33 = \mathbf{51.68\ kN}$$

Example 17.5 *A convergent-divergent nozzle has a throat area 500 mm² and an exit area of 1000 mm². Air enters the nozzle with a stagnation temperature of 360 K and a stagnation pressure of 1 MPa. Determine the maximum flow rate of air that the nozzle can pass, and the static pressure, static temperature, Mach number and velocity at the exit from the nozzle, if (a) the divergent section acts as a nozzle, and (b) the divergent section acts as a diffuser.* Ⓜ **[LO 17.5]**

Solution

$$\frac{A_2}{A^*} = \frac{1000}{500} = 2$$

From the isentropic flow tables, when $A_2/A^* = 2$ there are two values of the Mach number, one for supersonic flow when the divergent section acts as a nozzle, and the other for subsonic flow when the divergent section acts as a diffuser, which are $M_2 = 2.197, 0.308$ (Fig. 17.18).

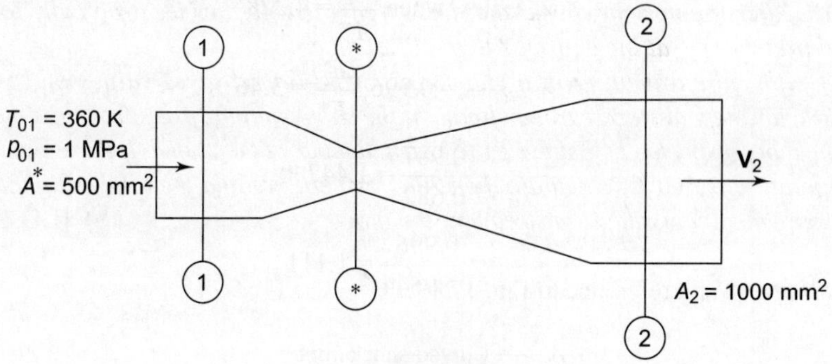

Fig. 17.18

(a) When $\quad M_2 = 2.197, \dfrac{p_2}{p_0} = 0.0939, \dfrac{T_2}{T_0} = 0.5089$

$\therefore$ $\quad p_2 = 0.0939 \times 1000 = 93.9 \text{ kPa}$

$$T_2 = 0.5089 \times 360 = 183.2 \text{ K}$$

$$c_2 = \sqrt{\gamma R T_2} = 20.045\sqrt{183.2}$$
$$= 271.2 \text{ m/s}$$

$\therefore$ $\quad V_2 = 271.2 \times 2.197 = \mathbf{596\ m/s}$

Mass flow rate

$$w = A^* \rho^* V^* = \rho_2 A_2 V_2 = \rho_1 A_1 V_1$$

For air $\frac{p^*}{p_0} = 0.528$ and $\frac{T^*}{T_0} = 0.833$

$$\rho^* = \frac{p^*}{RT^*} = \frac{0.528 \times 1000}{0.287 \times 0.833 \times 360} = 6.13 \text{ kg/m}^3$$

$$T^* = 360 \times 0.833 = 300 \text{ K}$$

$$\mathbf{V}^* = \sqrt{\gamma RT^*} = 20.045\sqrt{300} = 347.2 \text{ m/s}$$

$$\therefore \quad w = (500 \times 10^{-6}) \times 6.13 \times 347.2 = \mathbf{1.065 \text{ kg/s}}$$

(b) When $M = 0.308$, $\frac{p_2}{p_0} = 0.936$, $\frac{T_2}{T_0} = 0.9812$

$$p_2 = 0.936 \times 1000 = 936 \text{ kPa}$$

$$T_2 = 0.9812 \times 360 = 353.2 \text{ K}$$

$$c_2 = \sqrt{\gamma RT_2} = 20.045\sqrt{353.2} = 376.8 \text{ m/s}$$

$$\mathbf{V}_2 = 376.8 \times 0.308 = 116 \text{ m/s} \quad w = \mathbf{1.065 \text{ kg/s}}$$

Example 17.6 *When a Pitot tube is immersed in a supersonic stream, a curved shock wave is formed ahead of the Pitot tube mouth. Since the radius of the curvature of the shock is large, the shock may be assumed to be a normal shock. After the normal shock, the fluid stream decelerates isentropically to the total pressure p_{0y} at the entrance to the Pitot tube.*

A pitot tube travelling in a supersonic wind-tunnel gives values of 16 and 70 kPa for the static pressure upstream of the shock and the pressure at the mouth of the tube, respectively. Estimate the Mach number of the tunnel. If the stagnation temperature is 300°C, calculate the static temperature and the total (stagnation) pressure upstream and downstream of the tube. M **[LO 17.6]**

Solution With reference to Fig. 17.19,

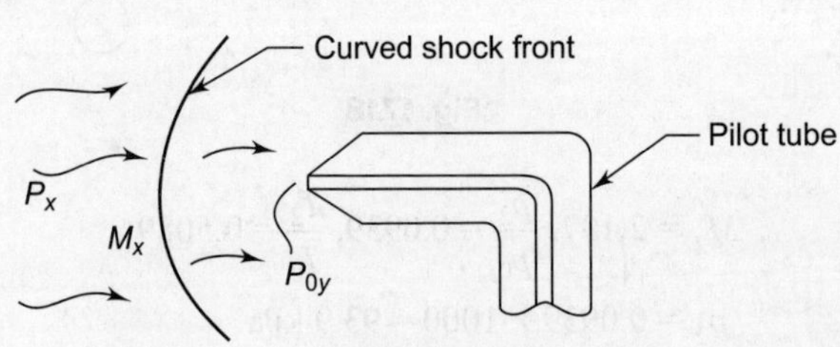

Fig. 17.19

$$p_x = 16 \text{ kPa}, p_{oy} = 70 \text{ kPa}$$

$$\therefore \quad \frac{p_{oy}}{p_x} = \frac{70}{16} = 4.375$$

From the gas tables for normal shock

When $\frac{p_{oy}}{p_x} = 4.375, M_x = 1.735, \frac{p_y}{p_x} = 3.34, \frac{\rho_y}{\rho_x} = 2.25$

$$\frac{T_y}{T_x} = 1.483, \frac{p_{oy}}{p_{ox}} = 0.84, M_y = 0.631$$

$$T_{ox} = T_{oy} = 573 \text{ K}$$

$$T_{ox} = \left(1 + \frac{\gamma - 1}{2} M_x^2\right) T_x = (1 + 0.2 \times 3) T_x = 1.6\, T_x$$

$$T_x = \frac{573}{1.6} = 358 \text{ K}$$

$$T_y = 358 \times 1.483 = 530 \text{ K} = 257°\text{C}$$

$$p_{ox} = \frac{p_{oy}}{0.84} = \frac{70}{0.84} = 83.3 \text{ kPa}$$

$$M_x = \mathbf{1.735}$$

Example 17.7 *A convergent-divergent nozzle operates at off-design condition while conducting air from a high pressure tank to a large container. A normal shock occurs in the divergent part of the nozzle at a section where the cross-sectional area is 18.75 cm². If the stagnation pressure and stagnation temperature at the inlet of the nozzle are 0.21 MPa and 36°C, respectively, and the throat area is 12.50 cm² and the exit area is 25 cm² estimate the exit Mach number, exit pressure, loss in stagnation pressure and entropy increase, during the flow between the tanks.* M [LO 17.6]

Solution With reference to Fig. 17.20, at shock section

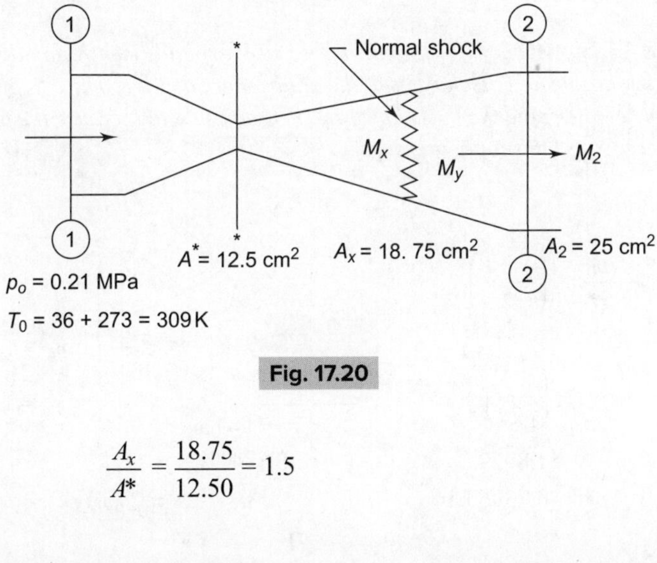

Fig. 17.20

$$\frac{A_x}{A^*} = \frac{18.75}{12.50} = 1.5$$

Up to the shock, the flow is isentropic. From the isentropic flow tables, when

$$\frac{A}{A^*} = 1.5, \quad M_x = 1.86 \quad \frac{p_x}{p_{ox}} = 0.159$$

∴ Static pressure upstream of the shock

$$p_x = 0.159 \times 0.21 \times 10^3 = 33.4 \text{ kPa}$$

From the gas tables on normal shocks

When $M_x = 1.86, M_y = 0.604, \frac{p_y}{p_x} = 3.87, \frac{p_{oy}}{p_x} = 4.95, \frac{p_{oy}}{p_{ox}} = 0.786$

∴ $$p_y = 3.87 \times 33.4 = 129.3 \text{ kPa}$$

$$p_{oy} = 4.95 \times 33.4 = 165.3 \text{ kPa}$$

From the shock section to the exit of the nozzle, the flow is again isentropic. When $M_y = 0.604$, from the isentropic flow tables

$$\frac{A_y}{A^*} = 1.183$$

∴ $$\frac{A_2}{A^*} = \frac{A_2 A_y}{A_y A^*} = \frac{25}{18.75} \times 1.183 = 1.582$$

When $A_2/A^* = 1.582$, from the isentropic flow tables, $M_2 = \mathbf{0.402}$

$$\frac{p_2}{p_{oy}} = 0.895$$

∴ $$p_2 = 0.895 \times 165.3 = \mathbf{147.94 \text{ kPa}}$$

Loss in stagnation pressure occurs only across the shock

$$p_{ox} - p_{oy} = 210 - 165.3 = \mathbf{44.7 \text{ kPa}}$$

Entropy increase, $s_y - s_x$

$$= -R \ln \frac{p_{oy}}{p_{ox}} = 0.287 \ln \frac{210}{165.3}$$

$$= 0.287 \times 0.239 = \mathbf{0.0686 \text{ kJ/kg K}}$$

Example 17.8 *Air at $p_0 = 1.4$ bar, $T_0 = 280$ K expands isentropically through a converging nozzle and discharges to the atmosphere at 1 bar. The exit plane area is 0.0013 m². Determine the mass flow rate. Is it possible to increase the mass flow rate by increasing the supply pressure?* E [LO 17.5]

Solution

$$\frac{p_0}{p^*} = \left(\frac{\gamma+1}{2}\right)^{\frac{\gamma}{\gamma-1}}$$

$$= \left(\frac{2.4}{2}\right)^{\frac{1.4}{0.4}}$$

$$= 1.8929$$

∴ $$p^* = 0.7396 \text{ bar}$$

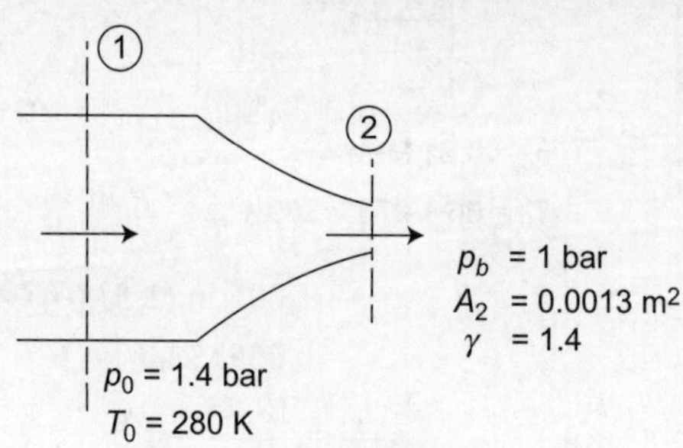

Fig. 17.21

The nozzle is not choked, since it is less than p_b (= 1 bar).
For isentropic expansion

$$T_2 = T_1 \left(\frac{p_2}{p_1}\right)^{\frac{\gamma-1}{\gamma}} = 254.3 \text{ K}$$

Now,

$$V_2 = \sqrt{2c_p(T_1 - T_2)}$$

$$= \sqrt{2 \times 1005(280 - 254.3)}$$

$$= 227.2 \text{ m/s}$$

$\therefore$ Mass flow rate, $\dot{m} = \dfrac{A_2 V_2 p_2}{RT_2} = \dfrac{0.0013 \times 227.2 \times 1000}{0.287 \times 254.3}$

$= \mathbf{0.4047 \text{ kg/s}}$

The mass flow rate can be increased by raising the supply pressure. However, when $p_0' = 1.8929$ bar, the nozzle will choke, and no further increase will occur.

Example 17.9 *Air as an ideal gas with $\gamma = 1.4$ undergoes a normal shock. The upstream conditions are $p_x = 0.5$ bar, $T_x = 280$ K and $M_x = 1.8$. Determine (a) the pressure p_y, (b) the stagnation pressure p_{ox}, (c) the stagnation temperature T_{ox} and (d) the change in specific entropy across the shock.* **E [LO 17.6]**

Solution (a) From Gas Table *A*.10

at $M_x = 1.8, \dfrac{p_y}{p_x} = 3.6133$

$\therefore \quad p_y = 3.6133 \times 0.5 = \mathbf{1.8066 \text{ bar}}$

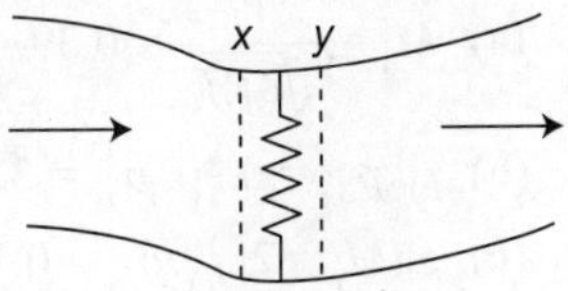

Fig. 17.22

(b) Also at $M_x = 1.8$, $p_x/p_{ox} = 0.17404$

$\therefore \quad p_{ox} = \dfrac{0.5}{0.17404} = \mathbf{2.873 \text{ bar}}$

(c) $\dfrac{T_x}{T_{ox}} = 0.60680$

$\therefore \quad T_{ox} = \dfrac{280}{0.60680} = \mathbf{461.4 \text{ K}}$

(d) $s_y - s_x = c_p \ln \dfrac{T_y}{T_x} - R \ln \dfrac{p_y}{p_x}$

$= 1.005 \ln (1.5316) - 0.287 \ln (3.6133)$

$= \mathbf{0.05935 \text{ kJ/kg K}}$

Example 17.10 *Air at 3.4 bar, 530 K and 180 m/s enters a converging-diverging nozzle operating at steady state. A normal shock stands in the diverging section at a location where the Mach number in $M_x = 2$. The flow is isentropic except where the shock stands. If the air behaves as an ideal gas, determine (a) the stagnation temperature T_{ox}, (b) the stagnation pressure p_{ox}, (c) the pressure p_x, (d) the pressure p_y, (e) the stagnation pressure p_{oy} and (f) the stagnation temperature T_{oy}.*

If the mass flow rate of air is 0.5 kg/s, and the exit plane pressure is 2.4 bar, determine the throat area and the exit plane area. M [LO 17.6]

Solution Using Gas Table *A*.10

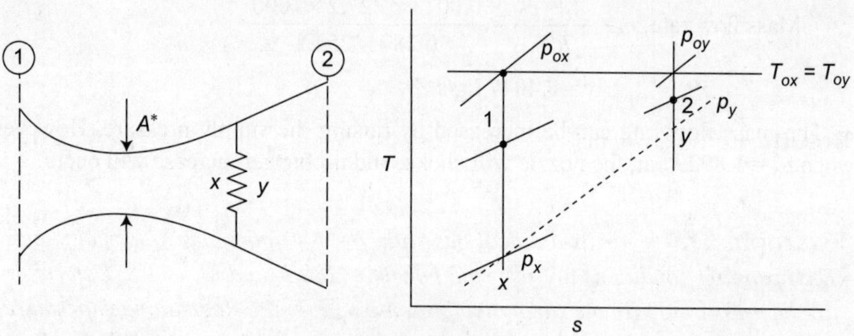

Fig. 17.23

(a) $M_1 = \dfrac{V_1}{\sqrt{8RT_1}} = 0.39$, $T_1/T_{ox} = 0.97032$, $T_{ox} = \mathbf{546.2\ K.}$

(b) $p_1/p_{ox} = 0.90$, $p_{ox} = \mathbf{3.778\ bar}$

(c) at $M_x = 2$, $p_x/p_{ox} = 0.1278$ $\therefore$ $p_x = \mathbf{0.4828\ bar}$

(d) at $M_x = 2$, $p_x/p_y = 4.5$, $p_y = \mathbf{2.1726\ bar}$

(e) $M_y = 0.57735$, $\dfrac{p_y}{p_{oy}} = 0.79737$, $p_{oy} = \mathbf{2.725\ bar}$

(f) $T_{oy} = T_{ox} = \mathbf{546.2\ K}$

To find A^*,

$$A_1 = \frac{\dot{m}RT_1}{V_1 p_1} = \frac{0.5 \times 0.287 \times 530}{180 \times 3.4 \times 100}$$

$$= 1.2427 \times 10^{-3}\ \text{m}^2$$

$$\text{At } M_1 = 0.39,\ A_1/A^* = 1.6346$$

$$\therefore \quad A^* = \mathbf{7.602 \times 10^{-4}\ m^2}$$

$$\text{At } M = 0.57735,\ \frac{A_2}{A^*} = 1.25$$

$$\therefore \quad A_2 = \text{Exit plane area} = 1.25 \times 7.602 \times 10^{-4}$$

$$= \mathbf{9.502 \times 10^{-4}\ m^2}$$

Example 17.11 *A frictionless converging nozzle is connected to a large tank of air where the pressure is 1.10 MPa (abs) and temperature is 110°C. The other side of the nozzle is connected to a constant area insulated duct as shown in Fig. 17.24. Air flows isentropically through the nozzle and the pressure at the nozzle exit is 150 kPa (abs). If the temperature at the duct exit is 70°C, find the pressure at the end of the duct. Also find the change in entropy.* M [LO 17.5]

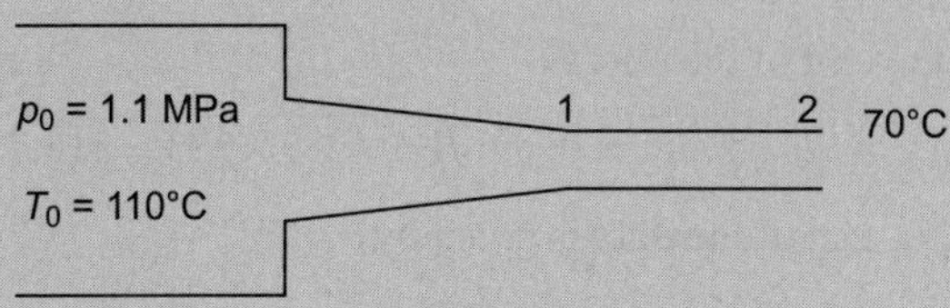

Fig. 17.24

Solution We know that

$$\frac{p_0}{p} = \left[1 + \frac{\gamma - 1}{2} M^2\right]^{\frac{\gamma}{\gamma-1}}$$

Mach number at the exit of the nozzle is

$$M_1 = \left[\frac{2}{\gamma - 1}\left\{\left(\frac{p_0}{p_1}\right)^{\frac{\gamma-1}{\gamma}} - 1\right\}\right]^{1/2}$$

$$= \left[\frac{2}{1.4 - 1}\left\{\left(\frac{1.1}{0.82}\right)^{\frac{1.4-1}{1.4}} - 1\right\}\right]^{1/2} = 0.66$$

Temperature of air at the exit of the nozzle is found as follows

$$\frac{T_0}{T_1} = 1 + \frac{\gamma - 1}{2} M_1^2$$

$$T_1 = \frac{T_0}{1 + \frac{\gamma - 1}{2} M_1^2} = \frac{383}{1 + \frac{1.4 - 1}{2}(0.66)^2} = 352.31 \text{ K}$$

Velocity of air at exit of the nozzle is

$$V_1 = M_1 c_1 = M_1 \sqrt{\gamma R T_1} = 0.66\sqrt{1.4 \times 287 \times 352.31} = 248.32 \text{ m/s}$$

The density of air at exit of the nozzle is

$$\rho_1 = \frac{p_1}{R T_1} = \frac{820}{0.287 \times 352.31} = 8.11 \text{ kg/m}^3$$

Mach number at the exit of the duct is then found to be

$$M_2 = \sqrt{\frac{2}{\gamma - 1}\left(\frac{T_0}{T_2} - 1\right)}$$

$$= \sqrt{\frac{2}{1.4-1}\left(\frac{383}{333} - 1\right)} = 0.866$$

Velocity of air at the exit of the duct is

$$V_2 = M_2 c_2 = M_2 \sqrt{\gamma R T_2} = 0.866\sqrt{1.4 \times 287 \times 333} = 316.77 \text{ m/s}$$

The density of air at exit of the duct is given by

$$w = \rho_1 A_1 V_1 = \rho_2 A_2 V_2$$

$$\rho_2 = \frac{\rho_1 V_1}{V_2} = \frac{8.11 \times 248.32}{316.77} = 6.36 \text{ kg/m}^3$$

The pressure at the exit of the duct will be

$$p_2 = \rho_2 R T_2 = 6.36 \times 287 \times 383 = 699097 \text{ N/m}^2 = \mathbf{699.097\ kPa}$$

The change in entropy is

$$s_2 - s_1 = c_p \ln\frac{T_2}{T_1} - R \ln\frac{p_2}{p_1}$$

$$= 1005 \ln\frac{383}{352.31} - 287 \ln\frac{699.097}{820} = \mathbf{129.72\ J/kgK}$$

SUMMARY

A compressible fluid is one whose density changes with the change in pressure or temperature. The velocity of a pressure pulse in such fluids, called *sonic or acoustic velocity* is given by

$$c = \sqrt{\left(\frac{\partial p}{\partial \rho}\right)_s}$$

For an ideal gas, $p = pRT$,

$$c = \sqrt{\gamma R T}$$

The Mach number M is the ratio of the actual fluid velocity, V, to the sonic velocity

$$M = \mathbf{V}/c.$$

The stagnation properties, designated with subscript zero, are those which refer to a compressible fluid when it is brought to rest adiabatically. Thus, the relations of stagnation and static properties of temperature and pressure are

$$\frac{T_o}{T} = 1 + \frac{\gamma - 1}{2} M^2 \text{ and } \frac{p_o}{p} = \left(1 + \frac{\gamma - 1}{2} M^2\right)^{\frac{\gamma}{\gamma - 1}}$$

For isentropic flow in a nozzle, the effect of area change is given by

$$\frac{dA}{A} = (M^2 - 1)\frac{d\mathbf{V}}{\mathbf{V}}$$

When $M < 1$, the flow is subsonic at inlet, as the area decreases, the velocity also increases, and if $M > 1$ at inlet, velocity increases as area increases. The discharge per unit area is

$$\frac{w}{A} = \sqrt{\frac{\gamma}{R}}\frac{p_o M}{\sqrt{T_o}}\frac{1}{\left(1+\frac{\gamma-1}{2}M^2\right)^{\frac{\gamma+1}{2(\gamma-1)}}}$$

The discharge is maximum when $M = 1$ or sonic velocity prevails at the throat who $dA = 0$. The properties are called *critical*, and designated by superscripts *.
Thus,

$$\frac{T_o^*}{T} = \frac{2}{\gamma+1} \text{ and } \frac{p_o^*}{p} = \left(\frac{2}{\gamma+1}\right)^{\frac{\gamma}{\gamma-1}}$$

For air, $\gamma = 1.4$, the critical pressure ratio is 0.528. The nozzle is said to be *choked* when critical condition is reached. The area ratio is given by

$$\frac{A^*}{A} = \left[\left(\frac{2}{\gamma+1}\right)\left(1+\frac{\gamma-1}{2}M^2\right)^{\frac{\gamma+1}{2(\gamma-1)}}\frac{1}{M}\right]$$

which shows that a subsonic nozzle is converging and a supersonic nozzle is diverging. The dimensionless velocity M^* is defined as

$$M^* = \frac{\mathbf{V}}{C^*} = \frac{\frac{\gamma+1}{2}M^2}{1+\frac{\gamma-1}{2}M^2}$$

The pressure and velocity of the fluid vary as the back pressure at nozzle exit is varied. When the back pressure $p_B = p_E$, the exit pressure, the flow remains isentropic. If $p_B > p_E$, the shock appears and the flow at exit adjusts to p_E. Shocks are highly irreversible and properties are discontinuous across them. Shocks appear only when the flow is supersonic and after the shocks, the flow is supersonic and after the shocks the flow becomes subsonic across the shocks, the upstream and downstream properties being designated by subscripts x and y, respectively,

Continuity equation: $\frac{w}{A} = P_x V_x = p_y V_y = G$ where G is the mass velocity (kg/m²s)

Momentum equation: $p_x + \rho_a V_x^2 = p_y + \rho_y V_y^2$ where F is called the *impulse function*

Energy equation: $h_x + \frac{V_x^2}{2} = h_y + \frac{V_y^2}{2} = h_o$

Second law gives $s_y - s_x > 0$

The energy equation can be written as

$$h = h_x + \frac{G^2}{2\rho_y^2}$$

The line passing through the points of the same G and h_o is called a *Fanno line*. Adiabatic flow in a duct with friction must follow a Fanno line.

The line representing the locus of states with the same *impulse pressure*, I (= F/A) and mass velocity is called the *Rayleigh line*. The normal shock end states *x* and *y* must lie at the intersections of the Fanno line and the Rayleigh line is also a model for flow in constant area duct with heat transfer but without friction.

REVIEW QUESTIONS

(E) **17.1** What is a compressible fluid? **[LO 17.1]**

(M) **17.2** What are the basic laws in compressible flow? **[LO 17.1]**

(E) **17.3** How is sonic velocity defined in terms of pressure and density of the fluid? **[LO 17.2]**

(M) **17.4** Show that the sonic velocity in an ideal gas depends on the temperature and the nature of the gas. **[LO 17.2]**

(E) **17.5** What is a Mach number? **[LO 17.2]**

(M) **17.6** What is stagnation state? What do you mean by stagnation properties? **[LO 17.3]**

(E) **17.7** What are a nozzle and a diffuser? **[LO 17.4]**

(M) **17.8** Explain the effect of area change in subsonic and supersonic flows. **[LO 17.4]**

(E) **17.9** What do you understand by choking in nozzle flows? **[LO 17.5]**

(D) **17.10** Show that the discharge through a nozzle is maximum when there is a sonic condition at its throat. **[LO 17.5]**

(M) **17.11** What do you understand by critical pressure ratio? What is its value for air? **[LO 17.5]**

(M) **17.12** Explain the effect of area ratio as a function of Mach number in an isentropic nozzle. **[LO 17.5]**

(M) **17.13** What is M^*? **[LO 17.5]**

(M) **17.14** What is a shock? Where does it occur in a nozzle? **[LO 17.5]**

(E) **17.15** What is the impulse function? **[LO 17.6]**

(D) **17.16** What is a Fanno line? Why do the end states of a normal shock lie on the Fanno line? **[LO 17.6]**

(D) **17.17** What is a Rayleigh line? Why do the end states of a normal shock also lie on the Rayleigh line? **[LO 17.6]**

(M) **17.18** Where does the local sound velocity occur on the Fanno line and on the Rayleigh line? **[LO 17.6]**

(M) **17.19** How is the strength of a shock defined? **[LO 17.6]**

(M) **17.20** Explain the occurrence of choking for adiabatic flow with friction and for diabatic flow without friction. **[LO 17.7]**

PROBLEMS WITH HINTS

(M) **17.1** A converging–diverging nozzle has an exit area to throat area ratio (A_2/A^*) of 2. Air enters the nozzle with a stagnation pressure of 1000 kPa and a stagnation temperature of 360 K. The throat area is 500 mm^2. Determine the mass flow rate of air, exit pressure, exit temperature, exit Mach number and the exit velocity if (a) the diverging section acts as a nozzle and (b) the diverging section acts as a diffuser. **[LO 17.5]**

Hint: From Gas Tables (A-10) in the Appendix for $A_2/A^* = 2$, there are two Mach numbers, one for supersonic flow when the divergent section acts as a nozzle, and the other for subsonic flow when the divergent section acts as a diffuser.

(a) $\frac{A_2}{A^*} = 2, M_2 = 2.197, \frac{p_2}{p_o} = 0.0939, \frac{T_2}{T_o} = 0.5089$

Therefore, find p_2 = **93.9 kPa**, T_2 = **183.2 K**,

$$c_2 = \sqrt{\gamma R T_2} = \mathbf{271.3\ m/s},\ V_2 = \mathbf{596.1\ m/s}$$

(b) $\frac{A_2}{A^*} = 2, M_2 = 0.308, \frac{p_2}{p_o} = 0.936, \frac{T_2}{T_o} = 0.9812$

$\therefore\ p_2$ = **936 kPa**, T_2 = **353.3 K**, V_e = **116 m/s**

(E) **17.2** Assume a normal shock appears in the exit plane of the supersonic nozzle of the above problem. Determine the static pressure and temperature and the stagnation pressure just downstream of the normal shock. **[LO 17.6]**

Hint: From Gas Tables A-10(b),

$$M_x = 2.197, M_y = 0.547, p_y/p_x = 5.46, \frac{T_y}{T_x} = 1.854,$$

$$\frac{p_{oy}}{p_{ox}} = 0.63. \text{ Find } p_y, T_y, P_{oy}.$$

PROBLEMS

(E) **17.1** Air in a reservoir has a temperature of 27°C and a pressure of 0.8 MPa. The air is allowed to escape through a channel at a rate of 2.5 kg/s. Assuming that the air velocity in the reservoir is negligible and that the flow through the channel is isentropic, find the Mach number, the velocity and the area at a section in the channel where the static pressure is 0.6 MPa. **[LO 17.2, 17.3]**

(M) **17.2** A supersonic wind tunnel nozzle is to be designed for $M = 2$, with a throat section, 0.11 m^2 in area. The supply pressure and temperature at the nozzle inlet, where the velocity is negligible, are 70 kPa and 37°C, respectively. Compute the mass flow rate, the exit area and the fluid properties at the throat and exit. Take $\gamma = 1.4$. **[LO 17.4, 17.5]**

(M) **17.3** An ideal gas flows into a convergent nozzle at a pressure of 0.565 MPa, a temperature of 280°C and negligible velocity. After reversible adiabatic expansion in the nozzle the gas flows directly into a large vessel. The gas in the vessel may be maintained at any specified state while the nozzle supply state is held constant. The exit area of the nozzle is 500 mm^2. For this gas $\gamma = 1.3$ and $c_p = 1.172$ kJ/kg K. Determine (a) the pressure of the gas leaving the nozzle when its temperature is 225°C and (b) the gas mass flow rate when the pressure in the vessel is 0.21 MPa. **[LO 17.4, 17.5]**

(E) **17.4** Air flows adiabatically through a pipe with a constant area. At point 1, the stagnation pressure is 0.35 MPa and the Mach number is 0.4. Further downstream the stagnation pressure is found to be 0.25 MPa. What is the Mach number at the second point for subsonic flow? **[LO 17.2, 17.3]**

(E) **17.5** The intake duct to an axial flow air compressor has a diameter of 0.3 m and compresses air at 10 kg/s. The static pressure inside the duct is 67 kPa and the stagnation temperature is 40°C. Calculate the Mach number in the duct. **[LO 17.2, 17.3]**

(M) **17.6** Show that for an ideal gas the fractional change in pressure across a small pressure pulse is given by

$$\frac{dp}{p} = \gamma \frac{d\mathbf{V}}{c}$$

and that the fractional change in absolute temperature is given by

$$\frac{dT}{T} = (\gamma - 1)\frac{d\mathbf{V}}{c}$$

[LO 17.6]

(E) **17.7** An airplane flies at an altitude of 13,000 m (temperature – 55°C, pressure 18.5 kPa) with a speed of 180 m/s. Neglecting frictional effects, calculate (a) the critical velocity of the air relative to the aircraft, and (b) the maximum possible velocity of the air relative to the aircraft. **[LO 17.2]**

(M) **17.8** A stream of air flowing in a duct is at a pressure of 150 kPa, has a Mach number of 0.6, and flows at a rate of 0.25 kg/s. The cross-sectional area of the duct is 625 mm^2. (a) Compute the stagnation temperature of the stream in°C. (b) What is the maximum percentage reduction in area which could be introduced without reducing the flow rate of the stream? (c) For the maximum area reduction of part (b), find the velocity and pressure at the minimum area, assuming no friction and heat transfer. **[LO 17.2, 17.3]**

(M) **17.9** A rocket combustion chamber is supplied with 12 kg/s of hydrogen and 38 kg/s of oxygen. Before entering the nozzle all the oxygen is consumed, the pressure is 2.3 MPa, and the temperature is 2800°C. Neglecting dissociation and friction, find the throat area of the nozzle required. Assume $\gamma = 1.25$. **[LO 17.4]**

(M) **17.10** A gas with specific heat ratio $\gamma = 1.4$ passes through a plane normal shock Immediately before the shock the Mach number is 2, and the stagnation pressure and temperature are 600 kPa and 400 K, respectively. Find the static pressure and static temperature immediately downstream of the shock. **[LO 17.6]**

(E) **17.11** Air flowing through a nozzle encounters a shock. The Mach number upstream of the shock is 1.8, and the static temperature downstream of the shock is 450 K. How much has the velocity changed across the shock? **[LO 17.6]**

(D) **17.12** The stagnation temperature and stagnation pressure of air in a reservoir supplying a convergent-divergent nozzle are 450 K and 400 kPa, respectively. The nozzle throat area is 625 mm^2 and the nozzle exit area is 1875 mm^2. A shock is noted at a position in the divergent section where the area is 1250 mm^2. (a)What are the exit pressure, temperature and velocity? (b) What value of the back pressure would cause the flow through the nozzle to be completely supersonic? (c) What value of back pressure would result in completely isentropic flow interior and exterior to the nozzle? **[LO 17.6]**

(E) **17.13** Assume that a flow through the nozzle has been established by setting the back pressure at 200 kPa. Because a back pressure greater than 150 kPa leads to a shock in the nozzle, find the area ratio where the shock occurs. **[LO 17.6]**

(M) **17.14** Air flows through a frictionless convergent-divergent nozzle. The area of the exit section is three times the area of the throat section, and the ratio of stagnation pressure at the entrance to the static pressure in the exit section is 2.5. Calculate (a) the Mach number M_x where the shock occurs, (b) the area A_x where the shock occurs and (c) the entropy increase. **[LO 17.6]**

(M) **17.15** A nozzle is designed assuming reversible adiabatic flow with an exit Mach number of 2.6. Air flows through it with a stagnation pressure and temperature of 2 MPa and 150°C, respectively. The mass rate of flow is 5 kg/s. (a) Determine the exit pressure, temperature, area and throat area. (b) If the back pressure at the nozzle exit is raised to 1.35 MPa, and the flow remains isentropic except for a normal shock wave, determine the exit Mach number and temperature, and the mass flow rate through the nozzle. **[LO 17.5, 17.6]**

(M) **17.16** A jet plane travels through the air with a speed of 1000 km per hour at an altitude of 6000 m, where the pressure is 40 kPa and the temperature is –12°C. Consider the diffuser of the engine. The air leaves the diffuser with a velocity of 100 m/s. Determine the pressure and temperature leaving the diffuser, and the ratio of the inlet to the exit area of the diffuser, assuming isentropic flow. **[LO 17.4]**

(M) **17.17** A normal shock occurs in the diverging section of a nozzle under steady flow conditions at a point where the air is at 150 kPa and 300 K travelling at 1000 m/s. What are the pressure and temperature on the subsonic side of the front? If the surroundings are at 20°C, what is the irreversibility caused by the shock process? **[LO 17.6]**

ANSWERS TO PROBLEMS

17.3 0.36 MPa, 0.48 kg/s

17.5 0.526

17.10 345 kPa, 376 K

17.11 378 m/s

17.12 (a) 234 kPa, 440 K, 141 m/s, (b) 150 kPa, (c) 18.8 kPa

17.16 61 kPa, 295 K, 0.487

17.17 1426.5 kPa, 762.1 K, 84.74 kJ/kg

18

Gas Compressors

Learning Objectives

LO 18.1 State function, types and applications of a gas compressor

LO 18.2 Explain adiabatic, polytropic and isothermal works for a reversible compression

LO 18.3 Demonstrate work required for single-stage reciprocating air compressor

LO 18.4 Discuss volumetric efficiency of a reciprocating air compressor

LO 18.5 Compare single-stage compression with multi-stage compression

LO 18.6 Discuss the *p-V* diagram of an air motor

LO 18.7 Compare different rotary compressors

LO 18.1

State function, types and applications of a gas compressor

A gas compressor is a device in which work is done on the gas to raise its pressure, with an appreciable increase in its density. The compression of gases is an important process in many power plants, refrigeration plants and industrial plants. Industrial uses occur in connection with compressed air motors for tools, air brakes for vehicles, servo-mechanisms, metallurgical and chemical processes, conveying of materials through ducts, transporting of natural gas and production of bottled gases.

Compressors can be (a) positive displacement machines like reciprocating compressors, Root's blower and vane-sealed machines and (b) turbine type like centrifugal and axial flow compressors.

18.1 COMPRESSION PROCESSES

A gas compression process may be either adiabatic or involving heat transfer, depending upon the purpose for which the gas is compressed. If the compressed gas is to be used promptly in an engine or in a combustion process, adiabatic compression is desirable so as to obtain the maximum energy in the gas at the end of compression. In many applications, however, the gas is not used promptly but is stored in a tank or receiver for use later when needed. The gas in the tank transfers heat to the surroundings so that when finally used it is at room temperature. In such cases, the objective of the compression and storage process is simply to increase the pressure of the gas without change of temperature. It will be shown below that if the gas is cooled during compression, the work required will be less than for adiabatic compression. Another advantage of cooling is the reduction in volume of gas and less pipe friction losses. Since cooling during the compression process is not very effective, *after-coolers* are often used to cool the gas leaving the compressor.

18.2 WORK OF COMPRESSION

LO 18.2

Explain adiabatic, polytropic and isothermal works for a reversible compression

The steady flow energy equation for a compressor (Fig. 18.1) gives

$$h_1 + Q = h_2 + W_x \tag{18.1}$$

neglecting the changes of potential and kinetic energy. From the property relation

$$Tds = dh - vdp \tag{18.2}$$

For a reversible process,

$$Q = \Delta h - \int vdp \tag{18.3}$$

Let us consider two particular idealised cases, namely, reversible adiabatic and reversible isothermal, as well as a general case of a reversible polytropic process (pv^n = constant). The paths of such processes are plotted in Fig. 18.2 for an ideal gas compressed from state 1 to state 2.

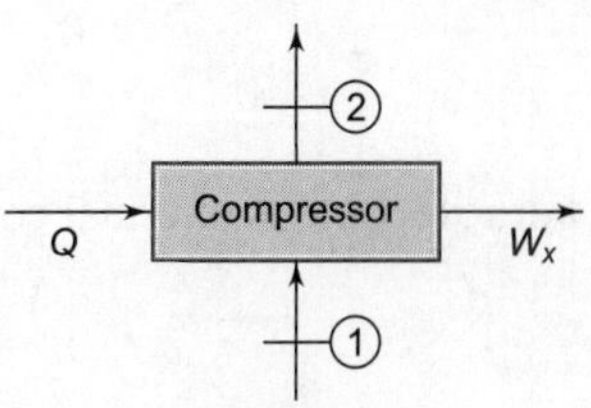

Fig. 18.1

As shown in Section 7.14 or from Eqs. (18.1 to 18.3) for adiabatic compression of unit mass,

$$W_x = -\int vdp \tag{18.4}$$

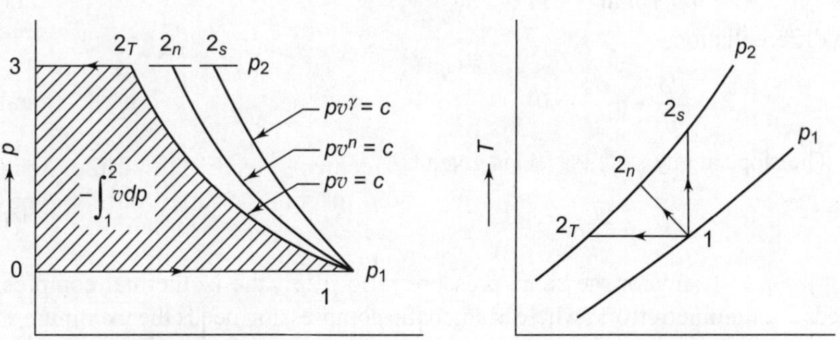

Fig. 18.2 Reversible compression processes

For a reversible adiabatic process, $pv^\gamma = C$,

$$v = \left(\frac{pv^\gamma}{p}\right)^{\frac{1}{\gamma}} = \frac{p_1^{1/\gamma} v_1}{p^{1/\gamma}} = \frac{p_2^{1/\gamma} v_2}{p^{1/\gamma}}$$

and
$$W_x = -\int_1^2 \frac{p_1^{1/\gamma} v_1}{p^{1/\gamma}} dp = -p_1^{1/\gamma} \cdot v_1 \cdot \frac{1}{1-\frac{1}{\gamma}} \left[p_2^{1-\frac{1}{\gamma}} - p_1^{1-\frac{1}{\gamma}} \right]$$

$$= -\frac{\gamma}{\gamma-1} p_1 v_1 \left[\left(\frac{p_2}{p_1}\right)^{\frac{\gamma-1}{\gamma}} - 1 \right]$$

The *work of compression* or the steady flow work input to the gas is the negative of the shaft work W_x, i.e. for reversible adiabatic compression,

$$W_s = \frac{\gamma}{\gamma-1} p_1 v_1 \left[\left(\frac{p_2}{p_1}\right)^{\frac{\gamma-1}{\gamma}} - 1 \right] \tag{18.5}$$

Similarly, for reversible polytropic compression

$$W_n = \frac{n}{n-1} p_1 v_1 \left[\left(\frac{p_2}{p_1}\right)^{\frac{n-1}{n}} - 1 \right] \tag{18.6}$$

For reversible isothermal compression of an ideal gas,

$$W_t = -\int_1^2 v dp = -\int_1^2 \frac{p_1 v_1}{p} dp = p_1 v_1 \ln \frac{p_2}{p_1} \tag{18.7}$$

In the *p-v* plot of Fig. 18.2 the work of compression for each type of process is represented by the area between the path of that process and the axis of pressures. For the general process $pv^n = C$, taking logarithm

$$\ln p + n \ln v = \ln C$$

On differentiation,

$$\frac{dp}{p} + n\frac{dv}{v} = 0$$

∴ The slope at any state (say 1) is given by

$$\frac{dp}{dv} = -n\frac{p_1}{v_1} \tag{18.8}$$

For $\gamma > n > 1$ and for the same pressure ratio p_2/p_1, the isothermal compression needs the minimum work, whereas adiabatic compression needs the maximum work, while the polytropic compression needing work in between the two. In isothermal compression, all the work done on the gas during the process is transformed into an internal energy increase where $Q - W = u_2 - u_1 = 0$ (for an ideal gas) or $W = Q$.

This heat is taken away from the gas by cooling. Thus in isothermal compression, considered to be ideal, no energy would be imparted to the gas, since its function is simply to raise the pressure of the gas (and not its temperature).

The efficiency of a compressor working in a steady flow process may be defined as

$$\eta_c = \frac{h_{2S} - h_1}{W_c} = \frac{W_s}{W_c} \tag{18.9}$$

where W_c is the shaft work supplied to the compressor per unit mass. For the idealised reversible isothermal process the compressor efficiency is sometimes defined as

$$\eta_c = \frac{W_t}{W_c} \tag{18.10}$$

where W_t is the work of reversible isothermal compression. The two efficiencies of Eqs. (18.9) and (18.10) are called, respectively, the *adiabatic efficiency* and *isothermal efficiency*. Because of the effects of cooling, the adiabatic efficiency of a real compressor may be greater than unity.

Many turbine-type compressors are essentially adiabatic machines due to their high speeds. For an adiabatic machine the work of compression is equal to the enthalpy rise of the gas,

$$W_c = h_2 - h_1$$

Then for an adiabatic compressor the efficiency is

$$\eta_c = \frac{h_{2S} - h_1}{h_2 - h_1} \tag{18.11}$$

For a reciprocating machine the compressor efficiency may be on an indicated work basis or a brake work basis, depending upon where the work input is measured (See Section 3.3).

18.3 SINGLE-STAGE RECIPROCATING AIR COMPRESSOR

LO 18.3

Demonstrate work required for single-stage reciprocating air compressor

Figure 18.3 shows the arrangement of a single-stage reciprocating air compressor, together with a typical indicator diagram. The compressor operates on a two-stroke cycle as follows:

Stroke 1 (*a-c*) The piston withdraws, causing the air in the clearance volume to expand, and when the pressure in the cylinder falls below atmospheric pressure (at *b*) the inlet valve opens and air is drawn into the cylinder for the remainder of the stroke.

Stroke 2 (*c-a*) The piston moves inwards, compressing the air in the cylinder, and the inlet valve closes when the cylinder pressure reaches atmospheric pressure. Further compression follows as the piston moves towards the top of its stroke until, when the pressure in the cylinder is more than that in the receiver, the delivery valve opens and air is delivered to the receiver for the remainder of the stroke.

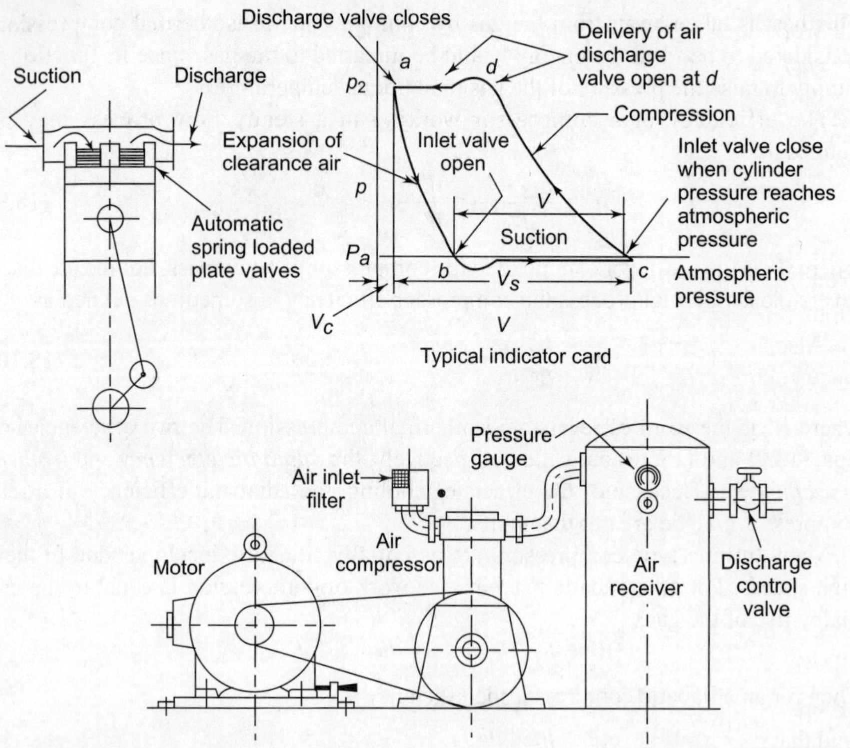

Fig. 18.3 Arrangement of single-stage reciprocating air compression

The intercept V on the indicator diagram represents the volume of air taken in per cycle. It is seen that the effect of clearance air is to reduce the quantity of air drawn in during the suction stroke, so that in practise the clearance space is made as small as possible.

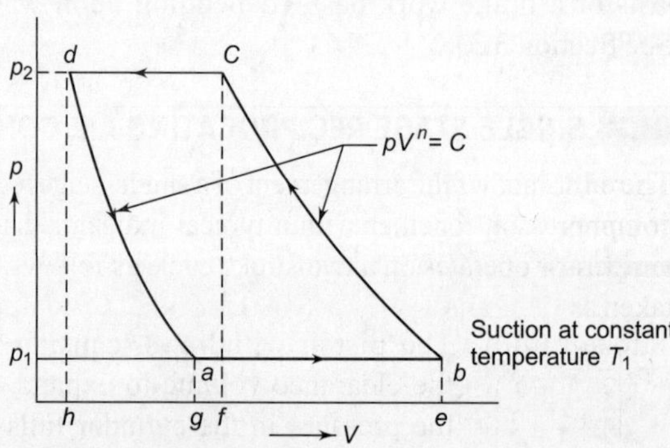

Fig. 18.4 Indicator diagram

The total area of the indicator diagram represents the actual work of the compressor on the gas or air. The areas above p_2 and below p_1 represent work done because of pressure drop through the valves and port passages; this work is called the valve loss. The idealised machine to which the actual machine is compared has an indicator diagram like Fig. 18.4. Both expansion and compression are supposed to follow the same law $pv^n = C$. The small quantity of high pressure air in the clearance volume expands to V_a and air drawn in during the suction stroke is $V_b - V_a$. From the diagram:

Work done on the air during the cycle = Enclosed area *abcd* = Area *dcfh* + Area *cbef* – Area *hgad* – Area *geba*

$$\therefore \quad W = p_2 (V_c - V_d) + \frac{p_2V_c - p_1V_b}{n-1} - \frac{p_2V_d - p_1V_a}{n-1} - p_1 (V_b - V_a)$$

$$= \frac{n}{n-1} [(p_2 V_c - p_1 V_b) - (p_2 V_d - p_1 V_a)]$$

Now, $p_2V_c = m_cRT_2$, $p_1V_b = m_bRT_1$ and $m_c - m_b$ = mass of air taken during compression.

Also, $p_2V_d = m_dRT_2$, $p_1V_a = m_aRT_1$ and $m_a - m_d$ = mass of air present in the clearance volume after delivery.

$$\therefore \quad \text{Work done/cycle} = \frac{n}{n-1} R[m_b (T_2 - T_1) - m_d (T_2 - T_1)]$$

$$= \frac{n}{n-1} (m_b - m_d) R(T_2 - T_1)$$

$$= \frac{n}{n-1}(m_b - m_d) RT_1 \left[\left(\frac{p_2}{p_1}\right)^{\frac{n-1}{n}} - 1\right] \tag{18.12}$$

where $m_b - m_d$ is the difference between the mass of air present at the end of suction and that present at the end of delivery.

Thus the expression *m* Eq. (18.12) is the same as obtained from the steady flow energy equation given in Eq. (18.6) where the clearance volume was neglected. Thus the mass of gas in the clearance volume does not have any effect on the work of compression.

18.4 VOLUMETRIC EFFICIENCY

LO 18.4

Discuss volumetric efficiency of a reciprocating air compressor

The amount of air dealt with in a given time by an air compressor is often referred to at free air conditions, i.e. the temperature and pressure of the environment, which may be taken as 15°C and 101.325 kPa, if not mentioned. It is known as free air delivery (FAD).

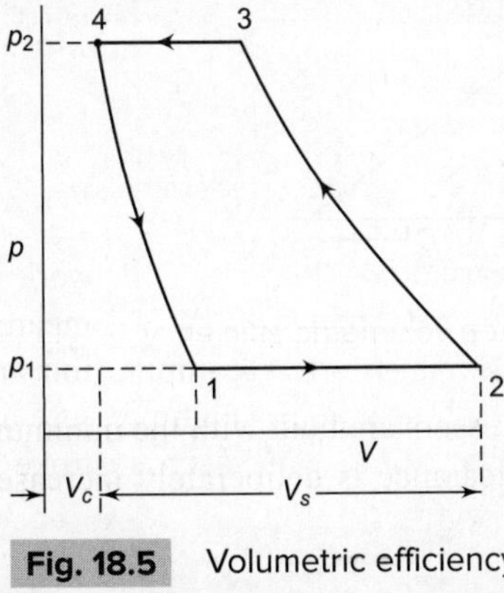

Fig. 18.5 Volumetric efficiency

The ratio of the actual volume of gas taken into the cylinder during suction stroke to the piston displacement volume (PD) or the swept volume (V_s) of the piston is called the *volumetric efficiency*, or

$$\eta_{\text{vol}} = \frac{\dot{m}v_1}{PD} = \frac{\dot{m}v_1}{V_s}$$

where $\dot{m}$ is the mass flow rate of the gas and v_1 is the specific volume of the gas at inlet to the compressor. With reference to Fig. 18.5

$$\eta_{vol} = \frac{V_2 - V_1}{V_s} = \frac{V_C + V_s - V_1}{V_s}$$

$$= 1 + \frac{V_c}{V_s} - \frac{V_1}{V_s}$$

Let C = clearance = $\dfrac{\text{Clearance volume}}{\text{PD or } V_s}$

$$= \frac{V_c}{V_s}$$

Since $p_1 V_1^n = p_2 V_4^n$, $\therefore V_1 = V_4 \left(\dfrac{p_2}{p_1}\right)^{\frac{1}{n}}$

$$= V_c \left(\frac{p_2}{p_1}\right)^{\frac{1}{n}}$$

$$\therefore \qquad \eta_{vol} = 1 + C - \frac{V_c}{V_s}\left(\frac{p_2}{p_1}\right)^{\frac{1}{n}}$$

$$= 1 + C - C\left(\frac{p_2}{p_1}\right)^{\frac{1}{n}} \qquad (18.13)$$

Equation (18.13) is plotted in Fig. 18.6. Since (p_2/p_1) is always greater than unity, it is evident that the volumetric efficiency decreases as the clearance increases and as the pressure ratio increases.

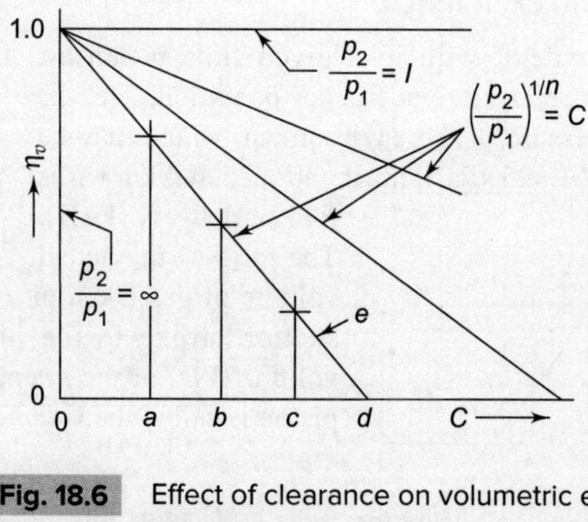

Fig. 18.6 Effect of clearance on volumetric efficiency

In order to get maximum flow capacity, compressors are built with the minimum practical clearance. Sometimes, however, the clearance is deliberately increased

$\left(\dot{m} = \dfrac{V_s n_v}{v_1}\right)$ as a means of controlling the flow through a compressor driven by a constant speed motor. The compressor cylinder of Fig. 18.7 is fitted with a clearance pocket which can be opened by a valve. Let us suppose that this machine is operating at conditions corresponding to line *e* in Fig. 18.6. If the clearance volume is at minimum value *a* the volumetric efficiency and the flow through the machine are maximum. If the clearance pocket is then opened to increase the clearance to *b*, the volumetric efficiency and the flow are reduced. By increasing the clearance in steps, as indicated by points *c* and *d*, the flow may be reduced in steps to zero. The work per kg of gas compressed is, however, not affected by the clearance volume in an idealised compressor.

For a given pressure ratio, η_{vol} is zero when the maximum clearance is

$$C_{max} = \frac{1}{\left(\dfrac{p_2}{p_1}\right)^{\frac{1}{n}} - 1} \tag{18.14}$$

18.4.1 Effect of Pressure Ratio on Volumetric Efficiency

It is evident from Fig. 18.6 that as the pressure ratio is increased the volumetric efficiency of a compressor of fixed clearance decreases, eventually becoming zero. This can also be seen in an indicator diagram, Fig. 18.8. As the discharge pressure is increased from p_{2a} to p_{2b}, the volume V_{Ia} taken at pressure p_1 decreases to V_{Ib}, and η_{vol} decreases. At some pressure p_{2c} the compression line intersects the line of clearance volume and there is no discharge of gas. An attempt to pump to p_{2c} (or any higher pressure) would result in compression and re-expansion of the same gas repeatedly, with no flow in or out.

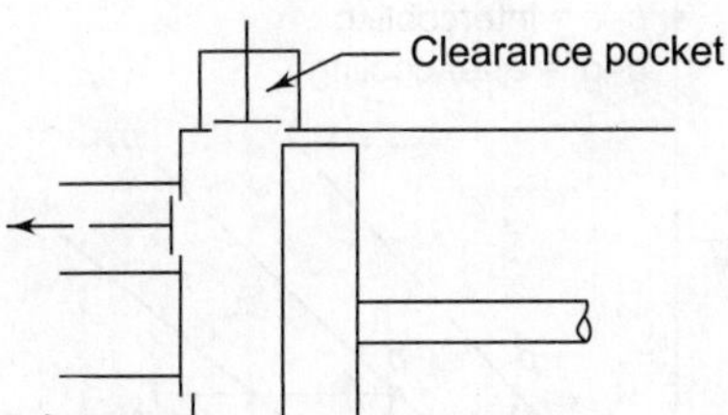

Fig. 18.7 Clearance pocket for capacity control

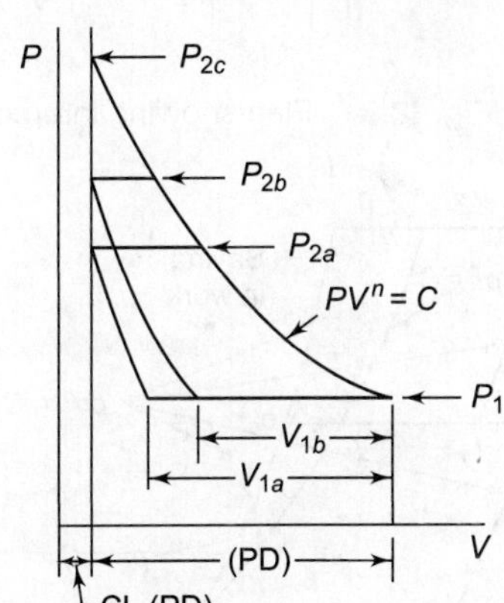

Fig. 18.8 Effect of pressure-ratio on capacity

The maximum pressure ratio attainable with a reciprocating compressor cylinder is then seen to be limited by the clearance, which is given from Eq. 18.13, as

$$\frac{p_{2\max}}{p_1} = \left(1+\frac{1}{C}\right)^n \tag{18.15}$$

18.5 MULTI-STAGE COMPRESSION

LO 18.5 Compare single-stage compression with multi-stage compression

When compressing a gas (or air) to high pressure it is advantageous to do it in stages. The condition for minimum work requires the compression to be isothermal. Since the temperature after compression is given by $T_2 = T_1 (p_2/p_1)^{(n-1)/n}$, the delivery temperature, T_2, increases with the pressure ratio. Also the volumetric efficiency as given by Eq. (18.13) decreases as the pressure ratio increases, as mentioned earlier.

The volumetric efficiency can be improved by carrying out the compression in two stages. After the first stage of compression (LP) from the state p_1, T_1 to the state p_x, T_x, the fluid is passed into a smaller cylinder (HP) in which the gas is compressed to the required final pressure p_2 at temperature T_2 (Fig. 18.9).

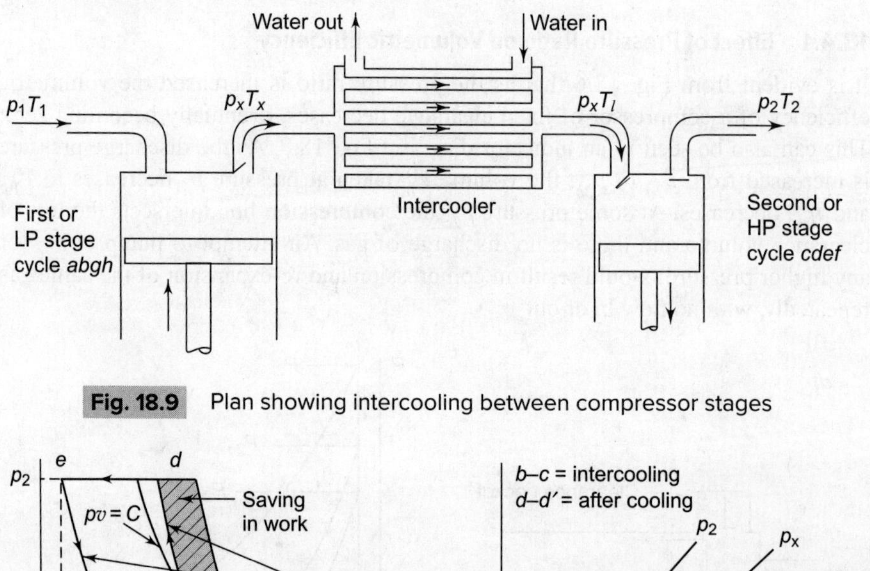

Fig. 18.9 Plan showing intercooling between compressor stages

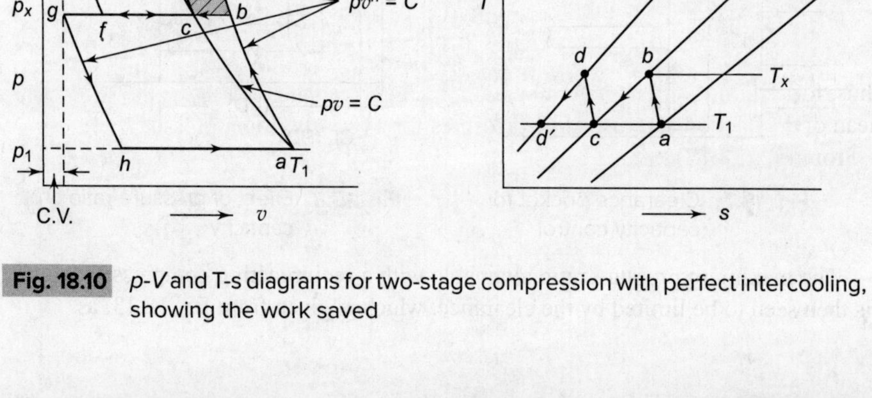

Fig. 18.10 *p-V* and T-s diagrams for two-stage compression with perfect intercooling, showing the work saved

The gas after being compressed in the LP (low pressure) cylinder ($a - b$) is passed on to an intercooler for getting cooled. After leaving the intercooler, the gas enters the HP (high pressure) cylinder for further compression (c–d). Figure 18.10 shows the p-V diagram for two-stage compression. Complete or perfect intercooling (b-c) means that the exiting gas from the intercooler at temperature T_x is cooled completely to the original (inlet) temperature T_1.

Similarly, perfect aftercooling (d–d') makes the gas, leaving the HP compressor, cooled also to the inlet temperature T_1. This aftercooling reduces the volume of the gas leaving, and thus the size of the receiver becomes smaller. The clearance volume in both the cylinders has here been assumed to be the same.

18.5.1 Ideal Intermediate Pressure

The intermediate pressure p_x (Fig. 18.10) has an optimum value for minimum work of compression. The total work per kg of gas is given by

$$W_c = \frac{n}{n-1} RT_1 \left[\left(\frac{p_x}{p_1}\right)^{\frac{n-1}{n}} - 1\right] + \frac{n}{n-1} RT_1 \left[\left(\frac{p_2}{p_x}\right)^{\frac{n-1}{n}} - 1\right]$$

$$= \frac{n}{n-1} RT_1 \left[\left(\frac{p_x}{p_1}\right)^{\frac{n-1}{n}} + \left(\frac{p_2}{p_x}\right)^{\frac{n-1}{n}} - 2\right] \quad (18.16)$$

Here p_1, T_1 and p_2 are fixed and p_x is the only variable. Differentiating Eq. (18.16) with respect to p_x and making it equal to zero,

$$\frac{dW_c}{dp_x} = \frac{n}{n-1} RT_1 \left[\frac{n-1}{n}\left(\frac{1}{p_1}\right)^{\frac{n-1}{n}} (p_x)^{1-\frac{1}{n}-1} + p_2^{\frac{n-1}{n}} \cdot p_x^{-1+\frac{1}{n}-1}\left(-\frac{n-1}{n}\right)\right] = 0$$

$$\therefore \quad p_x^{-\frac{1}{n}+2-\frac{1}{n}} = (p_2 p_1)^{\frac{n-1}{n}}$$

or

$$p_x^{2\left(\frac{n-1}{n}\right)} = (p_1 p_2)^{\frac{n-1}{n}}$$

$$\therefore \quad p_x = \sqrt{p_1 p_2} \quad (18.17)$$

Thus, for minimum work of compression, the intermediate pressure is the geometric mean of the suction and discharge pressures for a two-stage compressor.

From Eq. (18.17)

$$\frac{p_x}{p_1} = \frac{p_2}{p_x} = \left(\frac{p_2}{p_1}\right)^{\frac{1}{2}} \quad (18.18)$$

Pressure ratio in LP stage = Pressure ratio in HP stage

Also,
$$\frac{T_x}{T_1} = \left(\frac{p_x}{p_1}\right)^{\frac{n-1}{n}} \text{ and } \frac{T_2}{T_1} = \left(\frac{p_2}{p_x}\right)^{\frac{n-1}{n}}$$

$\therefore$
$$T_2 = T_x \tag{18.19}$$

From Fig. 18.10, it is seen that work required in LP compressor = Work required in HP compressor.

Thus, the intermediate pressure that produces minimum work will also result in equal pressure ratios in the two stages of compression, equal discharge temperatures and equal work for the two stages.

For two stage-compression, the minimum work, using Eq. (18.16) becomes

$$W_c = \frac{2nRT_1}{n-1}\left[\left(\frac{p_x}{p_1}\right)^{\frac{n-1}{n}} - 1\right] \tag{18.20}$$

where p_x/p_1 is the pressure ratio in each stage. In terms of overall pressure ratio, it becomes

$$W_c = \frac{2nRT_1}{n-1}\left[\left(\frac{p_2}{p_1}\right)^{\frac{n-1}{2n}} - 1\right] \tag{18.21}$$

Heat rejected in the intercooler, $Q_{bc} = c_p\,(T_x - T_1)$ kJ/kg

Let us now consider compression efficiencies and imperfect intercooling. As shown in Fig. 18.11, an ideal gas is compressed from the initial state p_1, T_1 to p_x. It is then cooled at constant pressure to T_y and then compressed from p_x, T_y to p_2. Given p_1, T_1, T_y and p_2, it is desired to find the value of p_x which gives minimum work. Let the adiabatic compression efficiencies of the two stages be, respectively, η_{c_1} and η_{c_2}; then the work of compression is

$$W_c = W_1 + W_2$$

$$= \frac{1}{\eta_{c_1}}\frac{\gamma}{\gamma-1}RT_1\left[\left(\frac{p_x}{p_1}\right)^{\frac{\gamma-1}{\gamma}} - 1\right] + \frac{1}{\eta_{c_2}}\frac{\gamma}{\gamma-1}RT_y\left[\left(\frac{p_2}{p_y}\right)^{\frac{\gamma-1}{\gamma}} - 1\right]$$

But
$$\left(\frac{p_x}{p_1}\right)^{\frac{\gamma-1}{\gamma}} = \frac{T_x'}{T_1} \text{ and } \left(\frac{p_2}{p_y}\right)^{\frac{\gamma-1}{\gamma}} = \left(\frac{p_2}{p_x}\right)^{\frac{\gamma-1}{\gamma}} = \frac{T_2'}{T_x'}$$

$\therefore$
$$W_c = \frac{\gamma R}{\gamma-1}\left[\frac{T_1}{\eta_{C_1}}\left(\frac{T_x'}{T_1} - 1\right) + \frac{T_y}{\eta_{C_2}}\left(\frac{T_2'}{T_x'} - 1\right)\right]$$

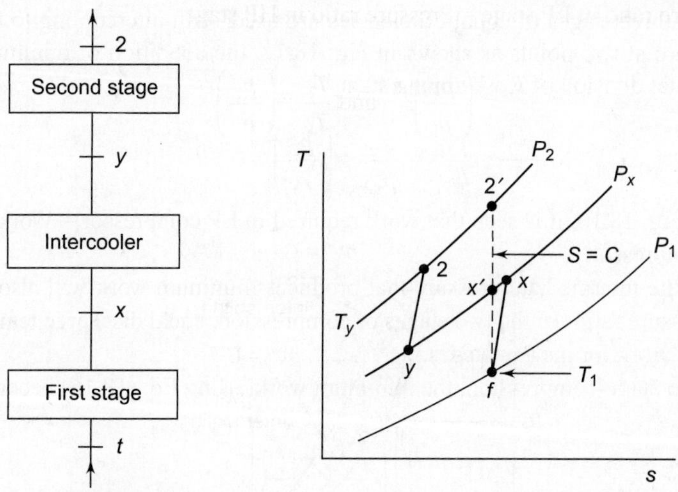

Fig. 18.11 *T-s* plot of two-stage compression process

Taking the derivative with respect to T'_x and setting it equal to zero (noting that T_1, T'_2 and T_y are constant),

$$\frac{dW_c}{dT'_x} = 0 = \frac{\gamma R}{\gamma - 1}\left[\frac{1}{\eta_{C_1}} + \frac{T_y T'_2}{\eta_{C_2}}\left(-\frac{1}{(T'_x)^2}\right)\right]$$

Then,
$$(T'_x)^2 = \frac{\eta_{C_1}}{\eta_{C_2}} T_y T'_2$$

and
$$\frac{T'_x}{T_1} = \sqrt{\frac{\eta_{C_1}}{\eta_{C_2}} \cdot \frac{T_y}{T_1} \cdot \frac{T'_2}{T_1}}$$

for minimum work. Now

$$\frac{T'_x}{T_1} = \left(\frac{p_x}{p_1}\right)^{\frac{\gamma-1}{\gamma}} \text{ and } \frac{T'_2}{T_1} = \left(\frac{p_2}{p_1}\right)^{\frac{\gamma-1}{\gamma}}$$

Therefore, for minimum work in two-stage compression of an ideal gas with inter-cooling to a fixed temperature T_y,

$$\frac{p_x}{p_1} = \sqrt{\left(\frac{\eta_{C_1}}{\eta_{C_2}} \frac{T_y}{T_1}\right)^{\frac{\gamma}{\gamma-1}} \cdot \frac{p_2}{p_1}} \tag{18.22}$$

For the special case of $T_y = T_1$ and $\eta_{c1} = \eta_{c2}$, which is often taken as a standard of comparison, the requirement for minimum work is as obtained earlier in Eq. (18.18).

$$\frac{p_x}{p_1} = \sqrt{\frac{p_2}{p_1}} \tag{18.23}$$

Also for this special case, the condition of minimum work is the condition of equal work in the two stages.

When, three stages of equal efficiencies are used, with intercooling to the initial temperature at two points as shown in Fig. 18.12, the condition of minimum work, and of equal division of work among stages is

$$\frac{p_{x1}}{p_1} = \frac{p_{x2}}{p_{x1}} = \frac{p_2}{p_{x2}} = \left(\frac{p_2}{p_1}\right)^{1/3} \tag{18.24}$$

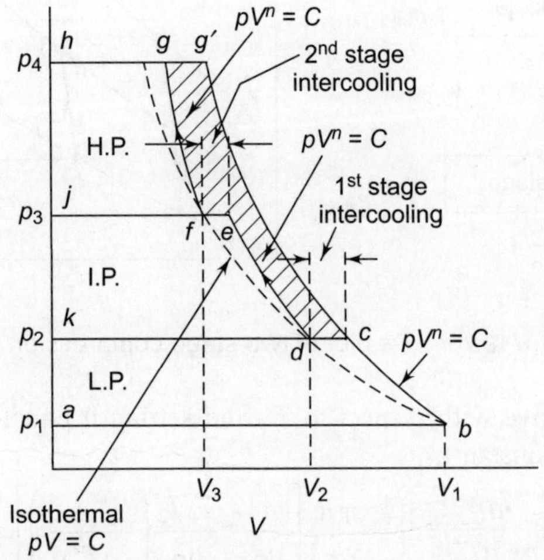

Fig. 18.12 Three-stage compression with perfect intercooling

Thus for three-stage compressor the optimum pressure ratio per stage can be written as

$$\frac{p_x}{p_1} = \left(\frac{p_d}{p_s}\right)^{\frac{1}{3}}$$

where p_d and p_s are the discharge and suction pressure, respectively.
For N-stage compressor, the optimum pressure ratio per stage is

$$\frac{p_x}{p_1} = \left(\frac{p_d}{p_s}\right)^{\frac{1}{N}} \tag{18.25}$$

The minimum work of compression for N-stages is then

$$W_c = \frac{NnRT_1}{n-1}\left[\left(\frac{p_d}{p_s}\right)^{\frac{n-1}{Nn}} - 1\right] \tag{18.26}$$

In the case of gas compression, the desirable idealised process is, however, a reversible isothermal process, and the isothermal efficiency is given by

$$\eta_t = \frac{W_t}{W_c} = \frac{p_1 v_1 \ln \frac{p_2}{p_1}}{W_c} \tag{18.27}$$

The indicator diagram for a three-stage compressor with perfect intercooling is shown in Fig. 18.12. The air delivered from the first stage at *c* is cooled to point *d* on the curve $pV = C$ (shown dotted) before admission to the second stage cylinder. Again the air is cooled from *e* to *f* after compression in the second stage and before admission to the third stage cylinder. The shaded area represents the saving of work resulting from such intercooling since without intercooling the compression curve would have followed the path bcg'.

Advantages of multi-stage compression are:

1. Improved overall volumetric efficiency. If all compression was done in one cylinder the gas in the clearance volume would expand to a large volume before the new intake could begin. This results in a very low volumetric efficiency. By cooling the gas between the stages a much higher efficiency can be obtained.
2. A reduction in work required per stroke, and therefore the total driving power.
3. Size and strength of cylinders can be adjusted to suit volume and pressure of gas.
4. Multi-cylinders give more uniform torque and better mechanical balance thus needing smaller flywheel.
5. Since the maximum temperature reached during the compression process is greatly reduced by intercooling lubrication difficulties and explosion hazards are lessened.
6. Leakage losses are reduced considerably.

   Practise appears to indicate that the economical value of pressure ratio per stage is in the range of 3 to 5. For compression of atmospheric air, single-stage machines are often used up to 550 kPa discharge pressure, two-stage from 350 kPa to 2.1 MPa, three stage from 2.1 MPa to 7.0 MPa and four or more stages for higher pressures.

18.6 AIR MOTORS

LO 18.6

Discuss the *p-V* diagram of an air motor

Compressed air is used in a wide variety of applications where the use of electric motors or SI engines is not permissible and where high safety is required to be met as in mining applications. Pneumatic breakers, picks, spades, rammers, vibrators, riveters, etc. form a range of hand tools having wide use in construction work.

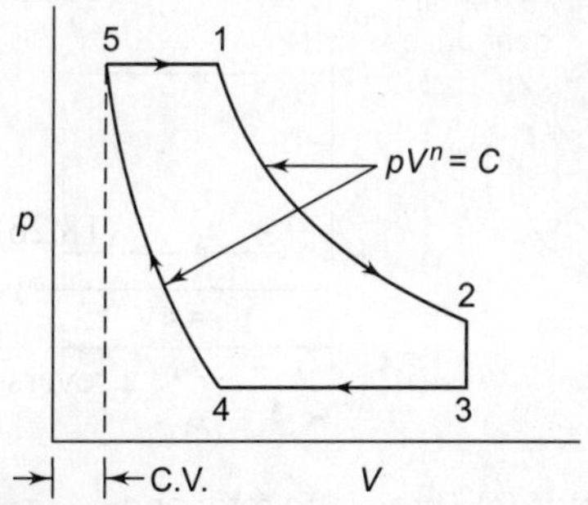

Fig. 18.13 *p-V* diagram for an air motor

Basically the cycle in the reciprocating expander is the reverse of that in the reciprocating compressor (Fig. 18.13). Air is supplied from an air receiver (process 5-1) in which air is at approximately ambient air temperature. The state 1 is the point of cut-off. Air expands in the cylinder according to the low pv^n = constant (where $n \cong 1.3$) pushing the piston out and doing work against the surroundings till the point of release at 2. There is blowdown of air from 2 to 3 and air is exhausted from 3 to 4. The compression of trapped or cushion air begins at 4 and ends at 5 according to pV^n = constant and the compressed air is again admitted to repeat the cycle.

18.7 ROTARY COMPRESSOR

LO 18.7

Compare different rotary compressors

Rotary compressors are used where large quantities of air or gas are needed at relatively low pressure. They may be classified into two main types

1. **Displacement compressors** in which air is compressed by being trapped in the reducing space formed by two sets of engaging surfaces,
2. **Steady-flow compressors** in which compression occurs by the transfer of kinetic energy from a rotor.

Rotary positive displacement machines are generally uncooled and the compression is largely adiabatic. There are two types: Roots blower and vane-type compressor.

Roots blower is an extension of the idea of a gear pump, popular in engines for pumping oil. There are two lobes on each rotor, and their shape is of cycloidal or involute form (Fig. 18.14). One of the lobes is connected to the drive and the second is gear driven from the first, the two rotating in opposite directions. Very small clearances between the lobes and between the casing and the lobes are provided to prevent leakages and to reduce wear. Four times the volume between the casing and one side of the rotor will be displaced in each revolution of the driving shaft. As each side of each lobe faces its side of the casing a volume of gas V, at pressure p_1, is displaced towards the delivery side at constant pressure. A further rotation of the rotor opens this volume to the receiver, and the gas flows back from the receiver, since this gas is at a higher pressure. The gas induced is compressed irreversibly by that from the receiver to the pressure p_2, and then delivery begins. This process is carried out four times per revolution of the driving shaft.

The p-V diagram for this machine is shown in Fig. 18.14(c), in which the pressure rises irreversibly from p_1 to p_2 at constant volume.

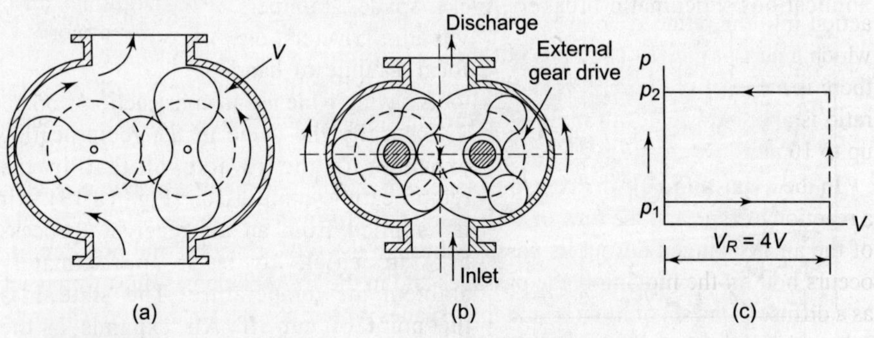

Fig. 18.14 Roots blower with (a) Two lobe rotor, (b) Three lobe rotor and (c) P-V diagram

For pressure ratios of 1.2, 1.6 and 2.0, the Roots blower efficiency becomes 0.945, 0.84 and 0.765, respectively, which shows that the efficiency decreases as the pressure ratio increases.

Vane-type compressor (Fig. 18.15(a)) consists of a rotor mounted eccentrically in the body and supported by ball-and-roller bearings at the ends. The rotor is slotted to take the blades which are of a non-metallic material, usually fibre or carbon. As each blade moves past the inlet passage, compression begins due to decreasing volume between the rotor and casing. Delivery begins with the arrival of each blade at the delivery passage. The *p-V* diagram is shown in Fig. 18.15 (b) where V_s is the induced volume at state p_1, T_1. Compression occurs to the pressure p_i isentropically. At this pressure the displaced gas is opened to the receiver and the gas flowing back from the receiver raises the pressure irreversibly to p_2. The work input is given by the sum of the areas *A* and *B*. Comparing the areas of Figs. 18.14(c) and 18.15(b) it is seen that for a given air flow and given pressure ratio the vane type requires less work input than the Roots blower.

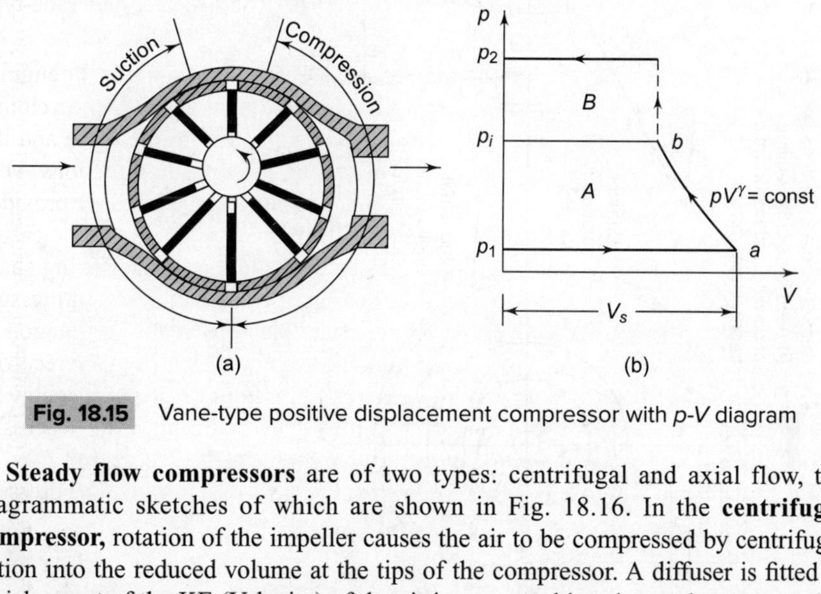

Fig. 18.15 Vane-type positive displacement compressor with *p-V* diagram

Steady flow compressors are of two types: centrifugal and axial flow, the diagrammatic sketches of which are shown in Fig. 18.16. In the **centrifugal compressor,** rotation of the impeller causes the air to be compressed by centrifugal action into the reduced volume at the tips of the compressor. A diffuser is fitted in which a part of the KE (Velocity) of the air is converted into internal energy so that there is a rise of pressure in both the impeller and the diffuser. The typical pressure ratio is about 1.4 to 1, but multi-stage arrangements can provide delivery pressures up to 10 atm.

In the **axial-flow compressor**, the blades are arranged in much the same way as in a reaction turbine and the flow of air is along the axis of the compressor. The velocity of the air is changed during its passage through the sets of blades, and pressure lift occurs both in the moving blade passages and in the fixed blade passages which act as a diffuser. One set of moving and fixed blades represents a stage and a typical pressure ratio per stage is about 1.2. Multi-stage arrangements achieve higher delivery pressures with high efficiency.

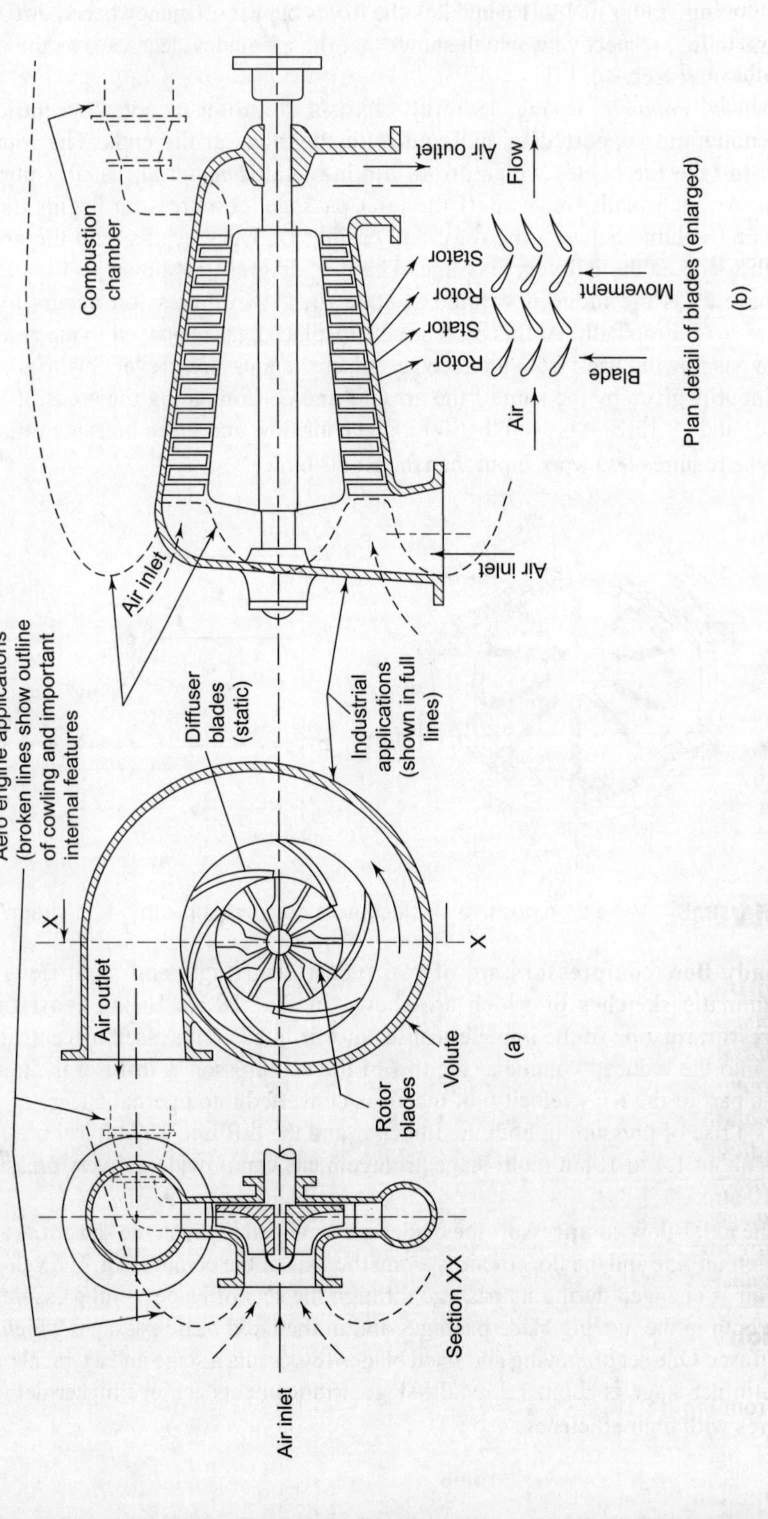

Fig. 18.16 (a) Centrifugal compressor, (b) Axial flow compressor

For uncooled rotary compressors having a necessarily high-speed action, the compression process is regarded as adiabatic while the ideal process is isentropic (reversible and adiabatic) (Fig. 18.17). In an irreversible adiabatic process the extra work which flows into the system to overcome friction increases the temperature and hence enthalpy of the gas. From SFEE neglecting PE and KE terms, the work input is

$$W_c = h_2 - h_1 = c_p (T_2 - T_1) \tag{18.28}$$

where T_2 is the final temperature for the actual compression and the isentropic efficiency η_s is given by

$$\eta_s = \frac{W_s}{W_c} = \frac{h_{2s} - h_1}{h_2 - h_1} = \frac{T_{2s} - T_1}{T_2 - T_1} \tag{18.29}$$

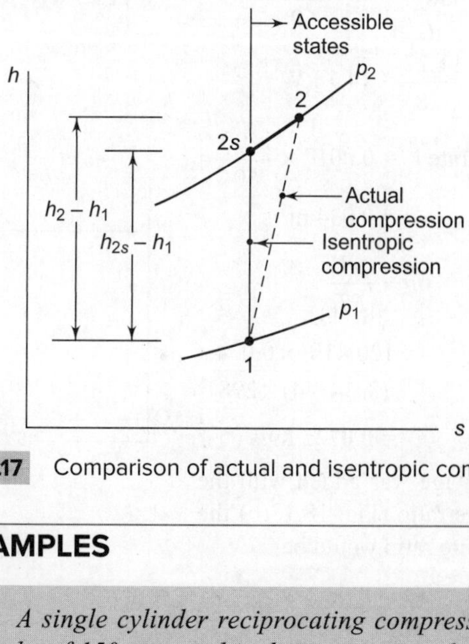

Fig. 18.17 Comparison of actual and isentropic compressions

SOLVED EXAMPLES

Example 18.1 *A single cylinder reciprocating compressor has a bore of 120 mm and a stroke of 150 mm, and is driven at a speed of 1200 rpm. It is compressing CO_2 gas from a pressure of 120 kPa and a temperature of 20°C to a temperature of 215°C. Assuming polytropic compression with n = 1.3, no clearance and volumetric efficiency of 100%, calculate (a) pressure ratio, (b) indicated power, (c) shaft power, with a mechanical efficiency of 80% and (d) mass flow rate.*

If a second stage of equal pressure ratio was added, calculate (e) the overall pressure ratio, and (f) the bore of the second stage cylinder, if the same stroke was maintained. M [LO 18.2]

Solution

(a) From Fig. 18.18(a) $\quad \dfrac{p_2}{p_1} = \left(\dfrac{T_2}{T_1}\right)^{\frac{n}{n-1}} = \left(\dfrac{488}{298}\right)^{\frac{1.3}{0.3}}$

$= \mathbf{8.48}$

(b) $$V_1 = V_S = \frac{\pi}{4}(0.12)^2 \times 0.15 = 0.0017 \text{ m}^3$$

$$W = \frac{n}{n-1} p_1 V_1 \left[\left(\frac{p_2}{p_1}\right)^{\frac{n-1}{n}} - 1\right]$$

$\therefore$ $$W = \frac{1.3}{0.3} \times 120 \times 10^3 \times 0.0017 \left[(8.48)^{\frac{0.3}{1.3}} - 1\right]$$

$$= 563.6 \text{ J}$$

$$\text{Indicated power} = \frac{563.6 \times 1200}{60} \times 10^{-3} = \mathbf{11.27 \text{ kW}}$$

(c) $$\text{Shaft power} = \frac{11.27}{0.8} = \mathbf{14.1 \text{ kW}}$$

(d) $$\text{Volume flow rate } \dot{V} = 0.0017 \times \frac{1200}{60}$$

$$= 0.034 \text{ m}^3/\text{s}$$

$\therefore$ $$\dot{m} = \frac{p_1 \dot{V}_1}{RT_1}$$

$$\frac{120 \times 10^3 \times 0.034}{(8314/44) \times 298}$$

$$= \mathbf{0.0725 \text{ kg/s}}$$

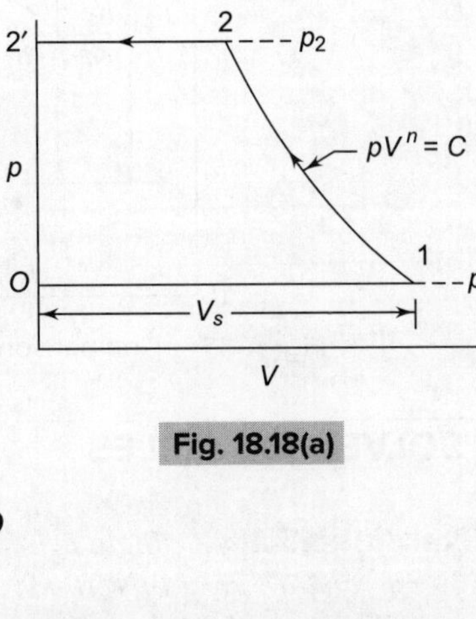

Fig. 18.18(a)

(e) If a second stage was added with the same pressure ratio (Fig. 18.18(b) the overall pressure ratio would be

$$\frac{p_3}{p_1} = \frac{p_3}{p_2} \times \frac{p_2}{p_1} = 8.48 \times 8.48 = \mathbf{71.9}$$

(f) Volume delivered per cycle is V_2

$$p_1 V_1^n = p_2 V_2^n$$

$\therefore$ $$V_2 = \left(\frac{p_1}{p_2}\right)^{\frac{1}{n}} \cdot V_1$$

$$= \left(\frac{1}{8.48}\right)^{\frac{1}{1.3}} \times 0.0017$$

$$= \mathbf{0.00033 \text{ m}^3}$$

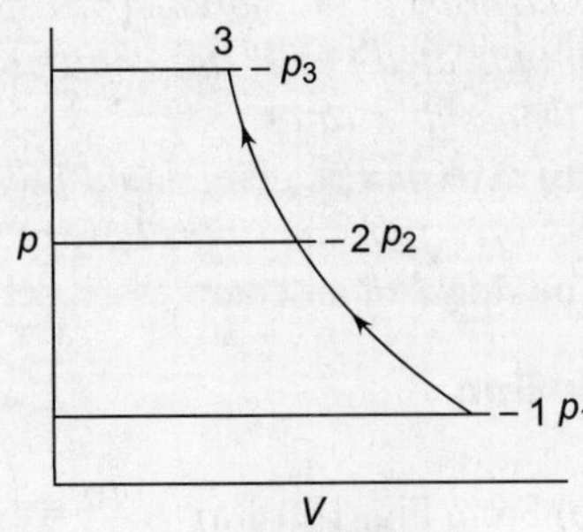

Fig. 18.18(b)

The second stage would have a swept volume of 0.00033 m^3. Since stroke is the same,

$$\frac{\pi}{4} d^2 \times 0.15 = 0.00033 \text{ m}^3$$

$$\therefore \quad d = 0.053 \text{ m} = \mathbf{53 \text{ mm}}$$

Example 18.2 *A single stage reciprocating air compressor has a swept volume of 2000 cm³ and runs at 800 rpm. It operates on a pressure ratio of 8, with a clearance of 5% of the swept volume. Assume NTP room conditions and at inlet (p = 101.3 kPa, t = 15°C), and polytropic compression and expansion with n = 1.25. Calculate (a) indicated power, (b) volumetric efficiency, (c) mass flow rate, (d) FAD, (e) isothermal efficiency and (f) the actual power needed to drive the compressor, if mechanical efficiency is 0.85.* Ⓜ **[LO 18.3]**

Solution

$$p_1 = 101.3 \text{ kPa}, p_2 = 8p_1 = 810.4 \text{ kPa}$$

$$T_1 = 288 \text{ K}, V_S = 2000 \text{ cm}^3$$

$$V_3 = V_c = 0.05\, V_S = 100 \text{ cm}^3$$

$$V_1 = V_c + V_s = 2100 \text{ cm}^3$$

$$p_3 V_3^n = p_4 V_4^n$$

$$\therefore \quad V_4 = \left(\frac{p_3}{p_4}\right)^{\frac{1}{n}} \cdot V_3 = (8)^{\frac{1}{1.25}} \times 100$$

$$= 528 \text{ cm}^3$$

$$V_1 - V_4 = 2100 - 528 = 1572 \text{ cm}^3$$

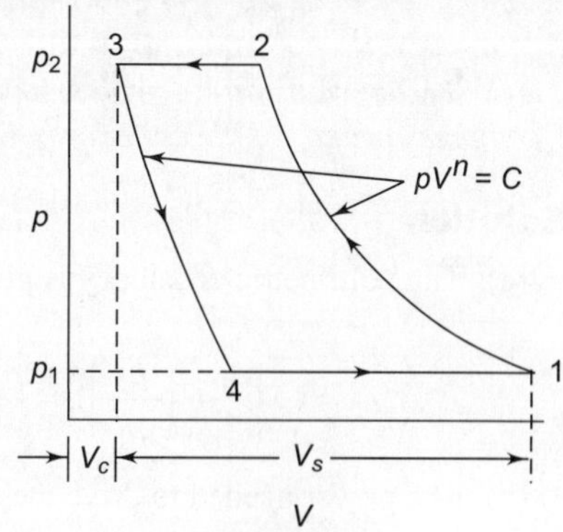

Fig. 18.19

$$W = \frac{n}{n-1} p_1 (V_1 - V_4) \left[\left(\frac{p_2}{p_1}\right)^{\frac{n-1}{n}} - 1\right]$$

$$= \frac{1.25}{0.25} \times 101.3 \times 10^3 \times 1572 \times 10^{-6} \left[(8)^{\frac{0.25}{1.25}} - 1\right] = 411 \text{ J}$$

(a) Indicated power $= \dfrac{411 \times 800 \times 10^{-3}}{60} = \mathbf{5.47 \text{ kW}}$

(b) Volumetric efficiency $= \dfrac{1572}{2000} \times 100 = \mathbf{78.6\%}$

(c) Mass of air compressed per cycle

$$m = \frac{pV}{RT} = \frac{101.3 \times 10^3 \times 1572 \times 10^{-6}}{287 \times 288} = 1.93 \times 10^{-3} \text{ kg}$$

$\therefore$ Mass flow rate $= 1.93 \times 10^{-3} \times 800 = \mathbf{1.54 \text{ kg/min}}$

(d) $\text{FAD} = \text{Free air delivery} = 1572 \times 10^{-6} \times 800 = \mathbf{1.26\ m^3/min}$

(e) $W_t = p_1 (V_1 - V_4) \ln \frac{p_2}{p_1} = 101.3 \times 10^3 \times 1572 \times 10^{-6} \ln 8$

$= 331 \text{ J}$

$$\eta_{\text{isothermal}} = \frac{0.331 \times 800}{5.47 \times 60} = \mathbf{80.7\%}$$

(f) $$\text{Input power} = \frac{5.47}{0.85} = \mathbf{6.44\ kW}$$

Example 18.3 *A single acting single cylinder reciprocating air compressor running at 450 rpm receives air at* 100 kPa, 20°C. *The compressor delivers air at* 600 kPa *at a flow rate of* 0.15 m³/s *measured at suction conditions. The clearance volume is 5% of the swept volume and the stroke/bore ratio is 1.2. The index of compression and expansion is 1.3. The mechanical efficiency of the compressor is 85%. Determine (a) the volumetric efficiency, (b) the power needed to drive the compressor and (c) the cylinder bore and stroke.* Ⓜ **[LO 18.3, 18.4]**

Solution

(a) The volumetric efficiency is given by

$$\eta_{vol} = 1 + C - C\left(\frac{p_2}{p_1}\right)^{\frac{1}{n}} = 1 + 0.05 - 0.05(6)^{\frac{1}{1.3}} = 0.8516 \text{ or } \mathbf{85.16\%}$$

(b) The power needed to drive the compressor is

$$W_{\text{comp}} = \frac{n}{n-1} p_1 (\text{FAD}) \left[\left(\frac{p_2}{p_1}\right)^{(n-1)/n} - 1\right] \frac{1}{\eta_{mech}}$$

$$= \frac{1.3}{1.3-1} \times 100 \times 0.15 \left[\left(\frac{600}{100}\right)^{(1.3-1)/1.3} - 1\right] \frac{1}{0.85}$$

$= \mathbf{39.16\ kW}$

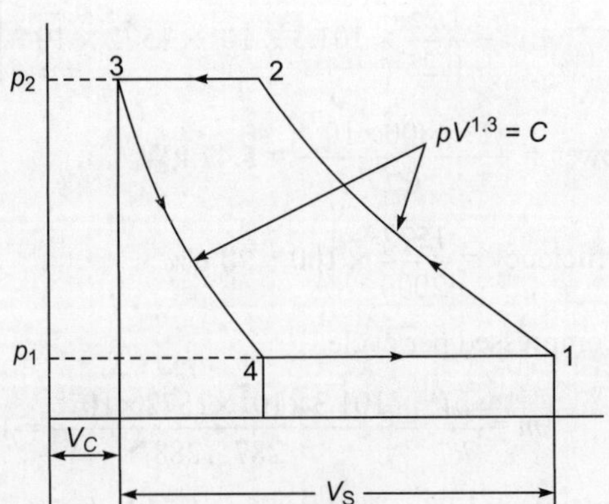

Fig. 18.20

(c) The volumetric efficiency can be expressed as (Fig. 18.20)

$$\eta_v = \frac{V_1 - V_4}{V_s}$$

or $$0.8516 = \frac{0.15}{V_s}$$

or $$V_s = 0.176 \text{ m}^2/\text{s} = \frac{0.176 \times 60}{450} = 0.02347 \text{ m}^3/\text{rev}$$

Now $$V_s = \frac{\pi}{4} D^2 L = 0.02347$$

or $$\frac{\pi}{4} D^2 \times 1.2D = 0.02347$$

or $$D = \mathbf{0.292\ m}$$

∴ $$L = 1.2 \times 0.292 = \mathbf{0.35\ m}$$

Example 18.4 *A single acting reciprocating air compressor deals with 0.01 m³/s of air at 101 kPa and 15°C. The compressor running at 150 rpm receives air at 98 kPa,* 30°C *and delivers at 750 kPa. The bore and stroke of the cylinder are 320 and 400 mm, respectively. The law of compression and expansion is* $pV^{1.25} = \text{constant}$. *The mechanical efficiency of the compressor is 82%. Determine (a) the power needed to drive the compressor (b) the isothermal efficiency (c) the volumetric efficiency and (d) the clearance of the cylinder.*

M **[LO 18.3, 18.4]**

Solution

(a) $$\frac{p_1 V_1}{T_1} = \frac{p_0 V_0}{T_0}$$

or $$\dot{V}_1 = \frac{p_0}{p_1}\frac{T_1}{T_0}\dot{V}_0 = \frac{101}{98} \times \frac{303}{288} \times 0.01 = 0.0108 \text{ m}^3/\text{s}$$

The power needed to drive the compressor is

$$W_{\text{comp}} = \frac{n}{n-1} p_1 \dot{V}_1 \left[\left(\frac{p_2}{p_1}\right)^{(n-1)/n} - 1\right] \frac{1}{\eta_{mech}}$$

$$= \frac{1.25}{1.25-1} \times 98 \times 0.0108 \left[\left(\frac{750}{98}\right)^{(1.25-1)/1.25} - 1\right] \frac{1}{0.82}$$

$$= \mathbf{3.24\ kW}$$

(b) The power requirement using the isothermal compression is

$$\dot{W}_t = p(V_1 - V_4) \ln \frac{p_2}{p_1} \frac{1}{\eta_{mech}} = 98 \times 0.0108 \ln \frac{750}{98} \times \frac{1}{0.82} = 2.627 \text{ kW}$$

Therefore isothermal efficiency is

$$\eta_{\text{isothermal}} = \frac{2.627}{3.24} = 0.8108 \text{ or } \mathbf{81.08\%}$$

(c) Swept volume of the cylinder is

$$V_s = \frac{\pi}{4} D^2 L = \frac{\pi}{4}(0.32)^2 \times 0.4 = 0.03217 \text{ m}^3$$

The volumetric efficiency is (Fig. 18.21)

$$\eta_v = \frac{V_1 - V_4}{V_s} = \frac{\dot{V}_1 \times \frac{N}{60}}{V_s} = \frac{0.0108 \times \frac{150}{60}}{0.03217} = 0.8393 \text{ or } \mathbf{83.93\%}$$

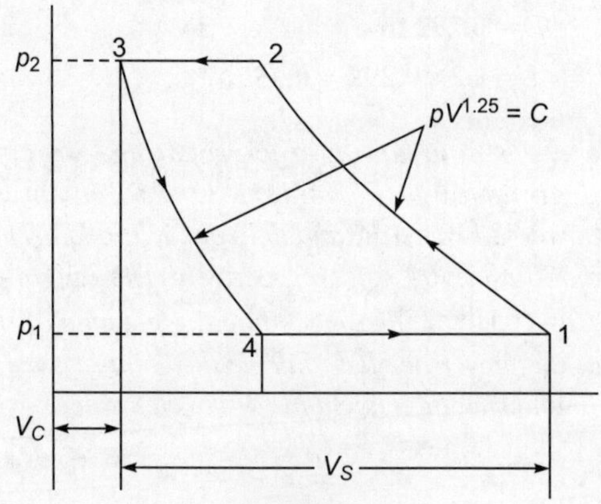

Fig. 18.21

(d) The volumetric efficiency can be expressed as

$$\eta_v = 1 + C - C\left(\frac{p_2}{p_1}\right)^{\frac{1}{n}} = 1 + 0.05 - 0.05(6)^{\frac{1}{1.3}} = 0.8516 \text{ or } 85.16\%$$

or $\quad 0.8393 = 1 + C\left[1 - \left(\frac{750}{98}\right)^{\frac{1}{1.25}}\right] = 1 - 4.094C$

or $\quad C = \mathbf{0.04}$

Therefore, the clearance is 4% of the swept volume.

Example 18.5 *A two-stage air compressor with perfect intercooling takes in air at 1 bar pressure and 27°C. The law of compression in both the stages is $pv^{1.3}$ = constant. The compressed air is delivered at 9 bar from the HP cylinder to an air receiver. Calculate per kilogram of air, the minimum work done, and the heat rejected to the intercooler.* E **[LO 18.5]**

Solution The minimum work required in a two-stage compressor is given by Eq. (18.20),

$$W_c = \frac{2nRT_1}{n-1}\left[\left(\frac{p_2}{p_1}\right)^{n-1/n} - 1\right]$$

$$= \frac{2\times1.3\times0.287\times300}{0.3}[(3)^{0.3/1.3} - 1]$$

$$= 26 \times 0.287 \times 100 \times 0.287 = \mathbf{214.16\ kJ/kg}$$

$$p_2 = \sqrt{p_1 p_3} = \sqrt{1\times9} = 3 \text{ bar}$$

$$\frac{T_2}{T_1} = \left(\frac{p_2}{p_1}\right)^{n-1/n} = 3^{0.3/1.3} = 1.28856$$

$$T_2 = 386.56 \text{ K}$$

Heat rejected to the intercooler

$$= 1.005\ (386.56 - 300) = \mathbf{86.99\ kJ/kg}$$

Example 18.6 *A single-acting two-stage air compressor deals with 4 m³/min of air at 1.013 bar and 15°C with a speed of 250 rpm. The delivery pressure is 80 bar. Assuming complete intercooling, find the minimum power required by the compressor and the bore and stroke of the compressor. Assume a piston speed of 3 m/s, mechanical efficiency of 75% and volumetric efficiency of 80% per stage. Assume the polytropic index of compression in both the stages to be n = 1.25 and neglect clearance.* M **[LO 18.5]**

Solution

$$p_2 = \sqrt{p_1 p_4} = \sqrt{1.013\times80} = 9 \text{ bar}$$

Minimum power required by the compressor

$$\dot{W} = \frac{2n}{n-1}p_1\dot{V}_1\left[\left(\frac{p_2}{p_1}\right)^{n-1/n} - 1\right]\times\frac{1}{\eta_{\text{mech}}}$$

$$= \frac{2\times1.25}{0.25}\times\frac{1.013\times100}{0.75}\times\frac{4}{60}\left[\left(\frac{9}{1.013}\right)^{0.25/1.25} - 1\right]$$

$$= \frac{1013\times4}{45}\times0.548 = \mathbf{49.34\ kW}$$

If L be the stroke length of the piston,

$$2L\frac{N}{60} = 3 \text{ m/s}$$

$$L = \frac{90\times100}{250} = \mathbf{36\ cm}$$

Effective LP swept volume = 4/250 = 0.016 m^3

$$\frac{\pi}{4}(D_{LP})^2 \times 0.36 \times \eta_{vol} = 0.016$$

$$\therefore \quad D_{LP} = \sqrt{\frac{0.016 \times 4}{\pi \times 0.36 \times 0.8}}$$

$$= 0.266 \text{ m} = \mathbf{26.6 \text{ cm}}$$

$$\frac{p_1 V_1}{T_1} = \frac{p_3 V_3}{T_3} \quad \therefore \quad \frac{V_3}{V_1} = \frac{p_1}{p_3}$$

$$\frac{\frac{\pi}{4} D_{HP}^2 L}{\frac{\pi}{4} D_{LP}^2 L} = \frac{1.013}{9}$$

$$D_{HP} = 0.266\sqrt{\frac{1.013}{9}} = 0.892 \text{ m} = \mathbf{8.92 \text{ cm}}$$

Example 18.7 *A reciprocating air compressor is to be designed to compress 6 kg/min of air from 100 kPa and 20°C. The compressor delivers compressed air at 1.8 MPa. The index of compression and expansion, n, is equal to 1.3. Calculate the savings in power consumption and gain in isothermal efficiency, when (a) a two-stage and (b) a three-stage compressor is used in place of a single stage compressor. For two- and three-stages assume there is perfect intercooling in the intermediate stages.* Ⓜ **[LO 18.3, 18.4, 18.5]**

Solution For a single-stage, the required power is

$$\dot{W} = \frac{n}{n-1}\dot{m}RT_1\left[\left(\frac{p_2}{p_1}\right)^{\frac{n-1}{n}} - 1\right]$$

$$= \frac{1.3}{1.3-1} \times \frac{6}{60} \times 0.287 \times 293\left[\left(\frac{1800}{100}\right)^{\frac{0.3}{1.3}} - 1\right] = 34.56 \text{ kW}$$

The power requirement using the isothermal compression is

$$\dot{W}_t = \dot{m}RT_1 \ln\frac{p_2}{p_1} = \frac{6}{60} \times 0.287 \times 293 \ln\frac{1800}{100} = 24.31 \text{ kW}$$

Therefore, isothermal efficiency is

$$\eta_{isothermal} = \frac{24.31}{34.56} = 0.7034 \text{ or } 70.34\%$$

(a) The power required for a two-stage compressor with perfect intercooler is

$$\dot{W}_{2\text{ stage}} = \frac{2n}{n-1}\dot{m}RT_1\left[\left(\frac{p_2}{p_1}\right)^{\frac{n-1}{2n}} - 1\right]$$

$$= \frac{2\times1.3}{1.3-1}\times\frac{6}{60}\times0.287\times293\left[\left(\frac{1800}{100}\right)^{\frac{0.3}{2\times1.3}} - 1\right] = 28.85 \text{ kW}$$

The savings in power consumption is $= \dfrac{34.56-28.85}{34.56} = 0.1652$ or **16.52%**

The power requirement using the isothermal compression for a two-stage compressor with perfect intercooler is

$$\dot{W}_t = \dot{m}RT_1 \ln\frac{p_2}{p_1} = \frac{6}{60}\times0.287\times293\ln\frac{1800}{100} = 24.31 \text{ kW}$$

Therefore, isothermal efficiency is

$$\eta_{\text{isothermal}} = \frac{24.31}{28.85} = 0.8426 \text{ or } 84.26\%$$

The gain in isothermal efficiency is $= \dfrac{0.8426-0.7034}{0.7034} = 0.1979$ or **19.79%**

(b) The power required for a three-stage compressor with perfect intercooler is

$$\dot{W}_{3\text{ stage}} = \frac{3n}{n-1}\dot{m}RT_1\left[\left(\frac{p_2}{p_1}\right)^{\frac{n-1}{3n}} - 1\right]$$

$$= \frac{3\times1.3}{1.3-1}\times\frac{6}{60}\times0.287\times293\left[\left(\frac{1800}{100}\right)^{\frac{0.3}{3\times1.3}} - 1\right] = 27.22 \text{ kW}$$

The savings in power consumption with respect to single-stage compressor is

$$= \frac{34.56-27.22}{34.56} = 0.2124 \text{ or } \mathbf{21.24\%}$$

The power requirement using the isothermal compression for a three-stage compressor with perfect intercooler is

$$\dot{W}_t = \dot{m}RT_1 \ln\frac{p_2}{p_1} = \frac{6}{60}\times0.287\times293\ln\frac{1800}{100} = 24.31 \text{ kW}$$

Therefore isothermal efficiency is

$$\eta_{\text{isothermal}} = \frac{24.31}{27.22} = 0.8931 \text{ or } 89.31\%$$

The gain in isothermal efficiency is $= \dfrac{0.8931 - 0.7034}{0.7034} = 0.2697$ or **26.97%**

Example 18.8 *The pressure and temperature of air supplied to an air engine are 700 kPa and 38°C, respectively. Cut-off takes place at 0.4 of the stroke. Expansion follows the law $pv^{1.3} = c$ (constant) to the release point, which is at the end of the outstroke. The pressure then falls to the constant back pressure of 112 kPa. Neglect the effect of clearance and assuming that the area of the actual indicator diagram is 0.85 of that outlined above, determine the indicated output if the air mass is 1.25 kg.* M **[LO 18.6]**

Solution From Fig. 18.22,

$$V_1 = 0.4\ V_2,\ \frac{T_1}{T_2} = \left(\frac{V_2}{V_1}\right)^{n-1}$$

$$= (2.5)^{0.3} = 1.316$$

$$\therefore \quad T_2 = \frac{273+38}{1.316} = 236.3 \text{ K}$$

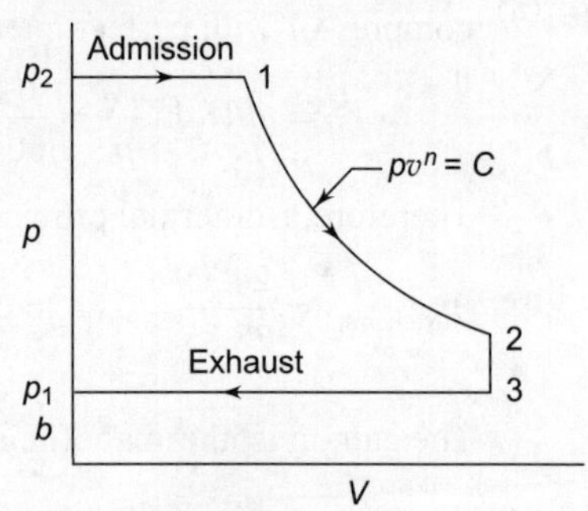

Fig. 18.22

$$p_1 V_1^{1.3} = p_2 V_2^{1.3}$$

$$\therefore \quad p_2 = p_1 \cdot \left(\frac{V_1}{V_2}\right)^{1.3} = 700(0.4)^{1.3}$$

$$= 212.7 \text{ kPa}$$

$$v_2 = \frac{mRT_2}{p_2} = \frac{1.25 \times 0.287 \times 236.3}{212.7} = 0.3986 \text{ m}^3$$

$$v_2 = \frac{V_2}{m} = \frac{0.3986}{1.25} = 0.3188 \text{ m}^3/\text{kg}$$

Area of indicator diagram

$$= p_1 v_1 + \frac{p_1 v_1 - p_2 v_2}{n-1} - p_3 v_3$$

$$= RT_1 + \frac{R(T_1 - T_2)}{n-1} - p_3 v_3$$

$$= 0.287 \times 311 + \frac{0.287(311 - 236.3)}{0.3} - 112 \times 0.3188$$

$$= 125 \text{ kJ/kg}$$

$$\text{Indicated output} = 125 \times 0.85 = 106.26 \text{ kJ/kg}$$

$$= 106.26 \times 1.25 = \mathbf{132.82\ kJ}$$

Example 18.9 *A three-stage single-acting reciprocating air compressor has a low-pressure (LP) cylinder of 450 mm bore and 300 mm stroke. The clearance volume of the LP cylinder is 5% of the swept volume. Intake pressure and temperature are 1 bar and 18°C, respectively, while the delivery pressure is 15 bar. Intermediate pressures are ideal and intercooling is perfect. The compression and expansion index can be taken as 1.3. Estimate (a) the intermediate pressures, (b) the effective swept volume of LP cylinder, (c) the temperature and volume of air delivered per stroke at 15 bar and (d) the work done per kg of air. Take R = 0.29 kJ/kg K.* D [LO 18.5]

Solution

(a) $$\frac{p_2}{p_1} = \frac{p_3}{p_2} = \frac{p_4}{p_3} = \left(\frac{p_4}{p_1}\right)^{1/3} = (15)^{1/3} = 2.466$$

$$p_2 = 2.466 \text{ bar}$$
$$p_3 = 2.466 \times 2.466 = \mathbf{6.081\ bar}$$

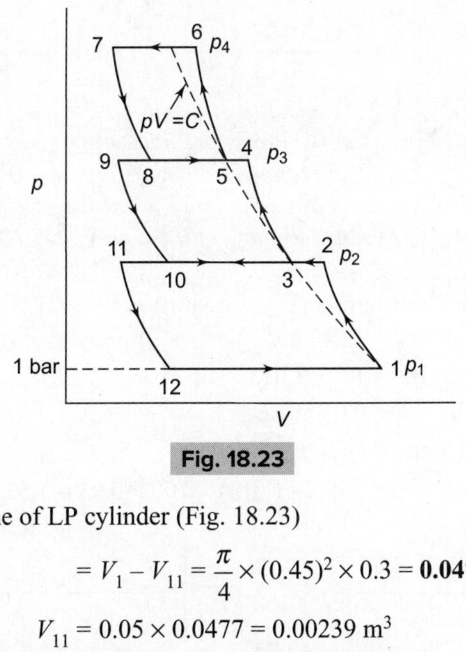

Fig. 18.23

(b) Swept volume of LP cylinder (Fig. 18.23)

$$= V_1 - V_{11} = \frac{\pi}{4} \times (0.45)^2 \times 0.3 = \mathbf{0.0477\ m^3}$$

$$V_{11} = 0.05 \times 0.0477 = 0.00239 \text{ m}^3$$

$$\therefore \quad V_1 = 0.0477 + 0.00239 = 0.05009 \text{ m}^3$$

$$p_{11}V_{11}^{1.3} = p_{12}V_{12}^{1.3}$$

$$\therefore \quad V_{12} = V_{11} \times \left(\frac{p_{11}}{p_{12}}\right)^{1/1.3} = 0.00239 \times (2.466)^{1/1.3} = 0.00478 \text{ m}^3$$

Effective swept volume of the LP cylinder is

$$V_1 - V_{12} = 0.05009 - 0.00478 = \mathbf{0.04531\ m^3}$$

(c) $T_5 = T_3 = T_1 = 291$ K

$$\frac{T_6}{T_5} = \left(\frac{p_4}{p_3}\right)^{\frac{n-1}{n}}$$

$$\therefore \quad T_6 = 291 \times (2.466)^{\frac{0.3}{1.3}} = 358.5 \text{ K}$$

$$\therefore \quad t_6 = \text{delivery temperature} = \mathbf{85.5°C}$$

$$\frac{p_4(V_6 - V_7)}{T_6} = \frac{p_1(V_1 - V_{12})}{T_1}$$

$$\therefore \quad V_6 - V_7 = \frac{p_1}{p_6} \cdot \frac{T_6}{T_1} \cdot (V_1 - V_{12}) = \frac{1}{15} \times \frac{358.5}{291} \times 0.04531$$

$$= \mathbf{0.00372\ m^3}$$

(d) Work per kg air

$$= \frac{3nRT_1}{n-1}\left[\left(\frac{p_2}{p_1}\right)^{\frac{n-1}{n}} - 1\right]$$

$$= \frac{3 \times 1.3 \times 0.29 \times 291}{0.3}\left[(2.466)^{\frac{0.3}{1.3}} - 1\right] = \mathbf{254.3\ kJ}$$

Example 18.10 *For a Roots blower, the inlet pressure is 1.013 bar and the pressure ratio is 1.5 to 1. The induced volume of air is 0.03 m³/rev. Estimate the work input. What would be the work input for a vane-type compressor if the internal compression takes place through half the pressure range.* M **[LO 18.7]**

Solution For the Roots blower (Fig. 18.24(a))

$$p_1 = 1.013 \text{ bar}$$

$$p_2 = 1.5 \times 1.013 = 1.52 \text{ bar}$$

$$V_s = 0.03 \text{ m}^3/\text{rev.}$$

$$\therefore \quad \text{Work done} = (p_2 - p_1)\, V_s$$

$$= (1.52 - 1.013) \times 100 \times 0.03 = \mathbf{1.52\ kJ/rev}$$

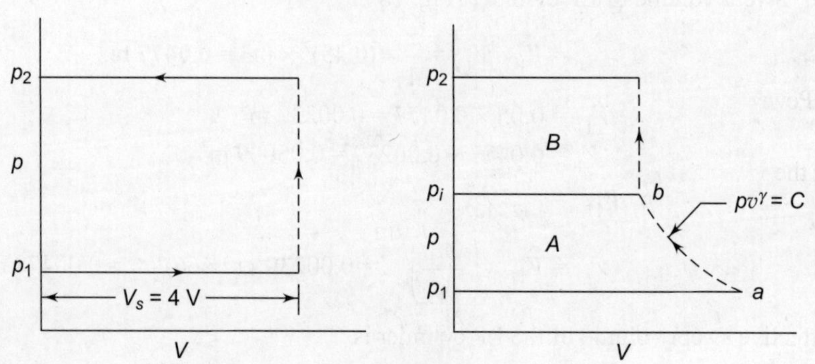

Fig. 18.24

For the vane-type compressor (Fig. 18.24(b))

$$p_i = \frac{(1.5 \times 1.013) + 1.013}{2} = 1.266 \text{ bar}$$

Work required = Area A + Area B

$$\text{Area } A = \frac{\gamma}{\gamma - 1} p_1 V_s \left[\left(\frac{p_i}{p_1} \right)^{\frac{\gamma-1}{\gamma}} - 1 \right]$$

$$= \frac{1.4}{0.4} \times 1.013 \times 100 \times 0.03 \left[\left(\frac{1.266}{1.013} \right)^{\frac{0.4}{1.4}} - 1 \right]$$

$$= 0.70 \text{ kJ/rev}$$

$$\text{Area } B = (p_2 - p_i)\, V_b$$

$$p_1\, V_a^\gamma = p_i\, V_b^\gamma$$

$$\therefore \quad V_b = V_a \cdot \left(\frac{p_1}{p_i} \right)^{\frac{1}{\gamma}} = 0.03 \left(\frac{1.013}{1.266} \right)^{\frac{1}{1.4}} = 0.0256 \text{ m}^3$$

$$\text{Area } B = (1.52 - 1.266) \times 100 \times 0.0256 = 0.65 \text{ kJ/rev}$$

$\therefore$ Work required = 0.70 + 0.65 = **1.35 kJ/rev.**

(Compared to 1.52 kJ/rev. for the Roots blower)

Example 18.11 *A Roots blower supplies air at the rate of 1 kg/s. The pressure ratio of the blower is 2:1 with an intake pressure and temperature of 1 bar and 70°C, respectively. Find the power required to drive the blower. Take R = 0.29 kJ/kg K.*

If a vane pump having the same air flow, pressure ratio and intake conditions has the volume reduced to 0.7 of the intake volume before delivering the air, estimate the power required. M [LO 18.7]

Solution

$$\dot{V} = \frac{\dot{m} R T_1}{p_1} = \frac{1 \times 0.29 \times 343}{100} = 0.995 \text{ m}^3\text{/s}$$

$\therefore$ Power required by the Roots blower

$$= \dot{V}\,(p_2 - p_1) = 0.995 \times 100 = \mathbf{99.5\ kW}$$

For the vane compressor,

$$p_1\, V_1^\gamma = p_2\, V_2^\gamma$$

$$\therefore \quad p_2 = p_1 \left(\frac{V_1}{V_2} \right)^\gamma = 1 \times \left(\frac{1}{0.7} \right)^{1.4} = 1.65 \text{ bar}$$

$$\dot{V}_2 = 0.7\, \dot{V}_1 = 0.7 \times 0.995 = 0.696 \text{ m}^3\text{/s}$$

$$\text{Power required} = \frac{\gamma}{\gamma-1} p_1\dot{V}_1\left[\left(\frac{p_2}{p_1}\right)^{\frac{\gamma-1}{\gamma}} - 1\right] + \dot{V}_2(p_3 - p_2)$$

$$= \frac{1.4}{0.4}\times 100\times 0.995\left[\left(\frac{1.65}{1}\right)^{\frac{0.4}{1.4}} - 1\right]$$

$$+ 0.696\,(2 - 1.65)\times 100 = \mathbf{78\ kW}$$

Example 18.12 *A gas turbine utilises a two-stage centrifugal compressor. The pressure ratios for the first and second stages are 2.5 to 1 and 2.1 to 1, respectively. The flow of air is 5 kg/s, this air being drawn at 1.013 bar, and 10°C. If the temperature drop in the intercooler is 50°C and the isentropic efficiency is 85% for each stage, calculate (a) the actual temperatures at the end of each stage, and (b) the total compressor power. Take* $\gamma = 1.4$ *and* $c_p = 1.005$ *kJ/kg K.*

M [LO 18.7]

Solution

(a) *Stage I* $\quad \dfrac{T_{2s}}{T_1} = \left(\dfrac{p_2}{p_1}\right)^{\frac{\gamma-1}{\gamma}}$ (Fig. 18.25)

$$T_{2s} = 283\,(2.5)^{0.4/1.4} = 367.7\text{ K}$$

$$T_2 - T_3 = 50°\text{C}$$

$$T_{2s} - T_1 = \eta_s\,(T_2 - T_1)$$

$$T_2 = T_1 + \frac{T_{2s} - T_1}{\eta_s}$$

$$= 283 + \frac{367.7 - 283}{0.85} = \mathbf{382.65\ K}$$

Stage II $\quad T_3 = T_2 - 50 = 382.65 - 50$

$$= 332.65\text{ K}$$

$$\frac{T_{4s}}{T_3} = \left(\frac{p_3}{p_2}\right)^{\frac{\gamma-1}{\gamma}}$$

$$\therefore \quad T_{4s} = 332.65\times(2.1)^{0.4/1.4}$$

$$= 411.19\text{ K}$$

$$T_4 - T_3 = \frac{T_{4s} - T_3}{\eta_s}$$

$$= \frac{411.19 - 322.65}{0.85}$$

$$= 104.17\text{ K}$$

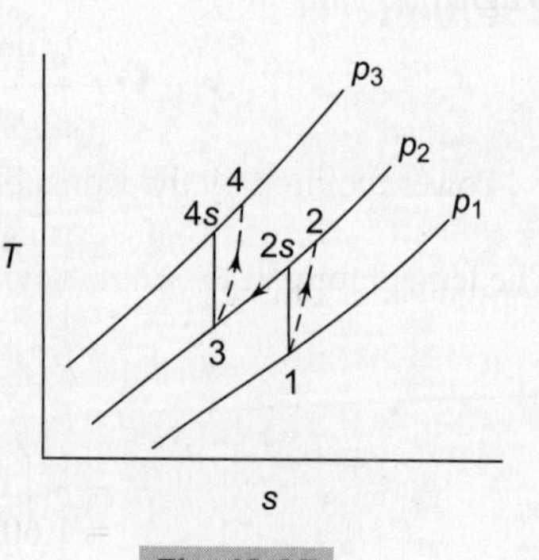

Fig. 18.25

$\therefore \qquad T_4 = 104.17 + 332.65 = \mathbf{436.82\ K}$

(b) Total compressor power $= \dot{m}_a c_p\,[(T_2 - T_1) + (T_4 - T_3)]$

$= 5 \times 1.005\,[(382.65 - 283) + (436.82 - 332.65)]$

$= \mathbf{1024.2\ kW}$

Example 18.13 *A rotary compressor is used to supercharge a petrol engine. The static pressure ratio across the rotor is 2.5:1. Static inlet pressure and temperature are 0.6 bar and 5°C, respectively. The air-fuel ratio is 13:1 and the engine consumes 0.04 kg fuel/s. For the air-fuel mixture take γ = 1.39 and c_p = 1.005 kJ/kg K. The isentropic efficiency of the compressor is 84%. Estimate the power required to drive the compressor. Taking the exit velocity from the compressor as 120 m/s and assuming that the mixture is adiabatically brought to rest in the engine cylinders, estimate the stagnation temperature and pressure of the mixture at the beginning of the compression stroke in the engine cylinder.*

Ⓜ **[LO 18.7]**

Solution

$$\frac{T_{2s}}{T_1} = \left(\frac{p_2}{p_1}\right)^{\frac{\gamma-1}{\gamma}} = (2.5)^{\frac{0.39}{1.39}} = 1.294$$

$$T_{2s} = 278 \times 1.294 = 359.7\ \text{K}$$

$$T_2 - T_1 = \frac{T_{2s} - T_1}{\eta_s} = \frac{359.7 - 278}{0.84} = 97.3\ \text{K}$$

$$\therefore \qquad T_2 = 97.3 + 278 = 375.3\ \text{K}$$

Mass of air-fuel mixture = 0.04 × (13 + 1) = 0.56 kg/s

Power to drive the compressor $= \dot{m}_g\, c_p\, (T_2 - T_1)$

$= 0.56 \times 1.005 \times 97.3$

$= \mathbf{54.8\ kW}$

Stagnation temperature

$$T_{02} = T_2 + \frac{V_2^2}{2c_p} = 375.3 + \frac{120^2}{2 \times 1.005 \times 1000}$$

$$= 382.46\ \text{K} = \mathbf{109.46°C}$$

The temperature in the engine cylinder is 109.46°C

$$p_{02} = p_2 \times \left(\frac{T_{02}}{T_2}\right)^{\frac{\gamma}{\gamma-1}} = 0.6 \times 2.5 \times \left(\frac{382.46}{375.3}\right)^{0.36}$$

$= 1.605$ bar or **160.5 kPa**

SUMMARY

A gas compressor is a device in which work is done on the gas to raise its pressure with an increase in its density. The adiabatic steady flow work input to the gas

$$W_s = -\int_1^2 v dp = \frac{\gamma}{\gamma - 1} p_1 v_1 \left[\left(\frac{p_2}{p_1} \right)^{\frac{\gamma-1}{\gamma}} - 1 \right]$$

For polytrophic process γ is replaced by n.
The efficiency of a compressor is given by

$$\eta = \frac{W_S}{W_c} = \frac{h_{2S} - h_i}{W_c}$$

where W_c is the shaft work input per unit mass.
For the same p_2/p_1, the isothermal compression needs the minimum work, whereas adiabatic compression needs the maximum work, while the polytrophic compression needing work in between the two. The isothermal efficiency of the compressor is

$$\eta_t = \frac{W_t}{W_c}$$

where W_t is the work of reversible isothermal compression. Many turbine type compressors are essentially adiabatic machine due to their high speeds and their efficiency is

$$\eta_c = \frac{W_S}{W_c} = \frac{h_{2S} - h_1}{h_2 - h_1}$$

The ratio of actual volume of gas taken into the cylinder during suction stroke to the piston displacement (PD) or the swept volume of the piston is called the *volumetric efficiency*.

$$\eta_{\text{vol}} = \frac{mv_1}{V_S}$$

The clearance volumetric efficiency is

$$\eta_{\text{vol}} = 1 + C - C \left(\frac{p_2}{p_1} \right)^{\frac{1}{n}}$$

where C is the *clearance*, the ratio of the clearance and swept volumes.
To reduce the work of compression and enhance its volumetric efficiency, the compression is staged with intercooling. For a two-stage compression process, the intermediate pressure p_x for minimum work

$$p_x = \sqrt{p_1 p_2} \text{ or}$$

$$\frac{p_x}{p_1} = \frac{p_2}{P_x} = \left(\frac{p_2}{p_1} \right)^{\frac{1}{2}}$$

Also, $T_2 = T_x$ and the minimum work is

$$W_x = \frac{2nRT_1}{n-1} \left[\left(\frac{p_x}{p_1} \right)^{\frac{n-1}{n}} - 1 \right] = \frac{2nRT_1}{n-1} \left[\left(\frac{p_2}{p_1} \right)^{\frac{n-1}{2n}} - 1 \right]$$

where $\frac{p_x}{p_1}$ is the pressure ratio in each stage.

For N–stages of compression

$$(W_x)_{\min} = \frac{NnRT_1}{n-1}\left[\left(\frac{p_2}{p_1}\right)^{\frac{n-1}{Nn}} - 1\right]$$

Air motors are used when safety is required as in mines. Pneumatic breakers, picks, spades, riveters, etc. use them in various construction works. Rotary compressors are used where large quantities of air are needed Roots blower is used often for pumping gas or air as a gear pump. Centrifugal and axial flow compressors are used in gas turbine power plants.

REVIEW QUESTIONS

(E) **18.1** What is the function of a compressor? What are the different types of compressors? **[LO 18.1]**

(E) **18.2** What are the main applications of compressors? **[LO 18.1]**

(D) **18.3** Derive the expressions for the reversible work of compression if the compression process is (a) adiabatic, (b) polytropic and (c) isothermal. **[LO 18.2]**

(E) **18.4** Which of the compression processes needs minimum work and which the maximum work? **[LO 18.2]**

(M) **18.5** Define (a) adiabatic efficiency and (b) isothermal efficiency. **[LO 18.2]**

(D) **18.6** Show that for a single-stage reciprocating compressor the work of compression remains the same irrespective of its derivation on the basis of (a) a steady flow system and (b) a closed system. **[LO 18.3]**

(M) **18.7** Define the volumetric efficiency of a compressor. On what factors does it depend? **[LO 18.4]**

(M) **18.8** Discuss how the volumetric efficiency varies with the clearance and the pressure ratio. For a given pressure ratio and the polytropic index, find the maximum clearance when the volumetric efficiency is reduced to zero. **[LO 18.4]**

(M) **18.9** What is the maximum pressure ratio attainable with a reciprocating compressor for a given clearance? **[LO 18.4]**

(E) **18.10** What is the need of staging the compression process? **[LO 18.5]**

(M) **18.11** Show that the optimum intermediate pressure of a two-stage reciprocating compressor for minimum work is the geometric mean of the suction and discharge pressures. **[LO 18.5]**

(D) **18.12** Explain how the use of intermediate pressure for minimum work results in equal pressure ratios in the two-stages of compression, equal discharge temperatures, and equal work for the two stages. **[LO 18.5]**

(M) **18.13** What do you mean by perfect intercooling? What is the amount of heat rejected in the intercooler? **[LO 18.5]**

(E) **18.14** What is the function of an aftercooler? **[LO 18.5]**

(E) **18.15** Explain the advantages of multistage compression. **[LO 18.5]**

(M) **18.16** Explain the operation of an air motor. **[LO 18.6]**

(D) **18.17** What are rotary compressors? How does the Roots blower operate? What is the power input to it? Define Roots blower efficiency. **[LO 18.7]**

(M) **18.18** What is a vane-type compressor? Briefly explain its operation. **[LO 18.7]**

(M) **18.19** How does the pressure rise in a centrifugal compressor? Where is it used? **[LO 18.7]**

(M) **18.20** How does an axial flow compressor operate? **[LO 18.7]**

PROBLEMS WITH HINTS

(E) **18.1** Air is compressed steadily by a reversible compressor from an inlet state of 100 kPa, 300 K to an exit pressure of 900 kPa. Determine the compressor work per unit mass for (a) isentropic compressor with $\gamma = 1.4$, (b) polytropic compression with $n = 1.3$, (c) isothermal compression and (d) ideal two-stage compression with intercooling and with $n = 1.3$. **[LO 18.2, 18.5]**

Hint:

(a) $W_s = \frac{\gamma RT_1}{\gamma - 1}\left[\left(\frac{p_2}{p_1}\right)^{\frac{\gamma-1}{\gamma}} - 1\right] = \mathbf{262.33\ kJ/kg}$

(b) $W_n = \frac{1.3 \times 0.287 \times 300}{1.3 - 1}\left[(9)^{\frac{1.3-1}{1.3}} - 1\right]$

$= \mathbf{246.4\ kJ/kg}$

(c) $W_T = RT \ln \frac{p_2}{p_1} = \mathbf{189.2\ kJ/kg}$

(d) $p_i = \sqrt{p_1 p_2} = 300$ kPa.

$\therefore \quad W_{2-5} = 2 \times$ compressor work for a single stage.

$$= \frac{2 \times 1.3 \times 0.287 \times 300}{1.3 - 1}\left[\left(\frac{300}{100}\right)^{\frac{1.3-1}{1.3}} - 1\right]$$

$$= \mathbf{215.3\ kJ/kg}$$

Isothermal compression requires minimum work while the isentropic compression requires maximum work. Compressor work is decreased with two stages of compression instead of one.

(E) **18.2** Air is compressed in a centrifugal compressor ($\gamma = 1.4$) from 110 to 300 kPa. The compression efficiency is 0.90. Determine the work per unit mass of air using the ideal gas low. **[LO 18.7]**

Hint:

$$\frac{T_{2S}}{T_1} = \left(\frac{p_2}{p_1}\right)^{\frac{\gamma-1}{\gamma}} \rightarrow T_{2S} = 410.62 \text{ K}$$

$$\eta = 0.9 = \frac{T_{2S} - T_1}{T_2 - T_1} \rightarrow T_2 = 422.91 \text{ K}$$

$$\therefore \quad W_c = c_p (T_2 - T_1) = \mathbf{123.5\ kJ/kg}$$

PROBLEMS

Ⓔ **18.1** A four-cylinder single stage air compressor has a bore of 200 mm and a stroke of 300 mm and runs at 400 rpm. At a working pressure of 620 kPa, it delivers 3.1 m^3 of air per min. at 270°C. Calculate (a) mass flow rate, (b) free air delivery (FAD), (c) effective swept volume and (d) volumetric efficiency. Take free air conditions at inlet as 101.3 kPa, 21°C. **[LO 18.3, 18.4]**

Ⓔ **18.2** 0.2 m^3 of air at 20°C and 100 kPa is compressed according to the relation $pv^{1.3}$ = constant by the piston in an engine cylinder that has a compression ratio of 6. Heat is then added while the pressure remains the same, until the piston returns to its original position. Calculate (a) mass of air, (b) pressure at the end of compression, (c) final temperature and (d) network transfer in the combined process. **[LO 18.2]**

Ⓔ **18.3** A single stage single-acting air compressor deals with 90 m^3/h of air at 101.325 kPa and 15°C. The pressure and temperature during the suction stroke remain constant at 98 kPa and 40°C, respectively, $n = 1.22$. The air is delivered at 735 kPa, $R_a = 0.287$ kJ/kg K. Find (a) the power needed to drive the compressor if the mechanical efficiency is 0.85, (b) the swept volume if the speed is 120 rpm. Take the volumetric efficiency as 0.78. **[LO 18.3, 18.4]**

Ⓔ **18.4** Find the stroke, piston diameter and indicated power of a single-acting air compressor, which operates under the following conditions : volume of FAD at 101.325 kPa and 15°C = 105 m^3/min; pressure and temperature at the beginning of compression, 98 kPa and 30°C; discharge pressure is 40 kPa; speed 220 rpm; $n = 1.25$; stroke = bore and the clearance volume is 6% of the swept volume. **[LO 18.3, 18.4]**

Ⓜ **18.5** A two-stage single-acting compressor compresses air from 101.325 kPa and 15°C to a pressure of 162 kPa. Calculate the work done per kg of air delivered, and the heat transferred to the intercooler if the intercooling is ideal.

If the volumetric efficiency is expected to be 85% for the LP stage and the speed is 400 rpm, calculate a suitable diameter for the LP cylinder for an air delivery of 2.5 kg/min if the stroke-bore ratio is unity. Assume the polytropic index to be 1.3 and state all the assumptions made. For air, $R = 0.287$ kJ/kg K and $c_p = 1.005$ kJ/kg K. **[LO 18.4, 18.5]**

Ⓜ **18.6** Air flows steadily into a compressor at a temperature of 17°C and a pressure of 1.05 bar and leaves at a temperature of 247°C and a pressure of 6.3 bar. There is no heat transfer to or from the air as it flows through the compressor; changes in elevation and velocity are negligible. Evaluate the external work done per kg of air, assuming air as an ideal gas for which $R = 0.287$ kJ/kg K and $\gamma = 1.4$. Evaluate the minimum external work required to compress the air adiabatically from the same initial state to the same final pressure and the isentropic efficiency of the compressor. **[LO 18.2]**

Ⓜ **18.7** A slow-speed reciprocating air compressor with a water jacket for cooling approximates a quasi-static compression process following a path $pv^{1.3}$ = const. If air enters at a temperature of 20°C and a pressure of 1 bar, and is compressed to 6 bar at a rate of 1000 kg/h, determine the discharge temperature of air, the power required and the heat transferred per kg. **[LO 18.2]**

Ⓔ **18.8** A single-acting two-stage reciprocating air compressor with complete intercooling delivers 6 kg/min at 15 bar pressure. Assume an intake condition of 1 bar and 15°C and that the compression and expansion processes are polytropic with $n = 1.3$. Calculate (a) the power required and (b) the isothermal efficiency. **[LO 18.2, 18.5]**

(M) **18.9** A two-stage air compressor receives 0.238 m^3/s of air at 1 bar and 27°C and discharges it at 10 bar. The polytropic index of compression is 1.35. Determine (a) the minimum power necessary for compression, (b) the power needed for single-stage compression to the same pressure, (c) the maximum temperature for (a) and (b) and (d) the heat removed in the intercooler. **[LO 18.5]**

(M) **18.10** A single-acting air compressor has a cylinder of bore 15 cm and the piston stroke is 25 cm. The crank speed is 600 rpm. Air taken from the atmosphere (1 atm, 27°C) is delivered at 11 bar. Assuming polytropic compression $pv^{1.25}$ = const., find the power required to drive the compressor, when its mechanical efficiency is 80%. The compressor has a clearance volume which is 1/20th of the stroke volume. How long will it take to deliver 1 m^3 of air at the compressor outlet conditions. Find the volumetric efficiency of the compressor. **[LO 18.3, 18.4]**

(M) **18.11** A multi-stage air compressor compresses air from 1 to 40 bar. The maximum temperature in any stage is not to exceed 400 K. (a) If the law of compression for all the stages is $pv^{1.3}$ = const., and the initial temperature is 300 K, find the number of stages for the minimum power input. (b) Find the intermediate pressures for optimum compression as well as the power needed. (c) What is the heat transfer in each of the intercooler? **[LO 18.5]**

(M) **18.12** A single stage reciprocating air compressor has a swept volume of 2000 cm^3 and runs at 800 rpm. It operates on a pressure ratio of 8, with a clearance of 5% of the swept volume. Assume NTP room conditions at inlet, and polytropic compression and expansion with n = 1.25. Calculate (a) the indicated power, (b) volumetric efficiency, (c) mass flow rate, (d) the free air delivery FAD, (e) isothermal efficiency and (f) actual power needed to drive the compressor, if the mechanical efficiency is 0.85. **[LO 18.3, 18.4]**

(M) **18.13** A two-stage single-acting reciprocating compressor takes in air at the rate of 0.2 m^3/s. Intake pressure and temperature are 0.1 MPa and 16°C, respectively. The air is compressed to a final pressure of 0.7 MPa. The intermediate pressure is ideal, and intercooling is perfect. The compression index is 1.25 and the compressor runs at 10 rps. Neglecting clearance, determine (a) the intermediate pressure, (b) the total volume of each cylinder, (c) the power required to drive the compressor and (d) the rate of heat absorption in the intercooler. **[LO 18.5]**

(M) **18.14** A three-stage single-acting air compressor running in an atmosphere at 1.013 bar and 15°C has a free air delivery of 2.83 m^3/min. The suction pressure and temperature are 0.98 bar and 32°C, respectively. The delivery pressure is to be 72 bar. Calculate the indicated power required, assuming complete intercooling, n = 1.3 and that the compressor is designed for minimum work. What will be the heat loss to the intercoolers? **[LO 18.5]**

(M) **18.15** A two-stage single-acting air compressor delivers 0.07 m^3 of free air per sec (free air condition 101.325 kN/m^2 and 15°C). Intake conditions are 95 kN/m^2 and 22°C. Delivery pressure from the compressor is 1300 kN/m^2. The intermediate pressure is ideal and there is perfect intercooling. The compression index is 1.25 in both cylinders. The overall mechanical and electrical efficiency is 75%. Neglecting clearance, determine (a) the energy input to the driving motor, (b) the heat transfer rate in the intercooler and (c) the percentage saving in work by using a two-stage intercooled compressor instead of a single-stage compressor. Take R = 0.287 kJ/kg K and c_p = 1.006 kJ/kg K. **[LO 18.5]**

(E) **18.16** Air at 1.013 bar and 15°C is to be compressed at the rate of 5.6 m^3/min to 1.75 bar. Two machines are considered: (a) the Roots blower and (b) a sliding vane rotary compressor. Compare the powers required, assuming for the vane type that internal compression takes place through 75% of the pressure rise before delivery takes place, and that the compressor is an ideal uncooled machine. **[LO 18.7]**

(M) **18.17** A rotary air compressor has an inlet static pressure and temperature of 100 kN/m^2 and 20°C, respectively. The compressor has an air mass flow rate of 2 kg/s through a pressure ratio of 5:1. The isentropic efficiency of compression is 85%. Exit velocity from the compressor is 150 m/s. Neglecting change of velocity through the compressor, determine the power required to drive the compressor. Estimate the stagnation temperature and pressure at exit from the compressor. **[LO 18.3, 18.7]**

(M) **18.18** The cylinder of an air motor has a bore of 63.5 mm and stroke of 114 mm. The supply pressure is 6.3 bar, the supply temperature 24°C and the exhaust pressure is 1.013 bar. The clearance volume is 5% of the swept volume and the cut-off ratio is 0.5. The air is compressed by the returning piston after it has travelled through 0.95 of its stroke. The law of compression and expansion is $pv^{1.3}$ = constant. Calculate (a) the temperature at the end of expansion, (b) the indicated power of the motor which runs at 300 rpm and (c) the air supplied per minute. **[LO 18.6]**

(M) **18.19** A four cylinder single-stage air compressor has a bore of 200 mm and a stroke of 300 mm and runs at 400 rpm. At a working pressure of 721.3 kPa it delivers 3.1 m^3 of air per min at 270°C. Calculate (a) the mass flow rate, (b) the free air delivery (FAD), (c) effective swept volume and (d) volumetric efficiency. Take the inlet condition as that of the free air at 101.3 kPa, 21°C. **[LO 18.3, 18.4]**

ANSWERS TO PROBLEMS

18.1. (a) 0.239 kg/s,
(b) 0.199 m^{3}s,
(c) 0.0299 m^3
(d) 79.2%

18.2 (a) 0.238 kg,
(b) 1.027 MPa,
(c) 2736°C,
(d) 123.7 kJ

18.6 – 225 kJ/kg, – 190 kJ/kg, 84.4%

18.7 443 K, 51.82 kW, 36 kJ/kg

18.8 (a) 26.15 kW
(b) 85.6%)

18.9 (a) 63.8 kW,
(b) 74.9 kW,
(c) 404.2 K, 544.9 K,
(d) 28.9 kW

18.10 12.25 kW, 3.55 min, 72%

18.11 (a) 3
(b) 3.48 bar, 12.1 bar, 373.1 kJ/kg
(c) 100.5 kJ/kg

18.12 (a) 5.47 kW
(b) 78.6%
(c) 1.54 kg/min
(d) 1.26 m^3/min,
(e) 80.7%
(f) 6.44 kW

18.13 (a) 0.264 MPa
(b) 0.0076 m^3 h.p. and 0.02 m^3 l.p. cylinder
(c) 42.8 kW
(d) 14.95 kW

18.14 25.568 kW, 13.78 kW

18.15 (a) 29.1 kW,
(b) 6.99 kW,
(c) 13%

18.16 6.88 kW, 5.71 kW

18.17 404 kW, 232.2°C, 541.5 kN/M^2

18.18 (a) – 28.4°C,
(b) 0.749 kW,
(c) 0.42 kg/min.

18.19 (a) 0.239 kg/s,
(b) 0.199 m^3/s,
(c) 0.299 m^3,
(d) 79.2%

19

Steam Turbine and Condenser

Learning Objectives

LO 19.1 Recall basic functions of steam turbine

LO 19.2 Explain classification of steam turbines based on principle of energy conversion

LO 19.3 Explain principle of operation of impulse turbines

LO 19.4 Analyze velocity diagram of steam turbines

LO 19.5 Examine optimum velocity ratio

LO 19.6 Demonstrate graphical construction of velocity diagram

LO 19.7 Explain basics of compounding in steam turbines

LO 19.8 Analyze pressure compounding or Rateau staging

LO 19.9 Analyze velocity compounding or Curtis staging

LO 19.10 Examine optimum velocity ratio for a Curtis stage

LO 19.11 Explain principle of operation of reaction turbines

LO 19.12 Explain carry-over efficiency

LO 19.13 Compare the optimum velocity ratios of simple impulse, Curtis and 50% reaction turbines

LO 19.14 Illustrate calculation of nozzle and blade heights

LO 19.15 Analyze nozzle efficiency

LO 19.16 Explain different losses in steam turbines

LO 19.17 Explain reheat factor and condition line

LO 19.18 Identify need of a condenser

LO 19.19 Discuss different types of condensers

LO 19.20 Explain air removal in condensers

19.1 INTRODUCTION

LO 19.1
Recall basic functions of steam turbine

A steam turbine is a prime mover which continuously converts the energy of high-pressure, high temperature steam supplied by a steam generator into shaft work with the low temperature steam exhausted to a condenser. This energy conversion essentially occurs in two steps:

1. The high-pressure, high-temperature steam first expands in nozzles and comes out at a high velocity.
2. The high velocity jets of steam coming out of the nozzles, impinge on the blades mounted on a wheel, get deflected by an angle and suffer a loss of momentum which is absorbed by the rotating wheel in producing torque.

A steam turbine is basically an assemblage of nozzles and blades.

The Greek inventor Hero of Alexandria built the first prototype of a steam turbine in 120 BC which operated on the reaction principles. A simple closed spherical vessel, mounted on bearings, carrying steam from a cauldron or boiler with four tangential pipes discharging steam is driven around by the reaction of the steam jets [Fig. 19.1].

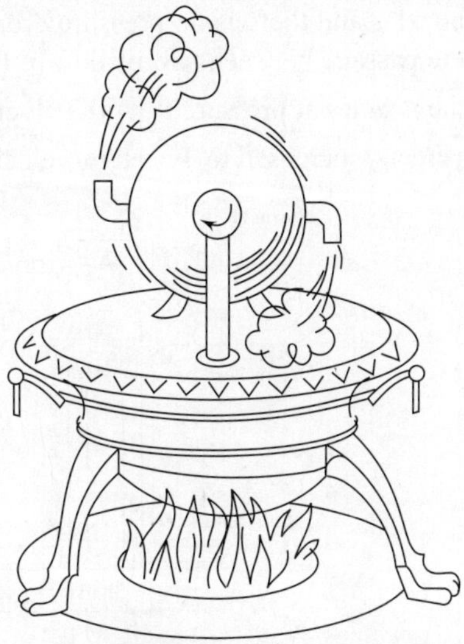

Fig. 19.1 Hero's turbine

It lacked blades, and the torque required for rotation was generated solely by the reaction force from steam exiting through strategically aligned nozzles. Although Hero's turbine was the first of its kind, it had no practical application. However, all of the steam was released directly into the atmosphere, making it highly not acceptable. Such a design is not feasible for modern power station turbines, which consume large quantities of steam. In real-world applications, turbines are designed to recycle steam after power generation to improve efficiency and sustainability.

19.2 CLASSIFICATION OF STEAM TURBINES

LO 19.2
Explain classification of steam turbines based on principle of energy conversion

Steam turbines can be classified based on various criteria, such as the direction of steam flow through the turbine, the method of steam admission and exhaust, number of cylinders and so on. However, the primary classification is based on the type of turbine blades, as this directly influences the process by which the steam's enthalpy is converted into mechanical energy. This classification and its implications are explained in the following sections.

19.2.1 Turbine Blading

Depending upon the types of blades used and the method of energy transfer from the fluid to the rotor wheel, the turbines may be of two types:

(a) Impulse turbines
(b) Reaction turbines

19.2.2 Impulse Turbines

LO 19.3
Explain principle of operation of impulse turbines

In impulse turbines, all pressure drops of steam occur in the nozzles and there is no pressure drop as steam flows through the passage between two blades. In Fig. 19.2(a), steam enters the nozzles at pressure p_0 with velocity $\overline{V}_0$, undergoes expansion to pressure p_1 with velocity increased to $\overline{V}_1$. High-velocity jets of steam impinge upon the blades with

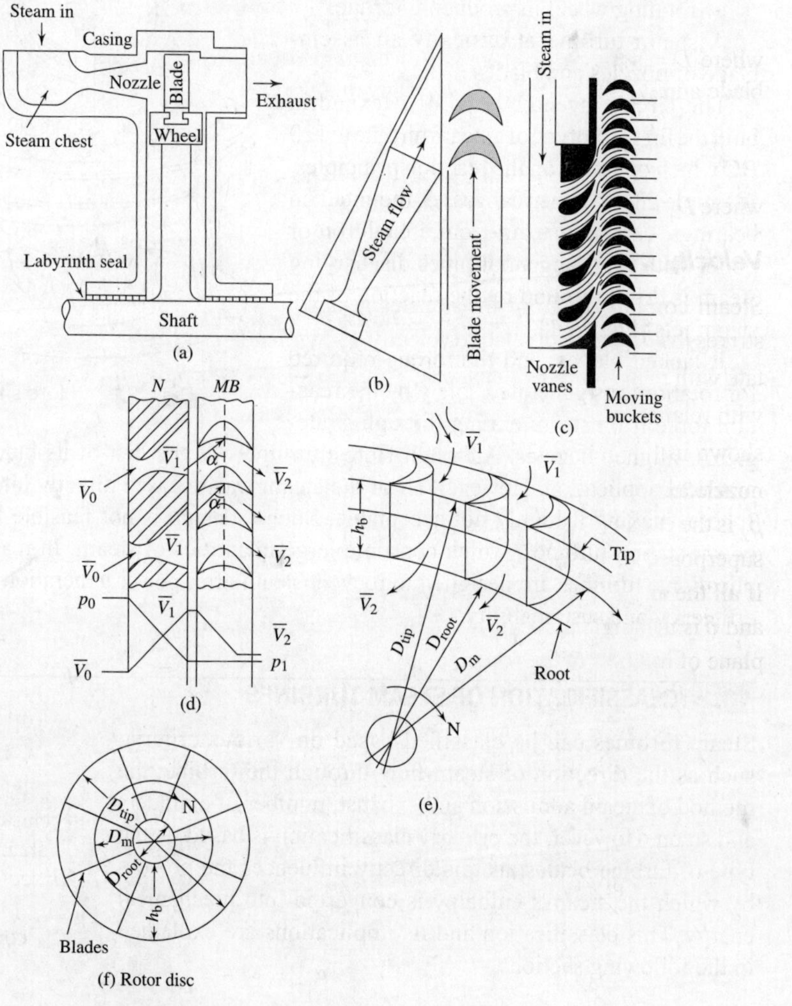

Fig. 19.2 Flow through impulse blades

velocity $\bar{V}_1$, get deflected by an angle, and come out at a lower velocity $\bar{V}_2$ impressing a torque on the blades. The pressure of steam p_1 remains essentially constant as steam flows through the blades.

From the principle of conservation of momentum,

Momentum of steam jets at inlet to the blades – momentum of jets at exit from the blades (both resolved in the direction of motion of the wheel) = momentum (angular) absorbed by the wheel in producing shaft work.

In Figs. 19.2(b) to 19.2(d), it is seen that the wheel rotates only due to the impulsive effect of the jets (i.e., the difference of momenta of the jets deflected by the blades). The blades of such a wheel are called impulse blades.

The mean peripheral velocity of the blades, also called the mean blade velocity, $\bar{V}_b$, is given by

$$\bar{V}_b = \frac{\pi D_m N}{60}$$

where D_m is the mean diameter of the wheel and N is its rpm. The area of flow or blade annulus [Fig. 19.2(e)], A_b, is given to be

$$A_b = \frac{\pi}{4}(D_2^2 - D_1^2) = \pi \frac{D_2 + D_1}{2}\frac{D_2 - D_1}{2} = \pi D_m h_b \tag{19.1}$$

where D_1 is the root diameter, D_2 is the tip diameter, and h_b is the height of the blades.

Velocity Diagrams, Diagram Work and Diagram Efficiency

LO 19.4
Analyze velocity diagram of steam turbines

Steam coming out from the nozzles at absolute velocity $\bar{V}_1$ strikes the blades with relative velocity $\bar{V}_{r_1}$ while the blades rotate with mean peripheral velocity $\bar{V}_b$. Steam leaves the blades with relative velocity $\bar{V}_{r_2}$ while its absolute velocity is $\bar{V}_2$. As shown in Fig. 19.3(a), α is the nozzle angle subtended by the nozzle axis with the direction of rotation of the wheel, β_1 is the inlet blade angle and β_2 is the exit blade angle. The inlet and exit velocity triangles of Fig. 19.3(a) have been superposed on a common $\bar{V}_b$ in Fig. 19.3(b) (for $\delta > 90°$) and 19.3(c) (for $\delta < 90°$). If all the angles are measured clockwise, then γ is the exit blade angle ($\gamma = 180 - \beta_2$) and δ is the angle made by absolute exit velocity of steam leaving the blades with the plane of rotation of the wheel.

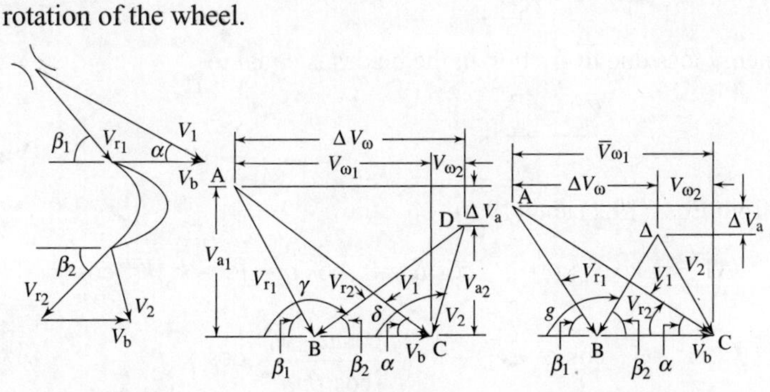

Fig. 19.3 Velocity diagrams for impulse blading

It is the difference of the cosine components of the velocities of steam which drives the wheel and produces the torque. It is called the change in the velocity of whirl, $\Delta\bar{V}_\omega$, which is given by

$$\Delta\bar{V}_\omega = \text{velocity of whirl at inlet-velocity of whirl at outlet}$$

$$= \bar{V}_{\omega_1} - \bar{V}_{\omega_2} = \bar{V}_1 \cos\alpha - \bar{V}_2 \cos\delta \tag{19.2}$$

If $\delta > 90°$, $\bar{V}_{\omega_1}$ and $\bar{V}_{\omega_2}$ become additive in estimating $\Delta\bar{V}_\omega$ [Fig. 19.3(b)]. If $\delta <$ 90°, $\bar{V}_{\omega_2}$ is to be subtracted from $\bar{V}_{\omega_1}$ to determine $\Delta\bar{V}_\omega$ [Fig. 19.3(c)].

From the inlet velocity triangle *ABC* [Fig. 19.3(b)],

$$\bar{V}_1 \cos\alpha - \bar{V}_{r_1} = \bar{V}_{r_1} \cos\beta_1$$

$$\bar{V}_1 \sin\alpha = \bar{V}_{r_1} \sin\beta_1$$

Therefore, on division,

$$\tan\beta_1 = \frac{\bar{V}_1 \sin\alpha}{\bar{V}_1 \cos\alpha - \bar{V}_b} \tag{19.3}$$

From the exit velocity triangle *DBC* [Fig. 19.3(b)],

$$\bar{V}_2 \cos(180 - \delta) + \bar{V}_b = \bar{V}_{r_2} \cos(180 - \gamma)$$

$$\bar{V}_b - \bar{V}_2 \cos\delta = -\bar{V}_{r_2} \cos\gamma$$

$$\therefore \quad \bar{V}_2 \cos\delta = \bar{V}_b + \bar{V}_{r_2} \cos\gamma \tag{19.4}$$

The same result can also be obtained from Fig. 19.3(c),

$$\bar{V}_b - \bar{V}_2 \cos\delta = \bar{V}_{r_2} \cos(180 - \gamma) = -\bar{V}_{r_2} \cos\gamma$$

$$\bar{V}_2 \cos\delta = \bar{V}_b + \bar{V}_{r_2} \cos\gamma \tag{19.5}$$

The ratio of relative velocity at exit to that at inlet is called "the blade friction factor", k_b,

$$k_b = \bar{V}_{r_2} / \bar{V}_{r_1} \tag{19.6}$$

The energy loss due to friction in the blades is equal to

$$\frac{\bar{V}_{r_1}^2 - \bar{V}_{r_2}^2}{2}$$

Now, from Eqs. (19.2) and (19.5),

$$\Delta\bar{V}_\omega = \bar{V}_1 \cos\alpha - \bar{V}_2 \cos\delta = \bar{V}_1 \cos\alpha - \bar{V}_b - k_b \bar{V}_{r_1} \cos\gamma$$

$$= \bar{V}_1 \cos\alpha - \bar{V}_b - k_b \frac{\bar{V}_1 \cos\alpha - \bar{V}_b}{\cos\beta_1} \cos\gamma$$

$$= (\bar{V}_1 \cos\alpha - \bar{V}_b)\left[1 - k_b \frac{\cos\gamma}{\cos\beta_1}\right] \tag{19.7}$$

Since $\gamma = 180 - \beta_1;\ \cos\gamma = -\cos\beta_1,$

$$\Delta\bar{V}_\omega = (\bar{V}_1 \cos\alpha - \bar{V}_b)(1 + k_b)$$

Also, $\Delta V_\omega = V_{r_1}\cos\beta_1 + V_{r_2}\cos\beta_2 = V_{r_1}\cos\beta_1 + k_b V_{r_2}\cos\beta_1$

$$= V_{r_1}\cos\beta_1(1 + k_b) = (V_1\cos\alpha - V_b)(1 + k_b) \tag{19.8}$$

Blades are said to be *symmetrical* if blade angles are equal, i.e. if $\beta_1 = \beta_2$. Impulse turbines mostly have symmetrical blades.

The tangential thrust impressed by the jets on the blades, P_t, is

$$P_t = \omega_s \cdot \Delta\bar{V}_\omega \tag{19.9}$$

where ω_s is the steam flow rate.

The axial thrust, i.e. the thrust produced by the difference in the axial components of the velocities (along the axis of the shaft),

$$P_a = \omega_s \cdot \Delta\bar{V}_a \tag{19.10}$$

where $\Delta\bar{V}_a = V_{r_1}\sin\beta_1 - V_{r_2}\sin\beta_2$ (19.11)

The rate at which work is done by the jets on the blades is called the *blading work or diagram work*, W_D, which is given by

$$W_D = P_t \times \bar{V}_b = \omega_s \cdot \Delta\bar{V}_\omega \bar{V}_b \tag{19.12}$$

The energy input to the blades is the kinetic energy of jets issuing out from the nozzles $= \omega_s \cdot (\bar{V}_1^2/2)$.

Therefore, the *blading or diagram efficiency*

$$\eta_D = \frac{\text{Rate of work done on the blades}}{\text{Rate of energy input to the blades}} = \frac{\omega_s \Delta\bar{V}_\omega \bar{V}_b}{\frac{1}{2}\omega_s \bar{V}_1^2}$$

or
$$\eta_D = \frac{2\Delta\bar{V}_\omega \bar{V}_b}{\bar{V}_1^2} \tag{19.13}$$

It is the fraction of kinetic energy of jets of steam issuing from nozzles which is converted to shaft work.

Optimum Velocity Ratio

LO 19.5
Examine optimum velocity ratio

Substituting $\Delta\bar{V}_\omega$ from Eq. (19.8) in Eq. (19.13), the diagram efficiency is given by

$$\eta_D = \frac{2(\bar{V}_1\cos\alpha - \bar{V}_b)(1 + k_b)\bar{V}_b}{\bar{V}_1^2} = \frac{2\bar{V}_b^2\left(\dfrac{\bar{V}_1\cos\alpha}{\bar{V}_b} - 1\right)(1 + k_b)}{\bar{V}_1^2}$$

Defining the *velocity ratio* ρ as $\overline{V}_b/\overline{V}_1$, the ratio of mean blade velocity V_b to the jet velocity $\overline{V}_1$,

$$\eta_D = 2\rho^2\left(\frac{\cos\alpha}{\rho} - 1\right)(1+k_b) = 2(\rho\cos\alpha - \rho^2)(1+k_b) \qquad (19.14)$$

There is a certain value of ρ which makes η_D the maximum. Differentiating η_D with respect to ρ and making it equal to zero,

$$\frac{d\eta_D}{d\rho} = 2(\cos\alpha - 2\rho)(1+k_b) = 0$$

Therefore, the optimum velocity ratio for impulse blading is

$$\rho_{opt} = \frac{\cos\alpha}{2} \qquad (19.15)$$

Substituting this optimum value of ρ in Eq. (19.14), the maximum blading efficiency is

$$(\eta_D)_{max} = 2\left[\frac{\cos^2\alpha}{2} - \frac{\cos^2\alpha}{4}\right](1+k_b) = \frac{1+k_b}{2}\cos^2\alpha \qquad (19.16)$$

If the energy loss due to friction in blades is small, $k_b \simeq 1$, then

$$(\eta_D)_{max} = \cos^2\alpha \qquad (19.17)$$

where α is the nozzle angle. The lower is the nozzle angle, the higher is the blading efficiency. However, too low a nozzle angle may cause energy loss at blade inlet. Therefore, the nozzle angle has to be maintained within a certain range, which varies from 16° to 22°.

Graphical Method

LO 19.6 Demonstrate graphical construction of velocity diagram

The diagram work and the diagram efficiency under a certain operating condition can also be estimated graphically by drawing to scale the inlet and exit velocity triangles. If the state of steam at inlet to the nozzles, the exit pressure and nozzle efficiency are known, then

$$\overline{V}_1 = 44.72[(h_0 - h_{1s})\eta_n]^{1/2}$$

can be estimated. Knowing D_m and N, $\overline{V}_b$ is found out from

$$\overline{V}_b = \frac{\pi D_m N}{60}$$

With α and k_b being given and with the blades being symmetrical ($\beta_1 = \beta_2$), a scale for the diagram is chosen, say, 1 cm = 50 m/s. A horizontal line AB is drawn, the length of which is proportional to $\overline{V}_b$ (Fig. 19.4). A straight line is drawn from B making an angle α and a length BC proportional to $\overline{V}_1$ is cut-off. CA is joined. The angle $\angle$CAD is measured. This is the inlet blade angle β_1. CA is measured, which represents $\overline{V}_{r_1}$. Again, a straight line AE is drawn from A making an angle β_1 (= β_2) and a length AE is cut off proportional to $\overline{V}_{r_2}$ $(= k_b\, V_{r_1})$. EB is joined. The velocity diagram is thus completed. From this diagram, $\Delta\overline{V}_\omega$, $\overline{V}_{a_1}$ and $\overline{V}_{a_2}$ are measured.

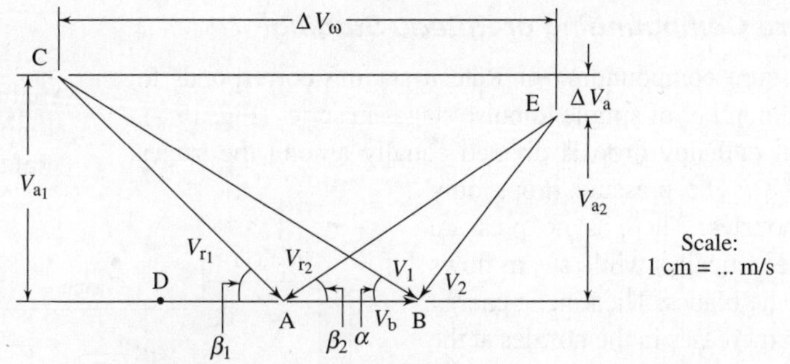

Fig. 19.4 Construction of velocity diagram

Then the desired particulars, $P_t = \omega_s \Delta \bar{V}_\omega$; $P_a = \omega_s (\bar{V}_{a_1} - \bar{V}_{a_2})$; $W_D = P_t \bar{V}_b$; $\eta_D = (2\Delta \bar{V}_\omega \bar{V}_b)/\bar{V}_1^2$ are estimated.

Compounding of Steam Turbines

> **LO 19.7**
> Explain basics of compounding in steam turbines

One row of nozzles followed by one row of blades is called a stage of a turbine. If steam is allowed to expand from boiler condition down to condenser vacuum in a single row of nozzles (Fig. 19.6), due to a large enthalpy drop, the velocity at exit from the nozzles ($\bar{V}_1$) is very large (Fig. 19.5). Since $\bar{V}_b/\bar{V}_1 = \cos \alpha/2$, the blade velocity ($\bar{V}_b$) also becomes very large. Again $\bar{V}_b$ is equal to $\pi D_m N/60$. Therefore, if D_m is fixed at some economic value, N becomes very large. Thus single-stage impulse turbines, also called de Laval turbines, have inherently very high rotative speeds (N).

Such high rotational speeds cannot be properly utilized. It entails large friction losses. The centrifugal stresses also become very large. Alternatively, if N is fixed, the diameter of the wheel becomes too large. Moreover, with a single stage, the velocity of steam at exit is sufficiently high and there is a considerable loss of kinetic energy with exiting steam. To obviate these difficulties, the turbines are compounded or staged, where steam, instead of expanding in a single stage, is made to expand in a number of stages, whereby the turbine speed is reduced while securing the same enthalpy drop of steam.

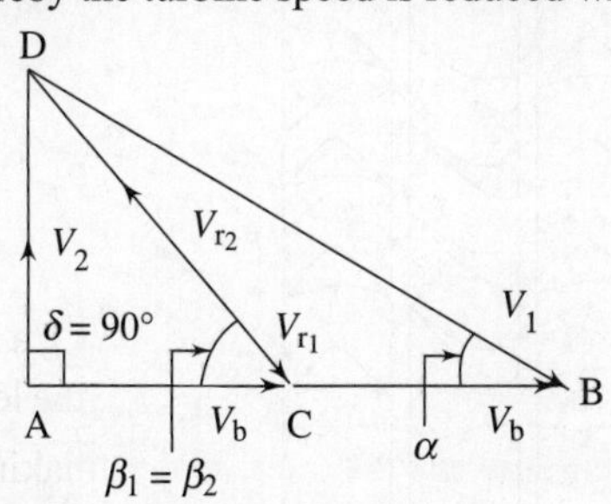

Fig. 19.5 Velocity diagram for frictionless symmetrical impulse blading with axial discharge

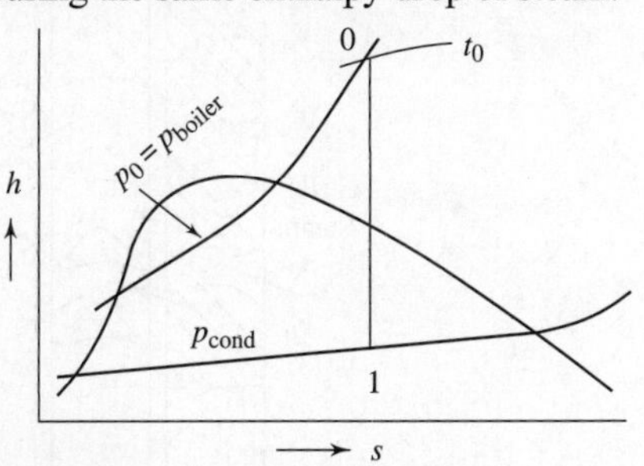

Fig. 19.6 Single stage expansion of steam

Basically, there are two ways of compounding steam turbines:

1. Pressure compounding or Rateau staging
2. Velocity compounding or Curtis staging

Pressure Compounding or Rateau Staging

LO 19.8
Analyze pressure compounding or Rateau staging

The pressure compounding or Rateau staging corresponds to putting a number of simple impulse stages in series (Fig. 19.7). The total enthalpy drop is divided equally among the stages (Fig. 19.8). The pressure drops only in the nozzles. There is no pressure drop (theoretically) while steam flows through the blades. The kinetic energy of steam increases in the nozzles at the expense of the pressure drop and it is absorbed (partially) by the blades in each stage, in producing torque. In Fig. 19.8, the total isentropic enthalpy drop of steam $(h_0 - h_4)$ is divided equally among the four stages of the turbine. In Mollier diagram, the enthalpy drop $(h_0 - h_4)$ is measured, then the enthalpy drop, $h_0 - h_1 = h_1 - h_2 = h_2 - h_3 = h_3 - h_4 = (h_0 - h_4)/4$ are computed and inserted on the isentropic line. The interstage pressures noted from the diagram are p_1, p_2 and p_3, i.e. the pressure after first stage is p_1, the pressure after second stage is p_2, and so on.

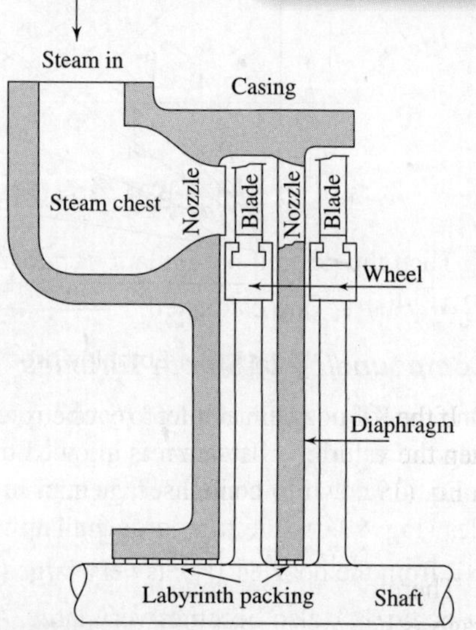

Fig. 19.7(a) Pressure-compounded impulse turbine

The velocity of steam at exit from the first row of nozzles is

$$\bar{V}_1 = 44.72[(h_0 - h_1)]^{1/2} = 44.72[(h_0 - h_4)/4]^{1/2}$$

or,
$$\bar{V}_1 = \frac{1}{2}(44.72)[(h_0 - h_4)]^{1/2} \tag{19.18}$$

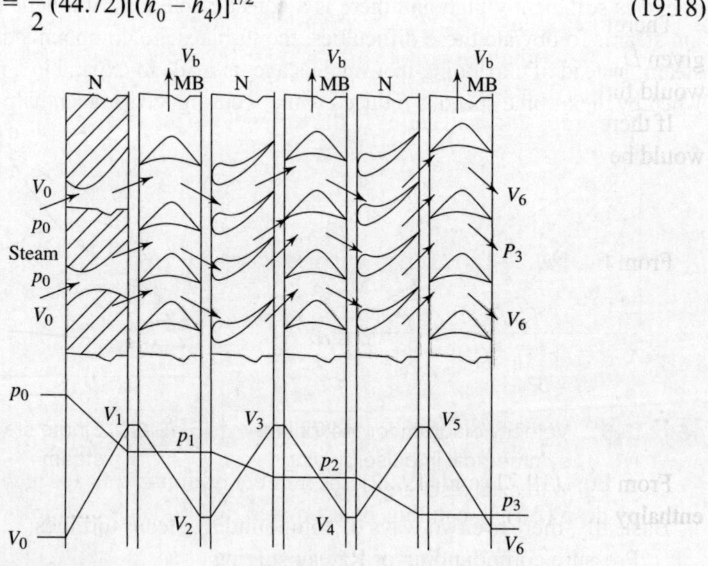

Fig. 19.7(b) Three pressure (or Rateau) stages in series

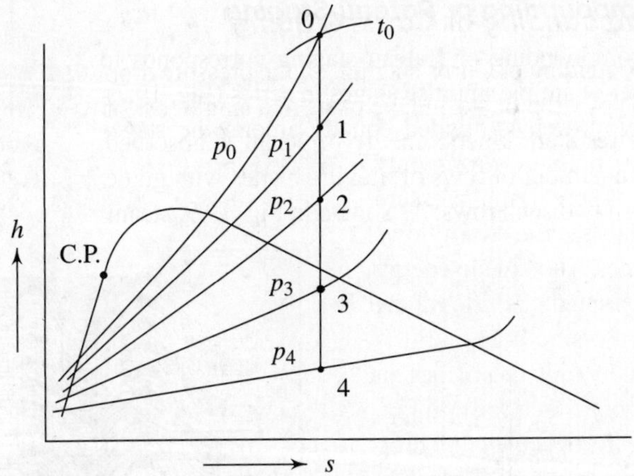

Fig. 19.8 Enthalpy drop per stage in a 4-stage turbine

If the KE of steam at inlet to each subsequent row of nozzles is small and neglected, then the velocity of steam leaving the nozzles in each row will be the same as given in Eq. (19.18). For a single-stage turbine, the jet velocity would have been

$$\overline{V}_1 = 44.72[h_0 - h_4]^{1/2} \tag{19.19}$$

Therefore, for a 4-stage turbine, the velocity of steam leaving the nozzles in each stage is half of that for a single-stage turbine. For a 9-stage turbine, it will be one-third.

For each impulse stage operating at its maximum blading efficiency, the blade velocity is given by

$$\overline{V}_b = \frac{\cos \alpha}{2} \overline{V}_1 \tag{19.20}$$

Therefore, along with $\overline{V}_1$, $\overline{V}_b$ also gets halved for a 4-stage turbine, and for a given D_m, N gets halved or for a given N, D_m gets halved. With more stages, N or D_m would further decrease.

If there are n stages in series in the turbine, the isentropic enthalpy drop per stage would be

$$(\Delta h)_{\text{stage}} = \frac{\Delta h_{\text{total}}}{n} \tag{19.21}$$

From Eq. (19.20), and for an impulse stage,

$$\overline{V}_1 = \frac{2}{\cos \alpha} \overline{V}_b = 44.72\, [\Delta h_{\text{stage}}]^{1/2}$$

$$\therefore \quad (\Delta h)_{\text{stage}} = \left[\frac{2\overline{V}_b}{44.72 \cos \alpha} \right]^2 = 4 \left[\frac{\overline{V}_b}{44.72 \cos \alpha} \right]^2 \tag{19.22}$$

From Eqs. (19.21) and (19.22), the number of impulse stages required for a certain enthalpy drop $(\Delta h)_{\text{total}}$ can thus be estimated under ideal condition, i.e.,

$$n = \frac{(\Delta h_s)_{\text{total}}}{(\Delta h_s)_{\text{stage}}}$$

Velocity Compounding or Curtis Staging

LO 19.9 Analyze velocity compounding or Curtis staging

In velocity compounding or Curtis staging, all the pressure drop and hence, enthalpy drop of steam take place in a single row of nozzles and the resultant kinetic energy of steam is absorbed by the wheel in a number of rows of moving blades with guide blades in between two such rows. As shown in Fig. 19.9, steam

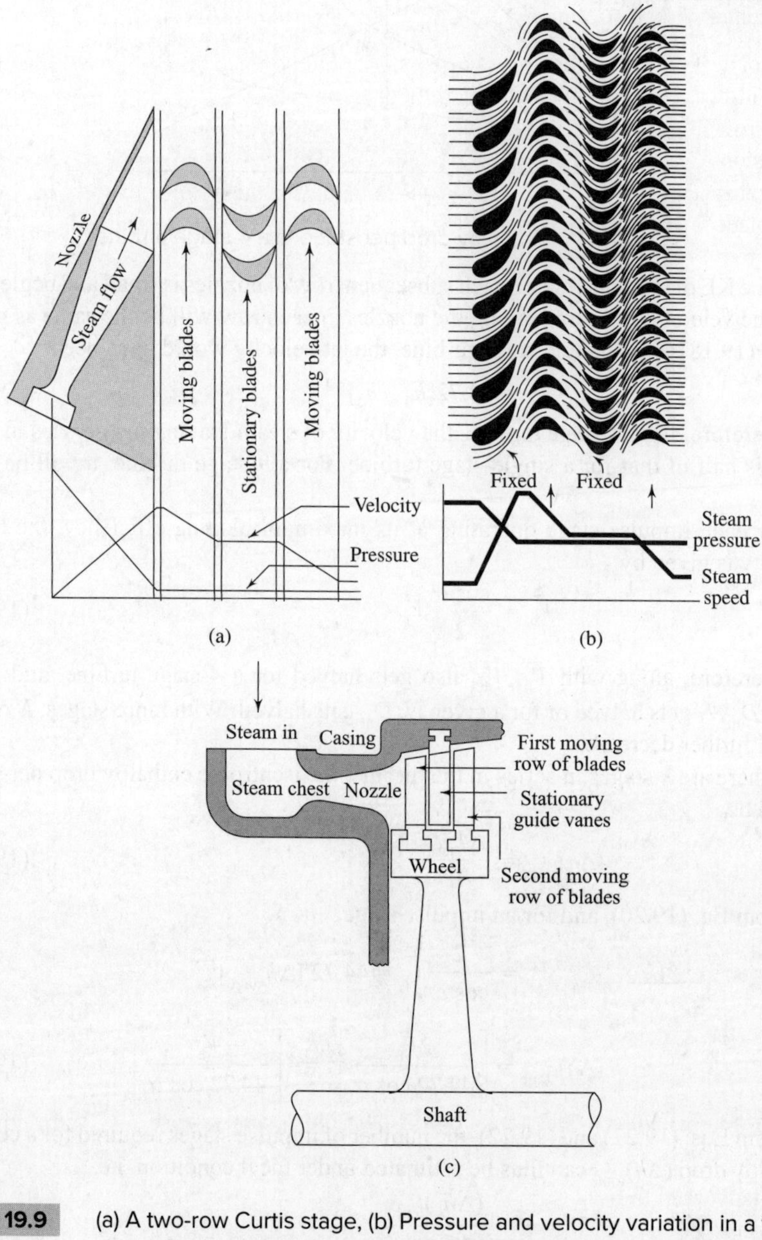

Fig. 19.9 (a) A two-row Curtis stage, (b) Pressure and velocity variation in a two-row Curtis stage, (c) Nozzles and blades in a two-row Curtis stage

expands in the single row of nozzles from p_0 to p_1 with velocity increasing from $\bar{V}_0$ to $\bar{V}_1$. The pressure p_1 remains essentially constant thereafter. The KE of steam jets $\left(\frac{1}{2}\omega_s \bar{V}_1^2\right)$ at nozzle exit is partially converted to shaft work in the first row of moving blades with velocity decreasing from $\bar{V}_1$ to $\bar{V}_2$. The exiting steam jets are then deflected by the stationary guide blades to the next row of moving blades where part of the remaining kinetic energy $\left(\frac{1}{2}\omega_s \bar{V}_2^2\right)$ is converted to shaft work. It is a two-row Curtis or velocity stage having two rows of moving blades with one row of guide blades in between. In a three-row Curtis stage, the two-row stage is followed by a second row of guide blades and then a third row or moving blades, so that the energy conversion from fluid to rotor takes place in three stages. Steam with high KE exiting the nozzles works on the first row of moving blades, gets deflected by the first row of guide blades, works on the second row of moving blades, again gets deflected by the second row of guide blades and finally does work on the third row of moving blades.

The velocity diagrams for the first row of moving blades and the second row of moving blades are shown respectively in (a) and (b) of Fig. 19.10. The blade friction factor k_b may be assumed the same for both moving and guide blades.

Here, $$\frac{\bar{V}_{r_2}}{\bar{V}_{r_1}} = \frac{\bar{V}_{r_3}}{\bar{V}_2} = \frac{\bar{V}_{r_4}}{\bar{V}_{r_3}} = k_b$$

α_1 = exit angle of the guide blades

β_1, β_2 = inlet and exit angles of the first row of moving blades

β_3, β_4 = inlet and exit angles of the second row of moving blades

$\Delta\bar{V}_{\omega_1}, \Delta\bar{V}_{\omega_2}$ = changes in the velocity of whirl in the first and second rows of moving blades

$\Delta V_{a_1}, \Delta V_{a_2}$ = changes in the axial components of velocity in the first and second rows of moving blades.

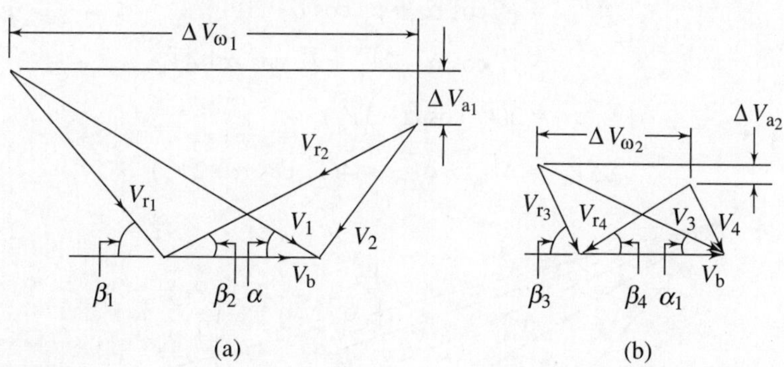

Fig. 19.10 Velocity diagrams for a 2-row Curtis stage: (a) first row of moving blades, (b) second row of moving blades

The tangential thrust,

$$P_t = \omega_s \Sigma \Delta\bar{V}_\omega = \omega_s \left[\Delta\bar{V}_{\omega_1} + \Delta\bar{V}_{\omega_2}\right]$$

The axial thrust,

$$P_a = \omega_s \Sigma\Delta\bar{V}_a = \omega_s\left[\Delta\bar{V}_{a_1} + \Delta\bar{V}_{a_2}\right]$$

Blading or diagram work, $W_D = P_t * \bar{V}_b$

Blading or diagram efficiency,

$$\eta_D = \frac{2\Sigma\Delta\bar{V}_\omega\bar{V}_b}{\bar{V}_1^2}$$

These parameters can be estimated either graphically by drawing to scale the velocity diagrams or trigonometrically.

For symmetrical blades ($\beta_1 = \beta_2$ and $\beta_3 = \beta_4$), the velocity triangles can also be drawn as shown in Fig. 19.11.

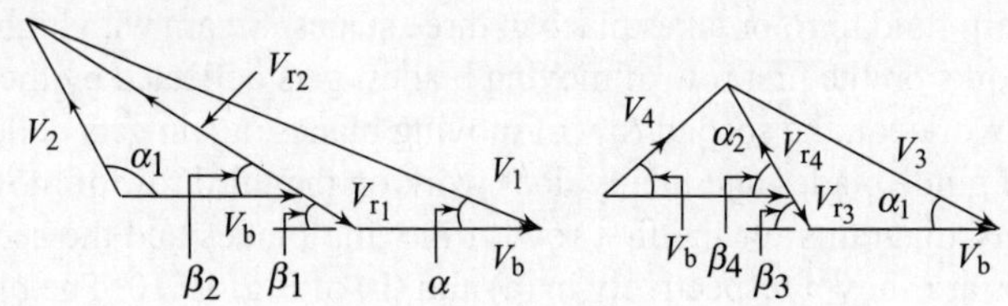

Fig. 19.11 Two-row Curtis stage with symmetrical blades: (a) first row of moving blades, (b) second row of moving blades

Optimum Velocity Ratio for a Curtis Stage

LO 19.10 Examine optimum velocity ratio for a Curtis stage

Let us again consider a two-row Curtis stage having symmetrical blading but with discharge not being axial. The velocity diagram is shown in Fig. 19.12.

$$\Delta\bar{V}_{\omega_1} = \bar{V}_1 \cos\alpha + \bar{V}_2 \cos\alpha_1 = \bar{V}_1 \cos\alpha + \bar{V}_1 \cos\alpha - 2\bar{V}_b$$

$$= 2(\bar{V}_1 \cos\alpha - \bar{V}_b)$$

$$\Delta\bar{V}_{\omega_2} = \bar{V}_3 \cos\alpha_1 + \bar{V}_4 \cos\alpha_2$$

$$= \bar{V}_1 \cos\alpha - 2\bar{V}_b + \bar{V}_1 \cos\alpha - 4\bar{V}_b$$

$$= 2(\bar{V}_1 \cos\alpha - 3\bar{V}_b)$$

$$\therefore \quad \Sigma\Delta\bar{V}_\omega = \Delta\bar{V}_{\omega_1} + \Delta\bar{V}_{\omega_2} = 4(\bar{V}_1 \cos\alpha - 2\bar{V}_b)$$

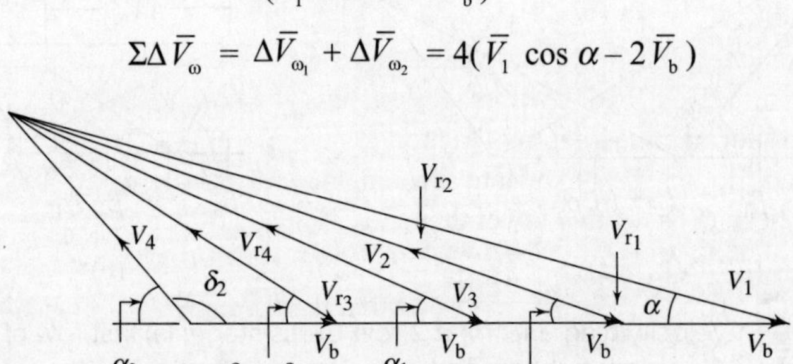

Fig. 19.12 Diagram of a two-row Curtis stage with frictionless symmetrical blading and non-axial discharge

The rate of energy transfer from fluid to rotor,

$$W_D = \omega_s \, \Delta \bar{V}_\omega \cdot \bar{V}_b$$

and the diagram efficiency (fraction of fluid energy converted to shaft work),

$$\eta_D = \frac{2\Sigma\Delta\bar{V}_\omega \bar{V}_b}{\bar{V}_1^2}$$

$$= \frac{2\times 4\,(\bar{V}_1 \cos\alpha - 2\bar{V}_b)\,\bar{V}_b}{\bar{V}_1^2} = \frac{8\bar{V}_b^2\left(\dfrac{\bar{V}_1 \cos\alpha}{\bar{V}_b} - 2\right)}{\bar{V}_1^2}$$

$$= 8\rho^2\left(\frac{\cos\alpha}{\rho} - 2\right) = 8(\rho\cos\alpha - 2\rho^2)$$

$$\frac{d\eta_D}{d\rho} = 8(\cos\alpha - 4\rho) = 0$$

$$\therefore \quad \rho_{opt} = \frac{\cos\alpha}{4} \tag{19.23}$$

For a three-row Curtis stage, it can similarly be shown that

$$\rho_{opt} = \frac{\cos\alpha}{6}$$

Therefore, for a Curtis stage having Z-rows of moving blades

$$\rho_{opt} = \frac{\cos\alpha}{2Z} \tag{19.24}$$

If $Z = 1$, i.e. for a simple impulse stage, $\rho_{opt} = \dfrac{\cos\alpha}{2}$ as obtained earlier.

For a two-row Curtis stage, $Z = 2$,

$$\rho_{opt} = \frac{\bar{V}_b}{\bar{V}_1} = \frac{\cos\alpha}{4} \quad \therefore \; \bar{V}_b = \frac{1}{2}\left(\frac{\cos\alpha}{2}\bar{V}_1\right)$$

Therefore, in a 2-row Curtis stage, the blade velocity is half of the value for a single-stage turbine. For a certain D_m, N gets halved. Substituting in the expression of diagram efficiency,

$$(\eta_D)_{max} = 8\left(\frac{\cos^2\alpha}{4} - \frac{\cos^2\alpha}{8}\right) = \cos^2\alpha \tag{19.25}$$

This holds good for the conditions as specified before, i.e. there is no friction and blades are symmetrical, and the discharge is axial. The same efficiency (i.e. $\cos^2\alpha$) was also obtained for symmetrical and frictionless simple impulse blading [Eq. (19.17)]. However, had friction been taken into account, the blading efficiency of the Curtis stage would have been considerably lower than the efficiency of the simple impulse blading.

$$\bar{V}_1 = \frac{4\bar{V}_b}{\cos\alpha} = 44.72\,[(\Delta h_s)_{Curtis}]^{1/2}$$

$$\therefore \quad (\Delta h_s)_{Curtis} = 16\left[\frac{\bar{V}_b}{44.72\cos\alpha}\right]^2 \tag{19.26}$$

This is the isentropic enthalpy drop in the 2-row Curtis stage.

19.2.3 Reaction Turbines

LO 19.11
Explain principle of operation of reaction turbines

In these turbines, pressure drop occurs both in the nozzles or the fixed row of blades and in the moving row of blades (Fig. 19.13), since the moving blade channels are also of the nozzle shape. Due to the expansion of steam while flowing through the blades, there is an increase in kinetic energy, which gives rise to reaction in the opposite direction (by Newton's third law of motion). Blades rotate due to both the impulse effect of the jets (due to change in their momentum) and the reaction force of the exiting jets impressed on the blades in the opposite direction. Such turbines are called impulse-reaction turbines, or to distinguish them from impulse turbines, simply reaction turbines. The degree of reaction (R) of these turbines is defined as

$$R = \frac{\Delta h_{\text{mb}}}{\Delta h_{\text{fb}} + \Delta h_{\text{mb}}} \tag{19.27}$$

where the subscripts "mb" and "fb" represent moving blades and fixed blades, respectively.

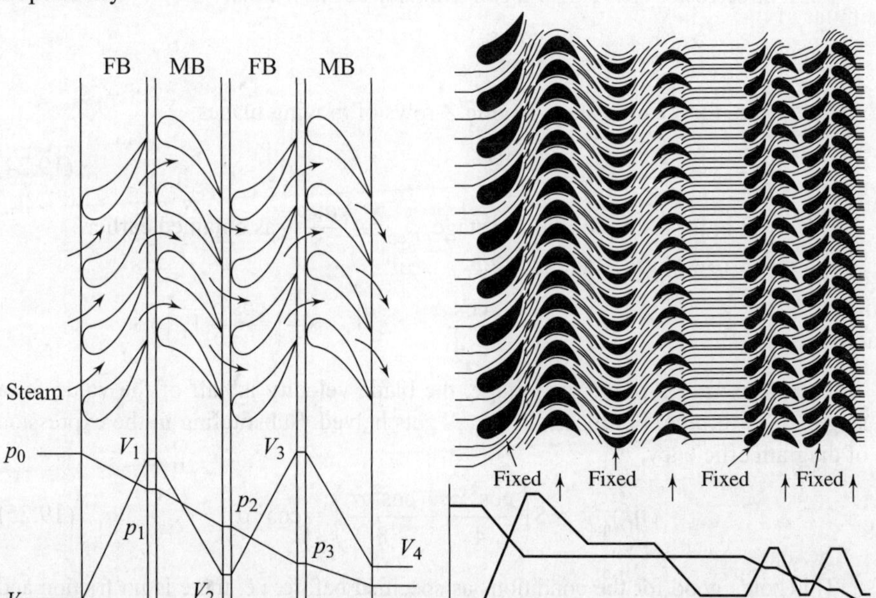

(a) (b)

Fig. 19.13 Variation of pressure and velocity in (a) two reaction stages, (b) in a two-row Curtis stage followed by two reaction stages

If $(\Delta h)_{\text{mb}} = 0$, $R = 0$, which is the case of pure impulse turbines where there is no enthalpy drop of steam in the moving blades, and all the enthalpy drops of the stage take place only in nozzles.

If $\Delta h_{\text{fb}} = 0$, $R = 1$, which is the case of a pure reaction (R = 100%) turbine, e.g. Hero's turbine.

If equal enthalpy drops occur in the fixed and moving blades, i.e. if $\Delta h_{fb} = \Delta h_{mb} = (\Delta h_{stage})/2$, $R = 1/2$ or 50%. Sometimes, 50% reaction turbines are also called *Parsons turbines*.

The velocity diagrams for the moving blades of a 50% reaction turbine are shown in Fig. 19.14. With a simple impulse type, the value of V_{r_2} $(= k_b V_{r_1})$ would be given by BE, but in the reaction turbine this velocity is increased to BC by further expansion of the steam in the nozzle-shape blade channels.

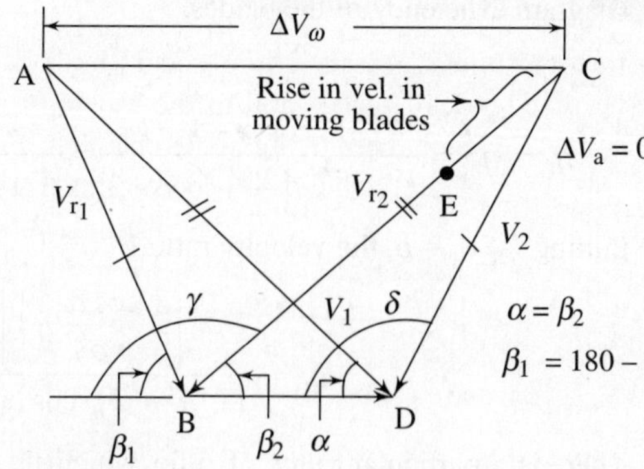

Fig. 19.14 Velocity diagram of a 50% reaction turbine

For manufacturing advantage, both fixed blades and moving blades are made similar in shape so that they can be extruded from the same set of dies.

Since $\qquad \Delta h_{fb} = \Delta h_{mb}, V_1 = V_{r_2}$

Again, for similar geometry, $\alpha = \beta_2 = 180° - \gamma$

The triangle ABD and DBC are similar.

$$\therefore \qquad V_{r_1} = V_2 \quad \text{and} \quad \beta_1 = 180° - \delta$$

Since $\beta_1 \neq \beta_2$, the blades are unsymmetrical. Again, since $\Delta V_a = 0$, there is no axial thrust imposed on the blades due to change in axial velocity in a 50% reaction turbine. However, there will be considerable axial thrust produced due to the pressure difference across the blades in each rotor disc since there is pressure drop of steam across the moving blades.

$$\Delta V_\omega = V_1 \cos\alpha - V_2 \cos\delta = V_1 \cos\alpha + V_{r_2} \cos\beta_2 - V_b \quad (\because \delta > 90°)$$
$$= 2V_1 \cos\alpha - V_b$$

Also, $\qquad \Delta V_\omega = V_{r_1} \cos\beta_1 + V_{r_2} \cos\beta_2 = V_1 \cos\alpha - V_b + V_1 \cos\alpha$

$$= 2V_1 \cos\alpha - V_b \qquad (19.28)$$

The diagram work per kg of steam,

$$W_D = \Delta V_\omega V_b = (2V_1 \cos\alpha - V_b)V_b \qquad (19.29)$$

Energy input to blades per kg of steam

$$= \frac{V_1^2}{2} + \frac{V_{r_2}^2 - V_{r_1}^2}{2} = \frac{V_1^2}{2} + \frac{V_1^2}{2} - \frac{V_{r_1}^2}{2} = V^2{}_1 - \frac{V_{r_1}^2}{2}$$

Now, $\qquad V_{r_2}^2 = V_1^2 + V_b^2 - 2V_1V_b \cos\alpha$

$$\therefore \qquad \text{Energy input} = V_1^2 - \frac{V_1^2 + V_b^2 - 2V_1V_b\cos\alpha}{2}$$

$$= \frac{V_1^2 - V_b^2 + 2V_1V_b\cos\alpha}{2} \qquad (19.30)$$

Diagram efficiency of the blades,

$$\eta_D = \eta_{b_1} = \frac{2(2V_1 \cos\alpha - V_b)V_b}{V_1^2 - V_b^2 + 2V_1V_b \cos\alpha} = \frac{2V_b^2\left[2\frac{V_1\cos\alpha}{V_b} - 1\right]}{V_1^2\,[1 - (V_b^2/V_1^2) + 2(V_b/V_1)\cos\alpha]}$$

Putting $V_b/V_1 = \rho$, the velocity ratio,

$$\eta_D = \frac{2\rho^2\left[2\frac{\cos\alpha}{\rho} - 1\right]}{1 - \rho^2 + 2\rho\cos\alpha} = \frac{2(2\rho\cos\alpha - \rho^2)}{1 - \rho^2 + 2\rho\cos\alpha} \quad (19.31)$$

There is a particular value of ρ for which η_D is a maximum. Differentiating η_D with respect to ρ and equating it zero,

$$\frac{d\eta_D}{d\rho} = \frac{(1-\rho^2 + 2\rho\cos\alpha)\,2(\cos\alpha - 2\rho) - 2\rho\,(2\cos\alpha - \rho)(-2\rho + 2\cos\alpha)}{(1-\rho^2+2\rho\cos\alpha)^2} = 0$$

$$4(1 - \rho^2 + 2\rho\cos\alpha)(\cos\alpha - \rho) - 4\rho(2\cos\alpha - \rho)(\cos\rho - \rho) = 0$$

$$4(\cos\alpha - \rho)(1 - \rho^2 + 2\rho\cos\alpha - 2\rho\cos\alpha + \rho^2) = 0$$

$$\therefore \quad \rho_{opt} = \cos\alpha \quad (19.32)$$

or $\quad V_b = V_1\cos\alpha$

Substituting the value of ρ_{opt} in Eq. (19.31),

$$(\eta_D)_{max} = \frac{2\cos^2\alpha}{1+\cos^2\alpha} \quad (19.33)$$

From Eqs. (19.29) and (19.32), the specific blading work corresponding to maximum blading efficiency.

$$W_D = (2V_b - V_b)V_b = V_b^2 \quad (19.34)$$

The velocity diagrams for a 50% reaction turbine operating with maximum blading efficiency are shown in Fig. 19.15.

$$\Delta V_\omega = V_1\cos\alpha = V_b$$

$$\therefore \quad W_D = \Delta V_\omega V_b = V_b^2$$

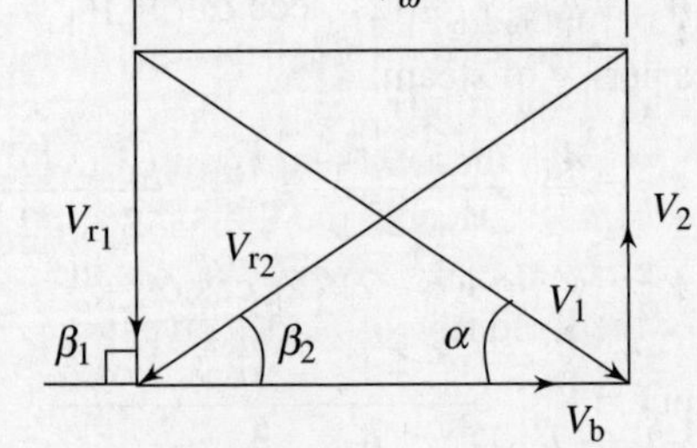

Fig. 19.15 Velocity diagram of a 50% reaction turbine operating with maximum blading efficiency

It is no longer convenient to talk of nozzles and blades, since in the reaction turbine both fixed and moving blades act as nozzles. It is usual to refer to the two sets of blades as the stator or fixed blades and the rotor or moving blades.

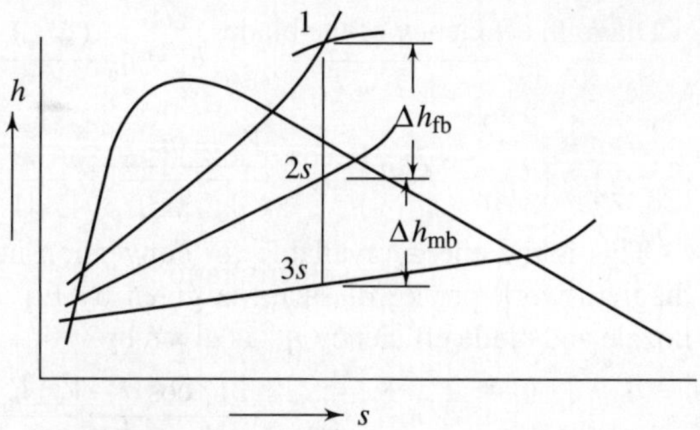

Fig. 19.16 Enthalpy drop in a 50% reaction stage

Since the isentropic enthalpy drop of the stage is equally divided between the fixed blades and moving blades (Fig. 19.16),

$$(\Delta h_s)_{fb} = h_1 - h_{2s} = \frac{V_1^2 - V_2^2}{2}$$

$$(\Delta h_s)_{mb} = h_{2s} - h_{3s} = \frac{V_{r_2}^2 - V_{r_1}^2}{2} = \frac{V_1^2 - V_2^2}{2}$$

$$\therefore \quad (\Delta h_s)_{fb} = (\Delta h_s)_{mb}$$

$$(\Delta h_s)_{stage} = h_1 - h_{3s} = 2(h_{2s} - h_{3s}) = 2(\Delta h_s)_{mb}$$

$$\therefore \quad R = \frac{h_{2s} - h_{3s}}{h_1 - h_{3s}} = \frac{\Delta h_{mb}}{2\,(\Delta h_{mb})} = \frac{1}{2}$$

Carry-over Efficiency

> **LO 19.12**
> Explain carry-over efficiency

For symmetrical staging (R = 50%), the isentropic enthalpy drop across the stage is evenly distributed between the stationary and moving rows of blades (Fig. 19.16) so that

$$(\Delta h_s)_{fb} = (\Delta h_s)_{mb} = \frac{(\Delta h_s)_{stage}}{2} = (\Delta h_s)_{row}$$

Assuming the same nozzle efficiency for both fixed and moving rows, the kinetic energy of fluid gained per row, fixed as well as moving, is

$$\eta_n \cdot \frac{(\Delta h_s)_{stage}}{2}$$

Now, the kinetic energy associated with the leaving velocity of the preceding stage is available to do work in the following stage. Of course, some loss is involved in the journey from one stage to the next and is taken into consideration by what is known as *carry-over efficiency*. Hence, the kinetic energy leaving one stage and available to the next is given by $\eta_{CO}\, V_2^2/2$, where η_{CO} is the carry-over efficiency.

The energy balance for the fixed row of blades gives

$$(V_1^2/2) + h_1 = \eta_{CO}\frac{V_2^2}{2} + h_2$$

$$\frac{V_1^2 - \eta_{CO} V_2^2}{2} = h_2 - h_1 = \eta_n \frac{(\Delta h_s)_{stage}}{2}$$

$$\therefore \qquad (\Delta h_s)_{stage} = \frac{V_1^2 - \eta_{CO} V_2^2}{\eta_n} \tag{19.35}$$

This is the energy available for conversion in one stage. The energy output or diagram work per kg of steam is given by Eq. (19.29). Therefore, the combined nozzle and blade efficiency η_{nb} is given by

$$\eta_{nb} = \frac{(2V_1 \cos\alpha - V_b) V_b}{(V_1^2 - \eta_{CO} V_2^2)/\eta_n}$$

or

$$\eta_{nb} = \eta_n \frac{V_b^2}{V_1^2} \frac{\left(\dfrac{2V_1 \cos\alpha}{V_b} - 1\right)}{1 - \eta_{CO}\left(\dfrac{V_2^2}{V_1^2}\right)} \tag{19.36}$$

Again,

$$V_2^2 = V_{r_2}^2 + V_b^2 - 2V_{r_2} V_b \cos\alpha$$

$$\therefore \qquad \frac{V_2^2}{V_1^2} = 1 + \rho^2 - 2\rho\cos\alpha, \text{ where } \rho = V_b/V_1$$

Substituting in Eq. (19.36),

$$\eta_{nb} = \eta_n \frac{\rho^2\left(\dfrac{2\cos\alpha}{\rho} - 1\right)}{1 - \eta_{CO}(1 + \rho^2 - 2\rho\cos\alpha)}$$

or

$$\eta_{nb} = \eta_n \frac{2\rho\cos\alpha - \rho^2}{1 - \eta_{CO}(1 + \rho^2 - 2\rho\cos\alpha)} \tag{19.37}$$

When $\eta_{CO} = 1$, $\eta_{nb} = \eta_n$, i.e. the losses are confined to the nozzle friction only.

When $\eta_{CO} = 0$, there is no carry-over of kinetic energy to the next stage (velocity of approach is zero), and

$$\eta_{nb} = \eta_n(2\rho\cos\alpha - \rho^2) \tag{19.38}$$

which is the efficiency of a single row of blades.

If $\eta_n = 1$,

$$\eta_{nb} = \eta_{b_1} = \eta_D = 2\rho\cos\alpha - \rho^2 \tag{19.39}$$

For maximum blading efficiency,

$$\frac{d\eta_D}{d\rho} = 2\cos\alpha - 2\rho = 0$$

$$\rho_{opt} = \cos\alpha$$

which is the same as Eq. (19.32).

Substituting in Eq. (19.39)

$$(\eta_{b1})_{max} = \cos^2\alpha \tag{19.40}$$

which is somewhat less than that obtained from Eq. (19.33).

A comparison of Eqs. (19.17), (19.25) and (19.40) reveals that the optimum efficiencies for simple impulse, Curtis and reaction blading are all equal. However, when friction is taken into account, the reaction stage is found to be the most efficient, followed by Rateau and Curtis staging in that order. The friction losses are less significant in the reaction stage since the flow velocities are lower.

LO 19.13

Compare the optimum velocity ratios of simple impulse, Curtis and 50% reaction turbines

Comparison of Enthalpy Drops in Various Stages

Assuming the blades operating at maximum efficiency, for a symmetric (50% R) reaction stage,

$$\rho_{opt} = \cos\alpha = V_b/V_1$$

$\therefore$ V_1 = velocity of steam exiting the nozzles or fixed blades

$$= V_b/\cos\alpha = 44.72\left[\frac{(\Delta h_s)_{stage}}{2}\right]^2$$

$$\therefore \quad (\Delta h_s)_{stage} = 2\left[\frac{V_b}{44.72\cos\alpha}\right]^2 \tag{19.41}$$

This is the isentropic enthalpy drop in a 50% reaction stage. If we compare this with the isentropic enthalpy drop for a simple impulse stage, Eq. (19.19) and that for a 2-row Curtis stage, Eq. (19.26),

$$(\Delta h_s)_{impulse\ stage} = 4\left[\frac{V_b}{44.72\cos\alpha}\right]^2$$

and $$(\Delta h_s)_{2\text{-row Curtis stage}} = 16\left[\frac{V_b}{44.72\cos\alpha}\right]^2$$

we find for the same values of $\bar{V}_b$ and α,

$$(\Delta h_s)_{50\%\ reaction\ stage} : (\Delta h_s)_{simple\ impulse\ stage} : (\Delta h_s)_{2\text{-row curtis stage}} = 1 : 2 : 8 \tag{19.42}$$

Therefore, a 2-row Curtis stage is equivalent (ideally) to four simple impulse stages and eight 50% reaction stages.

Since the number of stages required for a certain total isentropic enthalpy drop is

$$n = \frac{(\Delta h_s)_{total}}{(\Delta h_s)_{stage}}$$

the lower is the enthalpy drop per stage, the higher is the number of stages required for the same output.

For V_b = 320 m/s and α = 16°,

$$(\Delta h_s)_{impulse\ stage} = 221.64 \text{ kJ/kg}$$

$$(\Delta h_s)_{50\%\ reaction\ stage} = 110.82 \text{ kJ/kg}$$

$$(\Delta h_s)_{2\text{-row Curtis stage}} = 886.56 \text{ kJ/kg}$$

Thus, for the same output of a turbine the number of reaction (50%) stages required will be about twice the number of impulse or pressure (Rateau) stages and about 8 times the number of 2-row Curtis stages. So, reaction turbines are costlier than impulse turbines for the same output.

Again, for the same mean blade speed V_b,

$$V_1 = V_b/\cos\alpha \text{ for a 50\% reaction stage}$$
$$V_1 = 2V_b/\cos\alpha \text{ for a simple impulse stage, and}$$
$$V_1 = 4V_b/\cos\alpha \text{ for a 2-row Curtis stage.}$$

The energy loss due to friction is proportional to the square of the velocity of a fluid (steam). Since the fluid velocity is the highest for a 2-row Curtis stage and lowest in the 50% reaction stage, so the energy loss due to friction in the reaction stage is the least while that in the Curtis stage is the highest. The energy loss in the impulse stage will be in between these two values. Therefore, the efficiency of the reaction stage will be the highest and that of the Curtis stage will be the lowest, while that of impulse stage will lie in between.

$$\eta_{\text{reaction (50\%) stage}} > \eta_{\text{simple impulse stage}} > \eta_{\text{2-row Curtis}}$$

Figure 19.17 gives a comparison of diagram efficiencies with respect to the velocity ratio. Reaction turbines are more efficient but are more costly.

In modern turbines, a 2-row Curtis stage is generally used as the first stage. It is followed by a series of reaction stages. Sometimes impulse stages are also used after the Curtis stage. The use of the initial Curtis stage reduces the length of the rotor, i.e. the number of stages required (since a 2-row Curtis stage can replace about eight 50% reaction stages) and hence, the cost of the rotor significantly, at the expense of some loss in efficiency. It is often called the control stage, where steam at the highest pressure and temperature is immediately expanded down into a region of more moderate pressures and temperatures, such that leakage between stages is reduced, expansion problems are simplified and the overall length of the rotor is reduced. Thus, it is a good compromise to use one 2-row Curtis stage initially, sacrificing some loss in efficiency. A large drop in enthalpy occurs in this Curtis stage. The remaining enthalpy drop occurs in the subsequent stages, the number of which can be estimated from

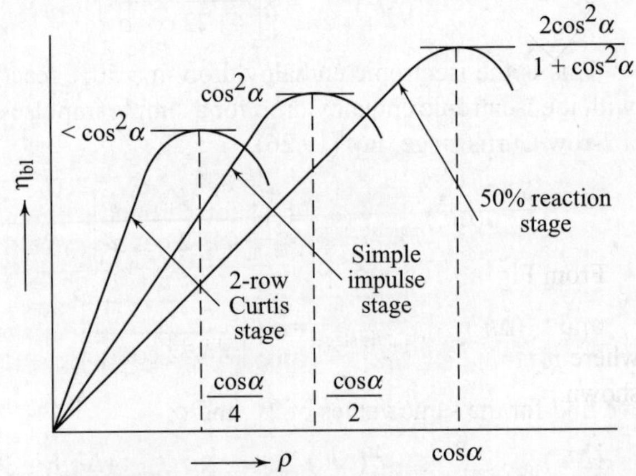

Fig. 19.17 Variation of diagram efficiency with speed ratio

$$n = \frac{(\Delta h_s)_{\text{total}} - (\Delta h_s)_{\text{2-row Curtis stage}}}{(\Delta h_s)_{\text{stage}}} \quad (19.43)$$

where Δh_s represents the isentropic enthalpy drop.

19.2.4 Nozzle and Blade Heights

LO 19.14 Illustrate calculation of nozzle and blade heights

Nozzle or blade height depends on the total annular area required to pass the desired flow of fluid. From Fig. 19.18, it can be seen that the area available for flow at exit of one nozzle passage is approximately

$$A = Oh_n$$

where O = width of flow passage at exit at mean nozzle height, and h_n = nozzle height.

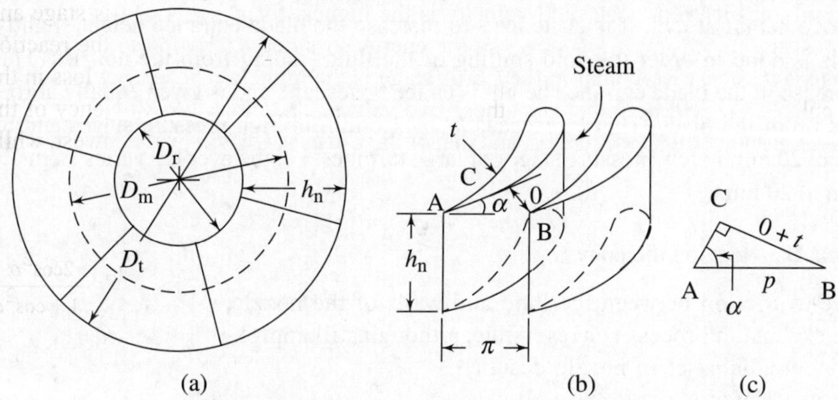

Fig. 19.18 Flow through a nozzle passage

From Fig. 19.18(c),

$$p \sin \alpha = 0 + t$$

where p = pitch of nozzles at mean nozzle height, and t = nozzle wall thickness, as shown.

$$\therefore \quad A = (p \sin \alpha - t)h_n$$

Assuming full peripheral admission, i.e. the nozzle diaphragm is completely occupied by nozzles, the number of nozzles, z, is given by

$$z = (\pi D_m)/p$$

Total nozzle area is

$$A_n = p \sin \alpha \left(1 - \frac{t}{p \sin \alpha}\right) h_n \frac{\pi D_m}{p}$$

or

$$A_n = pD_m h_n k_{tn} \sin a \tag{19.44}$$

where k_{tn} is the nozzle thickness factor.

Similarly, it can be shown that the area available for flow at the entrance to the blades is

$$A_b = \pi D_m h_b k_{tb} \sin \beta_1 \tag{19.45}$$

where k_{tb} is the edge thickness factor for blades.

From the continuity equation, it follows that

$$\omega_s = \frac{A_n V_1}{v_1} = \frac{A_b V_{r_1}}{v_1}$$

or $\quad \pi D_m h_n k_{tn} \sin \alpha \, V_1 = \pi D_m h_b k_{tb} \sin \beta_1 \, V_{r_1} \quad$ (19.46)

For all practical purposes, it can be assumed that $k_{tn} = k_{tb}$, so that

$$h_b/h_n = (V_1 \sin \alpha)/(V_{r_1} \sin \beta_1) = 1 \quad (19.47)$$

Although Eq. (19.46) indicates that the blade height at entrance is equal to the nozzle height at exit, it is customary to increase the blade entrance height slightly. This is done in order to avoid spilling of the fluid issuing from the nozzles. This increase in the blade entrance height is called "over lap" and is given equally at root and tip of the blade. This h_b exceeds h_n by 1.6 mm in high pressure stages and by about 20 mm in low pressure stages of large turbines, i.e. the over lap varies from 1.6 mm to 20 mm.

19.2.5 Nozzle Efficiency

LO 19.15
Analyze nozzle efficiency

Due to friction between the fluid and walls of the nozzle, the expansion process is irreversible, although still approximately adiabatic. In nozzle design, the usual practice is to base all calculations on isentropic flow and then to make an allowance for friction using a coefficient or efficiency.

The nozzle efficiency, η_n, is defined as the ratio of the actual enthalpy drop to the isentropic enthalpy drop (Fig. 19.19),

$$\eta_n = \frac{h_0 - h_1}{h_0 - h_{1s}} \quad (19.48)$$

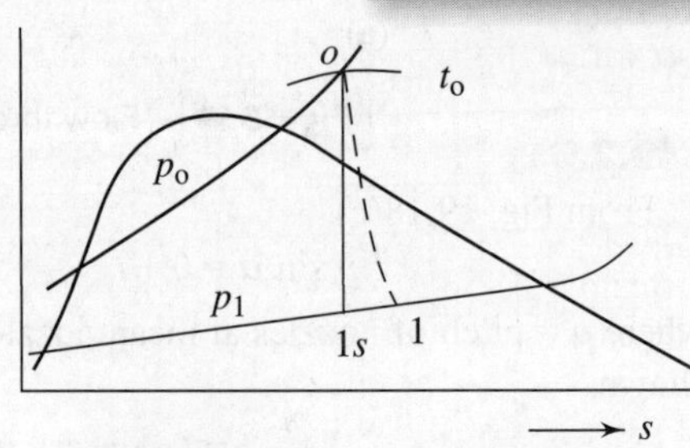

Fig. 19.19 Nozzle efficiency

For an ideal gas,

$$\eta_n = \frac{c_p (T_0 - T_1)}{c_p (T_0 - T_{1s})} = \frac{T_0 - T_1}{T_0 - T_{1s}} \quad (19.49)$$

If the actual velocity at exit from the nozzle is $\overline{V}_1$ and the velocity at exit when the flow is isentropic is $\overline{V}_{1s}$, then

$$\eta_n = \frac{\overline{V}_1^2 - \overline{V}_0^2}{\overline{V}_{1s}^2 - \overline{V}_0^2} \quad (19.50)$$

When the inlet velocity is very small, then

$$\eta_n = \frac{\overline{V}_1^2}{\overline{V}_{1s}^2} \quad (19.51)$$

Sometimes, a velocity coefficient ϕ is defined as the ratio of actual exit velocity to the exit velocity when the flow is isentropic between the same pressures,

$$\phi = \frac{\bar{V}_1}{\bar{V}_{1s}} \tag{19.52}$$

Therefore, $\phi = (\eta_n)^{1/2}$

Another coefficient which is frequently used is the ratio of the actual mass flow through the nozzle, ω, to the mass flow which would be passed if the flow were isentropic, ω_s. This is called the coefficient of discharge, C_d

$$C_d = \frac{\omega}{\omega_s} \tag{19.53}$$

The convergent part of a nozzle is usually sharp, while the divergent part is gradual. Most of the friction loss occurs in the divergent portion of the nozzle. If the semi-divergence angle α is large (Fig. 19.20), there will be flow separation from the wall with the formation of eddies, which entails energy loss. If the angle α is small, the length of the nozzle becomes large to provide the desired exit flow area, causing more energy loss due to friction. Usually, α varies from 5° to 8°.

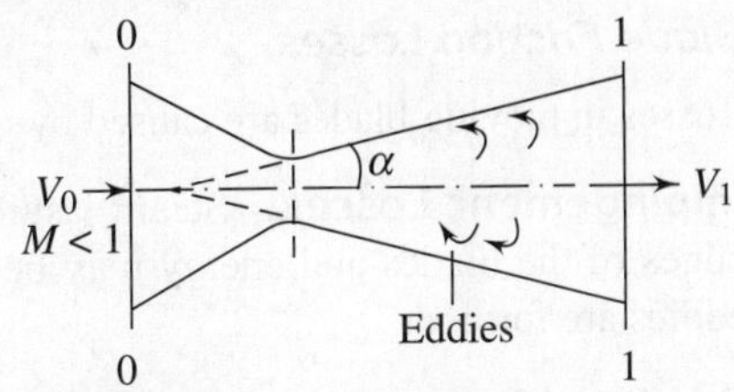

Fig. 19.20 Semi-divergence angle α

19.2.6 Losses in Steam Turbines

LO 19.16
Explain different losses in steam turbines

There are certain losses taking place during the expansion of steam within a turbine. These can be divided broadly into two groups.

1. Internal losses, which are connected with the flow of steam.
2. External losses, which occur outside the turbine casing.

The internal losses may be enumerated as follows:

(a) Losses in regulating valves
(b) Nozzle friction losses
(c) Blade friction losses
(d) Disc friction losses
(e) Partial admission losses
(f) Gland leakage losses
(g) Residual velocity losses
(h) Carry-over losses

Losses in Regulating Valves

Steam, before entering the turbine, passes through the main valve and the regulating valves, the flow through these being accompanied by pressure losses. Steam gets throttled adiabatically with constant enthalpy. However, the enthalpy drop in the turbine decreases yielding less specific output. Thus, some available energy of steam is lost due to the irreversible process of throttling. The pressure drop varies from 3 to 5% of the inlet steam pressure p_o.

Nozzle Friction Losses

The friction losses in nozzles were mentioned earlier. The effect of friction is taken care of by the nozzle efficiency. Losses are due to the growth of boundary layer and the formation of eddies in the wake, apart from the frictional resistance of walls, which varies with the height and length of passage. Losses are higher in a turbulent boundary layer than in a laminar one. In reaction turbine where pressure or enthalpy drop per stage is less due to lower velocity, the laminar condition persists over a greater length of passage. So, the friction loss is less than the impulse stage. However, due to the large number of stages, the total surface area exposed to flow is more, which increases the friction loss. Thus, the nozzle loss depends on its size, surface roughness, its length, roundness of entrance, divergence angle, space between nozzles, moisture and trailing edge.

Blade Friction Losses

Losses in moving blades are caused by various factors as enumerated below:

Impingement Losses Steam issuing out from the nozzles meets the leading edges of the blades and energy may be lost if the entry is not smooth enough and eddies are formed.

Frictional Losses Steam encounters these losses in the blade passages, which depends on the roughness of the blade surface.

Turning Losses These occur as the steam turns in the blade passage.

Wake Losses These occur at blade exit, depending on its shape and tip thickness.

The moving blade losses are taken care of by the blade friction coefficient ($k_b = V_{r_2}/V_{r_1}$) representing the reduction of relative velocity of steam from V_{r_1} to V_{r_2} due to friction.

Disc Friction Losses

When the turbine disc rotates in the viscous steam, there is surface friction loss due to relative motion between the disc and steam particles. Due to centrifugal force, steam is thrown radially outward. The moving disc surface exerts a drag on the steam, sets it in motion from root to tip, and produces a definite circulation. Some part of the kinetic energy of steam is lost due to this friction.

Partial Admission Loss

An impulse stage operating with partial admission, or an early stage in such a turbine with nozzles provided only over a part of the blade periphery, will have blades idle during part of the revolution [Fig. 19.21(a)]. Some portion of kinetic energy of the incoming steam is spent in clearing away the steam existing within the blade passage. These are called "scavenging losses" which together with disc friction losses are often referred to as "windage losses" in which some kinetic energy is imparted to the fluid at the expense of the kinetic energy of the blades. Since reaction turbines are designed for full peripheral admission, the windage loss as well as disc friction can be neglected.

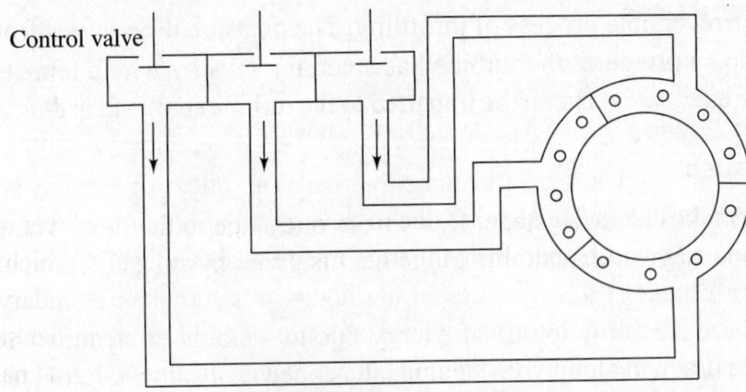

Fig. 19.21(a) Partial admission

Gland Leakage Losses

Leakage of steam can occur between stages and along the shaft at inlet and exit ends of the casing. Diaphragm leakage takes place in both impulse and reaction stages through the radial clearance between the stationary nozzle diaphragm and the shaft or drum. Tip leakage occurs in reaction stages through the clearance between the outer periphery of the moving blades and the casing because of the pressure difference existing across the blades. Shaft leakage occurs through the radial clearance between the shaft and casing at both high and low pressure ends of turbines. At the HP end, steam leaks out to the atmosphere, while at the LP end, the pressure being less than atmospheric, air leaks into the shell.

Since the leaked steam does not work on the blades, it represents energy loss. Both diaphragm and tip leakages can be minimized by reducing the radial clearances, but it must avoid rubbing or metal-to-metal contact. The clearance may be as low as 0.5 mm. However, proper balancing of the rotor, both static and dynamic, is a must to avoid any such rubbing. It is necessary to use seals or packing to further reduce the leakage flow. These seals may be labyrinths, carbon rings, water or steam seals, or gland leak-off. To prevent shaft leakage, labyrinths may be used with carbon rings and gland leak-off. Labyrinth seals consist of a series of thin strips fixed with the casing which maintain the smallest possible clearance with the shaft [Fig. 19.21(b)]. The small constrictions make the steam throttle to lower pressures many times, till only a very little quantity leaks out. Carbon ring seals, which consist of a ring of carbon divided into segments, have the rings fit snugly to the shaft by springs so as to prevent leakage, and may be used along with labyrinth glands in series in large turbines.

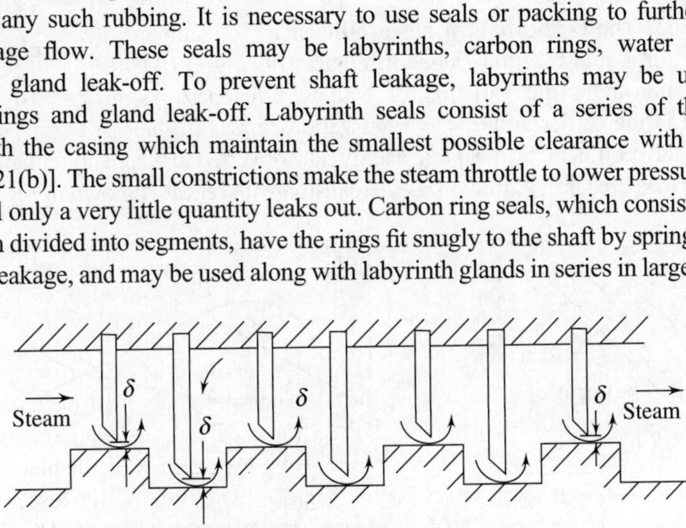

Fig. 19.21(b) Stepped labyrinth seals

Residual Velocity Loss

Steam leaving the last stage of the turbine has a certain velocity which represents an amount of kinetic energy that cannot be imparted to the turbine shaft and is thus wasted.

Carry-over Losses

Some energy loss takes place as steam flows from one stage to the next. The kinetic energy leaving one stage and available to the next is given by $\eta_{CO}(V_2^2/2)$, where η_{CO} is the carry-over efficiency.

In addition, there are some losses of energy due to wetness of steam (where the water particles are dragged along with steam at the expense of some K.E. of steam). If the quality of steam is less than 0.88, erosion and also corrosion can take place. Since the velocity of steam leaving the last stage of turbine is quite large (100–120 m/s), there will be energy losses due to friction in the exhaust hood of the turbine. Exhaust hoods to the condenser gradually increase in area like a diffusor and thus, there is a further decrease in velocity of steam and an increase in pressure as steam enters the condenser. Such hoods allow the turbine to operate down to a slightly lower pressure than that required by the condenser (depending on temperature and flow of cooling water, and air extraction from its shell), thus increasing the turbine work.

External Losses

There are some energy losses in the bearings and governing mechanisms which can be reduced by improving the lubrication systems. Some energy is consumed by oil pumps. Since the turbines are adequately insulated, the surface heat loss by radiation and convection is small. Modern large electric generators are hydrogen cooled, well designed, and very efficient, where the energy losses are within 2 to 3 per cent.

19.2.7 Reheat Factor and Condition Line

LO 19.17
Explain reheat factor and condition line

Figure 19.22 shows the typical stage efficiencies of simple impulse, Curtis and reaction stages in contrast to Fig. 19.31 which demonstrates the variation of blading efficiency. As discussed in the earlier article, there are various losses in the stage and the portion of the available energy not converted to work and remaining in the fluid is termed "reheat". A single-stage expansion with reheat is shown in Fig. 19.23.

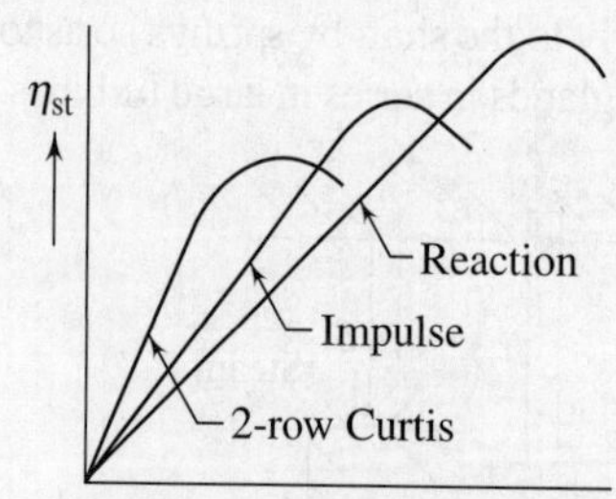

Fig. 19.22 Typical stage efficiencies

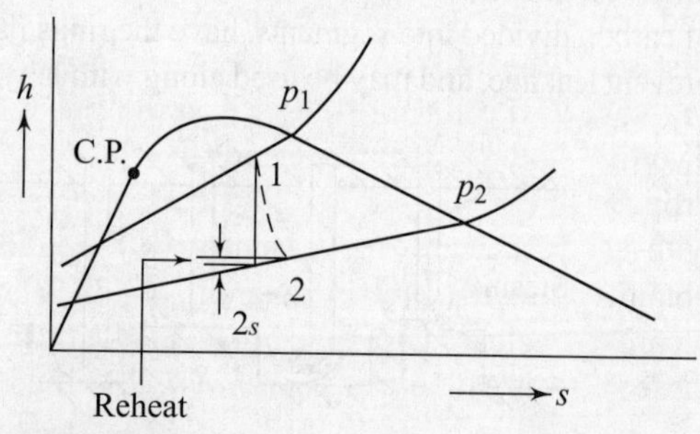

Fig. 19.23 Expansion in a single stage with reheat

$$\Delta h_s = h_1 - h_{2s} = \text{isentropic enthalpy drop}$$

$$\Delta h = h_1 - h_2 = \text{actual enthalpy drop}$$

$$\therefore \quad \text{Reheat} = h_2 - h_{2s}$$

Figure 19.24 shows the expansion in a 4-stage turbine, taking into account the effects of reheat. The following conclusions are drawn:

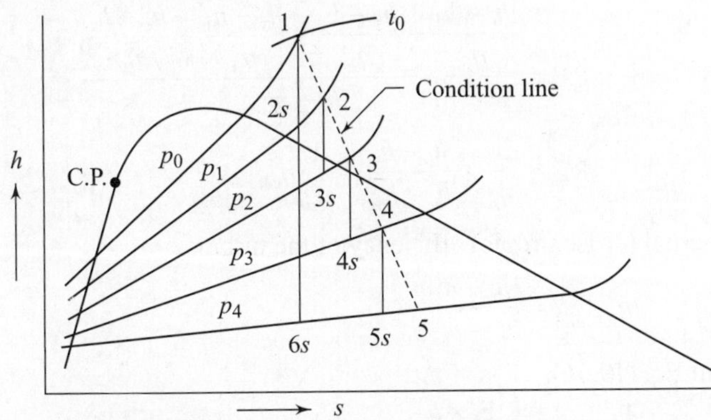

Fig. 19.24 Condition line for a four-stage turbine

1. Reheat takes place with increase in entropy.
2. The reheat in a given stage is available to do work in the succeeding stage except the last stage where the reheat is a loss.
3. The constant pressure lines diverge from one another, thereby increasing the enthalpy drop for the same pressure drop.
4. Because of (2) and (3), the sum of the available energies (isentropic enthalpy drops) for each stage is greater than the available energy (isentropic enthalpy drop) for the whole turbine.
5. The condition representing the actual expansion in the turbine is approximately the locus of points indicating the actual conditions of steam at the exit of each stage.

The effect of item 4 may be expressed, assuming equal available energy per stage, by a term called "*reheat factor*" (RF). The total isentropic enthalpy drop, $(h_1 - h_{6s})$, is divided equally into four parts in the four stages (Fig. 19.24) in the Mollier diagram, from which the interstage pressures p_1, p_2 and p_3 are noted, p_0 and p_4 being the boiler and condenser pressures, respectively. The isentropic enthalpy drop $(h_1 - h_{2s})$ in the first stage is multiplied by the stage efficiency to obtain the actual enthalpy drop $(h_1 - h_2)$, which is cut-off from 1 on the isobar at p_1 in the Mollier diagram. From 2, a vertical line is drawn which cuts the isobar at p_2 at state 3s. The isentropic enthalpy drop $(h_2 - h_{3s})$ is noted and the actual drop $(h_2 - h_3)$ is estimated by multiplying it with stage efficiency. In the same way, the enthalpy drops $(h_3 - h_{4s})$ and $(h_4 - h_{5s})$ are obtained from the Mollier diagram, and the state 5 is fixed. The states 2, 3, 4 and 5 record the actual conditions of steam at exit from each stage and the locus through these states is called the "*condition line*". The reheat factor is defined as

$$\text{RF} = \frac{(h_1 - h_{2s}) + (h_2 - h_{3s}) + (h_3 - h_{4s}) + (h_4 - h_{5s})}{h_1 - h_{6s}} \tag{19.54}$$

If the stage efficiency is assumed to be the same in all stages,

$$\eta_{st} = \frac{h_1 - h_2}{h_1 - h_{2s}} = \frac{h_2 - h_3}{h_2 - h_{3s}} = \frac{h_3 - h_4}{h_3 - h_{4s}} = \frac{h_4 - h_5}{h_4 - h_{5s}} \tag{19.55}$$

By substituting $h_1 - h_{2s} = (h_1 - h_2)/\eta_{st}$, $h_2 - h_{3s} = (h_2 - h_3)/\eta_{st}$ and so on in Eq. (19.54),

$$RF = \frac{\dfrac{h_1 - h_2}{\eta_{st}} + \dfrac{h_2 - h_3}{\eta_{st}} + \dfrac{h_3 - h_4}{\eta_{st}} + \dfrac{h_4 - h_5}{\eta_{st}}}{h_1 - h_{6s}}$$

or

$$RF = \frac{1}{\eta_{st}} \times \frac{h_1 - h_5}{h_1 - h_{6s}} = \frac{1}{\eta_{st}} \eta_{internal} \tag{19.56}$$

where the internal (or isentropic) efficiency of the turbine,

$$\eta_{internal} = \frac{h_1 - h_5}{h_1 - h_{6s}} \tag{19.57}$$

Thus, from Eq. (19.56),

$$\eta_{internal} = R.F. \times h_{stage} \tag{19.58}$$

Due to divergence of constant pressure lines, the reheat factor is greater than unity. The value of RF lies between 1.04 and 1.08. Therefore,

$$\eta_{internal} > \eta_{stage}$$

In other words, due to reheat, the expression $(h_2 - h_{2s}) + (h_3 - h_{3s}) + (h_4 - h_{4s}) + (h_5 - h_{5s})$ is greater than $(h_5 - h_{6s})$.

19.3 NEED OF A CONDENSER

LO 19.18
Identify need of a condenser

A condenser where the exhaust steam from the turbine is condensed operates at a pressure lower than atmosphere. There are two objects of using a condenser in a steam plant:

1. To reduce the turbine exhaust pressure so as to increase the specific output of the turbine (Fig. 19.25). If the circulating cooling water temperature in a condenser is low enough (say, 30 °C), it creates a low back pressure (vacuum) for the turbine. This

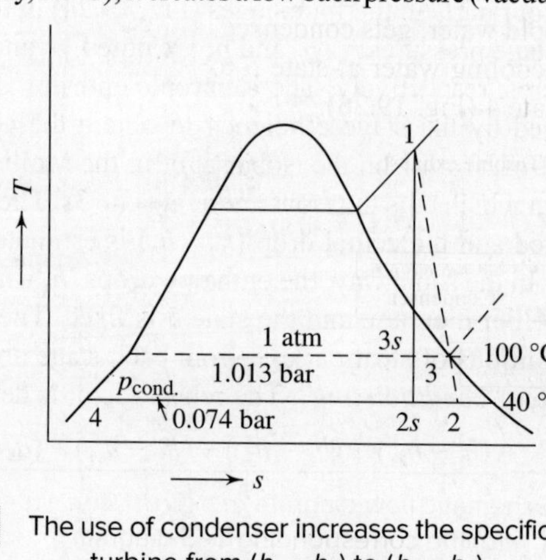

Fig. 19.25 The use of condenser increases the specific work output of turbine from $(h_1 - h_3)$ to $(h_1 - h_2)$

pressure is equal to the saturation pressure corresponding to the condensing steam temperature (say, 0.074 bar at 40 °C, Fig. 19.28), which, in turn, is a function of the cooling water temperature. It is known that the enthalpy drop or turbine work per unit pressure drop is much greater at the low pressure end than at the high pressure end of a turbine. A condenser, by lowering the back pressure, say, from 1.013 to 0.074 bar, thus increases the plant efficiency and reduces the steam flow for a given output. The lower the pressure, the greater the output and efficiency. Hence, it is important to use the lowest possible cooling water temperature. This restricts the temperature rise of cooling water in the condenser tubes to 5–8 °C so that the tube outer surface temperature remains low and, consequently, the condensing steam temperature is low and vacuum is high.

2. To recover high quality feedwater in the form of condensate and feed it back to the steam generator without any further treatment.

As a result, only the makeup water to replenish the water losses in the cyclic plant needs to be treated.

19.3.1 Types

LO 19.19
Discuss different types of condensers

There are two broad classes of condensers:

(a) *Direct-contact type condensers*, where the condensate and cooling water directly mix and come out as a single stream.

(b) *Surface condensers*, which are shell-and-tube heat exchangers where the two fluids do not come in direct contact and the heat released by the condensation of steam is transferred through the walls of the tubes into the cooling water continuously circulating inside them.

19.4 DIRECT-CONTACT CONDENSERS

Direct-contact condensers can be of three types:

(a) Spray condenser
(b) Barometric condenser
(c) Jet condenser

In a spray condenser, the cooling water is sprayed into the steam. Steam, by mixing directly with cold water, gets condensed. The exhaust steam from the turbine at state 2 mixes with cooling water at state 5 to produce saturated water at state 3, which is pumped to state 4 (Fig. 19.26).

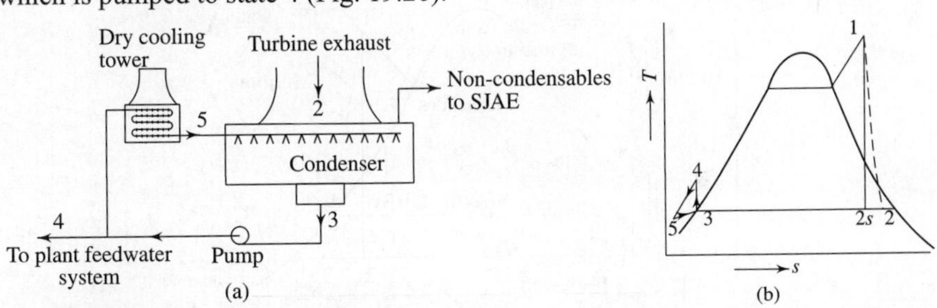

Fig. 19.26 (a) Schematic flow diagram of a direct-contact spray condenser, (b) the corresponding *T–S* diagram

Part of the condensate (ω_4), equal to the turbine exhaust flow (ω_2), is sent back to the plant as feedwater. The remainder is cooled in a dry cooling tower to state 5, and is then sprayed on to the turbine exhaust. Since the cooling water mixes with the steam and part of the condensate is used as feedwater, the water must be of high purity.

In a geothermal or OTEC (ocean thermal energy conversion) plant, only vacuum is required to be maintained in the condenser and no feedwater is needed. Hence, the mixture at state 4 is discarded.

For the spray condenser (Fig. 19.26), a mass balance and an energy balance give the following equations:

$$\omega_2 = \omega_4, \qquad \omega_3 = \omega_2 + \omega_5$$

$$\omega_2 h_2 + \omega_5 h_5 = \omega_3 h_3$$

from which
$$\frac{\mathrm{d}p}{c} = \left[\frac{\mathrm{d}p}{\mathrm{d}\rho}\right]^{1/2} \tag{19.59}$$

Since $h_2 - h_3$ is much greater than $h_3 - h_5$, the circulating water flow (ω_5) is much larger than the steam flow (ω_2).

In a barometric condenser [Fig. 19.27(a)], the cooling water is made to fall in a series of baffles to expose large surface area for the steam fed from below to come in direct contact. The steam condenses and the mixture falls in a tail pipe to the hot well below. By virtue of its static head, the tail pipe compresses the mixture to atmospheric pressure. Thus,

$$p_{\text{atm}} - p_{\text{cond}} + \Delta p_{\text{f}} = \rho g H \tag{19.60}$$

where ρ = density of mixture, H = height of tail pipe, and Δp_{f} is the pressure drop due to friction.

For low values of Δp_{f}, H is around 9.5 m. The higher is the value of H, the higher is the friction. Friction is lowered by increasing the tail pipe diameter, which results in a tall and heavy system.

In the jet condenser [Fig. 19.27(b)], the height of the tail pipe is reduced by replacing it with a diffuser. The diffuser helps raising the pressure in a short distance than a tail pipe.

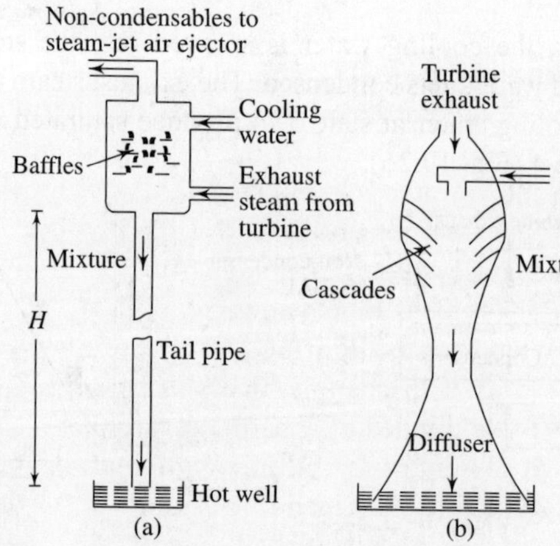

Fig. 19.27 Schematics of direct-contact condensers: (a) barometric, (b) diffuser or jet

In spray-type condensers, the non-condensable gases are usually removed with a steam jet air ejector (SJAE), as discussed later.

19.5 SURFACE CONDENSERS

Surface condensers are mostly used in power plants. They are essentially shell-and-tube heat exchangers. For the convenience of cleaning and maintenance, cooling water flows through the tubes and steam condenses outside the tubes. Figure 19.28 shows a surface condenser with two passes on the water side. It consists of a steel shell with water boxes on each side. The right water box is divided to allow for two water passes. At each end, there are tube sheets into which the water tubes are rolled. This prevents leakage of circulating water into the steam. An expansion joint allows for the different rates of expansion between the tubes and shell. There are vertical plates at intermediate points between the two tube sheets to provide support to the long tubes and to prevent tube vibration. The hot well acts as a reservoir of the condensate with a capacity equal to the total condensate flow for a certain period of time, say, 5 min.

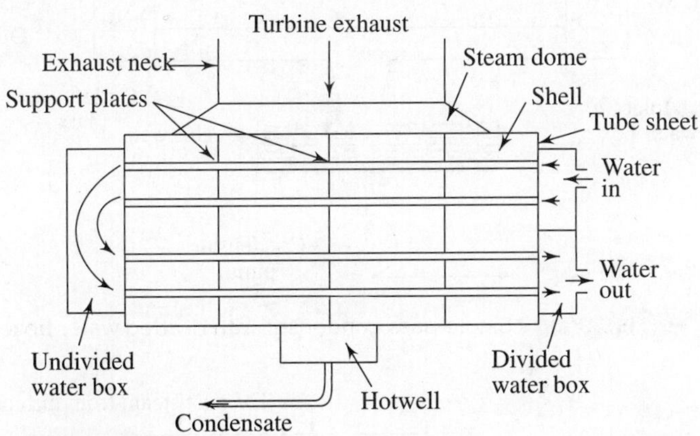

Fig. 19.28 Schematics of a two-pass surface condenser

In a single pass condenser (Fig. 19.29), cooling water flows through the tubes once, from one end to the other. Compared to a two-pass condenser, a single-pass condenser with the same number and size of the tubes and with the same water velocity requires twice as much water flow but results in half the water temperature rise and, therefore, lower condenser pressure. Thus, a single-pass condenser is good for overall plant efficiency and reduces thermal pollution, but requires more than twice the water flow and hence, four times the pumping power.

Apart from the division required to allow for a number of water passes, the water boxes are often further divided. For example, in a single-pass condenser, both the inlet and outlet water boxes are divided by a partition. This permits to take off half the condensing surface out of service for cleaning while water flows through the other half to keep the unit running at half load [Fig. 19.29(a)].

The tube arrangement in the condenser shell controls its effectiveness. Older designs tended to crowd a maximum number of tubes inside the shell to make available as much surface area as possible. Modern designs have steam lanes between tube banks to get maximum steam flow with least pressure drop and uniform distribution of steam in the shell. Tube lengths of 9–15 m are now quite common. Such long tubes result in a large rise in temperature of cooling water and the condensing ability decreases at the exit ends of the tubes. Thus, the tubes are closer at the cold end than at the hot end. To minimise the unequal distribution of steam flow from the turbine exhaust duct to condenser tubes, a well-tapered steam dome is added above the tube bundles. An expansion joint provided between the turbine exhaust and the condenser inlet permits the condenser to be rigidly mounted on the floor [Fig. 19.29(b)].

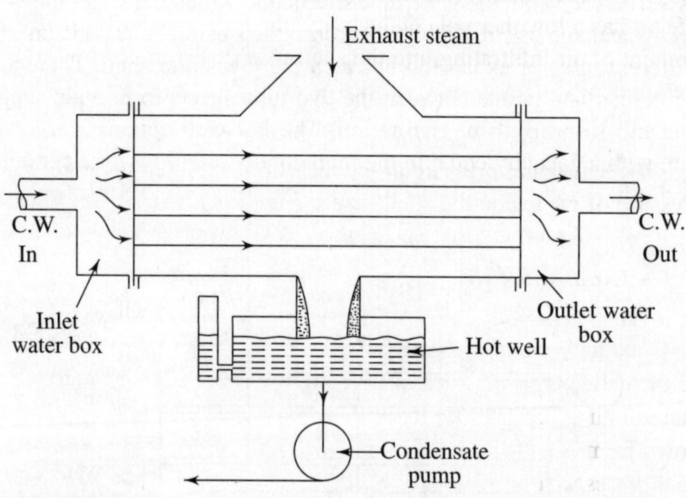

Fig. 19.29(a) Single-pass condenser with divided water boxes

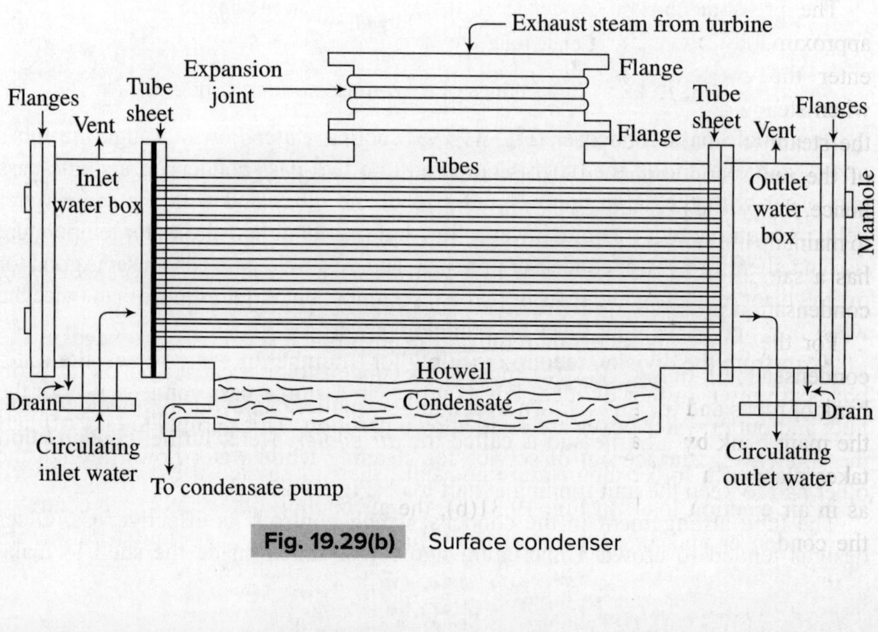

Fig. 19.29(b) Surface condenser

19.5.1 Air Removal

LO 19.20
Explain air removal in condensers

Air leaks into the condenser shell through flanges. Some air also comes along with steam, which has leaked into the exhaust end of the turbine along the shaft. This air affects the condenser performance badly because of the following reasons.

1. It reduces the heat transfer considerably.
2. It reduces the condenser vacuum and increases the turbine exhaust pressure, thus reducing the turbine output.

As air–water vapour mixture approaches the cold tube surface, water vapour condenses. Air, being non-condensable, forms an air film around the condensate film. Since air has a low thermal conductivity, the heat transfer is greatly reduced.

The amount of air infiltrating into the shell may be estimated by Dalton's law of partial pressures.

$$p_{sh} = p_{air} + p_{st} \tag{19.61}$$

where p_{st} is the saturation pressure at the measured shell temperature, t_{sh}, p_{air} is the partial pressure of air inside the shell, and p_{sh} is the measured total pressure of the shell (Fig. 19.30). Assuming that air behaves as an ideal gas at such a low pressure, we have

$$p_{air}\omega_s v_2 = \omega_a R_a (t_{sh} + 273) \tag{19.62}$$

where v_2 = specific volume of exhaust steam, ω_a = rate of air leakage, R_a = characteristic gas constant of air = 0.287 kJ/kg K. With p_{air} being known from Eq. (19.61), the rate of air leakage can be estimated from Eq. (19.62). This air has to be continuously removed from the condenser shell.

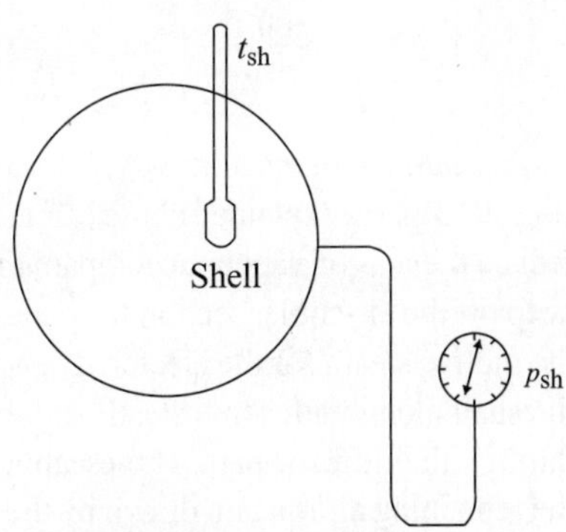

Fig. 19.30 Condenser shell pressure and temperature

The pressure in the condenser (p_{sh}) is approximately constant, and steam and air enter the condenser in fixed proportions when steady conditions prevail. As some of the steam is condensed, the partial pressure of the remaining steam (p_{st}) decreases and hence, the partial pressure of the air increases to maintain the same total pressure (Eq. 19.61). At reduced partial pressure, the steam has a saturation temperature which is below that of the incoming steam. Hence, condensation proceeds at progressively lower temperatures.

For the convenience of air removal, an air cooler section is provided in the condenser shell. In Fig. 19.31(a), most of the condensation is carried out on the main bank of tubes and the air is drawn over another smaller bank which is shielded from the main bank by a baffle and is called the *air cooler*. Here, further condensation takes place at a lower temperature and thus, there is saving in feedwater as well as in air ejection load. In Fig. 19.31(b), the air cooling tubes are in the centre of the condenser and air is removed from this section. The incoming steam passes all around the bank of tubes and some is drawn upward to the centre. In doing so, it meets the subcooled condensate falling below and thus reduces the amount of subcooling by heating it.

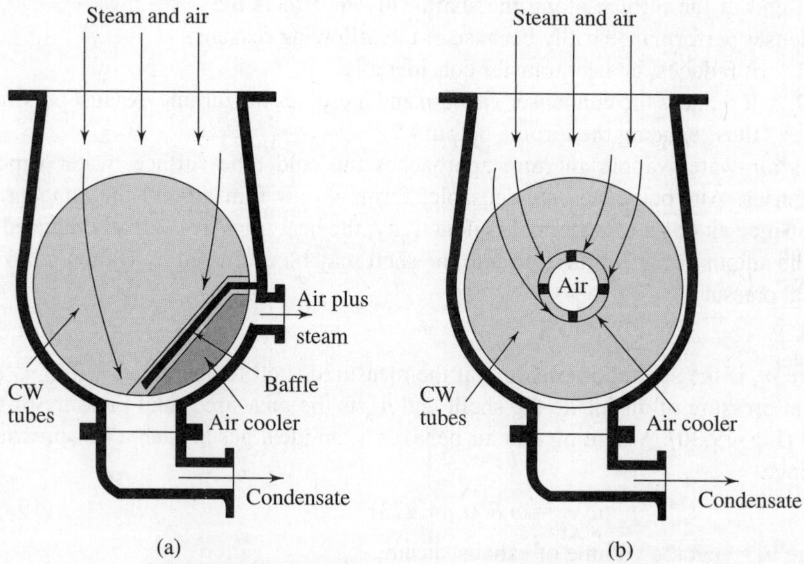

Fig. 19.31 Air cooler section in condensers

A steam jet air ejector, SJAE, is mostly used to remove air from the condenser shell (Fig. 19.32). As explained above, air is extracted from the coldest part of the condenser to reduce the loss of vapour accompanying it and hence, to reduce the ejector load. Baffles keep off the air-cooler section from the main shell. Steam bled from the turbine is used as the motive steam for the ejector. The vacuum created at the nozzle throat draws air from the shell along with some steam and the combined flow gets compressed while flowing through the diffusor part. The escaping vapour is condensed in the intercondenser and the remaining air is again drawn by the vacuum created at the nozzle throat of the second stage. The mixture of motive steam and air is further compressed in the diffusor. In the aftercondenser, the vapour is condensed and air, now at a pressure higher than atmosphere, is vented out. Cooling water flowing through both intercondenser and aftercondenser is the condensate from the hotwell prior to its flow through the feedwater heaters.

Sometimes, a term called 'vacuum efficiency' is often used in regard to a condenser. It is defined as

$$\text{Vacuum efficiency} = \frac{\text{Vacuum produced by steam at condenser inlet}}{\text{Barometric pressure} - \text{Saturation pressure at exhaust steam pressure}}$$

Another term called 'condenser efficiency' is also used sometimes, which is defined as

$$\text{Condenser efficiency} = \frac{\text{Actual temperature rise of cooling water}}{\text{Maximum temperature rise of cooling water}}$$

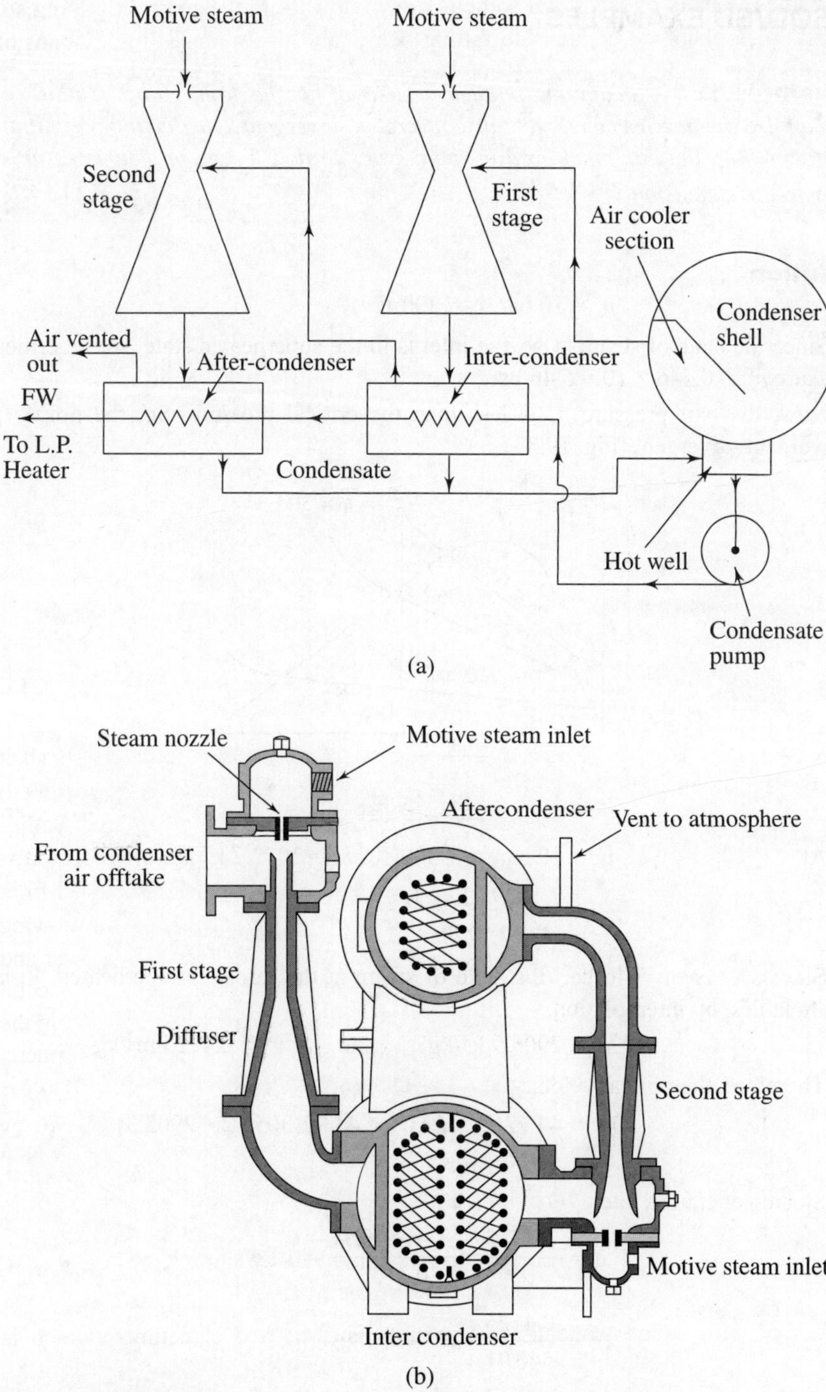

Fig. 19.32 Two-stage steam jet air ejector with inter-condenser and after condenser

SOLVED EXAMPLES

Example 19.1 *Steam is expanded in a set of nozzles from 10 bar, 300 °C to 2 bar. Are the nozzles convergent or convergent–divergent? Neglecting the initial velocity, find the minimum area of the nozzles to flow 1 kg/s of steam. Assume isentropic expansion.* E [LO 19.7]

Solution

$$p_0 = 10 \text{ bar}, t_0 = 300°\text{C}$$

Since the state of steam at nozzle inlet is in the superheated state, so the critical pressure $p^* = 0.546 \times 10 = 5.46$ bar.

Now, the exit pressure p_1 is less than the critical pressure. So, the nozzle is convergent–divergent (Fig. 19.33).

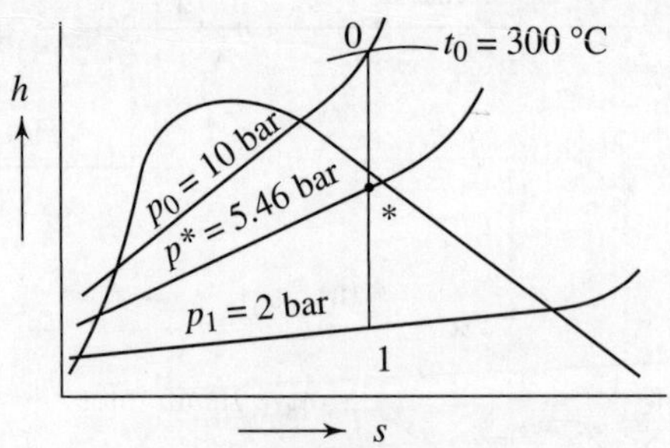

Fig. 19.33

At $p_0 = 10$ bar, $t_0 = 300$ °C, $h_0 = 3052.2$ kJ/kg,

$$s_0 = 7.1229 \text{ kJ/kg K}$$

$$s_0 = s^* = 7.1229$$

Since $s^* > s_g$ at 5.46 bar, the state of steam at the throat is superheated. From steam tables, by interpolation,

$$h^* = 2905.9 \text{ kJ/kg} \quad \text{and} \quad v^* = 0.4125 \text{ m}^3/\text{kg}$$

Therefore, the critical velocity

$$V^* = 44.72(h_0 - h^*)^{1/2} = 44.72(3052.2 - 2905.3)^{1/2}$$

$$= 540.91 \text{ m/s}$$

Steam generation rate,

$$\omega = \frac{A^* V^*}{v^*} = \frac{A^* \times 540.91}{0.4125} = 1 \text{ kg/s}$$

$$\therefore \quad A^* = \frac{0.4125}{540.91} \times 10^4 = 7.626 \text{ cm}^2$$

This is the minimum area of the nozzles. *Ans.*

Example 19.2 *Steam expands in a set of nozzles from 10 bar, 300 °C to 1 bar. The convergent parts of the nozzles are sharp and frictionless. In the divergent parts, the friction loss may be taken as 0.15 of the isentropic enthalpy drop. If the steam flow rate is 1 kg/s and the initial velocity of steam is negligible, find the minimum area of the nozzles. If the exit diameter of nozzles is 25 mm, find the number of nozzles.* Ⓜ [LO 19.7]

Solution

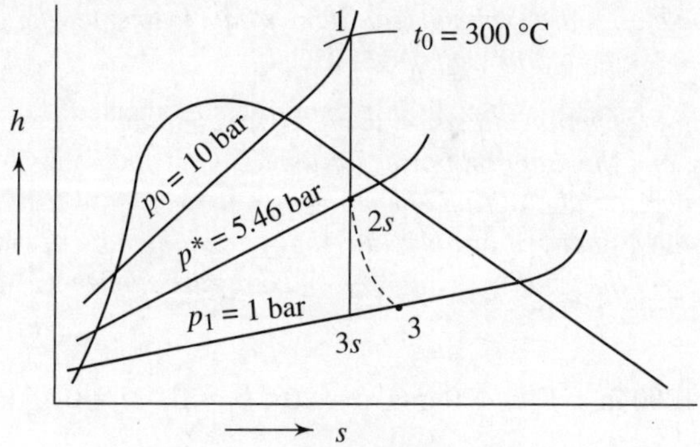

Fig. 19.34

The critical pressure at the throat,

$$p^* = 0.546 \times 10 = 5.46 \text{ bar (Fig. E19.34)}$$

$$h_1 = 3052.2 \text{ kJ/kg}, \qquad s_1 = 7.1276 \text{ kJ/kg K} = s_{2s}$$

$$h_{2s} = 2916.2 \text{ kJ/kg}$$

$$V^* = 44.72(h_1 - h_{2s})^{1/2} = 44.72(136)^{1/2} = 521.52 \text{ m/s}$$

$$h_{3s} = 2605 \text{ kJ/kg}$$

$$v_3 = 44.72[(h_1 - h_{2s}) + 0.85(h_{2s} - h_{3s})]^{1/2}$$

$$= 44.72[136 + 0.85(311)]^{1/2} = 894.79 \text{ m/s}$$

$$s_1 = s_{3s} = 7.1276 \text{ kJ/kg K} = 1.3025 + x_{3s} \times 6.0579$$

$$x_{3s} = 0.962$$

$$h_{3s} = 417.46 + 0.962 \times 2258.01 = 2589.67 \text{ kJ/kg}$$

$$h_{2s} - h_3 = 0.85(2916.2 - 2589.67) = 277.53 \text{ kJ/kg}$$

$$h_3 = 2638.67 \text{ kJ/kg} = 417.46 + x_3\, 2258.01$$

$$x_3 = 0.984$$

$$v_3 = 0.001043 + 0.984 \times 1.694 = 1.688 \text{ m}^3/\text{kg}$$

By interpolation (at 5.46 bar, 226 °C),

$$v^* = v_{2s} = 0.416 \text{ m}^3/\text{kg}$$

$$w_s = 1 \text{ kg/s} = (A^* V^*)/v^* = (A^* \times 521.52)/0.416$$

$\therefore$ $A^* = 7.98 \text{ cm}^2$ (minimum area) *Ans.*

Also, $$1 \text{ kg/s} = A_1 V_1 / v_3 = \frac{A_1 \times 894.79}{1.688}$$

$$\therefore \quad A_1 = 18.64 \text{ cm}^2 = n \times \frac{\pi}{4}(2.5)^2$$

$$n = 3.797 = 4$$ *Ans.*

Example 19.3 *The velocity of steam entering a simple impulse turbine is 1000 m/s, and the nozzle angle is 20°. The mean peripheral velocity of blades is 400 m/s and the blades are symmetrical. If the steam is to enter the blades without shock, what will be the blade angles?*

(a) Neglecting the friction effects on the blades, calculate the tangential force on the blades and the diagram power for a mass flow of 0.75 kg/s. Estimate also the axial thrust and diagram efficiency.

(b) If the relative velocity at exit is reduced by friction to 80% of that at inlet, estimate the axial thrust, diagram power and diagram efficiency. Ⓜ **[LO 19.4]**

Solution

Given: $V_1 = 1000$ m/s, $V_b = 400$ m/s, $\alpha = 20°$, $\beta_1 = \beta_2$, $\omega_s = 0.75$ kg/s (Fig. 19.35)

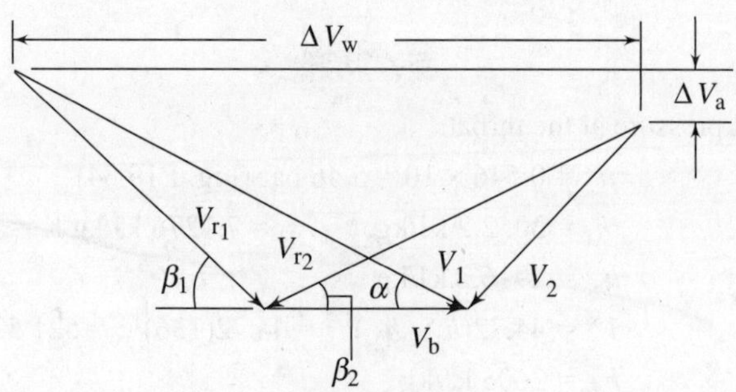

Fig. 19.35

(a)
$$k_b = 0$$

$$V_{r_1} \sin \beta_1 = V_1 \sin \alpha$$

$$V_{r_2} \cos \beta_1 = V_1 \cos \alpha - V_b$$

$$\beta_1 = \tan^{-1}\left[\frac{V_1 \sin \alpha}{V_1 \cos \alpha - V_b}\right] = \tan^{-1}\frac{1000 \sin 20°}{1000 \cos 20° - 400}$$

$$= \tan^{-1}\frac{342}{940 - 400} = 32.35° = \beta_2$$ *Ans.*

$$V_{r_1} \sin 32.35° = 342$$

$$\therefore \quad V_{r_1} = 639.25 \text{ m/s} = V_{r_2}$$

$$\Delta V_w = V_{r_1} \cos \beta_1 + V_{r_2} \cos \beta_2 = 2V_{r_1} \cos \beta_1$$

$$= 2 \times 639.25 \times \cos 20° = 1080.07 \text{ m/s}$$

$$\Delta V_a = V_{r_1} \sin \beta_1 - V_{r_2} \sin \beta_2 = 0$$

Tangential thrust is $P_t = \omega_s \Delta V_\omega = 0.75 \times 1080.07 = 810.05$ N

Diagram power, $\dot{W}_D = P_t \times V_b = 810.05 \times 400 = 324.02$ kW *Ans.*

$$\text{Diagram efficiency, } \eta_D = \frac{324.02 \text{ kW}}{\frac{1}{2} \times 0.75 \times 1000^2 \times 10^{-3} \text{ kW}}$$

$$= 0.864 \quad \text{or} \quad 86.4\%$$ *Ans.*

Axial thrust, $P_a = \omega_s \Delta V_a = 0$ *Ans.*

(b) $k_b = 0.8$

$$V_{r2} = 0.8\, V_{r1} = 0.8 \times 639.25 = 511.4 \text{ m/s}$$

$$\Delta V_w = 639.25 \cos 32.35° + 511.4 \cos 32.35° = 972.06 \text{ m/s}$$

Axial thrust, $P_a = \omega_s (V_{r_1} \sin \beta_1 - V_{r_2} \sin \beta_2)$

$$= 0.75 \times 127.85 \sin 32.35° = 51.3 \text{ N}$$ *Ans.*

Diagram power, $\dot{W}_D = 0.75 \times 972.06 \times 400 = 291.62$ kW *Ans.*

$$\text{Diagram efficiency, } \eta_D = \frac{291620}{\frac{1}{2} \times 0.75 \times 1000^2}$$

$$= 0.7776 \quad \text{or} \quad 77.76\%$$ *Ans.*

Example 19.4 *In a stage of an impulse turbine provided with a single row wheel, the mean diameter of the blade ring is 800 mm and the speed of rotation is 3000 rpm. The steam issues from the nozzles with a velocity of 300 m/s and the nozzle angle is 20°. The rotor blades are equiangular and the blade friction factor is 0.86. What is the power developed in the blading when the axial thrust on the blades is 140 newtons?* E [LO 19.4]

Solution

$$V_b = \frac{\pi D_m N}{60} = \frac{\pi \times 0.80 \times 3000}{60} = 125.6 \text{ m/s}$$

$$V_1 = 300 \text{ m/s}, \alpha = 20$$

$$\tan \beta_1 = \frac{V_1 \sin \alpha}{V_1 \cos \alpha - V_b} = \frac{300 \sin 20°}{300 \cos 20° - 125.6} = \frac{102.61}{281.91 - 125.6}$$

$$= 0.6565$$

$$\beta_1 = 33.3° = \beta_2$$

$$V_1 \sin \alpha = V_{r_1} \sin \beta_1$$

$$102.61 = V_{r_1} \sin 33.3°$$

$$\therefore \quad V_{r_1} = 187 \text{ m/s}$$

$$V_{r_2} = 0.86 \times 187 = 161 \text{ m/s}$$

Axial thrust, $P_a = \omega_s (V_{r_1} \sin \beta_1 - V_{r2} \sin \beta_2)$

$$= \omega_s V_{r_1} \sin \beta_1 (1 - k_b)$$

$$= \omega_s \times 187 \sin 33.3° (1 - 0.86)$$

$$= \omega_s \times 14.3654 = 140 \text{ N}$$

$$\therefore \quad \omega_s = 9.7456 \text{ kg/s}$$

$$\Delta V_w = V_{r_2} \cos \beta_2 + V_{r_1} \cos \beta_1 = V_{r_1} \cos \beta_1 (1 + k_b)$$

$$= 187 \cos 33.3° \times 1.86 = 290.71 \text{ m/s}$$

$$\therefore \quad \text{Power developed} = 9.7456 \times 290.71 \times 125.6 \times 10^{-3}$$

$$= 355.84 \text{ kW}$$

Ans.

Example 19.5 *The nozzles of the impulse stage of a turbine receive steam at 15 bar and 300 °C and discharge it at 10 bar. The nozzle efficiency is 95% and the nozzle angle is 20°. The blade speed is that required for maximum work, and the inlet angle of the blades is that required for entry of the steam without shock. The blade exit angle is 5° less than the inlet angle. The blade friction factor is 0.9. Calculate for a steam flow of 1350 kg/h (a) the axial thrust, (b) the diagram power, and (c) the diagram efficiency.* M **[LO 19.3, LO 19.4]**

Solution

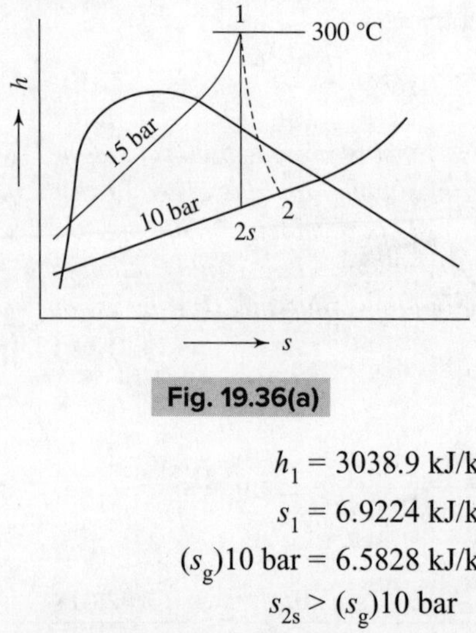

Fig. 19.36(a)

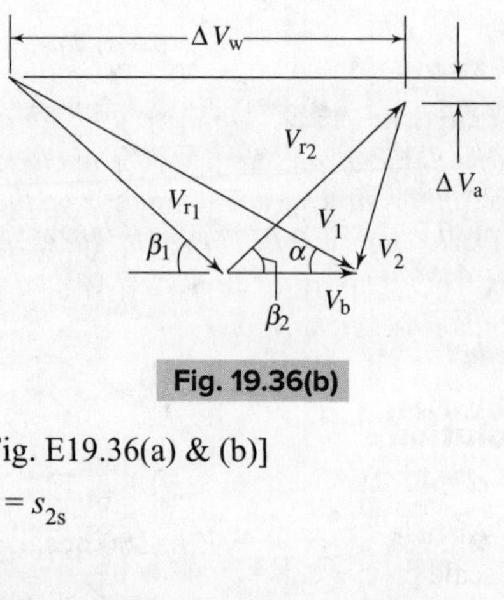

Fig. 19.36(b)

$$h_1 = 3038.9 \text{ kJ/kg [Fig. E19.36(a) \& (b)]}$$

$$s_1 = 6.9224 \text{ kJ/kg K} = s_{2s}$$

$$(s_g)10 \text{ bar} = 6.5828 \text{ kJ/kg K}$$

$$s_{2s} > (s_g)10 \text{ bar}$$

The state "2s" is in the superheated region

$$t_{2s} = 250\,°\text{C}, \; h_{2s} = 2943.1 \text{ kJ/kg}$$

$$V_1 = 44.72[(3038.9 - 2943.1)0.95]^{1/2} = 426.63 \text{ m/s}$$

$$\frac{V_b}{V_1} = \frac{\cos\alpha}{2} = \frac{\cos 20°}{2} = 0.4699$$

$$\therefore \quad V_b = 426.63 \times 0.4699 = 200.45 \text{ m/s}$$

$$\tan\beta_1 = \frac{V_1 \sin\alpha}{V_1 \cos\alpha - V_b} = \frac{145.916}{200.45} = 0.7279$$

$$\beta_1 = 36°$$

$$\therefore \quad \beta_2 = 36 - 5 = 31°$$

$$V_{r_1} = \frac{V_1 \sin\alpha}{\sin\beta_1} = 248.25 \text{ m/s}$$

$$V_{r_2} = k_b V_{r_1} = 223.42 \text{ m/s}$$

$$\Delta V_\omega = V_{r_1}\cos\beta_1 + V_{r_2}\cos\beta_2 = 392.35 \text{ m/s}$$

$$\Delta V_a = V_{r_1}\sin\beta_1 - V_{r_2}\sin\beta_2 = 30.85 \text{ m/s}$$

(a) Axial thrust, $P_a = \frac{1350}{3600} \times 30.85 = 11.57$ N *Ans.*

Tangential thrust, $P_t = \frac{1350}{3600} \times 392.35 = 147.13$ N

(b) Diagram power, $\dot{W}_D = 147.13 \times 200.45 \times 10^{-3}$

$= 29.492$ kW *Ans.*

(c) Diagram efficiency $\eta_D = \dfrac{29492}{\frac{1}{2} \times \frac{1350}{3600} \times (426.63)^2}$

$= 0.864$ or 86.4% *Ans.*

Example 19.6 *The following particulars apply to a two-row velocity compounded impulse stage of a turbine: nozzle angle 17°, mean blade speed 125 m/s; exit angles of the first row moving blades, the fixed blades, and the second row moving blades 22°, 26° and 30°, respectively; blade friction factor for each row 0.9. Assume that the absolute velocity of steam leaving the stage is in the axial direction. Draw the velocity diagrams for the stage and obtain (a) the absolute velocity of steam leaving the stage, (b) the diagram work, and (c) the diagram efficiency.* Ⓜ **[LO 19.4, LO 19.9]**

Solution

$\alpha = 17°$, $V_b = 125$ m/s, $\beta_2 = 22°$, $\alpha_1 = 26°$, $\beta_4 = 30°$, $k_b = 0.9$, $\delta_2 = 90°$

At first, the velocity diagrams for the second row of moving blades are drawn to scale (say, 1 cm = 50 m/s). Then those for the first row are drawn with the given particulars (Fig. 19.37).

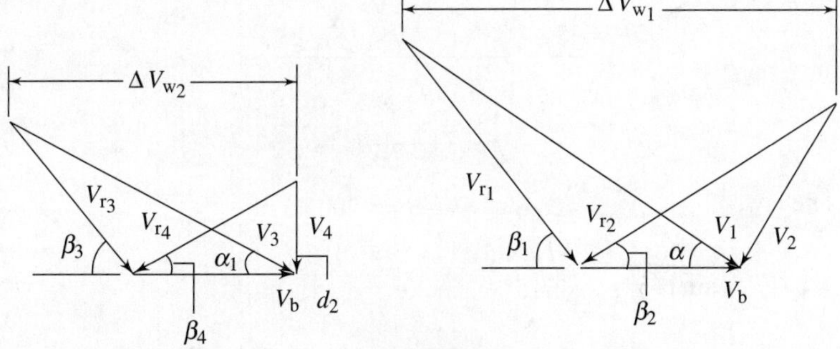

Fig. 19.37

ΔV_{w_1} and ΔV_{w_2} are measured from the diagrams.

$$\Delta V_{w_1} + \Delta V_{w_2} = (16 \text{ cm} + 4.8 \text{ cm}) \times 50 = 1040 \text{ m/s}$$

$$V_1 = 11.5 \text{ cm} \times 50 \frac{\text{m/s}}{\text{cm}} = 575 \text{ m/s}$$

(a) Velocity of steam exiting the stage, V_4 as read from the diagram

$$= 1.5 \text{ cm} \times 50 \frac{\text{m/s}}{\text{cm}} = 75 \text{ m/s}$$ *Ans.*

(b) Diagram work,

$$W_D = 1040 \times 1255 = 130{,}000 \text{ J/kg}$$
$$= 130 \text{ kJ/kg}$$ *Ans.*

(c) Diagram efficiency,

$$\eta_D = \frac{130{,}000}{\frac{1}{2} \times 575 \times 575}$$
$$= 0.7863 \quad \text{or} \quad 78.63\%$$ *Ans.*

Example 19.7 *The angles at inlet and discharge of the blading of a 50% reaction turbine are 35° and 20°, respectively. The speed of rotation is 1500 rpm and at a particular stage, the mean ring diameter is 0.67 m and the steam condition is at 1.5 bar, 0.96 dry. Estimate (a) the required height of blading to pass 3.6 kg/s of steam, and (b) the power developed by the ring.* E **[LO 19.4, LO 19.11]**

Solution

Given: $\beta_1 = 35° = \delta$, $\beta_2 = \alpha = 20°$ (Fig. E19.38)

$$V_b = \frac{\pi D_m N}{60} = \frac{\pi \times 0.67 \times 1500}{60} = 52.62 \text{ m/s}$$

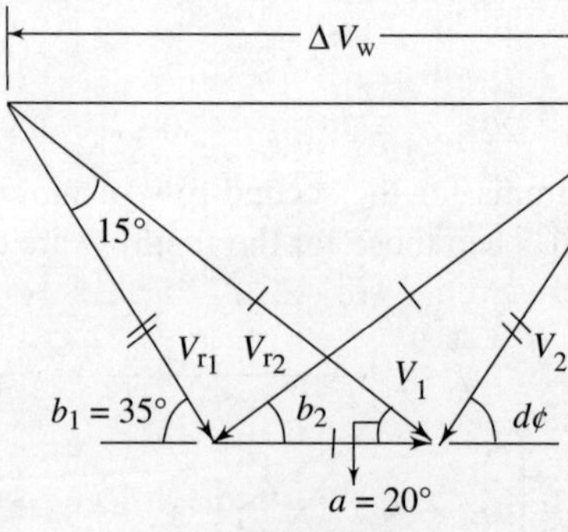

Fig. 19.38

By sine law,

$$\frac{V_1}{\sin 145°} = \frac{V_b}{\sin 15°} = \frac{V_{r_1}}{\sin 20°}$$

$$V_1 = 52.62 \times \frac{0.5736}{0.2588} = 116.63 \text{ m/s} = V_{r_2}$$

$$V_{r_1} = 52.62 \times \frac{0.342}{0.2588} = 69.54 \text{ m/s} = V_2$$

$$\Delta V_w = V_1 \cos\alpha + V_2 \cos\delta$$
$$= 116.63 \cos 20° + 69.54 \cos 35°$$
$$= 109.6 + 56.964 = 166.564 \text{ m/s}$$

$$v_1 = 0.001052 + 0.96 \times 1.15937 = 1.114 \text{ m}^3\text{/kg}$$

Now, $\omega_s = 3.6$ kg/s

$$= \frac{\pi D_m h_b V_1 \sin\alpha}{v_1} = \frac{\pi \times 0.67 \times h_b \times 116 \cdot 63 \sin 20°}{1.114}$$

$\therefore$ $h_b = 0.0478 \text{ m} = 47.8 \text{ mm}$ *Ans.*

Power developed by the ring

$$= \omega_s \, \Delta V_w \, V_b = 3.6 \times 166.564 \times 52.62 \times 10^{-3}$$
$$= 31.552 \text{ kW}$$ *Ans.*

Example 19.8 *A Parsons reaction (50%) turbine running at 400 rpm develops 5 MW using 6 kg/kWh of steam flow. The exit angle of the blades is 20° and the velocity of steam relative to the blades at exit is 1.35 times the mean blade speed. At a particular stage in the expansion, the pressure is 1.2 bar and the steam quality is 0.95. Calculate for this stage (a) a suitable blade height, assuming the ratio of D_m/h_b as 12, and (b) the diagram power.* E **[LO 19.4, LO 19.11]**

Solution

$V_{r_1} = V_2$, $V_1 = V_{r_2} = 1.35\, V_b$, $\beta_1 = \delta$, $\beta_2 = \alpha = 20°$ (Fig. 19.39)

$$\omega_s = 6 \text{ kg/kWh} = \frac{6 \times 5000}{3600} = \frac{25}{3} \text{ kg/s}$$

$$V_b = \frac{\pi D_m N}{60} = \frac{\pi \times 12 h_b \times 400}{60} = 80\, \pi h_b$$

$$V_1 = 1.35 \times 80\, \pi h_b = 108\, \pi h_b$$

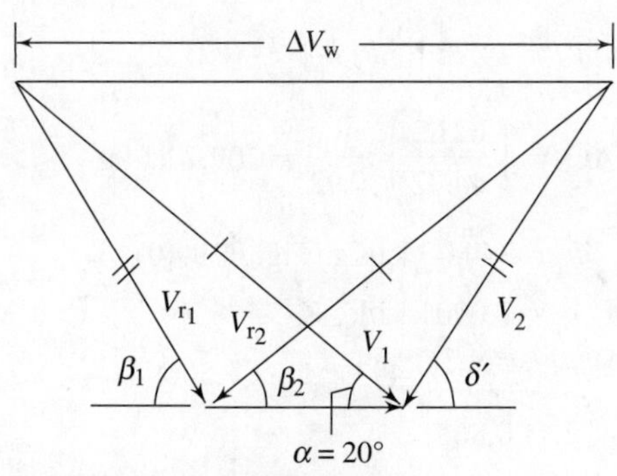

Fig. 19.39

v_1 at 1.2 bar, 0.95 dry = 0.0010468 + 0.95 (1.454) = 1.381 m^3/kg Volume flow of steam,

$$\omega_s v_1 = \pi D_m h_b V_1 \sin\alpha\, k_{tb}$$

$$\frac{25}{3} \times 1.381 = \pi(12h_b)h_b\,(108\pi h_b)\sin 20°$$

$$\therefore \quad h_b = 0.138 \text{ m} = 138 \text{ mm} \qquad \textit{Ans.} \text{ (a)}$$

$$V_b = 80\pi \times 0.138 = 34.67 \text{ m/s}$$

$$V_1 = 1.35 \times 34.67 = 46.8 \text{ m/s}$$

$$\Delta V_w = 2V_1 \cos\alpha - V_b = 2 \times 46.8 \cos 20° - 34.67 = 53.28 \text{ m/s}$$

Diagram power,

$$\dot{W}_D = \omega_s \Delta V_W V_b$$

$$= \frac{25}{3} \times 53.28 \times 34.67 \times 10^{-3} \text{ kW}$$

$$= 15.39 \text{ kW} \qquad \textit{Ans.} \text{ (b)}$$

Example 19.9 *A steam turbine is to develop 8 MW at 5000 rpm for driving a compressor. The steam enters at 40 bar, 500 °C and exhausts at 0.1 bar. The internal efficiency of the turbine is 0.85 and its mechanical efficiency is 0.96. Estimate (a) the number of impulse stages required if similar impulse stages are used throughout and (b) the nozzle height for the first stage with full admission. Assume nozzle efficiency as 0.92, nozzle angle 15°, limiting blade velocity 300 m/s, and the blades operating at maximum efficiency.* D **[LO 19.13]**

Solution

For maximum blading efficiency,

$$\frac{V_b}{V_1} = \cos a/2$$

$$V_1 = \frac{2 \times 300}{\cos 15°} = 621.2 \text{ m/s}$$

Now, $$V_1 = 44.72\,[(\Delta h)_s h_n]^{1/2} = 621.2 \text{ m/s}$$

$$\therefore \quad (\Delta h)_s = \left(\frac{621.2}{44.72}\right)^2 \frac{1}{0.92} = 209.7 \text{ kJ/kg}$$

$$h_1 = 3445.3 \text{ kJ/kg (Fig. E19.40)}$$

$$s_1 = 7.0901 \text{ kJ/kg K}$$

$$= s_2 = 0.6493 + x_{2s} \times 7.5009$$

$$x_2 = 0.86$$

$$h_{2s} = 191.83 + 0.86 \times 2392.8$$

$$= 2246.5 \text{ kJ/kg}$$

$$h_1 - h_{2s} = 1198.8 \text{ kJ/kg}$$

∴ Number of stages required

$$= \frac{h_1 - h_{2s}}{(\Delta h)_{1s}} = \frac{1198.8}{209.7} = 5.72 \quad \text{or} \quad 6 \text{ stages}$$ *Ans.* (a)

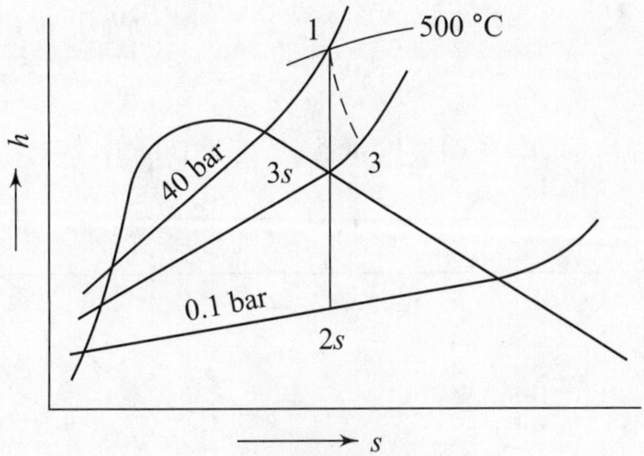

Fig. 19.40

(b) Isentropic output

$$= \frac{8000}{0.85 \times 0.96} = \omega_s \times 1198.8$$

$$\therefore \quad \omega_s = \frac{8000}{0.85 \times 0.96 \times 1198.8} = 8.178 \text{ kg/s} = 29.44 \text{ t/h}$$

$$h_1 - h_3 = 0.92 \times 209.7 = 193 \text{ kJ/kg}$$

$$h_3 = 3252.3 \text{ kJ/kg}$$

From Mollier chart,

$$v_3 = 0.17 \text{ m}^3/\text{kg}$$

$$\omega_s = \frac{(A_1 V_1)}{v_1} = \frac{A_1 \times 621.2}{0.17} = 8.178 \text{ kg/s}$$

$$A_1 = 0.002238 \text{ m}^2 = \pi D_m h_n \sin\alpha$$

$$V_b = 300 \text{ m/s} = \frac{\pi D_m \times 5000}{60}$$

$$\therefore \quad \pi D_m = 3.6 \text{ m}$$

$$\therefore \quad 3.6 \times h_n \times \sin 15° = 0.002238$$

$$h_n = 0.0024 \text{ m} = 2.4 \text{ mm}$$ *Ans.*

Example 19.10 *At a certain point in a 50% reaction turbine, the steam leaving a moving blade row is at 1.5 bar, 0.90 dry. The steam flow rate is 7 kg/s and the turbine speed is 3000 rpm. At entry to the moving blade row, the axial velocity of flow is 0.7 times and at exit from the row 0.75 times the mean blade velocity. The exit angles of both fixed and moving blades are 20°, measured from the plane of rotation, and the height of moving blades at exit is 1/10 of the mean diameter. Determine the height of the moving blades at exit and the power developed in the blade row.* **E [LO 19.4, LO 19.11]**

Solution

$\alpha = \beta_2 = 20°$, $V_1 \sin\alpha = 0.7\, V_b$, $V_2 \sin\delta = 0.75\, V_b$ (Fig. E19.41)

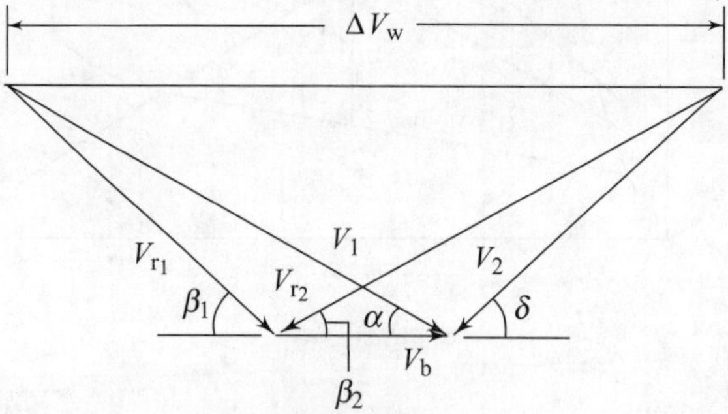

Fig. 19.41

$$\pi D_m h_b V_2 \sin\delta = \omega_s v_{exit}$$

$$v_{exit} = 0.001052 + 0.9 \times 1.15937 = 1.045 \text{ m}^3/\text{kg}$$

$$\pi D_m (0.1 D_m) 0.75 \frac{\pi D_m \times 3000}{60} = 7 \times 1.045$$

$$D_m^{\,3} = \frac{7 \times 1.045 \times 60}{\pi \times 0.1 \times 0.75 \times \pi \times 3000} = 0.19764$$

$$D_m = 0.582 \text{ m}$$

$$\therefore \quad h_b = 0.0582 \text{ m} = 58.2 \text{ mm} \qquad \textit{Ans.}$$

$$V_b = \frac{\pi \times 0.582 \times 3000}{60} = 91.42 \text{ m/s}$$

$$\Delta V_w = 2V_1 \cos\alpha - V_b = 2\frac{0.7\, V_b}{\sin\alpha} \cos\alpha - V_b$$

$$= \left(\frac{1.4}{\tan 20°} - 1\right) V_b = 260.22 \text{ m/s}$$

$$\text{Power developed} = \omega_s \Delta V_w V_b = 7 \times 260.22 \times 91.42 \times 10^{-3}$$

$$= 166.53 \text{ kW} \qquad \textit{Ans.}$$

Example 19.11 *Steam at 20 bar, 400 °C expands in a steam turbine to 0.1 bar. There are four stages in the turbine and the total enthalpy drop is divided equally among the stages. The stage efficiency is 75% and it is the same in all the stages. Determine the interstage pressures, the reheat factor and the turbine internal efficiency.* E [LO 19.17]

Solution

From Mollier chart (Fig. E19.42),

$$h_1 - h_{6s} = 3250 - 2282 = 968 \text{ kJ/kg}$$
$$h_1 - h_{2s} = 968/4 = 242 \text{ kJ/kg}$$

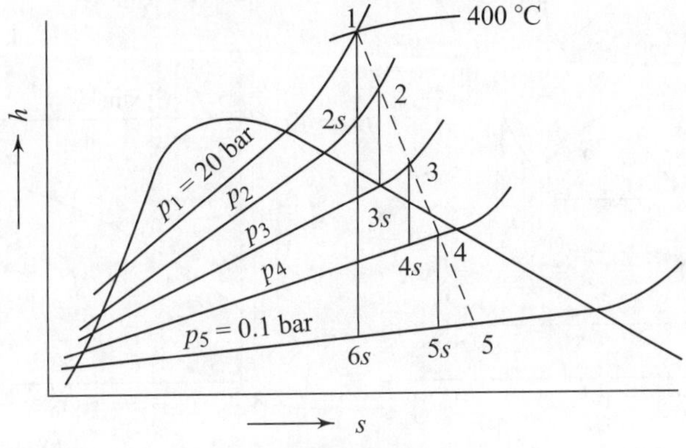

Fig. 19.42

The interestage pressures, p_2, p_3 and p_4 as read from the Mollier chart are 8 bar, 2.6 bar and 0.60 bar, respectively.

$$h_1 - h_2 = 0.75 \times 242 = 181.5 \text{ kJ/kg}$$
$$h_2 - h_{3s} = 3060 - 2800 = 260 \text{ kJ/kg}$$
$$h_2 - h_3 = 0.75 \times 260 = 195 \text{ kJ/kg}$$
$$h_3 - h_{4s} = 2870 - 2605 = 265 \text{ kJ/kg}$$
$$h_3 - h_4 = 0.75 \times 270 = 202.5 \text{ kJ/kg}$$
$$h_4 - h_{5s} = 2680 - 2410 = 270 \text{ kJ/kg}$$
$$h_4 - h_5 = 0.75 \times 270 = 202.5 \text{ kJ/kg}$$
$$h_5 = 2470 \text{ kJ/kg}, x_5 = 0.958$$

$$\text{Reheat factor} = \frac{(h_1 - h_{2s}) + (h_2 - h_{3s}) + (h_3 - h_{4s}) + (h_4 - h_{5s})}{h_1 - h_{6s}}$$

$$= \frac{242 + 260 + 265 + 270}{984} = \frac{1037}{968} = 1.071$$

$$\eta_{\text{internal}} = \frac{h_1 - h_5}{h_1 - h_{6s}} = \frac{780}{968} = 0.805 \quad \text{or} \quad 80.5\%$$

Also, $\eta_{\text{internal}} = h_{\text{stage}} \times \text{RF} = 0.75 \times 1.071 = 0.803$ or, 80.3% *Ans.*

Example 19.12 *A surface condenser receives 250 t/h of steam at 40 °C with 12% moisture. The cooling water enters at 32 °C and leaves at 38 °C. The pressure inside the condenser is found to be 0.078 bar. The velocity of circulating water is 1.8 m/s. The condenser tubes are of 25.4 mm od and 1.25 mm thickness. Taking the overall heat transfer coefficient as 2600 W/m²K, determine (a) the rate of flow of cooling water, (b) the rate of air leakage into the condenser shell, (c) the length of tubes, and (d) the number of tubes.* D [LO 19.19]

Solution

As shown in Fig. E19.43,

$$x_2 = 0.88,\ h_2 = h_f + 0.88h_{fg},\ h_3 = h_f$$

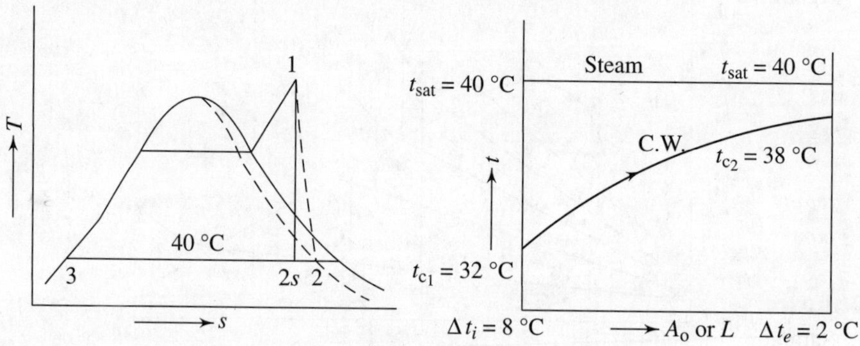

Fig. 19.43

At 40 °C, $h_{fg} = 2407$ kJ/kg

$$h_2 - h_3 = 0.88h_{fg} = 0.88 \times 2407 = 2118.16 \text{ kJ/kg}$$

By energy balance,

$$\omega_s(h_2 - h_3) = \omega_c c_{pc}(t_{c_2} - t_{c_1})$$

$$\frac{250 \times 1000}{3600} \times 2118.16 = \omega_c \times 4.187(38 - 32)$$

$$\omega_c = 5855.2 \text{ kg/s}$$ *Ans.* (a)

At 40 °C, $p_{sat} = 0.07375$ bar

$$p = p_{sat} + p_{air} = 0.078 \text{ bar}$$

$$p_{air} = 0.078 - 0.07375 = 0.00425 \text{ bar} = 0.425 \text{ kN/m}^2$$

$$v_2 = v_f + x_2 v_{fg} = 0.001008 + 0.88 \times 19.544 = 17.2 \text{ m}^3\text{/kg}$$

Now, $p_{air}\,\omega_s v_2 = \omega_{air} R_{air} T_{sh}$

$$0.425 \times \frac{250 \times 1000}{3600} \times 17.2 = \omega_{air} \times 0.287 \times 313$$

$$\omega_{air} = 5.651 \text{ kg/s}$$ *Ans.* (b)

$$\Delta t_{1.m} = \frac{\Delta t_i - \Delta t_e}{\ln(\Delta t_i / \Delta t_e)} = \frac{8 - 2}{\ln \frac{8}{2}} = 4.33\text{°C}$$

$$Q = U_0 A_0 \Delta t_{1.m} = \omega_s (h_2 - h_3)$$

$$2.6 \times A_0 \times 4.33 = \frac{250 \times 1000}{3600} \times 2118.16$$

$$A_0 = 13066 \text{ m}^2$$

$$\omega_c = \left(n \frac{\pi}{4} d_i^2 \right) \rho V = 5855.2 \text{ kg/s}$$

$$n \frac{\pi}{4} (25.4 - 2.5)^2 \times 10^{-6} \times 1000 \times 1.8 = 5855.2$$

$$n = \frac{5.8552 \times 4 \times 10^6}{524.41 \times \pi \times 1.8} = 7898$$ *Ans.* (d)

Again, $$A_0 = n\pi d_0 l = 13066 \text{ m}^2$$

$$l = \frac{13066}{7898 \times \pi \times 25.4 \times 10^{-3}} = 20.73 \text{ m}$$ *Ans.* (c)

Example 19.13 *Steam enters the condenser at 35°C. The condenser vacuum is 70 cm of mercury when the barometer reads 75.5 cm of Hg. Determine the vacuum efficiency. Estimate the mass of air present in the condenser per kg of steam.* E [LO 19.20]

Solution

Saturation pressure of steam at 35°C = 0.05622 bar and $v_g = 25.24$ m³/kg

1std. atm. pr = 76 cm Hg = 1.013 bar.

$$\therefore \quad p_{sat} = 0.05622 \text{ bar} = 4.2717 \text{ cm Hg}$$

$$P_{abs} = 75.5 - 70 = 5.5 \text{ cm Hg}$$

Vacuum gauge corrected to standard atmosphere

$$= 76 - 5.5 = 70.5 \text{ cm Hg}$$

$$\therefore \text{ Vacuum efficiency} = \frac{70.5}{76 - 4.2717} = 0.9829 \text{ or } 98.29\%$$ *Ans.*

Absolute pressure inside the condenser = 5.5 cm Hg

Partial pressure of steam at 35°C = 4.2717 cm Hg

∴ Partial pressure of air = 5.5 – 4.2717

= 1.2283 cm Hg

∴ Mass of air associated with 1 kg steam

$$= \frac{pv}{Rt} = \frac{1.2283}{76} \times \frac{1.013 \times 10^5 \times 25.24}{287 \times (273 + 35)}$$

= 0.467 kg air/kg steam *Ans.*

Example 19.14 *Exhaust steam having a quality of 0.9 enters a surface condenser at an absolute pressure of 0.13 bar and comes out as water at 45°C. The circulating water enters at 30°C and leaves at 40°C. Estimate the quantity of circulating water and the condenser efficiency.* E [LO 19.20]

Solution

At 0.13 bar, $t_{sat} = 51.06°C$

$$h_2 = h_f + x_2 h_{fg}$$
$$= 213 + 0.9 \times 2380.3$$
$$= 2355.97 \text{ kJ/kg}$$
$$h_f \text{ at } 45°C = 188.35 \text{ kJ/kg}$$

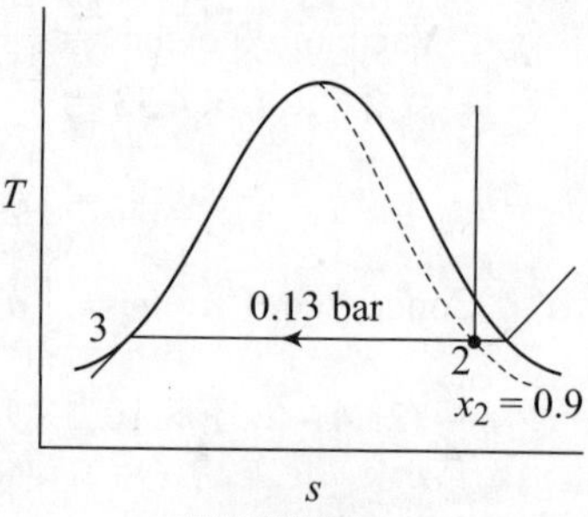

Fig. 19.44

By energy balance,

$$\dot{m}_w c_{pw}(t_{w_2} - t_{w_1}) = \dot{m}_s (h_2 - h_3)$$

$$\therefore \quad \dot{m}_w / \dot{m}_s = \frac{2355.97 - 188.35}{4.182 \times (40 - 30)}$$

$$= 51.77 \frac{\text{kg water}}{\text{kg steam}}$$ *Ans.*

$$\text{Condenser efficiency} = \frac{40 - 30}{51.06 - 30} = 0.475 \text{ or } 47.5\%$$ *Ans.*

Example 19.15 *The following readings were taken during a test on a surface condenser:*
Mean condenser temperature = 35°C, hot well temperature = 30°C, condenser vacuum = 69 cm Hg, barometer reading 76 cm Hg. Condensate collected 16 kg/min. Cooling water enters at 20°C and leaves at 32.5°C, flow rate being 37,500 kg/h. Calculate (a) the mass of air present per cubic metre of the condenser, (b) quality of steam at condenser inlet, (c) vacuum efficiency, and (d) condenser efficiency. M **[LO 19.20]**

Solution

Absolute pressure inside the condenser = 76 – 69 = 7 cm Hg = 0.0933 bar

$$p_{sat} \text{ at } 35°C = 0.05622 \text{ bar}$$

$$\therefore \quad \text{Partial pressure of air, } p_a = 0.0933 - 0.05622$$
$$= 0.03708 \text{ bar}$$

$$\therefore \text{Mass of air present, } m_a = \frac{pv}{RT}$$

or, $$m_a = \frac{0.03708 \times 10^5 \times 1}{287 \times (273 + 35)} = 0.042 \text{ kg/m}^3$$ *Ans.* (a)

Let x be the quality of steam at condenser inlet.

$$\dot{m}_s [x h_{fg} + c_p(t_{sat} - 30)] = \dot{m}_w c_{pw}(t_{w_2} - t_{w_1})$$

$$16 \times 60[x + 2418.8 + 4.182(35 - 30)] = 37{,}500 \times 4.182 \times (32.5 - 20)$$
$$= 1.96 \times 10^6$$

$$\therefore \quad x = 0.836$$ *Ans.* (b)

$$\text{Vacuum efficiency} = \frac{\text{Actual vacuum}}{\text{Ideal vacuum}}$$

$$= \frac{69}{79 - 4.22} = 096.13 \text{ or } 96.13\% \qquad \textit{Ans.} \text{ (c)}$$

$$\text{Condenser efficiency} = \frac{\text{Actual temperature rise}}{\text{Maximum temperature rise}}$$

$$= \frac{12.5}{35 - 20} = 0.833 \text{ or } 83.33\% \qquad \textit{Ans.} \text{ (d)}$$

Example 19.16 *A surface condenser receives 20 t/h of dry saturated steam at 40 °C. The air leakage to the condenser is estimated to be 0.35 kg per 1000 kg of steam. The condensate leaves at a temperature of 38 °C. Makeup water is supplied at 10 °C. The cooling water enters at 32 °C and leaves at 38 °C. A separate air extraction pump (SJAE) is added and from the air cooler section, air along with some steam leaves at 27 °C. The pressure in the condenser is assumed to remain constant. Calculate (a) the rate of saving of condensate and the rate of saving in the heat supply in the boiler due to separate air extraction pump, (b) the percentage reduction in air ejector load due to this separate air extraction method, and (c) the rate of cooling water flow.* D [LO 19.20]

Solution

At 40 °C, $p_{sat} = 0.07384$ bar and $v_g = 19.52\ \text{m}^3/\text{kg}$ [Fig. 19.45(a)].

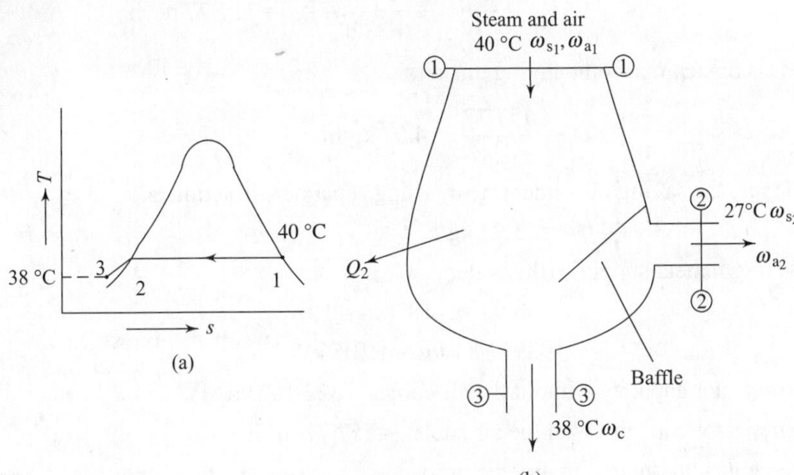

Fig. 19.45

From Eq. (19.58),

$$p_{air}\, \omega_s v_1 = \omega_a R_a (t_{sh} + 273)$$

$$p_{air} = \frac{0.35 \times 0.287 \times 313 \times 10^3}{1000 \times 19.52}\ \text{N/m}^2$$

$$= 1.61\ \text{N/m}^2 = 1.61 \times 10^{-5}\ \text{bar}$$

This is very small and can be neglected.

Condensate leaves at 38 °C. At 38 °C, $p_{sat} = 0.06624$ bar and $v_g = 21.63$ m³/kg.

The total pressure in the condenser is 0.07384 bar

$$p = p_{st} + p_{air}$$

$$0.07384 = 0.06624 + p_{air} \quad \text{or} \quad p_{air} = 0.00760 \text{ bar}$$

Mass of air removed per hour,

$$\omega_a = 20{,}000 \times \frac{0.35}{1000} = 7 \text{ kg/h}$$

Volume of air removed per hour,

$$V = \frac{\omega_a R_a (273 + 38)}{p_{air}} = \frac{7 \times 287 \times 311}{760} = 822.1 \text{ m}^3/\text{h}$$

The mass of steam accompanying this air is

$$\omega_s = \frac{822.1}{21.63} = 38 \text{ kg/h}$$

Separate air extraction pump: Air along with some steam leaves the air cooler section at 27 °C.

At 27°C, $p_{sat} = 0.03564$ bar and $v_g = 38.81$ m³/kg

$$p_{air} = 0.07384 - 0.03564 = 0.03820 \text{ bar}$$

Therefore, the volume of air removed,

$$V = \frac{\omega_a R_a T}{p_{air}} = \frac{7 \times 287 \times 300}{3820} = 157.77 \text{ m}^3/\text{h}$$

Mass of steam accompanying this air,

$$\omega_s = \frac{157.77}{38.81} = 4.07 \text{ kg/h}$$

Hence, the saving in condensate by using separate extraction is

$$38 - 4.07 = 33.93 \text{ kg/h} \qquad \textit{Ans. (a)}$$

Saving in heat supply in the boiler

$$= 33.98 \times 4.187(38 - 10)$$

$$= 3977.8 \text{ kJ/h} = 1.105 \text{ kW} \qquad \textit{Ans. (a)}$$

Air ejector capacity without the air cooler = 822.1 m³/h

Air ejector capacity with the air cooler = 157.77 m³/h

Percentage reduction in air ejector load

$$= \frac{822.1 - 157.77}{822.1} \times 100 = 80.8\% \qquad \textit{Ans. (b)}$$

By making an energy balance for the condenser [Fig. E19.25(b)], with subscripts s, a and c being used for steam, air and condensate, respectively,

$$Q_2 = \omega_{s_2} h_{s_1} + \omega_{a_1} h_{a_1} - (\omega_{s_2} h_{s_2} + \omega_{a_2} h_{a_2}) - \omega_c h_c$$

Here, $\omega_{a_1} = \omega_{a_2} = 7$ kg/h, $\omega_{s_2} = 4.07$ kg/h,

$$\omega_c = 20{,}000 - 4.07 = 20{,}000 \text{ kg/h} = \omega_{s_1}$$
$$h_c = h_f \text{ at } 38\,°\text{C} = 159.3 \text{ kJ/kg},$$
$$h_{s_1} = h_g \text{ at } 40\,°\text{C} = 2574.3 \text{ kJ/kg},$$
$$h_{s_2} = h_g \text{ at } 27\,°\text{C} = 2550.3 \text{ kJ/kg}$$
$$Q_2 = \omega_{s_1}(h_{s_1} - h_c) - \omega_{a_1}(h_{a_1} - h_{a_2}) - \omega_{s_2}h_{s_2}$$
$$= 20{,}000(2574.3 - 159.3) - 7 \times 1.005(40 - 27) - 4.07 \times 2550.3$$
$$= 48{,}300{,}000 - 91.5 - 10379.7$$
$$= 48389529 \text{ kJ/h} = 13441.5 \text{ kW}$$

Now,
$$Q_2 = \omega_c c_p(t_{c_2} - t_{c_1})$$
$$13441.5 = \omega_c \times 4.187 \times 6$$
$$\omega_c = 535 \text{ kg/s}$$ *Ans.* (c)

If we neglect the energy leaving with the flow to the ejector,
$$Q_2 = \omega_{s_1}(h_{s_1} - h_c) = \omega_c c_p(t_{c_2} - t_{c_1})$$
$$48{,}300{,}000 = \omega_c \times 4.187 \times 6$$
$$\omega_c = 534 \text{ kg/s}$$ *Ans.* (c)

SUMMARY

The steam turbine is the prime mover in a steam power plant, functioning as a thermodynamic heat engine that operates in a cycle to deliver mechanical power. Its main objective is to convert the enthalpy of steam - generated by the boiler - into mechanical energy at the turbine shaft.

This energy conversion occurs in two stages: first, the steam's enthalpy is transformed into high-velocity (kinetic energy) steam through nozzles; second, this kinetic energy is transferred to the turbine blades, producing shaft power. In impulse turbines, all enthalpy-to-kinetic energy conversion takes place in stationary nozzles, and the moving blades solely convert this kinetic energy into mechanical output. In contrast, reaction turbines have moving blades that also act as nozzles, enabling simultaneous conversion of steam enthalpy into both kinetic energy and shaft power.

Reaction turbines are generally more efficient than impulse turbines. The degree of reaction in a reaction turbine refers to the proportion of enthalpy drop occurring in the moving blades relative to the total enthalpy drop. A commonly used configuration in large power stations is the Parsons turbine, which has a 50% degree of reaction - indicating equal enthalpy drops in fixed and moving blades.

Turbine performance and efficiency are typically analyzed using velocity diagrams, which depict the relationships between absolute and relative steam velocities at the inlet and outlet of the moving blades, along with the blade velocity itself. These diagrams consist of inlet and outlet velocity triangles sharing a common base formed by the blade velocity vector.

The velocity ratio - defined as the ratio of blade speed to steam speed at the inlet of the moving blades -has an optimal value that varies by turbine type and is critical for achieving maximum efficiency as expressed below:

$$Popt = \frac{\cos\alpha}{2Z}$$ (for velocity compounded impulse turbine with Z rows of moving blades)

Hence, for simple impulse turbine with single row of moving blades, Z = 1 and

$$Popt = \frac{\cos\alpha}{2}$$

Also, for 50% degree of reaction or Parson's turbines,

$$Popt = \cos\alpha$$

Friction in both the nozzles and moving blades reduces the overall efficiency of a steam turbine. Nozzle efficiency is defined as the ratio of the actual enthalpy drop to the corresponding isentropic enthalpy drop within a nozzle. Similarly, friction in the moving blades leads to a reduction in the relative velocity of steam, decreasing the energy transfer to the shaft.

In multi-stage turbines, frictional losses result in the actual enthalpy drop in each stage being less than the corresponding isentropic enthalpy drop. To account for the cumulative impact of friction across all stages, the reheat factor is introduced. It is defined as the ratio of the sum of isentropic enthalpy drops across all stages to the total isentropic enthalpy drop for the same overall pressure drop.

Steam turbines experience various losses that reduce their overall efficiency. Major internal losses include blade and disc friction losses caused by steam rubbing against surfaces, leakage losses due to steam escaping through clearances, carryover losses from residual kinetic energy carried into the next stage, residual velocity losses from unutilized kinetic energy in exhaust steam. Partial admission losses occur specifically in impulse turbines, where only part of the blade row is exposed to steam at a time, unlike in reaction turbines which have continuous admission over the whole circumference. In addition to these internal losses, external losses such as bearing friction also contribute to efficiency reduction.

The condenser serves as the heat sink for rejecting waste heat in a steam power plant, a requirement for any cyclic heat engine as per the Kelvin-Planck statement of the second law of thermodynamics. By enabling heat rejection at vacuum pressures, the condenser effectively improves the efficiency of the steam turbine. It also facilitates water recycling within the cycle and plays a crucial role in removing non-condensable gases from steam. Cooling water in the condenser condenses the exhaust steam from the turbine. Although direct-contact (mixing) condensers are simpler and more efficient, surface or indirect-contact condensers are more commonly used in power plants to avoid contamination of boiler feedwater. To conserve cooling water, cooling towers are often employed; they use evaporative cooling to dissipate heat to the atmosphere and allow the cooling water to be recirculated. However, some loss of cooling water is inevitable due to evaporation in this process.

REVIEW QUESTIONS

(E) **19.1** What is the oldest known steam turbine? What were its limitations? **[LO 19.1]**

(M) **19.2** Explain the basic steps of energy conversion in a steam turbine. **[LO19.1]**

(E) **19.3** What are the two principal types of turbines based on energy conversion? **[LO 19.2]**

(M) **19.4** State the advantages and disadvantages of impulse turbines. **[LO 19.3]**

(D) **19.5** What is a velocity diagram? Why is it important? **[LO 19.4]**

(D) **19.6** What is diagram efficiency? Why is it named so? **[LO 19.4]**

(D) **19.7** What is velocity ratio? Why is it important? **[LO 19.5]**

(D) **19.8** What is the optimum velocity ratio for a steam turbine? Explain. **[LO 19.5]**

(D) **19.9** What is the expression for the optimum velocity ratio of a single-row impulse turbine? Derive the expression. **[LO 19.5]**

(M) **19.10** How does friction affect the maximum efficiency of an impulse turbine with a single row of blades? **[LO 19.5]**

(D) **19.11** Explain the method and advantages of the graphical construction of the velocity diagram. **[LO 19.6]**

(D) **19.12** What is compounding in steam turbines? How many types of compounding are there? **[LO 19.7]**

(M) **19.13** What are the advantages and disadvantages of pressure compounding? **[LO 19.8]**

(M) **19.14** What are the advantages and limitations of velocity compounding? **[LO 19.9]**

(D) **19.15** Show the pressure and velocity profiles in the axial direction of pressure-compounded and velocity-compounded turbines. **[LO 19.8 & LO 19.9]**

(D) **19.16** Derive the expression for the optimum velocity ratio for a Curtis turbine with multiple rows of blades. **[LO 19.10]**

(M) **19.17** How are impulse-reaction turbines different from impulse turbines? **[LO 19.11]**

(M) **19.18** Why are impulse-reaction turbines preferred over pure reaction turbines? **[LO 19.11]**

(M) **19.19** What is the degree of reaction in an impulse-reaction turbine? Explain. **[LO 19.11]**

(D) **19.20** What is the optimum velocity ratio for an impulse-reaction turbine with a degree of reaction of 50%? Derive it. Also derive the corresponding maximum efficiency **[LO 19.11]**

(M) **19.21** What is carryover efficiency? Explain. **[LO 19.12]**

(D) **19.22** Compare the optimum velocity ratios and corresponding maximum efficiencies of simple impulse, Curtis, and 50% reaction turbines. Explain the relative advantages between these three types with a plot of velocity ratio vs. maximum efficiencies. **[LO 19.13]**

(M) **19.23** Explain the principle of calculation of nozzle and blade heights. **[LO 19.14]**

(E) **19.24** How does friction affect the performance of a nozzle? **[LO 19.15]**

(M) **19.25** What is nozzle efficiency? Explain. **[LO 19.15]**

(E) **19.26** What do you understand by losses in steam turbines? **[LO 19.16]**

(D) **19.27** State and explain the different types of losses in steam turbines. **[LO 19.16]**

(M) **19.28** What type of loss is there in impulse turbines but not in reaction turbines? Why? **[LO 19.16]**

(D) **19.29** Explain the reheat factor and condition line. Why are they important in multi-stage turbines? **[LO 19.17]**

(M) **19.30** Why is a condenser required with a steam turbine? What are the different functions of a condenser? **[LO 19.18]**

(E) **19.31** Explain the different types of condensers and their relative advantages and disadvantages. **[LO 19]**

(E) **19.32** Why is the condenser ideal for air removal from steam? **[LO 19.20]**

(M) **19.33** What are vacuum efficiency and condenser efficiency? Explain. **[LO 19.20]**

PROBLEMS WITH HINTS

(M) **19.1** Steam expands in a turbine from 40 bar, 500 °C to 0.10 bar isentropically. Assuming ideal conditions, determine the mean diameter of the wheel if the turbine were of (a) single impulse stage, (b) single 50% reaction stage, (c) four pressure (or Rateau) stages, (d) one two-row Curtis stage, and (e) four 50% reaction stages. Take the nozzle angle as 16° and N as 300 rpm. **[LO 19.4, LO 19.13]**

Hints:

$$h_1 = 3445.3 \text{ kJ/kg}$$

$$s_1 = 7.0901 \text{ kJ/kg K} = s_2 = 0.6493 + x_{2s} \times 7.5009$$

$$x_{2s} = 0.86$$

$$h_{2s} = 191.83 + 0.86 \times 2392.8 = 2246.5 \text{ kJ/kg}$$

$$h_1 - h_{2s} = 1198.8 \text{ kJ/kg}$$

(a)

$$V_1 = 44.72(1198.8)^{1/2} = 1548.37 \text{ m/s}$$

$$\frac{V_b}{V_1} = \frac{\cos\alpha}{2} = \frac{\cos 16°}{2} = 0.4806$$

$$V_b = 744.2 \text{ m/s} = \frac{\pi D_m N}{60}$$

$$D_m = 4.73 \text{ m}$$ *Ans.*

(b)

$$V_1 = 44.72\left[\frac{\Delta h_{stage}}{2}\right]^{1/2} = 44.72\left[\frac{1198.8}{2}\right]^{1/2}$$

$$= 1094.86 \text{ m/s}$$

$$\frac{V_b}{V_1} = \cos\alpha = \cos 16°$$

$$V_b = 1052.4 \text{ m/s} = \frac{\pi D_m N}{60}$$

$$\therefore \quad D_m = 6.7 \text{ m}$$ *Ans.*

(c)

$$V_1 = 44.72\left[\frac{1198.8}{4}\right]^{1/2} = 774.19 \text{ m/s}$$

$$\frac{V_b}{V_1} = \frac{\cos\alpha}{2}, \; V_b = 372.07 \text{ m/s} = \frac{\pi D_m N}{60}$$

$$D_m = 2.37 \text{ m}$$ *Ans.*

(d)

$$\frac{V_b}{V_1} = \frac{\cos\alpha}{4} = 0.2403, \; V_1 = 1548.37 \text{ m/s}$$

$$V_b = 372.1 \text{ m/s} = \frac{\pi D_m N}{60}$$

$$D_m = 2.368 \text{ m}$$ *Ans.*

(e) Four 50% reaction stages

$$\therefore \quad \Delta h_{stage} = \frac{1198.8}{4} = 299.7 \text{ kJ/kg}$$

$$V_1 = 44.72\left[\frac{\Delta h_{stage}}{2}\right]^{1/2} = 547.43 \text{ m/s}$$

$$\frac{V_b}{V_1} = \cos a$$

$$\therefore \quad V_b = 526.22 = \frac{\pi D_m N}{60}$$

$$\therefore \quad D_m = 3.35 \text{ m}$$

Ans.

(M) **19.2** A 20-stage 50% reaction turbine develops a diagram power of 12 MW. The inlet condition of steam is at 15 bar, 350 °C, while the condensers pressure is 0.14 bar. The stage efficiency is 75% for each stage and the reheat factor is 1.04. At one stage the steam is at 1 bar, dry saturated. The exit angle of the blades is 20° and the blade velocity ratio is 0.7. The blade height may be taken as 1/12 of the mean blade diameter. Calculate: **[LO 19.4, LO 19.11]**

(a) the flow rate of steam required, assuming that all stages develop equal work,

(b) the mean blade diameter, and

(c) the speed of the rotor.

Hints:

$$\eta_{internal} = \eta_{st} \times RF = 0.75 \times 1.04 = 0.78$$

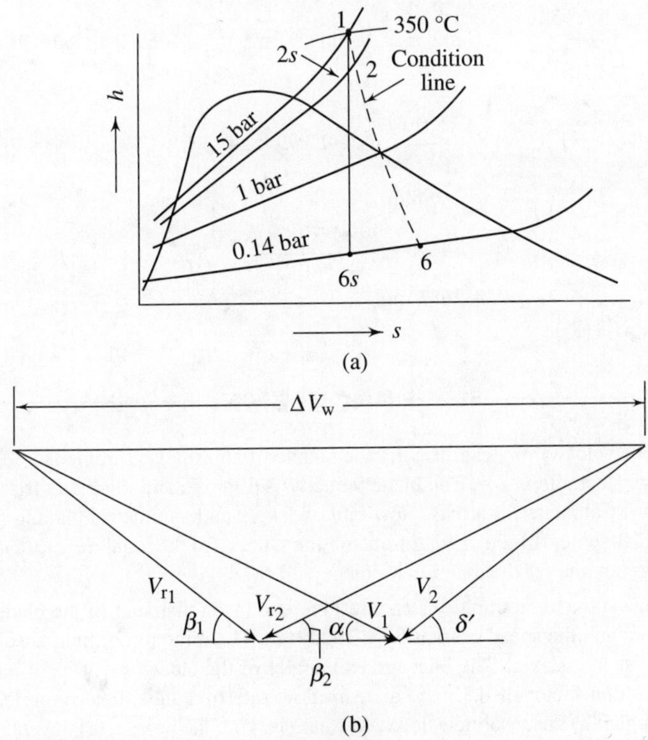

Fig. 19.46

From Mollier chart, $\Delta h_s = 855$ kJ/kg (Fig. 19.45)

$$\therefore \quad \Delta h_{act} = 0.78 \times 855 = 667 \text{ kJ/kg}$$

$$\omega_s(\Delta h)_{act} = 12 \text{ MW}$$

$$\therefore \quad \omega_s = \frac{12000}{667} = 17.99 \text{ kg/s} \qquad \textit{Ans.} \text{ (a)}$$

Work done per kg of steam flow,

$$W_D = V_b(2V_1 \cos\alpha - V_b)$$

Now, $$\frac{V_b}{V_1} = 0.7, \alpha = 20°$$

$$W_D = V_b\left(\frac{2V_b}{0.7}\cos 20° - V_b\right) = 1.6848\, V_b^2 \text{ J/kg}$$

Enthalpy drop per stage = work done per stage = $\frac{667}{20} = 33.35$ kJ/kg

$$\therefore \quad 1.6848 V_b^2 \times 10^{-3} = 33.35$$

$$\therefore \quad V_b = 141.4 \text{ m/s}$$

$$V_1 \sin\alpha = \frac{V_b}{0.7}\sin\alpha = \frac{141.4}{0.7}\sin 20° = 69.1 \text{ m/s}$$

Volume flow rate per sec at 1 bar

$$= \pi D_m h_b \times V_1 \sin\alpha = \pi D_m \frac{D_m}{12} \times 69.1 = 18.09\, D_m^2 \text{ m}^3\text{/s}$$

At 1 bar, $$v_g = 1.694 \text{ m}^3\text{/kg}$$

$$\omega_s = \frac{18.09\, D_m^2}{1.694} = 17.99 \text{ kg/s}$$

$$\therefore \quad D_m = 1.298 \text{ m} \qquad \textit{Ans.} \text{ (b)}$$

$$V_b = 141.4 = \frac{\pi \times 1.298 \times N}{60}$$

$$\therefore \quad N = 2081 \text{ rpm} \qquad \textit{Ans.} \text{ (c)}$$

PROBLEMS

(E) **19.1** The velocity of steam leaving the nozzle of an impulse turbine is 900 m/s and the nozzle angle is 20°. The blade velocity is 300 m/s and the blade friction factor is 0.7. Calculate for a mass flow rate of 1 kg/s and symmetric blading (a) the blade inlet angle, (b) the driving force on the wheel, (c) the axial thrust, (d) the diagram power, and (e) the diagram efficiency. **[LO 19.3]**

(M) **19.2** In a stage of an impulse steam turbine, the mean diameter of the blade ring is 800 mm and the speed of rotation is 3000 rpm. The direction of final absolute velocity of steam is axial. The inlet and exit angles of the blades are 30°. Assuming a blade friction factor of 0.85 and a steam flow rate of 1 kg/s, determine (a) the nozzle angle, (b) the absolute velocity of steam leaving the nozzle, (c) the enthalpy drop in the stage, (d) the tangential thrust, (e) the axial thrust, (f) the blading work, and (g) the blading efficiency. **[LO 19.4]**

(E) **19.3** A single-stage impulse turbine rotor has a mean blade ring diameter of 500 mm and rotates at a speed of 10,000 rpm. The nozzle angle is 20° and the steam leaves the nozzles with a velocity of 900 m/s. The blades are equiangular and the blade friction factor is 0.85. Construct the velocity diagrams for the blades and determine the inlet angle of the blades for shockless entry of steam. Determine (a) the diagram power for a steam flow of 750 kg/h, (b) the diagram efficiency, (c) the axial thrust, and (d) the loss of kinetic energy due to friction. **[LO 19.4]**

(D) **19.4** A stage of an impulse steam turbine operates close to the maximum blading efficiency. The blades are equiangular, and the friction effects in blades may be neglected. The mean blade velocity is 200 m/s and the steam flow rate is 0.75 kg/s. Find (a) the discharge angle at which the steam leaves the blades, and (b) the diagram power. **[LO 19.5]**

(D) **19.5** A single stage simple impulse turbine generator set is to operate under the following conditions:

Steam at inlet	: 8 bar, dry saturated
Exhaust	: 0.2 bar
Nozzle efficiency	: 0.90
Nozzle angle	: 15°
Internal efficiency	: 0.75
Mechanical efficiency	: 0.92
Generator efficiency	: 0.90

Find (a) the mean blade speed for maximum blading efficiency, and (b) the steam flow rate for developing 200 kW. **[LO 19.6]**

(D) **19.6** Deduce a general expression for the blade efficiency of a stage of an impulse turbine with single row wheels, assuming equiangular blades, a nozzle angle α, and a blade friction factor k_b. What is the maximum efficiency if $\alpha = 20°$ and $k_b = 0.83$? What is the velocity ratio? If the blade efficiency is 90% of the maximum value, what are the values of the velocity ratios? Draw the velocity diagrams for each case and state the blade angles. **[LO 19.6]**

(E) **19.7** In a four-stage pressure-compounded impulse turbine steam at 23 bar, 345 °C expands to 0.07 bar. The internal efficiency of the turbine is 0.72. Assuming that the work is shared equally among the stages and the condition line is straight, estimate the stage pressures, efficiency of each stage and reheat factor. **[LO 19.8]**

(M) **19.8** The following particulars refer to a two-row velocity-compounded impulse wheel which forms the first stage of a combination turbine:

Steam velocity at nozzle outlet	: 630 m/s
Mean blade velocity	: 125 m/s
Nozzle angle	: 16°
Outlet angle, first row of moving blades	: 18°
Outlet angle, fixed guide blades	: 22°
Outlet angle, second row of moving blades	: 36°
Steam flow rate	: 2.6 kg/s

The ratio of the relative velocity at outlet to that at inlet is 0.84 for all the blades. Determine (a) the velocity of whirl, (b) the tangential thrust on the blades, (c) the axial thrust on the blades, (d) the power developed, and (e) the blading efficiency. **[LO 19.9]**

(M) **19.9** Show that in a 50% reaction steam turbine stage, the maximum stage efficiency is

$$\frac{2\cos^2\alpha}{1+\cos^2\alpha}$$

where α is the nozzle angle.

In a particular stage the mean diameter is 500 mm and the blade height is 30 mm. The blade angles are 60° at inlet and 160° at outlet. The density of steam is 2.7 kg/m^3, the speed is 3000 rpm. Calculate the mass flow rate of steam, power developed and stage efficiency. **[LO 19.11]**

(E) **19.10** In a reaction stage of a steam turbine the nozzle angle is 20° and the absolute velocity of steam at inlet to the moving blades is 240 m/s. The blade velocity is 210 m/s. If the blading is designed for 50% reaction, determine (a) the blade height at inlet and exit, (b) the enthalpy drop per kg of steam in the moving blades and in the complete stage, (c) the diagram power for a steam flow of 1 kg/s, and (d) the diagram efficiency. **[LO 19.11]**

(M) **19.11** A stage of a 50% reaction turbine delivers dry saturated steam at 2.7 bar from the fixed blades at 90 m/s. The mean blade height is 40 mm, and the moving blade exit angle is 20°. The axial velocity of steam is 3/4 of the mean blade velocity. Steam flow rate is 9000 kg/h. The effect of blade tip thickness on the annulus area can be neglected. Calculate (a) the wheel speed in rpm, (b) the diagram power, (c) the diagram efficiency, and (d) the enthalpy drop of steam in this stage. **[LO 19.4, LO 19.11]**

(M) **19.12** A steam turbine is to operate between 140 bar, 560 °C and 0.075 bar. The maximum blade velocity is 320 m/s and the nozzle efficiency in all stages is 0.90. Nozzle angles will be 15° for impulse stages and 25° for reaction stages. All stages operate close to the maximum efficiency. Estimate the number of stages required for each of the following arrangements:

(a) all simple impulse stages,
(b) all 50% reaction stages,
(c) a two-row Curtis stage followed by simple impulse stages
(d) a two-row Curtis stage followed by 50% reaction stages. **[LO 19.13]**

(E) **19.13** Steam at 30 bar, 350 °C expands through a convergent–divergent nozzle. The exit plane pressure is 3 bar. The flow rate is 0.5 kg/s and the nozzle efficiency is 0.8. Assuming that the velocity at inlet is negligible, determine the throat and exit areas, steam velocity at the exit, and the quality of steam at the exit plane. The critical pressure ratio can be taken as 0.546. **[LO 19.14]**

(M) **19.14** Dry saturated steam at 10 bar is expanded in a convergent–divergent nozzle. The velocity of steam at exit is 685 m/s, the flow rate is 7 kg/s and the nozzle efficiency is 85%. Assume the flow to be isentropic up to the throat. The critical pressure ratio can be taken as 0.54. Determine the throat and exit areas of the nozzle and the pressure at exit. Neglect the velocity of steam at inlet to the nozzle. **[LO 19.14]**

(D) **19.15** A convergent nozzle receives steam at 4 bar, 150°C and negligible inlet velocity, and expands it into a space at atmospheric pressure. Assuming supersaturated expansion and a nozzle efficiency of 0.9, calculate the nozzle throat area required for a mass flow of 1.2 kg/s. **[LO 19.14]**

(D) **19.16** Steam at 20 bar, 400 °C expands in a 50% reaction turbine to a pressure 0.2 bar. The steam leaving the turbine is dry saturated. The reheat factor is 1.05 and the isentropic efficiency of each stage is the same throughout. There are 14 stages and the enthalpy drop is the same in each. All the blades have an exit angle of 22° and the mean value of blade velocity ratio is 0.82. The mass flow rate of steam is 34,000 kg/h. The turbine speed is 2400 rpm. Calculate (a) the stage efficiency, (b) the diagram power, (c) the drum diameter, (d) the blade height for the last row of moving blades, and (e) the pressure at the entry to the last stage. **[LO 19.11, LO 19.4]**

(M) **19.17** A 50% reaction turbine is supplied with steam at 60 bar, 600 °C. The condenser pressure is 0.07 bar. If the reheat factor is assumed to be 1.04 and the stage efficiency is constant throughout at 80%, calculate the steam flow required for a diagram power of 25 MW. **[LO 19.17]**

(M) **19.18** A five-stage steam turbine working between 20 bar, 350 °C and 2 bar has equal enthalpy drops in all stages, and the stage efficiency is 75% in all stages. Plot on the Mollier diagram the condition line. Find the interstage pressures, reheat factor, and turbine internal efficiency using properties from the diagram. **[LO 19.17]**

(M) **19.19** A condenser for a steam power plant receives 185 t/h of steam at 40 °C, 92% quality. Cooling water enters at 33 °C and leaves at 37 °C. The condensate leaves at 39 °C. The pressure inside the condenser is found to be 0.077 bar. Find (a) the cooling water flow required in m^3/s, and (b) the air leakage into the condenser in kg/h. **[LO 19.19]**

(D) **19.20** A surface condenser receives 20 t/h of dry saturated steam at 36.2 °C. The condensate leaves at 34.6 °C. The volume of air and vapour leaving the air cooler section at 29 °C is 3.8 m^3/min. Assuming constant pressure throughout the condenser and neglecting the partial pressure of air at inlet to the condenser, calculate (a) the mass of air removed per 10,000 kg of steam, (b) the mass of steam condensed in the air cooler per minute, and (c) the rate of heat removal by the cooling water. **[LO 19.20]**

ANSWERS TO PROBLEMS

19.1	(a) 29°24′, (b) 927.7 N, (c) 92.3 N, (d) 278.3 kW, (e) 68.7%
19.2	(a) 17.33°, (b) 72.56 m/s, (c) 41.06 kJ/kg, (d) 273.5 N, (e) 12.8 N, (f) 34.37 kW, (g) 83.7%
19.3	28°, (a) 56.8 kW, (b) 0.70, (c) 11.98 kW
19.4	(a) 90°, (b) 60 kW
19.5	(a) 490 m/s, (b) 0.564 kg/s
19.6	80.7%, 0.47, 0.32 and 0.62, 36° and 48°6
19.7	7.25 bar, 1.9 bar, 0.39 bar; 61.68%, 68.14%, 70.41%, 72.84%; 1.0574
19.8	(a) 1127.5 m/s, (b) 2.93 kN, (c) 188.5 N, (d) 366.4 kW, (e) 0.71
19.9	4.61 kg/s, 43.57 kW, 0.915
19.11	(a) 1824 rpm, (b) 13.14 kW, (c) 0.787 and (d) 5.26 kJ/kg
19.13	127.6 mm^2, 323.5 mm^2, 922 m/s, $x_2 = 0.985$
19.14	4870 mm^2, 7000 mm^2, 2.4 bar
19.15	2010 mm^2
19.16	(a) 67.7%, (b) 6025 kW, (c) 1.34 m, (d) 175 mm, (e) 0.32 bar
19.17	21 kg/s
19.20	(a) 2.63 kg, (b) 0.492 kg, (c) 13451 kW

Appendix A

A.1 Steam Tables*

Table A. 1 Saturated steam : Temperature table

Temp. °C T	Pressure kPa, MPa p	Specific Volume, m³/kg: Sat. Liquid v_f	Sat. Vapour v_g	Internal Energy, kJ/kg: Sat. Liquid u_f	Evap. u_{fg}	Sat. Vapour u_g	Enthalpy, kJ/kg: Sat. Liquid h_f	Evap. h_{fg}	Sat. Vapour h_g	Entropy, kJ/kg K: Sat. Liquid s_f	Evap. s_{fg}	Sat Vapour s_g
0.01	0.6113	0.001000	206.132	0.00	2375.3	2375.3	0.00	2501.3	2501.3	0.0000	9.1562	9.1562
5	0.8721	0.001000	147.118	20.97	2361.3	2382.2	20.98	2489.6	2510.5	0.0761	8.9496	9.0257
10	1.2276	0.001000	106.377	41.99	2347.2	2389.2	41.99	2477.7	2519.7	0.1510	8.7498	8.9007
15	1.7051	0.001001	77.925	62.98	2333.1	2396.0	62.98	2465.9	2528.9	0.2245	8.5569	8.7813
20	2.3385	0.001002	57.790	83.94	2319.0	2402.9	83.94	2454.1	2538.1	0.2966	8.3706	8.6671
25	3.1691	0.001003	43.359	104.86	2304.9	2409.8	104.87	2442.3	2547.2	0.3673	8.1905	8.5579
30	4.2461	0.001004	32.893	125.77	2290.8	2416.6	125.77	2430.5	2556.2	0.4369	8.0164	8.4533
35	5.6280	0.001006	25.216	146.65	2276.7	2423.4	146.66	2418.6	2565.3	0.5052	7.8478	8.3530
40	7.3837	0.001008	19.523	167.53	2262.6	2430.1	167.54	2406.7	2574.3	0.5724	7.6845	8.2569
45	9.5934	0.001010	15.258	188.41	2248.4	2436.8	188.42	2394.8	2583.2	0.6386	7.5261	8.1647
50	12.350	0.001012	12.032	209.30	2234.2	2443.5	209.31	2382.7	2592.1	0.7037	7.3725	8.0762
55	15.758	0.001015	9.568	230.19	2219.9	2450.1	230.20	2370.7	2600.9	0.7679	7.2234	7.9912
60	19.941	0.001017	7.671	251.09	2205.5	2456.6	251.11	2358.5	2609.6	0.8311	7.0784	7.9095
65	25.033	0.001020	6.197	272.00	2191.1	2463.1	272.03	2346.2	2618.2	0.8934	6.9375	7.8309
70	31.188	0.001023	5.042	292.93	2176.6	2469.5	292.96	2333.8	2626.8	0.9548	6.8004	7.7552
75	38.578	0.001026	4.131	313.87	2162.0	2475.9	313.91	2321.4	2635.3	1.0154	6.6670	7.6824
80	47.390	0.001029	3.407	334.84	2147.4	2482.2	334.88	2308.8	2643.7	1.0752	6.5369	7.6121
85	57.834	0.001032	2.828	355.82	2132.6	2488.4	355.88	2296.0	2651.9	1.1342	6.4102	7.5444
90	70.139	0.001036	2.361	376.82	2117.7	2494.5	376.90	2283.2	2660.1	1.1924	6.2866	7.4790
95	84.554	0.001040	1.982	397.86	2102.7	2500.6	397.94	2270.2	2668.1	1.2500	6.1659	7.4158
100	0.10135	0.001044	1.6729	418.91	2087.6	2506.5	419.02	2257.0	2676.0	1.3068	6.0480	7.3548

* Adapted from Joseph H. Keenan, Frederick G.Keyes, Philip G. Hill, and Joan G. Moore, *Steam Tables*, John Wiley and Sons, New York, 1969.

(Contd.)

Table A.1.1 (Contd.)

Temp. °C T	Specific Volume, m^3/kg Pressure kPa, MPa p	Sat Liquid v_f	Sat. Liquid v_g	Internal Energy, kJ/kg Sat. Liquid u_f	Evap. u_{fg}	Sat. Vapour u_g	Enthalpy, kJ/kg Sat. Liquid h_f	Evap. u_{fg}	Sat. Vapour h_g	Entropy, kJ/kg K Sat. Liquid s_f	Evap. s_{fg}	Sat. Vapour s_g
105	0.12082	0.001047	1.4194	440.00	2072.3	2512.3	440.13	2243.7	2683.8	1.3629	5.9328	7.2958
110	0.14328	0.001052	1.2102	461.12	2057.0	2518.1	461.27	2230.2	2691.5	1.4184	5.8202	7.2386
115	0.16906	0.001056	1.0366	482.28	2041.4	2523.7	482.46	2216.5	2699.0	1.4733	5.7100	7.1832
120	0.19853	0.001060	0.8919	503.48	2025.8	2529.2	503.69	2202.6	2706.3	1.5275	5.6020	7.1295
125	0.2321	0.001065	0.77059	524.72	2009.9	2534.6	524.96	2188.5	2713.5	1.5812	5.4962	7.0774
130	0.2701	0.001070	0.66850	546.00	1993.9	2539.9	546.29	2174.2	2720.5	1.6343	5.3925	7.0269
135	0.3130	0.001075	0.58217	567.34	1977.7	2545.0	567.67	2159.6	2727.3	1.6869	5.2907	6.9777
140	0.3613	0.001080	0.50885	588.72	1961.3	2550.0	589.11	2144.8	2733.9	1.7390	5.1908	6.9298
145	0.4154	0.001085	0.44632	610.16	1944.7	2554.9	610.61	2129.6	2740.3	1.7906	5.0926	6.8832
150	0.4759	0.001090	0.39278	631.66	1927.9	2559.5	632.18	2114.3	2746.4	1.8417	4.9960	6.8378
155	0.5431	0.001096	0.34676	653.23	1910.8	2564.0	653.82	2098.6	2752.4	1.8924	4.9010	6.7934
160	0.6178	0.001102	0.30706	674.85	1893.5	2568.4	675.53	2082.6	2758.1	1.9426	4.8075	6.7501
165	0.7005	0.001108	0.27269	696.55	1876.0	2572.5	697.32	2066.2	2763.5	1.9924	4.7153	6.7078
170	0.7917	0.001114	0.24283	718.31	1858.1	2576.5	719.20	2049.5	2768.7	2.0418	4.6244	6.6663
175	0.8920	0.001121	0.21680	740.16	1840.0	2580.2	741.16	2032.4	2773.6	2.0909	4.5347	6.6256
180	1.0022	0.001127	0.19405	762.08	1821.6	2583.7	763.21	2015.0	2778.2	2.1395	4.4461	6.5857
185	1.1227	0.001134	0.17409	784.08	1802.9	2587.0	785.36	1997.1	2782.4	2.1878	4.3586	6.5464
190	1.2544	0.001141	0.15654	806.17	1783.8	2590.0	807.61	1978.8	2786.4	2.2358	4.2720	6.5078
195	1.3978	0.001149	0.14105	828.36	1764.4	2592.8	829.96	1960.0	2790.0	2.2835	4.1863	6.4697
200	1.5538	0.001156	0.12736	850.64	1744.7	2595.3	852.43	1940.7	2793.2	2.3308	4.1014	6.4322
205	1.7230	0.001164	0.11521	873.02	1724.5	2597.5	875.03	1921.0	2796.0	2.3779	4.0172	6.3951
210	1.9063	0.001173	0.10441	895.51	1703.9	2599.4	897.75	1900.7	2798.5	2.4247	3.9337	6.3584
215	2.1042	0.001181	0.09479	918.12	1682.9	2601.1	920.61	1879.9	2800.5	2.4713	3.8507	6.3221
220	2.3178	0.001190	0.08619	940.85	1661.5	2602.3	943.61	1858.5	2802.1	2.5177	3.7683	6.2860

(Contd.)

Table A.1.1 (Contd.)

Temp. °C T	Specific Volume, m^3/kg Pressure kPa, MPa p	Sat. Liquid v_f	Sat. Vapour v_g	Internal Energy, kJ/kg Sat. Liquid u_f	Evap. u_{fg}	Sat. Vapour u_g	Enthalpy, kJ/kg Sat. Liquid h_f	Evap. u_{fg}	Sat. Vapour h_g	Entropy, kJ/kg K Sat. Liquid s_f	Evap. s_{fg}	Sat. Vapour s_g
225	2.5477	0.001199	0.7849	963.72	1639.6	2603.3	966.77	1836.5	2803.3	2.5639	3.6863	6.2502
230	2.7949	0.001209	0.07158	986.72	1617.2	2603.9	990.10	1813.8	2803.9	2.6099	3.6047	6.2146
235	3.0601	0.001219	0.06536	1009.88	1594.2	2604.1	1013.61	1790.5	2804.1	2.6557	3.5233	6.1791
240	3.3442	0.001229	0.05976	1033.19	1570.8	2603.9	1037.31	1766.5	2803.8	2.7015	3.4422	6.1436
245	3.6482	0.001240	0.05470	1056.69	1546.7	2603.4	1061.21	1741.7	2802.9	2.7471	3.3612	6.1083
250	3.9730	0.001251	0.05013	1080.37	1522.0	2602.4	1085.34	1716.2	2801.5	2.7927	3.2802	6.0729
255	4.3195	0.001263	0.04598	1104.26	1496.7	2600.9	1109.72	1689.8	2799.5	2.8382	3.1992	6.0374
260	4.6886	0.001276	0.04220	1128.37	1470.6	2599.0	1134.35	1662.5	2796.9	2.8837	3.1181	6.0018
265	5.0813	0.001289	0.03877	1152.72	1443.9	2596.6	1159.27	1634.3	2793.6	2.9293	3.0368	5.9661
270	5.4987	0.001302	0.03564	1177.33	1416.3	2593.7	1184.49	1605.2	2789.7	2.9750	2.9551	5.9301
275	5.9418	0.001317	0.03279	1202.23	1387.9	2590.2	1210.05	1574.9	2785.0	3.0208	2.8730	5.8937
280	6.4117	0.001332	0.03017	1227.43	1358.7	2586.1	1235.97	1543.6	2779.5	3.0667	2.7903	5.8570
285	6.9094	0.001348	0.02777	1252.98	1328.4	2581.4	1262.29	1511.0	2773.3	3.1129	2.7069	5.8198
290	7.4360	0.001366	0.02557	1278.89	1297.1	2576.0	1289.04	1477.1	2766.1	3.1593	2.6227	5.7821
295	7.9928	0.001384	0.02354	1305.21	1264.7	2569.9	1316.27	1441.8	2758.0	3.2061	2.5375	5.7436
300	8.5810	0.001404	0.02167	1331.97	1231.0	2563.0	1344.01	1404.9	2748.9	3.2533	2.4511	5.7044
305	9.2018	0.001425	0.01995	1359.22	1195.9	2555.2	1372.33	1366.4	2738.7	3.3009	2.3633	5.6642
310	9.8566	0.001447	0.01835	1387.03	1159.4	2546.4	1401.29	1326.0	2727.3	3.3492	2.2737	5.6229
315	10.547	0.001472	0.01687	1415.44	1121.1	2536.6	1430.97	1283.5	2714.4	3.3981	2.1821	5.5803
320	11.274	0.001499	0.01549	1444.55	1080.9	2525.5	1461.45	1238.6	2700.1	3.4479	2.0882	5.5361
330	12.845	0.001561	0.012996	1505.24	993.7	2498.9	1525.29	1140.6	2665.8	3.5506	1.8909	5.4416
340	14.586	0.001638	0.010797	1570.26	894.3	2464.5	1594.15	1027.9	2622.0	3.6593	1.6763	5.3356
350	16.514	0.001740	0.008813	1641.81	776.6	2418.4	1670.54	893.4	2563.9	3.7776	1.4336	5.2111
360	18.651	0.001892	0.006945	1725.19	626.3	2351.5	1760.48	720.5	2481.0	3.9146	1.1379	5.0525
370	21.028	0.002213	0.004926	1843.84	384.7	2228.5	1890.37	441.8	2332.1	4.1104	0.6868	4.7972
374.14	22.089	0.003155	0.003155	2029.58	0	2029.6	2099.26	0	2099.3	4.4297	0	4.4297

(Contd.)

Table. A. 1.2 Saturated water: Pressure table

Pressure MPa p	Tem. °C T	Specific Volume, m³/kg Sat. Liquid v_f	Sat. Vapour v_g	Internal Energy, kJ/kg Sat. Liquid u_f	Evap. u_{fg}	Sat. Vapour u_g	Entropy, kJ/kg Sat. Liquid h_f	Evap. h_{fg}	Sat. Vapour h_g	Entropy, kJ/kg K Sat. Liquid s_f	Evap. s_{fg}	Sat. Vapour s_g
0.6113	0.01	0.001000	206.132	0	2375.3	2375.3	0.00	2501.3	2501.3	0	9.1562	9.1562
1.0	6.98	0.001000	129.208	29.29	2355.7	2385.0	29.29	2484.9	2514.2	0.1059	8.8697	8.9756
1.5	13.03	0.001001	87.980	54.70	2338.6	2393.3	54.70	2470.6	2525.3	0.1956	8.6322	8.8278
2.0	17.50	0.001001	67.004	73.47	2326.0	2399.5	73.47	2460.0	2533.5	0.2607	8.4629	8.7236
2.5	21.08	0.001002	54.254	88.47	2315.9	2404.4	88.47	2451.6	2540.0	0.3120	8.3311	8.6431
3.0	24.08	0.001003	45.665	101.03	2307.5	2408.5	101.03	2444.5	2545.5	0.3545	8.2231	8.5775
4.0	28.96	0.001004	34.800	121.44	2293.7	2415.2	121.44	2432.9	2554.4	0.4226	8.0520	8.4746
5.0	32.88	0.001005	28.193	137.79	2282.7	2420.5	137.79	2423.7	2561.4	0.4763	7.9187	8.3950
7.5	40.29	0.001008	19.238	168.76	2261.7	2430.5	168.77	2406.0	2574.8	0.5763	7.6751	8.2514
10.0	45.81	0.001010	14.674	191.79	2246.1	2437.9	191.81	2392.8	2584.6	0.6492	7.5010	8.1501
15.0	53.97	0.001014	10.022	225.90	2222.8	2448.7	225.91	2373.1	2599.1	0.7548	7.2536	8.0084
20.0	60.06	0.001017	7.649	251.35	2205.4	2456.7	251.38	2358.3	2609.7	0.8319	7.0766	7.9085
25.0	64.97	0.001020	6.204	271.88	2191.2	2463.1	271.90	2346.3	2618.2	0.8930	6.9383	7.8313
30.0	69.10	0.001022	5.229	289.18	2179.2	2468.4	289.21	2336.1	2626.3	0.9439	6.8247	7.7686
40.0	75.87	0.001026	3.993	317.51	2159.5	2477.0	317.55	2319.2	2636.7	1.0258	6.6441	7.6700
50.0	81.33	0.001030	3.240	340.42	2143.4	2483.8	340.47	2305.4	2645.9	1.0910	6.5029	7.5939
75.0	91.77	0.001037	2.217	384.29	2112.4	2496.7	384.36	2278.6	2663.0	1.2129	6.2434	7.4563
MPa												
0.100	99.62	0.001043	1.6940	417.33	2088.7	2506.1	417.44	2258.0	2675.5	1.3025	6.0568	7.3593
0.125	105.99	0.001048	1.3749	444.16	2069.3	2513.5	444.30	2241.1	2685.3	1.3739	5.9104	7.2843
0.150	111.37	0.001053	1.1593	466.92	2052.7	2519.6	467.08	2226.5	2693.5	1.4335	5.7897	7.2232
0.175	116.06	0.001057	1.0036	486.78	2038.1	2524.9	486.97	2213.6	2700.5	1.4848	5.6868	7.1717
0.200	120.23	0.001061	0.8857	504.47	2025.0	2529.5	504.68	2202.0	2706.6	1.5300	5.5970	7.1271
0.225	124.00	0.001064	0.7933	520.45	2013.1	2533.6	520.69	2191.3	2712.0	1.5705	5.5173	7.0878
0.250	127.43	0.001067	0.7187	535.08	2002.1	2537.2	535.34	2181.5	2716.9	1.6072	5.4455	7.0526

(Contd.)

Table A.1.2 (Contd.)

Pressure MPa p	Temp. °C T	Specific Volume, m³/kg Sat. Liquid v_f	Sat. Vapour v_g	Internal Energy, kJ/kg Sat. Liquid u_f	Evap. u_{fg}	Sat. Vapour v_g	Enthalpy, kJ/kg Sat. Liquid h_f	Evap. h_{fg}	Sat. Vapour h_g	Entropy, kJ/kg K Sat. Liquid s_f	Evap. s_{fg}	Sat. Vapour s_g
0.275	130.60	0.001070	0.6573	548.57	1992.0	2540.5	548.87	2172.4	2721.3	1.6407	5.3801	7.0208
0.300	133.55	0.001073	0.6058	561.13	1982.4	2543.6	561.45	2163.9	2725.3	1.6717	5.3201	6.9918
0.350	138.88	0.001079	0.5243	583.93	1965.0	2548.9	584.31	2148.1	2732.4	1.7274	5.2130	6.9404
0.375	141.32	0.001081	0.4914	594.38	1956.9	2255.13	594.79	2140.8	2735.6	1.7527	5.1647	6.9174
0.40	143.63	0.001084	0.4625	604.29	1949.3	2553.6	604.73	2133.8	2738.5	1.7766	5.1193	6.8958
0.45	147.93	0.001088	0.4140	622.75	1934.9	2557.6	623.24	2120.7	2743.9	1.8206	5.0359	6.8565
0.50	151.86	0.001093	0.3749	639.66	1921.6	2561.2	640.21	2108.5	2748.7	1.8606	4.9606	6.8212
0.55	155.48	0.001097	0.3427	655.30	1909.2	2564.5	655.91	2097.0	2752.9	1.8972	4.8920	6.7892
0.60	158.85	0.001101	0.3157	669.88	1897.5	2567.4	670.54	2086.3	2756.8	1.9311	4.8289	6.7600
0.65	162.01	0.001104	0.2927	683.55	1886.5	2570.1	684.26	2076.0	2760.3	1.9627	4.7704	6.7330
0.70	164.97	0.001108	0.2729	696.43	1876.1	2572.5	697.20	2066.3	2763.5	1.9922	4.7158	6.7080
0.75	167.77	0.001111	0.2556	708.62	1866.1	2574.7	709.45	2057.0	2766.4	2.0199	4.6647	6.6846
0.80	170.43	0.001115	0.2404	720.20	1856.6	2576.8	721.10	2048.0	2769.1	2.0461	4.6166	6.6627
0.85	172.96	0.001118	0.2270	731.25	1847.4	2578.7	732.20	2039.4	2771.6	2.0709	4.5711	6.6421
0.90	175.38	0.001121	0.2150	741.81	1838.7	2580.5	742.82	2031.1	2773.9	2.0946	4.5280	6.6225
0.95	177.69	0.001124	0.2042	751.94	1830.2	2582.1	753.00	2023.1	2776.1	2.1171	4.4869	6.6040
1.00	179.91	0.001127	0.19444	761.67	1822.0	2583.6	762.79	2015.3	2778.1	2.1386	4.4478	6.5864
1.10	184.09	0.001133	0.17753	780.08	1806.3	2586.4	781.32	2000.4	2781.7	2.1791	4.3744	6.5535
1.20	187.99	0.001139	0.16333	797.27	1791.6	2588.8	798.64	1986.2	2784.8	2.2165	4.3067	6.5233
1.30	191.64	0.001144	0.15125	813.42	1777.5	2590.9	814.91	1972.7	2787.6	2.2514	4.2438	6.4953
1.40	195.07	0.001149	0.14084	828.68	1764.1	2592.8	830.29	1959.7	2790.0	2.2842	4.1850	6.4692
1.50	198.32	0.001154	0.13177	843.14	1751.3	2594.5	844.87	1947.3	2792.1	2.3150	4.2198	6.4448
1.75	205.76	0.001166	0.11349	876.44	1721.4	2597.8	878.48	1918.0	2796.4	2.3851	4.0044	6.3895
2.00	212.42	0.001177	0.09963	906.42	1693.8	2600.3	908.77	1890.7	2799.5	2.4473	3.8935	6.3408
2.25	218.45	0.001187	0.08875	933.81	1668.2	2602.0	936.48	1865.2	2801.7	2.5034	3.7938	6.2971

(Contd.)

Table A.1.2 (Contd.)

Pressure MPa p	Temp. °C T	Specific Volume, m³/kg Sat. Liquid v_f	Sat. Vapour v_g	Internal Energy, kJ/kg Sat. Liquid u_f	Evap. u_{fg}	Sat. Vapour u_g	Enthalpy, kJ/kg Sat. Liquid h_f	Evap. h_{fg}	Sat. Vapour h_g	Entropy, kJ/kg K Sat. Liquid s_f	Evap. s_{fg}	Sat. Vapour s_g
2.50	223.99	0.001197	0.07998	959.09	1644.0	2603.1	962.09	1841.0	2803.1	2.5546	3.7028	6.2574
2.75	229.12	0.001207	0.07275	982.65	1621.2	2603.8	985.97	1817.9	2803.9	2.6018	3.6190	6.2208
3.00	233.90	0.001216	0.06668	1004.76	1599.3	2604.1	1008.41	1795.7	2804.1	2.6456	3.5412	6.1869
3.25	238.38	0.001226	0.06152	1025.62	1578.4	2604.0	1029.60	1774.4	2804.0	2.6866	3.4685	6.1551
3.50	242.60	0.001235	0.05707	1045.41	1558.3	2603.7	1049.73	1753.7	2803.4	2.7252	3.4000	6.1252
4.0	250.40	0.001252	0.049778	1082.28	1520.0	2602.3	1087.29	1714.1	2801.4	2.7963	3.2737	6.0700
5.0	263.99	0.001286	0.039441	1147.78	1449.3	2597.1	1154.21	1640.1	2794.3	2.9201	3.0532	5.9733
6.0	275.64	0.001319	0.032440	1205.41	1384.3	2589.7	1213.32	1571.0	2784.3	3.0266	2.8625	5.8891
7.0	285.88	0.001351	0.027370	1257.51	1323.0	2580.5	1266.97	1505.1	2772.1	3.1210	2.6922	5.8132
8.0	295.06	0.001384	0.023518	1305.54	1264.3	2569.8	1316.61	1441.3	2757.9	3.2067	2.5365	5.7431
9.0	303.40	0.001418	0.020484	1350.47	1207.3	2557.8	1363.23	1378.9	2742.1	3.2857	2.3915	5.6771
10.0	311.06	0.001452	0.018026	1393.00	1151.4	2544.4	1407.53	1317.1	2724.7	3.3595	2.2545	5.6140
11.0	318.15	0.001489	0.015987	1433.68	1096.1	2529.7	1450.05	1255.5	2705.6	3.4294	2.1233	5.5527
12.0	324.75	0.001527	0.014263	1472.92	1040.8	2513.7	1491.24	1193.6	2684.8	3.4961	1.9962	5.4923
13.0	330.93	0.001567	0.012780	1511.09	985.0	2496.1	1531.46	1130.8	2662.2	3.5604	1.8718	5.4323
14.0	336.75	0.001611	0.011485	1548.53	928.2	2476.8	1571.08	1066.5	2637.5	3.6231	1.7485	5.3716
15.0	342.24	0.001658	0.010338	1585.58	869.8	2455.4	1610.45	1000.0	2610.5	3.6847	1.6250	5.3097
16.0	347.43	0.001711	0.009306	1622.63	809.1	2431.7	1650.00	930.6	2580.6	3.4760	1.4995	5.2454
17.0	352.37	0.001770	0.008365	1660.16	744.8	2405.0	1690.25	856.9	2547.2	3.8078	1.3698	5.1776
18.0	357.06	0.001840	0.007490	1698.86	675.4	2374.3	1731.97	777.1	2509.1	3.8713	1.2330	5.1044
19.0	361.54	0.001924	0.006657	1739.87	598.2	2338.1	1776.43	688.1	2464.5	3.9387	1.0841	5.0227
20.0	365.81	0.002035	0.005834	1785.47	507.6	2293.1	1826.18	583.6	2409.7	4.0137	0.9132	4.9269
21.0	369.89	0.002206	0.004953	1841.97	388.7	2230.7	1888.30	446.4	2334.7	4.1073	0.6942	4.8015
22.0	373.80	0.002808	0.003526	1973.16	108.2	2081.4	2034.92	124.0	2159.0	4.3307	0.1917	4.5224
22.09	374.14	0.003155	0.003155	2029.58	0	2029.6	2099.26	0	2099.3	4.4297	0	4.4297

Table A.1.3 Superheated vapour

T	p = 10 kPa (45.81)				p = 50 kPa, (81.33)				p = 100 kPa (99.62)			
	v	*u*	*h*	*s*	*v*	*u*	*h*	*s*	*v*	*u*	*h*	*s*
Sat.	14.674	2437.9	2584.6	8.1501	3.240	2483.8	2645.9	7.5939	1.6940	2506.1	2675.5	7.3593
50	14.869	2443.9	2592.6	8.1749	—	—	—	—	—	—	—	—
100	17.196	2515.5	2687.5	8.4479	3.418	2511.6	2682.5	7.6947	1.6958	2506.6	2676.2	7.3614
150	19.513	2587.9	2783.0	8.6881	3.889	2585.6	2780.1	7.9400	1.9364	2582.7	2776.4	7.6133
200	21.825	2661.3	2879.5	8.9037	4.356	2659.8	2877.6	8.1579	2.1723	2658.0	2875.3	7.8342
250	24.136	2736.0	2977.3	9.1002	4.821	2735.0	2976.0	8.3555	2.4060	2733.7	2974.3	8.0332
300	26.445	2812.1	3076.5	9.2812	5.284	2811.3	3075.5	8.5372	2.6388	2810.4	3074.3	8.2157
400	31.063	2968.9	3279.5	9.6076	6.209	2968.4	3278.9	8.8641	3.1026	2967.8	3278.1	8.5434
500	35.679	3132.3	3489.0	9.8977	7.134	3131.9	3488.6	9.1545	3.5655	3131.5	3488.1	8.8341
600	40.295	3302.5	3705.4	10.1608	8.058	3302.2	3705.1	9.4177	4.0278	3301.9	3704.7	9.0975
700	44.911	3479.6	3928.7	10.4028	8.981	3479.5	3928.5	9.6599	4.4899	3479.2	3928.2	9.3398
800	49.526	3663.8	4159.1	10.6281	9.904	3663.7	4158.9	9.8852	4.9517	3663.5	4158.7	9.5652
900	54.141	3855.0	4396.4	10.8395	10.828	3854.9	4396.3	10.0967	5.4135	3854.8	4396.1	9.7767
1000	58.757	4053.0	4640.6	11.0392	11.751	4052.9	4640.5	10.2964	5.8753	4052.8	4640.3	9.9764
1100	63.372	4257.5	4891.2	11.2287	12.674	4257.4	4891.1	10.4858	6.3370	4257.3	4890.9	10.1658
1200	67.987	4467.9	5147.8	11.4090	13.597	4467.8	5147.7	10.6662	6.7986	4467.7	5147.6	10.3462
1300	72.603	4683.7	4409.7	11.5810	14.521	4683.6	5409.6	10.8382	7.2603	4683.5	5409.5	10.5182
	p = 200 kPa (120.23)				p = 300 kPa (133.55)				p = 400 kPa (143.65)			
Sat.	0.88573	2529.5	2706.6	7.1271	0.60582	2543.6	2725.3	6.9918	0.46246	2553.6	2738.5	6.8958
150	0.95964	2576.9	2768.8	7.2795	0.63388	2570.8	2761.0	7.0778	0.47084	2564.5	2752.8	6.9299
200	1.08034	2654.4	2870.5	7.5066	0.71629	2650.7	2865.5	7.3115	0.53422	2646.8	2860.5	7.1706
250	1.19880	2731.2	2971.0	7.7085	0.79636	2728.7	2967.6	7.5165	0.59512	2726.1	2964.2	7.3788
300	1.31616	2808.6	3071.8	7.8926	0.87529	2806.7	3069.3	7.7022	0.65484	2804.8	3066.7	7.5661
400	1.54930	2966.7	3276.5	8.2217	1.03151	2965.5	3275.0	8.0329	0.77262	2964.4	3273.4	7.8984
500	1.78139	3130.7	3487.0	8.5132	1.18669	3130.0	3486.0	8.3250	0.88934	3129.2	3484.9	8.1912
600	2.01297	3301.4	3704.0	8.7769	1.34136	3300.8	3703.2	8.5892	1.00555	3300.2	3702.4	8.4557
700	2.24426	3478.8	3927.7	9.0194	1.49573	3478.4	3927.1	8.8319	1.12147	3477.9	3926.5	8.6987
800	2.47539	3663.2	4158.3	9.2450	1.64994	3662.9	4157.8	9.0575	1.23722	3662.5	4157.4	8.9244

(Contd.)

Table A.1.3 (Contd.)

T	p = 200 kPa (120.23) v	u	h	s	p = 300 kPa (133.55) v	u	h	s	p = 400 kPa (143.63) v	u	h	s
900	2.70643	3854.5	4395.8	9.4565	1.80406	3854.2	4395.4	9.2691	1.35288	3853.9	4395.1	9.1361
1000	2.93740	4052.5	4640.0	9.6563	1.95812	4052.3	4639.7	9.4689	1.46847	4052.0	4639.4	9.3360
1100	3.16834	4257.0	4890.7	9.8458	2.11214	4256.8	4890.4	9.6585	1.58404	4256.5	4890.1	9.5255
1200	3.39927	4467.5	5147.3	10.0262	2.26614	4467.2	5147.1	9.8389	1.69958	4467.0	5146.8	9.7059
1300	3.63018	4683.2	5409.3	10.1982	2.42013	4683.0	5409.0	10.0109	1.81511	4682.8	5408.8	9.8780
	p = 500 kPa (151.86)				**p = 600 kPa (158.85)**				**p = 800 kPa (170.43)**			
Sat.	0.37489	2561.2	2748.7	6.8212	0.31567	2567.4	2756.8	6.7600	0.24043	2576.8	2769.1	6.6627
200	0.42492	2642.9	2855.4	7.0592	0.35202	2638.9	2850.1	6.9665	0.26080	2630.6	2839.2	6.8158
250	0.47436	2723.5	2960.7	7.2708	0.39383	2720.9	2957.2	7.1816	0.29314	2715.5	2950.0	7.0384
300	0.52256	2802.9	3064.2	7.4598	0.43437	2801.0	3061.6	7.3723	0.32411	2797.1	3056.4	7.2372
350	0.57012	2882.6	3167.6	7.6328	0.47424	2881.1	3165.7	7.5463	0.35439	2878.2	3161.7	7.4088
400	0.61728	2963.2	3271.8	7.7937	0.51372	2962.0	3270.2	7.7078	0.38426	2959.7	3267.1	7.5715
500	0.71093	3128.4	3483.8	8.0872	0.59199	3127.6	3482.7	8.0020	0.44331	3125.9	3480.6	7.8672
600	0.80406	3299.6	3701.7	8.3521	0.66974	3299.1	3700.9	8.2673	0.50184	3297.9	3699.4	8.1332
700	0.89691	3477.5	3926.0	8.5952	0.74720	3477.1	3925.4	8.5107	0.56007	3476.2	3924.3	8.3770
800	0.98959	3662.2	4157.0	8.8211	0.82450	3661.8	4156.5	8.7367	0.61813	3661.1	4155.7	8.6033
900	1.08217	3853.6	4394.7	9.0329	0.90169	3853.3	4394.4	8.9485	0.67610	3852.8	4393.6	8.8153
1000	1.17469	4051.8	4639.1	9.2328	0.97883	4051.5	4638.8	9.1484	0.73401	4051.0	4638.2	9.0153
1100	1.26718	4256.3	4889.9	9.4224	1.05594	4256.1	4889.6	9.3381	0.79188	4255.6	4889.1	9.2049
1200	1.35964	4466.8	5146.6	9.6028	1.13302	4466.5	5146.3	9.5185	0.84974	4466.1	5145.8	9.3854
1300	1.45210	4682.5	5408.6	9.7749	1.21009	4682.3	5408.3	9.6906	0.90758	4681.8	5407.9	9.5575
	p = 1.00 MPa (179.91)				**p = 1.20 MPa (187.99)**				**p = 1.40 MPa (195.07)**			
Sat.	0.19444	2583.6	2778.1	6.5864	0.16333	2588.8	2784.8	6.5233	0.14084	2592.8	2790.0	6.4692
200	0.20596	2621.9	2827.9	6.6939	0.16930	2612.7	2815.9	6.5898	0.14302	2603.1	2803.3	6.4975
250	0.23268	2709.9	2942.6	6.9246	0.19235	2704.2	2935.0	6.8293	0.16350	2698.3	2927.2	6.7467
300	0.25794	2793.2	3051.2	7.1228	0.21382	2789.2	3045.8	7.0316	0.18228	2785.2	3040.4	6.9533
350	0.28247	2875.2	3157.7	7.3010	0.23452	2872.2	3153.6	7.2120	0.20026	2869.1	3149.5	7.1359
400	0.30659	2957.3	3263.9	7.4650	0.25480	2954.9	3260.7	7.3773	0.21780	2952.5	3257.4	7.3025
500	0.35411	3124.3	3478.4	7.7621	0.29463	3122.7	3476.3	7.6758	0.25215	3121.1	3474.1	7.6026

(Contd.)

Table A.1.3 (Contd.)

T	p = 1.00 MPa (179.91)				p = 1.20 MPa (187.99)				p = 1.40 MPa (195.07)			
	v	*u*	*h*	*s*	*v*	*u*	*h*	*s*	*v*	*u*	*h*	*s*
600	0.40109	3296.8	3697.9	8.0289	0.33393	3295.6	3696.3	7.9434	0.28596	3294.4	3694.8	7.8710
700	0.44779	3475.4	3923.1	8.2731	0.37294	3474.5	3922.0	8.1881	0.31947	3473.6	3920.9	8.1160
800	0.49432	3660.5	4154.8	8.4996	0.41177	3659.8	4153.9	8.4149	0.35281	3659.1	4153.0	8.3431
900	0.54075	3852.5	4392.9	8.7118	0.45051	3851.6	4392.2	8.6272	0.38606	3851.0	4391.5	8.5555
1000	0.58712	4050.5	4637.6	8.9119	0.48919	4050.0	4637.0	8.8274	0.41924	4049.5	4636.4	8.7558
1100	0.63345	4255.1	4888.5	9.1016	0.52783	4254.6	4888.0	9.0171	0.45239	4254.1	4887.5	8.9456
1200	0.67977	4465.6	5145.4	9.2821	0.56646	4465.1	5144.9	9.1977	0.48552	4464.6	5144.4	9.1262
1300	0.72608	4681.3	5407.4	9.4542	0.60507	4680.9	5406.9	9.3698	0.51864	4680.4	5406.5	9.2983
	p = 1.60 MPa (201.40)				p = 1.80 MPa (207.15)				p = 2.00 MPa (212.42)			
Sat.	0.12380	2595.9	2794.0	6.4217	0.11042	2598.4	2797.1	6.3793	0.09963	2600.3	2799.5	6.3408
225	0.13287	2644.6	2857.2	6.5518	0.11673	2636.6	2846.7	6.4807	0.10377	2628.3	2835.8	6.4146
250	0.14184	2692.3	2919.2	6.6732	0.12497	2686.0	2911.0	6.6066	0.11144	2679.6	2902.5	6.5452
300	0.15862	2781.0	3034.8	6.8844	0.14021	2776.8	3029.2	6.8226	0.12547	2772.6	3023.5	6.7663
350	0.17456	2866.0	3145.4	7.0693	0.15457	2862.9	3141.2	7.0099	0.13857	2859.8	3137.0	6.9562
400	0.19005	2950.1	3254.2	7.2373	0.16847	2947.7	3250.9	7.1793	0.15120	2945.2	3247.6	7.1270
500	0.22029	3119.5	3471.9	7.5389	0.19550	3117.8	3469.7	7.4824	0.17568	3116.2	3467.6	7.4316
600	0.24998	3293.3	3693.2	7.8080	0.22199	3292.1	3691.7	7.7523	0.19960	3290.9	3690.1	7.7023
700	0.27937	3472.7	3919.7	8.0535	0.24818	3471.9	3918.6	7.9983	0.22323	3471.0	3917.5	7.9487
800	0.30859	3658.4	4152.1	8.2808	0.27420	3657.7	4151.3	8.2258	0.24668	3657.0	4150.4	8.1766
900	0.33772	3850.5	4390.8	8.4934	0.30012	3849.9	4390.1	8.4386	0.27004	3849.3	4389.4	8.3895
1000	0.36678	4049.0	4635.8	8.6938	0.32598	4048.4	4635.2	8.6390	0.29333	4047.9	4634.6	8.5900
1100	0.39581	4253.7	4887.0	8.8837	0.35180	4253.2	4886.4	8.8290	0.31659	4252.7	4885.9	8.7800
1200	0.42482	4464.2	5143.9	9.0642	0.37761	4663.7	5143.4	9.0096	0.33984	4463.2	5142.9	8.9606
1300	0.45382	4679.9	5406.0	9.2364	0.40340	4679.4	5405.6	9.1817	0.36306	4679.0	5405.1	9.1382

Table A.1.3 (Contd.)

	p = 2.50 MPa (223.99)				p = 3.00 MPa (233.90)				p = 3.50 MPa (242.66)			
T	v	u	h	s	v	u	h	s	v	u	h	s
Sat.	0.07998	2603.1	2803.1	6.2574	0.06668	2604.1	2804.1	6.1869	0.05707	2603.7	2803.4	6.1252
225	0.08027	2605.6	2806.3	6.2638	—	—	—	—	—	—	—	—
250	0.08700	2662.5	2880.1	6.4084	0.07058	2644.0	2855.8	6.2871	0.05873	2623.7	2829.2	6.1748
300	0.09890	2761.6	3008.8	6.6437	0.08114	2750.0	2993.5	6.5389	0.06842	2738.0	2977.5	6.4460
350	0.10976	2851.8	3126.2	6.8402	0.09053	2843.7	3115.3	6.7427	0.07678	2835.3	3104.0	6.6578
400	0.12010	2939.0	3239.3	7.0147	0.09936	2932.7	3230.8	6.9211	0.08453	2926.4	3222.2	6.8404
450	0.13014	3025.4	3350.8	7.1745	0.10787	3020.4	3344.0	7.0833	0.09196	3015.3	3337.2	7.0051
500	0.13998	3112.1	3462.0	7.3233	0.11619	3107.9	3456.5	7.2337	0.09918	3103.7	3450.9	7.1571
600	0.15930	3288.0	3686.2	7.5960	0.13243	3285.0	3682.3	7.5084	0.11324	3282.1	3678.4	7.4338
700	0.17832	3468.8	3914.6	7.8435	0.14838	3466.0	3911.7	7.7571	0.12699	3464.4	3908.8	7.6837

(Contd.)

Table A.1.3 (Contd.)

T	p = 2.50 MPa (223.99)				p = 3.00 MPa (233.90)				p = 3.50 MPa (242.60)			
	v	u	h	s	v	u	h	s	v	u	h	s
800	0.19716	3655.3	4148.2	8.0720	0.16414	3653.6	4146.0	7.9862	0.14056	3651.8	4143.8	7.9135
900	0.21590	3847.9	4387.6	8.2853	0.17980	3846.5	4385.9	8.1999	0.15402	3845.0	4384.1	8.1275
1000	0.23458	4046.7	4633.1	8.4860	0.19541	4045.4	4631.6	8.4009	0.16743	4044.1	4630.1	8.3288
1100	0.25322	4251.5	4884.6	8.6761	0.21098	4250.3	4883.3	8.5911	0.18080	4249.1	4881.9	8.5191
1200	0.27185	4462.1	5141.7	8.8569	0.22652	4460.9	5140.5	8.7719	0.19415	4459.8	5139.3	8.7000
1300	0.29046	4677.8	5404.0	9.0291	0.24206	4676.6	5402.8	8.9442	0.20749	4675.5	5401.7	8.8723
	p = 4.00 MPa (250.40)				p = 4.50 MPa (257.48)				p = 5.00 MPa (263.99)			
Sat.	0.04978	2602.3	2801.4	6.0700	0.04406	2600.0	2798.3	6.0198	0.03944	2597.1	2794.3	5.9733
275	0.05457	2667.9	2886.2	6.2284	0.04730	2650.3	2863.1	6.1401	0.04141	2631.2	2838.3	6.0543
300	0.05884	2725.3	2960.7	6.3614	0.05135	2712.0	2943.1	6.2827	0.04532	2697.9	2924.5	6.2083
350	0.06645	2826.6	3092.4	6.5820	0.05840	2817.8	3080.6	6.5130	0.05194	2808.7	3068.4	6.4492
400	0.07341	2919.9	3213.5	6.7689	0.06475	2913.3	3204.7	6.7046	0.05781	2906.6	3195.6	6.6458
450	0.08003	3010.1	3330.2	6.9362	0.07074	3004.9	3323.2	6.8745	0.06330	2999.6	3316.1	6.8185
500	0.08643	3099.5	3445.2	7.0900	0.07651	3095.2	3439.5	7.0300	0.06857	3090.9	3433.8	6.9758
600	0.09885	3279.1	3674.4	7.3688	0.08765	3276.0	3670.5	7.3109	0.07869	3273.0	3666.5	7.2588
700	0.11095	3462.1	3905.9	7.6198	0.09847	3459.9	3903.0	7.5631	0.08849	3457.7	3900.1	7.5122
800	0.12287	3650.1	4141.6	7.8502	0.10911	3648.4	4139.4	7.7942	0.09811	3646.6	4137.2	7.7440
900	0.13469	3843.6	4382.3	8.0647	0.11965	3842.1	4380.6	8.0091	0.10762	3840.7	4378.8	7.9593
1000	0.14645	4042.9	4628.7	8.2661	0.13013	4041.6	4627.2	8.2108	0.11707	4040.3	4625.7	8.1612
1100	0.15817	4248.0	4880.6	8.4566	0.14056	4246.8	4879.3	8.4014	0.12648	4245.6	4878.0	8.3519
1200	0.16987	4458.6	5138.1	8.6376	0.15098	4457.4	5136.9	8.5824	0.13587	4456.3	5135.7	8.5330
1300	0.18156	4674.3	5400.5	8.8099	0.16139	4673.1	5399.4	8.7548	0.14526	4672.0	5398.2	8.7055

(Contd.)

Table A.1.3 (Contd.)

T	$p = 6.00$ MPa (275.64)				$p = 7.00$ MPa (285.88)				$p = 8.00$ MPa (295.06)			
	v	u	h	s	v	u	h	s	v	u	h	s
Sat.	0.03244	2589.7	2784.3	5.8891	0.02737	2580.5	2772.1	5.8132	0.02352	2569.8	2757.9	5.7431
300	0.03616	2667.2	2884.2	6.0673	0.02947	2632.1	2838.4	5.9304	0.02426	2590.9	2785.0	5.7905
350	0.04223	2789.6	3043.0	6.3334	0.03524	2769.3	3016.0	6.2282	0.02995	2747.7	2987.3	6.1300
400	0.04739	2892.8	3177.2	6.5407	0.03993	2878.6	3158.1	6.4477	0.03432	2863.8	3138.3	6.3633
450	0.05214	2988.9	3301.8	6.7192	0.04416	2977.9	3287.0	6.6326	0.03817	2966.7	3272.0	6.5550
500	0.05665	3082.2	3422.1	6.8802	0.04814	3073.3	3410.3	6.7974	0.04175	3064.3	3398.3	6.7239
550	0.06101	3174.6	3540.6	7.0287	0.05195	3167.2	3530.9	6.9486	0.04516	3159.8	3521.0	6.8778
600	0.06525	3266.9	3658.4	7.1676	0.05565	3260.7	3650.3	7.0894	0.04845	3254.4	3642.0	7.0205
700	0.07352	3453.2	3894.3	7.4234	0.06283	3448.6	3888.4	7.3476	0.05481	3444.0	3882.5	7.2812
800	0.08160	3643.1	4132.7	7.6566	0.06981	3639.6	4128.3	7.5822	0.06097	3636.1	4123.8	7.5173
900	0.08958	3837.8	4375.3	7.8727	0.07669	3835.0	4371.8	7.7991	0.06702	3832.1	4368.3	7.7350
1000	0.09749	4037.8	4622.7	8.0751	0.08350	4035.3	4619.8	8.0020	0.07301	4032.8	4616.9	7.9384
1100	0.10536	4243.3	4875.4	8.2661	0.09027	4240.9	4872.8	8.1933	0.07896	4238.6	4870.3	8.1299
1200	0.11321	4454.0	5133.3	8.4473	0.09703	4451.7	5130.9	8.3747	0.08489	4449.4	5128.5	8.3115
1300	0.12106	4669.6	5396.0	8.6199	0.10377	4667.3	5393.7	8.5472	0.09080	4665.0	5391.5	8.4842
	$p = 9.00$ MPa (303.40)				$p = 10.00$ MPa (311.06)				$p = 12.50$ MPa (327.89)			
Sat.	0.02048	2557.8	2742.1	5.6771	0.01803	2544.4	2724.7	5.6140	0.01350	2505.1	2673.8	5.4623
325	0.02327	2646.5	2855.9	5.8711	0.01986	2610.4	2809.0	5.7568	—	—	—	—
350	0.02580	2724.4	2956.5	6.0361	0.02242	2699.2	2923.4	5.9442	0.01613	2624.6	2826.2	5.7117
400	0.02993	2848.4	3117.8	6.2853	0.02641	2832.4	3096.5	6.2119	0.02000	2789.3	3039.3	6.0416
450	0.03350	2955.1	3256.6	6.4843	0.02975	2943.3	3240.8	6.4189	0.02299	3912.4	3199.8	6.2718
500	0.03677	3055.1	3386.1	6.6575	0.03279	3045.8	3373.6	6.5965	0.02560	3021.7	3341.7	6.4617
550	0.03987	3152.2	3511.0	6.8141	0.03564	3144.5	3500.9	6.7561	0.02801	3124.9	3475.1	6.6289
600	0.04285	3248.1	3633.7	6.9588	0.03837	3241.7	3625.3	6.9028	0.03029	3225.4	3604.0	6.7810
650	0.04574	3343.7	3755.3	7.0943	0.04101	3338.2	3748.3	7.0397	0.03248	3324.4	3730.4	6.9218
700	0.04857	3439.4	3876.5	7.2221	0.04358	3434.7	3870.5	7.1687	0.03460	3422.9	3855.4	7.0536

(Contd.)

Table A.1.3 (Contd.)

	$p = 9.00$ MPa (303.40)				$p = 10.00$ MPa (311.06)				$p = 12.50$ MPa (327.89)			
T	***v***	***u***	***h***	***s***	***v***	***u***	***h***	***s***	***v***	***u***	***h***	***s***
800	0.05409	3632.5	4199.4	7.4597	0.04859	3629.0	4114.9	7.4077	0.03869	3620.0	4103.7	7.2965
900	0.5950	3829.2	4364.7	7.6782	0.05349	3826.3	4361.2	7.6272	0.04267	3819.1	4352.5	7.5181
1000	0.06485	4030.3	4613.9	7.8821	0.05832	4027.8	4611.0	7.8315	0.04658	4021.6	4603.8	7.7237
1100	0.07016	4236.3	4867.7	8.0739	0.06312	4234.0	4865.1	8.0236	0.05045	4228.2	4858.8	7.9165
1200	0.07544	4447.2	5126.2	8.2556	0.06789	4444.9	5123.8	8.2054	0.05430	4439.3	5118.0	8.0987
1300	0.08072	4662.7	5389.2	8.4283	0.07265	4660.4	5387.0	8.3783	0.05813	4654.8	5381.4	8.2717
	$p = 15$ MPa (342.24)				$p = 17.5$ MPa (354.75)				$p = 20$ MPa (365.81)			
Sat.	.010338	2455.4	2610.5	5.3097	.0079204	2390.2	2528.8	5.1418	.0058342	2293.1	2409.7	4.9269
350	.011470	2520.4	2692.4	5.4420	.0017139	1632.0	1662.0	3.7612	.0016640	1612.3	1645.6	3.7280
400	.015649	2740.7	2975.4	5.8810	.0124477	2685.0	2902.8	5.7212	.0099423	2619.2	2818.1	5.5539
450	.018446	2879.5	3156.2	6.1403	.0151740	2844.2	3109.7	6.0182	.0126953	2806.2	3060.1	5.9016
500	.020800	2996.5	3308.5	6.3442	.0173585	2970.3	3274.0	6.2382	.0147683	2942.8	3238.2	6.1400
550	.022927	3104.7	3448.6	6.5198	.0192877	3083.8	3421.4	6.4229	.0165553	3062.3	3393.5	6.3347
600	.024911	3208.6	3582.3	6.6775	.0210640	3191.5	3560.1	6.5866	.0181781	3174.0	3537.6	6.5048
650	.026797	3310.4	3712.3	6.8223	.0227372	3296.0	3693.9	6.7356	.0196929	3281.5	3675.3	6.6582
700	.028612	3410.9	3840.1	6.9572	.0243365	3398.8	3824.7	6.8736	.0211311	3386.5	3809.1	6.7993
800	.032096	3611.0	4092.4	7.2040	.0273849	3601.9	4081.1	7.1245	.0238532	3592.7	4069.8	7.0544
900	.035457	3811.9	4343.8	7.4279	.0303071	3804.7	4335.1	7.3507	.0264463	3797.4	4326.4	7.2830
1000	.038748	4015.4	4596.6	7.6347	.0331580	4009.3	4589.5	7.5588	.0289666	4003.1	4582.5	7.4925
1100	.042001	4222.6	4852.6	7.8282	.0359695	4216.9	4846.4	7.7530	.0314471	4211.3	4840.2	7.6874
1200	.045233	4433.8	5112.3	8.0108	.0387605	4428.3	5106.6	7.9359	.0339071	4422.8	5101.0	7.8706
1300	.048455	4649.1	5375.9	8.1839	.0415417	4643.5	5370.5	8.1093	.0363574	4638.0	5365.1	8.0441

(Contd.)

Table A.1.3 (Contd.)

	p = 25 MPa				p = 30 MPa				p = 35 MPa			
T	v	u	h	s	v	u	h	s	v	u	h	s
375	.001973	1798.6	1847.9	4.0319	.001789	1737.8	1791.4	3.9303	.001700	1702.9	1762.4	3.8721
400	.006004	2430.1	2580.2	5.1418	.002790	2067.3	2151.0	4.4728	.002100	1914.0	1987.5	4.2124
425	.007882	2609.2	2806.3	5.4722	.005304	2455.1	2614.2	5.1503	.003428	2253.4	2373.4	4.7747
450	.009162	2720.7	2949.7	5.6743	.006735	2619.3	2821.4	5.4423	.004962	2498.7	2672.4	5.1962
500	.011124	2884.3	3162.4	5.9592	.008679	2820.7	3081.0	5.7904	.006927	2751.9	2994.3	5.6281
550	.012724	3017.5	3335.6	6.1764	.010168	2970.3	3275.4	6.0342	.008345	2920.9	3213.0	5.9025
600	.014138	3137.9	3491.4	6.3602	.011446	3100.5	3443.9	6.2330	.009527	3062.0	3395.5	6.1178
650	.015433	3251.6	3637.5	6.5229	.012596	3221.0	3598.9	6.4057	.010575	3189.8	3559.9	6.3010
700	.016647	3361.4	3777.6	6.6707	.013661	3335.8	3745.7	6.5606	.011533	3309.9	3713.5	6.4631
800	.018913	3574.3	4047.1	6.9345	.015623	3555.6	4024.3	6.8332	.013278	3536.8	4001.5	6.7450
900	.021045	3783.0	4309.1	7.1679	.017448	3768.5	4291.9	7.0717	.014883	3754.0	4274.9	6.9886
1000	.023102	3990.9	4568.5	7.3801	.019196	3978.8	4554.7	7.2867	.016410	3966.7	4541.1	7.2063
1100	.025119	4200.2	4828.2	7.5765	.020903	4189.2	4816.3	7.4845	.017895	4178.3	4804.6	7.4056
1200	.027115	4412.0	5089.9	7.7604	.022589	4410.3	5079.0	7.6691	.019360	4390.7	5068.4	7.5910
1300	.029101	4626.9	5354.4	7.9342	.024266	4616.0	5344.0	7.8432	.020815	4605.1	5333.6	7.7652
	p = 40 MPa				p = 50 MPa				p = 60 MPa			
375	.0016406	1677.1	1742.7	3.8289	.0015593	1638.6	1716.5	3.7638	.0015027	1609.3	1699.5	3.1740
400	.0019077	1854.5	1930.8	4.1134	.0017309	1788.0	1874.6	4.0030	.0016335	1745.3	1843.4	3.9317
425	.0025319	2096.8	2198.1	4.5028	.0020071	1959.6	2060.0	4.2733	.0018165	1892.7	2001.7	4.1625
450	.0036931	2365.1	2512.8	4.9459	.0024862	2159.6	2283.9	4.5883	.0020850	2053.9	2179.0	4.4119
500	.0056225	2678.4	2903.3	5.4699	.0038924	2525.5	2720.1	5.1725	.0029557	2390.5	2567.9	4.9320
600	.0080943	3022.6	3346.4	6.0113	.0061123	2942.0	3247.6	5.8177	.0048345	2861.1	3151.2	5.6451
650	.0090636	3158.0	3520.6	6.2054	.0069657	3093.6	3441.8	6.0342	.0055953	3028.8	3364.6	5.8829
700	.0099415	3283.6	3681.3	6.3750	.0077274	3230.5	3616.9	6.2189	.0062719	3177.3	3553.6	6.0824
800	.0115228	3517.9	3978.8	6.6662	.0090761	3479.8	3933.6	6.5290	.0074588	3441.6	3889.1	6.4110
900	.0129626	3739.4	4257.9	6.9150	.0102831	3710.3	4224.4	6.7882	.0085083	3681.0	4191.5	6.6805
1000	.0143238	3954.6	4527.6	7.1356	.0114113	3930.5	4501.1	7.0146	.0094800	3906.4	4475.2	6.9126
1100	.0156426	4167.4	4793.1	7.3364	.0124966	4145.7	4770.6	7.2183	.0104091	4124.1	4748.6	7.1194
1200	.0169403	4380.1	5057.7	7.5224	.0135606	4359.1	5037.2	7.4058	.0113167	4338.2	5017.2	7.3082
1300	.0182292	4594.3	5323.5	7.6969	.0146159	4572.8	5303.6	7.5807	.0122155	4551.4	5284.3	7.4837

Table A.1.4 Compressed liquid water

	p = 5.00 MPa (263.99)				p = 10.00 MPa (311.06)				p = 15.00 MPa (342.24)			
T	*v*	*u*	*h*	*s*	*v*	*u*	*h*	*s*	*v*	*u*	*h*	*s*
Sat.	.0012859	1147.78	1154.21	2.9201	.0014524	1393.00	1407.53	3.3595	.0016581	1585.58	1610.45	3.6847
0	.0009977	0.03	5.02	0.0001	.0009952	0.10	10.05	0.0003	.0009928	0.15	15.04	0.0004
20	.0009995	83.64	88.64	0.2955	.0009972	83.35	93.32	0.2945	.0009950	83.05	97.97	0.2934
40	.0010056	166.93	171.95	0.5706	.0010034	166.33	176.36	0.5685	.0010013	165.73	180.75	0.5665
60	.0010149	250.21	255.28	0.8284	.0010127	249.34	259.47	0.8258	.0010105	248.49	263.65	0.8231
80	.0010268	333.69	338.83	1.0719	.0010245	332.56	342.81	1.0687	.0010222	331.46	346.79	1.0655
100	.0010410	417.50	422.71	1.3030	.0010385	416.09	426.48	1.2992	.0010361	414.72	430.26	1.2954
120	.0010576	501.79	507.07	1.5232	.0010549	500.07	510.61	1.5188	.0010522	498.39	514.17	1.5144
140	.0010768	586.74	592.13	1.7342	.0010737	584.67	595.40	1.7291	.0010707	582.64	598.70	1.7241
160	.0010988	672.61	678.10	1.9374	.0010953	670.11	681.07	1.9316	.0010918	667.69	684.07	1.9259
180	.0011240	759.62	765.24	2.1341	.0011199	756.63	767.83	2.1274	.0011159	753.74	770.48	2.1209
200	.0011530	848.08	853.85	2.3254	.0011480	844.49	855.97	2.3178	.0011433	841.04	858.18	2.3103
220	.0011866	938.43	944.36	2.5128	.0011805	934.07	945.88	2.5038	.0011748	929.89	947.52	2.4952
240	.0012264	1031.34	1037.47	2.6978	.0012187	1025.94	1038.13	2.6872	.0012114	1020.82	1038.99	2.6770
260	.0012748	1127.92	1134.30	2.8829	.0012645	1121.03	1133.68	2.8698	.0012550	1114.59	1133.41	2.8575
280					.0013216	1220.90	1234.11	3.0547	.0013084	1212.47	1232.09	3.0392
300					.0013972	1328.34	1342.31	3.2468	.0013770	1316.58	1337.23	3.2259
320									.0014724	1431.05	1453.13	3.4246
340									.0016311	1567.42	1591.88	3.6545

(Contd.)

Table A.1.4 (Contd.)

T	p = 20 MPa (365.81)				p = 30 MPa				p = 50 MPa			
	v	u	h	s	v	u	h	s	v	u	h	s
Sat	.0020353	1785.47	1826.18	4.0137	—	—	—	—	—	—	—	—
0	.0009904	0.02	20.00	0.0004	.0009856	0.25	29.82	0.0001	.0009766	0.20	49.03	−0.0014
20	.0009928	82.75	102.61	0.2922	.0009886	82.16	111.82	0.2898	.0009804	80.98	130.00	0.2847
40	.0009992	165.15	185.14	0.5646	.0009951	164.01	193.87	0.5606	.0009872	161.84	211.20	0.5526
60	.0010084	247.66	267.82	0.8205	.0010042	246.03	276.16	0.8153	.0009962	242.96	292.77	0.8051
80	.0010199	330.38	350.78	1.0623	.0010156	328.28	358.75	1.0561	.0010073	324.32	374.68	1.0439
100	.0010337	413.37	434.04	1.2917	.0010290	410.76	441.63	1.2844	.0010201	405.86	456.87	1.2703
120	.0010496	496.75	517.74	1.5101	.0010445	493.58	524.91	1.5017	.0010348	487.63	539.37	1.4857
140	.0010678	580.67	602.03	1.7192	.0010621	576.86	608.27	1.7097	.0010515	569.76	622.33	1.6915
160	.0010885	665.34	687.11	1.9203	.0010821	660.81	693.27	1.9085	.0010703	652.39	705.91	1.8890
180	.0011120	750.94	773.18	2.1146	.0011047	745.57	778.71	2.1024	.0010912	735.68	790.24	2.0793
200	.0011387	837.70	860.47	2.3031	.0011302	831.34	865.24	2.2892	.0011146	819.73	875.46	2.2634
220	.0011693	925.89	949.27	2.4869	.0011590	918.32	953.09	2.4710	.0011408	904.67	961.71	2.4419
240	.0012046	1015.94	1040.04	2.6673	.0011920	1006.84	1042.60	2.6489	.0011702	990.69	1049.20	2.6158
260	.0012462	1108.53	1133.45	2.8459	.0012303	1097.38	1134.29	2.8242	.0012034	1078.06	1138.23	2.7860
280	.0012965	1204.69	1230.62	3.0248	.0012755	1190.69	1228.96	2.9985	.0012415	1167.19	1229.26	2.9536
300	.0013596	1306.10	1333.29	3.2071	.0013304	1287.89	1327.80	3.1740	.0012860	1258.66	1322.95	3.1200
320	.0014437	1415.66	1444.53	3.3978	.0013997	1390.64	1432.63	3.3538	.0013388	1353.23	1420.17	3.2867
340	.0015683	1539.64	1571.01	3.6074	.0014919	1501.71	1456.47	3.5425	.0014032	1451.91	1522.07	3.4556
360	.0018226	1702.78	1739.23	3.8770	.0016265	1626.57	1675.36	3.7492	.0014838	1555.97	1630.16	3.6290
380	—	—	—	—	.0018691	1781.35	1837.43	4.0010	.0015883	1667.13	1746.54	3.8100

A.2 Thermodynamic Properties of Refrigerant-12* (Dichlorodifluoromethane)

Table A.2.1 Saturated refrigerant-12

Temperature °C t	Presssure MPa p	Specific Volume: Sat. Liquid v_f Cm^3/g	Specific Volume: Sat. Vapour v_g m^3/kg	Enthalpy: Sat. Liquid h_f	Enthalpy: Evap. h_{fg} kJ/kg	Enthalpy: Sat. Vapour h_g	Entropy: Sat. Liquid s_f	Entropy: Sat. Vapour s_g kJ/kg K
– 90	0.0028	0.608	4.415545	– 43.243	189.618	146.375	– 0.2084	0.8268
– 85	0.0042	0.612	3.037316	– 38.968	187.608	148.640	– 0.1854	0.8116
– 80	0.0062	0.617	2.038345	– 34.688	185.612	150.924	– 0.1630	0.7979
– 75	0.0088	0.622	1.537651	– 30.401	183.625	153.224	– 0.1411	0.7855
– 70	0.0123	0.627	1.127280	– 26.103	181.640	155.536	– 0.1197	0.7744
– 65	0.0168	0.632	0.841166	– 21.793	179.651	157.857	– 0.0987	0.7643
– 60	0.0226	0.637	0.637910	– 17.469	177.653	160.184	– 0.0782	0.7552
– 55	0.0300	0.642	0.491000	– 13.129	175.641	162.512	– 0.0581	0.7470
– 50	0.0391	0.648	0.383105	– 8.772	173.611	164.840	– 0.0384	0.7396
– 45	0.0504	0.654	0.302682	– 4.396	171.558	167.163	– 0.0190	0.7329
– 40	0.0642	0.659	0.241910	– 0.000	169.479	169.479	– 0.0000	0.7269

(Contd.)

* Repeated from *Fundamentals of Classical Thermodynamics* by G.J. Van Wylen and R. Sonntag. John Wiley, New York 1976, P. 667–673 (with kind permission of the publishers, John Wiley & Sons, Inc, New York).

ble A.2.1 (Contd.)

T	p	v_f	v_g	h_f	h_{fg}	h_g	s_f	s_g
– 35	0.0807	0.666	0.195398	4.416	167.368	171.784	0.0187	0.7214
– 30	0.1004	0.672	0.159375	8.854	165.222	174.076	0.0371	0.7165
– 25	0.1237	0.679	0.131166	13.315	163.037	176.352	0.0552	0.7121
– 20	0.1509	0.685	0.108847	17.800	160.810	178.610	0.0730	0.7082
– 15	0.1826	0.693	0.091018	22.312	158.534	180.846	0.0906	0.7046
– 10	0.2191	0.700	0.076646	26.851	156.207	183.058	0.1079	0.7014
– 5	0.2610	0.708	0.064963	31.420	153.823	185.243	0.1250	0.6986
0	0.3086	0.716	0.055389	36.022	151.376	187.397	0.1418	0.6960
5	0.3626	0.724	0.047485	40.659	148.859	189.518	0.1585	0.6937
10	0.4233	0.733	0.040914	45.337	146.265	191.602	0.1750	0.6916
15	0.4914	0.743	0.035413	50.058	143.586	193.644	0.1914	0.6897
20	0.5673	0.752	0.030780	54.828	140.812	195.641	0.2076	0.6879
25	0.6516	0.763	0.026854	59.653	137.933	197.586	0.2237	0.6863
30	0.7449	0.774	0.023508	64.539	134.936	199.475	0.2397	0.6848
35	0.8477	0.786	0.020641	69.494	131.805	201.299	0.2557	0.6834
40	0.9607	0.798	0.018171	74.527	128.525	203.051	0.2716	0.6820
45	1.0843	0.811	0.016032	79.647	125.074	204.722	0.2875	0.6806
50	1.2193	0.826	0.014170	84.868	121.430	206.298	0.3034	0.6792
55	1.3663	0.841	0.012542	90.201	117.565	207.766	0.3194	0.6777
60	1.5259	0.858	0.01111	95.665	113.443	209.109	0.3355	0.6760
65	1.6988	0.877	0.009847	101.279	109.024	210.303	0.3518	0.6742
70	1.8858	0.897	0.008725	107.067	104.255	211.321	0.3683	0.6721
75	2.0874	0.920	0.007723	113.058	99.068	212.126	0.3851	0.6697
80	2.3046	0.946	0.006821	119.291	93.373	212.665	0.4023	0.6667
85	2.5380	0.976	0.006005	125.818	87.047	212.865	0.4201	0.6631
90	2.7885	1.012	0.005258	132.708	79.907	212.614	0.4385	0.6585
95	3.0569	1.056	0.004563	140.068	71.658	211.726	0.4579	0.6526
100	3.3440	1.113	0.003903	148.076	61.768	209.843	0.4788	0.6444
105	3.6509	1.197	0.003242	157.085	49.014	206.099	0.5023	0.6319
110	3.9784	1.364	0.002462	168.059	28.425	196.484	0.5322	0.6064

Table A.2.2 Saturated refrigerant-12

t °C	*v* m^3kg	*h* kJ/kg	*s* kJ/kg K	*v* m^3/kg	*h* kJ/kg	*s* kJ/kg K	*v* m/kg	*h* kJ/kg	*s* kJ/kg K
		0.05 MPa			***0.10 MPa***			***0.15 MPa***	
– 20.00	0.341857	181.042	0.7912	0.167701	179.861	0.7401			
– 10.00	0.356228	186.757	0.8133	0.175222	185.707	0.7628	0.114716	184.619	0.7318
0.00	0.370508	192.567	0.8350	0.182647	191.628	0.7849	0.119866	190.660	0.7543
10.00	0.384716	198.471	0.8562	0.189994	197.628	0.8064	0.124932	196.762	0.7763
20.00	0.398863	204.469	0.8770	0.197277	203.707	0.8275	0.129930	202.927	0.7977
30.00	0.412959	210.557	0.8974	0.204506	209.866	0.8482	0.134873	209.160	0.8186
40.00	0.427012	216.733	0.9175	0.211691	216.104	0.8684	0.139768	215.463	0.8390
50.00	0.441030	222.997	0.9372	0.218839	222.421	0.8883	0.144625	221.835	0.8591
60.00	0.455017	229.344	0.9565	0.225955	228.815	0.9078	0.149450	228.277	0.8787
70.00	0.458978	235.774	0.9755	0.233044	235.285	0.9269	0.154247	234.789	0.8980
80.00	0.482917	242.282	0.9942	0.240111	241.829	0.9457	0.159020	241.371	0.9169
90.00	0.496838	248.868	1.0126	0.247159	248.446	0.9642	0.163774	248.020	0.9354
		0.20 MPa			***0.25 MPa***			***0.30 Mpa***	
00.0	0.088608	189.669	0.7320	0.069752	188.644	0.7139	0.057150	187.583	0.6984
10.00	0.092550	195.878	0.7543	0.073024	194.969	0.7366	0.059984	194.034	0.7216
20.00	0.096418	202.135	0.7760	0.076218	201.322	0.7587	0.062734	200.490	0.7440
30.00	0.100228	208.446	0.7972	0.079350	207.715	0.7801	0.065418	206.969	0.7658
40.00	0.103989	214.814	0.8178	0.082431	214.153	0.8010	0.068049	213.480	0.7869
50.00	0.107710	221.243	0.8381	0.085470	220.642	0.8214	0.070635	220.030	0.8075
60.00	0.111397	227.735	0.8578	0.088474	227.185	0.8413	0.073185	226.627	0.8276
70.00	0.115055	134.291	0.8772	0.091449	233.785	0.8608	0.075750	233.273	0.8473
80.00	0.118690	240.910	0.8962	0.094398	240.443	0.8800	0.078200	239.271	0.8665
90.00	0.122304	247.593	0.9149	0.097327	247.160	0.8987	0.080673	246.723	0.8853

(Contd.)

Table A.2.2 (Contd.)

t	*v*	*h*	*s*	*v*	*h*	*s*	*v*	*h*	*s*
100.0	0.125901	254.339	0.9332	0.100238	253.936	0.9171	0.83127	253.530	0.9038
110.0	0.129483	261.147	0.9512	0.103134	260.770	0.9352	0.085566	260.391	0.9220
		0.40 MPa			***0.50 MPa***			***0.60 MPa***	
20.0	0.045836	198.762	0.7199	0.035646	196.935	0.6999			
30.0	0.047971	205.428	0.7423	0.037464	203.814	0.7230	0.030422	202.116	0.7063
40.0	0.050046	212.095	0.7639	0.039214	210.656	0.7452	0.031966	209.154	0.7291
50.0	0.052072	218.779	0.7849	0.040911	217.484	0.7667	0.033450	216.141	0.7511
60.0	0.054059	225.488	0.8054	0.042565	224.315	0.7875	0.034887	223.104	0.7723
70.0	0.056014	232.230	0.8253	0.044184	232.161	0.8077	0.036285	230.062	0.7929
80.0	0.057941	239.012	0.8448	0.045774	238.031	0.8275	0.037653	237.027	0.8129
90.0	0.059846	245.837	0.8638	0.047340	244.932	0.8467	0.038995	244.009	0.8324
100.0	0.061731	252.707	0.8825	0.048886	251.869	0.8656	0.040316	251.016	0.8514
110.0	0.063600	259.624	0.9008	0.050415	258.845	0.8840	0.041619	258.053	0.8700
		0.70 MPa			***0.80 MPa***			***0.90 MPa***	
40.0	0.026761	207.580	0.7148	0.022830	205.924	0.7016	0.019744	204.170	0.6982
50.0	0.028100	214.745	0.7373	0.024068	213.290	0.7248	0.020912	211.765	0.7131
60.0	0.029387	221.854	0.7590	0.025247	220.558	0.7469	0.022012	218.212	0.7358
70.0	0.030632	228.931	0.7799	0.026380	227.766	0.7682	0.023062	226.564	0.7575
80.0	0.031843	235.997	0.8002	0.027477	234.941	0.7888	0.024072	233.856	0.7785
90.0	0.033027	243.066	0.8199	0.028545	242.101	0.8088	0.025051	141.113	0.7987
100.0	0.034189	250.146	0.8392	0.029588	249.260	0.8283	0.026005	248.355	0.8184
110.0	0.035332	257.247	0.8579	0.030612	256.428	0.8472	0.026937	255.593	0.8376
		1.00 MPa			***1.20 MPa***			***1.40 MPa***	
50.0	0.018366	210.162	0.7021	0.014483	206.661	0.6812			
60.0	0.019410	217.810	0.7254	0.015463	214.805	0.7060	0.012579	211.457	0.6876

(Contd.)

Table A.2.2 (Contd.)

t	*v*	*h*	*s*	*v*	*h*	*s*	*v*	*h*	*s*
70.0	0.020397	225.319	0.7476	0.016368	222.687	0.7293	0.013448	219.822	0.7123
80.0	0.021341	232.739	0.7689	0.017221	230.398	0.7514	0.014247	227.891	0.7355
90.0	0.022251	240.101	0.7895	0.018032	237.995	0.7727	0.014997	235.766	0.7575
100.0	0.023133	247.430	0.8094	0.018812	245.518	0.7931	0.015710	243.512	0.7785
110.10	0.023993	254.743	0.8287	0.019567	252.993	0.8129	0.016393	251.170	0.7988
		1.60 MPa			***1.80 MPa***			***2.00 MPa***	
70.0	0.011208	216.650	0.6959	0.009406	213.049	0.6794			
80.0	0.011984	225.177	0.7204	0.010187	222.198	0.7057	0.008704	218.859	0.6909
90.0	0.012698	233.390	0.7433	0.010884	230.835	0.7298	0.009406	228.056	0.7166
100.0	0.013366	241.397	0.7651	0.011526	239.155	0.7524	0.010035	263.760	0.7402
110.0	0.014000	249.264	0.7859	0.012126	2467.264	0.7739	0.010615	245.154	0.7624

A.3 Thermodynamic Properties of Refrigerent-12 (Monochlorodifluoromethane)

Table A.3.1 Saturated refrigerant-22

Temp. °C	Abs. Press. MPa p	Specific volume (m^3/kg) Sat. Liquid v_f	Evap. v_{fg}	Sat. Vapour v_g	Enthalpy (kJ/kg) Sat. Liquid h_f	Evap. h_{fg}	Vapour h_g	Entropy (kJ/kg K) Sat. Liquid s_f	Evap. s_{fg}	Sat. Vapour s_g
70	0.0205	0.000670	0.940268	0.94093	– 30.607	249.425	218.180	– 0.1401	1.2277	1.0876
65	0.0280	0.000676	0.704796	0.705478	– 25.658	246.925	221.267	– 0.1161	1.1862	1.0701
60	0.0375	0.000682	0.536470	0.537152	– 20.652	244.354	223.702	– 0.0924	1.1463	1.0540
55	0.0495	0.000689	0.414138	0.414827	– 15.585	241.703	226.117	– 0.0689	1.1079	1.0390
50	0.0644	0.000695	0.323862	0.324557	– 10.456	238.965	228.509	– 0.0457	1.0708	1.0251
45	0.0827	0.000702	0.256288	0.256990	– 5.262	236.132	230.870	– 0.0227	1.0349	1.0122
40	0.1049	0.000709	0.205036	0.205745	0	233.198	233.197	0	1.0002	1.0002
35	0.1317	0.000717	0.165583	0.166400	5.328	230.156	235.484	0.0225	0.9664	0.9889
30	0.1635	0.000725	0.135120	0.135844	10.725	227.001	237.726	0.0449	0.9335	0.9784
25	0.2010	0.000733	0.111126	0.111859	16.191	223.727	239.918	0.0670	0.9015	0.9685
20	0.2448	0.000741	0.092102	0.092843	21.728	220.327	242.055	0.0890	0.8703	0.9593
15	0.2957	0.000750	0.076876	0.077625	27.334	216.798	244.132	0.1107	0.8398	0.9505
10	0.3543	0.000759	0.064581	0.065340	33.012	213.132	246.144	0.1324	0.8099	0.9422
5	0.4213	0.000768	0.054571	0.055339	38.762	209.323	248.085	0.1538	0.7806	0.9344
0	0.4976	0.000778	0.046357	0.047135	44.586	205.364	249.949	0.1751	0.7518	0.9269
5	0.5838	0.000789	0.039567	0.040356	50.485	201.246	251.731	0.1963	0.7235	0.9197
10	0.6807	0.000800	0.033914	0.034714	56.463	196.960	253.423	0.2173	0.6956	0.9129
15	0.7891	0.000812	0.029176	0.029987	62.523	192.495	255.018	0.2382	0.6680	0.9062
20	0.9099	0.000824	0.025179	0.026003	68.670	187.836	256.506	0.2590	0.6407	0.8997
25	1.0439	0.000838	0.021787	0.022624	74.910	182.968	257.877	0.2797	0.6137	0.8934
30	1.1919	0.000852	0.018890	0.019742	81.250	177.869	259.119	0.3004	0.5867	0.8871

(Contd.)

Table A.3.1 (Contd.)

Temp. °C	Abs. Press. MPa p	Specific volume (m^3/kg) Sat. Liquid v_f	Evap. v_{fg}	Sat. Vapour v_g	Enthalpy (kJ/kg) Sat. Liquid h_f	Evap. h_{fg}	Vapour h_g	Entropy (kJ/kg K) Sat. Liquid s_f	Evap. s_{fg}	Sat. Vapour s_g
35	1.3548	0.000867	0.016401	0.017269	87.700	172.516	260.216	0.3210	0.5598	0.8809
40	1.5335	0.000884	0.014251	0.015135	94.272	166.877	261.149	0.3417	0.5329	0.8746
45	1.7290	0.000902	0.012382	0.013284	100.982	160.914	261.896	0.3624	0.5058	0.8682
50	1.9423	0.000922	0.010747	0.011669	107.851	154.576	262.428	0.3832	0.4783	0.8615
55	2.1744	0.000944	0.009308	0.010252	114.905	147.800	262.705	0.4042	0.4504	0.8546
60	2.4266	0.000969	0.008032	0.009001	122.180	140.497	262.678	0.4255	0.4217	0.8472
65	2.6999	0.000997	0.006890	0.007887	129.729	132.547	262.276	0.4472	0.3920	0.8391
70	2.9959	0.001030	0.005859	0.006889	137.625	123.772	261.397	0.4695	0.3607	0.8302
75	3.3161	0.001069	0.004914	0.005983	145.986	113.902	259.888	0.4927	0.3272	0.8198
80	3.6623	0.001118	0.004031	0.005149	155.011	102.475	257.486	0.5173	0.2902	0.8075
85	4.0368	0.001183	0.003175	0.004358	165.092	88.598	253.690	0.5445	0.2474	0.7918
90	4.4425	0.001282	0.002282	0.003564	177.204	70.037	247.241	0.5767	0.1929	0.7695
95	4.8835	0.001521	0.001030	0.002551	196.359	34.925	231.284	0.6273	0.0949	0.7222
96.006	4.9773	0.001906	0	0.001906	212.546	0	212.546	0.6708	0	0.6708

ble A.3.2 Superheated refrigerant-22

Temp. °C	v m^3/kg	h kJ/kg	s. kJ/kg K	v m^3/kg	h kJ/kg	s. kJ/kg K	v m^3/kg	h kJ/kg	s kJ/kg K
		0.05 MPa			0.10 MPa			0.15 MPa	
– 40	0.440633	234.724	1.07616	0.216331	233.337	1.00523	—	—	—
– 30	0.460641	240.602	1.10084	0.226754	239.359	1.03052	0.148723	238.078	0.98773
– 20	0.480543	246.586	1.12495	0.237064	245.466	1.05513	0.155851	244.319	1.01288
– 10	0.500357	252.676	1.14855	0.247279	251.665	1.07914	0.162879	250.631	1.03733
0	0.520095	258.874	1.17166	0.257415	257.956	1.10261	0.169823	257.022	1.06116
10	0.539771	265.180	1.19433	0.267485	264.345	1.12558	0.176699	263.496	1.08444
20	0.559393	271.594	1.21659	0.277500	270.831	1.14809	0.183516	270.057	1.10721
30	0.578970	278.115	1.23846	0.287467	277.416	1.17017	0.190284	276.709	1.12952
40	0.598507	284.743	1.25998	0.297394	284.101	1.19187	0.197011	283.452	1.15140
50	0.618011	291.478	1.28114	0.307287	290.887	1.21320	0.203702	290.289	1.17289
60	0.637485	298.319	1.30199	0.317149	297.772	1.23418	0.210362	297.220	1.19402
70	0.656935	305.265	1.32253	0.326986	304.757	1.25484	0.216997	304.246	1.21479
80	0.676362	312.314	1.34278	0.336801	311.842	1.27519	0.223608	311.368	1.23525
90	0.695771	319.465	1.36275	0.346596	319.026	1.29524	0.230200	318.584	1.25540
		0.20 MPa			0.25 MPa			0.30 MPa	
– 20	0.115203	243.140	0.98184	—	—	—	—	—	—
– 10	0.120647	249.574	1.00676	0.95280	248.492	0.98231	0.078344	247.382	0.96170
0	0.126003	256.069	1.03098	0.099689	255.097	1.00695	0.082128	254.104	0.98677
10	0.131286	262.633	1.05458	0.104022	261.755	1.03089	0.085832	260.861	1.01106

(Contd.)

Table A.3.2 (Contd.)

Temp. °C	v m³/kg	h kJ/kg	s. kJ/kg K	v m³/kg	h kJ/kg	s kJ/kg K	v m³/kg	h kJ/kg	s kJ/kg K
		0.20 MPa			*0.25 MPa*			*0.30 MPa*	
20	0.136509	269.273	1.07763	0.108292	268.476	1.05421	0.089469	267.667	1.03468
30	0.141681	275.992	1.10016	0.112508	275.267	1.07699	0.093051	274.531	1.05771
40	0.146809	282.796	1.12224	0.116681	282.132	1.09927	0.096588	281.460	1.08019
50	0.151902	289.686	1.14390	0.120815	289.076	1.12109	0.100085	288.460	1.10220
60	0.156963	296.664	1.16516	0.124918	296.102	1.14250	0.103550	295.535	1.12376
70	0.161997	303.731	1.18607	0.128993	303.212	1.16353	0.106986	302.689	1.14491
80	0.167008	310.890	1.20663	0.133044	310.409	1.18420	0.110399	309.924	1.16569
90	0.171999	318.139	1.22687	0.137075	317.692	1.20454	0.113790	317.241	1.18612
100	0.176972	325.480	1.24681	0.141089	325.063	1.22456	0.117164	324.643	1.20623
110	0.181931	332.912	1.26646	0.145086	332.522	1.24428	0.120522	332.129	1.22603
		0.40 MPa			*0.50 MPa*			*0.60 MPa*	
0	0.060131	252.051	0.95359	—	—	—	—	—	—
10	0.063060	259.023	0.97866	0.049355	257.108	0.95223	0.040180	255.109	0.92945
20	0.065915	266.010	1.00291	0.051751	264.295	0.97717	0.042280	262.517	0.95517
30	0.068710	273.029	1.02646	0.054081	271.483	1.00128	0.044307	269.888	0.97989
40	0.071455	280.092	1.04938	0.056358	278.690	1.02467	0.046276	277.250	1.00378
50	0.074160	287.209	1.07175	0.058590	285.930	1.04743	0.048198	284.622	1.02695
60	0.076830	294.386	1.09362	0.060786	293.215	1.06963	0.050081	292.020	1.04950
70	0.079470	301.630	1.11504	0.062951	300.552	1.09133	0.051931	299.456	1.07149
80	0.082085	308.944	1.13605	0.065090	307.949	1.11257	0.053754	306.938	1.09298
90	0.084679	316.332	1.15668	0.067206	315.410	1.13340	0.055553	314.475	1.11403
100	0.087254	323.796	1.17695	0.069303	322.939	1.15386	0.057332	322.071	1.13466
110	0.089813	331.339	1.19690	0.071384	330.539	1.17395	0.059094	329.731	1.15492
120	0.092358	338.961	1.21654	0.073450	338.213	1.19373	0.060842	337.458	1.17482
130	0.094890	346.664	1.23588	0.075503	345.963	1.21319	0.062576	345.255	1.19441

(Contd.)

Table A.3.2 (Contd.)

Temp. °C	v m³/kg	h kJ/kg	s. kJ/kg K	v m³/kg	h kJ/kg	s kJ/kg K	v m³/kg	h kJ/kg	s kJ/kg K
	0.70 MPa			*0.80 MPa*			*0.90 MPa*		
20	0.035487	260.667	0.93565	0.030366	258.737	0.91787	0.026355	256.713	0.90132
30	0.037305	268.240	0.96105	0.032034	266.533	0.94402	0.027915	264.760	0.92831
40	0.039059	275.769	0.98549	0.033632	274.243	0.96905	0.029397	272.670	0.95398
50	0.040763	283.282	1.00910	0.035175	281.907	0.99314	0.030819	280.497	0.97859
60	0.042424	290.800	1.03201	0.036674	289.553	1.01644	0.032193	288.278	1.00230
70	0.044052	298.339	1.05431	0.038136	297.202	1.03906	0.033528	296.042	1.02526
80	0.045650	305.912	1.07606	0.039568	304.868	1.06108	0.034832	303.807	1.04757
90	0.047224	313.527	1.09732	0.040974	312.565	1.08257	0.036108	311.590	1.06930
100	0.048778	321.192	1.11815	0.042359	320.303	1.10359	0.037363	319.401	1.09052
110	0.050313	328.914	1.13856	0.043725	328.087	1.12417	0.038598	327.251	1.11128
120	0.051834	336.696	1.15861	0.045076	335.925	1.14437	0.039817	335.147	1.13162
130	0.053341	344.541	1.17832	0.046413	343.821	1.16420	0.041022	343.094	1.15158
140	0.054836	352.454	1.19770	0.047738	351.778	1.18369	0.042215	351.097	1.17119
150	0.056321	360.435	1.21679	0.049052	359.799	1.20288	0.043398	359.159	1.19047
	1.00 MPa			*1.20 MPa*			*1.40 MPa*		
30	0.024600	262.912	0.91358	—	—	—	—	—	—
40	0.025995	271.042	0.93996	0.020851	267.602	0.91411	0.017120	263.861	0.89010
50	0.027323	279.046	0.96512	0.022051	276.011	0.94055	0.018247	272.766	0.91809
60	0.028601	286.973	0.98928	0.023191	284.263	0.96570	0.019299	281.401	0.94441
70	0.029836	294.859	1.01260	0.024282	292.413	0.98981	0.020295	289.858	0.96942
80	0.031038	302.727	1.03520	0.025336	330.508	1.01305	0.021248	298.202	0.99339
90	0.032213	310.599	1.05718	0.026359	308.570	1.03556	0.022167	306.473	1.01649
100	0.033364	318.488	1.07861	0.027357	316.623	1.05744	0.023058	314.703	1.03884

(Contd.)

Table A.3.2 (Contd.)

Temp. °C	v m³/kg	h kJ/kg	s. kJ/kg K	v m³/kg	h kJ/k g	s. kJ/kg K	v m³/kg	h kJ/kg	s kJ/kg K
		1.00 MPa			*1.20 MPa*			*1.40 MPa*	
110	0.034495	326.405	1.09955	0.028334	324.682	1.07875	0.023926	322.916	1.06056
120	0.035609	334.360	1.12004	0.029292	332.762	1.09957	0.024775	331.128	1.08172
130	0.036709	342.360	1.14014	0.030236	340.871	1.11994	0.025608	339.354	1.10238
140	0.037797	350.410	1.15986	0.031166	349.019	1.13990	0.026426	347.603	1.12259
150	0.038873	358.514	1.17924	0.032084	357.210	1.15949	0.027233	355.885	1.14240
160	0.039940	366.677	1.19831	0.032993	365.450	1.17873	0.028029	364.206	1.16183
		1.60 MPa			*1.80 MPa*			*2.00 MPa*	
50	0.015351	269.262	0.89689	0.013052	265.423	0.87625	—	—	—
60	0.016351	278.358	0.92461	0.014028	275.097	0.90573	0.012135	271.563	0.88729
70	0.017284	287.171	0.95068	0.014921	284.331	0.93304	0.013008	281.310	0.91612
80	0.018167	295.797	0.97546	0.015755	293.282	0.95876	0.013811	290.640	0.94292
90	0.019011	304.301	0.99920	0.016546	302.046	0.98323	0.014563	299.697	0.96821
100	0.019825	312.725	1.02209	0.017303	310.683	1.00669	0.015277	308.571	0.99232
110	0.020614	321.103	1.04424	0.018032	319.239	1.02932	0.015960	317.322	1.01546
120	0.021382	329.457	1.06576	0.018738	327.745	1.05123	0.016619	325.991	1.03780
130	0.022133	337.805	1.08673	0.019427	336.224	1.07253	0.017258	334.610	1.05944
140	0.022869	346.162	1.10721	0.020099	344.695	1.09329	0.017881	343.201	1.08049
150	0.023592	354.540	1.12724	0.020759	353.172	1.11356	0.018490	351.783	1.10102
160	0.024305	362.945	1.14688	0.021407	361.666	1.13340	0.019087	360.369	1.12107
170	0.025008	371.386	1.16614	0.022045	370.186	1.15284	0.019673	368.970	1.14070
180	0.25703	379.869	1.18507	0.022675	378.738	1.17193	0.020251	377.595	1.15995
		2.50 MPa			*3.00 MPa*			*3.50 MPa*	
70	0.009459	272.677	0.87476	—	—	—	—	—	—
80	0.010243	283.332	0.90537	0.007747	274.530	0.86780	0.005765	262.739	0.82489
90	0.010948	293.338	0.93332	0.008465	286.042	0.89995	0.006597	277.268	0.86548

(Contd.)

Table A.3.2 (Contd.)

Temp. °C	v m³/kg	h kJ/kg	s. kJ/kg K	v m³/kg	h kJ/kg	s. kJ/kg K	v m³/kg	h kJ/kg	s kJ/kg K
	2.50 MPa			**3.00 MPa**			**3.50 MPa**		
100	0.011598	302.935	0.95939	0.009098	296.663	0.92881	0.007257	289.504	0.89872
110	0.012208	312.261	0.98405	0.009674	306.744	0.95547	0.007829	300.640	0.92818
120	0.012788	321.400	1.00760	0.010211	316.470	0.98053	0.008346	311.129	0.95520
130	0.013343	330.412	1.03023	0.010717	325.955	1.00435	0.008825	321.196	0.98047
140	0.013880	339.336	1.05210	0.011200	335.270	1.02718	0.009276	330.976	1.00445
150	0.014400	348.205	1.07331	0.011665	344.467	1.04918	0.009704	340.554	1.02736
160	0.014907	357.040	1.09395	0.012114	353.584	1.07047	0.010114	349.989	1.04940
170	0.015402	365.860	1.11408	0.012550	362.647	1.09116	0.010510	359.324	1.07071
180	0.015887	374.679	1.13376	0.012976	371.679	1.11131	0.010894	368.590	1.09138
190	0.016364	383.508	1.15303	0.013392	380.695	1.13099	0.011268	377.810	1.11151
200	0.016834	392.354	1.17912	0.013801	389.708	1.15024	0.011634	387.004	1.13115
	4.00 MPa			**5.00 MPa**			**6.00 MPa**		
90	0.005037	265.629	0.82544	—	—	—	—	—	—
100	0.005804	280.997	0.86721	0.003334	253.042	0.78005	—	—	—
110	0.006405	293.748	0.90094	0.004255	275.919	0.84064	0.002432	243.278	0.74674
120	0.006924	305.273	0.93064	0.004851	291.362	0.88045	0.003333	272.385	0.82185
130	0.007391	316.080	0.95778	0.005335	304.469	0.91337	0.003899	290.253	0.86675
140	0.007822	326.422	0.98312	0.005757	316.379	0.94256	0.004345	304.757	0.90230
150	0.008226	336.446	1.00710	0.006139	327.563	0.96931	0.004728	317.633	0.93310
160	0.008610	346.246	1.02999	0.006493	338.266	0.99431	0.005071	329.553	0.96094
170	0.008978	355.885	1.05199	0.006826	348.633	1.01797	0.005386	340.849	0.98673
180	0.009332	365.409	1.07324	0.007142	358.760	1.04057	0.005680	351.715	1.01098
190	0.009675	374.853	1.09386	0.007444	368.713	1.06230	0.005958	362.271	1.03402
200	0.010009	384.240	1.11391	0.007735	378.537	1.08328	0.006222	372.602	1.05609
210	0.010335	393.593	1.13347	0.008018	388.268	1.10363	0.006477	382.764	1.07734
220	0.010654	402.925	1.15259	0.008292	397.932	1.12343	0.006722	392.801	1.09790

(Contd.)

A.4 Thermodynamic Properties of Refrigerant-134a (1,1,1,1–tetrafluoroethane)

Table A.4.1 Saturated R-134a

Temp. °C	Abs. Press. MPa p	Specific Volume, m³/kg			Enthalpy, kJ/kg			Entropy, kJ/kg K		
		Sat. Liquid v_f	Evap. v_{fg}	Sat. Vapour v_g	Sat. Liquid h_f	Evap. h_{fg}	Sat. Vapour h_g	Sat. Liquid s_f	Evap. s_{fg}	Sat. Vapour s_g
– 33	0.0737	0.000718	0.25574	0.25646	157.417	220.491	377.908	0.8346	0.9181	1.7528
– 30	0.0851	0.000722	0.22330	0.22402	161.118	218.683	379.802	0.8499	0.8994	1.7493
26.25	0.1013	0.000728	0.18947	0.19020	165.802	216.360	382.162	0.8690	0.8763	1.7453
– 25	0.0173	0.000730	0.17956	0.18029	167.381	215.569	382.950	0.8754	0.8687	1.7441
– 20	0.1337	0.000738	0.14575	0.14649	173.744	212.340	386.083	0.9007	0.8388	1.7395
– 15	0.1650	0.000746	0.11932	0.12007	180.193	209.004	389.197	0.9258	0.8096	1.7354
– 10	0.2017	0.000755	0.098454	0.099209	186.721	205.564	392.285	0.9507	0.7812	1.7319
– 5	0.2445	0.000764	0.081812	0.082576	193.324	202.016	395.340	0.9755	0.7534	1.7288
0	0.2940	0.000773	0.068420	0.069193	200.000	198.356	398.356	1.0000	0.7262	1.7262
5	0.3509	0.000783	0.057551	0.058334	206.751	194.572	401.323	1.0243	0.6995	1.7239
10	0.4158	0.000794	0.048658	0.049451	213.580	190.652	404.233	1.0485	0.6733	1.7218
15	0.4895	0.000805	0.041326	0.042131	220.492	186.582	407.075	1.0725	0.6475	1.7200
20	0.5728	0.000817	0.035238	0.036055	227.493	182.345	409.838	1.0963	0.6220	1.7183
25	0.6663	0.000829	0.030148	0.030977	234.590	177.920	412.509	1.1201	0.5967	1.7168
30	0.7710	0.000843	0.025865	0.026707	241.790	173.285	415.075	1.1437	0.5716	1.7153
35	0.8876	0.000857	0.022237	0.023094	249.103	168.415	417.518	1.1673	0.5465	1.7139
40	1.0171	0.000873	0.019147	0.020020	256.539	163.282	419.821	1.1909	0.5214	1.7123
45	1.1602	0.000890	0.016499	0.017389	264.110	157.852	421.962	1.2145	0.4962	1.7106
50	1.3180	0.000908	0.014217	0.015124	271.830	152.085	423.915	1.2381	0.4706	1.7088
55	1.4915	0.000928	0.012237	0.013166	279.718	145.933	425.650	1.2619	0.4447	1.7066
60	1.6818	0.000951	0.010511	0.011462	287.794	139.336	427.130	1.2857	0.4182	1.7040

(Contd.)

Table A.4.1 (Contd.)

Temp. °C	Abs. Press. MPa p	Specific Volume, m^3/kg			Enthalpy, kJ/kg			Entropy, kJ/kg K		
		Sat. Liquid v_f	Evap. v_{fg}	Sat. Vapour v_g	Sat. Liquid h_f	Evap. h_{fg}	Sat. Vapour h_g	Sat. Liquid s_f	Evap. s_{fg}	Sat. Vapour s_g
65	1.8898	0.000976	0.008995	0.009970	296.088	132.216	428.305	1.3099	0.3910	1.7009
70	2.1169	0.001005	0.007653	0.008657	304.642	124.468	429.110	1.3343	0.3627	1.6970
75	2.3644	0.001038	0.006453	0.007491	313.513	115.939	429.451	1.3592	0.3330	1.6923
80	2.6337	0.001078	0.005368	0.006446	322.794	106.395	429.189	1.3849	0.3013	1.6862
85	2.9265	0.001128	0.004367	0.005495	332.644	95.440	428.084	1.4117	0.2665	1.6782
90	3.2448	0.001995	0.003412	0.004606	343.380	82.295	425.676	1.4404	0.2266	1.6670
95	3.5914	0.001297	0.002432	0.003729	355.834	64.984	420.818	1.4733	0.1765	1.6498
101.15	4.0640	0.001969	0	0.001969	390.977	0	390.977	1.5658	0	1.5658

Table A.4.2 Superheated R-134a

Temp. °C	v m³/kg	h kJ/kg	s kJ/kg K	v m³/kg	h kJ/kg	s kJ/kg K	v m³/kg	h kJ/kg	s kJ/kg K
	0.10 MPa			0.15 MPa			0.20 MPa		
– 25	0.19400	383.212	1.75058	—	—	—	—	—	—
– 20	0.19860	387.215	1.76655	—	—	—	—	—	—
– 10	0.20765	395.270	1.79775	0.13603	393.839	1.76058	0.10013	392.338	1.73276
0	0.21652	403.413	1.82813	0.14222	402.187	1.79171	0.10501	400.911	1.76474
10	0.22527	411.668	1.85780	0.14828	410.602	1.82197	0.10974	409.500	1.79562
20	0.23393	420.048	1.88689	0.15424	419.111	1.85150	0.11436	418.145	1.82563
30	0.24250	428.564	1.91545	0.16011	427.730	1.88041	0.11889	426.875	1.85491
40	0.25102	437.223	1.94355	0.16592	436.473	1.90879	0.12335	435.708	1.88357
50	0.25948	446.029	1.97123	0.17168	445.350	1.93669	0.12776	444.658	1.91171
60	0.26791	454.986	1.99853	0.17740	454.366	1.96416	0.13213	453.735	1.93937
70	0.27631	464.096	2.02547	0.18308	463.525	1.99125	0.13646	462.946	1.96661
80	0.28468	473.359	2.05208	0.18874	472.831	2.01798	0.14076	472.296	1.99346
90	0.29303	482.777	2.07837	0.19437	482.285	2.04438	0.14504	481.788	2.01997
100	0.30136	492.349	2.10437	0.19999	491.888	2.07046	0.14930	491.424	2.04614

(Contd.)

Table A.4.2 (Contd.)

Temp. °C	v m³/kg	h kJ/kg	s kJ/kg K	v m³/kg	h kJ/kg	s kJ/kg K	v m³/kg	h kJ/kg	s kJ/kg K
		0.25 MPa			0.30 MPa			0.40 MPa	
0	0.082637	399.579	1.74284	—	—	—	—	—	—
10	0.086584	408.357	1.77440	0.071110	407.171	1.75637	0.051681	404.651	1.72611
20	0.090408	417.151	1.80492	0.074415	416.124	1.78744	0.054362	413.965	1.75844
30	0.090139	425.997	1.83460	0.077620	425.096	1.81754	0.056926	423.216	1.78947
40	0.097798	434.925	1.86357	0.080748	434.124	1.84684	0.059402	432.465	1.81949
50	0.101401	443.953	1.89195	0.083816	443.234	1.87547	0.061812	441.751	1.84868
60	0.104958	453.094	1.91980	0.086838	452.442	1.90354	0.064169	451.104	1.87718
70	0.108480	462.359	1.94720	0.089821	461.763	1.93110	0.066484	460.545	1.90510
80	0.111972	471.754	1.97419	0.092774	471.206	1.95823	0.068767	470.088	1.93252
90	0.115440	481.285	2.00080	0.095702	480.777	1.98495	0.071022	479.745	1.95948
100	0.118888	490.955	2.02707	0.098609	490.482	2.01131	0.073254	489.523	1.98604
110	0.122318	500.766	2.05302	0.101498	500.324	2.03734	0.075468	499.428	2.01223
120	0.125734	510.720	2.07866	0.104371	510.304	2.06305	0.077665	509.464	2.03809

(Contd.)

Table A.4.2 (Contd.)

Temp. °C	v m³/kg	h kJ/kg	s kJ/kg K	v m³/kg	h kJ/kg	s kJ/kg K	v m³/kg	h kJ/kg	s kJ/kg K
	0.50 MPa			0.60 MPa			0.70 MPa		
20	0.042256	411.645	1.73420	—	—	—	—	—	—
30	0.044457	421.221	1.76632	0.036094	419.093	1.74610	0.030069	416.809	1.72770
40	0.046557	430.720	1.79715	0.037958	428.881	1.77786	0.031781	426.933	1.76056
50	0.048581	440.205	1.82696	0.039735	438.589	1.80838	0.033392	436.895	1.79187
60	0.050547	449.718	1.85596	0.041447	448.279	1.83791	0.034929	446.782	1.82201
70	0.052467	459.290	1.88426	0.043108	457.994	1.86664	0.036410	456.655	1.85121
80	0.054351	468.942	1.91199	0.044730	467.764	1.89471	0.037848	466.554	1.87964
90	0.056205	478.690	1.93921	0.046319	477.611	1.92220	0.039251	476.507	1.90743
100	0.058035	488.546	1.96598	0.047883	487.550	1.94920	0.040627	486.535	1.93467
110	0.059845	498.518	1.99235	0.049426	497.594	1.97576	0.041980	496.654	1.96143
120	0.061639	508.613	2.01836	0.050951	507.750	2.00193	0.043314	506.875	1.98777
130	0.063418	518.835	2.04403	0.052461	518.026	2.02774	0.044633	517.207	2.01372
140	0.065184	529.187	2.06940	0.053958	528.425	2.05322	0.045938	527.656	2.03932
	0.80 MPa			0.90 MPa			1.00 MPa		
40	0.027113	424.860	1.74457	0.023446	422.642	1.72943	0.020473	420.249	1.71479
50	0.028611	435.114	1.77680	0.024868	433.235	1.76273	0.021849	431.243	1.74936
60	0.030024	445.223	1.80761	0.026192	443.595	1.79431	0.023110	441.890	1.78181
70	0.031375	455.270	1.83732	0.027447	453.835	1.82459	0.024293	452.345	1.81273
80	0.032678	465.308	1.86616	0.028649	464.025	1.85387	0.025417	462.703	1.84248
90	0.033944	475.375	1.89427	0.029810	474.216	1.88232	0.026497	473.027	1.87131
100	0.035180	485.499	1.92177	0.030940	484.441	1.91010	0.027543	483.361	1.89938
110	0.036392	495.698	1.94874	0.032043	494.726	1.93730	0.028561	493.736	1.92682
120	0.037584	505.988	1.97525	0.033126	505.088	1.96399	0.029556	504.175	1.95371
130	0.038760	516.379	2.00135	0.034190	515.542	1.99025	0.030533	514.694	1.98013
140	0.039921	526.880	2.02708	0.035241	526.096	2.01611	0.031495	525.305	2.00613
150	0.041071	537.496	2.05247	0.036278	536.760	2.04161	0.032444	536.017	2.03175

(Contd.)

Table A.4.2 (Contd.)

Temp. °C	v m³/kg	h kJ/kg	s kJ/kg K	v m³/kg	h kJ/kg	s kJ/kg K	v m³/kg	h kJ/kg	s kJ/kg K
	1.20 MPa			1.40 MPa			1.60 MPa		
50	0.017243	426.485	1.72373	—	—	—	—	—	—
60	0.018439	438.210	1.75837	0.015032	434.079	1.73597	0.012392	429.322	1.71349
70	0.019530	449.179	1.79081	0.016083	445.720	1.77040	0.013449	441.888	1.75066
80	0.020548	459.925	1.82168	0.017040	456.944	1.80265	0.014378	453.722	1.78466
90	0.021512	470.551	1.85135	0.017931	467.931	1.83333	0.015225	465.145	1.81656
100	0.022436	481.128	1.88009	0.018775	478.790	1.86282	0.016015	476.333	1.84695
110	0.023329	491.702	1.90805	0.019583	489.589	1.89139	0.016763	487.390	1.87619
120	0.024197	502.307	1.93537	0.020362	500.379	1.91918	0.017479	498.387	1.90452
130	0.025044	512.965	1.96214	0.021118	511.192	1.94634	0.018169	509.371	1.93211
140	0.025874	523.697	1.98844	0.021856	522.054	1.97296	0.018840	520.376	1.95908
150	0.026691	534.514	2.01431	0.022579	532.984	1.99910	0.019493	531.427	1.98551
160	0.027495	545.426	2.03980	0.023289	543.994	2.02481	0.020133	542.542	2.01147
170	0.028289	556.443	2.06494	0.023988	555.097	2.05015	0.020761	553.735	2.03702
	1.80 MPa			2.0 MPa			2.50 MPa		
70	0.011341	437.562	1.73085	0.009581	432.531	1.71011	—	—	—
80	0.012273	450.202	1.76717	0.010550	446.304	1.74968	0.007221	433.797	1.70180
90	0.013099	462.164	1.80057	0.011374	458.951	1.78500	0.008157	449.499	1.74567
100	0.013854	473.741	1.83202	0.012111	470.996	1.81772	0.008907	463.279	1.78311
110	0.014560	485.095	1.86205	0.012789	482.693	1.84866	0.009558	476.129	1.81709
120	0.015230	496.325	1.89098	0.013424	494.187	1.87827	0.010148	488.457	1.84886
130	0.015871	507.498	1.91905	0.014028	505.569	1.90686	0.010694	500.474	1.87904
140	0.016490	518.659	1.94639	0.014608	516.900	1.93463	0.011208	512.307	1.90804
150	0.017091	529.841	1.97314	0.015168	528.224	1.96171	0.011698	524.037	1.93609
160	0.017677	541.068	1.99936	0.015712	539.571	1.98821	0.012169	535.722	1.96338
170	0.018251	552.357	2.02513	0.016242	550.963	2.01421	0.012624	547.399	1.99004
180	0.018814	563.724	2.05049	0.016762	562.418	2.03977	0.013066	559.098	2.01614
190	0.019369	575.177	2.07549	0.017272	573.950	2.06494	0.013498	570.841	2.04177

Table A.4.2 (Contd.)

Temp. °C	v m³/kg	h kJ/kg	s kJ/kg K	v m³/kg	h kJ/kg	s kJ/kg K	v m³/kg	h kJ/kg	s kJ/kg K
		3.0 MPa			3.50 MPa			4.0 MPa	
90	0.005755	436.193	1.69950	—	—	—	—	—	—
100	0.006653	453.731	1.74717	0.004839	440.433	1.70386	—	—	—
110	0.007339	468.500	1.78623	0.005667	459.211	1.75355	0.004277	446.844	1.71840
120	0.007924	482.043	1.82113	0.006289	474.697	1.79346	0.005005	465.987	1.76415
130	0.008446	494.915	1.85347	0.006813	488.771	1.82881	0.005559	481.865	1.80404
140	0.008926	507.388	1.88403	0.007279	502.079	1.86142	0.006027	496.295	1.83940
150	0.009375	519.618	1.91328	0.007706	514.928	1.89216	0.006444	509.925	1.87200
160	0.009801	531.704	1.94151	0.008103	527.496	1.92151	0.006825	523.072	1.90271
170	0.010208	543.713	1.96892	0.008480	539.890	1.94980	0.007181	535.917	1.93203
180	0.010601	555.690	1.99565	0.008839	552.185	1.97724	0.007517	548.573	1.96028
190	0.010982	567.670	2.02180	0.009185	564.430	2.00397	0.007837	561.117	1.98766
200	0.011353	579.678	2.04745	0.009519	576.665	2.03010	0.008145	573.601	2.01432

A.5 Thermodynamic Properties of Ammonia

Table A.5.1 Saturated ammonia table

T (°C)	p (kPa)	Specific volume (m³/kg)			Enthalpy (kJ/kg)			Entropy (kJ/kg K)	
		Sat. liquid vt	*t*	*Sat. vapour, vg*	*Sat. liquid, ht*	*Evap., htg*	*Sat. vapour, hg*	*Sat. liquid, st*	*Sat. vapour sg*
– 50	40.88	0.001424		2.6254	– 44.3	1416.7	1372.4	– 0.1942	6.1561
– 48	45.96	0.001429		2.3533	– 35.5	1411.3	1375.8	– 0.1547	6.1149
– 46	51.55	0.001434		2.1140	– 26.6	1405.8	1379.2	– 0.1156	6.0746
– 44	57.69	0.001439		1.9032	– 17.8	1400.3	1382.5	– 0.0768	6.0352
– 42	64.42	0.001444		1.7170	– 8.9	1394.7	1385.8	– 0.0382	5.9967
– 40	71.77	0.001449		1.5521	0.0	1389.0	1389.0	0.0000	5.9589
– 38	79.80	0.001454		1.4058	8.9	1383.3	1392.2	0.0380	5.9220
– 36	88.54	0.001460		1.2757	17.8	1377.6	1395.4	0.0757	5.8858
– 34	98.05	0.001465		1.1597	26.8	1371.8	1398.5	0.1132	5.8504
– 32	108.37	0.001470		1.0562	35.7	1365.9	1401.6	0.1504	5.8156
– 30	119.55	0.001476		0.9635	44.7	1360.0	1404.6	0.1873	5.7815
– 28	131.64	0.001481		0.8805	53.6	1354.0	1407.6	0.2240	5.7481
– 26	144.70	0.001487		0.8059	62.6	1347.9	1410.5	0.2605	5.7153
– 24	158.78	0.001492		0.7388	71.6	1341.8	1413.4	0.2967	5.6831
– 22	173.93	0.001498		0.6783	80.7	1335.6	1416.2	0.3327	5.6515
– 20	190.22	0.001504		0.6237	89.7	1329.3	1419.0	0.3684	5.6205
– 18	207.71	0.001510		0.5743	98.8	1322.9	1421.7	0.4040	5.5900

(Contd.)

Table A.5.1 (Contd.)

T (°C)	p (kPa)	Specific volume (m³/kg) Sat. liquid vt	t	Sat. vapour, vg	Enthalpy (kJ/kg) Sat. liquid, ht	Evap., htg	Sat. vapour, hg	Entropy (kJ/kg K) Sat. liquid, st	Sat. vapour sg
– 16	226.45	0.001515		0.5296	107.8	1316.5	1424.4	0.4393	5.5600
– 14	246.51	0.001521		0.4889	116.9	1310.0	1427.0	0.4744	5.5305
– 12	267.95	0.001528		0.4520	126.0	1303.5	1429.5	0.5093	5.5015
– 10	290.95	0.001534		0.4185	135.2	1296.8	1432.0	0.5440	5.4730
– 8	315.25	0.001540		0.3878	144.3	1290.1	1434.4	0.5785	5.4449
– 6	341.25	0.001546		0.3599	153.5	1283.3	1436.8	0.6128	5.4173
– 4	368.90	0.001553		0.3343	162.7	1276.4	1439.1	0.6469	5.3901
– 2	398.27	0.001559		0.3109	171.9	1269.4	1441.3	0.6808	5.3633
0	429.44	0.001566		0.2895	181.1	1262.4	1443.5	0.7145	5.3369
2	462.49	0.001573		0.2698	190.4	1255.2	1445.6	0.7481	5.3108
4	497.49	0.001580		0.2517	199.6	1243.0	1447.6	0.7815	5.2852
6	531.51	0.001587		0.2351	208.9	1240.6	1449.6	0.8148	5.2599
8	573.64	0.001594		0.2198	218.3	1233.2	1451.5	0.8479	5.2350
10	614.95	0.001601		0.2056	227.6	1225.7	1453.3	0.8808	5.2104
12	658.52	0.001608		0.1926	237.0	1218.1	1455.1	0.9136	5.1861
14	704.44	0.001616		0.1805	246.4	1210.4	1456.8	0.9463	5.1621
16	752.79	0.001623		0.1693	255.9	1202.6	1458.5	0.9788	5.1385
18	803.66	0.001631		0.1590	265.4	1194.7	1460.0	1.0112	5.1151
20	857.12	0.001639		0.1494	274.9	1186.7	1461.5	1.0434	5.0920
22	913.27	0.001647		0.1405	284.4	1178.5	1462.9	1.0755	5.0692
24	972.19	0.001655		0.1322	294.0	1170.3	1464.3	1.1075	5.0467
26	1033.97	0.001663		0.1245	303.6	1162.0	1465.6	1.1394	5.0244
28	1098.71	0.001671		0.1173	313.2	1153.6	1466.8	1.1711	5.0023
30	1166.49	0.001680		0.1106	322.9	1145.0	1467.9	1.2028	4.9805
32	1237.41	0.001689		0.1044	332.6	1136.4	1469.0	1.2343	4.9589
34	1311.55	0.001698		0.0986	342.3	1127.6	1469.9	1.2656	4.9374

(Contd.)

Table A.5.1 (Contd.)

T (°C)	p (kPa)	Specific volume (m^3/kg)			Enthalpy (kJ/kg)			Entropy (kJ/kg K)	
		Sat. liquid vt	t	Sat. vapour, vg	Sat. liquid, ht	Evap., htg	Sat. vapour, hg	Sat. liquid, st	Sat. vapour sg
36	1369.03	0.001707		0.0931	352.1	1118.7	1470.8	1.2969	4.9161
38	1469.92	0.001716		0.0880	361.9	1109.7	1471.5	1.3281	4.8950
40	1554.33	0.001726		0.0833	371.7	1100.5	1472.2	1.3591	4.8740
42	1642.35	0.001735		0.0788	381.6	1091.2	1472.8	1.3901	4.8530
44	1734.09	0.001745		0.0746	391.5	1081.7	1473.2	1.4209	4.8322
46	1829.65	0.001756		0.0707	401.5	1072.0	1473.5	1.4518	4.8113
48	1929.13	0.001766		0.0669	411.5	1062.2	1473.7	1.4826	4.7905
50	2032.62	0.001777		0.0635	421.7	1052.0	1473.7	1.5135	4.7696

Table A.5.2 Superheated ammonia table

Abs. Press. (kPa) (Sat. temp., °C)		*Temperature (°C)*											
		– 20	*– 10*	*0*	*10*	*20*	*30*	*40*	*50*	*60*	*70*	*80*	*100*
	v	2.4474	2.5481	2.6482	2.7479	2.8479	2.9464	3.0453	3.1441	3.2427	3.3413	3.4397	
50	*h*	1435.8	1457.0	1478.1	1499.2	1520.4	1541.7	1563.0	1584.5	1606.1	1627.8	1649.7	
(– 46.54)	*s*	6.3256	6.4077	6.4865	6.5625	6.6360	6.7073	6.7766	6.8441	6.9099	6.9743	7.0372	
	v	1.6233	1.6915	1.7591	1.8263	1.8932	1.9597	2.0261	2.0933	2.1584	2.2244	2.2903	
75	*h*	1433.0	1454.7	1476.1	1497.5	1518.9	1540.3	1561.8	1583.4	1605.1	1626.9	1648.9	
(– 39.18)	*s*	6.1190	6.2028	6.2828	6.3597	6.4339	6.5058	6.5756	6.6434	6.7096	6.7742	6.8373	
	v	1.2110	1.2631	1.3145	1.3654	1.4160	1.4664	1.5165	1.5664	1.6163	1.6659	1.7155	1.8145
100	*h*	1430.1	1452.2	1474.1	1495.7	1517.3	1538.9	1560.5	1582.2	1604.1	1626.0	1648.0	1692.6
(– 33.01)	*s*	5.9695	6.0552	6.1366	6.2144	6.2894	6.3618	6.4321	6.5003	0.5668	6.6316	6.6950	6.8177
	v	0.9635	1.0059	1.0476	1.0889	1.1297	1.1703	1.2107	1.2509	1.2909	1.3309	1.3707	1.4501
125	*h*	1427.2	1449.8	1472.0	1493.9	1515.7	1537.5	1559.3	1581.1	1603.0	1625.0	1647.2	1691.8
(– 29.08)	*s*	5.8512	5.9389	6.0217	6.1006	6.1763	6.2494	6.3201	6.3887	6.4555	6.5206	6.5842	6.7072
	v	0.7984	0.8344	0.8697	0.9045	6.9388	0.9729	1.0068	1.0405	1.0740	1.1074	1.1408	1.2072
150	*h*	1424.1	1447.3	1469.8	1492.1	1514.1	1536.1	1558.0	1580.0	1602.0	1624.1	1646.3	1691.1
(– 25.23)	*s*	5.7526	5.8424	5.9266	6.0066	6.0831	6.1568	6.2280	6.2970	6.3641	6.4295	6.4933	6.6167
	v		0.6199	0.6471	0.6738	0.7001	0.7261	0.7519	0.7774	0.8029	0.8282	0.8533	0.9035
200	*h*		1442.0	1465.5	1448.1	1510.9	1533.2	1555.5	1577.7	1599.9	1622.2	1644.6	1689.6
(– 18.86)	*s*		5.6863	5.7737	5.8559	5.9342	6.0091	6.0813	6.1512	6.2189	6.2849	6.3491	6.4732
	v		0.4910	0.5135	0.5354	0.5568	0.5780	0.5989	0.6196	0.6401	0.6605	0.6809	0.7212
250	*h*		1436.6	1461.0	1484.5	1507.6	1530.3	1552.9	1575.4	1597.8	1620.3	1642.8	1688.2
(– 13.67)	*s*		5.5609	5.6517	5.7465	5.8165	5.8928	5.9661	6.0368	6.1052	6.1717	6.2365	6.3613
	v			0.4243	0.4430	0.4613	0.4792	0.4968	0.5113	0.5316	0.5488	0.5658	0.5997
300	*h*			1456.3	1480.6	1504.2	1527.4	1550.3	1573.0	1595.7	1618.4	1641.1	1686.7
(– 9.23)	*s*			5.5193	5.6366	5.7186	5.7963	5.8707	5.9423	6.0114	6.0785	6.1437	6.2693
	v			0.3605	0.3770	0.3929	0.4086	0.4239	0.4391	0.4541	0.4689	0.4837	0.5129

(Contd.)

Table A.5.2 (Contd.)

Abs. Press. (kPa) (Sat. temp., °C)		Temperature (°C)											
		–20	–10	0	10	20	30	40	50	60	70	80	100
350	*h*			1451.5	1478.5	1590.7	1521.4	1547.6	1570.7	1593.6	1616.5	1639.3	1685.2
(–5.35)	*s*			5.4600	5.5502	5.6312	5.7135	5.7800	5.8615	5.9314	5.9990	6.0647	6.1910
	v			0.3125	0.3274	0.3417	0.3556	0.3692	0.3826	0.3959	0.4090	0.4220	0.4478
400	*h*			1446.5	1472.4	1497.2	1521.3	1544.9	1568.3	1591.5	1614.5	1637.6	1683.7
(–1.89)	*s*			5.3803	5.4735	5.5597	5.6405	5.7173	5.7907	5.8613	5.9296	5.9957	6.1228
	v			0.2752	0.2887	0.3017	0.3143	0.3266	0.3387	0.3506	0.3624	0.3740	0.3971
450	*h*			1441.3	1468.1	1493.6	1518.2	1542.2	1565.9	1589.3	1612.6	1635.8	1682.2
(1.26)	*s*			5.3078	5.4042	5492.6	5.5752	5.6532	5.7275	5.7989	5.8678	5.9345	6.0623
		20	**30**	**40**	**50**	**60**	**70**	**80**	**100**	**120**	**140**	**160**	**180**
	v	0.2698	0.2813	0.2926	0.3036	0.3144	0.3251	0.3357	0.3565	0.3771	0.3975		
500	*h*	1489.9	1515.0	1539.5	1563.4	1587.1	1610.6	1634.0	1680.7	1727.5	1774.7		
(4.14)	*s*	5.4314	5.5157	5.5950	5.6704	5.7425	5.8120	5.8793	6.0079	6.1301	6.2472		
	v	0.2217	0.2317	0.2414	0.2508	0.2600	0.2691	0.2781	0.2957	0.3130	0.3302		
600	*h*	1482.4	1508.6	1533.8	1558.5	1582.7	1606.6	1630.4	1677.7	1724.9	1772.4		
(9.29)	*s*	5.3222	5.4102	5.4923	5.5697	5.6436	5.7144	5.7826	5.9129	6.0363	6.1541		
	v	0.1874	0.1963	0.2048	0.2131	0.2212	0.2291	0.2369	0.2522	0.2672	0.2821		
700	*h*	1474.5	1501.9	1528.1	1553.4	1578.2	1602.6	1626.8	1674.6	1722.4	1770.2		
(13.81)	*s*	5.2259	5.3179	5.4029	5.4826	5.5582	5.6303	5.6997	5.8316	5.9562	6.0749		
	v	0.1615	0.1696	0.1773	0.1848	0.1920	0.1991	0.2060	0.2196	0.2329	0.2459	0.2589	
800	*h*	1466.3	1495.0	1522.2	1548.3	1573.7	1598.6	1623.1	1671.6	1719.8	1768.0	1816.4	
(17.86)	*s*	5.1387	5.2351	5.3232	5.4053	5.4827	5.5562	5.6268	5.7603	5.8861	6.0057	6.1202	
	v		0.1488	0.1559	0.1627	0.1693	0.1757	0.1820	0.1942	0.2061	0.2178	0.2294	
900	*h*		1488.0	1516.2	1543.0	1569.1	1594.4	1619.4	1668.5	1717.1	1765.7	1814.4	
(21.54)	*s*		5.1593	5.2508	5.3354	5.4147	5.4897	5.5614	5.6968	5.8237	5.9442	6.0594	
	v		0.1321	0.1388	0.1450	0.1511	0.1570	0.1627	0.1739	0.1847	0.1954	0.2058	0.2162

(Contd.)

Table A.5.2 (Contd.)

		20	30	40	50	60	70	80	100	120	140	160	180
1000	*h*		1480.6	1510.0	1537.7	1564.4	1590.3	1615.6	1665.4	1714.5	1763.4	1812.4	1861.7
(24.91)	*s*		5.0889	5.1840	5.2713	5.3525	5.4292	5.5021	5.6392	5.7674	5.8888	6.0047	6.1159
	v			0.1129	0.1185	0.1238	0.1289	0.1338	0.1434	0.1526	0.1616	0.1705	0.1792
1200	*h*			1497.1	1526.6	1554.7	1581.7	1608.0	1659.2	1709.2	1758.9	1808.5	1858.2
(30.96)	*s*			5.0629	5.1560	5.2416	5.3215	5.3970	5.5379	5.6687	5.7919	5.9091	6.0214
	v			0.0944	0.0995	0.1042	0.1088	0.1132	0.1216	0.1297	0.1376	0.1452	0.1528
1400	*h*			1483.4	1515.1	1544.7	1573.0	1600.2	1652.8	1703.9	1754.3	1804.5	1854.7
(36.28)	*s*			4.9534	5.0530	5.1434	5.2270	5.3053	5.4501	5.5836	5.7087	5.8273	5.9406
	v				0.0851	0.0895	0.0937	0.0977	0.1053	0.1125	0.1195	0.1263	0.1330
1600	*h*				1502.9	1534.4	1564.0	1592.3	1646.4	1698.5	1749.7	1800.5	1851.2
(14.05)	*s*				4.9584	5.0543	5.1419	5.2232	5.3722	5.5084	5.6355	5.7555	5.8699

Table A.6 Variation of $\bar{c}_p$ with temperature of various ideal gases

$$\frac{\bar{c}_p}{R} = \alpha + \beta T + \gamma T^2 + \delta T^3 + \epsilon T^4$$

T is in K, equations valid from 300 to 1000 K

Gas	α	$\beta \times 10^3$	$\gamma \times 10^6$	$\delta \times 10^9$	$\varepsilon \times 10^{12}$
CO	3.710	– 1.619	3.692	– 2.032	0.240
CO_2	2.401	8.735	– 6.607	2.002	0
H_2	3.057	2.677	– 5.180	5.521	– 1.812
H_2O	4.070	– 1.108	4.152	– 2.964	0.807
O_2	3.626	– 1.878	7.056	– 6.764	2.156
N_2	3.675	– 1.208	2.324	– 0.632	– 0.226
Air	3.653	– 1.337	3.294	– 1.913	0.2763
SO_2	3.267	5.324	0.684	– 5.281	2.559
CH_4	3.826	– 3.979	24.558	– 22.733	6.963
C_2H_2	1.410	19.057	.24.501	16.391	– 4.135
C_2H_4	1.426	11.383	7.989	– 16.254	6.749
Monatomic gases*	2.5	0	0	0	0

* For monatomic gases, such as He, Ne, and Ar, $\bar{c}_p$ is constant over a wide temperature range and is very nearly equal to 5/2 $\bar{R}$.

Source: Adapted from K. Wark, *Thermodynamics, 4th εd. McGraw-Hill,* New York, 1983, as based on NASA SP-273, U.S. Government Printing Office, Washington, DC, 1971.

Table A.7 Ideal gas specific heats of some common gases (kJ/kg.K)

Temp. K	Air c_p	Air c_v	Air k	Nitrogen, N_2 c_p	Nitrogen, N_2 c_v	Nitrogen, N_2 k	Oxygen, O_2 c_p	Oxygen, O_2 c_v	Oxygen, O_2 k	*Temp. K*
250	1.003	0.716	1.401	1.039	0.742	1.400	0.913	0.653	1.398	250
300	1.005	0.718	1.400	1.039	0.743	1.400	0.918	0.658	1.395	300
350	1.008	0.721	1.398	1.041	0.744	1.399	0.928	0.668	1.389	350
400	1.013	0.726	1.395	1.044	0.747	1.397	0.941	0.681	1.382	400
450	1.020	0.733	1.391	1.049	0.752	1.395	0.956	0.696	1.373	450
500	1.029	0.742	1.387	1.056	0.759	1.391	0.972	0.712	1.365	500
550	1.040	0.753	1.381	1.065	0.768	1.387	0.988	0.728	1.358	550
600	1.051	0.764	1.376	1.075	0.778	1.382	1.003	0.743	1.350	600
650	1.063	0.776	1.370	1.086	0.789	1.376	1.017	0.758	1.343	650
700	1.075	0.788	1.364	1.098	0.801	1.371	1.031	0.771	1.337	700
750	1.087	0.800	1.359	1.110	0.813	1.365	1.043	0.783	1.332	750
800	1.099	0.812	1.354	1.121	0.825	1.360	1.054	0.794	1.327	800
900	1.121	0.834	1.344	1.145	0.849	1.349	1.074	0.814	1.319	900
1000	1.142	0.855	1.336	1.167	0.870	1.341	1.090	0.830	1.313	1000

(Contd.)

Table A.7 (Contd.)

Temp. K	*Carbon dioxide CO_2*			*Carbon Monoxide, CO*			*Hydrogen, H_2*			*Temp. K*
250	0.791	0.602	1.314	1.039	0.743	1.400	14.051	9.927	1.416	250
300	0.846	0.657	1.288	1.040	0.744	1.399	14.307	10.183	1.045	300
350	0.895	0.706	1.268	1.043	0.746	1.398	14.427	10.302	1.400	350
400	0.939	0.750	1.252	1.047	0.751	1.395	14.476	10.352	1.398	400
450	0.978	0.790	1.239	1.054	0.757	1.392	14.501	10.377	1.398	450
500	1.014	0.825	1.229	1.063	0.767	1.387	14.513	10.389	1.397	500
550	1.046	0.857	1.220	1.075	0.778	1.382	14.530	10.405	1.396	550
600	1.075	0.886	1.213	1.087	0.790	1.376	14.546	10.422	1.396	600
650	1.102	0.913	1.207	1.100	0.803	1.370	14.571	10.447	1.395	650
700	1.126	0.937	1.202	1.113	0.816	1.364	14.604	10.480	1.394	700
750	1.148	0.959	1.197	1.126	0.829	1.358	14.645	10.521	1.392	750
800	1.169	0.980	1.193	1.139	0.842	1.353	14.695	10.570	1.390	800
900	1.204	1.015	1.186	1.163	0.866	1.343	14.822	10.698	1.385	900
1000	1.234	1.045	1.181	1.185	0.888	1.335	14.983	10.859	1.380	1000

Source: Adapted from K. Wark, *Thermodynamics, 4th ed.*, McGraw-Hill, New York, 1983, as based on "Tables of Thermal Properties of Gases," NBS Circular 564, 1955.

A.8 Logarithm to the Base 10 of the Equilibrium Constant K

Temp. K	log 10 K								Temp. °R
	$H_2 \rightleftharpoons 2H$	$O_2 \rightleftharpoons 2O$	$N_2 \rightleftharpoons 2N$	$\frac{1}{2}O_2 + \frac{1}{2}N_2 \rightleftharpoons NO$	$H_2O \rightleftharpoons H_2 + \frac{1}{2}O_2$	$H_2O \rightleftharpoons OH + \frac{1}{2}H_2$	$CO_2 \rightleftharpoons CO + \frac{1}{2}O$	$CO_2 + H_2 \rightleftharpoons CO + H_2O$	
298	– 71.224	– 81.208	– 159.600	– 15.171	– 40.048	– 46.054	– 45.066	– 5.018	537
500	– 40.316	– 45.880	– 92.672	– 8.783	– 22.886	– 26.130	– 25.025	– 2.139	900
1000	– 17.292	– 19.614	– 43.056	– 4.062	– 10.062	– 11.280	– 10.221	– 0.159	1800
1200	– 13.414	– 15.208	– 34.754	– 3.275	– 7.899	– 8.811	– 7.764	+ 0.135	2160
1400	– 10.630	– 12.054	– 28.812	– 2.712	– 6.347	– 7.021	– 6.014	+ 0.333	2520
1600	– 8.532	– 9.684	– 24.350	– 2.290	– 5.180	– 5.677	– 4.706	+ 0.474	2880
1700	– 7.666	– 8.706	– 22.512	– 2.116	– 4.699	– 5.124	– 4.169	+ 0.530	3060
1800	– 6.896	– 7.836	– 20.874	– 1.962	– 4.270	– 4.613	– 3.693	+ 0.577	3240
1900	– 6.204	– 7.058	– 19.410	– 1.823	– 3.886	– 4.190	– 3.267	+ 0.619	3420
2000	– 5.580	– 6.356	– 18.092	– 1.699	– 3.540	– 3.776	– 2.884	+ 0.656	3600
2100	– 5.016	– 5.720	– 16.898	– 1.586	– 3.227	– 3.434	– 2.539	+ 0.688	3780
2200	– 4.502	– 5.142	– 15.810	– 1.484	– 2.942	– 3.091	– 2.226	+ 0.716	3960
2300	– 4.032	– 4.614	– 14.818	– 1.391	– 2.682	– 2.809	– 1.940	+ 0.742	4140
2400	– 3.600	– 4.130	– 13.908	– 1.305	– 2.443	– 2.520	– 1.679	+ 0.764	4320
2500	– 3.202	– 3.684	– 13.070	– 1.227	– 2.224	– 2.270	– 1.440	+ 0.784	4500
2600	– 2.836	– 3.272	– 12.298	– 1.154	– 2.021	– 2.038	– 1.219	+ 0.802	4680
2700	– 2.494	– 2.892	– 11.580	– 1.087	– 1.833	– 1.823	– 1.015	+ 0.818	4860
2800	– 2.178	– 2.536	– 10.914	– 1.025	– 1.658	– 1.624	– 0.825	+ 0.833	5040
2900	– 1.882	– 2.206	– 10.294	– 0.967	– 1.495	– 1.438	– 0.649	+ 0.846	5220
3000	– 1.606	– 1.898	– 9.716	– 0.913	– 1.343	– 1.265	– 0.485	+ 0.858	5400
3100	– 1.348	– 1.610	– 9.174	– 0.863	– 1.201	– 1.103	– 0.322	+ 0.869	5580
3200	– 1.106	– 1.340	– 8.664	– 0.815	– 1.067	– 0.951	– 0.189	+ 0.878	5760
3300	– 0.878	– 1.086	– 8.186	– 0.771	– 0.942	– 0.809	– 0.054	+ 0.888	5940
3400	– 0.664	– 0.846	– 7.736	– 0.729	– 0.824	– 0.674	+ 0.071	+ 0.895	6120
3500	– 0.462	– 0.620	– 7.312	– 0.690	– 0.712	– 0.547	+ 0.190	+ 0.902	6300

Source: Based on data from the JANAF Thermochemical Tables, NSRDS–NBS–37, 1971.

Table A.9 Enthalpy of formation at 25°C, ideal gas enthalpy, and absolute entropy at 0.1 MPa (1 bar) Pressure*

Nitrogen, Diatomic (N_2) $(h°_f)_{298} = 0$ *kJ/kmol* $M = 28.013$		*Oxygen, Diatomic (O_2)* $(h°_f)_{298} = 0$ *kJ/kmol* $M = 31.999$		
Temp. K	$(\bar{h} - \bar{h}°_{298})$ **kJ/kmol**	$s°$ **kJ/kmol K**	$(\bar{h}° - \bar{h}°_{298})$ **kJ/kmol K**	$\bar{h}°$ **kJ/kmol K**
0	– 8669	0	– 8682	0
100	– 5770	159.813	– 5778	173.306
200	– 2858	179.988	– 2866	193.486
298	0	191.611	0	205.142
300	54	191.791	54	205.322
400	2971	200.180	3029	213.874
500	5912	206.740	6088	220.698
600	8891	212.175	9247	226.455
700	11937	216.866	12502	231.272
800	15046	221.016	15841	235.924
900	18221	224.757	19246	239.936
1000	21406	288.167	22707	243.585
1100	24757	231.309	26217	246.928
1200	28108	234.225	29765	250.016
1300	31501	236.941	33351	252.886
1400	34936	239.484	36966	255.564
1500	38405	241.878	40610	258.078
1600	41903	244.137	44279	260.446
1700	45430	246.275	47970	262.685
1800	48982	248.304	51689	264.810
1900	52551	250.237	55434	266.835
2000	56141	252.078	59199	268.764

* Adapted from Tables A. 11 (Pages 687 to 696) in *Fundamentals of Classical Thermodynamics* by G.J. Van Wylen and R. Sonntag, John Wiley, New York, 1976 (with the kind permission of the publishers, John Wiley & Sons, Inc., New York).

Table A.9 (Contd.)

Carbon Dioxide (CO_2) $(h^\circ_f)_{288} = -393522$ kJ/kmol, $M = 44.01$			*Carbon Monoxide (CO)* $(h^\circ_f)_{298} = -110529$ kJ/kmol, $M = 28.01$	
Temp. K	$(\bar{h}^\circ - \bar{h}^\circ)$ kJ/kmol	s° kJ/kmol K	$(\bar{h}^\circ - \bar{h}^\circ)$ kJ/kmol	$\bar{s}^\circ$ kJ/kmol K
0	– 9364	0	– 8669	0
100	– 6456	179.109	– 5770	165.850
200	– 3414	199.975	– 2858	186.025
298	0	213.795	0	197.653
300	67	214.025	54	197.833
400	4008	225.334	2975	206.234
500	8314	234.924	5929	212.828
600	12916	243.309	8941	218.313
700	17765	250.773	12021	223.062
800	22815	257.517	15175	227.271
900	28041	263.668	18397	231.006
1000	33405	269.325	21686	234.531
1100	38894	274.555	25033	237.719
1200	44484	279.417	28426	240.673
1300	50158	283.956	31865	243.426
1400	55907	288.216	35338	245.999
1500	61714	292.224	38848	248.421
1600	67580	296.010	42384	250.702
1700	73492	299.592	45940	252.861
1800	79442	302.993	49522	254.907
1900	85429	306.232	53124	256.852
2000	91450	309.320	56739	258.710

Table A.9 (Contd.)

	Water (H_2O) $(\bar{h}^\circ_f)_{298} = -241827$ kJ/kmol $M = 18.015$		Hydrogen, Diatomic (H_2) $(\bar{h}^\circ_f)_{298} = 0$ kJ/kmol $M = 2.016$	
Temp. K	$(\bar{h}^\circ - \bar{h}^\circ_f)$ kJ/kmol	$\bar{s}^\circ$ kJ/kmol K	$(\bar{h}^\circ - \bar{h}^\circ_f)$ kJ/kmol	K kJ/kmol K
0	– 9904	0	– 8468	0
100	– 6615	152.390	– 5293	102.145
200	– 3280	175.486	– 2770	119.437
298	0	188.833	0	130.684
300	63	189.038	54	130.864
400	3452	198.783	2958	139.215
500	6920	206.523	5883	145.738
600	10498	213.037	8812	151.077
700	14184	218.719	11749	155.608
800	17991	223.803	14703	159.549
900	21924	228.430	17682	163.060
1000	25978	232.706	20686	166.223
1100	30167	236.694	23723	169.118
1200	34476	240.443	26794	171.792
1300	38903	243.986	29907	174.281
1400	43447	247.350	33062	176.620
1500	48095	250.560	36267	178.833
1600	52844	253.622	39522	180.929
1700	57685	256.559	42815	182.929
1800	62609	259.372	46150	184.833
1900	67613	262.078	49522	186.657
2000	72689	264.681	52932	188.406

A.10 Gas Tables

Table A.10.1 Isentropic flow

M	M^*	$\frac{A}{A^*}$	$\frac{p}{p_0}$	$\frac{\rho}{\rho_0}$	$\frac{T}{T_0}$
0	0	∞	1.00000	1.00000	1.00000
0.10	0.10943	5.8218	0.99303	0.99502	0.99800
0.20	0.21822	2.9635	0.97250	0.98027	0.99206
0.30	0.32572	2.0351	0.93947	0.95638	0.98232
0.40	0.43133	1.5901	0.89562	0.92428	0.96899
0.50	0.53452	1.3398	0.84302	0.88517	0.95238
0.60	0.63480	1.1882	0.78400	0.84045	0.93284
0.70	0.73179	1.09437	0.72092	0.79158	0.91075
0.80	0.82514	1.03823	0.65602	0.74000	0.88652
0.90	0.91460	1.00886	0.59126	0.68704	0.86058
1.00	1.00000	1.00000	0.52828	0.63394	0.83333
1.10	1.08124	1.00793	0.46835	0.58169	0.80515
1.20	1.1583	1.03044	0.41238	0.53114	0.77640
1.30	1.2311	1.06631	0.36092	0.48291	0.74738
1.40	1.2999	1.1149	0.31424	0.43742	0.71839
1.50	1.3646	1.1762	0.27240	0.39498	0.68965
1.60	1.4254	1.2502	0.23527	0.35573	0.66138
1.70	1.4825	1.3376	0.20259	0.31969	0.63372
1.80	1.5360	1.4390	0.17404	0.28682	0.60680
1.90	1.5861	1.5552	0.14924	0.25699	0.58072
2.00	1.6330	1.6875	0.12780	0.23005	0.55556
2.10	1.6769	1.8369	0.10935	0.20580	0.53135
2.20	1.7179	2.0050	0.09352	0.18405	0.50813
2.30	1.7563	2.1931	0.07997	0.16458	0.48591
2.40	1.7922	2.4031	0.06840	0.14720	0.46468
2.50	1.8258	2.6367	0.05853	0.13169	0.44444
2.60	1.8572	2.8960	0.05012	0.11787	0.42517
2.70	1.8865	3.1830	0.04295	0.10557	0.40684
2.80	1.9140	3.5001	0.03685	0.09462	0.38941
2.90	1.9398	3.8498	0.03165	0.08489	0.37286
3.00	1.9640	4.2346	0.02722	0.07623	0.35714
3.50	2.0642	6.7896	0.01311	0.04523	0.28986
4.00	2.1381	10.719	0.00658	0.02766	0.23810
4.50	2.1936	16.562	0.00346	0.01745	0.19802
5.00	2.2361	25.000	$189(10)^{-5}$	0.01134	0.16667
6.00	2.2953	53.180	$633(10)^{-6}$	0.00519	0.12195
7.00	2.3333	104.143	$242(10)^{-6}$	0.00261	0.09259
9.00	2.3772	327.189	$474(10)^{-7}$	0.000815	0.05814
10.00	2.3904	535.938	$236(10)^{-7}$	0.000495	0.04762
∞	2.4495	∞	0	0	0

Table A.10.2 Normal shocks

M_x	M_y	$\frac{p_y}{p_x}$	$\frac{\rho_y}{\rho_x}$	$\frac{T_y}{T_x}$	$\frac{p_{ay}}{p_{ax}}$	$\frac{p_{ay}}{p_x}$
1.00	1.00000	1.00000	1.0000	1.0000	1.00000	1.8929
1.10	0.91177	1.2450	1.1691	1.0649	0.99892	2.1328
1.20	0.84217	1.5133	1.3416	1.1280	0.99280	2.4075
1.30	0.78596	1.8050	1.5157	1.1909	0.97935	2.7135
1.40	0.73971	2.1200	1.6896	1.2547	0.95819	3.0493
1.50	0.70109	2.4583	1.8621	1.3202	0.92978	3.4133
1.60	0.66844	2.8201	2.0317	1.3880	0.89520	3.8049
1.70	0.64055	3.2050	2.1977	0.4583	0.85573	4.2238
1.80	0.61650	3.6133	2.3592	1.5316	0.81268	4.6695
1.90	0.58562	4.0450	2.5157	1.6079	0.76635	5.1417
2.00	0.57735	4.5000	2.6666	1.6875	0.72088	5.6405
2.10	0.56128	4.9784	2.8119	1.7704	0.67422	6.1655
2.20	0.54706	5.4800	2.9512	1.8569	0.62812	6.7163
2.30	0.53441	6.0050	3.0846	1.9968	0.58331	7.2937
2.40	0.52312	6.5533	3.2119	2.0403	0.54015	7.8969
2.50	0.51299	7.1250	3.3333	2.1375	0.49902	8.5262
2.60	0.50387	7.7200	3.4489	2.2383	0.46012	9.1813
2.70	0.49563	8.3383	3.5590	2.3429	0.42359	9.8625
2.80	0.48817	8.9800	3.6635	2.4512	0.38946	10.569
2.90	0.48138	9.6450	3.7629	2.5632	0.35773	11.302
3.00	0.47519	10.333	3.8571	2.6790	0.32834	12.061
4.00	0.43496	18.500	4.5714	4.0469	0.13876	21.068
5.00	0.41523	29.000	5.0000	5.8000	0.06172	32.654
10.00	0.38757	116.50	5.7143	20.388	0.00304	129.217
∞	0.37796	∞	6.000	∞	0	∞

A.11 Critical point data*

Substance	*Formula*	*Molecular Weight*	*Temperature K*	*Pressure MPa*	*Volume m³/kmol*
Ammonia	NH_3	17.03	405.5	11.28	.0724
Argon	Ar	30.948	151	4.86	.0749
Bromine	Br_2	159.808	584	10.34	.1355
Carbon Dioxide	CO_3	44.01	304.2	7.39	.0943
Carbon Monoxide	CO	28.001	133	3.50	.0930
Chlorine	CL_2	70.906	417	7.71	.1242
Deuterium (Normal)	D_2	4.00	38.4	1.66	—
Helium	He	4.003	5.3	0.23	.0578
Helium[3]	He	3.00	3.3	0.12	—
Hydrogen (Normal)	H_2	2.106	33.3	1.30	.0649
Krypton	Kr	83.80	209.4	5.50	.0924
Neon	Ne	20.183	44.5	2.73	.0417
Nitrogen	N_2	28.013	126.2	3.39	.0899
Nitrous Oxide	N_2O	44.013	309.7	7.27	.0961
Oxygen	O_2	31.999	154.8	5.08	.0780
Sulfur Dioxide	SO_2	64.063	430.7	7.88	.1217
Water	H_2O	18.015	647.3	22.09	.0568
Xenon	Xe	131.30	289.8	5.88	.1186
Benzene	X_2H_6	78.115	562	4.92	.2603
n-Butane	C_4H_{10}	58.124	425.5	3.80	.2547
Carbon Tetra-chloride	CCL_4	153.82	556.4	4.56	.2759
Chloroform	$CHCL_3$	119.38	536.6	5.47	.2403
Dhichlorodi-fluoromethane	CCL_2F_2	120.91	384.7	4.01	.2179
Dichloro-fluoromethane	$CHCL_2F$	102.92	451.7	5.17	.1973
Ethane	C_2H_6	30.070	305.5	4.88	.1480
Ethyl Alcohol	C_2H_5OH	46.07	516	6.38	.1673
Ethylene	C_2H_{14}	28.054	282.4	5.12	.1242
n-Hexane	C_6H_4	86.178	507.9	3.03	.3677
Methane	CH_4	16.043	191.1	4.64	0.993
Methyl Alcohol	CH_3OH	32.042	513.2	7.95	.1180
Methyl Chloride	CH_3Cl	50.488	416.3	6.68	.1430
Propane	C_3H_8	44.097	370	4.26	.1998
Propene	C_3H_6	42.081	365	4.62	.1810
Propyne	C_3H_4	40.065	401	5.35	—
Trichloro-fluoromethane	CCL_3F	137.37	471.2	4.38	.2478

* K.A. Kobe and R.E. Lynn, Jr., ***Chem. Rev.***, 52: 117-236 (1953).

A.12 Properties of Air At Low Pressures (Si Units)

T (K)	h (kJ/kg)	p_r	u (kJ/kg)	v_r	ϕ (kJ/kg. K)
100	99.76	0.029 90	71.06	2230	1.4143
110	109.77	0.041 71	78.20	1758.4	1.5098
120	119.79	0.056 52	85.34	1415.7	1.5971
130	129.81	0.074 74	92.51	1159.8	1.6773
140	139.84	0.096 81	99.67	964.2	1.7515
150	149.86	0.123 18	106.81	812.0	1.8206
160	159.87	0.154 31	113.95	691.4	1.8853
170	169.89	0.190 68	121.11	594.5	1.9461
180	179.92	0.232 79	128.28	515.6	2.0033
190	189.94	0.281 14	135.40	450.6	2.0575
200	199.96	0.3363	142.56	396.6	2.1088
210	209.97	0.3987	149.70	351.2	2.1577
220	219.99	0.4690	156.84	312.8	2.2043
230	230.01	0.5477	163.98	280.0	2.2489
240	240.03	0.6355	171.15	251.8	2.2915
250	250.05	0.7329	178.29	227.45	2.3325
260	260.09	0.8405	185.45	206.26	2.3717
270	270.12	0.9590	192.59	187.74	2.4096
280	280.14	1.0889	199.78	171.45	2.4461
290	290.17	1.2311	206.92	157.07	2.4813
300	300.19	1.3860	214.09	144.32	2.5153
310	310.24	1.5546	221.27	132.96	2.5483
320	320.29	1.7375	228.45	122.81	2.5802
330	330.34	1.9352	235.65	113.70	2.6111
340	340.43	2.149	242.86	105.51	2.6412
350	350.48	2.379	250.05	98.11	2.6704
360	360.58	2.626	257.23	91.40	2.6987
370	370.67	2.892	264.47	85.31	2.7264
380	380.77	3.176	271.72	79.77	2.7534
390	390.88	3.481	278.96	74.71	2.7796
400	400.98	3.806	286.19	70.07	2.8052
410	411.12	4.153	293.45	65.83	2.8302

Table A.12 Contd.

T (K)	h (kJ/kg)	p_r	u (kJ/kg)	v_r	ϕ (kJ/kg. K)
420	421.26	4.522	300.73	61.93	2.8547
430	431.43	4.915	308.03	58.34	2.8786
440	441.61	5.332	315.34	55.02	2.9020
450	451.83	5.775	322.66	51.96	2.9249
460	462.01	6.245	329.99	49.11	2.9473
470	472.25	6.742	337.34	46.48	2.9693
480	482.48	7.268	344.74	44.04	2.9909
490	492.74	7.824	352.11	41.76	3.0120
500	503.02	8.411	359.53	39.64	3.0328
510	513.32	9.031	366.97	37.65	3.0532
520	523.63	9.684	374.39	35.80	3.0733
530	533.98	10.372	381.88	34.07	3.0930
540	544.35	11.097	389.40	32.45	3.1124
550	554.75	11.858	396.89	30.92	3.1314
560	565.17	12.659	404.44	29.50	3.1502
570	575.57	13.500	411.98	28.15	3.1686
580	586.04	14.382	419.56	26.89	3.1868
590	596.53	15.309	427.17	25.70	3.2047
600	607.02	16.278	434.80	24.58	3.2223
610	617.53	17.297	442.43	23.51	3.2397
620	628.07	18.360	450.13	22.52	3.2569
630	638.65	19.475	457.83	21.57	3.2738
640	649.21	20.64	465.55	20.674	3.2905
650	659.84	21.86	473.32	19.828	3.3069
660	670.47	23.13	481.06	19.026	3.3232
670	681.15	24.46	488.88	18.266	3.3392
680	691.82	25.85	496.65	17.543	3.3551
690	702.52	27.29	504.51	16.857	3.3707
700	713.27	28.80	512.37	16.205	3.3861
710	724.01	30.38	520.26	15.585	3.4014
720	734.20	31.92	527.72	15.027	3.4156
730	745.62	33.72	536.12	14.434	3.4314

Table A.12 Contd.

T (K)	h (kJ/kg)	p_r	u (kJ/kg)	v_r	ϕ (kJ/kg. K)
740	756.44	35.50	544.05	13.900	3.4461
750	767.30	37.35	552.05	13.391	3.4607
760	778.21	39.27	560.08	12.905	3.4751
770	789.10	41.27	568.10	12.440	3.4894
780	800.03	43.35	576.15	11.998	3.5035
790	810.98	45.51	584.22	11.575	3.5174
800	821.94	47.75	592.34	11.172	3.5312
810	832.96	50.08	600.46	10.785	3.5449
820	843.97	52.49	608.62	10.416	3.5584
830	855.01	55.00	616.79	10.062	3.5718
840	866.09	57.60	624.97	9.724	3.5850
850	877.16	60.29	633.21	9.400	3.5981
860	888.28	63.09	641.44	9.090	3.6111
870	899.42	65.98	649.70	8.792	3.6240
880	910.56	68.98	658.00	8.507	3.6367
890	921.75	72.08	666.31	8.233	3.6493
900	932.94	75.29	674.63	7.971	3.6619
910	944.15	78.61	682.98	7.718	3.6743
920	955.38	82.05	691.33	7.476	3.6865
930	966.64	85.60	699.73	7.244	3.6987
940	977.92	89.28	708.13	7.020	3.7108
950	989.22	93.08	716.57	6.805	3.7227
960	1000.53	97.00	725.01	6.599	3.7346
970	1011.88	101.06	733.48	6.400	3.7463
980	1023.25	105.24	741.99	6.209	3.7580
990	1034.63	109.57	750.48	6.025	3.7695
1000	1046.03	114.03	759.02	5.847	3.7810
1020	1068.89	123.12	775.67	5.521	3.8030
1040	1091.85	133.34	793.35	5.201	3.8259
1060	1114.85	143.91	810.61	4.911	3.8478
1080	1137.93	155.15	827.94	4.641	3.8694
1100	1161.07	167.07	845.34	4.390	3.8906

Table A.12 Contd.

T (K)	h (kJ/kg)	p_r	u (kJ/kg)	v_r	ϕ (kJ/kg. K)
1120	1184.28	179.71	862.85	4.156	3.9116
1140	1207.54	193.07	880.37	3.937	3.9322
1160	1230.90	207.24	897.98	3.732	3.9525
1180	1254.34	222.2	915.68	3.541	3.9725
1200	1277.79	238.0	933.40	3.362	3.9922
1220	1301.33	254.7	951.19	3.194	4.0117
1240	1324.89	272.3	969.01	3.037	4.0308
1260	1348.55	290.8	986.92	2.889	4.0497
1280	1372.25	310.4	1004.88	2.750	4.0684
1300	1395.97	330.9	1022.88	2.619	4.0868
1320	1419.77	352.5	1040.93	2.497	4.1049
1340	1433.61	375.3	1059.03	2.381	4.1229
1360	1467.50	399.1	1077.17	2.272	4.1406
1380	1491.43	424.2	1095.36	2.169	4.1580
1400	1515.41	450.5	1113.62	2.072	4.1753
1420	1539.44	478.0	1131.90	1.9808	4.1923
1440	1563.49	506.9	1150.23	1.8942	4.2092
1460	1587.61	537.1	1168.61	1.8124	4.2258
1480	1611.80	568.8	1187.03	1.7350	4.2422
1500	1635.99	601.9	1205.47	1.6617	4.2585
1550	1696.63	691.4	1251.78	1.4948	4.2983
1600	1757.55	791.2	1298.35	1.3485	4.3369
1650	1818.70	902.0	1345.17	1.2197	4.3745
1700	1907.39	1025.1	1392.18	1.1065	4.4112
1750	1941.63	1160.5	1439.38	1.0056	4.4469
1800	2003.36	1310.3	1486.76	0.9164	4.4817
1850	2065.27	1475.0	1534.33	0.8370	4.5156
1900	2127.37	1655.4	1582.09	0.7659	4.5487
1950	2182.72	1830.4	1624.91	0.7084	4.5777
2000	2252.06	2067.9	1678.07	0.6449	4.6127
2500	2883.91	5521	2166.41	0.3019	4.8946
3000	3526.54	12 490	2665.55	0.16015	5.1288
3500	4177.22	25 123	3172.71	0.09289	5.3294

Source: Adapted from J.H. Keenan and J. Kaye, *Gas Tables*, New York: John Wiley & Sons. 1945.

Appendix B

Figure B.1 shows the Mollier diagram for steam and pressure–enthalpy diagrams for refrigerants are 12, 11 and 134a given in Fig. B.2 to B.5. Psychrometric chart is shown in Fig. B.6.

(See extended sheet for figures).

Bibliography

1. Bejan, Adrian, Advanced Engineering Thermodynamics, John Wiley, 2016.
2. Bejan, Adrian, *Entropy Generation through Heat and Fluid Flow*, Wiley Int Science, New York, 1982.
3. Benson, R.S., *Advanced Engineering Thermodynamics*, Pergamon Press, London 1977.
4. Callen, H.B., *Introduction to Thermostatistics*, John Wiley, New York, 1985.
5. Cengel, Y.A. and M.A. Boles, *Thermodynamics*: *An Engineering Approach*, McGraw-Hill, N.Y., 2017.
6. De, S., Nag's Power Plant Engineering, McGraw Hill Education, 2021
7. Eastop, T.D. and Mcconkey, A., Applied Thermodynamics for Engineering and Technologists, Pearson Education, 1993.
8. Ganesan, V., Internal Combustion Engines, McGraw Hill, 2017
9. Gill, P.W., Smith, J.H. and Ziurys, E.J., Fundamentals of Internal Combustion Engines, Oxford and IBH Publications Ltd., 1959
10. Hatsopoulos, G.N. and J.H. Keenan, *Principles of General Thermodynamics*, John Wiley, New York, 1965.
11. Holman, Jack P., *Heat Transfer*, McGraw-Hill, New York, 2010.
12. Jones, J.B. and G.A. Hawkins, *Engineering Thermodynamics*, John Wiley, New York, 1986.
13. Keenan, J.H., *Thermodynamics*, John Wiley, New York, 1941.
14. Lee, J.F. and F.W. Sears, *Thermodynamics*, Addison Wesely, Cambridge, Mass, 1962.
15. Mooney, David A., *Mechanical Engineering Thermodynamics*, Prentice-Hall, New Jersey, 1953.
16. Michael J. Moran,, Howard N. Shapiro, Daisie D. Boettner and Margaret B. Bailey, *Fundamentals of Engineering Thermodynamics*, John Wiley, 2014.
17. Nag, P.K., *Heat and Mass Transfer*, McGraw-Hill, New Delhi, 2011.
18. Obert, E.F., *Concepts of Thermodynamics*, McGraw-Hill, New York, 1960.
19. Reynolds, W.C. and H.C. Perkins, *Engineering Thermodynamics*, McGraw-Hill, New York, 1979.
20. Saad, Michael A., *Thermodynamics for Engineers*, Prentice-Hall, New Delhi, 1966.
21. Shapiro, A.H., *The Dynamics and Thermodynamics of Compressible Fluid Flow*, The Ronald Press, New York, 1977.
22. Spalding, D.B. and E.H. Cole, *Engineering Thermodynamics*, Edward Arnold, London, 1973.
23. Smith, J.M., Van Ness, H. Abbott, M. and Swihart, M., Introduction to Chemical Engineering Thermodynamics, McGraw Hill, 2022.
24. Van Wylen, G.J. and R.E. Sonntag, *Fundamentals of Classical Thermodynamics*, John Wiley, New York, 1994.
25. Wallace, F.J. and W.A. Linning, *Basic Engineering Thermodynamics*, Pitman, London, 1970.
26. Wark, Kenneth, *Thermodynamics*, McGraw-Hill, New York, 1988.
27. Zemansky, M.W., *Heat and Thermodynamics*, McGraw-Hill, New York, 1997.

Index

T

U

V

W

Z